PARALLELOGRAM

Perimeter: $P = 2a + 2b$
Area: $A = bh$

CIRCLE

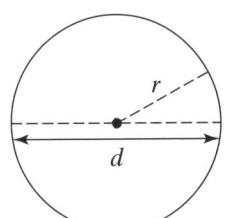

Circumference: $C = \pi d$
$\qquad\qquad\qquad C = 2\pi r$
Area: $A = \pi r^2$

RECTANGULAR SOLID

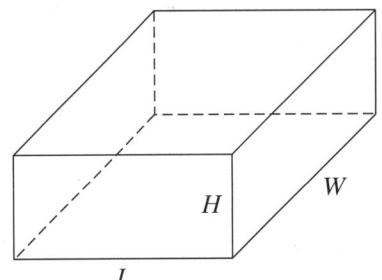

Volume: $V = LWH$
Surface Area: $A = 2HW + 2LW + 2LH$

CUBE

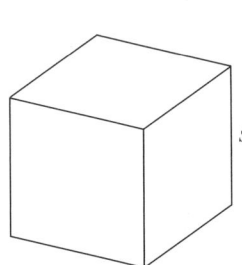

Volume: $V = s^3$
Surface Area: $A = 6s^2$

CONE

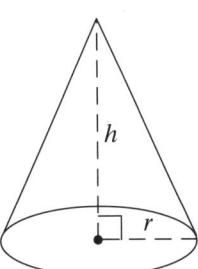

Volume: $V = \dfrac{1}{3}\pi r^2 h$
Lateral Surface Area: $A = \pi r \sqrt{r^2 + h^2}$

RIGHT CIRCULAR CYLINDER

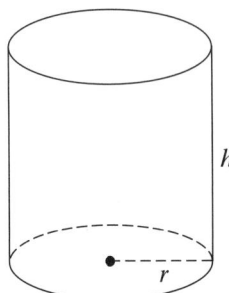

Volume: $V = \pi r^2 h$
Surface Area:
$A = 2\pi rh + 2\pi r^2$

OTHER FORMULAS

Distance: $d = rt$ (r = rate, t = time)

Temperature: $F = \dfrac{9}{5}C + 32 \qquad C = \dfrac{5}{9}(F - 32)$

Simple Interest: $I = Prt$
 (P = principal, r = annual interest rate, t = time in years)

Compound Interest: $A = P\left(1 + \dfrac{r}{n}\right)^{nt}$

 (P = principal, r = annual interest rate, t = time in years, n = number of compoundings per year)

Beginning and Intermediate Algebra

THIRD EDITION

K. Elayn Martin-Gay

University of New Orleans

PEARSON

Prentice
Hall

Upper Saddle River, New Jersey 07458

Library of Congress Cataloging-in-Publication Data
Martin-Gay, K. Elayn, 1955-
    Beginning and intermediate algebra/K. Elayn Martin-Gay.—3rd ed.
        p. cm.
    Includes index.
    ISBN (invalid) 0-13-144442-5
    1. Algebra. I. Title.

DEDICATION

In memory of Karin Elizabeth Wagner

I find the great thing in this world is not so much where we stand, as in what direction we are moving: To reach the port of heaven, we must sail sometimes with the wind and sometimes against it— but we must sail, and not drift, nor lie at anchor.

—Oliver Wendel Holmes

Editor in Chief: Christine Hoag
Senior Acquisitions Editor: Paul Murphy
Project Manager: Mary Beckwith
Assistant Editor: Christina Simoneau
Project Management: Elm Street Publishing Services, Inc.
Senior Managing Editor: Linda Mihatov Behrens
Executive Managing Editor: Kathleen Schiaparelli
**Assistant Vice President of Production and
    Manufacturing:** David W. Riccardi
Assistant Manufacturing Manager/Buyer: Michael Bell
Manufacturing Manager: Trudy Pisciotti
Art Editor: Tom Benfatti
Executive Marketing Manager: Eilish Collins Main
Marketing Assistant: Annett Uebel
Marketing Project Manager: Barbara Herbst
Development Editor: Elka Block
Editor in Chief, Development: Carol Trueheart
Art Director/Cover Designer: Maureen Eide
Cover Art: Geoffrey Cassar
Assistant to the Art Director: Dina Curro
Interior Designer: Donna Wickes
Creative Director: Carole Anson
Director of Creative Services: Paul Belfanti
Media Project Manager, Developmental Math: Audra J. Walsh

Composition: Pearson formatting:
    Manager, Electronic Composition: Jim Sullivan; Assistant
    Formatting Manager: Allyson Graesser; Electronic Production
    Specialists: Karen Noferi, Joanne Del Ben, Karen Stephens,
    Jacqueline Ambrosius, Julita Nazario, Vickie Croghan **and**
    Progressive Information Technologies
Director, Image Resource Center: Melinda Reo
Manager, Rights and Permissions: Zina Arabia
Interior Image Specialist: Beth Brenzel
Image Permission Coordinator: Craig Jones
Photo Researcher: Melinda Alexander
Art Studios: Artworks, Scientific Illustrators, Laserwords
  Artworks
    Managing Editor, AV Production & Management:
        Patricia Burns
    Production Manager: Ronda Whitson
    Production Technologies Manager: Matthew Haas
    Project Coordinator: Dan Missildine
    Art Supervisor: Kathryn Anderson
    Illustrators: Audrey Simonetti, Daniel Knopsnyder,
        Mark Landis, Nathan Storck, Ryan Currier, Stacy Smith, Scott
        Wieber
    Quality Supervisor: Pamela Taylor
    Quality Assurance: Cathy Shelly, Tim Nguyen, Ken Mooney

©2005, 2001, 1996 Pearson Education, Inc.
Pearson Prentice Hall
Pearson Education, Inc.
Upper Saddle River, New Jersey 07458

10 9 8 7 6 5 4 3 2

ISBN 0-13-144442-5

Pearson Education Ltd., London
Pearson Education Australia Pty. Limited, Sydney
Pearson Education Singapore Pte. Ltd.
Pearson Education North Asia, Ltd., Hong Kong
Pearson Education Canada, Ltd., Toronto
Pearson Educacion de Mexico, S.A., de C.V.
Pearson Education, Japan, Tokyo
Pearson Education Malaysia, Pte. Ltd.

CONTENTS

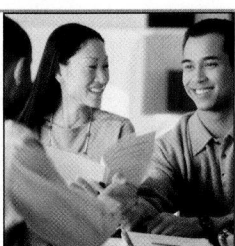

7 RATIONAL EXPRESSIONS   409

8 FUNCTIONS AND GRAPHS   477

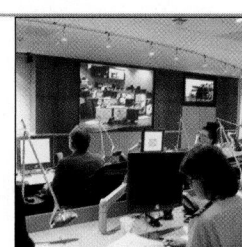

9 INEQUALITIES AND ABSOLUTE VALUE   521

13 CONIC SECTIONS 761

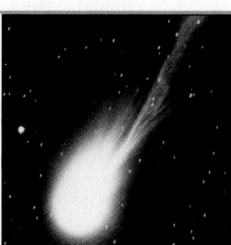

14 SEQUENCES, SERIES, AND THE BINOMIAL THEOREM 800

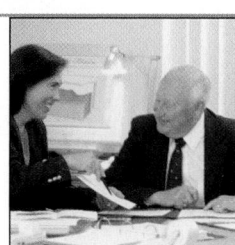

APPENDICES 839

PREFACE

ABOUT THE BOOK

Beginning and Intermediate Algebra, Third Edition was written to provide a **solid foundation in algebra**, as well as to develop students' problem-solving skills. Specific care has been taken to ensure that students have the most **up-to-date and relevant** text preparation for their next mathematics course, as well as to help students to succeed in nonmathematical courses that require a grasp of algebraic fundamentals. I have tried to achieve this by writing a user-friendly text that is keyed to objectives and contains many worked-out examples. The basic concepts of graphing and functions are introduced early, and problem solving techniques, real-life and real-data applications, data interpretation, appropriate use of technology, mental mathematics, number sense, critical thinking, decision-making, and geometric concepts are emphasized and integrated throughout the book.

The new edition includes an increased emphasis on study and test preparation skills. In addition, the third edition now includes a new resource, the Chapter Test Prep Video CDs. With these video CDs, students have instant access to video solutions for each of the chapter test questions contained in the text. It is designed to help them study efficiently.

The many factors that contributed to the success of the previous editions have been retained. In preparing this edition, I considered the comments and suggestions of colleagues throughout the country, students, and many users of the prior editions. The AMATYC Crossroads in Mathematics: Standards for Introductory College Mathematics before Calculus and the MAA and NCTM standards (plus Addenda), together with advances in technology, also influenced the writing of this text.

Beginning and Intermediate Algebra, Third Edition is **part of a series of texts** that can include *Basic College Mathematics, Second Edition* and *Prealgebra, Fourth Edition.* Also available are: *Beginning Algebra, Fourth Edition, Intermediate Algebra, Fourth Edition, Intermediate Algebra: A Graphing Approach, Third Edition*, and *Prealgebra and Introductory Algebra.* Throughout the series, pedagogical features are designed to develop student proficiency in algebra and problem solving, and to prepare students for future courses.

NEW FEATURES AND CHANGES IN THE NEW EDITION

The following new features have been added to the third edition.

INCREASED EMPHASIS ON STUDY SKILLS

New! Study Skills Reminders   Integrated throughout the text to help students hone their study skills and serve as a point-of-use support resource, reinforcing the skills covered in Section 1.1. Also, beginning in Section 2.3, and continued throughout the text, there are special Study Skills Reminders that help students organize an outline on how to recognize and how to solve various equations and inequalities. See the Study Skills Reminders in Sections 2.1, 2.4, 2.5, 2.6, and 2.7 for examples.

New! Integrated Reviews   serve as mid-chapter reviews and help students assimilate new skills and concepts they have learned separately over several sections. The reviews provide students with another opportunity to practice with *mixed* exercises as they master the topics.

Family of Faith Library

ENHANCED SECTION EXERCISE SETS

Mixed Practice exercises. Exercise sets have been organized to include mixed practice exercises where appropriate. They give students the chance to assimilate the concepts and skills covered in separate objectives. Students have the opportunity to practice the kind of decision making they will encounter on tests.

New! Concept Extension exercises have been added to the end of the section exercise sets. They extend the concepts and require students to combine several skills or concepts. They expose students to the way math ideas build upon each other and offer the opportunity for additional challenges.

MORE OPPORTUNITIES FOR STUDENTS TO CHECK THEIR UNDERSTANDING

New! Concept Checks are special exercises found in most sections following key examples. Working these will help students check their grasp of the concept being developed before moving to the next example.

INCREASED EMPHASIS ON IMPROVING TEST PREPARATION

New! Chapter Test Prep Video CDs packaged with each text and presented by Elayn Martin-Gay provide students with a resource to take and correct sample tests as they prepare for exams. Step-by-Step solutions are presented for every Chapter Test exercise contained in the text. Easy video navigation allows students to instantly access the solutions to the exact exercises they need help with.

KEY CONTINUING FEATURES

The following key features have been retained from previous editions.

Readability and Connections I have tried to make the writing style as clear as possible while still retaining the mathematical integrity of the content. When a new topic is presented, an effort has been made to **relate the new ideas to those that students may already know.** Constant reinforcement and connections within problem solving strategies, data interpretation, geometry, patterns, graphs, and situations from everyday life can help students gradually master both new and old information.

Problem Solving Process This is formally introduced in Chapter 2 with a **four-step process that is integrated throughout the text.** The four steps are Understand, Translate, Solve, and Interpret. The repeated use of these steps throughout the text in a variety of examples shows their wide applicability. Reinforcing the steps can increase students' confidence in tackling problems.

Applications and Connections Every effort was made to include as many accessible, interesting, and relevant real-life applications as possible throughout the text in both worked-out examples and exercise sets. The applications **strengthen students' understanding of mathematics in the real world** and help to motivate students. They show connections to a wide range of fields including agriculture, allied health, art, astronomy, automotive ownership, aviation, biology, business, chemistry, communication, computer technology, construction, consumer affairs, demographics, earth science, education, entertainment, environmental issues, finance and economics, food service, geography, government, history, hobbies, labor and career issues, life science, medicine, music, nutrition, physics, political science, population, recreation, sports, technology, transportation, travel, weather, and important related mathematical areas such as geometry and statistics. (See the Index of Applications on page xxvi.) Many of the applications are based on **recent and interesting real-life data.** Sources for data include newspapers, magazines, government publications, publicly held companies, special interest groups, research organizations, and reference books. Opportunities for obtaining your own real data are also included.

Helpful Hints   Helpful Hints, contain practical advice on applying mathematical concepts. These are found throughout the text and **strategically placed** where students are most likely to need immediate reinforcement. They are highlighted in a box for quick reference and, as appropriate, an indicator line is used to precisely identify the particular part of a problem or concept being discussed. For instance, see pages 94 and 96.

Visual Reinforcement of Concepts   The text contains numerous graphics, models, and illustrations to visually clarify and reinforce concepts. These include **new and updated** bar graphs, circle graphs in two and three dimensions, line graphs, calculator screens, application illustrations, photographs, and geometric figures.

Real World Chapter Openers   The chapter openers focus on how math is used in a specific career, and reference a "Spotlight on Decision Making" feature within the chapter for further exploration of the **career and the relevance of algebra**.

Student Resource Icons   At the beginning of each exercise set, videotape, tutorial software CD-Rom, Student Solutions Manual, Study Guide, and tutor center icons are displayed. These icons help reinforce that these learning aids are available should students wish to use them to review concepts and skills at their own pace. These items have **direct correlation to the text** and emphasize the text's methods of solution.

Chapter Highlights   Found at the end of each chapter, the Chapter Highlights contain key definitions, concepts, *and* examples to **help students understand and retain** what they have learned.

Chapter Project   This feature occurs at the end of each chapter, often serving as a chapter wrap-up. For **individual or group completion**, the multi-part Chapter Project, usually hands-on or data based, allows students to problem solve, make interpretations, and to think and write about algebra.

EXERCISE SETS

Each text section ends with an exercise set. The exercises are carefully graded in level of difficulty. In the beginning, **exercises are keyed** to at least one worked example in the text. Once a student has gained confidence, **the latter exercises are not keyed to examples**. Exercises and examples marked with a video icon (🔒) have been worked out step-by-step by the author in the lecture videos that accompany this text.

Throughout the text exercises there is an emphasis on data and graphical interpretation via tables, charts, and graphs. The ability to interpret data and read and create a variety of types of graphs is developed gradually so students become comfortable with it. Similarly, throughout the text there is integration of geometric concepts, such as perimeter and area. Exercises and examples marked with a geometry icon (△) have been identified for convenience.

Each exercise set contains one or more of the following features.

Mental Math   These problems are found at the beginning of many exercise sets. They are mental warm-ups that **reinforce concepts** found in the accompanying section and increase students' confidence before they tackle an exercise set. By relying on their own mental skills, students increase not only their confidence in themselves, but also their number sense and estimation ability. This edition includes a greater number of Mental Math exercises.

Writing Exercises   These exercises are found in almost every exercise set and are marked with the icon (✎). They require students to **assimilate information** and provide a written response to explain concepts or justify their thinking. Guidelines recommended by the American Mathematical Association of Two Year Colleges

(AMATYC) and other professional groups recommend incorporating writing in mathematics courses to reinforce concepts.

Mixed Practice exercises. Exercise sets have been organized to include mixed practice exercises where appropriate. They give students the chance to assimilate the concepts and skills covered in separate objectives. Students have the opportunity to practice the kind of decision making they will encounter on tests.

New! Concept Extension exercises have been added to the end of the section exercise sets. They extend the concepts and require students to combine several skills or concepts. They expose students to the way math ideas build upon each other and offer the opportunity for additional challenges.

Data and Graphical Interpretation Throughout the text there is an emphasis on data interpretation in exercises via tables, bar charts, line graphs, or circle graphs. The ability to interpret data and read and create a variety of graphs is **developed gradually** so students become comfortable with it.

Calculator Explorations and Exercises These optional explorations offer guided instruction, through examples and exercises, on the proper use of **scientific and graphing calculators or computer graphing utilities as tools in the mathematical problem-solving process**. Placed appropriately throughout the text, these explorations reinforce concepts or motivate discovery learning.

Additional exercises building on the skills developed in the Explorations may be found in exercise sets throughout the text, and are marked with the icon 🖩 for graphing calculator use.

Review and Preview These exercises occur in each exercise set (except for those in Chapter 1). These problems are **keyed to earlier sections** and review concepts learned earlier in the text that are needed in the next section or in the next chapter. These exercises show the **links between earlier topics and later material**.

Vocabulary Checks Vocabulary checks, provide an opportunity for students to become more familiar with the use of mathematical terms as they strengthen verbal skills. They are found at the end of each chapter.

Chapter Review and Chapter Test The end of each chapter contains a review of topics introduced in the chapter. The review problems are keyed to sections. The Chapter Test is not keyed to sections. The Chapter Test Prep Video CDs provide solutions by the author to every Chapter Test exercise in the text.

Cumulative Review Each chapter after the first contains a **cumulative review of all chapters beginning with the first** up through the chapter at hand. The odd problems contained in the cumulative reviews are actually earlier worked examples in the text. The even problems are keyed to sections where students can go to review the material in an exercise.

KEY CONTENT CHANGES IN THE THIRD EDITION

The following changes to content are included.

- **Chapter 3: Graphing** has been reorganized to offer greater flexibility, improved progression of topics, and a more thorough, better-paced treatment of graphing and functions. The chapter now includes **dedicated sections 3.5 The Slope-Intercept Form and 3.6 The Point-Slope Form**, which were formerly covered as objectives. Students are now exposed to a more in-depth treatment of these topics. To streamline coverage, Reading Graphs, formerly Section 1.9, has been integrated into Section 3.1 Reading Graphs and the Rectangular Coordinate System. Functions are now introduced in Section 3.7 (formerly 3.3).

To provide a more thorough treatment on functions, and improve progression, sections covering more advanced concepts of graphing and writing linear and nonlinear functions have been moved to a **new Chapter 8: Functions and Graphs**. Graphing Linear Inequalities (formerly Section 3.7) has been moved to a **new Chapter 9: Inequalities and Absolute Value**.

- **New Chapter 8: Functions and Graphs**. This new chapter provides a more focused, better-paced treatment on functions. Contents include: **8.1 Graphing and Writing Linear Functions; 8.2 Graphing Nonlinear Functions; 8.3 The Algebra of Composite Functions**; and **8.4 Variation and Problem Solving**.
- **Former Chapter 8: Transitions to Intermediate Algebra**, has been removed from the Third Edition. Topics have been distributed to other chapters and the appendices.
- **New Chapter 9: Inequalities and Absolute Value**. This new chapter has been reorganized from a portion of the former Chapter 8, and included here to follow the work on Functions and Graphs now covered in Chapter 8. Contents include: **9.1 Compound Inequalities** (formerly 8.3); **9.2 Absolute Value Inequalities** (formerly 8.4); **9.3 Absolute Value Inequalities** (formerly 8.5); and **9.4 Graphing Linear Inequalities in Two Variables and Systems of Equations** (formerly 3.7).
- **Chapter 4: Systems of Linear Equations** now includes Section 4.4 Solving Systems of Linear Equations in Three Variables (formerly 8.6).

Solving Systems of Equations by Matrices (former Section 8.7) and Solving Systems of Equations by Determinants (former Section 8.8) have been moved to the Appendices.

- **New! Appendix C Review of Beginning Algebra Topics** has been added to help students review and will benefit those who may join the course in the intermediate portion, without having taken a beginning algebra course. Contents include: **Equations and Inequalities; Problem Solving; Graphing; Polynomials and Factoring; and Rational Expressions**.
- **Chapter 5**. To improve the progression of topics, Negative Exponents and Scientific Notation are now covered in Section 5.5 (formerly 5.2).
- **Chapter 7**. To improve progression, Simplifying Complex Fractions is now covered in Section 7.7 (formerly 7.5). To streamline the sections on problem solving, former Sections 7.7 and 7.8 have been combined into **7.6 Proportion and Problem Solving with Rational Equations**.
- Conic Sections are now Chapter 13; Sequences, Series, and the Binomial Theorem are Chapter 14.

Specific Discussions have been enhanced, including:

- **Increased coverage on factoring trinomials by grouping**. See Section 6.3.
- **Increased discussion about slope as a rate of change**. See Section 3.4.

Examples, and Helpful Hints: In response to reviewer feedback, discussions have been revised to include additional examples. More Helpful Hints and Point of Use Helpful Hints facilitate student understanding.

Exercise Sets: Revised exercise sets are better organized and contain an improved progression from easier to more difficult exercises. All applications have been thoroughly updated. The Exercise Sets contain these enhancements:

- **Mixed Practice:** A new organization of the exercises is designed to make it easy for instructors to assign homework problems and tailor the difficulty level of the course.

- **New! Concept Extension** exercises allow instructors to pick and choose more difficult and/or more conceptual types of exercises. These are included at the end of the section exercise sets.
- **More real-life data exercises**, found throughout the text, incorporate the relevance of math to students' daily lives. See the Applications Index.
- Overall, the types of the exercises available for students has increased, so there are now more choices for exercises at increased level of difficulty.
- More allied health examples and exercises.
- **Cumulative Reviews** have an increased number of exercises, with the odds paired with new evens so instructors have expanded options for assignments.

INSTRUCTOR AND STUDENT RESOURCES

The third edition is supported by a comprehensive resource for instructors and students.

INSTRUCTOR RESOURCES—PRINT

Annotated Instructor's Edition (ISBN 0-13-144443-3)

- Answers to exercises on the text page or in Graphing Answer Section
- Graphing Answer Section contains answers to exercises requiring graphical solutions, chapter projects, and Spotlight on Decision Making exercises
- Teaching Tips throughout the text placed at key points in the margin, found in places where students historically need extra help together with ideas on how to help students through these concepts, as well as placed appropriately to provide ideas for expanding upon a certain concept, other ways to present a concept, or ideas for classroom activities
- New Classroom Examples have been added. Each Classroom Example parallels the text example for an added resource during lecture

Instructor's Solutions Manual (ISBN 0-13-144503-0)

- Detailed step-by-step solutions to even-numbered section exercises
- Solutions to every Spotlight on Decision Making exercise
- Solutions to every Calculator Exploration exercise
- Solutions to every Chapter Test and Chapter Review exercise
- Solution methods reflect those emphasized in the textbook

Instructor's Resource Manual with Tests (ISBN 0-13-144787-4)

- Notes to the Instructor that include new suggested assignments for each exercise set
- Eight Chapter Tests per chapter (5 free response, 3 multiple choice)
- Two Cumulative Review Tests (one free response, one multiple choice)
- Eight Final Exams (4 free response, 4 multiple choice)
- Twenty additional exercises per section for added test exercises or worksheets, if needed
- Group Activities (on average of two per chapter; providing short group activities in a convenient ready-to-use handout format)
- Answers to all items

INSTRUCTOR RESOURCES—MEDIA

TestGen with QuizMaster   enables instructors to build, edit, print, and administer tests using a computerized bank of questions developed to cover all the objectives of the text. Instructors can modify test bank questions or add new questions by using the built-in question editor, which allows users to create graphs, import graphics, and insert math notation, variable numbers, or text. Tests can be printed or administered online via the Internet or another network. TestGen comes packaged with QuizMaster, which allows students to take tests on a local area network. The software is available on a dual-platform Windows/Macintosh CD-ROM.

"Instructor to Instructor" Videos   authored by Elayn Martin-Gay these videos offer topical and teaching technique instruction to new instructors and adjuncts to enhance effective classroom communication, and provide seasoned faculty with additional teaching ideas and approaches. They also provide suggestions for presenting the topics in class, alternative approaches, time saving strategies, classroom activities, and much more.

New MyMathLab® (instructor)   is a series of text-specific, easily customizable online courses for Prentice Hall mathematics textbooks. MyMathLab is powered by CourseCompass_—Pearson Education's online teaching and learning environment—and by MathXL®—our online homework, tutorial, and assessment system. MyMathLab gives you the tools you need to deliver all or a portion of your course online, whether your students are in a lab setting or working from home.

MyMathLab provides a rich and flexible set of course materials, featuring free-response exercises that are algorithmically generated for unlimited practice and mastery. Students can also use online tools such as video lectures, animations, and a multimedia textbook to independently improve their understanding and performance. Instructors can use MyMathLab's homework and test managers to select and assign online exercises correlated directly to the textbook, and they can also import TestGen tests into MyMathLab for added flexibility. MyMathLab's online gradebook—designed specifically for mathematics—automatically tracks students' homework and test results and gives the instructor control over how to calculate final grades.

MyMathLab is available to qualified adopters. For more information, visit our website at www.mymathlab.com or contact your Prentice Hall sales representative for a product demonstration.

MathXL®   is a powerful online homework, tutorial, and assessment system that accompanies your Prentice Hall mathematics textbook. With MathXL, instructors can create, edit, and assign online homework and tests using algorithmically generated exercises correlated at the objective level to your textbook. All student work is tracked in MathXL's online gradebook. Students can take chapter tests in MathXL and receive personalized study plans based on their test results. The study plan diagnoses weaknesses and links students directly to tutorial exercises for the objectives they need to study and retest. Students can also access supplemental animations and video clips directly from selected exercises. MathXL is available to qualified adopters. For more information, visit our website at www.mathxl.com, or contact your Prentice Hall sales representative for a product demonstration.

MathXL® Tutorials on CD   This interactive tutorial CD-ROM provides algorithmically generated practice exercises that are correlated at the objective level to the exercises in the textbook. Every practice exercise is accompanied by an example and a guided solution designed to involve students in the solution process. Selected exercises may also include a video clip to help students visualize concepts. The software tracks student activity and scores and can generate printed summaries of students' progress.

STUDENT RESOURCES—PRINT

Student Solutions Manual (ISBN 0-13-144504-9)

- Detailed step-by-step solutions to odd-numbered section exercises
- Solutions to every (odd and even) Mental Math exercise
- Solutions to odd-numbered Calculator Exploration exercises
- Solutions to every (odd and even) exercise found in the Chapter Reviews and Chapter Tests
- Solution methods reflect those emphasized in the textbook
- Ask your bookstore about ordering

Study Guide (ISBN 0-13-144505-7)

- Additional step-by-step worked out examples and exercises
- Practice tests and final examination
- Includes Study Skills and Note-taking suggestions
- Includes Hints and warnings section
- Solutions to all exercises, tests, and final examination
- Solution methods reflect those emphasized in the text
- Ask your bookstore about ordering

STUDENT RESOURCES—MEDIA

New! Chapter Test Prep Video CDs   Provides a step-by-step video solution to each problem in the textbook Chapter Test, presented by Elayn Martin-Gay.

MyMathLab® (student)   is a complete online course designed to help students succeed in learning and understanding mathematics. MyMathLab contains an online version of your textbook with links to multimedia resources—such as video clips, practice exercises, and animations—that are correlated to the examples and exercises in the text. MyMathLab also provides students with online homework and tests and generates a personalized study plan based on their test results. The study plan links directly to unlimited tutorial exercises for the areas students need to study and re-test, so they can practice until they have mastered the skills and concepts in the textbook. All of the online homework, tests, and tutorial work students do is tracked in their MyMathLab gradebook.

MathXL®   is a powerful online homework, tutorial, and assessment system that accompanies your Prentice Hall mathematics textbook. With MathXL, instructors can create, edit, and assign online homework and tests using algorithmically generated exercises correlated at the objective level to your textbook. All student work is tracked in MathXL's online gradebook. Students can take chapter tests in MathXL and receive personalized study plans based on their test results. The study plan diagnoses weaknesses and links students directly to tutorial exercises for the objectives they need to study and retest. Students can also access supplemental animations and video clips directly from selected exercises. MathXL is available to qualified adopters. For more information, visit our website at www.mathxl.com, or contact your Prentice Hall sales representative for a product demonstration.

MathXL® Tutorials on CD   This interactive tutorial CD-ROM provides algorithmically generated practice exercises that are correlated at the objective level to the exercises in the textbook. Every practice exercise is accompanied by an example and a guided solution designed to involve students in the solution process. Selected exercises may also include a video clip to help students visualize concepts. The software tracks student activity and scores and can generate printed summaries of students' progress.

Lecture Series Videos   Digitized on CD-ROM (0-13-144508-1) and on VHS Tape (0-13-144507-3)

- Keyed to each section of the text
- Step-by-step solutions to exercises from each section of the text
- In-text exercises marked with a video icon appear on the videos
- Digitized videos offer convenient anytime access to video tutorial support when shrinkwrapped with the text

PH Tutor Center (0-13-064604-0)

- Free tutorial support via phone, fax or email
- Available Sunday—Thursday 5pm e.s.t. to midnight—5 days a week, 7 hours a day
- Staffed by developmental math faculty
- Accessed through a registration number that may be bundled with a new text or purchased separately with a used book.
- See www.prenhall.com/tutorcenter for FAQ

ACKNOWLEDGMENTS

First, as usual, I would like to thank my husband, Clayton, for his constant encouragement. I would also like to thank my children, Eric and Bryan, for continuing to eat my burnt meals. Thankfully, they have started to cook a little themselves.

I would also like to thank my extended family for their invaluable help and wonderful sense of humor. Their contributions are too numerous to list. They are Rod and Karen Pasch; Peter, Michael, Christopher, Matthew, and Jessica Callac; Stuart, Earline, Melissa, Mandy, Bailey, and Ethan Martin; Mark, Sabrina, and Madison Martin; Leo and Barbara Miller; and Jewett Gay.

I would like to thank the following reviewers for their input and suggestions:

Jeff Bonn, *Michigan Technical University*
Gail Brewer, *Amarillo College*
Cynthia Broughton, *Arizona Western*
Joe Bruce, *College of the Redwoods*
Edie Carter, *Amarillo College*
Dr. Marshall Cates, *California State Los Angeles*
Roy Cavanaugh, *Arizona Western*
Jackie Coomes, *Eastern Washington University*
Janice Ervin, *Central Piedmont Community College*
Theresa Evans, *Odessa College*
Pamela Gerszewski, *College of Albemarle*
Jerry Graveman, *Gainesville College*
Alan Hayashi, *Oxnard College*
Eric Hsu, *San Francisco State University*
Gary Hughes, *Iowa Western Community College*
Kathy Johnson, *Volunteer State Community College*
Sonya Johnson, *Central Piedmont Community College*
Fran Kowalewski, *Hood College*
Lisa Lindloff, *McLennan Community College*
Michelle Moravec, *McLennan Community College*
Kim Peppard, *West Texas A & M*
Christopher Reisch, *Jamestown Community College*
Jill Shirley, *California State Polytech Pomona*

Verlyn Stahlecker, *St. Philip's College*
Ann Thrower, *Kilgore College*
Matthew Utz, *University of Arkansas Fort Smith*
Ray Watson, *University of Arkansas Fort Smith*
Alma Wlazlinski, *McLennan Community College*
Susan Yellott, *Kilgore College*
Stephanie Zwyghuizen, *Jamestown Community College*

There were many people who helped me develop this text and I will attempt to thank some of them here. Lauri Semarne was invaluable for contributing to the overall accuracy of this text. Elka Block, Chris Callac, and Miriam Daunis were invaluable for their many suggestions and contributions during the development and writing of this third edition. Brandi Nelson provided guidance throughout the production process. I thank Richard Semmler for all his work on the solutions, text, and accuracy.

Sadly, executive editor of this project, Karin Wagner, passed away this year. She will be dearly remembered by me and the rest of the staff at Prentice Hall for her integrity, wisdom, commitment, and especially her sense of humor and infectious laugh.

A very special thank you to my project manager, Mary Beckwith, for taking over during a difficult period for all of us.

Lastly, my thanks to the staff at Prentice Hall for all their support: Linda Behrens, Mike Bell, Patty Burns, Tom Benfatti, Paul Belfanti, Maureen Eide, Eilish Main, Patrice Jones, Chris Hoag, Paul Corey, Tim Bozik, Jim Sullivan, Allyson Graesser, Karen Noferi, Joanne Del Ben, Karen Stephens, Jacqueline Ambrosius, Julita Nazario, and Vickie Croghan.

K. Elayn Martin-Gay

ABOUT THE AUTHOR

K. Elayn Martin-Gay has taught mathematics at the University of New Orleans for 25 years. Her numerous teaching awards include the local University Alumni Association's Award for Excellence in Teaching, and Outstanding Developmental Educator at University of New Orleans, presented by the Louisiana Association of Developmental Educators.

Prior to writing textbooks, K. Elayn Martin-Gay developed an acclaimed series of lecture videos to support developmental mathematics students in their quest for success. These highly successful videos originally served as the foundation material for her texts. Today the tapes specifically support each book in the Martin-Gay series.

Elayn is the author of over ten published textbooks as well as multimedia interactive mathematics, all specializing in developmental mathematics courses such as basic mathematics, prealgebra, beginning and intermediate algebra. She has provided author participation across the broadest range of materials: textbook, videos, tutorial software, and Interactive Math courseware. All the components are designed to work together. This offers an opportunity of various combinations for an integrated teaching and learning package offering great consistency and comfort for the student.

Every Student Can Succeed

Beginning & Intermediate Algebra, Third Edition has been written and designed to help you succeed in this course. Special care has been taken to ensure students have the most up-to-date and relevant text features, and as many real-world applications as possible to provide you with a solid foundation in algebra and prepare you for future courses.

Good study skills are essential for success in mathematics. This edition provides an increased emphasis on study and test preparation skills. Take a few minutes to examine the features and resources that have been incorporated into *Beginning & Intermediate Algebra, Third Edition* to help students excel.

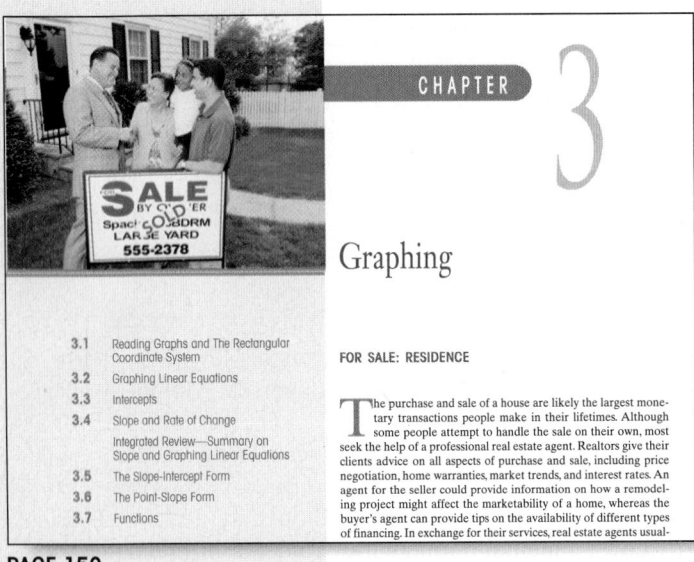

PAGE 150

◀ Real-World Chapter Openers

Real-world chapter openers focus on how algebraic concepts relate to the world around you. They also reference a **Spotlight on Decision Making** feature within the chapter for further exploration.

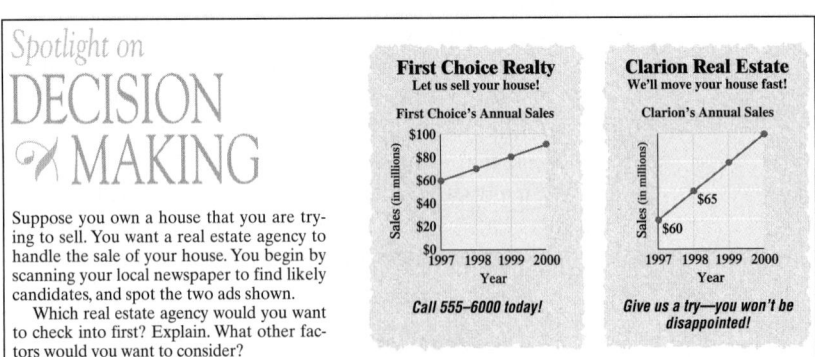

PAGE 189

▲ Spotlight on Decision Making

These unique applications encourage students to develop decision making and problem solving abilities. Primarily workplace or career-related situations are highlighted in each feature.

Become a Confident Problem Solver!

A goal of this text is to help you develop problem-solving abilities.

EXAMPLE 5

FINDING THE INVESTMENT AMOUNT

Rajiv Puri invested part of his $20,000 inheritance in a mutual funds account that pays 7% simple interest yearly and the rest in a certificate of deposit that pays 9% simple interest yearly. At the end of one year, Rajiv's investments earned $1550. Find the amount he invested at each rate.

Solution
1. UNDERSTAND. Read and reread the problem. Next, guess a solution. Suppose that Rajiv invested $8000 in the 7% fund and the rest, $12,000, in the fund paying 9%. To check, find his interest after one year. Recall the formula, $I = PRT$, so the interest from the 7% fund = $8000(0.07)(1) = $560. The interest from the 9% fund = $12,000(0.09)(1) = $1080. The sum of the interests is $560 + $1080 = $1640. Our guess is incorrect, since the sum of the interests is not $1550, but we now have a better understanding of the problem.

Let

$$x = \text{amount of money in the account paying 7\%.}$$

The rest of the money is $20,000 less x or

$$20,000 - x = \text{amount of money in the account paying 9\%.}$$

2. TRANSLATE. We apply the simple interest formula $I = PRT$ and organize our information in the following chart. Since there are two different rates of interest and two different amounts invested, we apply the formula twice.

	Principal ·	Rate ·	Time =	Interest
7% Fund	x	0.07	1	$x(0.07)(1)$ or $0.07x$
9% Fund	$20,000 - x$	0.09	1	$(20,000 - x)(0.09)(1)$ or $0.09(20,000 - x)$
Total	20,000			1550

The total interest earned, $1550, is the sum of the interest earned at 7% and the interest earned at 9%.

In words:	interest at 7%	+	interest at 9%	=	total interest
Translate:	$0.07x$	+	$0.09(20,000 - x)$	=	1550

3. SOLVE.

$$0.07x + 0.09(20,000 - x) = 1550$$

$0.07x + 1800 - 0.09x = 1550$ Apply the distributive property.
$1800 - 0.02x = 1550$ Combine like terms.
$-0.02x = -250$ Subtract 1800 from both sides.
$x = 12,500$ Divide both sides by -0.02.

4. INTERPRET.

Check: If $x = 12,500$, then $20,000 - x = 20,000 - 12,500$ or 7500. These solutions are reasonable, since their sum is $20,000 as required. The annual interest on $12,500 at 7% is $875; the annual interest on $7500 at 9% is $675, and $875 + $675 = $1550.
State: The amount invested at 7% is $12,500. The amount invested at 9% is $7500.

◀ General Strategy for Problem Solving

Save time by having a plan. The organization of this text can help you. Note the outlined problem-solving steps: *Understand, Translate, Solve,* and *Interpret.* Problem solving is introduced early, emphasized, and integrated throughout the chapters. The problem-solving procedure is illustrated step-by-step in the in-text examples.

Geometry ▶

Geometric concepts are integrated throughout the text in examples and exercises and are identified with a triangle icon. The inside front cover contains *Geometric Formulas* for convenient reference.

△ **57.** The CART Fed Ex Championship Series is an open-wheeled race car competition based in the United States. A CART car has a maximum length of 199 inches, a maximum width of 78.5 inches, and a maximum height of 33 inches. When the CART series travels to another country for a grand prix, teams must ship their cars. Find the volume of the smallest shipping crate needed to ship a CART car of maximum dimensions. (*Source:* Championship Auto Racing Teams, Inc.)

CART Racing Car

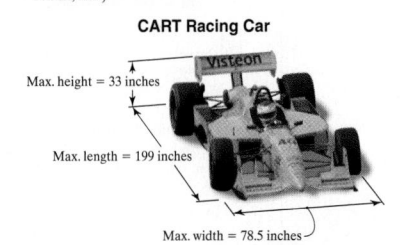

Max. height = 33 inches

Max. length = 199 inches

Max. width = 78.5 inches

Learn Better—Master and Apply Skills and Concepts

K. Elayn Martin-Gay provides thorough explanations of concepts and enlivens the content by integrating successful and innovative instructional tools. These features have been included to enhance your understanding of mathematical concepts.

 CONCEPT CHECK

Write the equations of any three parallel lines.

3 The slope-intercept form can also be used to write the equation of a line given its slope and *y*-intercept.

PAGE 198

◀ New Concept Checks

Concept Checks are special exercises found in most sections following key examples. Work these to help measure your grasp of the concept being explained before moving to the next example. Answers appear at the bottom of the text page.

Mixed Practice Exercises ▶

Mixed Practice exercises have been added to the exercise sets to give students the chance to assimilate the concepts and skills that have been covered in separate objectives.

MIXED PRACTICE

Graph each linear equation. See Examples 2 through 5.

9. $x + y = 4$ **10.** $x + y = 7$
11. $x - y = -2$ **12.** $-x + y = 6$
13. $x - 2y = 6$ **14.** $-x + 5y = 5$
15. $y = 6x + 3$ **16.** $y = -2x + 7$
17. $x - 2y = -6$ **18.** $-x + 2y = 5$
19. $y = 6x$ **20.** $x = -2y$
21. $3y - 10 = 5x$ **22.** $-2x + 7 = 2y$
23. $x + 3y = 9$ **24.** $2x + y = -2$
25. $y - x = -1$ **26.** $x - y = 5$
27. $x = -3y$ **28.** $y = -x$
29. $5x - y = 10$ **30.** $7x - y = 2$
31. $y = \frac{1}{2}x + 2$ **32.** $y = -\frac{1}{5}x - 1$

PAGE 172

Concept Extensions

Solve.

59. American Automobile Association (AAA) membership was 21 million in 1980. In 2002, their membership was 45 million. (*Source*: AAA)
   a. Write two ordered pairs of the form (years after 1980, membership in millions).
   b. The relationship between the years after 1980 and membership is approximately linear over this period. Use the ordered pairs from part (a) to write an equation of the line relating year to AAA membership.

60. In 1998, 1.7 million electronic bill statements were delivered and payments occurred. In 2002, that number rose to 9.9 million. (*Source*: USA Today)
   a. Write two ordered pairs of the form (years after 1998, millions of electronic bills).
   b. The relationship between years after 1998 and 2002 is approximately linear over this period. Use the ordered pairs from part a to write an equation of the line relating to electronic bills.
   c. Use the linear equation from part (b) to estimate the electronic bills to be delivered and paid in 2006.

◀ New Concept Extensions

Concept Extension exercises have been added to the exercise sets. They extend the concepts and require students to combine several skills or concepts. These exercises expose students to the way math ideas build upon each other.

PAGE 203

Study Better—Build Confidence and Develop Study Skills

Several features of this text can be helpful in building your confidence and mathematical competence. They will also help improve your study skills.

Tips for Success ▶

Coverage of study skills in Section 1.1 reinforces this important component to success in this course.

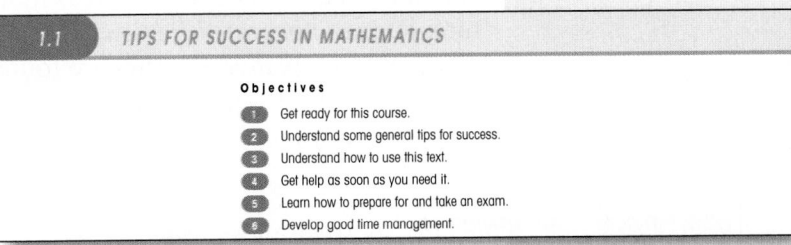

1.1   TIPS FOR SUCCESS IN MATHEMATICS

Objectives
1. Get ready for this course.
2. Understand some general tips for success.
3. Understand how to use this text.
4. Get help as soon as you need it.
5. Learn how to prepare for and take an exam.
6. Develop good time management.

PAGE 2

STUDY SKILLS REMINDER

Tips for Studying for an Exam

To prepare for an exam, try the following study techniques.

▶ Start the study process days before your exam.
▶ Make sure that you are current and up-to-date on your assignments.
▶ If there is a topic that you are unsure of, use one of the many resources that are available to you. For example,
  See your instructor.
  Visit a learning resource center on campus where math tutors are available.
  Read the textbook material and examples on the topic.
  View a videotape on the topic.
▶ Reread your notes and carefully review the Chapter Highlights at the end of the chapter.
▶ Work the review exercises at the end of the chapter and check your answers. Make sure that you correct any missed exercises. If you have trouble on a topic, use a resource listed above.
▶ Find a quiet place to take the Chapter Test found at the end of the chapter. Do not use any resources when taking this sample test. This way you will have a clear indication of how prepared you are for your exam. Check your answers and make sure that you correct any missed exercises.
▶ Get lots of rest the night before the exam. It's hard to show how well you know the material if your brain is foggy from lack of sleep.

Good luck and keep a positive attitude.

PAGE 201

Study Skills Reminders

◀ *Study Skills Reminders* are integrated throughout the text to reinforce Section 1.1 and encourage the development of strong study skills. A special type of reminder is included to help students develop a cumulative outline that will help them learn to recognize and solve equations and inequalities.

Mental Math ▼

Mental Math warm-up exercises reinforce concepts found in the accompanying section and can increase your confidence before beginning an execise set.

MENTAL MATH

The graph of each equation below is a line. Use the equation to identify the slope and a point of the line.

1. $y - 8 = 3(x - 4)$ 2. $y - 1 = 5(x - 2)$
3. $y + 3 = -2(x - 10)$ 4. $y + 6 = -7(x - 2)$
5. $y = \frac{2}{5}(x + 1)$ 6. $y = \frac{3}{7}(x + 4)$

PAGE 208

Helpful Hint

If a graph crosses the x-axis at $(-3, 0)$ and the y-axis at $(0, 7)$, then

$$\underbrace{(-3, 0)}_{x\text{-intercept}} \qquad \underbrace{(0, 7)}_{y\text{-intercept}}$$

Notice that for the y-intercept, the x-value is 0 and for the x-intercept, the y-value is 0.

Note: Sometimes in mathematics, you may see just the number 7 stated as the y-intercept, and -3 stated as the x-intercept.

PAGE 174

◀ Helpful Hints

Found throughout the text, these features contain practical advice on applying mathematical concepts. They are strategically placed where students are most likely to need immediate reinforcement.

Vocabulary Checks ▶

Vocabulary Checks help students strengthen verbal skills and answer questions about a chapter's content by filling in the blank with the correct word from the vocabulary list.

3   CHAPTER VOCABULARY CHECK

Fill in each blank with one of the words listed below.

relation	function	domain	range	standard	slope-intercept
y-axis	x-axis	solution	linear	slope	point-slope
x-intercept	y-intercept	y	x		

1. An ordered pair is a _____ of an equation in two variables if replacing the variables by the coordinates of the ordered pair results in a true statement.
2. The vertical number line in the rectangular coordinate system is called the _____.
3. A _____ equation can be written in the form $Ax + By = C$.
4. A(n) _____ is a point of the graph where the graph crosses the x-axis.
5. The form $Ax + By = C$ is called _____ form.
6. A(n) _____ is a point of the graph where the graph crosses the y-axis.

PAGE 221

Test Better—Test Yourself and Check Your Understanding

Good exercise sets and an abundance of worked-out examples are essential for building confidence. The exercises in this text are intended to help you build skills and understand concepts as well as motivate and challenge you. In addition, features such as *Integrated Reviews, Chapter Highlights, Chapter Reviews, Chapter Tests,* and *Cumulative Reviews* are found in each chapter to help you study and organize your notes. A new resource—*Chapter Test Prep Video CDs*—will help students study more effectively than ever before.

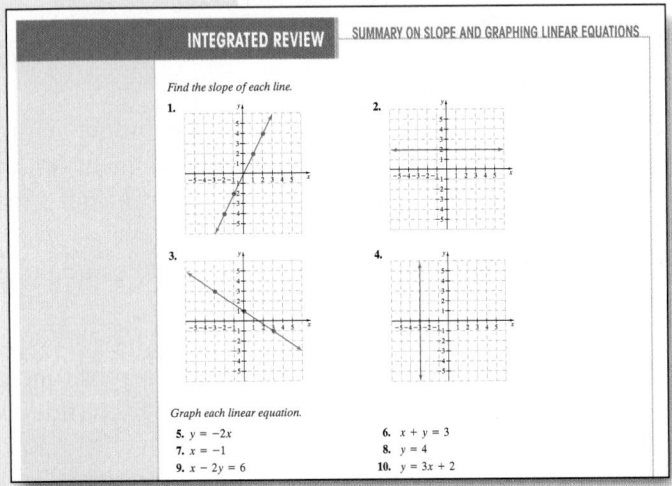

PAGE 195

◄ New Integrated Reviews

Integrated Reviews serve as mid-chapter reviews and help you learn the new skills you have been studying over several sections. This allows students to practice the type of decision making that will prepare them for tests.

Chapter Highlights ►

Each chapter ends with *Chapter Highlights.* They contain key definitions, concepts, and examples to help you understand and retain what you have learned in the chapter.

PAGE 221

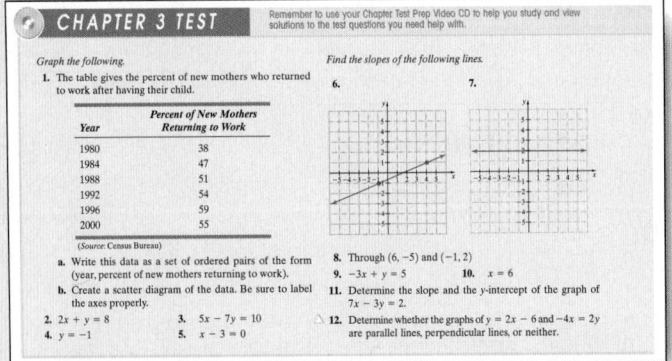

PAGE 229

◄ Chapter Tests

Take the *Chapter Test* at the end of each chapter to prepare for a class exam. With the new *Chapter Test Prep Video CDs* (packaged with your text), you can instantly view the solution to the exact exercises you're working as part of your studying. See the next page for more details.

Begin Studying More Effectively Today!

New Chapter Test Prep Video CDs

Good study skills are essential for success in mathematics. Use the student resources that accompany *Beginning & Intermediate Algebra, Third Edition* to make the most of your valuable study time.

The *Chapter Test Prep Video* CDs, presented by K. Elayn Martin-Gay, provide students with instant access to step-by-step video solutions for each of the Chapter Test questions in the text. To make the most of this resource when studying for a test, follow these three steps:

1. Take the Chapter Test at the end of each chapter.
2. Check your answers in the back of the text.
3. Use the *Chapter Test Prep Video* to review **every step** of the worked-out solution for those specific questions you answered incorrectly or didn't understand on the test.

The Video CDs' easy navigation is designed to help students study efficiently. Students select the exact test questions they missed or don't understand and instantly view the solution worked out by the author.

Previous and Next buttons allow easy navigation within tests.

The Chapter Test menu allows students to select the test questions they wish to view.

Rewind, Pause, and Play buttons let students view the solutions at their own pace.

The Video CDs includes an Introduction with instructions on how to use the *Chapter Test Prep Video CDs* as well as test-taking study skills.

To begin studying and preparing for your test, see the Video CDs and Note to Students in your text. The Video CDs are also available through your campus bookstore.

Get Motivated!

The third edition strongly emphasizes visualization. Graphing and functions are introduced early and intuitively. Knowing how to read and use graphs is a valuable skill in the workplace as well as in this and other courses. In addition, this edition includes a wealth of real-world applications to show the relevance of math in everyday life.

Real-World Applications ▼

Many applications are included often based on real data drawn from current and familiar sources.

PAGE 162

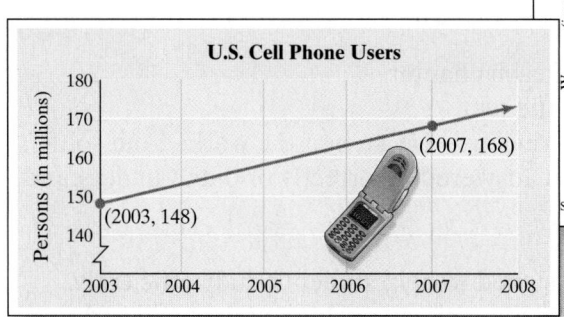

PAGE 193

Scientific and Graphing ▶ Calculator Explorations

These exploration features contain examples and exercises to reinforce concepts, help interpret graphs, and motivate discovery learning. Exercises can also be found in exercise sets.

PAGE 200

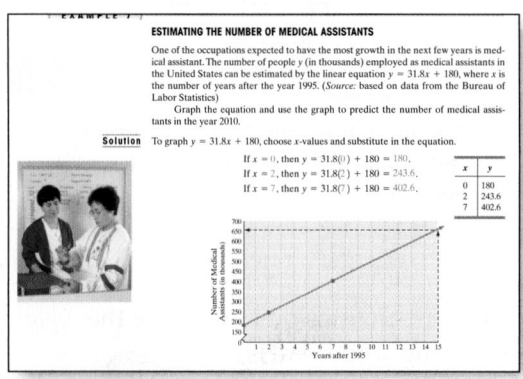

PAGE 170

▲ Visualization of Topics

Many illustrations, models, photographs, tables, charts, and graphs provide visual reinforcement of concepts and opportunities for data interpretation.

Resources for Student Success

Beginning & Intermediate Algebra, Third Edition is supported by comprehensive resource packages for both instructors and students. Please refer to the Preface for detailed descriptions.

Highlights of the packages include the following:

Chapter Test Prep Video CDs—included in every student text and annotated instructor's edition. They provide step-by-step worked out solutions to every question contained on the chapter tests in the text. Hosted by K. Elayn Martin-Gay.

MyMathLab®—a series of text-specific, easily customizable online courses for Prentice Hall mathematics texts.

MathXL®—a powerful online homework, tutorial, and assessment system.

MathXL® tutorials on CD—provides algorithmically generated practice exercises that are correlated at the objective level to the exercises in the text.

CD Lecture Series—text-specific videos available on CD. They are hosted by K. Elayn Martin-Gay and cover each objective in every chapter section of the text.

Prentice Hall Tutor Center—provides text-specific tutoring via phone, fax, and email.

Student Solutions Manual and Student Study Guide

FOR INSTRUCTORS

Annotated Instructor's Edition—now includes one additional classroom example for each example in the text, Teaching Tips, and answers to all text exercises.

Instructor's Solutions Manual and Instructor's Resource Manual with Tests

TestGen—enables instructors to build, edit, print, and administer tests using a computerized bank of questions developed to cover all objectives of the text.

Instructor-to-Instructor Videos—provide suggestions for presenting topics, time-saving tips, alternative strategies, and much more.

Review of Real Numbers

The power of mathematics is its flexibility. We apply numbers to almost every aspect of our lives, from an ordinary trip to the grocery store to a rocket launched into space. The power of algebra is its generality. Using letters to represent numbers, we tie together the trip to the grocery store and the launched rocket.

In this chapter we review the basic symbols and words—the language—of arithmetic and introduce using variables in place of numbers. This is our starting place in the study of algebra.

HEALTH CARE PROFESSIONALS IN DEMAND

During a visit to the dentist's office for a regular checkup, patients usually spend most of their time with the dental hygienist. Hygienists are professional health care providers who examine and clean a patient's gums and teeth. They remove plaque from teeth, apply fluoride for cavity prevention, floss teeth, and take and develop dental X rays. Frequently, hygienists teach their patients, especially children, how to practice good oral hygiene.

According to the U.S. Department of Labor, the demand for dental hygienists in the workplace is expected to grow through the year 2010. This increase in demand is due to many factors, including population growth and greater retention of natural teeth.

In the Spotlight on Decision Making feature on page 44, you will have the opportunity to make a decision concerning the health of a patient's gums as a dental hygienist.
Link: American Dental Hygienists' Association—www.adha.org
Source of text: http://www.bls.gov/oco/ocos097.htm

1.1 TIPS FOR SUCCESS IN MATHEMATICS

Objectives

1 Get ready for this course.

2 Understand some general tips for success.

3 Understand how to use this text.

4 Get help as soon as you need it.

5 Learn how to prepare for and take an exam.

6 Develop good time management.

Before reading this section, remember that your instructor is your best source for information. Please see your instructor for any additional help or information.

1 **Getting Ready for This Course**

Now that you have decided to take this course, remember that a *positive attitude* will make all the difference in the world. Your belief that you can succeed is just as important as your commitment to this course. Make sure that you are ready for this course by having the time and positive attitude that it takes to succeed.

Next, make sure that you have scheduled your math course at a time that will give you the best chance for success. For example, if you are also working, you may want to check with your employer to make sure that your work hours will not conflict with your course schedule. Also, schedule your class during a time of day when you are more attentive and do your best work.

On the day of your first class period, double-check your schedule and allow yourself extra time to arrive in case of traffic problems or difficulty locating your classroom. Make sure that you bring at least your textbook, paper, and a writing instrument. Are you required to have a lab manual, graph paper, calculator, or some other supply besides this text? If so, also bring this material with you.

2 **General Tips for Success**

Below are some general tips that will increase your chance for success in a mathematics class. Many of these tips will also help you in other courses you may be taking.

Exchange names and phone numbers with at least one other person in class. This contact person can be a great help if you miss an assignment or want to discuss math concepts or exercises that you find difficult.

Choose to attend all class periods and be on time. If possible, sit near the front of the classroom. This way, you will see and hear the presentation better. It may also be easier for you to participate in classroom activities.

Do your homework. You've probably heard the phrase "practice makes perfect" in relation to music and sports. It also applies to mathematics. You will find that the more time you spend solving mathematics problems, the easier the process becomes. Be sure to schedule enough time to complete your assignments before the next class period.

Check your work. Review the steps you made while working a problem. Learn to check your answers to the original problems. You may also compare your answers with the answers to selected exercises section in the back of the book. If you have made a mistake, try to figure out what went wrong. Then correct your mistake. If you can't find what went wrong, don't erase your work or throw it away. Bring your work to your instructor, a tutor in a math lab, or a classmate. It is easier for someone to find where you had trouble if they look at your original work.

Learn from your mistakes and be patient with yourself. Everyone, even your instructor, makes mistakes. (That definitely includes me—Elayn Martin-Gay.) Use your errors to learn and to become a better math student. The key is finding and understanding your errors.

Was your mistake a careless one, or did you make it because you can't read your own math writing? If so, try to work more slowly or write more neatly and make a conscious effort to carefully check your work.

Did you make a mistake because you don't understand a concept? Take the time to review the concept or ask questions to better understand it.

Did you skip too many steps? Skipping steps or trying to do too many steps mentally may lead to preventable mistakes.

Know how to get help if you need it. It's all right to ask for help. In fact, it's a good idea to ask for help whenever there is something that you don't understand. Make sure you know when your instructor has office hours and how to find his or her office. Find out whether math tutoring services are available on your campus. Check out the hours, location, and requirements of the tutoring service. Videotapes and software are available with this text. Learn how to access these resources.

Organize your class materials, including homework assignments, graded quizzes and tests, and notes from your class or lab. All of these items will be valuable references throughout your course especially when studying for upcoming tests and the final exam. Make sure that you can locate these materials when you need them.

Read your textbook before class. Reading a mathematics textbook is unlike leisure reading such as reading a novel or newspaper. Your pace will be much slower. It is helpful to have paper and a pencil with you when you read. Try to work out examples on your own as you encounter them in your text. You should also write down any questions that you want to ask in class. When you read a mathematics textbook, some of the information in a section may be unclear. But when you hear a lecture or watch a videotape on that section, you will understand it much more easily than if you had not read your text beforehand.

Don't be afraid to ask questions. Instructors are not mind readers. Many times we do not know a concept is unclear until a student asks a question. You are not the only person in class with questions. Other students are normally grateful that someone has spoken up.

Hand in assignments on time. This way you can be sure that you will not lose points for being late. Show every step of a problem and be neat and organized. Also be sure that you understand which problems are assigned for homework. You can always double-check this assignment with another student in your class.

3 **Using This Text**

There are many helpful resources that are available to you in this text. It is important that you become familiar with and use these resources. They should increase your chances for success in this course.

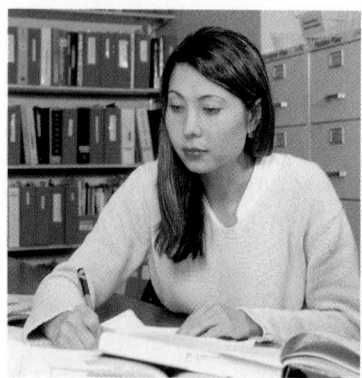

- The main section of exercises in each exercise set is referenced by an example(s). Use this referencing if you have trouble completing an assignment from the exercise set.

- If you need extra help in a particular section, look at the beginning of the section to see what videotapes and software are available.

- Make sure that you understand the meaning of the icons that are beside many exercises. The video icon 📹 tells you that the corresponding exercise may be viewed on the videotape that corresponds to that section. The pencil icon ✎ tells you that this exercise is a writing exercise in which you should answer in complete sentences. The △ icon tells you that the exercise involves geometry.

- Integrated Reviews in each chapter offer you a chance to practice—in one place—the many concepts that you have learned separately over several sections.

- There are many opportunities at the end of each chapter to help you understand the concepts of the chapter.

  Chapter Highlights contain chapter summaries and examples.

  Chapter Reviews contain review problems organized by section.

  Chapter Tests are sample tests to help you prepare for an exam.

  Cumulative Reviews are reviews consisting of material from the beginning of the book to the end of that particular chapter.

See the preface at the beginning of this text for a more thorough explanation of the features of this text.

④ Getting Help

If you have trouble completing assignments or understanding the mathematics, get help as soon as you need it! This tip is presented as an objective on its own because it is so important. In mathematics, usually the material presented in one section builds on your understanding of the previous section. This means that if you don't understand the concepts covered during a class period, there is a good chance that you will not understand the concepts covered during the next class period. If this happens to you, get help as soon as you can.

Where can you get help? Many suggestions have been made in the section on where to get help, and now it is up to you to do it. Try your instructor, a tutoring center, or a math lab, or you may want to form a study group with fellow classmates. If you do decide to see your instructor or go to a tutoring center, make sure that you have a neat notebook and are ready with your questions.

⑤ Preparing for and Taking an Exam

Make sure that you allow yourself plenty of time to prepare for a test. If you think that you are a little "math anxious," it may be that you are not preparing for a test in a way that will ensure success. The way that you prepare for a test in mathematics is important. To prepare for a test,

1. Review your previous homework assignments. You may also want to rework some of them.

2. Review any notes from class and section-level quizzes you have taken. (If this is a final exam, also review chapter tests you have taken.)

3. Review concepts and definitions by reading the Highlights at the end of each chapter.

4. Practice working out exercises by completing the Chapter Review found at the end of each chapter. (If this is a final exam, go through a Cumulative

Review. There is one found at the end of each chapter except Chapter 1. Choose the review found at the end of the latest chapter that you have covered in your course.) *Don't stop here!*

5. It is important that you place yourself in conditions similar to test conditions to find out how you will perform. In other words, as soon as you feel that you know the material, get a few blank sheets of paper and take a sample test. There is a Chapter Test available at the end of each chapter. During this sample test, do not use your notes or your textbook. Once you complete the Chapter Test, check your answers in the back of the book. If any answer is incorrect, there is a CD available with each exercise of each chapter test worked. Use this CD or your instructor to correct your sample test. Your instructor may also provide you with a review sheet. If you are not satisfied with the results, study the areas that you are weak in and try again.

6. Get a good night's sleep before the exam.

7. On the day of the actual test, allow yourself plenty of time to arrive at where you will be taking your exam.

When taking your test,

1. Read the directions on the test carefully.

2. Read each problem carefully as you take the test. Make sure that you answer the question asked.

3. Pace yourself by first completing the problems you are most confident with. Then work toward the problems you are least confident with. Watch your time so you do not spend too much time on one particular problem.

4. If you have time, check your work and answers.

5. Do not turn your test in early. If you have extra time, spend it double-checking your work.

6   **Managing Your Time**

As a college student, you know the demands that classes, homework, work, and family place on your time. Some days you probably wonder how you'll ever get everything done. One key to managing your time is developing a schedule. Here are some hints for making a schedule:

1. Make a list of all of your weekly commitments for the term. Include classes, work, regular meetings, extracurricular activities, etc. You may also find it helpful to list such things as laundry, regular workouts, grocery shopping, etc.

2. Next, estimate the time needed for each item on the list. Also make a note of how often you will need to do each item. Don't forget to include time estimates for reading, studying, and homework you do outside of your classes. You may want to ask your instructor for help estimating the time needed.

3. In the following exercise set, you are asked to block out a typical week on the schedule grid given. Start with items with fixed time slots like classes and work.

4. Next, include the items on your list with flexible time slots. Think carefully about how best to schedule some items such as study time.

5. Don't fill up every time slot on the schedule. Remember that you need to allow time for eating, sleeping, and relaxing! You should also allow a little extra time in case some items take longer than planned.

6. If you find that your weekly schedule is too full for you to handle, you may need to make some changes in your workload, classload, or in other areas of your life. You may want to talk to your advisor, manager or supervisor at work, or someone in your college's academic counseling center for help with such decisions.

Note: In this chapter, we begin a feature called Study Skills Reminder. The purpose of this feature is to remind you of some of the information given in this section and to further expand on some topics in this section.

EXERCISE SET 1.1

| STUDY | CD/ | PH MATH | MathXL®Tutorials | MathXL® | MyMathLab® |
| GUIDE/SSM | VIDEO | TUTOR CENTER | ON CD | | |

1. What is your instructor's name?
2. What are your instructor's office location and office hours?
3. What is the best way to contact your instructor?
4. What does the ＼ icon mean?
5. What does the 🔒 icon mean?
6. What does the △ icon mean?
7. Where are answers located in this text?
8. What Exercise Set answers are available to you in the answers section?
9. What Chapter Review, Chapter Test, and Cumulative Review answers are available to you in the answer section?
10. Are there worked-out solutions to exercises in this text?
11. If the answer to Exercise 10 is yes, what worked-out solutions are available to you in this text?
12. Go to the Highlights section at the end of this chapter. Describe how this section may be helpful to you when preparing for a test.

13. Do you have the name and contact information of at least one other student in class?
14. Will your instructor allow you to use a calculator in this class?
15. Are videotapes, CDs, and/or tutorial software available to you? If so, where?
16. Is there a tutoring service available? If so, what are its hours?
17. Have you attempted this course before? If so, write down ways that you might improve your chance of success during this second attempt.
18. List some steps that you can take if you begin having trouble understanding the material or completing an assignment.
19. Read or reread objective 6 and fill out the schedule grid below.
20. Study your filled-out grid from Exercise 19. Decide whether you have the time necessary to successfully complete this course and any other courses you may be registered for.

	Monday	*Tuesday*	*Wednesday*	*Thursday*	*Friday*	*Saturday*	*Sunday*
7:00 A.M.							
8:00 A.M.							
9:00 A.M.							
10:00 A.M.							
11:00 A.M.							
12:00 A.M.							
1:00 P.M.							
2:00 P.M.							
3:00 P.M.							
4:00 P.M.							
5:00 P.M.							
6:00 P.M.							
7:00 P.M.							
8:00 P.M.							
9:00 P.M.							

1.2   *SYMBOLS AND SETS OF NUMBERS*

Objectives

1 Use a number line to order numbers.

2 Translate sentences into mathematical statements.

3 Identify natural numbers, whole numbers, integers, rational numbers, irrational numbers, and real numbers.

4 Find the absolute value of a real number.

1   We begin with a review of the set of natural numbers and the set of whole numbers and how we use symbols to compare these numbers. A **set** is a collection of objects, each of which is called a **member** or **element** of the set. A pair of brace symbols { } encloses the list of elements and is translated as "the set of" or "the set containing."

Natural Numbers

The set of **natural numbers** is $\{1, 2, 3, 4, 5, 6, \ldots\}$.

Whole Numbers

The set of **whole numbers** is $\{0, 1, 2, 3, 4, \ldots\}$.

The three dots (an ellipsis) at the end of the list of elements of a set means that the list continues in the same manner indefinitely.

These numbers can be pictured on a **number line**. We will use number lines often to help us visualize distance and relationships between numbers. Visualizing mathematical concepts is an important skill and tool, and later we will develop and explore other visualizing tools.

To draw a number line, first draw a line. Choose a point on the line and label it 0. To the right of 0, label any other point 1. Being careful to use the same distance as from 0 to 1, mark off equally spaced distances. Label these points 2, 3, 4, 5, and so on. Since the whole numbers continue indefinitely, it is not possible to show every whole number on this number line. The arrow at the right end of the line indicates that the pattern continues indefinitely.

Picturing whole numbers on a number line helps us to see the order of the numbers. Symbols can be used to describe concisely in writing the order that we see.

The **equal symbol** $=$ means "is equal to."

The symbol $\neq$ means "is not equal to."

These symbols may be used to form a **mathematical statement**. The statement might be true or it might be false. The two statements below are both true.

$2 = 2$ states that "two is equal to two"

$2 \neq 6$ states that "two is not equal to six"

If two numbers are not equal, then one number is larger than the other. The symbol $>$ means "is greater than." The symbol $<$ means "is less than." For example,

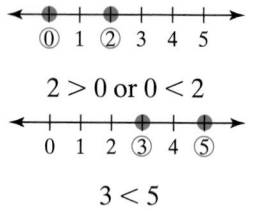

2 > 0 or 0 < 2

3 < 5

2 > 0   states that "two is greater than zero"

3 < 5   states that "three is less than five"

On a number line, we see that a number **to the right of** another number is **larger**. Similarly, a number **to the left of** another number is smaller. For example, 3 is to the left of 5 on a number line, which means that 3 is less than 5, or 3 < 5. Similarly, 2 is to the right of 0 on a number line, which means 2 is greater than 0, or 2 > 0. Since 0 is to the left of 2, we can also say that 0 is less than 2, or 0 < 2.

The symbols ≠, <, and > are called **inequality symbols**.

> ### Helpful Hint
>
> Notice that 2 > 0 has exactly the same meaning as 0 < 2. Switching the order of the numbers and reversing the "direction of the inequality symbol" does not change the meaning of the statement.
>
> 5 > 3 has the same meaning as 3 < 5.
>
> Also notice that, when the statement is true, the inequality arrow points to the smaller number.

EXAMPLE 1

Insert <, >, or = in the space between each pair of numbers to make each statement true.

a. 2   3   **b.** 7   4   **c.** 72   27

Solution   **a.** 2 < 3 since 2 is to the left of 3 on the number line.

b. 7 > 4 since 7 is to the right of 4 on the number line.

c. 72 > 27 since 72 is to the right of 27 on the number line.

Two other symbols are used to compare numbers. The symbol ≤ means "is less than or equal to." The symbol ≥ means "is greater than or equal to." For example,

7 ≤ 10 states that "seven is less than or equal to ten"

This statement is true since 7 < 10 is true. If either 7 < 10 or 7 = 10 is true, then 7 ≤ 10 is true.

3 ≥ 3 states that "three is greater than or equal to three"

This statement is true since 3 = 3 is true. If either 3 > 3 or 3 = 3 is true, then 3 ≥ 3 is true.

The statement 6 ≥ 10 is false since neither 6 > 10 nor 6 = 10 is true. The symbols ≤ and ≥ are also called **inequality symbols**.

EXAMPLE 2

Tell whether each statement is true or false.

a. $8 \geq 8$ **b.** $8 \leq 8$ **c.** $23 \leq 0$ **d.** $23 \geq 0$

Solution   **a.** True, since $8 = 8$ is true.   **b.** True, since $8 = 8$ is true.

c. False, since neither $23 < 0$ nor $23 = 0$ is true.   **d.** True, since $23 > 0$ is true.

2   Now, let's use the symbols discussed above to translate sentences into mathematical statements.

EXAMPLE 3

Translate each sentence into a mathematical statement.

a. Nine is less than or equal to eleven.

b. Eight is greater than one.

c. Three is not equal to four.

Solution

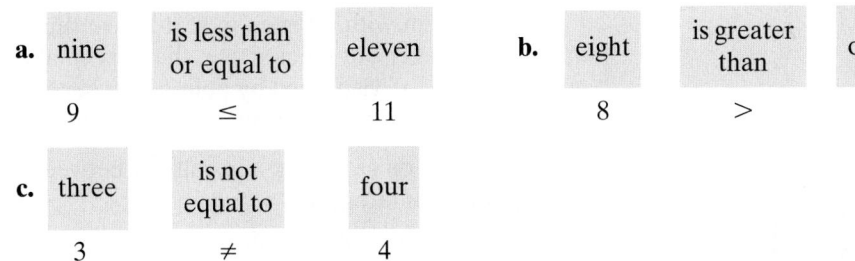

nine	is less than or equal to	eleven		eight	is greater than	one
9	$\leq$	11		8	$>$	1

a. **b.**

three	is not equal to	four
3	$\neq$	4

c.

3   Whole numbers are not sufficient to describe many situations in the real world. For example, quantities smaller than zero must sometimes be represented, such as temperatures less than 0 degrees.

We can picture numbers less than zero on a number line as follows:

Numbers less than 0 are to the left of 0 and are labeled $-1, -2, -3$, and so on. A $-$ sign, such as the one in -1, tells us that the number is to the left of 0 on a number line. In words, -1 is read "negative one." A $+$ sign or no sign tells us that a number lies to the right of 0 on the number line. For example, 3 and $+3$ both mean positive three.

The numbers we have pictured are called the set of **integers**. Integers to the left of 0 are called **negative integers**; integers to the right of 0 are called **positive integers**. The integer **0 is neither positive nor negative**.

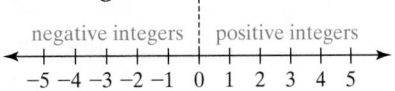

Integers

The set of **integers** is $\{ \ldots, -3, -2, -1, 0, 1, 2, 3, \ldots \}$.

Notice the ellipses (three dots) to the left and to the right of the list for the integers. This indicates that the positive integers and the negative integers continue indefinitely.

EXAMPLE 4

Use an integer to express the number in the following. "Pole of Inaccessibility, Antarctica, is the coldest location in the world, with an average annual temperature of 72 degrees below zero." (*Source: The Guinness Book of Records*)

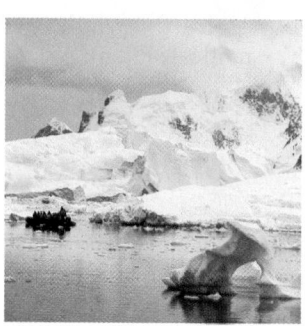

Solution   The integer -72 represents 72 degrees below zero.

A problem with integers in real-life settings arises when quantities are smaller than some integer but greater than the next smallest integer. On a number line, these quantities may be visualized by points between integers. Some of these quantities between integers can be represented as a quotient of integers. For example,

The point on a number line halfway between 0 and 1 can be represented by $\frac{1}{2}$, a quotient of integers.

The point on a number line halfway between 0 and -1 can be represented by $-\frac{1}{2}$. Other quotients of integers and their graphs are shown.

The set numbers, each of which can be represented as a quotient of integers, is called the set of **rational numbers**. Notice that every integer is also a rational number since each integer can be expressed as a quotient of integers. For example, the integer 5 is also a rational number since $5 = \frac{5}{1}$.

Rational Numbers

The set of **rational numbers** is the set of all numbers that can be expressed as a quotient of integers with denominator not zero.

The number line also contains points that cannot be expressed as quotients of integers. These numbers are called **irrational numbers** because they cannot be represented by rational numbers. For example, $\sqrt{2}$ and π are irrational numbers.

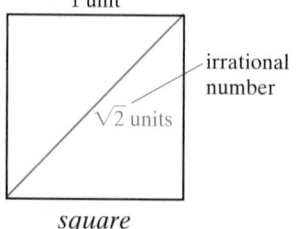

square

Irrational Numbers

The set of **irrational numbers** is the set of all numbers that correspond to points on the number line but that are not rational numbers. That is, an irrational number is a number that cannot be expressed as a quotient of integers.

Rational numbers and irrational numbers can be written as decimal numbers. The decimal equivalent of a rational number will either terminate or repeat in a pattern. For

example, upon dividing we find that

$$\frac{3}{4} = 0.75 \text{ (decimal number terminates or ends) and}$$

$$\frac{2}{3} = 0.66666\ldots \text{ (decimal number repeats in a pattern)}$$

The decimal representation of an irrational number will neither terminate nor repeat. For example, the decimal representations of irrational numbers $\sqrt{2}$ and π are

$$\sqrt{2} = 1.414213562\ldots \text{ (decimal number does not terminate or repeat in a pattern)}$$

$$\pi = 3.141592653\ldots \text{ (decimal number does not terminate or repeat in a pattern)}$$

(For further review of decimals, see the Appendix.)

Combining the natural numbers with the irrational numbers gives the set of **real numbers**. One and only one point on a number line corresponds to each real number.

Real Numbers

The set of **real numbers** is the set of all numbers each of which corresponds to a point on a number line.

On the following number line, we see that real numbers can be positive, negative, or 0. Numbers to the left of 0 are called **negative numbers**; numbers to the right of 0 are called **positive numbers**. Positive and negative numbers are also called **signed numbers**.

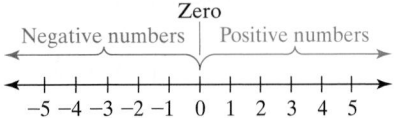

Several different sets of numbers have been discussed in this section. The following diagram shows the relationships among these sets of real numbers.

Common Sets of Numbers

STUDY SKILLS REMINDER

Drawing a diagram, like the one shown on the previous page, for yourself in your study notes will help you understand and remember mathematical relationships.

EXAMPLE 5

Given the set $\left\{-2, 0, \frac{1}{4}, -1.5, 112, -3, 11, \sqrt{2}\right\}$, list the numbers in this set that belong to the set of:

a. Natural numbers   **b.** Whole numbers   **c.** Integers

d. Rational numbers   **e.** Irrational numbers   **f.** Real numbers

Solution   **a.** The natural numbers are 11 and 112.

b. The whole numbers are 0, 11, and 112.

c. The integers are $-3, -2, 0, 11,$ and 112.

d. Recall that integers are rational numbers also. The rational numbers are $-3, -2, -1.5, 0, \frac{1}{4}, 11,$ and 112.

e. The irrational number is $\sqrt{2}$.

f. The real numbers are all numbers in the given set.

We can now extend the meaning and use of inequality symbols such as $<$ and $>$ to apply to all real numbers.

Order Property for Real Numbers

Given any two real numbers a and b, $a < b$ if a is to the left of b on a number line. Similarly, $a > b$ if a is to the right of b on a number line.

EXAMPLE 6

Insert $<$, $>$, or $=$ in the appropriate space to make each statement true.

a. $-1 \quad 0$ **b.** $7 \quad \frac{14}{2}$ **c.** $-5 \quad -6$

Solution   **a.** $-1 < 0$ since -1 is to the left of 0 on a number line.

$$-1 < 0$$

b. $7 = \frac{14}{2}$ since $\frac{14}{2}$ simplifies to 7.

c. $-5 > -6$ since -5 is to the right of -6 on the number line.

$$-5 > -6$$

4   A number line not only gives us a picture of the real numbers, it also helps us visualize the distance between numbers. The distance between a real number a and 0 is given a special name called the **absolute value** of a. "The absolute value of a" is written in symbols as $|a|$.

Absolute Value

The absolute value of a real number a, denoted by $|a|$, is the distance between a and 0 on a number line.

For example, $|3| = 3$ and $|-3| = 3$ since both 3 and -3 are a distance of 3 units from 0 on a number line.

> ### Helpful Hint
> Since $|a|$ is a distance, $|a|$ is always either positive or 0, never negative. That is, **for any real number a, $|a| \geq 0$.**

  EXAMPLE 7

Find the absolute value of each number.

a. $|4|$ **b.** $|-5|$ **c.** $|0|$ **d.** $\left|-\frac{1}{2}\right|$ **e.** $|5.6|$

Solution   **a.** $|4| = 4$ since 4 is 4 units from 0 on a number line.

b. $|-5| = 5$ since -5 is 5 units from 0 on a number line.

c. $|0| = 0$ since 0 is 0 units from 0 on a number line.

d. $\left|-\frac{1}{2}\right| = \frac{1}{2}$ since $-\frac{1}{2}$ is $\frac{1}{2}$ units from 0 on a number line.

e. $|5.6| = 5.6$ since 5.6 is 5.6 units from 0 on a number line.

EXAMPLE 8

Insert $<$, $>$, or $=$ in the appropriate space to make each statement true.

a. $|0|$ 2   **b.** $|-5|$ 5   **c.** $|-3|$ $|-2|$ **d.** $|5|$ $|6|$ **e.** $|-7|$ $|6|$

Solution   **a.** $|0| < 2$ since $|0| = 0$ and $0 < 2$.

b. $|-5| = 5$ since $5 = 5$.

c. $|-3| > |-2|$ since $3 > 2$.

d. $|5| < |6|$ since $5 < 6$.

e. $|-7| > |6|$ since $7 > 6$.

Spotlight on

DECISION
⌐ MAKING

Suppose you are a quality control engineering technician in a factory that makes machine screws. You have just helped to install programmable machinery on the production line that measures the length of each screw. If a screw's length is greater than 4.05 centimeters or less than or equal to 3.98 centimeters, the machinery is programmed to discard the screw. To check that the machinery works properly, you test six screws with known lengths. The results of the test are displayed. Is the new machinery working properly? Explain.

TEST RESULTS

Test Screw	Actual Length of Test Screw (cm)	Machine Action on Test Screw
A	4.03	Accept
B	3.96	Reject
C	4.05	Accept
D	4.08	Reject
E	3.98	Reject
F	4.01	Accept

EXERCISE SET 1.2

| STUDY GUIDE/SSM | CD/ VIDEO | PH MATH TUTOR CENTER | MathXL®Tutorials ON CD | MathXL® | MyMathLab® |

Insert <, >, or = in the appropriate space to make the statement true. See Example 1.

 1. 7   3

2. 9   15

3. 6.26   6.26

4. 2.13   1.13

5. 0   7

6. 20   0

7. −2   2

8. −4   −6

9. The freezing point of water is 32° Fahrenheit. The boiling point of water is 212° Fahrenheit. Write an inequality statement using < or > comparing the numbers 32 and 212.

10. The freezing point of water is 0° Celsius. The boiling point of water is 100° Celsius. Write an inequality statement using < or > comparing the numbers 0 and 100.

11. The average salary in the United States for an experienced registered nurse is $44,300. The average salary for a drafter is $34,611. Write an inequality statement using < or > comparing the numbers 44,300 and 34,611. (*Source*: U.S. Department of Labor)

12. The state of New York is home to 312 institutions of higher learning. California claims a total of 384 colleges and universities. Write an inequality statement using < or > comparing the numbers 312 and 384. (*Source*: U.S. Department of Education)

Are the following statements true or false? See Example 2.

 13. $11 \leq 11$

14. $4 \geq 7$

15. $10 > 11$

16. $17 > 16$

17. $3 + 8 \geq 3(8)$

18. $8 \cdot 8 \leq 8 \cdot 7$

19. $7 > 0$

20. $4 < 7$

21. An angle measuring 30° is shown and an angle measuring 45° is shown. Use the inequality symbol $\leq$ or $\geq$ to write a statement comparing the numbers 30 and 45.

22. The sum of the measures of the angles of a triangle is 180°. The sum of the measures of the angles of a parallelogram is 360°. Use the inequality symbol $\leq$ or $\geq$ to write a statement comparing the numbers 360 and 180.

Write each sentence as a mathematical statement. See Example 3.

23. Eight is less than twelve.

24. Fifteen is greater than five.

25. Five is greater than or equal to four.

26. Negative ten is less than or equal to thirty-seven.

27. Fifteen is not equal to negative two.

28. Negative seven is not equal to seven.

Use integers to represent the values in each statement. See Example 4.

29. Driskill Mountain, in Louisiana, has an altitude of 535 feet. New Orleans, Louisiana, lies 8 feet below sea level. (*Source*: U.S. Geological Survey)

30. During a Green Bay Packers football game, the team gained 23 yards and then lost 12 yards on consecutive plays.

31. From 1990 to 2000, the population of Washington, D.C. decreased by 34,841. (*Source: 2003 World Almanac*)

32. From 1990 to 2000, the population of the state of Alaska rose by 406,513. (*Source: 2003 World Almanac*)

33. Aaron Miller deposited $350 in his savings account. He later withdrew $126.

34. Aris Peña was deep-sea diving. During her dive, she ascended 30 feet and later descended 50 feet.

The graph below is called a bar graph. This particular graph shows the average monthly home mortgage payment in the United States. Each bar represents a different year and the height of the bar represents the mortgage payment (principal and interest).

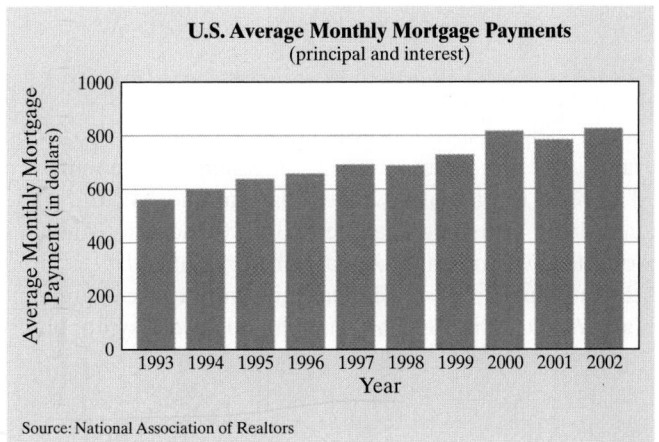

U.S. Average Monthly Mortgage Payments
(principal and interest)

Source: National Association of Realtors

35. In which year was the average mortgage payment the least?

36. What is the greatest mortgage payment shown?

37. In what years was the average mortgage payment less than $600?

38. In what years was the average mortgage payment greater than $800?

39. Write an inequality statement using $\leq$ or $\geq$ comparing the average mortgage payment for 2002 and 2000.

40. Do you notice any trends shown by this bar graph?

Tell which set or sets each number belongs to: natural numbers, whole numbers, integers, rational numbers, irrational numbers, and real numbers. See Example 5.

41. 0   **42.** $\frac{1}{4}$ **43.** -2 **44.** $-\frac{1}{2}$

45. 6   **46.** 5   **47.** $\frac{2}{3}$ **48.** $\sqrt{3}$

49. $-\sqrt{5}$ **50.** $-1\frac{5}{9}$

Tell whether each statement is true or false.

51. Every rational number is also an integer.

52. Every negative number is also a rational number.

53. Every natural number is positive.

54. Every rational number is also a real number.

55. 0 is a real number.

56. Every real number is also a rational number.

57. Every whole number is an integer.

58. $\frac{1}{2}$ is an integer.

59. A number can be both rational and irrational.

60. Every whole number is positive.

Insert $<$, $>$, or $=$ in the appropriate space to make a true statement. See Examples 6 through 8.

61. -10 -100 **62.** -200 -20 **63.** 32 5.2

64. 7 -7 **65.** $\frac{18}{3}$ $\frac{24}{3}$ **66.** $\frac{8}{2}$ $\frac{12}{3}$

67. -51 -50 **68.** $|-20|$ -200 **69.** $|-5|$ -4

70. 0 $|0|$ **71.** $|-1|$ $|1|$ **72.** $\left|\frac{2}{5}\right|$ $\left|-\frac{2}{5}\right|$

73. $|-2|$ $|-3|$ **74.** -500 $|-50|$ **75.** $|0|$ $|-8|$

76. $|-12|$ $\frac{24}{2}$

Concept Extensions

The apparent magnitude of a star is the measure of its brightness as seen by someone on Earth. The smaller the apparent magnitude, the brighter the star. Use the apparent magnitudes in the table to answer Exercises 77 through 82.

Star	Apparent Magnitude	Star	Apparent Magnitude
Arcturus	-0.04	Spica	0.98
Sirius	-1.46	Rigel	0.12
Vega	0.03	Regulus	1.35
Antares	0.96	Canopus	-0.72
Sun	-26.7	Hadar	0.61

(*Source: Norton's 2000: Star Atlas and Reference Handbook*, 18th ed., Longman Group, UK, 1989)

77. The apparent magnitude of the sun is −26.7. The apparent magnitude of the star Arcturus is −0.04. Write an inequality statement comparing the numbers −0.04 and −26.7.

78. The apparent magnitude of Antares is 0.96. The apparent magnitude of Spica is 0.98. Write an inequality statement comparing the numbers 0.96 and 0.98.

79. Which is brighter, the sun or Arcturus?

80. Which is dimmer, Antares or Spica?

81. Which star listed is the brightest?

82. Which star listed is the dimmest?

Rewrite the following inequalities so that the inequality symbol points in the opposite direction and the resulting statement has the same meaning as the given one.

83. $25 \geq 20$

84. $-13 \leq 13$

85. $0 < 6$

86. $5 > 3$

87. $-10 > -12$

88. $-4 < -2$

89. In your own words, explain how to find the absolute value of a number.

90. Give an example of a real-life situation that can be described with integers but not with whole numbers.

1.3 FRACTIONS

Objectives

1 Write fractions in simplest form.

2 Multiply and divide fractions.

3 Add and subtract fractions.

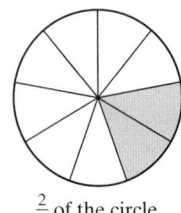

$\frac{2}{9}$ of the circle is shaded.

1   A quotient of two numbers such as $\frac{2}{9}$ is called a **fraction**. In the fraction $\frac{2}{9}$, the top number, 2, is called the **numerator** and the bottom number, 9, is called the **denominator**.

A fraction may be used to refer to part of a whole. For example, $\frac{2}{9}$ of the circle below is shaded. The denominator 9 tells us how many equal parts the whole circle is divided into and the numerator 2 tells us how many equal parts are shaded. To simplify fractions, we can factor the numerator and the denominator. In the statement $3 \cdot 5 = 15$, 3 and 5 are called **factors** and 15 is the **product**. (The raised dot symbol indicates multiplication.)

$$3 \quad \cdot \quad 5 \quad = \quad 15$$
$$\uparrow \qquad\qquad \uparrow \qquad\qquad\quad \uparrow$$
$$\text{factor} \qquad \text{factor} \qquad \text{product}$$

To **factor** 15 means to write it as a product. The number 15 can be factored as $3 \cdot 5$ or as $1 \cdot 15$.

A fraction is said to be **simplified** or in **lowest terms** when the numerator and the denominator have no factors in common other than 1. For example, the fraction $\frac{5}{11}$ is in lowest terms since 5 and 11 have no common factors other than 1.

To help us simplify fractions, we write the numerator and the denominator as a product of **prime numbers**.

Prime Number

A prime number is a whole number, other than 1, whose only factors are 1 and itself. The first few prime numbers are

$$2, 3, 5, 7, 11, 13, 17, 19, 23, 29, \text{ and so on.}$$

A natural number, other than 1, that is not a prime number is called a **composite number**. Every composite number can be written as a product of prime numbers. We call this product of prime numbers the prime factorization of the composite number.

EXAMPLE 1

Write each of the following numbers as a product of primes.

a. 40   **b.** 63

Solution   **a.** First, write 40 as the product of any two whole numbers, other than 1.

$$40 = 4 \cdot 10$$

Next, factor each of these numbers. Continue this process until all of the factors are prime numbers.

$$40 = \quad 4 \quad \cdot \quad 10$$
$$= 2 \cdot 2 \cdot 2 \cdot 5$$

All the factors are now prime numbers. Then 40 written as a product of primes is

$$40 = 2 \cdot 2 \cdot 2 \cdot 5$$

b. $63 = \quad 9 \cdot 7$
$$= 3 \cdot 3 \cdot 7$$

To use prime factors to write a fraction in lowest terms, apply the fundamental principle of fractions.

Fundamental Principle of Fractions

If $\dfrac{a}{b}$ is a fraction and c is a nonzero real number, then

$$\frac{a \cdot c}{b \cdot c} = \frac{a}{b}$$

To understand why this is true, we use the fact that since c is not zero, then $\dfrac{c}{c} = 1$.

$$\frac{a \cdot c}{b \cdot c} = \frac{a}{b} \cdot \frac{c}{c} = \frac{a}{b} \cdot 1 = \frac{a}{b}$$

We will call this process dividing out the common factor of c.

EXAMPLE 2

Write each fraction in lowest terms.

a. $\dfrac{42}{49}$ **b.** $\dfrac{11}{27}$ **c.** $\dfrac{88}{20}$

Solution   **a.** Write the numerator and the denominator as products of primes; then apply the fundamental principle to the common factor 7.

$$\frac{42}{49} = \frac{2 \cdot 3 \cdot 7}{7 \cdot 7} = \frac{2 \cdot 3}{7} = \frac{6}{7}$$

b. $\dfrac{11}{27} = \dfrac{11}{3 \cdot 3 \cdot 3}$

There are no common factors other than 1, so $\dfrac{11}{27}$ is already in lowest terms.

c. $\dfrac{88}{20} = \dfrac{2 \cdot 2 \cdot 2 \cdot 11}{2 \cdot 2 \cdot 5} = \dfrac{22}{5}$

✔ **CONCEPT CHECK**

Explain the error in the following steps.

a. $\dfrac{15}{55} = \dfrac{1\,5}{5\,5} = \dfrac{1}{5}$ **b.** $\dfrac{6}{7} = \dfrac{5+1}{5+2} = \dfrac{1}{2}$

2   To multiply two fractions, multiply numerator times numerator to obtain the numerator of the product; multiply denominator times denominator to obtain the denominator of the product.

Multiplying Fractions

$$\frac{a}{b} \cdot \frac{c}{d} = \frac{a \cdot c}{b \cdot d}, \qquad \text{if } b \neq 0 \text{ and } d \neq 0$$

EXAMPLE 3

Multiply $\dfrac{2}{15}$ and $\dfrac{5}{13}$. Write the product in lowest terms.

Solution

$$\frac{2}{15} \cdot \frac{5}{13} = \frac{2 \cdot 5}{15 \cdot 13} \qquad \text{Multiply numerators.}$$
$$\text{Multiply denominators.}$$

Next, simplify the product by dividing the numerator and the denominator by any common factors.

$$= \frac{2 \cdot 5}{3 \cdot 5 \cdot 13}$$
$$= \frac{2}{39}$$

Before dividing fractions, we first define **reciprocals**. Two fractions are reciprocals of each other if their product is 1. For example $\frac{2}{3}$ and $\frac{3}{2}$ are reciprocals since $\frac{2}{3} \cdot \frac{3}{2} = 1$. Also, the reciprocal of 5 is $\frac{1}{5}$ since $5 \cdot \frac{1}{5} = \frac{5}{1} \cdot \frac{1}{5} = 1$.

To divide fractions, multiply the first fraction by the reciprocal of the second fraction.

Dividing Fractions

$$\frac{a}{b} \div \frac{c}{d} = \frac{a}{b} \cdot \frac{d}{c}, \qquad \text{if } b \neq 0, d \neq 0, \text{ and } c \neq 0$$

EXAMPLE 4

Divide. Write all quotients in lowest terms.

a. $\dfrac{4}{5} \div \dfrac{5}{16}$ **b.** $\dfrac{7}{10} \div 14$ **c.** $\dfrac{3}{8} \div \dfrac{3}{10}$

Solution   **a.** $\dfrac{4}{5} \div \dfrac{5}{16} = \dfrac{4}{5} \cdot \dfrac{16}{5} = \dfrac{4 \cdot 16}{5 \cdot 5} = \dfrac{64}{25}$

b. $\dfrac{7}{10} \div 14 = \dfrac{7}{10} \div \dfrac{14}{1} = \dfrac{7}{10} \cdot \dfrac{1}{14} = \dfrac{7 \cdot 1}{2 \cdot 5 \cdot 2 \cdot 7} = \dfrac{1}{20}$.

c. $\dfrac{3}{8} \div \dfrac{3}{10} = \dfrac{3}{8} \cdot \dfrac{10}{3} = \dfrac{3 \cdot 2 \cdot 5}{2 \cdot 2 \cdot 2 \cdot 3} = \dfrac{5}{4}$

3   To add or subtract fractions with the same denominator, combine numerators and place the sum or difference over the common denominator.

Adding and Subtracting Fractions with the Same Denominator

$$\frac{a}{b} + \frac{c}{b} = \frac{a + c}{b}, \qquad \text{if } b \neq 0$$

$$\frac{a}{b} - \frac{c}{b} = \frac{a - c}{b}, \qquad \text{if } b \neq 0$$

EXAMPLE 5

Add or subtract as indicated. Write each result in lowest terms.

a. $\dfrac{2}{7} + \dfrac{4}{7}$ **b.** $\dfrac{3}{10} + \dfrac{2}{10}$ **c.** $\dfrac{9}{7} - \dfrac{2}{7}$ **d.** $\dfrac{5}{3} - \dfrac{1}{3}$

Solution   **a.** $\dfrac{2}{7} + \dfrac{4}{7} = \dfrac{2 + 4}{7} = \dfrac{6}{7}$ **b.** $\dfrac{3}{10} + \dfrac{2}{10} = \dfrac{3 + 2}{10} = \dfrac{5}{10} = \dfrac{5}{2 \cdot 5} = \dfrac{1}{2}$

c. $\dfrac{9}{7} - \dfrac{2}{7} = \dfrac{9 - 2}{7} = \dfrac{7}{7} = 1$ **d.** $\dfrac{5}{3} - \dfrac{1}{3} = \dfrac{5 - 1}{3} = \dfrac{4}{3}$

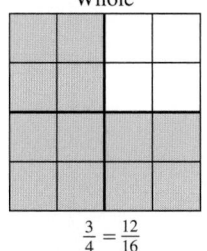

Whole

$\dfrac{3}{4} = \dfrac{12}{16}$

To add or subtract fractions without the same denominator, first write the fractions as **equivalent fractions** with a common denominator. Equivalent fractions are fractions that represent the same quantity. For example, $\frac{3}{4}$ and $\frac{12}{16}$ are equivalent fractions since they represent the same portion of a whole, as the diagram shows. Count the larger squares and the shaded portion is $\frac{3}{4}$. Count the smaller squares and the shaded portion is $\frac{12}{16}$. Thus, $\frac{3}{4} = \frac{12}{16}$.

We can write equivalent fractions by multiplying a given fraction by 1, as shown in the next example. Multiplying a fraction by 1 does not change the value of the fraction.

EXAMPLE 6

Write $\dfrac{2}{5}$ as an equivalent fraction with a denominator of 20.

Solution   Since $5 \cdot 4 = 20$, multiply the fraction by $\dfrac{4}{4}$. Multiplying by $\dfrac{4}{4} = 1$ does not change the value of the fraction.

Multiply by $\dfrac{4}{4}$ or **1**.

$$\frac{2}{5} = \frac{2}{5} \cdot \frac{4}{4} = \frac{2 \cdot 4}{5 \cdot 4} = \frac{8}{20}$$

EXAMPLE 7

Add or subtract as indicated. Write each answer in lowest terms.

a. $\dfrac{2}{5} + \dfrac{1}{4}$ **b.** $\dfrac{1}{2} + \dfrac{17}{22} - \dfrac{2}{11}$ **c.** $3\dfrac{1}{6} - 1\dfrac{11}{12}$

Solution   **a.** Fractions must have a common denominator before they can be added or subtracted. Since 20 is the smallest number that both 5 and 4 divide into evenly, 20 is the **least common denominator**. Write both fractions as equivalent fractions with denominators of 20. Since

$$\frac{2}{5} \cdot \frac{4}{4} = \frac{2 \cdot 4}{5 \cdot 4} = \frac{8}{20} \quad \text{and} \quad \frac{1}{4} \cdot \frac{5}{5} = \frac{1 \cdot 5}{4 \cdot 5} = \frac{5}{20}$$

then

$$\frac{2}{5} + \frac{1}{4} = \frac{8}{20} + \frac{5}{20} = \frac{13}{20}$$

b. The least common denominator for denominators 2, 22, and 11 is 22. First, write each fraction as an equivalent fraction with a denominator of 22. Then add or subtract from left to right.

$$\frac{1}{2} = \frac{1}{2} \cdot \frac{11}{11} = \frac{11}{22}, \quad \frac{17}{22} = \frac{17}{22}, \quad \text{and} \quad \frac{2}{11} = \frac{2}{11} \cdot \frac{2}{2} = \frac{4}{22}$$

Then

$$\frac{1}{2} + \frac{17}{22} - \frac{2}{11} = \frac{11}{22} + \frac{17}{22} - \frac{4}{22} = \frac{24}{22} = \frac{12}{11}$$

c. To find $3\dfrac{1}{6} - 1\dfrac{11}{12}$, lets use a vertical format.

$$
\begin{array}{ccccc}
 & & & \overset{\displaystyle 2+1\frac{2}{12}}{\frown} & \\
3\dfrac{1}{6} & = & 3\dfrac{2}{12} & = & 2\dfrac{14}{12} \\
-1\dfrac{11}{12} & = & -1\dfrac{11}{12} & = & -1\dfrac{11}{12} \\
\hline
 & & \underset{\substack{\uparrow \\ \text{Need to} \\ \text{borrow}}}{} & & 1\dfrac{3}{12} \text{ or } 1\dfrac{1}{4}
\end{array}
$$

Spotlight on

DECISION MAKING

Suppose you are fishing on a freshwater lake in Canada. You catch a whitefish weighing $14\dfrac{5}{32}$ pounds. According to the International Game Fish Association, the world's record for largest lake whitefish ever caught is $14\dfrac{3}{8}$ pounds. Did you set a new world's record? Explain. By how much did you beat or miss the existing world record?

MENTAL MATH

Represent the shaded part of each geometric figure by a fraction.

1.   **2.**   **3.**   **4.**

For Exercises 5 and 6, fill in the blank.

5. In the fraction $\frac{3}{5}$, 3 is called the _____ and 5 is called the _____.

6. The reciprocal of $\frac{7}{11}$ is ___.

EXERCISE SET 1.3

STUDY   CD/   PH MATH   MathXL®Tutorials   MathXL®   MyMathLab®
GUIDE/SSM   VIDEO   TUTOR CENTER   ON CD

Write each number as a product of primes. See Example 1.
1. 33   **2.** 60   **3.** 98   **4.** 27   **5.** 20
6. 56   **7.** 75   **8.** 32   **9.** 45   **10.** 24

Write the fraction in lowest terms. See Example 2.
11. $\frac{2}{4}$ **12.** $\frac{3}{6}$ **13.** $\frac{10}{15}$ **14.** $\frac{15}{20}$
15. $\frac{3}{7}$ **16.** $\frac{5}{9}$ **17.** $\frac{18}{30}$ **18.** $\frac{42}{45}$

Multiply or divide as indicated. Write the answer in lowest terms. See Examples 3 and 4.
19. $\frac{1}{2} \cdot \frac{3}{4}$ **20.** $\frac{10}{6} \cdot \frac{3}{5}$ **21.** $\frac{2}{3} \cdot \frac{3}{4}$
22. $\frac{7}{8} \cdot \frac{3}{21}$ **23.** $\frac{1}{2} \div \frac{7}{12}$ **24.** $\frac{7}{12} \div \frac{1}{2}$
25. $\frac{3}{4} \div \frac{1}{20}$ **26.** $\frac{3}{5} \div \frac{9}{10}$ **27.** $\frac{7}{10} \cdot \frac{5}{21}$
28. $\frac{3}{35} \cdot \frac{10}{63}$ **29.** $2\frac{7}{9} \cdot \frac{1}{3}$ **30.** $\frac{1}{4} \cdot 5\frac{5}{6}$

The area of a plane figure is a measure of the amount of surface of the figure. Find the area of each figure below. (The area of a rectangle is the product of its length and width. The area of a triangle is $\frac{1}{2}$ the product of its base and height.)

△ **31.**   △ **32.**

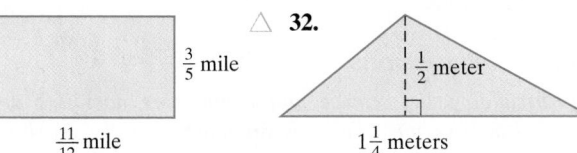

Add or subtract as indicated. Write the answer in lowest terms. See Example 5.
33. $\frac{4}{5} - \frac{1}{5}$ **34.** $\frac{6}{7} - \frac{1}{7}$ **35.** $\frac{4}{5} + \frac{1}{5}$
36. $\frac{6}{7} + \frac{1}{7}$ **37.** $\frac{17}{21} - \frac{10}{21}$ **38.** $\frac{18}{35} - \frac{11}{35}$

39. $\frac{23}{105} + \frac{4}{105}$ **40.** $\frac{13}{132} + \frac{35}{132}$

Write each fraction as an equivalent fraction with the given denominator. See Example 6.
41. $\frac{7}{10}$ with a denominator of 30
42. $\frac{2}{3}$ with a denominator of 9
43. $\frac{2}{9}$ with a denominator of 18
44. $\frac{8}{7}$ with a denominator of 56
45. $\frac{4}{5}$ with a denominator of 20
46. $\frac{4}{5}$ with a denominator of 25

Add or subtract as indicated. Write the answer in lowest terms. See Example 7.
47. $\frac{2}{3} + \frac{3}{7}$ **48.** $\frac{3}{4} + \frac{1}{6}$ **49.** $2\frac{13}{15} - 1\frac{1}{5}$
50. $5\frac{2}{9} - 3\frac{1}{6}$ **51.** $\frac{5}{22} - \frac{5}{33}$ **52.** $\frac{7}{10} - \frac{8}{15}$
53. $\frac{12}{5} - 1$ **54.** $2 - \frac{3}{8}$

Each circle below represents a whole, or 1. Use subtraction to determine the unknown part of the circle.

55.   **56.**

57.

58.

59.

60.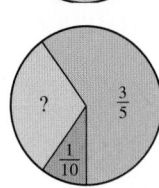

MIXED PRACTICE

Perform the following operations. Write answers in lowest terms.

61. $\dfrac{10}{21} + \dfrac{5}{21}$

62. $\dfrac{11}{35} + \dfrac{3}{35}$

 63. $\dfrac{10}{3} - \dfrac{5}{21}$

64. $\dfrac{11}{7} - \dfrac{3}{35}$

65. $\dfrac{2}{3} \cdot \dfrac{3}{5}$

66. $\dfrac{2}{3} \div \dfrac{3}{4}$

67. $\dfrac{3}{4} \div \dfrac{7}{12}$

68. $\dfrac{3}{5} + \dfrac{2}{3}$

69. $\dfrac{5}{12} + \dfrac{4}{12}$

70. $\dfrac{2}{7} + \dfrac{4}{7}$

71. $5 + \dfrac{2}{3}$

72. $7 + \dfrac{1}{10}$

73. $\dfrac{7}{8} \div 3\dfrac{1}{4}$

74. $3 \div \dfrac{3}{4}$

75. $\dfrac{7}{18} \div \dfrac{14}{36}$

76. $4\dfrac{3}{7} \div \dfrac{31}{7}$

77. $\dfrac{23}{105} - \dfrac{2}{105}$

78. $\dfrac{57}{132} - \dfrac{13}{132}$

 79. $1\dfrac{1}{2} + 3\dfrac{2}{3}$

80. $2\dfrac{3}{5} + 4\dfrac{7}{10}$

81. $\dfrac{2}{3} - \dfrac{5}{9} + \dfrac{5}{6}$

82. $\dfrac{8}{11} - \dfrac{1}{4} + \dfrac{1}{2}$

The perimeter of a plane figure is the total distance around the figure. Find the perimeter of each figure in Exercises 83 and 84.

△ **83.**

△ **84.**

85. Svetlana Feofanova currently holds the Women's Indoor Pole Vault world record at $4\dfrac{3}{4}$ meters. The Men's Indoor Pole Vault world record is currently held by Sergei Bubka, at $1\dfrac{2}{5}$ meters higher than the Women's record. What is the current Men's Indoor Pole Vault Record? (*Source: 2003 World Almanac*)

86. In March 1999, a proposal to increase the size of rectangular escape vents in lobster traps was brought before the Maine state legislature. The proposed change would add $\dfrac{1}{16}$ of an inch to the current vent height of $1\dfrac{7}{8}$ inches. What would be the new vent height under the proposal? (*Source: The Boston Sunday Globe*, April 4, 1999)

87. In your own words, explain how to add two fractions with different denominators.

88. In your own words, explain how to multiply two fractions.

The following trail chart is given to visitors at the Lakeview Forest Preserve.

Trail Name	Distance (miles)
Robin Path	$3\dfrac{1}{2}$
Red Falls	$5\dfrac{1}{2}$
Green Way	$2\dfrac{1}{8}$
Autumn Walk	$1\dfrac{3}{4}$

89. How much longer is Red Falls Trail than Green Way Trail?

90. Find the total distance traveled by someone who hiked along all four trails.

Concept Extensions

The breakdown of science and engineering doctorate degrees awarded in the United States is summarized in the graph on the next page, called a circle graph or a pie chart. Use the graph to answer the questions. (Source: National Science Foundation)

91. What fraction of science and engineering doctorates are awarded in the physical sciences?

92. Engineering doctorates make up what fraction of all science and engineering doctorates awarded in the United States?

Science and Engineering Doctorates Awarded, by Field of Study

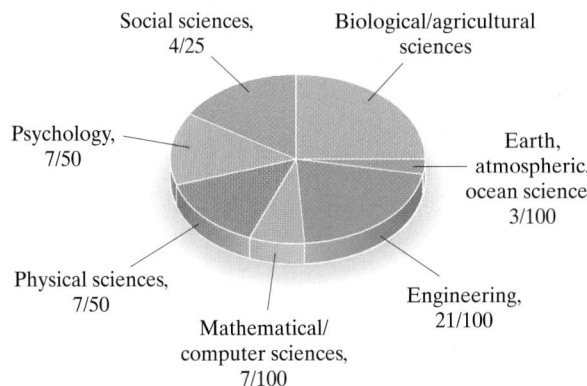

95. What fraction of Gap-brand stores were Old Navy stores? Simplify this fraction.

96. What fraction of Gap-brand stores were either domestic or international Gap stores? Simplify this fraction.

The area of a plane figure is a measure of the amount of surface of the figure. Find the area of each figure. (The area of a triangle is $\frac{1}{2}$ the product of its base and height. The area of a rectangle is the product of its length and width. Recall that area is measured in square units.)

△ **97.**

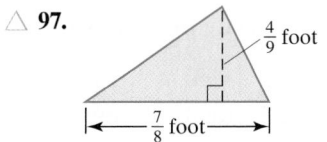

93. What fraction of all science and engineering doctorates are awarded in the biological and agricultural sciences?

94. Social sciences and psychology doctorates together make up what fraction of all science and engineering doctorates awarded in the United States?

△ **98.**

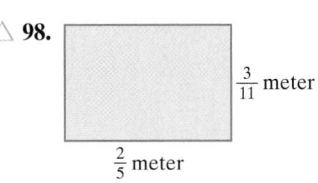

As of February 2001, Gap Inc. operated a total of 3678 stores worldwide. The following chart shows the store breakdown by brand. (Source: Gap Inc.)

Brand	Number of Stores
Gap (Domestic)	2079
Gap (International)	529
Banana Republic	404
Old Navy	666
Total	3678

1.4   INTRODUCTION TO VARIABLE EXPRESSIONS AND EQUATIONS

Objectives

1. Define and use exponents and the order of operations.
2. Evaluate algebraic expressions, given replacement values for variables.
3. Determine whether a number is a solution of a given equation.
4. Translate phrases into expressions and sentences into equations.

1   Frequently in algebra, products occur that contain repeated multiplication of the same factor. For example, the volume of a cube whose sides each measure 2 centimeters is $(2 \cdot 2 \cdot 2)$ cubic centimeters. We may use **exponential notation** to write such products in a more compact form. For example,

$$2 \cdot 2 \cdot 2 \quad \textit{may be written as} \quad 2^3.$$

Volume is $(2 \cdot 2 \cdot 2)$
cubic centimeters.

The 2 in 2^3 is called the **base**; it is the repeated factor. The 3 in 2^3 is called the **exponent** and is the number of times the base is used as a factor. The expression 2^3 is called an **exponential expression**.

$$2^3 = 2 \cdot 2 \cdot 2 = 8$$

exponent

base — 2 is a factor 3 times

EXAMPLE 1

Evaluate the following:

a. 3^2 [read as "3 squared" or as "3 to second power"]
b. 5^3 [read as "5 cubed" or as "5 to the third power"]
c. 2^4 [read as "2 to the fourth power"]
d. 7^1 **e.** $\left(\frac{3}{7}\right)^2$

Solution   **a.** $3^2 = 3 \cdot 3 = 9$ **b.** $5^3 = 5 \cdot 5 \cdot 5 = 125$
c. $2^4 = 2 \cdot 2 \cdot 2 \cdot 2 = 16$ **d.** $7^1 = 7$
e. $\left(\frac{3}{7}\right)^2 = \left(\frac{3}{7}\right)\left(\frac{3}{7}\right) = \frac{9}{49}$

Helpful Hint
$2^3 \neq 2 \cdot 3$ since 2^3 indicates repeated **multiplication** of the same factor.
$2^3 = 2 \cdot 2 \cdot 2 = 8$, whereas $2 \cdot 3 = 6$.

Using symbols for mathematical operations is a great convenience. However, the more operation symbols presented in an expression, the more careful we must be when performing the indicated operation. For example, in the expression $2 + 3 \cdot 7$, do we add first or multiply first? To eliminate confusion, **grouping symbols** are used. Examples of grouping symbols are parentheses ( ), brackets [ ], braces { }, and the fraction bar. If we wish $2 + 3 \cdot 7$ to be simplified by adding first, we enclose $2 + 3$ in parentheses.

$$(2 + 3) \cdot 7 = 5 \cdot 7 = 35$$

If we wish to multiply first, $3 \cdot 7$ may be enclosed in parentheses.

$$2 + (3 \cdot 7) = 2 + 21 = 23$$

To eliminate confusion when no grouping symbols are present, use the following agreed upon order of operations.

Order of Operations

Simplify expressions using the order below. If grouping symbols such as parentheses are present, simplify expressions within those first, starting with the innermost set. If fraction bars are present, simplify the numerator and the denominator separately.
1. Evaluate exponential expressions.
2. Perform multiplications or divisions in order from left to right.
3. Perform additions or subtractions in order from left to right.

Now simplify $2 + 3 \cdot 7$. There are no grouping symbols and no exponents, so we multiply and then add.

$$2 + 3 \cdot 7 = 2 + 21 \qquad \text{Multiply.}$$
$$= 23 \qquad \text{Add.}$$

EXAMPLE 2

Simplify each expression.

a. $6 \div 3 + 5^2$ **b.** $\dfrac{2(12 + 3)}{|-15|}$ **c.** $3 \cdot 10 - 7 \div 7$ **d.** $3 \cdot 4^2$ **e.** $\dfrac{3}{2} \cdot \dfrac{1}{2} - \dfrac{1}{2}$

Solution **a.** Evaluate 5^2 first.

$$6 \div 3 + 5^2 = 6 \div 3 + 25$$

Next divide, then add.

$$= 2 + 25 \qquad \text{Divide.}$$
$$= 27 \qquad \text{Add.}$$

b. First, simplify the numerator and the denominator separately.

$$\frac{2(12 + 3)}{|-15|} = \frac{2(15)}{15} \qquad \text{Simplify numerator and}$$
$$\qquad\qquad\qquad\qquad \text{denominator separately.}$$
$$= \frac{30}{15}$$
$$= 2 \qquad \text{Simplify.}$$

c. Multiply and divide from left to right. Then subtract.

$$3 \cdot 10 - 7 \div 7 = 30 - 1$$
$$= 29 \qquad \text{Subtract.}$$

d. In this example, only the 4 is squared. The factor of 3 is not part of the base because no grouping symbol includes it as part of the base.

$$3 \cdot 4^2 = 3 \cdot 16 \qquad \text{Evaluate the exponential expression.}$$
$$= 48 \qquad \text{Multiply.}$$

e. The order of operations applies to operations with fractions in exactly the same way as it applies to operations with whole numbers.

$$\frac{3}{2} \cdot \frac{1}{2} - \frac{1}{2} = \frac{3}{4} - \frac{1}{2} \qquad \text{Multiply.}$$
$$= \frac{3}{4} - \frac{2}{4} \qquad \text{The least common denominator is 4.}$$
$$= \frac{1}{4} \qquad \text{Subtract.}$$

> **Helpful Hint**
>
> Be careful when evaluating an exponential expression. In $3 \cdot 4^2$, the exponent 2 applies only to the base 4. In $(3 \cdot 4)^2$, we multiply first because of parentheses, so the exponent 2 applies to the product $3 \cdot 4$.
>
> $$3 \cdot 4^2 = 3 \cdot 16 = 48 \qquad (3 \cdot 4)^2 = (12)^2 = 144$$

Expressions that include many grouping symbols can be confusing. When simplifying these expressions, keep in mind that grouping symbols separate the expression into distinct parts. Each is then simplified separately.

EXAMPLE 3

Simplify $\dfrac{3 + |4 - 3| + 2^2}{6 - 3}$.

Solution The fraction bar serves as a grouping symbol and separates the numerator and denominator. Simplify each separately. Also, the absolute value bars here serve as a grouping symbol. We begin in the numerator by simplifying within the absolute value bars.

$$\frac{3 + |4 - 3| + 2^2}{6 - 3} = \frac{3 + |1| + 2^2}{6 - 3}$$ Simplify the expression inside the absolute value bars.

$$= \frac{3 + 1 + 2^2}{3}$$ Find the absolute value and simplify the denominator.

$$= \frac{3 + 1 + 4}{3}$$ Evaluate the exponential expression.

$$= \frac{8}{3}$$ Simplify the numerator.

STUDY SKILLS REMINDER

Make a practice of neatly writing down enough steps so that you are comfortable with your computations.

EXAMPLE 4

Simplify $3[4 + 2(10 - 1)]$.

Solution Notice that both parentheses and brackets are used as grouping symbols. Start with the innermost set of grouping symbols.

Helpful Hint

Be sure to follow order of operations and resist the temptation to incorrectly add 4 and 2 first.

$$3[4 + 2(10 - 1)] = 3[4 + 2(9)]$$ Simplify the expression in parentheses.

$$= 3[4 + 18]$$ Multiply.

$$= 3[22]$$ Add.

$$= 66$$ Multiply.

EXAMPLE 5

Simplify $\dfrac{8 + 2 \cdot 3}{2^2 - 1}$.

Solution

$$\frac{8 + 2 \cdot 3}{2^2 - 1} = \frac{8 + 6}{4 - 1} = \frac{14}{3}$$

2   In algebra, we use symbols, usually letters such as x, y, or z, to represent unknown numbers. A symbol that is used to represent a number is called a **variable**. An **algebraic expression** is a collection of numbers, variables, operation symbols, and grouping symbols. For example,

$$2x, \qquad -3, \qquad 2x + 10, \qquad 5(p^2 + 1), \qquad \text{and} \qquad \frac{3y^2 - 6y + 1}{5}$$

are algebraic expressions. The expression $2x$ means $2 \cdot x$. Also, $5(p^2 + 1)$ means $5 \cdot (p^2 + 1)$ and $3y^2$ means $3 \cdot y^2$. If we give a specific value to a variable, we can **evaluate an algebraic expression**. To evaluate an algebraic expression means to find its numerical value once we know the values of the variables.

Algebraic expressions often occur during problem solving. For example, the expression

$$16t^2$$

gives the distance in feet (neglecting air resistance) that an object will fall in t seconds. (See Exercise 63 in this section.)

EXAMPLE 6

Evaluate each expression if $x = 3$ and $y = 2$.

a. $2x - y$ **b.** $\dfrac{3x}{2y}$ **c.** $\dfrac{x}{y} + \dfrac{y}{2}$ **d.** $x^2 - y^2$

Solution   **a.** Replace x with 3 and y with 2.

$$
\begin{aligned}
2x - y &= 2(3) - 2 &&\text{Let } x = 3 \text{ and } y = 2. \\
&= 6 - 2 &&\text{Multiply.} \\
&= 4 &&\text{Subtract.}
\end{aligned}
$$

b. $\dfrac{3x}{2y} = \dfrac{3 \cdot 3}{2 \cdot 2} = \dfrac{9}{4}$ Let $x = 3$ and $y = 2$.

c. Replace x with 3 and y with 2. Then simplify.

$$\frac{x}{y} + \frac{y}{2} = \frac{3}{2} + \frac{2}{2} = \frac{5}{2}$$

d. Replace x with 3 and y with 2.

$$x^2 - y^2 = 3^2 - 2^2 = 9 - 4 = 5$$

3   Many times a problem-solving situation is modeled by an equation. An **equation** is a mathematical statement that two expressions have equal value. The equal symbol "$=$" is used to equate the two expressions. For example, $3 + 2 = 5$, $7x = 35$, $\dfrac{2(x - 1)}{3} = 0$, and $I = PRT$ are all equations.

> ▶ **Helpful Hint**
>
> An equation contains the equal symbol "=". An algebraic expression does not.

✔ **CONCEPT CHECK**

Which of the following are equations? Which are expressions?

a. $5x = 8$ **b.** $5x - 8$ **c.** $12y + 3x$ **d.** $12y = 3x$

When an equation contains a variable, deciding which values of the variable make an equation a true statement is called **solving** an equation for the variable. A **solution** of an equation is a value for the variable that makes the equation true. For example, 3 is a solution of the equation $x + 4 = 7$, because if x is replaced with 3 the statement is true.

$$x + 4 = 7$$
$$\downarrow$$
$$3 + 4 = 7 \qquad \text{Replace } x \text{ with 3.}$$
$$7 = 7 \qquad \text{True}$$

Similarly, 1 is not a solution of the equation $x + 4 = 7$, because $1 + 4 = 7$ is **not** a true statement.

EXAMPLE 7

Decide whether 2 is a solution of $3x + 10 = 8x$.

Solution   Replace x with 2 and see if a true statement results.

$$3x + 10 = 8x \qquad \text{Original equation}$$
$$3(2) + 10 \stackrel{?}{=} 8(2) \qquad \text{Replace } x \text{ with 2.}$$
$$6 + 10 \stackrel{?}{=} 16 \qquad \text{Simplify each side.}$$
$$16 = 16 \qquad \text{True}$$

Since we arrived at a true statement after replacing x with 2 and simplifying both sides of the equation, 2 is a solution of the equation.

④   Now that we know how to represent an unknown number by a variable, let's practice translating phrases into algebraic expressions and sentences into equations. Oftentimes solving problems requires the ability to translate word phrases and sentences into symbols. Below is a list of some key words and phrases to help us translate.

Addition (+)	Subtraction (−)	Multiplication (•)	Division (÷)	Equality (=)
Sum	Difference of	Product	Quotient	Equals
Plus	Minus	Times	Divide	Gives
Added to	Subtracted from	Multiply	Into	Is/was/ should be
More than	Less than	Twice	Ratio	Yields
Increased by	Decreased by	Of	Divided by	Amounts to
Total	Less			Represents
				Is the same as

EXAMPLE 8

Write an algebraic expression that represents each phrase. Let the variable x represent the unknown number.

   a. The sum of a number and 3

   b. The product of 3 and a number

   c. Twice a number

   d. 10 decreased by a number

   e. 5 times a number, increased by 7

Solution   **a.** $x + 3$ since "sum" means to add

   b. $3 \cdot x$ and $3x$ are both ways to denote the product of 3 and x

   c. $2 \cdot x$ or $2x$

   d. $10 - x$ because "decreased by" means to subtract

   e. $\underbrace{5x}_{\text{5 times a number}} + 7$

> **Helpful Hint**
>
> Make sure you understand the difference when translating phrases containing "decreased by," "subtracted from," and "less than."
>
> | Phrase | Translation | |
> |---|---|---|
> | A number decreased by 10 | $x - 10$ | ⎫ |
> | A number subtracted from 10 | $10 - x$ | ⎬ Notice the order. |
> | 10 less than a number | $x - 10$ | |
> | A number less 10 | $x - 10$ | ⎭ |

Now let's practice translating sentences into equations.

EXAMPLE 9

Write each sentence as an equation. Let x represent the unknown number.

   a. The quotient of 15 and a number is 4.

   b. Three subtracted from 12 is a number.

   c. Four times a number, added to 17, is 21.

Solution   **a.** In words:

$$\frac{15}{x} = 4$$

   b. In words:

three subtracted **from** 12	is	a number
↓	↓	↓

Translate: $12 - 3$ = x

Care must be taken when the operation is subtraction. The expression $3 - 12$ would be incorrect. Notice that $3 - 12 \neq 12 - 3$.

c. In words:

four times a number	added to	17	is	21
↓	↓	↓	↓	↓

Translate: $4x$ $+$ 17 $=$ 21

Calculator Explorations

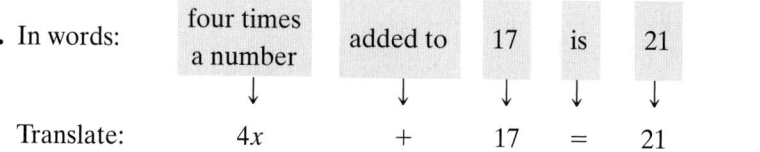

Exponents

To evaluate exponential expressions on a scientific calculator, find the key marked $\boxed{y^x}$ or $\boxed{\wedge}$. To evaluate, for example, 3^5, press the following keys: $\boxed{3}$ $\boxed{y^x}$ $\boxed{5}$ $\boxed{=}$ or $\boxed{3}$ $\boxed{\wedge}$ $\boxed{5}$ $\boxed{=}$.

↕ or
$\boxed{\text{ENTER}}$

The display should read $\boxed{\qquad 243}$ or $\boxed{\begin{array}{l} 3 \wedge 5 \\ \qquad 243 \end{array}}$

Order of Operations

Some calculators follow the order of operations, and others do not. To see whether or not your calculator has the order of operations built in, use your calculator to find $2 + 3 \cdot 4$. To do this, press the following sequence of keys:

$\boxed{2}$ $\boxed{+}$ $\boxed{3}$ $\boxed{\times}$ $\boxed{4}$ $\boxed{=}$.

↕ or
$\boxed{\text{ENTER}}$

The correct answer is 14 because the order of operations is to multiply before we add. If the calculator displays $\boxed{\qquad 14}$, then it has the order of operations built in.

Even if the order of operations is built in, parentheses must sometimes be inserted. For example, to simplify $\dfrac{5}{12 - 7}$, press the keys

$\boxed{5}$ $\boxed{\div}$ $\boxed{(}$ $\boxed{1}$ $\boxed{2}$ $\boxed{-}$ $\boxed{7}$ $\boxed{)}$ $\boxed{=}$.

↕ or
$\boxed{\text{ENTER}}$

The display should read $\boxed{\qquad 1}$ or $\boxed{\begin{array}{l} 5/(12 - 7) \\ \qquad \qquad 1 \end{array}}$

Use a calculator to evaluate each expression.

1. 5^4 **2.** 7^4

3. 9^5 **4.** 8^6

5. $2(20 - 5)$ **6.** $3(14 - 7) + 21$

7. $24(862 - 455) + 89$ **8.** $99 + (401 + 962)$

9. $\dfrac{4623 + 129}{36 - 34}$ **10.** $\dfrac{956 - 452}{89 - 86}$

Spotlight on
DECISION *of* MAKING

Suppose you are a local area network (LAN) administrator for a small college and you are configuring a new LAN for the mathematics department. The department would like a network of 20 computers so that each user can transmit data over the network at a speed of 0.25 megabits per second. The collective speed for a LAN is given by the expression rn, where r is the data transmission speed needed by each of the n computers on the LAN. You know that the network will drastically lose its efficiency if the collective speed of the network exceeds 8 megabits per second. Decide whether the LAN requested by the math department will operate efficiently. Explain your reasoning.

MENTAL MATH

Fill in the blank with add, subtract, multiply, or divide.

1. To simplify the expression $1 + 3 \cdot 6$, first _____.
2. To simplify the expression $(1 + 3) \cdot 6$, first _____.
3. To simplify the expression $(20 - 4) \cdot 2$, first _____.
4. To simplify the expression $20 - 4 \div 2$, first _____.

EXERCISE SET 1.4

STUDY GUIDE/SSM   CD/VIDEO   PH MATH TUTOR CENTER   MathXL®Tutorials ON CD   MathXL®   MyMathLab®

Evaluate. See Example 1.

1. 3^5
2. 5^3
 3. 3^3
4. 4^4
5. 1^5
6. 1^8
7. 5^1
8. 8^1
9. $\left(\dfrac{1}{5}\right)^3$
10. $\left(\dfrac{6}{11}\right)^2$
 11. $\left(\dfrac{2}{3}\right)^4$
12. $\left(\dfrac{1}{2}\right)^5$
13. 7^2
14. 9^2
15. 4^2
16. 2^4
17. $(1.2)^2$
18. $(0.07)^2$

Simplify each expression. See Examples 2 through 5.

 19. $5 + 6 \cdot 2$
20. $8 + 5 \cdot 3$
21. $4 \cdot 8 - 6 \cdot 2$
22. $12 \cdot 5 - 3 \cdot 6$
23. $2(8 - 3)$
24. $5(6 - 2)$
25. $2 + (5 - 2) + 4^2$
26. $6 - 2 \cdot 2 + 2^5$
27. $5 \cdot 3^2$
28. $2 \cdot 5^2$
29. $\dfrac{1}{4} \cdot \dfrac{2}{3} - \dfrac{1}{6}$
30. $\dfrac{3}{4} \cdot \dfrac{1}{2} + \dfrac{2}{3}$
31. $\dfrac{6 - 4}{9 - 2}$
32. $\dfrac{8 - 5}{24 - 20}$
33. $2[5 + 2(8 - 3)]$
34. $3[4 + 3(6 - 4)]$
35. $\dfrac{19 - 3 \cdot 5}{6 - 4}$
36. $\dfrac{4 \cdot 3 + 2}{4 + 3 \cdot 2}$
 37. $\dfrac{|6 - 2| + 3}{8 + 2 \cdot 5}$
38. $\dfrac{15 - |3 - 1|}{12 - 3 \cdot 2}$

39. $\dfrac{3 + 3(5 + 3)}{3^2 + 1}$
40. $\dfrac{3 + 6(8 - 5)}{4^2 + 2}$
41. $\dfrac{6 + |8 - 2| + 3^2}{18 - 3}$
42. $\dfrac{16 + |13 - 5| + 4^2}{17 - 5}$

43. Are parentheses necessary in the expression $2 + (3 \cdot 5)$? Explain your answer.
44. Are parentheses necessary in the expression $(2 + 3) \cdot 5$? Explain your answer.

For Exercises 45 and 46, match each expression in the first column with its value in the second column.

45.
a. $(6 + 2) \cdot (5 + 3)$	19
b. $(6 + 2) \cdot 5 + 3$	22
c. $6 + 2 \cdot 5 + 3$	64
d. $6 + 2 \cdot (5 + 3)$	43

46.
a. $(1 + 4) \cdot 6 - 3$	15
b. $1 + 4 \cdot (6 - 3)$	13
c. $1 + 4 \cdot 6 - 3$	27
d. $(1 + 4) \cdot (6 - 3)$	22

Evaluate each expression when $x = 1$, $y = 3$, and $z = 5$. See Example 6.

47. $3y$
48. $4x$
49. $\dfrac{z}{5x}$
50. $\dfrac{y}{2z}$
51. $3x - 2$
52. $6y - 8$

53. $|2x + 3y|$ **54.** $|5z - 2y|$

55. $5y^2$ **56.** $2z^2$

Evaluate each expression if $x = 12$, $y = 8$, *and* $z = 4$. *See Example 6.*

57. $\dfrac{x}{z} + 3y$ **58.** $\dfrac{y}{z} + 8x$ **59.** $x^2 - 3y + x$

60. $y^2 - 3x + y$ **61.** $\dfrac{x^2 + z}{y^2 + 2z}$ **62.** $\dfrac{y^2 + x}{x^2 + 3y}$

Neglecting air resistance, the expression $16t^2$ *gives the distance in feet an object will fall in t seconds.*

63. Complete the chart below. To evaluate $16t^2$, remember to first find t^2, then multiply by 16.

Time t (in seconds)	Distance $16t^2$ (in feet)
1	
2	
3	
4	

64. Does an object fall the same distance *during* each second? Why or why not? (See Exercise 63.)

Decide whether the given number is a solution of the given equation. See Example 7.

65. Is 5 a solution of $3x - 6 = 9$?

66. Is 6 a solution of $2x + 7 = 3x$?

67. Is 0 a solution of $2x + 6 = 5x - 1$?

68. Is 2 a solution of $4x + 2 = x + 8$?

69. Is 8 a solution of $2x - 5 = 5$?

70. Is 6 a solution of $3x - 10 = 8$?

71. Is 2 a solution of $x + 6 = x + 6$?

72. Is 10 a solution of $x + 6 = x + 6$?

73. Is 0 a solution of $x = 5x + 15$?

74. Is 1 a solution of $4 = 1 - x$?

Write each phrase as an algebraic expression. Let x represent the unknown number. See Example 8.

75. Fifteen more than a number

76. One-half times a number

77. Five subtracted from a number

78. The quotient of a number and 9

79. Three times a number, increased by 22

80. The product of 8 and a number

Write each sentence as an equation or inequality. Use x to represent any unknown number. See Example 9.

81. One increased by two equals the quotient of nine and three.

82. Four subtracted from eight is equal to two squared.

83. Three is not equal to four divided by two.

84. The difference of sixteen and four is greater than ten.

85. The sum of 5 and a number is 20.

86. Twice a number is 17.

87. Thirteen minus three times a number is 13.

88. Seven subtracted from a number is 0.

89. The quotient of 12 and a number is $\dfrac{1}{2}$.

90. The sum of 8 and twice a number is 42.

91. In your own words, explain the difference between an expression and an equation.

92. Determine whether each is an expression or an equation.

   a. $3x^2 - 26$ **b.** $3x^2 - 26 = 1$

   c. $2x - 5 = 7x - 5$ **d.** $9y + x - 8$

Concept Extensions

93. Insert parentheses so that the following expression simplifies to 32.

$$20 - 4 \cdot 4 \div 2$$

94. Insert parentheses so that the following expression simplifies to 28.

$$2 \cdot 5 + 3^2$$

Solve the following.

95. The perimeter of a figure is the distance around the figure. The expression $2l + 2w$ represents the perimeter of a rectangle when l is its length and w is its width. Find the perimeter of the following rectangle by substituting 8 for l and 6 for w.

96. The expression $a + b + c$ represents the perimeter of a triangle when a, b, and c are the lengths of its sides. Find the perimeter of the following triangle.

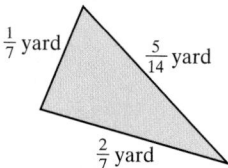

97. The area of a figure is the total enclosed surface of the figure. Area is measured in square units. The expression lw

represents the area of a rectangle when l is its length and w is its width. Find the area of the following rectangular-shaped lot.

100 feet

120 feet

△ **98.** A trapezoid is a four-sided figure with exactly one pair of parallel sides. The expression $\frac{1}{2}h(B + b)$ represents its area, when B and b are the lengths of the two parallel sides and h is the height between these sides. Find the area if $B = 15$ inches, $b = 7$ inches, and $h = 5$ inches.

7 inches

5 inches

15 inches

99. The expression $\frac{I}{PT}$ represents the rate of interest being charged if a loan of P dollars for T years required I dollars in interest to be paid. Find the interest rate if a \$650 loan for 3 years to buy a used IBM personal computer requires \$126.75 in interest to be paid.

100. The expression $\frac{d}{t}$ represents the average speed r in miles per hour if a distance of d miles is traveled in t hours. Find the rate to the nearest whole number if the distance between Dallas, Texas, and Kaw City, Oklahoma, is 432 miles, and it takes Peter Callac 8.5 hours to drive the distance.

101. Sprint Communications Company offers a long-distance telephone plan called Sprint Simple 7 that charges \$4.00 per month and \$0.07 per minute of calling. The expression $4.00 + 0.07m$ represents the monthly long-distance bill for a customer who makes m minutes of long-distance calling on this plan. Find the monthly bill for a customer who makes 228 minutes of long-distance calls on the Sprint Simple 7 plan.

102. In forensics, the density of a substance is used to help identify it. The expression $\frac{M}{V}$ represents the density of an object with a mass of M grams and a volume of V milliliters. Find the density of an object having a mass of 29.76 grams and a volume of 12 milliliters.

1.5   ADDING REAL NUMBERS

Objectives

1. Add real numbers with the same sign.
2. Add real numbers with unlike signs.
3. Solve problems that involve addition of real numbers.
4. Find the opposite of a number.

1   Real numbers can be added, subtracted, multiplied, divided, and raised to powers, just as whole numbers can. We use a number line to help picture the addition of real numbers.

EXAMPLE 1

Add: $3 + 2$

Solution   Recall that 3 and 2 are called addends. We start at 0 on a number line, and draw an arrow representing the addend 3. This arrow is three units long and points to the right since 3 is positive. From the tip of this arrow, we draw another arrow representing the addend 2. The number below the tip of this arrow is the sum, 5.

$$3 + 2 = 5$$

EXAMPLE 2

Add: $-1 + (-2)$

Solution   Here, -1 and -2 are addends. We start at 0 on a number line, and draw an arrow representing -1. This arrow is one unit long and points to the left since -1 is negative. From the tip of this arrow, we draw another arrow representing -2. The number below the tip of this arrow is the sum, -3.

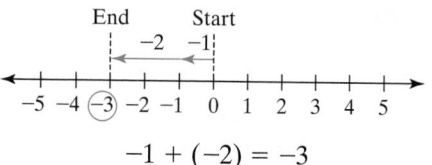

$$-1 + (-2) = -3$$

Thinking of signed numbers as money earned or lost might help make addition more meaningful. Earnings can be thought of as positive numbers. If \$1 is earned and later another \$3 is earned, the total amount earned is \$4. In other words, $1 + 3 = 4$.

On the other hand, losses can be thought of as negative numbers. If \$1 is lost and later another \$3 is lost, a total of \$4 is lost. In other words, $(-1) + (-3) = -4$.

Using a number line each time we add two numbers can be time consuming. Instead, we can notice patterns in the previous examples and write rules for adding signed numbers. When adding two numbers with the same sign, notice that the sign of the sum is the same as the sign of the addends.

Adding Two Numbers with the Same Sign

Add their absolute values. Use their common sign as the sign of the sum.

EXAMPLE 3

Add.

a. $-3 + (-7)$ **b.** $-1 + (-20)$ **c.** $-2 + (-10)$

Solution   Notice that each time, we are adding numbers with the same sign.

a. $-3 + (-7) = -10$ ← Add their absolute values: $3 + 7 = 10$.
         └────── Use their common sign.

b. $-1 + (-20) = -21$ ← Add their absolute values: $1 + 20 = 21$.
         └───── Common sign.

c. $-2 + (-10) = -12$ ← Add their absolute values.
         └───── Common sign.

2   Adding numbers whose signs are not the same can also be pictured on a number line.

EXAMPLE 4

Add: $-4 + 6$

Solution

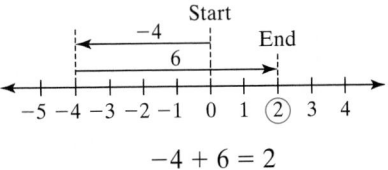

$$-4 + 6 = 2$$

Using temperature as an example, if the thermometer registers 4 degrees below 0 degrees and then rises 6 degrees, the new temperature is 2 degrees above 0 degrees. Thus, it is reasonable that $-4 + 6 = 2$.

Once again, we can observe a pattern: when adding two numbers with different signs, the sign of the sum is the same as the sign of the addend whose absolute value is larger.

> ### Adding Two Numbers with Different Signs
>
> Subtract the smaller absolute value from the larger absolute value. Use the sign of the number whose absolute value is larger as the sign of the sum.

EXAMPLE 5

Add.

a. $3 + (-7)$ **b.** $-2 + 10$ **c.** $0.2 + (-0.5)$

Solution   Notice that each time, we are adding numbers with different signs.

a. $3 + (-7) = -4$ ← Subtract their absolute values: $7 - 3 = 4$.

    └── The negative number, -7, has the larger absolute value so the sum is negative.

b. $-2 + 10 = 8$ ← Subtract their absolute values: $10 - 2 = 8$.

    └── The positive number, 10, has the larger absolute value so the sum is positive.

c. $0.2 + (-0.5) = -0.3$ ← Subtract their absolute values: $0.5 - 0.2 = 0.3$.

    └── The negative number, -0.5, has the larger absolute value so the sum is negative.

EXAMPLE 6

Add.

a. $-8 + (-11)$ **b.** $-5 + 35$ **c.** $0.6 + (-1.1)$

d. $-\dfrac{7}{10} + \left(-\dfrac{1}{10}\right)$ **e.** $11.4 + (-4.7)$ **f.** $-\dfrac{3}{8} + \dfrac{2}{5}$

Solution **a.** $-8 + (-11) = -19$ *Same sign. Add absolute values and use the common sign.*

b. $-5 + 35 = 30$ *Different signs. Subtract absolute values and use the sign of the number with the larger absolute value.*

c. $0.6 + (-1.1) = -0.5$ *Different signs.*

d. $-\dfrac{7}{10} + \left(-\dfrac{1}{10}\right) = -\dfrac{8}{10} = -\dfrac{4}{5}$ *Same sign.*

e. $11.4 + (-4.7) = 6.7$

f. $-\dfrac{3}{8} + \dfrac{2}{5} = -\dfrac{15}{40} + \dfrac{16}{40} = \dfrac{1}{40}$

> **Helpful Hint**
>
> Don't forget that a common denominator is needed when adding or subtracting fractions. The common denominator here is 40.

EXAMPLE 7

Add.

a. $3 + (-7) + (-8)$

b. $[7 + (-10)] + [-2 + |-4|]$

Solution **a.** Perform the additions from left to right.

$$3 + (-7) + (-8) = -4 + (-8) \quad \text{\textit{Adding numbers with different signs.}}$$
$$= -12 \quad \text{\textit{Adding numbers with like signs.}}$$

b. Simplify inside brackets first.

$$[7 + (-10)] + [-2 + |-4|] = [-3] + [-2 + 4]$$
$$= [-3] + [2]$$
$$= -1 \quad \text{\textit{Add.}}$$

> **Helpful Hint**
>
> Don't forget that brackets are grouping symbols. We simplify within them first.

3   Positive and negative numbers are often used in everyday life. Stock market returns show gains and losses as positive and negative numbers. Temperatures in cold climates often dip into the negative range, commonly referred to as "below zero" temperatures. Bank statements report deposits and withdrawals as positive and negative numbers.

EXAMPLE 8

FINDING THE GAIN OR LOSS OF A STOCK

During a three-day period, a share of Lamplighter's International stock recorded the following gains and losses:

Monday	**Tuesday**	**Wednesday**
a gain of $2	a loss of $1	a loss of $3

Find the overall gain or loss for the stock for the three days.

Solution Gains can be represented by positive numbers. Losses can be represented by negative numbers. The overall gain or loss is the sum of the gains and losses.

In words: gain   plus   loss   plus   loss
              ↓   ↓   ↓   ↓   ↓

Translate: 2   + (-1) + $(-3) = -2$

The overall loss is $2.

4   To help us subtract real numbers in the next section, we first review the concept of opposites. The graphs of 4 and −4 are shown on a number line below.

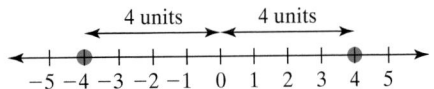

Notice that 4 and −4 lie on opposite sides of 0, and each is 4 units away from 0.

This relationship between −4 and +4 is an important one. Such numbers are known as **opposites** or **additive inverses** of each other.

Opposites or Additive Inverses

Two numbers that are the same distance from 0 but lie on opposite sides of 0 are called opposites or additive inverses of each other.

Let's discover another characteristic about opposites. Notice that the sum of a number and its opposite is 0.

$$10 + (-10) = 0$$
$$-3 + 3 = 0$$
$$\frac{1}{2} + \left(-\frac{1}{2}\right) = 0$$

In general, we can write the following:

The sum of a number a and its opposite $-a$ is 0.

$$a + (-a) = 0$$

This is why opposites are also called additive inverses. Notice that this also means that the opposite of 0 is then 0 since $0 + 0 = 0$.

EXAMPLE 9

Find the opposite or additive inverse of each number.

a. 5   **b.** −6   **c.** $\frac{1}{2}$ **d.** −4.5

Solution   **a.** The opposite of 5 is −5. Notice that 5 and −5 are on opposite sides of 0 when plotted on a number line and are equal distances away.

b. The opposite of −6 is 6.

c. The opposite of $\frac{1}{2}$ is $-\frac{1}{2}$.

d. The opposite of −4.5 is 4.5.

We use the symbol "−" to represent the phrase "the opposite of" or "the additive inverse of." In general, if *a* is a number, we write the opposite or additive inverse of *a* as −*a*. We know that the opposite of −3 is 3. Notice that this translates as

the opposite of   −3   is   3
$$- \qquad (-3) \qquad = \qquad 3$$

This is true in general.

> If *a* is a number, then $-(-a) = a$.

EXAMPLE 10

Simplify each expression.

a. $-(-10)$ **b.** $-\left(-\dfrac{1}{2}\right)$ **c.** $-(-2x)$ **d.** $-|-6|$

Solution   **a.** $-(-10) = 10$ **b.** $-\left(-\dfrac{1}{2}\right) = \dfrac{1}{2}$ **c.** $-(-2x) = 2x$

d. Since $|-6| = 6$, then $-|-6| = -6$.

MENTAL MATH

Tell whether the sum is a positive number, a negative number, or 0. Do not actually find the sum.

1. $-80 + (-127)$

2. $-162 + 164$

3. $-162 + 162$

4. $-1.26 + (-8.3)$

5. $-3.68 + 0.27$

6. $-\dfrac{2}{3} + \dfrac{2}{3}$

EXERCISE SET 1.5

STUDY GUIDE/SSM   CD/ VIDEO   PH MATH TUTOR CENTER   MathXL®Tutorials ON CD   MathXL®   MyMathLab®

MIXED PRACTICE

Add. See Examples 1 through 7.

1. $6 + 3$

2. $9 + (-12)$

 3. $-6 + (-8)$

4. $-6 + (-14)$

5. $8 + (-7)$

6. $6 + (-4)$

7. $-14 + 2$

8. $-10 + 5$

9. $-2 + (-3)$

10. $-7 + (-4)$

 11. $-9 + (-3)$

12. $7 + (-5)$

13. $-7 + 3$

14. $-5 + 9$

15. $10 + (-3)$

16. $8 + (-6)$

17. $5 + (-7)$

18. $3 + (-6)$

19. $-16 + 16$

20. $23 + (-23)$

21. $27 + (-46)$

22. $53 + (-37)$

23. $-18 + 49$

24. $-26 + 14$

25. $-33 + (-14)$

26. $-18 + (-26)$

27. $6.3 + (-8.4)$

28. $9.2 + (-11.4)$

29. $|-8| + (-16)$

30. $|-6| + (-61)$

31. $117 + (-79)$

32. $144 + (-88)$

33. $-9.6 + (-3.5)$

34. $-6.7 + (-7.6)$

35. $-\frac{3}{8} + \frac{5}{8}$

36. $-\frac{5}{12} + \frac{7}{12}$

37. $-\frac{7}{16} + \frac{1}{4}$

38. $-\frac{5}{9} + \frac{1}{3}$

39. $-\frac{7}{10} + \left(-\frac{3}{5}\right)$

40. $-\frac{5}{6} + \left(-\frac{2}{3}\right)$

41. $-15 + 9 + (-2)$

42. $-9 + 15 + (-5)$

43. $-21 + (-16) + (-22)$

44. $-18 + (-6) + (-40)$

45. $-23 + 16 + (-2)$

46. $-14 + (-3) + 11$

47. $|5 + (-10)|$

48. $|7 + (-17)|$

49. $6 + (-4) + 9$

50. $8 + (-2) + 7$

51. $[-17 + (-4)] + [-12 + 15]$

52. $[-2 + (-7)] + [-11 + 22]$

53. $|9 + (-12)| + |-16|$

54. $|43 + (-73)| + |-20|$

55. $-1.3 + [0.5 + (-0.3) + 0.4]$

56. $-3.7 + [0.1 + (-0.6) + 8.1]$

Solve. See Example 8.

57. The low temperature in Anoka, Minnesota, was $-15°$ last night. During the day it rose only $9°$. Find the high temperature for the day.

58. On January 2, 1943, the temperature was $-4°$ at 7:30 A.M. in Spearfish, South Dakota. Incredibly, it got $49°$ warmer in the next 2 minutes. To what temperature did it rise by 7:32?

59. The lowest elevation on Earth is -1349 feet (that is, 1349 feet below sea level) at the Dead Sea. If you are standing 695 feet above the Dead Sea, what is your elevation? (*Source:* National Geographic Society)

60. The lowest point in Africa is -512 feet at Lake Assal in Djibouti. If you are standing at a point 658 feet above Lake Assal, what is your elevation? (*Source:* Microsoft Encarta)

A negative net income results when a company's expenses are more than the money it brings in.

61. The table below shows net incomes for Amazon.com for the years 2000, 2001, and 2002. Find the total net income for three years.

Year	Net Income (in millions)
2000	$-\$1411$
2001	$-\$567$
2002	$-\$149$

(*Source:* amazon.com)

62. The table below shows net incomes for Continental Airlines for the years 2000, 2001, and 2002. Find the total net income for these years.

Year	Net Income (in millions)
2000	$\$342$
2001	$-\$95$
2002	$-\$451$

(*Source:* Continental Airlines)

In golf, scores that are under par for the entire round are shown as negative scores; positive scores are shown for scores that are over par, and 0 is par.

63. Candie Kung was the winner of the 2003 LPGA Takefuji Classic. Her scores were $5, -5, -2$. What was her overall score? (*Source:* Ladies Professional Golf Association)

64. During the 2003 LPGA Kraft Nabisco Championship, Patricia Meunier-Lebouc won with scores of $-2, -4, -2, 1$. What was her overall score? (*Source:* Ladies Professional Golf Association)

Find each additive inverse or opposite. See Example 9.

65. 6

66. 4

67. -2

68. -8

69. 0

70. $-\frac{1}{4}$

71. $|-6|$

72. $|-11|$

73. In your own words, explain how to find the opposite of a number.

74. In your own words, explain why 0 is the only number that is its own opposite.

Simplify each of the following. See Example 10.

75. $-|-2|$

76. $-(-3)$

77. $-|0|$

78. $\left|-\frac{2}{3}\right|$

79. $-\left|-\frac{2}{3}\right|$

80. $-(-7)$

81. Explain why adding a negative number to another negative number always gives a negative sum.

82. When a positive and a negative number are added, sometimes the sum is positive, sometimes it is zero, and sometimes it is negative. Explain why and when this happens.

Decide whether the given number is a solution of the given equation.

83. Is -4 a solution of $x + 9 = 5$?

84. Is 10 a solution of $7 = -x + 3$?

85. Is -1 a solution of $y + (-3) = -7$?

86. Is -6 a solution of $1 = y + 7$?

Concept Extensions

The following bar graph shows each month's average daily low temperature in degrees Fahrenheit for Barrow, Alaska. Use this graph to answer Exercises 87 through 92.

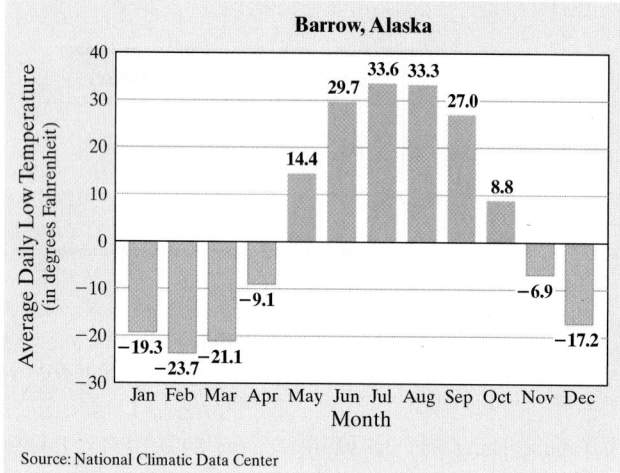

Barrow, Alaska

Source: National Climatic Data Center

87. For what month is the graphed temperature the highest?

88. For what month is the graphed temperature the lowest?

89. For what month is the graphed temperature positive *and* closest to 0°?

90. For what month is the graphed temperature negative *and* closest to 0°?

91. Find the average of the temperatures shown for the months of April, May, and October. (To find the average of three temperatures, find their sum and divide by 3.)

92. Find the average of the temperatures shown for the months of January, September, and October.

If a is a positive number and b is a negative number, fill in the blanks with the words positive or negative.

93. $-a$ is a _____.

94. $-b$ is a _____.

95. $a + a$ is a _____.

96. $b + b$ is a _____.

1.6   SUBTRACTING REAL NUMBERS

Objectives

1 Subtract real numbers.

2 Add and subtract real numbers.

3 Evaluate algebraic expressions using real numbers.

4 Solve problems that involve subtraction of real numbers.

1   Now that addition of signed numbers has been discussed, we can explore subtraction. We know that $9 - 7 = 2$. Notice that $9 + (-7) = 2$, also. This means that

$$9 - 7 = 9 + (-7)$$

Notice that the difference of 9 and 7 is the same as the sum of 9 and the opposite of 7. In general, we have the following.

> ### Subtracting Two Real Numbers
> If a and b are real numbers, then $a - b = a + (-b)$.

In other words, to find the difference of two numbers, add the first number to the opposite of the second number.

EXAMPLE 1

Subtract.

a. $-13 - 4$ **b.** $5 - (-6)$ **c.** $3 - 6$ **d.** $-1 - (-7)$

Solution

a. $-13 - 4 = -13 + (-4)$ Add -13 to the opposite of $+4$, which is -4.
$= -17$

b. $5 - (-6) = 5 + (6)$ Add 5 to the opposite of -6, which is 6.
$= 11$

c. $3 - 6 = 3 + (-6)$ Add 3 to the opposite of 6, which is -6.
$= -3$

d. $-1 - (-7) = -1 + (7) = 6$

> **Helpful Hint**
> Study the patterns indicated.
>
> No change ──── Change to addition.
> ──── Change to opposite.
>
> $5 - 11 = 5 + (-11) = -6$
> $-3 - 4 = -3 + (-4) = -7$
> $7 - (-1) = 7 + (1) = 8$

EXAMPLE 2

Subtract.

a. $5.3 - (-4.6)$ **b.** $-\dfrac{3}{10} - \dfrac{5}{10}$ **c.** $-\dfrac{2}{3} - \left(-\dfrac{4}{5}\right)$

Solution   **a.** $5.3 - (-4.6) = 5.3 + (4.6) = 9.9$

b. $-\dfrac{3}{10} - \dfrac{5}{10} = -\dfrac{3}{10} + \left(-\dfrac{5}{10}\right) = -\dfrac{8}{10} = -\dfrac{4}{5}$

c. $-\dfrac{2}{3} - \left(-\dfrac{4}{5}\right) = -\dfrac{2}{3} + \left(\dfrac{4}{5}\right) = -\dfrac{10}{15} + \dfrac{12}{15} = \dfrac{2}{15}$ The common denominator is 15.

EXAMPLE 3

Subtract 8 from -4.

Solution Be careful when interpreting this: The order of numbers in subtraction is important. 8 is to be subtracted **from** -4.

$$-4 - 8 = -4 + (-8) = -12$$

2 If an expression contains additions and subtractions, just write the subtractions as equivalent additions. Then simplify from left to right.

EXAMPLE 4

Simplify each expression.

a. $-14 - 8 + 10 - (-6)$ **b.** $1.6 - (-10.3) + (-5.6)$

Solution **a.** $-14 - 8 + 10 - (-6) = -14 + (-8) + 10 + 6$
$$= -6$$

b. $1.6 - (-10.3) + (-5.6) = 1.6 + 10.3 + (-5.6)$
$$= 6.3$$

When an expression contains parentheses and brackets, remember the order of operations. Start with the innermost set of parentheses or brackets and work your way outward.

EXAMPLE 5

Simplify each expression.

a. $-3 + [(-2 - 5) - 2]$ **b.** $2^3 - |10| + [-6 - (-5)]$

Solution **a.** Start with the innermost sets of parentheses. Rewrite $-2 - 5$ as a sum.

$$
\begin{aligned}
-3 + [(-2 - 5) - 2] &= -3 + [(-2 + (-5)) - 2] \\
&= -3 + [(-7) - 2] &&\text{Add: } -2 + (-5). \\
&= -3 + [-7 + (-2)] &&\text{Write } -7 - 2 \text{ as a sum.} \\
&= -3 + [-9] &&\text{Add.} \\
&= -12 &&\text{Add.}
\end{aligned}
$$

b. Start simplifying the expression inside the brackets by writing $-6 - (-5)$ as a sum.

$$
\begin{aligned}
2^3 - |10| + [-6 - (-5)] &= 2^3 - |10| + [-6 + 5] \\
&= 2^3 - |10| + [-1] &&\text{Add.} \\
&= 8 - 10 + (-1) &&\text{Evaluate } 2^3 \text{ and } |10|. \\
&= 8 + (-10) + (-1) &&\text{Write } 8 - 10 \text{ as a sum.} \\
&= -2 + (-1) &&\text{Add.} \\
&= -3 &&\text{Add.}
\end{aligned}
$$

3 Knowing how to evaluate expressions for given replacement values is helpful when checking solutions of equations and when solving problems whose unknowns satisfy given expressions. The next example illustrates this.

EXAMPLE 6

Find the value of each expression when $x = 2$ and $y = -5$.

a. $\dfrac{x - y}{12 + x}$ **b.** $x^2 - y$

Solution   **a.** Replace x with 2 and y with -5. Be sure to put parentheses around -5 to separate signs. Then simplify the resulting expression.

$$\frac{x - y}{12 + x} = \frac{2 - (-5)}{12 + 2}$$
$$= \frac{2 + 5}{14}$$
$$= \frac{7}{14} = \frac{1}{2}$$

b. Replace the x with 2 and y with -5 and simplify.

$$x^2 - y = 2^2 - (-5)$$
$$= 4 - (-5)$$
$$= 4 + 5$$
$$= 9$$

4   One use of positive and negative numbers is in recording altitudes above and below sea level, as shown in the next example.

EXAMPLE 7

FINDING THE DIFFERENCE IN ELEVATIONS

The lowest point on the surface of the Earth is the Dead Sea, at an elevation of 1349 feet below sea level. The highest point is Mt. Everest, at an elevation of 29,035 feet. How much of a variation in elevation is there between these two world extremes? (*Source*: National Geographic Society)

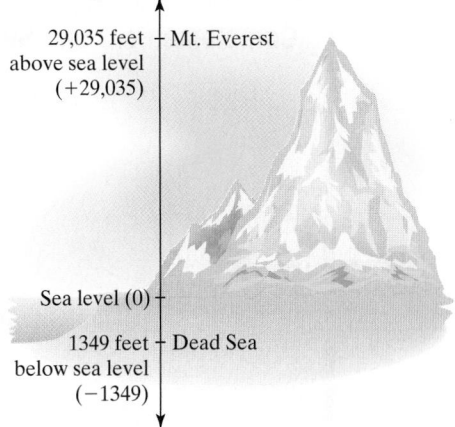

29,035 feet Mt. Everest
above sea level
(+29,035)

Sea level (0)

1349 feet Dead Sea
below sea level
(−1349)

Solution   To find the variation in elevation between the two heights, find the difference of the high point and the low point.

In words:	high point	minus	low point	
	↓	↓	↓	
Translate:	29,035	−	(−1349)	= 29,035 + 1349
				= 30,384 feet

Thus, the variation in elevation is 30,384 feet.

Spotlight on

DECISION of MAKING

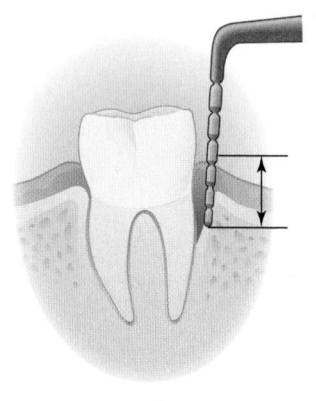

Suppose you are a dental hygienist. As part of a new patient assessment, you measure the depth of the gum tissue pocket around the patient's teeth with a dental probe and record the results. If these pockets deepen over time, this could indicate a problem with gum health or be an indication of gum disease. Now, a year later, you measure the patient's gum tissue pocket depth again to compare to the intial measurement. Based on these findings, would you alert the dentist to a problem with the health of the patien's gums? Explain.

Dental Chart

	Gum Tissue Pocket Depth *(millimeters)*					
Tooth:	22	23	24	25	26	27
Initial	2	3	3	2	4	2
Current	2	2	4	5	6	5

A knowledge of geometric concepts is needed by many professionals, such as doctors, carpenters, electronic technicians, gardeners, machinists, and pilots, just to name a few. With this in mind, we review the geometric concepts of **complementary** and **supplementary angles**.

Complementary and Supplementary Angles

Two angles are **complementary** if their sum is 90°.

$x + y = 90°$

Two angles are **supplementary** if their sum is 180°.

$x + y = 180°$

EXAMPLE 8

Find each unknown complementary or supplementary angle.

a.

b.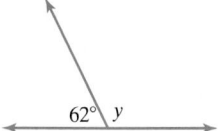

Solution   **a.** These angles are complementary, so their sum is $90°$. This means that x is $90° - 38°$.

$$x = 90° - 38° = 52°$$

b. These angles are supplementary, so their sum is $180°$. This means that y is $180° - 62°$.

$$y = 180° - 62° = 118°$$

EXERCISE SET 1.6

STUDY GUIDE/SSM   CD/VIDEO   PH MATH TUTOR CENTER   MathXL®Tutorials ON CD   MathXL®   MyMathLab®

MIXED PRACTICE

Subtract. See Examples 1 through 5.

1. $-6 - 4$
2. $-12 - 8$
3. $4 - 9$
4. $8 - 11$
5. $16 - (-3)$
6. $12 - (-5)$
7. $\dfrac{1}{2} - \dfrac{1}{3}$
8. $\dfrac{3}{4} - \dfrac{7}{8}$
9. $-16 - (-18)$
10. $-20 - (-48)$
11. $-6 - 5$
12. $-8 - 4$
13. $7 - (-4)$
14. $3 - (-6)$
15. $-6 - (-11)$
16. $-4 - (-16)$
17. $16 - (-21)$
18. $15 - (-33)$
19. $9.7 - 16.1$
20. $8.3 - 11.2$
21. $-44 - 27$
22. $-36 - 51$
23. $-21 - (-21)$
24. $-17 - (-17)$
25. $-2.6 - (-6.7)$
26. $-6.1 - (-5.3)$
27. $-\dfrac{3}{11} - \left(-\dfrac{5}{11}\right)$
28. $-\dfrac{4}{7} - \left(-\dfrac{1}{7}\right)$
29. $-\dfrac{1}{6} - \dfrac{3}{4}$
30. $-\dfrac{1}{10} - \dfrac{7}{8}$
31. $8.3 - (-0.62)$
32. $4.3 - (-0.87)$

Perform the operation. See Example 3.

33. Subtract -5 from 8.
34. Subtract 3 from -2.
35. Subtract -1 from -6.
36. Subtract 17 from 1.
37. Subtract 8 from 7.
38. Subtract 9 from -4.
39. Decrease -8 by 15.

40. Decrease 11 by -14.
41. In your own words, explain why $5 - 8$ simplifies to a negative number.
42. Explain why $6 - 11$ is the same as $6 + (-11)$.

Simplify each expression. (Remember the order of operations.) See Examples 4 and 5.

43. $-10 - (-8) + (-4) - 20$
44. $-16 - (-3) + (-11) - 14$
45. $5 - 9 + (-4) - 8 - 8$
46. $7 - 12 + (-5) - 2 + (-2)$
47. $-6 - (2 - 11)$ **48.** $-9 - (3 - 8)$
49. $3^3 - 8 \cdot 9$ **50.** $2^3 - 6 \cdot 3$
51. $2 - 3(8 - 6)$ **52.** $4 - 6(7 - 3)$
53. $(3 - 6) + 4^2$ **54.** $(2 - 3) + 5^2$
55. $-2 + [(8 - 11) - (-2 - 9)]$
56. $-5 + [(4 - 15) - (-6) - 8]$
57. $|-3| + 2^2 + [-4 - (-6)]$
58. $|-2| + 6^2 + (-3 - 8)$

Evaluate each expression when $x = -5$, $y = 4$, and $t = 10$. See Example 6.

59. $x - y$ **60.** $y - x$
61. $|x| + 2t - 8y$ **62.** $|x + t - 7y|$
63. $\dfrac{9 - x}{y + 6}$ **64.** $\dfrac{15 - x}{y + 2}$
65. $y^2 - x$ **66.** $t^2 - x$
67. $\dfrac{|x - (-10)|}{2t}$ **68.** $\dfrac{|5y - x|}{6t}$

Solve. See Example 7.

69. Within 24 hours in 1916, the temperature in Browning, Montana, fell from 44 degrees to −56 degrees. How large a drop in temperature was this?

70. Much of New Orleans is below sea level. If George descends 12 feet from an elevation of 5 feet above sea level, what is his new elevation?

71. In a series of plays, the San Francisco 49ers gain 2 yards, lose 5 yards, and then lose another 20 yards. What is their total gain or loss of yardage?

72. In some card games, it is possible to have a negative score. Lavonne Schultz currently has a score of 15 points. She then loses 24 points. What is her new score?

73. Pythagoras died in the year −475 (or 475 B.C.). When was he born, if he was 94 years old when he died?

74. The Greek astronomer and mathematician Geminus died in 60 A.D. at the age of 70. When was he born?

75. A commercial jet liner hits an air pocket and drops 250 feet. After climbing 120 feet, it drops another 178 feet. What is its overall vertical change?

76. Tyson Industries stock posted a loss of 1.625 points yesterday. If it drops another 0.75 point today, find its overall change for the two days.

77. The highest point in Africa is Mt. Kilimanjaro, Tanzania, at an elevation of 19,340 feet. The lowest point is Lake Assal, Djibouti, at 512 feet below sea level. How much higher is Mt. Kilimanjaro than Lake Assal? (*Source:* National Geographic Society)

78. The airport in Bishop, California, is at an elevation of 4101 feet above sea level. The nearby Furnace Creek Airport in Death Valley, California, is at an elevation of 226 feet below sea level. How much higher in elevation is the Bishop Air-

port than the Furnace Creek Airport? (*Source:* National Climatic Data Center)

Find each unknown complementary or supplementary angle. See Example 8.

△ **79.**   △ **80.**

△ **81.**   △ **82.**

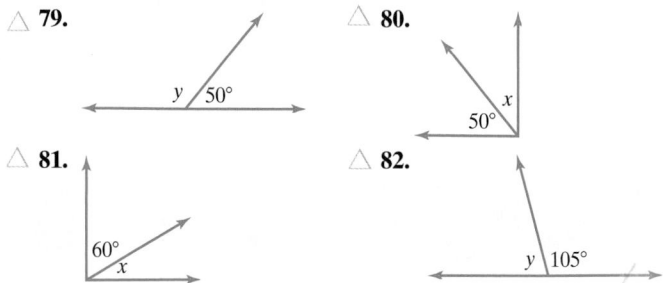

Decide whether the given number is a solution of the given equation.

83. Is −4 a solution of $x - 9 = 5$?
84. Is 3 a solution of $x - 10 = -7$?
85. Is −2 a solution of $-x + 6 = -x - 1$?
86. Is −10 a solution of $-x - 6 = -x - 1$?
87. Is 2 a solution of $-x - 13 = -15$?
88. Is 5 a solution of $4 = 1 - x$?

Concept Extensions

Recall from the last section the bar graph below that shows each month's average daily low temperature in degrees Fahrenheit for Barrow, Alaska. Use this graph to answer Exercises 89 through 91.

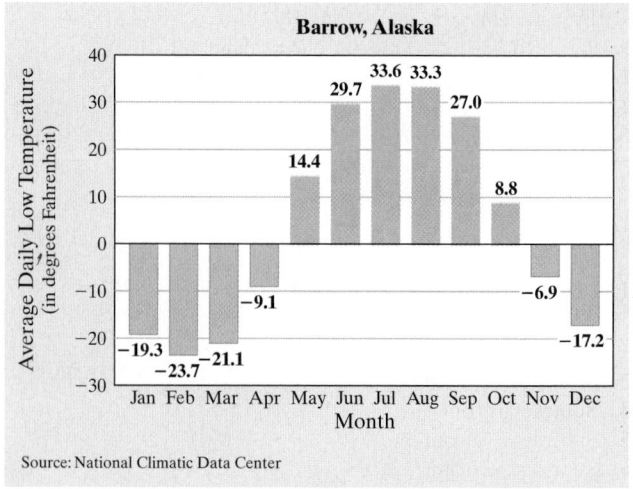

89. Record the monthly increases and decreases in the low temperature from the previous month.

Month	Monthly Increase or Decrease
February	
March	
April	
May	
June	
July	
August	
September	
October	
November	
December	

90. Which month had the greatest increase in temperature?

91. Which month had the greatest decrease in temperature?

If a is a positive number and b is a negative number, determine whether each statement is true or false.

92. $a - b$ is always a positive number.

93. $b - a$ is always a negative number.

94. $|b| - |a|$ is always a positive number.

95. $|b - a|$ is always a positive number.

Without calculating, determine whether each answer is positive or negative. Then use a calculator to find the exact difference.

96. $56,875 - 87,262$ **97.** $4.362 - 7.0086$

1.7   MULTIPLYING AND DIVIDING REAL NUMBERS

Objectives

1 Multiply and divide real numbers.

2 Evaluate algebraic expressions using real numbers.

1   In this section, we discover patterns for multiplying and dividing real numbers. To discover sign rules for multiplication, recall that multiplication is repeated addition. Thus $3 \cdot 2$ means that 2 is an addend 3 times. That is,

$$2 + 2 + 2 = 3 \cdot 2$$

which equals 6. Similarly, $3 \cdot (-2)$ means -2 is an addend 3 times. That is,

$$(-2) + (-2) + (-2) = 3 \cdot (-2)$$

Since $(-2) + (-2) + (-2) = -6$, then $3 \cdot (-2) = -6$. This suggests that the product of a positive number and a negative number is a negative number.

What about the product of two negative numbers? To find out, consider the following pattern.

Factor decreases by **1** each time

$$\left.\begin{array}{l} -3 \cdot 2 = -6 \\ -3 \cdot 1 = -3 \\ -3 \cdot 0 = 0 \end{array}\right\} \text{Product increases by } \mathbf{3} \text{ each time.}$$

This pattern continues as

Factor decreases by **1** each time

$$\left.\begin{array}{l} -3 \cdot \quad = 3 \\ -3 \cdot -2 = 6 \end{array}\right\} \text{Product increases by } \mathbf{3} \text{ each time.}$$

This suggests that the product of two negative numbers is a positive number.

> ## Multiplying Real Numbers
>
> **1.** The product of two numbers with the *same* sign is a positive number.
> **2.** The product of two numbers with *different* signs is a negative number.

EXAMPLE 1

Multiply.

a. $(-6)(4)$ **b.** $2(-1)$ **c.** $(-5)(-10)$

Solution   **a.** $(-6)(4) = -24$ **b.** $2(-1) = -2$ **c.** $(-5)(-10) = 50$

We know that every whole number multiplied by zero equals zero. This remains true for real numbers.

> ## Zero as a Factor
>
> If b is a real number, then $b \cdot 0 = 0$. Also, $0 \cdot b = 0$.

EXAMPLE 2

Perform the indicated operations.

a. $(7)(0)(-6)$ **b.** $(-2)(-3)(-4)$ **c.** $(-1)(5)(-9)$ **d.** $(-4)(-11) - (5)(-2)$

Solution   **a.** By the order of operations, we multiply from left to right. Notice that, because one of the factors is 0, the product is 0.

$$(7)(0)(-6) = 0(-6) = 0$$

b. Multiply two factors at a time, from left to right.

$$(-2)(-3)(-4) = (6)(-4) \qquad \text{Multiply } (-2)(-3).$$
$$= -24$$

c. Multiply from left to right.

$$(-1)(5)(-9) = (-5)(-9) \qquad \text{Multiply } (-1)(5).$$
$$= 45$$

d. Follow the rules for order of operation.

$$(-4)(-11) - (5)(-2) = 44 - (-10) \qquad \text{Find each product.}$$
$$= 44 + 10 \qquad \text{Add } \mathbf{44} \text{ to the opposite of } -\mathbf{10}.$$
$$= 54 \qquad \text{Add.}$$

> ## ▶ Helpful Hint
>
> You may have noticed from the example that if we multiply:
>
> • an *even* number of negative numbers, the product is *positive*.
> • an *odd* number of negative numbers, the product is *negative*.

Multiplying signed decimals or fractions is carried out exactly the same way as multiplying by integers.

EXAMPLE 3

Multiply.

a. $(-1.2)(0.05)$ **b.** $\dfrac{2}{3} \cdot \left(-\dfrac{7}{10}\right)$ **c.** $\left(-\dfrac{4}{5}\right)(-20)$

Solution   **a.** The product of two numbers with different signs is negative.

$$(-1.2)(0.05) = -[(1.2)(0.05)]$$
$$= -0.06$$

b. $\dfrac{2}{3} \cdot \left(-\dfrac{7}{10}\right) = -\dfrac{2 \cdot 7}{3 \cdot 10} = -\dfrac{2 \cdot 7}{3 \cdot 2 \cdot 5} = -\dfrac{7}{15}$

c. $\left(-\dfrac{4}{5}\right)(-20) = \dfrac{4 \cdot 20}{5 \cdot 1} = \dfrac{4 \cdot 4 \cdot 5}{5 \cdot 1} = \dfrac{16}{1}$ or 16

Now that we know how to multiply positive and negative numbers, let's see how we find the values of $(-4)^2$ and -4^2, for example. Although these two expressions look similar, the difference between the two is the parentheses. In $(-4)^2$, the parentheses tell us that the base, or repeated factor, is -4. In -4^2, only 4 is the base. Thus,

$$(-4)^2 = (-4)(-4) = 16 \qquad \textit{The base is } -\textbf{4.}$$
$$-4^2 = -(4 \cdot 4) = -16 \qquad \textit{The base is } \textbf{4.}$$

EXAMPLE 4

Evaluate.

a. $(-2)^3$ **b.** -2^3 **c.** $(-3)^2$ **d.** -3^2

Solution   **a.** $(-2)^3 = (-2)(-2)(-2) = -8$ *The base is* $-$**2.**
b. $-2^3 = -(2 \cdot 2 \cdot 2) = -8$ *The base is* **2.**
c. $(-3)^2 = (-3)(-3) = 9$ *The base is* $-$**3.**
d. $-3^2 = -(3 \cdot 3) = -9$ *The base is* **3.**

> **Helpful Hint**
> Be careful when identifying the base of an exponential expression.
>
> $\qquad\qquad (-3)^2 \qquad\qquad\qquad\qquad\qquad -3^2$
> $\qquad\qquad$ Base is -3 $\qquad\qquad\qquad\qquad\quad$ Base is 3
> $\qquad (-3)^2 = (-3)(-3) = 9 \qquad\quad -3^2 = -(3 \cdot 3) = -9$

Just as every difference of two numbers $a - b$ can be written as the sum $a + (-b)$, so too every quotient of two numbers can be written as a product. For example, the quotient $6 \div 3$ can be written as $6 \cdot \frac{1}{3}$. Recall that the pair of numbers 3 and $\frac{1}{3}$ has a special relationship. Their product is 1 and they are called reciprocals or **multiplicative inverses** of each other.

Reciprocals or Multiplicative Inverses

Two numbers whose product is 1 are called reciprocals or multiplicative inverses of each other.

Notice that **0 has no multiplicative inverse** since 0 multiplied by any number is never 1 but always 0.

EXAMPLE 5

Find the reciprocal of each number.

a. 22   **b.** $\dfrac{3}{16}$ **c.** -10 **d.** $-\dfrac{9}{13}$

Solution   **a.** The reciprocal of 22 is $\frac{1}{22}$ since $22 \cdot \frac{1}{22} = 1$.
b. The reciprocal of $\frac{3}{16}$ is $\frac{16}{3}$ since $\frac{3}{16} \cdot \frac{16}{3} = 1$.
c. The reciprocal of -10 is $-\frac{1}{10}$.
d. The reciprocal of $-\frac{9}{13}$ is $-\frac{13}{9}$.

We may now write a quotient as an equivalent product.

Quotient of Two Real Numbers

If a and b are real numbers and b is not 0, then

$$a \div b = \frac{a}{b} = a \cdot \frac{1}{b}$$

In other words, the quotient of two real numbers is the product of the first number and the multiplicative inverse or reciprocal of the second number.

EXAMPLE 6

Use the definition of the quotient of two numbers to divide.

a. $-18 \div 3$ **b.** $\dfrac{-14}{-2}$ **c.** $\dfrac{20}{-4}$

Solution   **a.** $-18 \div 3 = -18 \cdot \dfrac{1}{3} = -6$ **b.** $\dfrac{-14}{-2} = -14 \cdot -\dfrac{1}{2} = 7$

c. $\dfrac{20}{-4} = 20 \cdot -\dfrac{1}{4} = -5$

Since the quotient $a \div b$ can be written as the product $a \cdot \frac{1}{b}$, it follows that sign patterns for dividing two real numbers are the same as sign patterns for multiplying two real numbers.

Multiplying and Dividing Real Numbers

1. The product or quotient of two numbers with the *same* sign is a positive number.

2. The product or quotient of two numbers with *different* signs is a negative number.

EXAMPLE 7

Divide.

a. $\dfrac{-24}{-4}$ **b.** $\dfrac{-36}{3}$ **c.** $\dfrac{2}{3} \div \left(-\dfrac{5}{4}\right)$ **d.** $-\dfrac{3}{2} \div 9$

Solution   **a.** $\dfrac{-24}{-4} = 6$ **b.** $\dfrac{-36}{3} = -12$ **c.** $\dfrac{2}{3} \div \left(-\dfrac{5}{4}\right) = \dfrac{2}{3} \cdot \left(-\dfrac{4}{5}\right) = -\dfrac{8}{15}$

d. $-\dfrac{3}{2} \div 9 = -\dfrac{3}{2} \cdot \dfrac{1}{9} = -\dfrac{3 \cdot 1}{2 \cdot 9} = -\dfrac{3 \cdot 1}{2 \cdot 3 \cdot 3} = -\dfrac{1}{6}$

The definition of the quotient of two real numbers does not allow for division by 0 because 0 does not have a multiplicative inverse. There is no number we can multiply 0 by to get 1. How then do we interpret $\frac{3}{0}$? We say that division by 0 is not allowed or not defined and that $\frac{3}{0}$ does not represent a real number. The denominator of a fraction can never be 0.

Can the numerator of a fraction be 0? Can we divide 0 by a number? Yes. For example,

$$\frac{0}{3} = 0 \cdot \frac{1}{3} = 0$$

In general, the quotient of 0 and any nonzero number is 0.

> ### Zero as a Divisor or Dividend
>
> **1.** The quotient of any nonzero real number and 0 is undefined. In symbols, if
> $$a \neq 0, \frac{a}{0} \text{ is } \textbf{undefined}.$$
>
> **2.** The quotient of 0 and any real number except 0 is 0. In symbols, if
> $$a \neq 0, \frac{0}{a} = 0.$$

EXAMPLE 8

Perform the indicated operations.

a. $\dfrac{1}{0}$ **b.** $\dfrac{0}{-3}$ **c.** $\dfrac{0(-8)}{2}$

Solution   **a.** $\dfrac{1}{0}$ is undefined   **b.** $\dfrac{0}{-3} = 0$ **c.** $\dfrac{0(-8)}{2} = \dfrac{0}{2} = 0$

Notice that $\dfrac{12}{-2} = -6$, $-\dfrac{12}{2} = -6$, and $\dfrac{-12}{2} = -6$. This means that

$$\frac{12}{-2} = -\frac{12}{2} = \frac{-12}{2}$$

In words, a single negative sign in a fraction can be written in the denominator, in the numerator, or in front of the fraction without changing the value of the fraction. Thus,

$$\frac{1}{-7} = \frac{-1}{7} = -\frac{1}{7}$$

In general, if a and b are real numbers, $b \neq 0$, $\dfrac{a}{-b} = \dfrac{-a}{b} = -\dfrac{a}{b}$.

Examples combining basic arithmetic operations along with the principles of order of operations help us to review these concepts.

EXAMPLE 9

Simplify each expression.

a. $\dfrac{(-12)(-3) + 3}{-7 - (-2)}$ **b.** $\dfrac{2(-3)^2 - 20}{-5 + 4}$

Solution **a.** First, simplify the numerator and denominator separately, then divide.

$$\frac{(-12)(-3) + 3}{-7 - (-2)} = \frac{36 + 3}{-7 + 2}$$

$$= \frac{39}{-5} \text{ or } -\frac{39}{5}$$

b. Simplify the numerator and denominator separately, then divide.

$$\frac{2(-3)^2 - 20}{-5 + 4} = \frac{2 \cdot 9 - 20}{-5 + 4} = \frac{18 - 20}{-5 + 4} = \frac{-2}{-1} = 2$$

2 Using what we have learned about multiplying and dividing real numbers, we continue to practice evaluating algebraic expressions.

EXAMPLE 10

If $x = -2$ and $y = -4$, evaluate each expression.

a. $5x - y$ **b.** $x^4 - y^2$ **c.** $\dfrac{3x}{2y}$

Solution **a.** Replace x with -2 and y with -4 and simplify.

$$5x - y = 5(-2) - (-4) = -10 - (-4) = -10 + 4 = -6$$

b. Replace x with -2 and y with -4.

$$x^4 - y^2 = (-2)^4 - (-4)^2 \quad \text{Substitute the given values for the variables.}$$

$$= 16 - (16) \quad \text{Evaluate exponential expressions.}$$

$$= 0 \quad \text{Subtract.}$$

c. Replace x with -2 and y with -4 and simplify.

$$\frac{3x}{2y} = \frac{3(-2)}{2(-4)} = \frac{-6}{-8} = \frac{3}{4}$$

Calculator Explorations

Entering Negative Numbers on a Scientific Calculator

To enter a negative number on a scientific calculator, find a key marked $\boxed{+/-}$. (On some calculators, this key is marked $\boxed{\text{CHS}}$ for "change sign.") To enter -8, for example, press the keys $\boxed{8}$ $\boxed{+/-}$. The display will read $\boxed{-8}$. (*continued*)

Entering Negative Numbers on a Graphing Calculator

To enter a negative number on a graphing calculator, find a key marked $\boxed{(-)}$.
Do not confuse this key with the key $\boxed{-}$, which is used for subtraction. To enter -8, for example, press the keys $\boxed{(-)}$ $\boxed{8}$. The display will read $\boxed{-8}$.

Operations with Real Numbers

To evaluate $-2(7 - 9) - 20$ on a calculator, press the keys

$\boxed{2}$ $\boxed{+/-}$ $\boxed{\times}$ $\boxed{(}$ $\boxed{7}$ $\boxed{-}$ $\boxed{9}$ $\boxed{)}$ $\boxed{-}$ $\boxed{2}$ $\boxed{0}$ $\boxed{=}$, or

$\boxed{(-)}$ $\boxed{2}$ $\boxed{(}$ $\boxed{7}$ $\boxed{-}$ $\boxed{9}$ $\boxed{)}$ $\boxed{-}$ $\boxed{2}$ $\boxed{0}$ $\boxed{\text{ENTER}}$.

The display will read $\boxed{-16}$ or $\boxed{\begin{array}{c} -2(7 - 9) - 20 \\ -16 \end{array}}$.

Use a calculator to simplify each expression.

1. $-38(26 - 27)$

2. $-59(-8) + 1726$

3. $134 + 25(68 - 91)$

4. $45(32) - 8(218)$

5. $\dfrac{-50(294)}{175 - 265}$

6. $\dfrac{-444 - 444.8}{-181 - 324}$

7. $9^5 - 4550$

8. $5^8 - 6259$

9. $(-125)^2$ (Be careful.)

10. -125^2 (Be careful.)

MENTAL MATH

Answer the following with positive or negative.

1. The product of two negative numbers is a _____ number.
2. The quotient of two negative numbers is a _____ number.
3. The quotient of a positive number and a negative number is a _____ number.
4. The product of a positive number and a negative number is a _____ number.
5. The reciprocal of a positive number is a _____ number.
6. The opposite of a positive number is a _____ number.

EXERCISE SET 1.7

STUDY GUIDE/SSM   CD/ VIDEO   PH MATH TUTOR CENTER   MathXL®Tutorials ON CD   MathXL®   MyMathLab®

Multiply. See Examples 1 through 3.

 1. $-6(4)$

2. $-8(5)$

3. $2(-1)$

4. $7(-4)$

5. $-5(-10)$

6. $-6(-11)$

7. $-3 \cdot 4$

8. $-2 \cdot 8$

9. $-7 \cdot 0$

10. $-6 \cdot 0$

11. $2(-9)$

12. $3(-5)$

13. $-\dfrac{1}{2}\left(-\dfrac{3}{5}\right)$

14. $-\dfrac{1}{8}\left(-\dfrac{1}{3}\right)$

15. $-\dfrac{3}{4}\left(-\dfrac{8}{9}\right)$

16. $-\dfrac{5}{6}\left(-\dfrac{3}{10}\right)$

17. $5(-1.4)$

18. $6(-2.5)$

19. $-0.2(-0.7)$

20. $-0.5(-0.3)$

21. $-10(80)$

22. $-20(60)$

23. $4(-7)$

24. $5(-9)$

25. $(-5)(-5)$

26. $(-7)(-7)$

27. $\dfrac{2}{3}\left(-\dfrac{4}{9}\right)$

28. $\dfrac{2}{7}\left(-\dfrac{2}{11}\right)$

29. $-11(11)$

30. $-12(12)$

31. $-\dfrac{20}{25}\left(\dfrac{5}{16}\right)$

32. $-\dfrac{25}{36}\left(\dfrac{6}{15}\right)$

33. $(-1)(2)(-3)(-5)$ **34.** $(-2)(-3)(-4)(-2)$

Perform the indicated operation. See Example 2.

35. $(-2)(5) - (-11)(3)$ **36.** $8(-3) - 4(-5)$
37. $(-6)(-1)(-2) - (-5)$
38. $20 - (-4)(3)(-2)$

Decide whether each statement is true or false.

39. The product of three negative integers is negative.
40. The product of three positive integers is positive.
41. The product of four negative integers is negative.
42. The product of four positive integers is positive.

Evaluate. See Example 4.

43. $(-2)^4$ **44.** -2^4
45. -1^5 **46.** $(-1)^5$
47. $(-5)^2$ **48.** -5^2
49. -7^2 **50.** $(-7)^2$

Find each reciprocal or multiplicative inverse. See Example 5.

51. 9 **52.** 100 **53.** $\dfrac{2}{3}$

54. $\dfrac{1}{7}$ **55.** -14 **56.** -8

57. $-\dfrac{3}{11}$ **58.** $-\dfrac{6}{13}$ **59.** 0.2

60. 1.5 **61.** $\dfrac{1}{-6.3}$ **62.** $\dfrac{1}{-8.9}$

Divide. See Examples 6 through 8.

63. $\dfrac{18}{-2}$ **64.** $\dfrac{20}{-10}$ **65.** $\dfrac{-16}{-4}$

66. $\dfrac{-18}{-6}$ **67.** $\dfrac{-48}{12}$ **68.** $\dfrac{-60}{5}$

69. $\dfrac{0}{-4}$ **70.** $\dfrac{0}{-9}$ **71.** $-\dfrac{15}{3}$

72. $-\dfrac{24}{8}$ **73.** $\dfrac{5}{0}$ **74.** $\dfrac{3}{0}$

75. $\dfrac{-12}{-4}$ **76.** $\dfrac{-45}{-9}$ **77.** $\dfrac{30}{-2}$

78. $\dfrac{14}{-2}$ **79.** $\dfrac{6}{7} \div \left(-\dfrac{1}{3}\right)$ **80.** $\dfrac{4}{5} \div \left(-\dfrac{1}{2}\right)$

81. $-\dfrac{5}{9} \div \left(-\dfrac{3}{4}\right)$ **82.** $-\dfrac{1}{10} \div \left(-\dfrac{8}{11}\right)$

83. $-\dfrac{4}{9} \div \dfrac{4}{9}$ **84.** $-\dfrac{5}{12} \div \dfrac{5}{12}$

MIXED PRACTICE

Simplify. See Example 9.

85. $\dfrac{-9(-3)}{-6}$ **86.** $\dfrac{-6(-3)}{-4}$

87. $\dfrac{12}{9 - 12}$ **88.** $\dfrac{-15}{1 - 4}$

89. $\dfrac{-6^2 + 4}{-2}$ **90.** $\dfrac{3^2 + 4}{5}$

91. $\dfrac{8 + (-4)^2}{4 - 12}$ **92.** $\dfrac{6 + (-2)^2}{4 - 9}$

93. $\dfrac{22 + (3)(-2)}{-5 - 2}$ **94.** $\dfrac{-20 + (-4)(3)}{1 - 5}$

95. $\dfrac{-3 - 5^2}{2(-7)}$ **96.** $\dfrac{-2 - 4^2}{3(-6)}$

97. $\dfrac{6 - 2(-3)}{4 - 3(-2)}$ **98.** $\dfrac{8 - 3(-2)}{2 - 5(-4)}$

99. $\dfrac{-3 - 2(-9)}{-15 - 3(-4)}$ **100.** $\dfrac{-4 - 8(-2)}{-9 - 2(-3)}$

101. $\dfrac{|5 - 9| + |10 - 15|}{|2(-3)|}$ **102.** $\dfrac{|-3 + 6| + |-2 + 7|}{|-2 \cdot 2|}$

If $x = -5$ and $y = -3$, evaluate each expression. See Example 10.

103. $3x + 2y$ **104.** $4x + 5y$
105. $2x^2 - y^2$ **106.** $x^2 - 2y^2$
107. $x^3 + 3y$ **108.** $y^3 + 3x$

109. $\dfrac{2x - 5}{y - 2}$ **110.** $\dfrac{2y - 12}{x - 4}$

111. $\dfrac{-3 - y}{x - 4}$ **112.** $\dfrac{4 - 2x}{y + 3}$

113. At the end of 2002, Delta Airlines posted a net loss of $1272 million, which we will write as $-\$1272$ million. If this continued, what would Delta's income be after four years? (*Source*: Delta Airlines)

114. Gateway sells personal computers through mail order and retail stores. In 2002, Gateway posted a net income of $-\$298$ million. If this continued, what would Gateway's income be after three years? (*Source*: Gateway)

Decide whether the given number is a solution of the given equation.

115. Is 7 a solution of $-5x = -35$?

116. Is -4 a solution of $2x = x - 1$?

117. Is -20 a solution of $\dfrac{x}{10} = 2$?

118. Is -3 a solution of $\dfrac{45}{x} = -15$?

119. Is 5 a solution of $-3x - 5 = -20$?

120. Is -4 a solution of $2x + 4 = x + 8$?

Concept Extensions

121. Explain why the product of an even number of negative numbers is a positive number.

122. If a and b are any real numbers, is the statement $a \cdot b = b \cdot a$ always true? Why or why not?

123. Find any real numbers that are their own reciprocal.

124. Explain why 0 has no reciprocal.

If q is a negative number, r is a negative number, and t is a positive number, determine whether each expression simplifies to a positive or negative number. If it is not possible to determine, state so.

125. $\dfrac{q}{r \cdot t}$

126. $q^2 \cdot r \cdot t$

127. $q + t$

128. $t + r$

129. $t(q + r)$

130. $r(q - t)$

Write each of the following as an expression and evaluate.

131. The sum of -2 and the quotient of -15 and 3

132. The sum of 1 and the product of -8 and -5

133. Twice the sum of -5 and -3

134. 7 subtracted from the quotient of 0 and 5

OPERATIONS ON REAL NUMBERS

INTEGRATED REVIEW

Answer the following with positive or negative.

1. The product of two negative numbers is a _____ number.

2. The quotient of two negative numbers is a _____ number.

3. The quotient of a positive number and a negative number is a _____ number.

4. The product of a positive number and a negative number is a _____ number.

5. The reciprocal of a positive number is a _____ number.

6. The opposite of a positive number is a _____ number.

7. The sum of two negative numbers is a _____ number.

8. The absolute value of a negative number is a _____ number.

Perform each indicated operation and simplify.

9. $5(-7)$

10. $-3(-10)$

11. $\dfrac{-20}{-4}$

12. $\dfrac{30}{-6}$

13. $7 - (-3)$

14. $-8 - 10$

15. $-14 - (-12)$

16. $-3 - (-1)$

17. $-\dfrac{1}{2}\left(-\dfrac{3}{4}\right)$

18. $-\dfrac{2}{7}\left(\dfrac{11}{12}\right)$

19. $\dfrac{-12}{0.2}$

20. $\dfrac{-3.8}{-2}$

21. $-19 + (-23)$

22. $18 + (-25)$

23. $-15 + 17$

24. $-2 + (-37)$

25. $(-8)^2$

26. -9^2

27. -3^3

28. $(-2)^4$

29. -1^{10}

30. $(-1)^{10}$

31. $(-2)^5$

32. -2^5

33. $(2)(-8)(-3)$

34. $3(-2)(5)$

35. $-6(2) + 20 \div 2 - 4$

36. $-4(-3) + 9 \div 3 - 6$

37. $-3^2 - [6 + 5|-2 - 1|]$

38. $-5^2 - [4 + 3|-3 - 2|]$

39. $2(19 - 17)^3 - 3(7 - 9)^2$

40. $3(10 - 9)^2 - 6(20 - 19)^3$

41. $\dfrac{19 - 25}{3(-1)}$

42. $\dfrac{8(-4)}{-2}$

43. $\dfrac{-2(3 - 6) - 6(10 - 9)}{-6 - (-5)}$

44. $\dfrac{5(7 - 9) - 3(100 - 97)}{4 - 5}$

45. $\dfrac{-4(8 - 10)^3}{-2 - 1 - 12}$

46. $\dfrac{6(7 - 10)^2}{6 - (-1) - 2}$

1.8   *PROPERTIES OF REAL NUMBERS*

Objectives

1   Use the commutative and associative properties.

2   Use the distributive property.

3   Use the identity and inverse properties.

1   In this section we give names to properties of real numbers with which we are already familiar. Throughout this section, the variables *a*, *b*, and *c* represent real numbers.

We know that order does not matter when adding numbers. For example, we know that $7 + 5$ is the same as $5 + 7$. This property is given a special name—the **commutative property of addition**. We also know that order does not matter when multiplying numbers. For example, we know that $-5(6) = 6(-5)$. This property means that multiplication is commutative also and is called the **commutative property of multiplication**.

Commutative Properties

Addition: $a + b = b + a$

Multiplication: $a \cdot b = b \cdot a$

These properties state that the *order* in which any two real numbers are added or multiplied does not change their sum or product. For example, if we let $a = 3$ and $b = 5$, then the commutative properties guarantee that

$$3 + 5 = 5 + 3 \qquad \text{and} \qquad 3 \cdot 5 = 5 \cdot 3$$

Helpful Hint

Is subtraction also commutative? Try an example. Is $3 - 2 = 2 - 3$? **No!** The left side of this statement equals 1; the right side equals -1. There is no commutative property of subtraction. Similarly, there is no commutative property for division. For example, $10 \div 2$ does not equal $2 \div 10$.

STUDY SKILLS REMINDER

You may want to start a list in your notes describing the situations in which you should take extra caution, such as

1. cannot divide by 0

2. subtraction is not commutative—order matters

EXAMPLE 1

Use a commutative property to complete each statement.

a. $x + 5 =$ _____   **b.** $3 \cdot x =$ _____

Solution **a.** $x + 5 = 5 + x$ By the commutative property of addition
b. $3 \cdot x = x \cdot 3$ By the commutative property of multiplication

✔ **CONCEPT CHECK**

Which of the following pairs of actions are commutative?

a. "raking the leaves" and "bagging the leaves"
b. "putting on your left glove" and "putting on your right glove"
c. "putting on your coat" and "putting on your shirt"
d. "reading a novel" and "reading a newspaper"

Let's now discuss grouping numbers. We know that when we add three numbers, the way in which they are grouped or associated does not change their sum. For example, we know that $2 + (3 + 4) = 2 + 7 = 9$. This result is the same if we group the numbers differently. In other words, $(2 + 3) + 4 = 5 + 4 = 9$, also. Thus, $2 + (3 + 4) = (2 + 3) + 4$. This property is called the **associative property of addition**.

We also know that changing the grouping of numbers when multiplying does not change their product. For example, $2 \cdot (3 \cdot 4) = (2 \cdot 3) \cdot 4$ (check it). This is the **associative property of multiplication**.

Associative Properties

Addition: $(a + b) + c = a + (b + c)$
Multiplication: $(a \cdot b) \cdot c = a \cdot (b \cdot c)$

These properties state that the way in which three numbers are *grouped* does not change their sum or their product.

EXAMPLE 2

Use an associative property to complete each statement.

a. $5 + (4 + 6) =$ _____   **b.** $(-1 \cdot 2) \cdot 5 =$ _____

Solution **a.** $5 + (4 + 6) = (5 + 4) + 6$ By the associative property of addition
b. $(-1 \cdot 2) \cdot 5 = -1 \cdot (2 \cdot 5)$ By the associative property of multiplication

---

▶ **Helpful Hint**
Remember the difference between the commutative properties and the associative properties. The commutative properties have to do with the *order* of numbers, and the associative properties have to do with the *grouping* of numbers.

Concept Check Answer:
b, d

Let's now illustrate how these properties can help us simplify expressions.

EXAMPLE 3

Simplify each expression.

a. $10 + (x + 12)$ **b.** $-3(7x)$

Solution   **a.** $10 + (x + 12) = 10 + (12 + x)$ *By the commutative property of addition*

$$= (10 + 12) + x$$ *By the associative property of addition*

$$= 22 + x$$ *Add.*

b. $-3(7x) = (-3 \cdot 7)x$ *By the associative property of multiplication*

$$= -21x$$ *Multiply.*

2   The **distributive property of multiplication over addition** is used repeatedly throughout algebra. It is useful because it allows us to write a product as a sum or a sum as a product.

We know that $7(2 + 4) = 7(6) = 42$. Compare that with $7(2) + 7(4) = 14 + 28 = 42$. Since both original expressions equal 42, they must equal each other, or

$$7(2 + 4) = 7(2) + 7(4)$$

This is an example of the distributive property. The product on the left side of the equal sign is equal to the sum on the right side. We can think of the 7 as being distributed to each number inside the parentheses.

Distributive Property of Multiplication Over Addition

$$a(b + c) = ab + ac$$

Since multiplication is commutative, this property can also be written as

$$(b + c)a = ba + ca$$

The distributive property can also be extended to more than two numbers inside the parentheses. For example,

$$3(x + y + z) = 3(x) + 3(y) + 3(z)$$

$$= 3x + 3y + 3z$$

Since we define subtraction in terms of addition, the distributive property is also true for subtraction. For example

$$2(x - y) = 2(x) - 2(y)$$

$$= 2x - 2y$$

EXAMPLE 4

Use the distributive property to write each expression without parentheses. Then simplify if possible.

a. $2(x + y)$ **b.** $-5(-3 + 2z)$ **c.** $5(x + 3y - z)$

d. $-1(2 - y)$ **e.** $-(3 + x - w)$ **f.** $4(3x + 7) + 10$

Solution   **a.** $2(x + y) = 2 \cdot x + 2 \cdot y$

$\qquad\qquad = 2x + 2y$

b. $-5(-3 + 2z) = -5(-3) + (-5)(2z)$

$\qquad\qquad\qquad = 15 - 10z$

c. $5(x + 3y - z) = 5(x) + 5(3y) - 5(z)$

$\qquad\qquad\qquad = 5x + 15y - 5z$

d. $-1(2 - y) = (-1)(2) - (-1)(y)$

$\qquad\qquad\quad = -2 + y$

e. $-(3 + x - w) = -1(3 + x - w)$

$\qquad\qquad\qquad = (-1)(3) + (-1)(x) - (-1)(w)$

$\qquad\qquad\qquad = -3 - x + w$

> **Helpful Hint**
>
> Notice in part **e** that $-(3 + x - w)$ is first rewritten as $-1(3 + x - w)$.

f. $4(3x + 7) + 10 = 4(3x) + 4(7) + 10$ Apply the distributive property.

$\qquad\qquad\qquad = 12x + 28 + 10$ Multiply.

$\qquad\qquad\qquad = 12x + 38$ Add.

We can use the distributive property in reverse to write a sum as a product.

EXAMPLE 5

Use the distributive property to write each sum as a product.

a. $8 \cdot 2 + 8 \cdot x$ **b.** $7s + 7t$

Solution   **a.** $8 \cdot 2 + 8 \cdot x = 8(2 + x)$ **b.** $7s + 7t = 7(s + t)$

3   Next, we look at the **identity properties**.

The number 0 is called the identity for addition because when 0 is added to any real number, the result is the same real number. In other words, the *identity* of the real number is not changed.

The number 1 is called the identity for multiplication because when a real number is multiplied by 1, the result is the same real number. In other words, the *identity* of the real number is not changed.

> ## Identities for Addition and Multiplication
>
> 0 is the identity element for addition.
>
> $$a + 0 = a \qquad \text{and} \qquad 0 + a = a$$
>
> 1 is the identity element for multiplication.
>
> $$a \cdot 1 = a \qquad \text{and} \qquad 1 \cdot a = a$$

Notice that 0 is the *only* number that can be added to any real number with the result that the sum is the same real number. Also, 1 is the *only* number that can be multiplied by any real number with the result that the product is the same real number.

Additive inverses or **opposites** were introduced in Section 1.5. Two numbers are called additive inverses or opposites if their sum is 0. The additive inverse or opposite of 6 is −6 because $6 + (-6) = 0$. The additive inverse or opposite of −5 is 5 because $-5 + 5 = 0$.

Reciprocals or **multiplicative inverses** were introduced in Section 1.3. Two nonzero numbers are called reciprocals or multiplicative inverses if their product is 1. The reciprocal or multiplicative inverse of $\frac{2}{3}$ is $\frac{3}{2}$ because $\frac{2}{3} \cdot \frac{3}{2} = 1$. Likewise, the reciprocal of −5 is $-\frac{1}{5}$ because $-5\left(-\frac{1}{5}\right) = 1$.

CONCEPT CHECK

Which of the following is the

a. opposite of $-\frac{3}{10}$, and which is the

b. reciprocal of $-\frac{3}{10}$?

$$1, -\frac{10}{3}, \frac{3}{10}, 0, \frac{10}{3}, -\frac{3}{10}$$

Additive or Multiplicative Inverses

The numbers a and $-a$ are additive inverses or opposites of each other because their sum is 0; that is,

$$a + (-a) = 0$$

The numbers b and $\frac{1}{b}$ (for $b \neq 0$) are reciprocals or multiplicative inverses of each other because their product is 1; that is,

$$b \cdot \frac{1}{b} = 1$$

EXAMPLE 6

Name the property or properties illustrated by each true statement.

Solution
a. $3 \cdot y = y \cdot 3$ — Commutative property of multiplication (order changed)
b. $(x + 7) + 9 = x + (7 + 9)$ — Associative property of addition (grouping changed)
c. $(b + 0) + 3 = b + 3$ — Identity element for addition
d. $0.2 \cdot (z \cdot 5) = 0.2 \cdot (5 \cdot z)$ — Commutative property of multiplication (order changed)
e. $-2 \cdot \left(-\frac{1}{2}\right) = 1$ — Multiplicative inverse property
f. $-2 + 2 = 0$ — Additive inverse property
g. $-6 \cdot (y \cdot 2) = (-6 \cdot 2) \cdot y$ — Commutative and associative properties of multiplication (order and grouping changed)

Concept Check Answer:
a. $\frac{3}{10}$ **b.** $-\frac{10}{3}$

EXERCISE SET 1.8

STUDY   CD/   PH MATH   MathXL®Tutorials   MathXL®   MyMathLab®
GUIDE/SSM   VIDEO   TUTOR CENTER   ON CD

Use a commutative property to complete each statement. See Examples 1 and 3.

 1. $x + 16 =$ _____   **2.** $4 + y =$ _____

3. $-4 \cdot y =$ _____   **4.** $-2 \cdot x =$ _____

 5. $xy =$ __   **6.** $ab =$ ____

7. $2x + 13 =$ _____   **8.** $19 + 3y =$ _____

Use an associative property to complete each statement. See Examples 2 and 3.

9. $(xy) \cdot z =$ _____   **10.** $3 \cdot (xy) =$ _____

11. $2 + (a + b) =$ _____   **12.** $(y + 4) + z =$ _____

13. $4 \cdot (ab) =$ _____   **14.** $(-3y) \cdot z =$ _____

15. $(a + b) + c =$ _____

16. $6 + (r + s) =$ _____

Use the commutative and associative properties to simplify each expression. See Example 3.

 17. $8 + (9 + b)$ **18.** $(r + 3) + 11$

19. $4(6y)$ **20.** $2(42x)$

21. $\frac{1}{5}(5y)$ **22.** $\frac{1}{8}(8z)$

23. $(13 + a) + 13$ **24.** $7 + (x + 4)$

25. $-9(8x)$ **26.** $-3(12y)$

27. $\frac{3}{4}\left(\frac{4}{3}s\right)$ **28.** $\frac{2}{7}\left(\frac{7}{2}r\right)$

29. Write an example that shows that division is not commutative.

30. Write an example that shows that subtraction is not commutative.

Use the distributive property to write each expression without parentheses. Then simplify the result. See Example 4.

31. $4(x + y)$ **32.** $7(a + b)$

33. $9(x - 6)$ **34.** $11(y - 4)$

35. $2(3x + 5)$ **36.** $5(7 + 8y)$

37. $7(4x - 3)$ **38.** $3(8x - 1)$

 39. $3(6 + x)$ **40.** $2(x + 5)$

41. $-2(y - z)$ **42.** $-3(z - y)$

43. $-7(3y + 5)$ **44.** $-5(2r + 11)$

45. $5(x + 4m + 2)$ **46.** $8(3y + z - 6)$

47. $-4(1 - 2m + n)$ **48.** $-4(4 + 2p + 5)$

49. $-(5x + 2)$ **50.** $-(9r + 5)$

 51. $-(r - 3 - 7p)$ **52.** $-(q - 2 + 6r)$

53. $\frac{1}{2}(6x + 8)$ **54.** $\frac{1}{4}(4x - 2)$

55. $-\frac{1}{3}(3x - 9y)$ **56.** $-\frac{1}{5}(10a - 25b)$

57. $3(2r + 5) - 7$ **58.** $10(4s + 6) - 40$

59. $-9(4x + 8) + 2$ **60.** $-11(5x + 3) + 10$

61. $-4(4x + 5) - 5$ **62.** $-6(2x + 1) - 1$

Use the distributive property to write each sum as a product. See Example 5.

63. $4 \cdot 1 + 4 \cdot y$ **64.** $14 \cdot z + 14 \cdot 5$

65. $11x + 11y$ **66.** $9a + 9b$

67. $(-1) \cdot 5 + (-1) \cdot x$ **68.** $(-3)a + (-3)b$

69. $30a + 30b$ **70.** $25x + 25y$

Name the properties illustrated by each true statement. See Example 6.

71. $3 \cdot 5 = 5 \cdot 3$

72. $4(3 + 8) = 4 \cdot 3 + 4 \cdot 8$

73. $2 + (x + 5) = (2 + x) + 5$

74. $(x + 9) + 3 = (9 + x) + 3$

75. $9(3 + 7) = 9 \cdot 3 + 9 \cdot 7$

76. $1 \cdot 9 = 9$

77. $(4 \cdot y) \cdot 9 = 4 \cdot (y \cdot 9)$

78. $6 \cdot \frac{1}{6} = 1$

79. $0 + 6 = 6$

80. $(a + 9) + 6 = a + (9 + 6)$

81. $-4(y + 7) = -4 \cdot y + (-4) \cdot 7$

82. $(11 + r) + 8 = (r + 11) + 8$

83. $-4 \cdot (8 \cdot 3) = (8 \cdot -4) \cdot 3$

84. $r + 0 = r$

Concept Extensions

Fill in the table with the opposite (additive inverse), and the reciprocal (multiplicative inverse). Assume that the value of each expression is not 0.

	Expression	Opposite	Reciprocal
85.	8		
86.	$-\frac{2}{3}$		
87.	x		
88.	$4y$		
89.			$\frac{1}{2x}$
90.		$7x$	

Determine which pairs of actions are commutative.

91. "taking a test" and "studying for the test"

92. "putting on your shoes" and "putting on your socks"

93. "putting on your left shoe" and "putting on your right shoe"

94. "reading the sports section" and "reading the comics section"

95. Explain why 0 is called the identity element for addition.

96. Explain why 1 is called the identity element for multiplication.

CHAPTER 1 PROJECT

Creating and Interpreting Graphs

Companies often rely on market research to investigate the types of consumers who buy their products and their competitors' products. One way of gathering this type of data is through the use of surveys. The raw data collected from surveys can be difficult to interpret without some organization. Graphs organize data visually. They also allow a user to interpret the data quickly.

In this project, you will conduct a brief survey and create tables and graphs to represent the results. This project may be completed by working in groups or individually.

1. Begin by conducting this survey with fellow students, either in this class or another class.

2. For each survey question, tally the results for each response category. Present the results in a table.

3. For each survey question, find the fraction of the total number of responses that fall in each response category.

4. For each survey question, decide which type of graph would best represent the data: bar graph, line graph, or circle graph. Then create an appropriate graph for each set of responses to the questions.

5. Use the data for the two Internet questions and the two newspaper questions to record the increase or decrease in the number of people who participated in these activities. Complete the table in the next column.

6. Study the tables and graphs. What may you conclude from them? What do they tell you about your survey respondents? Write a paragraph summarizing your findings.

Activity	Increase or Decrease from 3 Years Ago
Using the Internet more than 4 hours per week	
Reading a newspaper at least once per week	

Survey

What is your age?

Under 20   20s   30s   40s   50s   60 and older

What is your gender?

Female   Male

Do you currently use the Internet more than 4 hours per week?

Yes   No

Three years ago, did you use the Internet more than 4 hours per week?

Yes   No

Do you currently read a newspaper at least once per week?

Yes   No

Three years ago, did you read a newspaper at least once per week?

Yes   No

CHAPTER VOCABULARY CHECK

Fill in each blank with one of the words or phrases listed below.

set	inequality symbols	opposites	absolute value	numerator
denominator	grouping symbols	exponent	base	reciprocals
variable	equation	solution		

1. The symbols ≠, <, and > are called _____.

2. A mathematical statement that two expressions are equal is called an _____.

3. The _____ of a number is the distance between that number and 0 on the number line.

4. A symbol used to represent a number is called a _____.

5. Two numbers that are the same distance from 0 but lie on opposite sides of 0 are called _____.

6. The number in a fraction above the fraction bar is called the _____.

7. A _____ of an equation is a value for the variable that makes the equation a true statement.

8. Two numbers whose product is 1 are called _____.

9. In 2^3, the 2 is called the ____ and the 3 is called the _____.

10. The number in a fraction below the fraction bar is called the _____.

11. Parentheses and brackets are examples of _____.

12. A ___ is a collection of objects.

CHAPTER 1 HIGHLIGHTS

Definitions and Concepts	Examples

Section 1.2 Symbols and Sets of Numbers

A **set** is a collection of objects, called **elements**, enclosed in braces.

$\{a, c, e\}$

Natural Numbers: $\{1, 2, 3, 4, \dots\}$
Whole Numbers: $\{0, 1, 2, 3, 4, \dots\}$

Integers: $\{\dots, -3, -2, -1, 0, 1, 2, 3, \dots\}$

Rational Numbers: {real numbers that can be expressed as a quotient of integers}

Irrational Numbers: {real numbers that cannot be expressed as a quotient of integers}

Real Numbers: {all numbers that correspond to a point on the number line}

A line used to picture numbers is called a **number line**.

The **absolute value** of a real number a, denoted by $|a|$, is the distance between a and 0 on the number line.

Symbols: $=$ is equal to
$\neq$ is not equal to
$>$ is greater than
$<$ is less than
$\leq$ is less than or equal to
$\geq$ is greater than or equal to

Order Property for Real Numbers
For any two real numbers a and b, a is less than b if a is to the left of b on a number line.

Given the set $\{-3.4, \sqrt{3}, 0, \frac{2}{3}, 5, -4\}$, list the numbers that belong to the set of

Natural numbers: 5

Whole numbers: 0, 5

Integers: $-4, 0, 5$

Rational numbers: $-4, -3.4, 0, \frac{2}{3}, 5$

Irrational Numbers: $\sqrt{3}$

Real numbers: $-4, -3.4, 0, \frac{2}{3}, \sqrt{3}, 5$

$|5| = 5 \qquad |0| = 0 \qquad |-2| = 2$

$-7 = -7$

$3 \neq -3$

$4 > 1$

$1 < 4$

$6 \leq 6$

$18 \geq -\dfrac{1}{3}$

$-3 < 0 \qquad 0 > -3 \qquad 0 < 2.5 \qquad 2.5 > 0$

Definitions and Concepts	**Examples**

Section 1.3 Fractions

A quotient of two integers is called a **fraction**. The **numerator** of a fraction is the top number. The **denominator** of a fraction is the bottom number.

$$\frac{13}{17} \begin{matrix} \leftarrow \text{numerator} \\ \leftarrow \text{denominator} \end{matrix}$$

If $a \cdot b = c$, then a and b are **factors** and c is the **product**.

$$7 \quad \cdot \quad 9 \quad = \quad 63$$
$$\downarrow \qquad \quad \downarrow \qquad \qquad \downarrow$$
$$\text{factor} \qquad \text{factor} \qquad \text{product}$$

A fraction is in **lowest terms** when the numerator and the denominator have no factors in common other than 1.

$\frac{13}{17}$ is in lowest terms.

To write a fraction in lowest terms, factor the numerator and the denominator; then apply the fundamental property.

Write in lowest terms.

$$\frac{6}{14} = \frac{2 \cdot 3}{2 \cdot 7} = \frac{3}{7}$$

Two fractions are **reciprocals** if their product is 1. The reciprocal of $\frac{a}{b}$ is $\frac{b}{a}$.

The reciprocal of $\frac{6}{25}$ is $\frac{25}{6}$

To multiply fractions, numerator times numerator is the numerator of the product and denominator times denominator is the denominator of the product.

Perform the indicated operations.

$$\frac{2}{5} \cdot \frac{3}{7} = \frac{6}{35}$$

To divide fractions, multiply the first fraction by the reciprocal of the second fraction.

$$\frac{5}{9} \div \frac{2}{7} = \frac{5}{9} \cdot \frac{7}{2} = \frac{35}{18}$$

To add fractions with the same denominator, add the numerators and place the sum over the common denominator.

$$\frac{5}{11} + \frac{3}{11} = \frac{8}{11}$$

To subtract fractions with the same denominator, subtract the numerators and place the difference over the common denominator.
Fractions that represent the same quantity are called **equivalent fractions**.

$$\frac{13}{15} - \frac{3}{15} = \frac{10}{15} = \frac{2}{3}$$

$$\frac{1}{5} = \frac{1 \cdot 4}{5 \cdot 4} = \frac{4}{20}$$

$\frac{1}{5}$ and $\frac{4}{20}$ are equivalent fractions.

Section 1.4 Introduction to Variable Expressions and Equations

The expression a^n is an **exponential expression**. The number a is called the **base**; it is the repeated factor. The number n is called the **exponent**; it is the number of times that the base is a factor.

$$4^3 = 4 \cdot 4 \cdot 4 = 64$$
$$7^2 = 7 \cdot 7 = 49$$

Order of Operations
Simplify expressions in the following order. If grouping symbols are present, simplify expressions within those first, starting with the innermost set. Also, simplify the numerator and the denominator of a fraction separately.

$$\frac{8^2 + 5(7-3)}{3 \cdot 7} = \frac{8^2 + 5(4)}{21}$$
$$= \frac{64 + 5(4)}{21}$$
$$= \frac{64 + 20}{21}$$
$$= \frac{84}{21}$$
$$= 4$$

1. Simplify exponential expressions.
2. Multiply or divide in order from left to right.
3. Add or subtract in order from left to right.

Definitions and Concepts	**Examples**

Section 1.4 Introduction to Variable Expressions and Equations

A symbol used to represent a number is called a **variable**.

An **algebraic expression** is a collection of numbers, variables, operation symbols, and grouping symbols.

To evaluate an algebraic expression containing a variable, substitute a given number for the variable and simplify.

A mathematical statement that two expressions are equal is called an **equation**.

A **solution** or **root** of an equation is a value for the variable that makes the equation a true statement.

Examples of variables are:
$$q, x, z$$
Examples of algebraic expressions are:
$$5x, 2(y - 6), \frac{q^2 - 3q + 1}{6}$$

Evaluate $x^2 - y^2$ if $x = 5$ and $y = 3$.
$$x^2 - y^2 = (5)^2 - 3^2$$
$$= 25 - 9$$
$$= 16$$

Equations:
$$3x - 9 = 20$$
$$A = \pi r^2$$

Determine whether 4 is a solution of $5x + 7 = 27$.
$$5x + 7 = 27$$
$$5(4) + 7 \stackrel{?}{=} 27$$
$$20 + 7 \stackrel{?}{=} 27$$
$$27 = 27 \quad \text{True}$$
4 is a solution.

Section 1.5 Adding Real Numbers

To Add Two Numbers with the Same Sign
1. Add their absolute values.
2. Use their common sign as the sign of the sum.

To Add Two Numbers with Different Signs
1. Subtract their absolute values.
2. Use the sign of the number whose absolute value is larger as the sign of the sum.

Two numbers that are the same distance from 0 but lie on opposite sides of 0 are called **opposites** or **additive inverses**. The opposite of a number a is denoted by $-a$.

The sum of a number a and its opposite, $-a$, is 0.
$$a + (-a) = 0$$
If a is a number, then $-(-a) = a$.

Add.
$$10 + 7 = 17$$
$$-3 + (-8) = -11$$

$$-25 + 5 = -20$$
$$14 + (-9) = 5$$

The opposite of -7 is 7.
The opposite of 123 is -123.

$$-4 + 4 = 0$$
$$12 + (-12) = 0$$
$$-(-8) = 8$$
$$-(-14) = 14$$

Definitions and Concepts	Examples

Section 1.6 Subtracting Real Numbers

To subtract two numbers a and b, add the first number a to the opposite of the second number b. $$a - b = a + (-b)$$	Subtract. $$3 - (-44) = 3 + 44 = 47$$ $$-5 - 22 = -5 + (-22) = -27$$ $$-30 - (-30) = -30 + 30 = 0$$

Section 1.7 Multiplying and Dividing Real Numbers

Quotient of two real numbers $$\frac{a}{b} = a \cdot \frac{1}{b}$$	Multiply or divide. $$\frac{42}{2} = 42 \cdot \frac{1}{2} = 21$$
Multiplying and Dividing Real Numbers The product or quotient of two numbers with the same sign is a positive number. The product or quotient of two numbers with different signs is a negative number.	$$7 \cdot 8 = 56 \quad -7 \cdot (-8) = 56$$ $$-2 \cdot 4 = -8 \quad 2 \cdot (-4) = -8$$ $$\frac{90}{10} = 9 \qquad \frac{-90}{-10} = 9$$ $$\frac{42}{-6} = -7 \qquad \frac{-42}{6} = -7$$
Products and Quotients Involving Zero The product of 0 and any number is 0. $$b \cdot 0 = 0 \quad \text{and} \quad 0 \cdot b = 0$$	$$-4 \cdot 0 = 0 \qquad 0 \cdot \left(-\frac{3}{4}\right) = 0$$
The quotient of a nonzero number and 0 is undefined. $\frac{b}{0}$ is undefined.	$\frac{-85}{0}$ is undefined.
The quotient of 0 and any nonzero number is 0. $$\frac{0}{b} = 0$$	$$\frac{0}{18} = 0 \qquad \frac{0}{-47} = 0$$

Section 1.8 Properties of Real Numbers

Commutative Properties Addition: $a + b = b + a$ Multiplication: $a \cdot b = b \cdot a$	$$3 + (-7) = -7 + 3$$ $$-8 \cdot 5 = 5 \cdot (-8)$$
Associative Properties Addition: $(a + b) + c = a + (b + c)$ Multiplication: $(a \cdot b) \cdot c = a \cdot (b \cdot c)$	$$(5 + 10) + 20 = 5 + (10 + 20)$$ $$(-3 \cdot 2) \cdot 11 = -3 \cdot (2 \cdot 11)$$
Two numbers whose product is 1 are called **multiplicative inverses** or **reciprocals**. The reciprocal of a nonzero number a is $\frac{1}{a}$ because $a \cdot \frac{1}{a} = 1$.	The reciprocal of 3 is $\frac{1}{3}$. The reciprocal of $-\frac{2}{5}$ is $-\frac{5}{2}$.

Definitions and Concepts	Examples
Section 1.8 Properties of Real Numbers	

Definitions and Concepts	Examples
Distributive Property $a(b + c) = a \cdot b + a \cdot c$	$5(6 + 10) = 5 \cdot 6 + 5 \cdot 10$ $-2(3 + x) = -2 \cdot 3 + (-2)(x)$
Identities $\quad a + 0 = a \qquad 0 + a = a$ $\qquad\qquad\quad a \cdot 1 = a \qquad\; 1 \cdot a = a$	$5 + 0 = 5 \qquad 0 + (-2) = -2$ $-14 \cdot 1 = -14 \qquad 1 \cdot 27 = 27$
Inverses Addition or opposite: $\;a + (-a) = 0$ Multiplication or reciprocal: $\;b \cdot \dfrac{1}{b} = 1$	$7 + (-7) = 0$ $3 \cdot \dfrac{1}{3} = 1$

STUDY SKILLS REMINDER

Are You Preparing for a Test on Chapter 1?

Below I have listed some *common trouble areas* for topics covered in Chapter 1. After studying for your test—but before taking your test—read these.

▶ Do you know the difference between $|-3|$, $-|-3|$, and $-(-3)$?

$|-3| = 3; \quad -|-3| = -3; \quad$ and $\quad -(-3) = 3 \qquad$ (Section 1.2)

▶ Evaluate $x - y$ if $x = 7$ and $y = -3$.

$x - y = 7 - (-3) = 10 \qquad$ (Section 1.4)

▶ Make sure you are familiar with order of operations. Sometimes the simplest-looking expressions can give you the most trouble.

$1 + 2(3 + 6) = 1 + 2(9) = 1 + 18 = 19 \qquad$ (Section 1.4)

▶ Do you know the difference between $(-3)^2$ and -3^2?

$(-3)^2 = 9 \quad$ and $\quad -3^2 = -9 \qquad$ (Section 1.7)

▶ Do you know that these fractions are equivalent?

$-\dfrac{1}{3} = \dfrac{-1}{3} = \dfrac{1}{-3} \qquad$ (Section 1.8)

(continued)

▶ Do you know the difference between opposite and reciprocal? If not, study the table below.

Number	Opposite	Reciprocal
5	-5	$\dfrac{1}{5}$
$-\dfrac{4}{7}$	$\dfrac{4}{7}$	$-\dfrac{7}{4}$
$-\dfrac{1}{3}$	$\dfrac{1}{3}$	-3

(Sections 1.5 and 1.7)

Remember: This is simply a checklist of selected topics given to check your understanding. For a review of Chapter 1 in the text, see the material at the end of Chapter 1.

CHAPTER REVIEW

(1.2) *Insert* $<, >,$ *or* $=$ *in the appropriate space to make the following statements true.*

1. 8 ___ 10

2. 7 ___ 2

3. -4 ___ -5

4. $\dfrac{12}{2}$ ___ -8

5. $|-7|$ ___ $|-8|$

6. $|-9|$ ___ -9

7. $-|-1|$ ___ -1

8. $|-14|$ ___ $-(-14)$

9. 1.2 ___ 1.02

10. $-\dfrac{3}{2}$ ___ $-\dfrac{3}{4}$

Translate each statement into symbols.

11. Four is greater than or equal to negative three.

12. Six is not equal to five.

13. 0.03 is less than 0.3.

14. Lions and hyenas were featured in the Disney film *The Lion King*. For short distances, lions can run at a rate of 50 miles per hour whereas hyenas can run at a rate of 40 miles per hour. Write an inequality statement comparing the numbers 50 and 40.

Given the following sets of numbers, list the numbers in each set that also belong to the set of:

a. Natural numbers

b. Whole numbers

c. Integers

d. Rational numbers

e. Irrational numbers

f. Real numbers

15. $\left\{ -6, 0, 1, 1\dfrac{1}{2}, 3, \pi, 9.62 \right\}$

16. $\left\{ -3, -1.6, 2, 5, \dfrac{11}{2}, 15.1, \sqrt{5}, 2\pi \right\}$

The following chart shows the gains and losses in dollars of Density Oil and Gas stock for a particular week.

Day	Gain or Loss in Dollars
Monday	$+1$
Tuesday	-2
Wednesday	$+5$
Thursday	$+1$
Friday	-4

17. Which day showed the greatest loss?

18. Which day showed the greatest gain?

(1.3) *Write the number as a product of prime factors.*

19. 36

20. 120

Perform the indicated operations. Write results in lowest terms.

21. $\dfrac{8}{15} \cdot \dfrac{27}{30}$

22. $\dfrac{7}{8} \div \dfrac{21}{32}$

23. $\dfrac{7}{15} + \dfrac{5}{6}$

24. $\dfrac{3}{4} - \dfrac{3}{20}$

25. $2\dfrac{3}{4} + 6\dfrac{5}{8}$

26. $7\dfrac{1}{6} - 2\dfrac{2}{3}$

27. $5 \div \dfrac{1}{3}$ **28.** $2 \cdot 8\dfrac{3}{4}$

29. Determine the unknown part of the given circle.

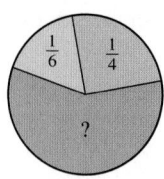

Find the area and the perimeter of each figure.

△ **30.**

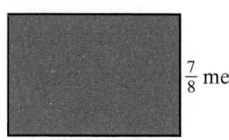

$\dfrac{7}{8}$ meter

$1\dfrac{1}{3}$ meter

△ **31.**

$\dfrac{5}{11}$ in. $\dfrac{3}{11}$ in. $\dfrac{3}{11}$ in. $\dfrac{5}{11}$ in.

32. A trim carpenter needs a piece of quarter round molding $6\dfrac{1}{8}$ feet long for a bathroom. She finds a piece $7\dfrac{1}{2}$ feet long. How long a piece does she need to cut from the $7\dfrac{1}{2}$-foot-long molding in order to use it in the bathroom?

In December 1998, Nkem Chukwu gave birth to the world's first surviving octuplets in Houston, Texas. The following chart gives the octuplets' birthweights. The babies are listed in order of birth.

Baby's Name	Gender	Birthweight (pounds)
Ebuka	girl	$1\dfrac{1}{2}$
Chidi	girl	$1\dfrac{11}{16}$
Echerem	girl	$1\dfrac{3}{4}$
Chima	girl	$1\dfrac{5}{8}$
Odera	girl	$\dfrac{11}{16}$
Ikem	boy	$1\dfrac{1}{8}$
Jioke	boy	$1\dfrac{13}{16}$
Gorom	girl	$1\dfrac{1}{8}$

(*Source*: Texas Children's Hospital, Houston, Texas)

33. What was the total weight of the boy octuplets?
34. What was the total weight of the girl octuplets?
35. Find the combined weight of all eight octuplets.
36. Which baby weighed the most?
37. Which baby weighed the least?
38. How much more did the heaviest baby weigh than the lightest baby?
39. By March 1999, Chima weighed $5\dfrac{1}{2}$ pounds. How much weight had she gained since birth?

40. By March 1999, Ikem weighed $4\dfrac{5}{32}$ pounds. How much weight had he gained since birth?

(1.4) Simplify each expression.

41. 2^4 **42.** 5^2

43. $\left(\dfrac{2}{7}\right)^2$ **44.** $\left(\dfrac{3}{4}\right)^3$

45. $6 \cdot 3^2 + 2 \cdot 8$ **46.** $68 - 5 \cdot 2^3$

47. $3(1 + 2 \cdot 5) + 4$ **48.** $8 + 3(2 \cdot 6 - 1)$

49. $\dfrac{4 + |6 - 2| + 8^2}{4 + 6 \cdot 4}$ **50.** $5[3(2 + 5) - 5]$

Translate each word statement to symbols.

51. The difference of twenty and twelve is equal to the product of two and four.

52. The quotient of nine and two is greater than negative five.

Evaluate each expression if $x = 6$, $y = 2$, and $z = 8$.

53. $2x + 3y$ **54.** $x(y + 2z)$

55. $\dfrac{x}{y} + \dfrac{z}{2y}$ **56.** $x^2 - 3y^2$

△ **57.** The expression $180 - a - b$ represents the measure of the unknown angle of the given triangle. Replace a with 37 and b with 80 to find the measure of the unknown angle.

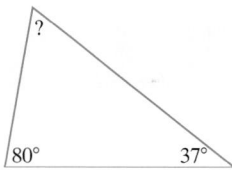

Decide whether the given number is a solution to the given equation.

58. Is $x = 3$ a solution of $7x - 3 = 18$?
59. Is $x = 1$ a solution of $3x^2 + 4 = x - 1$?

(1.5) Find the additive inverse or the opposite.

60. -9 **61.** $\dfrac{2}{3}$

62. $|-2|$ **63.** $-|-7|$

Find the following sums.

64. $-15 + 4$ **65.** $-6 + (-11)$

66. $\dfrac{1}{16} + \left(-\dfrac{1}{4}\right)$ **67.** $-8 + |-3|$

68. $-4.6 + (-9.3)$ **69.** $-2.8 + 6.7$

70. The lowest elevation in North America is -282 feet at Death Valley in California. If you are standing at a point 728 feet above Death Valley, what is your elevation? (*Source*: National Geographic Society)

(1.6) Perform the indicated operations.

71. $6 - 20$

72. $-3.1 - 8.4$

73. $-6 - (-11)$

74. $4 - 15$

75. $-21 - 16 + 3(8 - 2)$

76. $\dfrac{11 - (-9) + 6(8 - 2)}{2 + 3 \cdot 4}$

If $x = 3$, $y = -6$, and $z = -9$, evaluate each expression.

77. $2x^2 - y + z$

78. $\dfrac{y - x + 5x}{2x}$

(1.7) Find the multiplicative inverse or reciprocal.

79. -6

80. $\dfrac{3}{5}$

Simplify each expression.

81. $6(-8)$

82. $(-2)(-14)$

83. $\dfrac{-18}{-6}$

84. $\dfrac{42}{-3}$

85. $\dfrac{4(-3) + (-8)}{2 + (-2)}$

86. $\dfrac{3(-2)^2 - 5}{-14}$

87. $\dfrac{-6}{0}$

88. $\dfrac{0}{-2}$

89. $-4^2 - (-3 + 5) \div (-1) \cdot 2$

90. $-5^2 - (2 - 20) \div (-3) \cdot 3$

If $x = -5$ and $y = -2$, evaluate each expression.

91. $x^2 - y^4$

92. $x^2 - y^3$

93. During the 1999 LPGA Sara Lee Classic, Michelle McGann had scores of -9, -7, and $+1$ in three rounds of golf. Find her average score per round. (*Source:* Ladies Professional Golf Association)

94. During the 1999 PGA Masters Tournament, Bob Estes had scores of -1, 0, -3, and 0 in four rounds of golf. Find his average score per round. (*Source*: Professional Golf Association)

(1.8) Name the property illustrated.

95. $-6 + 5 = 5 + (-6)$

96. $6 \cdot 1 = 6$

97. $3(8 - 5) = 3 \cdot 8 + 3 \cdot (-5)$

98. $4 + (-4) = 0$

99. $2 + (3 + 9) = (2 + 3) + 9$

100. $2 \cdot 8 = 8 \cdot 2$

101. $6(8 + 5) = 6 \cdot 8 + 6 \cdot 5$

102. $(3 \cdot 8) \cdot 4 = 3 \cdot (8 \cdot 4)$

103. $4 \cdot \dfrac{1}{4} = 1$

104. $8 + 0 = 8$

105. $4(8 + 3) = 4(3 + 8)$

CHAPTER 1 TEST

Remember to use your Chapter Test Prep Video CD to help you study and view solutions to the test questions you need help with.

Translate the statement into symbols.

1. The absolute value of negative seven is greater than five.

2. The sum of nine and five is greater than or equal to four.

Simplify the expression.

3. $-13 + 8$

4. $-13 - (-2)$

5. $12 \div 4 \cdot 3 - 6 \cdot 2$

6. $(13)(-3)$

7. $(-6)(-2)$

8. $\dfrac{|-16|}{-8}$

9. $\dfrac{-8}{0}$

10. $\dfrac{|-6| + 2}{5 - 6}$

11. $\dfrac{1}{2} - \dfrac{5}{6}$

12. $-1\dfrac{1}{8} + 5\dfrac{3}{4}$

13. $(2 - 6) \div \dfrac{-2 - 6}{-3 - 1} - \dfrac{1}{2}$

14. $3(-4)^2 - 80$

15. $6[5 + 2(3 - 8) - 3]$

16. $\dfrac{-12 + 3 \cdot 8}{4}$

17. $\dfrac{(-2)(0)(-3)}{-6}$

Insert $<$, $>$, or $=$ in the appropriate space to make each of the following statements true.

18. -3 $\quad$ -7

19. 4 $\quad$ -8

20. 2 $\quad$ $|-3|$

21. $|-2|$ $\quad$ $-1 - (-3)$

22. In the state of Massachusetts, there are 2221 licensed child care centers and 10,993 licensed home-based child care providers. Write an inequality statement comparing the numbers 2221 and 10,993.(*Source*: Children's Foundation)

23. Given $\{-5, -1, 0, \frac{1}{4}, 1, 7, 11.6, \sqrt{7}, 3\pi\}$, list the numbers in this set that also belong to the set of:

a. Natural numbers

b. Whole numbers

c. Integers

d. Rational numbers

e. Irrational numbers

f. Real numbers

If $x = 6$, $y = -2$, and $z = -3$, evaluate each expression.

24. $x^2 + y^2$

25. $x + yz$

26. $2 + 3x - y$

27. $\dfrac{y + z - 1}{x}$

Identify the property illustrated by each expression.

28. $8 + (9 + 3) = (8 + 9) + 3$

29. $6 \cdot 8 = 8 \cdot 6$

30. $-6(2 + 4) = -6 \cdot 2 + (-6) \cdot 4$

31. $\frac{1}{6}(6) = 1$

32. Find the opposite of -9.

33. Find the reciprocal of $-\frac{1}{3}$.

The New Orleans Saints were 22 yards from the goal when the following series of gains and losses occurred.

Gains and Losses in Yards	
First Down	5
Second Down	−10
Third Down	−2
Fourth Down	29

34. During which down did the greatest loss of yardage occur?

35. Was a touchdown scored?

36. The temperature at the Winter Olympics was a frigid 14 degrees below zero in the morning, but by noon it had risen 31 degrees. What was the temperature at noon?

37. United HealthCare is a health insurance provider. In 3 consecutive recent years, it had net incomes of $356 million, $460 million, and −$166 million. What was United HealthCare's total net income for these three years? (*Source*: United HealthCare Corp.)

38. Jean Avarez decided to sell 280 shares of stock, which decreased in value by $1.50 per share yesterday. How much money did she lose?

Equations, Inequalities, and Problem Solving

Much of mathematics relates to deciding which statements are true and which are false. When a statement, such as an equation, contains variables, it is usually not possible to decide whether the equation is true or false until the variable has been replaced by a value. For example, the statement $x + 7 = 15$ is an equation stating that the sum $x + 7$ has the same value as 15. Is this statement true or false? It is false for some values of x and true for just one value of x, namely 8. Our purpose in this chapter is to learn ways of deciding which values make an equation or an inequality true.

CALLING ALL ACCOUNTANTS

There is an old saying: Nothing is certain, except death and taxes. And with ever-changing tax laws, is it any wonder that income tax preparation has become a multimillion-dollar industry? Companies such as H&R Block and Jackson-Hewitt employ accountants as both full-time and seasonal workers to prepare taxes and advise clients on financial strategies.

But accountants do so much more! Certified public accountants, or CPAs, also work for private companies or the federal government as auditors, consultants, and managers. Because they like to work with figures, their methodical approach to problem solving makes them valuable employees at all levels of business.

In the Spotlight on Decision Making feature on page 126, you will have the opportunity to make a decision involving itemized deductions as a personal income tax preparer.

Link: American Institute of Certified Public Accountants—
www.aicpa.org
Source of text: American Institute of Certified Public Accountants—
www.aicpa.org

SIMPLIFYING ALGEBRAIC EXPRESSIONS

Objectives

1   Identify terms, like terms, and unlike terms.

2   Combine like terms.

3   Use the distributive property to remove parentheses.

4   Write word phrases as algebraic expressions.

As we explore in this section, an expression such as $3x + 2x$ is not as simple as possible, because—even without replacing x by a value—we can perform the indicated addition.

1   Before we practice simplifying expressions, some new language of algebra is presented. A **term** is a number or the product of a number and variables raised to powers.

Terms

$$-y, \quad 2x^3, \quad -5, \quad 3xz^2, \quad \frac{2}{y}, \quad 0.8z$$

The **numerical coefficient** (sometimes also simply called the **coefficient**) of a term is the numerical factor. The numerical coefficient of $3x$ is 3. Recall that $3x$ means $3 \cdot x$.

Term	Numerical Coefficient	
$3x$	3	
$\dfrac{y^3}{5}$	$\dfrac{1}{5}$	*since* $\dfrac{y^3}{5}$ *means* $\dfrac{1}{5} \cdot y^3$
$0.7ab^3c^5$	0.7	
z	1	
$-y$	-1	
-5	-5	

> **Helpful Hint**
>
> The term $-y$ means $-1y$ and thus has a numerical coefficient of -1. The term z means $1z$ and thus has a numerical coefficient of 1.

EXAMPLE 1

Identify the numerical coefficient in each term.

a. $-3y$ **b.** $22z^4$ **c.** y **d.** $-x$ **e.** $\dfrac{x}{7}$

Solution   **a.** The numerical coefficient of $-3y$ is -3.

b. The numerical coefficient of $22z^4$ is 22.

c. The numerical coefficient of y is 1, since y is $1y$.

d. The numerical coefficient of $-x$ is -1, since $-x$ is $-1x$.

e. The numerical coefficient of $\dfrac{x}{7}$ is $\dfrac{1}{7}$, since $\dfrac{x}{7}$ means $\dfrac{1}{7} \cdot x$.

Terms with the same variables raised to exactly the same powers are called **like terms**. Terms that aren't like terms are called **unlike terms**.

Like Terms	Unlike Terms	
$3x, 2x$	$5x, 5x^2$	Why? Same variable x, but different powers x and x^2
$-6x^2y, 2x^2y, 4x^2y$	$7y, 3z, 8x^2$	Why? Different variables
$2ab^2c^3, ac^3b^2$	$6abc^3, 6ab^2$	Why? Different variables and different powers

> **Helpful Hint**
>
> In like terms, each variable and its exponent must match exactly, but these factors don't need to be in the same order.
>
> $$2x^2y \text{ and } 3yx^2 \text{ are like terms.}$$

EXAMPLE 2

Determine whether the terms are like or unlike.

a. $2x, 3x^2$ **b.** $4x^2y, x^2y, -2x^2y$ **c.** $-2yz, -3zy$ **d.** $-x^4, x^4$

Solution   **a.** Unlike terms, since the exponents on x are not the same.

b. Like terms, since each variable and its exponent match.

c. Like terms, since $zy = yz$ by the commutative property.

d. Like terms.

2   An algebraic expression containing the sum or difference of like terms can be simplified by applying the distributive property. For example, by the distributive property, we rewrite the sum of the like terms $3x + 2x$ as

$$3x + 2x = (3 + 2)x = 5x$$

Also,

$$-y^2 + 5y^2 = (-1 + 5)y^2 = 4y^2$$

Simplifying the sum or difference of like terms is called **combining like terms**.

EXAMPLE 3

Simplify each expression by combining like terms.

a. $7x - 3x$ **b.** $10y^2 + y^2$ **c.** $8x^2 + 2x - 3x$

Solution   **a.** $7x - 3x = (7 - 3)x = 4x$

b. $10y^2 + y^2 = 10y^2 + 1y^2 = (10 + 1)y^2 = 11y^2$

c. $8x^2 + 2x - 3x = 8x^2 + (2 - 3)x = 8x^2 - x$

EXAMPLE 4

Simplify each expression by combining like terms.

a. $2x + 3x + 5 + 2$ **b.** $-5a - 3 + a + 2$ **c.** $4y - 3y^2$

d. $2.3x + 5x - 6$ **e.** $-\dfrac{1}{2}b + b$

Solution   Use the distributive property to combine the numerical coefficients of like terms.

a. $2x + 3x + 5 + 2 = (2 + 3)x + (5 + 2)$
$$= 5x + 7$$

b. $-5a - 3 + a + 2 = -5a + 1a + (-3 + 2)$
$$= (-5 + 1)a + (-3 + 2)$$
$$= -4a - 1$$

c. $4y - 3y^2$ These two terms cannot be combined because they are unlike terms.

d. $2.3x + 5x - 6 = (2.3 + 5)x - 6$
$$= 7.3x - 6$$

e. $-\dfrac{1}{2}b + b = \left(-\dfrac{1}{2} + 1\right)b = \dfrac{1}{2}b$

The examples above suggest the following:

> ## Combining Like Terms
>
> To **combine like terms**, add the numerical coefficients and multiply the result by the common variable factors.

3 Simplifying expressions makes frequent use of the distributive property to also remove parentheses.

EXAMPLE 5

Find each product by using the distributive property to remove parentheses.

a. $5(x + 2)$ **b.** $-2(y + 0.3z - 1)$ **c.** $-(x + y - 2z + 6)$

Solution   **a.** $5(x + 2) = 5 \cdot x + 5 \cdot 2$ Apply the distributive property.
$$= 5x + 10$$ Multiply.

b. $-2(y + 0.3z - 1) = -2(y) + (-2)(0.3z) + (-2)(-1)$ Apply the distributive property.
$$= -2y - 0.6z + 2$$ Multiply.

c. $-(x + y - 2z + 6) = -1(x + y - 2z + 6)$ Distribute -1 over each term.
$$= -1(x) - 1(y) - 1(-2z) - 1(6)$$
$$= -x - y + 2z - 6$$

> ## ▶ Helpful Hint
>
> If a "$-$" sign precedes parentheses, the sign of each term inside the parentheses is changed when the distributive property is applied to remove parentheses.
>
> **Examples:**
>
> $-(2x + 1) = -2x - 1$ $-(-5x + y - z) = 5x - y + z$
>
> $-(x - 2y) = -x + 2y$ $-(-3x - 4y - 1) = 3x + 4y + 1$

When simplifying an expression containing parentheses, we often use the distributive property in both directions—first to remove parentheses and then again to combine any like terms.

EXAMPLE 6

Simplify the following expressions.

a. $3(2x - 5) + 1$ **b.** $-2(4x + 7) - (3x - 1)$ **c.** $9 + 3(4x + 10)$

Solution   **a.** $3(2x - 5) + 1 = 6x - 15 + 1$ Apply the distributive property.

$$= 6x - 14$$ Combine like terms.

b. $-2(4x + 7) - (3x - 1) = -8x - 14 - 3x + 1$ Apply the distributive property.

$$= -11x - 13$$ Combine like terms.

c. $9 - 3(4x + 10) = 9 - 12x - 30$ Apply the distributive property.

$$= -21 - 12x$$ Combine like terms.

> ### Helpful Hint
> Don't forget to use the distributive property to multiply before adding or subtracting like terms.

EXAMPLE 7

Write the phrase below as an algebraic expression. Then simplify if possible. Subtract $4x - 2$ from $2x - 3$.

Solution   "Subtract $4x - 2$ **from** $2x - 3$" translates to $(2x - 3) - (4x - 2)$. Next, simplify the algebraic expression.

$$(2x - 3) - (4x - 2) = 2x - 3 - 4x + 2$$ Apply the distributive property.
$$= -2x - 1$$ Combine like terms.

4   Next, we practice writing word phrases as algebraic expressions.

EXAMPLE 8

Write the following phrases as algebraic expressions and simplify if possible. Let x represent the unknown number.

a. Twice a number, added to 6
b. The difference of a number and 4, divided by 7
c. Five added to 3 times the sum of a number and 1
d. The sum of twice a number, 3 times the number, and 5 times the number

Solution   **a.** In words:

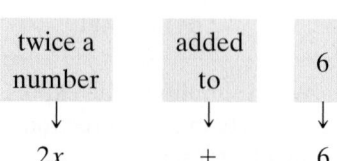

twice a number	added to	6
↓	↓	↓

Translate: $2x$ +   6

b. In words:

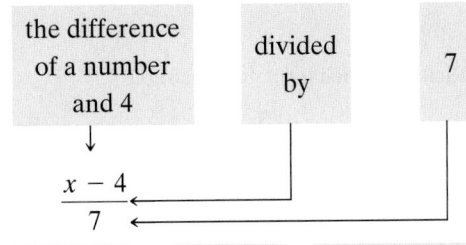

Translate:

$$\frac{x-4}{7}$$

c. In words:

five	added to	3 times	the sum of a number and 1
↓	↓	↓	↓

Translate: 5   +   3· $(x + 1)$

Next, we simplify this expression.

$$5 + 3(x + 1) = 5 + 3x + 3 \qquad \text{Use the distributive property.}$$
$$= 8 + 3x \qquad \text{Combine like terms.}$$

d. The phrase "the sum of" means that we add.

In words:

twice a number	added to	3 times the number	added to	5 times the number
↓	↓	↓	↓	↓

Translate: $2x$ + $3x$ + $5x$

Now let's simplify.

$$2x + 3x + 5x = 10x \qquad \text{Combine like terms.}$$

STUDY SKILLS REMINDER

Are You Getting all the Mathematics Help that You Need?

Remember that in addition to your instructor, there are many places to get help with your mathematics course. For example, see the list below.

▶ There is an accompanying video lesson for every section in this text.

▶ The back of this book contains answers to odd-numbered exercises as well as answers to every exercise in the Integrated Reviews, Chapter Reviews, Chapter Tests, and Cumulative Reviews.

▶ MathPro is available with this text. It is a tutorial software program with lessons corresponding to each section in the text.

▶ There is a student solutions manual available that contains worked-out solutions to odd-numbered exercises as well as solutions to every exercise in the Integrated Reviews, Chapter Reviews, Chapter Tests, and Cumulative Reviews.

▶ Check with your instructor for other local resources available to you, such as the tutoring center.

MENTAL MATH

Identify the numerical coefficient of each term. See Example 1.

1. $-7y$ **2.** $3x$ **3.** x **4.** $-y$ **5.** $17x^2y$

6. $1.2xyz$ **7.** $\dfrac{p}{8}$ **8.** $-\dfrac{5y}{3}$ **9.** $-\dfrac{2}{3}z$

Indicate whether the following lists of terms are like or unlike. See Example 2.

10. $5y, -y$ **11.** $-2x^2y, 6xy$ **12.** $2z, 3z^2$

13. $ab^2, -7ab^2$ **14.** $8wz, \dfrac{1}{7}zw$ **15.** $7.4p^3q^2, 6.2p^3q^2r$

EXERCISE SET 2.1

STUDY GUIDE/SSM   CD/VIDEO   PH MATH TUTOR CENTER   MathXL®Tutorials ON CD   MathXL®   MyMathLab®

Simplify each expression by combining any like terms. See Examples 3 and 4.

1. $7y + 8y$
2. $3x + 2x$
3. $-9n - 6n$
4. $-12p + 3p$
5. $3.5t - 4.5t$
6. $8.6y - 7.6y$
7. $8w - w + 6w$
8. $c - 7c + 2c$
9. $3b - 5 - 10b - 4$
10. $6g + 5 - 3g - 7$
11. $m - 4m + 2m - 6$
12. $a + 3a - 2 - 7a$

Simplify each expression. First use the distributive property to remove any parentheses. See Examples 5 and 6.

13. $5(y - 4)$
14. $7(r - 3)$
15. $7(d - 3) + 10$
16. $9(z + 7) - 15$
17. $-(3x - 2y + 1)$
18. $-(y + 5z - 7)$
19. $5(x + 2) - (3x - 4)$
20. $4(2x - 3) - 2(x + 1)$

21. In your own words, explain how to combine like terms.
22. Do like terms contain the same numerical coefficients? Explain your answer.

Write each of the following as an algebraic expression. Simplify if possible. See Example 7.

23. Add $6x + 7$ to $4x - 10$.
24. Add $3y - 5$ to $y + 16$.
25. Subtract $7x + 1$ from $3x - 8$.
26. Subtract $4x - 7$ from $12 + x$.

MIXED PRACTICE
Simplify each expression.

27. $7x^2 + 8x^2 - 10x^2$
28. $8x + x - 11x$
29. $6x - 5x + x - 3 + 2x$
30. $8h + 13h - 6 + 7h - h$
31. $-5 + 8(x - 6)$
32. $-6 + 5(r - 10)$

33. $6.2x - 4 + x - 1.2$
34. $7.9y - 0.7 - y + 0.2$
35. $2k - k - 6$
36. $7c - 8 - c$
37. $0.5(m + 2) + 0.4m$
38. $0.2(k + 8) - 0.1k$
39. $-4(3y - 4)$
40. $-3(2x + 5)$
41. $3(2x - 5) - 5(x - 4)$
42. $2(6x - 1) - (x - 7)$
43. $6x + 0.5 - 4.3x - 0.4x + 3$
44. $0.4y - 6.7 + y - 0.3 - 2.6y$
45. $-2(3x - 4) + 7x - 6$
46. $8y - 2 - 3(y + 4)$
47. $-9x + 4x + 18 - 10x$
48. $5y - 14 + 7y - 20y$
49. $5k - (3k - 10)$
50. $-11c - (4 - 2c)$
51. $(3x + 4) - (6x - 1)$
52. $(8 - 5y) - (4 + 3y)$

Write each of the following phrases as an algebraic expression and simplify if possible. Let x represent the unknown number. See Examples 7 and 8.

53. Twice a number, decreased by four
54. The difference of a number and two, divided by five
55. Three-fourths of a number, increased by twelve
56. Eight more than triple a number
57. The sum of 5 times a number and -2, added to 7 times a number
58. The sum of 3 times a number and 10, **subtracted from** 9 times a number
59. Subtract $5m - 6$ from $m - 9$.
60. Subtract $m - 3$ from $2m - 6$.
61. Eight times the sum of a number and six
62. Five, subtracted from four times a number
63. Double a number, minus the sum of the number and ten
64. Half a number, minus the product of the number and eight
65. Seven, multiplied by the quotient of a number and six
66. The product of a number and ten, less twenty
67. The sum of 2, three times a number, -9, and four times a number

68. The sum of twice a number, -1, five times a number, and -12

△ **69.** Recall that the perimeter of a figure is the total distance around the figure. Given the following rectangle, express the perimeter as an algebraic expression containing the variable x.

$5x$ feet

$(4x - 1)$ feet

$(4x - 1)$ feet

$5x$ feet

△ **70.** Given the following triangle, express its perimeter as an algebraic expression containing the variable x.

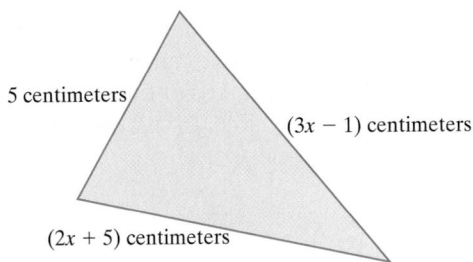

5 centimeters

$(3x - 1)$ centimeters

$(2x + 5)$ centimeters

REVIEW AND PREVIEW

Evaluate the following expressions for the given values. See Section 1.7.

71. If $x = -1$ and $y = 3$, find $y - x^2$.

72. If $g = 0$ and $h = -4$, find $gh - h^2$.

73. If $a = 2$ and $b = -5$, find $a - b^2$.

74. If $x = -3$, find $x^3 - x^2 + 4$.

75. If $y = -5$ and $z = 0$, find $yz - y^2$.

76. If $x = -2$, find $x^3 - x^2 - x$.

Concept Extensions

Given the following, determine whether each scale is balanced or not.

1 cone balances 1 cube

1 cylinder balances 2 cubes

77.

78.

79.

80.

81. To convert from feet to inches, we multiply by 12. For example, the number of inches in 2 feet is $12 \cdot 2$ inches. If one board has a length of $(x + 2)$ *feet* and a second board has a length of $(3x - 1)$ *inches*, express their total length in inches as an algebraic expression.

82. The value of 7 nickels is $5 \cdot 7$ cents. Likewise, the value of x nickels is $5x$ cents. If the money box in a drink machine contains x *nickels*, $3x$ *dimes*, and $(30x - 1)$ *quarters*, express their total value in cents as an algebraic expression.

For Exercises 83 through 88, see the example below.

Example

Simplify $-3xy + 2x^2y - (2xy - 1)$.

Solution

$$-3xy + 2x^2y - (2xy - 1)$$
$$= -3xy + 2x^2y - 2xy + 1 = -5xy + 2x^2y + 1$$

Simplify each expression.

83. $5b^2c^3 + 8b^3c^2 - 7b^3c^2$ **84.** $4m^4p^2 + m^4p^2 - 5m^2p^4$

85. $3x - (2x^2 - 6x) + 7x^2$ **86.** $9y^2 - (6xy^2 - 5y^2) - 8xy^2$

87. $-(2x^2y + 3z) + 3z - 5x^2y$

88. $-(7c^3d - 8c) - 5c - 4c^3d$

2.2   *THE ADDITION AND MULTIPLICATION PROPERTIES OF EQUALITY*

Objectives

1. Define linear equation in one variable and equivalent equations.
2. Use the addition property of equality to solve linear equations.
3. Use the multiplication property of equality to solve linear equations.
4. Use both properties of equality to solve linear equations.
5. Write word phrases as algebraic expressions.

1   Recall from Section 1.4 that an equation is a statement that two expressions have the same value. Also, a value of the variable that makes an equation a true statement is called a solution or root of the equation. The process of finding the solution of an equation is called **solving** the equation for the variable. In this section we concentrate on solving **linear equations** in one variable.

Linear Equation in One Variable

A linear equation in one variable can be written in the form

$$ax + b = c$$

where a, b, and c are real numbers and $a \neq 0$.

Evaluating a linear equation for a given value of the variable, as we did in Section 1.4, can tell us whether that value is a solution, but we can't rely on evaluating an equation as our method of solving it.

Instead, to solve a linear equation in x, we write a series of simpler equations, all *equivalent* to the original equation, so that the final equation has the form

$$x = \textbf{number} \qquad \textbf{or} \qquad \textbf{number} = x$$

Equivalent equations are equations that have the same solution. This means that the "number" above is the solution to the orginal equation.

2   The first property of equality that helps us write simpler equivalent equations is the **addition property of equality**.

Addition Property of Equality

If a, b, and c are real numbers, then

$$a = b \qquad \text{and} \qquad a + c = b + c$$

are equivalent equations.

This property guarantees that adding the same number to both sides of an equation does not change the solution of the equation. Since subtraction is defined in terms of addition, we may also **subtract the same number from both sides** without changing the solution.

A good way to picture a true equation is as a balanced scale. Since it is balanced, each side of the scale weighs the same amount.

If the same weight is added to or subtracted from each side, the scale remains balanced.

We use the addition property of equality to write equivalent equations until the variable is by itself on one side of the equation, and the equation looks like "$x = $ number" or "number $= x$."

EXAMPLE 1

Solve $x - 7 = 10$ for x.

Solution To solve for x, we want x alone on one side of the equation. To do this, we add 7 to both sides of the equation.

$$x - 7 = 10$$
$$x - 7 + 7 = 10 + 7 \qquad \text{Add 7 to both sides.}$$
$$x = 17 \qquad \text{Simplify.}$$

The solution of the equation $x = 17$ is obviously 17. Since we are writing equivalent equations, the solution of the equation $x - 7 = 10$ is also 17.

To check, replace x with 17 in the original equation.

$$x - 7 = 10$$
$$17 - 7 \stackrel{?}{=} 10 \qquad \text{Replace } x \text{ with 17 in the original equation.}$$
$$10 = 10 \qquad \text{True}$$

Since the statement is true, 17 is the solution.

CONCEPT CHECK

Use the addition property to fill in the blank so that the middle equation simplifies to the last equation.

$$x - 5 = 3$$
$$x - 5 + \underline{} = 3 + \underline{}$$
$$x = 8$$

Concept Check Answer:
5

EXAMPLE 2

Solve $y + 0.6 = -1.0$ for y.

Solution   To get y alone on one side of the equation, subtract 0.6 from both sides of the equation.

$$y + 0.6 = -1.0$$
$$y + 0.6 - 0.6 = -1.0 - 0.6 \qquad \text{Subtract 0.6 from both sides.}$$
$$y = -1.6 \qquad \text{Combine like terms.}$$

To check the proposed solution, -1.6, replace y with -1.6 in the original equation.

Check

$$y + 0.6 = -1.0$$
$$-1.6 + 0.6 \stackrel{?}{=} -1.0 \qquad \text{Replace } y \text{ with } -1.6 \text{ in the} $$
$$\text{original equation.}$$
$$-1.0 = -1.0 \qquad \text{True}$$

The solution is -1.6.

Many times, it is best to simplify one or both sides of an equation before applying the addition property of equality.

EXAMPLE 3

Solve: $2x + 3x - 5 + 7 = 10x + 3 - 6x - 4$

Solution   First we simplify both sides of the equation.

$$2x + 3x - 5 + 7 = 10x + 3 - 6x - 4$$
$$5x + 2 = 4x - 1 \qquad \text{Combine like terms on each side}$$
$$\text{of the equation.}$$

Next, we want all terms with a variable on one side of the equation and all numbers on the other side.

$$5x + 2 - 4x = 4x - 1 - 4x \qquad \text{Subtract } 4x \text{ from both sides.}$$
$$x + 2 = -1 \qquad \text{Combine like terms.}$$
$$x + 2 - 2 = -1 - 2 \qquad \text{Subtract 2 from both sides}$$
$$\text{to get } x \text{ alone.}$$
$$x = -3 \qquad \text{Combine like terms.}$$

Check

$$2x + 3x - 5 + 7 = 10x + 3 - 6x - 4 \qquad \text{Original equation.}$$
$$2(-3) + 3(-3) - 5 + 7 \stackrel{?}{=} 10(-3) + 3 - 6(-3) - 4 \qquad \text{Replace } x \text{ with } -3.$$
$$-6 - 9 - 5 + 7 \stackrel{?}{=} -30 + 3 + 18 - 4 \qquad \text{Multiply.}$$
$$-13 = -13 \qquad \text{True}$$

The solution is -3.

If an equation contains parentheses, we use the distributive property to remove them.

EXAMPLE 4

Solve: $7 = -5(2a - 1) - (-11a + 6)$.

Solution

$$7 = -5(2a - 1) - (-11a + 6)$$
$$7 = -10a + 5 + 11a - 6 \qquad \text{Apply the distributive property.}$$
$$7 = a - 1 \qquad \text{Combine like terms.}$$
$$7 + 1 = a - 1 + 1 \qquad \text{Add } \mathbf{1} \text{ to both sides to get } a \text{ alone.}$$
$$8 = a \qquad \text{Combine like terms.}$$

Check to see that 8 is the solution.

> **Helpful Hint**
> We may solve an equation so that the variable is alone on either side of the equation. For example, $8 = a$ is equivalent to $a = 8$.

When solving equations, we may sometimes encounter an equation such as

$$-x = 5.$$

This equation is not solved for x because x is not isolated. One way to solve this equation for x is to recall that

"$-$" can be read as "the opposite of."

We can read the equation $-x = 5$ then as "the opposite of $x = 5$." If the opposite of x is 5, this means that x is the opposite of 5 or -5.

In summary,

$$-x = 5 \quad \text{and} \quad x = -5$$

are equivalent equations and $x = -5$ is solved for x.

3   As useful as the addition property of equality is, it cannot help us solve every type of linear equation in one variable. For example, adding or subtracting a value on both sides of the equation does not help solve

$$\frac{5}{2}x = 15.$$

Instead, we apply another important property of equality, the **multiplication property of equality**.

> **Multiplication Property of Equality**
> If a, b, and c, are real numbers and $c \neq 0$, then
> $$a = b \quad \text{and} \quad ac = bc$$
> are equivalent equations.

This property guarantees that multiplying both sides of an equation by the same nonzero number does not change the solution of the equation. Since division is defined in terms of multiplication, we may also **divide both sides of the equation by the same nonzero number** without changing the solution.

EXAMPLE 5

Solve: $\dfrac{5}{2}x = 15$

Solution To get x alone, multiply both sides of the equation by the reciprocal of $\dfrac{5}{2}$, which is $\dfrac{2}{5}$.

Helpful Hint

Don't forget to multiply *both sides* by $\dfrac{2}{5}$.

$$\frac{5}{2}x = 15$$

$$\frac{2}{5} \cdot \frac{5}{2}x = \frac{2}{5} \cdot 15 \qquad \text{Multiply both sides by } \frac{2}{5}.$$

$$\left(\frac{2}{5} \cdot \frac{5}{2}\right)x = \frac{2}{5} \cdot 15 \qquad \text{Apply the associative property.}$$

$$1x = 6 \qquad \text{Simplify.}$$

or

$$x = 6$$

Check Replace x with 6 in the original equation.

$$\frac{5}{2}x = 15 \qquad \text{Original equation}$$

$$\frac{5}{2}(6) \stackrel{?}{=} 15 \qquad \text{Replace } x \text{ with 6.}$$

$$15 = 15 \qquad \text{True}$$

The solution is 6.

In the equation $\dfrac{5}{2}x = 15$, $\dfrac{5}{2}$ is the coefficient of x. When the coefficient of x is a *fraction*, we will get x alone by multiplying by the reciprocal. When the coefficient of x is an integer or a decimal, it is usually more convenient to divide both sides by the coefficient. (Dividing by a number is, of course, the same as multiplying by the reciprocal of the number.)

EXAMPLE 6

Solve: $-3x = 33$

Solution Recall that $-3x$ means $-3 \cdot x$. To get x alone, we divide both sides by the coefficient of x, that is, -3.

$$-3x = 33$$

$$\frac{-3x}{-3} = \frac{33}{-3} \qquad \text{Divide both sides by } -3.$$

$$1x = -11 \qquad \text{Simplify.}$$

$$x = -11$$

Check

$$-3x = 33 \qquad \text{Original equation}$$

$$-3(-11) \stackrel{?}{=} 33 \qquad \text{Replace } x \text{ with } -11.$$

$$33 = 33 \qquad \text{True}$$

The solution is -11.

EXAMPLE 7

Solve: $\dfrac{y}{7} = 20$

Solution Recall that $\dfrac{y}{7} = \dfrac{1}{7}y$. To get y alone, we multiply both sides of the equation by 7, the reciprocal of $\dfrac{1}{7}$.

$$\dfrac{y}{7} = 20$$

$$\dfrac{1}{7}y = 20$$

$$7 \cdot \dfrac{1}{7}y = 7 \cdot 20 \qquad \text{Multiply both sides by 7.}$$

$$1y = 140 \qquad \text{Simplify.}$$

$$y = 140$$

Check

$$\dfrac{y}{7} = 20 \qquad \text{Original equation}$$

$$\dfrac{140}{7} \overset{?}{=} 20 \qquad \text{Replace } y \text{ with } \mathbf{140}.$$

$$20 = 20 \qquad \text{True}$$

The solution is 140.

4   Next, we practice solving equations using both properties.

EXAMPLE 8

Solve: $12a - 8a = 10 + 2a - 13 - 7$

Solution First, simplify both sides of the equation by combining like terms.

$$12a - 8a = 10 + 2a - 13 - 7$$

$$4a = 2a - 10 \qquad \text{Combine like terms.}$$

To get all terms containing a variable on one side, subtract $2a$ from both sides.

$$4a - 2a = 2a - 10 - 2a \qquad \text{Subtract } \mathbf{2a} \text{ from both sides.}$$

$$2a = -10 \qquad \text{Simplify.}$$

$$\dfrac{2a}{2} = \dfrac{-10}{2} \qquad \text{Divide both sides by 2.}$$

$$a = -5 \qquad \text{Simplify.}$$

Check   Check by replacing a with -5 in the original equation. The solution is -5.

Next, we practice writing algebraic expressions.

EXAMPLE 9

a. The sum of two numbers is 8. If one number is 3, find the other number.

b. The sum of two numbers is 8. If one number is x, write an expression representing the other number.

Solution   a. If the sum of two numbers is 8 and one number is 3, we find the other number by subtracting 3 from 8. The other number is $8 - 3$ or 5.

b. If the sum of two numbers is 8 and one number is x, we find the other number by subtracting x from 8. The other number is represented by $8 - x$.

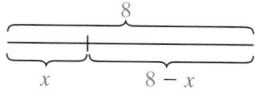

EXAMPLE 10

If x is the first of three consecutive integers, express the sum of the three integers in terms of x. Simplify if possible.

Solution   An example of three consecutive integers is

The second consecutive integer is always 1 more than the first, and the third consecutive integer is 2 more than the first. If x is the first of three consecutive integers, the three consecutive integers are

Their sum is

In words:

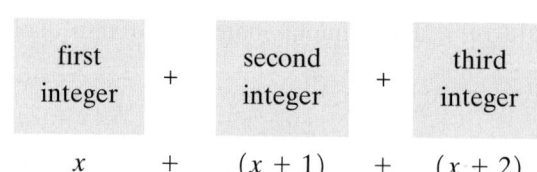

first integer	+	second integer	+	third integer

Translate: x + $(x + 1)$ + $(x + 2)$

which simplifies to $3x + 3$.

Below are examples of consecutive even and odd integers.

Even integers:

Odd integers:

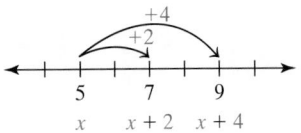

Helpful Hint

If x is an odd integer, then $x + 2$ is the next odd integer. This 2 simply means that odd integers are always 2 units from each other. (The same is true for even integers. They are always 2 units from each other.)

STUDY SKILLS REMINDER

Have You Decided to Successfully Complete this Course?

Ask yourself if one of your current goals is to successfully complete this course.

If it is not a goal of yours, ask yourself why not. One common reason is fear of failure. Amazingly enough, fear of failure alone can be strong enough to keep many of us from doing our best in any endeavor. Another common reason is that you simply haven't taken the time to make successfully completing this course one of your goals.

If you are taking this mathematics course, then successfully completing this course probably should be one of your goals. To make it a goal, start by writing this goal in your mathematics notebook. Then read or reread Section 1.1 and make a commitment to try the suggestions in this section.

If successfully completing this course is already a goal of yours, also read or reread Section 1.1 and try some of the suggestions in this section so that you are actively working toward your goal.

Good luck and don't forget that a positive attitude will make a big difference, also.

MENTAL MATH

Solve each equation mentally. See Examples 1 and 2.

1. $x + 4 = 6$ **2.** $x + 7 = 10$ **3.** $n + 18 = 30$

4. $z + 22 = 40$ **5.** $b - 11 = 6$ **6.** $d - 16 = 5$

7. $3a = 27$ **8.** $9c = 54$ **9.** $5b = 10$

10. $7t = 14$ **11.** $6x = -30$ **12.** $8r = -64$

EXERCISE SET 2.2

STUDY   CD/   PH MATH   MathXL®Tutorials   MathXL®   MyMathLab®
GUIDE/SSM   VIDEO   TUTOR CENTER   ON CD

Solve each equation. Check each solution. See Examples 1 and 2.

1. $x + 7 = 10$

2. $x + 14 = 25$

3. $x - 2 = -4$

4. $y - 9 = 1$

5. $3 + x = -11$

6. $8 + z = -8$

7. $r - 8.6 = -8.1$

8. $t - 9.2 = -6.8$

9. $8x = 7x - 3$

10. $2x = x - 5$

11. $5b - 0.7 = 6b$

12. $9x + 5.5 = 10x$

13. $7x - 3 = 6x$

14. $18x - 9 = 19x$

Solve each equation. See Examples 3 and 4.

15. $3x - 6 = 2x + 5$

16. $7y + 2 = 6y + 2$

17. $3t - t - 7 = t - 7$

18. $4c + 8 - c = 8 + 2c$

19. $7x + 2x = 8x - 3$

20. $3n + 2n = 7 + 4n$

21. $-2(x + 1) + 3x = 14$

22. $10 = 8(3y - 4) - 23y + 20$

Solve each equation. See Example 6.

23. $-5x = 20$

24. $-7x = -49$

25. $3x = 0$

26. $-2x = 0$

27. $-x = -12$

28. $-y = 8$

29. $3x + 2x = 50$

30. $-y + 4y = 33$

Solve each equation. See Examples 5 and 7.

31. $\frac{2}{3}x = -8$

32. $\frac{3}{4}n = -15$

33. $\frac{1}{6}d = \frac{1}{2}$

34. $\frac{1}{8}v = \frac{1}{4}$

35. $\frac{a}{-2} = 1$

36. $\frac{d}{15} = 2$

37. $\frac{k}{7} = 0$

38. $\frac{f}{-5} = 0$

39. In your own words, explain the addition property of equality.

40. In your own words, explain the multiplication property of equality.

Solve each equation. Check each solution. See Example 8.

41. $2x - 4 = 16$

42. $3x - 1 = 26$

43. $-x + 2 = 22$

44. $-x + 4 = -24$

45. $6a + 3 = 3$

46. $8t + 5 = 5$

47. $6x + 10 = -20$

48. $-10y + 15 = 5$

49. $5 - 0.3k = 5$

50. $2 + 0.4p = 2$

51. $-2x + \frac{1}{2} = \frac{7}{2}$

52. $-3n - \frac{1}{3} = \frac{8}{3}$

53. $\frac{x}{3} + 2 = -5$

54. $\frac{b}{4} - 1 = -7$

55. $10 = 2x - 1$

56. $12 = 3j - 4$

57. $6z - 8 - z + 3 = 0$

58. $4a + 1 + a - 11 = 0$

59. $10 - 3x - 6 - 9x = 7$

60. $12x + 30 + 8x - 6 = 10$

61. $\frac{5}{6}x = 10$

62. $-\frac{3}{4}x = 9$

63. $1 = 0.4x - 0.6x - 5$

64. $19 = 0.4x - 0.9x - 6$

65. $z - 5z = 7z - 9 - z$

66. $t - 6t = -13 + t - 3t$

67. $0.4x - 0.6x - 5 = 1$

68. $0.4x - 0.9x - 6 = 19$

69. $6 - 2x + 8 = 10$

70. $-5 - 6y + 6 = 19$

71. $-3a + 6 + 5a = 7a - 8a$

72. $4b - 8 - b = 10b - 3b$

73. $20 = -3(2x + 1) + 7x$

74. $-3 = -5(4x + 3) + 21x$

See Example 9.

75. Two numbers have a sum of 20. If one number is p, express the other number in terms of p.

76. Two numbers have a sum of 13. If one number is y, express the other number in terms of y.

77. A 10-foot board is cut into two pieces. If one piece is x feet long, express the other length in terms of x.

78. A 5-foot piece of string is cut into two pieces. If one piece is x feet long, express the other length in terms of x.

79. Two angles are *supplementary* if their sum is 180°. If one angle measures $x°$, express the measure of its supplement in terms of x.

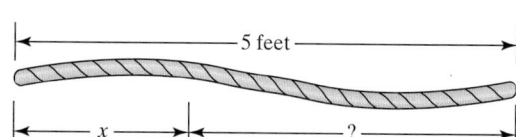

80. Two angles are *complementary* if their sum is 90°. If one angle measures $x°$, express the measure of its complement in terms of x.

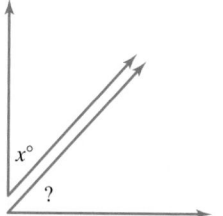

81. In a mayoral election, April Catarella received 284 more votes than Charles Pecot. If Charles received n votes, how many votes did April receive?

82. The length of the top of a computer desk is $1\frac{1}{2}$ feet longer than its width. If its width measures m feet, express its length as an algebraic expression in m.

83. The Verrazano-Narrows Bridge in New York City is the longest suspension bridge in North America. The Golden Gate Bridge in San Francisco is 60 feet shorter than the Verrazano-Narrows Bridge. If the length of the Verrazano-Narrows Bridge is m feet, express the length of the Golden Gate Bridge as an algebraic expression in m. (*Source: World Almanac*, 2000)

84. The longest interstate highway in the U.S. is I-90, which connects Seattle, Washington, and Boston, Massachusetts. The second longest interstate highway, I-80 (connecting San Francisco, California, and Teaneck, New Jersey), is 178.5 miles shorter than I-90. If the length of I-80 is m miles, express the length of I-90 as an algebraic expression in m. (*Source:* U.S. Department of Transportation–Federal Highway Administration)

85. In a recent election, Pat Ahumada ran against Solomon P. Ortiz for one of Texas's seats in the U.S. House of Representatives. Ahumada received 47,628 fewer votes than Ortiz. If Ahumada received n votes, how many did Ortiz receive? (*Source:* Voter News Service)

86. In a recent U.S. Senate race in Maine, Susan M. Collins received 30,898 more votes than Joseph E. Brennan. If Joseph received n votes, how many did Susan receive? (*Source:* Voter News Service)

87. The area of the Sahara Desert in Africa is 7 times the area of the Gobi Desert in Asia. If the area of the Gobi Desert is x square miles, express the area of the Sahara Desert as an algebraic expression in x.

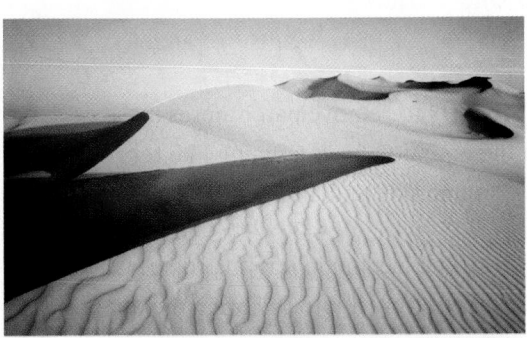

88. The largest meteorite in the world is the Hoba West located in Namibia. Its weight is 3 times the weight of the Armanty meteorite located in Outer Mongolia. If the weight of the Armanty meteorite is y kilograms, express the weight of the Hoba West meteorite as an algebraic expression in y.

Write each algebraic expression described. Simplify if possible. See Example 10.

89. If x represents the first of two consecutive odd integers, express the sum of the two integers in terms of x.

90. If x is the first of four consecutive even integers, write their sum as an algebraic expression in x.

91. If x is the first of three consecutive integers, express the sum of the first integer and the third integer as an algebraic expression containing the variable x.

92. If x is the first of two consecutive integers, express the sum of 20 and the second consecutive integer as an algebraic expression containing the variable x.

REVIEW AND PREVIEW

Simplify each expression. See Section 2.1.

93. $5x + 2(x - 6)$ **94.** $-7y + 2y - 3(y + 1)$

95. $-(x - 1) + x$ **96.** $-(3a - 3) + 2a - 6$

Insert $<$, $>$, or $=$ in the appropriate space to make each statement true. See Sections 1.2 and 1.7.

97. $(-3)^2$ -3^2

98. $(-2)^4$ -2^4

99. $(-2)^3$ -2^3

100. $(-4)^3$ -4^3

Concept Extensions

101. The sum of the angles of a triangle is $180°$. If one angle of a triangle measures $x°$ and a second angle measures $(2x + 7)°$, express the measure of the third angle in terms of x. Simplify the expression.

△ **102.** A quadrilateral is a four-sided figure like the one shown below whose angle sum is 360°. If one angle measures $x°$, a second angle measures $3x°$, and a third angle measures $5x°$, express the measure of the fourth angle in terms of x. Simplify the expression.

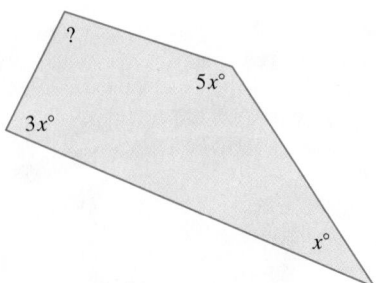

103. A licensed nurse practitioner is instructed to give a patient 2100 milligrams of an antibiotic over a period of 36 hours. If the antibiotic is to be given every 4 hours starting immediately, how much antibiotic should be given in each dose? To answer this question, solve the equation $9x = 2100$.

104. Suppose you are a pharmacist and a customer asks you the following question. His child is to receive 13.5 milliliters of a nausea medicine over a period of 54 hours. If the nausea medicine is to be administered every 6 hours starting immediately, how much medicine should be given in each dose?

105. A nurse's aide recorded the following fluid intakes for a patient on her night shift: 200 ml, 150 ml, 400 ml. If the patient's doctor requested that a total of 1000 ml of fluid be taken by the patient overnight, how much more fluid must the nurse give the patient? To solve this problem, solve the equation $200 + 150 + 400 + x = 1000$.

106. Let $x = 1$ and then $x = 2$ in the equation $x + 5 = x + 6$. Is either number a solution? How many solutions do you think this equation has? Explain your answer.

107. Let $x = 1$ and then $x = 2$ in the equation $x + 3 = x + 3$. Is either number a solution? How many solutions do you think this equation has? Explain your answer.

Solve each equation.

108. $-3.6x = 10.62$

109. $4.95y = -31.185$

110. $7x - 5.06 = -4.92$

111. $0.06y + 2.63 = 2.5562$

2.3 SOLVING LINEAR EQUATIONS

Objectives

1. Apply the general strategy for solving a linear equation.
2. Solve equations containing fractions.
3. Solve equations containing decimals.
4. Recognize identities and equations with no solution.
5. Write sentences as equations and solve.

1 We now present a general strategy for solving linear equations. One new piece of strategy is a suggestion to "clear an equation of fractions" as a first step. Doing so makes the equation more manageable, since operating on integers is more convenient than operating on fractions.

> **Solving Linear Equations in One Variable**
>
> **Step 1.** Multiply on both sides by the LCD to clear the equation of fractions if they occur.
>
> **Step 2.** Use the distributive property to remove parentheses if they occur.
>
> **Step 3.** Simplify each side of the equation by combining like terms.
>
> **Step 4.** Get all variable terms on one side and all numbers on the other side by using the addition property of equality.
>
> **Step 5.** Get the variable alone by using the multiplication property of equality.
>
> **Step 6.** Check the solution by substituting it into the original equation.

EXAMPLE 1

Solve: $4(2x - 3) + 7 = 3x + 5$

Solution   There are no fractions, so we begin with Step 2.

$$4(2x - 3) + 7 = 3x + 5$$

Step 2. $8x - 12 + 7 = 3x + 5$ Apply the distributive property.

Step 3. $8x - 5 = 3x + 5$ Combine like terms.

Step 4. Get all variable terms on the same side of the equation by subtracting $3x$ from both sides, then adding 5 to both sides.

$$8x - 5 - 3x = 3x + 5 - 3x$$ Subtract **3x** from both sides.
$$5x - 5 = 5$$ Simplify.
$$5x - 5 + 5 = 5 + 5$$ Add **5** to both sides.
$$5x = 10$$ Simplify.

Step 5. Use the multiplication property of equality to get x alone.

$$\frac{5x}{5} = \frac{10}{5}$$ Divide both sides by **5**.
$$x = 2$$ Simplify.

Step 6. Check.

$$4(2x - 3) + 7 = 3x + 5$$ Original equation
$$4[2(2) - 3] + 7 \stackrel{?}{=} 3(2) + 5$$ Replace x with **2**.
$$4(4 - 3) + 7 \stackrel{?}{=} 6 + 5$$
$$4(1) + 7 \stackrel{?}{=} 11$$
$$4 + 7 \stackrel{?}{=} 11$$
$$11 = 11$$ True

The solution is 2.

> **Helpful Hint**
> When checking solutions, use the original written equation.

EXAMPLE 2

Solve: $8(2 - t) = -5t$

Solution   First, we apply the distributive property.

$$8(2 - t) = -5t$$

Step 2. $16 - 8t = -5t$ Use the distributive property.

Step 4. $16 - 8t + 8t = -5t + 8t$ To get variable terms on one side, add **8t** to both sides.

$$16 = 3t$$ Combine like terms.

Step 5. $\frac{16}{3} = \frac{3t}{3}$ Divide both sides by **3**.

$$\frac{16}{3} = t$$ Simplify.

Step 6. Check.

$$8(2 - t) = -5t \qquad \text{Original equation}$$

$$8\left(2 - \frac{16}{3}\right) \stackrel{?}{=} -5\left(\frac{16}{3}\right) \qquad \text{Replace } t \text{ with } \frac{16}{3}.$$

$$8\left(\frac{6}{3} - \frac{16}{3}\right) \stackrel{?}{=} -\frac{80}{3} \qquad \text{The LCD is 3.}$$

$$8\left(-\frac{10}{3}\right) \stackrel{?}{=} -\frac{80}{3} \qquad \text{Subtract fractions.}$$

$$-\frac{80}{3} = -\frac{80}{3} \qquad \text{True}$$

The solution is $\frac{16}{3}$.

2   If an equation contains fractions, we can clear the equation of fractions by multiplying both sides by the LCD of all denominators. By doing this, we avoid working with time-consuming fractions.

EXAMPLE 3

Solve: $\dfrac{x}{2} - 1 = \dfrac{2}{3}x - 3$

Solution   We begin by clearing fractions. To do this, we multiply both sides of the equation by the LCD of 2 and 3, which is 6.

$$\frac{x}{2} - 1 = \frac{2}{3}x - 3$$

Step 1. $6\left(\dfrac{x}{2} - 1\right) = 6\left(\dfrac{2}{3}x - 3\right)$ Multiply both sides by the LCD, **6**.

Step 2. $6\left(\dfrac{x}{2}\right) - 6(1) = 6\left(\dfrac{2}{3}x\right) - 6(3)$ Apply the distributive property.

> **Helpful Hint**
>
> Don't forget to multiply *each* term by the LCD.

$$3x - 6 = 4x - 18 \qquad \text{Simplify.}$$

There are no longer grouping symbols and no like terms on either side of the equation, so we continue with Step 4.

$$3x - 6 = 4x - 18$$

Step 4. $3x - 6 - 3x = 4x - 18 - 3x$ To get variable terms on one side, subtract **3x** from both sides.

$$-6 = x - 18 \qquad \text{Simplify.}$$

$$-6 + 18 = x - 18 + 18 \qquad \text{Add 18 to both sides.}$$

$$12 = x \qquad \text{Simplify.}$$

Step 5.   The variable is now alone, so there is no need to apply the multiplication property of equality.

Step 6. Check.

$$\frac{x}{2} - 1 = \frac{2}{3}x - 3 \qquad \text{Original equation}$$

$$\frac{12}{2} - 1 \stackrel{?}{=} \frac{2}{3} \cdot 12 - 3 \qquad \text{Replace } x \text{ with } \mathbf{12}.$$

$$6 - 1 \stackrel{?}{=} 8 - 3 \qquad \text{Simplify.}$$

$$5 = 5 \qquad \text{True}$$

The solution is 12.

EXAMPLE 4

Solve: $\dfrac{2(a + 3)}{3} = 6a + 2$

Solution   We clear the equation of fractions first.

$$\frac{2(a + 3)}{3} = 6a + 2$$

Step 1.
$$3 \cdot \frac{2(a + 3)}{3} = 3(6a + 2) \qquad \begin{array}{l}\text{Clear the fraction by multiplying}\\ \text{both sides by the LCD, } \mathbf{3}.\end{array}$$

$$2(a + 3) = 3(6a + 2)$$

Step 2. Next, we use the distributive property and remove parentheses.

$$2a + 6 = 18a + 6 \qquad \text{Apply the distributive property.}$$

Step 4.
$$2a + 6 - 6 = 18a + 6 - 6 \qquad \text{Subtract } \mathbf{6} \text{ from both sides.}$$

$$2a = 18a$$

$$2a - 18a = 18a - 18a \qquad \text{Subtract } \mathbf{18a} \text{ from both sides.}$$

$$-16a = 0$$

Step 5.
$$\frac{-16a}{-16} = \frac{0}{-16} \qquad \text{Divide both sides by } -\mathbf{16}.$$

$$a = 0 \qquad \text{Write the fraction in simplest form.}$$

Step 6. To check, replace a with 0 in the original equation. The solution is 0.

---

3   When solving a problem about money, you may need to solve an equation containing decimals. If you choose, you may multiply to clear the equation of decimals.

EXAMPLE 5

Solve: $0.25x + 0.10(x - 3) = 0.05(22)$

Solution   First we clear this equation of decimals by multiplying both sides of the equation by 100. Recall that multiplying a decimal number by 100 has the effect of moving the decimal point 2 places to the right.

$$0.25x + 0.10(x - 3) = 0.05(22)$$

Helpful Hint

By the distributive property, 0.10 is multiplied by x and -3. Thus to multiply each term here by 100, we only need to multiply 0.10 by 100.

Step 1.

$$0.25x + 0.10(x - 3) = 0.05(22) \quad \text{Multiply both sides by 100.}$$
$$25x + 10(x - 3) = 5(22)$$

Step 2. $25x + 10x - 30 = 110 \quad$ Apply the distributive property.

Step 3. $35x - 30 = 110 \quad$ Combine like terms.

Step 4. $35x - 30 + 30 = 110 + 30 \quad$ Add **30** to both sides.

$35x = 140 \quad$ Combine like terms.

Step 5. $\dfrac{35x}{35} = \dfrac{140}{35} \quad$ Divide both sides by **35**.

$x = 4$

Step 6. To check, replace x with 4 in the original equation. The solution is 4.

4 So far, each equation that we have solved has had a single solution. However, not every equation in one variable has a single solution. Some equations have no solution, while others have an infinite number of solutions. For example,

$$x + 5 = x + 7$$

has no solution since no matter which **real number** we replace x with, the equation is false.

real number $+ 5 =$ same **real number** $+ 7$ **FALSE**

On the other hand,

$$x + 6 = x + 6$$

has infinitely many solutions since x can be replaced by any real number and the equation is always true.

real number $+ 6 =$ same **real number** $+ 6$ **TRUE**

The equation $x + 6 = x + 6$ is called an **identity**. The next few examples illustrate special equations like these.

EXAMPLE 6

Solve: $-2(x - 5) + 10 = -3(x + 2) + x$

Solution

$$-2(x - 5) + 10 = -3(x + 2) + x$$
$$-2x + 10 + 10 = -3x - 6 + x \quad \text{Apply the distributive property on both sides.}$$
$$-2x + 20 = -2x - 6 \quad \text{Combine like terms.}$$
$$-2x + 20 + 2x = -2x - 6 + 2x \quad \text{Add 2x to both sides.}$$
$$20 = -6 \quad \text{Combine like terms.}$$

The final equation contains no variable terms, and there is no value for x that makes $20 = -6$ a true equation. We conclude that there is **no solution** to this equation. In set notation, we can write the empty set, { }, or use the empty set or null set symbol, $\emptyset$. In this chapter, we will simply write *no solution*.

EXAMPLE 7

Solve: $3(x - 4) = 3x - 12$

Solution

$$3(x - 4) = 3x - 12$$
$$3x - 12 = 3x - 12 \qquad \text{Apply the distributive property.}$$

The left side of the equation is now identical to the right side. Every real number may be substituted for x and a true statement will result. We arrive at the same conclusion if we continue.

$$3x - 12 = 3x - 12$$
$$3x - 12 + 12 = 3x - 12 + 12 \qquad \text{Add 12 to both sides.}$$
$$3x = 3x \qquad \text{Combine like terms.}$$
$$3x - 3x = 3x - 3x \qquad \text{Subtract 3x from both sides.}$$
$$0 = 0$$

Again, one side of the equation is identical to the other side. Thus, $3(x - 4) = 3x - 12$ is an **identity** and **all real numbers** are solutions. In set notation, this is {all real numbers}.

✔ **CONCEPT CHECK**

Suppose you have simplified several equations and obtain the following results. What can you conclude about the solutions to the original equation?

    a. $7 = 7$ **b.** $x = 0$ **c.** $7 = -4$

5   We can apply our equation-solving skills to solving problems written in words. Many times, writing an equation that describes or models a problem involves a direct translation from a word sentence to an equation.

EXAMPLE 8

FINDING AN UNKNOWN NUMBER

Twice a number, added to seven, is the same as three subtracted from the number. Find the number.

Solution   Translate the sentence into an equation and solve.

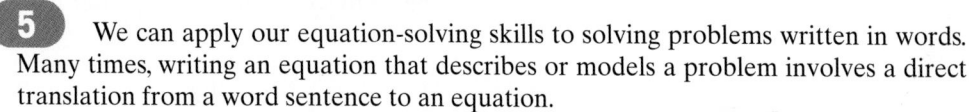

In words:	twice a number	added to	seven	is the same as	three subtracted from the number
Translate:	$2x$	$+$	7	$=$	$x - 3$

To solve, begin by subtracting x from both sides to isolate the variable term.

$$2x + 7 = x - 3$$
$$2x + 7 - x = x - 3 - x \qquad \text{Subtract } x \text{ from both sides.}$$
$$x + 7 = -3 \qquad \text{Combine like terms.}$$
$$x + 7 - 7 = -3 - 7 \qquad \text{Subtract 7 from both sides.}$$
$$x = -10 \qquad \text{Combine like terms.}$$

Concept Check Answer:

a. Every real number is a solution.
b. The solution is 0.
c. There is no solution.

Check the solution in the problem as it was originally stated. To do so, replace "number" in the sentence with −10. Twice "−10" added to 7 is the same as 3 subtracted from "−10."

$$2(-10) + 7 = -10 - 3$$
$$-13 = -13$$

The unknown number is −10.

> **Helpful Hint**
> When checking solutions, go back to the original stated problem, rather than to your equation in case errors have been made in translating to an equation.

STUDY SKILLS REMINDER

This is a special reminder that will be repeated and expanded throughout this text. It is very important for you to be able to recognize and solve different types of equations and inequalities. To help you do this, we will begin an outline below and continually expand this outline as different equations and inequalities are introduced. Although suggestions will be given, this outline should be in your own words and you should include at least "how to recognize" and "how to begin to solve" under each letter heading.

For example:

Solving Equations and Inequalities

I. Equations
   A. **Linear equations**—Recognize: *power on variable is 1 when there are no variables in denominator*—Solve: simplify (if fractions, multiply by LCD) and move variable terms to one side of the equation, constants to the other side.

II. Inequalities

See Appendix A for exercises.

Calculator Explorations

Checking Equations

We can use a calculator to check possible solutions of equations. To do this, replace the variable by the possible solution and evaluate both sides of the equation separately.

Equation: $3x - 4 = 2(x + 6)$ *Solution:* $x = 16$

$3x - 4 = 2(x + 6)$ Original equation

$3(16) - 4 \stackrel{?}{=} 2(16 + 6)$ Replace x with 16.

Now evaluate each side with your calculator.

(continued)

Evaluate left side:

$\boxed{3}$ $\boxed{\times}$ $\boxed{16}$ $\boxed{-}$ $\boxed{4}$ $\boxed{=}$ or $\boxed{\text{ENTER}}$ Display: $\boxed{44}$ or $\boxed{\begin{array}{l}3*16-4\\ \hspace{3em}44\end{array}}$

Evaluate right side:

$\boxed{2}$ $\boxed{(}$ $\boxed{16}$ $\boxed{+}$ $\boxed{6}$ $\boxed{)}$ $\boxed{=}$ or $\boxed{\text{ENTER}}$ Display: $\boxed{44}$ or $\boxed{\begin{array}{l}2(16+6)\\ \hspace{3em}44\end{array}}$

Since the left side equals the right side, the equation checks.

Use a calculator to check the possible solutions to each equation.

1. $2x = 48 + 6x$; $x = -12$ **2.** $-3x - 7 = 3x - 1$; $x = -1$

3. $5x - 2.6 = 2(x + 0.8)$; $x = 4.4$ **4.** $-1.6x - 3.9 = -6.9x - 25.6$; $x = 5$

5. $\dfrac{564x}{4} = 200x - 11(649)$; $x = 121$ **6.** $20(x - 39) = 5x - 432$; $x = 23.2$

EXERCISE SET 2.3

STUDY   CD/   PH MATH   MathXL®Tutorials   MathXL®   MyMathLab®
GUIDE/SSM   VIDEO   TUTOR CENTER   ON CD

Solve each equation. See Examples 1 and 2.

1. $-2(3x - 4) = 2x$ **2.** $-(5x - 1) = 9$

3. $4(2n - 1) = (6n + 4) + 1$

4. $3(4y + 2) = 2(1 + 5y) + 8$

 5. $5(2x - 1) - 2(3x) = 1$

6. $3(2 - 5x) + 4(6x) = 12$

7. $6(x - 3) + 10 = -8$ **8.** $-4(2 + n) + 9 = 1$

Solve each equation. See Examples 3 through 5.

9. $\dfrac{3}{4}x - \dfrac{1}{2} = 1$ **10.** $\dfrac{2}{3}x + \dfrac{5}{3} = \dfrac{5}{3}$

11. $x + \dfrac{5}{4} = \dfrac{3}{4}x$ **12.** $\dfrac{7}{8}x + \dfrac{1}{4} = \dfrac{3}{4}x$

13. $\dfrac{x}{2} - 1 = \dfrac{x}{5} + 2$ **14.** $\dfrac{x}{5} - 2 = \dfrac{x}{3}$

15. $\dfrac{6(3 - z)}{5} = -z$ **16.** $\dfrac{4(5 - w)}{3} = -w$

17. $\dfrac{2(x + 1)}{4} = 3x - 2$ **18.** $\dfrac{3(y + 3)}{5} = 2y + 6$

19. $0.50x + 0.15(70) = 0.25(142)$

20. $0.40x + 0.06(30) = 0.20(49)$

21. $0.12(y - 6) + 0.06y = 0.08y - 0.07(10)$

22. $0.60(z - 300) + 0.05z = 0.70z - 0.41(500)$

Solve each equation. See Examples 6 and 7.

23. $5x - 5 = 2(x + 1) + 3x - 7$

24. $3(2x - 1) + 5 = 6x + 2$

25. $\dfrac{x}{4} + 1 = \dfrac{x}{4}$ **26.** $\dfrac{x}{3} - 2 = \dfrac{x}{3}$

27. $3x - 7 = 3(x + 1)$ **28.** $2(x - 5) = 2x + 10$

29. Explain the difference between simplifying an expression and solving an equation.

30. When solving an equation, if the final equivalent equation is $0 = 5$, what can we conclude? If the final equivalent equation is $-2 = -2$, what can we conclude?

31. On your own, construct an equation for which every real number is a solution.

32. On your own, construct an equation that has no solution.

MIXED PRACTICE

Solve each equation.

33. $4x + 3 = 2x + 11$ **34.** $6y - 8 = 3y + 7$

35. $-2y - 10 = 5y + 18$ **36.** $7n + 5 = 10n - 10$

37. $0.6x - 0.1 = 0.5x + 0.2$ **38.** $0.2x - 0.1 = 0.6x - 2.1$

39. $2y + 2 = y$ **40.** $7y + 4 = -3$

41. $3(5c - 1) - 2 = 13c + 3$

42. $4(3t + 4) - 20 = 3 + 5t$

43. $x + \dfrac{7}{6} = 2x - \dfrac{7}{6}$ **44.** $\dfrac{5}{2}x - 1 = x + \dfrac{1}{4}$

45. $2(x - 5) = 7 + 2x$

46. $-3(1 - 3x) = 9x - 3$

47. $\dfrac{2(z + 3)}{3} = 5 - z$

48. $\dfrac{3(w + 2)}{4} = 2w + 3$

49. $\dfrac{4(y-1)}{5} = -3y$

50. $\dfrac{5(1-x)}{6} = -4x$

51. $8 - 2(a-1) = 7 + a$

52. $5 - 6(2+b) = b - 14$

53. $2(x+3) - 5 = 5x - 3(1+x)$

54. $4(2+x) + 1 = 7x - 3(x-2)$

55. $\dfrac{5x-7}{3} = x$

56. $\dfrac{7n+3}{5} = -n$

57. $\dfrac{9+5v}{2} = 2v - 4$

58. $\dfrac{6-c}{2} = 5c - 8$

59. $-3(t-5) + 2t = 5t - 4$

60. $-(4a-7) - 5a = 10 + a$

61. $0.02(6t-3) = 0.12(t-2) + 0.18$

62. $0.03(2m+7) = 0.06(5+m) - 0.09$

63. $0.06 - 0.01(x+1) = -0.02(2-x)$

64. $-0.01(5x+4) = 0.04 - 0.01(x+4)$

65. $\dfrac{3(x-5)}{2} = \dfrac{2(x+5)}{3}$

66. $\dfrac{5(x-1)}{4} = \dfrac{3(x+1)}{2}$

Write each of the following as equations. Then solve. See Example 8.

67. The sum of twice a number and 7 is equal to the sum of a number and 6. Find the number.

68. The difference of three times a number and 1 is the same as twice a number. Find the number.

69. Three times a number, minus 6, is equal to two times a number, plus 8. Find the number.

70. The sum of 4 times a number and −2 is equal to the sum of 5 times a number and −2. Find the number.

71. One-third of a number is five-sixths. Find the number.

72. Seven-eighths of a number is one-half. Find the number.

73. If the quotient of a number and 4 is added to $\frac{1}{2}$, the result is $\frac{3}{4}$. Find the number.

74. If $\frac{3}{4}$ is added to three times a number, the result is $\frac{1}{2}$ subtracted from twice the number. Find the number.

75. Five times a number, subtracted from ten, is triple the number. Find the number.

76. Nine is equal to ten subtracted from double a number. Find the number.

77. The perimeter of a geometric figure is the sum of the lengths of its sides. If the perimeter of the following pentagon (five-sided figure) is 28 centimeters, find the length of each side.

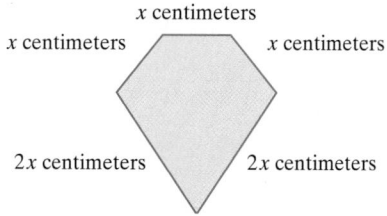

x centimeters
x centimeters x centimeters
$2x$ centimeters $2x$ centimeters

78. The perimeter of the following triangle is 35 meters. Find the length of each side.

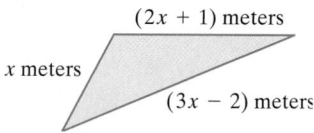

$(2x+1)$ meters
x meters
$(3x-2)$ meters

REVIEW AND PREVIEW

Evaluate. See Section 1.7.

79. $|2^3 - 3^2| - |5 - 7|$

80. $|5^2 - 2^2| + |9 + (-3)|$

81. $\dfrac{5}{4 + 3 \cdot 7}$

82. $\dfrac{8}{24 - 8 \cdot 2}$

See Section 2.1.

83. A plot of land is in the shape of a triangle. If one side is x meters, a second side is $(2x-3)$ meters and a third side is $(3x-5)$ meters, express the perimeter of the lot as a simplified expression in x.

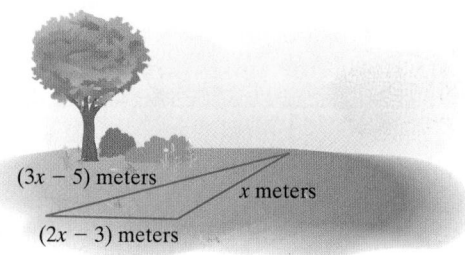

$(3x-5)$ meters
x meters
$(2x-3)$ meters

84. A portion of a board has length x feet. The other part has length $(7x-9)$ feet. Express the total length of the board as a simplified expression in x.

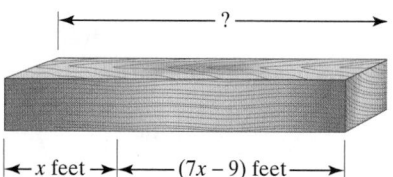

x feet $(7x-9)$ feet

Concept Extensions

The bar graph below shows the five most common names of cities, towns, or villages in the United States.

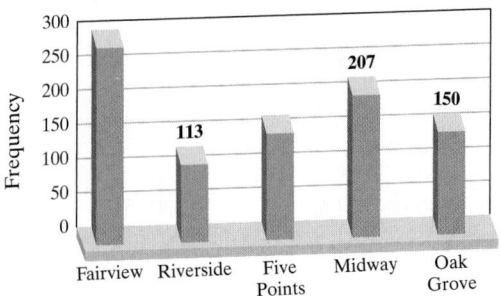

Source: Interior Department

85. What is the most popular name of a city, town, or village in the United States?

86. How many more cities, towns, or villages are named Oak Grove than named Riverside?

87. Let x represent "the number of cities, towns, or villages named Five Points" and use the information given to determine the unknown number. "The number of cities, towns, or villages named Five Points" added to 65 is equal to twice "the number of cities, towns, or villages named Five Points" minus 90. Check your answer by noticing the height of the bar representing Five Points. Is your answer reasonable?

88. Let x represent "the number of cities, towns, or villages named Fairview" and use the information given to determine the unknown number. Three times "the number of cities, towns, or villages named Fairview" added to 119 is equal to 168 subtracted from 4 times "the number of cities, towns, or villages named Fairview." Check your answer by noticing the height of the bar representing Fairview. Is your answer reasonable?

89. $1000(7x - 10) = 50(412 + 100x)$

90. $10,000(x + 4) = 100(16 + 75x)$

91. $0.035x + 5.112 = 0.010x + 5.107$

92. $0.127x - 2.685 = 0.027x - 2.38$

For Exercises 93 through 96, see the example below.

Example
Solve: $t(t + 4) = t^2 - 2t + 6$.

Solution
$$t(t + 4) = t^2 - 2t + 6$$
$$t^2 + 4t = t^2 - 2t + 6$$
$$t^2 + 4t - t^2 = t^2 - 2t + 6 - t^2$$
$$4t = -2t + 6$$
$$4t + 2t = -2t + 6 + 2t$$
$$6t = 6$$
$$t = 1$$

Solve each equation.

93. $x(x - 3) = x^2 + 5x + 7$

94. $t^2 - 6t = t(8 + t)$

95. $2z(z + 6) = 2z^2 + 12z - 8$

96. $y^2 - 4y + 10 = y(y - 5)$

INTEGRATED REVIEW   SOLVING LINEAR EQUATIONS

Solve. Feel free to use the steps given in Section 2.3.

1. $x - 10 = -4$

2. $y + 14 = -3$

3. $9y = 108$

4. $-3x = 78$

5. $-6x + 7 = 25$

6. $5y - 42 = -47$

7. $\dfrac{2}{3}x = 9$

8. $\dfrac{4}{5}z = 10$

9. $\dfrac{r}{-4} = -2$

10. $\dfrac{y}{-8} = 8$

11. $6 - 2x + 8 = 10$

12. $-5 - 6y + 6 = 19$

13. $2x - 7 = 2x - 27$

14. $3 + 8y = 8y - 2$

15. $-3a + 6 + 5a = 7a - 8a$

16. $4b - 8 - b = 10b - 3b$

17. $-\dfrac{2}{3}x = \dfrac{5}{9}$

18. $-\dfrac{3}{8}y = -\dfrac{1}{16}$

19. $10 = -6n + 16$

20. $-5 = -2m + 7$

21. $3(5c - 1) - 2 = 13c + 3$

22. $4(3t + 4) - 20 = 3 + 5t$

23. $\dfrac{2(z + 3)}{3} = 5 - z$

24. $\dfrac{3(w + 2)}{4} = 2w + 3$

25. $-2(2x - 5) = -3x + 7 - x + 3$

26. $-4(5x - 2) = -12x + 4 - 8x + 4$

27. $0.02(6t - 3) = 0.04(t - 2) + 0.02$

28. $0.03(m + 7) = 0.02(5 - m) + 0.03$

29. $-3y = \dfrac{4(y - 1)}{5}$

30. $-4x = \dfrac{5(1 - x)}{6}$

31. $\dfrac{5}{3}x - \dfrac{7}{3} = x$

32. $\dfrac{7}{5}n + \dfrac{3}{5} = -n$

2.4   AN INTRODUCTION TO PROBLEM SOLVING

Objective

1   Apply the steps for problem solving.

1   In previous sections, you practiced writing word phrases and sentences as algebraic expressions and equations to help prepare for problem solving. We now use these translations to help write equations that model a problem. The problem-solving steps given next may be helpful.

General Strategy for Problem Solving

1. UNDERSTAND the problem. During this step, become comfortable with the problem. Some ways of doing this are:

   Read and reread the problem.
   Choose a variable to represent the unknown.
   Construct a drawing, whenever possible.
   Propose a solution and check. Pay careful attention to how you check your proposed solution. This will help when writing an equation to model the problem.

2. TRANSLATE the problem into an equation.
3. SOLVE the equation.
4. INTERPRET the results: *Check* the proposed solution in the stated problem and state your conclusion.

Much of problem solving involves a direct translation from a sentence to an equation. Although we have been practicing these translations in previous sections, this section will often have translations that require a placement of parentheses.

EXAMPLE 1

FINDING AN UNKNOWN NUMBER

Twice the sum of a number and 4 is the same as four times the number decreased by 12. Find the number.

Solution   1. UNDERSTAND. Read and reread the problem. If we let

$$x = \text{the unknown number, then}$$

"the sum of a number and 4" translates to "$x + 4$" and "four times the number" translates to "$4x$."

2. TRANSLATE.

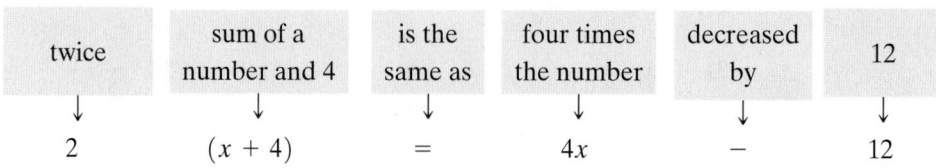

twice	sum of a number and 4	is the same as	four times the number	decreased by	12
↓	↓	↓	↓	↓	↓
2	$(x + 4)$	=	$4x$	−	12

3. SOLVE.

$$2(x + 4) = 4x - 12$$

$$2x + 8 = 4x - 12 \qquad \text{Apply the distributive property.}$$

$$2x + 8 - 4x = 4x - 12 - 4x \qquad \text{Subtract } 4x \text{ from both sides.}$$

$$-2x + 8 = -12$$

$$-2x + 8 - 8 = -12 - 8 \qquad \text{Subtract } 8 \text{ from both sides.}$$

$$-2x = -20$$

$$\frac{-2x}{-2} = \frac{-20}{-2} \qquad \text{Divide both sides by } -2.$$

$$x = 10$$

4. INTERPRET.

Check: Check this solution in the problem as it was originally stated. To do so, replace "number" with 10. Twice the sum of "10" and 4 is 28, which is the same as 4 times "10" decreased by 12.

State: The number is 10.

Next, we continue to review consecutive integers.

EXAMPLE 2

Some states have a single area code for the entire state. Two such states have area codes that are consecutive odd integers. If the sum of these integers is 1208, find the two area codes. (*Source: World Almanac*, 2003)

Solution   **1.** UNDERSTAND. Read and reread the problem. If we let

$$x = \text{the first odd integer, then}$$

$$x + 2 = \text{the next odd integer}$$

> **Helpful Hint**
>
> Remember, the 2 here means that odd integers are 2 units apart, for example, the odd integers 13 and 13 + 2 = 15.

2. TRANSLATE.

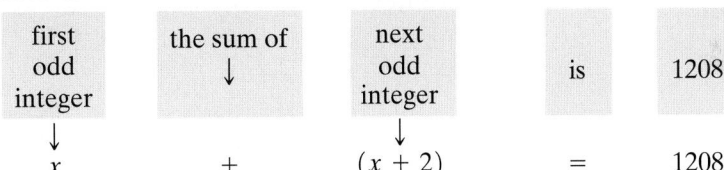

first odd integer	the sum of	next odd integer	is	1208
↓	↓	↓		
x	$+$	$(x + 2)$	$=$	1208

3. SOLVE.

$$x + x + 2 = 1208$$

$$2x + 2 = 1208$$

$$2x + 2 - 2 = 1208 - 2$$

$$2x = 1206$$

$$\frac{2x}{2} = \frac{1206}{2}$$

$$x = 603$$

4. INTERPRET.

Check: If $x = 603$, then the next odd integer $x + 2 = 603 + 2 = 605$. Notice their sum, $603 + 605 = 1208$, as needed.

State: The area codes are 603 and 605.

Note: New Hampshire's area code is 603 and South Dakota's area code is 605.

EXAMPLE 3

FINDING THE LENGTH OF A BOARD

A 10-foot board is to be cut into two pieces so that the longer piece is 4 times the shorter. Find the length of each piece.

Solution 1. UNDERSTAND the problem. To do so, read and reread the problem. You may also want to propose a solution. For example, if 3 feet represents the length of the shorter piece, then $4(3) = 12$ feet is the length of the longer piece, since it is 4 times the length of the shorter piece. This guess gives a total board length of 3 feet + 12 feet = 15 feet, too long. However, the purpose of proposing a solution is not to guess correctly, but to help better understand the problem and how to model it.

Since the length of the longer piece is given in terms of the length of the shorter piece, let's let

$$x = \text{length of shorter piece, then}$$
$$4x = \text{length of longer piece}$$

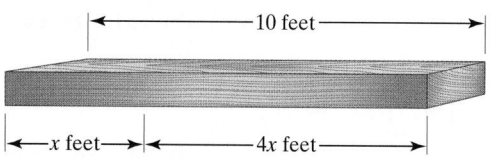

2. TRANSLATE the problem. First, we write the equation in words.

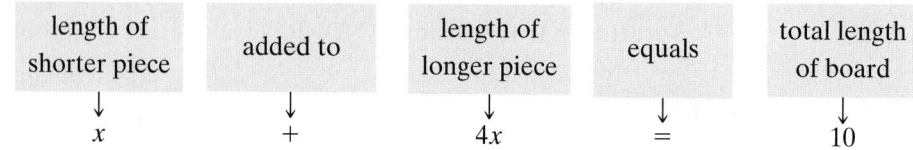

length of shorter piece	added to	length of longer piece	equals	total length of board
↓	↓	↓	↓	↓
x	$+$	$4x$	$=$	10

3. SOLVE.

$$
\begin{aligned}
x + 4x &= 10 \\
5x &= 10 \qquad \text{Combine like terms.} \\
\frac{5x}{5} &= \frac{10}{5} \qquad \text{Divide both sides by 5.} \\
x &= 2
\end{aligned}
$$

4. INTERPRET.

Check: Check the solution in the stated problem. If the shorter piece of board is 2 feet, the longer piece is $4 \cdot (2 \text{ feet}) = 8$ feet and the sum of the two pieces is 2 feet + 8 feet = 10 feet.

State: The shorter piece of board is 2 feet and the longer piece of board is 8 feet.

> **Helpful Hint**
> Make sure that units are included in your answer, if appropriate.

EXAMPLE 4

FINDING THE NUMBER OF REPUBLICAN AND DEMOCRATIC SENATORS

In a recent year, the U.S. House of Representatives had a total of 431 Democrats and Republicans. There were 15 more Republican representatives than

Democratic representatives. Find the number of representatives from each party. (*Source:* Office of the Clerk of the U.S. House of Representatives)

Solution   1. UNDERSTAND the problem. Read and reread the problem. Let's suppose that there were 200 Democratic representatives. Since there were 15 more Republicans than Democrats, there must have been $200 + 15 = 215$ Republicans. The total number of Democrats and Republicans was then $200 + 215 = 415$. This is incorrect since the total should be 431, but we now have a better understanding of the problem.

In general, if we let

$$x = \text{number of Democrats, then}$$

$$x + 15 = \text{number of Republicans}$$

2. TRANSLATE the problem. First, we write the equation in words.

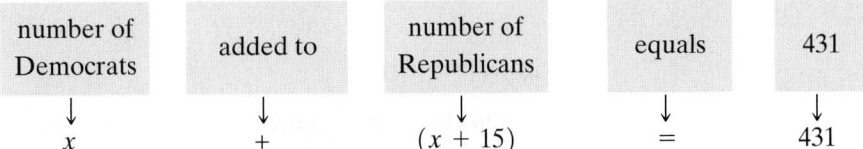

number of Democrats	added to	number of Republicans	equals	431
↓	↓	↓	↓	↓
x	$+$	$(x + 15)$	$=$	431

3. SOLVE.

$$x + (x + 15) = 431$$
$$2x + 15 = 431 \qquad \text{Combine like terms.}$$
$$2x + 15 - 15 = 431 - 15 \qquad \text{Subtract \textbf{15} from both sides.}$$
$$2x = 416$$
$$\frac{2x}{2} = \frac{416}{2} \qquad \text{Divide both sides by \textbf{2}.}$$
$$x = 208$$

4. INTERPRET.

Check: If there were 208 Democratic representatives, then there were $208 + 15 = 223$ Republican representatives. The total number of representatives is then $208 + 223 = 431$. The results check.

State: There were 208 Democratic and 223 Republican representatives in Congress.

EXAMPLE 5

CALCULATING HOURS ON JOB

A computer science major at a local university has a part-time job working on computers for his clients. He charges $20 to come to your home or office and then $25 per hour. During one month he visited 10 homes or offices and his total income was $575. How many hours did he spend working on computers?

Solution   1. UNDERSTAND. Read and reread the problem. Let's propose that the student spent 20 hours working on computers. Pay careful attention as to how his income is calculated. For 20 hours and 10 visits, his income is $20(\$25) + 10(\$20) = \$700$, more than $575. We now have a better understanding of the problem and know that the time working on computers is less than 20 hours.

Let's let

$$x = \text{hours working on computers. Then}$$
$$25x = \text{amount of money made while working on computers}$$

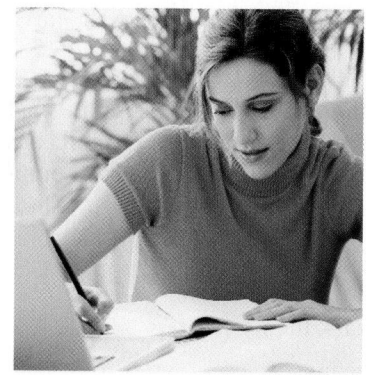

2. TRANSLATE.

money made while working on computers	plus	money made for visits	is equal to	575
↓	↓	↓	↓	↓
$25x$	$+$	$10(20)$	$=$	575

3. SOLVE.

$$25x + 200 = 575$$
$$25x + 200 - 200 = 575 - 200 \quad \text{Subtract 200 from both sides.}$$
$$25x = 375 \quad \text{Simplify.}$$
$$\frac{25x}{25} = \frac{375}{25} \quad \text{Divide both sides by 25.}$$
$$x = 15 \quad \text{Simplify.}$$

4. INTERPRET.

Check: If the student works 15 hours and makes 10 visits, his income is 15($25) + 10($20) = $575.

State: The student spent 15 hours working on computers.

△ (**EXAMPLE 6**)

FINDING ANGLE MEASURES

If the two walls of the Vietnam Veterans Memorial in Washington D.C. were connected, an isosceles triangle would be formed. The measure of the third angle is 97.5° more than the measure of either of the other two equal angles. Find the measure of the third angle. (*Source:* National Park Service)

Solution

1. UNDERSTAND. Read and reread the problem. We then draw a diagram (recall that an isosceles triangle has two angles with the same measure) and let

$$x = \text{degree measure of one angle}$$
$$x = \text{degree measure of the second equal angle}$$
$$x + 97.5 = \text{degree measure of the third angle}$$

2. TRANSLATE. Recall that the sum of the measures of the angles of a triangle equals 180.

measure of first angle		measure of second angle		measure of third angle	equals	180
↓		↓		↓	↓	↓
x	$+$	x	$+$	$(x + 97.5)$	$=$	180

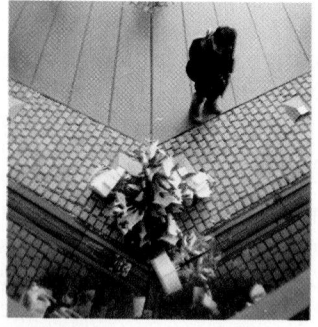

3. SOLVE.

$$x + x + (x + 97.5) = 180$$
$$3x + 97.5 = 180 \quad \text{Combine like terms.}$$
$$3x + 97.5 - 97.5 = 180 - 97.5 \quad \text{Subtract 97.5 from both sides.}$$
$$3x = 82.5$$
$$\frac{3x}{3} = \frac{82.5}{3} \quad \text{Divide both sides by 3.}$$
$$x = 27.5$$

4. INTERPRET.

Check: If $x = 27.5$, then the measure of the third angle is $x + 97.5 = 125$. The sum of the angles is then $27.5 + 27.5 + 125 = 180$, the correct sum.

State: The third angle measures 125°.*

(* The two walls actually meet at an angle of 125 degrees 12 minutes. The measurement of 97.5° given in the problem is an approximation.)

STUDY SKILLS REMINDER

How Are Your Homework Assignments Going?

It is so important in mathematics to keep up with homework. Why? Many concepts build on each other. Oftentimes, your understanding of a day's lecture in mathematics depends on an understanding of the previous day's material.

Remember that completing your homework assignment involves a lot more than attempting a few of the problems assigned.

To complete a homework assignment, remember these four things:

1. Attempt all of it.
2. Check it.
3. Correct it.
4. If needed, ask questions about it.

EXERCISE SET 2.4

STUDY GUIDE/SSM   CD/VIDEO   PH MATH TUTOR CENTER   MathXL®Tutorials ON CD   MathXL®   MyMathLab®

Solve. See Example 1.

1. Twice the difference of a number and 8 is equal to three times the sum of the number and 3. Find the number.

2. Five times the sum of a number and −1 is the same as 6 times the number. Find the number.

3. The product of twice a number and three is the same as the difference of five times the number and $\frac{3}{4}$. Find the number.

4. If the difference of a number and four is doubled, the result is $\frac{1}{4}$ less than the number. Find the number.

Solve. See Example 2.

5. The left and right page numbers of an open book are two consecutive integers whose sum is 469. Find these page numbers.

6. The room numbers of two adjacent classrooms are two consecutive even numbers. If their sum is 654, find the classroom numbers.

7. To make an international telephone call, you need the code for the country you are calling. The codes for Belgium, France, and Spain are three consecutive integers whose sum is 99. Find the code for each country. (*Source: The World Almanac and Book of Facts,* 2003)

8. To make an international telephone call, you need the code for the country you are calling. The codes for Mali Republic, Côte d'Ivoire, and Niger are three consecutive odd integers whose sum is 675. Find the code for each country.

Solve. See Examples 3 and 4.

9. A 40-inch board is to be cut into three pieces so that the second piece is twice as long as the first piece and the third piece is 5 times as long as the first piece. If x represents the length of the first piece, find the lengths of all three pieces.

10. A 21-foot beam is to be divided so that the longer piece is 1 foot more than 3 times the shorter piece. If x represents the length of the shorter piece, find the lengths of both pieces.

Solve. See Examples 2 through 5.

11. The governor of Washington state makes twice as much money as the governor of Nebraska. If the total of their salaries is $195,000, find the salary of each.

12. In the 2002 Winter Olympics, Germany won 2 more gold medals than the United States. If the total number of gold medals for both is 22, find the number of gold medals that each team won. (*Source: World Almanac*, 2003)

Solve. See Example 5.

13. A car rental agency advertised renting a Buick Century for $24.95 per day and $0.29 per mile. If you rent this car for 2 days, how many whole miles can you drive on a $100 budget?

14. A plumber gave an estimate for the renovation of a kitchen. Her hourly pay is $27 per hour and the plumber's parts will cost $80. If her total estimate is $404, how many hours does she expect this job to take?

15. In one U.S. city, the taxi cost is $3 plus $0.80 per mile. If you are traveling from the airport, there is an additional charge of $4.50 for tolls. How far can you travel from the airport by taxi for $27.50?

16. A professional carpet cleaning service charges $30 plus $25.50 per hour to come to your home. If your total bill from this company is $119.25 before taxes, how many hours were you charged?

Solve. See Example 6.

17. The flag of Equatorial Guinea contains an isosceles triangle. (Recall that an isosceles triangle contains two angles with the same measure.) If the measure of the third angle of the triangle is 30° more than twice the measure of either of the other two angles, find the measure of each angle of the triangle. (*Hint:* Recall that the sum of the measures of the angles of a triangle is 180°.)

18. The flag of Brazil contains a parallelogram. One angle of the parallelogram is 15° less than twice the measure of the angle next to it. Find the measure of each angle of the parallelogram. (*Hint:* Recall that opposite angles of a parallelogram have the same measure and that the sum of the measures of the angles is 360°.)

19. The sum of the measures of the angles of a parallelogram is 360°. In the parallelogram below, angles A and D have the same measure as well as angles C and B. If the measure of angle C is twice the measure of angle A, find the measure of each angle.

20. Recall that the sum of the measures of the angles of a triangle is 180°. In the triangle below, angle C has the same measure as angle B, and angle A measures 42° less than angle B. Find the measure of each angle.

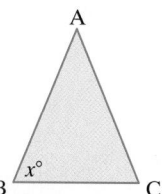

MIXED PRACTICE

21. A 17-foot piece of string is cut into two pieces so that one piece is 2 feet longer than twice the shorter piece. If the shorter piece is x feet long, find the lengths of both pieces.

22. An 18-foot wire is to be cut so that the longer piece is 5 times longer than the shorter piece. Find the length of each piece.

23. From 1997 to 2001, the number of prescriptions written for ADHD drugs increased by 6.6 million. If the sum of the number of prescriptions for these two years is 35.7 million, find the number of prescriptions for each year. Check to see that your results agree with the heights of the bars in the graph.

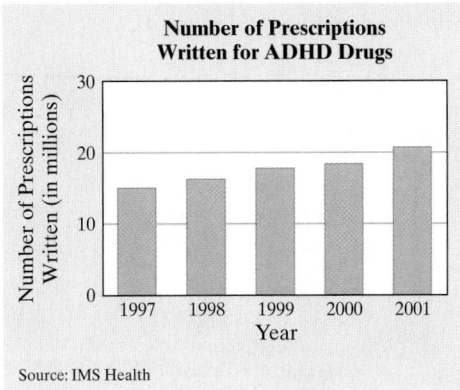

Source: IMS Health

24. The Pentagon Building in Washington, D.C., is the headquarters for the U.S. Department of Defense. The Pentagon is also the world's largest office building in terms of ground space with a floor area of over 6.5 million square feet. This is three times the floor area of the Empire State Building. About how much floor space does the Empire State Building have? Round to the nearest tenth.

25. Two angles are supplementary if their sum is 180°. One angle measures three times the measure of a smaller angle. If x represents the measure of the smaller angle and these two angles are supplementary, find the measure of each angle.

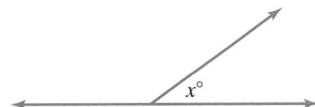

26. Two angles are complementary if their sum is 90°. Given the measures of the complementary angles shown, find the measure of each angle.

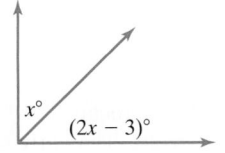

27. The measures of the angles of a triangle are 3 consecutive even integers. Find the measure of each angle.

28. A quadrilateral is a polygon with 4 sides. The sum of the measures of the 4 angles in a quadrilateral is 360°. If the measures of the angles of a quadrilateral are consecutive odd integers, find the measures.

29. Hertz Car Rental charges a daily rate of $39 plus $0.20 per mile for a certain car. Suppose that you rent that car for a day and your bill (before taxes) is $95. How many miles did you drive?

30. A woman's $15,000 estate is to be divided so that her husband receives twice as much as her son. If x represents the amount of money that her son receives, find the amount of money that her husband receives and the amount of money that her son receives.

31. In SuperBowl XXXVII in San Diego, California, the Tampa Bay Buccaneers won over the Oakland Raiders with a 27 point lead. If the total of the two scores was 69, find the individual team scores. (*Source: World Almanac*, 2003)

32. During the 2002 Houston Bowl, Oklahoma State beat Southern Miss by 10 points. If their combined scores total 56, find the individual team scores. (*Source*: ESPN)

33. In the 2000 Summer Olympics, China won more medals than Australia, who won more medals than Germany. If the number of medals won by each country is three consecutive integers whose sum is 174, how many medals did each country win? (*Source: The World Almanac and Book of Facts 2001*)

34. The number of counties in California and the number of counties in Montana are consecutive even integers whose sum is 114. If California has more counties than Montana, how many counties does each state have? (*Source: The World Almanac and Book of Facts 2001*)

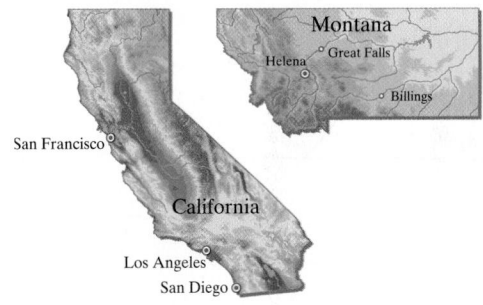

35. Starting in 2004, the state of California will have 21 more electoral votes for president than the state of Texas. If the total electoral votes for these two states is 89, find the number of electoral votes for each state.

36. After a recent election, there were 2 more Republican governors than Democratic governors in the United States. How many Democrats and how many Republicans held governor's offices after this election? (*Source: The World Almanac and Book of Facts, 2003*)

37. Over the past few years the satellite Voyager II has passed by the planets Saturn, Uranus, and Neptune, continually updating information about these planets, including the number of moons for each. Uranus is now believed to have 13 more moons than Neptune. Also, Saturn is now believed to have 2 more than twice the number of moons of Neptune. If the total number of moons for these planets is 47, find the number of moons for each planet. (*Source*: National Space Science Data Center)

38. On April 7, 2001, the Mars Odyssey spacecraft was launched, beginning a multi-year mission to observe and map the planet Mars. Mars Odyssey was launched on Boeing's Delta II 7925 launch vehicle using nine strap-on solid rocket motors. Each solid rocket motor has a height that is 8 meters more than 5 times its diameter. If the sum of the height and the diameter for a single solid rocket motor is 14 meters, find each dimension. (*Source*: NASA)

39. If the sum of a number and five is tripled, the result is one less than twice the number. Find the number.

40. Twice the sum of a number and six equals three times the sum of the number and four. Find the number.

41. In a recent election in Florida for a seat in the United States House of Representatives, Corrine Brown received 13,288 more votes than Bill Randall. If the total number of votes was 119,436, find the number of votes for each candidate.

42. In a recent election in Texas for a seat in the United States House of Representatives, Max Sandlin received 25,557 more votes than opponent Dennis Boerner. If the total number of votes was 135,821, find the number of votes for each candidate.(*Source*: Voter News Service)

The graph below shows the states with the highest tourism budgets. Use the graph for Exercises 43 through 47.

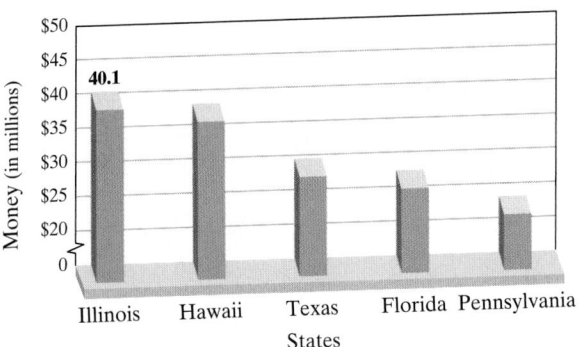

Source: Travel Industry Association of America

43. Which state spends the most money on tourism?

44. Which states spend between $25 and $30 million on tourism?

45. The states of Texas and Florida spend a total of $56.6 million for tourism. The state of Texas spends $2.2 million more than the state of Florida. Find the amount that each state spends on tourism.

46. The states of Hawaii and Pennsylvania spend a total of $60.9 million for tourism. The state of Hawaii spends $8.1 million less than twice the amount of money that the state of Pennsylvania spends. Find the amount that each state spends on tourism.

47. Compare the heights of the bars in the graph with your results of Exercises 45 and 46. Are your answers reasonable?

REVIEW AND PREVIEW

Perform the indicated operations. See Sections 1.5 and 1.6.

48. $3 + (-7)$

49. $-2 + (-8)$

50. $4 - 10$

51. $-11 + 2$

52. $-5 - (-1)$

53. $-12 - 3$

Concept Extensions

△ **54.** The golden rectangle is a rectangle whose length is approximately 1.6 times its width. The early Greeks thought that a rectangle with these dimensions was the most pleasing to the eye and examples of the golden rectangle are found in many early works of art. For example, the Parthenon in Athens contains many examples of golden rectangles. Mike Hallahan would like to plant a rectangular garden in the shape of a golden rectangle. If he has 78 feet of fencing available, find the dimensions of the garden.

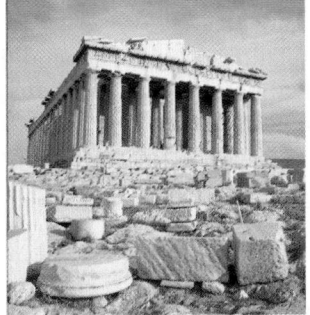

The length-width rectangle approximates the golden rectangle as well as the width-height rectangle.

55. Dr. Dorothy Smith gave the students in her geometry class at the University of New Orleans the following question. Is it possible to construct a triangle such that the second angle of the triangle has a measure that is twice the measure of the first angle and the measure of the third angle is 3 times the measure of the first? If so, find the measure of each angle. (*Hint:* Recall that the sum of the measures of the angles of a triangle is 180°.)

56. The human eye blinks once every 5 seconds on average. How many times does the average eye blink in one hour? In one 16-hour day while awake? In one year?

57. Give an example of how you recently solved a problem using mathematics.

58. In your own words, explain why a solution of a word problem should be checked using the original wording of the problem and not the equation written from the wording.

Recall from Exercise 54 that the golden rectangle is a rectangle whose length is approximately 1.6 times its width.

△ **59.** It is thought that for about 75% of adults, a rectangle in the shape of the golden rectangle is the most pleasing to the eye.

Draw 3 rectangles, one in the shape of the golden rectangle, and poll your class. Do the results agree with the percentage given above?

△ **60.** Examples of golden rectangles can be found today in architecture and manufacturing packaging. Find an example of a golden rectangle in your home. A few suggestions: the front face of a book, the floor of a room, the front of a box of food.

△ **61.** Measure the dimensions of each rectangle and decide which one best approximates the shape of the golden rectangle.

a.

b.

c.

2.5   FORMULAS AND PROBLEM SOLVING

Objectives

1. Use formulas to solve problems.
2. Solve a formula or equation for one of its variables.

1   An equation that describes a known relationship among quantities, such as distance, time, volume, weight, and money is called a **formula**. These quantities are represented by letters and are thus variables of the formula. Here are some common formulas and their meanings.

$A = lw$
Area of a rectangle = length · width

$I = PRT$
Simple Interest = Principal · Rate · Time

$P = a + b + c$
Perimeter of a triangle = side a + side b + side c

$d = rt$
distance = rate · time

$V = lwh$
Volume of a rectangular solid = length · width · height

$F = \left(\dfrac{9}{5}\right)C + 32$

degrees Fahrenheit = $\left(\dfrac{9}{5}\right)$ · degrees Celsius + 32

Formulas are valuable tools because they allow us to calculate measurements as long as we know certain other measurements. For example, if we know we traveled a distance of 100 miles at a rate of 40 miles per hour, we can replace the variables d and r in the formula $d = rt$ and find our time, t.

$$d = rt \qquad \text{Formula.}$$
$$100 = 40t \qquad \text{Replace } d \text{ with } \mathbf{100} \text{ and } r \text{ with } \mathbf{40}.$$

This is a linear equation in one variable, t. To solve for t, divide both sides of the equation by 40.

$$\frac{100}{40} = \frac{40t}{40} \qquad \text{Divide both sides by } \mathbf{40}.$$
$$\frac{5}{2} = t \qquad \text{Simplify.}$$

The time traveled is $\frac{5}{2}$ hours or $2\frac{1}{2}$ hours.

In this section we solve problems that can be modeled by known formulas. We use the same problem-solving steps that were introduced in the previous section. These steps have been slightly revised to include formulas.

EXAMPLE 1

FINDING TIME GIVEN RATE AND DISTANCE

A glacier is a giant mass of rocks and ice that flows downhill like a river. Portage Glacier in Alaska is about 6 miles, or 31,680 *feet*, long and moves 400 *feet* per year. Icebergs are created when the front end of the glacier flows into Portage Lake. How long does it take for ice at the head (beginning) of the glacier to reach the lake?

Solution   1. UNDERSTAND.   Read and reread the problem. The appropriate formula needed to solve this problem is the distance formula, $d = rt$. To become familiar with this formula, let's find the distance that ice traveling at a rate of 400 feet per year travels in 100 years. To do so, we let time t be 100 years and rate r be the given 400 feet per year, and substitute these values into the formula $d = rt$. We then have that distance $d = 400(100) = 40,000$ feet. Since we are interested in finding how long it takes ice to travel 31,680 feet, we now know that it is less than 100 years.

Since we are using the formula $d = rt$, we let

t = the time in years for ice to reach the lake

r = rate or speed of ice

d = distance from beginning of glacier to lake

2. TRANSLATE.   To translate to an equation, we use the formula $d = rt$ and let distance $d = 31,680$ feet and rate $r = 400$ feet per year.

$$d = r \cdot t$$
$$31,680 = 400 \cdot t \qquad \text{Let } d = \textbf{31,680} \text{ and } r = \textbf{400.}$$

3. SOLVE. Solve the equation for t. To solve for t, divide both sides by 400.

$$\frac{31,680}{400} = \frac{400 \cdot t}{400} \qquad \text{Divide both sides by \textbf{400.}}$$
$$79.2 = t \qquad \text{Simplify.}$$

> ▶ **Helpful Hint**
> Don't forget to include units, if appropriate.

4. INTERPRET.

Check: To check, substitute 79.2 for t and 400 for r in the distance formula and check to see that the distance is 31,680 feet.

State: It takes 79.2 years for the ice at the head of Portage Glacier to reach the lake.

EXAMPLE 2

CALCULATING THE LENGTH OF A GARDEN

Charles Pecot can afford enough fencing to enclose a rectangular garden with a perimeter of 140 feet. If the width of his garden must be 30 feet, find the length.

$w = 30$ feet

l

Solution
1. UNDERSTAND. Read and reread the problem. The formula needed to solve this problem is the formula for the perimeter of a rectangle, $P = 2l + 2w$. Before continuing, let's become familiar with this formula.

l = the length of the rectangular garden

w = the width of the rectangular garden

P = perimeter of the garden

2. TRANSLATE. To translate to an equation, we use the formula $P = 2l + 2w$ and let perimeter $P = 140$ feet and width $w = 30$ feet.

$$P = 2l + 2w$$
$$\downarrow \qquad \searrow$$
$$140 = 2l + 2(30) \qquad \text{Let } P = 140 \text{ and } w = 30.$$

3. SOLVE.

$$140 = 2l + 2(30)$$
$$140 = 2l + 60 \qquad \qquad \text{Multiply } 2(30).$$
$$140 - 60 = 2l + 60 - 60 \qquad \text{Subtract } 60 \text{ from both sides.}$$
$$80 = 2l \qquad \qquad \text{Combine like terms.}$$
$$40 = l \qquad \qquad \text{Divide both sides by } 2.$$

4. INTERPRET.

Check: Substitute 40 for l and 30 for w in the perimeter formula and check to see that the perimeter is 140 feet.

State: The length of the rectangular garden is 40 feet.

EXAMPLE 3

FINDING AN EQUIVALENT TEMPERATURE

The average maximum temperature for January in Algerias, Algeria, is 59° Fahrenheit. Find the equivalent temperature in degrees Celsius.

Solution
1. UNDERSTAND. Read and reread the problem. A formula that can be used to solve this problem is the formula for converting degrees Celsius to degrees Fahrenheit, $F = \frac{9}{5}C + 32$. Before continuing, become familiar with this formula. Using

this formula, we let

$$C = \text{temperature in degrees Celsius, and}$$
$$F = \text{temperature in degrees Fahrenheit.}$$

2. TRANSLATE. To translate to an equation, we use the formula $F = \dfrac{9}{5}C + 32$ and let degrees Fahrenheit $F = 59$.

Formula: $$F = \dfrac{9}{5}C + 32$$

Substitute: $$59 = \dfrac{9}{5}C + 32 \qquad \text{Let } F = 59.$$

3. SOLVE.

$$59 = \dfrac{9}{5}C + 32$$

$$59 - 32 = \dfrac{9}{5}C + 32 - 32 \qquad \text{Subtract \textbf{32} from both sides.}$$

$$27 = \dfrac{9}{5}C \qquad \text{Combine like terms.}$$

$$\dfrac{5}{9} \cdot 27 = \dfrac{5}{9} \cdot \dfrac{9}{5}C \qquad \text{Multiply both sides by } \tfrac{5}{9}.$$

$$15 = C \qquad \text{Simplify.}$$

4. INTERPRET.

Check: To check, replace C with 15 and F with 59 in the formula and see that a true statement results.

State: Thus, 59° Fahrenheit is equivalent to 15° Celsius.

In the next example, we again use the formula for perimeter of a rectangle as in Example 2. In Example 2, we knew the width of the rectangle. In this example, both the length and width are unknown.

EXAMPLE 4

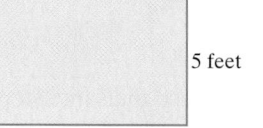

5 feet

13 feet

FINDING ROAD SIGN DIMENSIONS

The length of a rectangular road sign is 2 feet less than three times its width. Find the dimensions if the perimeter is 28 feet.

WESTVIEW	92
RIVERSIDE	162
SUNDALE	205

Solution   **1.** UNDERSTAND. Read and reread the problem. Recall that the formula for the perimeter of a rectangle is $P = 2l + 2w$. Draw a rectangle and guess the solution. If the width of the rectangular sign is 5 feet, its length is 2 feet less than 3 times the width or $3(5 \text{ feet}) - 2 \text{ feet} = 13 \text{ feet}$. The perimeter P of the rectangle is then $2(13 \text{ feet}) + 2(5 \text{ feet}) = 36 \text{ feet}$, too much. We now know that the width is less than 5 feet.

w

$3w - 2$

Let

w = the width of the rectangular sign; then

$3w - 2$ = the length of the sign.

Draw a rectangle and label it with the assigned variables.

2. TRANSLATE.

Formula: $P = 2l + 2w$ or

Substitute: $28 = 2(3w - 2) + 2w.$

3. SOLVE.

$$28 = 2(3w - 2) + 2w$$
$$28 = 6w - 4 + 2w \qquad \text{Apply the distributive property.}$$
$$28 = 8w - 4$$
$$28 + 4 = 8w - 4 + 4 \qquad \text{Add 4 to both sides.}$$
$$32 = 8w$$
$$\frac{32}{8} = \frac{8w}{8} \qquad \text{Divide both sides by 8.}$$
$$4 = w$$

4. INTERPRET.

Check: If the width of the sign is 4 feet, the length of the sign is $3(4 \text{ feet})$ $- 2$ feet $= 10$ feet. This gives a perimeter of $P = 2(4 \text{ feet}) + 2(10 \text{ feet}) = 28$ feet, the correct perimeter.

State: The width of the sign is 4 feet and the length of the sign is 10 feet.

2   We say that the formula $F = \frac{9}{5}C + 32$ is solved for F because F is alone on one side of the equation and the other side of the equation contains no F's. Suppose that we need to convert many Fahrenheit temperatures to equivalent degrees Celsius. In this case, it is easier to perform this task by solving the formula $F = \frac{9}{5}C + 32$ for C. (See Example 8.) For this reason, it is important to be able to solve an equation for any one of its specified variables. For example, the formula $d = rt$ is solved for d in terms of r and t. We can also solve $d = rt$ for t in terms of d and r. To solve for t, divide both sides of the equation by r.

$$d = rt$$
$$\frac{d}{r} = \frac{rt}{r} \qquad \text{Divide both sides by } r.$$
$$\frac{d}{r} = t \qquad \text{Simplify.}$$

To solve a formula or an equation for a specified variable, we use the same steps as for solving a linear equation. These steps are listed next.

Solving Equations for a Specified Variable

Step 1. Multiply on both sides to clear the equation of fractions if they occur.

Step 2. Use the distributive property to remove parentheses if they occur.

Step 3. Simplify each side of the equation by combining like terms.

Step 4. Get all terms containing the specified variable on one side and all other terms on the other side by using the addition property of equality.

Step 5. Get the specified variable alone by using the multiplication property of equality.

EXAMPLE 5

Solve $V = lwh$ for l.

Solution

This formula is used to find the volume of a box. To solve for l, divide both sides by wh.

$$V = lwh$$

$$\frac{V}{wh} = \frac{lwh}{wh} \qquad \text{Divide both sides by } wh.$$

$$\frac{V}{wh} = l \qquad \text{Simplify.}$$

Since we have l alone on one side of the equation, we have solved for l in terms of V, w, and h. Remember that it does not matter on which side of the equation we isolate the variable.

EXAMPLE 6

Solve $y = mx + b$ for x.

Solution

The term containing the variable we are solving for, mx, is on the right side of the equation. Get mx alone by subtracting b from both sides.

$$y = mx + b$$

$$y - b = mx + b - b \qquad \text{Subtract } b \text{ from both sides.}$$

$$y - b = mx \qquad \text{Combine like terms.}$$

Next, solve for x by dividing both sides by m.

$$\frac{y - b}{m} = \frac{mx}{m}$$

$$\frac{y - b}{m} = x \qquad \text{Simplify.}$$

EXAMPLE 7

Solve $P = 2l + 2w$ for w.

Solution

This formula relates the perimeter of a rectangle to its length and width. Find the term containing the variable w. To get this term, $2w$, alone subtract $2l$ from both sides.

Helpful Hint

The 2's may *not* be divided out here. Although 2 is a factor of the denominator, 2 is *not* a factor of the numerator since it is not a factor of both terms in the numerator.

$$P = 2l + 2w$$

$$P - 2l = 2l + 2w - 2l \qquad \text{Subtract } 2l \text{ from both sides.}$$

$$P - 2l = 2w \qquad \text{Combine like terms.}$$

$$\frac{P - 2l}{2} = \frac{2w}{2} \qquad \text{Divide both sides by 2.}$$

$$\frac{P - 2l}{2} = w \qquad \text{Simplify.}$$

The next example has an equation containing a fraction. We will first clear the equation of fractions and then solve for the specified variable.

EXAMPLE 8

Solve $F = \frac{9}{5}C + 32$ for C.

Solution

$$F = \frac{9}{5}C + 32$$

$$5(F) = 5\left(\frac{9}{5}C + 32\right) \quad \text{Clear the fraction by multiplying both sides by the LCD.}$$

$$5F = 9C + 160 \quad \text{Distribute the 5.}$$

$$5F - 160 = 9C + 160 - 160 \quad \text{To get the term containing the variable } C \text{ alone, subtract 160 from both sides.}$$

$$5F - 160 = 9C \quad \text{Combine like terms.}$$

$$\frac{5F - 160}{9} = \frac{9C}{9} \quad \text{Divide both sides by 9.}$$

$$\frac{5F - 160}{9} = C \quad \text{Simplify.}$$

Spotlight on

DECISION
MAKING

Suppose you have just purchased a house. You are required by your mortgage company to buy home-owner's insurance. In choosing a policy, you must also choose a deductible level. A deductible is how much you, the home-owner, pay to repair or replace a loss before the insurance company starts paying. Typically, deductibles start at $250. However, other levels such as $500 or $1000 are also available. One way to save money when purchasing your homeowner's policy is to raise the deductible amount. However, doing so means you will have to pay more out of pocket in case of theft or fire. Study the accompanying chart. Which amount of deductible would you choose? Why?

Increase Your Deductible to:	
$500	save up to 12%
$1000	save up to 24%
$2500	save up to 30%
$5000	save up to 37%

on the cost of a homeowner's policy, depending on your insurance company.

Source: Insurance Information Institute

EXERCISE SET 2.5

STUDY GUIDE/SSM   CD/VIDEO   PH MATH TUTOR CENTER   MathXL®Tutorials ON CD   MathXL®   MyMathLab®

Substitute the given values into each given formula and solve for the unknown variable. If necessary, round to one decimal place. See Examples 1 through 3.

△ **1.** $A = bh$; $A = 45, b = 15$ (Area of a parallelogram)

2. $d = rt$; $d = 195, t = 3$ (Distance formula)

△ **3.** $S = 4lw + 2wh$; $S = 102, l = 7, w = 3$ (Surface area of a special rectangular box)

△ **4.** $V = lwh$; $l = 14, w = 8, h = 3$ (Volume of a rectangular box)

△ **5.** $A = \frac{1}{2}h(B + b)$; $A = 180, B = 11, b = 7$ (Area of a trapezoid)

△ **6.** $A = \frac{1}{2}h(B + b)$; $A = 60, B = 7, b = 3$ (Area of a trapezoid)

7. $P = a + b + c$; $P = 30, a = 8, b = 10$ (Perimeter of a triangle)

 8. $V = \frac{1}{3}Ah$; $V = 45, h = 5$ (Volume of a pyramid)

9. $C = 2\pi r$; $C = 15.7$ (use the approximation 3.14 or a calculator approximation for π) (Circumference of a circle)

10. $A = \pi r^2$; $r = 4.5$ (use the approximation 3.14 or a calculator approximation for π)(Area of a circle)

11. $I = PRT$; $I = 3750, P = 25,000, R = 0.05$ (Simple interest formula)

12. $I = PRT$; $I = 1,056,000, R = 0.055, T = 6$ (Simple interest formula)

13. $V = \frac{1}{3}\pi r^2 h$; $V = 565.2, r = 6$ (use a calculator approximation for π)(Volume of a cone)

14. $V = \frac{4}{3}\pi r^3$; $r = 3$ (use a calculator approximation for π) (Volume of a sphere)

Solve each formula for the specified variable. See Examples 5 through 8.

15. $f = 5gh$ for h

16. $C = 2\pi r$ for r

 17. $V = lwh$ for w

18. $T = mnr$ for n

19. $3x + y = 7$ for y

20. $-x + y = 13$ for y

21. $A = P + PRT$ for R

22. $A = P + PRT$ for T

23. $V = \frac{1}{3}Ah$ for A

24. $D = \frac{1}{4}fk$ for k

25. $P = a + b + c$ for a

26. $PR = s_1 + s_2 + s_3 + s_4$ for s_3

27. $S = 2\pi rh + 2\pi r^2$ for h **28.** $S = 4lw + 2wh$ for h

Solve. See Examples 1 through 4.

29. The world's largest sign for Coca-Cola is located in Arica, Chile. The rectangular sign has a length of 400 feet and has an area of 52,400 square feet. Find the width of the sign. (*Source: Fabulous Facts about Coca-Cola, Atlanta, GA*)

30. The length of a rectangular garden is 6 meters. If 21 meters of fencing are required to fence the garden, find its width.

6 meters

31. Convert Nome, Alaska's 14°F high temperature to Celsius.

32. Convert Paris, France's low temperature of −5°C to Fahrenheit.

33. The X-30 is a "space plane" that skims the edge of space at 4000 miles per hour. Neglecting altitude, if the circumference of the Earth is approximately 25,000 miles, how long will it take for the X-30 to travel around the Earth?

34. In the United States, a notable hang glider flight was a 303-mile, $8\frac{1}{2}$ hour flight from New Mexico to Kansas. What was the average rate during this flight?

35. An architect designs a rectangular flower garden such that the width is exactly two-thirds of the length. If 260 feet of antique picket fencing are to be used, find the dimensions of the garden.

$?$

← x feet →

36. If the length of a rectangular parking lot is 10 meters less than twice its width, and the perimeter is 400 meters, find the length of the parking lot.

x meters

$?$

37. A flower bed is in the shape of a triangle with one side twice the length of the shortest side, and the third side is 30 feet more than the length of the shortest side. Find the dimensions if the perimeter is 102 feet.

38. The perimeter of a yield sign in the shape of an isosceles triangle is 22 feet. If the shortest side is 2 feet less than the other two sides, find the length of the shortest side. (*Hint:* An isosceles triangle has two sides the same length.)

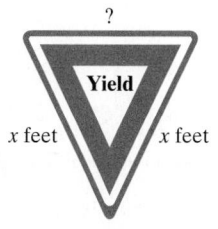

$?$

Yield

x feet x feet

39. The Cat is a high-speed catamaran auto ferry that operates between Bar Harbor, Maine, and Yarmouth, Nova Scotia. The Cat can make the trip in about $2\frac{1}{2}$ hours at a speed of 55 mph. About how far apart are Bar Harbor and Yarmouth? (*Source:* Bay Ferries)

40. A family is planning their vacation to Disney World. They will drive from a small town outside New Orleans, Louisiana, to Orlando, Florida, a distance of 700 miles. They plan to average a rate of 55 mph. How long will this trip take?

41. Piranha fish require 1.5 cubic feet of water per fish to maintain a healthy environment. Find the maximum number of piranhas you could put in a tank measuring 8 feet by 3 feet by 6 feet.

42. Find how many goldfish you can put in a cylindrical tank whose diameter is 8 meters and whose height is 3 meters, if each goldfish needs 2 cubic meters of water.

43. A lawn is in the shape of a trapezoid with a height of 60 feet and bases of 70 feet and 130 feet. How many bags of fertilizer must be purchased to cover the lawn if each bag covers 4000 square feet?

44. If the area of a right-triangularly shaped sail is 20 square feet and its base is 5 feet, find the height of the sail.

45. Maria's Pizza sells one 16-inch cheese pizza or two 10-inch cheese pizzas for $9.99. Determine which size gives more pizza.

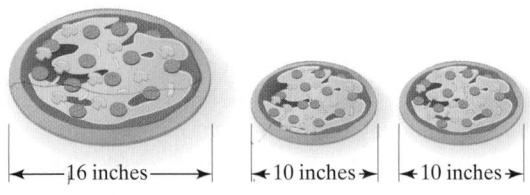

46. Find how much rope is needed to wrap around the Earth at the equator, if the radius of the Earth is 4000 miles. (*Hint:* Use 3.14 for π and the formula for circumference.)

47. A Japanese "bullet" train set a new world record for train speed at 552 kilometers per hour during a manned test run on the Yamanashi Maglev Test Line in April 1999. The Yamanashi Maglev Test Line is 42.8 kilometers long. How many *minutes* would a test run on the Yamanashi Line last at this record-setting speed? Round to the nearest hundredth of a minute. (*Source:* Japan Railways Central Co.)

48. In 1983, the Hawaiian volcano Kilauea began erupting in a series of episodes still occurring at the time of this writing. At times, the lava flows advanced at speeds of up to 0.5 kilometer per hour. In 1983 and 1984 lava flows destroyed 16 homes in the Royal Gardens subdivision, about 6 km away from the eruption site. Roughly how long did it take the lava to reach Royal Gardens? (*Source:* U.S. Geological Survey Hawaiian Volcano Observatory)

49. The perimeter of an equilateral triangle is 7 inches more than the perimeter of a square, and the side of the triangle is 5 inches longer than the side of the square. Find the side of the triangle.

50. A square animal pen and a pen shaped like an equilateral triangle have equal perimeters. Find the length of the sides of each pen if the sides of the triangular pen are fifteen less than twice a side of the square pen. (*Hint:* An equilateral triangle has three sides the same length.)

51. Find how long it takes Tran Nguyen to drive 135 miles on I-10 if he merges onto I-10 at 10 A.M. and drives nonstop with his cruise control set on 60 mph.

52. Beaumont, Texas, is about 150 miles from Toledo Bend. If Leo Miller leaves Beaumont at 4 A.M. and averages 45 mph, when should he arrive at Toledo Bend?

53. The longest runway at Los Angeles International Airport has the shape of a rectangle and an area of 1,813,500 square feet. This runway is 150 feet wide. How long is the runway? (*Source:* Los Angeles World Airports)

54. Bolts of lightning can travel at the speed of 270,000 miles per second. How many times can a lightning bolt travel around the world in one second? (See Exercise 46. Round to the nearest tenth.)

55. The highest temperature ever recorded in Europe was 122°F in Seville, Spain, in August of 1881. Convert this record high temperature to Celsius. (*Source:* National Climatic Data Center)

56. The lowest temperature ever recorded in Oceania was −10°C at the Haleakala Summit in Maui, Hawaii, in January 1961. Convert this record low temperature to Fahrenheit. (*Source:* National Climatic Data Center)

△ **57.** The CART Fed Ex Championship Series is an open-wheeled race car competition based in the United States. A CART car has a maximum length of 199 inches, a maximum width of 78.5 inches, and a maximum height of 33 inches. When the CART series travels to another country for a grand prix, teams must ship their cars. Find the volume of the smallest shipping crate needed to ship a CART car of maximum dimensions. (*Source:* Championship Auto Racing Teams, Inc.)

CART Racing Car

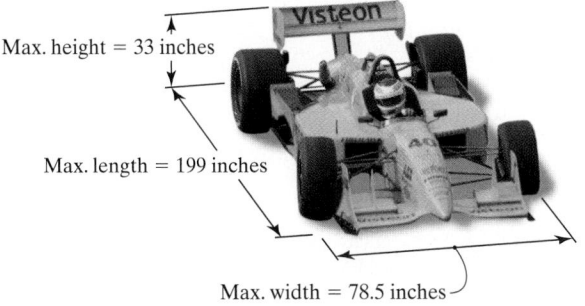

Max. height = 33 inches
Max. length = 199 inches
Max. width = 78.5 inches

58. On a road course, a CART car's speed can average up to around 105 mph. Based on this speed, how long would it take a CART driver to travel from Los Angeles to New York City, a distance of about 2810 miles by road, without stopping? Round to the nearest tenth of an hour.

△ **59.** The Hoberman Sphere is a toy ball that expands and contracts. When it is completely closed, it has a diameter of 9.5 inches. Find the volume of the Hoberman Sphere when it is completely closed. Use 3.14 for π. Round to the nearest whole cubic inch. (*Source:* Hoberman Designs, Inc.)

△ **60.** When the Hoberman Sphere (see Exercise 59) is completely expanded, its diameter is 30 inches. Find the volume of the Hoberman Sphere when it is completely expanded. Use 3.14 for π. Round to the nearest whole cubic inch. (*Source:* Hoberman Designs, Inc.)

See Exercise 60.

61. The average temperature on the planet Mercury is 167°C. Convert this temperature to degrees Fahrenheit. (*Source:* National Space Science Data Center)

62. The average temperature on the planet Jupiter is −227°F. Convert this temperature to degrees Celsius. Round to the nearest degree. (*Source:* National Space Science Data Center)

REVIEW AND PREVIEW

Write the following phrases as algebraic expressions. See Section 2.1.

63. Nine divided by the sum of a number and 5

64. Half the product of a number and five

65. Three times the sum of a number and four

66. Double the sum of ten and four times a number

67. Triple the difference of a number and twelve

68. A number minus the sum of the number and six

Concept Extensions

69. Dry ice is a name given to solidified carbon dioxide. At −78.5° Celsius it changes directly from a solid to a gas. Convert this temperature to Fahrenheit.

70. Lightning bolts can reach a temperature of 50,000° Fahrenheit. Convert this temperature to Celsius.

71. The distance from the sun to the Earth is approximately 93,000,000 miles. If light travels at a rate of 186,000 miles per second, how long does it take light from the sun to reach us?

72. Light travels at a rate of 186,000 miles per second. If our moon is 238,860 miles from the Earth, how long does it take light from the moon to reach us? (Round to the nearest tenth of a second.)

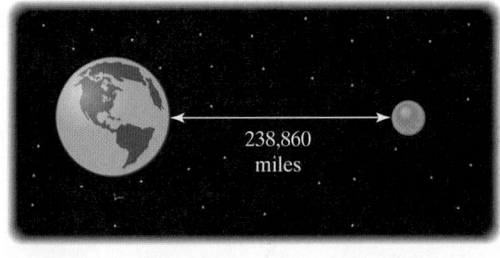

238,860 miles

73. A glacier is a giant mass of rocks and ice that flows downhill like a river. Exit Glacier, near Seward, Alaska, moves at a rate of 20 inches a day. Find the distance in feet the glacier moves in a year. (Assume 365 days a year. Round to 2 decimal places.)

74. Flying fish do not *actually* fly, but glide. They have been known to travel a distance of 1300 feet at a rate of 20 miles per hour. How many seconds did it take to travel this distance? (*Hint:* First convert miles per hour to feet per second. Recall that 1 mile = 5280 feet. Round to the nearest tenth of a second.)

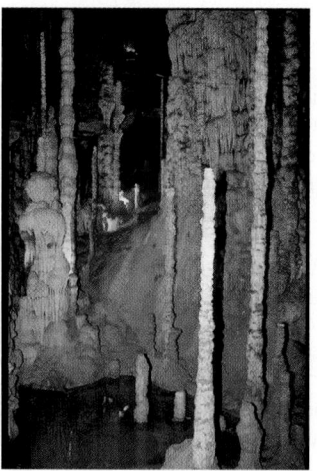

76. Find the temperature at which the Celsius measurement and Fahrenheit measurement are the same number.

77. Normal room temperature is about 78°F. Convert this temperature to Celsius.

△ **75.** Stalactites join stalagmites to form columns. A column found at Natural Bridge Caverns near San Antonio, Texas, rises 15 feet and has a *diameter* of only 2 inches. Find the volume of this column in cubic inches. (*Hint:* Use the formula for volume of a cylinder and use a calculator approximation for π. Round your answer to the nearest tenth.)

△ **78.** The formula $V = LWH$ is used to find the volume of a box. If the length of a box is doubled, the width is doubled, and the height is doubled, how does this affect the volume?

△ **79.** The formula $A = bh$ is used to find the area of a parallelogram. If the base of a parallelogram is doubled and its height is doubled, how does this affect the area?

2.6   *FURTHER PROBLEM SOLVING*

Objectives

1   Solve problems involving percents.

2   Solve problems involving distance.

3   Solve problems involving mixtures.

4   Solve problems involving interest.

This section is devoted to solving problems in the categories listed. The same problem-solving steps used in previous sections are also followed in this section. They are listed below for review.

> ### General Strategy for Problem Solving
>
> **1.** UNDERSTAND the problem. During this step, become comfortable with the problem. Some ways of doing this are:
>
> Read and reread the problem.
>
> Choose a variable to represent the unknown.
>
> Construct a drawing, whenever possible.
>
> Propose a solution and check. Pay careful attention to how you check your proposed solution. This will help writing an equation to model the problem.
>
> **2.** TRANSLATE the problem into an equation.
>
> **3.** SOLVE the equation.
>
> **4.** INTERPRET the results: *Check* the proposed solution in the stated problem and *state* your conclusion.

1   The first two examples involve percents.

Percent increase or percent decrease is a common way to describe how some measurement has increased or decreased. For example, crime increased by 8%, teachers received a 5.5% increase in salary, or a company decreased its employees by 10%. The next example is a review of percent increase.

EXAMPLE 1

CALCULATING THE COST OF ATTENDING COLLEGE

The cost of attending a public college rose from $5324 in 1990 to $8086 in 2000. Find the percent increase. (*Source:* U.S. Department of Education. *Note:* These costs include room and board.)

Solution   **1.** UNDERSTAND. Read and reread the problem. Let's guess that the percent increase is 20%. To see if this is the case, we find 20% of $5324 to find the *increase* in cost. Then we add this increase to $5324 to find the *new cost*. In other words, 20% ($5324) = 0.20($5324) = $1064.80, the *increase* in cost. The new cost then would be $5324 + $1064.80 = $6388.80, less than the actual new cost of $8086. We now know that the increase is greater than 20% and we know how to check our proposed solution.

Let x = the percent increase.

2. TRANSLATE. First, find the **increase**, and then the **percent increase**. The increase in cost is found by:

In words:	increase	=	new cost	−	old cost	or
Translate:	increase	=	$8086	−	$5324	
		=	$2762			

Next, find the percent increase. The percent increase or percent decrease is always a percent of the original number or in this case, the old cost.

In words:	increase	is	what percent increase	of	old cost
Translate:	$2762	=	x	·	$5324

3. SOLVE.

$$2762 = 5324x$$
$$0.519 \approx x$$
$$51.9\% \approx x$$

Divide both sides by **5324** and round to **3** decimal places.

Write as a percent.

4. INTERPRET.

Check: Check the proposed solution.
State: The percent increase in cost is approximately 51.9%.

EXAMPLE 2

Most of the movie screens in the United States project analog films, but the number of cinemas using digital are increasing. Find the number of digital screens last year if after a 175% increase, the number this year is 124. Round to the nearest whole.

Solution

1. UNDERSTAND. Read and reread the problem. Let's guess a solution and see how we would check our guess. If the number of digital screens last year was 50, we would see if 50 plus the increase is 124; that is,

$$50 + 175\% \cdot 50 = 50 + 1.75(50) = 2.75(50) = 137.50$$

Since 137.5 is too large, we know that our guess of 50 is too large. We also have a better understanding of the problem. Let

$$x = \text{number of digital screens last year}$$

2. TRANSLATE. To translate to an equation, we remember that

In words:

number of digital screen last year	plus	increase	equals	number of digital screens this year
Translate: x	$+$	$1.75x$	$=$	124

3. SOLVE.

$$2.75x = 124$$
$$x = \frac{124}{2.75}$$
$$x \approx 45$$

4. INTERPRET.

Check: Recall that x represents the number of digital screens last year. If this number is approximately 45, let's see if 45 plus the increase is close to 124. (We use the word "close" since 45 is rounded.)

$$45 + 175\%(45) = 45 + 1.75(45) = 2.75(45) = 123.75,$$

which is close to 124.

State: There were approximately 45 digital screens last year.

2 The next example involves distance.

EXAMPLE 3

FINDING TIME GIVEN RATE AND DISTANCE

Marie Antonio, a bicycling enthusiast, rode her 21-speed at an average speed of 18 miles per hour on level roads and then slowed down to an average of 10 mph on the hilly roads of the trip. If she covered a distance of 98 miles, how long did the entire trip take if traveling the level roads took the same time as traveling the hilly roads?

Solution

1. UNDERSTAND the problem. To do so, read and reread the problem. The formula $d = r \cdot t$ is needed. At this time, let's guess a solution. Suppose that she spent 2 hours traveling on the level roads. This means that she also spent 2 hours traveling on the hilly roads, since the times spent were the same. What is her total distance? Her distance on the level road is rate · time = 18(2) = 36 miles. Her distance on the hilly roads is rate · time = 10(2) = 20 miles. This gives a total distance of 36 miles + 20 miles = 56 miles, not the correct distance of 98 miles. Remember that the purpose of guessing a solution is not to guess correctly (although this may happen) but to help better understand the problem and how to model it with an equation.

We are looking for the length of the entire trip, so we begin by letting

$$x = \text{the time spent on level roads.}$$

Because the same amount of time is spent on hilly roads, then also

$$x = \text{the time spent on hilly roads.}$$

2. TRANSLATE. To help us translate to an equation, we now summarize the information from the problem on the following chart. Fill in the rates given, the variables used to represent the times, and use the formula $d = r \cdot t$ to fill in the distance column.

	Rate ·	Time =	Distance
Level	18	x	$18x$
Hilly	10	x	$10x$

Since the entire trip covered 98 miles, we have that

In words:	total distance	=	level distance	+	hilly distance
Translate:	98	=	$18x$	+	$10x$

3. SOLVE.

$$98 = 28x \qquad \text{Add like terms.}$$
$$\frac{98}{28} = \frac{28x}{28} \qquad \text{Divide both sides by 28.}$$
$$3.5 = x$$

4. INTERPRET the results.

Check: Recall that x represents the time spent on the level portion of the trip and also the time spent on the hilly portion. If Marie rides for 3.5 hours at 18 mph, her distance is 18(3.5) = 63 miles. If Marie rides for 3.5 hours at 10 mph, her distance is 10(3.5) = 35 miles. The total distance is 63 miles + 35 miles = 98 miles, the required distance.

State: The time of the entire trip is then 3.5 hours + 3.5 hours or 7 hours.

3   Mixture problems involve two or more different quantities being combined to form a new mixture. These applications range from Dow Chemical's need to form a chemical mixture of a required strength to Planter's Peanut Company's need to find the correct mixture of peanuts and cashews, given taste and price constraints.

EXAMPLE 4

CALCULATING PERCENT FOR A LAB EXPERIMENT

A chemist working on his doctoral degree at Massachusetts Institute of Technology needs 12 liters of a 50% acid solution for a lab experiment. The stockroom has only

40% and 70% solutions. How much of each solution should be mixed together to form 12 liters of a 50% solution?

Solution

1. UNDERSTAND. First, read and reread the problem a few times. Next, guess a solution. Suppose that we need 7 liters of the 40% solution. Then we need $12 - 7 = 5$ liters of the 70% solution. To see if this is indeed the solution, find the amount of pure acid in 7 liters of the 40% solution, in 5 liters of the 70% solution, and in 12 liters of a 50% solution, the required amount and strength.

number of liters	×	acid strength	=	amount of pure acid
7 liters	×	40%	=	7(0.40) or 2.8 liters
5 liters	×	70%	=	5(0.70) or 3.5 liters
12 liters	×	50%	=	12(0.50) or 6 liters

Since 2.8 liters + 3.5 liters = 6.3 liters and not 6, our guess is incorrect, but we have gained some valuable insight into how to model and check this problem.

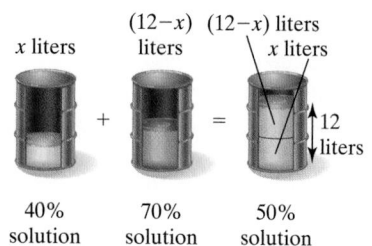

x liters · (12−x) liters · (12−x) liters · x liters · 12 liters

40% solution · 70% solution · 50% solution

Let

$$x = \text{number of liters of 40\% solution; then}$$
$$12 - x = \text{number of liters of 70\% solution.}$$

2. TRANSLATE. To help us translate to an equation, the following table summarizes the information given. Recall that the amount of acid in each solution is found by multiplying the acid strength of each solution by the number of liters.

	No. of Liters	·	*Acid* Strength	=	Amount of Acid
40% Solution	x		40%		$0.40x$
70% Solution	$12 - x$		70%		$0.70(12 - x)$
50% Solution Needed	12		50%		$0.50(12)$

The amount of acid in the final solution is the sum of the amounts of acid in the two beginning solutions.

In words: acid in 40% solution + acid in 70% solution = acid in 50% mixture

Translate: $0.40x$ + $0.70(12 - x)$ = $0.50(12)$

3. SOLVE.

$$0.40x + 0.70(12 - x) = 0.50(12)$$
$$0.4x + 8.4 - 0.7x = 6 \qquad \text{Apply the distributive property.}$$
$$-0.3x + 8.4 = 6 \qquad \text{Combine like terms.}$$
$$-0.3x = -2.4 \qquad \text{Subtract 8.4 from both sides.}$$
$$x = 8 \qquad \text{Divide both sides by } -0.3.$$

4. INTERPRET.

Check: To check, recall how we checked our guess.
State: If 8 liters of the 40% solution are mixed with $12 - 8$ or 4 liters of the 70% solution, the result is 12 liters of a 50% solution.

4   The next example is an investment problem.

EXAMPLE 5

FINDING THE INVESTMENT AMOUNT

Rajiv Puri invested part of his $20,000 inheritance in a mutual funds account that pays 7% simple interest yearly and the rest in a certificate of deposit that pays 9% simple interest yearly. At the end of one year, Rajiv's investments earned $1550. Find the amount he invested at each rate.

Solution   1. **UNDERSTAND.** Read and reread the problem. Next, guess a solution. Suppose that Rajiv invested $8000 in the 7% fund and the rest, $12,000, in the fund paying 9%. To check, find his interest after one year. Recall the formula, $I = PRT$, so the interest from the 7% fund = $8000(0.07)(1) = $560. The interest from the 9% fund = $12,000(0.09)(1) = $1080. The sum of the interests is $560 + $1080 = $1640. Our guess is incorrect, since the sum of the interests is not $1550, but we now have a better understanding of the problem.

Let
$$x = \text{amount of money in the account paying } 7\%.$$

The rest of the money is $20,000 less x or

$$20{,}000 - x = \text{amount of money in the account paying } 9\%.$$

2. **TRANSLATE.** We apply the simple interest formula $I = PRT$ and organize our information in the following chart. Since there are two different rates of interest and two different amounts invested, we apply the formula twice.

	Principal ·	Rate ·	Time ·	=	Interest
7% Fund	x	0.07	1		$x(0.07)(1)$ or $0.07x$
9% Fund	$20{,}000 - x$	0.09	1		$(20{,}000 - x)(0.09)(1)$ or $0.09(20{,}000 - x)$
Total	20,000				1550

The total interest earned, $1550, is the sum of the interest earned at 7% and the interest earned at 9%.

In words: | interest at 7% | + | interest at 9% | = | total interest |

Translate: $0.07x$ + $0.09(20{,}000 - x)$ = 1550

3. **SOLVE.**

$$0.07x + 0.09(20{,}000 - x) = 1550$$
$$0.07x + 1800 - 0.09x = 1550 \qquad \text{Apply the distributive property.}$$
$$1800 - 0.02x = 1550 \qquad \text{Combine like terms.}$$
$$-0.02x = -250 \qquad \text{Subtract } \mathbf{1800} \text{ from both sides.}$$
$$x = 12{,}500 \qquad \text{Divide both sides by } -\mathbf{0.02}.$$

4. **INTERPRET.**

Check: If $x = 12{,}500$, then $20{,}000 - x = 20{,}000 - 12{,}500$ or 7500. These solutions are reasonable, since their sum is $20,000 as required. The annual interest on $12,500 at 7% is $875; the annual interest on $7500 at 9% is $675, and $875 + $675 = $1550.
State: The amount invested at 7% is $12,500. The amount invested at 9% is $7500.

Spotlight on

DECISION ❧ MAKING

Suppose you are a personal income tax preparer. Your clients, Jose and Felicia Fernandez, are filing jointly Form 1040 as their individual income tax return. You know that medical expenses may be written off as an itemized deduction if the expenses exceed 7.5% of their adjusted gross income. Furthermore, only the portion of medical expenses that exceed 7.5% of their adjusted gross income can be deducted. Is the Fernandez family eligible to deduct their medical expenses? Explain.

Internal Revenue Service
Form 1040

Jose Fernandez
Felicia Fernandez

Adjusted Gross Income . . . $33,650

Fernandez Family Deductible Medical Expenses	
Medical bills	$1025
Dental bills	$ 325
Prescription drugs	$ 360
Medical Insurance premiums	$1200

EXERCISE SET 2.6

STUDY GUIDE/SSM · CD/VIDEO · PH MATH TUTOR CENTER · MathXL®Tutorials ON CD · MathXL® · MyMathLab®

Solve. See Examples 1 and 2.

1. Nordstrom's advertised a 25% off sale. If a London Fog coat originally sold for $256, find the decrease in price and the sale price.

2. Time Saver increased the price of a $0.95 cola by 15%. Find the increase in price and the new price.

3. Find the original price of a pair of shoes if the sale price is $78 after a 25% discount.

4. Find last year's salary if after a 4% pay raise, this year's salary is $33,800.

5. The number of fraud complaints (usually ID theft) rose from 220,000 in 2001 to 380,000 in 2002. Find the percent increase. Round to the nearest whole percent.

6. The number of text messages rose from 996 million in June to 1100 million in December. Find the percent increase. Round to the nearest whole percent.

Solve. See Example 3.

7. A jet plane traveling at 500 mph overtakes a propeller plane traveling at 200 mph that had a 2-hour head start. How far from the starting point are the planes?

8. How long will it take a bus traveling at 60 miles per hour to overtake a car traveling at 40 mph if the car had a 1.5-hour head start?

9. The Jones family drove to Disneyland at 50 miles per hour and returned on the same route at 40 mph. Find the distance to Disneyland if the total driving time was 7.2 hours.

10. A bus traveled on a level road for 3 hours at an average speed 20 miles per hour faster than it traveled on a winding road. The time spent on the winding road was 4 hours. Find the average speed on the level road if the entire trip was 305 miles.

Solve. See Example 4.

11. How much pure acid should be mixed with 2 gallons of a 40% acid solution in order to get a 70% acid solution?

12. How many cubic centimeters of a 25% antibiotic solution should be added to 10 cubic centimeters of a 60% antibiotic solution in order to get a 30% antibiotic solution?

13. Planter's Peanut Company wants to mix 20 pounds of peanuts worth $3 a pound with cashews worth $5 a pound in order to make an experimental mix worth $3.50 a pound. How many pounds of cashews should be added to the peanuts?

14. Community Coffee Company wants a new flavor of Cajun coffee. How many pounds of coffee worth $7 a pound should be added to 14 pounds of coffee worth $4 a pound to get a mixture worth $5 a pound?

15. Is it possible to mix a 30% antifreeze solution with a 50% antifreeze solution and obtain a 70% antifreeze solution? Why or why not?

16. A trail mix is made by combining peanuts worth $3 a pound, raisins worth $2 a pound, and M & M's worth $4 a pound. Would it make good business sense to sell the trail mix for $1.98 a pound? Why or why not?

Solve. See Example 5.

17. Zoya Lon invested part of her $25,000 advance at 8% annual simple interest and the rest at 9% annual simple interest. If her total yearly interest from both accounts was $2135, find the amount invested at each rate.

18. Karen Waugtal invested some money at 9% annual simple interest and $250 more than that amount at 10% annual simple interest. If her total yearly interest was $101, how much was invested at each rate?

19. Sam Mathius invested part of his $10,000 bonus in a fund that paid an 11% profit and invested the rest in stock that suffered a 4% loss. Find the amount of each investment if his overall net profit was $650.

20. Bruce Blossum invested a sum of money at 10% annual simple interest and invested twice that amount at 12% annual simple interest. If his total yearly income from both investments was $2890, how much was invested at each rate?

MIXED PRACTICE

Solve.

21. A student at the University of New Orleans makes money by buying and selling used cars. Charles bought a used car and later sold it for a 20% profit. If he sold it for $4680, how much did Charles pay for the car?

22. After a downturn in the economy, a company downsized by 35%. If there are still 78 employees, how many employees were there prior to the layoffs?

23. How can $54,000 be invested, part at 8% annual simple interest and the remainder at 10% annual simple interest, so that the interest earned by the two accounts will be equal?

24. Ms. Mills invested her $20,000 bonus in two accounts. She took a 4% loss on one investment and made a 12% profit on another investment, but ended up breaking even. How much was invested in each account?

25. Kathleen and Cade Williams leave simultaneously from the same point hiking in opposite directions, Kathleen walking at 4 miles per hour and Cade at 5 mph. How long can they talk on their walkie-talkies if the walkie-talkies have a 20-mile radius?

26. Alan and Dave Schaferkötter leave from the same point driving in opposite directions, Alan driving at 55 miles per hour and Dave at 65 mph. Alan has a one-hour head start. How long will they be able to talk on their car phones if the phones have a 250-mile range?

27. How much of an alloy that is 20% copper should be mixed with 200 ounces of an alloy that is 50% copper in order to get an alloy that is 30% copper?

28. How much water should be added to 30 gallons of a solution that is 70% antifreeze in order to get a mixture that is 60% antifreeze?

29. Iceberg lettuce is grown and shipped to stores for about 40 cents a head, and consumers purchase it for about 70 cents a head. Find the percent increase.

30. The lettuce consumption per capita in 1968 was about 21.5 pounds, and in 2000 the consumption rose to 24.3 pounds. Find the percent increase. (Round to the nearest tenth of a percent.)

31. If $3000 is invested at 6% annual simple interest, how much should be invested at 9% annual simple interest so that the total yearly income from both investments is $585?

32. Trudy Waterbury, a financial planner, invested a certain amount of money at 9% annual simple interest, twice that amount at 10% annual simple interest, and three times that amount at 11% annual simple interest. Find the amount invested at each rate if her total yearly income from the investments was $2790.

33. Smart Cards (cards with an embedded computer chip) have been growing in popularity in recent years. In 2006, 500 million Smart Cards are expected to be issued. This represents a 117% increase from the number of cards that were issued in 2001. How many Smart Cards were issued in 2001? Round to the nearest million. (*Source:* The Freedonia Group)

34. April Thrower spent $32.25 to take her daughter's birthday party guests to the movies. Adult tickets cost $5.75 and children tickets cost $3.00. If 8 persons were at the party, how many adult tickets were bought?

35. Two hikers are 11 miles apart and walking toward each other. They meet in 2 hours. Find the rate of each hiker if one hiker walks 1.1 mph faster than the other.

36. On a 255-mile trip, Gary Alessandrini traveled at an average speed of 70 mph, got a speeding ticket, and then traveled at 60 mph for the remainder of the trip. If the entire trip took 4.5 hours and the speeding ticket stop took 30 minutes, how long did Gary speed before getting stopped?

37. The ACT Assessment is a college entrance exam taken by about 60% of college-bound students. The national average score was 20.7 in 1993 and rose to 21.0 in 2001. Find the percent increase. (Round to the nearest hundredth of a percent.)

38. The first Barbie doll was introduced in March 1959 and cost $3. This same 1959 Barbie doll now costs up to $5000. Find the percent increase rounded to the nearest whole percent.

39. Nedra and Latonya Dominguez are 12 miles apart hiking toward each other. How long will it take them to meet if Nedra walks at 3 mph and Latonya walks 1 mph faster?

40. Mark Martin can row upstream at 5 mph and downstream at 11 mph. If Mark starts rowing upstream until he gets tired and then rows downstream to his starting point, how far did Mark row if the entire trip took 4 hours?

REVIEW AND PREVIEW

Perform the indicated operations. See Section 1.6.

41. $3 + (-7)$

42. $(-2) + (-8)$

43. $\dfrac{3}{4} - \dfrac{3}{16}$

44. $-11 + 2.9$

45. $-5 - (-1)$

46. $-12 - 3$

Place $<, >,$ *or* $=$ *in the appropriate space to make each a true statement. See Sections 1.1 and 1.7.*

47. $-5 \quad\quad -7$

48. $\dfrac{12}{3} \quad\quad 2^2$

49. $|-5| \quad\quad -(-5)$

50. $-3^3 \quad\quad (-3)^3$

Concept Extensions

To "break even" in a manufacturing business, revenue R (income) ***must equal*** *the cost C of production, or R = C.*

51. The cost C to produce x number of skateboards is given by $C = 100 + 20x$. The skateboards are sold wholesale for $24 each, so revenue R is given by $R = 24x$. Find how many skateboards the manufacturer needs to produce and sell to break even. (*Hint*: Set the expression for R equal to the expression for C, then solve for x.)

52. The revenue R from selling x number of computer boards is given by $R = 60x$, and the cost C of producing them is given by $C = 50x + 5000$. Find how many boards must be sold to break even. Find how much money is needed to produce the break-even number of boards.

53. The cost C of producing x number of paperback books is given by $C = 4.50x + 2400$. Income R from these books is given by $R = 7.50x$. Find how many books should be produced and sold to break even.

54. Find the break-even quantity for a company that makes x number of computer monitors at a cost C given by $C = 870 + 70x$ and receives revenue R given by $R = 105x$.

55. Problems 51 through 54 involve finding the break-even point for manufacturing. Discuss what happens if a company makes and sells fewer products than the break-even point. Discuss what happens if more products than the break-even point are made and sold.

Standardized nutrition labels like the ones below have been displayed on food items since 1994. The percent column on the right shows the percent of daily values based on a 2000-calorie diet shown at the bottom of the label. For example, a serving of this food contains 4 grams of total fat, where the recommended daily fat based on a 2000-calorie diet is 65 grams of fat. This means that $\frac{4}{65}$ or approximately 6% (as shown) of your daily recommended fat is taken in by eating a serving of this food.

56. Based on a 2000-calorie diet, what percent of the daily values of sodium is contained in a serving of this food? In other words, find x. (Round to the nearest tenth of a percent.)

57. Based on a 2000-calorie diet, what percent of the daily value of total carbohydrate is contained in a serving of this food? In other words, find y. (Round to the nearest tenth of a percent.)

58. Notice on the nutrition label that one serving of this food contains 130 calories and 35 of these calories are from fat. Find the percent of calories from fat. (Round to the nearest tenth of a percent.) It is recommended that no more than 30% of calorie intake come from fat. Does this food satisfy this recommendation?

Below is a nutrition label for a particular food.

NUTRITIONAL INFORMATION PER SERVING

Serving Size: 9.8 oz.	Servings Per Container: 1
Calories................280	Polyunsaturated Fat..........1g
Protein..................12g	Saturated Fat...............3g
Carbohydrate.............45g	Cholesterol..............20mg
Fat......................6g	Sodium...............520mg
Percent of Calories from Fat......?	Potassium.............220mg

59. If fat contains approximately 9 calories per gram, find the percent of calories from fat in one serving of this food. (Round to the nearest tenth of a percent.)

60. If protein contains approximately 4 calories per gram, find the percent of calories from protein from one serving of this food. (Round to the nearest tenth of a percent.)

61. Find a food that contains more than 30% of its calories per serving from fat. Analyze the nutrition label and verify that the percents shown are correct.

Nutrition Facts

Serving Size 18 crackers (31g)
Servings Per Container About 9

Amount Per Serving

Calories 130 Calories from Fat 35

	% Daily Value*
Total Fat 4g	**6%**
Saturated Fat 0.5g	**3%**
Polyunsaturated Fat 0g	
Monounsaturated Fat 1.5g	
Cholesterol 0mg	**0%**
Sodium 230mg	***x***
Total Carbohydrate 23g	***y***
Dietary Fiber 2g	**8%**
Sugars 3g	
Protein 2g	
Vitamin A 0% • Vitamin C 0%	
Calcium 2% • Iron 6%	

*Percent Daily Values are based on a 2,000 calorie diet. Your daily values may be higher or lower depending on your calorie needs.

		Calories:	2,000	2,500
Total Fat	Less than		65g	80g
Sat Fat	Less than		20g	25g
Cholesterol	Less than		300mg	300mg
Sodium	Less than		2400mg	2400mg
Total Carbohydrate			300g	375g
Dietary Fiber			25g	30g

2.7   *SOLVING LINEAR INEQUALITIES*

Objectives

1   Define linear inequality in one variable.

2   Graph solution sets on a number line and use interval notation.

3   Solve linear inequalities.

4   Solve compound inequalities.

5   Solve inequality applications.

1   In Chapter 1, we reviewed these inequality symbols and their meanings:

$<$ means "is less than" $\leq$ means "is less than or equal to"

$>$ means "is greater than" $\geq$ means "is greater than or equal to"

Equations	Inequalities
$x = 3$	$x \leq 3$
$5n - 6 = 14$	$5n - 6 > 14$
$12 = 7 - 3y$	$12 \leq 7 - 3y$
$\dfrac{x}{4} - 6 = 1$	$\dfrac{x}{4} - 6 > 1$

A linear inequality is similar to a linear equation except that the equality symbol is replaced with an inequality symbol.

Linear Inequality in One Variable

A **linear inequality in one variable** is an inequality that can be written in the form

$$ax + b < c$$

where a, b, and c are real numbers and a is not 0.

This definition and all other definitions, properties, and steps in this section also hold true for the inequality symbols, $>$, $\geq$, and $\leq$.

2   A **solution of an inequality** is a value of the variable that makes the inequality a true statement. The solution set is the set of all solutions. For the inequality $x < 3$, replacing x with any number less than 3, that is, to the left of 3 on a number line, makes the resulting inequality true. This means that any number less than 3 is a solution of the inequality $x < 3$.

Since there are infinitely many such numbers, we cannot list all the solutions of the inequality. We *can* use set notation and write

$$\{x \qquad | \qquad x < 3\}.$$ Recall that this is read

the   such

set of   that x is less than 3.

all x

We can also picture the solutions on a number line. If we use open/closed-circle notation, the graph of $\{x|x < 3\}$ looks like the following.

In this text, a convenient notation, called **interval notation**, will be used to write solution sets of inequalities. To help us understand this notation, a different graphing notation will be used. Instead of an open circle, we use a parenthesis; instead of a closed circle, we use a bracket. With this new notation, the graph of $\{x|x < 3\}$ now looks like

and can be represented in interval notation as $(-\infty, 3)$. The symbol $-\infty$, read as "negative infinity," does not indicate a number, but does indicate that the shaded arrow to the left never ends. In other words, the interval $(-\infty, 3)$ includes *all* numbers less than 3.

Picturing the solutions of an inequality on a number line is called **graphing** the solutions or graphing the inequality, and the picture is called the **graph** of the inequality.

To graph $\{x|x \leq 3\}$ or simply $x \leq 3$, shade the numbers to the left of 3 and place a bracket at 3 on the number line. The bracket indicates that 3 **is** a solution: 3 **is** less than or equal to 3. In interval notation, we write $(-\infty, 3]$.

> **Helpful Hint**
>
> When writing an inequality in interval notation, it may be easier to first graph the inequality, then write it in interval notation. To help, think of the number line as approaching $-\infty$ to the left and $+\infty$ or ∞ to the right. Then simply write the interval notation by following your shading from left to right.
>
>

EXAMPLE 1

Graph $x \geq -1$. Then write the solutions in interval notation.

Solution   We place a bracket at -1 since the inequality symbol is $\geq$ and -1 is greater than or equal to -1. Then we shade to the right of -1.

In interval notation, this is $[-1, \infty)$.

3   When solutions of a linear inequality are not immediately obvious, they are found through a process similar to the one used to solve a linear equation. Our goal is to get the variable alone, and we use properties of inequality similar to properties of equality.

Addition Property of Inequality

If a, b, and c are real numbers, then

$$a < b \quad \text{and} \quad a + c < b + c$$

are equivalent inequalities.

This property also holds true for subtracting values, since subtraction is defined in terms of addition. In other words, adding or subtracting the same quantity from both sides of an inequality does not change the solution of the inequality.

EXAMPLE 2

Solve $x + 4 \le -6$ for x. Graph the solution set and write it in interval notation.

Solution   To solve for x, subtract 4 from both sides of the inequality.

$$
\begin{array}{ll}
x + 4 \le -6 & \text{Original inequality.} \\
x + 4 - 4 \le -6 - 4 & \text{Subtract 4 from both sides.} \\
x \le -10 & \text{Simplify.}
\end{array}
$$

The solution set is $(-\infty, -10]$.

▶ **Helpful Hint**

Notice that any number less than or equal to -10 is a solution to $x \le -10$. For example, solutions include

$$-10, \ -200, \ -11\frac{1}{2}, \ -7\pi, \ -\sqrt{130}, \ -50.3$$

An important difference between linear equations and linear inequalities is shown when we multiply or divide both sides of an inequality by a nonzero real number. For example, start with the true statement $6 < 8$ and multiply both sides by 2. As we see below, the resulting inequality is also true.

$$
\begin{array}{ll}
6 < 8 & \text{True} \\
2(6) < 2(8) & \text{Multiply both sides by 2.} \\
12 < 16 & \text{True}
\end{array}
$$

But if we start with the same true statement $6 < 8$ and multiply both sides by -2, the resulting inequality is not a true statement.

$$
\begin{array}{ll}
6 < 8 & \text{True} \\
-2(6) < -2(8) & \text{Multiply both sides by } -2. \\
-12 < -16 & \text{False}
\end{array}
$$

Notice, however, that if we reverse the direction of the inequality symbol, the resulting inequality is true.

$$
\begin{array}{ll}
-12 < -16 & \text{False} \\
-12 > -16 & \text{True}
\end{array}
$$

This demonstrates the multiplication property of inequality.

> ### Multiplication Property of Inequality
> 1. If a, b, and c are real numbers, and c is **positive**, then
> $$a < b \quad \text{and} \quad ac < bc$$
> are equivalent inequalities.
> 2. If a, b, and c are real numbers, and c is **negative**, then
> $$a < b \quad \text{and} \quad ac > bc$$
> are equivalent inequalities.

Because division is defined in terms of multiplication, this property also holds true when dividing both sides of an inequality by a nonzero number. If we multiply or divide both sides of an inequality by a negative number, **the direction of the inequality sign must be reversed for the inequalities to remain equivalent**.

> ### ▶ Helpful Hint
> Whenever both sides of an inequality are multiplied or divided by a negative number, the direction of the inequality symbol **must be** reversed to form an equivalent inequality.

EXAMPLE 3

Solve $-2x \le -4$. Graph the solution set and write it in interval notation.

Solution   Remember to reverse the direction of the inequality symbol when dividing by a negative number.

> **▶ Helpful Hint**
> Don't forget to reverse the direction of the inequality sign.

$$-2x \le -4$$
$$\frac{-2x}{-2} \ge \frac{-4}{-2} \qquad \text{Divide both sides by } -2$$
$$\text{and reverse the direction of the inequality sign.}$$
$$x \ge 2 \qquad \text{Simplify.}$$

The solution set $[2, \infty)$ is graphed as shown.

EXAMPLE 4

Solve $2x < -4$. Graph the solution set and write it in interval notation.

Solution

> **▶ Helpful Hint**
> Do not reverse the inequality sign.

$$2x < -4$$
$$\frac{2x}{2} < \frac{-4}{2} \qquad \text{Divide both sides by 2.}$$
$$\text{Do not reverse the direction of the inequality sign.}$$
$$x < -2 \qquad \text{Simplify.}$$

The solution set $(-\infty, -2)$ is graphed as shown.

✔ **CONCEPT CHECK**

Fill in the blank with $<, >, \leq,$ or $\geq$.

a. Since $-8 < -4$, then $3(-8)$ ____ $3(-4)$.

b. Since $5 \geq -2$, then $\dfrac{5}{-7}$ ____ $\dfrac{-2}{-7}$.

c. If $a < b$, then $2a$ ____ $2b$.

d. If $a \geq b$, then $\dfrac{a}{-3}$ ____ $\dfrac{b}{-3}$.

The following steps may be helpful when solving inequalities. Notice that these steps are similar to the ones given in Section 2.3 for solving equations.

Solving Linear Inequalities in One Variable

Step 1. Clear the inequality of fractions by multiplying both sides of the inequality by the lowest common denominator (LCD) of all fractions in the inequality.

Step 2. Remove grouping symbols such as parentheses by using the distributive property.

Step 3. Simplify each side of the inequality by combining like terms.

Step 4. Write the inequality with variable terms on one side and numbers on the other side by using the addition property of inequality.

Step 5. Get the variable alone by using the multiplication property of inequality.

▶ **Helpful Hint**

Don't forget that if both sides of an inequality are multiplied or divided by a negative number, the direction of the inequality sign must be reversed.

EXAMPLE 5

Solve $-4x + 7 \geq -9$. Graph the solution set and write it in interval notation.

Solution

$$-4x + 7 \geq -9$$
$$-4x + 7 - 7 \geq -9 - 7 \qquad \text{Subtract 7 from both sides.}$$
$$-4x \geq -16 \qquad \text{Simplify.}$$
$$\frac{-4x}{-4} \leq \frac{-16}{-4} \qquad \text{Divide both sides by } -4 \text{ and reverse the direction of the inequality sign.}$$
$$x \leq 4 \qquad \text{Simplify.}$$

The solution set $(-\infty, 4]$ is graphed as shown.

EXAMPLE 6

Solve $2x + 7 \le x - 11$. Graph the solution set and write it in interval notation.

Solution

$$2x + 7 \le x - 11$$
$$2x + 7 - x \le x - 11 - x \qquad \text{Subtract } x \text{ from both sides.}$$
$$x + 7 \le -11 \qquad \text{Combine like terms.}$$
$$x + 7 - 7 \le -11 - 7 \qquad \text{Subtract } 7 \text{ from both sides.}$$
$$x \le -18 \qquad \text{Combine like terms.}$$

The graph of the solution set $(-\infty, -18]$ is shown.

$$-20 \ -19 \ -18 \ -17 \ -16 \ -15 \ -14$$

EXAMPLE 7

Solve $-5x + 7 < 2(x - 3)$. Graph the solution set and write it in interval notation.

Solution

$$-5x + 7 < 2(x - 3)$$
$$-5x + 7 < 2x - 6 \qquad \text{Apply the distributive property.}$$
$$-5x + 7 - 2x < 2x - 6 - 2x \qquad \text{Subtract } 2x \text{ from both sides.}$$
$$-7x + 7 < -6 \qquad \text{Combine like terms.}$$
$$-7x + 7 - 7 < -6 - 7 \qquad \text{Subtract } 7 \text{ from both sides.}$$
$$-7x < -13 \qquad \text{Combine like terms.}$$
$$\frac{-7x}{-7} > \frac{-13}{-7} \qquad \text{Divide both sides by } -7 \text{ and \textit{reverse}}$$
$$\qquad\qquad\qquad \textit{the direction of the inequality sign.}$$
$$x > \frac{13}{7} \qquad \text{Simplify.}$$

The graph of the solution set $\left(\frac{13}{7}, \infty\right)$ is shown.

$$\frac{13}{7}$$
$$-2 \ -1 \ 0 \ 1 \ 2 \ 3 \ 4$$

EXAMPLE 8

Solve $2(x - 3) - 5 \le 3(x + 2) - 18$. Graph the solution set and write it in interval notation.

Solution

$$2(x - 3) - 5 \le 3(x + 2) - 18$$
$$2x - 6 - 5 \le 3x + 6 - 18 \qquad \text{Apply the distributive property.}$$
$$2x - 11 \le 3x - 12 \qquad \text{Combine like terms.}$$
$$-x - 11 \le -12 \qquad \text{Subtract } 3x \text{ from both sides.}$$
$$-x \le -1 \qquad \text{Add } 11 \text{ to both sides.}$$

$$\frac{-x}{-1} \geq \frac{-1}{-1}$$ Divide both sides by -1 and reverse the direction of the inequality sign.

$$x \geq 1$$ Simplify.

The graph of the solution set $[1, \infty)$ is shown.

4   Inequalities containing one inequality symbol are called **simple inequalities**, while inequalities containing two inequality symbols are called **compound inequalities**. A compound inequality is really two simple inequalities in one. The compound inequality

$$3 < x < 5 \quad \text{means} \quad 3 < x \text{ and } x < 5$$

This can be read "x is greater than 3 and less than 5."

A solution of a compound inequality is a value that is a solution of both of the simple inequalities that make up the compound inequality. For example,

$$4\frac{1}{2} \text{ is a solution of } 3 < x < 5 \text{ since } 3 < 4\frac{1}{2} \text{ and } 4\frac{1}{2} < 5.$$

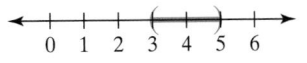

To graph $3 < x < 5$, place parentheses at both 3 and 5 and shade between.

EXAMPLE 9

Graph $2 < x \leq 4$. Write the solution in interval notation.

Solution   Graph all numbers greater than 2 and less than or equal to 4. Place a parenthesis at 2, a bracket at 4, and shade between.

When we solve a simple inequality, we isolate the variable on one side of the inequality. When we solve a compound inequality, we isolate the variable in the middle part of the inequality. Also, when solving a compound inequality, we must perform the same operation to all **three** parts of the inequality: left, middle, and right.

EXAMPLE 10

Solve $-1 \leq 2x - 3 < 5$. Graph the solution set and write it in interval notation.

Solution

$$-1 \leq 2x - 3 < 5$$
$$-1 + 3 \leq 2x - 3 + 3 < 5 + 3 \qquad \text{Add 3 to all three parts.}$$
$$2 \leq 2x < 8 \qquad \text{Combine like terms.}$$
$$\frac{2}{2} \leq \frac{2x}{2} < \frac{8}{2} \qquad \text{Divide all three parts by 2.}$$
$$1 \leq x < 4 \qquad \text{Simplify.}$$

The solution set $[1, 4)$ is graphed.

EXAMPLE 11

Solve $3 \le \dfrac{3x}{2} + 4 \le 5$. Graph the solution set and write it in interval notation.

Solution

$$3 \le \dfrac{3x}{2} + 4 \le 5$$

$$2(3) \le 2\left(\dfrac{3x}{2} + 4\right) \le 2(5) \qquad \text{Multiply all three parts by } \mathbf{2} \text{ to clear the fraction.}$$

$$6 \le 3x + 8 \le 10 \qquad \text{Distribute.}$$

$$-2 \le 3x \le 2 \qquad \text{Subtract } \mathbf{8} \text{ from all three parts.}$$

$$\dfrac{-2}{3} \le \dfrac{3x}{3} \le \dfrac{2}{3} \qquad \text{Divide all three parts by } \mathbf{3}.$$

$$-\dfrac{2}{3} \le x \le \dfrac{2}{3} \qquad \text{Simplify.}$$

The graph of the solution set $\left[-\frac{2}{3}, \frac{2}{3}\right]$ is shown.

5 Problems containing words such as "at least," "at most," "between," "no more than," and "no less than" usually indicate that an inequality should be solved instead of an equation. In solving applications involving linear inequalities, use the same procedure you use to solve applications involving linear equations.

EXAMPLE 12

STAYING WITHIN BUDGET

Marie Chase and Jonathan Edwards are having their wedding reception at the Gallery Reception Hall. They may spend at most $2000 for the reception. If the reception hall charges a $100 cleanup fee plus $36 per person, find the greatest number of people that they can invite and still stay within their budget.

Solution

1. UNDERSTAND. Read and reread the problem. Next, guess a solution. If 40 people attend the reception, the cost is $100 + $36(40) = $100 + $1440 = $1540.
Let x = the number of people who attend the reception.

2. TRANSLATE.

In words:	cleanup fee	+	cost per person	must be less than or equal to	$2000
Translate:	100	+	36x	≤	2000

3. SOLVE.

$$100 + 36x \le 2000$$
$$36x \le 1900 \qquad \text{Subtract } \mathbf{100} \text{ from both sides.}$$
$$x \le 52\dfrac{7}{9} \qquad \text{Divide both sides by } \mathbf{36}.$$

4. INTERPRET.

Check: Since x represents the number of people, we round down to the nearest whole, or 52. Notice that if 52 people attend, the cost is

$100 + $36(52) = $1972. If 53 people attend, the cost is
$100 + $36(53) = $2008, which is more than the given $2000.

State: Marie Chase and Jonathan Edwards can invite at most 52 people to the reception.

STUDY SKILLS REMINDER

Continue your outline started in Section 2.3. Write how to recognize and how to solve linear inequalities in your own words. For example:

Solving Equations and Inequalities

I. Equations
   A. Linear equations—(Section 2.3)

II. Inequalities
   A. **Linear inequalities**—Recognize: *inequality sign and power on x is 1 when there are no variables in denominator*—Solve: simplify and move variable terms to one side of the inequality, constants to the other side—don't forget, if multiply or divide by a negative number, reverse the direction of the inequality sign.

See Appendix A for summary exercises.

MENTAL MATH

Solve each of the following inequalities.

1. $5x > 10$

2. $4x < 20$

3. $2x \geq 16$

4. $9x \leq 63$

EXERCISE SET 2.7

STUDY GUIDE/SSM · CD/VIDEO · PH MATH TUTOR CENTER · MathXL®Tutorials ON CD · MathXL® · MyMathLab®

Graph each set of numbers given in interval notation. Then write an inequality statement in x describing the numbers graphed.

1. $[2, \infty)$

2. $(-3, \infty)$

3. $(-\infty, -5)$

4. $(-\infty, 4]$

7. $x < \dfrac{1}{2}$

8. $z < -\dfrac{2}{3}$

9. $y \geq 5$

10. $x > 3$

Graph each inequality on a number line. Then write the solutions in interval notation. See Example 1.

5. $x \leq -1$

6. $y < 0$

Solve each inequality. Graph the solution set and write it in interval notation. See Examples 2 through 4.

11. $2x < -6$

12. $3x > -9$

13. $x - 2 \geq -7$

14. $x + 4 \leq 1$

 15. $-8x \leq 16$

16. $-5x < 20$

Solve each inequality. Graph the solution set and write it in interval notation. See Examples 5 and 6.

17. $3x - 5 > 2x - 8$

18. $3 - 7x \geq 10 - 8x$

19. $4x - 1 \leq 5x - 2x$

20. $7x + 3 < 9x - 3x$

Solve each inequality. Graph the solution set and write it in interval notation. See Examples 7 and 8.

21. $x - 7 < 3(x + 1)$

22. $3x + 9 \geq 5(x - 1)$

23. $-6x + 2 \geq 2(5 - x)$

24. $-7x + 4 > 3(4 - x)$

25. $4(3x - 1) \leq 5(2x - 4)$

26. $3(5x - 4) \leq 4(3x - 2)$

 27. $3(x + 2) - 6 > -2(x - 3) + 14$

28. $7(x - 2) + x \leq -4(5 - x) - 12$

MIXED PRACTICE

Solve the following inequalities. Graph each solution set and write it in interval notation.

29. $-2x \leq -40$

30. $-7x > 21$

31. $-9 + x > 7$

32. $y - 4 \leq 1$

33. $3x - 7 < 6x + 2$

34. $2x - 1 \geq 4x - 5$

35. $5x - 7x \geq x + 2$

36. $4 - x < 8x + 2x$

37. $\dfrac{3}{4}x > 2$

38. $\dfrac{5}{6}x \geq -8$

39. $3(x - 5) < 2(2x - 1)$

40. $5(x + 4) < 4(2x + 3)$

41. $4(2x + 1) < 4$

42. $6(2 - x) \geq 12$

43. $-5x + 4 \geq -4(x - 1)$

44. $-6x + 2 < -3(x + 4)$

45. $-2(x - 4) - 3x < -(4x + 1) + 2x$

46. $-5(1 - x) - x \leq -(6 - 2x) + 6$

47. $-3x + 6 \geq 2x + 6$

48. $-(x - 4) < 4$

49. Explain how solving a linear inequality is similar to solving a linear equation.

50. Explain how solving a linear inequality is different from solving a linear equation.

Graph each inequality. Then write the solutions in interval notation. See Example 9.

51. $-1 < x < 3$

52. $2 \leq y \leq 3$

53. $0 \leq y < 2$

54. $-1 \leq x \leq 4$

Solve each inequality. Graph the solution set and write it in interval notation. See Examples 10 and 11.

55. $-3 < 3x < 6$

56. $-5 < 2x < -2$

57. $2 \leq 3x - 10 \leq 5$

58. $4 \leq 5x - 6 \leq 19$

59. $-4 < 2(x - 3) \leq 4$

60. $0 < 4(x + 5) \leq 8$

61. $-2 < 3x - 5 < 7$

62. $1 < 4 + 2x \leq 7$

 63. $-6 < 3(x - 2) \leq 8$

64. $-5 \leq 2(x + 4) < 8$

65. Explain how solving a linear inequality is different from solving a compound inequality.

66. Explain how solving a linear inequality is similar to solving a compound inequality.

67. Six more than twice a number is greater than negative fourteen. Find all numbers that make this statement true.

68. Five times a number, increased by one, is less than or equal to ten. Find all such numbers.

69. Dennis and Nancy Wood are celebrating their 30th wedding anniversary by having a reception at Tiffany Oaks reception hall. They have budgeted $3000 for their reception. If the reception hall charges a $50.00 cleanup fee plus $34 per person, find the greatest number of people that they may invite and still stay within their budget.

70. A surprise retirement party is being planned for Pratep Puri. A total of $860 has been collected for the event, which is to be held at a local reception hall. This reception hall charges a cleanup fee of $40 and $15 per person for drinks and light snacks. Find the greatest number of people that may be invited and still stay within $860.

△ **71.** Find the values for x so that the perimeter of this rectangle is no greater than 100 centimeters.

△ **72.** Find the values for x so that the perimeter of this triangle is no longer than 87 inches.

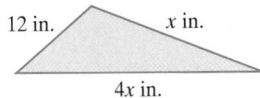

73. A financial planner has a client with $15,000 to invest. If he invests $10,000 in a certificate of deposit paying 11% annual simple interest, at what rate does the remainder of the money need to be invested so that the two investments together yield at least $1600 in yearly interest?

74. Alex earns $600 per month plus 4% of all his sales over $1000. Find the minimum sales that will allow Alex to earn at least $3000 per month.

75. Ben Holladay bowled 146 and 201 in his first two games. What must he bowl in his third game to have an average of at least 180?

76. On an NBA team the two forwards measure 6′8″ and 6′6″ and the two guards measure 6′0″ and 5′9″ tall. How tall a center should they hire if they wish to have a starting team average height of at least 6′5″?

Solve the following. See Example 12.

77. High blood cholesterol levels increase the risk of heart disease in adults. Doctors recommend that total blood cholesterol be less than 200 milligrams per deciliter. Total cholesterol levels from 200 up to 240 milligrams per deciliter are considered borderline. Any total cholesterol reading above 240 milligrams per deciliter is considered high.

Letting x represent a patient's total blood cholesterol level, write a series of three inequalities that describe the ranges corresponding to recommended, borderline, and high levels of total blood cholesterol.

78. T. Theodore Fujita created the Fujita Scale (or F-Scale), which uses ratings from 0 to 5 to classify tornadoes. An F-0 tornado has wind speeds between 40 and 72 mph, inclusive. The winds in an F-1 tornado range from 73 to 112 mph. In an

F-2 tornado, winds are from 113 to 157 mph. An F-3 tornado has wind speeds ranging from 158 to 206 mph. Wind speeds in an F-4 tornado are clocked at 207 to 260 mph. The most violent tornadoes are ranked at F-5, with wind speeds of at least 261 mph. (*Source:* National Weather Service)

Letting y represent a tornado's wind speed, write a series of six inequalities that describe the wind speed ranges corresponding to each Fujita Scale rank.

79. Twice a number, increased by one, is between negative five and seven. Find all such numbers.

80. Half a number, decreased by four, is between two and three. Find all such numbers.

81. The temperatures in Ohio range from −39°C to 45°C. Use a compound inequality to convert these temperatures to Fahrenheit temperatures. (*Hint:* Use $C = \frac{5}{9}(F - 32)$.)

82. Mario Lipco has scores of 85, 95, and 92 on his algebra tests. Use a compound inequality to find the range of scores he can make on his final exam in order to receive an A in the course. The final exam counts as three tests, and an A is received if the final course average is from 90 to 100. (*Hint:* The average of a list of numbers is their sum divided by the number of numbers in the list.)

REVIEW AND PREVIEW

Evaluate the following. See Section 1.3.

83. $(2)^3$ **84.** $(3)^3$ **85.** $(1)^{12}$

86. 0^5 **87.** $\left(\dfrac{4}{7}\right)^2$ **88.** $\left(\dfrac{2}{3}\right)^3$

This broken line graph shows the average cell phone prices for the given years. Use this for Exercises 89 through 92.

89. What was the average cell phone price in 1986?

90. What was the average cell phone price in 2002?

91. Which period had the greatest drop in cell phone prices?

92. What years had cell phone prices over $500?

Concept Extensions

△ **93.** The formula $C = 3.14d$ can be used to approximate the circumference of a circle given its diameter. Waldo Manufacturing manufactures and sells a certain washer with an outside circumference of 3 centimeters. The company has decided that a washer whose actual circumference is in the interval $2.9 \le C \le 3.1$ centimeters is acceptable. Use a compound inequality and find the corresponding interval for diameters of these washers. (Round to 3 decimal places.)

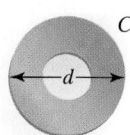

△ **94.** Bunnie Supplies manufactures plastic Easter eggs that open. The company has determined that if the circumference of the opening of each part of the egg is in the interval $118 \le C \le 122$ millimeters, the eggs will open and close comfortably. Use a compound inequality and find the corresponding interval for diameters of these openings. (Round to 2 decimal places.)

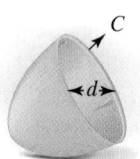

For Exercises 95 through 98, see the example below.

Solve $x(x - 6) > x^2 - 5x + 6$. Graph the solution set and write it in interval notation.

Solution

$$x(x - 6) > x^2 - 5x + 6$$
$$x^2 - 6x > x^2 - 5x + 6$$
$$x^2 - 6x - x^2 > x^2 - 5x + 6 - x^2$$
$$-6x > -5x + 6$$
$$-x > 6$$
$$\frac{-x}{-1} < \frac{6}{-1}$$
$$x < -6$$

The solution set $(-\infty, -6)$ is graphed as shown.

$$\xleftarrow{\quad\qquad} \underset{-7 \ -6 \ -5 \ -4 \ -3 \ -2 \ -1}{+\!\!+\!\!+\!\!+\!\!+\!\!+\!\!+\!\!+} \xrightarrow{\quad\qquad}$$

Solve each inequality. Graph the solution set and write it in interval notation.

95. $x(x + 4) > x^2 - 2x + 6$

96. $x(x - 3) \ge x^2 - 5x - 8$

97. $x^2 + 6x - 10 < x(x - 10)$

98. $x^2 - 4x + 8 < x(x + 8)$

CHAPTER 2 PROJECT

Developing a Budget

Whether you are rolling in dough or pinching every penny, a budget can help put you in control of your personal finances. Listing your income lets you see where your money comes from, and by tracking your expenses you can analyze your spending practices. Putting these together in a budget helps you make informed decisions about what you spend.

In this project, you will have the opportunity to develop a personal budget. This project may be completed by working in groups or individually.

1. Keep track of your spending for a week or two. Be sure to write down *everything* that you spend during that time—no matter how small—from a daily newspaper or a can of soda to your tuition bill for the term.

2. Decide on a timeframe for your budget: a month, a term, a school year, or the entire year. How many weeks are in your timeframe?

3. Make a list of all of your income for your timeframe of choice. Be sure to include sources like grants or scholarships, portion of savings earmarked for college expenses for this timeframe, take-home pay from a job, gifts, financial support from family, and interest. Total your income from all sources.

4. Make a list of all of your expected expenses for your timeframe of choice. Be sure to include major expenses such as car

payments, auto and health insurance, room and board or rent/mortgage payment, and tuition. Try to estimate more variable expenses like telephone and utilities, books and supplies, groceries and/or restaurant meals, entertainment, transportation or parking, personal care, clothing, dues and/or lab fees, etc. Don't forget to take into account small or irregular purchases. Use the expense record you made in Question 1 to help you gauge your levels of spending and identify spending categories. Total your expected expenses in all categories.

5. In the equation $I = E + wx$, I represents your income from Question 3, E represents your expected expenses from Question 4, w represents the number of weeks in your timeframe of choice (from Question 2), and x represents the weekly shortage or surplus in your budget. If x is positive, this is the extra amount you have in your budget each week to save or spend as "pocket money." If x is negative, this is the amount by which you fall short each week. To keep your budget balanced, you should try to reduce nonessential expenses by this amount each week.

Substitute for I, E, and w in the equation and solve for x. Interpret your result.

6. If you have a weekly shortage, which expenses could you try to reduce to balance your budget? If you have a weekly surplus, what would you do with this extra money?

2 CHAPTER VOCABULARY CHECK

Fill in each blank with one of the words or phrases listed below.

like terms   numerical coefficient   linear inequality in one variable
equivalent equations   formula   compound inequalities
linear equation in one variable

1. Terms with the same variables raised to exactly the same powers are called _____.

2. A _____ can be written in the form $ax + b = c$.

3. Equations that have the same solution are called _____.

4. Inequalities containing two inequality symbols are called _____.

5. An equation that describes a known relationship among quantities is called a _____.

6. A _____ can be written in the form $ax + b < c$, (or $>, \leq, \geq$).

7. The _____ of a term is its numerical factor.

CHAPTER 2   HIGHLIGHTS

Definitions and Concepts	Examples

Section 2.1 Simplifying Algebraic Expressions

Definitions and Concepts	Examples
The **numerical coefficient** of a **term** is its numerical factor.	**Term** **Numerical Coefficient** $-7y$ -7 x 1 $\frac{1}{5}a^2b$ $\frac{1}{5}$
Terms with the same variables raised to exactly the same powers are **like terms**.	**Like Terms** **Unlike Terms** $12x, -x$ $3y, 3y^2$ $-2xy, 5yx$ $7a^2b, -2ab^2$
To combine like terms, add the numerical coefficients and multiply the result by the common variable factor. To remove parentheses, apply the distributive property.	$9y + 3y = 12y$ $-4z^2 + 5z^2 - 6z^2 = -5z^2$ $-4(x + 7) + 10(3x - 1)$ $= -4x - 28 + 30x - 10$ $= 26x - 38$

Definitions and Concepts	Examples

Section 2.2 The Addition and Multiplication Properties of Equality

A **linear equation in one variable** can be written in the form $ax + b = c$ where a, b, and c are real numbers and $a \neq 0$.

Linear Equations
$$-3x + 7 = 2$$
$$3(x - 1) = -8(x + 5) + 4$$

Equivalent equations are equations that have the same solution.

$x - 7 = 10$ and $x = 17$
are equivalent equations.

Addition Property of Equality

Adding the same number to or subtracting the same number from both sides of an equation does not change its solution.

$$y + 9 = 3$$
$$y + 9 - 9 = 3 - 9$$
$$y = -6$$

Multiplication Property of Equality

Multiplying both sides or dividing both sides of an equation by the same nonzero number does not change its solution.

$$\frac{2}{3}a = 18$$
$$\frac{3}{2}\left(\frac{2}{3}a\right) = \frac{3}{2}(18)$$
$$a = 27$$

Section 2.3 Solving Linear Equations

To Solve Linear Equations

Solve: $\dfrac{5(-2x + 9)}{6} + 3 = \dfrac{1}{2}$

1. Clear the equation of fractions.

   1. $6 \cdot \dfrac{5(-2x + 9)}{6} + 6 \cdot 3 = 6 \cdot \dfrac{1}{2}$
   $$5(-2x + 9) + 18 = 3$$

2. Remove any grouping symbols such as parentheses.

   2. $-10x + 45 + 18 = 3$ Distributive property

3. Simplify each side by combining like terms.

   3. $-10x + 63 = 3$ Combine like terms.

4. Write variable terms on one side and numbers on the other side using the addition property of equality.

   4. $-10x + 63 - 63 = 3 - 63$ Subtract 63.
   $$-10x = -60$$

5. Get the variable alone using the multiplication property of equality.

   5. $\dfrac{-10x}{-10} = \dfrac{-60}{-10}$ Divide by -10.
   $$x = 6$$

6. Check by substituting in the original equation.

   6. $\dfrac{5(-2x + 9)}{6} + 3 = \dfrac{1}{2}$
   $$\dfrac{5(-2 \cdot 6 + 9)}{6} + 3 \overset{?}{=} \dfrac{1}{2}$$
   $$\dfrac{5(-3)}{6} + 3 \overset{?}{=} \dfrac{1}{2}$$
   $$-\dfrac{5}{2} + \dfrac{6}{2} \overset{?}{=} \dfrac{1}{2}$$
   $$\dfrac{1}{2} = \dfrac{1}{2} \quad \text{True}$$

Definitions and Concepts	**Examples**

Section 2.4 An Introduction to Problem Solving

Problem-Solving Steps

The height of the Hudson volcano in Chili is twice the height of the Kiska volcano in the Aleutian Islands. If the sum of their heights is 12,870 feet, find the height of each.

1. UNDERSTAND the problem.

1. Read and reread the problem. Guess a solution and check your guess.

Let x be the height of the Kiska volcano. Then $2x$ is the height of the Hudson volcano.

x⌐ $2x$⌐
Kiska   Hudson

2. TRANSLATE the problem.

2. In words:

height of Kiska	added to	height of Hudson	is	12,870
Translate: x	$+$	$2x$	$=$	12,870

3. SOLVE.

3.

$$x + 2x = 12{,}870$$
$$3x = 12{,}870$$
$$x = 4290$$

4. INTERPRET the results.

4. *Check*: If x is 4290 then $2x$ is 2(4290) or 8580. Their sum is $4290 + 8580$ or 12,870, the required amount.

State: Kiska volcano is 4290 feet high and Hudson volcano is 8580 feet high.

Section 2.5 Formulas and Problem Solving

An equation that describes a known relationship among quantities is called a **formula**.

To solve a formula for a specified variable, use the same steps as for solving a linear equation. Treat the specified variable as the only variable of the equation.

Formulas

$A = lw$ (area of a rectangle)
$I = PRT$ (simple interest)

Solve $P = 2l + 2w$ for l.

$$P = 2l + 2w$$
$$P - 2w = 2l + 2w - 2w \qquad \text{Subtract } 2w.$$
$$P - 2w = 2l$$
$$\frac{P - 2w}{2} = \frac{2l}{2} \qquad \text{Divide by 2.}$$
$$\frac{P - 2w}{2} = l \qquad \text{Simplify.}$$

If all values for the variables in a formula are known except for one, this unknown value may be found by substituting in the known values and solving.

If $d = 182$ miles and $r = 52$ miles per hour in the formula $d = r \cdot t$, find t.

$$d = r \cdot t$$
$$182 = 52 \cdot t \qquad \text{Let } d = 182 \text{ and } r = 52.$$
$$3.5 = t$$

The time is 3.5 hours.

Definitions and Concepts	**Examples**

Section 2.6 Further Problem Solving

	How many liters of a 20% acid solution must be mixed with a 50% acid solution in order to obtain 12 liters of a 30% solution?
1. UNDERSTAND.	1. Read and reread. Guess a solution and check. Let x = number of liters of 20% solution. Then $12 - x$ = number of liters of 50% solution.

2. TRANSLATE.

	No. of Liters $\cdot$ *Strength* = *Amount of Acid*

	No. of Liters	*Strength*	*Amount of Acid*
20% Solution	x	20%	$0.20x$
50% Solution	$12 - x$	50%	$0.50(12 - x)$
30% Solution Needed	12	30%	$0.30(12)$

In words:

acid in 20% solution	+	acid in 50% solution	=	acid in 30% solution

Translate: $0.20x$ + $0.50(12 - x)$ = $0.30(12)$

3. SOLVE.

3. Solve $0.20x + 0.50(12 - x) = 0.30(12)$
$0.20x + 6 - 0.50x = 3.6$ Apply the distributive
$-0.30x + 6 = 3.6$ property.
$-0.30x = -2.4$ Subtract 6.
$x = 8$ Divide by -0.30.

4. INTERPRET.

4. *Check, then state.*
If 8 liters of a 20% acid solution are mixed with $12 - 8$ or 4 liters of a 50% acid solution, the result is 12 liters of a 30% solution.

Section 2.7 Solving Linear Inequalities

A **linear inequality in one variable** is an inequality that can be written in one of the forms:

$$ax + b < c \qquad ax + b \le c$$
$$ax + b > c \qquad ax + b \ge c$$

where a, b, and c are real numbers and a is not 0.

Linear Inequalities

$$2x + 3 < 6 \qquad\qquad 5(x - 6) \ge 10$$

$$\frac{x - 2}{5} > \frac{5x + 7}{2} \qquad \frac{-(x + 8)}{9} \le \frac{-2x}{11}$$

Addition Property of Inequality

Adding the same number to or subtracting the same number from both sides of an inequality does not change the solutions.

$$y + 4 \le -1$$
$$y + 4 - 4 \le -1 - 4 \quad \text{Subtract 4.}$$
$$y \le -5$$

Definitions and Concepts	**Examples**

Multiplication Property of Inequality

Multiplying or dividing both sides of an inequality by the same positive number does not change its solutions.

$$\frac{1}{3}x > -2$$

$$3\left(\frac{1}{3}x\right) > 3 \cdot -2 \quad \text{Multiply by 3.}$$

$$x > -6$$

Multiplying or dividing both sides of an inequality by the same **negative number and reversing the direction of the inequality sign** does not change its solutions.

$$-2x \le 4$$

$$\frac{-2x}{-2} \ge \frac{4}{-2} \quad \text{Divide by } -2, \text{reverse inequality sign.}$$

$$x \ge -2$$

To Solve Linear Inequalities

1. Clear the equation of fractions.
2. Remove grouping symbols.
3. Simplify each side by combining like terms.
4. Write variable terms on one side and numbers on the other side using the addition property of inequality.

5. Get the variable alone using the multiplication property of inequality.

Solve: $3(x + 2) \le -2 + 8$

1. No fractions to clear. $3(x + 2) \le -2 + 8$
2. $3x + 6 \le -2 + 8$ Distributive property
3. $3x + 6 \le 6$ Combine like terms.
4. $3x + 6 - 6 \le 6 - 6$ Subtract 6.
   $3x \le 0$

5. $\dfrac{3x}{3} \le \dfrac{0}{3}$ Divide by 3.
   $x \le 0$

Inequalities containing two inequality symbols are called **compound inequalities**.

To solve a compound inequality, isolate the variable in the middle part of the inequality. Perform the same operation to all three parts of the inequality: left, middle, right.

Compound Inequalities

$$-2 < x < 6$$

$$5 \le 3(x - 6) < \frac{20}{3}$$

Solve: one only $-2 < 3x + 1 < 7$

$$-2 - 1 < 3x + 1 - 1 < 7 - 1 \quad \text{Subtract 1.}$$

$$-3 < 3x < 6$$

$$\frac{-3}{3} < \frac{3x}{3} < \frac{6}{3} \quad \text{Divide by 3.}$$

$$-1 < x < 2$$

2 CHAPTER REVIEW

(2.1) *Simplify the following expressions.*

1. $5x - x + 2x$
2. $0.2z - 4.6x - 7.4z$
3. $\frac{1}{2}x + 3 + \frac{7}{2}x - 5$
4. $\frac{4}{5}y + 1 + \frac{6}{5}y + 2$
5. $2(n - 4) + n - 10$
6. $3(w + 2) - (12 - w)$

7. Subtract $7x - 2$ from $x + 5$.

8. Subtract $1.4y - 3$ from $y - 0.7$.

Write each of the following as algebraic expressions.

9. Three times a number decreased by 7

10. Twice the sum of a number and 2.8 added to 3 times a number

(2.2) *Solve the following.*

11. $8x + 4 = 9x$
12. $5y - 3 = 6y$
13. $3x - 5 = 4x + 1$
14. $2x - 6 = x - 6$
15. $4(x + 3) = 3(1 + x)$
16. $6(3 + n) = 5(n - 1)$

Use the addition property to fill in the blank so that the middle equation simplifies to the last equation.

17.
$$x - 5 = 3$$
$$x - 5 + \underline{\quad} = 3 + \underline{\quad}$$
$$x = 8$$

18.
$$x + 9 = -2$$
$$x + 9 - \underline{\quad} = -2 - \underline{\quad}$$
$$x = -11$$

Write each as an algebraic expression.

19. The sum of two numbers is 10. If one number is x, express the other number in terms of x.

20. Mandy is 5 inches taller than Melissa. If x inches represents the height of Mandy, express Melissa's height in terms of x.

△ **21.** If one angle measures $(x + 5)°$, express the measure of its supplement in terms of x.

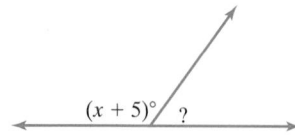

Solve each equation.

22. $\frac{3}{4}x = -9$
23. $\frac{x}{6} = \frac{2}{3}$
24. $-3x + 1 = 19$
25. $5x + 25 = 20$
26. $5x + x = 9 + 4x - 1 + 6$
27. $-y + 4y = 7 - y - 3 - 8$

28. Express the sum of three even consecutive integers as an expression in x. Let x be the first even integer.

(2.3) *Solve the following.*

29. $\frac{2}{7}x - \frac{5}{7} = 1$
30. $\frac{5}{3}x + 4 = \frac{2}{3}x$

31. $-(5x + 1) = -7x + 3$
32. $-4(2x + 1) = -5x + 5$
33. $-6(2x - 5) = -3(9 + 4x)$
34. $3(8y - 1) = 6(5 + 4y)$
35. $\frac{3(2 - z)}{5} = z$
36. $\frac{4(n + 2)}{5} = -n$
37. $5(2n - 3) - 1 = 4(6 + 2n)$
38. $-2(4y - 3) + 4 = 3(5 - y)$
39. $9z - z + 1 = 6(z - 1) + 7$
40. $5t - 3 - t = 3(t + 4) - 15$
41. $-n + 10 = 2(3n - 5)$
42. $-9 - 5a = 3(6a - 1)$
43. $\frac{5(c + 1)}{6} = 2c - 3$
44. $\frac{2(8 - a)}{3} = 4 - 4a$
45. $200(70x - 3560) = -179(150x - 19,300)$
46. $1.72y - 0.04y = 0.42$

Solve.

47. The quotient of a number and 3 is the same as the difference of the number and two. Find the number.

48. Double the sum of a number and six is the opposite of the number. Find the number.

(2.4) *Solve each of the following.*

49. The height of the Eiffel Tower is 68 feet more than three times a side of its square base. If the sum of these two dimensions is 1380 feet, find the height of the Eiffel Tower.

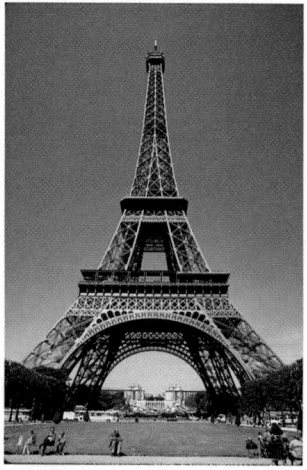

50. A 12-foot board is to be divided into two pieces so that one piece is twice as long as the other. If x represents the length of the shorter piece, find the length of each piece.

51. One area code in Ohio is 34 more than three times another area code used in Ohio. If the sum of these area codes is 1262, find the two area codes.

52. Find three consecutive even integers whose sum is -114.

(2.5) Substitute the given values into the given formulas and solve for the unknown variable.

△ **53.** $P = 2l + 2w; P = 46, l = 14$

△ **54.** $V = lwh; V = 192, l = 8, w = 6$

Solve each of the following for the indicated variable.

55. $y = mx + b$ for m **56.** $r = vst - 9$ for s

57. $2y - 5x = 7$ for x **58.** $3x - 6y = -2$ for y

△ **59.** $C = \pi D$ for π △ **60.** $C = 2\pi r$ for π

△ **61.** A swimming pool holds 900 cubic meters of water. If its length is 20 meters and its height is 3 meters, find its width.

62. The highest temperature on record in Rome, Italy, is $104°$ Fahrenheit. Convert this temperature to Celsius.

63. A charity 10K race is given annually to benefit a local hospice organization. How long will it take to run/walk a 10K race (10 kilometers or 10,000 meters) if your average pace is 125 **meters** per minute?

(2.6) Solve each of the following.

64. A $50,000 retirement pension is to be invested into two accounts: a money market fund that pays 8.5% and a certifi-

cate of deposit that pays 10.5%. How much should be invested at each rate in order to provide a yearly interest income of $4550?

65. A pay phone is holding its maximum number of 500 coins consisting of nickels, dimes, and quarters. The number of quarters is twice the number of dimes. If the value of all the coins is $88.00, how many nickels were in the pay phone?

66. How long will it take an Amtrak passenger train to catch up to a freight train if their speeds are 60 and 45 mph and the freight train had an hour and a half head start?

67. In 2003, Lance Armstrong became the 4th cyclist in history to win the Tour de France 5 years in a row. Suppose he rides a bicycle up a mountain trail at 8 mph and down the same trail at 12 mph. Find the round-trip distance traveled if the total travel time was 5 hours.

(2.7) Solve and graph the solution of each of the following inequalities.

68. $x \le -2$ **69.** $x > 0$

70. $-1 < x < 1$ **71.** $0.5 \le y < 1.5$

72. $-2x \ge -20$ **73.** $-3x > 12$

74. $5x - 7 > 8x + 5$ **75.** $x + 4 \ge 6x - 16$

76. $2 \le 3x - 4 < 6$ **77.** $-3 < 4x - 1 < 2$

78. $-2(x - 5) > 2(3x - 2)$ **79.** $4(2x - 5) \le 5x - 1$

80. Tina earns $175 per week plus a 5% commission on all her sales. Find the minimum amount of sales to ensure that she earns at least $300 per week.

81. Ellen Catarella shot rounds of 76, 82, and 79 golfing. What must she shoot on her next round so that her average will be below 80?

● **CHAPTER 2 TEST**   Remember to use your Chapter Test Prep Video CD to help you study and view solutions to the test questions you need help with.

Simplify each of the following expressions.

1. $2y - 6 - y - 4$

2. $2.7x + 6.1 + 3.2x - 4.9$

3. $4(x - 2) - 3(2x - 6)$

4. $7 + 2(5y - 3)$

Solve each of the following equations.

5. $-\dfrac{4}{5}x = 4$ **6.** $4(n - 5) = -(4 - 2n)$

7. $5y - 7 + y = -(y + 3y)$

8. $4z + 1 - z = 1 + z$

9. $\dfrac{2(x + 6)}{3} = x - 5$ **10.** $\dfrac{1}{2} - x + \dfrac{3}{2} = x - 4$

11. $-0.3(x - 4) + x = 0.5(3 - x)$

12. $-4(a + 1) - 3a = -7(2a - 3)$

13. $-2(x - 3) = x + 5 - 3x$

Solve each of the following applications.

14. A number increased by two-thirds of the number is 35. Find the number.

△ **15.** A gallon of water seal covers 200 square feet. How many gallons are needed to paint two coats of water seal on a deck that measures 20 feet by 35 feet?

20 feet

35 feet

16. Sedric Angell invested an amount of money in Amoxil stock that earned an annual 10% return, and then he invested twice the original amount in IBM stock that earned an annual 12% return. If his total return from both investments was $2890, find how much he invested in each stock.

17. Two trains leave Los Angeles simultaneously traveling on the same track in opposite directions at speeds of 50 and 64 mph. How long will it take before they are 285 miles apart?

18. Find the value of x if $y = -14$, $m = -2$, and $b = -2$ in the formula $y = mx + b$.

Solve each of the following equations for the indicated variable.

△ **19.** $V = \pi r^2 h$ for h

20. $3x - 4y = 10$ for y

Solve and graph each of the following inequalities.

21. $3x - 5 \geq 7x + 3$

22. $x + 6 > 4x - 6$

23. $-2 < 3x + 1 < 8$

24. $\dfrac{2(5x + 1)}{3} > 2$

CHAPTER CUMULATIVE REVIEW

1. Given the set $\{-2, 0, \frac{1}{4}, 112, -3, 11, \sqrt{2}\}$, list the numbers in this set that belong to the set of:
  a. Natural numbers
  b. Whole numbers
  c. Integers
  d. Rational numbers
  e. Irrational numbers
  f. Real numbers

2. Given the set $\{7, 2, -\frac{1}{5}, 0, \sqrt{3}, -185, 8\}$, list the numbers in this set that belong to the set of:
  a. Natural numbers
  b. Whole numbers
  c. Integers
  d. Rational numbers
  e. Irrational numbers
  f. Real numbers

3. Find the absolute value of each number.
  a. $|4|$ **b.** $|-5|$ **c.** $|0|$

4. Find the absolute value of each number.
  a. $|5|$ **b.** $|-8|$ **c.** $\left|-\frac{2}{3}\right|$

5. Write each of the following numbers as a product of primes.
  a. 40   **b.** 63

6. Write each number as a product of primes.
  a. 44   **b.** 90

7. Write $\dfrac{2}{5}$ as an equivalent fraction with a denominator of 20.

8. Write $\dfrac{2}{3}$ as an equivalent fraction with a denominator of 24.

9. Simplify $3[4(5 + 2) - 10]$.

10. Simplify $5[16 - 4(2 + 1)]$.

11. Decide whether 2 is a solution of $3x + 10 = 8x$.

12. Decide whether 3 is a solution of $5x - 2 = 4x$.

Add.

13. $-1 + (-2)$

14. $(-2) + (-8)$

15. $-4 + 6$

16. $-3 + 10$

17. Simplify each expression.
  a. $-(-10)$ **b.** $-\left(-\dfrac{1}{2}\right)$
  c. $-(-2x)$ **d.** $-|-6|$

18. Simplify each expression.
  a. $-(-5)$ **b.** $-\left(-\dfrac{2}{3}\right)$
  c. $-(-a)$ **d.** $-|-3|$

19. Subtract.
  a. $5.3 - (-4.6)$ **b.** $-\dfrac{3}{10} - \dfrac{5}{10}$
  c. $-\dfrac{2}{3} - \left(-\dfrac{4}{5}\right)$

20. Subtract.
  a. $-2.7 - 8.4$ **b.** $-\dfrac{4}{5} - \left(-\dfrac{3}{5}\right)$
  c. $\dfrac{1}{4} - \left(-\dfrac{1}{2}\right)$

21. Find each unknown complementary or supplementary angle.

a.

b.

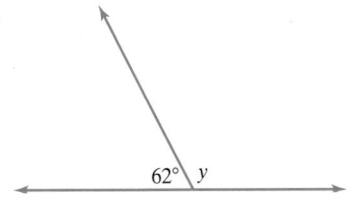

22. Find each unknown complementary or supplementary angle.

a. 　　**b.**

23. Find each product.

   a. $(-1.2)(0.05)$

   b. $\dfrac{2}{3} \cdot -\dfrac{7}{10}$

24. Find each product.

   a. $(4.5)(-0.08)$　　**b.** $\dfrac{-3}{4} \cdot \dfrac{-8}{17}$

25. Find each quotient.

   a. $\dfrac{-24}{-4}$　　**b.** $\dfrac{-36}{3}$

   c. $\dfrac{2}{3} \div \left(-\dfrac{5}{4}\right)$

26. Find each quotient.

   a. $\dfrac{-32}{8}$　　**b.** $\dfrac{-108}{-12}$

   c. $\dfrac{-5}{7} \div \left(\dfrac{-9}{2}\right)$

27. Use a commutative property to complete each statement.

   a. $x + 5 = $ _____　　**b.** $3 \cdot x = $ _____

28. Use a commutative property to complete each statement.

   a. $y + 1 = $ _____　　**b.** $y \cdot 4 = $ _____

29. Use the distributive property to write each sum as a product.

   a. $8 \cdot 2 + 8 \cdot x$　　**b.** $7s + 7t$

30. Use the distributive property to write each sum as a product.

   a. $4 \cdot y + 4 \cdot \dfrac{1}{3}$　　**b.** $0.10x + 0.10y$

31. Subtract $4x - 2$ from $2x - 3$.

32. Subtract $10x + 3$ from $-5x + 1$.

Solve.

33. $y + 0.6 = -1.0$　　**34.** $z - 0.4 = 1.5$

35. $7 = -5(2a - 1) - (-11a + 6)$

36. $-3x + 1 - (-4x - 6) = 10$

37. $\dfrac{y}{7} = 20$　　　　　　　**38.** $\dfrac{x}{4} = 18$

39. $4(2x - 3) + 7 = 3x + 5$

40. $6x + 5 = 4(x + 4) - 1$

41. Twice the sum of a number and 4 is the same as four times the number decreased by 12. Find the number.

42. A number increased by 4 is the same as 3 times the number decreased by 8. Find the number.

43. Solve $V = lwh$ for l.

44. Solve $C = 2\pi r$ for r.

45. Solve $x + 4 \leq -6$ for x. Graph the solution set and write it in interval notation.

46. Solve $x - 3 > 2$ for x. Graph the solution set and write it in interval notation.

Graphing

In the previous chapter we learned to solve and graph the solutions of linear equations and inequalities in one variable. Now we define and present techniques for solving and graphing linear equations.

FOR SALE: RESIDENCE

The purchase and sale of a house are likely the largest monetary transactions people make in their lifetimes. Although some people attempt to handle the sale on their own, most seek the help of a professional real estate agent. Realtors give their clients advice on all aspects of purchase and sale, including price negotiation, home warranties, market trends, and interest rates. An agent for the seller could provide information on how a remodeling project might affect the marketability of a home, whereas the buyer's agent can provide tips on the availability of different types of financing. In exchange for their services, real estate agents usually collect a commission based on a percent of the selling price of the home.

In the Spotlight on Decision Making feature on page 189, you will have the opportunity to choose a real estate agency as the seller of a home.
*Link: National Association of Realtors—*www.realtor.org

3.1   *READING GRAPHS AND THE RECTANGULAR COORDINATE SYSTEM*

Objectives

1   Read bar and line graphs.

2   Define the rectangular coordinate system and plot ordered pairs of numbers.

3   Graph paired data to create a scatter diagram.

4   Determine whether an ordered pair is a solution of an equation in two variables.

5   Find the missing coordinate of an ordered pair solution, given one coordinate of the pair.

In today's world, where the exchange of information must be fast and entertaining, graphs are becoming increasingly popular. They provide a quick way of making comparisons, drawing conclusions, and approximating quantities. Bar and line graphs are common in previous sections, but in this section, we do not label the heights of the bars or points.

1   A **bar graph** consists of a series of bars arranged vertically or horizontally. The bar graph in Example 1 shows a comparison of the costs of the tests normally used in higher education. The names of the tests are listed horizontally and a bar is shown for each test. Corresponding to the height of the bar for each test is a number along a vertical axis. These vertical numbers are dollars charged for administering the test.

EXAMPLE 1

The following bar graph shows the cost charged for taking different tests often used in higher education.

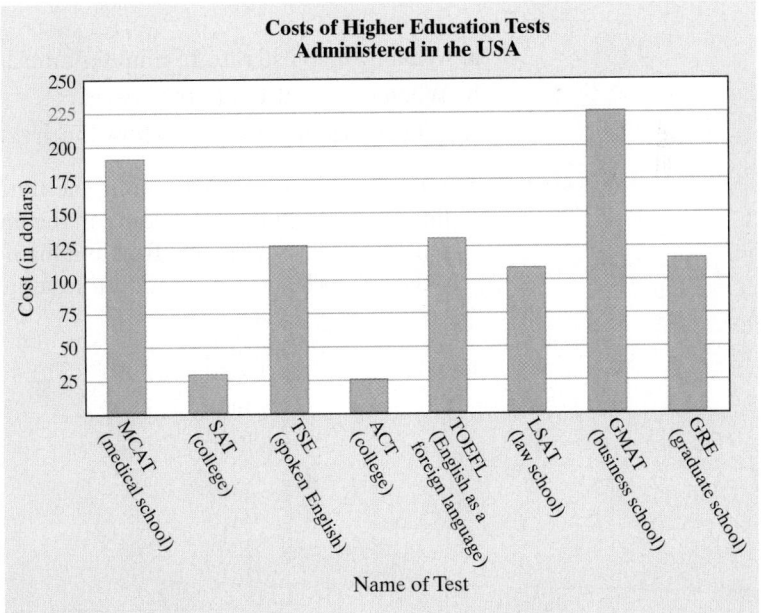

a. Which test costs the most to take? Approximate the cost to take this test.

b. Which test costs the least to take?

c. Find the cost to take the GRE test.

Solution   **a.** The tallest bar corresponds to the most expensive test. The GMAT test is the most expensive. To find its cost, go to the top of the bar for this test. From the top of the bar, move horizontally to the left until the vertical axis is reached. The cost, $225, is shown in red on the vertical axis.

b. The shortest bar corresponds to the least expensive test. The ACT test is the least expensive.

c. Go to the top of the bar for the GRE test. From the top of the bar, move horizontally to the left until the vertical axis is reached. The height of the bar is between the 100 and 125 marks and a little closer to the 125 mark, so we approximate the cost of the GRE test to be $115.

A **line graph** consists of a series of points connected by a line (or sometimes line segments). Example is a line graph, but it's also sometimes called a **broken line graph**.

EXAMPLE 2

The line graph shows the relationship between time spent smoking a cigarette and pulse rate. Time is recorded along the horizontal axis in minutes, with 0 minutes being the moment a smoker lights a cigarette. Pulse is recorded along the vertical axis in heartbeats per minute.

a. What is the pulse rate 15 minutes after lighting a cigarette?
b. When is the pulse rate the lowest?
c. When does the pulse rate show the greatest change?

Solution

a. We locate the number 15 along the time axis and move vertically upward until the line is reached. From this point on the line, we move horizontally to the left until the pulse rate axis is reached. Reading the number of beats per minute, we find that the pulse rate is 80 beats per minute 15 minutes after lighting a cigarette.

b. We find the lowest point of the line graph, which represents the lowest pulse rate. From this point, we move vertically downward to the time axis. We find that the pulse rate is the lowest at −5 minutes, which means 5 minutes *before* lighting a cigarette.

c. The pulse rate shows the greatest change during the 5 minutes between 0 and 5. Notice that the line graph is *steepest* between 0 and 5 minutes.

2   Notice in the previous graph that there are two numbers associated with each point of the graph. For example, we discussed earlier that 15 minutes after "lighting up," the pulse rate is 80 beats per minute. If we agree to write the time first and the pulse rate second, we can say there is a point on the graph corresponding to the **ordered pair** of numbers $(15, 80)$. A few more ordered pairs are listed alongside their corresponding points.

In general, we use this same ordered pair idea to describe the location of a point in a plane (such as a piece of paper). We start with a horizontal and a vertical axis. Each axis is a number line, and for the sake of consistency we construct our axes to intersect at the 0 coordinate of both. This point of intersection is called the **origin**. Notice that these two number lines or axes divide the plane into four regions called **quadrants**. The quadrants are usually numbered with Roman numerals as shown. The axes are not considered to be in any quadrant.

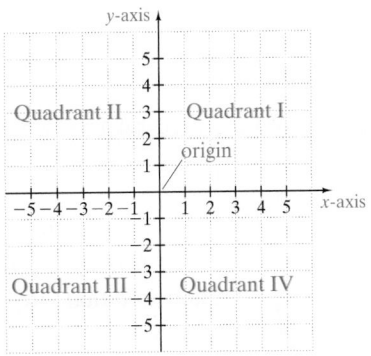

It is helpful to label axes, so we label the horizontal axis the **x-axis** and the vertical axis the **y-axis**. We call the system described above the **rectangular coordinate system**.

Just as with the pulse rate graph, we can then describe the locations of points by ordered pairs of numbers. We list the horizontal **x-axis** measurement first and the vertical **y-axis** measurement second.

To plot or graph the point corresponding to the ordered pair

$$(a, b)$$

we start at the origin. We then move a units left or right (right if a is positive, left if a is negative). From there, we move b units up or down (up if b is positive, down if b is negative). For example, to plot the point corresponding to the ordered pair $(3, 2)$, we start at the origin, move 3 units right, and from there move 2 units up. (See the figure on the next page.) The x-value, 3, is called the **x-coordinate** and the y-value, 2, is called the **y-coordinate**. From now on, we will call the point with coordinates $(3, 2)$ simply the point $(3, 2)$. The point $(-2, 5)$ is graphed also.

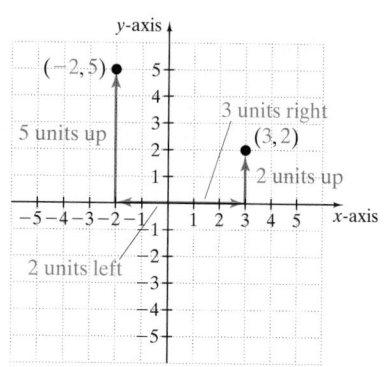

Does the order in which the coordinates are listed matter? Yes! Notice that the point corresponding to the ordered pair $(2, 3)$ is in a different location than the point corresponding to $(3, 2)$. These two ordered pairs of numbers describe two different points of the plane.

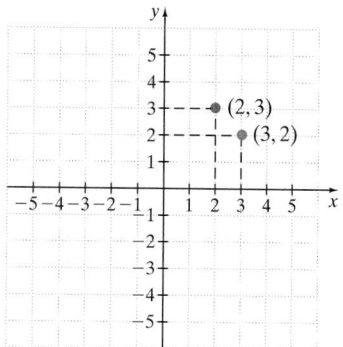

✔ **CONCEPT CHECK**

Is the graph of the point $(-5, 1)$ in the same location as the graph of the point $(1, -5)$? Explain.

> **Helpful Hint**
>
> Don't forget that **each ordered pair corresponds to exactly one point in the plane and that each point in the plane corresponds to exactly one ordered pair.**

EXAMPLE 3

On a single coordinate system, plot each ordered pair. State in which quadrant, if any, each point lies.

a. $(5, 3)$ **b.** $(-5, 3)$ **c.** $(-2, -4)$ **d.** $(1, -2)$

e. $(0, 0)$ **f.** $(0, 2)$ **g.** $(-5, 0)$ **h.** $\left(0, -5\frac{1}{2}\right)$

Solution

Point $(5, 3)$ lies in quadrant I.

Point $(-5, 3)$ lies in quadrant II.

Point $(-2, -4)$ lies in quadrant III.

Point $(1, -2)$ lies in quadrant IV.

Points $(0, 0), (0, 2), (-5, 0),$ and $\left(0, -5\frac{1}{2}\right)$ lie on axes, so they are not in any quadrant.

Concept Check Answer:

The graph of point $(-5, 1)$ lies in quadrant II and the graph of point $(1, -5)$ lies in quadrant IV. They are *not* in the same location.

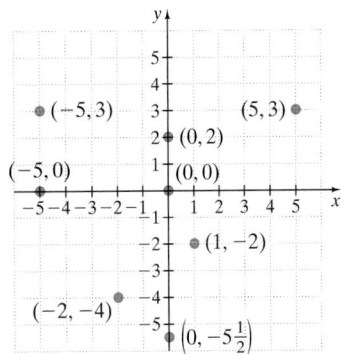

From Example 3, notice that the *y*-coordinate of any point on the *x*-axis is 0. For example, the point $(-5, 0)$ lies on the *x*-axis. Also, the *x*-coordinate of any point on the *y*-axis is 0. For example, the point $(0, 2)$ lies on the *y*-axis.

✔ CONCEPT CHECK

For each description of a point in the rectangular coordinate system, write an ordered pair that represents it.

a. Point A is located three units to the left of the *y*-axis and five units above the *x*-axis.

b. Point B is located six units below the origin.

3   Data that can be represented as an ordered pair is called **paired data**. Many types of data collected from the real world are paired data. For instance, the annual measurement of a child's height can be written as an ordered pair of the form (year, height in inches) and is paired data. The graph of paired data as points in the rectangular coordinate system is called a **scatter diagram**. Scatter diagrams can be used to look for patterns and trends in paired data.

EXAMPLE 4

The table gives the annual net sales for Wal-Mart Stores for the years shown. (*Source: Wal-Mart Stores, Inc.*)

Year	Wal-Mart Net Sales (in billions of dollars)
1997	105
1998	118
1999	138
2000	165
2001	191
2002	218
2003	245

a. Write this paired data as a set of ordered pairs of the form (year, sales in billions of dollars).

b. Create a scatter diagram of the paired data.

c. What trend in the paired data does the scatter diagram show?

Solution   **a.** The ordered pairs are (1997, 105), (1998, 118), (1999, 138), (2000, 165), (2001, 191), (2002, 218), and (2003, 245).

b. We begin by plotting the ordered pairs. Because the x-coordinate in each ordered pair is a year, we label the x-axis "Year" and mark the horizontal axis with the years given. Then we label the y-axis or vertical axis "Net Sales (in billions of dollars)." In this case it is convenient to mark the vertical axis in multiples of 20. Since no net sale is less than 100, we use the notation $\lessgtr$ to skip to 100, then proceed by multiples of 20.

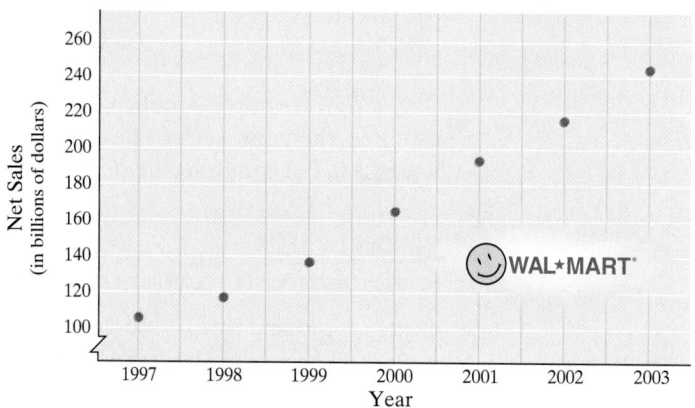

c. The scatter diagram shows that Wal-Mart net sales steadily increased over the years 1997–2003.

4   Let's see how we can use ordered pairs to record solutions of equations containing two variables. An equation in one variable such as $x + 1 = 5$ has one solution, which is 4: the number 4 is the value of the variable x that makes the equation true.

An equation in two variables, such as $2x + y = 8$, has solutions consisting of two values, one for x and one for y. For example, $x = 3$ and $y = 2$ is a solution of $2x + y = 8$ because, if x is replaced with 3 and y with 2, we get a true statement.

$$2x + y = 8$$
$$2(3) + 2 = 8$$
$$8 = 8 \quad \text{True}$$

The solution $x = 3$ and $y = 2$ can be written as $(3, 2)$, an **ordered pair** of numbers. The first number, 3, is the x-value and the second number, 2, is the y-value.

In general, an ordered pair is a **solution** of an equation in two variables if replacing the variables by the values of the ordered pair results in a true statement.

EXAMPLE 5

Determine whether each ordered pair is a solution of the equation $x - 2y = 6$.

a. $(6, 0)$ **b.** $(0, 3)$ **c.** $\left(1, -\dfrac{5}{2} \right)$

Solution   **a.** Let $x = 6$ and $y = 0$ in the equation $x - 2y = 6$.

$$x - 2y = 6$$
$$6 - 2(0) = 6 \quad \text{Replace } x \text{ with 6 and } y \text{ with 0.}$$
$$6 - 0 = 6 \quad \text{Simplify.}$$
$$6 = 6 \quad \text{True}$$

$(6, 0)$ is a solution, since $6 = 6$ is a true statement.

b. Let $x = 0$ and $y = 3$.

$$x - 2y = 6$$
$$0 - 2(3) = 6 \qquad \text{Replace } x \text{ with } 0 \text{ and } y \text{ with } 3.$$
$$0 - 6 = 6$$
$$-6 = 6 \qquad \text{False}$$

$(0, 3)$ is *not* a solution, since $-6 = 6$ is a false statement.

c. Let $x = 1$ and $y = -\frac{5}{2}$ in the equation.

$$x - 2y = 6$$
$$1 - 2\left(-\frac{5}{2}\right) = 6 \qquad \text{Replace } x \text{ with } 1 \text{ and } y \text{ with } -\frac{5}{2}.$$
$$1 + 5 = 6$$
$$6 = 6 \qquad \text{True}$$

$\left(1, -\frac{5}{2}\right)$ is a solution, since $6 = 6$ is a true statement.

5   If one value of an ordered pair solution of an equation is known, the other value can be determined. To find the unknown value, replace one variable in the equation by its known value. Doing so results in an equation with just one variable that can be solved for the variable using the methods of Chapter 2.

EXAMPLE 6

Complete the following ordered pair solutions for the equation $3x + y = 12$.

a. $(0, \ )$ **b.** $(\ , 6)$ **c.** $(-1, \ )$

Solution   **a.** In the ordered pair $(0, \ )$, the x-value is 0. Let $x = 0$ in the equation and solve for y.

$$3x + y = 12$$
$$3(0) + y = 12 \qquad \text{Replace } x \text{ with } 0.$$
$$0 + y = 12$$
$$y = 12$$

The completed ordered pair is $(0, 12)$.

b. In the ordered pair $(\ , 6)$, the y-value is 6. Let $y = 6$ in the equation and solve for x.

$$3x + y = 12$$
$$3x + 6 = 12 \qquad \text{Replace } y \text{ with } 6.$$
$$3x = 6 \qquad \text{Subtract 6 from both sides.}$$
$$x = 2 \qquad \text{Divide both sides by 3.}$$

The ordered pair is $(2, 6)$.

c. In the ordered pair $(-1, \ )$, the x-value is -1. Let $x = -1$ in the equation and solve for y.

$$3x + y = 12$$
$$3(-1) + y = 12 \qquad \text{Replace } x \text{ with } -1.$$
$$-3 + y = 12$$
$$y = 15 \qquad \text{Add 3 to both sides.}$$

The ordered pair is $(-1, 15)$.

Solutions of equations in two variables can also be recorded in a **table of values**, as shown in the next example.

EXAMPLE 7

Complete the table for the equation $y = 3x$.

	x	y
a.	-1	
b.		0
c.		-9

Solution **a.** Replace x with -1 in the equation and solve for y.

$$y = 3x$$
$$y = 3(-1) \quad \text{Let } x = -1.$$
$$y = -3$$

The ordered pair is $(-1, -3)$.

b. Replace y with 0 in the equation and solve for x.

$$y = 3x$$
$$0 = 3x \quad \text{Let } y = 0.$$
$$0 = x \quad \text{Divide both sides by 3.}$$

The ordered pair is $(0, 0)$.

c. Replace y with -9 in the equation and solve for x.

$$y = 3x$$
$$-9 = 3x \quad \text{Let } y = -9.$$
$$-3 = x \quad \text{Divide both sides by 3.}$$

x	y
-1	-3
0	0
-3	-9

The ordered pair is $(-3, -9)$. The completed table is shown to the left.

EXAMPLE 8

Complete the table for the equation $y = 3$.

x	y
-2	
0	
-5	

Solution The equation $y = 3$ is the same as $0x + y = 3$. Replace x with -2 and we have $0(-2) + y = 3$ or $y = 3$. Notice that no matter what value we replace x by, y always equals 3. The completed table is shown on the right.

x	y
-2	3
0	3
-5	3

By now, you have noticed that equations in two variables often have more than one solution. We discuss this more in the next section.

A table showing ordered pair solutions may be written vertically or horizontally as shown in the next example.

EXAMPLE 9

FINDING THE VALUE OF A COMPUTER

A computer was recently purchased for a small business for $2000. The business manager predicts that the computer will be used for 5 years and the value in dollars y of the computer in x years is $y = -300x + 2000$. Complete the table.

x	0	1	2	3	4	5
y						

Solution

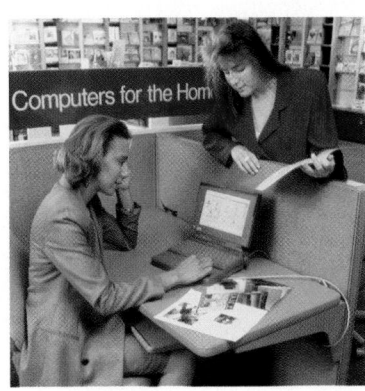

To find the value of y when x is 0, replace x with 0 in the equation. We use this same procedure to find y when x is 1 and when x is 2.

When $x = 0$,	**When $x = 1$,**	**When $x = 2$,**
$y = -300x + 2000$	$y = -300x + 2000$	$y = -300x + 2000$
$y = -300 \cdot 0 + 2000$	$y = -300 \cdot 1 + 2000$	$y = -300 \cdot 2 + 2000$
$y = 0 + 2000$	$y = -300 + 2000$	$y = -600 + 2000$
$y = 2000$	$y = 1700$	$y = 1400$

We have the ordered pairs (0, 2000), (1, 1700), and (2, 1400). This means that in 0 years the value of the computer is $2000, in 1 year the value of the computer is $1700, and in 2 years the value is $1400. To complete the table of values, we continue the procedure for $x = 3$, $x = 4$, and $x = 5$.

When $x = 3$,	**When $x = 4$,**	**When $x = 5$,**
$y = -300x + 2000$	$y = -300x + 2000$	$y = -300x + 2000$
$y = -300 \cdot 3 + 2000$	$y = -300 \cdot 4 + 2000$	$y = -300 \cdot 5 + 2000$
$y = -900 + 2000$	$y = -1200 + 2000$	$y = -1500 + 2000$
$y = 1100$	$y = 800$	$y = 500$

The completed table is

x	0	1	2	3	4	5
y	2000	1700	1400	1100	800	500

The ordered pair solutions recorded in the completed table for the example above are graphed on the next page. Notice that the graph gives a visual picture of the decrease in value of the computer.

Computer Value

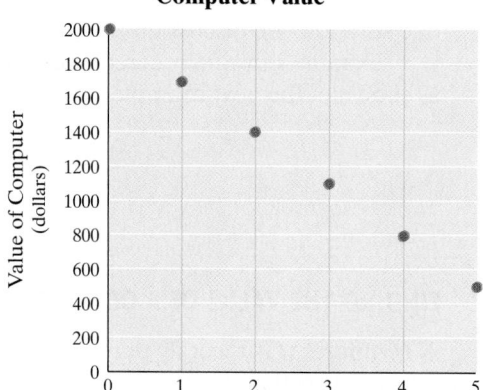

x	y
0	2000
1	1700
2	1400
3	1100
4	800
5	500

Spotlight on

DECISION

of MAKING

Suppose you own a small mail-order business. One of your employees brings you this graph showing the expected retail revenue from on-line shopping sales on the Internet. Should you consider expanding your business to include taking orders over the Internet? Explain your reasoning. What other factors would you want to consider?

Source: Forrester Research Inc.

STUDY SKILLS REMINDER

Are You Satisfied with Your Performance on a Particular Quiz or Exam?

If not, analyze your quiz or exam like you would a good mystery novel. Look for common themes in your errors.

Were most of your errors a result of

- *Carelessness?* If your errors were careless, did you turn in your work before the allotted time expired? If so, resolve to use the entire time allotted next time. Any extra time can be spent checking your work.

- *Running out of time?* If so, make a point to better manage your time on your next exam. A few suggestions are to work any questions that you are unsure of last and to check your work after all questions have been answered.

- *Not understanding a concept?* If so, review that concept and correct your work. Remember next time to make sure that all concepts on a quiz or exam are understood before the exam.

MENTAL MATH

Give two ordered pair solutions for each of the following linear equations.

1. $x + y = 10$　　　　**2.** $x + y = 6$　　　　**3.** $x = 3$　　　　**4.** $y = -2$

EXERCISE SET 3.1

STUDY   CD/   PH MATH   MathXL®Tutorials   MathXL®   MyMathLab®
GUIDE/SSM   VIDEO   TUTOR CENTER   ON CD

The following bar graph shows the top 10 tourist destinations and the number of tourists that visit each country per year. Use this graph to answer Exercises 1 through 6. See Example 1.

The line graph below shows the number of students per computer in U.S. public schools. Use this graph for Exercises 7 through 12. See Example 2.

Source: World Tourism Organization

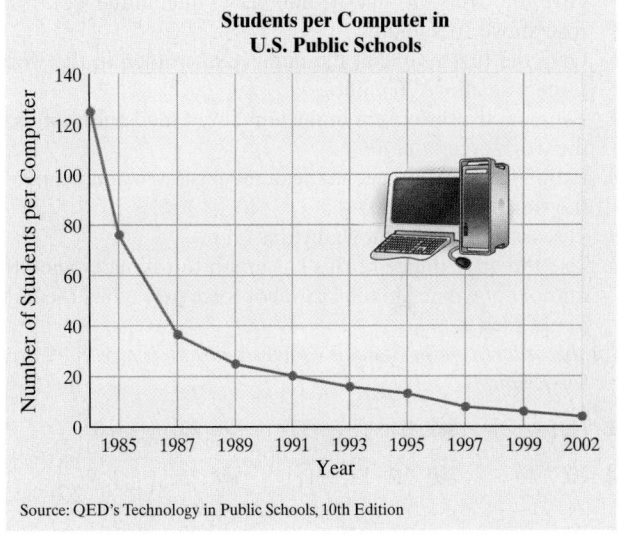

Source: QED's Technology in Public Schools, 10th Edition

1. Which country is the most popular tourist destination?

2. Which country shown is the least popular tourist destination?

3. Which countries have more than 40 million tourists per year?

4. Which countries shown have fewer than 20 million tourists per year?

5. Estimate the number of tourists per year whose destination is Italy.

6. Estimate the number of tourists per year whose destination is France.

7. Approximate the number of students per computer in 1991.

8. Approximate the number of students per computer in 2002.

9. During what year was the greatest decrease in number of students per computer?

10. What was the first year that the number of students per computer fell below 20?

11. What was the first year shown that the number of students per computer fell below 10?

12. Discuss any trends shown by this line graph.

The special bar graph shown is called a double bar graph. This double bar graph is used to compare men and women in the U.S. labor force per year. Use this graph for Exercises 13 through 20.

U.S. Labor Force

Source: U.S. Bureau of the Census

13. Estimate the number of women in the workforce in 2000.

14. Estimate the number of men in the workforce in 2000.

15. Give the first year that the number of men in the workforce rose above 20 million.

16. Give the first year that the number of women in the workforce rose above 20 million.

17. Estimate the difference in the number of men and women in the workforce in 1940.

18. Estimate the difference in the number of men and women in the workforce in 2000.

19. Discuss any trends shown by this graph.

20. Use the appearance of this bar graph to estimate when the number of women in the U.S. labor force may equal the number of men.

Plot the ordered pairs. State in which quadrant, if any, each point lies. See Example 3.

21. $(1, 5)$

22. $(-5, -2)$

23. $(-3, 0)$

24. $(0, -1)$

25. $(2, -4)$

26. $\left(-1, 4\frac{1}{2}\right)$

27. $\left(4\frac{3}{4}, 0\right)$

28. $\left(0, \frac{7}{8}\right)$

29. $(0, 0)$

30. $(5, 0)$

31. $(0, 4)$

32. $(-3, -3)$

Find the x- and y-coordinates of the following labeled points.

33.

34.

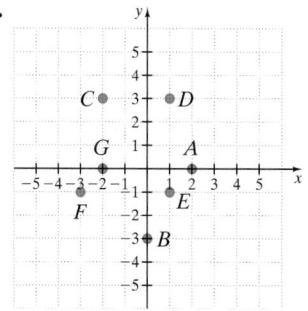

35. Plot the points $(-1, 5), (3, 5), (3, -4)$, and $(-1, -4)$ on a rectangular coordinate system. Find the perimeter of the rectangle whose vertices are these points.

36. Plot the points $(5, 2), (5, -6), (0, -6)$, and $(0, 2)$ on a rectangular coordinate system. Find the area of the rectangle whose vertices are these points.

Solve. See Example 4.

37. The table shows the distance from the equator (in miles) and the average annual snowfall (in inches) for each of eight selected U.S. cities. (*Sources*: National Climatic Data Center, Wake Forest University Albatross Project)

City	Distance from Equator (in miles)	Average Annual Snowfall (in inches)
1. Atlanta, GA	2313	2
2. Austin, TX	2085	1
3. Baltimore, MD	2711	21
4. Chicago, IL	2869	39
5. Detroit, MI	2920	42
6. Juncau, AK	4038	99
7. Miami, FL	1783	0
8. Winston-Salem, NC	2493	9

a. Write this paired data as a set of ordered pairs of the form (distance from equator, average annual snowfall).

b. Create a scatter diagram of the paired data. Be sure to label the axes appropriately.

Average Annual Snowfall for Selected U.S. Cities

c. What trend in the paired data does the scatter diagram show?

38. The table shows the average farm size (in acres) in the United States during the years shown. (*Source*: National Agricultural Statistics Service)

Year	Average Farm Size (in acres)
1995	438
1996	438
1997	436
1998	435
1999	432
2000	434
2001	436

a. Write this paired data as a set of ordered pairs of the form (year, average farm size).

b. Create a scatter diagram of the paired data. Be sure to label the axes appropriately.

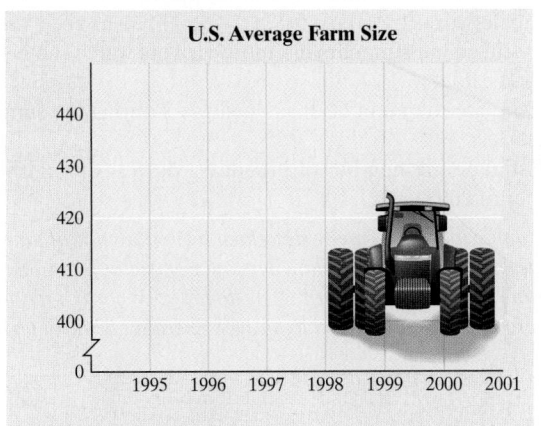

Determine whether each ordered pair is a solution of the given linear equation. See Example 5.

39. $y = -5x$; $(-1, -5), (0, 0), (2, -10)$

40. $x = 2y$; $(0, 0), (2, 1), (-2, -1)$

41. $x = 5$; $(4, 5), (5, 4), (5, 0)$

42. $y = -2$; $(-2, 2), (2, -2), (0, -2)$

43. $2x - y = 11$; $(3, -4), (9, 8)$

44. $x - 4y = 14$; $(2, -3), (14, 6)$

45. $x = \frac{1}{3}y$; $(0, 0), (3, 9)$

46. $y = -\frac{1}{2}x$; $(0, 0), (4, 2)$

Complete each ordered pair so that it is a solution of the given linear equation. See Examples 6 through 8.

47. $x - 4y = 4$; $(\quad, -2), (4, \quad)$

48. $x - 5y = -1$; $(\quad, -2), (4, \quad)$

49. $y = -7$; $(11, \quad), (\quad, -7)$

50. $x = \frac{1}{2}$; $(\quad, 0), \left(\frac{1}{2}, \quad\right)$

51. $-2x + 7y = -3$ $(\quad, 1) (1, \quad)$

52. $-3x + 4y = -1$ $(\quad, 2) (-2, \quad)$

Complete the table of values for each given equation; then plot each solution. Use a single coordinate system for each equation. See Examples 6 through 8.

53. $2x + 7y = 5$

x	y
0	
	0
	1

54. $x - 6y = 3$

x	y
0	
1	
	-1

55. $x = 3$

x	y
	0
	-0.5
	$\frac{1}{4}$

56. $y = -1$

x	y
-2	
0	
-1	

57. $x = -5y$

x	y
	0
	1
10	

58. $y = -3x$

x	y
0	
-2	
	9

59. Discuss any similarities in the graphs of the ordered pair solutions for Exercises 53–58.

60. Explain why equations in two variables have more than one solution.

Solve. See Example 9.

61. The cost in dollars y of producing x computer desks is given by $y = 80x + 5000$.

a. Complete the following table and graph the results.

x	100	200	300
y			

b. Find the number of computer desks that can be produced for $8600. (*Hint:* Find x when $y = 8600$.)

62. The hourly wage y of an employee at a certain production company is given by $y = 0.25x + 9$ where x is the number of units produced in an hour.

a. Complete the table and graph the results.

x	0	1	5	10
y				

b. Find the number of units that must be produced each hour to earn an hourly wage of $12.25. (*Hint:* Find x when $y = 12.25$.)

The graph below shows the number of Target stores for each year. Use this graph to answer Exercises 63–68.

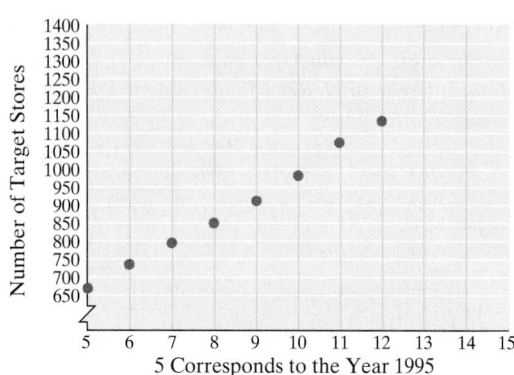

Source: *USA Today* 1/14/99

63. The ordered pair $(5, 670)$ is a point of the graph. Write a sentence describing the meaning of this ordered pair.

64. The ordered pair $(10, 984)$ is a point of the graph. Write a sentence describing the meaning of this ordered pair.

65. Estimate the increase in Target stores for years 6, 7, and 8.

66. Use a straightedge or ruler and this graph to predict the number of Target stores in the year 2009.

67. When is the graph of the ordered pair (a, b) the same as the graph of the ordered pair (b, a)?

68. In your own words, describe how to plot an ordered pair.

REVIEW AND PREVIEW

Solve each equation for y. See Section 2.6.

69. $10x = -5y$

70. $4y = -8x$

71. $x - 3y = 6$

72. $2x - 9y = -20$

Concept Extensions

Determine the quadrant or quadrants in which the points described below lie.

73. The first coordinate is positive and the second coordinate is negative.

74. Both coordinates are negative.

75. The first coordinate is negative.

76. The second coordinate is positive.

77. Three vertices of a rectangle are $(-2, -3)$, $(-7, -3)$, and $(-7, 6)$.
   a. Find the coordinates of the fourth vertex of a rectangle.
   b. Find the perimeter of the rectangle.
   c. Find the area of the rectangle.

78. Three vertices of a square are $(-4, -1)$, $(-4, 8)$, and $(5, 8)$.
   a. Find the coordinates of the fourth vertex of the square.
   b. Find the perimeter of the square.
   c. Find the area of the square.

This special line graph is a double line graph. Here, we can compare the number of viewers who watch cable TV during the summer and the number of viewers who watch broadcast networks during the summer.

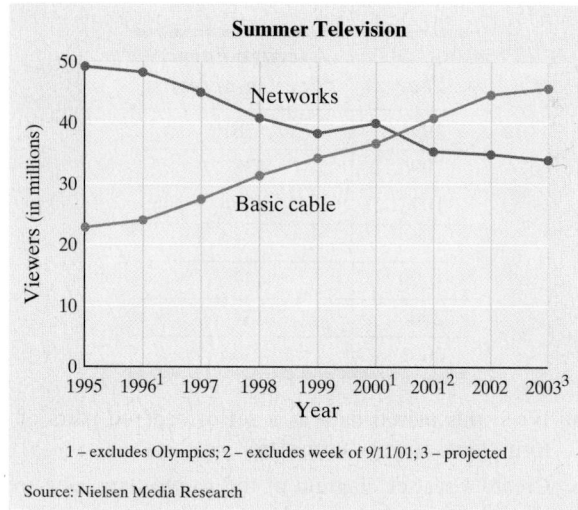

1 – excludes Olympics; 2 – excludes week of 9/11/01; 3 – projected

Source: Nielsen Media Research

79. During what summers did more viewers watch cable than the networks?

80. During what summers did more viewers watch the networks than cable?

81. Estimate the number of cable viewers during the summer of 2003.

82. Estimate the number of broadcast viewers during the summer of 2003.

Geographic locations can be described by a gridwork of lines called latitudes and longitudes, as shown below. For example, the location of Houston, Texas, can be described by latitude 30° north and longitude 95° west. Use the map shown to answer Exercises 83 through 86.

83. Using latitude and longitude, describe the location of New Orleans, Louisiana.

84. Use an atlas and describe the location of your hometown.

85. Using latitude and longitude, describe the location of Denver, Colorado.

86. Give another name for 0° latitude.

3.2   *GRAPHING LINEAR EQUATIONS*

Objectives

1. Identify linear equations.
2. Graph a linear equation by finding and plotting ordered pair solutions.

1. In the previous section, we found that equations in two variables may have more than one solution. For example, both $(6, 0)$ and $(2, -2)$ are solutions of the equation $x - 2y = 6$. In fact, this equation has an infinite number of solutions. Other solutions include $(0, -3)$, $(4, -1)$, and $(-2, -4)$. If we graph these solutions, notice that a pattern appears.

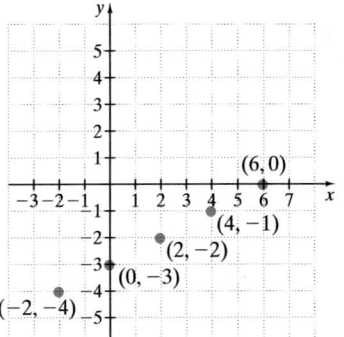

These solutions all appear to lie on the same line, which has been filled in below. It can be shown that every ordered pair solution of the equation corresponds to a point on this line, and every point on this line corresponds to an ordered pair solution. Thus, we say that this line is the graph of the equation $x - 2y = 6$.

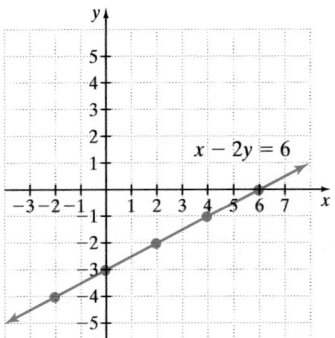

The equation $x - 2y = 6$ is called a **linear equation in two variables** and **the graph of every linear equation in two variables is a line.**

Linear Equation in Two Variables

A linear equation in two variables is an equation that can be written in the form

$$Ax + By = C$$

where A, B, and C are real numbers and A and B are not both 0. **The graph of a linear equation in two variables is a straight line.**

The form $Ax + By = C$ is called **standard form**.

> **Helpful Hint**
> Notice in the form $Ax + By = C$, the understood exponent on both x and y is 1.

Examples of Linear Equations in Two Variables

$$2x + y = 8 \qquad -2x = 7y \qquad y = \frac{1}{3}x + 2 \qquad y = 7$$

Before we graph linear equations in two variables, let's practice identifying these equations.

EXAMPLE 1

Identify the linear equations in two variables.

a. $x - 1.5y = -1.6$ **b.** $y = -2x$ **c.** $x + y^2 = 9$ **d.** $x = 5$

Solution **a.** This is a linear equation in two variables because it is written in the form $Ax + By = C$ with $A = 1$, $B = -1.5$, and $C = -1.6$.

b. This is a linear equation in two variables because it can be written in the form $Ax + By = C$.

$$y = -2x$$
$$2x + y = 0 \qquad \text{Add } 2x \text{ to both sides.}$$

c. This is *not* a linear equation in two variables because y is squared.

d. This is a linear equation in two variables because it can be written in the form $Ax + By = C$.

$$x = 5$$
$$x + 0y = 5 \qquad \text{Add } 0 \cdot y.$$

2 From geometry, we know that a straight line is determined by just two points. Graphing a linear equation in two variables, then, requires that we find just two of its infinitely many solutions. Once we do so, we plot the solution points and draw the line connecting the points. Usually, we find a third solution as well, as a check.

EXAMPLE 2

Graph the linear equation $2x + y = 5$.

Solution Find three ordered pair solutions of $2x + y = 5$. To do this, choose a value for one variable, x or y, and solve for the other variable. For example, let $x = 1$. Then $2x + y = 5$ becomes

$$2x + y = 5$$
$$2(1) + y = 5 \qquad \text{Replace } x \text{ with } 1.$$
$$2 + y = 5 \qquad \text{Multiply.}$$
$$y = 3 \qquad \text{Subtract } 2 \text{ from both sides.}$$

Since $y = 3$ when $x = 1$, the ordered pair $(1, 3)$ is a solution of $2x + y = 5$. Next, let $x = 0$.

$$2x + y = 5$$
$$2(0) + y = 5 \qquad \text{Replace } x \text{ with } \mathbf{0}.$$
$$0 + y = 5$$
$$y = 5$$

The ordered pair $(0, 5)$ is a second solution.

The two solutions found so far allow us to draw the straight line that is the graph of all solutions of $2x + y = 5$. However, we find a third ordered pair as a check. Let $y = -1$.

$$2x + y = 5$$
$$2x + (-1) = 5 \qquad \text{Replace } y \text{ with } -\mathbf{1}.$$
$$2x - 1 = 5$$
$$2x = 6 \qquad \text{Add } \mathbf{1} \text{ to both sides.}$$
$$x = 3 \qquad \text{Divide both sides by } \mathbf{2}.$$

The third solution is $(3, -1)$. These three ordered pair solutions are listed in table form as shown. The graph of $2x + y = 5$ is the line through the three points.

x	y
1	3
0	5
3	-1

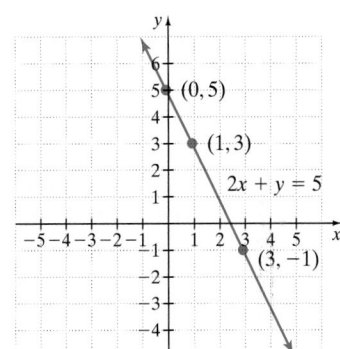

> **Helpful Hint**
>
> All three points should fall on the same straight line. If not, check your ordered pair solutions for a mistake.

EXAMPLE 3

Graph the linear equation $-5x + 3y = 15$.

Solution   Find three ordered pair solutions of $-5x + 3y = 15$.

Let $x = 0$.	**Let $y = 0$.**	**Let $x = -2$.**
$-5x + 3y = 15$	$-5x + 3y = 15$	$-5x + 3y = 15$
$-5 \cdot 0 + 3y = 15$	$-5x + 3 \cdot 0 = 15$	$-5(-2) + 3y = 15$
$0 + 3y = 15$	$-5x + 0 = 15$	$10 + 3y = 15$
$3y = 15$	$-5x = 15$	$3y = 5$
$y = 5$	$x = -3$	$y = \frac{5}{3}$

The ordered pairs are $(0, 5)$, $(-3, 0)$, and $(-2, \frac{5}{3})$. The graph of $-5x + 3y = 15$ is the line through the three points.

x	y
0	5
-3	0
-2	$\frac{5}{3} = 1\frac{2}{3}$

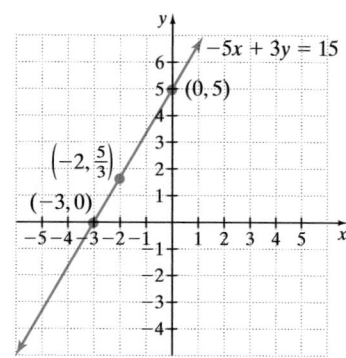

EXAMPLE 4

Graph the linear equation $y = 3x$.

Solution   To graph this linear equation, we find three ordered pair solutions. Since this equation is solved for y, choose three x values.

If $x = 2$, $y = 3 \cdot 2 = 6$.

If $x = 0$, $y = 3 \cdot 0 = 0$.

If $x = -1$, $y = 3 \cdot -1 = -3$.

x	y
2	6
0	0
-1	-3

Next, graph the ordered pair solutions listed in the table above and draw a line through the plotted points as shown on the next page. The line is the graph of $y = 3x$. Every point on the graph represents an ordered pair solution of the equation and every ordered pair solution is a point on this line.

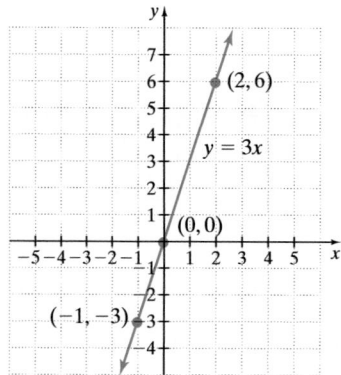

EXAMPLE 5

Graph the linear equation $y = -\frac{1}{3}x$.

Solution   Find three ordered pair solutions, graph the solutions, and draw a line through the plotted solutions. To avoid fractions, choose x values that are multiples of 3 to substitute in the equation. When a multiple of 3 is multiplied by $-\frac{1}{3}$, the result is an integer. See the calculations shown to the right of the table on the top of the next page.

If $x = 6$, then $y = -\dfrac{1}{3} \cdot 6 = -2$.

If $x = 0$, then $y = -\dfrac{1}{3} \cdot 0 = 0$.

If $x = -3$, then $y = -\dfrac{1}{3} \cdot -3 = 1$.

x	y
6	−2
0	0
−3	1

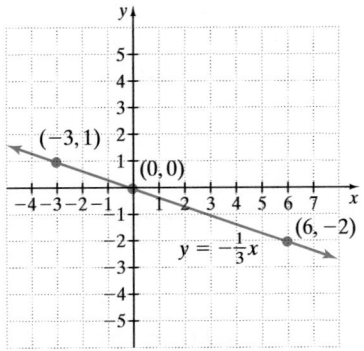

Let's compare the graphs in Examples 4 and 5. The graph of $y = 3x$ tilts upward (as we follow the line from left to right) and the graph of $y = -\frac{1}{3}x$ tilts downward (as we follow the line from left to right). Also notice that both lines go through the origin or that $(0, 0)$ is an ordered pair solution of both equations. In general, the graphs of $y = 3x$ and $y = -\frac{1}{3}x$ are of the form $y = mx$ where m is a constant. The graph of an equation in this form always goes through the origin $(0, 0)$ because when x is 0, $y = mx$ becomes $y = m \cdot 0 = 0$.

EXAMPLE 6

Graph the linear equation $y = 3x + 6$ and compare this graph with the graph of $y = 3x$ in Example 4.

Solution Find ordered pair solutions, graph the solutions, and draw a line through the plotted solutions. We choose x values and substitute in the equation $y = 3x + 6$.

If $x = -3$, then $y = 3(-3) + 6 = -3$.

If $x = 0$, then $y = 3(0) + 6 = 6$.

If $x = 1$, then $y = 3(1) + 6 = 9$.

x	y
−3	−3
0	6
1	9

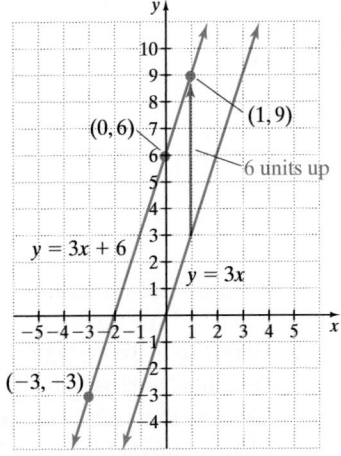

The most startling similarity is that both graphs appear to have the same upward tilt as we move from left to right. Also, the graph of $y = 3x$ crosses the y-axis at the origin, while the graph of $y = 3x + 6$ crosses the y-axis at 6. In fact, the graph of $y = 3x + 6$ is the same as the graph of $y = 3x$ moved vertically upward 6 units.

Notice that the graph of $y = 3x + 6$ crosses the y-axis at 6. This happens because when $x = 0$, $y = 3x + 6$ becomes $y = 3 \cdot 0 + 6 = 6$. The graph contains the point $(0, 6)$, which is on the y-axis.

In general, if a linear equation in two variables is solved for y, we say that it is written in the form $y = mx + b$. The graph of this equation contains the point $(0, b)$ because when $x = 0$, $y = mx + b$ is $y = m \cdot 0 + b = b$.

> The graph of $y = mx + b$ crosses the y-axis at $(0, b)$.

We will review this again in Section 3.5.

Linear equations are often used to model real data as seen in the next example.

EXAMPLE 7

ESTIMATING THE NUMBER OF MEDICAL ASSISTANTS

One of the occupations expected to have the most growth in the next few years is medical assistant. The number of people y (in thousands) employed as medical assistants in the United States can be estimated by the linear equation $y = 31.8x + 180$, where x is the number of years after the year 1995. (*Source:* based on data from the Bureau of Labor Statistics)

Graph the equation and use the graph to predict the number of medical assistants in the year 2010.

Solution To graph $y = 31.8x + 180$, choose x-values and substitute in the equation.

If $x = 0$, then $y = 31.8(0) + 180 = 180$.
If $x = 2$, then $y = 31.8(2) + 180 = 243.6$.
If $x = 7$, then $y = 31.8(7) + 180 = 402.6$.

x	y
0	180
2	243.6
7	402.6

To use the graph to *predict* the number of medical assistants in the year 2010, we need to find the y-coordinate that corresponds to $x = 15$. (15 years after 1995 is the year 2010.) To do so, find 15 on the x-axis. Move vertically upward to the graphed line and then horizontally to the left. We approximate the number on the y-axis to be 655. Thus in the year 2010, we predict that there will be 655 thousand medical assistants. (The actual value, using 15 for x, is 657.)

Graphing Calculator Explorations

In this section, we begin an optional study of graphing calculators and graphing software packages for computers. These graphers use the same point plotting technique that was introduced in this section. The advantage of this graphing technology is, of course, that graphing calculators and computers can find and plot ordered pair solutions much faster than we can. Note, however, that the features described in these boxes may not be available on all graphing calculators.

The rectangular screen where a portion of the rectangular coordinate system is displayed is called a **window**. We call it a **standard window** for graphing when both the x- and y-axes show coordinates between -10 and 10. This information is often displayed in the window menu on a graphing calculator as

$$Xmin = -10$$
$$Xmax = 10$$
$$Xscl = 1 \qquad \text{The scale on the } x\text{-axis is one unit per tick mark.}$$
$$Ymin = -10$$
$$Ymax = 10$$
$$Yscl = 1 \qquad \text{The scale on the } y\text{-axis is one unit per tick mark.}$$

To use a graphing calculator to graph the equation $y = 2x + 3$, press the $\boxed{\text{Y=}}$ key and enter the keystrokes $\boxed{2}$ $\boxed{x}$ $\boxed{+}$ $\boxed{3}$. The top row should now read $Y_1 = 2x + 3$. Next press the $\boxed{\text{GRAPH}}$ key, and the display should look like this:

Use a standard window and graph the following linear equations. (Unless otherwise stated, use a standard window when graphing.)

1. $y = -3x + 7$ **2.** $y = -x + 5$

3. $y = 2.5x - 7.9$ **4.** $y = -1.3x + 5.2$

5. $y = -\dfrac{3}{10}x + \dfrac{32}{5}$ **6.** $y = \dfrac{2}{9}x - \dfrac{22}{3}$

EXERCISE SET 3.2

|STUDY
GUIDE/SSM|CD/
VIDEO|PH MATH
TUTOR CENTER|MathXL®Tutorials
ON CD|MathXL®|MyMathLab®|

Determine whether each equation is a linear equation in two variables. See Example 1.

1. $-x = 3y + 10$ **2.** $y = x - 15$

3. $x = y$ **4.** $x = y^3$

5. $x^2 + 2y = 0$ **6.** $0.01x - 0.2y = 8.8$

7. $y = -1$ **8.** $x = 25$

MIXED PRACTICE

Graph each linear equation. See Examples 2 through 5.

9. $x + y = 4$ **10.** $x + y = 7$

11. $x - y = -2$ **12.** $-x + y = 6$

13. $x - 2y = 6$ **14.** $-x + 5y = 5$

15. $y = 6x + 3$ **16.** $y = -2x + 7$

17. $x - 2y = -6$ **18.** $-x + 2y = 5$

19. $y = 6x$ **20.** $x = -2y$

21. $3y - 10 = 5x$ **22.** $-2x + 7 = 2y$

23. $x + 3y = 9$ **24.** $2x + y = -2$

25. $y - x = -1$ **26.** $x - y = 5$

27. $x = -3y$ **28.** $y = -x$

29. $5x - y = 10$ **30.** $7x - y = 2$

31. $y = \dfrac{1}{2}x + 2$ **32.** $y = -\dfrac{1}{5}x - 1$

Graph each pair of linear equations on the same set of axes. Discuss how the graphs are similar and how they are different. See Example 6.

33. $y = 5x;\ y = 5x + 4$ **34.** $y = 2x;\ y = 2x + 5$

35. $y = -2x;\ y = -2x - 3$ **36.** $y = x;\ y = x - 7$

37. $y = \dfrac{1}{2}x;\ y = \dfrac{1}{2}x + 2$ **38.** $y = -\dfrac{1}{4}x;\ y = -\dfrac{1}{4}x + 3$

The graph of $y = 5x$ is given below as well as Figures a–d. For Exercises 39 through 42, match each equation with its graph. Hint: Recall that if an equation is written in the form $y = mx + b$, its graph crosses the y-axis at b.

a.

b.

c.

d.

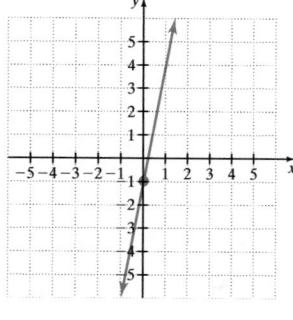

39. $y = 5x + 5$ **40.** $y = 5x - 4$

41. $y = 5x - 1$ **42.** $y = 5x + 2$

Solve. See Example 7.

43. Skateboarding has been growing in popularity in recent years, and the sales of skateboarding apparel and hardware have been increasing at a steady rate. The sales of skateboard apparel and hardware (in millions of dollars) from the years 1996 to 2001 are given by the equation $y = 180x + 450$ where x is the number of years after 1996. If this trend continues, what will the total sales be in 2009? (*Source*: Transworld SKATEboarding Magazine)

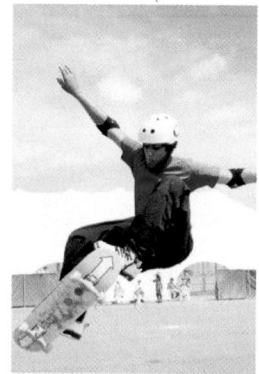

44. The revenue y (in billions of dollars) for Home Depot stores during the years 1998 through 2001 is given by the equation $y = 8x + 30$, where x is the number of years after 1998. Graph this equation and use the graph to predict the revenue for Home Depot Stores in the year 2010. (*Source*: Based on data from The Home Depot Inc.)

45. The average weekly earnings (in dollars) by U.S. production workers can be approximated by the equation $y = 16x + 144$ from 1998 to 2001 where x is the number of years after 1998. Graph this equation and use it to predict the average weekly earnings in 2007. (*Source*: U.S. Bureau of Labor Statistics)

46. The U.S. silver production (in metric tons) from 1998 to 2001 has been steadily dropping, and can be approximated by the equation $y = -105x + 2060$ where x is the number of years after 1998. If this current trend continues, use the equation to estimate the U.S. silver production in 2008. (*Source*: U.S. Geological Survey)

REVIEW AND PREVIEW

△ **47.** The coordinates of three vertices of a rectangle are $(-2,5)$, $(4,5)$, and $(-2,-1)$. Find the coordinates of the fourth vertex. See Section 3.1.

△ **48.** The coordinates of two vertices of a square are $(-3,-1)$ and $(2,-1)$. Find the coordinates of two pairs of points possible for the third and fourth vertices. See Section 3.1.

Solve the following equations. See Section 2.3.

49. $3(x - 2) + 5x = 6x - 16$ **50.** $5 + 7(x + 1) = 12 + 10x$

51. $3x + \dfrac{2}{5} = \dfrac{1}{10}$ **52.** $\dfrac{1}{6} + 2x = \dfrac{2}{3}$

Complete each table. See Section 3.1.

53. $x - y = -3$

x	y
0	
	0

54. $y - x = 5$

x	y
0	
	0

55. $y = 2x$

x	y
0	
	0

56. $x = -3y$

x	y
0	
	0

Concept Extensions

Write each statement as an equation in two variables. Then graph the equation.

57. The y-value is 5 more than the x-value.

58. The y-value is twice the x-value.

59. Two times the x-value, added to three times the y-value is 6.

60. Five times the x-value, added to twice the y-value is -10.

61. Explain how to find ordered pair solutions of linear equations in two variables.

△ **62.** The perimeter of the trapezoid below is 22 centimeters. Write a linear equation in two variables for the perimeter. Find y if x is 3 cm.

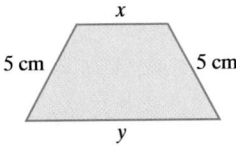

63. If (a, b) is an ordered pair solution of $x + y = 5$, is (b, a) also a solution? Explain why or why not.

△ **64.** The perimeter of the rectangle below is 50 miles. Write a linear equation in two variables for this perimeter. Use this equation to find x when y is 20.

65. Graph the nonlinear equation $y = x^2$ by completing the table shown. Plot the ordered pairs and connect them with a smooth curve.

x	y
0	
1	
-1	
2	
-2	

66. Graph the nonlinear equation $y = |x|$ by completing the table shown. Plot the ordered pairs and connect them. This curve is "V" shaped.

$$y = |x|$$

x	y
0	
1	
-1	
2	
-2	

3.3 INTERCEPTS

Objectives

1. Identify intercepts of a graph.
2. Graph a linear equation by finding and plotting intercepts.
3. Identify and graph vertical and horizontal lines.

1 In this section, we graph linear equations in two variables by identifying intercepts. For example, the graph of $y = 4x - 8$ is shown below. Notice that this graph crosses the y-axis at the point $(0, -8)$. This point is called the **y-intercept**. Likewise, the graph crosses the x-axis at $(2, 0)$, and this point is called the **x-intercept**.

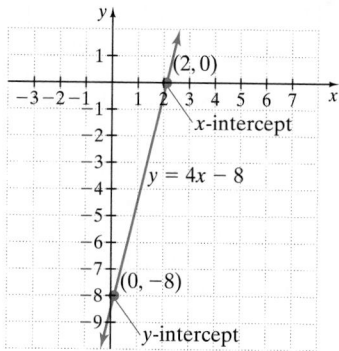

Helpful Hint

If a graph crosses the x-axis at $(-3, 0)$ and the y-axis at $(0, 7)$, then

$$\underbrace{(-3, 0)}_{\text{x-intercept}} \qquad \underbrace{(0, 7)}_{\text{y-intercept}}$$

Notice that for the y-intercept, the x-value is 0 and for the x-intercept, the y-value is 0.

Note: Sometimes in mathematics, you may see just the number 7 stated as the y-intercept, and -3 stated as the x-intercept.

EXAMPLE 1

Identify the x- and y-intercepts.

a.

b.

c.

d.

e.

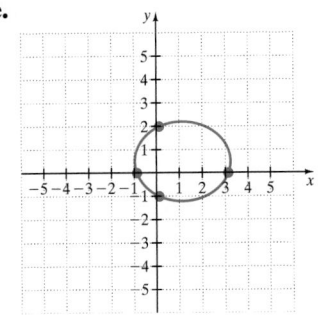

> ## Helpful Hint
>
> Notice that any time $(0,0)$ is a point of a graph, then it is an x-intercept and a y-intercept.

Solution

a. The graph crosses the x-axis at -3, so the x-intercept is $(-3, 0)$. The graph crosses the y-axis at 2, so the y-intercept is $(0, 2)$.

b. The graph crosses the x-axis at -4 and -1, so the x-intercepts are $(-4, 0)$ and $(-1, 0)$. The graph crosses the y-axis at 1, so the y-intercept is $(0, 1)$.

c. The x-intercept and the y-intercept are both $(0, 0)$.

d. The x-intercept is $(2, 0)$. There is no y-intercept.

e. The x-intercepts are $(-1, 0)$ and $(3, 0)$. The y-intercepts are $(0, -1)$, and $(0, 2)$.

2   Given the equation of a line, intercepts are usually easy to find since one coordinate is 0.

One way to find the y-intercept of a line, given its equation, is to let $x = 0$, since a point on the y-axis has an x-coordinate of 0. To find the x-intercept of a line, let $y = 0$, since a point on the x-axis has a y-coordinate of 0.

> ## Finding x- and y-intercepts
>
> To find the x-intercept, let $y = 0$ and solve for x.
> To find the y-intercept, let $x = 0$ and solve for y.

EXAMPLE 2

Graph $x - 3y = 6$ by finding and plotting intercepts.

Solution   Let $y = 0$ to find the x-intercept and let $x = 0$ to find the y-intercept.

$$\begin{aligned}
&\text{Let } y = 0 && \text{Let } x = 0 \\
&x - 3y = 6 && x - 3y = 6 \\
&x - 3(0) = 6 && 0 - 3y = 6 \\
&x - 0 = 6 && -3y = 6 \\
&x = 6 && y = -2
\end{aligned}$$

The x-intercept is $(6, 0)$ and the y-intercept is $(0, -2)$. We find a third ordered pair solution to check our work. If we let $y = -1$, then $x = 3$. Plot the points $(6, 0)$, $(0, -2)$, and $(3, -1)$. The graph of $x - 3y = 6$ is the line drawn through these points, as shown.

x	y
6	0
0	-2
3	-1

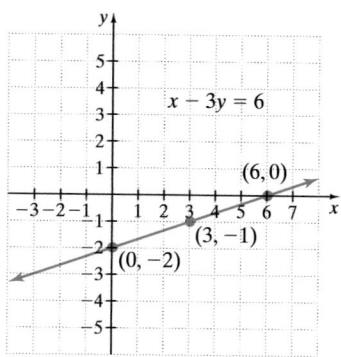

EXAMPLE 3

Graph $x = -2y$ by plotting intercepts.

Solution   Let $y = 0$ to find the x-intercept and $x = 0$ to find the y-intercept.

$$
\begin{array}{ll}
\text{Let } y = 0 & \text{Let } x = 0 \\
x = -2y & x = -2y \\
x = -2(0) & 0 = -2y \\
x = 0 & 0 = y
\end{array}
$$

Both the x-intercept and y-intercept are $(0, 0)$. In other words, when $x = 0$, then $y = 0$, which gives the ordered pair $(0, 0)$. Also, when $y = 0$, then $x = 0$, which gives the same ordered pair $(0, 0)$. This happens when the graph passes through the origin. Since two points are needed to determine a line, we must find at least one more ordered pair that satisfies $x = -2y$. Let $y = -1$ to find a second ordered pair solution and let $y = 1$ as a checkpoint.

$$
\begin{array}{ll}
\text{Let } y = -1 & \text{Let } y = 1 \\
x = -2(-1) & x = -2(1) \\
x = 2 & x = -2
\end{array}
$$

The ordered pairs are $(0, 0)$, $(2, -1)$, and $(-2, 1)$. Plot these points to graph $x = -2y$.

x	y
0	0
2	-1
-2	1

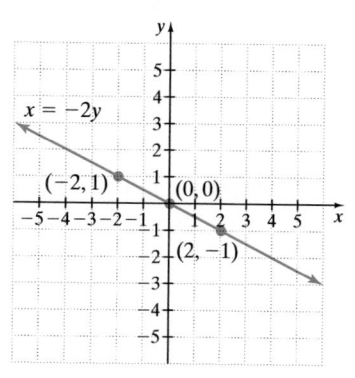

EXAMPLE 4

Graph $4x = 3y - 9$.

Solution   Find the x- and y-intercepts, and then choose $x = 2$ to find a third checkpoint.

Let $y = 0$

$$4x = 3(0) - 9$$
$$4x = -9$$

Solve for x.

$$x = -\frac{9}{4} \text{ or } -2\frac{1}{4}$$

Let $x = 0$

$$4 \cdot 0 = 3y - 9$$
$$9 = 3y$$

Solve for y.

$$3 = y$$

Let $x = 2$

$$4(2) = 3y - 9$$
$$8 = 3y - 9$$

Solve for y.

$$17 = 3y$$
$$\frac{17}{3} = y \text{ or } y = 5\frac{2}{3}$$

The ordered pairs are $\left(-2\frac{1}{4}, 0\right)$, $(0, 3)$, and $\left(2, 5\frac{2}{3}\right)$. The equation $4x = 3y - 9$ is graphed as follows.

x	y
$-2\frac{1}{4}$	0
0	3
2	$5\frac{2}{3}$

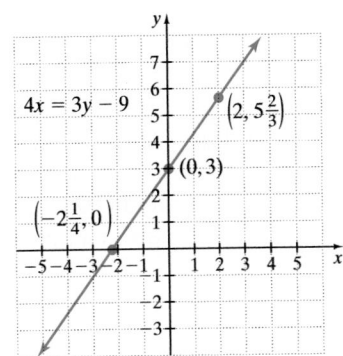

3   The equation $x = c$, where c is a real number constant, is a linear equation in two variables because it can be written in the form $x + 0y = c$. The graph of this equation is a vertical line as shown in the next example.

EXAMPLE 5

Graph $x = 2$.

Solution   The equation $x = 2$ can be written as $x + 0y = 2$. For any y-value chosen, notice that x is 2. No other value for x satisfies $x + 0y = 2$. Any ordered pair whose x-coordinate is 2 is a solution of $x + 0y = 2$. We will use the ordered pair solutions $(2, 3)$, $(2, 0)$, and $(2, -3)$ to graph $x = 2$.

x	y
2	3
2	0
2	-3

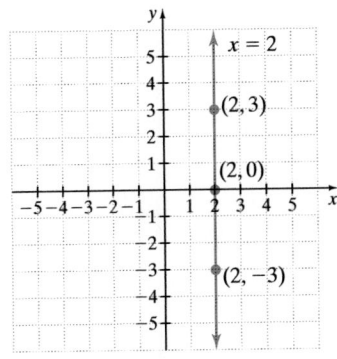

The graph is a vertical line with x-intercept $(2, 0)$. Note that this graph has no y-intercept because x is never 0.

Vertical Lines

The graph of $x = c$, where c is a real number, is a vertical line with x-intercept $(c, 0)$.

EXAMPLE 6

Graph $y = -3$.

Solution The equation $y = -3$ can be written as $0x + y = -3$. For any x-value chosen, y is -3. If we choose 4, 1, and -2 as x-values, the ordered pair solutions are $(4, -3)$, $(1, -3)$, and $(-2, -3)$. Use these ordered pairs to graph $y = -3$. The graph is a horizontal line with y-intercept $(0, -3)$ and no x-intercept.

x	y
4	-3
1	-3
-2	-3

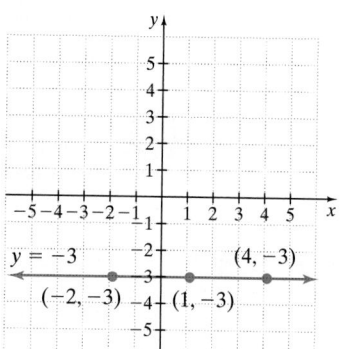

Horizontal Lines

The graph of $y = c$, where c is a real number, is a horizontal line with y-intercept $(0, c)$.

Graphing Calculator Explorations

You may have noticed that to use the $\boxed{Y=}$ key on a grapher to graph an equation, the equation must be solved for y. For example, to graph $2x + 3y = 7$, we solve this equation for y.

$$2x + 3y = 7$$
$$3y = -2x + 7 \qquad \text{Subtract } 2x \text{ from both sides.}$$
$$\frac{3y}{3} = -\frac{2x}{3} + \frac{7}{3} \qquad \text{Divide both sides by 3.}$$
$$y = -\frac{2}{3}x + \frac{7}{3} \qquad \text{Simplify.}$$

To graph $2x + 3y = 7$ or $y = -\frac{2}{3}x + \frac{7}{3}$, press the $\boxed{Y=}$ key and enter

$$Y_1 = -\frac{2}{3}x + \frac{7}{3}$$

Graph each linear equation.

1. $x = 3.78y$

2. $-2.61y = x$

3. $3x + 7y = 21$

4. $-4x + 6y = 12$

5. $-2.2x + 6.8y = 15.5$

6. $5.9x - 0.8y = -10.4$

MENTAL MATH

Answer the following true or false.

1. The graph of $x = 2$ is a horizontal line.

2. All lines have an x-intercept *and* a y-intercept.

3. The graph of $y = 4x$ contains the point $(0, 0)$.

4. The graph of $x + y = 5$ has an x-intercept of $(5, 0)$ and a y-intercept of $(0, 5)$.

5. The graph of $y = 5x$ contains the point $(5, 1)$.

6. The graph of $y = 5$ is a horizontal line.

EXERCISE SET 3.3

STUDY GUIDE/SSM CD/VIDEO PH MATH TUTOR CENTER MathXL®Tutorials ON CD MathXL® MyMathLab®

Identify the intercepts. See Example 1.

 1.

2.

3.

4.

5.

6.

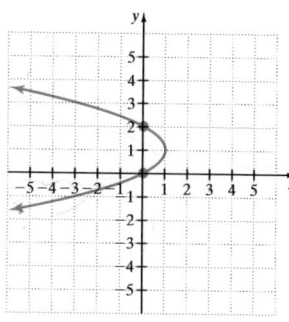

7. What is the greatest number of intercepts for a line?

8. What is the least number of intercepts for a line?

9. What is the least number of intercepts for a circle?

10. What is the greatest number of intercepts for a circle?

Graph each linear equation by finding x- and y-intercepts. See Examples 2 through 4.

11. $x - y = 3$

12. $x - y = -4$

13. $x = 5y$

14. $2x = y$

 15. $-x + 2y = 6$

16. $x - 2y = -8$

17. $2x - 4y = 8$

18. $2x + 3y = 6$

Graph each linear equation. See Examples 5 and 6.

19. $x = -1$

20. $y = 5$

21. $y = 0$

22. $x = 0$

23. $y + 7 = 0$

24. $x - 2 = 0$

MIXED PRACTICE

Graph each linear equation.

25. $x + 2y = 8$

26. $x - 3y = 3$

27. $x - 7 = 3y$

28. $y - 3x = 2$

29. $x = -3$

30. $y = 3$

31. $3x + 5y = 7$

32. $3x - 2y = 5$

33. $x = y$

34. $x = -y$

35. $x + 8y = 8$

36. $x - 3y = 9$

37. $5 = 6x - y$

38. $4 = x - 3y$

39. $-x + 10y = 11$

40. $-x + 9 = -y$

41. $y = 4.5$

42. $x = -\dfrac{1}{2}$

43. $y = \dfrac{1}{2}x$

44. $y = -2x$

45. $x + 4 = 0$

46. $y - 5.2 = 0$

47. $3x - 4y = -12$

48. $-2x + 5y = 10$

49. $2x + 3y = 6$

50. $4x + y = 5$

For Exercises 51 through 56, match each equation with its graph.

A.

B.

C.

D.

E.

F.

51. $y = 3$
53. $x = -1$
55. $y = 2x + 3$

52. $y = 2x + 2$
54. $x = 3$
56. $y = -2x$

REVIEW AND PREVIEW

Simplify.

57. $\dfrac{-6 - 3}{2 - 8}$

58. $\dfrac{4 - 5}{-1 - 0}$

59. $\dfrac{-8 - (-2)}{-3 - (-2)}$

60. $\dfrac{12 - 3}{10 - 9}$

61. $\dfrac{0 - 6}{5 - 0}$

62. $\dfrac{2 - 2}{3 - 5}$

Concept Extensions

63. The average price of a digital camera (y) can be modeled by the linear equation $y = -78.1x + 569.9$ where x represents the number of years after 1999. (*Source*: NPDTechworld)

   a. Find the y-intercept of this equation.

   b. What does this y-intercept mean?

64. Since the summer of 1999, the number of G- and PG-rated movies has been steadily increasing. The number of G and PG movies released each year can be estimated by the equation $y = 4x + 49$ where x represents the number of years after 1999. (*Source*: Nielsen EDI)

   a. Find the x-intercept of this equation.

   b. What does this x-intercept mean?

   c. Use part b to comment on the limitations of using equations modeling real data.

65. The number of music cassettes y (in millions) shipped to retailers in the United States from 1995 through 2000 can be modeled by the equation $y = -37.2x + 264.4$, where x represents the number of years after 1995. (*Source*: Recording Industry Association of America)

   a. Find the x-intercept of this equation. (Round to the nearest tenth.)

   b. What does this x-intercept mean?

   c. Use part b to comment on the limitations of using equations to model real data.

66. The number of Disney Stores y for the years 1996–2000 can be modeled by the equation $y = 51.6x + 560.2$, where x represents the number of years after 1996. (*Source*: The Walt Disney Company Fact Book 2000)

   a. Find the y-intercept of this equation.

   b. What does this y-intercept mean?

67. The production supervisor at Alexandra's Office Products finds that it takes 3 hours to manufacture a particular office chair and 6 hours to manufacture an office desk. A total of 1200 hours is available to produce office chairs and desks of this style. The linear equation that models this situation is $3x + 6y = 1200$, where x represents the number of chairs produced and y the number of desks manufactured.

   a. Complete the ordered pair solution $(0, \quad)$ of this equation. Describe the manufacturing situation that corresponds to this solution.

   b. Complete the ordered pair solution $(\quad, 0)$ of this equation. Describe the manufacturing situation that corresponds to this solution.

   c. Use the ordered pairs found above and graph the equation $3x + 6y = 1200$.

   d. If 50 desks are manufactured, find the greatest number of chairs that they can make.

Two lines in the same plane that do not intersect are called **parallel lines.**

68. Draw a line parallel to the line $x = 5$ that intersects the x-axis at $(1, 0)$. What is the equation of this line?

69. Draw a line parallel to the line $y = -1$ that intersects the y-axis at $(0, -4)$. What is the equation of this line?

70. Discuss whether a vertical line ever has a y-intercept.

71. Explain why it is a good idea to use three points to graph a linear equation.

72. Discuss whether a horizontal line ever has an x-intercept.

73. Explain how to find intercepts.

3.4   SLOPE AND RATE OF CHANGE

Objectives

1. Find the slope of a line given two points of the line.
2. Find the slopes of horizontal and vertical lines.
3. Compare the slopes of parallel and perpendicular lines.
4. Solve applications of slope.

1   Thus far, much of this chapter has been devoted to graphing lines. You have probably noticed by now that a key feature of a line is its slant or steepness. In mathematics, the slant or steepness of a line is formally known as its **slope**. We measure the slope of a line by the ratio of vertical change to the corresponding horizontal change as we move along the line.

On the line below, for example, suppose that we begin at the point $(1, 2)$ and move to the point $(4, 6)$. The vertical change is the change in y-coordinates: $6 - 2$ or 4 units. The corresponding horizontal change is the change in x-coordinates: $4 - 1 = 3$ units. The ratio of these changes is

$$\text{slope} = \frac{\text{change in } y \text{ (vertical change)}}{\text{change in } x \text{ (horizontal change)}} = \frac{4}{3}$$

The slope of this line, then, is $\frac{4}{3}$: for every 4 units of change in y-coordinates, there is a corresponding change of 3 units in x-coordinates.

> ▶ **Helpful Hint**
>
> It makes no difference what two points of a line are chosen to find its slope. The slope of a line is the same everywhere on the line.

To find the slope of a line, then, choose two points of the line. Label the two x-coordinates of two points, x_1 and x_2 (read "x sub one" and "x sub two"), and label the corresponding y-coordinates y_1 and y_2.

The vertical change or **rise** between these points is the difference in the y-coordinates: $y_2 - y_1$. The horizontal change or **run** between the points is the difference of the x-coordinates: $x_2 - x_1$. The slope of the line is the ratio of $y_2 - y_1$ to $x_2 - x_1$, and we traditionally use the letter m to denote slope $m = \dfrac{y_2 - y_1}{x_2 - x_1}$.

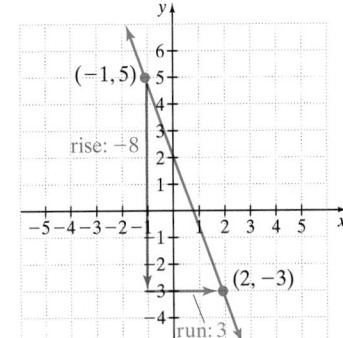

Slope of a Line

The slope m of the line containing the points (x_1, y_1) and (x_2, y_2) is given by

$$m = \frac{\text{rise}}{\text{run}} = \frac{\text{change in } y}{\text{change in } x} = \frac{y_2 - y_1}{x_2 - x_1}, \qquad \text{as long as } x_2 \neq x_1$$

EXAMPLE 1

Find the slope of the line through $(-1, 5)$ and $(2, -3)$. Graph the line.

Solution   If we let (x_1, y_1) be $(-1, 5)$, then $x_1 = -1$ and $y_1 = 5$. Also, let (x_2, y_2) be $(2, -3)$ so that $x_2 = 2$ and $y_2 = -3$. Then, by the definition of slope,

$$m = \frac{y_2 - y_1}{x_2 - x_1}$$

$$= \frac{-3 - 5}{2 - (-1)}$$

$$= \frac{-8}{3} = -\frac{8}{3}$$

The slope of the line is $-\dfrac{8}{3}$.

✔ CONCEPT CHECK

The points $(-2, -5)$, $(0, -2)$, $(4, 4)$, and $(10, 13)$ all lie on the same line. Work with a partner and verify that the slope is the same no matter which points are used to find slope.

> ### Helpful Hint
>
> When finding slope, it makes no difference which point is identified as (x_1, y_1) and which is identified as (x_2, y_2). Just remember that whatever y-value is first in the numerator, its corresponding x-value is first in the denominator. Another way to calculate the slope in Example 1 is:
>
> $$m = \frac{y_2 - y_1}{x_2 - x_1} = \frac{5 - (-3)}{-1 - 2} = \frac{8}{-3} \quad \text{or} \quad -\frac{8}{3} \leftarrow \text{Same slope as found in Example 1.}$$

Concept Check Answer:
$m = \dfrac{3}{2}$

EXAMPLE 2

Find the slope of the line through $(-1, -2)$ and $(2, 4)$. Graph the line.

Solution

y-value
$$m = \frac{-2 - 4}{-1 - 2} = \frac{-6}{-3} = 2$$
corresponding *x*-value

The slope is the same if we begin with the other *y*-value.

y-value
$$m = \frac{4 - (-2)}{2 - (-1)} = \frac{6}{3} = 2$$
corresponding *x*-value

The slope is 2.

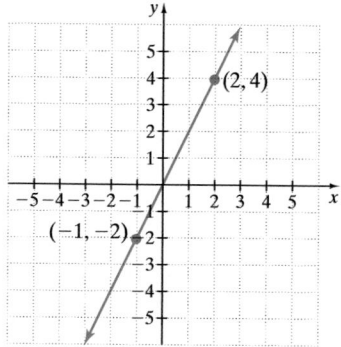

✔ CONCEPT CHECK

What is wrong with the following slope calculation for the points $(3, 5)$ and $(-2, 6)$?

$$m = \frac{5 - 6}{-2 - 3} = \frac{-1}{-5} = \frac{1}{5}$$

Notice that the slope of the line in Example 1 is negative, whereas the slope of the line in Example 2 is positive. Let your eye follow the line with negative slope from left to right and notice that the line "goes down." Following the line with positive slope from left to right, notice that the line "goes up." This is true in general.

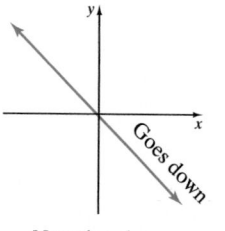

Negative slope

Positive slope

> **Helpful Hint**
>
> To decide whether a line "goes up" or "goes down," always follow the line from left to right.

Concept Check Answer:
The order in which the *x*- and *y*-values are used must be the same.
$$m = \frac{5 - 6}{3 - (-2)} = \frac{-1}{5} = -\frac{1}{5}$$

2 If a line tilts upward from left to right, its slope is positive. If a line tilts downward from left to right, its slope is negative. Let's now find the slopes of two special lines, horizontal and vertical lines.

EXAMPLE 3

Find the slope of the line $y = -1$.

Solution Recall that $y = -1$ is a horizontal line with y-intercept $(0, -1)$. To find the slope, find two ordered pair solutions of $y = -1$. Solutions of $y = -1$ must have a y-value of -1. Let's use points $(2, -1)$ and $(-3, -1)$, which are on the line.

$$m = \frac{y_2 - y_1}{x_2 - x_1} = \frac{-1 - (-1)}{-3 - 2} = \frac{0}{-5} = 0$$

The slope of the line $y = -1$ is 0. The graph of $y = -1$ is given below.

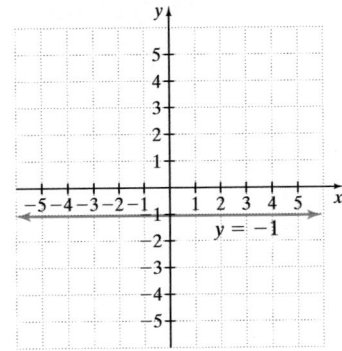

Any two points of a horizontal line will have the same y-values. This means that the y-values will always have a difference of 0 for all horizontal lines. Thus, **all horizontal lines have a slope 0**.

EXAMPLE 4

Find the slope of the line $x = 5$.

Solution Recall that the graph of $x = 5$ is a vertical line with x-intercept $(5, 0)$.

To find the slope, find two ordered pair solutions of $x = 5$. Solutions of $x = 5$ must have an x-value of 5. Let's use points $(5, 0)$ and $(5, 4)$, which are on the line.

$$m = \frac{y_2 - y_1}{x_2 - x_1} = \frac{4 - 0}{5 - 5} = \frac{4}{0}$$

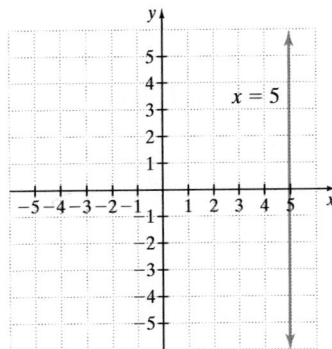

Since $\frac{4}{0}$ is undefined, we say the slope of the vertical line $x = 5$ is undefined.

Any two points of a vertical line will have the same x-values. This means that the x-values will always have a difference of 0 for all vertical lines. Thus **all vertical lines have undefined slope**.

> ### Helpful Hint
> Slope of 0 and undefined slope are not the same. Vertical lines have undefined slope or no slope, while horizontal lines have a slope of 0.

Here is a general review of slope.

Summary of Slope

Slope m of the line through (x_1, y_1) and (x_2, y_2) is given by the equation

$$m = \frac{y_2 - y_1}{x_2 - x_1}.$$

Upward line

Downward line

Positive slope: $m > 0$

Negative slope: $m < 0$

Horizontal line
$y = c$

Zero slope: $m = 0$

Vertical line
$x = c$

Undefined slope or no slope

3   Two lines in the same plane are **parallel** if they do not intersect. Slopes of lines can help us determine whether lines are parallel. Parallel lines have the same steepness, so it follows that they have the same slope.

Parallel Lines

Nonvertical parallel lines have the same slope.

How do the slopes of perpendicular lines compare? Two lines that intersect at right angles are said to be **perpendicular**. The product of the slopes of two perpendicular lines is -1.

Perpendicular Lines

If the product of the slopes of two lines is -1, the lines are perpendicular.

(Two nonvertical lines are perpendicular if the slope of one is the negative reciprocal of the slope of the other.)

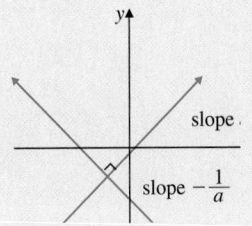

Helpful Hint

Here are examples of numbers that are negative (opposite) reciprocals.

Number	Negative Reciprocal	Their product is -1.
$\dfrac{2}{3}$	$-\dfrac{3}{2}$	$\dfrac{2}{3} \cdot -\dfrac{3}{2} = -\dfrac{6}{6} = -1$
-5 or $-\dfrac{5}{1}$	$\dfrac{1}{5}$	$-5 \cdot \dfrac{1}{5} = -\dfrac{5}{5} = -1$

Helpful Hint

Here are a few important facts about vertical and horizontal lines.

- Two distinct vertical lines are parallel.
- Two distinct horizontal lines are parallel.
- A horizontal line and a vertical line are always perpendicular.

EXAMPLE 5

Is the line passing through the points $(-6, 0)$ and $(-2, 3)$ parallel to the line passing through the points $(5, 4)$ and $(9, 7)$?

Solution   To see if these lines are parallel, we find and compare slopes. The line passing through the points $(-6, 0)$ and $(-2, 3)$ has slope

$$m = \frac{3 - 0}{-2 - (-6)} = \frac{3}{4}$$

The line passing through the points $(5, 4)$ and $(9, 7)$ has slope

$$m = \frac{7 - 4}{9 - 5} = \frac{3}{4}$$

Since the slopes are the same, these lines are parallel.

EXAMPLE 6

Find the slope of a line perpendicular to the line passing through the points $(-1, 7)$ and $(2, 2)$.

Solution   First, let's find the slope of the line through $(-1, 7)$ and $(2, 2)$. This line has slope

$$m = \frac{2 - 7}{2 - (-1)} = \frac{-5}{3}$$

The slope of every line perpendicular to the given line has a slope equal to the negative reciprocal of

$$-\frac{5}{3} \quad \text{or} \quad -\left(-\frac{3}{5}\right) = \frac{3}{5}$$

↑   ↑
negative   reciprocal

The slope of a line perpendicular to the given line has slope $\frac{3}{5}$.

4   Slope can also be interpreted as a rate of change. In other words, slope tells us how fast y is changing with respect to x. To see this, let's look at a few of the many real-world applications of slope. For example, the pitch of a roof, used by builders and architects, is its slope. The pitch of the roof on the left is $\frac{7}{10}$ $\left(\frac{\text{rise}}{\text{run}}\right)$. This means that the roof rises vertically 7 feet for every horizontal 10 feet. The rate of change for the roof is 7 vertical feet (y) per 10 horizontal feet (x).

The grade of a road is its slope written as a percent. A 7% grade, as shown below, means that the road rises (or falls) 7 feet for every horizontal 100 feet. (Recall that $7\% = \frac{7}{100}$.) Here, the slope of $\frac{7}{100}$ gives us the rate of change. The road rises (in our diagram) 7 vertical feet (y) for every 100 horizontal feet (x).

$\frac{7}{10}$ pitch

$\frac{7}{100} = 7\%$ grade

7 feet

100 feet

EXAMPLE 7

FINDING THE GRADE OF A ROAD

At one part of the road to the summit of Pikes Peak, the road rises at a rate of 15 vertical feet for a horizontal distance of 250 feet. Find the grade of the road.

Solution   Recall that the grade of a road is its slope written as a percent.

$$\text{grade} = \frac{\text{rise}}{\text{run}} = \frac{15}{250} = 0.06 = 6\%$$

15 feet

250 feet

The grade is 6%.

EXAMPLE 8

FINDING THE SLOPE OF A LINE

The following graph shows the cost y (in cents) of an in-state long-distance telephone call in Massachusetts where x is the length of the call in minutes. Find the slope of the line and attach the proper units for the rate of change.

Solution   Use $(2, 48)$ and $(5, 81)$ to calculate slope.

$$m = \frac{81 - 48}{5 - 2} = \frac{33}{3} = \frac{11}{1} \frac{\text{cents}}{\text{minute}}$$

This means that the rate of change of a phone call is 11 cents per 1 minute or the cost of the phone call increases 11 cents per minute.

Spotlight on
DECISION
MAKING

Suppose you own a house that you are trying to sell. You want a real estate agency to handle the sale of your house. You begin by scanning your local newspaper to find likely candidates, and spot the two ads shown.

Which real estate agency would you want to check into first? Explain. What other factors would you want to consider?

Graphing Calculator Explorations

It is possible to use a grapher to sketch the graph of more than one equation on the same set of axes. This feature can be used to confirm our findings from Section 3.2 when we learned that the graph of an equation written in the form $y = mx + b$ has a y-intercept of b. For example, graph the equations $y = \frac{2}{5}x$, $y = \frac{2}{5}x + 7$, and $y = \frac{2}{5}x - 4$ on the same set of axes. To do so, press the $\boxed{Y=}$ key and enter the equations on the first three lines.

$$Y_1 = \left(\frac{2}{5}\right)x$$

$$Y_2 = \left(\frac{2}{5}\right)x + 7$$

$$Y_3 = \left(\frac{2}{5}\right)x - 4$$

The screen should look like:

Notice that all three graphs appear to have the same positive slope. The graph of $y = \frac{2}{5}x + 7$ is the graph of $y = \frac{2}{5}x$ moved 7 units upward with a y-intercept of 7. Also, the graph of $y = \frac{2}{5}x - 4$ is the graph of $y = \frac{2}{5}x$ moved 4 units downward with a y-intercept of -4.

Graph the equations on the same set of axes. Describe the similarities and differences in their graphs.

1. $y = 3.8x$, $y = 3.8x - 3$, $y = 3.8x + 9$

2. $y = -4.9x$, $y = -4.9x + 1$, $y = -4.9x + 8$

3. $y = \frac{1}{4}x$; $y = \frac{1}{4}x + 5$, $y = \frac{1}{4}x - 8$

4. $y = -\frac{3}{4}x$, $y = -\frac{3}{4}x - 5$, $y = -\frac{3}{4}x + 6$

MENTAL MATH

Decide whether a line with the given slope is upward, downward, horizontal, or vertical.

1. $m = \frac{7}{6}$ **2.** $m = -3$ **3.** $m = 0$ **4.** m is undefined.

EXERCISE SET 3.4

STUDY   CD/   PH MATH   MathXL®Tutorials   MathXL®   MyMathLab®
GUIDE/SSM   VIDEO   TUTOR CENTER   ON CD

Find the slope of each line if it exists. See Example 1.

1.

2.

3.

4.

5.

6.

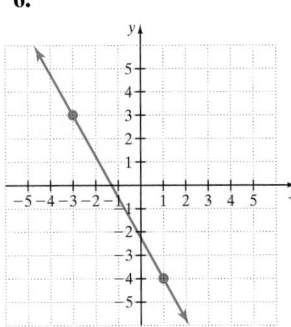

Find the slope of the line that goes through the given points. See Examples 1 and 2.

7. $(0,0)$ and $(7,8)$

8. $(-1,5)$ and $(0,0)$

 9. $(-1,5)$ and $(6,-2)$

10. $(-1,9)$ and $(-3,4)$

11. $(1,4)$ and $(5,3)$

12. $(3,1)$ and $(2,6)$

13. $(-4,3)$ and $(-4,5)$

14. $(6,-6)$ and $(6,2)$

15. $(-2,8)$ and $(1,6)$

16. $(4,-3)$ and $(2,2)$

17. $(1,0)$ and $(1,1)$

18. $(0,13)$ and $(-4,13)$

19. $(5,1)$ and $(-2,1)$

20. $(5,4)$ and $(0,5)$

For each graph, determine which line has the greater slope.

21.

22.

23.

24.

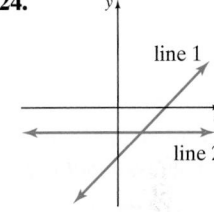

In Exercises 25–30, match each line with its slope.

A. $m = 0$ **B.** undefined slope   **C.** $m = 3$

D. $m = 1$ **E.** $m = -\dfrac{1}{2}$ **F.** $m = -\dfrac{3}{4}$

25.

26.

27.

28.

29.

30.

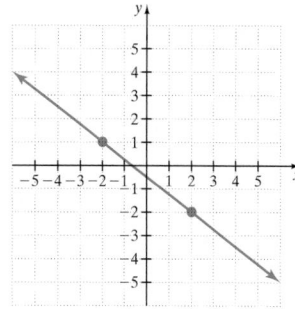

Find the slope of each line. See Examples 3 and 4.

31. $x = 6$

32. $y = 4$

33. $y = -4$

34. $x = 2$

35. $x = -3$

36. $x = -11$

37. $y = 0$

38. $x = 0$

△ *Find the slope of the line **a.** parallel and **b.** perpendicular to the line through each pair of points. See Examples 5 and 6.*

39. $(-3, -3)$ and $(0, 0)$

40. $(6, -2)$ and $(1, 4)$

41. $(-8, -4)$ and $(3, 5)$

42. $(6, -1)$ and $(-4, -10)$

△ *Determine whether the lines through each pair of points are parallel, perpendicular, or neither. See Examples 5 and 6.*

43. $(0, 6)$ and $(-2, 0)$
$(0, 5)$ and $(1, 8)$

44. $(1, -1)$ and $(-1, -11)$
$(-2, -8)$ and $(0, 2)$

45. $(2, 6)$ and $(-2, 8)$
$(0, 3)$ and $(1, 5)$

46. $(-1, 7)$ and $(1, 10)$
$(0, 3)$ and $(1, 5)$

47. $(3, 6)$ and $(7, 8)$
$(0, 6)$ and $(2, 7)$

48. $(4, 2)$ and $(6, 6)$
$(0, -2)$ and $(1, 0)$

49. $(2, -3)$ and $(6, -5)$
$(5, -2)$ and $(-3, -4)$

50. $(-1, -5)$ and $(4, 4)$
$(10, 8)$ and $(-7, 4)$

51. $(-4, -3)$ and $(-1, 0)$
$(4, -4)$ and $(0, 0)$

52. $(2, -2)$ and $(4, -8)$
$(0, 7)$ and $(3, 8)$

△ *Solve. See Examples 5 and 6.*

53. Find the slope of a line parallel to the line passing through $(-7, -5)$ and $(-2, -6)$.

54. Find the slope of a line parallel to the line passing through the origin and $(-2, 10)$.

55. Find the slope of a line perpendicular to the line passing through the origin and $(1, -3)$.

56. Find the slope of the line perpendicular to the line passing through $(-1, 3)$ and $(2, -8)$.

57. Find the slope of a line parallel to the line passing through $(3, 3)$ and $(-3, -3)$.

58. Find the slope of a line parallel to the line passing through $(0, 3)$ and $(6, 0)$.

The pitch of a roof is its slope. Find the pitch of each roof shown. See Example 7.

59.

60.

The grade of a road is its slope written as a percent. Find the grade of each road shown. See Example 7.

61.

62.

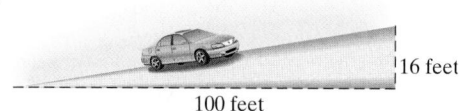

63. One of Japan's superconducting "bullet" trains is researched and tested at the Yamanashi Maglev Test Line near Otsuki City. The steepest section of the track has a rise of 2580 meters for a horizontal distance of 6450 meters. What is the grade for this section of track? (*Source:* Japan Railways Central Co.)

64. The steepest street is Baldwin Street in Dunedin, New Zealand. It has a maximum rise of 10 meters for a horizontal distance of 12.66 meters. Find the grade for this section of road. Round to the nearest whole percent. (*Source: The Guinness Book of Records*)

65. Professional plumbers suggest that a sewer pipe should rise 0.25 inch for every horizontal foot. Find the recommended slope for a sewer pipe. Round to the nearest hundredth.

0.25 inch

12 inches

66. According to federal regulations, a wheel chair ramp should rise no more than 1 foot for a horizontal distance of 12 feet. Write the slope as a grade. Round to the nearest tenth of a percent.

Find the slope of each line and write the slope as a rate of change. Don't forget to attach the proper units. See Example 8.

67. This graph shows a projection of the increase of cell phone users *y* (in millions) for year *x*.

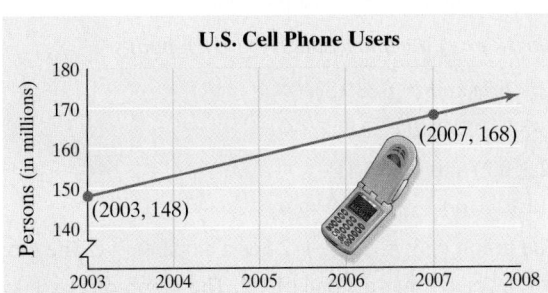

U.S. Cell Phone Users

(2007, 168)

(2003, 148)

Persons (in millions)

2003   2004   2005   2006   2007   2008

68. This graph approximates the total number of cosmetic surgeons for year *x*.

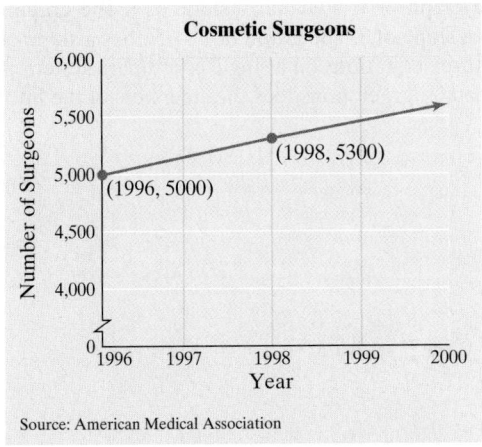

Cosmetic Surgeons

(1998, 5300)

(1996, 5000)

Number of Surgeons

1996   1997   1998   1999   2000

Year

Source: American Medical Association

69. The graph below shows the total cost *y* (in dollars) of owning and operating a compact car where *x* is the number of miles driven.

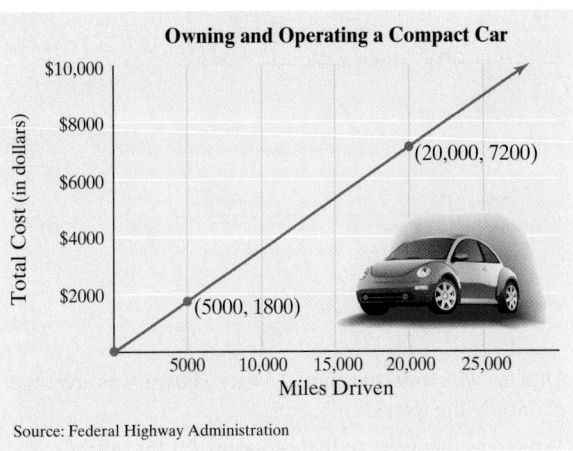

Owning and Operating a Compact Car

Total Cost (in dollars)

(20,000, 7200)

(5000, 1800)

5000   10,000   15,000   20,000   25,000

Miles Driven

Source: Federal Highway Administration

70. The graph below shows the total cost *y* (in dollars) of owning and operating a full-size pickup truck, where *x* is the number of miles driven.

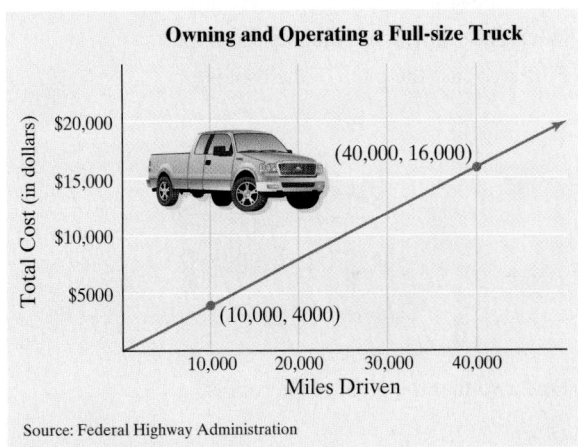

Owning and Operating a Full-size Truck

Total Cost (in dollars)

(40,000, 16,000)

(10,000, 4000)

10,000   20,000   30,000   40,000

Miles Driven

Source: Federal Highway Administration

REVIEW AND PREVIEW

Solve each equation for y. See Section 2.6.

71. $x + y = 10$

72. $-x + y = -7$

73. $x + 2y = -12$

74. $x - 3y = 15$

75. $5x - y = 17$

76. $4x - y = 20$

Concept Extensions

The line graph on the next page shows the average fuel economy (in miles per gallon) by passenger automobiles produced during each of the model years shown. Use this graph to answer Exercises 77 through 82.

77. What was the average fuel economy (in miles per gallon) for automobiles produced during 1995?

78. Find the decrease in average fuel economy for automobiles for the years 1991 to 1992.

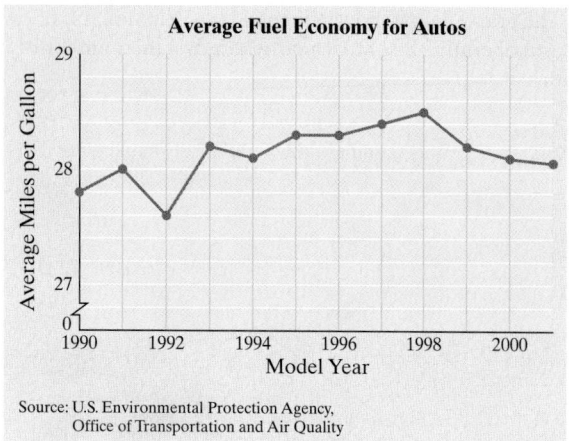

Average Fuel Economy for Autos

Source: U.S. Environmental Protection Agency,
Office of Transportation and Air Quality

79. During which of the model years shown was average fuel economy the lowest?

What was the average fuel economy for that year?

80. During which of the model years shown was average fuel economy the highest?

What was the average fuel economy for that year?

81. What line segment has the greatest slope?

82. What line segment has the smallest slope?

83. Find x so that the pitch of the roof is $\frac{1}{3}$.

84. Find x so that the pitch of the roof is $\frac{2}{5}$.

85. The average price of an acre of U.S. farmland was $782 in 1994. In 2001, the price of an acre rose to approximately $1132. (*Source:* National Agricultural Statistics Service)

  a. Write two ordered pairs of the form (year, price of acre).
  b. Find the slope of the line through the two points.
  c. Write a sentence explaining the meaning of the slope as a rate of change.

86. There were approximately 13,290 kidney transplants performed in the United States in 2000. In 2002, the number of kidney transplants performed in the United States rose to 14,774. (*Source:* Organ Procurement and Transplantation Network)

  a. Write two ordered pairs of the form (year, number of kidney transplants).
  b. Find the slope of the line between the two points.
  c. Write a sentence explaining the meaning of the slope as a rate of change.

△ **87.** Show that a triangle with vertices at the points $(1, 1)$, $(-4, 4)$, and $(-3, 0)$ is a right triangle.

△ **88.** Show that the quadrilateral with vertices $(1, 3)$, $(2, 1)$, $(-4, 0)$, and $(-3, -2)$ is a parallelogram.

Find the slope of the line through the given points.

89. $(2.1, 6.7)$ and $(-8.3, 9.3)$

90. $(-3.8, 1.2)$ and $(-2.2, 4.5)$

91. $(2.3, 0.2)$ and $(7.9, 5.1)$

92. $(14.3, -10.1)$ and $(9.8, -2.9)$

93. The graph of $y = -\frac{1}{3}x + 2$ has a slope of $-\frac{1}{3}$. The graph of $y = -2x + 2$ has a slope of -2. The graph of $y = -4x + 2$ has a slope of -4. Graph all three equations on a single coordinate system. As the absolute value of the slope becomes larger, how does the steepness of the line change?

94. The graph of $y = \frac{1}{2}x$ has a slope of $\frac{1}{2}$. The graph of $y = 3x$ has a slope of 3. The graph of $y = 5x$ has a slope of 5. Graph all three equations on a single coordinate system. As slope becomes larger, how does the steepness of the line change?

INTEGRATED REVIEW   SUMMARY ON SLOPE AND GRAPHING LINEAR EQUATIONS

Find the slope of each line.

1.

2.

3.

4.

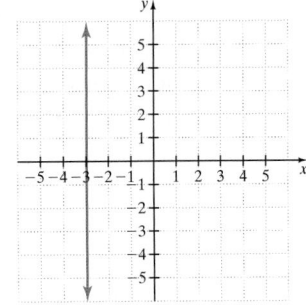

Graph each linear equation.

5. $y = -2x$ **6.** $x + y = 3$

7. $x = -1$ **8.** $y = 4$

9. $x - 2y = 6$ **10.** $y = 3x + 2$

11. $5x + 3y = 15$ **12.** $2x - 4y = 8$

Determine whether the lines through the points are parallel, perpendicular, or neither.

13. $(-1, 3)$ and $(1, -3)$ **14.** $(-6, -6)$ and $(-1, -2)$
$(2, -1)$ and $(4, -7)$ $(-4, 3)$ and $(3, -3)$

15. In the years 2000 through 2002 (the number of admissions to movie theaters in the United States can be modeled by the linear equation $y = 110x + 1407$ where x is years after 2000 and y is admissions in millions. (*Source:* Motion Picture Assn. of America)

a. Find the y-intercept of this line.

b. Write a sentence explaining the meaning of this intercept.

c. Find the slope of this line.

d. Write a sentence explaining the meaning of the slope as a rate of change.

3.5 THE SLOPE-INTERCEPT FORM

Objectives

△ **1** Use the slope-intercept form to find the slope and the y-intercept of a line.

△ **2** Use the slope-intercept form to determine whether two lines are parallel, perpendicular, or neither.

3 Use the slope-intercept form to write an equation of a line.

4 Use the slope-intercept form to graph a linear equation.

1 In Section 3.4, we learned that the slant of a line is called its slope. Recall that the slope of a line m is the ratio of vertical change (change in y) to the corresponding horizontal change (change in x) as we move along the line.

> **Helpful Hint**
> Don't forget that the slope of a line is the same no matter which ordered pairs of the line are used to calculate slope.

EXAMPLE 1

Find the slope of the line whose equation is $y = \frac{3}{4}x + 6$.

Solution To find the slope of this line, find any two ordered pair solutions of the equation. Let's find and use intercept points as our two ordered pair solutions. Recall from Section 3.2 that the graph of an equation of the form $y = mx + b$ has y-intercept $(0, b)$.

This means that the graph of $y = \frac{3}{4}x + 6$ has a y-intercept of $(0, 6)$ as shown below.

To find the y-intercept, let $x = 0$.

Then $y = \frac{3}{4}x + 6$ becomes

$$y = \frac{3}{4} \cdot 0 + 6 \text{ or}$$

$$y = 6$$

To find the x-intercept, let $y = 0$.

Then $y = \frac{3}{4}x + 6$ becomes

$$0 = \frac{3}{4}x + 6 \text{ or}$$

$$0 = 3x + 24 \quad \text{Multiply both sides by 4.}$$

$$-24 = 3x \quad \text{Subtract 24 from both sides.}$$

$$-8 = x \quad \text{Divide both sides by 3.}$$

The y-intercept is $(0, 6)$, as expected. The x-intercept is $(-8, 0)$.

Use the points $(0, 6)$ and $(-8, 0)$ to find the slope. Then

$$m = \frac{\text{change in } y}{\text{change in } x} = \frac{0 - 6}{-8 - 0} = \frac{-6}{-8} = \frac{3}{4}$$

The slope of the line is $\frac{3}{4}$.

Analyzing the results of Example 1, you may notice a striking pattern:

The slope of $y = \frac{3}{4}x + 6$ is $\frac{3}{4}$, the same as the coefficient of x.

Also, as mentioned earlier, the y-intercept is $(0, 6)$. Notice that the y-value 6 is the same as the constant term.

When a linear equation is written in the form $y = mx + b$, not only is $(0, b)$ the y-intercept of the line, but m is its slope. The form $y = mx + b$ is appropriately called the **slope-intercept form**.

$$\uparrow \quad \uparrow$$
$$\text{slope} \quad y\text{-intercept}$$
$$(\mathbf{0, b})$$

Slope-Intercept Form

When a linear equation in two variables is written in slope-intercept form,

$$y = mx + b$$

m is the slope of the line and $(0, b)$ is the y-intercept of the line.

EXAMPLE 2

Find the slope and the y-intercept of the line whose equation is $5x + y = 2$.

Solution   Write the equation in slope-intercept form by solving the equation for y.

$$5x + y = 2$$
$$y = -5x + 2 \quad \text{Subtract } 5x \text{ from both sides.}$$

The coefficient of x, -5, is the slope and the constant term, 2, is the y-value of the y-intercept, $(0, 2)$.

EXAMPLE 3

Find the slope and the y-intercept of the line whose equation is $3x - 4y = 4$.

Solution   Write the equation in slope-intercept form by solving for y.

$$3x - 4y = 4$$
$$-4y = -3x + 4 \quad \text{Subtract } 3x \text{ from both sides.}$$
$$\frac{-4y}{-4} = \frac{-3x}{-4} + \frac{4}{-4} \quad \text{Divide both sides by } -4.$$
$$y = \frac{3}{4}x - 1 \quad \text{Simplify.}$$

The coefficient of x, $\frac{3}{4}$, is the slope, and the y-intercept is $(0, -1)$.

2   The slope-intercept form can be used to determine whether two lines are parallel or perpendicular. Recall that nonvertical parallel lines have the same slope and different y-intercepts. Also, nonvertical perpendicular lines have slopes whose product is -1.

EXAMPLE 4

Determine whether the graphs of $y = -\frac{1}{5}x + 1$ and $2x + 10y = 30$ are parallel lines, perpendicular lines, or neither.

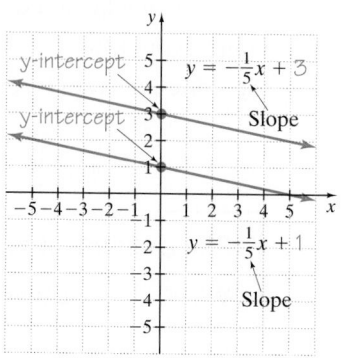

Solution   The graph of $y = -\frac{1}{5}x + 1$ is a line with slope $-\frac{1}{5}$ and with y-intercept $(0, 1)$. To find the slope and the y-intercept of the graph of $2x + 10y = 30$, write this equation in slope-intercept form. To do this, solve the equation for y.

$$2x + 10y = 30$$

$$10y = -2x + 30 \qquad \text{Subtract } 2x \text{ from both sides.}$$

$$y = -\frac{1}{5}x + 3 \qquad \text{Divide both sides by } 10.$$

The graph of this equation is a line with slope $-\frac{1}{5}$ and y-intercept $(0, 3)$. Because both lines have the *same* slope but different y-intercepts, these lines are parallel.

✔ **CONCEPT CHECK**

Write the equations of any three parallel lines.

3   The slope-intercept form can also be used to write the equation of a line given its slope and y-intercept.

EXAMPLE 5

Find an equation of the line with y-intercept $(0, -3)$ and slope of $\frac{1}{4}$.

Solution   We are given the slope and the y-intercept. Let $m = \frac{1}{4}$ and $b = -3$, and write the equation in slope-intercept form, $y = mx + b$.

Concept Check Answer:
For example,
$y = 2x - 3, y = 2x - 1, y = 2x$

$$y = mx + b$$

$$y = \frac{1}{4}x + (-3) \qquad \text{Let } m = \frac{1}{4} \text{ and } b = -3.$$

$$y = \frac{1}{4}x - 3 \qquad \text{Simplify.}$$

4   We now use the slope-intercept form of the equation of a line to graph a linear equation.

EXAMPLE 6

Graph: $y = \frac{3}{5}x - 2$.

Solution   Since the equation $y = \frac{3}{5}x - 2$ is written in slope-intercept form $y = mx + b$, the slope of its graph is $\frac{3}{5}$ and the y-intercept is $(0, -2)$. To graph, begin by plotting the intercept $(0, -2)$. From this point, find another point of the graph by using the slope $\frac{3}{5}$ and recalling that slope is $\frac{\text{rise}}{\text{run}}$. Start at the intercept point and move 3 units up since the numerator of the slope is 3; then move 5 units to the right since the denominator of the slope is 5. Stop at the point $(5, 1)$. The line through $(0, -2)$ and $(5, 1)$ is the graph of $y = \frac{3}{5}x - 2$.

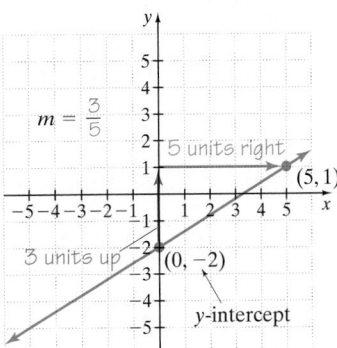

EXAMPLE 7

Use the slope-intercept form to graph the equation $4x + y = 1$.

Solution   First, write the given equation in slope-intercept form.

$$4x + y = 1$$
$$y = -4x + 1$$

The graph of this equation will have slope -4 and y-intercept $(0, 1)$. To graph this line, first plot the intercept point $(0, 1)$. To find another point of the graph, use the slope -4, which can be written as $\frac{-4}{1}$. Start at the point $(0, 1)$ and move 4 units down (since the numerator of the slope is -4); then move 1 unit to the right (since the denominator of the slope is 1). We arrive at the point $(1, -3)$. The line through $(0, 1)$ and $(1, -3)$ is the graph of $4x + y = 1$.

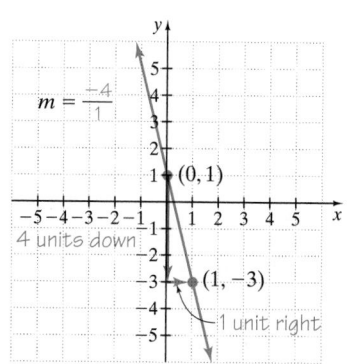

Helpful Hint

In Example 7, if we interpret the slope of -4 as $\dfrac{4}{-1}$, we arrive at $(-1, 5)$ for a second point. Notice that this point is also on the line.

Graphing Calculator Explorations

A grapher is a very useful tool for discovering patterns. To discover the change in the graph of a linear equation caused by a change in slope, try the following. Use a standard window and graph a linear equation in the form $y = mx + b$. Recall that the graph of such an equation will have slope m and y-intercept b.

First graph $y = x + 3$. To do so, press the $\boxed{Y=}$ key and enter $Y_1 = x + 3$. Notice that this graph has slope 1 and that the y-intercept is 3. Next, on the same set of axes, graph $y = 2x + 3$ and $y = 3x + 3$ by pressing $\boxed{Y=}$ and entering $Y_2 = 2x + 3$ and $Y_3 = 3x + 3$.

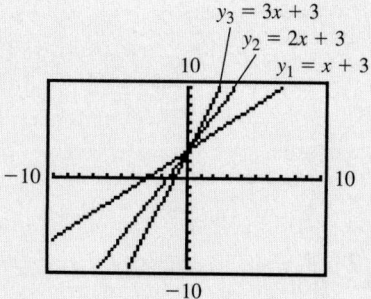

Notice the difference in the graph of each equation as the slope changes from 1 to 2 to 3. How would the graph of $y = 5x + 3$ appear? To see the change in the graph caused by a change in negative slope, try graphing $y = -x + 3$, $y = -2x + 3$, and $y = -3x + 3$ on the same set of axes.

Use a grapher to graph the following equations. For each exercise, graph the first equation and use its graph to predict the appearance of the other equations. Then graph the other equations on the same set of axes and check your prediction.

1. $y = x$; $y = 6x$, $y = -6x$

2. $y = -x$; $y = -5x$, $y = -10x$

3. $y = \dfrac{1}{2}x + 2$; $y = \dfrac{3}{4}x + 2$, $y = x + 2$

4. $y = x + 1$; $y = \dfrac{5}{4}x + 1$, $y = \dfrac{5}{2}x - 1$

5. $y = -7x + 5$; $y = 7x + 5$

6. $y = 3x - 1$; $y = -3x - 1$

STUDY SKILLS REMINDER

Tips for Studying for an Exam

To prepare for an exam, try the following study techniques.

▶ Start the study process days before your exam.

▶ Make sure that you are current and up-to-date on your assignments.

▶ If there is a topic that you are unsure of, use one of the many resources that are available to you. For example,

   See your instructor.

   Visit a learning resource center on campus where math tutors are available.

   Read the textbook material and examples on the topic.

   View a videotape on the topic.

▶ Reread your notes and carefully review the Chapter Highlights at the end of the chapter.

▶ Work the review exercises at the end of the chapter and check your answers. Make sure that you correct any missed exercises. If you have trouble on a topic, use a resource listed above.

▶ Find a quiet place to take the Chapter Test found at the end of the chapter. Do not use any resources when taking this sample test. This way you will have a clear indication of how prepared you are for your exam. Check your answers and make sure that you correct any missed exercises.

▶ Get lots of rest the night before the exam. It's hard to show how well you know the material if your brain is foggy from lack of sleep.

Good luck and keep a positive attitude.

MENTAL MATH

Identify the slope and the *y*-intercept of the graph of each equation.

1. $y = 2x - 1$

2. $y = -7x + 3$

3. $y = x + \dfrac{1}{3}$

4. $y = -x - \dfrac{2}{9}$

5. $y = \dfrac{5}{7}x - 4$

6. $y = -\dfrac{1}{4}x + \dfrac{3}{5}$

EXERCISE SET 3.5

STUDY   CD/   PH MATH   MathXL®Tutorials   MathXL®   MyMathLab®
GUIDE/SSM   VIDEO   TUTOR CENTER   ON CD

Determine the slope and the y-intercept of the graph of each equation. See Examples 1 through 3.

1. $2x + y = 4$

2. $-3x + y = -5$

 3. $x + 9y = 1$

4. $x + 7y = 6$

5. $4x - 3y = 12$

6. $2x - 6y = 12$

7. $x + y = 0$

8. $x - y = 0$

9. $y = -3$

10. $y + 2 = 0$

11. $-x + 5y = 20$

12. $-8x + 3y = 11$

In Exercises 13–16, match each linear equation with its graph.

13. $y = 2x + 1$

14. $y = -x + 1$

15. $y = -3x - 2$

16. $y = \dfrac{5}{3}x - 2$

A.

B.

C.

D.

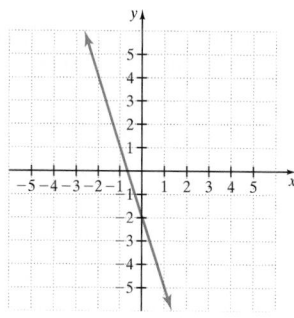

Determine whether the lines are parallel, perpendicular, or neither. See Example 4.

△ **17.** $x - 3y = -6$
   $3x - y = 0$

△ **18.** $-5x + y = -6$
   $x + 5y = 5$

△ **19.** $2x - 7y = 1$
   $2y = 7x - 2$

△ **20.** $y = 4x - 2$
   $4x + y = 5$

🔒 **21.** $10 + 3x = 5y$
   $5x + 3y = 1$

△ **22.** $-x + 2y = -2$
   $2x = 4y + 3$

△ **23.** $6x = 5y + 1$
   $-12x + 10y = 1$

△ **24.** $11x + 10y = 3$
   $11y = 10x + 5$

25. Explain how to write the equation of a line if you are given the slope of the line and its y-intercept.

26. Explain why the graphs of $y = x$ and $y = -x$ are perpendicular lines.
△

Use the slope-intercept form of the linear equation to write an equation of each line with given slope and y-intercept. See Example 5.

27. Slope -1; y-intercept $(0, 1)$

28. Slope $\dfrac{1}{2}$; y-intercept $(0, -6)$

🔒 **29.** Slope 2; y-intercept $\left(0, \dfrac{3}{4}\right)$

30. Slope -3; y-intercept $\left(0, -\dfrac{1}{5}\right)$

31. Slope $\dfrac{2}{7}$; y-intercept $(0, 0)$

32. Slope $-\dfrac{4}{5}$; y-intercept $(0, 0)$

Use the slope-intercept form to graph each equation. See Examples 6 and 7.

33. $y = \dfrac{2}{3}x + 5$

34. $y = \dfrac{1}{4}x - 3$

35. $y = -\dfrac{3}{5}x - 2$

36. $y = -\dfrac{5}{7}x + 5$

37. $y = 2x + 1$

38. $y = -4x - 1$

🔒 **39.** $y = -5x$

40. $y = 6x$

41. $4x + y = 6$

42. $-3x + y = 2$

43. $x - y = -2$

44. $x + y = 3$

45. $3x + 5y = 10$

46. $2x + 3y = 5$

🔒 **47.** $4x - 7y = -14$

48. $3x - 4y = 4$

REVIEW AND PREVIEW

Solve each equation for y. See Section 2.6.

49. $y - (-6) = 2(x - 4)$

50. $y - 7 = -9(x - 6)$

51. $y - 1 = -6(x - (-2))$

52. $y - (-3) = 4(x - (-5))$

In Exercises 53–58, match each line with its slope.

A. $m = 0$ **B.** undefined slope   **C.** $m = 3$

D. $m = 1$ **E.** $m = -\dfrac{1}{2}$ **F.** $m = -\dfrac{3}{4}$

53.

54.

55.

56.

57.

58.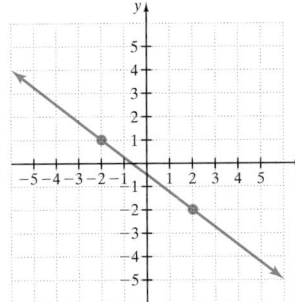

Concept Extensions

Solve.

59. American Automobile Association (AAA) membership was 21 million in 1980. In 2002, their membership was 45 million. (*Source*: AAA)

   a. Write two ordered pairs of the form (years after 1980, membership in millions).

   b. The relationship between the years after 1980 and membership is approximately linear over this period. Use the ordered pairs from part (a) to write an equation of the line relating year to AAA membership.

60. In 1998, 1.7 million electronic bill statements were delivered and payments occurred. In 2002, that number rose to 9.9 million. (*Source: USA Today*)

   a. Write two ordered pairs of the form (years after 1998, millions of electronic bills).

   b. The relationship between years after 1998 and 2002 is approximately linear over this period. Use the ordered pairs from part a to write an equation of the line relating to electronic bills.

   c. Use the linear equation from part (b) to estimate the electronic bills to be delivered and paid in 2006.

61. Write an equation of the line parallel to $2y + 4x = 12$ that has a y-intercept of $(0, 5)$.

62. Write an equation of the line perpendicular to $3y - 9x = 15$ that has a y-intercept of $(0, -2)$.

63. The equation relating Celsius temperature and Fahrenheit temperature is linear and its graph is shown below. This line contains the point $(0, 32)$, which means that $0°$ Celsius is equivalent to $32°$ Fahrenheit. This line also passes through the point $(100, 212)$.

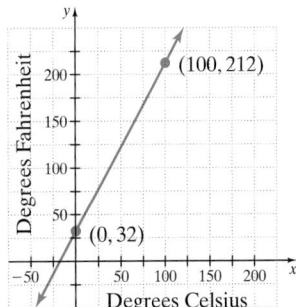

   a. Write the meaning of the ordered pair $(100, 212)$ in words.

   b. Use the graph and approximate the Fahrenheit temperature equivalent to $20°$ Celsius.

   c. Use the graph and approximate the Celsius temperature equivalent to $80°$ Fahrenheit.

   d. Use the ordered pairs $(0, 32)$ and $(100, 212)$ and the slope-intercept form and write an equation of the line relating Celsius temperatures and Fahrenheit temperatures. (*Hint:* Use both ordered pairs to find slope. Then substitute the slope and y-intercept in slope-intercept form.)

   e. Use the linear equation found in part (d) to check your approximations from (b) and (c).

64. The equation relating temperatures on the Kelvin scale and the Celsius scale is linear. If $0°$ Celsius is equivalent to 273 Kelvin and $100°$ Celsius is equivalent to 373 Kelvin, find the linear equation that models the relationship between these two temperature scales. (*Hint:* See Exercise 63 and use the ordered pairs $(0, 273)$ and $(100, 373)$.)

65. Show that the slope of a line whose equation is $y = mx + b$ is m. To do so, find two ordered pair solutions of this equation and use the slope formula. For example, if $x = 0$, then $y = 0 \cdot x + b$ or $y = b$. Thus, one ordered pair solution is $(0, b)$. Find one more ordered pair solution and find the slope.

3.6   THE POINT-SLOPE FORM

Objectives

1. Use the point-slope form to find an equation of a line given its slope and a point of the line.
2. Use the point-slope form to find an equation of a line given two points of the line.
3. Find equations of vertical and horizontal lines.
4. Use the point-slope form to solve problems.

1 From the last section, we know that if the y-intercept of a line and its slope are given, then the line can be graphed and an equation of this line can be found. Our goal in this section is to answer the following: If any two points of a line are given, can an equation of the line be found? To answer this question, first let's see if we can find an equation of a line given one point (not necessarily the y-intercept) and the slope of the line.

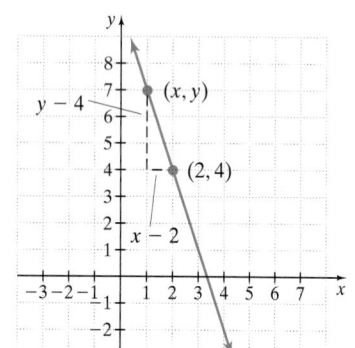

Suppose that the slope of a line is -3 and the line contains the point $(2, 4)$. For any other point with ordered pair (x, y) to be on the line, its coordinates must satisfy the slope equation

$$\frac{y - 4}{x - 2} = -3$$

Now multiply both sides of this equation by $x - 2$.

$$y - 4 = -3(x - 2)$$

This equation is a linear equation whose graph is a line that contains the point $(2, 4)$ and has slope -3. This form of a linear equation is called the **point-slope form.**

In general, when the slope of a line and any point on the line are known, the equation of the line can be found. To do this, use the slope formula to write the slope of a line that passes through points (x, y) and (x_1, y_1). We have

$$\frac{y - y_1}{x - x_1} = m$$

Multiply both sides of this equation by $x - x_1$ to obtain

$$y - y_1 = m(x - x_1)$$
$$\uparrow$$
$$slope$$

Point-Slope Form of the Equation of a Line

The point-slope form of the equation of a line is $y - y_1 = m(x - x_1)$, where m is the slope of the line and (x_1, y_1) is a point on the line.

EXAMPLE 1

Find an equation of the line passing through $(-1, 5)$ with slope -2. Write the equation in standard form: $Ax + By = C$.

Solution   Since the slope and a point on the line are given, use point-slope form $y - y_1 = m(x - x_1)$ to write the equation. Let $m = -2$ and $(-1, 5) = (x_1, y_1)$.

$$y - y_1 = m(x - x_1)$$
$$y - 5 = -2[x - (-1)] \quad \text{Let } m = -2 \text{ and } (x_1, y_1) = (-1, 5).$$
$$y - 5 = -2(x + 1) \quad \text{Simplify.}$$
$$y - 5 = -2x - 2 \quad \text{Use the distributive property.}$$
$$y = -2x + 3 \quad \text{Add 5 to both sides.}$$
$$2x + y = 3 \quad \text{Add } 2x \text{ to both sides.}$$

In standard form, the equation is $2x + y = 3$.

2   We may also find the equation of a line given any two points of the line, as seen in the next example.

EXAMPLE 2

Find an equation of the line through $(2, 5)$ and $(-3, 4)$. Write the equation in standard form.

Solution   First, use the two given points to find the slope of the line.

$$m = \frac{4 - 5}{-3 - 2} = \frac{-1}{-5} = \frac{1}{5}$$

Next, use the slope $\frac{1}{5}$ and either one of the given points to write the equation in point-slope form. We use $(2, 5)$. Let $x_1 = 2$, $y_1 = 5$, and $m = \frac{1}{5}$.

$$y - y_1 = m(x - x_1) \quad \text{Use point-slope form.}$$
$$y - 5 = \frac{1}{5}(x - 2) \quad \text{Let } x_1 = 2, y_1 = 5, \text{ and } m = \frac{1}{5}.$$
$$5(y - 5) = 5 \cdot \frac{1}{5}(x - 2) \quad \text{Multiply both sides by 5 to clear fractions.}$$
$$5y - 25 = x - 2 \quad \text{Use the distributive property and simplify.}$$
$$-x + 5y - 25 = -2 \quad \text{Subtract } x \text{ from both sides.}$$
$$-x + 5y = 23 \quad \text{Add 25 to both sides.}$$

Helpful Hint
Multiply both sides of the equation $-x + 5y = 23$ by -1, and it becomes $x - 5y = -23$. Both $-x + 5y = 23$ and $x - 5y = -23$ are in standard form, and they are equations of the same line.

3   Recall from Section 3.3 that:

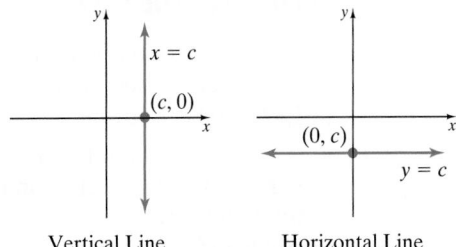

Vertical Line   Horizontal Line

EXAMPLE 3

Find an equation of the vertical line through $(-1, 5)$.

Solution The equation of a vertical line can be written in the form $x = c$, so an equation for a vertical line passing through $(-1, 5)$ is $x = -1$.

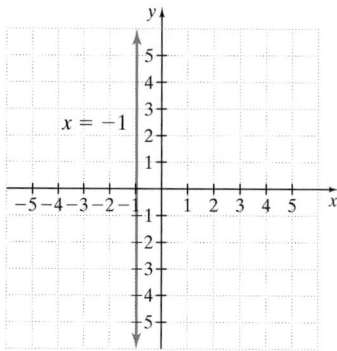

EXAMPLE 4

Find an equation of the line parallel to the line $y = 5$ and passing through $(-2, -3)$.

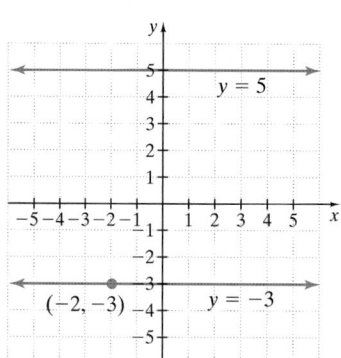

Solution Since the graph of $y = 5$ is a horizontal line, any line parallel to it is also horizontal. The equation of a horizontal line can be written in the form $y = c$. An equation for the horizontal line passing through $(-2, -3)$ is $y = -3$.

4 Problems occurring in many fields can be modeled by linear equations in two variables. The next example is from the field of marketing and shows how consumer demand of a product depends on the price of the product.

EXAMPLE 5

PREDICTING THE SALES OF FRISBEES

The Whammo Company has learned that by pricing a newly released Frisbee at $6, sales will reach 2000 Frisbees per day. Raising the price to $8 will cause the sales to fall to 1500 Frisbees per day.

a. Assume that the relationship between sales price and number of Frisbees sold is linear and write an equation describing this relationship. Write the equation in slope-intercept form.

b. Predict the daily sales of Frisbees if the price is $7.50.

Solution

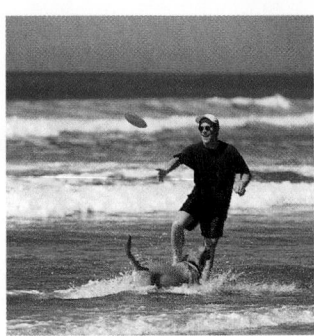

a. First, use the given information and write two ordered pairs. Ordered pairs will be in the form (sales price, number sold) so that our ordered pairs are (6, 2000) and (8, 1500). Use the point-slope form to write an equation. To do so, we find the slope of the line that contains these points.

$$m = \frac{2000 - 1500}{6 - 8} = \frac{500}{-2} = -250$$

Next, use the slope and either one of the points to write the equation in point-slope form. We use (6, 2000).

$$y - y_1 = m(x - x_1) \quad \text{Use point-slope form.}$$
$$y - 2000 = -250(x - 6) \quad \text{Let } x_1 = 6, y_1 = 2000, \text{ and } m = -250.$$
$$y - 2000 = -250x + 1500 \quad \text{Use the distributive property.}$$
$$y = -250x + 3500 \quad \text{Write in slope-intercept form.}$$

b. To predict the sales if the price is $7.50, we find y when $x = 7.50$.

$$y = -250x + 3500$$
$$y = -250(7.50) + 3500 \quad \text{Let } x = 7.50.$$
$$y = -1875 + 3500$$
$$y = 1625$$

If the price is $7.50, sales will reach 1625 Frisbees per day.

The preceding example may also be solved by using ordered pairs of the form (number sold, sales price).

Forms of Linear Equations

$Ax + By = C$	**Standard form** of a linear equation.
	A and B are not both 0.
$y = mx + b$	**Slope-intercept form** of a linear equation.
	The slope is m and the y-intercept is $(0, b)$.
$y - y_1 = m(x - x_1)$	**Point-slope form** of a linear equation.
	The slope is m and (x_1, y_1) is a point on the line.
$y = c$	**Horizontal line**
	The slope is 0 and the y-intercept is $(0, c)$.
$x = c$	**Vertical line**
	The slope is undefined and the x-intercept is $(c, 0)$.

Parallel and Perpendicular Lines

Nonvertical parallel lines have the same slope.
The product of the slopes of two nonvertical perpendicular lines is -1.

Spotlight on
DECISION
of MAKING

Suppose you are a physical therapist. At weekly sessions, you are administering treadmill treatment to a patient. In the first phase of the treatment plan, the patient is to walk on a treadmill at a speed of 2 mph until fatigue sets in. Once the patient is able to walk 10 minutes at this speed, he will be ready for the next phase of treatment. However, you may change the treatment plan if it looks like it will take longer than 10 weeks to build up to 10 minutes of walking at 2 mph.

You record and plot the patient's progress on his chart for 5 weeks. Do you think you may need to change the first phase of treatment? Explain your reasoning. If not, when do you think the patient will be ready for the next phase of treatment?

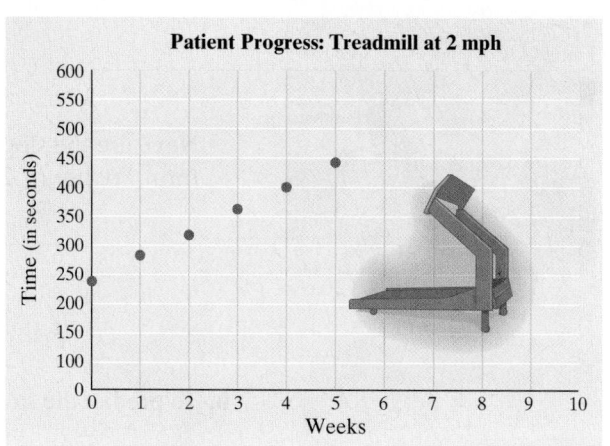

Patient Progress: Treadmill at 2 mph

MENTAL MATH

The graph of each equation below is a line. Use the equation to identify the slope and a point of the line.

1. $y - 8 = 3(x - 4)$

2. $y - 1 = 5(x - 2)$

3. $y + 3 = -2(x - 10)$

4. $y + 6 = -7(x - 2)$

5. $y = \dfrac{2}{5}(x + 1)$

6. $y = \dfrac{3}{7}(x + 4)$

EXERCISE SET 3.6

STUDY GUIDE/SSM   CD/VIDEO   PH MATH TUTOR CENTER   MathXL®Tutorials ON CD   MathXL®   MyMathLab®

Use the point-slope form of the linear equation to find an equation of each line with the given slope and passing through the given point. Then write the equation in standard form. See Example 1.

1. Slope 6; through $(2, 2)$

2. Slope 4; through $(1, 3)$

3. Slope -8; through $(-1, -5)$

4. Slope -2; through $(-11, -12)$

5. Slope $\dfrac{1}{2}$; through $(5, -6)$

6. Slope $\dfrac{2}{3}$; through $(-8, 9)$

Find an equation of the line through the given points. Write the equation in standard form. See Example 2.

7. Through $(3, 2)$ and $(5, 6)$

8. Through $(6, 2)$ and $(8, 8)$

9. Through $(-1, 3)$ and $(-2, -5)$

10. Through $(-4, 0)$ and $(6, -1)$

11. Through $(2, 3)$ and $(-1, -1)$

12. Through $(0, 0)$ and $\left(\dfrac{1}{2}, \dfrac{1}{3}\right)$

Find an equation of each line. See Example 3.

13. Vertical line through $(0, 2)$

14. Horizontal line through $(1, 4)$

15. Horizontal line through $(-1, 3)$

16. Vertical line through $(-1, 3)$

17. Vertical line through $\left(-\dfrac{7}{3}, -\dfrac{2}{5}\right)$

18. Horizontal line through $\left(\dfrac{2}{7}, 0\right)$

Find an equation of each line. See Example 4.

19. Parallel to $y = 5$, through $(1, 2)$

20. Perpendicular to $y = 5$, through $(1, 2)$

△ **21.** Perpendicular to $x = -3$, through $(-2, 5)$

△ **22.** Parallel to $y = -4$, through $(0, -3)$

△ **23.** Parallel to $x = 0$, through $(6, -8)$

△ **24.** Perpendicular to $x = 7$, through $(-5, 0)$

MIXED PRACTICE

Find an equation of each line described. Write each equation in standard form.

25. With slope $-\frac{1}{2}$, through $\left(0, \frac{5}{3}\right)$

26. With slope $\frac{5}{7}$, through $(0, -3)$

27. Slope 1, through $(-7, 9)$

28. Slope 5, through $(6, -8)$

29. Through $(10, 7)$ and $(7, 10)$

30. Through $(5, -6)$ and $(-6, 5)$

△ **31.** Through $(6, 7)$, parallel to the x-axis

△ **32.** Through $(0, -5)$, parallel to the y-axis

33. Slope $-\frac{4}{7}$, through $(-1, -2)$

34. Slope $-\frac{3}{5}$, through $(4, 4)$

35. Through $(-8, 1)$ and $(0, 0)$

36. Through $(2, 3)$ and $(0, 0)$

37. Through $(0, 0)$ with slope 3

38. Through $(0, -2)$ with slope -1

39. Through $(-6, -6)$ and $(0, 0)$

40. Through $(0, 0)$ and $(4, 4)$

41. Slope -5, y-intercept 7

42. Slope -2; y-intercept -4

43. Through $(-1, 5)$ and $(0, -6)$

44. Through $(4, 0)$ and $(0, -5)$

45. With undefined slope, through $\left(-\frac{3}{4}, 1\right)$

46. With slope 0, through $(6.7, 12.1)$

△ **47.** Through $(-2, -3)$, perpendicular to the y-axis

△ **48.** Through $(0, 12)$, perpendicular to the x-axis

49. With slope 7, through $(1, 3)$

50. With slope -10, through $(5, -1)$

Solve. Assume that each exercise describes a linear relationship. See Example 5.

51. In 2001 there were 6680 electric-powered vehicles in use in the United States. In 1998 only 4760 electric vehicles were being used. (*Source*: U.S. Energy Information Administration)

    a. Write an equation describing the relationship between time and number of electric-powered vehicles. Use ordered pairs of the form (years past 1998, number of vehicles).

b. Use this equation to predict the number of electric-powered vehicles in use in 2008.

52. In 1996 there were 457 thousand eating establishments in the United States. In 2000, there were 483 thousand eating establishments.

    a. Write an equation describing the relationship between time and number of eating establishments. Use ordered pairs of the form (years past 1996, number of eating establishments in thousands).

    b. Use this equation to predict the number of eating establishments in 2009.

53. A rock is dropped from the top of a 400-foot cliff. After 1 second, the rock is traveling 32 feet per second. After 3 seconds, the rock is traveling 96 feet per second.

    a. Assume that the relationship between time and speed is linear and write an equation describing this relationship. Use ordered pairs of the form (time, speed).

    b. Use this equation to determine the speed of the rock 4 seconds after it was dropped.

54. A Hawaiian fruit company is studying the sales of a pineapple sauce to see if this product is to be continued. At the end of its first year, profits on this product amounted to $30,000. At the end of the fourth year, profits are $66,000.

    a. Assume that the relationship between years on the market and profit is linear and write an equation describing this relationship. Use ordered pairs of the form (years on the market, profit).

b. Use this equation to predict the profit at the end of 7 years.

55. In 2000, the U.S. population per square mile of land area was 79.6. In 1990, the person per square mile population was 70.3.

    a. Write an equation describing the relationship between year and persons per square mile. Use ordered pairs of the form (years past 1990, persons per square mile).

    b. Use this equation to predict the person per square mile population in 2007.

56. In 1996, there were 135 thousand apparel and accessory stores in the United States. In 2001, there were a total of 152 thousand apparel and accessory stores. (*Source*: U.S. Bureau of the Census. *County Business Patterns*, annual)

    a. Write an equation describing this relationship. Use ordered pairs of the form (years past 1996, number of stores in thousands).

    b. Use this equation to predict the number of apparel and accessory stores in 2009.

57. In 1995, the sales of battery-operated ride-on toys in the United States were $191 million. In 2000, the sales of these ride-on toys had risen to $260 million. (*Source*: Toy Manufacturers of America, Inc.)

    a. Write two ordered pairs of the form (years after 1995, sales of battery-operated ride-on toys) for this situation.

    b. The relationship between years after 1995 and sales of battery-operated ride-on toys is linear over this period. Use the ordered pairs from part (a) to write an equation of the line relating year to sales of battery-operated ride-on toys.

    c. Use the linear equation from part (b) to estimate the sales of ride-on toys in 1999.

58. In 2000, crude oil production by OPEC countries was 29.2 million barrels per day. In 2002, OPEC crude oil production had risen to about 28.7 million barrels per day. (*Source*: Energy Information Administration)

    a. Write two ordered pairs of the form (years after 2000, crude oil production) for this situation.

    b. Assume that the relationship between years after 1997 and crude oil production is linear over this period. Use the ordered pairs from part (a) to write an equation of the line relating year to crude oil production.

    c. Use the linear equation from part (b) to estimate the crude oil production by OPEC countries in 2001.

△ **59.** A linear equation can be written that relates the radius of a circle to its circumference. A circle with a radius of 10 centimeters has a circumference of approximately 63 centimeters. A circle with a radius of 15 centimeters has a circumference of approximately 94 centimeters. Use the ordered pairs (10, 63) and (15, 94) to write a linear equation in standard form that approximates the relationship between the radius of the circle and its circumference.

60. The value of a computer bought in 1998 depreciates or decreases as time passes. Two years after the computer was bought, it was worth $2600 and 5 years after it was bought, it was worth $2000.

    a. If the relationship between number of years past 1998 and value of the computer is linear, write an equation describing this relationship. Use ordered pairs of the form (years past 1998, value of computer).

    b. Use this equation to estimate the value of the computer in the year 2008.

61. The Pool Fun Company has learned that, by pricing a newly released Fun Noodle at $3, sales will reach 10,000 Fun Noodles per day during the summer. Raising the price to $5 will cause the sales to fall to 8000 Fun Noodles per day.

    a. Assume that the relationship between sales price and number of Fun Noodles sold is linear and write an equation describing this relationship. Use ordered pairs of the form (sales price, number sold).

    b. Predict the daily sales of Fun Noodles if the price is $3.50.

62. The value of a building bought in 1985 may be depreciated (or decreased) as time passes for income tax purposes. Seven years after the building was bought, this value was $165,000 and 12 years after it was bought, this value was $140,000.

    a. If the relationship between number of years past 1985 and the depreciated value of the building is linear, write an equation describing this relationship. Use ordered pairs of the form (years past 1985, value of building).

    b. Use this equation to estimate the depreciated value of the building in 2005.

REVIEW AND PREVIEW

Find the value of $x^2 - 3x + 1$ for each given value of x. See Section 1.7.

63. 2   **64.** 5   **65.** -1 **66.** -3

For each graph, determine whether any x-values correspond to two or more y-values. See Section 3.7.

67.

68.

69.

70.

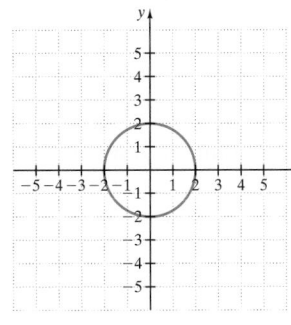

Concept Extensions

71. Given the equation of a nonvertical line, explain how to find the slope without finding two points on the line.

72. Given two points on a nonvertical line, explain how to use the point-slope form to find the equation of the line.

△ **73.** Write an equation in standard form of the line that contains the point $(-1, 2)$ and is

    a. parallel to the line $y = 3x - 1$.

    b. perpendicular to the line $y = 3x - 1$.

△ **74.** Write an equation in standard form of the line that contains the point $(4, 0)$ and is

    a. parallel to the line $y = -2x + 3$.

    b. perpendicular to the line $y = -2x + 3$.

△ **75.** Write an equation in standard form of the line that contains the point $(3, -5)$ and is

    a. parallel to the line $3x + 2y = 7$.

    b. perpendicular to the line $3x + 2y = 7$.

△ **76.** Write an equation in standard form of the line that contains the point $(-2, 4)$ and is

    a. parallel to the line $x + 3y = 6$.

    b. perpendicular to the line $x + 3y = 6$.

3.7   **FUNCTIONS**

Objectives

  1 Identify relations, domains, and ranges.

  2 Identify functions.

  3 Use the vertical line test.

  4 Use function notation.

1 In previous sections, we have discussed the relationships between two quantities. For example, the relationship between the length of the side of a square x and its area y is described by the equation $y = x^2$. These variables x and y are related in the following way: for any given value of x, we can find the corresponding value of y by squaring the x-value. Ordered pairs can be used to write down solutions of this equation. For example, $(2, 4)$ is a solution of $y = x^2$, and this notation tells us that the x-value 2 is related to the y-value 4 for this equation. In other words, when the length of the side of a square is 2 units, its area is 4 square units.

A set of ordered pairs is called a **relation**. The set of all x-coordinates is called the **domain** of a relation, and the set of all y-coordinates is called the **range** of a relation. Equations such as $y = x^2$ are also called relations since equations in two variables define a set of ordered pair solutions.

EXAMPLE 1

Find the domain and the range of the relation $\{(0, 2), (3, 3), (-1, 0), (3, -2)\}$.

Solution   The domain is the set of all x-values or $\{-1, 0, 3\}$, and the range is the set of all y-values, or $\{-2, 0, 2, 3\}$.

2   Some relations are also functions.

> ### Function
> A function is a set of ordered pairs that assigns to each x-value exactly one y-value.

EXAMPLE 2

Which of the following relations are also functions?

a. $\{(-1, 1), (2, 3), (7, 3), (8, 6)\}$ **b.** $\{(0, -2), (1, 5), (0, 3), (7, 7)\}$

Solution   **a.** Although the ordered pairs $(2, 3)$ and $(7, 3)$ have the same y-value, each x-value is assigned to only one y-value so this set of ordered pairs is a function.

b. The x-value 0 is assigned to two y-values, -2 and 3, so this set of ordered pairs is not a function.

Relations and functions can be described by a graph of their ordered pairs.

EXAMPLE 3

Which graph is the graph of a function?

a.   **b.**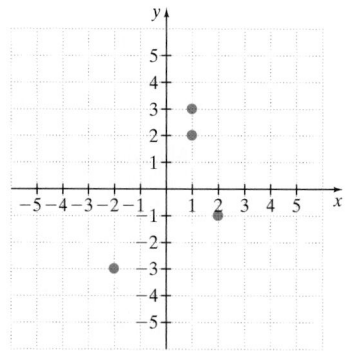

Solution   **a.** This is the graph of the relation $\{(-4, -2), (-2, -1)(-1, -1), (1, 2)\}$. Each x-coordinate has exactly one y-coordinate, so this is the graph of a function.

b. This is the graph of the relation $\{(-2, -3), (1, 2), (1, 3), (2, -1)\}$. The x-coordinate 1 is paired with two y-coordinates, 2 and 3, so this is not the graph of a function.

3   The graph in Example 3(b) was not the graph of a function because the x-coordinate 1 was paired with two y-coordinates, 2 and 3. Notice that when an x-coordinate is paired with more than one y-coordinate, a vertical line can be drawn that will intersect the graph at more than one point. We can use this fact to determine whether a relation is also a function. We call this the **vertical line test**.

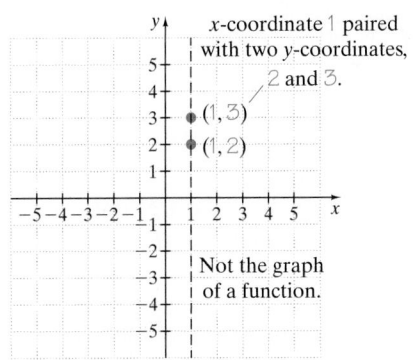

Vertical Line Test

If a vertical line can be drawn so that it intersects a graph more than once, the graph is not the graph of a function.

This vertical line test works for all types of graphs on the rectangular coordinate system.

EXAMPLE 4

Use the vertical line test to determine whether each graph is the graph of a function.

a.   **b.**   **c.**   **d.**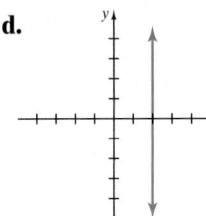

Solution   **a.** This graph is the graph of a function since no vertical line will intersect this graph more than once.

b. This graph is also the graph of a function; no vertical line will intersect it more than once.

c. This graph is not the graph of a function. Vertical lines can be drawn that intersect the graph in two points. An example of one is shown.

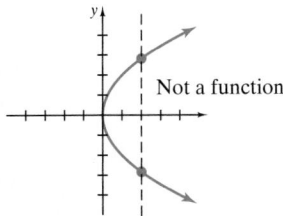

d. This graph is not the graph of a function. A vertical line can be drawn that intersects this line at every point.

Recall that the graph of a linear equation is a line, and a line that is not vertical will pass the vertical line test. **Thus, all linear equations are functions except those of the form $x = c$, which are vertical lines.**

EXAMPLE 5

Which of the following linear equations are functions?

a. $y = x$ **b.** $y = 2x + 1$ **c.** $y = 5$ **d.** $x = -1$

Solution **a, b,** and **c** are functions because their graphs are nonvertical lines. **d** is not a function because its graph is a vertical line.

Examples of functions can often be found in magazines, newspapers, books, and other printed material in the form of tables or graphs such as that in Example 6.

EXAMPLE 6

The graph shows the sunrise time for Indianapolis, Indiana, for the year. Use this graph to answer the questions.

a. Approximate the time of sunrise on February 1.
b. Approximately when does the sun rise at 5 A.M.?
c. Is this the graph of a function?

Solution **a.** To approximate the time of sunrise on February 1, we find the mark on the horizontal axis that corresponds to February 1. From this mark, we move vertically upward until the graph is reached. From that point on the graph, we move horizontally to the left until the vertical axis is reached. The vertical axis there reads 7 A.M.

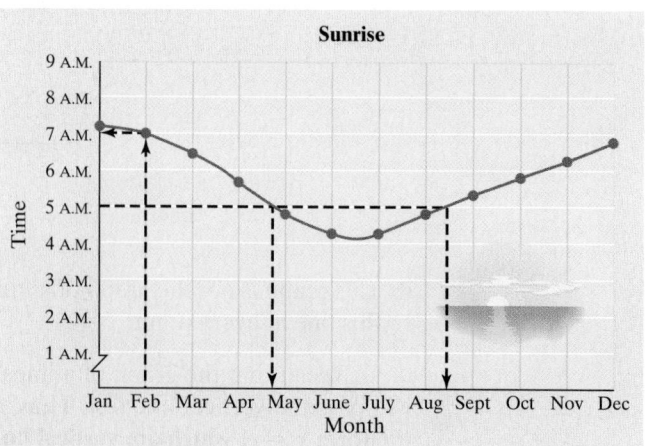

b. To approximate when the sun rises at 5 A.M., we find 5 A.M. on the time axis and move horizontally to the right. Notice that we will reach the graph twice, corresponding to two dates for which the sun rises at 5 A.M. We follow both points on the graph vertically downward until the horizontal axis is reached. The sun rises at 5 A.M. at approximately the end of the month of April and the middle of the month of August.

c. The graph is the graph of a function since it passes the vertical line test. In other words, for every day of the year in Indianapolis, there is exactly one sunrise time.

4   The graph of the linear equation $y = 2x + 1$ passes the vertical line test, so we say that $y = 2x + 1$ is a function. In other words, $y = 2x + 1$ gives us a rule for writing ordered pairs where every x-coordinate is paired with one y-coordinate.

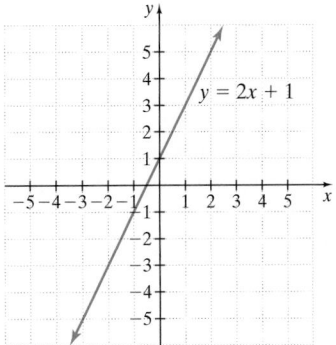

We often use letters such as f, g, and h to name functions. For example, the symbol $f(x)$ means *function of x* and is read "f of x." This notation is called **function notation**. The equation $y = 2x + 1$ can be written as $f(x) = 2x + 1$ using function notation, and these equations mean the same thing. In other words, $y = f(x)$.

The notation $f(1)$ means to replace x with 1 and find the resulting y or function value. Since

$$f(x) = 2x + 1$$

then

$$f(1) = 2(1) + 1 = 3$$

This means that, when $x = 1$, y or $f(x) = 3$, and we have the ordered pair $(1, 3)$. Now let's find $f(2), f(0)$, and $f(-1)$.

$$
\begin{array}{lll}
f(x) = 2x + 1 & f(x) = 2x + 1 & f(x) = 2x + 1 \\
f(2) = 2(2) + 1 & f(0) = 2(0) + 1 & f(-1) = 2(-1) + 1 \\
\quad = 4 + 1 & \quad = 0 + 1 & \quad = -2 + 1 \\
\quad = 5 & \quad = 1 & \quad = -1
\end{array}
$$

Ordered
Pair: $(2, 5)$ $(0, 1)$ $(-1, -1)$

> **Helpful Hint**
>
> Note that $f(x)$ is a special symbol in mathematics used to denote a function. The symbol $f(x)$ is read "f of x." It does **not** mean $f \cdot x$ (f times x).

EXAMPLE 7

Given $g(x) = x^2 - 3$, find the following. Then write down the corresponding ordered pairs generated.

a. $g(2)$ **b.** $g(-2)$ **c.** $g(0)$

Solution

a. $g(x) = x^2 - 3$
$g(2) = 2^2 - 3$
$= 4 - 3$
$= 1$

b. $g(x) = x^2 - 3$
$g(-2) = (-2)^2 - 3$
$= 4 - 3$
$= 1$

c. $g(x) = x^2 - 3$
$g(0) = 0^2 - 3$
$= 0 - 3$
$= -3$

Ordered Pair:
$g(2) = 1$ gives $(2, 1)$
$g(-2) = 1$ gives $(-2, 1)$
$g(0) = -3$ gives $(0, -3)$

We now practice finding the domain and the range of a function. The domain of our functions will be the set of all possible real numbers that x can be replaced by. The range is the set of corresponding y-values.

EXAMPLE 8

Find the domain of each function.

a. $g(x) = \dfrac{1}{x}$ **b.** $f(x) = 2x + 1$

Solution

a. Recall that we cannot divide by 0 so that the domain of $g(x)$ is the set of all real numbers except 0. In interval notation, we can write $(-\infty, 0) \cup (0, \infty)$.

b. In this function, x can be any real number. The domain of $f(x)$ is the set of all real numbers, or $(-\infty, \infty)$ in interval notation.

✔ **CONCEPT CHECK**

Suppose that the value of f is -7 when the function is evaluated at 2. Write this situation in function notation.

EXAMPLE 9

Find the domain and the range of each function graphed. Use interval notation.

a.

b.

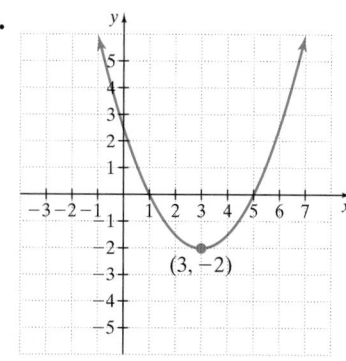

Concept Check Answer:
$f(2) = -7$

Solution   a.

b.

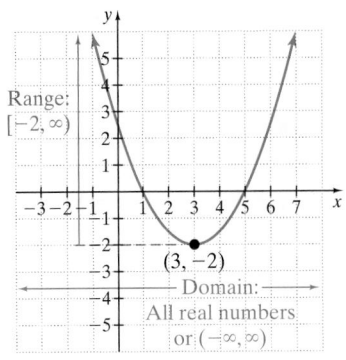

Spotlight on DECISION MAKING

Suppose you are the property and grounds manager for a small company in Portland, Oregon. A recent heat wave has caused employees to ask for air-conditioning in the building. As you begin to research air-conditioning installation, you discover this graph of average high temperatures for Portland. Because the building is surrounded by trees, the current ventilation system (without air-conditioning) is able to keep the temperature 10°F cooler than the outside temperature on a warm day. A comfortable temperature range for working is 68°F to 74°F. Would you recommend installing air-conditioning in the building? Explain your reasoning. What other factors would you want to consider?

Source: The Weather Channel Enterprises, Inc.

EXERCISE SET 3.7

STUDY GUIDE/SSM   CD/ VIDEO   PH MATH TUTOR CENTER   MathXL®Tutorials ON CD   MathXL®   MyMathLab®

Find the domain and the range of each relation. See Example 1.

1. $\{(2, 4), (0, 0), (-7, 10), (10, -7)\}$

2. $\{(3, -6), (1, 4), (-2, -2)\}$

3. $\{(0, -2), (1, -2)(5, -2)\}$

4. $\{(5, 0), (5, -3), (5, 4), (5, 3)\}$

Determine which relations are also functions. See Example 2.

5. $\{(1, 1), (2, 2), (-3, -3), (0, 0)\}$

6. $\{(1, 2), (3, 2), (4, 2)\}$

7. $\{(-1, 0), (-1, 6), (-1, 8)\}$

8. $\{(11, 6), (-1, -2), (0, 0), (3, -2)\}$

Use the vertical line test to determine whether each graph is the graph of a function. See Example 4.

9.

10.

11.

12.

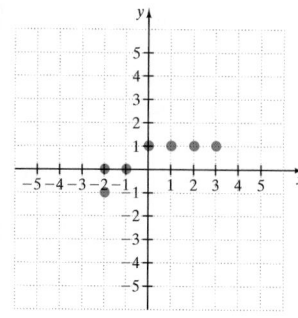

Decide whether the equation describes a function. See Example 5.

21. $y = x + 1$

22. $y = x - 1$

23. $y - x = 7$

24. $2x - 3y = 9$

25. $y = 6$

26. $x = 3$

27. $x = -2$

28. $y = -9$

29. $x = y^2$

30. $y = x^2 - 3$

Use the graph in Example 6 to answer Exercises 31 through 34.

31. Approximate the time of sunrise on September 1 in Indianapolis.

32. Approximate the date(s) when the sun rises in Indianapolis at 7 A.M.

33. Describe the change in sunrise over the year for Indianapolis.

34. When, in Indianapolis, is the earliest sunrise? What point on the graph does this correspond to?

13.

14.

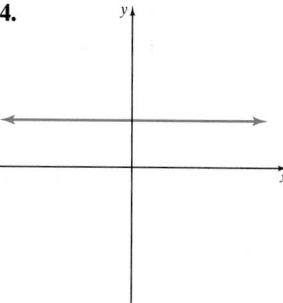

This graph shows the U.S. hourly minimum wage at the beginning of each year shown. Use this graph to answer Exercises 35 through 40.

15.

16.

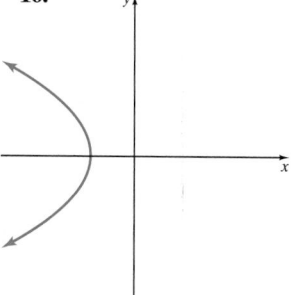

35. Approximate the minimum wage at the beginning of 1997.

36. Approximate the minimum wage at the beginning of 1999.

37. Approximate the year when the minimum wage increased to over $4.50 per hour.

38. Approximate the year when the minimum wage increased to over $5.00 per hour.

17.

18.

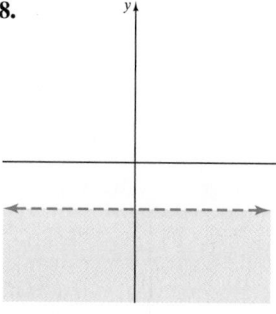

39. Is this graph the graph of a function? Why or why not?

40. Do you think that a similar graph of your hourly wage on January 1 of every year (whether you are working or not) will be the graph of a function? Why or why not?

Given the following functions, find $f(-2)$, $f(0)$, and $f(3)$. See Example 7.

19.

20.

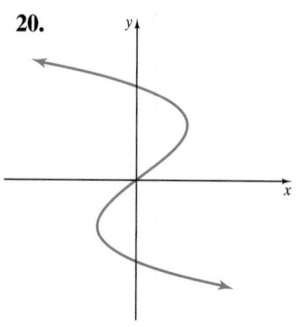

41. $f(x) = 2x - 5$

42. $f(x) = 3 - 7x$

43. $f(x) = x^2 + 2$

44. $f(x) = x^2 - 4$

45. $f(x) = x^3$

46. $f(x) = -x^3$

47. $f(x) = |x|$

48. $f(x) = |2 - x|$

Given the following functions, find h(−1), h(0), and h(4). Then write the corresponding ordered pair. See Example 7.

49. $h(x) = 5x$

50. $h(x) = -3x$

51. $h(x) = 2x^2 + 3$

52. $h(x) = -x^2$

53. $h(x) = -x^2 - 2x + 3$

54. $h(x) = -x^2 + 4x - 3$

55. $h(x) = 6$

56. $h(x) = -12$

Find the domain of each function. See Example 8.

57. $f(x) = 3x - 7$

58. $g(x) = 5 - 2x$

59. $h(x) = \dfrac{1}{x + 5}$

60. $f(x) = \dfrac{1}{x - 6}$

61. $g(x) = |x + 1|$

62. $h(x) = |2x|$

Find the domain and the range of each function graphed. See Example 9.

 63.

64.

(1, 5)

65.

66.

67.

68.

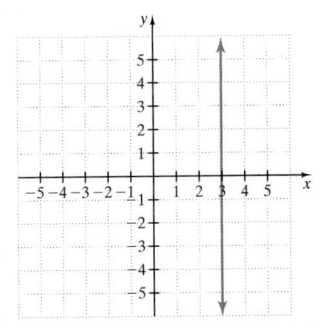

REVIEW AND PREVIEW

Find the coordinates of the point of intersection. See Section 3.1.

69.

70.

71.

72.

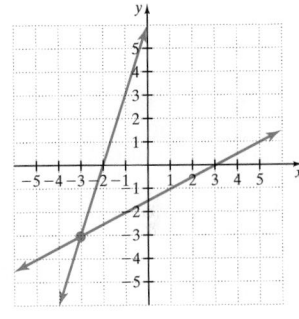

Concept Extensions

73. Forensic scientists use the function

$$H(x) = 2.59x + 47.24$$

to estimate the height of a woman in centimeters given the length x of her femur bone.

a. Estimate the height of a woman whose femur measures 46 centimeters.

b. Estimate the height of a woman whose femur measures 39 centimeters.

74. The dosage in milligrams D of Ivermectin, a heartworm preventive for a dog who weighs x pounds, is given by the function

$$D(x) = \dfrac{136}{25}x$$

a. Find the proper dosage for a dog that weighs 35 pounds.

b. Find the proper dosage for a dog that weighs 70 pounds.

75. In your own words define **(a)** function; **(b)** domain; **(c)** range.

76. Explain the vertical line test and how it is used.

77. Since $y = x + 7$ is a function, rewrite the equation using function notation.

See the example below for Exercises 78 through 81.

Example

If $f(x) = x^2 + 2x + 1$, find $f(\pi)$.

Solution:

$f(x) = x^2 + 2x + 1$
$f(\pi) = \pi^2 + 2\pi + 1$

Given the following functions, find the indicated values.

78. $f(x) = 2x + 7$

    a. $f(2)$ **b.** $f(a)$

79. $g(x) = -3x + 12$

    a. $g(s)$ **b.** $g(r)$

80. $h(x) = x^2 + 7$

    a. $h(3)$ **b.** $h(a)$

81. $f(x) = x^2 - 12$

    a. $f(12)$ **b.** $f(a)$

CHAPTER 3 PROJECT

Financial Analysis

Investment analysts investigate a company's sales, net profit, debt, and assets to decide whether investing in it is a wise choice. One way to analyze this data is to graph it and look for trends over time. Another way is to find algebraically the rate at which the data change over time.

    The table below gives the net profits in millions of dollars for the leading U.S. businesses in the pharmaceutical industry for the years 1997 and 1998. In this project, you will analyze the performances of these companies and, based on this information alone, make an investment recommendation. This project may be completed by working in groups or individually.

PHARMACEUTICAL INDUSTRY NET PROFITS (IN MILLIONS OF DOLLARS)

Company	2000	2001
Merck	$40,363	$47,716
Pfizer	$29,574	$32,259
Johnson & Johnson	$29,139	$33,004
Bristol-Myers Squibb	$21,331	$21,717
Pharmacia	$18,150	$19,299
Abbott Laboratories	$13,746	$16,285
Eli Lilly	$10,862	$11,543
Schering-Plough	$ 9815	$ 9802

(*Source: Fortune Magazine*)

1. Scan the table. Did any of the companies have a loss during the years shown? If so, which company and when? What does this mean?

2. Write the data for each company as two ordered pairs of the form (year, net profit). Assuming that the trends in net profit are linear, use graph paper to graph the line represented by the ordered pairs for each company. Describe the trend shown by each graph.

3. Find the slope of the line for each company.

4. Which of the lines, if any, have positive slopes? What does that mean in this context? Which of the lines, if any, have negative slopes? What does that mean in this context?

Of these pharmaceutical companies, which one(s) would you recommend as an investment choice? Why?

Do you think it is wise to make a decision after looking at only 2 years of net profits? What other factors do you think should be taken into consideration when making an investment choice?

(Optional) Use financial magazines, company annual reports, or online investing information to find net profit information for two different years for two to four companies in the same industry. Analyze the net profits and make an investment recommendation.

3 CHAPTER VOCABULARY CHECK

Fill in each blank with one of the words listed below.

relation	function	domain	range	standard	slope-intercept
y-axis	*x*-axis	solution	linear	slope	point-slope
x-intercept	*y*-intercept	*y*	*x*		

1. An ordered pair is a _____ of an equation in two variables if replacing the variables by the coordinates of the ordered pair results in a true statement.
2. The vertical number line in the rectangular coordinate system is called the _____.
3. A _____ equation can be written in the form $Ax + By = C$.
4. A(n) _____ is a point of the graph where the graph crosses the *x*-axis.
5. The form $Ax + By = C$ is called _____ form.
6. A(n) _____ is a point of the graph where the graph crosses the *y*-axis.
7. The equation $y = 7x - 5$ is written in _____ form.
8. The equation $y + 1 = 7(x - 2)$ is written in _____ form.
9. To find an *x*-intercept of a graph, let ___ = 0.
10. The horizontal number line in the rectangular coordinate system is called the _____.
11. To find a *y*-intercept of a graph, let ___ = 0.
12. The _____ of a line measures the steepness or tilt of a line.
13. A set of ordered pairs that assigns to each *x*-value exactly one *y*-value is called a _____.
14. The set of all *x*-coordinates of a relation is called the _____ of the relation.
15. The set of all *y*-coordinates of a relation is called the _____ of the relation.
16. A set of ordered pairs is called a _____.

CHAPTER 3   HIGHLIGHTS

Definitions and Concepts	**Examples**
Section 3.1 Reading Graphs and the Rectangular Coordinate System	

The **rectangular coordinate system** consists of a plane and a vertical and a horizontal number line intersecting at their 0 coordinates. The vertical number line is called the **y-axis** and the horizontal number line is called the **x-axis**. The point of intersection of the axes is called the **origin**.	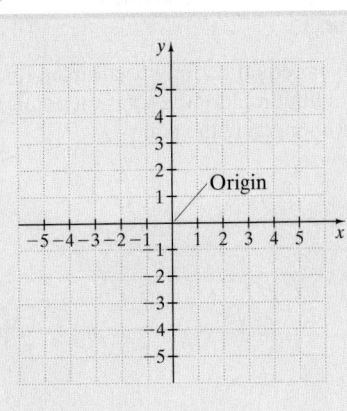

(continued)

| **Definitions and Concepts** | **Examples** |

Section 3.1   Reading Graphs and the Rectangular Coordinate System

To **plot** or **graph** an ordered pair means to find its corresponding point on a rectangular coordinate system.

To plot or graph an ordered pair such as $(3, -2)$, start at the origin. Move 3 units to the right and from there, 2 units down.

To plot or graph $(-3, 4)$ start at the origin. Move 3 units to the left and from there, 4 units up.

An ordered pair is a **solution** of an equation in two variables if replacing the variables by the coordinates of the ordered pair results in a true statement.

Determine whether $(-1, 5)$ is a solution of $2x + 3y = 13$.

$$2x + 3y = 13$$
$$2(-1) + 3 \cdot 5 = 13 \quad \text{Let } x = -1, y = 5$$
$$-2 + 15 = 13$$
$$13 = 13 \quad \text{True}$$

If one coordinate of an ordered pair solution is known, the other value can be determined by substitution.

Complete the ordered pair solution $(0, \quad)$ for the equation $x - 6y = 12$.

$$x - 6y = 12$$
$$0 - 6y = 12 \quad \text{Let } x = 0.$$
$$\frac{-6y}{-6} = \frac{12}{-6} \quad \text{Divide by } -6.$$
$$y = -2$$

The ordered pair solution is $(0, -2)$.

Section 3.2   Graphing Linear Equations

A **linear equation in two variables** is an equation that can be written in the form $Ax + By = C$ where A and B are not both 0. The form $Ax + By = C$ is called **standard form**.

To graph a linear equation in two variables, find three ordered pair solutions. Plot the solution points and draw the line connecting the points.

Linear Equations

$$3x + 2y = -6 \qquad x = -5$$
$$y = 3 \qquad y = -x + 10$$

$x + y = 10$ is in standard form.

Graph $x - 2y = 5$.

x	y
5	0
1	-2
-1	-3

(continued)

Definitions and Concepts	**Examples**

Section 3.3   Intercepts

An **intercept** of a graph is a point where the graph intersects an axis. If a graph intersects the x-axis at a, then $(a, 0)$ is the **x-intercept**. If a graph intersects the y-axis at b, then $(0, b)$ is the **y-intercept**.

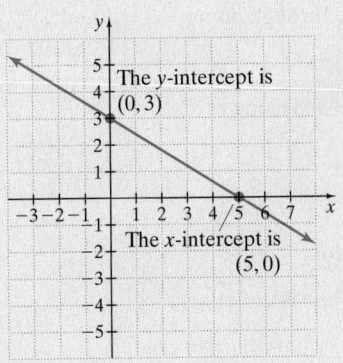

To find the x-intercept, let $y = 0$ and solve for x.

To find the y-intercept, let $x = 0$ and solve for y.

Graph $2x - 5y = -10$ by finding intercepts.

$$\begin{array}{ll}
\text{If } y = 0, \text{ then} & \text{If } x = 0, \text{ then} \\
2x - 5 \cdot 0 = -10 & 2 \cdot 0 - 5y = -10 \\
2x = -10 & -5y = -10 \\
\dfrac{2x}{2} = \dfrac{-10}{2} & \dfrac{-5y}{-5} = \dfrac{-10}{-5} \\
x = -5 & y = 2
\end{array}$$

The x-intercept is $(-5, 0)$. The y-intercept is $(0, 2)$.

The graph of $x = c$ is a vertical line with x-intercept $(c, 0)$.

The graph of $y = c$ is a horizontal line with y-intercept $(0, c)$.

(continued)

Definitions and Concepts	**Examples**

Section 3.4 Slope and Rate of Change

The **slope** m of the line through points (x_1, y_1) and (x_2, y_2) is given by

$$m = \frac{y_2 - y_1}{x_2 - x_1} \quad \text{as long as } x_2 \neq x_1$$

A horizontal line has slope 0.

The slope of a vertical line is undefined.

Nonvertical parallel lines have the same slope.

Two nonvertical lines are perpendicular if the slope of one is the negative reciprocal of the slope of the other.

The slope of the line through points $(-1, 6)$ and $(-5, 8)$ is

$$m = \frac{y_2 - y_1}{x_2 - x_1} = \frac{8 - 6}{-5 - (-1)} = \frac{2}{-4} = -\frac{1}{2}$$

The slope of the line $y = -5$ is 0.
The line $x = 3$ has undefined slope.

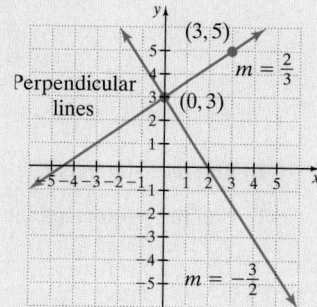

Section 3.5 The Slope-Intercept Form

Slope-Intercept Form

$$y = mx + b$$

m is the slope of the line.
$(0, b)$ is the y-intercept.

Find the slope and the y-intercept of the line whose equation is $2x + 3y = 6$.

Solve for y:

$$2x + 3y = 6$$
$$3y = -2x + 6 \qquad \text{Subtract } 2x.$$
$$y = -\frac{2}{3}x + 2 \qquad \text{Divide by 3.}$$

The slope of the line is $-\frac{2}{3}$ and the y-intercept is $(0, 2)$.

Find an equation of the line with slope 3 and y-intercept $(0, -1)$.

The equation is $y = 3x - 1$.

Section 3.6 The Point-Slope Form

Point-Slope Form

$$y - y_1 = m(x - x_1)$$

m is the slope.
(x_1, y_1) is a point on the line.

Find an equation of the line with slope $\frac{3}{4}$ that contains the point $(-1, 5)$.

$$y - 5 = \frac{3}{4}[x - (-1)]$$

$$4(y - 5) = 3(x + 1) \qquad \text{Multiply by 4.}$$
$$4y - 20 = 3x + 3 \qquad \text{Distribute.}$$
$$-3x + 4y = 23 \qquad \text{Subtract } 3x \text{ and add 20.}$$

(continued)

Definitions and Concepts	Examples

Section 3.7 Functions

A set of ordered pairs is a **relation**. The set of all *x*-coordinates is called the **domain** of the relation and the set of all *y*-coordinates is called the **range** of the relation.

A **function** is a set of ordered pairs that assigns to each *x*-value exactly one *y*-value.

Vertical Line Test

If a vertical line can be drawn so that it intersects a graph more than once, the graph is not the graph of a function.

The symbol $f(x)$ means **function of *x***. This notation is called **function** notation.

The domain of the relation $\{(0, 5), (2, 5), (4, 5), (5, -2)\}$ is $\{0, 2, 4, 5\}$. The range is $\{-2, 5\}$.

Which are graphs of functions?

 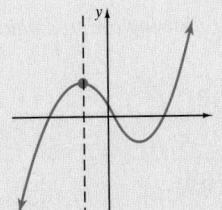

This graph is not the graph of a function.

This graph is the graph of a function.

If $f(x) = 2x^2 + 6x - 1$, find $f(3)$.

$$f(3) = 2(3)^2 + 6 \cdot 3 - 1$$
$$= 2 \cdot 9 + 18 - 1$$
$$= 18 + 18 - 1$$
$$= 35$$

CHAPTER REVIEW

(3.1) *Use the following graph to answer Exercises 1 through 4.*

Source: Cellular Telecommunications Industry Association Semi-Annual Wireless Survey

1. Approximate the number of cellular phone subscribers in 2001.

2. Approximate the increase in cellular phone subscribers in 2001.

3. What year shows the greatest number of subscribers?

4. What trend is shown by this graph?

Use the following graph to answer Exercises 5 through 8.

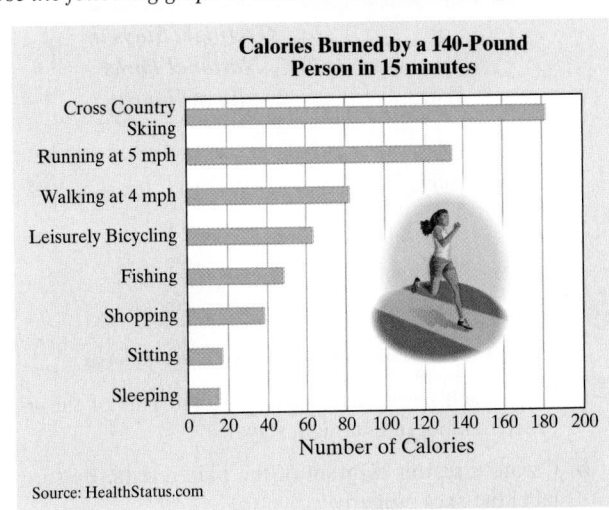

Source: HealthStatus.com

5. Which activity burns the most calories? Estimate the number of calories.

6. Which activity burns the fewest calories? Estimate the number of calories.

7. Estimate how many more calories running at 5mph burns than walking at 4mph.

8. Estimate how many more calories fishing burns than shopping.

Plot the following ordered pairs on a Cartesian coordinate system.

9. $(-7, 0)$

10. $\left(0, 4\frac{4}{5}\right)$

11. $(-2, -5)$

12. $(1, -3)$

13. $(0.7, 0.7)$

14. $(-6, 4)$

15. A local lumberyard uses quantity pricing. The table shows the price per board for different amounts of lumber purchased.

Price per Board (in dollars)	Number of Boards Purchased
8.00	1
7.50	10
6.50	25
5.00	50
2.00	100

a. Write each paired data as an ordered pair of the form (price per board, number of boards purchased).

b. Create a scatter diagram of the paired data. Be sure to label the axes appropriately.

16. The table shows the annual overnight stays in national parks (*Source*: National Park Service)

Year	Overnight Stays in National Parks (in millions)
1996	10.5
1997	10
1998	9.8
1999	9.9
2000	9.6
2001	9.8

a. Write each paired data as an ordered pair of the form (year, number of overnight stays).

b. Create a scatter diagram of the paired data. Be sure to label the axes properly.

Determine whether each ordered pair is a solution of the given equation.

17. $7x - 8y = 56$; $(0, 56), (8, 0)$

18. $-2x + 5y = 10$; $(-5, 0), (1, 1)$

19. $x = 13$; $(13, 5), (13, 13)$

20. $y = 2$; $(7, 2), (2, 7)$

Complete the ordered pairs so that each is a solution of the given equation.

21. $-2 + y = 6x$; $(7, \quad)$

22. $y = 3x + 5$; $(\quad, -8)$

Complete the table of values for each given equation; then plot the ordered pairs. Use a single coordinate system for each exercise.

23. $9 = -3x + 4y$

x	y
	0
	3
9	

24. $y = 5$

x	y
7	
-7	
0	

25. $x = 2y$

x	y
	0
	5
	-5

26. The cost in dollars of producing x compact disk holders is given by $y = 5x + 2000$.

a. Complete the following table.

x	y
1	
100	
1000	

b. Find the number of compact disk holders that can be produced for $6430.

(3.2) *Graph each linear equation.*

27. $x - 3y = 12$

28. $5x - y = -8$

29. $x = 3y$

30. $y = -2x$

31. $2x - 3y = 6$

32. $4x - 3y = 12$

33. The projected U.S. long-distance revenue (in billions of dollars) from 1999 to 2004 is given by the equation, $y = 3x + 111$ where x is the number of years after 1999. Graph this equation and use it to estimate the amount of long-distance revenue in 2007. (*Source*: Giga Information Group)

(3.3) Identify the intercepts.

34.

35.

36.

37.

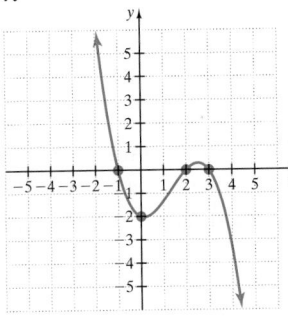

Graph each linear equation by finding its intercepts.

38. $x - 3y = 12$

39. $-4x + y = 8$

40. $y = -3$

41. $x = 5$

42. $y = -3x$

43. $x = 5y$

(3.4) Find the slope of each line.

44.

45.

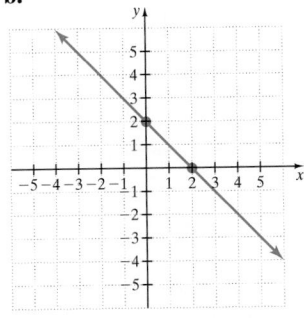

In Exercises 46–50, match each line with its slope.

a.

b.

c.

d.

e.

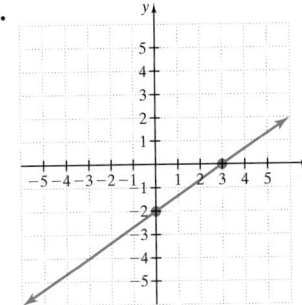

46. $m = 0$

47. $m = -1$

48. undefined slope

49. $m = 3$

50. $m = \dfrac{2}{3}$

Find the slope of the line that goes through the given points.

51. $(2, 5)$ and $(6, 8)$

52. $(4, 7)$ and $(1, 2)$

53. $(1, 3)$ and $(-2, -9)$

54. $(-4, 1)$ and $(3, -6)$

Find the slope of each line.

55. $x = 5$

56. $y = -1$

57. $y = -2$

58. $x = 0$

Find the slope of each line and write the slope as a rate of change. Don't forget to attach the proper units.

59. The graph below shows the average monthly day care cost for a 3-year-old attending 8 hours a day, 5 days a week.

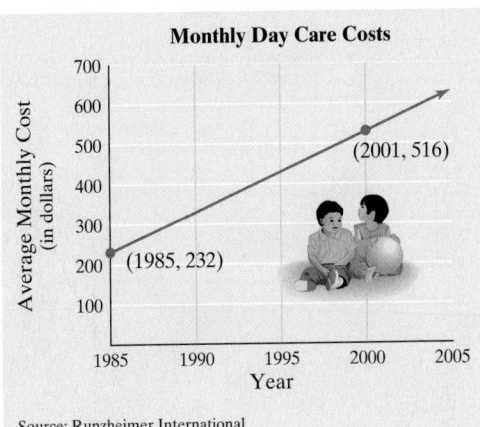

Monthly Day Care Costs

(2001, 516)

(1985, 232)

Source: Runzheimer International

60. The graph below shows the U.S. government's projected spending (in billions of dollars) on technology.

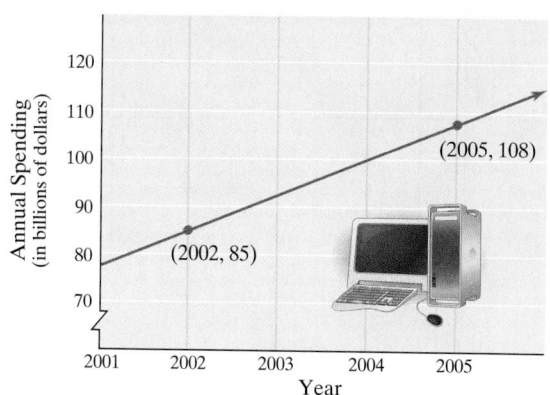

(3.5) Determine the slope and the y-intercept of the graph of each equation.

61. $3x + y = 7$

62. $x - 6y = -1$

63. $y = 2$

64. $x = -5$

Determine whether the lines are parallel, perpendicular, or neither.

△ **65.** $x - y = -6$
$x + y = 3$

△ **66.** $3x + y = 7$
$-3x - y = 10$

△ **67.** $y = 4x + \dfrac{1}{2}$
$4x + 2y = 1$

Write an equation of each line in slope-intercept form.

68. slope -5; y-intercept $\dfrac{1}{2}$

69. slope $\dfrac{2}{3}$; y-intercept 6

Use the slope-intercept form to graph each equation.

70. $y = -3x$

71. $y = 3x - 1$

72. $-x + 2y = 8$

73. $5x - 3y = 15$

In Exercises 74–77, match each equation with its graph.

74. $y = -2x + 1$

75. $y = -4x$

76. $y = 2x$

77. $y = 2x - 1$

a.

b.

c.

d.

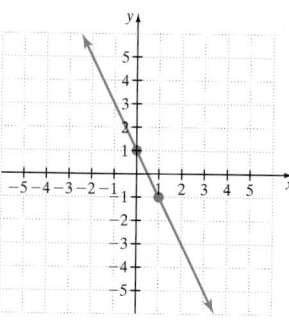

(3.6) Write an equation of each line in standard form.

78. With slope 4, through $(2, 0)$

79. With slope -3, through $(0, -5)$

80. With slope $\dfrac{1}{2}$, through $\left(0, -\dfrac{7}{2}\right)$

81. With slope 0, through $(-2, -3)$

82. With 0 slope, through the origin

83. With slope -6, through $(2, -1)$

84. With slope 12, through $\left(\dfrac{1}{2}, 5\right)$

85. Through $(0, 6)$ and $(6, 0)$

86. Through $(0, -4)$ and $(-8, 0)$

87. Vertical line, through $(5, 7)$

88. Horizontal line, through $(-6, 8)$

△ **89.** Through $(6, 0)$, perpendicular to $y = 8$

△ **90.** Through $(10, 12)$, perpendicular to $x = -2$

△ **91.** Write an equation in standard form of the line that contains $(5, 0)$ and is
  a. parallel to the line $y = -3x + 7$.
  b. perpendicular to the line $y = -3x + 7$.

(3.7) Determine which of the following are functions

92. $\{(7, 1), (7, 5), (2, 6)\}$

93. $\{(0, -1), (5, -1), (2, 2)\}$

94. $7x - 6y = 1$

95. $y = 7$

96. $x = 2$

97. $y = x^3$

98. $x = y^2$

99.

100.

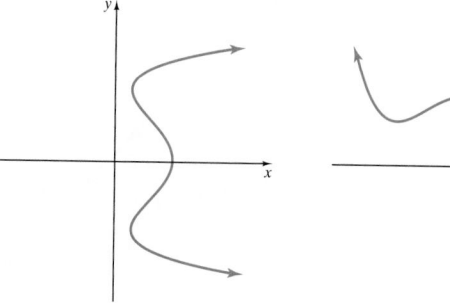

Given the following functions, find the indicated function values.

101. Given $f(x) = -2x + 6$, find

   a. $f(0)$ **b.** $f(-2)$ **c.** $f\left(\dfrac{1}{2}\right)$

102. Given $h(x) = -5 - 3x$, find

   a. $h(2)$ **b.** $h(-3)$ **a.** $h(0)$

103. Given $g(x) = x^2 + 12x$, find

   a. $g(3)$ **b.** $g(-5)$ **a.** $g(0)$

104. Given $h(x) = 6 - |x|$, find

   a. $h(-1)$ **b.** $h(1)$ **c.** $h(-4)$

Find the domain of each function.

105. $f(x) = 2x + 7$ **106.** $g(x) = \dfrac{7}{x - 2}$

Find the domain and the range of each function graphed.

107.

108.

109.

110.

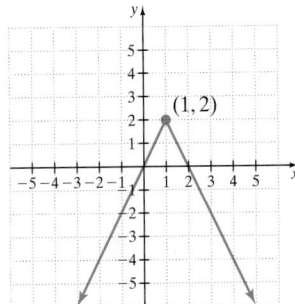

CHAPTER 3 TEST

Remember to use your Chapter Test Prep Video CD to help you study and view solutions to the test questions you need help with.

Graph the following.

1. The table gives the percent of new mothers who returned to work after having their child.

Year	Percent of New Mothers Returning to Work
1980	38
1984	47
1988	51
1992	54
1996	59
2000	55

(*Source*: Census Bureau)

   a. Write this data as a set of ordered pairs of the form (year, percent of new mothers returning to work).

   b. Create a scatter diagram of the data. Be sure to label the axes properly.

2. $2x + y = 8$ **3.** $5x - 7y = 10$

4. $y = -1$ **5.** $x - 3 = 0$

Find the slopes of the following lines.

6.

7.

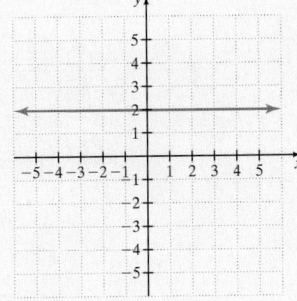

8. Through $(6, -5)$ and $(-1, 2)$

9. $-3x + y = 5$ **10.** $x = 6$

11. Determine the slope and the y-intercept of the graph of $7x - 3y = 2$.

△ **12.** Determine whether the graphs of $y = 2x - 6$ and $-4x = 2y$ are parallel lines, perpendicular lines, or neither.

Find equations of the following lines. Write the equation in standard form.

13. With slope of $-\frac{1}{4}$, through $(2, 2)$

14. Through the origin and $(6, -7)$

15. Through $(2, -5)$ and $(1, 3)$

△ **16.** Through $(-5, -1)$ and parallel to $x = 7$

17. With slope $\frac{1}{8}$ and y-intercept $(0, 12)$

Which of the following are functions?

18. **19.**

Given the following functions, find the indicated function values.

20. $h(x) = x^3 - x$

    a. $h(-1)$ **b.** $h(0)$ **c.** $h(4)$

21. Find the domain of $y = \dfrac{1}{x + 1}$.

Find the domain and the range of each function graphed.

22.

23.

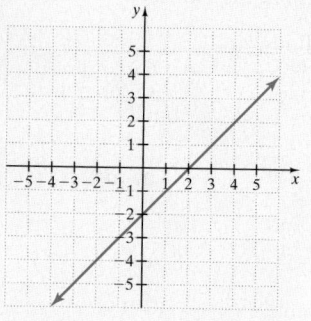

This graph shows the sunset times for Seward, Alaska. Use this graph to answer Exercises 24 through 29.

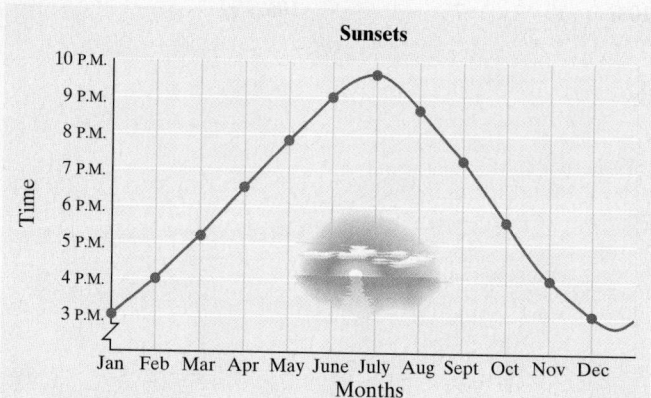

24. Approximate the time of sunset on June 1.

25. Approximate the time of sunset on November 1.

26. Approximate the date(s) when the sunset is 3 P.M.

27. Approximate the date(s) when the sunset is 9 P.M.

28. Is this graph the graph of a function? Why or why not?

29. Do you think a graph of sunset times for any location will always be a function? Why or why not?

CHAPTER CUMULATIVE REVIEW

1. Insert $<, >$, or $=$ in the space between each pair of numbers to make each statement true.

   a. 2   3   **b.** 7   4   **c.** 72   27

2. Write the fraction $\dfrac{56}{64}$ in lowest terms.

3. Multiply $\dfrac{2}{15}$ and $\dfrac{5}{13}$. Write the product in lowest terms

4. Add: $\dfrac{10}{3} + \dfrac{5}{21}$

5. Simplify: $\dfrac{3 + |4 - 3| + 2^2}{6 - 3}$

6. Simplify: $16 - 3 \cdot 3 + 2^4$

7. Add.

   a. $-8 + (-11)$ **b.** $-5 + 35$

   c. $0.6 + (-1.1)$ **d.** $-\dfrac{7}{10} + \left(-\dfrac{1}{10}\right)$

   e. $11.4 + (-4.7)$ **f.** $-\dfrac{3}{8} + \dfrac{2}{5}$

8. Simplify: $|9 + (-20)| + |-10|$

9. Simplify each expression.

   a. $-14 - 8 + 10 - (-6)$

   b. $1.6 - (-10.3) + (-5.6)$

10. Simplify: $-9 - (3 - 8)$

11. If $x = -2$ and $y = -4$, evaluate each expression.

   a. $5x - y$ **b.** $x^4 - y^2$

   c. $\dfrac{3x}{2y}$

12. Is -20 a solution of $\dfrac{x}{-10} = 2$?

13. Simplify each expression.

   a. $10 + (x + 12)$ **b.** $-3(7x)$

14. Simplify: $(12 + x) - (4x - 7)$

15. Identify the numerical coefficient in each term.

   a. $-3y$ **b.** $22z^4$

   c. y **d.** $-x$

   e. $\dfrac{x}{7}$

16. Multiply: $-5(x - 7)$

17. Solve $y + 0.6 = -1.0$ for y.

18. Solve: $5(3 + z) - (8z + 9) = -4$

19. Solve: $\dfrac{5}{2}x = 15$

20. Solve: $\dfrac{x}{4} - 1 = -7$

21. If x is the first of three consecutive integers, express the sum of the three integers in terms of x. Simplify if possible.

22. Solve: $\dfrac{x}{3} - 2 = \dfrac{x}{3}$

23. Solve: $\dfrac{2(a + 3)}{3} = 6a + 2$

24. Solve: $x + 2y = 6$ for y.

25. In a recent year, the U.S. House of Representatives had a total of 431 Democrats and Republicans. There were 15 more Republican representatives than Democratic. Find the number of representatives from each party.

26. Solve $5(x + 4) \geq 4(2x + 3)$. Write the solution set in interval notation.

27. Charles Pecot can afford enough fencing to enclose a rectangular garden with a perimeter of 140 feet. If the width of his garden is to be 30 feet, find the length.

28. Solve $-3 < 4x - 1 \leq 2$. Write the solution set in interval notation.

29. Solve $y = mx + b$ for x.

30. Complete the table for $y = -5x$.

x	y
0	
-1	
	-10

31. A chemist working on his doctoral degree at Massachusetts Institute of Technology needs 12 liters of a 50% acid solution for a lab experiment. The stockroom has only 40% and 70% solutions. How much of each solution should be mixed together to form 12 liters of a 50% solution?

32. Graph: $y = -3x + 5$

33. Graph $x \geq -1$.

34. Find the x- and y-intercepts of $2x + 4y = -8$.

35. Solve $-1 \leq 2x - 3 < 5$, and graph the solution set.

36. Graph $x = 2$ on   a   rectangular   coordinate   system.

37. Determine whether each ordered pair is a solution of the equation $x - 2y = 6$.

   a. $(6, 0)$

   b. $(0, 3)$

   c. $\left(1, -\dfrac{5}{2}\right)$

38. Find the slope of the line through $(0, 5)$ and $(-5, 4)$.

39. Identify the linear equations in two variables.

   a. $x - 1.5y = -1.6$

   b. $y = -2x$

   c. $x + y^2 = 9$

   d. $x = 5$

40. Find the slope of $x = -10$.

41. Find the slope of the line $y = -1$.

42. Find the slope and y-intercept of the line whose equation is $2x - 5y = 10$.

43. Find the slope of a line perpendicular to the line passing through the points $(-1, 7)$ and $(2, 2)$.

44. Write an equation of the line through $(2, 3)$ and $(0, 0)$. Write the equation in standard form.

CHAPTER

Systems of Linear Equations

LARGEST HEALTH CARE OCCUPATION IN THE UNITED STATES

O ver 2.5 million people work as nurses in settings as varied as hospitals, private homes, corporate offices, nursing homes, overnight camps, doctors' offices, and community health centers. The U.S. Bureau of Labor Statistics predicts that the demand for nurses will continue to grow rapidly as the American population ages, requiring more long-term health care and home health care.

Registered nurses must be licensed in the state in which they work. About 25% of registered nurses hold a diploma from a hospital program, 35% hold an associate's degree, 30% hold a bachelor's degree, and 10% hold a higher degree. Nurses use math skills nearly every day in taking and comparing vital signs, administering medications, and tracking fluid intake and output.

In the Spotlight on Decision Making feature on page 239, you will have the opportunity, as a registered nurse, to make a decision concerning a patient's blood pressure.

In Chapter 3, we graphed equations containing two variables. Equations like these are often needed to represent relationships between two different values. For example, an economist attempts to predict what effects a price change will have on the sales prospects of calculators. There are many real-life opportunities to compare and contrast two such equations, called a system of equations. This chapter presents linear systems and ways we solve these systems and apply them to real-life situations.

4.1   SOLVING SYSTEMS OF LINEAR EQUATIONS BY GRAPHING

Objectives

1 Determine if an ordered pair is a solution of a system of equations in two variables.

2 Solve a system of linear equations by graphing.

3 Without graphing, determine the number of solutions of a system.

1 A **system of linear equations** consists of two or more linear equations. In this section, we focus on solving systems of linear equations containing two equations in two variables. Examples of such linear systems are

$$\begin{cases} 3x - 3y = 0 \\ x = 2y \end{cases} \qquad \begin{cases} x - y = 0 \\ 2x + y = 10 \end{cases} \qquad \begin{cases} y = 7x - 1 \\ y = 4 \end{cases}$$

A **solution** of a system of two equations in two variables is an ordered pair of numbers that is a solution of both equations in the system.

EXAMPLE 1

Which of the following ordered pairs is a solution of the given system?

$$\begin{cases} 2x - 3y = 6 & \text{First equation} \\ x = 2y & \text{Second equation} \end{cases}$$

a. $(12, 6)$ **b.** $(0, -2)$

Solution   If an ordered pair is a solution of both equations, it is a solution of the system.

a. Replace x with 12 and y with 6 in both equations.

$$2x - 3y = 6 \quad \text{First equation} \qquad\qquad x = 2y \quad \text{Second equation}$$
$$2(12) - 3(6) \overset{?}{=} 6 \quad \text{Let } x = 12 \text{ and } y = 6. \qquad 12 \overset{?}{=} 2(6) \quad \text{Let } x = 12 \text{ and } y = 6.$$
$$24 - 18 \overset{?}{=} 6 \quad \text{Simplify.} \qquad\qquad\qquad 12 = 12 \quad \text{True}$$
$$6 = 6 \quad \text{True}$$

Since $(12, 6)$ is a solution of both equations, it is a solution of the system.

b. Start by replacing x with 0 and y with -2 in both equations.

$$2x - 3y = 6 \quad \text{First equation} \qquad\qquad x = 2y \quad \text{Second equation}$$
$$2(0) - 3(-2) \overset{?}{=} 6 \quad \text{Let } x = 0 \text{ and } y = -2. \qquad 0 \overset{?}{=} 2(-2) \quad \text{Let } x = 0 \text{ and } y = -2.$$
$$0 + 6 \overset{?}{=} 6 \quad \text{Simplify.} \qquad\qquad\qquad 0 = -4 \quad \text{False}$$
$$6 = 6 \quad \text{True}$$

While $(0, -2)$ is a solution of the first equation, it is not a solution of the second equation, so it is **not** a solution of the system.

2 Since a solution of a system of two equations in two variables is a solution common to both equations, it is also a point common to the graphs of both equations. Let's practice finding solutions of both equations in a system—that is, solutions of a system—by graphing and identifying points of intersection.

EXAMPLE 2

Solve the system of equations by graphing.

$$\begin{cases} -x + 3y = 10 \\ x + y = 2 \end{cases}$$

Solution   On a single set of axes, graph each linear equation.

$-x + 3y = 10$

x	y
0	$\dfrac{10}{3}$
-4	2
2	4

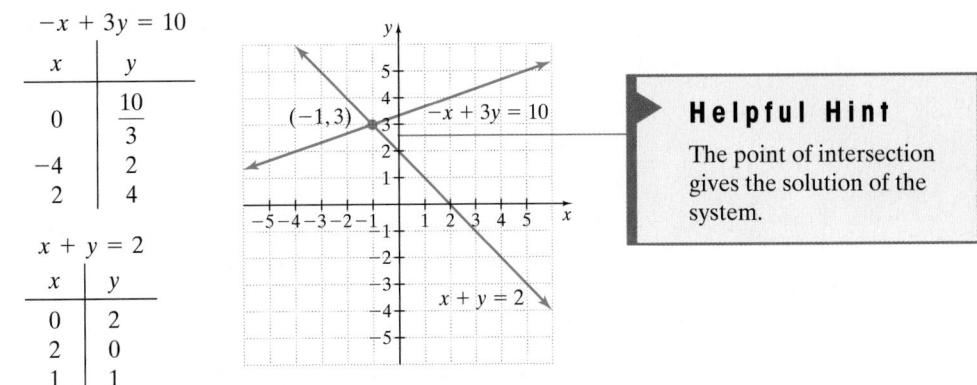

$x + y = 2$

x	y
0	2
2	0
1	1

Helpful Hint

The point of intersection gives the solution of the system.

The two lines appear to intersect at the point $(-1, 3)$. To check, we replace x with -1 and y with 3 in both equations.

$-x + 3y = 10$ First equation

$-(-1) + 3(3) \overset{?}{=} 10$ Let $x = -1$ and $y = 3$.

$1 + 9 \overset{?}{=} 10$ Simplify.

$10 = 10$ True

$x + y = 2$ Second equation

$-1 + 3 \overset{?}{=} 2$ Let $x = -1$ and $y = 3$.

$2 = 2$ True

$(-1, 3)$ checks, so it is the solution of the system.

Helpful Hint

Neatly drawn graphs can help when you are estimating the solution of a system of linear equations by graphing.

In the example above, notice that the two lines intersected in a point. This means that the system has 1 solution.

A system of equations that has at least one solution as in Example 2 is said to be a **consistent system**. A system that has no solution is said to be an **inconsistent system**.

EXAMPLE 3

Solve the following system of equations by graphing.

$$\begin{cases} 2x + y = 7 \\ 2y = -4x \end{cases}$$

Solution Graph the two lines in the system.

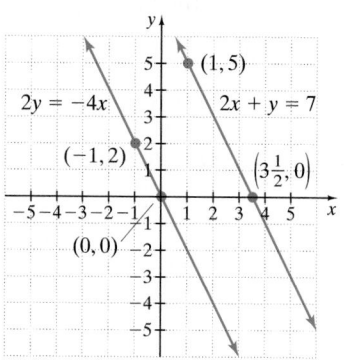

The lines **appear** to be parallel. To confirm this, write both equations in slope-intercept form by solving each equation for y.

$2x + y = 7$	First equation	$2y = -4x$	Second equation
$y = -2x + 7$	Subtract $2x$ from both sides.	$\dfrac{2y}{2} = \dfrac{-4x}{2}$	Divide both sides by 2.
		$y = -2x$	

Recall that when an equation is written in slope-intercept form, the coefficient of x is the slope. Since both equations have the same slope, -2, but different y-intercepts, the lines are parallel and have no points in common. Thus, there is no solution of the system and the system is inconsistent.

In Examples 2 and 3, the graphs of the two linear equations of each system are different. When this happens, we call these equations **independent equations**. If the graphs of the two equations in a system are identical, we call the equations **dependent equations**.

 EXAMPLE 4

Solve the system of equations by graphing.

$$\begin{cases} x - y = 3 \\ -x + y = -3 \end{cases}$$

Solution Graph each line.

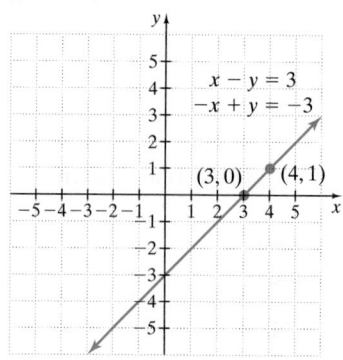

These graphs **appear** to be identical. To confirm this, write each equation in slope-intercept form.

$x - y = 3$ First equation $-x + y = -3$ Second equation

$-y = -x + 3$ Subtract x from both sides. $y = x - 3$ Add x to both sides.

$\dfrac{-y}{-1} = \dfrac{-x}{-1} + \dfrac{3}{-1}$ Divide both sides by -1.

$y = x - 3$

The equations are identical and so must be their graphs. The lines have an infinite number of points in common. Thus, there is an infinite number of solutions of the system and this is a consistent system. The equations are dependent equations.

As we have seen, three different situations can occur when graphing the two lines associated with the equations in a linear system:

One point of intersection: one solution

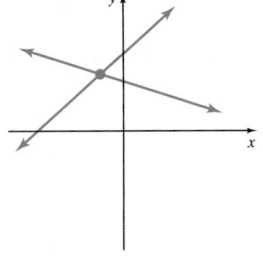

Consistent system
(at least one solution)
Independent equations
(graphs of equations differ)

Parallel lines: no solution

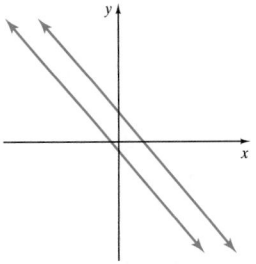

Inconsistent system
(no solution)
Independent equations
(graphs of equations differ)

Same line: infinite number of solutions

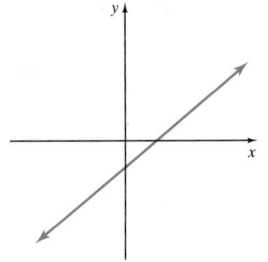

Consistent system
(at least one solution)
Dependent equations
(graphs of equations identical)

3   You may have suspected by now that graphing alone is not an accurate way to solve a system of linear equations. For example, a solution of $\left(\frac{1}{2}, \frac{2}{9}\right)$ is unlikely to be read correctly from a graph. The next two sections present two accurate methods of solving these systems. In the meantime, we can decide how many solutions a system has by writing each equation in the slope-intercept form.

EXAMPLE 5

Without graphing, determine the number of solutions of the system.

$$\begin{cases} \dfrac{1}{2}x - y = 2 \\ \quad x = 2y + 5 \end{cases}$$

Solution First write each equation in slope-intercept form.

$$\frac{1}{2}x - y = 2 \qquad \text{First equation} \qquad\qquad\qquad x = 2y + 5 \qquad \text{Second equation}$$

$$\frac{1}{2}x = y + 2 \qquad \text{Add } y \text{ to both sides.} \qquad\qquad x - 5 = 2y \qquad \text{Subtract 5 from both sides.}$$

$$\frac{x}{2} - \frac{5}{2} = \frac{2y}{2} \qquad \text{Divide both sides by 2.}$$

$$\frac{1}{2}x - 2 = y \qquad \text{Subtract 2 from both sides.} \qquad\qquad \frac{1}{2}x - \frac{5}{2} = y \qquad \text{Simplify.}$$

The slope of each line is $\frac{1}{2}$, but they have different y-intercepts. This tells us that the lines representing these equations are parallel. Since the lines are parallel, the system has no solution and is inconsistent.

EXAMPLE 6

Determine the number of solutions of the system.

$$\begin{cases} 3x - y = 4 \\ x + 2y = 8 \end{cases}$$

Solution Once again, the slope-intercept form helps determine how many solutions this system has.

$$3x - y = 4 \qquad \text{First equation} \qquad\qquad\qquad x + 2y = 8 \qquad \text{Second equation}$$

$$3x = y + 4 \qquad \text{Add } y \text{ to both sides.} \qquad\qquad x = -2y + 8 \qquad \text{Subtract 2y from both sides.}$$

$$x - 8 = -2y \qquad \text{Subtract 8 from both sides.}$$

$$3x - 4 = y \qquad \text{Subtract 4 from both sides.} \qquad\qquad \frac{x}{-2} - \frac{8}{-2} = \frac{-2y}{-2} \qquad \text{Divide both sides by } -2.$$

$$-\frac{1}{2}x + 4 = y \qquad \text{Simplify.}$$

The slope of the second line is $-\frac{1}{2}$, whereas the slope of the first line is 3. Since the slopes are not equal, the two lines are neither parallel nor identical and must intersect. Therefore, this system has one solution and is consistent.

Spotlight on
DECISION
⁊ MAKING

Suppose you are a registered nurse. Today you are working at a health fair providing free blood pressure screenings. You measure an attendee's blood pressure as 168/82 (read as "168 over 82," where the systolic blood pressure is listed first and the diastolic blood pressure is listed second). What would you recommend that this health fair attendee do? Explain.

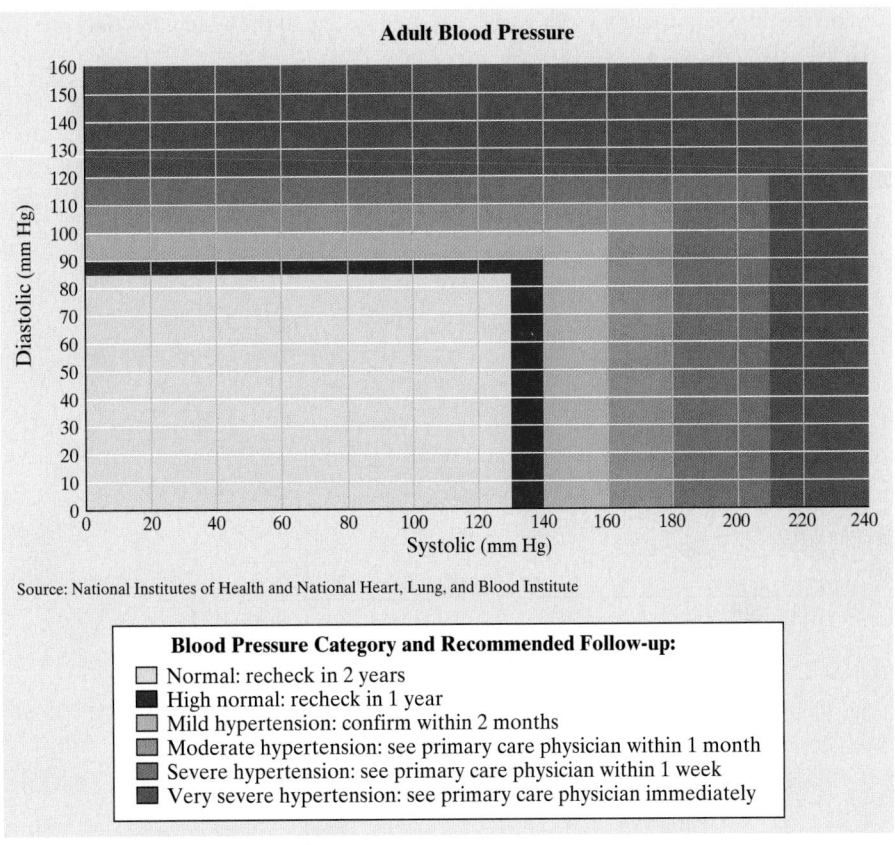

Source: National Institutes of Health and National Heart, Lung, and Blood Institute

Blood Pressure Category and Recommended Follow-up:
- ☐ Normal: recheck in 2 years
- ■ High normal: recheck in 1 year
- ■ Mild hypertension: confirm within 2 months
- ■ Moderate hypertension: see primary care physician within 1 month
- ■ Severe hypertension: see primary care physician within 1 week
- ■ Very severe hypertension: see primary care physician immediately

Graphing Calculator Explorations

A graphing calculator may be used to approximate solutions of systems of equations. For example, to approximate the solution of the system

$$\begin{cases} y = -3.14x - 1.35 \\ y = 4.88x + 5.25, \end{cases}$$

first graph each equation on the same set of axes. Then use the intersect feature of your calculator to approximate the point of intersection.

The approximate point of intersection is $(-0.82, 1.23)$.

Solve each system of equations. Approximate the solutions to two decimal places.

1. $\begin{cases} y = -2.68x + 1.21 \\ y = 5.22x - 1.68 \end{cases}$

2. $\begin{cases} y = 4.25x + 3.89 \\ y = -1.88x + 3.21 \end{cases}$

3. $\begin{cases} 4.3x - 2.9y = 5.6 \\ 8.1x + 7.6y = -14.1 \end{cases}$

4. $\begin{cases} -3.6x - 8.6y = 10 \\ -4.5x + 9.6y = -7.7 \end{cases}$

MENTAL MATH

Each rectangular coordinate system shows the graph of the equations in a system of equations. Use each graph to determine the number of solutions for each associated system. If the system has only one solution, give its coordinates. (The coordinates will be integers.)

1.

2.

3.

4.

5.

6.

7.

8.

EXERCISE SET 4.1

STUDY   CD/   PH MATH   MathXL®Tutorials   MathXL®   MyMathLab®
GUIDE/SSM   VIDEO   TUTOR CENTER   ON CD

Determine whether the given ordered pairs satisfy the system of the linear equations. See Example 1.

1. $\begin{cases} x + y = 8 \\ 3x + 2y = 21 \end{cases}$

   a. $(2, 4)$
   b. $(5, 3)$
   c. $(1, 9)$

2. $\begin{cases} 2x + y = 5 \\ x + 3y = 5 \end{cases}$

   a. $(5, 0)$
   b. $(1, 2)$
   c. $(2, 1)$

3. $\begin{cases} 3x - y = 5 \\ x + 2y = 11 \end{cases}$

   a. $(2, -1)$
   b. $(3, 4)$
   c. $(0, -5)$

4. $\begin{cases} 2x - 3y = 8 \\ x - 2y = 6 \end{cases}$

   a. $(4, 0)$
   b. $(-2, -4)$
   c. $(7, 2)$

5. $\begin{cases} 2y = 4x \\ 2x - y = 0 \end{cases}$

   a. $(-3, -6)$

   b. $(0, 0)$

   c. $(1, 2)$

6. $\begin{cases} 4x = 1 - y \\ x - 3y = -8 \end{cases}$

   a. $(0, 1)$

   b. $(1, -3)$

   c. $(-2, 2)$

7. Construct a system of two linear equations that has $(2, 5)$ as a solution.

8. Construct a system of two linear equations that has $(0, 1)$ as a solution.

MIXED PRACTICE

Solve each system of equations by graphing the equations on the same set of axes. Tell whether the system is consistent or inconsistent and whether the equations are dependent or independent. See Examples 2 through 4.

9. $\begin{cases} y = x + 1 \\ y = 2x - 1 \end{cases}$

10. $\begin{cases} y = 3x - 4 \\ y = x + 2 \end{cases}$

11. $\begin{cases} 2x + y = 0 \\ 3x + y = 1 \end{cases}$

12. $\begin{cases} 2x + y = 1 \\ 3x + y = 0 \end{cases}$

13. $\begin{cases} y = -x - 1 \\ y = 2x + 5 \end{cases}$

14. $\begin{cases} y = x - 1 \\ y = -3x - 5 \end{cases}$

15. $\begin{cases} 2x - y = 6 \\ \quad\quad y = 2 \end{cases}$

16. $\begin{cases} x + y = 5 \\ \quad\quad x = 4 \end{cases}$

17. $\begin{cases} x + y = 5 \\ x + y = 6 \end{cases}$

18. $\begin{cases} 2x + y = 4 \\ \quad x + y = 2 \end{cases}$

19. $\begin{cases} y - 3x = -2 \\ 6x - 2y = 4 \end{cases}$

20. $\begin{cases} y + 2x = 3 \\ 4x = 2 - 2y \end{cases}$

21. $\begin{cases} x - 2y = 2 \\ 3x + 2y = -2 \end{cases}$

22. $\begin{cases} x + 3y = 7 \\ 2x - 3y = -4 \end{cases}$

23. $\begin{cases} \dfrac{1}{2}x + y = -1 \\ \quad\quad x = 4 \end{cases}$

24. $\begin{cases} x + \dfrac{3}{4}y = 2 \\ \quad\quad x = -1 \end{cases}$

25. $\begin{cases} y = x - 2 \\ y = 2x + 3 \end{cases}$

26. $\begin{cases} y = x + 5 \\ y = -2x - 4 \end{cases}$

27. $\begin{cases} x + y = 7 \\ x - y = 3 \end{cases}$

28. $\begin{cases} x + y = -4 \\ x - y = 2 \end{cases}$

29. Explain how to use a graph to determine the number of solutions of a system.

30. The ordered pair $(-2, 3)$ is a solution of all three independent equations:

$$x + y = 1$$
$$2x - y = -7$$
$$x + 3y = 7$$

Describe the graph of all three equations on the same axes.

Without graphing, decide.

   a. Are the graphs of the equations identical lines, parallel lines, or lines intersecting at a single point?

   b. How many solutions does the system have? See Examples 5 and 6.

31. $\begin{cases} 4x + y = 24 \\ x + 2y = 2 \end{cases}$

32. $\begin{cases} 3x + y = 1 \\ 3x + 2y = 6 \end{cases}$

33. $\begin{cases} 2x + y = 0 \\ 2y = 6 - 4x \end{cases}$

34. $\begin{cases} 3x + y = 0 \\ 2y = -6x \end{cases}$

35. $\begin{cases} 6x - y = 4 \\ \dfrac{1}{2}y = -2 + 3x \end{cases}$

36. $\begin{cases} 3x - y = 2 \\ \dfrac{1}{3}y = -2 + 3x \end{cases}$

37. $\begin{cases} x = 5 \\ y = -2 \end{cases}$

38. $\begin{cases} y = 3 \\ x = -4 \end{cases}$

39. $\begin{cases} 3y - 2x = 3 \\ x + 2y = 9 \end{cases}$

40. $\begin{cases} 2y = x + 2 \\ y + 2x = 3 \end{cases}$

41. $\begin{cases} 6y + 4x = 6 \\ 3y - 3 = -2x \end{cases}$

42. $\begin{cases} 8y + 6x = 4 \\ 4y - 2 = 3x \end{cases}$

43. $\begin{cases} x + y = 4 \\ x + y = 3 \end{cases}$

44. $\begin{cases} 2x + y = 0 \\ y = -2x + 1 \end{cases}$

REVIEW AND PREVIEW

Solve each equation. See Section 2.3.

45. $5(x - 3) + 3x = 1$

46. $-2x + 3(x + 6) = 17$

47. $4\left(\dfrac{y + 1}{2}\right) + 3y = 0$

48. $-y + 12\left(\dfrac{y - 1}{4}\right) = 3$

49. $8a - 2(3a - 1) = 6$

50. $3z - (4z - 2) = 9$

Concept Extensions

51. Explain how writing each equation in a linear system in the point-slope form helps determine the number of solutions of a system.

52. Is it possible for a system of two linear equations in two variables to be inconsistent, but with dependent equations? Why or why not?

The double line graph below shows the number of pounds of fishery products from U.S. domestic catch and from imports. Use this graph to answer Exercises 53 and 54.

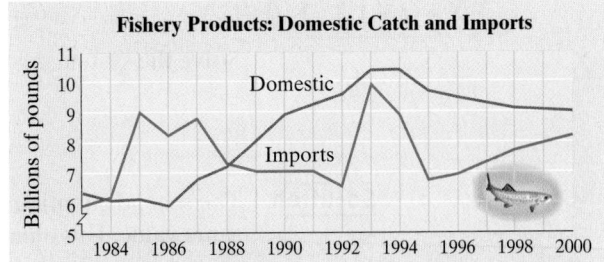

Fishery Products: Domestic Catch and Imports

Source: U.S. Bureau of the Census, Statistical Abstract of the United States: 1998, 115th ed., Washington, DC, 1995.

53. In what year(s) was the number of pounds of fishery products imported equal to the number of pounds of domestic catch?

54. In what year(s) was the number of pounds of fishery products imported greater than the number of pounds of domestic catch?

The double line graph below shows the number of Kmart stores vs. the number of Wal-Mart and Wal-Mart Supercenter stores. Use this graph to answer Exercises 55 through 58.

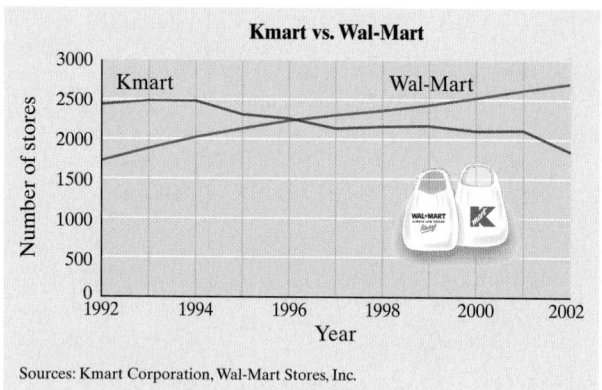

Kmart vs. Wal-Mart

Sources: Kmart Corporation, Wal-Mart Stores, Inc.

55. In what year was the number of Kmart stores approximately equal to the number of Wal-Mart stores?

56. In what years was the number of Wal-Mart stores greater than the number of Kmart stores?

57. Describe any trends you see in this graph.

58. Approximate how many more stores Wal-Mart had than Kmart in 2002.

59. Construct a system of two linear equations that has $(1, 3)$ as a solution.

60. Construct a system of two linear equations that has $(0, 7)$ as a solution.

61. Below are two tables of values for two linear equations. Using the tables,

   a. find a solution of the corresponding system.

   b. graph several ordered pairs from each table and sketch the two lines.

   Does your graph confirm the solution from part a?

x	y	x	y
1	3	1	6
2	5	2	7
3	7	3	8
4	9	4	9
5	11	5	10

4.2   SOLVING SYSTEMS OF LINEAR EQUATIONS BY SUBSTITUTION

Objective

1   Use the substitution method to solve a system of linear equations.

1   As we stated in the preceding section, graphing alone is not an accurate way to solve a system of linear equations. In this section, we discuss a second, more accurate method for solving systems of equations. This method is called the **substitution method** and is introduced in the next example.

EXAMPLE 1

Solve the system:

$$\begin{cases} 2x + y = 10 & \text{First equation} \\ x = y + 2 & \text{Second equation} \end{cases}$$

Solution   The second equation in this system is $x = y + 2$. This tells us that x and $y + 2$ have the same value. This means that we may substitute $y + 2$ for x in the first equation.

$$2x + y = 10 \qquad \text{First equation}$$

$$2\,(y + 2) + y = 10 \qquad \text{Substitute } y + 2 \text{ for } x \text{ since } x = y + 2.$$

Notice that this equation now has one variable, y. Let's now solve this equation for y.

$$2(y + 2) + y = 10$$

$$2y + 4 + y = 10 \quad \text{Use the distributive property.}$$

$$3y + 4 = 10 \quad \text{Combine like terms.}$$

$$3y = 6 \quad \text{Subtract 4 from both sides.}$$

$$y = 2 \quad \text{Divide both sides by 3.}$$

Now we know that the y-value of the ordered pair solution of the system is 2. To find the corresponding x-value, we replace y with 2 in the equation $x = y + 2$ and solve for x.

$$x = y + 2$$

$$x = 2 + 2 \quad \text{Let } y = 2.$$

$$x = 4$$

The solution of the system is the ordered pair $(4, 2)$. Since an ordered pair solution must satisfy both linear equations in the system, we could have chosen the equation $2x + y = 10$ to find the corresponding x-value. The resulting x-value is the same.

Check   We check to see that $(4, 2)$ satisfies both equations of the original system.

First Equation	*Second Equation*
$2x + y = 10$	$x = y + 2$
$2(4) + 2 \overset{?}{=} 10$	$4 \overset{?}{=} 2 + 2 \quad \text{Let } x = 4 \text{ and } y = 2.$
$10 = 10 \quad \text{True}$	$4 = 4 \quad \text{True}$

The solution of the system is $(4, 2)$.

A graph of the two equations shows the two lines intersecting at the point $(4, 2)$.

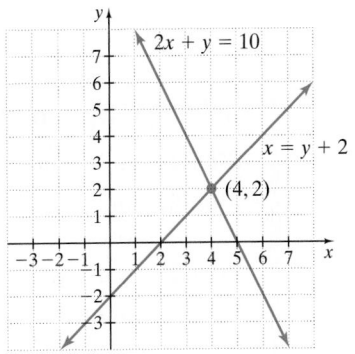

To solve a system of equations by substitution, we first need an equation solved for one of its variables.

EXAMPLE 2

Solve the system:

$$\begin{cases} x + 2y = 7 \\ 2x + 2y = 13 \end{cases}$$

Solution   We choose one of the equations and solve for x or y. We will solve the first equation for x by subtracting $2y$ from both sides.

$$x + 2y = 7 \qquad \text{First equation}$$
$$x = 7 - 2y \qquad \text{Subtract } 2y \text{ from both sides.}$$

Since $x = 7 - 2y$, we now substitute $7 - 2y$ for x in the second equation and solve for y.

$$2x + 2y = 13 \qquad \text{Second equation}$$
$$2(7 - 2y) + 2y = 13 \qquad \text{Let } x = 7 - 2y.$$

$$14 - 4y + 2y = 13 \qquad \text{Use the distributive property.}$$
$$14 - 2y = 13 \qquad \text{Simplify.}$$
$$-2y = -1 \qquad \text{Subtract 14 from both sides.}$$
$$y = \frac{1}{2} \qquad \text{Divide both sides by } -2.$$

> **Helpful Hint**
> Don't forget to insert parentheses when substituting $7 - 2y$ for x.

To find x, we let $y = \frac{1}{2}$ in the equation $x = 7 - 2y$.

$$x = 7 - 2y$$
$$x = 7 - 2\left(\frac{1}{2}\right) \qquad \text{Let } y = \frac{1}{2}.$$
$$x = 7 - 1$$
$$x = 6$$

The solution is $\left(6, \frac{1}{2}\right)$. Check the solution in both equations of the original system.

The following steps may be used to solve a system of equations by the substitution method.

> **Solving a System of Two Linear Equations by the Substitution Method**
> **Step 1.** Solve one of the equations for one of its variables.
> **Step 2.** Substitute the expression for the variable found in Step 1 into the other equation.
> **Step 3.** Solve the equation from Step 2 to find the value of one variable.
> **Step 4.** Substitute the value found in Step 3 in any equation containing both variables to find the value of the other variable.
> **Step 5.** Check the proposed solution in the original system.

✔ **CONCEPT CHECK**
As you solve the system $\begin{cases} 2x + y = -5 \\ x - y = 5 \end{cases}$ you find that $y = -5$. Is this the solution of the system?

EXAMPLE 3

Solve the system: $\begin{cases} 7x - 3y = -14 \\ -3x + y = 6 \end{cases}$

Solution   To avoid introducing fractions, we will solve the second equation for y.

$$-3x + y = 6 \qquad \text{\textit{Second equation}}$$
$$y = 3x + 6$$

Next, substitute $3x + 6$ for y in the first equation.

$$7x - 3y = -14 \qquad \text{\textit{First equation}}$$
$$7x - 3(3x + 6) = -14$$
$$7x - 9x - 18 = -14$$
$$-2x - 18 = -14$$
$$-2x = 4$$
$$\frac{-2x}{-2} = \frac{4}{-2}$$
$$x = -2$$

To find the corresponding y-value, substitute -2 for x in the equation $y = 3x + 6$. Then $y = 3(-2) + 6$ or $y = 0$. The solution of the system is $(-2, 0)$. Check this solution in both equations of the system.

> **Helpful Hint**
>
> When solving a system of equations by the substitution method, begin by solving an equation for one of its variables. If possible, solve for a variable that has a coefficient of 1 or -1. This way, we avoid working with time-consuming fractions.

EXAMPLE 4

Solve the system: $\begin{cases} \dfrac{1}{2}x - y = 3 \\ x = 6 + 2y \end{cases}$

Solution   The second equation is already solved for x in terms of y. Thus we substitute $6 + 2y$ for x in the first equation and solve for y.

$$\frac{1}{2}x - y = 3 \qquad \text{\textit{First equation}}$$
$$\frac{1}{2}(6 + 2y) - y = 3 \qquad \text{\textit{Let } } x = 6 + 2y.$$
$$3 + y - y = 3$$
$$3 = 3$$

Arriving at a true statement such as $3 = 3$ indicates that the two linear equations in the original system are equivalent. This means that their graphs are identical and there are an infinite number of solutions of the system. Any solution of one equation is also a solution of the other.

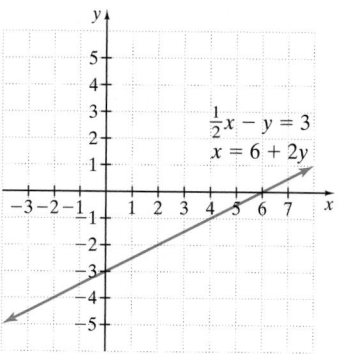

EXAMPLE 5

Use substitution to solve the system.

$$\begin{cases} 6x + 12y = 5 \\ -4x - 8y = 0 \end{cases}$$

Solution Choose the second equation and solve for y.

$-4x - 8y = 0$	Second equation
$-8y = 4x$	Add $4x$ to both sides.
$\dfrac{-8y}{-8} = \dfrac{4x}{-8}$	Divide both sides by -8.
$y = -\dfrac{1}{2}x$	Simplify.

Now replace y with $-\dfrac{1}{2}x$ in the first equation.

$6x + 12y = 5$	First equation
$6x + 12\left(-\dfrac{1}{2}x\right) = 5$	Let $y = -\dfrac{1}{2}x$.
$6x + (-6x) = 5$	Simplify.
$0 = 5$	Combine like terms.

The false statement $0 = 5$ indicates that this system has no solution and is inconsistent. The graph of the linear equations in the system is a pair of parallel lines.

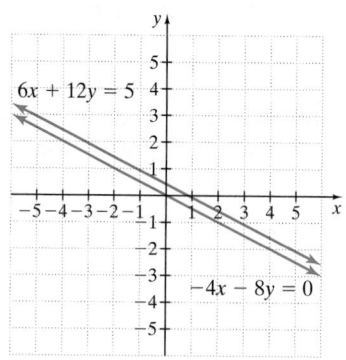

✔ **CONCEPT CHECK**

Describe how the graphs of the equations in a system appear if the system has

a. no solution   **b.** one solution   **c.** an infinite number of solutions

STUDY SKILLS REMINDER

Are You Satisfied with Your Performance on a Particular Quiz or Exam?

If not, don't forget to analyze your quiz or exam and look for common errors.

Were most of your errors a result of

▶ *Carelessness?* If your errors were careless, did you turn in your work before the allotted time expired? If so, resolve to use the entire time allotted next time. Any extra time can be spent checking your work.

▶ *Running out of time?* If so, make a point to better manage your time on your next exam. A few suggestions are to work any questions that you are unsure of last and to check your work after all questions have been answered.

▶ *Not understanding a concept?* If so, review that concept and correct your work. Remember next time to make sure that all concepts on a quiz or exam are understood before the exam.

MENTAL MATH

Give the solution of each system. If the system has no solution or an infinite number of solutions, say so. If the system has 1 solution, find it.

1. $\begin{cases} y = 4x \\ -3x + y = 1 \end{cases}$

When solving, you obtain $x = 1$

2. $\begin{cases} 4x - y = 17 \\ -8x + 2y = 0 \end{cases}$

When solving, you obtain $0 = 34$

3. $\begin{cases} 4x - y = 17 \\ -8x + 2y = -34 \end{cases}$

When solving, you obtain $0 = 0$

4. $\begin{cases} 5x + 2y = 25 \\ x = y + 5 \end{cases}$

When solving, you obtain $y = 0$

5. $\begin{cases} x + y = 0 \\ 7x - 7y = 0 \end{cases}$

When solving, you obtain $x = 0$

6. $\begin{cases} y = -2x + 5 \\ 4x + 2y = 10 \end{cases}$

When solving, you obtain $0 = 0$

EXERCISE SET 4.2

| STUDY GUIDE/SSM | CD/ VIDEO | PH MATH TUTOR CENTER | MathXL®Tutorials ON CD | MathXL® | MyMathLab® |

MIXED PRACTICE

Solve each system of equations by the substitution method. See Examples 1 through 5.

1. $\begin{cases} x + y = 3 \\ x = 2y \end{cases}$

2. $\begin{cases} x + y = 20 \\ x = 3y \end{cases}$

3. $\begin{cases} x + y = 6 \\ y = -3x \end{cases}$

4. $\begin{cases} x + y = 6 \\ y = -4x \end{cases}$

5. $\begin{cases} 3x + 2y = 16 \\ x = 3y - 2 \end{cases}$

6. $\begin{cases} 2x + 3y = 18 \\ x = 2y - 5 \end{cases}$

7. $\begin{cases} 3x - 4y = 10 \\ x = 2y \end{cases}$

8. $\begin{cases} 3x - 4y = 10 \\ y = 2x \end{cases}$

9. $\begin{cases} y = 3x + 1 \\ 4y - 8x = 12 \end{cases}$

10. $\begin{cases} y = 2x + 3 \\ 5y - 7x = 18 \end{cases}$

11. $\begin{cases} x + 2y = 6 \\ 2x + 3y = 8 \end{cases}$

12. $\begin{cases} x + 3y = -5 \\ 2x + 2y = 6 \end{cases}$

13. $\begin{cases} 2x - 5y = 1 \\ 3x + y = -7 \end{cases}$

14. $\begin{cases} 4x + 2y = 5 \\ 2x + y = -4 \end{cases}$

15. $\begin{cases} 2y = x + 2 \\ 6x - 12y = 0 \end{cases}$

16. $\begin{cases} 3y = x + 6 \\ 4x + 12y = 0 \end{cases}$

17. $\begin{cases} \dfrac{1}{3}x - y = 2 \\ x - 3y = 6 \end{cases}$

18. $\begin{cases} \dfrac{1}{4}x - 2y = 1 \\ x - 8y = 4 \end{cases}$

19. $\begin{cases} 4x + y = 11 \\ 2x + 5y = 1 \end{cases}$

20. $\begin{cases} 3x + y = -14 \\ 4x + 3y = -22 \end{cases}$

21. $\begin{cases} 2x - 3y = -9 \\ 3x = y + 4 \end{cases}$

22. $\begin{cases} 8x - 3y = -4 \\ 7x = y + 3 \end{cases}$

23. $\begin{cases} 6x - 3y = 5 \\ x + 2y = 0 \end{cases}$

24. $\begin{cases} 10x - 5y = -21 \\ x + 3y = 0 \end{cases}$

25. $\begin{cases} 3x - y = 1 \\ 2x - 3y = 10 \end{cases}$

26. $\begin{cases} 2x - y = -7 \\ 4x - 3y = -11 \end{cases}$

27. $\begin{cases} -x + 2y = 10 \\ -2x + 3y = 18 \end{cases}$

28. $\begin{cases} -x + 3y = 18 \\ -3x + 2y = 19 \end{cases}$

29. $\begin{cases} 5x + 10y = 20 \\ 2x + 6y = 10 \end{cases}$

30. $\begin{cases} 2x + 4y = 6 \\ 5x + 10y = 15 \end{cases}$

31. $\begin{cases} 3x + 6y = 9 \\ 4x + 8y = 16 \end{cases}$

32. $\begin{cases} 6x + 3y = 12 \\ 9x + 6y = 15 \end{cases}$

33. $\begin{cases} y = 2x + 9 \\ y = 7x + 10 \end{cases}$

34. $\begin{cases} y = 5x - 3 \\ y = 8x + 4 \end{cases}$

35. Explain how to identify an inconsistent system (no solution) when using the substitution method.

36. Occasionally, when using the substitution method, the equation $0 = 0$ is obtained. Explain how this result indicates that the equations are dependent.

Solve each system by the substitution method. First simplify each equation by combining like terms.

37. $\begin{cases} -5y + 6y = 3x + 2(x - 5) - 3x + 5 \\ 4(x + y) - x + y = -12 \end{cases}$

38. $\begin{cases} 5x + 2y - 4x - 2y = 2(2y + 6) - 7 \\ 3(2x - y) - 4x = 1 + 9 \end{cases}$

REVIEW AND PREVIEW

Write equivalent equations by multiplying both sides of the given equation by the given nonzero number. See Section 2.3.

39. $3x + 2y = 6; -2$

40. $-x + y = 10; 5$

41. $-4x + y = 3; 3$

42. $5a - 7b = -4; -4$

Add the binomials. See Section 2.1.

43. $\begin{aligned} 3n + 6m \\ 2n - 6m \end{aligned}$

44. $\begin{aligned} -2x + 5y \\ 2x + 11y \end{aligned}$

45. $\begin{aligned} -5a - 7b \\ 5a - 8b \end{aligned}$

46. $\begin{aligned} 9q + p \\ -9q - p \end{aligned}$

Concept Extensions

47. The number of men and women receiving bachelor's degrees each year has been steadily increasing. For the years 1970 through the projection of 2010, the number of men receiving degrees (in thousands) is given by the equation $y = 2.5x + 450$, and for women, the equation is $y = 11.85x + 337$ where x is the number of years after 1970. (*Source:* National Center for Education Statistics)

   a. Use the substitution method to solve this system of equations. (Round your final results to the nearest whole numbers.)

   b. Explain the meaning of your answer to part (a).

   c. Sketch a graph of the system of equations. Write a sentence describing the trends for men and women receiving bachelor degrees.

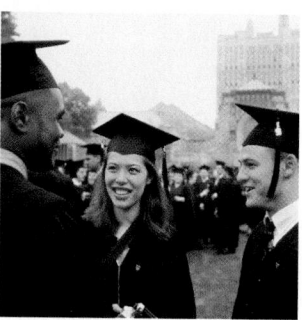

48. The number of country music stations in the United States from 1995–2002 is given by the equation $y = -82.18x + 2612.39$ where x is the number of years after 1995. The number of Top 40 music stations is given by the equation $y = 27.89x + 306.46$ for the same time period. (*Source:* M Street Corporation)

   a. Use the substitution method to solve this system of equations. (Round your final results to the nearest tenth.)

   b. Explain the meaning of your answer to part (a).

   c. Sketch a graph of the system of equations. Write a sentence describing the trends in the popularity of these two types of music formats.

Solve each system by substitution. When necessary, round answers to the nearest hundredth.

49. $\begin{cases} y = 5.1x + 14.56 \\ y = -2x - 3.9 \end{cases}$

50. $\begin{cases} y = 3.1x - 16.35 \\ y = -9.7x + 28.45 \end{cases}$

51. $\begin{cases} 3x + 2y = 14.05 \\ 5x + y = 18.5 \end{cases}$

52. $\begin{cases} x + y = -15.2 \\ -2x + 5y = -19.3 \end{cases}$

4.3   SOLVING SYSTEMS OF LINEAR EQUATIONS BY ADDITION

Objective

1   Use the addition method to solve a system of linear equations.

1   We have seen that substitution is an accurate way to solve a linear system. Another method for solving a system of equations accurately is the **addition** or **elimination method**. The addition method is based on the addition property of equality: adding equal quantities to both sides of an equation does not change the solution of the equation. In symbols,

$$\text{if } A = B \text{ and } C = D, \text{ then } A + C = B + D.$$

EXAMPLE 1

Solve the system: $\begin{cases} x + y = 7 \\ x - y = 5 \end{cases}$

Solution   Since the left side of each equation is equal to the right side, we add equal quantities by adding the left sides of the equations together and the right sides of the equations together. If we choose wisely, this adding gives us an equation in one variable, x, which we can solve for x.

$$
\begin{array}{ll}
x + y = 7 & \text{First equation} \\
\underline{x - y = 5} & \text{Second equation} \\
2x \quad\quad = 12 & \text{Add the equations.} \\
\quad\quad x = 6 & \text{Divide both sides by 2.}
\end{array}
$$

The x-value of the solution is 6. To find the corresponding y-value, let $x = 6$ in either equation of the system. We will use the first equation.

$$
\begin{array}{ll}
x + y = 7 & \text{First equation} \\
6 + y = 7 & \text{Let } x = 6. \\
y = 7 - 6 & \text{Solve for } y. \\
y = 1 & \text{Simplify.}
\end{array}
$$

The solution is $(6, 1)$. Check this in both equations.

First Equation	***Second Equation***
$x + y = 7$	$x - y = 5$
$6 + 1 \stackrel{?}{=} 7$	$6 - 1 \stackrel{?}{=} 5$ Let $x = 6$ and $y = 1$.
$7 = 7$ True	$5 = 5$ True

Thus, the solution of the system is $(6, 1)$ and the graphs of the two equations intersect at the point $(6, 1)$ as shown on the top of the next page.

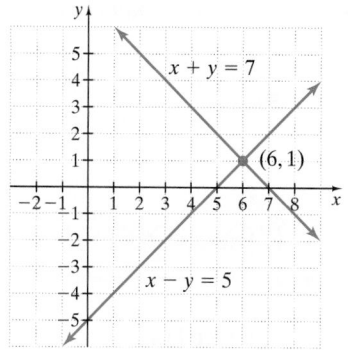

EXAMPLE 2

Solve the system: $\begin{cases} -2x + y = 2 \\ -x + 3y = -4 \end{cases}$

Solution If we simply add the two equations, the result is still an equation in two variables. However, our goal is to eliminate one of the variables. Notice what happens if we multiply *both sides* of the first equation by -3, which we are allowed to do by the multiplication property of equality. The system

$$\begin{cases} -3(-2x + y) = -3(2) \\ -x + 3y = -4 \end{cases} \quad \text{simplifies to} \quad \begin{cases} 6x - 3y = -6 \\ -x + 3y = -4 \end{cases}$$

Now add the resulting equations and the y-variable is eliminated.

$$\begin{array}{rcl} 6x - 3y &=& -6 \\ \underline{-x + 3y} &=& \underline{-4} \\ 5x &=& -10 \qquad \text{Add.} \\ x &=& -2 \qquad \text{Divide both sides by 5.} \end{array}$$

To find the corresponding y-value, let $x = -2$ in any of the preceding equations containing both variables. We use the first equation of the original system.

$$\begin{array}{ll} -2x + y = 2 & \text{First equation} \\ -2(-2) + y = 2 & \text{Let } x = -2. \\ 4 + y = 2 & \\ y = -2 & \end{array}$$

The solution is $(-2, -2)$. Check this ordered pair in both equations of the original system.

In Example 2, the decision to multiply the first equation by -3 was no accident. **To eliminate a variable** when adding two equations, **the coefficient of the variable in one equation must be the opposite of its coefficient in the other equation.**

> **Helpful Hint**
>
> Be sure to multiply *both sides* of an equation by a chosen number when solving by the addition method. A common mistake is to multiply only the side containing the variables.

EXAMPLE 3

Solve the system: $\begin{cases} 2x - y = 7 \\ 8x - 4y = 1 \end{cases}$

Solution   Multiply both sides of the first equation by -4 and the resulting coefficient of x is -8, the opposite of 8, the coefficient of x in the second equation. The system becomes

> **Helpful Hint**
>
> Don't forget to multiply both sides by -4.

$$\begin{cases} -4(2x - y) = -4(7) \\ 8x - 4y = 1 \end{cases} \text{ simplifies to } \begin{cases} -8x + 4y = -28 \\ 8x - 4y = 1 \end{cases}$$

Now add the resulting equations.

$$\begin{array}{r} -8x + 4y = -28 \\ 8x - 4y = 1 \\ \hline 0 = -27 \quad \text{False} \end{array}$$

When we add the equations, both variables are eliminated and we have $0 = -27$, a false statement. This means that the system has no solution. The graphs of these equations are parallel lines.

EXAMPLE 4

Solve the system: $\begin{cases} 3x - 2y = 2 \\ -9x + 6y = -6 \end{cases}$

Solution   First we multiply both sides of the first equation by 3, then we add the resulting equations.

$$\begin{cases} 3(3x - 2y) = 3(2) \\ -9x + 6y = -6 \end{cases} \text{ simplifies to } \begin{cases} 9x - 6y = 6 \\ -9x + 6y = -6 \quad \text{Add the equations.} \\ 0 = 0 \quad \text{True} \end{cases}$$

Both variables are eliminated and we have $0 = 0$, a true statement. Whenever you eliminate a variable and get the equation $0 = 0$, the system has an infinite number of solutions. The graphs of these equations are identical.

✔ **CONCEPT CHECK**

Suppose you are solving the system

$$\begin{cases} 3x + 8y = -5 \\ 2x - 4y = 3 \end{cases}$$

You decide to use the addition method by multiplying both sides of the second equation by 2. In which of the following was the multiplication performed correctly? Explain.

a. $4x - 8y = 3$ **b.** $4x - 8y = 6$

EXAMPLE 5

Solve the system: $\begin{cases} 3x + 4y = 13 \\ 5x - 9y = 6 \end{cases}$

Solution   We can eliminate the variable y by multiplying the first equation by 9 and the second equation by 4.

$$\begin{cases} 9(3x + 4y) = 9(13) \\ 4(5x - 9y) = 4(6) \end{cases} \quad \text{simplifies to} \quad \begin{cases} 27x + 36y = 117 \\ 20x - 36y = 24 \end{cases}$$

$$\begin{aligned} 47x &= 141 \qquad \text{Add the equations.} \\ x &= 3 \end{aligned}$$

To find the corresponding y-value, we let $x = 3$ in any equation in this example containing two variables. Doing so in any of these equations will give $y = 1$. The solution to this system is $(3, 1)$. Check to see that $(3, 1)$ satisfies each equation in the original system.

If we had decided to eliminate x instead of y in Example 5, the first equation could have been multiplied by 5 and the second by -3. Try solving the original system this way to check that the solution is $(3, 1)$.

The following steps summarize how to solve a system of linear equations by the addition method.

Solving a System of Two Linear Equations by the Addition Method

Step 1. Rewrite each equation in standard form $Ax + By = C$.

Step 2. If necessary, multiply one or both equations by a nonzero number so that the coefficients of a chosen variable in the system are opposites.

Step 3. Add the equations.

Step 4. Find the value of one variable by solving the resulting equation from Step 3.

Step 5. Find the value of the second variable by substituting the value found in Step 4 into either of the original equations.

Step 6. Check the proposed solution in the original system.

✔ **CONCEPT CHECK**

Suppose you are solving the system

$$\begin{cases} -4x + 7y = 6 \\ x + 2y = 5 \end{cases}$$

by the addition method.

a. What step(s) should you take if you wish to eliminate x when adding the equations?
b. What step(s) should you take if you wish to eliminate y when adding the equations?

EXAMPLE 6

Concept Check Answer:

a. multiply the second equation by 4
b. possible answer: multiply the first equation by -2 and the second equation by 7

Solve the system: $\begin{cases} -x - \dfrac{y}{2} = \dfrac{5}{2} \\ -\dfrac{x}{2} + \dfrac{y}{4} = 0 \end{cases}$

Solution We begin by clearing each equation of fractions. To do so, we multiply both sides of the first equation by the LCD 2 and both sides of the second equation by the LCD 4. Then

the system

$$\begin{cases} 2\left(-x - \dfrac{y}{2}\right) = 2\left(\dfrac{5}{2}\right) \\ 4\left(-\dfrac{x}{2} + \dfrac{y}{4}\right) = 4(0) \end{cases}$$ simplifies to $\begin{cases} -2x - y = 5 \\ -2x + y = 0 \end{cases}$

Now we add the resulting equations in the simplified system.

$$\begin{array}{r} -2x - y = 5 \\ \underline{-2x + y = 0} \\ -4x \quad\quad = 5 \end{array} \qquad \text{Add.}$$

$$x = -\frac{5}{4}$$

To find y, we could replace x with $-\dfrac{5}{4}$ in one of the equations with two variables.

Instead, let's go back to the simplified system and multiply by appropriate factors to eliminate the variable x and solve for y. To do this, we multiply the first equation in the simplified system by -1. Then the system

$$\begin{cases} -1(-2x - y) = -1(5) \\ -2x + y = 0 \end{cases} \quad \text{simplifies to} \quad \begin{array}{r} 2x + y = -5 \\ \underline{-2x + y = \ \ 0} \\ 2y = -5 \end{array} \quad \text{Add.}$$

$$y = -\frac{5}{2}$$

Check the ordered pair $\left(-\dfrac{5}{4}, -\dfrac{5}{2}\right)$ in both equations of the original system. The solution is $\left(-\dfrac{5}{4}, -\dfrac{5}{2}\right)$.

EXERCISE SET 4.3

| STUDY GUIDE/SSM | CD/ VIDEO | PH MATH TUTOR CENTER | MathXL®Tutorials ON CD | MathXL® | MyMathLab® |

Solve each system of equations by the addition method. See Examples 1 through 5.

1. $\begin{cases} 3x + y = 5 \\ 6x - y = 4 \end{cases}$

2. $\begin{cases} 4x + y = 13 \\ 2x - y = 5 \end{cases}$

3. $\begin{cases} x - 2y = 8 \\ -x + 5y = -17 \end{cases}$

4. $\begin{cases} x - 2y = -11 \\ -x + 5y = 23 \end{cases}$

5. $\begin{cases} x + y = 6 \\ x - y = 6 \end{cases}$

6. $\begin{cases} x - y = 1 \\ -x + 2y = 0 \end{cases}$

7. $\begin{cases} 3x + y = 4 \\ 9x + 3y = 6 \end{cases}$

8. $\begin{cases} 2x + y = 6 \\ 4x + 2y = 12 \end{cases}$

9. $\begin{cases} 3x - 2y = 7 \\ 5x + 4y = 8 \end{cases}$

10. $\begin{cases} 6x - 5y = 25 \\ 4x + 15y = 13 \end{cases}$

11. $\begin{cases} \dfrac{2}{3}x + 4y = -4 \\ 5x + 6y = 18 \end{cases}$

12. $\begin{cases} \dfrac{3}{2}x + 4y = 1 \\ 9x + 24y = 5 \end{cases}$

13. $\begin{cases} 4x - 6y = 8 \\ 6x - 9y = 12 \end{cases}$

14. $\begin{cases} 9x - 3y = 12 \\ 12x - 4y = 18 \end{cases}$

15. $\begin{cases} 3x + y = -11 \\ 6x - 2y = -2 \end{cases}$

16. $\begin{cases} 4x + y = -13 \\ 6x - 3y = -15 \end{cases}$

17. $\begin{cases} 3x + 2y = 11 \\ 5x - 2y = 29 \end{cases}$

18. $\begin{cases} 4x + 2y = 2 \\ 3x - 2y = 12 \end{cases}$

19. $\begin{cases} x + 5y = 18 \\ 3x + 2y = -11 \end{cases}$

20. $\begin{cases} x + 4y = 14 \\ 5x + 3y = 2 \end{cases}$

21. $\begin{cases} 2x - 5y = 4 \\ 3x - 2y = 4 \end{cases}$

22. $\begin{cases} 6x - 5y = 7 \\ 4x - 6y = 7 \end{cases}$

23. $\begin{cases} 2x + 3y = 0 \\ 4x + 6y = 3 \end{cases}$

24. $\begin{cases} -x + 5y = -1 \\ 3x - 15y = 3 \end{cases}$

Solve each system of equations by the addition method. See Example 6.

25. $\begin{cases} \dfrac{x}{3} + \dfrac{y}{6} = 1 \\ \dfrac{x}{2} - \dfrac{y}{4} = 0 \end{cases}$

26. $\begin{cases} \dfrac{x}{2} + \dfrac{y}{8} = 3 \\ x - \dfrac{y}{4} = 0 \end{cases}$

27. $\begin{cases} x - \dfrac{y}{3} = -1 \\ -\dfrac{x}{2} + \dfrac{y}{8} = \dfrac{1}{4} \end{cases}$

28. $\begin{cases} 2x - \dfrac{3y}{4} = -3 \\ x + \dfrac{y}{9} = \dfrac{13}{3} \end{cases}$

29. $\begin{cases} \dfrac{x}{3} - y = 2 \\ -\dfrac{x}{2} + \dfrac{3y}{2} = -3 \end{cases}$

30. $\begin{cases} \dfrac{x}{2} + \dfrac{y}{4} = 1 \\ -\dfrac{x}{4} - \dfrac{y}{8} = 1 \end{cases}$

31. $\begin{cases} 8x = -11y - 16 \\ 2x + 3y = -4 \end{cases}$

32. $\begin{cases} 10x + 3y = -12 \\ 5x = -4y - 16 \end{cases}$

33. When solving a system of equations by the addition method, how do we know when the system has no solution?

34. To solve the system $\begin{cases} 2x - 3y = 5 \\ 5x + 2y = 6 \end{cases}$, explain why the addition method might be preferred rather than the substitution method.

MIXED PRACTICE

Solve each system by either the addition method or the substitution method.

35. $\begin{cases} 2x - 3y = -11 \\ y = 4x - 3 \end{cases}$

36. $\begin{cases} 4x - 5y = 6 \\ y = 3x - 10 \end{cases}$

37. $\begin{cases} x + 2y = 1 \\ 3x + 4y = -1 \end{cases}$

38. $\begin{cases} x + 3y = 5 \\ 5x + 6y = -2 \end{cases}$

39. $\begin{cases} 2y = x + 6 \\ 3x - 2y = -6 \end{cases}$

40. $\begin{cases} 3y = x + 14 \\ 2x - 3y = -16 \end{cases}$

41. $\begin{cases} y = 2x - 3 \\ y = 5x - 18 \end{cases}$

42. $\begin{cases} y = 6x - 5 \\ y = 4x - 11 \end{cases}$

43. $\begin{cases} x + \dfrac{1}{6}y = \dfrac{1}{2} \\ 3x + 2y = 3 \end{cases}$

44. $\begin{cases} x + \dfrac{1}{3}y = \dfrac{5}{12} \\ 8x + 3y = 4 \end{cases}$

45. $\begin{cases} \dfrac{x + 2}{2} = \dfrac{y + 11}{3} \\ \dfrac{x}{2} = \dfrac{2y + 16}{6} \end{cases}$

46. $\begin{cases} \dfrac{x + 5}{2} = \dfrac{y + 14}{4} \\ \dfrac{x}{3} = \dfrac{2y + 2}{6} \end{cases}$

47. $\begin{cases} 2x + 3y = 14 \\ 3x - 4y = -69.1 \end{cases}$

48. $\begin{cases} 5x - 2y = -19.8 \\ -3x + 5y = -3.7 \end{cases}$

REVIEW AND PREVIEW

Rewrite the following sentences using mathematical symbols. Do not solve the equations. See Sections 2.2 and 2.3.

49. Twice a number, added to 6, is 3 less than the number.

50. The sum of three consecutive integers is 66.

51. Three times a number, subtracted from 20, is 2.

52. Twice the sum of 8 and a number is the difference of the number and 20.

53. The product of 4 and the sum of a number and 6 is twice a number.

54. The quotient of twice a number and 7 is subtracted from the reciprocal of the number.

Concept Extensions

55. In recent years, the number of Americans (in millions) who went downhill skiing more than once in a year has been decreasing, while the number who went snowboarding has been increasing. The number of downhill skiers who went more than once in a year from 1996 to 2001 is given by $0.35x + y = 9.3$ and the number of snowboarders can be given by the equation $0.56x - y = -2.5$. For both equations, x is the number of years after 1996. (*Source*: National Ski & Snowboard Association)

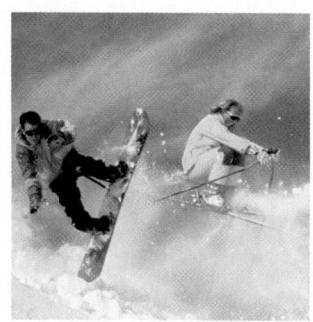

a. Suppose the trend continues. Use the addition method to predict the year in which the number of downhill skiers will equal the number of snowboarders. (Round up to the next whole number.)

b. Use the equations to determine the number of skiers and the number of snowboarders during the year found in part a. Are the numbers the same? Why or why not?

56. Commercial broadcast television stations can be divided into VHF stations (channels 2 through 13) and UHF stations (channels 14 through 83). The number y of VHF stations in the United States from 1980 through 2001 is given by the equation $3x - y = -514$, where x is the number of years after 1980. The number y of UHF stations in the United States from 1980 through 2001 is given by the equation $-23x + y = 244$, where x is the number of years after 1980. (*Source:* Based on data from the Television Bureau of Advertising, Inc.)

a. Use the addition method to solve this system of equations.

b. Interpret your solution from part a.

c. During which years were there more UHF commercial television stations than VHF stations?

57. Use the system of linear equations below to answer the questions.

$$\begin{cases} x + y = 5 \\ 3x + 3y = b \end{cases}$$

a. Find the value of b so that the equations are dependent and the system has an infinite number of solutions.

b. Find a value of b so that the system is inconsistent and there are no solutions to the system.

58. Use the system of linear equations below to answer the questions.

$$\begin{cases} x + y = 4 \\ 2x + by = 8 \end{cases}$$

a. Find the value of b so that the equations are dependent and the system has an infinite number of solutions.

b. Find a value of b so that the system is consistent and the system has a single solution.

59. Suppose you are solving the system

$$\begin{cases} -4x + 7y = 6 \\ x + 2y = 5 \end{cases}$$

by the addition method.

a. What step(s) should you take if you wish to eliminate x when adding the equations?

b. What step(s) should you take if you wish to eliminate y when adding the equations?

60. Suppose you are solving the system

$$\begin{cases} 3x + 8y = -5 \\ 2x - 4y = 3 \end{cases}$$

You decide to use the addition method by multiplying both sides of the second equation by 2. In which of the following was the multiplication performed correctly? Explain.

a. $4x - 8y = 3$

b. $4x - 8y = 6$

INTEGRATED REVIEW   SOLVING SYSTEMS OF EQUATIONS

The graphs of various systems of equations are shown. Match each graph with the solution of its corresponding system.

A   B   C   D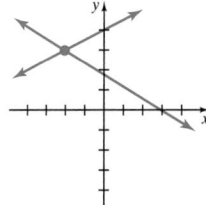

1. Solution: $(1, 2)$ **2.** Solution: $(-2, 3)$

3. No solution   **4.** Infinite number of solutions

Solve each system by either the addition method or the substitution method.

5. $\begin{cases} 2x - 3y = -11 \\ y = 4x - 3 \end{cases}$ **6.** $\begin{cases} 4x - 5y = 6 \\ y = 3x - 10 \end{cases}$ **7.** $\begin{cases} x + y = 3 \\ x - y = 7 \end{cases}$

8. $\begin{cases} x - y = 20 \\ x + y = -8 \end{cases}$

9. $\begin{cases} x + 2y = 1 \\ 3x + 4y = -1 \end{cases}$

10. $\begin{cases} x + 3y = 5 \\ 5x + 6y = -2 \end{cases}$

11. $\begin{cases} y = x + 3 \\ 3x - 2y = -6 \end{cases}$

12. $\begin{cases} y = -2x \\ 2x - 3y = -16 \end{cases}$

13. $\begin{cases} y = 2x - 3 \\ y = 5x - 18 \end{cases}$

14. $\begin{cases} y = 6x - 5 \\ y = 4x - 11 \end{cases}$

15. $\begin{cases} x + \dfrac{1}{6}y = \dfrac{1}{2} \\ 3x + 2y = 3 \end{cases}$

16. $\begin{cases} x + \dfrac{1}{3}y = \dfrac{5}{12} \\ 8x + 3y = 4 \end{cases}$

17. $\begin{cases} x - 5y = 1 \\ -2x + 10y = 3 \end{cases}$

18. $\begin{cases} -x + 2y = 3 \\ 3x - 6y = -9 \end{cases}$

19. $\begin{cases} 0.2x - 0.3y = -0.95 \\ 0.4x + 0.1y = 0.55 \end{cases}$

20. $\begin{cases} 0.08x - 0.04y = -0.11 \\ 0.02x - 0.06y = -0.09 \end{cases}$

21. $\begin{cases} x = 3y - 7 \\ 2x - 6y = -14 \end{cases}$

22. $\begin{cases} y = \dfrac{x}{2} - 3 \\ 2x - 4y = 0 \end{cases}$

23. Which method, substitution or addition, would you prefer to use to solve the system below? Explain your reasoning.

$$\begin{cases} 3x + 2y = -2 \\ y = -2x \end{cases}$$

24. Which method, substitution or addition, would you prefer to use to solve the system below? Explain your reasoning.

$$\begin{cases} 3x - 2y = -3 \\ 6x + 2y = 12 \end{cases}$$

4.4 SOLVING SYSTEMS OF LINEAR EQUATIONS IN THREE VARIABLES

Objective

1 Solve a system of three linear equations in three variables.

In this section, the algebraic methods of solving systems of two linear equations in two variables are extended to systems of three linear equations in three variables. We call the equation $3x - y + z = -15$, for example, a **linear equation in three variables** since there are three variables and each variable is raised only to the power 1. A solution of this equation is an **ordered triple (x, y, z)** that makes the equation a true statement. For example, the ordered triple $(2, 0, -21)$ is a solution of $3x - y + z = -15$ since replacing x with 2, y with 0, and z with -21 yields the true statement $3(2) - 0 + (-21) = -15$. The graph of this equation is a plane in three-dimensional space, just as the graph of a linear equation in two variables is a line in two-dimensional space.

Although we will not discuss the techniques for graphing equations in three variables, visualizing the possible patterns of intersecting planes gives us insight into the possible patterns of solutions of a system of three three-variable linear equations. There are four possible patterns.

1. Three planes have a single point in common. This point represents the single solution of the system. This system is **consistent**.

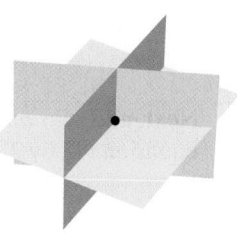

2. Three planes intersect at no point common to all three. This system has no solution. A few ways that this can occur are shown. This system is **inconsistent**.

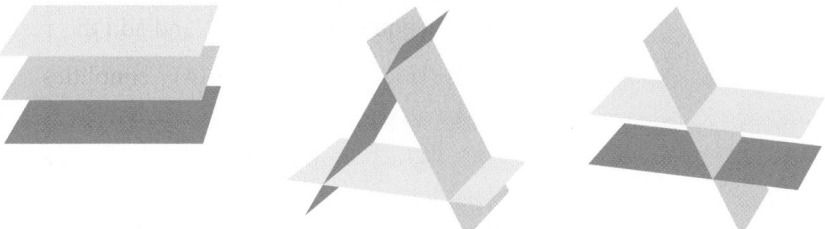

3. Three planes intersect at all the points of a single line. The system has infinitely many solutions. This system is **consistent**.

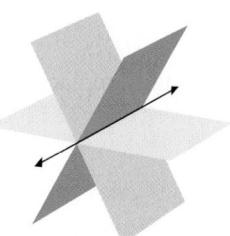

4. Three planes coincide at all points on the plane. The system is consistent, and the equations are **dependent**.

1   Just as with systems of two equations in two variables, we can use the elimination or substitution method to solve a system of three equations in three variables. To use the elimination method, we eliminate a variable and obtain a system of two equations in two variables. Then we use the methods we learned in the previous section to solve the system of two equations.

EXAMPLE 1

Solve the system.

$$\begin{cases} 3x - y + z = -15 & \text{Equation (1)} \\ x + 2y - z = 1 & \text{Equation (2)} \\ 2x + 3y - 2z = 0 & \text{Equation (3)} \end{cases}$$

Solution   Add equations (1) and (2) to eliminate z.

$$\begin{aligned} 3x - y + z &= -15 \\ x + 2y - z &= 1 \\ \hline 4x + y &= -14 \quad \text{Equation (4)} \end{aligned}$$

> **Helpful Hint**
>
> Don't forget to add two other equations besides equations (1) and (2) *and* to **eliminate the same variable.**

Next, add two *other* equations and *eliminate z again*. To do so, multiply both sides of equation (1) by 2 and add this resulting equation to equation (3). Then

$$\begin{cases} 2(3x - y + z) = 2(-15) \\ 2x + 3y - 2z = 0 \end{cases} \quad \begin{matrix}\text{simplifies} \\ \text{to} \end{matrix} \quad \begin{cases} 6x - 2y + 2z = -30 \\ 2x + 3y - 2z = 0 \\ \hline 8x + y = -30 \quad \text{Equation (5)} \end{cases}$$

Now solve equations (4) and (5) for x and y. To solve by elimination, multiply both sides of equation (4) by -1 and add this resulting equation to equation (5). Then

$$\begin{cases} -1(4x + y) = -1(-14) \\ 8x + y = -30 \end{cases} \quad \begin{matrix}\text{simplifies} \\ \text{to} \end{matrix} \quad \begin{cases} -4x - y = 14 \\ 8x + y = -30 \\ \hline 4x = -16 \quad \text{Add the equations.} \\ x = -4 \quad \text{Solve for } x. \end{cases}$$

Replace x with -4 in equation (4) or (5).

$$\begin{aligned} 4x + y &= -14 \quad \text{Equation (4)} \\ 4(-4) + y &= -14 \quad \text{Let } x = -4. \\ y &= 2 \quad \text{Solve for } y. \end{aligned}$$

Finally, replace x with -4 and y with 2 in equation (1), (2), or (3).

$$\begin{aligned} x + 2y - z &= 1 \quad \text{Equation (2)} \\ -4 + 2(2) - z &= 1 \quad \text{Let } x = -4 \text{ and } y = 2. \\ -4 + 4 - z &= 1 \\ -z &= 1 \\ z &= -1 \end{aligned}$$

The solution is $(-4, 2, -1)$. To check, let $x = -4$, $y = 2$, and $z = -1$ in all three original equations of the system.

Equation (1)	*Equation (2)*	*Equation (3)*
$3x - y + z = -15$	$x + 2y - z = 1$	$2x + 3y - 2z = 0$
$3(-4) - 2 + (-1) \overset{?}{=} -15$	$-4 + 2(2) - (-1) \overset{?}{=} 1$	$2(-4) + 3(2) - 2(-1) \overset{?}{=} 0$
$-12 - 2 - 1 \overset{?}{=} -15$	$-4 + 4 + 1 \overset{?}{=} 1$	$-8 + 6 + 2 \overset{?}{=} 0$
$-15 = -15$	$1 = 1$	$0 = 0$
True	True	True

All three statements are true, so the solution is $(-4, 2, -1)$.

EXAMPLE 2

Solve the system.

$$\begin{cases} 2x - 4y + 8z = 2 & (1) \\ -x - 3y + z = 11 & (2) \\ x - 2y + 4z = 0 & (3) \end{cases}$$

Solution   Add equations (2) and (3) to eliminate x, and the new equation is

$$-5y + 5z = 11 \quad (4)$$

To eliminate x again, multiply both sides of equation (2) by 2, and add the resulting equation to equation (1). Then

$$\begin{cases} 2x - 4y + 8z = 2 \\ 2(-x - 3y + z) = 2(11) \end{cases} \quad \begin{array}{c} \text{simplifies} \\ \text{to} \end{array} \quad \begin{cases} 2x - 4y + 8z = 2 \\ \underline{-2x - 6y + 2z = 22} \\ -10y + 10z = 24 \quad (5) \end{cases}$$

Next, solve for y and z using equations (4) and (5). Multiply both sides of equation (4) by -2, and add the resulting equation to equation (5).

$$\begin{cases} -2(-5y + 5z) = -2(11) \\ -10y + 10z = 24 \end{cases} \quad \begin{array}{c} \text{simplifies} \\ \text{to} \end{array} \quad \begin{cases} 10y - 10z = -22 \\ \underline{-10y + 10z = 24} \\ 0 = 2 \quad \text{False} \end{cases}$$

Since the statement is false, this system is inconsistent and has no solution. The solution set is the empty set $\{\ \}$ or $\varnothing$.

The elimination method is summarized next.

Solving a System of Three Linear Equations by the Elimination Method

Step 1: Write each equation in standard form $Ax + By + Cz = D$.

Step 2: Choose a pair of equations and use the equations to eliminate a variable.

Step 3: Choose any other pair of equations and eliminate the **same variable** as in Step 2.

Step 4: Two equations in two variables should be obtained from Step 2 and Step 3. Use methods from Section 4.1 to solve this system for both variables.

Step 5: To solve for the third variable, substitute the values of the variables found in Step 4 into any of the original equations containing the third variable.

Step 6: Check the ordered triple solution in *all three* original equations.

> **Helpful Hint**
>
> Make sure you read closely and follow Step 3.

✔ **CONCEPT CHECK**

In the system

$$\begin{cases} x + y + z = 6 & \text{Equation (1)} \\ 2x - y + z = 3 & \text{Equation (2)} \\ x + 2y + 3z = 14 & \text{Equation (3)} \end{cases}$$

equations (1) and (2) are used to eliminate y. Which action could be used to best finish solving? Why?

a. Use (1) and (2) to eliminate z.   **b.** Use (2) and (3) to eliminate y.
c. Use (1) and (3) to eliminate x.

EXAMPLE 3

Solve the system.

$$\begin{cases} 2x + 4y = 1 & (1) \\ 4x - 4z = -1 & (2) \\ y - 4z = -3 & (3) \end{cases}$$

Concept Check Answer:
b

Solution Notice that equation (2) has no term containing the variable y. Let us eliminate y using equations (1) and (3). Multiply both sides of equation (3) by -4, and add the resulting equation to equation (1). Then

$$\begin{cases} 2x + 4y = 1 \\ -4(y - 4z) = -4(-3) \end{cases} \quad \text{simplifies} \atop \text{to} \quad \begin{cases} 2x + 4y = 1 \\ \quad\ -4y + 16z = 12 \\ \overline{2x \qquad + 16z = 13} \end{cases} \quad \text{(4)}$$

Next, solve for z using equations (4) and (2). Multiply both sides of equation (4) by -2 and add the resulting equation to equation (2).

$$\begin{cases} (2x + 16z) = (13) \\ 4x - 4z = -1 \end{cases} \quad \text{simplifies} \atop \text{to} \quad \begin{cases} -4x - 32z = -26 \\ \underline{4x - 4z = -1} \\ \qquad\ -36z = -27 \\ \qquad\qquad z = \dfrac{3}{4} \end{cases}$$

Replace z with $\dfrac{3}{4}$ in equation (3) and solve for y.

$$y - 4\left(\frac{3}{4}\right) = -3 \qquad \text{Let } z = \frac{3}{4} \text{ in equation (3).}$$

$$y - 3 = -3$$
$$y = 0$$

Replace y with 0 in equation (1) and solve for x.

$$2x + 4(0) = 1$$
$$2x = 1$$
$$x = \frac{1}{2}$$

The solution is $\left(\dfrac{1}{2}, 0, \dfrac{3}{4}\right)$. Check to see that this solution satisfies all three equations of the system.

EXAMPLE 4

Solve the system.

$$\begin{cases} x - 5y - 2z = 6 & \text{(1)} \\ -2x + 10y + 4z = -12 & \text{(2)} \\ \dfrac{1}{2}x - \dfrac{5}{2}y - z = 3 & \text{(3)} \end{cases}$$

Solution Multiply both sides of equation (3) by 2 to eliminate fractions, and multiply both sides of equation (2) by $-\dfrac{1}{2}$ so that the coefficient of x is 1. The resulting system is then

$$\begin{cases} x - 5y - 2z = 6 & \text{(1)} \\ x - 5y - 2z = 6 & \text{Multiply (2) by } -\frac{1}{2}. \\ x - 5y - 2z = 6 & \text{Multiply (3) by 2.} \end{cases}$$

All three equations are identical, and therefore equations (1), (2), and (3) are all equivalent. There are infinitely many solutions of this system. The equations are dependent. The solution set can be written as $\{(x, y, z) | x - 5y - 2z = 6\}$. ⬭

As mentioned earlier, we can also use the substitution method to solve a system of linear equations in three variables.

EXAMPLE 5

Solve the system:

$$\begin{cases} x - 4y - 5z = 35 & (1) \\ x - 3y = 0 & (2) \\ -y + z = -25 & (3) \end{cases}$$

Solution Notice in equations (2) and (3) that a variable is missing. Also notice that both equations contain the variable y. Let's use the substitution method by solving equation (2) for x and equation (3) for z and substituting the results in equation (1).

$$x - 3y = 0 \qquad (2)$$
$$x = 3y \qquad \text{Solve equation (2) for } x.$$
$$-y + z = -55 \qquad (3)$$
$$z = y - 55 \qquad \text{Solve equation (3) for } z.$$

Now substitute $3y$ for x and $y - 55$ for z in equation (1).

Helpful Hint

Do not forget to distribute.

$$x - 4y - 5z = 35 \qquad (1)$$
$$3y - 4y - 5(y - 55) = 35 \qquad \text{Let } x = 3y \text{ and } z = y - 55.$$
$$3y - 4y - 5y + 275 = 35 \qquad \text{Use the distributive law and multiply.}$$
$$-6y + 275 = 35 \qquad \text{Combine like terms.}$$
$$-6y = -240 \qquad \text{Subtract 275 from both sides.}$$
$$y = 40 \qquad \text{Solve.}$$

To find x, recall that $x = 3y$ and substitute 40 for y. Then $x = 3y$ becomes $x = 3 \cdot 40 = 120$. To find z, recall that $z = y - 55$ and substitute 40 for y, also. Then $z = y - 55$ becomes $z = 40 - 55 = -15$. The solution is $(120, 40, -15)$. ⬭

STUDY SKILLS REMINDER

How Well Do You Know Your Textbook?

See if you can answer the questions below.

1. What does the 🖥 icon mean?

2. What does the \ icon mean?

3. What does the △ icon mean?

4. Where can you find a review for each chapter? What answers to this review can be found in the back of your text?

5. Each chapter contains an overview of the chapter along with examples. What is this feature called?

6. Does this text contain any solutions to exercises? If so, where?

EXERCISE SET 4.4

STUDY GUIDE/SSM   CD/VIDEO   PH MATH TUTOR CENTER   MathXL®Tutorials ON CD   MathXL®   MyMathLab®

Solve.

1. Choose the equation(s) that has $(-1, 3, 1)$ as a solution.

a. $x + y + z = 3$ **b.** $-x + y + z = 5$

c. $-x + y + 2z = 0$ **d.** $x + 2y - 3z = 2$

2. Choose the equation(s) that has $(2, 1, -4)$ as a solution.

a. $x + y + z = -1$ **b.** $x - y - z = -3$

c. $2x - y + z = -1$ **d.** $-x - 3y - z = -1$

3. Use the result of Exercise 1 to determine whether $(-1, 3, 1)$ is a solution of the system below. Explain your answer.

$$\begin{cases} x + y + z = 3 \\ -x + y + z = 5 \\ x + 2y - 3z = 2 \end{cases}$$

4. Use the result of Exercise 2 to determine whether $(2, 1, -4)$ is a solution of the system below. Explain your answer.

$$\begin{cases} x + y + z = -1 \\ x - y - z = -3 \\ 2x - y + z = -1 \end{cases}$$

MIXED PRACTICE

Solve each system. See Examples 1 through 5.

5. $\begin{cases} x - y + z = -4 \\ 3x + 2y - z = 5 \\ -2x + 3y - z = 15 \end{cases}$

6. $\begin{cases} x + y - z = -1 \\ -4x - y + 2z = -7 \\ 2x - 2y - 5z = 7 \end{cases}$

7. $\begin{cases} x + y = 3 \\ 2y = 10 \\ 3x + 2y - 3z = 1 \end{cases}$

8. $\begin{cases} 5x = 5 \\ 2x + y = 4 \\ 3x + y - 4z = -15 \end{cases}$

9. $\begin{cases} 2x + 2y + z = 1 \\ -x + y + 2z = 3 \\ x + 2y + 4z = 0 \end{cases}$

10. $\begin{cases} 2x - 3y + z = 5 \\ x + y + z = 0 \\ 4x + 2y + 4z = 4 \end{cases}$

11. $\begin{cases} x - 2y + z = -5 \\ -3x + 6y - 3z = 15 \\ 2x - 4y + 2z = -10 \end{cases}$

12. $\begin{cases} 3x + y - 2z = 2 \\ -6x - 2y + 4z = -2 \\ 9x + 3y - 6z = 6 \end{cases}$

13. $\begin{cases} 4x - y + 2z = 5 \\ 2y + z = 4 \\ 4x + y + 3z = 10 \end{cases}$

14. $\begin{cases} 5y - 7z = 14 \\ 2x + y + 4z = 10 \\ 2x + 6y - 3z = 30 \end{cases}$

15. $\begin{cases} x + 5z = 0 \\ 5x + y = 0 \\ y - 3z = 0 \end{cases}$

16. $\begin{cases} x - 5y = 0 \\ x - z = 0 \\ -x + 5z = 0 \end{cases}$

17. $\begin{cases} 6x - 5z = 17 \\ 5x - y + 3z = -1 \\ 2x + y = -41 \end{cases}$

18. $\begin{cases} x + 2y = 6 \\ 7x + 3y + z = -33 \\ x - z = 16 \end{cases}$

19. $\begin{cases} x + y + z = 8 \\ 2x - y - z = 10 \\ x - 2y - 3z = 22 \end{cases}$

20. $\begin{cases} 5x + y + 3z = 1 \\ x - y + 3z = -7 \\ -x + y = 1 \end{cases}$

21. $\begin{cases} x + 2y - z = 5 \\ 6x + y + z = 7 \\ 2x + 4y - 2z = 5 \end{cases}$

22. $\begin{cases} 4x - y + 3z = 10 \\ x + y - z = 5 \\ 8x - 2y + 6z = 10 \end{cases}$

🖥 **23.** $\begin{cases} 2x - 3y + z = 2 \\ x - 5y + 5z = 3 \\ 3x + y - 3z = 5 \end{cases}$

24. $\begin{cases} 4x + y - z = 8 \\ x - y + 2z = 3 \\ 3x - y + z = 6 \end{cases}$

25. $\begin{cases} -2x - 4y + 6z = -8 \\ x + 2y - 3z = 4 \\ 4x + 8y - 12z = 16 \end{cases}$

26. $\begin{cases} -6x + 12y + 3z = -6 \\ 2x - 4y - z = 2 \\ -x + 2y + \dfrac{z}{2} = -1 \end{cases}$

27. $\begin{cases} 2x + 2y - 3z = 1 \\ y + 2z = -14 \\ 3x - 2y = -1 \end{cases}$

28. $\begin{cases} 7x + 4y = 10 \\ x - 4y + 2z = 6 \\ y - 2z = -1 \end{cases}$

29. $\begin{cases} x + 2y - z = 5 \\ -3x - 2y - 3z = 11 \\ 4x + 4y + 5z = -18 \end{cases}$

30. $\begin{cases} 3x - 3y + z = -1 \\ 3x - y - z = 3 \\ -6x + 2y + 2z = -6 \end{cases}$

31. $\begin{cases} \dfrac{3}{4}x - \dfrac{1}{3}y + \dfrac{1}{2}z = 9 \\ \dfrac{1}{6}x + \dfrac{1}{3}y - \dfrac{1}{2}z = 2 \\ \dfrac{1}{2}x - y + \dfrac{1}{2}z = 2 \end{cases}$

32. $\begin{cases} \dfrac{1}{3}x - \dfrac{1}{4}y + z = -9 \\ \dfrac{1}{2}x - \dfrac{1}{3}y - \dfrac{1}{4}z = -6 \\ x - \dfrac{1}{2}y - z = -8 \end{cases}$

REVIEW AND PREVIEW

Solve. See Section 2.5.

33. The sum of two numbers is 45 and one number is twice the other. Find the numbers.

34. The difference between two numbers is 5. Twice the smaller number added to five times the larger number is 53. Find the numbers.

Solve. See Section 2.3.

35. $2(x - 1) - 3x = x - 12$

36. $7(2x - 1) + 4 = 11(3x - 2)$

37. $-y - 5(y + 5) = 3y - 10$

38. $z - 3(z + 7) = 6(2z + 1)$

39. Write a single linear equation in three variables that has $(-1, 2, -4)$ as a solution. (There are many possibilities.) Explain the process you used to write an equation.

40. Write a system of three linear equations in three variables that has $(2, 1, 5)$ as a solution. (There are many possibilities.) Explain the process you used to write an equation.

41. Write a system of linear equations in three variables that has the solution $(-1, 2, -4)$. Explain the process you used to write your system.

42. When solving a system of three equation in three unknowns, explain how to determine that a system has no solution.

43. The fraction $\dfrac{1}{24}$ can be written as the following sum:

$$\frac{1}{24} = \frac{x}{8} + \frac{y}{4} + \frac{z}{3}$$

where the numbers x, y, and z are solutions of

$$\begin{cases} x + y + z = 1 \\ 2x - y + z = 0 \\ -x + 2y + 2z = -1 \end{cases}$$

Solve the system and see that the sum of the fractions is $\dfrac{1}{24}$.

44. The fraction $\dfrac{1}{18}$ can be written as the following sum:

$$\frac{1}{18} = \frac{x}{2} + \frac{y}{3} + \frac{z}{9}$$

where the numbers x, y, and z are solutions of

$$\begin{cases} x + 3y + z = -3 \\ -x + y + 2z = -14 \\ 3x + 2y - z = 12 \end{cases}$$

Solve the system and see that the sum of the fractions is $\dfrac{1}{18}$.

Solving systems involving more than three variables can be accomplished with methods similar to those encountered in this section. Apply what you already know to solve each system of equations in four variables.

45. $\begin{cases} x + y - w = 0 \\ y + 2z + w = 3 \\ x - z = 1 \\ 2x - y - w = -1 \end{cases}$

46. $\begin{cases} 5x + 4y = 29 \\ y + z - w = -2 \\ 5x + z = 23 \\ y - z + w = 4 \end{cases}$

47. $\begin{cases} x + y + z + w = 5 \\ 2x + y + z + w = 6 \\ x + y + z = 2 \\ x + y = 0 \end{cases}$

48. $\begin{cases} 2x - z = -1 \\ y + z + w = 9 \\ y - 2w = -6 \\ x + y = 3 \end{cases}$

49. Write a system of three linear equations in three variables that are dependent equations.

50. What is the solution to the system in Exercise 49?

4.5   *SYSTEMS OF LINEAR EQUATIONS AND PROBLEM SOLVING*

Objectives

1   Solve problems that can be modeled by a system of two linear equations.

2   Solve problems with cost and revenue functions.

3   Solve problems that can be modeled by a system of three linear equations.

1   Thus far, we have solved problems by writing one-variable equations and solving for the variable. Some of these problems can be solved, perhaps more easily, by writing a system of equations, as illustrated in this section.

EXAMPLE 1

PREDICTING EQUAL CONSUMPTION OF RED MEAT AND POULTRY

America's consumption of red meat has decreased most years since 1980 while consumption of poultry has increased. The function $y = -0.71x + 125.6$ approximates the annual pounds of red meat consumed per capita, where x is the number of years since 1980. The function $y = 1.56x + 39.7$ approximates the annual pounds of poultry consumed per capita, where x is also the number of years since 1980. If this trend continues, determine the year when the annual consumption of red meat and poultry are equal.

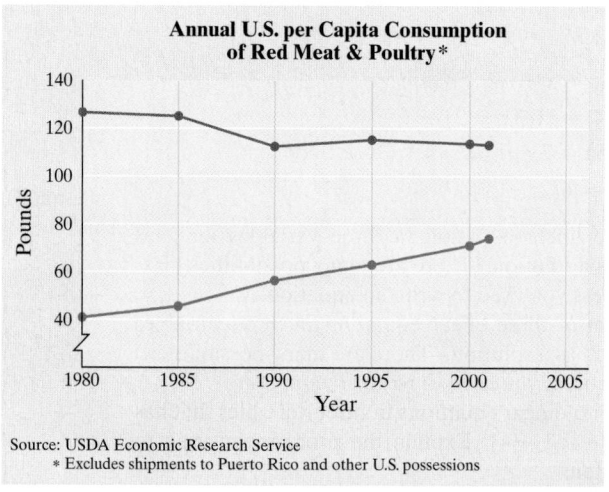

Source: USDA Economic Research Service
* Excludes shipments to Puerto Rico and other U.S. possessions

Solution   **1.** UNDERSTAND. Read and reread the problem and guess a year. Let's guess the year 2010. This year is 30 years since 1980, so $x = 30$. Now let $x = 30$ in each given function.

Red meat: $y = -0.71x + 125.6 = -0.71(30) + 125.6 = 104.3$ pounds
Poultry: $y = 1.56x + 39.7\quad = 1.56(30) + 39.7\quad = 86.5$ pounds

Since the projected pounds in 2010 for red meat and poultry are not the same, we guessed incorrectly, but we do have a better understanding of the problem. We also know that the year will be later than 2010 since projected consumption of red meat is still greater than poultry that year.

2. TRANSLATE. We are already given the system of equations.

3. SOLVE. We want to know the year x in which pounds y are the same, so we solve the system:

$$\begin{cases} y = -0.71x + 125.6 \\ y = 1.56x + 39.7 \end{cases}$$

Since both equations are solved for y, one way to solve is to use the substitution method.

$$y = -0.71x + 125.6 \qquad \text{First equation}$$

$$1.56x + 39.7 = -0.71x + 125.6 \qquad \text{Let } y = 1.56x + 39.7.$$

$$2.27x = 85.9$$

$$x = \frac{85.9}{2.27} \approx 37.84$$

4. INTERPRET. Since we are only asked to find the year, we need only solve for x.
Check: To check, see whether $x \approx 37.84$ gives approximately the same number of pounds of red meat and poultry.

Red meat: $y = -0.71x + 125.6 = -0.71(37.84) + 125.6 \approx 98.73$ pounds

Poultry: $y = 1.56x + 39.7 = 1.56(37.84) + 39.7 \approx 98.73$ pounds

Since we rounded the number of years, the number of pounds do differ slightly. They differ only by 0.0032, so we can assume that we solved correctly.

State: The consumption of red meat and poultry will be the same about 37.84 years after 1980, or 2017.84. Thus, in the year 2017, we predict the consumption will be the same.

EXAMPLE 2

FINDING UNKNOWN NUMBERS

A first number is 4 less than a second number. Four times the first number is 6 more than twice the second. Find the numbers.

Solution   **1.** UNDERSTAND. Read and reread the problem and guess a solution. If a first number is 10 and this is 4 less than a second number, the second number is 14. Four times the first number is 4(10), or 40. This is not equal to 6 more than twice the second number, which is 2(14) + 6 or 34. Although we guessed incorrectly, we now have a better understanding of the problem.

Since we are looking for two numbers, we will let

$$x = \text{first number}$$
$$y = \text{second number}$$

2. TRANSLATE. Since we have assigned two variables to this problem, we will translate the given facts into two equations. For the first statement we have

In words:	the first number	is	4 less than the second number
	$\downarrow$	$\downarrow$	$\downarrow$
Translate:	x	$=$	$y - 4$

Next we translate the second statement into an equation.

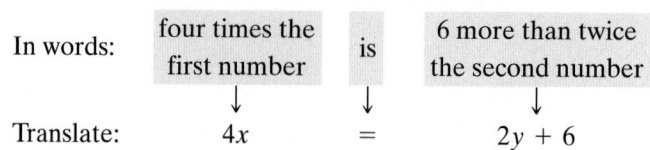

In words: four times the first number | is | 6 more than twice the second number

Translate: $4x$ $=$ $2y + 6$

3. SOLVE. Here we solve the system

$$\begin{cases} x = y - 4 \\ 4x = 2y + 6 \end{cases}$$

Since the first equation expresses x in terms of y, we will use substitution. We substitute $y - 4$ for x in the second equation and solve for y.

$$4x = 2y + 6 \qquad \textit{Second equation}$$

$$4\,(y - 4) = 2y + 6$$

$$4y - 16 = 2y + 6 \qquad \textit{Let } x = y - 4.$$

$$2y = 22$$

$$y = 11$$

Now we replace y with 11 in the equation $x = y - 4$ and solve for x. Then $x = y - 4$ becomes $x = 11 - 4 = 7$. The ordered pair solution of the system is $(7, 11)$.

4. INTERPRET. Since the solution of the system is $(7, 11)$, then the first number we are looking for is 7 and the second number is 11.
Check: Notice that 7 *is* 4 less than 11, and 4 times 7 *is* 6 more than twice 11. The proposed numbers, 7 and 11, are correct.
State: The numbers are 7 and 11.

EXAMPLE 3

FINDING THE RATE OF SPEED

Two cars leave Indianapolis, one traveling east and the other west. After 3 hours they are 297 miles apart. If one car is traveling 5 mph faster than the other, what is the speed of each?

Solution **1.** UNDERSTAND. Read and reread the problem. Let's guess a solution and use the formula $d = rt$ to check. Suppose that one car is traveling at a rate of 55 miles per hour. This means that the other car is traveling at a rate of 50 miles per hour since we are told that one car is traveling 5 mph faster than the other. To find the distance apart after 3 hours, we will first find the distance traveled by each car. One car's distance is rate · time $= 55(3) = 165$ miles. The other car's distance is rate · time $= 50(3) = 150$ miles. Since one car is traveling east and the other west, their distance apart is the sum of their distances, or 165 miles $+$ 150 miles $=$ 315 miles. Although this distance apart is not the required distance of 297 miles, we now have a better understanding of the problem.

$50(3) = 150$ mi $55(3) = 165$ mi

$150 + 165 = 315$ mi

Let's model the problem with a system of equations. We will let

$$x = \text{speed of one car}$$
$$y = \text{speed of the other car}$$

We summarize the information on the following chart. Both cars have traveled 3 hours. Since distance $=$ rate $\cdot$ time, their distances are $3x$ and $3y$ miles, respectively.

	Rate $\cdot$	*Time* $=$	*Distance*
One Car	x	3	$3x$
Other Car	y	3	$3y$

2. TRANSLATE. We can now translate the stated conditions into two equations.

In words: one car's distance   added to   the other car's distance   is   297

Translate: $3x$ $+$ $3y$ $=$ 297

In words: one car's speed   is   5 mph faster than the other

Translate: x $=$ $y + 5$

3. SOLVE. Here we solve the system

$$\begin{cases} 3x + 3y = 297 \\ x = y + 5 \end{cases}$$

Again, the substitution method is appropriate. We replace x with $y + 5$ in the first equation and solve for y.

$$3x + 3y = 297 \quad \text{First equation}$$
$$3(y + 5) + 3y = 297 \quad \text{Let } x = y + 5.$$
$$3y + 15 + 3y = 297$$
$$6y = 282$$
$$y = 47$$

To find x, we replace y with 47 in the equation $x = y + 5$. Then $x = 47 + 5 = 52$. The ordered pair solution of the system is $(52, 47)$.

4. INTERPRET. The solution $(52, 47)$ means that the cars are traveling at 52 mph and 47 mph, respectively.

Check: Notice that one car is traveling 5 mph faster than the other. Also, if one car travels 52 mph for 3 hours, the distance is $3(52) = 156$ miles. The other car traveling for 3 hours at 47 mph travels a distance of $3(47) = 141$ miles. The sum of the distances $156 + 141$ is 297 miles, the required distance.

State: The cars are traveling at 52 mph and 47 mph.

> ▶ **Helpful Hint**
>
> Don't forget to attach units, if appropriate.

EXAMPLE 4

MIXING SOLUTIONS

Lynn Pike, a pharmacist, needs 70 liters of a 50% alcohol solution. She has available a 30% alcohol solution and an 80% alcohol solution. How many liters of each solution should she mix to obtain 70 liters of a 50% alcohol solution?

Solution   1. UNDERSTAND. Read and reread the problem. Next, guess the solution. Suppose that we need 20 liters of the 30% solution. Then we need $70 - 20 = 50$ liters of the 80% solution. To see if this gives us 70 liters of a 50% alcohol solution, let's find the amount of pure alcohol in each solution.

number of liters	×	alcohol strength	=	amount of pure alcohol
↓		↓		↓
20 liters	×	0.30	=	6 liters
50 liters	×	0.80	=	40 liters
70 liters	×	0.50	=	35 liters

Since 6 liters + 40 liters = 46 liters and not 35 liters, our guess is incorrect, but we have gained some insight as to how to model and check this problem.
We will let

$$x = \text{amount of 30\% solution, in liters}$$
$$y = \text{amount of 80\% solution, in liters}$$

and use a table to organize the given data.

	Number of Liters	Alcohol Strength	Amount of Pure Alcohol
30% Solution	x	30%	$0.30x$
80% Solution	y	80%	$0.80y$
50% Solution Needed	70	50%	$(0.50)(70)$

2. TRANSLATE. We translate the stated conditions into two equations.

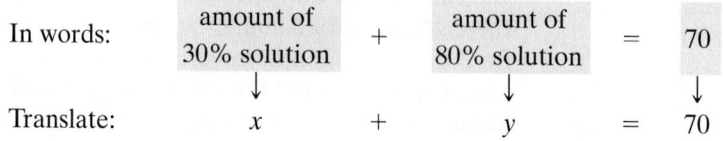

	amount of pure alcohol in 30% solution		amount of pure alcohol in 80% solution		amount of pure alcohol in 50% solution
In words:	amount of pure alcohol in 30% solution	+	amount of pure alcohol in 80% solution	=	amount of pure alcohol in 50% solution
	↓		↓		↓
Translate:	$0.30x$	+	$0.80y$	=	$(0.50)(70)$

3. SOLVE. Here we solve the system

$$\begin{cases} x + y = 70 \\ 0.30x + 0.80y = (0.50)(70) \end{cases}$$

To solve this system, we use the elimination method. We multiply both sides of the first equation by -3 and both sides of the second equation by 10. Then

$$\begin{cases} -3(x + y) = -3(70) \\ 10(0.30x + 0.80y) = 10(0.50)(70) \end{cases} \quad \begin{matrix} \text{simplifies} \\ \text{to} \end{matrix} \quad \begin{cases} -3x - 3y = -210 \\ \underline{3x + 8y = 350} \\ 5y = 140 \\ y = 28 \end{cases}$$

Now we replace y with 28 in the equation $x + y = 70$ and find that $x + 28 = 70$, or $x = 42$.
The ordered pair solution of the system is $(42, 28)$.

4. INTERPRET.
  Check: Check the solution in the same way that we checked our guess.
  State: The pharmacist needs to mix 42 liters of 30% solution and 28 liters of 80% solution to obtain 70 liters of 50% solution.

✔ **CONCEPT CHECK**

Suppose you mix an amount of 25% acid solution with an amount of 60% acid solution. You then calculate the acid strength of the resulting acid mixture. For which of the following results should you suspect an error in your calculation? Why?

  a. 14%   **b.** 32%   **c.** 55%

2   Recall that businesses are often computing cost and revenue functions or equations to predict sales, to determine whether prices need to be adjusted, and to see whether the company is making or losing money. Recall also that the value at which revenue equals cost is called the break-even point. When revenue is less than cost, the company is losing money; when revenue is greater than cost, the company is making money.

EXAMPLE 5

FINDING A BREAK-EVEN POINT

A manufacturing company recently purchased $3000 worth of new equipment to offer new personalized stationery to its customers. The cost of producing a package of personalized stationery is $3.00, and it is sold for $5.50. Find the number of packages that must be sold for the company to break even.

Solution   **1.** UNDERSTAND. Read and reread the problem. Notice that the cost to the company will include a one-time cost of $3000 for the equipment and then $3.00 per package produced. The revenue will be $5.50 per package sold.

To model this problem, we will let

$$x = \text{number of packages of personalized stationery}$$
$$C(x) = \text{total cost for producing } x \text{ packages of stationery}$$
$$R(x) = \text{total revenue for selling } x \text{ packages of stationery}$$

2. TRANSLATE. The revenue equation is

In words:

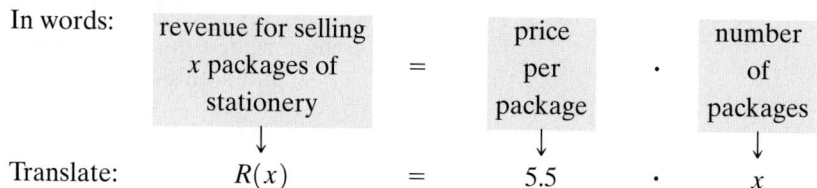

Translate: $R(x)$ $=$ 5.5 $\cdot$ x

The cost equation is

In words:

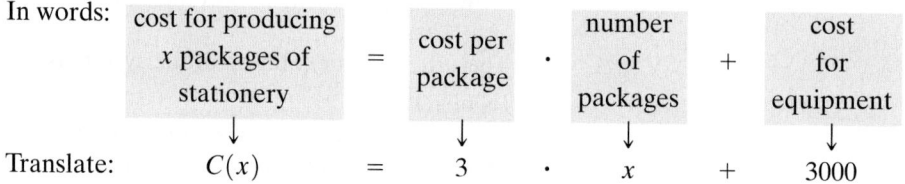

Translate: $C(x)$ $=$ 3 $\cdot$ x $+$ 3000

Since the break-even point is when $R(x) = C(x)$, we solve the equation

$$5.5x = 3x + 3000$$

3. SOLVE.

$$5.5x = 3x + 3000$$
$$2.5x = 3000 \qquad \text{Subtract } 3x \text{ from both sides.}$$
$$x = 1200 \qquad \text{Divide both sides by 2.5.}$$

4. INTERPRET.

Check: To see whether the break-even point occurs when 1200 packages are produced and sold, see if revenue equals cost when $x = 1200$. When $x = 1200$, $R(x) = 5.5x = 5.5(1200) = 6600$ and $C(x) = 3x + 3000 = 3(1200) + 3000 = 6600$. Since $R(1200) = C(1200) = 6600$, the break-even point is 1200.

State: The company must sell 1200 packages of stationery to break even. The graph of this system is shown.

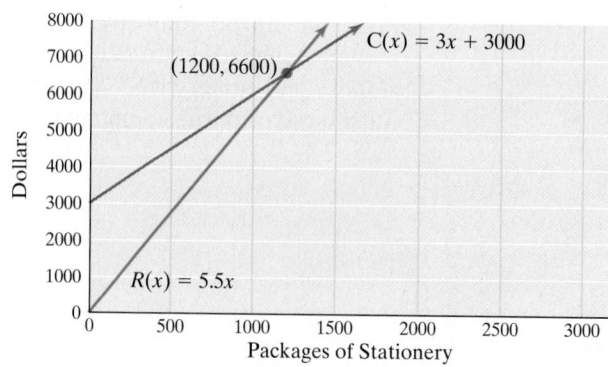

3   To introduce problem solving by writing a system of three linear equations in three variables, we solve a problem about triangles.

EXAMPLE 6

FINDING ANGLE MEASURES

The measure of the largest angle of a triangle is 80° more than the measure of the smallest angle, and the measure of the remaining angle is 10° more than the measure of the smallest angle. Find the measure of each angle.

Solution

1. UNDERSTAND. Read and reread the problem. Recall that the sum of the measures of the angles of a triangle is 180°. Then guess a solution. If the smallest angle measures 20°, the measure of the largest angle is 80° more, or $20° + 80° = 100°$. The measure of the remaining angle is 10° more than the measure of the smallest angle, or $20° + 10° = 30°$. The sum of these three angles is $20° + 100° + 30° = 150°$, not the required 180°. We now know that the measure of the smallest angle is greater than 20°.

To model this problem we will let

$$x = \text{degree measure of the smallest angle}$$
$$y = \text{degree measure of the largest angle}$$
$$z = \text{degree measure of the remaining angle}$$

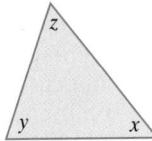

2. TRANSLATE. We translate the given information into three equations.

In words: the sum of the measures   =   180

Translate: $x + y + z$ =   180

In words: the largest angle   is   80 more than the smallest angle

Translate y = $x + 80$

In words: the remaining angle   is   10 more than the smallest angle

Translate: z = $x + 10$

3. SOLVE. We solve the system

$$\begin{cases} x + y + z = 180 \\ \qquad\quad y = x + 80 \\ \qquad\quad z = x + 10 \end{cases}$$

Since y and z are both expressed in terms of x, we will solve using the substitution method. We substitute $y = x + 80$ and $z = x + 10$ in the first equation. Then

$$x + y + z = 180 \qquad \text{First equation}$$

$$x + \overbrace{(x + 80)} + \overbrace{(x + 10)} = 180 \qquad \text{Let } y = x + 80 \text{ and } z = x + 10.$$

$$3x + 90 = 180$$

$$3x = 90$$

$$x = 30$$

Then $y = x + 80 = 30 + 80 = 110$, and $z = x + 10 = 30 + 10 = 40$. The ordered triple solution is $(30, 110, 40)$.

4. INTERPRET.
   Check: Notice that $30° + 40° + 110° = 180°$. Also, the measure of the largest angle, $110°$, is $80°$ more than the measure of the smallest angle, $30°$. The measure of the remaining angle, $40°$, is $10°$ more than the measure of the smallest angle, $30°$.
   State: The angles measure $30°$, $110°$, and $40°$.

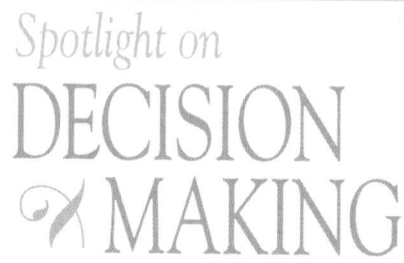

Suppose you are choosing a long-distance telephone plan. You have narrowed your choices to the One Rate® 7¢ Plan offered by AT&T and the Qwest Communicator plan offered by Qwest Communications International, Inc. Under the AT&T One Rate® 7¢ Plan, calls made any time cost $0.07 per minute and users are charged a $5.95 monthly fee. Under the Qwest Communicator plan, calls made any time cost $0.05 per minute and users are charged a $9.95 monthly fee. Which long-distance plan would you choose? Why? What other factors would you want to consider? Would you change your choice if you knew that your long-distance calls averaged 4 hours per month? Explain your reasoning.

EXERCISE SET 4.5

| STUDY GUIDE/SSM | CD/ VIDEO | PH MATH TUTOR CENTER | MathXL®Tutorials ON CD | MathXL® | MyMathLab® |

Solve. See Examples 1 through 4.

1. One number is two more than a second number. Twice the first is 4 less than 3 times the second. Find the numbers.
2. Three times one number minus a second is 8, and the sum of the numbers is 12. Find the numbers.
3. The United States has the world's only "large deck" aircraft carriers which can hold up to 72 aircraft. The Enterprise class carrier is longest in length while the Nimitz class carrier is the second longest. The total length of these two carriers is 2193 feet while the difference of their lengths is only 9 feet. (*Source:* U.S.A. Today, May, 2001)
   a. Find the length of each class carrier.

b. If a football field has a length of 100 yards, determine the length of the Enterprise class carrier in terms of number of football fields.

4. The rate of growth of participation (age 6 and older) in sports featured in the X-Games is surpassing that for older sports such as football and baseball. The most popular X-Game sport is in-line roller skating, followed by skateboarding. In 2000, the total number of participants in both sports was 40.6 million. If the number of participants in roller skating was 5.8 million more than twice the number of participants in skateboarding, find the number of participants in each sport. (*Source:* January 2001 Sporting Goods Manufacturers Association Sports Participation Topline Report)

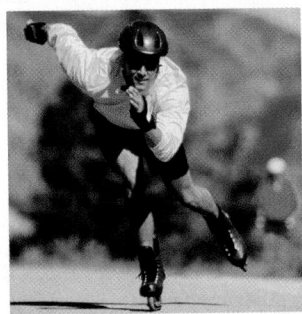

5. A Delta 727 traveled 560 mph with the wind and 480 mph against the wind. Find the speed of the plane in still air and the speed of the wind.

6. Terry Watkins can row about 10.6 kilometers in 1 hour downstream and 6.8 kilometers upstream in 1 hour. Find how fast he can row in still water, and find the speed of the current.

7. Find how many quarts of 4% butterfat milk and 1% butterfat milk should be mixed to yield 60 quarts of 2% butterfat milk.

8. A pharmacist needs 500 milliliters of a 20% phenobarbital solution but has only 5% and 25% phenobarbital solutions available. Find how many milliliters of each he should mix to get the desired solution.

9. In a recent year, the United Kingdom was the most popular host country for U.S. students traveling abroad to study. Spain was the second most popular destination. A total of 40,012 students visited one of the two countries. If 15,428 more U.S. students studied in the United Kingdom than in Spain, find how many students studied abroad in each country. (*Source:* Institute of International Education, Open Doors 2000)

10. In 2000, Washington D.C. had the most conventions while Orlando was the second most popular convention location. A total of 353 conventions were held in one of the two destinations. If Washington D.C. had 27 more conventions than Orlando, find the number of conventions held in each location. (*Source: Successful Meetings* magazine)

11. Karen Karlin bought some large frames for $15 each and some small frames for $8 each at a closeout sale. If she bought 22 frames for $239, find how many of each type she bought.

12. Hilton University Drama Club sold 311 tickets for a play. Student tickets cost 50 cents each; nonstudent tickets cost $1.50. If total receipts were $385.50, find how many tickets of each type were sold.

13. One number is two less than a second number. Twice the first is 4 more than 3 times the second. Find the numbers.

14. Twice one number plus a second number is 42, and the first number minus the second number is −6. Find the numbers.

15. In the United States, the percent of women using the Internet is increasing faster than the percent of men. For the years 1996–2001, the function $y = 7x + 18.7$ can be used to estimate the percent of females using the Internet while the function $y = 6x + 27.7$ can be used to estimate the percent of males. For both functions, x is the number of years since 1996. If this trend continues, predict the year in which the percent of females using the Internet is equal to the percent of males. (*Source:* Pew Internet & American Life Project)

16. The percent of car-vehicle sales is decreasing while the percent of light-truck (pickups, sport-utility vans and minivans) vehicle sales is increasing. For the years 1997–2000, the function $y = -x + 54.5$ can be used to estimate the percent of vehicle sales being cars while the function $y = x + 45.5$ can be used to estimate the percent of vehicle sales being light trucks. For both functions, x is the number of years since 1997.

   a. If this trend continues, predict the year in which the percent of car sales equals the percent of light-truck sales.

   b. Before the actual 2001 vehicle sales data was published, *USA Today* predicted that light-truck sales would likely be greater than car sales in the year 2001. Does your prediction from part a agree with this statement? (*Source: USA Today* and Autodata)

17. An office supply store in San Diego sells 7 writing tablets and 4 pens for $6.40. Also, 2 tablets and 19 pens cost $5.40. Find the price of each.

18. A Candy Barrel shop manager mixes M&M's worth $2.00 per pound with trail mix worth $1.50 per pound. Find how many pounds of each she should use to get 50 pounds of a party mix worth $1.80 per pound.

19. An airplane takes 3 hours to travel a distance of 2160 miles with the wind. The return trip takes 4 hours against the wind. Find the speed of the plane in still air and the speed of the wind.

20. Two cyclists start at the same point and travel in opposite directions. One travels 4 mph faster than the other. In 4 hours they are 112 miles apart. Find how fast each is traveling.

21. The percent of men 65 years of age or older in the labor force has decreased most years since 1900 while the percent of women 65 years of age or older has slightly increased. For the years 1900–2000, the function $y = -0.52x + 64.5$ approximates the percent of men 65 years of age or older in the labor force while the function $y = 0.02x + 7.2$ approximates the percent of women 65 years of age or older in the labor force. For both functions, x is the number of years after 1900. (*Source: The World Almanac,* 2002)

   a. Explain how the decrease of men described can be verified by their given function while the slight increase of women described can be verified by their given function.

b. If this trend continues, determine the year when the percent of men and woman 65 years of age or older in the labor force are the same.

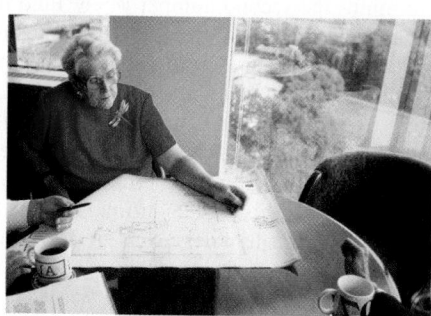

22. The annual U.S. per capita consumption of butter has remained about the same since 1980 while consumption of margarine has decreased. For the years 1995–1999, the function $y = 0.05x + 4.3$ approximates the annual U.S. per capita consumption of butter in pounds and the function $y = -0.31x + 9.3$ approximates the annual U.S. per capita consumption of margarine in pounds. For both functions, x is the number of years after 1995. If this trend continues, determine the year when the pounds of butter consumed equals the pounds of margarine consumed.

△ **23.** The perimeter of a triangle is 93 centimeters. If two sides are equally long and the third side is 9 centimeters longer than the others, find the lengths of the three sides.

24. Jack Reinholt, a car salesman, has a choice of two pay arrangements: a weekly salary of $200 plus 5% commission on sales, or a straight 15% commission. Find the amount of weekly sales for which Jack's earnings are the same regardless of the pay arrangement.

25. Hertz car rental agency charges $25 daily plus 10 cents per mile. Budget charges $20 daily plus 25 cents per mile. Find the daily mileage for which the Budget charge for the day is twice that of the Hertz charge for the day.

26. Carroll Blakemore, a drafting student, bought three templates and a pencil one day for $6.45. Another day he bought two pads of paper and four pencils for $7.50. If the price of a pad of paper is three times the price of a pencil, find the price of each type of item.

△ **27.** In the figure, line l and line m are parallel lines cut by transversal t. Find the values of x and y.

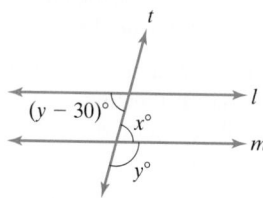

△ **28.** Find the values of x and y in the following isosceles triangle.

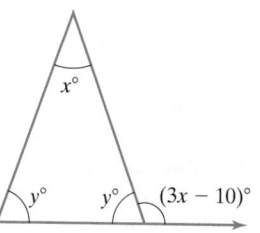

Given the cost function $C(x)$ and the revenue function $R(x)$, find the number of units x that must be sold to break even. See Example 5.

29. $C(x) = 30x + 10,000$ $R(x) = 46x$

30. $C(x) = 12x + 15,000$ $R(x) = 32x$

31. $C(x) = 1.2x + 1500$ $R(x) = 1.7x$

32. $C(x) = 0.8x + 900$ $R(x) = 2x$

33. $C(x) = 75x + 160,000$ $R(x) = 200x$

34. $C(x) = 105x + 70,000$ $R(x) = 245x$

35. The planning department of Abstract Office Supplies has been asked to determine whether the company should introduce a new computer desk next year. The department estimates that $6000 of new manufacturing equipment will need to be purchased and that the cost of constructing each desk will be $200. The department also estimates that the revenue from each desk will be $450.

  a. Determine the revenue function $R(x)$ from the sale of x desks.

  b. Determine the cost function $C(x)$ for manufacturing x desks.

  c. Find the break-even point.

36. Baskets, Inc., is planning to introduce a new woven basket. The company estimates that $500 worth of new equipment will be needed to manufacture this new type of basket and that it will cost $15 per basket to manufacture. The company also estimates that the revenue from each basket will be $31.

  a. Determine the revenue function $R(x)$ from the sale of x baskets.

  b. Determine the cost function $C(x)$ for manufacturing x baskets.

  c. Find the break-even point.

Solve. See Example 6.

37. Rabbits in a lab are to be kept on a strict daily diet that includes 30 grams of protein, 16 grams of fat, and 24 grams of carbohydrates. The scientist has only three food mixes available with the following grams of nutrients per unit.

	Protein	*Fat*	*Carbohydrate*
Mix A	4	6	3
Mix B	6	1	2
Mix C	4	1	12

Find how many units of each mix are needed daily to meet each rabbit's dietary need.

38. Gerry Gundersen mixes different solutions with concentrations of 25%, 40%, and 50% to get 200 liters of a 32% solution. If he uses twice as much of the 25% solution as of the 40% solution, find how many liters of each kind he uses.

△ **39.** The perimeter of a quadrilateral (four-sided polygon) is 29 inches. The longest side is twice as long as the shortest side. The other two sides are equally long and are 2 inches longer than the shortest side. Find the length of all four sides.

△ **40.** The measure of the largest angle of a triangle is 90° more than the measure of the smallest angle, and the measure of the remaining angle is 30° more than the measure of the smallest angle. Find the measure of each angle.

41. The sum of three numbers is 40. One number is five more than a second number. It is also twice the third. Find the numbers.

42. The sum of the digits of a three-digit number is 15. The tens-place digit is twice the hundreds-place digit, and the ones-place digit is 1 less than the hundreds-place digit. Find the three-digit number.

43. In 2003, the WNBA's top scorer was Lauren Jackson of the Seattle Storm. She scored a total of 698 points during the regular season. The number of two-point field goals Jackson made was 19 less than six times the number of three-point field goals she made. The number of free throws (each worth one point) she made was 64 fewer than the number of two-point field goals she made. Find how many free throws, two-point field goals, and three-point field goals Lauren Jackson made during the 2003 regular season. (*Source:* Women's National Basketball Association)

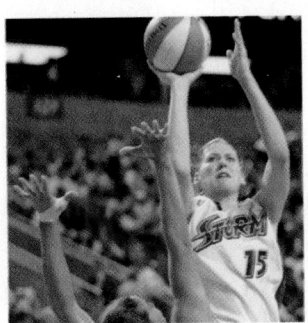

44. During the 2003 NBA playoffs, the top-scoring player was Kobe Bryant of the Los Angeles Lakers . Bryant scored a total of 385 points during the playoffs. The number of free throws (each worth one point) he made was 11 more than three times the number of three-point field goals he made. Bryant also made 26 fewer free throws than two-point field goals. How many free throws, two-point field goals, and three-point field goals did Kobe Bryant make during the 2003 playoffs? (*Source:* National Basketball Association)

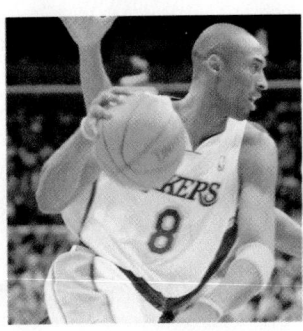

△ **45.** Find the values of x, y, and z in the following triangle.

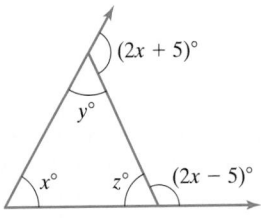

△ **46.** The sum of the measures of the angles of a quadrilateral is 360°. Find the value of x, y, and z in the following quadrilateral.

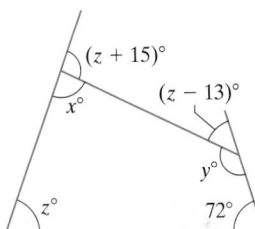

REVIEW AND PREVIEW

Multiply both sides of equation (1) by 2, and add the resulting equation to equation (2). See Section 4.4.

47. $3x - y + z = 2$ (1)
$-x + 2y + 3z = 6$ (2)

48. $2x + y + 3z = 7$ (1)
$-4x + y + 2z = 4$ (2)

Multiply both sides of equation (1) by -3, and add the resulting equation to equation (2). See Section 4.4.

49. $x + 2y - z = 0$ (1)
$3x + y - z = 2$ (2)

50. $2x - 3y + 2z = 5$ (1)
$x - 9y + z = -1$ (2)

Concept Extensions

51. The number of personal bankruptcy petitions filed in the United States has been on the rise since the early 1980s. In 2001, the number of petitions filed was 200,000 more than four times the number of petitions filed in 1980. This is equivalent to an increase of 1,100,000 petitions filed from 1980 to 2001. Find how many personal bankruptcy petitions were filed in each year. (*Source:* Based on data from the Administrative Office of the United States Courts)

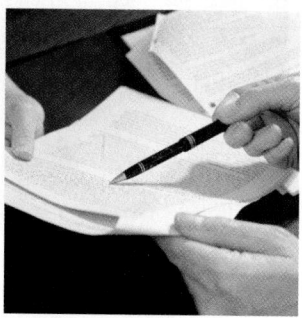

52. In 2000, the median weekly earnings for male bus drivers in the United States was $100 more than the median weekly earnings for female bus drivers. The median weekly earnings for female bus drivers was 0.75 times that of their male counterparts. Also in 2000, the median weekly earnings for female financial managers in the United States was $400 less than the median weekly earnings for male financial managers. The median weekly earnings for male financial managers was 1.5 times that of their female counterparts. (*Source:* Based on data from the U.S. Bureau of Labor Statistics)

   a. Find the median weekly earnings for female bus drivers in the United States in 2000.

   b. Find the median weekly earnings for female financial managers in the United States in 2000.

c. Of the four groups of workers described in the problem, which group makes the greatest weekly earnings? Which group makes the least weekly earnings?

53. Find the values of a, b, and c such that the equation $y = ax^2 + bx + c$ has ordered pair solutions $(1, 6)$, $(-1, -2)$, and $(0, -1)$. To do so, substitute each ordered pair solution into the equation. Each time, the result is an equation in three unknowns: a, b, and c. Then solve the resulting system of three linear equations in three unknowns, a, b, and c.

54. Find the values of a, b, and c such that the equation $y = ax^2 + bx + c$ has ordered pair solutions $(1, 2)$, $(2, 3)$ and $(-1, 6)$. (*Hint:* See Exercise 53.)

55. Data (x, y) for the total number y (in thousands) of college-bound students who took the ACT assessment in the year x are $(0, 1065)$, $(1, 1070)$ and $(3, 1175)$, where $x = 0$ represents 2000 and $x = 1$ represents 2001. Find the values of a, b, and c such that the equation $y = ax^2 + bx + c$ models this data. According to your model, how many students will take the ACT in 2009? (*Source:* ACT, Inc.)

56. Monthly normal rainfall data (x, y) for Portland, Oregon, are $(4, 2.47)$, $(7, 0.6)$, $(8, 1.1)$, where x represents time in months (with $x = 1$ representing January) and y represents rainfall in inches. Find the values of a, b, and c rounded to 2 decimal places such that the equation $y = ax^2 + bx + c$ models this data. According to your model, how much rain should Portland expect during September? (*Source:* National Climatic Data Center)

CHAPTER 4 PROJECT

Analyzing the Courses of Ships

From overhead photographs or satellite imagery of ships on the ocean, defense analysts can tell a lot about a ship's immediate course by looking at its wake. Assuming that two ships will maintain their present courses, it is possible to extend their paths, based on the wakes visible in the photograph. This can be used to find possible points of collision for the two ships.

In this project, you will investigate the courses and possibility of collision of the two ships shown in the figure. This project may be completed by working in groups or individually.

1. Using each ship's wake as a guide, use a straight-edge to extend the paths of the ships on the figure. Estimate the coordinates of the point of intersection of the ships' courses from the grid. If the ships continue in these courses, they could possibly collide at the point of intersection of their paths.

2. What factor will govern whether or not the ships actually collide at the point found in Question 1?

Scale: $\frac{1}{4}$ inch = 10 miles

3. Using the coordinates labeled on each ship's wake, find a linear equation that describes each path.

4. (Optional) Use a graphing calculator to graph both equations in the same window. Use the Intersect or Trace feature to estimate the point of intersection of the two paths. Compare this estimate to your estimate in Question 1.

5. Solve the system of two linear equations using one of the methods in this chapter. The solution is the point of intersection of the two paths. Compare your answer to your estimates from Questions 1 and 4.

6. Plot the point of intersection you found in Question 5 on the figure. Use the figure's scale to find each ship's distance from this point of collision by measuring from the bow (tip) of each ship with a ruler. Suppose that the speed of Ship A is r_1 and the speed of ship B is r_2. Given the present positions and courses of the two ships, find a relationship between their speeds that would ensure their collision.

4   CHAPTER VOCABULARY CHECK

Fill in each blank with one of the words or phrases listed below.

system of linear equations	solution	consistent	independent
dependent	inconsistent	substitution	addition

1. In a system of linear equations in two variables, if the graphs of the equations are the same, the equations are _____ equations.

2. Two or more linear equations are called a _____ .

3. A system of equations that has at least one solution is called a(n) _____ system.

4. A _____ of a system of two equations in two variables is an ordered pair of numbers that is a solution of both equations in the system.

5. Two algebraic methods for solving systems of equations are _____ and _____ .

6. A system of equations that has no solution is called a(n) _____ system.

7. In a system of linear equations in two variables, if the graphs of the equations are different, the equations are _____ equations.

STUDY SKILLS REMINDER

Are You Preparing for a Test on Chapter 4?

Below I have listed some common trouble areas for topics covered in Chapter 4. After studying for your test—but before taking your test—read these.

▶ If you are having trouble drawing a neat graph, remember to ask your instructor if you can use graph paper on your test. This will save you time and keep your graphs neat.

▶ Do you remember how to check solutions of systems of equations? If $(-1, 5)$ is a solution of the system

$$\begin{cases} 3x - y = -8 \\ -x + y = 6 \end{cases},$$

then the ordered pair will make *both* equations a true statement.

$$
\begin{array}{ll}
3x - y = -8 & -x + y = 6 \\
3(-1) - 5 = -8 & -(-1) + 5 = 6 \quad \text{Let } x = -1 \\
& \text{and } y = 5. \\[6pt]
-8 = -8 \quad \text{True} & 6 = 6 \quad \text{True}
\end{array}
$$

Remember: This is simply a list of a few common trouble areas. For a review of Chapter 4, see the Highlights and Chapter Review at the end of this chapter.

CHAPTER 4 HIGHLIGHTS

Definitions and Concepts	Examples

Section 4.1 Solving Systems of Linear Equations by Graphing

A **solution** of a system of two equations in two variables is an ordered pair of numbers that is a solution of both equations in the system.

Determine whether $(-1, 3)$ is a solution of the system:

$$\begin{cases} 2x - y = -5 \\ x = 3y - 10 \end{cases}$$

Replace x with -1 and y with 3 in both equations.

$$
\begin{array}{ll}
2x - y = -5 & x = 3y - 10 \\
2(-1) - 3 \stackrel{?}{=} -5 & -1 \stackrel{?}{=} 3 \cdot 3 - 10 \\
-5 = -5 \quad \text{True} & -1 = -1 \qquad \text{True}
\end{array}
$$

$(-1, 3)$ is a solution of the system.

(continued)

Definitions and Concepts	**Examples**

Section 4.1   Solving Systems of Linear Equations by Graphing

Graphically, a solution of a system is a point common to the graphs of both equations.

Solve by graphing. $\begin{cases} 3x - 2y = -3 \\ x + y = 4 \end{cases}$

A system of equations with at least one solution is a **consistent system**. A system that has no solution is an **inconsistent system**.

If the graphs of two linear equations are identical, the equations are **dependent**. If their graphs are different, the equations are **independent**.

Consistent
and independent

Consistent
and dependent

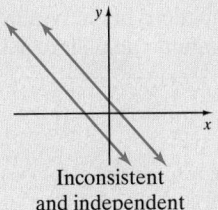

Inconsistent
and independent

Section 4.2   Solving Systems of Linear Equations by Substitution

To solve a system of linear equations by the substitution method

Step 1. Solve one equation for a variable.

Step 2. Substitute the expression for the variable into the other equation.

Step 3. Solve the equation from Step 2 to find the value of one variable.

Step 4. Substitute the value from Step 3 in either original equation to find the value of the other variable.

Step 5. Check the solution in both equations.

Solve by substitution.

$$\begin{cases} 3x + 2y = 1 \\ x = y - 3 \end{cases}$$

Substitute $y - 3$ for x in the first equation.

$$3x + 2y = 1$$
$$3(y - 3) + 2y = 1$$
$$3y - 9 + 2y = 1$$
$$5y = 10$$
$$y = 2 \qquad \text{Divide by 5.}$$

To find x, substitute 2 for y in $x = y - 3$ so that $x = 2 - 3$ or -1. The solution $(-1, 2)$ checks.   *(continued)*

Definitions and Concepts	**Examples**

Section 4.3 Solving Systems of Linear Equations by Addition

To solve a system of linear equations by the addition method	Solve by addition.
Step 1. Rewrite each equation in standard form $Ax + By = C$.	$$\begin{cases} x - 2y = 8 \\ 3x + y = -4 \end{cases}$$
Step 2. Multiply one or both equations by a nonzero number so that the coefficients of a variable are opposites.	Multiply both sides of the first equation by -3.
Step 3. Add the equations.	$$\begin{cases} -3x + 6y = -24 \\ \underline{3x + y = -4} \end{cases}$$
Step 4. Find the value of one variable by solving the resulting equation.	$7y = -28$ Add. $y = -4$ Divide by 7.

To find x, let $y = -4$ in an original equation.

$$x - 2(-4) = 8 \quad \text{First equation}$$
$$x + 8 = 8$$
$$x = 0$$

Step 5. Substitute the value from Step 4 into either original equation to find the value of the other variable.

The solution $(0, -4)$ checks.

Step 6. Check the solution in both equations.

Solve: $\begin{cases} 2x - 6y = -2 \\ x = 3y - 1 \end{cases}$

If solving a system of linear equations by substitution or addition yields a true statement such as $-2 = -2$, then the graphs of the equations in the system are identical and there is an infinite number of solutions of the system.

Substitute $3y - 1$ for x in the first equation.

$$2(3y - 1) - 6y = -2$$
$$6y - 2 - 6y = -2$$
$$-2 = -2 \quad \text{True}$$

The system has an infinite number of solutions.

Section 4.4 Solving Systems of Linear Equations in Three Variables

A **solution** of an equation in three variables x, y, and z is an **ordered triple** (x, y, z) that makes the equation a true statement.	Verify that $(-2, 1, 3)$ is a solution of $2x + 3y - 2z = -7$. Replace x with -2, y with 1, and z with 3.

$$2(-2) + 3(1) - 2(3) = -7$$
$$-4 + 3 - 6 = -7$$
$$-7 = -7 \quad \text{True}$$

$(-2, 1, 3)$ is a solution.

To solve a system of three linear equations by the elimination method:

Solve the system.

Step 1: Write each equation in standard form, $Ax + By + Cz = D$.

$$\begin{cases} 2x + y - z = 0 & (1) \\ x - y - 2z = -6 & (2) \\ -3x - 2y + 3z = -22 & (3) \end{cases}$$

Step 2: Choose a pair of equations and use the equations to eliminate a variable.

1. Each equation is written in standard form.

2.
$$\begin{array}{r} 2x + y - z = 0 \quad \textbf{(1)} \\ \underline{x - y - 2z = -6} \quad \textbf{(2)} \quad \text{Add.} \\ 3x - 3z = -6 \quad \textbf{(4)} \end{array}$$

(continued)

Definitions and Concepts	Examples

Step 3: Choose any other pair of equations and eliminate the same variable.

Step 4: Solve the system of two equations in two variables from Steps 1 and 2.

Step 5: Solve for the third variable by substituting the values of the variables from Step 4 into any of the original equations.

3. Eliminate y from equations (1) and (3) also.

$$\begin{array}{rll} 4x + 2y - 2z = \ \ 0 & \text{Multiply equation} \\ \underline{-3 - 2y + 3z = -22} & \text{(3) (1) by 2.} \\ x \ \ \ \ \ \ \ \ + \ z = -22 & \text{(5) Add.} \end{array}$$

4. Solve

$$\begin{cases} 3x - 3z = \ \ -6 & (4) \\ x + \ \ z = -22 & (5) \end{cases}$$

$$\begin{array}{rll} x - \ z = \ \ -2 & \\ \underline{x + \ z = -22} & \text{(5)} \quad \text{Divide equation} \\ 2x \ \ \ \ \ \ = -24 & \text{(4) by 3.} \\ x \ \ \ \ \ \ = -12 & \end{array}$$

To find z, use equation (5).

$$\begin{array}{r} x + z = -22 \\ -12 + z = -22 \\ z = -10 \end{array}$$

5. To find y, use equation (1).

$$\begin{array}{r} 2x + y - z = 0 \\ 2(-12) + y - (-10) = 0 \\ -24 + y + 10 = 0 \\ y = 14 \end{array}$$

The solution is $(-12, 14, -10)$.

Problem-solving steps
1. UNDERSTAND. Read and reread the problem.

Two angles are supplementary if their sum is 180°. The larger of two supplementary angles is three times the smaller, decreased by twelve. Find the measure of each angle. Let

$$x = \text{measure of smaller angle}$$
$$y = \text{measure of larger angle}$$

2. TRANSLATE.

In words:

the sum of supplementary angles	is	180°
↓	↓	↓
Translate: $x + y$	=	180

(continued)

Definitions and Concepts	**Examples**

Section 4.5 Systems of Linear Equations and Problem Solving

	In words:
	larger angle · is · 3 times smaller · decreased by · 12
	↓ ↓ ↓ ↓ ↓
	Translate: y = $3x$ − 12

3. SOLVE.

Solve the system:

$$\begin{cases} x + y = 180 \\ y = 3x - 12 \end{cases}$$

Use the substitution method and replace y with $3x - 12$ in the first equation.

$$x + y = 180$$
$$x + (3x - 12) = 180$$
$$4x = 192$$
$$x = 48$$

Since $y = 3x - 12$, then $y = 3 \cdot 48 - 12$ or 132.

4. INTERPRET.

The solution checks. The smaller angle measures 48° and the larger angle measures 132°.

4 CHAPTER REVIEW

(4.1) *Determine whether any of the following ordered pairs satisfy the system of linear equations.*

1. $\begin{cases} 2x - 3y = 12 \\ 3x + 4y = 1 \end{cases}$

  a. $(12, 4)$

  b. $(3, -2)$

  c. $(-3, 6)$

2. $\begin{cases} 4x + y = 0 \\ -8x - 5y = 9 \end{cases}$

  a. $\left(\dfrac{3}{4}, -3\right)$

  b. $(-2, 8)$

  c. $\left(\dfrac{1}{2}, -2\right)$

3. $\begin{cases} 5x - 6y = 18 \\ 2y - x = -4 \end{cases}$

  a. $(-6, -8)$

  b. $\left(3, \dfrac{5}{2}\right)$

  c. $\left(3, -\dfrac{1}{2}\right)$

4. $\begin{cases} 2x + 3y = 1 \\ 3y - x = 4 \end{cases}$

  a. $(2, 2)$

  b. $(-1, 1)$

  c. $(2, -1)$

Solve each system of equations by graphing.

5. $\begin{cases} 2x + y = 5 \\ \quad\; 3y = -x \end{cases}$

6. $\begin{cases} 3x + y = -2 \\ 2x - y = -3 \end{cases}$

7. $\begin{cases} y - 2x = 4 \\ \;\; x + y = -5 \end{cases}$

8. $\begin{cases} y - 3x = 0 \\ 2y - 3 = 6x \end{cases}$

9. $\begin{cases} 3x + y = 2 \\ 3x - 6 = -9y \end{cases}$

10. $\begin{cases} 2y + x = 2 \\ \;\; x - y = 5 \end{cases}$

*Without graphing, **(a)** decide whether the graphs of the system are identical lines, parallel lines, or lines intersecting at a single point and **(b)** determine the number of solutions for each system.*

11. $\begin{cases} 2x - y = 3 \\ \quad\;\; y = 3x + 1 \end{cases}$

12. $\begin{cases} 3x + y = 4 \\ \quad\;\; y = -3x + 1 \end{cases}$

13. $\begin{cases} \dfrac{2}{3}x + \dfrac{1}{6}y = 0 \\ \qquad\quad y = -4x \end{cases}$

14. $\begin{cases} \dfrac{1}{4}x + \dfrac{1}{8}y = 0 \\ \qquad\quad y = -6x \end{cases}$

(4.2) *Solve the following systems of equations by the substitution method. If there is a single solution, give the ordered pair. If not, state whether the system is inconsistent or whether the equations are dependent.*

15. $\begin{cases} \qquad\quad y = 2x + 6 \\ 3x - 2y = -11 \end{cases}$

16. $\begin{cases} \qquad\quad y = 3x - 7 \\ 2x - 3y = 7 \end{cases}$

17. $\begin{cases} x + 3y = -3 \\ 2x + y = 4 \end{cases}$

18. $\begin{cases} 3x + y = 11 \\ x + 2y = 12 \end{cases}$

19. $\begin{cases} 4y = 2x - 3 \\ x - 2y = 4 \end{cases}$

20. $\begin{cases} 2x = 3y - 18 \\ x + 4y = 2 \end{cases}$

21. $\begin{cases} 2(3x - y) = 7x - 5 \\ 3(x - y) = 4x - 6 \end{cases}$

22. $\begin{cases} 4(x - 3y) = 3x - 1 \\ 3(4y - 3x) = 1 - 8x \end{cases}$

23. $\begin{cases} \frac{3}{4}x + \frac{2}{3}y = 2 \\ 3x + y = 18 \end{cases}$

24. $\begin{cases} \frac{2}{5}x + \frac{3}{4}y = 1 \\ x + 3y = -2 \end{cases}$

(4.3) *Solve the following systems of equations by the addition method. If there is a single solution, give the ordered pair. If not, state whether the system is inconsistent or whether the equations are dependent.*

25. $\begin{cases} 2x + 3y = -6 \\ x - 3y = -12 \end{cases}$

26. $\begin{cases} 4x + y = 15 \\ -4x + 3y = -19 \end{cases}$

27. $\begin{cases} 2x - 3y = -15 \\ x + 4y = 31 \end{cases}$

28. $\begin{cases} x - 5y = -22 \\ 4x + 3y = 4 \end{cases}$

29. $\begin{cases} 2x = 6y - 1 \\ \frac{1}{3}x - y = \frac{-1}{6} \end{cases}$

30. $\begin{cases} 8x = 3y - 2 \\ \frac{4}{7}x - y = \frac{-5}{2} \end{cases}$

31. $\begin{cases} 5x = 6y + 25 \\ -2y = 7x - 9 \end{cases}$

32. $\begin{cases} -4x = 8 + 6y \\ -3y = 2x - 3 \end{cases}$

33. $\begin{cases} 3(x - 4) = -2y \\ 2x = 3(y - 19) \end{cases}$

34. $\begin{cases} 4(x + 5) = -3y \\ 3x - 2(y + 18) = 0 \end{cases}$

35. $\begin{cases} \frac{2x + 9}{3} = \frac{y + 1}{2} \\ \frac{x}{3} = \frac{y - 7}{6} \end{cases}$

36. $\begin{cases} \frac{2 - 5x}{4} = \frac{2y - 4}{2} \\ \frac{x + 5}{3} = \frac{y}{5} \end{cases}$

(4.4) *Solve each system of equations in three variables.*

37. $\begin{cases} x + z = 4 \\ 2x - y = 4 \\ x + y - z = 0 \end{cases}$

38. $\begin{cases} 2x + 5y = 4 \\ x - 5y + z = -1 \\ 4x - z = 11 \end{cases}$

39. $\begin{cases} 4y + 2z = 5 \\ 2x + 8y = 5 \\ 6x + 4z = 1 \end{cases}$

40. $\begin{cases} 5x + 7y = 9 \\ 14y - z = 28 \\ 4x + 2z = -4 \end{cases}$

41. $\begin{cases} 3x - 2y + 2z = 5 \\ -x + 6y + z = 4 \\ 3x + 14y + 7z = 20 \end{cases}$

42. $\begin{cases} x + 2y + 3z = 11 \\ y + 2z = 3 \\ 2x + 2z = 10 \end{cases}$

43. $\begin{cases} 7x - 3y + 2z = 0 \\ 4x - 4y - z = 2 \\ 5x + 2y + 3z = 1 \end{cases}$

44. $\begin{cases} x - 3y - 5z = -5 \\ 4x - 2y + 3z = 13 \\ 5x + 3y + 4z = 22 \end{cases}$

(4.5) *Solve by writing and solving a system of two linear equations in two variables.*

45. The sum of two numbers is 16. Three times the larger number decreased by the smaller number is 72. Find the two numbers.

46. The Forrest Theater can seat a total of 360 people. They take in $15,150 when every seat is sold. If orchestra section tickets cost $45 and balcony tickets cost $35, find the number of people that can be seated in the orchestra section.

47. A riverboat can head 340 miles upriver in 19 hours, but the return trip takes only 14 hours. Find the current of the river and find the speed of the ship in still water to the nearest tenth of a mile.

	d	$=$	r	$\cdot$	t
Upriver	340		$x - y$		19
Downriver	340		$x + y$		14

48. Sam Abney invested $9000 one year ago. Part of the money was invested at 6%, the rest at 10%. If the total interest earned in one year was $652.80, find how much was invested at each rate.

49. Find the amount of 6% acid solution and the amount of 14% acid solution Pat should combine to prepare 50 cc (cubic centimeters) of a 12% solution.

50. The Deli charges $3.80 for a breakfast of 3 eggs and 4 strips of bacon. The charge is $2.75 for 2 eggs and 3 strips of bacon. Find the cost of each egg and the cost of each strip of bacon.

51. An exercise enthusiast alternates between jogging and walking. He traveled 15 miles during the past 3 hours. He jogs at a rate of 7.5 miles per hour and walks at a rate of 4 miles per hour. Find how much time, to the nearest hundredth of an hour, he actually spent jogging.

△ 52. The perimeter of an isosceles (two sides equal) triangle is 73 centimeters. If two sides are of equal length and the third side is 7 centimeters longer than the others, find the lengths of the three sides.

53. The sum of three numbers is 98. The sum of the first and second is two more than the third number, and the second is four times the first. Find the numbers.

54. Chris Kringler has $2.77 in his coin jar—all in pennies, nickels, and dimes. If he has 53 coins in all and four more nickels than dimes, find how many of each type of coin he has.

55. An employee at a See's Candy Store needs a special mixture of candy. She has creme-filled chocolates that sell for $3.00 per pound, chocolate-covered nuts that sell for $2.70 per pound, and chocolate-covered raisins that sell for $2.25 per pound. She wants to have twice as many raisins as nuts in the mixture. Find how many pounds of each she should use to make 45 pounds worth $2.80 per pound.

56. The sum of three numbers is 295. One number is five more than a second and twice the third. Find the numbers.

CHAPTER 4 TEST

Answer each question true or false.

1. A system of two linear equations in two variables can have exactly two solutions.

2. Although $(1, 4)$ is not a solution of $x + 2y = 6$, it can still be a solution of the system $\begin{cases} x + 2y = 6 \\ x + y = 5 \end{cases}$.

3. If the two equations in a system of linear equations are added and the result is $3 = 0$, the system has no solution.

4. If the two equations in a system of linear equations are added and the result is $3x = 0$, the system has no solution.

Is the ordered pair a solution of the given linear system?

5. $\begin{cases} 2x - 3y = 5 \\ 6x + y = 1 \end{cases}; (1, -1)$ **6.** $\begin{cases} 4x - 3y = 24 \\ 4x + 5y = -8 \end{cases}; (3, -4)$

7. Use graphing to find the solutions of the system $\begin{cases} y - x = 6 \\ y + 2x = -6 \end{cases}$

8. Use the substitution method to solve the system $\begin{cases} 3x - 2y = -14 \\ x + 3y = -1 \end{cases}$

9. Use the substitution method to solve the system

$$\begin{cases} \dfrac{1}{2}x + 2y = -\dfrac{15}{4} \\ 4x = -y \end{cases}$$

10. Use the addition method to solve the system $\begin{cases} 3x + 5y = 2 \\ 2x - 3y = 14 \end{cases}$

11. Use the addition method to solve the system $\begin{cases} 4x - 6y = 7 \\ -2x + 3y = 0 \end{cases}$

Solve each system using the substitution method or the addition method.

12. $\begin{cases} 3x + y = 7 \\ 4x + 3y = 1 \end{cases}$

13. $\begin{cases} 3(2x + y) = 4x + 20 \\ x - 2y = 3 \end{cases}$

14. $\begin{cases} \dfrac{x - 3}{2} = \dfrac{2 - y}{4} \\ \dfrac{7 - 2x}{3} = \dfrac{y}{2} \end{cases}$

15. Find the amount of a 12% saline solution a lab assistant should add to 80 cc (cubic centimeters) of a 22% saline solution in order to have a 16% solution.

16. Don has invested $4000, part at 5% simple annual interest and the rest at 9%. Find how much he invested at each rate if the total interest after 1 year is $311.

17. Although the number of farms in the U.S. is still decreasing, small farms are making a comeback. Texas and Missouri are the states with the most number of farms. Texas has 116 thousand more farms than Missouri and the total number of farms for these two states is 336 thousand. Find the number of farms for each state.

18. $\begin{cases} 2x - 3y = 4 \\ 3y + 2z = 2 \\ x - z = -5 \end{cases}$ **19.** $\begin{cases} 3x - 2y - z = -1 \\ 2x - 2y = 4 \\ 2x - 2z = -12 \end{cases}$

20. The measure of the largest angle of a triangle is three less than 5 times the measure of the smallest angle. The measure of the remaining angle is 1 less than twice the measure of the smallest angle. Find the measure of each angle.

4 CHAPTER CUMULATIVE REVIEW

1. Insert $<$, $>$, or $=$ in the space between the paired numbers to make each statement true.
   a. $-1 \quad 0$ **b.** $7 \quad \frac{14}{2}$ **c.** $-5 \quad -6$

2. Evaluate.
   a. 5^2 **b.** 2^5

3. Name the property or properties illustrated by each true statement.
   a. $3 \cdot y = y \cdot 3$
   b. $(x + 7) + 9 = x + (7 + 9)$
   c. $(b + 0) + 3 = b + 3$
   d. $0.2 \cdot (z \cdot 5) = 0.2 \cdot (5 \cdot z)$
   e. $-2 \cdot \left(-\frac{1}{2}\right) = 1$
   f. $-2 + 2 = 0$
   g. $-6 \cdot (y \cdot 2) = (-6 \cdot 2) \cdot y$

4. Evaluate $y^2 - 3x$ for $x = 8$ and $y = 5$.

5. Subtract $4x - 2$ from $2x - 3$.

6. Simplify: $7 - 12 + (-5) - 2 + (-2)$

7. Solve: $7 = -5(2a - 1) - (-11a + 6)$.

8. Evaluate $2y^2 - x^2$ for $x = -7$ and $y = -3$.

9. Solve: $\frac{5}{2}x = 15$

10. Simplify: $0.4y - 6.7 + y - 0.3 - 2.6y$

11. Solve: $\frac{x}{2} - 1 = \frac{2}{3}x - 3$

12. Solve: $7(x - 2) - 6(x + 1) = 20$

13. Twice the sum of a number and 4 is the same as four times the number, decreased 12. Find the number.

14. Solve: $5(y - 5) = 5y + 10$

15. Solve $y = mx + b$ for x.

16. Five times the sum of a number and -1 is the same as 6 times the number. Find the number.

17. Solve $-2x \leq -4$. Write the solution set in interval notation.

18. Solve $P = a + b + c$ for b.

19. Graph $x = -2y$ by plotting intercepts.

20. Solve $3x + 7 \geq x - 9$. Write the solution set in interval notation.

21. Find the slope of the line through $(-1, 5)$ and $(2, -3)$.

22. Complete the table of values for $x - 3y = 3$

x	y
0	
	0
	2

23. Find the slope of the line whose equation is $y = \frac{3}{4}x + 6$.

24. Find the slope of the line parallel to the line passing through $(-1, 3)$ and $(2, -8)$.

25. Find the slope and the y-intercept of the line whose equation is $3x - 4y = 4$.

26. Find the slope and y-intercept of the line whose equation is $y = 7x$.

27. Find an equation of the line passing through $(-1, 5)$ with slope -2. Write the equation in standard form: $Ax + By = C$.

28. Determine whether the lines are parallel, perpendicular or neither.
$$y = 4x - 5$$
$$-4x + y = 7$$

29. Find an equation of the vertical line through $(-1, 5)$.

30. Write an equation of the line with slope -5, through $(-2, 3)$.

31. Find the domain and the range of the relation $\{(0, 2), (3, 3), (-1, 0), (3, -2)\}$.

32. If $f(x) = 5x^2 - 6$, find $f(0)$ and $f(-2)$.

33. Which of the following relations are also functions?
   a. $\{(-1, 1), (2, 3), (7, 3), (8, 6)\}$
   b. $\{(0, -2), (1, 5), (0, 3), (7, 7)\}$

34. Determine which graph(s) are graphs of functions.

a.   **b.**   **c.**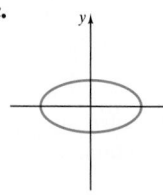

35. Determine the number of solutions of the system.

$$\begin{cases} 3x - y = 4 \\ x + 2y = 8 \end{cases}$$

36. Determine whether any ordered pairs satisfy the given system.

$$\begin{cases} 2x - y = 6 \\ 3x + 2y = -5 \end{cases}$$

a. $(1, -4)$ **b.** $(0, 6)$ **c.** $(3, 0)$

Solve each system.

37. $\begin{cases} x + 2y = 7 \\ 2x + 2y = 13 \end{cases}$ **38.** $\begin{cases} 3x - 4y = 10 \\ y = 2x \end{cases}$

39. $\begin{cases} x + y = 7 \\ x - y = 5 \end{cases}$ **40.** $\begin{cases} x = 5y - 3 \\ x = 8y + 4 \end{cases}$

41. Solve the system.

$$\begin{cases} 3x - y + z = -15 \\ x + 2y - z = 1 \\ 2x + 3y - 2z = 0 \end{cases}$$

42. Solve the system.

$$\begin{cases} x - 2y + z = 0 \\ 3x - y - 2z = -15 \\ 2x - 3y + 3z = 7 \end{cases}$$

43. A first number is 4 less than a second number. Four times the first number is 6 more than twice the second. Find the numbers.

44. Find two numbers whose sum is 37 and whose difference is 21.

CHAPTER 5

Exponents and Polynomials

Recall from Chapter 1 that an exponent is a shorthand notation for repeated factors. This chapter explores additional concepts about exponents and exponential expressions. An especially useful type of exponential expression is a polynomial. Polynomials model many real-world phenomena. This chapter will focus on operations on polynomials.

People who have been stuck in rush hour traffic know that there are millions of automobiles on the road today. Keeping those cars in good working order is the job of a professional auto technician. Auto technicians enjoy problem solving and, due to technological advances in automobile design, they dabble in computer science, electronics, and mathematics. Many technicians are also small business owners, and college courses in business and finance are helpful.

Auto technicians can be certified by the National Institute for Automobile Service Excellence (ASE). ASE offers certification exams in over 40 areas of specialization and requires its certified technicians to participate in continuing education programs. Employment in this field is expected to increase through the year 2010, according to the U.S. Bureau of Labor Statistics.

In the Spotlight on Decision Making feature on page 295, you will have the opportunity to make a decision concerning the modification of a car engine as an auto mechanic.

Link: ASE—www.asecert.org
Sources of text: (ASE) www.asecert.org
(BLS) http://www.bls.gov/oco/ocos181.htm

5.1 EXPONENTS

Objectives

1. Evaluate exponential expressions.
2. Use the product rule for exponents.
3. Use the power rule for exponents.
4. Use the power rules for products and quotients.
5. Use the quotient rule for exponents, and define a number raised to the 0 power.

1 As we reviewed in Section 1.4, an exponent is a shorthand notation for repeated factors. For example, $2 \cdot 2 \cdot 2 \cdot 2 \cdot 2$ can be written as 2^5. The expression 2^5 is called an **exponential expression**. It is also called the fifth **power** of 2, or we say that 2 is **raised** to the fifth power.

$$5^6 = \underbrace{5 \cdot 5 \cdot 5 \cdot 5 \cdot 5 \cdot 5}_{6 \text{ factors; each factor is } 5} \qquad \text{and} \qquad (-3)^4 = \underbrace{(-3) \cdot (-3) \cdot (-3) \cdot (-3)}_{4 \text{ factors; each factor is } -3}$$

The **base** of an exponential expression is the repeated factor. The **exponent** is the number of times that the base is used as a factor.

$$5^6 \overset{\curvearrowleft}{\underset{\curvearrowleft}{}} \begin{matrix} \text{exponent} \\ \text{base} \end{matrix} \qquad (-3)^4 \overset{\curvearrowleft}{\underset{\curvearrowleft}{}} \begin{matrix} \text{exponent} \\ \text{base} \end{matrix}$$

EXAMPLE 1

Evaluate each expression.

a. 2^3 **b.** 3^1 **c.** $(-4)^2$ **d.** -4^2 **e.** $\left(\dfrac{1}{2}\right)^4$ **f.** $(0.5)^3$ **g.** $4 \cdot 3^2$

Solution
a. $2^3 = 2 \cdot 2 \cdot 2 = 8$

b. To raise 3 to the first power means to use 3 as a factor only once. Therefore, $3^1 = 3$. Also, when no exponent is shown, the exponent is assumed to be 1.

c. $(-4)^2 = (-4)(-4) = 16$ **d.** $-4^2 = -(4 \cdot 4) = -16$

e. $\left(\dfrac{1}{2}\right)^4 = \dfrac{1}{2} \cdot \dfrac{1}{2} \cdot \dfrac{1}{2} \cdot \dfrac{1}{2} = \dfrac{1}{16}$ **f.** $(0.5)^3 = (0.5)(0.5)(0.5) = 0.125$

g. $4 \cdot 3^2 = 4 \cdot 9 = 36$

Notice how similar -4^2 is to $(-4)^2$ in the example above. The difference between the two is the parentheses. In $(-4)^2$, the parentheses tell us that the base, or repeated factor, is -4. In -4^2, only 4 is the base.

> **Helpful Hint**
>
> Be careful when identifying the base of an exponential expression. Pay close attention to the use of parentheses.
>
>
>
> $(-3)^2$ -3^2 $2 \cdot 3^2$
> The base is -3.   The base is 3.   The base is 3.
> $(-3)^2 = (-3)(-3) = 9$ $-3^2 = -(3 \cdot 3) = -9$ $2 \cdot 3^2 = 2 \cdot 3 \cdot 3 = 18$

2   An exponent has the same meaning whether the base is a number or a variable. If x is a real number and n is a positive integer, then x^n is the product of n factors, each of which is x.

$$x^n = \underbrace{x \cdot x \cdot x \cdot x \cdot x \cdot \ldots \cdot x}_{n \text{ factors of } x}$$

EXAMPLE 2

Evaluate each expression for the given value of x.

a. $2x^3$; x is 5

b. $\dfrac{9}{x^2}$; x is -3

Solution **a.** If x is 5, $2x^3 = 2 \cdot (5)^3$

$= 2 \cdot (5 \cdot 5 \cdot 5)$

$= 2 \cdot 125$

$= 250$

b. If x is -3, $\dfrac{9}{x^2} = \dfrac{9}{(-3)^2}$

$= \dfrac{9}{(-3)(-3)}$

$= \dfrac{9}{9} = 1$

Exponential expressions can be multiplied, divided, added, subtracted, and themselves raised to powers. By our definition of an exponent,

$$5^4 \cdot 5^3 = \underbrace{(5 \cdot 5 \cdot 5 \cdot 5)}_{4 \text{ factors of } 5} \cdot \underbrace{(5 \cdot 5 \cdot 5)}_{3 \text{ factors of } 5}$$

$$= \underbrace{5 \cdot 5 \cdot 5 \cdot 5 \cdot 5 \cdot 5 \cdot 5}_{7 \text{ factors of } 5}$$

$$= 5^7$$

Also,

$$x^2 \cdot x^3 = (x \cdot x) \cdot (x \cdot x \cdot x)$$

$$= x \cdot x \cdot x \cdot x \cdot x$$

$$= x^5$$

In both cases, notice that the result is exactly the same if the exponents are added.

$$5^4 \cdot 5^3 = 5^{4+3} = 5^7 \quad \text{and} \quad x^2 \cdot x^3 = x^{2+3} = x^5$$

This suggests the following rule.

Product Rule for Exponents

If m and n are positive integers and a is a real number, then

$$a^m \cdot a^n = a^{m+n}$$

For example, $3^5 \cdot 3^7 = 3^{5+7} = 3^{12}$.

In other words, to multiply two exponential expressions with a **common base**, keep the base and add the exponents. We call this simplifying the exponential expression.

EXAMPLE 3

Use the product rule to simplify.

a. $4^2 \cdot 4^5$ **b.** $x^4 \cdot x^6$ **c.** $y^3 \cdot y$ **d.** $y^3 \cdot y^2 \cdot y^7$ **e.** $(-5)^7 \cdot (-5)^8$ **f.** $a^2 \cdot b^2$

Solution   **a.** $4^2 \cdot 4^5 = 4^{2+5} = 4^7$

b. $x^4 \cdot x^6 = x^{4+6} = x^{10}$

c. $y^3 \cdot y = y^3 \cdot y^1$
$= y^{3+1}$
$= y^4$

> **Helpful Hint**
>
> Don't forget that if no exponent is written, it is assumed to be 1.

d. $y^3 \cdot y^2 \cdot y^7 = y^{3+2+7} = y^{12}$

e. $(-5)^7 \cdot (-5)^8 = (-5)^{7+8} = (-5)^{15}$

f. $a^2 \cdot b^2$ Cannot be simplified because a and b are different bases.

EXAMPLE 4

Use the product rule to simplify $(2x^2)(-3x^5)$.

Solution   Recall that $2x^2$ means $2 \cdot x^2$ and $-3x^5$ means $-3 \cdot x^5$.

$$(2x^2)(-3x^5) = 2 \cdot x^2 \cdot -3 \cdot x^5 \qquad \text{Remove parentheses.}$$
$$= 2 \cdot -3 \cdot x^2 \cdot x^5 \qquad \text{Group factors with common bases.}$$
$$= -6x^7 \qquad \text{Simplify.}$$

> **Helpful Hint**
>
> These examples will remind you of the difference between adding and multiplying terms.
>
> **Addition**
>
> $$5x^3 + 3x^3 = (5 + 3)x^3 = 8x^3 \qquad \text{By the distributive property.}$$
> $$7x + 4x^2 = 7x + 4x^2 \qquad \text{Cannot be combined.}$$
>
> **Multiplication**
>
> $$(5x^3)(3x^3) = 5 \cdot 3 \cdot x^3 \cdot x^3 = 15x^{3+3} = 15x^6 \qquad \text{By the product rule.}$$
> $$(7x)(4x^2) = 7 \cdot 4 \cdot x \cdot x^2 = 28x^{1+2} = 28x^3 \qquad \text{By the product rule.}$$

3   Exponential expressions can themselves be raised to powers. Let's try to discover a rule that simplifies an expression like $(x^2)^3$. By definition,

$$(x^2)^3 = \underbrace{(x^2)(x^2)(x^2)}_{\text{3 factors of } x^2}$$

which can be simplified by the product rule for exponents.

$$(x^2)^3 = (x^2)(x^2)(x^2) = x^{2+2+2} = x^6$$

Notice that the result is exactly the same if we multiply the exponents.

$$(x^2)^3 = x^{2 \cdot 3} = x^6$$

The following property states this result.

Power Rule for Exponents

If m and n are positive integers and a is a real number, then
$$(a^m)^n = a^{mn}$$

For example, $(7^2)^5 = 7^{2 \cdot 5} = 7^{10}$.

To raise a power to a power, keep the base and multiply the exponents.

EXAMPLE 5

Use the power rule to simplify.

a. $(x^2)^5$ **b.** $(y^8)^2$ **c.** $[(-5)^3]^7$

Solution   **a.** $(x^2)^5 = x^{2 \cdot 5} = x^{10}$ **b.** $(y^8)^2 = y^{8 \cdot 2} = y^{16}$ **c.** $[(-5)^3]^7 = (-5)^{21}$

Helpful Hint

Take a moment to make sure that you understand when to apply the product rule and when to apply the power rule.

Product Rule → *Add Exponents*	*Power Rule* → *Multiply Exponents*
$x^5 \cdot x^7 = x^{5+7} = x^{12}$	$(x^5)^7 = x^{5 \cdot 7} = x^{35}$
$y^6 \cdot y^2 = y^{6+2} = y^8$	$(y^6)^2 = y^{6 \cdot 2} = y^{12}$

---

4   When the base of an exponential expression is a product, the definition of x^n still applies. To simplify $(xy)^3$, for example,

$$(xy)^3 = (xy)(xy)(xy) \qquad (xy)^3 \text{ means 3 factors of } (xy).$$
$$= x \cdot x \cdot x \cdot y \cdot y \cdot y \qquad \text{Group factors with common bases.}$$
$$= x^3 y^3 \qquad \text{Simplify.}$$

Notice that to simplify the expression $(xy)^3$, we raise each factor within the parentheses to a power of 3.

$$(xy)^3 = x^3 y^3$$

In general, we have the following rule.

Power of a Product Rule

If n is a positive integer and a and b are real numbers, then
$$(ab)^n = a^n b^n$$

For example, $(3x)^5 = 3^5 x^5$.

In other words, to raise a product to a power, we raise each factor to the power.

EXAMPLE 6

Simplify each expression.

a. $(st)^4$ **b.** $(2a)^3$ **c.** $\left(\dfrac{1}{3}mn^3\right)^2$ **d.** $(-5x^2y^3z)^2$

Solution
a. $(st)^4 = s^4 \cdot t^4 = s^4 t^4$ Use the power of a product rule.

b. $(2a)^3 = 2^3 \cdot a^3 = 8a^3$ Use the power of a product rule.

c. $\left(\dfrac{1}{3}mn^3\right)^2 = \left(\dfrac{1}{3}\right)^2 \cdot (m)^2 \cdot (n^3)^2 = \dfrac{1}{9}m^2 n^6$

d. $(-5x^2y^3z)^2 = (-5)^2 \cdot (x^2)^2 \cdot (y^3)^2 \cdot (z^1)^2$ Use the power of a product rule.
$\qquad\qquad = 25x^4y^6z^2$ Use the power rule for exponents.

Let's see what happens when we raise a quotient to a power. To simplify $\left(\dfrac{x}{y}\right)^3$, for example,

$$\left(\dfrac{x}{y}\right)^3 = \left(\dfrac{x}{y}\right)\left(\dfrac{x}{y}\right)\left(\dfrac{x}{y}\right)$$ $\left(\dfrac{x}{y}\right)^3$ means 3 factors of $\left(\dfrac{x}{y}\right)$.

$$= \dfrac{x \cdot x \cdot x}{y \cdot y \cdot y}$$ Multiply fractions.

$$= \dfrac{x^3}{y^3}$$ Simplify.

Notice that to simplify the expression $\left(\dfrac{x}{y}\right)^3$, we raise both the numerator and the denominator to a power of 3.

$$\left(\dfrac{x}{y}\right)^3 = \dfrac{x^3}{y^3}$$

In general, we have the following.

Power of a Quotient Rule

If n is a positive integer and a and c are real numbers, then

$$\left(\dfrac{a}{c}\right)^n = \dfrac{a^n}{c^n}, \quad c \neq 0$$

For example, $\left(\dfrac{y}{7}\right)^4 = \dfrac{y^4}{7^4}$.

In other words, to raise a quotient to a power, we raise both the numerator and the denominator to the power.

EXAMPLE 7

Simplify each expression.

a. $\left(\dfrac{m}{n}\right)^7$ **b.** $\left(\dfrac{x^3}{3y^5}\right)^4$

Solution   **a.** $\left(\dfrac{m}{n}\right)^7 = \dfrac{m^7}{n^7}, n \neq 0$ *Use the power of a quotient rule.*

b. $\left(\dfrac{x^3}{3y^5}\right)^4 = \dfrac{(x^3)^4}{3^4 \cdot (y^5)^4}, y \neq 0$ *Use the power of a product or quotient rule.*

$= \dfrac{x^{12}}{81y^{20}}$ *Use the power rule for exponents.*

5   Another pattern for simplifying exponential expressions involves quotients.

To simplify an expression like $\dfrac{x^5}{x^3}$, in which the numerator and the denominator have a common base, we can apply the fundamental principle of fractions and divide the numerator and the denominator by the common base factors. Assume for the remainder of this section that denominators are not 0.

$$\dfrac{x^5}{x^3} = \dfrac{x \cdot x \cdot x \cdot x \cdot x}{x \cdot x \cdot x}$$

$$= \dfrac{x \cdot x \cdot x \cdot x \cdot x}{x \cdot x \cdot x}$$

$$= x \cdot x$$

$$= x^2$$

Notice that the result is exactly the same if we subtract exponents of the common bases.

$$\dfrac{x^5}{x^3} = x^{5-3} = x^2$$

The quotient rule for exponents states this result in a general way.

Quotient Rule for Exponents

If m and n are positive integers and a is a real number, then

$$\dfrac{a^m}{a^n} = a^{m-n}$$

as long as a is not 0.

For example, $\dfrac{x^6}{x^2} = x^{6-2} = x^4$.

In other words, to divide one exponential expression by another with a common base, keep the base and subtract exponents.

EXAMPLE 8

Simplify each quotient.

a. $\dfrac{x^5}{x^2}$ **b.** $\dfrac{4^7}{4^3}$ **c.** $\dfrac{(-3)^5}{(-3)^2}$ **d.** $\dfrac{s^2}{t^3}$ **e.** $\dfrac{2x^5y^2}{xy}$

Solution   **a.** $\dfrac{x^5}{x^2} = x^{5-2} = x^3$ *Use the quotient rule.*

b. $\dfrac{4^7}{4^3} = 4^{7-3} = 4^4 = 256$ *Use the quotient rule.*

c. $\dfrac{(-3)^5}{(-3)^2} = (-3)^3 = -27$

d. $\dfrac{s^2}{t^3}$ *Cannot be simplified because s and t are different bases.*

e. Begin by grouping common bases.

$$\frac{2x^5y^2}{xy} = 2 \cdot \frac{x^5}{x^1} \cdot \frac{y^2}{y^1}$$
$$= 2 \cdot (x^{5-1}) \cdot (y^{2-1}) \qquad \text{Use the quotient rule.}$$
$$= 2x^4y^1 \quad \text{or} \quad 2x^4y$$

✔ **CONCEPT CHECK**

Suppose you are simplifying each expression. Tell whether you would *add* the exponents, *subtract* the exponents, *multiply* the exponents, *divide* the exponents, or *none of these*.

a. $\left(x^{63}\right)^{21}$ **b.** $\dfrac{y^{15}}{y^3}$ **c.** $z^{16} + z^8$ **d.** $w^{45} \cdot w^9$

Let's now give meaning to an expression such as x^0. To do so, we will simplify $\dfrac{x^3}{x^3}$ in two ways and compare the results.

$$\frac{x^3}{x^3} = x^{3-3} = x^0 \qquad \text{Apply the quotient rule.}$$

$$\frac{x^3}{x^3} = \frac{x \cdot x \cdot x}{x \cdot x \cdot x} = 1 \qquad \text{Apply the fundamental principle for fractions.}$$

Since $\dfrac{x^3}{x^3} = x^0$ and $\dfrac{x^3}{x^3} = 1$, we define that $x^0 = 1$ as long as x is not 0.

> ### Zero Exponent
>
> $a^0 = 1$, as long as a is not 0.

In other words, any base raised to the 0 power is 1, as long as the base is not 0.

EXAMPLE 9

Simplify the following expressions.

a. 3^0 **b.** $(ab)^0$ **c.** $(-5)^0$ **d.** -5^0 **e.** $\left(\dfrac{3}{100}\right)^0$

Solution   **a.** $3^0 = 1$

b. Assume that neither a nor b is zero.

$$(ab)^0 = a^0 \cdot b^0 = 1 \cdot 1 = 1$$

c. $(-5)^0 = 1$

d. $-5^0 = -1 \cdot 5^0 = -1 \cdot 1 = -1$

e. $\left(\dfrac{3}{100}\right)^0 = 1$

Concept Check Answer:

a. multiply   **b.** subtract
c. none of these   **d.** add

In the next example, exponential expressions are simplified using two or more of the exponent rules presented in this section.

EXAMPLE 10

Simplify the following.

a. $\left(\dfrac{-5x^2}{y^3}\right)^2$ **b.** $\dfrac{(x^3)^4 x}{x^7}$ **c.** $\dfrac{(2x)^5}{x^3}$ **d.** $\dfrac{(a^2b)^3}{a^3b^2}$

Solution   **a.** Use the power of a product or quotient rule; then use the power rule for exponents.

$$\left(\frac{-5x^2}{y^3}\right)^2 = \frac{(-5)^2(x^2)^2}{(y^3)^2} = \frac{25x^4}{y^6}$$

b. $\dfrac{(x^3)^4 x}{x^7} = \dfrac{x^{12} \cdot x}{x^7} = \dfrac{x^{12+1}}{x^7} = \dfrac{x^{13}}{x^7} = x^{13-7} = x^6$

c. Use the power of a product rule; then use the quotient rule.

$$\frac{(2x)^5}{x^3} = \frac{2^5 \cdot x^5}{x^3} = 2^5 \cdot x^{5-3} = 32x^2$$

d. Begin by applying the power of a product rule to the numerator.

$$\frac{(a^2b)^3}{a^3b^2} = \frac{(a^2)^3 \cdot b^3}{a^3 \cdot b^2}$$

$$= \frac{a^6b^3}{a^3b^2} \qquad \text{Use the power rule for exponents.}$$

$$= a^{6-3}b^{3-2} \qquad \text{Use the quotient rule.}$$

$$= a^3b^1 \quad \text{or} \quad a^3b$$

Spotlight on

DECISION ✗ MAKING

Suppose you are an auto mechanic and amateur racing enthusiast. You have been modifying the engine in your car and would like to enter a local amateur race. The racing classes depend on the size, or displacement, of the engine.

Engine displacement can be calculated using the formula $d = \dfrac{\pi}{4}b^2sc$. In the formula,

BRENTWOOD AMATEUR RACING CLUB	
Racing Class	*Displacement Limit*
A	up to 2000 cc
B	up to 2400 cc
C	up to 2650 cc
D	up to 3000 cc

d = the engine displacement in cubic centimeters (cc)

b = the bore or engine cylinder diameter in centimeters

s = the stroke or distance the piston travels up or down within the cylinder in centimeters

c = the number of cylinders the engine has.

You have made the following measurements on your modified engine: 8.4 cm bore, 7.6 cm stroke, and 6 cylinders. In which racing class would you enter your car? Explain.

STUDY SKILLS REMINDER

What Should You Do on the Day of an Exam?

On the day of an exam, try the following:

▶ Allow yourself plenty of time to arrive at your classroom.

▶ Read the directions on the test carefully.

▶ Read each problem carefully as you take your test. Make sure that you answer the question asked.

▶ Watch your time and pace yourself. Work the problems that you are most confident with first.

▶ If you have time, check your work and answers.

▶ Do not turn your test in early. If you have extra time, spend it double-checking your work.

Good luck!

MENTAL MATH

State the bases and the exponents for each of the following expressions.

1. 3^2 **2.** 5^4 **3.** $(-3)^6$ **4.** -3^7 **5.** -4^2

6. $(-4)^3$ **7.** $5 \cdot 3^4$ **8.** $9 \cdot 7^6$ **9.** $5x^2$ **10.** $(5x)^2$

EXERCISE SET 5.1

STUDY GUIDE/SSM   CD/VIDEO   PH MATH TUTOR CENTER   MathXL®Tutorials ON CD   MathXL®   MyMathLab®

Evaluate each expression. See Example 1.

 1. 7^2 **2.** -3^2 **3.** $(-5)^1$

4. $(-3)^2$ **5.** -2^4 **6.** -4^3

 7. $(-2)^4$ **8.** $(-4)^3$ **9.** $(0.1)^5$

10. $(0.2)^5$ **11.** $\left(\dfrac{1}{3}\right)^4$ **12.** $\left(-\dfrac{1}{9}\right)^2$

13. $7 \cdot 2^5$ **14.** $9 \cdot 1^7$ **15.** $-2 \cdot 5^3$

16. $-4 \cdot 3^3$

17. Explain why $(-5)^4 = 625$, while $-5^4 = -625$.

18. Explain why $5 \cdot 4^2 = 80$, while $(5 \cdot 4)^2 = 400$.

Evaluate each expression given the replacement values for x. See Example 2.

19. x^2; $x = -2$ **20.** x^3; $x = -2$

21. $5x^3$; $x = 3$ **22.** $4x^2$; $x = -1$

23. $2xy^2$; $x = 3$ and $y = 5$ **24.** $-4x^2y^3$; $x = 2$ and $y = -1$

25. $\dfrac{2z^4}{5}$; $z = -2$ **26.** $\dfrac{10}{3y^3}$; $y = 5$

27. The formula $V = x^3$ can be used to find the volume V of a cube with side length x. Find the volume of a cube with side length 7 meters. (Volume is measured in cubic units.)

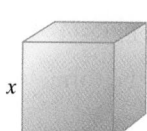

28. The formula $S = 6x^2$ can be used to find the surface area S of a cube with side length x. Find the surface area of the cube with side length 5 meters. (Surface area is measured in square units.)

29. To find the amount of water that a swimming pool in the shape of a cube can hold, do we use the formula for volume of the cube or surface area of the cube? (See Exercises 27 and 28.)

30. To find the amount of material needed to cover an ottoman in the shape of a cube, do we use the formula for volume of the cube or surface area of the cube? (See Exercises 27 and 28.)

Use the product rule to simplify each expression. Write the results using exponents. See Examples 3 and 4.

 31. $x^2 \cdot x^8$

32. $y^2 \cdot y$

33. $(-3)^3 \cdot (-3)^9$

34. $(-5)^7 \cdot (-5)^6$

35. $(5y^4)(3y)$

36. $(-2z^3)(-2z^2)$

37. $(4z^{10})(-6z^7)(z^3)$

38. $(12x^5)(-x^6)(x^4)$

Use the power rule and the power of a product or quotient rule to simplify each expression. See Examples 5 through 7.

39. $(pq)^7$

40. $(4s)^3$

41. $\left(\dfrac{m}{n}\right)^9$

42. $\left(\dfrac{xy}{7}\right)^2$

 43. $(x^2y^3)^5$

44. $(a^4b)^7$

45. $\left(\dfrac{-2xz}{y^5}\right)^2$

46. $\left(\dfrac{y^4}{-3z^3}\right)^3$

Use the quotient rule and simplify each expression. See Example 8.

47. $\dfrac{x^3}{x}$

48. $\dfrac{y^{10}}{y^9}$

49. $\dfrac{(-2)^5}{(-2)^3}$

50. $\dfrac{(-5)^{14}}{(-5)^{11}}$

51. $\dfrac{p^7q^{20}}{pq^{15}}$

52. $\dfrac{x^8y^6}{y^5}$

53. $\dfrac{7x^2y^6}{14x^2y^3}$

54. $\dfrac{9a^4b^7}{3ab^2}$

Simplify the following. See Example 9.

55. $(2x)^0$

56. $-4x^0$

57. $-2x^0$

58. $(4y)^0$

59. $5^0 + y^0$

60. $-3^0 + 4^0$

Simplify the following. See Example 10.

61. $\left(\dfrac{-3a^2}{b^3}\right)^3$

62. $\left(\dfrac{q^7}{-2p^5}\right)^5$

63. $\dfrac{(x^5)^7 \cdot x^8}{x^4}$

64. $\dfrac{y^{20}}{(y^2)^3 \cdot y^9}$

65. $\dfrac{(z^3)^6}{(5z)^4}$

66. $\dfrac{(3x)^4}{(x^2)^2}$

67. $\dfrac{(6mn)^5}{mn^2}$

68. $\dfrac{(6xy)^2}{9x^2y^2}$

MIXED PRACTICE

Simplify the following.

69. -5^2

70. $(-5)^2$

71. $\left(\dfrac{1}{4}\right)^3$

72. $\left(\dfrac{2}{3}\right)^3$

73. $(9xy)^2$

74. $(2ab)^5$

75. $(6b)^0$

76. $(5ab)^0$

77. $2^3 + 2^5$

78. $7^2 - 7^0$

79. b^4b^2

80. y^4y^1

81. $a^2a^3a^4$

82. $x^2x^{15}x^9$

 83. $(2x^3)(-8x^4)$

84. $(3y^4)(-5y)$

85. $(4a)^3$

86. $(2ab)^4$

87. $(-6xyz^3)^2$

88. $(-3xy^2a^3b)^3$

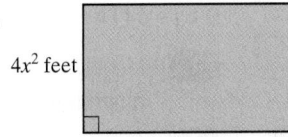 **89.** $\left(\dfrac{3y^5}{6x^4}\right)^3$

90. $\left(\dfrac{2ab}{6yz}\right)^4$

91. $\dfrac{x^5}{x^4}$

92. $\dfrac{5x^9}{x^3}$

93. $\dfrac{2x^3y^2z}{xyz}$

94. $\dfrac{x^{12}y^{13}}{x^5y^7}$

95. $\dfrac{(3x^2y^5)^5}{x^3y}$

96. $\dfrac{(4a^2)^4}{a^4b}$

97. In your own words, explain why $5^0 = 1$.

98. In your own words, explain when $(-3)^n$ is positive and when it is negative.

REVIEW AND PREVIEW

Simplify each expression by combining any like terms. Use the distributive property to remove any parentheses. See Section 2.1.

99. $y - 10 + y$

100. $-6z + 20 - 3z$

101. $7x + 2 - 8x - 6$

102. $10y - 14 - y - 14$

103. $2(x - 5) + 3(5 - x)$

104. $-3(w + 7) + 5(w + 1)$

Concept Extensions

Find the area of each figure. See the geometric formulas on the inside covers. If the formula contains π, do not approximate.

105.

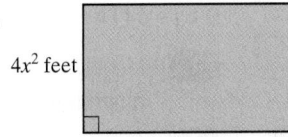

$4x^2$ feet

$5x^3$ feet

106.

$2y^{10}$ meters

$9y^7$ meters

107.

$5y$ centimeters

108.

$8z^5$ decimeters

Find the volume of each figure. See the geometric formulas in the inside covers. If the formula includes π, do not approximate.

△ **109.**

3y⁴ feet

$3y^4$ feet

$3y^4$ feet

△ **110.**

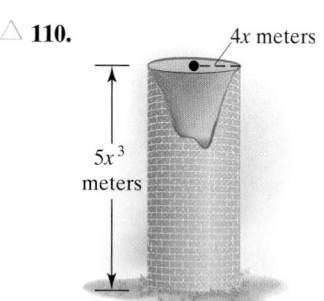

4x meters

$5x^3$ meters

Simplify each expression. Assume that variables represent positive integers.

111. $x^{5a}x^{4a}$

112. $b^{9a}b^{4a}$

113. $\dfrac{x^{9a}}{x^{4a}}$

114. $\dfrac{y^{15b}}{y^{6b}}$

115. $(x^a y^b z^c)^{5a}$

116. $(9a^2 b^3 c^4 d^5)^{ab}$

117. Suppose you borrow money for 6 months. If the interest rate is compounded monthly, the formula $A = P\left(1 + \dfrac{r}{12}\right)^6$ gives the total amount A to be repaid at the end of 6 months. For a loan of $P = \$1000$ and interest rate of 9% ($r = 0.09$), how much money will you need to pay off the loan?

118. On April 18, 2001, the Federal Reserve discount rate was set at 4%. (*Source*: Federal Reserve Board) The discount rate is the interest rate at which banks can borrow money from the Federal Reserve System. Suppose a bank needs to borrow money from the Federal Reserve System for 3 months. If the interest is compounded monthly, the formula $A = P\left(1 + \dfrac{r}{12}\right)^3$ gives the total amount A to be repaid at the end of 3 months. For a loan of $P = \$500,000$ and interest rate of $r = 0.04$, how much money will the bank repay to the Federal Reserve at the end of 3 months? Round to the nearest dollar.

5.2 POLYNOMIAL FUNCTIONS AND ADDING AND SUBTRACTING POLYNOMIALS

Objectives

1 Identify term, constant, polynomial, monomial, binomial, trinomial, and the degree of a term and of a polynomial.

2 Define polynomial functions.

3 Review combining like terms.

4 Add polynomials.

5 Subtract polynomials.

1   In this section, we introduce a special algebraic expression called a polynomial. Let's first review some definitions presented in Section 2.1.

Recall that a term is a number or the product of a number and variables raised to powers. The terms of the expression $4x^2 + 3x$ are $4x^2$ and $3x$. The terms of the expression $9x^4 - 7x - 1$ are $9x^4$, $-7x$, and -1.

Expression	Terms
$4x^2 + 3x$	$4x^2, 3x$
$9x^4 - 7x - 1$	$9x^4, -7x, -1$
$7y^3$	$7y^3$
5	5

The **numerical coefficient** of a term, or simply the **coefficient**, is the numerical factor of each term. If no numerical factor appears in the term, then the coefficient is understood to be 1. If the term is a number only, it is called a **constant** term or simply a constant.

Term	Coefficient
x^5	1
$3x^2$	3
$-4x$	-4
$-x^2y$	-1
3 (constant)	3

Polynomial

A **polynomial in x** is a finite sum of terms of the form ax^n, where a is a real number and n is a whole number.

For example,

$$x^5 - 3x^3 + 2x^2 - 5x + 1$$

is a polynomial in x. Notice that this polynomial is written in **descending powers** of x because the powers of x decrease from left to right. (Recall that the term 1 can be thought of as $1x^0$.)

On the other hand,

$$x^{-5} + 2x - 3$$

is **not** a polynomial because it contains an exponent, -5, that is not a whole number. (We study negative exponents in Section 5 of this chapter.)

A **polynomial** is a finite sum of terms in which all variables are raised to nonnegative integer powers and no variables appear in any denominator.

Polynomials	Not Polynomials	
$4x^5y + 7xz$	$5x^{-3} + 2x$	Negative integer exponent
$-5x^3 + 2x + \dfrac{2}{3}$	$\dfrac{6}{x^2} - 5x + 1$	Variable in denominator

Some polynomials are given special names.

A **monomial** is a polynomial with exactly one term.

A **binomial** is a polynomial with exactly two terms.

A **trinomial** is a polynomial with exactly three terms.

The following are examples of monomials, binomials, and trinomials. Each of these examples is also a polynomial.

POLYNOMIALS

Monomials	Binomials	Trinomials	None of These
ax^2	$x + y$	$x^2 + 4xy + y^2$	$5x^3 - 6x^2 + 3x - 6$
$-3z$	$3p + 2$	$x^5 + 7x^2 - x$	$-y^5 + y^4 - 3y^3 - y^2 + y$
4	$4x^2 - 7$	$-q^4 + q^3 - 2q$	$x^6 + x^4 - x^3 + 1$

Each term of a polynomial has a **degree**.

> ### Degree of a Term
> The degree of a term is the sum of the exponents on the variables contained in the term.

EXAMPLE 1

Find the degree of each term.

a. $3x^2$ **b.** -2^3x^5 **c.** y **d.** $12x^2yz^3$ **e.** 5

Solution
a. The exponent on x is 2, so the degree of the term is 2.
b. The exponent on x is 5, so the degree of the term is 5. (Recall that the degree is the sum of the exponents on only the *variables*.)
c. The degree of y, or y^1, is 1.
d. The degree is the sum of the exponents on the variables, or $2 + 1 + 3 = 6$.
e. The degree of 5, which can be written as $5x^0$, is 0.

From the preceding example, we can say that the degree of a constant is 0. Also, the term 0 has no degree.
Each polynomial also has a degree.

> ### Degree of a Polynomial
> The **degree of a polynomial** is the largest degree of all its terms.

EXAMPLE 2

Find the degree of each polynomial and indicate whether the polynomial is also a monomial, binomial, or trinomial.

	Polynomial	*Degree*	*Classification*
a.	$7x^3 - 3x + 2$	3	Trinomial
b.	$-xyz$	$1 + 1 + 1 = 3$	Monomial
c.	$x^4 - 16$	4	Binomial

EXAMPLE 3

Find the degree of the polynomial
$$3xy + x^2y^2 - 5x^2 - 6.$$

Solution The degree of each term is
$$3xy + x^2y^2 - 5x^2 - 6$$

Degree: 2 4 2 0

The largest degree of any term is 4, so the degree of this polynomial is 4.

2   At times, it is convenient to use function notation to represent polynomials. For example, we may write $P(x)$ to represent the polynomial $3x^2 - 2x - 5$. In symbols, this is

$$P(x) = 3x^2 - 2x - 5$$

This function is called a **polynomial function** because the expression $3x^2 - 2x - 5$ is a polynomial.

> **Helpful Hint**
>
> Recall that the symbol $P(x)$ **does not mean** P times x. It is a special symbol used to denote a function.

EXAMPLE 4

If $P(x) = 3x^2 - 2x - 5$, find the following.

a. $P(1)$ **b.** $P(-2)$

Solution   **a.** Substitute 1 for x in $P(x) = 3x^2 - 2x - 5$ and simplify.

$$P(x) = 3x^2 - 2x - 5$$
$$P(1) = 3(1)^2 - 2(1) - 5 = -4$$

b. Substitute -2 for x in $P(x) = 3x^2 - 2x - 5$ and simplify.

$$P(x) = 3x^2 - 2x - 5$$
$$P(-2) = 3(-2)^2 - 2(-2) - 5 = 11$$

Many real-world phenomena are modeled by polynomial functions. If the polynomial function model is given, we can often find the solution of a problem by evaluating the function at a certain value.

EXAMPLE 5

FINDING THE HEIGHT OF AN OBJECT

The world's highest bridge, Royal Gorge suspension bridge in Colorado, is 1053 feet above the Arkansas River. An object is dropped from the top of this bridge. Neglecting air resistance, the height of the object at time t seconds is given by the polynomial function $P(t) = -16t^2 + 1053$. Find the height of the object when $t = 1$ second and when $t = 8$ seconds.

Solution   To find the height of the object at 1 second, we find $P(1)$.

$$P(t) = -16t^2 + 1053$$
$$P(1) = -16(1)^2 + 1053$$
$$P(1) = 1037$$

When $t = 1$ second, the height of the object is 1037 feet.

To find the height of the object at 8 seconds, we find $P(8)$.

$$P(t) = -16t^2 + 1053$$
$$P(8) = -16(8)^2 + 1053$$
$$P(8) = -1024 + 1053$$
$$P(8) = 29$$

When $t = 8$ seconds, the height of the object is 29 feet. Notice that as time t increases, the height of the object decreases.

3 Before we add polynomials, recall that terms are considered to be **like terms** if they contain exactly the same variables raised to exactly the same powers.

Like Terms	Unlike Terms
$-5x^2, -x^2$	$4x^2, 3x$
$7xy^3z, -2xzy^3$	$12x^2y^3, -2xy^3$

To simplify a polynomial, **combine like terms** by using the distributive property. For example, by the distributive property,

$$5x + 7x = (5 + 7)x = 12x$$

EXAMPLE 6

Simplify by combining like terms.

a. $-12x^2 + 7x^2 - 6x$ **b.** $3xy - 2x + 5xy - x$

Solution   By the distributive property,

a. $-12x^2 + 7x^2 - 6x = (-12 + 7)x^2 - 6x = -5x^2 - 6x$

b. Use the associative and commutative properties to group together like terms; then combine.

$$3xy - 2x + 5xy - x = 3xy + 5xy - 2x - x$$
$$= (3 + 5)xy + (-2 - 1)x$$
$$= 8xy - 3x$$

4 Now we have reviewed the necessary skills to add polynomials.

Adding Polynomials

Combine all like terms.

EXAMPLE 7

Add.

a. $(7x^3y - xy^3 + 11) + (6x^3y - 4)$ **b.** $(3a^3 - b + 2a - 5) + (a + b + 5)$

Solution   **a.** To add, remove the parentheses and group like terms.

$$(7x^3y - xy^3 + 11) + (6x^3y - 4)$$
$$= 7x^3y - xy^3 + 11 + 6x^3y - 4$$
$$= 7x^3y + 6x^3y - xy^3 + 11 - 4 \quad \text{Group like terms.}$$
$$= 13x^3y - xy^3 + 7 \quad \text{Combine like terms.}$$

b. $(3a^3 - b + 2a - 5) + (a + b + 5)$

$$= 3a^3 - b + 2a - 5 + a + b + 5$$

$$= 3a^3 - b + b + 2a + a - 5 + 5 \quad \text{Group like terms.}$$

$$= 3a^3 + 3a \quad\quad\quad\quad\quad\quad\quad \text{Combine like terms.}$$

EXAMPLE 8

Add $11x^3 - 12x^2 + x - 3$ and $x^3 - 10x + 5$.

Solution $(11x^3 - 12x^2 + x - 3) + (x^3 - 10x + 5)$

$$= 11x^3 + x^3 - 12x^2 + x - 10x - 3 + 5 \quad \text{Group like terms.}$$

$$= 12x^3 - 12x^2 - 9x + 2 \quad\quad\quad\quad \text{Combine like terms.}$$

Sometimes it is more convenient to add polynomials vertically. To do this, line up like terms beneath one another and add like terms.

5   The definition of subtraction of real numbers can be extended to apply to polynomials. To subtract a number, we add its opposite.

$$a - b = a + (-b)$$

Likewise, to subtract a polynomial, we add its opposite. In other words, if P and Q are polynomials, then

$$P - Q = P + (-Q)$$

The polynomial $-Q$ is the **opposite**, or **additive inverse**, of the polynomial Q. We can find $-Q$ by writing the opposite of each term of Q.

> ## Subtracting Polynomials
>
> To subtract a polynomial, add its opposite.

For example,

To subtract, add its opposite (found by writing the opposite of each term).

$$(3x^2 + 4x - 7) - (3x^2 - 2x - 5) = (3x^2 + 4x - 7) + (-3x^2 + 2x + 5)$$

$$= 3x^2 + 4x - 7 - 3x^2 + 2x + 5$$

$$= 6x - 2 \quad \text{Combine like terms.}$$

✔ **CONCEPT CHECK**

Which polynomial is the opposite of $16x^3 - 5x + 7$?

a. $-16x^3 - 5x + 7$ **b.** $-16x^3 + 5x - 7$

c. $16x^3 + 5x + 7$ **d.** $-16x^3 + 5x + 7$

EXAMPLE 9

Subtract $(12z^5 - 12z^3 + z) - (-3z^4 + z^3 + 12z)$.

Solution   To subtract, add the opposite of the second polynomial to the first polynomial.

$$(12z^5 - 12z^3 + z) - (-3z^4 + z^3 + 12z)$$
$$= 12z^5 - 12z^3 + z + 3z^4 - z^3 - 12z)$$ *Add the opposite of the polynomial being subtracted.*
$$= 12z^5 + 3z^4 - 12z^3 - z^3 + z - 12z$$ *Group like terms.*
$$= 12z^5 + 3z^4 - 13z^3 - 11z$$ *Combine like terms.*

✔ **CONCEPT CHECK**

Why is the following subtraction incorrect?

$$(7z - 5) - (3z - 4)$$
$$= 7z - 5 - 3z - 4$$
$$= 4z - 9$$

EXAMPLE 10

Subtract $4x^3y^2 - 3x^2y^2 + 2y^2$ from $10x^3y^2 - 7x^2y^2$.

Solution   If we subtract 2 from 8, the difference is $8 - 2 = 6$. Notice the order of the numbers, and then write "Subtract $4x^3y^2 - 3x^2y^2 + 2y^2$ from $10x^3y^2 - 7x^2y^2$" as a mathematical expression.

$$(10x^3y^2 - 7x^2y^2) - (4x^3y^2 - 3x^2y^2 + 2y^2)$$
$$= 10x^3y^2 - 7x^2y^2 - 4x^3y^2 + 3x^2y^2 - 2y^2$$ *Remove parentheses.*
$$= 6x^3y^2 - 4x^2y^2 - 2y^2$$ *Combine like terms.*

To add or subtract polynomials vertically, just remember to line up like terms. For example, perform the subtraction $(10x^3y^2 - 7x^2y^2) - (4x^3y^2 - 3x^2y^2 + 2y^2)$ vertically. Add the opposite of the second polynomial.

$$\begin{array}{l} 10x^3y^2 - 7x^2y^2 \\ \underline{-(4x^3y^2 - 3x^2y^2 + 2y^2)} \end{array} \quad \text{is equivalent to} \quad \begin{array}{l} 10x^3y^2 - 7x^2y^2 \\ \underline{-4x^3y^2 + 3x^2y^2 - 2y^2} \\ 6x^3y^2 - 4x^2y^2 - 2y^2 \end{array}$$

Polynomial functions, like polynomials, can be added, subtracted, multiplied, and divided. For example, if

$$P(x) = x^2 + x + 1$$

then

$$2P(x) = 2(x^2 + x + 1) = 2x^2 + 2x + 2$$ *Use the distributive property.*

Also, if $Q(x) = 5x^2 - 1$, then $P(x) + Q(x) = (x^2 + x + 1) + (5x^2 - 1)$
$$= 6x^2 + x.$$

Concept Check Answer:
With parentheses removed, the expression should be
$$7z - 5 - 3z + 4 = 4z - 1$$

A useful business and economics application of subtracting polynomial functions is finding the profit function $P(x)$ when given a revenue function $R(x)$ and a cost function $C(x)$. In business, it is true that

$$\text{profit} = \text{revenue} - \text{cost, or}$$
$$P(x) = R(x) - C(x)$$

For example, if the revenue function is $R(x) = 7x$ and the cost function is $C(x) = 2x + 5000$, then the profit function is

$$P(x) = R(x) - C(x)$$

or

$$P(x) = 7x - (2x + 5000) \quad \text{Substitute } R(x) = 7x$$
$$P(x) = 5x - 5000 \qquad\qquad \text{and } C(x) = 2x + 5000.$$

Problem-solving exercises involving profit are in the exercise set.

Graphing Calculator Explorations

A graphing calculator may be used to visualize addition and subtraction of polynomials in one variable. For example, to visualize the following polynomial subtraction statement

$$(3x^2 - 6x + 9) - (x^2 - 5x + 6) = 2x^2 - x + 3$$

graph both

$$Y_1 = (3x^2 - 6x + 9) - (x^2 - 5x + 6) \quad \text{Left side of equation}$$

and

$$Y_2 = 2x^2 - x + 3 \qquad\qquad \text{Right side of equation}$$

on the same screen and see that their graphs coincide. (*Note:* If the graphs do not coincide, we can be sure that a mistake has been made in combining polynomials or in calculator keystrokes. If the graphs appear to coincide, we cannot be sure that our work is correct. This is because it is possible for the graphs to differ so slightly that we do not notice it.)

The graphs of Y_1 and Y_2 are shown. The graphs appear to coincide, so the subtraction statement

$$(3x^2 - 6x + 9) - (x^2 - 5x + 6) = 2x^2 - x + 3$$

appears to be correct.

Perform the indicated operations. Then visualize by using the procedure described above.

1. $(2x^2 + 7x + 6) + (x^3 - 6x^2 - 14)$
2. $(-14x^3 - x + 2) + (-x^3 + 3x^2 + 4x)$
3. $(1.8x^2 - 6.8x - 1.7) - (3.9x^2 - 3.6x)$
4. $(-4.8x^2 + 12.5x - 7.8) - (3.1x^2 - 7.8x)$
5. $(1.29x - 5.68) + (7.69x^2 - 2.55x + 10.98)$
6. $(-0.98x^2 - 1.56x + 5.57) + (4.36x - 3.71)$

EXERCISE SET 5.2

STUDY   CD/   PH MATH   MathXL®Tutorials   MathXL®   MyMathLab®
GUIDE/SSM   VIDEO   TUTOR CENTER   ON CD

Find the degree of each term. See Example 1.

1. 4

2. 7

 3. $5x^2$

4. $-z^3$

 5. $-3xy^2$

6. $12x^3z$

Find the degree of each polynomial and indicate whether the polynomial is a monomial, binomial, trinomial, or none of these. See Examples 2 and 3.

7. $6x + 3$

8. $7x - 8$

9. $3x^2 - 2x + 5$

10. $5x^2 - 3x^2y - 2x^3$

11. $-xyz$

12. -9

13. $x^2y - 4xy^2 + 5x + y$

14. $-2x^2y - 3y^2 + 4x + y^5$

15. In your own words, describe how to find the degree of a term.

16. In your own words, describe how to find the degree of a polynomial.

If $P(x) = x^2 + x + 1$ and $Q(x) = 5x^2 - 1$, find the following. See Example 4.

17. $P(7)$

18. $Q(4)$

19. $Q(-10)$

20. $P(-4)$

21. $P(0)$

22. $Q(0)$

23. $Q\left(\frac{1}{4}\right)$

24. $P\left(\frac{1}{2}\right)$

Refer to Example 5 for Exercises 25 through 28.

25. Find the height of the object at $t = 2$ seconds.

26. Find the height of the object at $t = 4$ seconds.

27. Find the height of the object at $t = 6$ seconds.

28. Approximate (to the nearest second) how long it takes before the object hits the ground. (*Hint:* The object hits the ground when $P(x) = 0$.)

Simplify by combining like terms. See Example 6.

29. $5y + y$

30. $-x + 3x$

31. $4x + 7x - 3$

32. $-8y + 9y + 4y^2$

33. $4xy + 2x - 3xy - 1$

34. $-8xy^2 + 4x - x + 2xy^2$

35. $7x^2 - 2xy + 5y^2 - x^2 + xy + 11y^2$

36. $-a^2 + 18ab - 2b^2 + 14a^2 - 12ab - b^2$

MIXED PRACTICE

Perform the indicated operations. See Examples 7 through 10.

37. $(9y^2 - 8) + (9y^2 - 9)$

38. $(x^2 + 4x - 7) + (8x^2 + 9x - 7)$

 39. Add $(x^2 + xy - y^2)$ and $(2x^2 - 4xy + 7y^2)$.

40. Add $(4x^3 - 6x^2 + 5x + 7)$ and $(2x^2 + 6x - 3)$.

41. $x^2 - 6x + 3$
$\underline{+\ (2x + 5)}$

42. $-2x^2 + 3x - 9$
$\underline{+\quad\ (2x - 3)}$

43. $(9y^2 - 7y + 5) - (8y^2 - 7y + 2)$

44. $(2x^2 + 3x + 12) - (5x - 7)$

45. Subtract $(6x^2 - 3x)$ from $(4x^2 + 2x)$.

46. Subtract $(xy + x - y)$ from $(xy + x - 3)$.

47. $3x^2 - 4x + 8$
$\underline{-\ (5x^2 - 7)}$

48. $-3x^2 - 4x + 8$
$\underline{-\quad\ (5x + 12)}$

49. $(5x - 11) + (-x - 2)$

50. $(3x^2 - 2x) + (5x^2 - 9x)$

51. $(7x^2 + x + 1) - (6x^2 + x - 1)$

52. $(4x - 4) - (-x - 4)$

53. $(7x^3 - 4x + 8) + (5x^3 + 4x + 8x)$

54. $(9xyz + 4x - y) + (-9xyz - 3x + y + 2)$

55. $(9x^3 - 2x^2 + 4x - 7) - (2x^3 - 6x^2 - 4x + 3)$

56. $(3x^2 + 6xy + 3y^2) - (8x^2 - 6xy - y^2)$

57. Add $(y^2 + 4yx + 7)$ and $(-19y^2 + 7yx + 7)$.

58. Subtract $(x - 4)$ from $(3x^2 - 4x + 5)$.

59. $(3x^3 - b + 2a - 6) + (-4x^3 + b + 6a - 6)$

60. $(5x^2 - 6) + (2x^2 - 4x + 8)$

61. $(4x^2 - 6x + 2) - (-x^2 + 3x + 5)$

62. $(5x^2 + x + 9) - (2x^2 - 9)$

63. $(-3x + 8) + (-3x^2 + 3x - 5)$

64. $(5y^2 - 2y + 4) + (3y + 7)$

65. $(-3 + 4x^2 + 7xy^2) + (2x^3 - x^2 + xy^2)$

66. $(-3x^2y + 4) - (-7x^2y - 8y)$

67. $6y^2 - 6y + 4$
$\underline{-(-y^2 - 6y + 7)}$

68. $-4x^3 + 4x^2 - 4x$
$\underline{-(2x^3 - 2x^2 + 3x)}$

69. $3x^2 + 15x + 8$
$\underline{+(2x^2 + 7x + 8)}$

70. $9x^2 + 9x - 4$
$\underline{+(7x^2 - 3x - 4)}$

71. $\left(\frac{1}{2}x^2 - \frac{1}{3}x^2y + 2y^3\right) + \left(\frac{1}{4}x^2 - \frac{8}{3}x^2y^2 - \frac{1}{2}y^3\right)$

72. $\left(\frac{2}{5}a^2 - ab + \frac{4}{3}b^2\right) + \left(\frac{1}{5}a^2b - ab + \frac{5}{6}b^2\right)$

73. Find the sum of $(5q^4 - 2q^2 - 3q)$ and $(-6q^4 + 3q^2 + 5)$.

74. Find the sum of $(5y^4 - 7y^2 + x^2 - 3)$ and $(-3y^4 + 2y^2 + 4)$.

75. Subtract $(3x + 7)$ from the sum of $(7x^2 + 4x + 9)$ and $(8x^2 + 7x - 8)$.

76. Subtract $(9x + 8)$ from the sum of $(3x^2 - 2x - x^3 + 2)$ and $(5x^2 - 8x - x^3 + 4)$.

77. Find the sum of $(4x^4 - 7x^2 + 3)$ and $(2 - 3x^4)$.

78. Find the sum of $(8x^4 - 14x^2 + 6)$ and $(-12x^6 - 21x^4 - 9x^2)$.

79. $\left(\frac{2}{3}x^2 - \frac{1}{6}x + \frac{5}{6}\right) - \left(\frac{1}{3}x^2 + \frac{5}{6}x - \frac{1}{6}\right)$

80. $\left(\frac{3}{16}x^2 + \frac{5}{8}x - \frac{1}{4}\right) - \left(\frac{5}{16}x^2 - \frac{3}{8}x + \frac{3}{4}\right)$

Solve. See Example 5.

The surface area of a rectangular box is given by the polynomial

$$2HL + 2LW + 2HW$$

and is measured in square units. In business, surface area is often calculated to help determine cost of materials.

△ **81.** A rectangular box is to be constructed to hold a new camcorder. The box is to have dimensions 5 inches by 4 inches by 9 inches. Find the surface area of the box.

△ **82.** Suppose it has been determined that a box of dimensions 4 inches by 4 inches by 8.5 inches can be used to contain the camcorder in Exercise 81. Find the surface area of this box and calculate the square inches of material saved by using this box instead of the box in Exercise 81.

83. A projectile is fired upward from the ground with an initial velocity of 300 feet per second. Neglecting air resistance, the height of the projectile at any time t can be described by the polynomial function $P(t) = -16t^2 + 300t$. Find the height of the projectile at each given time.

a. $t = 1$ second **b.** $t = 2$ seconds
c. $t = 3$ seconds **d.** $t = 4$ seconds
e. Explain why the height increases and then decreases as time passes.

f. Approximate (to the nearest second) how long before the object hits the ground.

84. An object is thrown upward with an initial velocity of 25 feet per second from the top of the 984-foot-high Eiffel Tower in Paris, France. The height of the object at any time t can be described by the polynomial function $P(t) = -16t^2 + 25t + 984$. Find the height of the projectile at each given time.(*Source:* Council on Tall Buildings and Urban Habitat, Lehigh University)

a. $t = 1$ second
b. $t = 3$ seconds
c. $t = 5$ seconds
d. Approximate (to the nearest second) how long before the object hits the ground.

85. The polynomial function $P(x) = 45x - 100,000$ models the relationship between the number of computer briefcases x that a company sells and the profit the company makes, $P(x)$. Find $P(4000)$, the profit from selling 4000 computer briefcases.

86. The total cost (in dollars) for MCD, Inc., Manufacturing Company to produce x blank audiocassette tapes per week is given by the polynomial function $C(x) = 0.8x + 10,000$. Find the total cost of producing 20,000 tapes per week.

87. The total revenues (in dollars) for MCD, Inc., Manufacturing Company to sell x blank audiocasette tapes per week is given by the polynomial function $R(x) = 2x$. Find the total revenue from selling 20,000 tapes per week.

88. In business, profit equals revenue minus cost, or $P(x) = R(x) - C(x)$. Find the profit function for MCD, Inc. by subtracting the given functions in Exercises 86 and 87.

REVIEW AND PREVIEW

Multiply. See Section 5.1.

89. $x \cdot x$

90. $y \cdot y^2$

91. $x^2 \cdot x^2$

92. $z^3 \cdot z$

Multiply. See Section 1.8.

93. $5(3x - 2)$

94. $-7(2z - 6y)$

95. $-2(x^2 - 5x + 6)$

96. $5(-3y^2 - 2y + 7)$

Concept Extensions

Perform indicated operations.

97. $(4x^{2a} - 3x^a + 0.5) - (x^{2a} - 5x^a - 0.2)$

98. $(9y^{5a} - 4y^{3a} + 1.5y) - (6y^{5a} - y^{3a} + 4.7y)$

99. $(8x^{2y} - 7x^y + 3) + (-4x^{2y} + 9x^y - 14)$

100. $(14z^{5x} + 3z^{2x} + z) - (2z^{5x} - 10z^{2x} + 3z)$

Find each perimeter.

△ **101.**

$(x + 5y)$ units

$(3x^2 - x + 2y)$ units

△ **102.**

$(z + 2)$ units

$(2z^2 + z)$ units

$(z^3 - 4z + 1)$ units

If $P(x) = 3x + 3$, $Q(x) = 4x^2 - 6x + 3$, and $R(x) = 5x^2 - 7$, find the following.

103. $P(x) + Q(x)$

104. $R(x) + P(x)$

105. $Q(x) - R(x)$

106. $P(x) - Q(x)$

107. $2[Q(x)] - R(x)$

108. $-5[P(x)] - Q(x)$

109. $3[R(x)] + 4[P(x)]$

110. $2[Q(x)] + 7[R(x)]$

*If $P(x)$ is the polynomial given, find **a.** $P(a)$, **b.** $P(-x)$, and **c.** $P(x + h)$.*

111. $P(x) = 2x - 3$

112. $P(x) = 8x + 3$

113. $P(x) = 4x$

114. $P(x) = -4x$

115. $P(x) = 4x - 1$

116. $P(x) = 3x - 2$

117. The function $f(x) = -246.7x^2 + 1887.9x + 1016.9$ can be used to approximate the increasing number of radio stations on the Internet during the years 1998–2003 where x is the number of years after 1998 and $f(x)$ is the number of stations. Round answers to the nearest whole. (*Source:* BRS Media, Inc.)

   a. Approximate the number of radio stations on the Internet in 1998.

   b. Approximate the number of radio stations on the Internet in 2000.

   c. Use the function to predict the number of radio stations on the Internet in 2006.

   d. From parts (a), (b), and (c), determine whether the number of radio stations on the Internet is increasing at a steady rate. Explain why or why not.

118. The function $f(x) = 0.21x^2 - 0.55x + 8.57$ can be used to approximate the number of Americans enrolled in health maintenance organizations (HMOs) during the period 1980–2001 where x is the number of years after 1980 and $f(x)$ is the number of millions of Americans. Round answers to the nearest tenth of a million. (*Source:* Based on data from *Health, United States, 2001*, National Center for Health Statistics)

   a. Approximate the number of Americans enrolled in HMOs in 1995.

   b. Approximate the number of Americans enrolled in HMOs in 2000.

   c. Use the function to predict the number of Americans enrolled in HMOs in 2005.

   d. From parts (a), (b), and (c), determine whether the number of Americans enrolled in HMOs is changing at a steady rate. Explain why or why not.

119. Sport utility vehicle (SUV) sales in the United States. have increased since 1992. The function $f(x) = 0.014x^2 + 0.12x + 0.85$ can be used to approximate the number of SUV sales during the years 1990–2000 where x is the number of years after 1990 and $f(x)$ is the SUV sales (in millions). Round answers to the nearest tenth of a million. (*Source:* Wards Communications)

   a. Approximate the number of SUVs sold in 1999.

   b. Use the function to predict the number of SUVs sold in 2005.

120. Digital camera sales have increased since 1996. The function $f(x) = 34.7x^2 + 68.2x + 377.3$ can be used to approximate the revenue from selling digital cameras for the years 1996–2001 where x is the number of years after 1996 and $f(x)$ is the sales revenue (in million of dollars). Round answers to the nearest whole million. (*Source:* International Data Corporation)

   a. Approximate the revenue from digital camera sales in 2000.

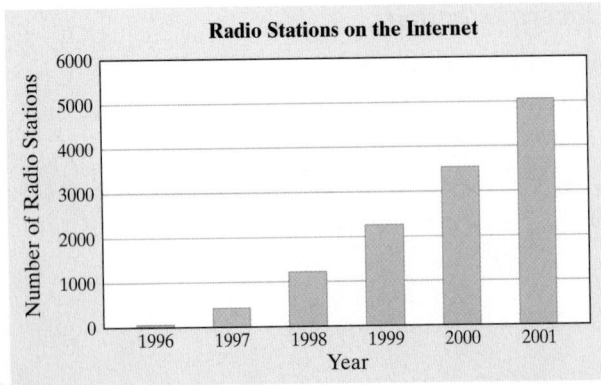

Radio Stations on the Internet

(bar graph: Number of Radio Stations vs. Year, with years 1996, 1997, 1998, 1999, 2000, 2001)

b. Use the function to predict the revenue from digital camera sales in 2004.

121. The function $f(x) = 1.4x^2 + 129.6x + 939$ can be used to approximate spending for health care in the United States, where x is the number of years since 1980 and $f(x)$ is the amount of money spent per capita. (*Source:* U.S. Health Care Financing Administration)

a. Approximate the amount of money spent on health care per capita in the year 1985.

b. Approximate the amount of money spent on health care per capita in the year 1995.

c. Use the given function to predict the amount of money that will be spent on health care per capita in the year 2010.

d. From parts **a, b,** and **c,** is the amount of money spent rising at a steady rate? Why or why not?

122. The function $f(x) = 904x^2 - 10{,}311x + 70{,}479$ can be used to approximate the number of AIDS cases reported in the United States from 1995 to 2002, where x is the number of years since 1995. (*Source:* Based on data from the U.S. Centers for Disease Control and Prevention)

a. Approximate the number of AIDS cases reported in the United States in 1995.

b. Approximate the number of AIDS cases reported in the United States in 2002.

c. Approximate the number of AIDS cases reported in the United States in 2000.

d. Describe the trend in the number of AIDS cases reported during the period covered by the model.

5.3   *MULTIPLYING POLYNOMIALS*

Objectives

1 Use the distributive property to multiply polynomials.

2 Multiply polynomials vertically.

1   To multiply polynomials, we apply our knowledge of the rules and definitions of exponents.

To multiply two monomials such as $(-5x^3)$ and $(-2x^4)$, use the associative and commutative properties and regroup. Remember that to multiply exponential expressions with a common base we add exponents.

$$(-5x^3)(-2x^4) = (-5)(-2)(x^3)(x^4) = 10x^7$$

EXAMPLES

Multiply.

1. $6x \cdot 4x = (6 \cdot 4)(x \cdot x)$ *Use the commutative and associative properties.*

$= 24x^2$ *Multiply.*

2. $-7x^2 \cdot 2x^5 = (-7 \cdot 2)(x^2 \cdot x^5)$

$= -14x^7$

3. $(-12x^5)(-x) = (-12x^5)(-1x)$

$= (-12)(-1)(x^5 \cdot x)$

$= 12x^6$

To multiply polynomials that are not monomials, use the distributive property.

EXAMPLE 4

Use the distributive property to find each product.

a. $5x(2x^3 + 6)$ **b.** $-3x^2(5x^2 + 6x - 1)$

Solution **a.** $5x(2x^3 + 6) = 5x(2x^3) + 5x(6)$ Use the distributive property.

$= 10x^4 + 30x$ Multiply.

b. $-3x^2(5x^2 + 6x - 1)$

$= (-3x^2)(5x^2) + (-3x^2)(6x) + (-3x^2)(-1)$ Use the distributive property.

$= -15x^4 - 18x^3 + 3x^2$ Multiply.

We also use the distributive property to multiply two binomials. To multiply $(x + 3)$ by $(x + 1)$, distribute the factor $(x + 3)$ first.

$(x + 3)(x + 1) = x(x + 1) + 3(x + 1)$ Distribute $(x + 3)$.

$= x(x) + x(1) + 3(x) + 3(1)$ Apply distributive property a second time.

$= x^2 + x + 3x + 3$ Multiply.

$= x^2 + 4x + 3$ Combine like terms.

This idea can be expanded so that we can multiply any two polynomials.

To Multiply Two Polynomials

Multiply each term of the first polynomial by each term of the second polynomial, and then combine like terms.

EXAMPLE 5

Multiply $(3x + 2)(2x - 5)$.

Solution Multiply each term of the first binomial by each term of the second.

$(3x + 2)(2x - 5) = 3x(2x) + 3x(-5) + 2(2x) + 2(-5)$

$= 6x^2 - 15x + 4x - 10$ Multiply.

$= 6x^2 - 11x - 10$ Combine like terms.

EXAMPLE 6

Multiply $(2x - y)^2$.

Solution Recall that $a^2 = a \cdot a$, so $(2x - y)^2 = (2x - y)(2x - y)$. Multiply each term of the first polynomial by each term of the second.

$(2x - y)(2x - y) = 2x(2x) + 2x(-y) + (-y)(2x) + (-y)(-y)$

$= 4x^2 - 2xy - 2xy + y^2$ Multiply.

$= 4x^2 - 4xy + y^2$ Combine like terms.

EXAMPLE 7

Multiply $(t + 2)$ by $(3t^2 - 4t + 2)$.

Solution   Multiply each term of the first polynomial by each term of the second.

$$(t + 2)(3t^2 - 4t + 2) = t(3t^2) + t(-4t) + t(2) + 2(3t^2) + 2(-4t) + 2(2)$$
$$= 3t^3 - 4t^2 + 2t + 6t^2 - 8t + 4$$
$$= 3t^3 + 2t^2 - 6t + 4 \qquad \text{Combine like terms.}$$

EXAMPLE 8

Multiply $(3a + b)^3$.

Solution   Write $(3a + b)^3$ as $(3a + b)(3a + b)(3a + b)$.

$$(3a + b)(3a + b)(3a + b) = (9a^2 + 3ab + 3ab + b^2)(3a + b)$$
$$= (9a^2 + 6ab + b^2)(3a + b)$$
$$= (9a^2 + 6ab + b^2)3a + (9a^2 + 6ab + b^2)b$$
$$= 27a^3 + 18a^2b + 3ab^2 + 9a^2b + 6ab^2 + b^3$$
$$= 27a^3 + 27a^2b + 9ab^2 + b^3$$

2   Another convenient method for multiplying polynomials is to use a vertical format similar to the format used to multiply real numbers. We demonstrate this method by multiplying $(3y^2 - 4y + 1)$ by $(y + 2)$.

EXAMPLE 9

Multiply $(3y^2 - 4y + 1)(y + 2)$. Use a vertical format.

Solution   **Step 1.** Multiply 2 by each term of the top polynomial. Write the first **partial product** below the line.

$$
\begin{array}{r}
3y^2 - 4y + 1 \\
\times \qquad y + 2 \\
\hline
6y^2 - 8y + 2 \quad \text{Partial product}
\end{array}
$$

Step 2. Multiply y by each term of the top polynomial. Write this partial product underneath the previous one, being careful to line up like terms.

$$
\begin{array}{r}
3y^2 - 4y + 1 \\
\times \qquad y + 2 \\
\hline
6y^2 - 8y + 2 \quad \text{Partial product} \\
3y^3 - 4y^2 + \ y \qquad \text{Partial product}
\end{array}
$$

Step 3. Combine like terms of the partial products.

$$
\begin{array}{r}
3y^2 - 4y + 1 \\
\times \qquad y + 2 \\
\hline
6y^2 - 8y + 2 \quad \text{Combine like terms.} \\
3y^3 - 4y^2 + \ y \\
\hline
3y^3 + 2y^2 - 7y + 2
\end{array}
$$

Thus, $(y + 2)(3y^2 - 4y + 1) = 3y^3 + 2y^2 - 7y + 2.$

When multiplying vertically, be careful if a power is missing, you may want to leave space in the partial products and take care that like terms are lined up.

EXAMPLE 10

Multiply $(2x^3 - 3x + 4)(x^2 + 1)$. Use a vertical format.

Solution

$$
\begin{array}{r}
2x^3 - 3x + 4 \\
\times \qquad x^2 + 1 \\
\hline
2x^3 \qquad - 3x + 4 \\
2x^5 - 3x^3 + 4x^2 \\
\hline
2x^5 - x^3 + 4x^2 - 3x + 4
\end{array}
$$

Leave space for missing powers of x.

Combine like terms.

MENTAL MATH

Find the following products mentally. See Examples 1 through 3.

1. $5x(2y)$ **2.** $7a(4b)$ **3.** $x^2 \cdot x^5$ **4.** $z \cdot z^4$

5. $6x(3x^2)$ **6.** $5a^2(3a^2)$ **7.** $-9x^3 \cdot 3x^2$ **8.** $-8x(-4x^7)$

Simplify, if possible.

9. $a^2 \cdot a^5$ **10.** $(a^2)^5$ **11.** $a^2 + a^5$ **12.** $\dfrac{a^5}{a^2}$

EXERCISE SET 5.3

STUDY GUIDE/SSM   CD/ VIDEO   PH MATH TUTOR CENTER   MathXL®Tutorials ON CD   MathXL®   MyMathLab®

Find the following products. See Example 4.

1. $2a(2a - 4)$ **2.** $3a(2a + 7)$

3. $7x(x^2 + 2x - 1)$ **4.** $-5y(y^2 + y - 10)$

5. $3x^2(2x^2 - x)$ **6.** $-4y^2(5y - 6y^2)$

△ **7.** The area of the larger rectangle below is $x(x + 3)$. Find another expression for this area by finding the sum of the areas of the smaller rectangles.

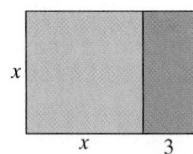

△ **8.** Write an expression for the area of the larger rectangle below in two different ways.

Find the following products. See Examples 5 and 6.

9. $(a + 7)(a - 2)$ **10.** $(y + 5)(y + 7)$

11. $(2y - 4)^2$ **12.** $(6x - 7)^2$

13. $(5x - 9y)(6x - 5y)$ **14.** $(3x - 7y)(7x + 2y)$

15. $(2x^2 - 5)^2$ **16.** $(x^2 - 4)^2$

△ **17.** The area of the figure below is $(x + 2)(x + 3)$. Find another expression for this area by finding the sum of the areas of the smaller rectangles.

△ **18.** Write an expression for the area of the figure below in two different ways.

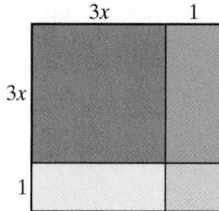

Find the following products. See Example 7.

19. $(x - 2)(x^2 - 3x + 7)$ **20.** $(x + 3)(x^2 + 5x - 8)$

21. $(x + 5)(x^3 - 3x + 4)$ **22.** $(a + 2)(a^3 - 3a^2 + 7)$

23. $(2a - 3)(5a^2 - 6a + 4)$ **24.** $(3 + b)(2 - 5b - 3b^2)$

Find the following products. See Example 8.

25. $(x + 2)^3$ **26.** $(y - 1)^3$

27. $(2y - 3)^3$ **28.** $(3x + 4)^3$

Find the following products. Use the vertical multiplication method. See Examples 9 and 10.

29. $(x + 3)(2x^2 + 4x - 1)$ **30.** $(2x - 5)(3x^2 - 4x + 7)$

31. $(x^3 + 5x - 7)(x^2 - 9)$ **32.** $(3x^4 - x^2 + 2)(2x^3 + 1)$

33. Evaluate each of the following.

    a. $(2 + 3)^2; 2^2 + 3^2$ **b.** $(8 + 10)^2; 8^2 + 10^2$

    c. Does $(a + b)^2 = a^2 + b^2$ no matter what the values of a and b are? Why or why not?

34. Perform the indicated operations. Explain the difference between the two expressions.

    a. $(3x + 5) + (3x + 7)$ **b.** $(3x + 5)(3x + 7)$

MIXED PRACTICE

Find the following products.

35. $2a(a + 4)$ **36.** $-3a(2a + 7)$

37. $3x(2x^2 - 3x + 4)$ **38.** $-4x(5x^2 - 6x - 10)$

39. $(5x + 9y)(3x + 2y)$ **40.** $(5x - 5y)(2x - y)$

41. $(x + 2)(x^2 + 5x + 6)$ **42.** $(x - 7)(x^2 - 15x + 56)$

43. $(7x + 4)^2$ **44.** $(3x - 2)^2$

45. $-2a^2(3a^2 - 2a + 3)$ **46.** $-4b^2(3b^3 - 12b^2 - 6)$

47. $(x + 3)(x^2 + 7x + 12)$ **48.** $(n + 1)(n^2 - 7n - 9)$

49. $(a + 1)^3$ **50.** $(x - y)^3$

51. $(x + y)(x + y)$ **52.** $(x + 3)(7x + 1)$

53. $(x - 7)(x - 6)$ **54.** $(4x + 5)(-3x + 2)$

55. $3a(a^2 + 2)$ **56.** $x^3(x + 12)$

57. $-4y(y^2 + 3y - 11)$ **58.** $-2x(5x^2 - 6x + 1)$

59. $(5x + 1)(5x - 1)$ **60.** $(2x + y)(3x - y)$

61. $(5x + 4)(x^2 - x + 4)$ **62.** $(x - 2)(x^2 - x + 3)$

63. $(2x - 5)^3$ **64.** $(3y - 1)^3$

65. $(4x + 5)(8x^2 + 2x - 4)$ **66.** $(x + 7)(x^2 - 7x - 8)$

67. $(7xy - y)^2$ **68.** $(x + y)^2$

69. $(5y^2 - y + 3)(y^2 - 3y - 2)$

70. $(2x^2 + x - 1)(x^2 + 3x - 4)$

71. $(3x^2 + 2x - 4)(2x^2 - 4x + 3)$

72. $(a^2 + 3a - 2)(2a^2 - 5a - 1)$

REVIEW AND PREVIEW

Perform the indicated operation. See Section 5.1.

73. $(5x)^2$ **74.** $(4p)^2$

75. $(-3y^3)^2$ **76.** $(-7m^2)^2$

*For income tax purposes, Rob Calcutta, the owner of Copy Services, uses a method called **straight-line depreciation** to show the depreciated (or decreased) value of a copy machine he recently purchased. Rob assumes that he can use the machine for 7 years. The graph below shows the depreciated value of the machine over the years. Use this graph to answer Exercises 77 through 82. See Section and 3.1.*

77. What was the purchase price of the copy machine?

78. What is the depreciated value of the machine in 7 years?

79. What loss in value occurred during the first year?

80. What loss in value occurred during the second year?

81. Why do you think this method of depreciating is called straight-line depreciation?

82. Why is the line tilted downward?

Concept Extensions

Express each of the following as polynomials.

△ **83.** Find the area of the rectangle.

(2x + 5) yards

(2x − 5) yards

△ **84.** Find the area of the square field.

(x + 4) feet

△ **85.** Find the area of the triangle.

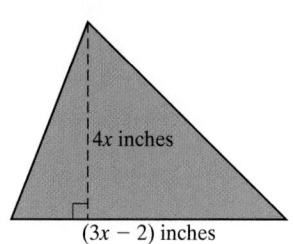

4x inches

(3x − 2) inches

△ **86.** Find the volume of the cube.

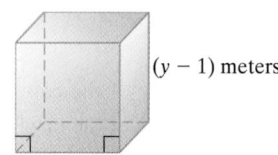

(y − 1) meters

△ **87.** Write a polynomial that describes the area of the shaded region.

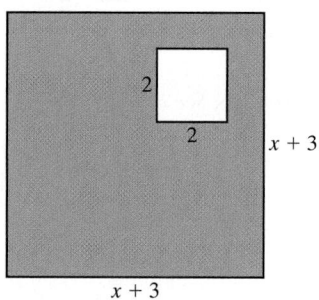

2

2

x + 3

x + 3

△ **88.** Write a polynomial that describes the area of the shaded region.

x + 4

x + 1

x

89. Multiply the following polynomials.

a. $(a + b)(a - b)$ b. $(2x + 3y)(2x - 3y)$

c. $(4x + 7)(4x - 7)$

d. Can you make a general statement about all products of the form $(x + y)(x - y)$?

5.4   SPECIAL PRODUCTS

Objectives

1 Multiply two binomials using the FOIL method.

2 Square a binomial.

3 Multiply the sum and difference of two terms.

1   In this section, we multiply binomials using special products. First, a special order for multiplying binomials called the FOIL order or method is introduced. This method is demonstrated by multiplying $(3x + 1)$ by $(2x + 5)$.

F stands for the product of the **First** terms. $(3x + 1)(2x + 5)$
$$(3x)(2x) = 6x^2 \quad \textbf{F}$$
O stands for the product of the **Outer** terms. $(3x + 1)(2x + 5)$
$$(3x)(5) = 15x \quad \textbf{O}$$
I stands for the product of the **Inner** terms. $(3x + 1)(2x + 5)$
$$(1)(2x) = 2x \quad \textbf{I}$$
L stands for the product of the **Last** terms. $(3x + 1)(3x + 5)$
$$(1)(5) = 5 \quad \textbf{L}$$

$$\begin{array}{cccc} \text{F} & \text{O} & \text{I} & \text{L} \end{array}$$
$$(3x + 1)(2x + 5) = 6x^2 + 15x + 2x + 5$$
$$= 6x^2 + 17x + 5 \qquad \textit{Combine like terms.}$$

✔ CONCEPT CHECK

Multiply $(3x + 1)(2x + 5)$ using methods from the last section. Show that the product is still $6x^2 + 17x + 5$.

EXAMPLE 1

Multiply $(x - 3)(x + 4)$ by the FOIL method.

Solution

$$\begin{array}{cccc} \text{F} & \text{O} & \text{I} & \text{L} \end{array}$$
$$(x - 3)(x + 4) = (x)(x) + (x)(4) + (-3)(x) + (-3)(4)$$

$$= x^2 + 4x - 3x - 12$$
$$= x^2 + x - 12 \qquad \textit{Combine like terms.}$$

EXAMPLE 2

Multiply $(5x - 7)(x - 2)$ by the FOIL method.

Solution

$$\begin{array}{cccc} \text{F} & \text{O} & \text{I} & \text{L} \end{array}$$
$$(5x - 7)(x - 2) = 5x(x) + 5x(-2) + (-7)(x) + (-7)(-2)$$

$$= 5x^2 - 10x - 7x + 14$$
$$= 5x^2 - 17x + 14 \qquad \textit{Combine like terms.}$$

Concept Check Answer:
Multiply and simplify:
$3x(2x + 5) + 1(2x + 5)$

EXAMPLE 3

Multiply $2(y + 6)(2y - 1)$.

Solution
$$\begin{array}{cccc} \text{F} & \text{O} & \text{I} & \text{L} \end{array}$$
$$2(y + 6)(2y - 1) = 2(2y^2 - 1y + 12y - 6)$$
$$= 2(2y^2 + 11y - 6)$$
$$= 4y^2 + 22y - 12 \qquad \textit{Now use the distributive property.}$$

2 Now, try squaring a binomial using the FOIL method.

EXAMPLE 4

Multiply $(3y + 1)^2$.

Solution

$$(3y + 1)^2 = (3y + 1)(3y + 1)$$

$$\qquad\qquad\quad \text{F} \qquad \text{O} \qquad \text{I} \qquad \text{L}$$

$$= (3y)(3y) + (3y)(1) + 1(3y) + 1(1)$$

$$= 9y^2 + 3y + 3y + 1$$

$$= 9y^2 + 6y + 1$$

Notice the pattern that appears in Example 4.

$(3y + 1)^2 = 9y^2 + 6y + 1$ $9y^2$ is the first term of the binomial squared. $(3y)^2 = 9y^2$.

6y is 2 times the product of both terms of the binomial. $(2)(3y)(1) = 6y$.

1 is the second term of the binomial squared. $(1)^2 = 1$.

This pattern leads to the following, which can be used when squaring a binomial. We call these **special products**.

Squaring a Binomial

A binomial squared is equal to the square of the first term plus or minus twice the product of both terms plus the square of the second term.

$$(a + b)^2 = a^2 + 2ab + b^2$$
$$(a - b)^2 = a^2 - 2ab + b^2$$

This product can be visualized geometrically.

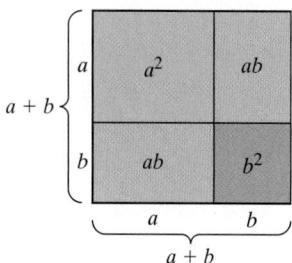

The area of the large square is side · side.

Area $= (a + b)(a + b) = (a + b)^2$

The area of the large square is also the sum of the areas of the smaller rectangles.

Area $= a^2 + ab + ab + b^2 = a^2 + 2ab + b^2$

Thus, $(a + b)^2 = a^2 + 2ab + b^2$.

EXAMPLE 5

Use a special product to square each binomial.

a. $(t + 2)^2$ **b.** $(p - q)^2$ **c.** $(2x + 5)^2$ **d.** $(x^2 - 7y)^2$

Solution

	first term squared	plus or minus	twice the product of the terms	plus	second term squared

a. $\quad (t + 2)^2 \;=\; t^2 \;+\; 2(t)(2) \;+\; 2^2 = t^2 + 4t + 4$

b. $\quad (p - q)^2 \;=\; p^2 \;-\; 2(p)(q) \;+\; q^2 = p^2 - 2pq + q^2$

c. $\quad (2x + 5)^2 \;=\; (2x)^2 \;+\; 2(2x)(5) \;+\; 5^2 = 4x^2 + 20x + 25$

d. $\quad (x^2 - 7y)^2 \;=\; (x^2)^2 \;-\; 2(x^2)(7y) \;+\; (7y^2) = x^4 - 14x^2y + 49y^2$

Helpful Hint

Notice that

$$(a + b)^2 \neq a^2 + b^2 \qquad \text{The middle term } \textbf{2ab} \text{ is missing.}$$

$$(a + b)^2 = (a + b)(a + b) = a^2 + 2ab + b^2$$

Likewise,

$$(a - b)^2 \neq a^2 - b^2$$

$$(a - b)^2 = (a - b)(a - b) = a^2 - 2ab + b^2$$

Another special product is the product of the sum and difference of the same two terms, such as $(x + y)(x - y)$. Finding this product by the FOIL method, we see a pattern emerge.

$$\text{F O I L}$$
$$(x + y)(x - y) = x^2 - xy + xy - y^2$$
$$= x^2 - y^2$$

Notice that the middle two terms subtract out. This is because the **O**uter product is the opposite of the **I**nner product. Only the **difference of squares** remains.

3

Multiplying the Sum and Difference of Two Terms

The product of the sum and difference of two terms is the square of the first term minus the square of the second term.

$$(a + b)(a - b) = a^2 - b^2$$

EXAMPLE 6

Use a special product to multiply.

a. $4(x + 4)(x - 4)$ 　　　b. $(6t + 7)(6t - 7)$ 　　　c. $\left(x - \dfrac{1}{4}\right)\left(x + \dfrac{1}{4}\right)$

d. $(2p - q)(2p + q)$ 　　　e. $(3x^2 - 5y)(3x^2 + 5y)$

Solution

	first term squared	minus	second term squared

a. $4(x + 4)(x - 4) = 4(x^2 \quad - \quad 4^2) = 4(x^2 - 16) = 4x^2 - 64$

b. $(6t + 7)(6t - 7) = (6t)^2 \quad - \quad 7^2 = 36t^2 - 49$

c. $\left(x - \dfrac{1}{4}\right)\left(x + \dfrac{1}{4}\right) = x^2 - \left(\dfrac{1}{4}\right)^2 = x^2 - \dfrac{1}{16}$

d. $(2p - q)(2p + q) = (2p)^2 - q^2 = 4p^2 - q^2$

e. $(3x^2 - 5y)(3x^2 + 5y) = (3x^2)^2 - (5y)^2 = 9x^4 - 25y^2$

✔ **CONCEPT CHECK**

Match each expression below to the equivalent expression or expressions in the list below.

$$(a + b)^2$$
$$(a + b)(a - b)$$

a. $(a + b)(a + b)$ **b.** $a^2 - b^2$ **c.** $a^2 + b^2$ **d.** $a^2 - 2ab + b^2$ **e.** $a^2 + 2ab + b^2$

Let's now practice multiplying polynomials in general. If possible, use a special product.

EXAMPLE 7

Multiply.

a. $(x - 5)(3x + 4)$ **b.** $(7x + 4)^2$ **c.** $(y - 0.6)(y + 0.6)$
d. $(y^4 + 2)(3y^2 - 1)$ **e.** $(a - 3)(a^2 + 2a - 1)$

Solution

a. $(x - 5)(3x + 4) = 3x^2 + 4x - 15x - 20$ FOIL.
$\qquad\qquad\qquad\quad = 3x^2 - 11x - 20$

b. $(7x + 4)^2 = (7x)^2 + 2(7x)(4) + 4^2$ Squaring a binomial.
$\qquad\qquad\quad = 49x^2 + 56x + 16$

c. $(y - 0.6)(y + 0.6) = y^2 - (0.6)^2 = y^2 - 0.36$ Multiplying the sum and difference of 2 terms.

d. $(y^4 + 2)(3y^2 - 1) = 3y^6 - y^4 + 6y^2 - 2$

e. I've inserted this product as a reminder that since it is not a binomial times a binomial, the FOIL order may not be used.

$(a - 3)(a^2 + 2a - 1) = a(a^2 + 2a - 1) - 3(a^2 + 2a - 1)$ Multiplying each term
$\qquad\qquad\qquad\qquad\quad = a^3 + 2a^2 - a - 3a^2 - 6a + 3$ of the binomial by
$\qquad\qquad\qquad\qquad\quad = a^3 - a^2 - 7a + 3$ each term of the trinomial.

Concept Check Answer:
a or e, b

MENTAL MATH

Answer each exercise true or false.

1. $(x + 4)^2 = x^2 + 16$

2. For $(x + 6)(2x - 1)$ the product of the first terms is $2x^2$.

3. $(x + 4)(x - 4) = x^2 + 16$

4. The product $(x - 1)(x^3 + 3x - 1)$ is a polynomial of degree 5.

EXERCISE SET 5.4

STUDY
GUIDE/SSM

CD/
VIDEO

PH MATH
TUTOR CENTER

MathXL®Tutorials
ON CD

MathXL®

MyMathLab®

Find each product using the FOIL method. See Examples 1 through 3.

1. $(x + 3)(x + 4)$

2. $(x + 5)(x - 1)$

3. $(x - 5)(x + 10)$

4. $(y - 12)(y + 4)$

5. $(5x - 6)(x + 2)$

6. $(3y - 5)(2y - 7)$

7. $(y - 6)(4y - 1)$

8. $(2x - 9)(x - 11)$

9. $(2x + 5)(3x - 1)$

10. $(6x + 2)(x - 2)$

Find each product. See Examples 4 and 5.

11. $(x - 2)^2$

12. $(x + 7)^2$

13. $(2x - 1)^2$

14. $(7x - 3)^2$

15. $(3a - 5)^2$

16. $(5a + 2)^2$

17. $(5x + 9)^2$

18. $(6s - 2)^2$

19. Using your own words, explain how to square a binomial such as $(a + b)^2$.

20. Explain how to find the product of two binomials using the FOIL method.

Find each product. See Example 6.

21. $(a - 7)(a + 7)$

22. $(b + 3)(b - 3)$

23. $(3x - 1)(3x + 1)$

24. $(4x - 5)(4x + 5)$

25. $\left(3x - \dfrac{1}{2}\right)\left(3x + \dfrac{1}{2}\right)$

26. $\left(10x + \dfrac{2}{7}\right)\left(10x - \dfrac{2}{7}\right)$

27. $(9x + y)(9x - y)$

28. $(2x - y)(2x + y)$

29. $(2x + 0.1)(2x - 0.1)$

30. $(5x - 1.3)(5x + 1.3)$

MIXED PRACTICE

Find each product. See Example 7.

31. $(a + 5)(a + 4)$

32. $(a - 5)(a - 7)$

33. $(a + 7)^2$

34. $(b - 2)^2$

35. $(4a + 1)(3a - 1)$

36. $(6a + 7)(6a + 5)$

37. $(x + 2)(x - 2)$

38. $(x - 10)(x + 10)$

39. $(3a + 1)^2$

40. $(4a - 2)^2$

41. $(x^2 + y)(4x - y^4)$

42. $(x^3 - 2)(5x + y)$

43. $(x + 3)(x^2 - 6x + 1)$

44. $(x - 2)(x^2 - 4x + 2)$

45. $(2a - 3)^2$

46. $(5b - 4x)^2$

47. $(5x - 6z)(5x + 6z)$

48. $(11x - 7y)(11x + 7y)$

49. $(x^5 - 3)(x^5 - 5)$

50. $(a^4 + 5)(a^4 + 6)$

51. $\left(x - \dfrac{1}{3}\right)\left(x + \dfrac{1}{3}\right)$

52. $\left(3x + \dfrac{1}{5}\right)\left(3x - \dfrac{1}{5}\right)$

53. $(a^3 + 11)(a^4 - 3)$

54. $(x^5 + 5)(x^2 - 8)$

55. $3(x - 2)^2$

56. $2(3b + 7)^2$

57. $(3b + 7)(2b - 5)$

58. $(3y - 13)(y - 3)$

59. $(7p - 8)(7p + 8)$

60. $(3s - 4)(3s + 4)$

61. $\left(\dfrac{1}{3}a^2 - 7\right)\left(\dfrac{1}{3}a^2 + 7\right)$

62. $\left(\dfrac{2}{3}a - b^2\right)\left(\dfrac{2}{3}a - b^2\right)$

63. $5x^2(3x^2 - x + 2)$

64. $4x^3(2x^2 + 5x - 1)$

65. $(2r - 3s)(2r + 3s)$

66. $(6r - 2x)(6r + 2x)$

67. $(3x - 7y)^2$

68. $(4s - 2y)^2$

69. $(4x + 5)(4x - 5)$

70. $(3x + 5)(3x - 5)$

71. $(8x + 4)^2$

72. $(3x + 2)^2$

73. $\left(a - \dfrac{1}{2}y\right)\left(a + \dfrac{1}{2}y\right)$

74. $\left(\dfrac{a}{2} + 4y\right)\left(\dfrac{a}{2} - 4y\right)$

75. $\left(\dfrac{1}{5}x - y\right)\left(\dfrac{1}{5}x + y\right)$

76. $\left(\dfrac{y}{6} - 8\right)\left(\dfrac{y}{6} + 8\right)$

77. $(a + 1)(3a^2 - a + 1)$

78. $(b + 3)(2b^2 + b - 3)$

REVIEW AND PREVIEW

Simplify each expression. See Section 5.1.

79. $\dfrac{50b^{10}}{70b^5}$ **80.** $\dfrac{x^3y^6}{xy^2}$ **81.** $\dfrac{8a^{17}b^{15}}{-4a^7b^{10}}$

82. $\dfrac{-6a^8y}{3a^4y}$ **83.** $\dfrac{2x^4y^{12}}{3x^4y^4}$ **84.** $\dfrac{-48ab^6}{32ab^3}$

Find the slope of each line. See Section 3.4.

85.   **86.**

87.   **88.**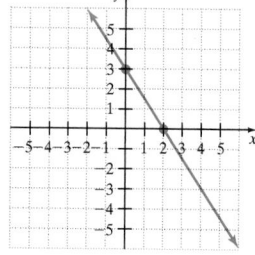

Express each of the following as a polynomial in x.

 89. Find the area of the square rug shown if its side is $(2x + 1)$ feet.

 90. Find the area of the rectangular canvas if its length is $(3x - 2)$ inches and its width is $(x - 4)$ inches.

Concept Extensions

Find the area of each shaded region.

△ **91.**

△ **92.**

△ **93.**

△ **94.**

△ **95.**   △ **96.**

97. When is the product of two binomials a binomial? To help you answer this question, find the product $(x + 2)(x + 2)$ and then find the product $(x + 2)(x - 2)$.

Find each product. For example,

$$[(a + b) - 2][(a + b) + 2] = (a + b)^2 - 2^2$$
$$= a^2 + 2ab + b^2 - 4$$

98. $[(x + y) - 3][(x + y) + 3]$

99. $[(a + c) - 5][(a + c) + 5]$

100. $[(a - 3) + b][(a - 3) - b]$

101. $[(x - 2) + y][(x - 2) - y]$

INTEGRATED REVIEW   EXPONENTS AND OPERATIONS ON POLYNOMIALS

Perform the indicated operations and simplify.

1. $(5x^2)(7x^3)$

2. $(4y^2)(8y^7)$

3. -4^2

4. $(-4)^2$

5. $(x - 5)(2x + 1)$

6. $(3x - 2)(x + 5)$

7. $(x - 5) + (2x + 1)$

8. $(3x - 2) + (x + 5)$

9. $\dfrac{7x^9y^{12}}{x^3y^{10}}$

10. $\dfrac{20a^2b^8}{14a^2b^2}$

11. $(12m^7n^6)^2$

12. $(4y^9z^{10})^3$

13. $3(4y - 3)(4y + 3)$

14. $2(7x - 1)(7x + 1)$

15. $(x^7y^5)^9$

16. $(3^1x^9)^3$

17. $(7x^2 - 2x + 3) - (5x^2 + 9)$

18. $(10x^2 + 7x - 9) - (4x^2 - 6x + 2)$

19. $0.7y^2 - 1.2 + 1.8y^2 - 6y + 1$

20. $7.8x^2 - 6.8x + 3.3 + 0.6x^2 - 9$

21. $(x + 4y)^2$

22. $(y - 9z)^2$

23. $(x + 4y) + (x + 4y)$

24. $(y - 9z) + (y - 9z)$

25. $7x^2 - 6xy + 4(y^2 - xy)$

26. $5a^2 - 3ab + 6(b^2 - a^2)$

27. $(x - 3)(x^2 + 5x - 1)$

28. $(x + 1)(x^2 - 3x - 2)$

29. $(2x^3 - 7)(3x^2 + 10)$

30. $(5x^3 - 1)(4x^4 + 5)$

31. $(2x - 7)(x^2 - 6x + 1)$

32. $(5x - 1)(x^2 + 2x - 3)$

Perform exercises and simplify, if possible.

33. $5x^3 + 5y^3$

34. $(5x^3)(5y^3)$

35. $(5x^3)^3$

36. $\dfrac{5x^3}{5y^3}$

37. $x + x$

38. $x \cdot x$

5.5   *NEGATIVE EXPONENTS AND SCIENTIFIC NOTATION*

Objectives

1 Evaluate numbers raised to negative integer powers.

2 Use all the rules and definitions for exponents to simplify exponential expressions.

3 Write numbers in scientific notation.

4 Convert numbers from scientific notation to standard form.

1 Our work with exponential expressions so far has been limited to exponents that are positive integers or 0. Here we expand to give meaning to an expression like x^{-3}.

Suppose that we wish to simplify the expression $\dfrac{x^2}{x^5}$. If we use the quotient rule for exponents, we subtract exponents:

$$\frac{x^2}{x^5} = x^{2-5} = x^{-3}, \quad x \neq 0$$

But what does x^{-3} mean? Let's simplify $\dfrac{x^2}{x^5}$ using the definition of x^n.

$$\frac{x^2}{x^5} = \frac{x \cdot x}{x \cdot x \cdot x \cdot x \cdot x}$$

$$= \frac{x \cdot x}{x \cdot x \cdot x \cdot x \cdot x} \qquad \text{Divide numerator and denominator by common factors by applying the fundamental principle for fractions.}$$

$$= \frac{1}{x^3}$$

If the quotient rule is to hold true for negative exponents, then x^{-3} must equal $\dfrac{1}{x^3}$.

From this example, we state the definition for negative exponents.

Negative Exponents

If a is a real number other than 0 and n is an integer, then

$$a^{-n} = \frac{1}{a^n}$$

For example, $x^{-3} = \dfrac{1}{x^3}$.

In other words, another way to write a^{-n} is to take its reciprocal and change the sign of its exponent.

EXAMPLE 1

Simplify by writing each expression with positive exponents only.

a. 3^{-2} **b.** $2x^{-3}$ **c.** $2^{-1} + 4^{-1}$ **d.** $(-2)^{-4}$ **e.** $\dfrac{1}{y^{-4}}$ **f.** $\dfrac{1}{7^{-2}}$

Solution   **a.** $3^{-2} = \dfrac{1}{3^2} = \dfrac{1}{9}$ *Use the definition of negative exponents.*

b. $2x^{-3} = 2 \cdot \dfrac{1}{x^3} = \dfrac{2}{x^3}$ *Use the definition of negative exponents.*

> ### Helpful Hint
> Don't forget that since there are no parentheses, only x is the base for the exponent -3.

c. $2^{-1} + 4^{-1} = \dfrac{1}{2} + \dfrac{1}{4} = \dfrac{2}{4} + \dfrac{1}{4} = \dfrac{3}{4}$

d. $(-2)^{-4} = \dfrac{1}{(-2)^4} = \dfrac{1}{(-2)(-2)(-2)(-2)} = \dfrac{1}{16}$

e. $\dfrac{1}{y^{-4}} = \dfrac{1}{\frac{1}{y^4}} = y^4$ **f.** $\dfrac{1}{7^{-2}} = \dfrac{1}{\frac{1}{7^2}} = \dfrac{7^2}{1}$ or 49

> **Helpful Hint**
>
> From Example 1, we see that
>
> $$\frac{1}{y^{-4}} = \frac{y^4}{1} \text{ or } y^4.$$
>
> Also, notice that a negative exponent *does not affect* the sign of its base. The key word to remember when working with negative exponents is reciprocal.
>
> $$x^{-2} = \frac{1}{x^2} \qquad 2^{-3} = \frac{1}{2^3} = \frac{1}{8}$$
>
> $$\frac{1}{5^{-2}} = \frac{5^2}{1} \text{ or } 25 \qquad \frac{y^{-7}}{x^{-11}} = \frac{x^{11}}{y^7}$$

EXAMPLE 2

Simplify each expression. Write results using positive exponents only.

a. $\dfrac{1}{x^{-3}}$ **b.** $\dfrac{1}{3^{-4}}$ **c.** $\dfrac{p^{-4}}{q^{-9}}$ **d.** $\dfrac{5^{-3}}{2^{-5}}$

Solution **a.** $\dfrac{1}{x^{-3}} = \dfrac{x^3}{1} = x^3$ **b.** $\dfrac{1}{3^{-4}} = \dfrac{3^4}{1} = 81$ **c.** $\dfrac{p^{-4}}{q^{-9}} = \dfrac{q^9}{p^4}$ **d.** $\dfrac{5^{-3}}{2^{-5}} = \dfrac{2^5}{5^3} = \dfrac{32}{125}$

EXAMPLE 3

Simplify each expression. Write answers with positive exponents.

a. $\dfrac{y}{y^{-2}}$ **b.** $\dfrac{3}{x^{-4}}$ **c.** $\dfrac{x^{-5}}{x^7}$

Solution **a.** $\dfrac{y}{y^{-2}} = \dfrac{y^1}{y^{-2}} = y^{1-(-2)} = y^3$ Remember that $\dfrac{a^m}{a^n} = a^{m-n}$.

b. $\dfrac{3}{x^{-4}} = 3 \cdot \dfrac{1}{x^{-4}} = 3 \cdot x^4$ or $3x^4$ **c.** $\dfrac{x^{-5}}{x^7} = x^{-5-7} = x^{-12} = \dfrac{1}{x^{12}}$

2   All the previously stated rules for exponents apply for negative exponents also. Here is a summary of the rules and definitions for exponents.

> ## Summary of Exponent Rules
>
> If m and n are integers and a, b, and c are real numbers, then:
>
> Product rule for exponents: $a^m \cdot a^n = a^{m+n}$
>
> Power rule for exponents: $(a^m)^n = a^{m \cdot n}$
>
> Power of a product: $(ab)^n = a^n b^n$
>
> Power of a quotient: $\left(\dfrac{a}{c}\right)^n = \dfrac{a^n}{c^n}, \quad c \neq 0$
>
> Quotient rule for exponents: $\dfrac{a^m}{a^n} = a^{m-n}, \quad a \neq 0$
>
> Zero exponent: $a^0 = 1, \quad a \neq 0$
>
> Negative exponent: $a^{-n} = \dfrac{1}{a^n}, \quad a \neq 0$

EXAMPLE 4

Simplify the following expressions. Write each result using positive exponents only.

a. $\left(\dfrac{2}{3}\right)^{-3}$ **b.** $\dfrac{(x^3)^4 x}{x^7}$ **c.** $\left(\dfrac{3a^2}{b}\right)^{-3}$

d. $\dfrac{4^{-1}x^{-3}y}{4^{-3}x^2 y^{-6}}$ **e.** $(y^{-3}z^6)^{-6}$ **f.** $\left(\dfrac{-2x^3 y}{xy^{-1}}\right)^3$

Solution

a. $\left(\dfrac{2}{3}\right)^{-3} = \dfrac{2^{-3}}{3^{-3}} = \dfrac{3^3}{2^3} = \dfrac{27}{8}$

b. $\dfrac{(x^3)^4 x}{x^7} = \dfrac{x^{12} \cdot x}{x^7} = \dfrac{x^{12+1}}{x^7} = \dfrac{x^{13}}{x^7} = x^{13-7} = x^6$ Use the power rule.

c. $\left(\dfrac{3a^2}{b}\right)^{-3} = \dfrac{3^{-3}(a^2)^{-3}}{b^{-3}}$ Raise each factor in the numerator and the denominator to the -3 power.

$= \dfrac{3^{-3}a^{-6}}{b^{-3}}$ Use the power rule.

$= \dfrac{b^3}{3^3 a^6}$ Use the negative exponent rule.

$= \dfrac{b^3}{27a^6}$ Write 3^3 as **27**.

d. $\dfrac{4^{-1}x^{-3}y}{4^{-3}x^2 y^{-6}} = 4^{-1-(-3)}x^{-3-2}y^{1-(-6)} = 4^2 x^{-5} y^7 = \dfrac{4^2 y^7}{x^5} = \dfrac{16y^7}{x^5}$

e. $(y^{-3}z^6)^{-6} = y^{18} \cdot z^{-36} = \dfrac{y^{18}}{z^{36}}$

f. $\left(\dfrac{-2x^3 y}{xy^{-1}}\right)^3 = \dfrac{(-2)^3 x^9 y^3}{x^3 y^{-3}} = \dfrac{-8x^9 y^3}{x^3 y^{-3}} = -8x^{9-3}y^{3-(-3)} = -8x^6 y^6$

3   Both very large and very small numbers frequently occur in many fields of science. For example, the distance between the sun and the planet Pluto is approximately 5,906,000,000 kilometers, and the mass of a proton is approximately 0.000000000000000000000165 gram. It can be tedious to write these numbers in this standard decimal notation, so **scientific notation** is used as a convenient shorthand for expressing very large and very small numbers.

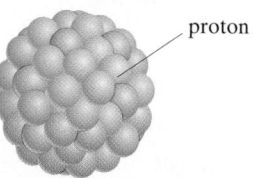

proton

Mass of proton is approximately
0.000 000 000 000 000 000 000 001 65 gram

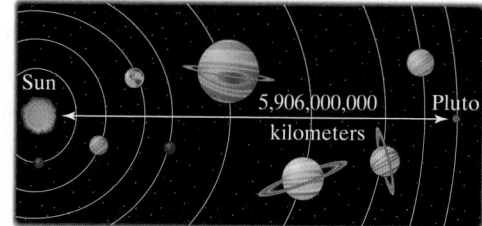

Sun   5,906,000,000 kilometers   Pluto

Scientific Notation

A positive number is written in scientific notation if it is written as the product of a number a, where $1 \le a < 10$, and an integer power r of 10:

$$a \times 10^r$$

The numbers below are written in scientific notation. The $\times$ sign for multiplication is used as part of the notation.

2.03×10^2 7.362×10^7 5.906×10^9 (Distance between the sun and Pluto)

1×10^{-3} 8.1×10^{-5} 1.65×10^{-24} (Mass of a proton)

The following steps are useful when writing numbers in scientific notation.

To Write a Number in Scientific Notation

Step 1. Move the decimal point in the original number to the left or right so that the new number has a value between 1 and 10.

Step 2. Count the number of decimal places the decimal point is moved in Step 1. If the original number is 10 or greater, the count is positive. If the original number is less than 1, the count is negative.

Step 3. Multiply the new number in Step 1 by 10 raised to an exponent equal to the count found in Step 2.

EXAMPLE 5

Write each number in scientific notation.

a. 367,000,000   **b.** 0.000003   **c.** 20,520,000,000   **d.** 0.00085

Solution   **a. Step 1.** Move the decimal point until the number is between 1 and 10.

367,000,000

8 places

Step 2. The decimal point is moved 8 places, and the original number is 10 or greater, so the count is positive 8.

Step 3. $367,000,000 = 3.67 \times 10^8$.

b. Step 1. Move the decimal point until the number is between 1 and 10.

0.000003

6 places

Step 2. The decimal point is moved 6 places, and the original number is less than 1, so the count is -6.

Step 3. $0.000003 = 3.0 \times 10^{-6}$

c. $20,520,000,000 = 2.052 \times 10^{10}$

d. $0.00085 = 8.5 \times 10^{-4}$

4   A number written in scientific notation can be rewritten in standard form. For example, to write 8.63×10^3 in standard form, recall that $10^3 = 1000$.

$$8.63 \times 10^3 = 8.63(1000) = 8630$$

Notice that the exponent on the 10 is positive 3, and we moved the decimal point 3 places to the right.

To write 7.29×10^{-3} in standard form, recall that $10^{-3} = \dfrac{1}{10^3} = \dfrac{1}{1000}$.

$$7.29 \times 10^{-3} = 7.29\left(\frac{1}{1000}\right) = \frac{7.29}{1000} = 0.00729$$

The exponent on the 10 is negative 3, and we moved the decimal to the left 3 places.

In general, **to write a scientific notation number in standard form**, move the decimal point the same number of places as the exponent on 10. If the exponent is positive, move the decimal point to the right; if the exponent is negative, move the decimal point to the left.

EXAMPLE 6

Write each number in standard notation, without exponents.

a. 1.02×10^5 **b.** 7.358×10^{-3} **c.** 8.4×10^7 **d.** 3.007×10^{-5}

Solution   **a.** Move the decimal point 5 places to the right.

$$1.02 \times 10^5 = 102{,}000.$$

b. Move the decimal point 3 places to the left.

$$7.358 \times 10^{-3} = 0.007358.$$

c. $8.4 \times 10^7 = 84{,}000{,}000.$ *7 places to the right*

d. $3.007 \times 10^{-5} = 0.00003007$ *5 places to the left*

✔ **CONCEPT CHECK**

Which number in each pair is larger?

a. 7.8×10^3 or 2.1×10^5 **b.** 9.2×10^{-2} or 2.7×10^4 **c.** 5.6×10^{-4} or 6.3×10^{-5}

Performing operations on numbers written in scientific notation makes use of the rules and definitions for exponents.

EXAMPLE 7

Perform each indicated operation. Write each result in standard decimal notation.

a. $(8 \times 10^{-6})(7 \times 10^3)$ **b.** $\dfrac{12 \times 10^2}{6 \times 10^{-3}}$

Solution   **a.** $(8 \times 10^{-6})(7 \times 10^3) = (8 \cdot 7) \times (10^{-6} \cdot 10^3)$

$$= 56 \times 10^{-3}$$
$$= 0.056$$

Concept Check Answer:

a. 2.1×10^5 **b.** 2.7×10^4
c. 5.6×10^{-4}

b. $\dfrac{12 \times 10^2}{6 \times 10^{-3}} = \dfrac{12}{6} \times 10^{2-(-3)} = 2 \times 10^5 = 200{,}000$

Calculator Explorations

Scientific Notation

To enter a number written in scientific notation on a scientific calculator, locate the scientific notation key, which may be marked $\boxed{\text{EE}}$ or $\boxed{\text{EXP}}$. To enter 3.1×10^7, press $\boxed{3.1}$ $\boxed{\text{EE}}$ $\boxed{7}$. The display should read $\boxed{3.1 \quad 07}$.

(continued)

Enter each number written in scientific notation on your calculator.

1. 5.31×10^3

2. -4.8×10^{14}

3. 6.6×10^{-9}

4. -9.9811×10^{-2}

Multiply each of the following on your calculator. Notice the form of the result.

5. $3,000,000 \times 5,000,000$

6. $230,000 \times 1000$

Multiply each of the following on your calculator. Write the product in scientific notation.

7. $(3.26 \times 10^6)(2.5 \times 10^{13})$

8. $(8.76 \times 10^{-4})(1.237 \times 10^9)$

Spotlight on
DECISION
MAKING

Suppose you are a paralegal for a law firm. You are investigating the facts of a mineral rights case. Drilling on property adjacent to the client's has struck a natural gas reserve. The client believes that a portion of this reserve lies within her own property boundaries and she is, therefore, entitled to a portion of the proceeds from selling the natural gas. As part of your investigation, you contact two different experts, who fax you the following estimates for the size of the natural gas reserve:

Expert A	**Expert B**
Estimate of entire reserve: 4.6×10^7 cubic feet	Estimate of entire reserve: 6.7×10^6 cubic feet
Estimate of size of reserve on client's property 1.84×10^7 cubic feet	Estimate of size of reserve on client's property: 2.68×10^6 cubic feet

How, if at all, would you use these estimates in the case? Explain.

MENTAL MATH

State each expression using positive exponents only.

1. $5x^{-2}$

2. $3x^{-3}$

3. $\dfrac{1}{y^{-6}}$

4. $\dfrac{1}{x^{-3}}$

5. $\dfrac{4}{y^{-3}}$

6. $\dfrac{16}{y^{-7}}$

EXERCISE SET 5.5

STUDY GUIDE/SSM   CD/ VIDEO   PH MATH TUTOR CENTER   MathXL®Tutorials ON CD   MathXL®   MyMathLab®

Simplify each expression. Write each result using positive exponents only. See Examples 1 through 3.

1. 4^{-3}

2. 6^{-2}

3. $(-2)^{-4}$

4. $(-3)^{-5}$

5. $7x^{-3}$

6. $(7x)^{-3}$

7. $\left(\dfrac{1}{2}\right)^{-5}$

8. $\left(\dfrac{1}{8}\right)^{-2}$

9. $\left(-\dfrac{1}{4}\right)^{-3}$

10. $\left(-\dfrac{1}{8}\right)^{-2}$

11. $3^{-1} + 2^{-1}$

12. $4^{-1} + 4^{-2}$

13. $\dfrac{1}{p^{-3}}$

14. $\dfrac{1}{q^{-5}}$

15. $\dfrac{p^{-5}}{q^{-4}}$

16. $\dfrac{r^{-5}}{s^{-2}}$

17. $\dfrac{x^{-2}}{x}$

18. $\dfrac{y}{y^{-3}}$

19. $2^0 + 3^{-1}$ **20.** $4^{-2} - 4^{-3}$ **21.** $\dfrac{-1}{p^{-4}}$

22. $\dfrac{-1}{y^{-6}}$ **23.** $-2^0 - 3^0$ **24.** $5^0 + (-5)^0$

Simplify each expression. Write each result using positive exponents only. See Example 4.

25. $\dfrac{x^2 x^5}{x^3}$ **26.** $\dfrac{y^4 y^5}{y^6}$ **27.** $\dfrac{p^2 p}{p^{-1}}$

28. $\dfrac{y^3 y}{y^{-2}}$ **29.** $\dfrac{(m^5)^4 m}{m^{10}}$ **30.** $\dfrac{(x^2)^8 x}{x^9}$

31. $\dfrac{r}{r^{-3} r^{-2}}$ **32.** $\dfrac{p}{p^{-3} q^{-5}}$ **33.** $(x^5 y^3)^{-3}$

34. $(z^5 x^5)^{-3}$ **35.** $\dfrac{(x^2)^3}{x^{10}}$ **36.** $\dfrac{(y^4)^2}{y^{12}}$

37. $\dfrac{(a^5)^2}{(a^3)^4}$ **38.** $\dfrac{(x^2)^5}{(x^4)^3}$ **39.** $\dfrac{8k^4}{2k}$

40. $\dfrac{27r^4}{3r^6}$ **41.** $\dfrac{-6m^4}{-2m^3}$ **42.** $\dfrac{15a^4}{-15a^5}$

43. $\dfrac{-24a^6 b}{6ab^2}$ **44.** $\dfrac{-5x^4 y^5}{15x^4 y^2}$

45. $(-2x^3 y^{-4})(3x^{-1} y)$

46. $(-5a^4 b^{-7})(-a^{-4} b^3)$

47. $(a^{-5} b^2)^{-6}$ **48.** $(4^{-1} x^5)^{-2}$

49. $\left(\dfrac{x^{-2} y^4}{x^3 y^7}\right)^2$ **50.** $\left(\dfrac{a^5 b}{a^7 b^{-2}}\right)^{-3}$

51. $\dfrac{4^2 z^{-3}}{4^3 z^{-5}}$ **52.** $\dfrac{3^{-1} x^4}{3^3 x^{-7}}$

53. $\dfrac{2^{-3} x^{-4}}{2^2 x}$ **54.** $\dfrac{5^{-1} z^7}{5^{-2} z^9}$

55. $\dfrac{7ab^{-4}}{7^{-1} a^{-3} b^2}$ **56.** $\dfrac{6^{-5} x^{-1} y^2}{6^{-2} x^{-4} y^4}$

57. $\left(\dfrac{a^{-5} b}{ab^3}\right)^{-4}$ **58.** $\left(\dfrac{r^{-2} s^{-3}}{r^{-4} s^{-3}}\right)^{-3}$

59. $\dfrac{(xy^3)^5}{(xy)^{-4}}$ **60.** $\dfrac{(rs)^{-3}}{(r^2 s^3)^2}$

61. $\dfrac{(-2xy^{-3})^{-3}}{(xy^{-1})^{-1}}$ **62.** $\dfrac{(-3x^2 y^2)^{-2}}{(xyz)^{-2}}$

63. $\dfrac{6x^2 y^3}{-7xy^5}$ **64.** $\dfrac{-8xa^2 b}{-5xa^5 b}$

Write each number in scientific notation. See Example 5.

65. 78,000   **66.** 9,300,000,000

67. 0.00000167   **68.** 0.00000017

69. 0.00635   **70.** 0.00194

71. 1,160,000   **72.** 700,000

73. The temperature at the interior of the Earth is 20,000,000 degrees Celsius. Write 20,000,000 in scientific notation.

74. The half-life of a carbon isotope is 5000 years. Write 5000 in scientific notation.

75. The temperature of the Sun at its core is 15,600,000 degrees Kelvin. Write 15,600,000 in scientific notation. (*Source:* Students for the Exploration and Development of Space)

76. Google.com is an Internet search engine that allows users to search over 1,346,966,000 Web pages. Write 1,346,966,000 in scientific notation. (*Source:* Google, Inc.)

77. At this writing, the world's largest optical telescopes are the twin Keck Telescopes located near the summit of Mauna Kea in Hawaii. The elevation of the Keck Telescopes is about 13,600 feet above sea level. Write 13,600 in scientific notation. (*Source:* W.M. Keck Observatory)

78. More than 2,000,000,000 pencils are manufactured in the United States annually. Write this number in scientific notation. (*Source:* AbsoluteTrivia.com)

79. In 2003, the population of the United States was roughly 292,000,000. Write 292,000,000 in scientific notation. (*Source:* U.S. Census Bureau)

80. Pioneer 10 became the first spacecraft to leave the solar system eleven years after it was launched in 1972. When it was contacted in April 2001 to verify that it could still transmit a radio signal, Pioneer 10 was 7,290,000,000 miles from Earth. Write 7,290,000,000 in scientific notation. (*Source:* NASA Ames Research Center)

Write each number in standard notation. See Example 6.

81. 8.673×10^{-10}

82. 9.056×10^{-4}

83. 3.3×10^{-2}

84. 4.8×10^{-6}

85. 2.032×10^{4}

86. 9.07×10^{10}

87. One coulomb of electricity is 6.25×10^{18} electrons. Write this number in standard notation.

88. The mass of a hydrogen atom is 1.7×10^{-24} grams. Write this number in standard notation.

89. The distance light travels in 1 year is 9.460×10^{12} kilometers. Write this number in standard notation.

90. The population of the world is 6.067×10^{9}. Write this number in standard notation. (*Source:* U.S. Bureau of the Census)

Evaluate each expression using exponential rules. Write each result in standard notation. See Example 7.

91. $(1.2 \times 10^{-3})(3 \times 10^{-2})$

92. $(2.5 \times 10^{6})(2 \times 10^{-6})$

93. $(4 \times 10^{-10})(7 \times 10^{-9})$

94. $(5 \times 10^{6})(4 \times 10^{-8})$

95. $\dfrac{8 \times 10^{-1}}{16 \times 10^{5}}$

96. $\dfrac{25 \times 10^{-4}}{5 \times 10^{-9}}$

97. $\dfrac{1.4 \times 10^{-2}}{7 \times 10^{-8}}$

98. $\dfrac{0.4 \times 10^{5}}{0.2 \times 10^{11}}$

REVIEW AND PREVIEW

Simplify the following. See Section 5.1.

99. $\dfrac{5x^{7}}{3x^{4}}$

100. $\dfrac{27y^{14}}{3y^{7}}$

101. $\dfrac{15z^{4}y^{3}}{21zy}$

102. $\dfrac{18a^{7}b^{17}}{30a^{7}b}$

Use the distributive property and multiply. See Sections 5.3 and 5.5.

103. $\dfrac{1}{y}(5y^{2} - 6y + 5)$

104. $\dfrac{2}{x}(3x^{5} + x^{4} - 2)$

△ **105.** Find the volume of the cube.

$\dfrac{3x^{-2}}{z}$ inches

△ **106.** Find the area of the triangle.

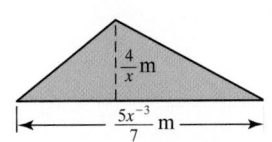

$\dfrac{4}{x}$ m

$\dfrac{5x^{-3}}{7}$ m

Concept Extensions

Simplify each expression. Write each result in standard notation.

107. $(2.63 \times 10^{12})(-1.5 \times 10^{-10})$

108. $(6.785 \times 10^{-4})(4.68 \times 10^{10})$

Light travels at a rate of 1.86×10^{5} miles per second. Use this information and the distance formula $d = r \cdot t$ to answer Exercises 109 and 110.

109. If the distance from the moon to the Earth is 238,857 miles, find how long it takes the reflected light of the moon to reach the Earth. (Round to the nearest tenth of a second.)

110. If the distance from the sun to the Earth is 93,000,000 miles, find how long it takes the light of the sun to reach the Earth. (Round to the nearest tenth of a second.)

111. It was stated earlier that for an integer n,

$$x^{-n} = \frac{1}{x^{n}}, \quad x \neq 0$$

Explain why x may not equal 0.

112. Determine whether each statement is true or false.

a. $5^{-1} < 5^{-2}$

b. $\left(\dfrac{1}{5}\right)^{-1} < \left(\dfrac{1}{5}\right)^{-2}$

c. $a^{-1} < a^{-2}$ for all nonzero numbers.

113. Although the actual amount varies by season and time of day, the average volume of water that flows over Niagara Falls (the American and Canadian falls combined) each second is 7.5×10^{5} gallons. How much water flows over Niagara Falls in an hour? Write the result in scientific notation. (*Hint:* 1 hour equals 3600 seconds) (*Source:* niagarafallslive.com)

114. A beam of light travels 9.460×10^{12} kilometers per year. How far does light travel in 10,000 years? Write the result in scientific notation.

115. Explain why $(a^{-1})^{3}$ has the same value as $(a^{3})^{-1}$.

116. If $a = \dfrac{1}{10}$, then find the value of a^{-2}.

Simplify each expression. Assume that variables represent positive integers.

117. $a^{-4m} \cdot a^{5m}$

118. $(x^{-3s})^{3}$

119. $\dfrac{y^{4a}}{y^{-a}}$

120. $\dfrac{y^{-6a}}{zy^{6a}}$

121. $(z^{3a+2})^{-2}$

122. $(a^{4x-1})^{-1}$

5.6 DIVISION OF POLYNOMIALS

Objectives

1 Divide a polynomial by a monomial.

2 Use long division to divide a polynomial by another polynomial.

1 Now that we know how to add, subtract, and multiply polynomials, we practice dividing polynomials.

To divide a polynomial by a monomial, recall addition of fractions. Fractions that have a common denominator are added by adding the numerators:

$$\frac{a}{c} + \frac{b}{c} = \frac{a + b}{c}$$

If we read this equation from right to left and let a, b, and c be monomials, $c \neq 0$, we have the following:

Dividing a Polynomial by a Monomial

Divide each term of the polynomial by the monomial.

$$\frac{a + b}{c} = \frac{a}{c} + \frac{b}{c}, \quad c \neq 0$$

Throughout this section, we assume that denominators are not 0.

EXAMPLE 1

Divide $6m^2 + 2m$ by $2m$.

Solution We begin by writing the quotient in fraction form. Then we divide each term of the polynomial $6m^2 + 2m$ by the monomial $2m$.

$$\frac{6m^2 + 2m}{2m} = \frac{6m^2}{2m} + \frac{2m}{2m}$$
$$= 3m + 1 \qquad \text{Simplify.}$$

Check We know that if $\dfrac{6m^2 + 2m}{2m} = 3m + 1$, then $2m \cdot (3m + 1)$ must equal $6m^2 + 2m$. Thus, to check, we multiply.

$$2m(3m + 1) = 2m(3m) + 2m(1) = 6m^2 + 2m$$

The quotient $3m + 1$ checks.

EXAMPLE 2

Divide $\dfrac{9x^5 - 12x^2 + 3x}{3x^2}$.

Solution

$$\frac{9x^5 - 12x^2 + 3x}{3x^2} = \frac{9x^5}{3x^2} - \frac{12x^2}{3x^2} + \frac{3x}{3x^2} \qquad \text{Divide each term by } 3x^2.$$
$$= 3x^3 - 4 + \frac{1}{x} \qquad \text{Simplify.}$$

Notice that the quotient is not a polynomial because of the term $\frac{1}{x}$. This expression is called a rational expression—we will study rational expressions further in Chapter 10. Although the quotient of two polynomials is not always a polynomial, we may still check by multiplying.

Check

$$3x^2\left(3x^3 - 4 + \frac{1}{x}\right) = 3x^2(3x^3) - 3x^2(4) + 3x^2\left(\frac{1}{x}\right)$$

$$= 9x^5 - 12x^2 + 3x$$

EXAMPLE 3

Divide $\dfrac{8x^2y^2 - 16xy + 2x}{4xy}$.

Solution

$$\frac{8x^2y^2 - 16xy + 2x}{4xy} = \frac{8x^2y^2}{4xy} - \frac{16xy}{4xy} + \frac{2x}{4xy} \qquad \text{Divide each term by } 4xy.$$

$$= 2xy - 4 + \frac{1}{2y} \qquad \text{Simplify.}$$

Check

$$4xy\left(2xy - 4 + \frac{1}{2y}\right) = 4xy(2xy) - 4xy(4) + 4xy\left(\frac{1}{2y}\right)$$

$$= 8x^2y^2 - 16xy + 2x$$

✔ **CONCEPT CHECK**

In which of the following is $\dfrac{x+5}{5}$ simplified correctly?

a. $\dfrac{x}{5} + 1$ **b.** x **c.** $x + 1$

2   To divide a polynomial by a polynomial other than a monomial, we use a process known as long division. Polynomial long division is similar to number long division, so we review long division by dividing 13 into 3660.

> **Helpful Hint**
> Recall that 3660 is called the dividend.

$$
\begin{array}{r}
281 \\
13\overline{)3660} \\
\end{array}
$$

$\underline{26}{\downarrow}$	$2 \cdot 13 = 26$
106	Subtract and bring down the next digit in the dividend.
$\underline{104}{\downarrow}$	$8 \cdot 13 = 104$
20	Subtract and bring down the next digit in the dividend.
$\underline{13}$	$1 \cdot 13 = 13$
7	Subtract. There are no more digits to bring down, so the remainder is **7**.

The quotient is 281 R 7, which can be written as $281\dfrac{7}{13}$ $\begin{array}{l}\leftarrow \text{remainder} \\ \leftarrow \text{divisor}\end{array}$

Recall that division can be checked by multiplication. To check a division problem such as this one, we see that

$$13 \cdot 281 + 7 = 3660$$

Now we demonstrate long division of polynomials.

EXAMPLE 4

Divide $x^2 + 7x + 12$ by $x + 3$ using long division.

Solution To subtract, change the signs of these terms and add.

$$
\begin{array}{r}
x \\
x + 3 \overline{)\, x^2 + 7x + 12} \\
\underline{x^2 + 3x} \downarrow \\
4x + 12
\end{array}
$$

How many times does x divide x^2? $\dfrac{x^2}{x} = x$.

Multiply: $x(x + 3)$.

Subtract and bring down the next term.

Now we repeat this process.

$$
\begin{array}{r}
x + 4 \\
x + 3 \overline{)\, x^2 + 7x + 12} \\
\underline{x^2 + 3x} \\
4x + 12 \\
\underline{4x + 12} \\
0
\end{array}
$$

How many times does x divide $4x$? $\dfrac{4x}{x} = 4$.

To subtract, change the signs of these terms and add.

Multiply: $4(x + 3)$.

Subtract. The remainder is 0.

The quotient is $x + 4$.

Check We check by multiplying.

divisor	·	quotient	+	remainder	=	dividend
↓		↓		↓		↓

or $(x + 3)$ · $(x + 4)$ + 0 = $x^2 + 7x + 12$

The quotient checks.

EXAMPLE 5

Divide $6x^2 + 10x - 5$ by $3x - 1$ using long division.

Solution

$$
\begin{array}{r}
2x + 4 \\
3x - 1 \overline{)\, 6x^2 + 10x - 5} \\
\underline{6x^2 - 2x} \downarrow \\
12x - 5 \\
\underline{12x - 4} \\
-1
\end{array}
$$

$\dfrac{6x^2}{3x} = 2x$, so $2x$ is a term of the quotient.

Multiply $2x(3x - 1)$.

Subtract and bring down the next term.

$\dfrac{12x}{3x} = 4$, $4(3x - 1)$

Subtract. The remainder is -1.

Thus $(6x^2 + 10x - 5)$ divided by $(3x - 1)$ is $(2x + 4)$ with a remainder of -1. This can be written as

$$
\frac{6x^2 + 10x - 5}{3x - 1} = 2x + 4 + \frac{-1}{3x - 1} \quad \begin{array}{l} \leftarrow \text{remainder} \\ \leftarrow \text{divisor} \end{array}
$$

Check To check, we multiply $(3x - 1)(2x + 4)$. Then we add the remainder, -1, to this product.

$$
(3x - 1)(2x + 4) + (-1) = (6x^2 + 12x - 2x - 4) - 1
$$
$$
= 6x^2 + 10x - 5
$$

The quotient checks.

In Example 5, the degree of the divisor, $3x - 1$, is 1 and the degree of the remainder, -1, is 0. The division process is continued until the degree of the remainder polynomial is less than the degree of the divisor polynomial.

EXAMPLE 6

Divide $\dfrac{4x^2 + 7 + 8x^3}{2x + 3}$.

Solution   Before we begin the division process, we rewrite $4x^2 + 7 + 8x^3$ as $8x^3 + 4x^2 + 0x + 7$. Notice that we have written the polynomial in descending order and have represented the missing x term by $0x$.

$$
\begin{array}{r}
4x^2 - 4x + 6 \\
2x+3\overline{)8x^3 + 4x^2 + 0x + 7} \\
\underline{8x^3 + 12x^2} \\
-8x^2 + 0x \\
\underline{+8x^2 + 12x} \\
12x + 7 \\
\underline{12x + 18} \\
-11 \quad \text{remainder}
\end{array}
$$

Thus, $\dfrac{4x^2 + 7 + 8x^3}{2x + 3} = 4x^2 - 4x + 6 + \dfrac{-11}{2x + 3}$.

EXAMPLE 7

Divide $\dfrac{2x^4 - x^3 + 3x^2 + x - 1}{x^2 + 1}$.

Solution   Before dividing, rewrite the divisor polynomial $(x^2 + 1)$ as $(x^2 + 0x + 1)$. The $0x$ term represents the missing x^1 term in the divisor.

$$
\begin{array}{r}
2x^2 - x + 1 \\
x^2+0x+1\overline{)2x^4 - x^3 + 3x^2 + x - 1} \\
\underline{2x^4 + 0x^3 + 2x^2} \\
-x^3 + x^2 + x \\
\underline{+x^3 + 0x^2 + x} \\
x^2 + 2x - 1 \\
\underline{x^2 + 0x + 1} \\
2x - 2 \quad \text{remainder}
\end{array}
$$

Thus, $\dfrac{2x^4 - x^3 + 3x^2 + x - 1}{x^2 + 1} = 2x^2 - x + 1 + \dfrac{2x - 2}{x^2 + 1}$.

MENTAL MATH

Simplify each expression mentally.

1. $\dfrac{a^6}{a^4}$ 2. $\dfrac{y^2}{y}$ 3. $\dfrac{a^3}{a}$ 4. $\dfrac{p^8}{p^3}$ 5. $\dfrac{k^5}{k^2}$

6. $\dfrac{k^7}{k^5}$ 7. $\dfrac{p^8}{p^3}$ 8. $\dfrac{k^5}{k^2}$ 9. $\dfrac{k^7}{k^5}$

EXERCISE SET 5.6

STUDY GUIDE/SSM   CD/VIDEO   PH MATH TUTOR CENTER   MathXL®Tutorials ON CD   MathXL®   MyMathLab®

Perform each division. See Examples 1 through 3.

1. $\dfrac{15p^3 + 18p^2}{3p}$

2. $\dfrac{14m^2 - 27m^3}{7m}$

3. $\dfrac{-9x^4 + 18x^5}{6x^5}$

4. $\dfrac{6x^5 + 3x^4}{3x^4}$

 5. $\dfrac{-9x^5 + 3x^4 - 12}{3x^3}$

6. $\dfrac{6a^2 - 4a + 12}{2a^2}$

7. $\dfrac{4x^4 - 6x^3 + 7}{-4x^4}$

8. $\dfrac{-12a^3 + 36a - 15}{3a}$

9. $\dfrac{25x^5 - 15x^3 + 5}{5x^2}$

10. $\dfrac{-4y^2 + 4y + 6}{2y}$

△ **11.** The perimeter of a square is $(12x^3 + 4x - 16)$ feet. Find the length of its side.

Perimeter is
$(12x^3 + 4x - 16)$
feet

△ **12.** The volume of the swimming pool shown is $(36x^5 - 12x^3 + 6x^2)$ cubic feet. If its height is $2x$ feet and its width is $3x$ feet, find its length.

3x feet

2x feet

Perform each division. See Examples 4 through 7.

 13. $\dfrac{x^2 + 4x + 3}{x + 3}$

14. $\dfrac{x^2 + 7x + 10}{x + 5}$

15. $\dfrac{2x^2 + 13x + 15}{x + 5}$

16. $\dfrac{3x^2 + 8x + 4}{x + 2}$

17. $\dfrac{2x^2 - 7x + 3}{x - 4}$

18. $\dfrac{3x^2 - x - 4}{x - 1}$

19. $\dfrac{8x^2 + 6x - 27}{2x - 3}$

20. $\dfrac{18w^2 + 18w - 8}{3w + 4}$

21. $\dfrac{9a^3 - 3a^2 - 3a + 4}{3a + 2}$

22. $\dfrac{-x^3 - 6x^2 + 2x - 3}{x - 1}$

 23. $\dfrac{2b^3 + 9b^2 + 6b - 4}{b + 4}$

24. $\dfrac{2x^3 + 3x^2 - 3x + 4}{x + 2}$

25. Explain how to check a polynomial long division result when the remainder is 0.

26. Explain how to check a polynomial long division result when the remainder is not 0.

△ **27.** The area of the following parallelogram is $(10x^2 + 31x + 15)$ square meters. If its base is $(5x + 3)$ meters, find its height.

?

$(5x + 3)$ meters

△ **28.** The area of the top of the Ping-Pong table is $(49x^2 + 70x - 200)$ square inches. If its length is $(7x + 20)$ inches, find its width.

?

$(7x + 20)$ inches

MIXED PRACTICE

Perform each division.

29. $\dfrac{20x^2 + 5x + 9}{5x^3}$

30. $\dfrac{8x^3 - 4x^2 + 6x + 2}{2x^2}$

31. $\dfrac{5x^2 + 28x - 10}{x + 6}$

32. $\dfrac{2x^2 + x - 15}{x + 3}$

33. $\dfrac{10x^3 - 24x^2 - 10x}{10x}$

34. $\dfrac{2x^3 + 12x^2 + 16}{4x^2}$

35. $\dfrac{6x^2 + 17x - 4}{x + 3}$

36. $\dfrac{2x^2 - 9x + 15}{x - 6}$

37. $\dfrac{12x^4 + 3x^2}{3x^2}$

38. $\dfrac{15x^2 - 9x^5}{9x^5}$

39. $\dfrac{2x^3 + 2x^2 - 17x + 8}{x - 2}$

40. $\dfrac{4x^3 + 11x^2 - 8x - 10}{x + 3}$

41. $\dfrac{30x^2 - 17x + 2}{5x - 2}$

42. $\dfrac{4x^2 - 13x - 12}{4x + 3}$

43. $\dfrac{3x^4 - 9x^3 + 12}{-3x}$

44. $\dfrac{8y^6 - 3y^2 - 4y}{4y}$

45. $\dfrac{8x^2 + 10x + 1}{2x + 1}$

46. $\dfrac{3x^2 + 17x + 7}{3x + 2}$

47. $\dfrac{4x^2 - 81}{2x - 9}$

48. $\dfrac{16x^2 - 36}{4x + 6}$

49. $\dfrac{4x^3 + 12x^2 + x - 12}{2x + 3}$

50. $\dfrac{6x^2 + 11x - 10}{3x - 2}$

51. $\dfrac{x^3 - 27}{x - 3}$

52. $\dfrac{x^3 + 64}{x + 4}$

53. $\dfrac{x^3 + 1}{x + 1}$

54. $\dfrac{x^5 + x^2}{x^2 + x}$

55. $\dfrac{1 - 3x^2}{x + 2}$

56. $\dfrac{7 - 5x^2}{x + 3}$

57. $\dfrac{-4b + 4b^2 - 5}{2b - 1}$

58. $\dfrac{-3y + 2y^2 - 15}{2y + 5}$

REVIEW AND PREVIEW

Multiply each expression. See Section 5.3.

59. $2a(a^2 + 1)$

60. $-4a(3a^2 - 4)$

61. $2x(x^2 + 7x - 5)$

62. $4y(y^2 - 8y - 4)$

63. $-3xy(xy^2 + 7x^2y + 8)$

64. $-9xy(4xyz + 7xy^2z + 2)$

65. $9ab(ab^2c + 4bc - 8)$

66. $-7sr(6s^2r + 9sr^2 + 9rs + 8)$

Use the bar graph below to answer Exercises 67 through 70. See Section 3.1.

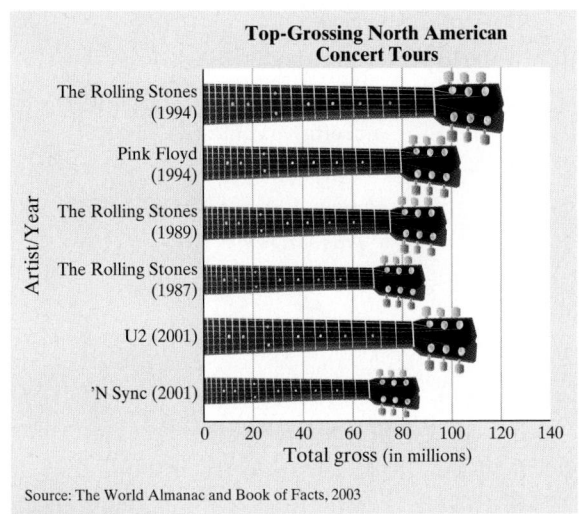

Top-Grossing North American Concert Tours

Source: The World Almanac and Book of Facts, 2003

67. Which artist has grossed the most money on one tour?

68. Estimate the amount of money made by the 1994 concert tour of Pink Floyd.

69. Estimate the amount of money made by the 2001 concert tour of U2.

70. Which artist shown has grossed the least amount of money on a tour?

Concept Extensions

Divide.

71. $\dfrac{x^7 + x^4}{x^4 + x^3}$

72. $\dfrac{x^6 - x^4}{x^3 + 1}$

73. $(18x^{10a} - 12x^{8a} + 14x^{5a} - 2x^{3a}) \div 2x^{3a}$

74. $(25y^{11b} + 5y^{6b} - 20y^{3b} + 100y^b) \div 5y^b$

5.7   SYNTHETIC DIVISION AND THE REMAINDER THEOREM

Objectives

1   Use synthetic division to divide a polynomial by a binomial.

2   Use the remainder theorem to evaluate polynomials.

1   When a polynomial is to be divided by a binomial of the form $x - c$, a shortcut process called **synthetic division** may be used. On the left is an example of long division, and on the right, the same example showing the coefficients of the variables only.

$$\begin{array}{r} 2x^2 + 5x + 2 \\ x - 3)\overline{2x^3 - x^2 - 13x + 1} \\ \underline{2x^3 - 6x^2} \\ 5x^2 - 13x \\ \underline{5x^2 - 15x} \\ 2x + 1 \\ \underline{2x - 6} \\ 7 \end{array} \qquad \begin{array}{r} 2 \quad 5 \quad 2 \\ 1 - 3)\overline{2 - 1 - 13 + 1} \\ \underline{2 - 6} \\ 5 - 13 \\ \underline{5 - 15} \\ 2 + 1 \\ \underline{2 - 6} \\ 7 \end{array}$$

Notice that as long as we keep coefficients of powers of x in the same column, we can perform division of polynomials by performing algebraic operations on the coefficients only. This shortcut process of dividing with coefficients only in a special format is called synthetic division. To find $(2x^3 - x^2 - 13x + 1) \div (x - 3)$ by synthetic division, follow the next example.

EXAMPLE 1

Use synthetic division to divide $2x^3 - x^2 - 13x + 1$ by $x - 3$.

Solution To use synthetic division, the divisor must be in the form $x - c$. Since we are dividing by $x - 3$, c is 3. Write down 3 and the coefficients of the dividend.

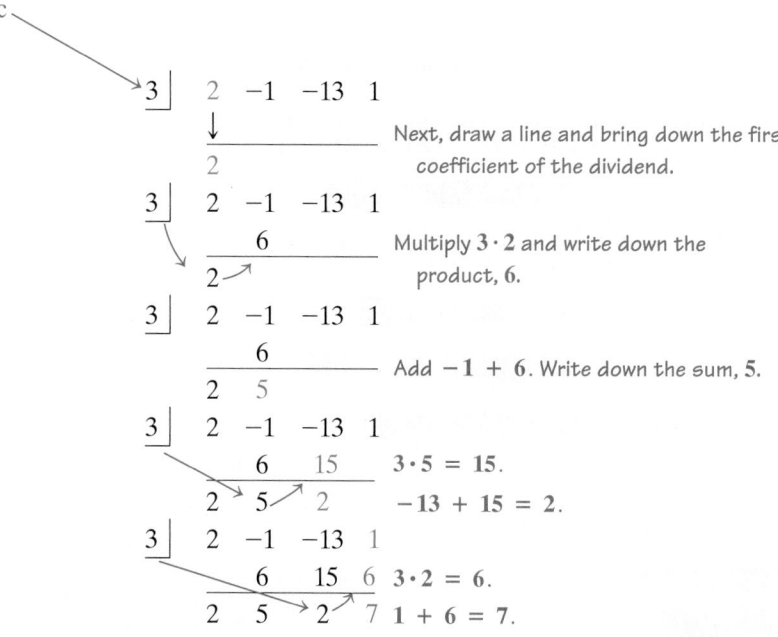

The quotient is found in the bottom row. The numbers 2, 5, and 2 are the coefficients of the quotient polynomial, and the number 7 is the remainder. The degree of the quotient polynomial is one less than the degree of the dividend. In our example, the degree of the dividend is 3, so the degree of the quotient polynomial is 2. As we found when we performed the long division, the quotient is

$$2x^2 + 5x + 2, \quad \text{remainder } 7$$

or

$$2x^2 + 5x + 2 + \frac{7}{x - 3}.$$

EXAMPLE 2

Use synthetic division to divide $x^4 - 2x^3 - 11x^2 + 5x + 34$ by $x + 2$.

Solution   The divisor is $x + 2$, which we write in the form $x - c$ as $x - (-2)$. Thus, c is -2. The dividend coefficients are $1, -2, -11, 5,$ and 34.

$$
c \longrightarrow \begin{array}{r|rrrrr}
-2 & 1 & -2 & -11 & 5 & 34 \\
   &   & -2 & 8   & 6 & -22 \\
\hline
   & 1 & -4 & -3 & 11 & 12
\end{array}
$$

The dividend is a fourth-degree polynomial, so the quotient polynomial is a third-degree polynomial. The quotient is $x^3 - 4x^2 - 3x + 11$ with a remainder of 12. Thus,

$$
\frac{x^4 - 2x^3 - 11x^2 + 5x + 34}{x + 2} = x^3 - 4x^2 - 3x + 11 + \frac{12}{x + 2}.
$$

✔ **CONCEPT CHECK**

Which division problems are candidates for the synthetic division process?

  a. $(3x^2 + 5) \div (x + 4)$

  b. $(x^3 - x^2 + 2) \div (3x^3 - 2)$

  c. $(y^4 + y - 3) \div (x^2 + 1)$

  d. $x^5 \div (x - 5)$

> **Helpful Hint**
> Before dividing by synthetic division, write the dividend in descending order of variable exponents. Any "missing powers" of the variable should be represented by 0 times the variable raised to the missing power.

EXAMPLE 3

If $P(x) = 2x^3 - 4x^2 + 5$

  a. Find $P(2)$ by substitution.
  b. Use synthetic division to find the remainder when $P(x)$ is divided by $x - 2$.

Solution   **a.** $P(x) = 2x^3 - 4x^2 + 5$

  $P(2) = 2(2)^3 - 4(2)^2 + 5$

  $\qquad = 2(8) - 4(4) + 5 = 16 - 16 + 5 = 5$

  Thus, $P(2) = 5$.

  b. The coefficients of $P(x)$ are $2, -4, 0,$ and 5. The number 0 is a coefficient of the missing power of x^1. The divisor is $x - 2$, so c is 2.

$$
c \longrightarrow \begin{array}{r|rrrr}
2 & 2 & -4 & 0 & 5 \\
  &   & 4 & 0 & 0 \\
\hline
  & 2 & 0 & 0 & 5 \quad \text{remainder}
\end{array}
$$

The remainder when $P(x)$ is divided by $x - 2$ is 5.

2   Notice in the preceding example that $P(2) = 5$ and that the remainder when $P(x)$ is divided by $x - 2$ is 5. This is no accident. This illustrates the **remainder theorem.**

> ### Remainder Theorem
>
> If a polynomial $P(x)$ is divided by $x - c$, then the remainder is $P(c)$.

EXAMPLE 4

Use the remainder theorem and synthetic division to find $P(4)$ if

$$P(x) = 4x^6 - 25x^5 + 35x^4 + 17x^2.$$

Solution   To find $P(4)$ by the remainder theorem, we divide $P(x)$ by $x - 4$. The coefficients of $P(x)$ are $4, -25, 35, 0, 17, 0,$ and 0. Also, c is 4.

$$
\begin{array}{r|rrrrrrr}
 c \longrightarrow & & & & & & & \\
 4 & 4 & -25 & 35 & 0 & 17 & 0 & 0 \\
 & & 16 & -36 & -4 & -16 & 4 & 16 \\
 \hline
 & 4 & -9 & -1 & -4 & 1 & 4 & 16 \quad \text{remainder}
\end{array}
$$

Thus, $P(4) = 16$, the remainder.

EXERCISE SET 5.7

    Math XL

STUDY   CD/   PH MATH   MathXL®Tutorials   MathXL®   MyMathLab®
GUIDE/SSM   VIDEO   TUTOR CENTER   ON CD

Use synthetic division to divide. See Examples 1 and 2.

1. $(x^2 + 3x - 40) \div (x - 5)$

2. $(x^2 - 14x + 24) \div (x - 2)$

3. $(x^2 + 5x - 6) \div (x + 6)$

4. $(x^2 + 12x + 32) \div (x + 4)$

5. $(x^3 - 7x^2 - 13x + 5) \div (x - 2)$

6. $(x^3 + 6x^2 + 4x - 7) \div (x + 5)$

7. $(4x^2 - 9) \div (x - 2)$ **8.** $(3x^2 - 4) \div (x - 1)$

*For the given polynomial $P(x)$ and the given c, find $P(c)$ by **(a)** direct substitution and **(b)** the remainder theorem. See Examples 3 and 4.*

9. $P(x) = 3x^2 - 4x - 1; P(2)$

10. $P(x) = x^2 - x + 3; P(5)$

11. $P(x) = 4x^4 + 7x^2 + 9x - 1; P(-2)$

12. $P(x) = 8x^5 + 7x + 4; P(-3)$

13. $P(x) = x^5 + 3x^4 + 3x - 7; P(-1)$

14. $P(x) = 5x^4 - 4x^3 + 2x - 1; P(-1)$

MIXED PRACTICE

Use synthetic division to divide.

15. $(x^3 - 3x^2 + 2) \div (x - 3)$

16. $(x^2 + 12) \div (x + 2)$

17. $(6x^2 + 13x + 8) \div (x + 1)$

18. $(x^3 - 5x^2 + 7x - 4) \div (x - 3)$

19. $(2x^4 - 13x^3 + 16x^2 - 9x + 20) \div (x - 5)$

20. $(3x^4 + 5x^3 - x^2 + x - 2) \div (x + 2)$

21. $(3x^2 - 15) \div (x + 3)$

22. $(3x^2 + 7x - 6) \div (x + 4)$

23. $(3x^3 - 6x^2 + 4x + 5) \div \left(x - \dfrac{1}{2}\right)$

24. $(8x^3 - 6x^2 - 5x + 3) \div \left(x + \dfrac{3}{4}\right)$

25. $(3x^3 + 2x^2 - 4x + 1) \div \left(x - \dfrac{1}{3}\right)$

26. $(9y^3 + 9y^2 - y + 2) \div \left(y + \dfrac{2}{3}\right)$

27. $(7x^2 - 4x + 12 + 3x^3) \div (x + 1)$

28. $(x^4 + 4x^3 - x^2 - 16x - 4) \div (x - 2)$

29. $(x^3 - 1) \div (x - 1)$

30. $(y^3 - 8) \div (y - 2)$

31. $(x^2 - 36) \div (x + 6)$

32. $(4x^3 + 12x^2 + x - 12) \div (x + 3)$

For the given polynomial $P(x)$ and the given c, use the remainder theorem to find $P(c)$.

33. $P(x) = x^3 + 3x^2 - 7x + 4; 1$

34. $P(x) = x^3 + 5x^2 - 4x - 6; 2$

35. $P(x) = 3x^3 - 7x^2 - 2x + 5; -3$

36. $P(x) = 4x^3 + 5x^2 - 6x - 4; -2$

37. $P(x) = 4x^4 + x^2 - 2; -1$

38. $P(x) = x^4 - 3x^2 - 2x + 5; -2$

39. $P(x) = 2x^4 - 3x^2 - 2; \dfrac{1}{3}$

40. $P(x) = 4x^4 - 2x^3 + x^2 - x - 4; \dfrac{1}{2}$

41. $P(x) = x^5 + x^4 - x^3 + 3; \dfrac{1}{2}$

42. $P(x) = x^5 - 2x^3 + 4x^2 - 5x + 6; \dfrac{2}{3}$

43. Explain an advantage of using the remainder theorem instead of direct substitution.

44. Explain an advantage of using synthetic division instead of long division.

REVIEW AND PREVIEW

Solve each equation for x. See Sections 2.3.

45. $7x + 2 = x - 3$

46. $4 - 2x = 17 - 5x$

47. $\dfrac{x}{3} - 5 = 13$

48. $\dfrac{2x}{9} + 1 = \dfrac{7}{9}$

Evaluate. See Section 1.3.

49. 2^3

50. 3^4

51. $(-2)^5$

52. -2^5

53. $3 \cdot 4^2$

54. $4 \cdot 3^3$

Evaluate each expression for the given replacement value. See Section 1.7.

55. x^2 if x is -5

56. x^3 if x is -5

57. $2x^3$ if x is -1

58. $3x^2$ if x is -1

Concept Extensions

We say that 2 is a factor of 8 because 2 divides 8 evenly, or with a remainder of 0. In the same manner, the polynomial $x - 2$ is a factor of the polynomial $x^3 - 14x^2 + 24x$ because the remainder is 0 when $x^3 - 14x^2 + 24x$ is divided by $x - 2$. Use this information for Exercises 59 through 61.

59. Use synthetic division to show that $x + 3$ is a factor of $x^3 + 3x^2 + 4x + 12$.

60. Use synthetic division to show that $x - 2$ is a factor of $x^3 - 2x^2 - 3x + 6$.

61. From the remainder theorem, the polynomial $x - c$ is a factor of a polynomial function $P(x)$ if $P(c)$ is what value?

62. If a polynomial is divided by $x - 5$, the quotient is $2x^2 + 5x - 6$ and the remainder is 3. Find the original polynomial.

63. If a polynomial is divided by $x + 3$, the quotient is $x^2 - x + 10$ and the remainder is -2. Find the original polynomial.

64. If the area of a parallelogram is $(x^4 - 23x^2 + 9x - 5)$ square centimeters and its base is $(x + 5)$ centimeters, find its height.

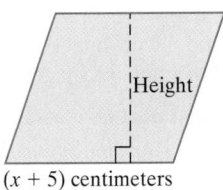

Height

$(x + 5)$ centimeters

65. If the volume of a box is $(x^4 + 6x^3 - 7x^2)$ cubic meters, its height is x^2 meters, and its length is $(x + 7)$ meters, find its width.

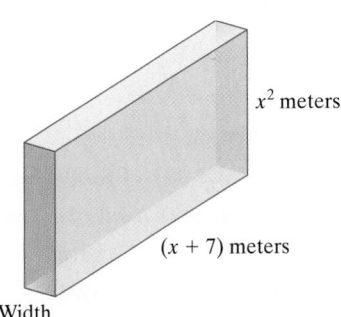

x^2 meters

$(x + 7)$ meters

Width

CHAPTER 5 PROJECT

Modeling with Polynomials

The polynomial model $1.65x^2 + 14.77x + 364.27$ dollars represents consumer spending per person per year on all U.S. media from 1990 to 1996. This includes spending on subscription TV services, recorded music, newspapers, magazines, books, home video, theater movies, video games, and educational software. The polynomial model $0.81x^2 + 3.41x + 88.98$ dollars represents consumer spending per person per year on subscription TV services alone during this same period. In both models, x is the number of years after 1990. (*Source:* Based on data from Veronis, Suhler & Associates, New York, NY, *Communications Industry Report*, annual)

In this project, you will have the opportunity to investigate these polynomial models numerically, algebraically, and graphically. This project may be completed by working in groups or individually.

1. Use the polynomials to complete the following table showing the annual consumer spending per person over the period 1990–1996 by evaluating each polynomial at the given values of x. Then subtract each value in the fourth column from the corresponding value in the third column. Record the result in the last column, "Difference." What do you think these values represent? What trends do you notice in the data?

Year	x	Consumer Spending per Person per Year on ALL U.S. Media	Consumer spending per Person per Year on Subscription TV	Difference
1990	0			
1992	2			
1994	4			
1996	6			

2. Use the polynomial models to find a new polynomial model representing the amount of consumer spending per person on U.S. media other than subscription TV services (such as recorded music, newspapers, magazines, books, home video, theater movies, video games, and educational software). Then use this new polynomial model to complete the following table.

Year	x	Consumer Spending per Person per Year on Media Other Than Subscription TV
1990	0	
1992	2	
1994	4	
1996	6	

3. Compare the values in the last column of the table in Question 1 to the values in the last column of the table in Question 2. What do you notice? What can you conclude?

4. Use the polynomial models to estimate consumer spending on (a) all U.S. media, (b) subscription TV, and (c) media other than subscription TV for the year 1998.

5. Use the polynomial models to estimate consumer spending on (a) all U.S. media, (b) subscription TV, and (c) media other than subscription TV for the year 2000.

6. Create a bar graph that represents the data for consumer spending on all U.S. media in the years 1990, 1992, 1994, and 1996 along with your estimates for 1998 and 2000. Study your bar graph. Discuss what the graph implies about the future.

CHAPTER VOCABULARY CHECK

Fill in each blank with one of the words or phrases listed below.

term	coefficient	monomial	binomial	trinomial
polynomials	degree of a term	degree of a polynomial		FOIL

1. A _____ is a number or the product of numbers and variables raised to powers.
2. The _____ method may be used when multiplying two binomials.

3. A polynomial with exactly 3 terms is called a _____.

4. The _____ is the greatest degree of any term of the polynomial.

5. A polynomial with exactly 2 terms is called a _____.

6. The _____ of a term is its numerical factor.

7. The _____ is the sum of the exponents on the variables in the term.

8. A polynomial with exactly 1 term is called a _____.

9. Monomials, binomials, and trinomials are all examples of _____.

STUDY SKILLS REMINDER

Are You Preparing for a Test on Chapter 5?

Below is a list of some *common trouble areas* for topics covered in Chapter 5. After studying for your test—but before taking your test—read these.

▶ Do you know that a negative exponent does not make the base a negative number? For example,

$$3^{-2} = \frac{1}{3^2} = \frac{1}{9}$$

▶ Make sure you remember that x has an understood coefficient of 1 and an understood exponent of 1. For example,

$$2x + x = 2x + 1x = 3x; \quad x^5 \cdot x = x^5 \cdot x^1 = x^6$$

▶ Do you know the difference between $5x^2$ and $(5x)^2$?

$$5x^2 \text{ is } 5 \cdot x^2; \quad (5x)^2 = 5^2 \cdot x^2 \text{ or } 25 \cdot x^2$$

▶ Can you evaluate $x^2 - x$ when $x = -2$?

$$x^2 - x = (-2)^2 - (-2) = 4 - (-2) = 4 + 2 = 6$$

▶ Can you subtract $5x^2 + 1$ from $3x^2 - 6$?

$$(3x^2 - 6) - (5x^2 + 1) = 3x^2 - 6 - 5x^2 - 1 = -2x^2 - 7$$

▶ Make sure you are familiar with squaring a binomial and other special products.

$$(3x - 4)^2 = (3x)^2 - 2(3x)(4) + 4^2 = 9x^2 - 24x + 16$$

or

$$(3x - 4)^2 = (3x - 4)(3x - 4) = 9x^2 - 24x + 16$$
$$(2x^2 + 1)(2x^2 - 1) = (2x^2)^2 - 1^2 = 4x^4 - 1$$

Remember: This is simply a checklist of common trouble areas. For a review of Chapter 5, see the Highlights and Chapter Review.

CHAPTER 5 HIGHLIGHTS

Definitions and Concepts	**Examples**

Section 5.1 Exponents

a^n means the product of n factors, each of which is a.

$$3^2 = 3 \cdot 3 = 9$$
$$(-5)^3 = (-5)(-5)(-5) = -125$$
$$\left(\frac{1}{2}\right)^4 = \frac{1}{2} \cdot \frac{1}{2} \cdot \frac{1}{2} \cdot \frac{1}{2} = \frac{1}{16}$$

If m and n are integers and no denominators are 0,

Product Rule: $a^m \cdot a^n = a^{m+n}$

Power Rule: $(a^m)^n = a^{mn}$

Power of a Product Rule: $(ab)^n = a^n b^n$

Power of a Quotient Rule: $\left(\dfrac{a}{b}\right)^n = \dfrac{a^n}{b^n}$

Quotient Rule: $\dfrac{a^m}{a^n} = a^{m-n}$

Zero Exponent: $a^0 = 1, a \neq 0$.

$$x^2 \cdot x^7 = x^{2+7} = x^9$$
$$(5^3)^8 = 5^{3 \cdot 8} = 5^{24}$$
$$(7y)^4 = 7^4 y^4$$
$$\left(\frac{x}{8}\right)^3 = \frac{x^3}{8^3}$$
$$\frac{x^9}{x^4} = x^{9-4} = x^5$$
$$5^0 = 1, x^0 = 1, x \neq 0$$

Section 5.2 Polynomial Functions and Adding and Subtracting Polynomials

A **polynomial** is a finite sum of terms in which all variables have exponents raised to nonnegative integer powers and no variables appear in the denominator.

Polynomials

$$1.3x^2 \quad \text{(monomial)}$$
$$-\frac{1}{3}y + 5 \quad \text{(binomial)}$$
$$6z^2 - 5z + 7 \quad \text{(trinomial)}$$

A function P is a **polynomial function** if $P(x)$ is a polynomial.

For the polynomial function

$$P(x) = -x^2 + 6x - 12, \text{ find } P(-2)$$
$$P(-2) = -(-2)^2 + 6(-2) - 12 = -28.$$

To add polynomials, combine all like terms.

Add

$$(3y^2x - 2yx + 11) + (-5y^2x - 7)$$
$$= -2y^2x - 2yx + 4$$

To subtract polynomials, change the signs of the terms of the polynomial being subtracted, then add.

Subtract

$$(-2z^3 - z + 1) - (3z^3 + z - 6)$$
$$= -2z^3 - z + 1 - 3z^3 - z + 6$$
$$= -5z^3 - 2z + 7$$

Definitions and Concepts	**Examples**

Section 5.3 Multiplying Polynomials

| **To multiply two polynomials**, multiply each term of one polynomial by each term of the other polynomial, and then combine like terms. | Multiply:
$(2x + 1)(5x^2 - 6x + 2)$

$= 2x(5x^2 - 6x + 2) + 1(5x^2 - 6x + 2)$
$= 10x^3 - 12x^2 + 4x + 5x^2 - 6x + 2$
$= 10x^3 - 7x^2 - 2x + 2$ |

Section 5.4 Special Products

The **FOIL method** may be used when multiplying two binomials.	Multiply: $(5x - 3)(2x + 3)$ $\begin{array}{cccc} & F & O & I & L \\ (5x - 3)(2x + 3) = (5x)(2x) + (5x)(3) + (-3)(2x) + (-3)(3) \end{array}$ $= 10x^2 + 15x - 6x - 9$ $= 10x^2 + 9x - 9$
Squaring a Binomial $(a + b)^2 = a^2 + 2ab + b^2$ $(a - b)^2 = a^2 - 2ab + b^2$	Square each binomial. $(x + 5)^2 = x^2 + 2(x)(5) + 5^2$ $= x^2 + 10x + 25$ $(3x - 2y)^2 = (3x)^2 - 2(3x)(2y) + (2y)^2$ $= 9x^2 - 12xy + 4y^2$
Multiplying the Sum and Difference of Two Terms $(a + b)(a - b) = a^2 - b^2$	Multiply. $(6y + 5)(6y - 5) = (6y)^2 - 5^2$ $= 36y^2 - 25$

Section 5.5 Negative Exponents and Scientific Notation

| If $a \neq 0$ and n is an integer,

$a^{-n} = \dfrac{1}{a^n}$

Rules for exponents are true for positive and negative integers. | $3^{-2} = \dfrac{1}{3^2} = \dfrac{1}{9}; 5x^{-2} = \dfrac{5}{x^2}$

Simplify: $\left(\dfrac{x^{-2}y}{x^5}\right)^{-2} = \dfrac{x^4 y^{-2}}{x^{-10}}$
$= x^{4-(-10)}y^{-2}$
$= \dfrac{x^{14}}{y^2}$ |

(continued)

Definitions and Concepts	**Examples**

Section 5.5 Negative Exponents and Scientific Notation

A positive number is written in scientific notation if it is as the product of a number a, $1 \leq a < 10$, and an integer power r of 10. $$a \times 10^r$$	Write each number in scientific notation. $$12{,}000 = 1.2 \times 10^4$$ $$0.00000568 = 5.68 \times 10^{-6}$$

Section 5.6 Division of Polynomials

To divide a polynomial by a monomial: $$\frac{a + b}{c} = \frac{a}{c} + \frac{b}{c}$$ To divide a polynomial by a polynomial other than a monomial, use long division.	Divide: $$\frac{15x^5 - 10x^3 + 5x^2 - 2x}{5x^2} = \frac{15x^5}{5x^2} - \frac{10x^3}{5x^2} + \frac{5x^2}{5x^2} - \frac{2x}{5x^2}$$ $$= 3x^3 - 2x + 1 - \frac{2}{5x}$$

$$5x - 1 + \dfrac{-4}{2x + 3}$$

$$
\begin{array}{r}
2x + 3 \overline{\smash{)}\, 10x^2 + 13x - 7} \\
\underline{10x^2 + 15x} \\
-2x - 7 \\
\underline{-2x - 3} \\
-4
\end{array}
$$

Section 5.7 Synthetic Division and the Remainder Theorem

A shortcut method called **synthetic division** may be used to divide a polynomial by a binomial of the form $x - c$.	Use synthetic division to divide $2x^3 - x^2 - 8x - 1$ by $x - 2$.

$$
\begin{array}{r|rrrr}
2 & 2 & -1 & -8 & -1 \\
  & \downarrow & 4 & 6 & -4 \\
\hline
  & 2 & 3 & -2 & -5
\end{array}
$$

The quotient is $2x^2 + 3x - 2 - \dfrac{5}{x - 2}$.

5 CHAPTER REVIEW

(5.1) *State the base and the exponent for each expression.*

1. 3^2

2. $(-5)^4$

3. -5^4

Evaluate each expression.

4. 8^3

5. $(-6)^2$

6. -6^2

7. $-4^3 - 4^0$

8. $(3b)^0$

9. $\dfrac{8b}{8b}$

Simplify each expression.

10. $5b^3b^5a^6$

11. $2^3 \cdot x^0$

12. $[(-3)^2]^3$

13. $(2x^3)(-5x^2)$

14. $\left(\dfrac{mn}{q}\right)^2 \cdot \left(\dfrac{mn}{q}\right)$

15. $\left(\dfrac{3ab^2}{6ab}\right)^4$

16. $\dfrac{x^9}{x^4}$

17. $\dfrac{2x^7y^8}{8xy^2}$

18. $\dfrac{3x^4y^{10}}{12xy^6}$

19. $5a^7(2a^4)^3$

20. $(2x)^2(9x)$

21. $\dfrac{(-4)^2(3^3)}{(4^5)(3^2)}$

22. $\dfrac{(-7)^2(3^5)}{(-7)^3(3^4)}$

23. $\dfrac{(2x)^0(-4)^2}{16x}$

24. $\dfrac{(8xy)(3xy)}{18x^2y^2}$

25. $m^0 + p^0 + 3q^0$

26. $(-5a)^0 + 7^0 + 8^0$

27. $(3xy^2 + 8x + 9)^0$

28. $8x^0 + 9^0$

29. $6(a^2b^3)^3$

30. $\dfrac{(x^3z)^a}{x^2z^2}$

(5.2) *Find the degree of each term.*

31. $-5x^4y^3$

32. $95xyz$

Find the degree of each polynomial.

33. $-14x^2y - 28x^2y^3 - 42x^2y^2$

34. $6x^2y^2z^2 + 5x^2y^3 - 12xyz$

35. a. Complete the table for the polynomial
$3a^2b - 2a^2 + ab - b^2 - 6$.

Term	Numerical Coefficient	Degree of Term
$3a^2b$		
$-2a^2$		
ab		
$-b^2$		
-6		

b. What is the degree of the polynomial?

36. a. Complete the table for the polynomial
$x^2y^2 + 5x^2 - 7y^2 + 11xy - 1$.

Term	Numerical Coefficient	Degree of Term
x^2y^2		
$5x^2$		
$-7y^2$		
$11xy$		
-1		

b. What is the degree of the polynomial?

△ **37.** The surface area of a box with a square base and a height of 5 units is given by the polynomial $2x^2 + 20x$. Fill in the table below by evaluating $2x^2 + 20x$ for the given values of x.

x	1	3	5.1	10
$2x^2 + 20x$				

Simplify by combining like terms.

38. $4x + 8x - 6x^2 - 6x^2y$

39. $-8xy^3 + 4xy^3 - 3x^3y$

Add or subtract as indicated.

40. $(3x + 7y) + (4x^2 - 3x + 7) + (y - 1)$

41. $(4x^2 - 6xy + 9y^2) - (8x^2 - 6xy - y^2)$

42. $(3x^2 - 4b + 28) + (9x^2 - 30) - (4x^2 - 6b + 20)$

43. Add $(9xy + 4x^2 + 18)$ and $(7xy - 4x^3 - 9x)$.

44. Subtract $(x - 7)$ from the sum of $(3x^2y - 7xy - 4)$ and $(9x^2y + x)$.

45. $\begin{array}{r} x^2 - 5x + 7 \\ -(x + 4) \\ \hline \end{array}$

46. $\begin{array}{r} x^3 \quad + 2xy^2 - y \\ + (x - 4xy^2 \ -7) \\ \hline \end{array}$

If $P(x) = 9x^2 - 7x + 8$, find the following.

47. $P(6)$ **48.** $P(-2)$

49. $P(-3)$

If $P(x) = 2x - 1$ and $Q(x) = x^2 + 2x - 5$, find the following.

50. $P(x) + Q(x)$

51. $2[P(x)] - Q(x)$

52. Find the perimeter of the rectangle.

$$x^2y + 5 \text{ cm}$$

$$2x^2y - 6x + 1 \text{ cm}$$

53. With the popularity of the Internet growing rapidly, the number of new trademarks registered each year has also been increasing. The number of trademark registrations (in thousands) from 1990 to 2000 can be given by the polynomial $72.5x^2 - 17.5x + 120$, where x is the number of years since 1990. Use this model to predict the number of trademark registrations in 2010. (*Source*: International Trademark Association)

(5.3) Multiply each expression.

54. $9x(x^2y)$ **55.** $-7(8xz^2)$

56. $(6xa^2)(xya^3)$ **57.** $(4xy)(-3xa^2y^3)$

58. $6(x + 5)$ **59.** $9(x - 7)$

60. $4(2a + 7)$ **61.** $9(6a - 3)$

62. $-7x(x^2 + 5)$ **63.** $-8y(4y^2 - 6)$

64. $-2(x^3 - 9x^2 + x)$ **65.** $-3a(a^2b + ab + b^2)$

66. $(3a^3 - 4a + 1)(-2a)$ **67.** $(6b^3 - 4b + 2)(7b)$

68. $(2x + 5)(x - 7)$ **69.** $(2x - 5)(3x + 2)$

70. $(4a - 1)(a + 7)$ **71.** $(6a - 1)(7a + 3)$

72. $(x + 7)(x^3 + 4x - 5)$ **73.** $(x + 2)(x^5 + x + 1)$

74. $(x^2 + 2x + 4)(x^2 + 2x - 4)$

75. $(x^3 + 4x + 4)(x^3 + 4x - 4)$

76. $(x + 7)^3$ **77.** $(2x - 5)^3$

(5.4) Multiply.

78. $2x(3x^2 - 7x + 1)$ **79.** $3y(5y^2 - y + 2)$

80. $(6x^5 - 1)(4x^2 + 3)$ **81.** $(4a^3 - 1)(3a^2 + 7)$

82. $(x^2 + 7y)^2$ **83.** $(x^3 - 5y)^2$

84. $(3x - 7)^2$ **85.** $(4x + 2)^2$

86. $(y + 1)(y^2 - 6y - 5)$ **87.** $(x - 2)(x^2 - x - 2)$

88. $(5x - 9)^2$ **89.** $(5x + 1)(5x - 1)$

90. $(7x + 4)(7x - 4)$ **91.** $(a + 2b)(a - 2b)$

92. $(2x - 6)(2x + 6)$ **93.** $(4a^2 - 2b)(4a^2 + 2b)$

(5.5) Simplify each expression.

94. 7^{-2} **95.** -7^{-2}

96. $2x^{-4}$ **97.** $(2x)^{-4}$

98. $\left(\dfrac{1}{5}\right)^{-3}$ **99.** $\left(\dfrac{-2}{3}\right)^{-2}$

100. $2^0 + 2^{-4}$ **101.** $6^{-1} - 7^{-1}$

Simplify each expression. Assume that variables in an exponent represent positive integers only. Write each answer using positive exponents.

102. $\dfrac{1}{(2q)^{-3}}$ **103.** $\dfrac{-1}{(qr)^{-3}}$

104. $\dfrac{r^{-3}}{s^{-4}}$ **105.** $\dfrac{rs^{-3}}{r^{-4}}$

106. $(-x^{-3}y^5)(5xy^{-2})$ **107.** $(-3x^5y^{-2})(-4x^{-5}y)$

108. $(2x^{-5})^{-3}$ **109.** $(3y^{-6})^{-1}$

110. $(3a^{-1}b^{-1}c^{-2})^{-2}$ **111.** $(4x^{-2}y^{-3}z)^{-3}$

112. $\dfrac{5^{-2}x^8}{5^{-3}x^{11}}$ **113.** $\dfrac{7^5y^{-2}}{7^7y^{-10}}$

114. $\left(\dfrac{bc^{-2}}{bc^{-3}}\right)^4$ **115.** $\left(\dfrac{x^{-3}y^{-4}}{x^{-2}y^{-5}}\right)^{-3}$

116. $\dfrac{x^{-4}y^{-6}z^3}{x^2y^7z^3}$ **117.** $\dfrac{a^5b^{-5}c^4}{a^{-5}b^5c^4}$

118. $-2^0 + 2^{-4}$ **119.** $-3^{-2} - 3^{-3}$

120. $a^{6m}a^{5m}$ **121.** $\dfrac{(x^{5+h})^3}{x^5}$

122. $(3xy^{2z})^3$ **123.** $a^{m+2}a^{m+3}$

Write each number in scientific notation.

124. 0.00027 **125.** 0.8868

126. $80,800,000$ **127.** $868,000$

128. In a recent year, the United States imported approximately 109,379,000 kilograms of coffee. Write this number in scientific notation. (*Source:* International Coffee Organization)

129. The approximate diameter of the Milky Way galaxy is 150,000 light years. Write this number in scientific notation. (*Source:* NASA IMAGE/POETRY Education and Public Outreach Program)

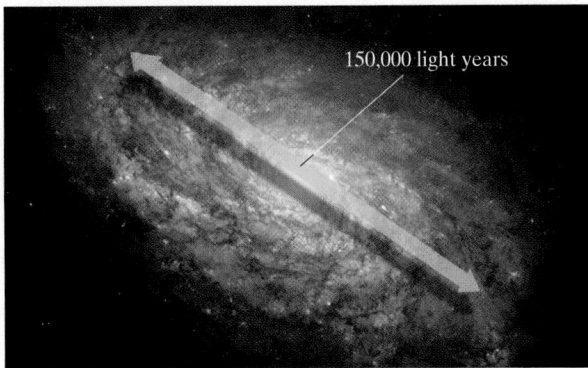

150,000 light years

Write each number in standard form.

130. 8.67×10^5

131. 3.86×10^{-3}

132. 8.6×10^{-4}

133. 8.936×10^5

134. The volume of the planet Jupiter is 1.43128×10^{15} cubic kilometers. Write this number in standard notation. (*Source:* National Space Science Data Center)

135. An angstrom is a unit of measure, equal to 1×10^{-10} meter, used for measuring wavelengths or the diameters of atoms. Write this number in standard notation. (*Source:* National Institute of Standards and Technology)

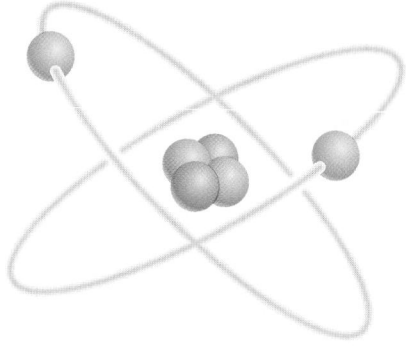

Simplify. Express each result in standard form.

136. $(8 \times 10^4)(2 \times 10^{-7})$

137. $\dfrac{8 \times 10^4}{2 \times 10^{-7}}$

(5.6) *Perform each division.*

138. $\dfrac{x^2 + 21x + 49}{7x^2}$

139. $\dfrac{5a^3b - 15ab^2 + 20ab}{-5ab}$

140. $(a^2 - a + 4) \div (a - 2)$

141. $(4x^2 + 20x + 7) \div (x + 5)$

142. $\dfrac{a^3 + a^2 + 2a + 6}{a - 2}$

143. $\dfrac{9b^3 - 18b^2 + 8b - 1}{3b - 2}$

144. $\dfrac{4x^4 - 4x^3 + x^2 + 4x - 3}{2x - 1}$

145. $\dfrac{-10x^2 - x^3 - 21x + 18}{x - 6}$

(5.7) *Use synthetic division to find each quotient.*

146. $(3x^3 + 12x - 4) \div (x - 2)$

147. $(3x^3 + 2x^2 - 4x - 1) \div \left(x + \dfrac{3}{2}\right)$

148. $(x^5 - 1) \div (x + 1)$

149. $(x^3 - 81) \div (x - 3)$

150. $(x^3 - x^2 + 3x^4 - 2) \div (x - 4)$

151. $(3x^4 - 2x^2 + 10) \div (x + 2)$

If $P(x) = 3x^5 - 9x + 7$, use the remainder theorem to find the following.

152. $P(4)$

153. $P(-5)$

CHAPTER 5 TEST

Remember to use your Chapter Test Prep Video CD to help you study and view solutions to the test questions you need help with.

Evaluate each expression.

1. 2^5 **2.** $(-3)^4$ **3.** -3^4 **4.** 4^{-3}

Simplify each exponential expression. Write the result using only positive exponents.

5. $(3x^2)(-5x^9)$ **6.** $\dfrac{y^7}{y^2}$

7. $\dfrac{r^{-8}}{r^{-3}}$ **8.** $\left(\dfrac{x^2y^3}{x^3y^{-4}}\right)^2$ **9.** $\dfrac{6^2x^{-4}y^{-1}}{6^3x^{-3}y^7}$

Express each number in scientific notation.

10. 563,000   **11.** 0.0000863

Write each number in standard form.

12. 1.5×10^{-3} **13.** 6.23×10^4

14. Simplify. Write the answer in standard form.
$$(1.2 \times 10^5)(3 \times 10^{-7})$$

15. a. Complete the table for the polynomial $4xy^2 + 7xy + x^3y - 2$.

Term	Numerical Coefficient	Degree of Term
$4xy^2$		
$7xy$		
x^3y		
-2		

  b. What is the degree of the polynomial?

16. Simplify by combining like terms.
$$5x^2 + 4xy - 7x^2 + 11 + 8xy$$

Perform each indicated operation.

17. $(8x^3 + 7x^2 + 4x - 7) + (8x^3 - 7x - 6)$

18. $\begin{array}{r} 5x^3 + \;\;x^2 + 5x - 2 \\ -(8x^3 - 4x^2 + \;\;x - 7) \end{array}$

19. Subtract $(4x + 2)$ from the sum of $(8x^2 + 7x + 5)$ and $(x^3 - 8)$.

Multiply.

20. $(3x + 7)(x^2 + 5x + 2)$

21. $3x^2(2x^2 - 3x + 7)$

22. $(x + 7)(3x - 5)$

23. $(4x - 2)^2$

24. $(x^2 - 9b)(x^2 + 9b)$

25. The height of the Bank of China in Hong Kong is 1001 feet. Neglecting air resistance, the height of an object dropped from this building at time t seconds is given by the polynomial $-16t^2 + 1001$. Find the height of the object at the given times below.

t	0 seconds	1 second	3 seconds	5 seconds
$-16t^2 + 1001$				

Divide.

26. $\dfrac{4x^2 + 24xy - 7x}{8xy}$ **27.** $(x^2 + 7x + 10) \div (x + 5)$

28. $\dfrac{27x^3 - 8}{3x + 2}$

29. A pebble is hurled upward from the top of the Canada Trust Tower, which is 880 feet tall, with an initial velocity of 96 feet per second. Neglecting air resistance, the height $h(t)$ of the pebble after t seconds is given by the polynomial function
$$h(t) = -16t^2 + 96t + 880$$

  a. Find the height of the pebble when $t = 1$.

  b. Find the height of the pebble when $t = 5.1$.

  c. When will the pebble hit the ground?

30. Use synthetic division to divide $(4x^4 - 3x^3 - x - 1)$ by $(x + 3)$.

31. If $P(x) = 4x^4 + 7x^2 - 2x - 5$, use the remainder theorem to find $P(-2)$.

5 CHAPTER CUMULATIVE REVIEW

1. Tell whether each statement is true or false.
   a. $8 \geq 8$ b. $8 \leq 8$
   c. $23 \leq 0$ d. $23 \geq 0$

2. Find the absolute value of each number.
   a. $|-7.2|$ b. $|0|$ c. $\left|-\dfrac{1}{2}\right|$

3. Divide. Write all quotients in lowest terms.
   a. $\dfrac{4}{5} \div \dfrac{5}{16}$ b. $\dfrac{7}{10} \div 14$
   c. $\dfrac{3}{8} \div \dfrac{3}{10}$

4. Multiply. Write products in lowest terms.
   a. $\dfrac{3}{4} \cdot \dfrac{7}{21}$ b. $\dfrac{1}{2} \cdot 4\dfrac{5}{6}$

5. Evaluate the following:
   a. 3^2 b. 5^3
   c. 2^4 d. 7^1
   e. $\left(\dfrac{3}{7}\right)^2$

6. Evaluate $\dfrac{2x - 7y}{x^2}$ for $x = 5$ and $y = 1$.

7. Add.
   a. $-3 + (-7)$ b. $-1 + (-20)$
   c. $-2 + (-10)$

8. Simplify: $8 + 3(2 \cdot 6 - 1)$

9. Subtract 8 from -4.

10. Is $x = 1$ a solution of $5x^2 + 2 = x - 8$.

11. Find the reciprocal of each number.
    a. 22 b. $\dfrac{3}{16}$
    c. -10 d. $-\dfrac{9}{13}$

12. Subtract:
    a. $7 - 40$ b. $-5 - (-10)$

13. Use an associative property to complete each statement.
    a. $5 + (4 + 6) = \underline{\qquad}$
    b. $(-1 \cdot 2) \cdot 5 = \underline{\qquad}$

14. Simplify: $\dfrac{4(-3) + (-8)}{5 + (-5)}$

15. Use the distributive property to write each expression without parentheses. Then simplify if possible.
    a. $2(x + y)$ b. $-5(-3 + 2z)$ c. $5(x + 3y - z)$
    d. $-1(2 - y)$ e. $-(3 + x - w)$ f. $4(3x + 7) + 10$

16. Use the distributive property to write $-2(x + 3y - z)$ without parentheses.

17. Find each product by using the distributive property to remove parentheses.
    a. $5(x + 2)$
    b. $-2(y + 0.3z - 1)$
    c. $-(x + y - 2z + 6)$

18. Simplify: $2(6x - 1) - (x - 7)$

19. Solve $x - 7 = 10$ for x.

20. Write the phrase as an algebraic expression: double a number, subtracted from the sum of a number and seven.

21. Solve: $\dfrac{5}{2}x = 15$

22. Solve: $2x + \dfrac{1}{8} = x - \dfrac{3}{8}$

23. Twice a number, added to seven, is the same as three subtracted from the number. Find the number.

24. Solve: $10 = 5j - 2$

25. Twice the sum of a number and 4 is the same as four times the number decreased by 12. Find the number.

26. Solve: $\dfrac{7x + 5}{3} = x + 3$

27. The length of a rectangular road sign is 2 feet less than three times its width. Find the dimensions if the perimeter is 28 feet.

28. Graph $x < 5$ and write in interval notation.

29. Solve $F = \dfrac{9}{5}C + 32$ for C.

30. Find the slope of each line.
    a. $x = -1$ b. $y = 7$

31. Graph $2 < x \leq 4$.

32. Recall that the grade of a road is its slope written as a percent. Find the grade of the road shown.

20 feet
2 feet

33. Complete the following ordered-pair solutions for the equation $3x + y = 12$.
    a. $(0, \ )$ b. $(\ , 6)$
    c. $(-1, \ )$

34. Solve the system: $\begin{cases} 3x + 2y = -8 \\ 2x - 6y = -9 \end{cases}$

35. Graph the linear equation $2x + y = 5$.

36. Solve the system: $\begin{cases} x = -3y + 3 \\ 2x + 9y = 5 \end{cases}$

37. Graph $x = 2$.

38. Evaluate.

   a. $(-5)^2$ **b.** -5^2 **c.** $2 \cdot 5^2$

39. Find the slope of the line $x = 5$.

40. Simplify: $\dfrac{(z^2)^3 \cdot z^7}{z^9}$

41. Subtract $(12z^5 - 12z^3 + z) - (-3z^4 + z^3 + 12z)$.

42. Subtract $(5y^2 - 6) - (y^2 + 2)$.

43. Use the product rule to simplify $(2x^2)(-3x^5)$.

44. Find the value of $-x^2$ when

   a. $x = 2$ **b.** $x = -2$

45. If $P(x) = 3x^2 - 2x - 5$, find

   a. $P(1)$ **b.** $P(-2)$

46. Multiply $(10x^2 - 3)(10x^2 + 3)$.

47. Multiply $(2x - y)^2$.

48. Multiply $(10x^2 + 3)^2$.

49. Divide $6m^2 + 2m$ by $2m$.

50. Evaluate.

   a. 5^{-1} **b.** 7^{-2}

6

Factoring Polynomials

In Chapter 5, you learned how to multiply polynomials. This chapter deals with an operation that is the reverse process of multiplying, called *factoring*. Factoring is an important algebraic skill because this process allows us to write a sum as a product.

At the end of this chapter, we use factoring to help us solve equations other than linear equations, and in Chapter 7 we use factoring to simplify and perform arithmetic operations on rational expressions.

Results from a national survey conducted by the Nellie Mae Corporation show that graduating college students in 2002 had an average credit card debt of $3000. The National Foundation for Credit Counseling (NFCC) offers college students the following advice on the responsible use of credit cards:

- Compare credit cards and shop around for the best interest rates.
- Be careful of hidden credit card fees.
- Avoid stress by shopping wisely and buying only what you can afford to repay immediately.
- Closely guard your credit card to protect yourself from identity theft.

For further help, contact a certified credit counselor through the NFCC. These counselors offer their services at little or no cost to the public.

In the Spotlight on Decision Making feature on page 379, you will have the opportunity to make a decision about the best choice of three credit card offers.

Link: National Foundation for Credit Counseling: www.nfcc.org
Source of text: (same)

6.1   THE GREATEST COMMON FACTOR AND FACTORING BY GROUPING

Objectives

1 Find the greatest common factor of a list of integers.

2 Find the greatest common factor of a list of terms.

3 Factor out the greatest common factor from a polynomial.

4 Factor a polynomial by grouping.

When an integer is written as the product of two or more other integers, each of these integers is called a **factor** of the product. This is true for polynomials, also. When a polynomial is written as the product of two or more other polynomials, each of these polynomials is called a factor of the product.

The process of writing a polynomial as a product is called **factoring** the polynomial.

$$\underset{\substack{\uparrow \\ factor}}{2} \cdot \underset{\substack{\uparrow \\ factor}}{3} = \underset{\substack{\uparrow \\ product}}{6}$$

$$\underset{\substack{\uparrow \\ factor}}{x^2} \cdot \underset{\substack{\uparrow \\ factor}}{x^3} = \underset{\substack{\uparrow \\ product}}{x^5}$$

$$\underset{\substack{\uparrow \\ factor}}{(x + 2)}\underset{\substack{\uparrow \\ factor}}{(x + 3)} = \underset{\substack{\uparrow \\ product}}{x^2 + 5x + 6}$$

Notice that factoring is the reverse process of multiplying.

$$\overset{\text{factoring}}{x^2 + 5x + 6 = (x + 2)(x + 3)}$$
$$\underset{\text{multiplying}}{}$$

✔ **CONCEPT CHECK**

Multiply: $2(x - 4)$

What do you think the result of factoring $2x - 8$ would be? Why?

The first step in factoring a polynomial is to see whether the terms of the polynomial have a common factor. If there is one, we can write the polynomial as a product by **factoring out** the common factor. We will usually factor out the **greatest common factor (GCF)**.

1   The GCF of a list of integers is the largest integer that is a factor of all the integers in the list. For example, the GCF of 12 and 20 is 4 because 4 is the largest integer that is a factor of both 12 and 20. With large integers, the GCF may not be easily found by inspection. When this happens, use the following steps.

Finding the GCF of a List of Integers

Step 1. Write each number as a product of prime numbers.

Step 2. Identify the common prime factors.

Step 3. The product of all common prime factors found in step 2 is the greatest common factor. If there are no common prime factors, the greatest common factor is 1.

Recall from Section 1.3 that a prime number is a whole number other than 1, whose only factors are 1 and itself.

EXAMPLE 1

Find the GCF of each list of numbers.

a. 28 and 40   **b.** 55 and 21   **c.** 15, 18, and 66

Solution   **a.** Write each number as a product of primes.

$$28 = 2 \cdot 2 \cdot 7 = 2^2 \cdot 7$$
$$40 = 2 \cdot 2 \cdot 2 \cdot 5 = 2^3 \cdot 5$$

There are two common factors, each of which is 2, so the GCF is

$$\text{GCF} = 2 \cdot 2 = 4$$

b. $55 = 5 \cdot 11$
$21 = 3 \cdot 7$

There are no common prime factors; thus, the GCF is 1.

c. $15 = 3 \cdot 5$
$18 = 2 \cdot 3 \cdot 3 = 2 \cdot 3^2$
$66 = 2 \cdot 3 \cdot 11$

The only prime factor common to all three numbers is 3, so the GCF is

$$\text{GCF} = 3$$

2   The greatest common factor of a list of variables raised to powers is found in a similar way. For example, the GCF of x^2, x^3, and x^5 is x^2 because each term contains a factor of x^2 and no higher power of x is a factor of each term.

$$x^2 = x \cdot x$$
$$x^3 = x \cdot x \cdot x$$
$$x^5 = x \cdot x \cdot x \cdot x \cdot x$$

There are two common factors, each of which is x, so the GCF $= x \cdot x$ or x^2. From this example, we see that **the GCF of a list of common variables raised to powers is the variable raised to the smallest exponent in the list**.

EXAMPLE 2

Find the GCF of each list of terms.

a. x^3, x^7, and x^5 **b.** y, y^4, and y^7

Solution   **a.** The GCF is x^3, since 3 is the smallest exponent to which x is raised.
b. The GCF is y^1 or y, since 1 is the smallest exponent on y.

In general, the **greatest common factor (GCF) of a list of terms** is the product of the GCF of the numerical coefficients and the GCF of the variable factors.

EXAMPLE 3

Find the GCF of each list of terms.

a. $6x^2$, $10x^3$, and $-8x$ **b.** $8y^2$, y^3, and y^5 **c.** a^3b^2, a^5b, and a^6b^2

Solution   **a.** The GCF of the numerical coefficients 6, 10, and -8 is 2.
The GCF of variable factors x^2, x^3, and x is x.
Thus, the GCF of $6x^2$, $10x^3$, and $-8x$ is $2x$.

b. The GCF of the numerical coefficients 8, 1, and 1 is 1.
The GCF of variable factors y^2, y^3, and y^5 is y^2.
Thus, the GCF of $8y^2$, y^3, and y^5 is $1y^2$ or y^2.

c. The GCF of a^3, a^5, and a^6 is a^3.

The GCF of b^2, b, and b^2 is b. Thus, the GCF of the terms is a^3b.

3 The first step in factoring a polynomial is to find the GCF of its terms. Once we do so, we can write the polynomial as a product by **factoring out** the GCF.

The polynomial $8x + 14$, for example, contains two terms: $8x$ and 14. The GCF of these terms is 2. We factor out 2 from each term by writing each term as a product of 2 and the term's remaining factors.

$$8x + 14 = 2 \cdot 4x + 2 \cdot 7$$

Using the distributive property, we can write

$$8x + 14 = 2 \cdot 4x + 2 \cdot 7$$
$$= 2(4x + 7)$$

Thus, a factored form of $8x + 14$ is $2(4x + 7)$.

> **Helpful Hint**
>
> A factored form of $8x + 14$ is *not*
>
> $$2 \cdot 4x + 2 \cdot 7$$
>
> Although the *terms* have been factored (written as a product), the *polynomial* $8x + 14$ has not been factored (written as a product). A factored form of $8x + 14$ is the *product* $2(4x + 7)$.

✔ **CONCEPT CHECK**

Which of the following is/are factored form(s) of $7t + 21$?

a. 7   **b.** $7 \cdot t + 7 \cdot 3$ **c.** $7(t + 3)$ **d.** $7(t + 21)$

EXAMPLE 4

Factor each polynomial by factoring out the GCF.

a. $6t + 18$ **b.** $y^5 - y^7$

Solution   **a.** The GCF of terms $6t$ and 18 is 6.

$$6t + 18 = 6 \cdot t + 6 \cdot 3$$
$$= 6(t + 3) \qquad \text{Apply the distributive property.}$$

Our work can be checked by multiplying 6 and $(t + 3)$.

$$6(t + 3) = 6 \cdot t + 6 \cdot 3 = 6t + 18, \text{ the original polynomial.}$$

b. The GCF of y^5 and y^7 is y^5. Thus,

> **Helpful Hint**
>
> Don't forget the 1.

$$y^5 - y^7 = (y^5)1 - (y^5)y^2$$
$$= y^5(1 - y^2)$$

Concept Check Answer:
c

EXAMPLE 5

Factor $-9a^5 + 18a^2 - 3a$.

Solution $-9a^5 + 18a^2 - 3a = (3a)(-3a^4) + (3a)(6a) + (3a)(-1)$
$$= 3a(-3a^4 + 6a - 1)$$

> **Helpful Hint**
> Don't forget the -1.

In Example 5 we could have chosen to factor out a $-3a$ instead of $3a$. If we factor out a $-3a$, we have

$$-9a^5 + 18a^2 - 3a = (-3a)(3a^4) + (-3a)(-6a) + (-3a)(1)$$
$$= -3a(3a^4 - 6a + 1)$$

> **Helpful Hint**
> Notice the changes in signs when factoring out $-3a$.

EXAMPLE 6

Factor $25x^4z + 15x^3z + 5x^2z$.

Solution   The greatest common factor is $5x^2z$.

$$25x^4z + 15x^3z + 5x^2z = 5x^2z(5x^2 + 3x + 1)$$

> **Helpful Hint**
> Be careful when the GCF of the terms is the same as one of the terms in the polynomial. The greatest common factor of the terms of $8x^2 - 6x^3 + 2x$ is $2x$. When factoring out $2x$ from the terms of $8x^2 - 6x^3 + 2x$, don't forget a term of 1.
> $$8x^2 - 6x^3 + 2x = 2x(4x) - 2x(3x^2) + 2x(1)$$
> $$= 2x(4x - 3x^2 + 1)$$
> Check by multiplying.
> $$2x(4x - 3x^2 + 1) = 8x^2 - 6x^3 + 2x$$

EXAMPLE 7

Factor $5(x + 3) + y(x + 3)$.

Solution   The binomial $(x + 3)$ is the greatest common factor. Use the distributive property to factor out $(x + 3)$.

$$5(x + 3) + y(x + 3) = (x + 3)(5 + y)$$

EXAMPLE 8

Factor $3m^2n(a + b) - (a + b)$.

Solution   The greatest common factor is $(a + b)$.

$$3m^2n(a + b) - 1(a + b) = (a + b)(3m^2n - 1)$$

4 Once the GCF is factored out, we can often continue to factor the polynomial, using a variety of techniques. We discuss here a technique for factoring polynomials called **grouping**.

EXAMPLE 9

Factor $xy + 2x + 3y + 6$ by grouping. Check by multiplying.

Solution The GCF of the first two terms is x, and the GCF of the last two terms is 3.

$$xy + 2x + 3y + 6 = \underline{x(y + 2) + 3(y + 2)}$$

> **Helpful Hint**
> Notice that this is *not* a factored form of the original polynomial. It is a sum, not a product.

Next, factor out the common binomial factor of $(y + 2)$.

$$x(y + 2) + 3(y + 2) = (y + 2)(x + 3)$$

To check, multiply $(y + 2)$ by $(x + 3)$.

$$(y + 2)(x + 3) = xy + 2x + 3y + 6, \text{ the original polynomial.}$$

Thus, the factored form of $xy + 2x + 3y + 6$ is $(y + 2)(x + 3)$.

Factoring a Four-term Polynomial by Grouping

Step 1. Arrange the terms so that the first two terms have a common factor and the last two terms have a common factor.

Step 2. For each pair of terms, use the distributive property to factor out the pair's greatest common factor.

Step 3. If there is now a common binomial factor, factor it out.

Step 4. If there is no common binomial factor in Step 3, begin again, rearranging the terms differently. If no rearrangement leads to a common binomial factor, the polynomial cannot be factored.

EXAMPLE 10

Factor $3x^2 + 4xy - 3x - 4y$ by grouping.

Solution The first two terms have a common factor x. Factor -1 from the last two terms so that the common binomial factor of $(3x + 4y)$ appears.

$$3x^2 + 4xy - 3x - 4y = x(3x + 4y) - 1(3x + 4y)$$

Next, factor out the common factor $(3x + 4y)$.

$$= (3x + 4y)(x - 1)$$

> **Helpful Hint**
>
> One more reminder: When **factoring** a polynomial, make sure the polynomial is written as a **product**. For example, it is true that
>
> $$3x^2 + 4xy - 3x - 4y = x(3x + 4y) - 1(3x + 4y)$$
>
> but $x(3x + 4y) - 1(3x + 4y)$ is not a **factored form** of the original polynomial since it is a **sum (difference)**, not a **product**. The factored form of $3x^2 + 4xy - 3x - 4y$ is $(3x + 4y)(x - 1)$.

Factoring out a greatest common factor first makes factoring by any method easier, as we see in the next example.

EXAMPLE 11

Factor $4ax - 4ab - 2bx + 2b^2$.

Solution   First, factor out the common factor 2 from all four terms.

$$
\begin{aligned}
4ax - 4ab - 2bx + 2b^2 \\
&= 2(2ax - 2ab - bx + b^2) && \text{Factor out \textbf{2} from all four terms.} \\
&= 2[2a(x - b) - b(x - b)] && \text{Factor out common factors from each pair of terms.} \\
&= 2(x - b)(2a - b) && \text{Factor out the common binomial.}
\end{aligned}
$$

Notice that we factored out $-b$ instead of b from the second pair of terms so that the binomial factor of each pair is the same.

STUDY SKILLS REMINDER

How Well Do You Know Your Textbook?

See if you can answer the questions below.

1. What does the 🔒 icon mean?
2. What does the ↘ icon mean?
3. What does the △ icon mean?
4. Where can you find a review for each chapter? What answers to this review can be found in the back of your text?
5. Each chapter contains an overview of the chapter along with examples. What is this feature called?
6. Does this text contain any solutions to exercises? If so, where?

MENTAL MATH

Find the prime factorization of the following integers.

1. 14   **2.** 15   **3.** 10   **4.** 70

Find the GCF of the following pairs of integers.

5. 6, 15   **6.** 20, 15   **7.** 3, 18   **8.** 14, 35

EXERCISE SET 6.1

STUDY GUIDE/SSM   CD/VIDEO   PH MATH TUTOR CENTER   MathXL®Tutorials ON CD   MathXL®   MyMathLab®

Find the GCF for each list. See Examples 1 through 3.

1. $32, 36$
2. $36, 90$
3. $12, 18, 36$
4. $24, 14, 21$
5. y^2, y^4, y^7
6. x^3, x^2, x^3
7. $x^{10}y^2, xy^2, x^3y^3$
8. p^7q, p^8q^2, p^9q^3
9. $8x, 4$
10. $9y, y$
 11. $12y^4, 20y^3$
12. $32x, 18x^2$
13. $12x^3, 6x^4, 3x^5$
14. $15y^2, 5y^7, 20y^3$
15. $18x^2y, 9x^3y^3, 36x^3y$
16. $7x, 21x^2y^2, 14xy$

Factor out the GCF from each polynomial. See Examples 4 through 6.

17. $30x - 15$
18. $42x - 7$
19. $24cd^3 - 18c^2d$
20. $25x^4y^3 - 15x^2y^2$
21. $-24a^4x + 18a^3x$
22. $-15a^2x + 9ax$
23. $12x^3 + 16x^2 - 8x$
24. $6x^3 - 9x^2 + 12x$
25. $5x^3y - 15x^2y + 10xy$
26. $14x^3y + 7x^2y - 7xy$
27. Construct a binomial whose greatest common factor is $5a^3$.
28. Construct a trinomial whose greatest common factor is $2x^2$.

Factor out the GCF from each polynomial. See Examples 7 and 8.

29. $y(x + 2) + 3(x + 2)$
30. $z(y + 4) + 3(y + 4)$
31. $x(y - 3) - 4(y - 3)$
32. $6(x + 2) - y(x + 2)$
33. $2x(x + y) - (x + y)$
34. $xy(y + 1) - (y + 1)$

Factor the following four-term polynomials by grouping. See Examples 9 through 11.

35. $5x + 15 + xy + 3y$
36. $xy + y + 2x + 2$
37. $2y - 8 + xy - 4x$
38. $6x - 42 + xy - 7y$
39. $3xy - 6x + 8y - 16$
40. $xy - 2yz + 5x - 10z$
41. $y^3 + 3y^2 + y + 3$
42. $x^3 + 4x + x^2 + 4$

Write an expression for the area of each shaded region. Then write the expression as a factored polynomial.

△ **43.**

△ **44.**

△ **45.**

△ **46.**

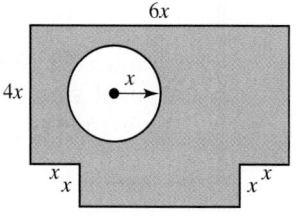

MIXED PRACTICE

Factor the following polynomials.

47. $3x - 6$
48. $4x - 16$
49. $32xy - 18x^2$
50. $10xy - 15x^2$
51. $4x - 8y + 4$
52. $7x + 21y - 7$
53. $8(x + 2) - y(x + 2)$
54. $x(y^2 + 1) - 3(y^2 + 1)$
55. $-40x^8y^6 - 16x^9y^5$
56. $-21x^3y - 49x^2y^2$
57. $-3x + 12$
58. $-10x + 20$
59. $18x^3y^3 - 12x^3y^2 + 6x^5y^2$
60. $32x^3y^3 - 24x^2y^3 + 8x^2y^4$

Factor.

61. $y^2(x - 2) + (x - 2)$
62. $x(y + 4) + (y + 4)$
63. $5xy + 15x + 6y + 18$
64. $2x^3 + x^2 + 8x + 4$
65. $4x^2 - 8xy - 3x + 6y$
66. $2x^3 - x^2 - 10x + 5$
67. $126x^3yz + 210y^4z^3$
68. $231x^3y^2z - 143yz^2$
69. $3y - 5x + 15 - xy$
70. $2x - 9y + 18 - xy$
71. $12x^2y - 42x^2 - 4y + 14$
72. $90 + 15y^2 - 18x - 3xy^2$

Fill in the chart by finding two numbers that have the given product and sum. The first row is filled in for you.

	Two Numbers	Their Product	Their Sum
	4, 7	28	11
73.		12	8
74.		20	9
75.		8	−9
76.		16	−10
77.		−10	3
78.		−9	0
79.		−24	−5
80.		−36	−5

REVIEW AND PREVIEW

Multiply. See Section 5.4.

81. $(x + 2)(x + 5)$

82. $(y + 3)(y + 6)$

83. $(a - 7)(a - 8)$

84. $(z - 4)(z - 4)$

Concept Extensions

85. Explain how you can tell whether a polynomial is written in factored form.

86. Construct a 4-term polynomial that can be factored by grouping.

Which of the following expressions are factored?

87. $(a + 6)(b - 2)$

88. $(x + 5)(x + y)$

89. $5(2y + z) - b(2y + z)$

90. $3x(a + 2b) + 2(a + 2b)$

Write an expression for the length of each rectangle.

91.

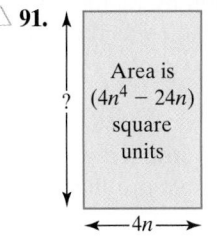

Area is $(4n^4 - 24n)$ square units

?

4n units

92.

Area is $(5x^5 - 5x^2)$ square units

$5x^2$ units

?

Factor each polynomial by grouping.

93. $x^{2n} + 2x^n + 3x^n + 6$ (*Hint:* Don't forget that $x^{2n} = x^n \cdot x^n$.)

94. $x^{2n} + 6x^n + 10x^n + 60$

95. $3x^{2n} + 21x^n - 5x^n - 35$

96. $12x^{2n} - 10x^n - 30x^n + 25$

97. The number (in millions) of CDs shipped annually in the United States each year during 1999–2001 can be modeled by the polynomial $8x^2 - 56x + 124$, where x is the number of years since 1999. (*Source:* Recording Industry Association of America)

a. Find the number of CDs sold in 2001. To do so, let $x = 2$ and evaluate $8x^2 - 56x + 124$.

b. Use this expression to predict the number of CDs sold in 2006.

c. Factor the polynomial $8x^2 - 56x + 124$.

98. The number (in thousands) of students who graduated from U.S. high schools each year during 1998–2000 can be modeled by $12x^2 + 21x + 2457$, where x is the number of years since 1998. (*Source:* U.S. Bureau of the Census)

a. Find the number of students who graduated from U.S. high schools in 2000. To do so, let $x = 2$ and evaluate $12x^2 + 21x + 2457$.

b. Use this expression to predict the number of students who will graduate from U.S. high schools in 2007.

c. Factor the polynomial $12x^2 + 21x + 2457$.

6.2 FACTORING TRINOMIALS OF THE FORM $x^2 + bx + c$

Objectives

1 Factor trinomials of the form $x^2 + bx + c$.

2 Factor out the greatest common factor and then factor a trinomial of the form $x^2 + bx + c$.

1 In this section, we factor trinomials of the form $x^2 + bx + c$, such as

$$x^2 + 4x + 3, \quad x^2 - 8x + 15, \quad x^2 + 4x - 12, \quad r^2 - r - 42$$

Notice that for these trinomials, the coefficient of the squared variable is 1.

Recall that factoring means to write as a product and that factoring and multiplying are reverse processes. Using the FOIL method of multiplying binomials, we have that

$$
\overset{\text{F} \qquad \text{O} \qquad \text{I} \qquad \text{L}}{(x + 3)(x + 1) = x^2 + 1x + 3x + 3}
$$
$$
= x^2 + 4x + 3
$$

Thus, a factored form of $x^2 + 4x + 3$ is $(x + 3)(x + 1)$.

Notice that the product of the first terms of the binomials is $x \cdot x = x^2$, the first term of the trinomial. Also, the product of the last two terms of the binomials is $3 \cdot 1 = 3$, the third term of the trinomial. The sum of these same terms is $3 + 1 = 4$, the coefficient of the middle term, x, of the trinomial.

The product of these numbers is **3**.

$$
x^2 + 4x + 3 = (x + 3)(x + 1)
$$

The sum of these numbers is **4**.

Many trinomials, such as the one above, factor into two binomials. To factor $x^2 + 7x + 10$, let's assume that it factors into two binomials and begin by writing two pairs of parentheses. The first term of the trinomial is x^2, so we use x and x as the first terms of the binomial factors.

$$
x^2 + 7x + 10 = (x + \quad)(x + \quad)
$$

To determine the last term of each binomial factor, we look for two integers whose product is 10 and whose sum is 7. Since our numbers must have a positive product and a positive sum, we list pairs of positive integer factors of 10 only.

Positive Factors of 10	Sum of Factors
1, 10	$1 + 10 = 11$
2, 5	$2 + 5 = 7$

The correct pair of numbers is 2 and 5 because their product is 10 and their sum is 7. Now we can fill in the last terms of the binomial factors.

$$
x^2 + 7x + 10 = (x + 2)(x + 5)
$$

To see if we have factored correctly, multiply.

$$
(x + 2)(x + 5) = x^2 + 5x + 2x + 10
$$
$$
= x^2 + 7x + 10 \qquad \text{Combine like terms.}
$$

Helpful Hint

Since multiplication is commutative, the factored form of $x^2 + 7x + 10$ can be written as either $(x + 2)(x + 5)$ or $(x + 5)(x + 2)$.

Factoring a Trinomial of the Form $x^2 + bx + c$

To factor a trinomial of the form $x^2 + bx + c$, look for two numbers whose product is c and whose sum is b. The factored form of $x^2 + bx + c$ is

product is c

$$
(x + \text{one number})(x + \text{other number})
$$

sum is b

EXAMPLE 1

Factor $x^2 + 7x + 12$.

Solution   Begin by writing the first terms of the binomial factors.

$$(x + \quad)(x + \quad)$$

Next, look for two numbers whose product is 12 and whose sum is 7. Since our numbers must have a positive product and a positive sum, we look at positive pairs of factors of 12 only.

Positive Factors of 12	Sum of Factors
1, 12	$1 + 12 = 13$
2, 6	$2 + 6 = 8$
3, 4	$3 + 4 = 7$

The correct pair of numbers is 3 and 4 because their product is 12 and their sum is 7. Use these numbers as the last terms of the binomial factors.

$$x^2 + 7x + 12 = (x + 3)(x + 4)$$

To check, multiply $(x + 3)$ by $(x + 4)$.

EXAMPLE 2

Factor $x^2 - 8x + 15$.

Solution   Begin by writing the first terms of the binomials.

$$(x + \quad)(x + \quad)$$

Now look for two numbers whose product is 15 and whose sum is -8. Since our numbers must have a positive product and a negative sum, we look at negative factors of 15 only.

Negative Factors of 15	Sum of Factors
$-1, -15$	$-1 + (-15) = -16$
$-3, -5$	$-3 + (-5) = -8$

The correct pair of numbers is -3 and -5 because their product is 15 and their sum is -8. Then

$$x^2 - 8x + 15 = (x - 3)(x - 5)$$

EXAMPLE 3

Factor $x^2 + 4x - 12$.

Solution $x^2 + 4x - 12 = (x + \quad)(x + \quad)$

Look for two numbers whose product is -12 and whose sum is 4. Since our numbers have a negative product, their signs must be different.

Factors of -12	Sum of Factors
$-1, 12$	$-1 + 12 = 11$
$1, -12$	$1 + (-12) = -11$
$-2, 6$	$-2 + 6 = 4$
$2, -6$	$2 + (-6) = -4$
$-3, 4$	$-3 + 4 = 1$
$3, -4$	$3 + (-4) = -1$

The correct pair of numbers is -2 and 6 since their product is -12 and their sum is 4. Hence

$$x^2 + 4x - 12 = (x - 2)(x + 6)$$

EXAMPLE 4

Factor $r^2 - r - 42$.

Solution   Because the variable in this trinomial is r, the first term of each binomial factor is r.

$$r^2 - r - 42 = (r + \quad)(r + \quad)$$

Find two numbers whose product is -42 and whose sum is -1, the numerical coefficient of r. The numbers are 6 and -7. Therefore,

$$r^2 - r - 42 = (r + 6)(r - 7)$$

EXAMPLE 5

Factor $a^2 + 2a + 10$.

Solution   Look for two numbers whose product is 10 and whose sum is 2. Neither 1 and 10 nor 2 and 5 give the required sum, 2. We conclude that $a^2 + 2a + 10$ is not factorable with integers. The polynomial $a^2 + 2a + 10$ is called a **prime polynomial**.

EXAMPLE 6

Factor $x^2 + 5xy + 6y^2$.

Solution $x^2 + 5xy + 6y^2 = (x + \quad)(x + \quad)$

Look for two terms whose product is $6y^2$ and whose sum is $5y$, the coefficient of x in the middle term of the trinomial. The terms are $2y$ and $3y$ because $2y \cdot 3y = 6y^2$ and $2y + 3y = 5y$. Therefore,

$$x^2 + 5xy + 6y^2 = (x + 2y)(x + 3y)$$

The following sign patterns may be useful when factoring trinomials.

> **Helpful Hint—Sign Patterns**
>
> A positive constant in a trinomial tells us to look for two numbers with the same sign. The sign of the coefficient of the middle term tells us whether the signs are both positive or both negative.
>
> both   same
> positive sign
> $\downarrow$ $\downarrow$
> $$x^2 + 10x + 16 = (x + 2)(x + 8)$$
>
> both   same
> negative sign
> $\downarrow$ $\downarrow$
> $$x^2 - 10x + 16 = (x - 2)(x - 8)$$
>
> *(continued)*

A negative constant in a trinomial tells us to look for two numbers with opposite signs.

opposite
signs
↓

opposite
signs
↓

$$x^2 + 6x - 16 = (x + 8)(x - 2) \qquad x^2 - 6x - 16 = (x - 8)(x + 2)$$

2   Remember that the first step in factoring any polynomial is to factor out the greatest common factor (if there is one other than 1 or -1).

EXAMPLE 7

Factor $3m^2 - 24m - 60$.

Solution   First factor out the greatest common factor, 3, from each term.

$$3m^2 - 24m - 60 = 3(m^2 - 8m - 20)$$

Next, factor $m^2 - 8m - 20$ by looking for two factors of -20 whose sum is -8. The factors are -10 and 2.

$$3m^2 - 24m - 60 = 3(m + 2)(m - 10)$$

Remember to write the common factor 3 as part of the answer. Check by multiplying.

$$3(m + 2)(m - 10) = 3(m^2 - 8m - 20)$$
$$= 3m^2 - 24m - 60$$

Helpful Hint

When factoring a polynomial, remember that factored out common factors are part of the final factored form. For example,

$$5x^2 - 15x - 50 = 5(x^2 - 3x - 10)$$
$$= 5(x + 2)(x - 5)$$

Thus, $5x^2 - 15x - 50$ **factored completely** is $5(x + 2)(x - 5)$.

MENTAL MATH

Complete the following.

1. $x^2 + 9x + 20 = (x + 4)(x \quad)$ **2.** $x^2 + 12x + 35 = (x + 5)(x \quad)$ **3.** $x^2 - 7x + 12 = (x - 4)(x \quad)$

4. $x^2 - 13x + 22 = (x - 2)(x \quad)$ **5.** $x^2 + 4x + 4 = (x + 2)(x \quad)$ **6.** $x^2 + 10x + 24 = (x + 6)(x \quad)$

EXERCISE SET 6.2

STUDY GUIDE/SSM CD/VIDEO PH MATH TUTOR CENTER MathXL®Tutorials ON CD MathXL® MyMathLab®

Factor each trinomial completely. If the polynomial can't be factored, write prime. See Examples 1 through 5.

1. $x^2 + 7x + 6$
2. $x^2 + 6x + 8$
3. $x^2 + x - 20$
4. $x^2 + 7x - 30$
5. $x^2 - 8x + 15$
6. $x^2 - 9x + 14$
7. $x^2 - 10x + 9$
8. $x^2 - 6x + 9$
9. $x^2 - 15x + 5$
10. $x^2 - 13x + 30$
11. $x^2 - 3x - 18$
12. $x^2 - x - 30$
13. $x^2 + 5x + 2$
14. $x^2 - 7x + 5$

Factor each trinomial completely. See Example 6.

15. $x^2 + 8xy + 15y^2$
16. $x^2 + 6xy + 8y^2$
17. $x^2 - 2xy + y^2$
18. $x^2 - 11xy + 30y^2$
19. $x^2 - 3xy - 4y^2$
20. $x^2 - 4xy - 77y^2$

Factor each trinomial completely. See Example 7.

21. $2z^2 + 20z + 32$
22. $3x^2 + 30x + 63$
23. $2x^3 - 18x^2 + 40x$
24. $x^3 - x^2 - 56x$
25. $7x^2 + 14xy - 21y^2$
26. $3r^2 - 3rs - 6s^2$

27. To factor $x^2 + 13x + 42$, think of two numbers whose _____ is 42 and whose ____ is 13.

28. Write a polynomial that factors as $(x - 3)(x + 8)$.

MIXED PRACTICE

Factor each trinomial completely.

29. $x^2 + 15x + 36$
30. $x^2 + 19x + 60$
31. $x^2 - x - 2$
32. $x^2 - 5x - 14$
33. $r^2 - 16r + 48$
34. $r^2 - 10r + 21$
35. $x^2 - 4x - 21$
36. $x^2 - 4x - 32$
37. $x^2 + 7xy + 10y^2$
38. $x^2 + 5xy + 4y^2$
39. $r^2 - 3r + 6$
40. $x^2 + 4x - 10$
41. $2t^2 + 24t + 64$
42. $2t^2 + 20t + 50$
43. $x^3 - 2x^2 - 24x$
44. $x^3 - 3x^2 - 28x$
45. $x^2 - 16x + 63$
46. $x^2 - 19x + 88$
47. $x^2 + xy - 2y^2$
48. $x^2 - xy - 6y^2$
49. $3x^2 - 60x + 108$
50. $2x^2 - 24x + 70$
51. $x^2 - 18x - 144$
52. $x^2 + x - 42$
53. $6x^3 + 54x^2 + 120x$
54. $3x^3 + 3x^2 - 126x$
55. $2t^5 - 14t^4 + 24t^3$
56. $3x^6 + 30x^5 + 72x^4$
57. $5x^3y - 25x^2y^2 - 120xy^3$
58. $3x^2 - 6xy - 72y^2$
59. $4x^2y + 4xy - 12y$
60. $3x^2y - 9xy + 45y$
61. $2a^2b - 20ab^2 + 42b^3$
62. $-x^2z + 14xz^2 - 28z^3$

REVIEW AND PREVIEW

Multiply. See Section 5.4.

63. $(2x + 1)(x + 5)$
64. $(3x + 2)(x + 4)$

65. $(5y - 4)(3y - 1)$
66. $(4z - 7)(7z - 1)$
67. $(a + 3)(9a - 4)$
68. $(y - 5)(6y + 5)$

Concept Extensions

Find a positive value of b so that each trinomial is factorable.

69. $x^2 + bx + 15$
70. $y^2 + by + 20$
71. $m^2 + bm - 27$
72. $x^2 + bx - 14$

Find a positive value of c so that each trinomial is factorable.

73. $x^2 + 6x + c$
74. $t^2 + 8t + c$
75. $y^2 - 4y + c$
76. $n^2 - 16n + c$

Write the perimeter of each rectangle as a simplified polynomial. Then factor the polynomial.

77.

$4x + 33$

$x^2 + 10x$

78.

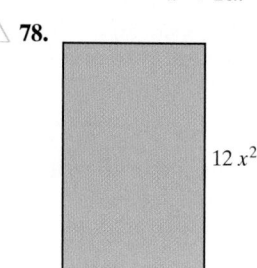

$12\,x^2$

$2x^3 + 16x$

Complete the following sentences in your own words.

79. If $x^2 + bx + c$ is factorable and c is negative, then the signs of the last term factors of the binomial are opposite because....

80. If $x^2 + bx + c$ is factorable and c is positive, then the signs of the last term factors of the binomials are the same because....

Factor each trinomial completely. Don't forget to first factor out the greatest common factor.

81. $2x^2y + 30xy + 100y$
82. $3x^2z^2 + 9xz^2 + 6z^2$
83. $-12x^2y^3 - 24xy^3 - 36y^3$
84. $-4x^2t^4 + 4xt^4 + 24t^4$
85. $y^2(x + 1) - 2y(x + 1) - 15(x + 1)$
86. $z^2(x + 1) - 3z(x + 1) - 70(x + 1)$

Factor each trinomial. (Hint: Notice that $x^{2n} + 4x^n + 3$ factors as $(x^n + 1)(x^n + 3)$.)

87. $x^{2n} + 5x^n + 6$
88. $x^{2n} + 8x^n - 20$

6.3   FACTORING TRINOMIALS OF THE FORM $ax^2 + bx + c$

Objectives

1 Factor trinomials of the form $ax^2 + bx + c$.

2 Factor out a GCF before factoring a trinomial of the form $ax^2 + bx + c$.

3 Factor perfect square trinomials.

4 Factor trinomials of the form $ax^2 + bx + c$ by grouping.

1   In this section, we factor trinomials of the form $ax^2 + bx + c$, such as

$$3x^2 + 11x + 6, \qquad 8x^2 - 22x + 5, \qquad 2x^2 + 13x - 7$$

Notice that the coefficient of the squared variable in these trinomials is a number other than 1. We will factor these trinomials using a trial-and-check method based on FOIL and our work in the last section.

To begin, let's review the relationship between the numerical coefficients of the trinomial and the numerical coefficients of its factored form. For example, since $(2x + 1)(x + 6) = 2x^2 + 13x + 6$, the factored form of $2x^2 + 13x + 6$ is

$$2x^2 + 13x + 6 = (2x + 1)(x + 6)$$

Notice that $2x$ and x are factors of $2x^2$, the first term of the trinomial. Also, 6 and 1 are factors of 6, the last term of the trinomial, as shown:

$$2x^2 + 13x + 6 = (2x + 1)(x + 6)$$

Also notice that $13x$, the middle term, is the sum of the following products:

$$2x^2 + 13x + 6 = (2x + 1)(x + 6)$$

$$1x$$
$$\underline{+12x}$$
$$13x \qquad \text{Middle term}$$

Let's use this pattern to factor $5x^2 + 7x + 2$. First, we find factors of $5x^2$. Since all numerical coefficients in this trinomial are positive, we will use factors with positive numerical coefficients only. Thus, the factors of $5x^2$ are $5x$ and x. Let's try these factors as first terms of the binomials. Thus far, we have

$$5x^2 + 7x + 2 = (5x + \quad)(x + \quad)$$

Next, we need to find positive factors of 2. Positive factors of 2 are 1 and 2. Now we try possible combinations of these factors as second terms of the binomials until we obtain a middle term of $7x$.

$$(5x + 1)(x + 2) = 5x^2 + 11x + 2$$

$$1x$$
$$\underline{+10x}$$
$$11x \longrightarrow \text{Incorrect middle term}$$

Let's try switching factors 2 and 1.

$$(5x + 2)(x + 1) = 5x^2 + 7x + 2$$

$$2x$$
$$+5x$$
$$7x \longrightarrow \text{Correct middle term}$$

Thus the factored form of $5x^2 + 7x + 2$ is $(5x + 2)(x + 1)$. To check, we multiply $(5x + 2)$ and $(x + 1)$. The product is $5x^2 + 7x + 2$.

EXAMPLE 1

Factor $3x^2 + 11x + 6$.

Solution Since all numerical coefficients are positive, we use factors with positive numerical coefficients. We first find factors of $3x^2$.

Factors of $3x^2$: $3x^2 = 3x \cdot x$

If factorable, the trinomial will be of the form

$$3x^2 + 11x + 6 = (3x + \quad)(x + \quad)$$

Next we factor 6.

Factors of 6: $6 = 1 \cdot 6$, $6 = 2 \cdot 3$

Now we try combinations of factors of 6 until a middle term of $11x$ is obtained. Let's try 1 and 6 first.

$$(3x + 1)(x + 6) = 3x^2 + 19x + 6$$

$$1x$$
$$+18x$$
$$19x \longrightarrow \text{Incorrect middle term}$$

Now let's next try 6 and 1.

$$(3x + 6)(x + 1)$$

Before multiplying, notice that the terms of the factor $3x + 6$ have a common factor of 3. The terms of the original trinomial $3x^2 + 11x + 6$ have no common factor other than 1, so the terms of the factored form of $3x^2 + 11x + 6$ can contain no common factor other than 1. This means that $(3x + 6)(x + 1)$ is not a factored form.

Next let's try 2 and 3 as last terms.

$$(3x + 2)(x + 3) = 3x^2 + 11x + 6$$

$$2x$$
$$+9x$$
$$11x \longrightarrow \text{Correct middle term}$$

Thus the factored form of $3x^2 + 11x + 6$ is $(3x + 2)(x + 3)$.

Helpful Hint

If the terms of a trinomial have no common factor (other than 1), then the terms of neither of its binomial factors will contain a common factor (other than 1).

(continued)

For example,

$$\underbrace{3x^2 + 11x + 6}_{\text{no common factor}} \qquad (3x + 3)(x + 2)$$

↑ ↑
Common factor of **3**,
so cannot be the
factorization of
$3x^2 + 11x + 6$

✔ **CONCEPT CHECK**

Do the terms of $3x^2 + 29x + 18$ have a common factor? Without multiplying, decide which of the following factored forms could not be a factored form of $3x^2 + 29x + 18$.

a. $(3x + 18)(x + 1)$ **b.** $(3x + 2)(x + 9)$ **c.** $(3x + 6)(x + 3)$ **d.** $(3x + 9)(x + 2)$

EXAMPLE 2

Factor $8x^2 - 22x + 5$.

Solution Factors of $8x^2$: $8x^2 = 8x \cdot x$, $8x^2 = 4x \cdot 2x$

We'll try $8x$ and x.

$$8x^2 - 22x + 5 = (8x + \quad)(x + \quad)$$

Since the middle term, $-22x$, has a negative numerical coefficient, we factor 5 into negative factors.

Factors of 5: $5 = -1 \cdot -5$

Let's try -1 and -5.

$$(8x - 1)(x - 5) = 8x^2 - 41x + 5$$

$$\underbrace{\substack{-1x \\ +(-40x)}}$$
$$-41x \longrightarrow \text{Incorrect middle term}$$

Now let's try -5 and -1.

$$(8x - 5)(x - 1) = 8x^2 - 13x + 5$$

$$\underbrace{\substack{-5x \\ +(-8x)}}$$
$$-13x \longrightarrow \text{Incorrect middle term}$$

Don't give up yet! We can still try other factors of $8x^2$. Let's try $4x$ and $2x$ with -1 and -5.

$$(4x - 1)(2x - 5) = 8x^2 - 22x + 5$$

$$\underbrace{\substack{-2x \\ +(-20x)}}$$
$$-22x \longrightarrow \text{Correct middle term}$$

The factored form of $8x^2 - 22x + 5$ is $(4x - 1)(2x - 5)$.

EXAMPLE 3

Factor: $2x^2 + 13x - 7$

Solution Factors of $2x^2$: $2x^2 = 2x \cdot x$

Factors of -7: $-7 = -1 \cdot 7$, $-7 = 1 \cdot -7$

Concept Check Answer:
no; a, c, d

We try possible combinations of these factors:

$$(2x + 1)(x - 7) = 2x^2 - 13x - 7 \qquad \text{\textit{Incorrect middle term}}$$
$$(2x - 1)(x + 7) = 2x^2 + 13x - 7 \qquad \text{\textit{Correct middle term}}$$

The factored form of $2x^2 + 13x - 7$ is $(2x - 1)(x + 7)$.

EXAMPLE 4

Factor $10x^2 - 13xy - 3y^2$.

Solution Factors of $10x^2$: $10x^2 = 10x \cdot x$, $10x^2 = 2x \cdot 5x$

Factors of $-3y^2$: $-3y^2 = -3y \cdot y$, $-3y^2 = 3y \cdot -y$

We try some combinations of these factors:

$$(10x - 3y)(x + y) = 10x^2 + 7xy - 3y^2$$
$$(x + 3y)(10x - y) = 10x^2 + 29xy - 3y^2$$
$$(5x + 3y)(2x - y) = 10x^2 + xy - 3y^2$$
$$(2x - 3y)(5x + y) = 10x^2 - 13xy - 3y^2 \qquad \text{\textit{Correct middle term}}$$

The factored form of $10x^2 - 13xy - 3y^2$ is $(2x - 3y)(5x + y)$.

② Don't forget that the best first step in factoring any polynomial is to look for a common factor to factor out.

EXAMPLE 5

Factor $24x^4 + 40x^3 + 6x^2$.

Solution Notice that all three terms have a common factor of $2x^2$. First, factor out $2x^2$.

$$24x^4 + 40x^3 + 6x^2 = 2x^2(12x^2 + 20x + 3)$$

Next, factor $12x^2 + 20x + 3$.

Factors of $12x^2$: $12x^2 = 6x \cdot 2x$, $12x^2 = 4x \cdot 3x$, $12x^2 = 12x \cdot x$

Since all terms in the trinomial have positive numerical coefficients, factor 3 using positive factors only.

Factors of 3: $3 = 1 \cdot 3$

We try some combinations of the factors.

$$2x^2(4x + 3)(3x + 1) = 2x^2(12x^2 + 13x + 3)$$
$$2x^2(12x + 1)(x + 3) = 2x^2(12x^2 + 37x + 3)$$
$$2x^2(2x + 3)(6x + 1) = 2x^2(12x^2 + 20x + 3) \qquad \text{\textit{Correct middle term}}$$

The factored form of $24x^4 + 40x^3 + 6x^2$ is $2x^2(2x + 3)(6x + 1)$.

> ▶ **Helpful Hint**
>
> Don't forget to include the common factor in the factored form.

EXAMPLE 6

Factor $4x^2 - 12x + 9$.

Solution Factors of $4x^2$: $4x^2 = 2x \cdot 2x$, $4x^2 = 4x \cdot x$

Since the middle term $-12x$ has a negative numerical coefficient, factor 9 into negative factors only.

Factors of 9: $9 = -3 \cdot -3, \quad 9 = -1 \cdot -9$

The correct combination is

$$(2x - 3)(2x - 3) = 4x^2 - 12x + 9$$

$$\underbrace{}_{\substack{-6x \\ +(-6x)}}$$

$$-12x \longrightarrow \text{Correct middle term}$$

Thus, $4x^2 - 12x + 9 = (2x - 3)(2x - 3)$, which can also be written as $(2x - 3)^2$.

3   Notice in Example 6 that $4x^2 - 12x + 9 = (2x - 3)^2$. The trinomial $4x^2 - 12x + 9$ is called a **perfect square trinomial** since it is the square of the binomial $2x - 3$.

In the last chapter, we learned a shortcut special product for squaring a binomial, recognizing that

$$(a + b)^2 = a^2 + 2ab + b^2$$

The trinomial $a^2 + 2ab + b^2$ is a perfect square trinomial, since it is the square of the binomial $a + b$. We can use this pattern to help us factor perfect square trinomials. To use this pattern, we must first be able to recognize a perfect square trinomial. A trinomial is a perfect square when its first term is the square of some expression a, its last term is the square of some expression b, and its middle term is twice the product of the expressions a and b. When a trinomial fits this description, its factored form is $(a + b)^2$.

Perfect Square Trinomials

$$a^2 + 2ab + b^2 = (a + b)^2$$
$$a^2 - 2ab + b^2 = (a - b)^2$$

EXAMPLE 7

Factor $x^2 + 12x + 36$.

Solution   This trinomial is a perfect square trinomial since:

1. The first term is the square of x: $x^2 = (x)^2$.
2. The last term is the square of 6: $36 = (6)^2$.
3. The middle term is twice the product of x and 6: $12x = 2 \cdot x \cdot 6$.
Thus, $x^2 + 12x + 36 = (x + 6)^2$.

EXAMPLE 8

Factor $25x^2 + 25xy + 4y^2$.

Solution   Determine whether or not this trinomial is a perfect square by considering the same three questions.

1. Is the first term a square? Yes, $25x^2 = (5x)^2$.
2. Is the last term a square? Yes, $4y^2 = (2y)^2$.
3. Is the middle term twice the product of $5x$ and $2y$? **No.** $2 \cdot 5x \cdot 2y = 20xy$, not $25xy$.

Therefore, $25x^2 + 25xy + 4y^2$ is not a perfect square trinomial. It is factorable, though. Using earlier techniques, we find that $25x^2 + 25xy + 4y^2 = (5x + 4y)(5x + y)$.

> **Helpful Hint**
>
> A perfect square trinomial that is not recognized as such can be factored by other methods.

EXAMPLE 9

Factor $4m^2 - 4m + 1$.

Solution This is a perfect square trinomial since $4m^2 = (2m)^2$, $1 = (1)^2$, and $4m = 2 \cdot 2m \cdot 1$.

$$4m^2 - 4m + 1 = (2m - 1)^2$$

4   If we extend our work from Section 6.1, grouping can also be used to factor trinomials of the form $ax^2 + bx + c$. To use this method, write the trinomial as a four-term polynomial. For example, to factor $2x^2 + 11x + 12$ using grouping, find two numbers whose product is $2 \cdot 12 = 24$ and whose sum is 11. Since we want a positive product and a positive sum, we consider positive pairs of factors of 24 only.

Factors of 24	*Sum of Factors*
1, 24	$1 + 24 = 25$
2, 12	$2 + 12 = 14$
3, 8	$3 + 8 = 11$

The factors are 3 and 8. Use these factors to write the middle term $11x$ as $3x + 8x$. Replace $11x$ with $3x + 8x$ in the original trinomial and factor by grouping.

$$
\begin{aligned}
2x^2 + 11x + 12 &= 2x^2 + 3x + 8x + 12 \\
&= (2x^2 + 3x) + (8x + 12) \\
&= x(2x + 3) + 4(2x + 3) \\
&= (2x + 3)(x + 4)
\end{aligned}
$$

In general, we have the following:

> **Factoring Trinomials of the Form $ax^2 + bx + c$ by Grouping**
>
> **Step 1.** Find two numbers whose product is $a \cdot c$ and whose sum is b.
>
> **Step 2.** Write the middle term, bx, using the factors found in Step 1.
>
> **Step 3.** Factor by grouping.

 EXAMPLE 10

Factor $8x^2 - 14x + 5$ by grouping.

Solution This trinomial is of the form $ax^2 + bx + c$ with $a = 8$, $b = -14$, and $c = 5$.

Step 1. Find two numbers whose product is $a \cdot c$ or $8 \cdot 5 = 40$, and whose sum is b or -14. The numbers are -4 and -10.

Step 2. Write $-14x$ as $-4x - 10x$ so that

$$8x^2 - 14x + 5 = 8x^2 - 4x - 10x + 5$$

Step 3. Factor by grouping.

$$
\begin{aligned}
8x^2 - 4x - 10x + 5 &= 4x(2x - 1) - 5(2x - 1) \\
&= (2x - 1)(4x - 5)
\end{aligned}
$$

EXAMPLE 11

Factor $3x^2 - x - 10$ by grouping.

Solution   In $3x^2 - x - 10$, $a = 3$, $b = -1$, and $c = -10$.

Step 1. Find two numbers whose product is $a \cdot c$ or $3(-10) = -30$ and whose sum is b or -1. The numbers are -6 and 5.

Step 2. $3x^2 - x - 10 = 3x^2 - 6x + 5x - 10$

Step 3. $= 3x(x - 2) + 5(x - 2)$

        $= (x - 2)(3x + 5)$

EXAMPLE 12

Factor $4x^2 + 11x - 3$ by grouping.

Solution   In $4x^2 + 11x - 3$, $a = 4$, $b = 11$, and $c = -3$.

Step 1. Find two numbers whose product is $a \cdot c$ or $4(-3) = -12$ and whose sum is b or 11. The numbers are -1 and 12.

Step 2. $4x^2 + 11x - 3 = 4x^2 - 1x + 12x - 3$

        $= x(4x - 1) + 3(4x - 1)$

        $= (4x - 1)(x + 3)$

STUDY SKILLS REMINDER

Are You Satisfied with Your Performance in This Course Thus Far?

If not, ask yourself the following questions:

▶ Am I attending all class periods and arriving on time?

▶ Am I working and checking my homework assignments?

▶ Am I getting help when I need it?

▶ In addition to my instructor, am I using the supplements to this text that could help me? For example, the tutorial video lessons? MathPro, the tutorial software?

▶ Am I satisfied with my performance on quizzes and tests?

     If you answered no to *any* of these questions, read or reread Section 1.1 for suggestions in these areas. Also, you may want to contact your instructor for additional feedback.

MENTAL MATH

State whether or not each trinomial is a perfect trinomial square.

1. $x^2 + 14x + 49$ **2.** $9x^2 - 12x + 4$ **3.** $y^2 + 2y + 4$

4. $x^2 - 4x + 2$ **5.** $9y^2 + 6y + 1$ **6.** $y^2 - 16y + 64$

EXERCISE SET 6.3

STUDY GUIDE/SSM | CD/VIDEO | PH MATH TUTOR CENTER | MathXL®Tutorials ON CD | MathXL® | MyMathLab®

Factor completely. See Examples 1 through 5 or 10 through 12.

1. $2x^2 + 13x + 15$

2. $3x^2 + 8x + 4$

3. $2x^2 - 9x - 5$

4. $3x^2 + 20x - 63$

5. $2y^2 - y - 6$

6. $8y^2 - 17y + 9$

7. $16a^2 - 24a + 9$

8. $25x^2 + 20x + 4$

9. $36r^2 - 5r - 24$

10. $20r^2 + 27r - 8$

11. $10x^2 + 17x + 3$

12. $21x^2 - 41x + 10$

13. $21x^2 - 48x - 45$

14. $12x^2 - 14x - 10$

15. $12x^2 - 14x - 6$

16. $20x^2 - 2x + 6$

17. $4x^3 - 9x^2 - 9x$

18. $6x^3 - 31x^2 + 5x$

Factor the following perfect square trinomials. See Examples 6 through 9.

19. $x^2 + 22x + 121$

20. $x^2 + 18x + 81$

21. $x^2 - 16x + 64$

22. $x^2 - 12x + 36$

23. $16y^2 - 40y + 25$

24. $9y^2 + 48y + 64$

25. $x^2y^2 - 10xy + 25$

26. $4x^2y^2 - 28xy + 49$

27. Describe a perfect square trinomial.

28. Write a perfect square trinomial that factors as $(x + 3y)^2$.

MIXED PRACTICE
Factor the following completely.

29. $2x^2 - 7x - 99$

30. $2x^2 + 7x - 72$

31. $4x^2 - 8x - 21$

32. $6x^2 - 11x - 10$

33. $30x^2 - 53x + 21$

34. $21x^2 - 6x - 30$

35. $24x^2 - 58x + 9$

36. $36x^2 + 55x - 14$

37. $9x^2 - 24xy + 16y^2$

38. $25x^2 + 60xy + 36y^2$

39. $x^2 - 14xy + 49y^2$

40. $x^2 + 10xy + 25y^2$

41. $2x^2 + 7x + 5$

42. $2x^2 + 7x + 3$

43. $3x^2 - 5x + 1$

44. $3x^2 - 7x + 6$

45. $-2y^2 + y + 10$

46. $-4x^2 - 23x + 6$

47. $16x^2 + 24xy + 9y^2$

48. $4x^2 - 36xy + 81y^2$

49. $8x^2y + 34xy - 84y$

50. $6x^2y^2 - 2xy^2 - 60y^2$

51. $3x^2 + x - 2$

52. $8y^2 + y - 9$

53. $x^2y^2 + 4xy + 4$

54. $x^2y^2 - 6xy + 9$

55. $49y^2 + 42xy + 9x^2$

56. $16x^2 - 8xy + y^2$

57. $3x^2 - 42x + 63$

58. $5x^2 - 75x + 60$

59. $42a^2 - 43a + 6$

60. $54a^2 + 39ab - 8b^2$

61. $18x^2 - 9x - 14$

62. $8x^2 + 6x - 27$

63. $25p^2 - 70pq + 49q^2$

64. $36p^2 - 18pq + 9q^2$

65. $15x^2 - 16x - 15$

66. $12x^2 + 7x - 12$

67. $-27t + 7t^2 - 4$

68. $4t^2 - 7 - 3t$

The area of the largest square in the figure is $(a + b)^2$. Use this figure to answer Exercises 69 through 70.

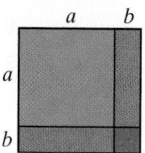

69. Write the area of the largest square as the sum of the areas of the smaller squares and rectangles.

70. What factoring formula from this section is visually represented by this square?

REVIEW AND REVIEW
Multiply the following. See Section 5.4.

71. $(x - 2)(x + 2)$

72. $(y^2 + 3)(y^2 - 3)$

73. $(a + 3)(a^2 - 3a + 9)$

74. $(z - 2)(z^2 + 2z + 4)$

75. $(y - 5)(y^2 + 5y + 25)$

76. $(m + 7)(m^2 - 7m + 49)$

As of 2001, approximately 66% of U.S. households had a personal computer. The following graph shows the percent of selected households having a computer grouped according to household income. See Section 3.1.

Source: *The World Almanac and Book of Facts*, 2000

77. Which range of household income corresponds to the greatest percent of households having a personal computer?

78. Which range of household income corresponds to the greatest increase in percent of households having a personal computer?

79. Describe any trend you notice from this graph.

80. Why don't the percents shown in the graph add to 100%?

Concept Extensions

Find a positive value of b so that each trinomial is factorable.

81. $3x^2 + bx - 5$ **82.** $2y^2 + by + 3$

83. $2z^2 + bz - 7$ **84.** $5x^2 + bx - 1$

Find a positive value of c so that each trinomial is factorable.

85. $5x^2 + 7x + c$ **86.** $7x^2 + 22x + c$

87. $3x^2 - 8x + c$ **88.** $11y^2 - 40y + c$

Factor completely. Don't forget to first factor out the greatest common factor.

89. $-12x^3y^2 + 3x^2y^2 + 15xy^2$ **90.** $-12r^3x^2 + 38r^2x^2 + 14rx^2$

91. $-30p^3q + 88p^2q^2 + 6pq^3$ **92.** $3x^3y^2 + 3x^2y^3 - 18xy^4$

93. $4x^2(y-1)^2 + 10x(y-1)^2 + 25(y-1)^2$

94. $3x^2(a+3)^3 - 28x(a+3)^3 + 25(a+3)^3$

Factor.

95. $3x^{2n} + 17x^n + 10$ **96.** $2x^{2n} + 5x^n - 12$

6.4   FACTORING BINOMIALS

Objectives

1 Factor the difference of two squares.

2 Factor the sum or difference of two cubes.

1 When learning to multiply binomials in Chapter 5, we studied a special product, the product of the sum and difference of two terms, a and b:

$$(a + b)(a - b) = a^2 - b^2$$

For example, the product of $x + 3$ and $x - 3$ is

$$(x + 3)(x - 3) = x^2 - 9$$

The binomial $x^2 - 9$ is called a **difference of squares**. In this section, we use the pattern for the product of a sum and difference to factor the binomial difference of squares.

To use this pattern to help us factor, we must be able to recognize a difference of squares. A binomial is a difference of squares when it is the difference of the square of some expression a and the square of some expression b.

Difference of Two Squares

$$a^2 - b^2 = (a + b)(a - b)$$

EXAMPLE 1

Factor $x^2 - 25$.

Solution $x^2 - 25$ is the difference of two squares since $x^2 - 25 = x^2 - 5^2$. Therefore,

$$x^2 - 25 = x^2 - 5^2 = (x + 5)(x - 5)$$

Multiply to check.

EXAMPLE 2

Factor each difference of squares.

    a. $4x^2 - 1$ **b.** $25a^2 - 9b^2$

Solution   **a.** $4x^2 - 1 = (2x)^2 - 1^2 = (2x + 1)(2x - 1)$
              b. $25a^2 - 9b^2 = (5a)^2 - (3b)^2 = (5a + 3b)(5a - 3b)$

EXAMPLE 3

Factor $x^4 - y^6$.

Solution   Write x^4 as $(x^2)^2$ and y^6 as $(y^3)^2$.

$$x^4 - y^6 = (x^2)^2 - (y^3)^2 = (x^2 + y^3)(x^2 - y^3)$$

EXAMPLE 4

Factor $9x^2 - 36$.

Solution   Remember when factoring always to check first for common factors. If there are common factors, factor out the GCF and then factor the resulting polynomial.

$$9x^2 - 36 = 9(x^2 - 4) \qquad \text{\small Factor out the GCF 9.}$$
$$= 9(x^2 - 2^2)$$
$$= 9(x + 2)(x - 2)$$

In this example, if we forget to factor out the GCF first, we still have the difference of two squares.

$$9x^2 - 36 = (3x)^2 - (6)^2 = (3x + 6)(3x - 6)$$

This binomial has not been factored completely since both terms of both binomial factors have a common factor of 3.

$$3x + 6 = 3(x + 2) \quad \text{and} \quad 3x - 6 = 3(x - 2)$$

Then

$$9x^2 - 36 = (3x + 6)(3x - 6) = 3(x + 2)3(x - 2) = 9(x + 2)(x - 2)$$

Factoring is easier if the GCF is factored out first before using other methods.

EXAMPLE 5

Factor $x^2 + 4$.

Solution   The binomial $x^2 + 4$ is the **sum** of squares since we can write $x^2 + 4$ as $x^2 + 2^2$. We might try to factor using $(x + 2)(x + 2)$ or $(x - 2)(x - 2)$. But when multiplying to check, neither factoring is correct.

$$(x + 2)(x + 2) = x^2 + 4x + 4$$
$$(x - 2)(x - 2) = x^2 - 4x + 4$$

In both cases, the product is a trinomial, not the required binomial. In fact, $x^2 + 4$ is a **prime polynomial**.

> **Helpful Hint**
> After the greatest common factor has been removed, the *sum* of two squares cannot be factored further using real numbers.

2   Although the sum of two squares usually does not factor, the sum or difference of two cubes can be factored and reveals factoring patterns. The pattern for the sum of cubes is illustrated by multiplying the binomial $x + y$ and the trinomial $x^2 - xy + y^2$.

$$x^2 - xy + y^2$$
$$\underline{x + y}$$
$$x^2y - xy^2 + y^3$$
$$\underline{x^3 - x^2y + xy^2}$$
$$x^3 \qquad\qquad + y^3$$
$$(x + y)(x^2 - xy + y^2) = x^3 + y^3 \qquad \text{Sum of cubes}$$

The pattern for the difference of two cubes is illustrated by multiplying the binomial $x - y$ by the trinomial $x^2 + xy + y^2$. The result is

$$(x - y)(x^2 + xy + y^2) = x^3 - y^3 \qquad \text{Difference of cubes}$$

> **Sum or Difference of Two Cubes**
> $$a^3 + b^3 = (a + b)(a^2 - ab + b^2)$$
> $$a^3 - b^3 = (a - b)(a^2 + ab + b^2)$$

EXAMPLE 6

Factor $x^3 + 8$.

Solution   First, write the binomial in the form $a^3 + b^3$.

$$x^3 + 8 = x^3 + 2^3 \qquad \text{Write in the form } a^3 + b^3.$$

If we replace a with x and b with 2 in the formula above, we have

$$x^3 + 2^3 = (x + 2)[x^2 - (x)(2) + 2^2]$$
$$= (x + 2)(x^2 - 2x + 4)$$

> **Helpful Hint**
> When factoring sums or differences of cubes, notice the sign patterns.
>
> $$x^3 + y^3 = (x + y)(x^2 - xy + y^2)$$
> same sign, opposite signs, always positive
>
> $$x^3 - y^3 = (x - y)(x^2 + xy + y^2)$$
> same sign, opposite signs, always positive

EXAMPLE 7

Factor $y^3 - 27$.

Solution

$$y^3 - 27 = y^3 - 3^3 \qquad \text{Write in the form } a^3 - b^3.$$
$$= (y - 3)[y^2 + (y)(3) + 3^2]$$
$$= (y - 3)(y^2 + 3y + 9)$$

EXAMPLE 8

Factor $64x^3 + 1$.

Solution

$$64x^3 + 1 = (4x)^3 + 1^3$$
$$= (4x + 1)[(4x)^2 - (4x)(1) + 1^2]$$
$$= (4x + 1)(16x^2 - 4x + 1)$$

 EXAMPLE 9

Factor $54a^3 - 16b^3$.

Solution

Remember to factor out common factors first before using other factoring methods.

$$54a^3 - 16b^3 = 2(27a^3 - 8b^3) \qquad \text{Factor out the GCF 2.}$$
$$= 2[(3a)^3 - (2b)^3] \qquad \text{Difference of two cubes}$$
$$= 2(3a - 2b)[(3a)^2 + (3a)(2b) + (2b)^2]$$
$$= 2(3a - 2b)(9a^2 + 6ab + 4b^2)$$

Graphing Calculator Explorations

$$\boxed{\begin{array}{l} 2 \rightarrow X \\ \qquad\qquad 2 \\ X^2 - 6X \\ \qquad\qquad -8 \end{array}}$$

A graphing calculator is a convenient tool for evaluating an expression at a given replacement value. For example, let's evaluate $x^2 - 6x$ when $x = 2$. To do so, store the value 2 in the variable x and then enter and evaluate the algebraic expression. The value of $x^2 - 6x$ when $x = 2$ is -8. You may want to use this method for evaluating expressions as you explore the following.

We can use a graphing calculator to explore factoring patterns numerically. Use your calculator to evaluate $x^2 - 2x + 1$, $x^2 - 2x - 1$, and $(x - 1)^2$ for each value of x given in the table. What do you observe?

	$x^2 - 2x + 1$	$x^2 - 2x - 1$	$(x - 1)^2$
$x = 5$			
$x = -3$			
$x = 2.7$			
$x = -12.1$			
$x = 0$			

(continued)

Notice in each case that $x^2 - 2x - 1 \neq (x - 1)^2$. Because for each x in the table the value of $x^2 - 2x + 1$ and the value of $(x - 1)^2$ are the same, we might guess that $x^2 - 2x + 1 = (x - 1)^2$. We can verify our guess algebraically with multiplication:

$$(x - 1)(x - 1) = x^2 - x - x + 1 = x^2 - 2x + 1$$

MENTAL MATH

Write each number as a square.

1. 1   **2.** 25   **3.** 81   **4.** 64   **5.** 9   **6.** 100

Write each number as a cube.

7. 1   **8.** 64   **9.** 8   **10.** 27

EXERCISE SET 6.4

STUDY GUIDE/SSM   CD/VIDEO   PH MATH TUTOR CENTER   MathXL®Tutorials ON CD   MathXL®   MyMathLab®

Factor the difference of two squares. See Examples 1 through 4.

1. $x^2 - 4$

2. $y^2 - 81$

3. $y^2 - 49$

4. $x^2 - 100$

5. $25y^2 - 9$

6. $49a^2 - 16$

7. $121 - 100x^2$

8. $144 - 81x^2$

9. $12x^2 - 27$

10. $36x^2 - 64$

11. $169a^2 - 49b^2$

12. $225a^2 - 81b^2$

13. $x^2y^2 - 1$

14. $16 - a^2b^2$

15. $x^4 - 9$

16. $y^4 - 25$

17. $49a^4 - 16$

18. $49b^4 - 1$

19. $x^4 - y^{10}$

20. $x^{14} - y^4$

21. What binomial multiplied by $(x - 6)$ gives the difference of two squares?

22. What binomial multiplied by $(5 + y)$ gives the difference of two squares?

Factor the sum or difference of two cubes. See Examples 6 through 9.

23. $a^3 + 27$

24. $b^3 - 8$

25. $8a^3 + 1$

26. $64x^3 - 1$

27. $5k^3 + 40$

28. $6r^3 - 162$

29. $x^3y^3 - 64$

30. $8x^3 - y^3$

31. $x^3 + 125$

32. $a^3 - 216$

33. $24x^4 - 81xy^3$

34. $375y^6 - 24y^3$

35. What binomial multiplied by $(4x^2 - 2xy + y^2)$ gives the sum or difference of two cubes?

36. What binomial multiplied by $(1 + 4y + 16y^2)$ gives the sum or difference of two cubes?

MIXED PRACTICE

Factor the binomials completely.

37. $x^2 - 121$

38. $x^2 - 36$

39. $81 - p^2$

40. $100 - t^2$

41. $4r^2 - 1$

42. $9t^2 - 1$

43. $9x^2 - 16$

44. $36y^2 - 25$

45. $16r^2 + 1$

46. $49y^2 + 1$

47. $27 - t^3$

48. $125 + r^3$

49. $8r^3 - 64$

50. $54r^3 + 2$

51. $t^3 - 343$

52. $s^3 + 216$

53. $x^2 - 169y^2$

54. $x^2 - 225y^2$

55. $x^2y^2 - z^2$

56. $x^3y^3 - z^3$

57. $x^3y^3 + 1$

58. $x^2y^2 + z^2$

🔒 **59.** $s^3 - 64t^3$

60. $8t^3 + s^3$

61. $18r^2 - 8$

62. $32t^2 - 50$

63. $9xy^2 - 4x$

64. $16xy^2 - 64x$

65. $25y^4 - 100y^2$

🔒 **66.** $xy^3 - 9xyz^2$

67. $x^3y - 4xy^3$

68. $12s^3t^3 + 192s^5t$

69. $8s^6t^3 + 100s^3t^6$

70. $25x^5y + 121x^3y$

71. $27x^2y^3 - xy^2$

72. $8x^3y^3 + x^3y$

REVIEW AND PREVIEW

Divide the following. See Section 5.6.

73. $\dfrac{8x^4 + 4x^3 - 2x + 6}{2x}$

74. $\dfrac{3y^4 + 9y^2 - 6y + 1}{3y^2}$

Use long division to divide the following. See Section 5.6.

75. $\dfrac{2x^2 - 3x - 2}{x - 2}$

76. $\dfrac{4x^2 - 21x + 21}{x - 3}$

77. $\dfrac{3x^2 + 13x + 10}{x + 3}$

78. $\dfrac{5x^2 + 14x + 12}{x + 2}$

Concept Extensions

Factor each expression completely.

79. $(x + 2)^2 - y^2$

80. $(y - 6)^2 - z^2$

81. $a^2(b - 4) - 16(b - 4)$

82. $m^2(n + 8) - 9(n + 8)$

83. $(x^2 + 6x + 9) - 4y^2$ (*Hint:* Factor the trinomial in parentheses first.)

84. $(x^2 + 2x + 1) - 36y^2$

85. $x^{2n} - 100$

86. $x^{2n} - 81$

87. An object is dropped from the top of Pittsburgh's USX Towers, which is 841 feet tall. (*Source: World Almanac* research) The height of the object after t seconds is given by the expression $841 - 16t^2$.

   a. Find the height of the object after 2 seconds.

   b. Find the height of the object after 5 seconds.

   c. To the nearest whole second, estimate when the object hits the ground.

   d. Factor $841 - 16t^2$.

841 feet

88. A worker on the top of the Aetna Life Building in San Francisco accidentally drops a bolt. The Aetna Life Building is 529 feet tall. (*Source: World Almanac* research) The height of the bolt after t seconds is given by the expression $529 - 16t^2$.

   a. Find the height of the bolt after 1 second.

   b. Find the height of the bolt after 4 seconds.

   c. To the nearest whole second, estimate when the bolt hits the ground.

   d. Factor $529 - 16t^2$.

529 feet

89. At this writing, the world's tallest building is the Petronas Twin Towers in Kuala Lumpur, Malaysia, at a height of 1483 feet. (*Source:* Council on Tall Buildings and Urban Habitat) Suppose a worker is suspended 39 feet below the tip of the pinnacle atop one of the towers, at a height of 1444 feet above the ground. If the worker accidentally drops a bolt, the height of the bolt after t seconds is given by the expression $1444 - 16t^2$.

   a. Find the height of the bolt after 3 seconds.

   b. Find the height of the bolt after 7 seconds.

   c. To the nearest whole second, estimate when the bolt hits the ground.

   d. Factor $1444 - 16t^2$.

90. A performer with the Moscow Circus is planning a stunt involving a free fall from the top of the Moscow State University building, which is 784 feet tall. (*Source:* Council on Tall Buildings and Urban Habitat) Neglecting air resistance, the performer's height above gigantic cushions positioned at ground level after t seconds is given by the expression $784 - 16t^2$.

c. To the nearest whole second, estimate when the performer reaches the cushions positioned at ground level.

d. Factor $784 - 16t^2$.

91. In your own words, explain how to tell whether a binomial is a difference of squares. Then explain how to factor a difference of squares.

a. Find the performer's height after 2 seconds.

b. Find the performer's height after 5 seconds.

92. In your own words, explain how to tell whether a binomial is a sum of cubes. Then explain how to factor a sum of cubes.

Spotlight on DECISION & MAKING

Suppose you are shopping for a credit card and have received the following credit card offers in the mail.

Credit Card Offer No.1

We offer a low 9.8% APR interest rate, coupled with a generous 20-day grace period and low $35 annual fee. We can offer you a $2000 credit limit.

In addition, you'll earn one frequent flier mile for every dollar charged on your card. And every time you use your card to pay for airline tickets, we'll even give you $250,000 in flight insurance—absolutely free!

Credit Card Offer No. 2

With our card, you'll get a 14.5% APR interest rate, absolutely *no* annual fee, a 25-day grace period, and a $2500 credit limit.

And, only with our card, you can receive a year-end cash bonus of up to 1% of the total value of purchases made to your card during the year! That's our way of saying "Thank you" for choosing our card.

Credit Card Offer No. 3

Just say "yes" to this offer, and you can get our lowest 17% APR interest rate, along with a 30-day grace period and easy-to-swallow $20 annual fee. You also qualify for a $3000 credit limit.

That's not all: you earn gift certificates for a local mall with every purchase! Plus, you'll receive $100,000 of flight insurance whenever you charge airline tickets.

You construct a decision grid to help make your choice. In the decision grid, give each of the decision criteria a rank reflecting its importance to you, with 1 being not important to 10 being very important. Then for each card offer, decide how well the criteria are supported, assigning a rating of 1 for poor support to a rating of 10 for excellent support. For Credit Card Offer #1, fill in the Score column by multiplying rank by rating for each criteria. Repeat for each credit card offer. Finally, total the scores for each credit card offer. The offer with the highest score is likely to be the best choice for you.

Based on your decision-grid analysis, which credit card would you choose? Explain.

(continued)

Spotlight on **DECISION ⚘ MAKING**

Criteria	Rank	Credit Card Offer #1		Credit Card Offer #2		Credit Card Offer #3	
		Rating	*Score*	*Rating*	*Score*	*Rating*	*Score*
Interest rate							
Annual fee							
Grace period							
Credit limit							
Automatic flight insurance							
Rebates/ incentives/bonuses							
TOTAL							

INTEGRATED REVIEW   CHOOSING A FACTORING STRATEGY

The following steps may be helpful when factoring polynomials.

Factoring a Polynomial

Step 1. Are there any common factors? If so, factor out the GCF.

Step 2. How many terms are in the polynomial?

  a. If there are **two** terms, decide if one of the following can be applied.

   i. Difference of two squares: $a^2 - b^2 = (a + b)(a - b)$.

   ii. Difference of two cubes: $a^3 - b^3 = (a - b)(a^2 + ab + b^2)$.

   iii. Sum of two cubes: $a^3 + b^3 = (a + b)(a^2 - ab + b^2)$.

  b. If there are **three** terms, try one of the following.

   i. Perfect square trinomial: $a^2 + 2ab + b^2 = (a + b)^2$.

   ii. If not a perfect square trinomial, factor using the methods presented in Sections 6.2 and 6.3.

  c. If there are **four** or more terms, try factoring by grouping.

Step 3. See if any factors in the factored polynomial can be factored further.

Step 4. Check by multiplying.

Study the next five examples to help you use the steps on the previous page.

EXAMPLE 1

Factor $10t^2 - 17t + 3$.

Solution   **Step 1.**   The terms of this polynomial have no common factor (other than 1).

Step 2.   There are three terms, so this polynomial is a trinomial. This trinomial is not a perfect square trinomial, so factor using methods from earlier sections.

Factors of $10t^2$: $10t^2 = 2t \cdot 5t$, $10t^2 = t \cdot 10t$

Since the middle term, $-17t$, has a negative numerical coefficient, find negative factors of 3.

Factors of 3: $3 = -1 \cdot -3$

Try different combinations of these factors. The correct combination is

$$(2t - 3)(5t - 1) = 10t^2 - 17t + 3$$

$$\underbrace{}$$
$$-15t$$
$$\underline{-2t}$$
$$-17t \qquad \text{\textit{Correct middle term}}$$

Step 3.   No factor can be factored further, so we have factored completely.

Step 4.   To check, multiply $2t - 3$ and $5t - 1$.

$$(2t - 3)(5t - 1) = 10t^2 - 2t - 15t + 3 = 10t^2 - 17t + 3$$

The factored form of $10t^2 - 17t + 3$ is $(2t - 3)(5t - 1)$.

EXAMPLE 2

Factor $2x^3 + 3x^2 - 2x - 3$.

Solution   **Step 1.**   There are no factors common to all terms.

Step 2.   Try factoring by grouping since this polynomial has four terms.

$$2x^3 + 3x^2 - 2x - 3 = x^2(2x + 3) - 1(2x + 3) \quad \textit{Factor out the greatest common factor for each pair of terms.}$$

$$= (2x + 3)(x^2 - 1) \quad \textit{Factor out } 2x + 3.$$

Step 3.   The binomial $x^2 - 1$ can be factored further. It is the difference of two squares.

$$= (2x + 3)(x + 1)(x - 1) \quad \textit{Factor } x^2 - 1 \textit{ as a difference of squares.}$$

Step 4.   Check by finding the product of the three binomials.

The polynomial factored completely is $(2x + 3)(x + 1)(x - 1)$.

EXAMPLE 3

Factor $12m^2 - 3n^2$.

Solution   **Step 1.**   The terms of this binomial contain a greatest common factor of 3.

$$12m^2 - 3n^2 = 3(4m^2 - n^2) \quad \textit{Factor out the greatest common factor.}$$

382 CHAPTER 6 FACTORING POLYNOMIALS

Step 2. The binomial $4m^2 - n^2$ is a difference of squares.

$$= 3(2m + n)(2m - n) \quad \text{Factor the difference of squares.}$$

Step 3. No factor can be factored further.

Step 4. We check by multiplying.

$$3(2m + n)(2m - n) = 3(4m^2 - n^2) = 12m^2 - 3n^2$$

The factored form of $12m^2 - 3n^2$ is $3(2m + n)(2m - n)$. ⬭

EXAMPLE 4

Factor $x^3 + 27y^3$.

Solution **Step 1.** The terms of this binomial contain no common factor (other than 1).

Step 2. This binomial is the sum of two cubes.

$$x^3 + 27y^3 = (x)^3 + (3y)^3$$
$$= (x + 3y)[x^2 - x(3y) + (3y)^2]$$
$$= (x + 3y)(x^2 - 3xy + 9y^2)$$

Step 3. No factor can be factored further.

Step 4. We check by multiplying.

$$(x + 3y)(x^2 - 3xy + 9y^2) = x(x^2 - 3xy + 9y^2) + 3y(x^2 - 3xy + 9y^2)$$
$$= x^3 - 3x^2y + 9xy^2 + 3x^2y - 9xy^2 + 27y^3$$
$$= x^3 + 27y^3$$

Thus, $x^3 + 27y^3$ factored completely is $(x + 3y)(x^2 - 3xy + 9y^2)$. ⬭

EXAMPLE 5

Factor $30a^2b^3 + 55a^2b^2 - 35a^2b$.

Solution **Step 1.** $30a^2b^3 + 55a^2b^2 - 35a^2b = 5a^2b(6b^2 + 11b - 7)$ Factor out the GCF.

Step 2. $= 5a^2b(2b - 1)(3b + 7)$ Factor the resulting trinomial.

Step 3. No factor can be factored further.

Step 4. Check by multiplying.

The trinomial factored completely is $5a^2b(2b - 1)(3b + 7)$. ⬭

Factor the following completely.

1. $a^2 + 2ab + b^2$ **2.** $a^2 - 2ab + b^2$ **3.** $a^2 + a - 12$

4. $a^2 - 7a + 10$ **5.** $a^2 - a - 6$ **6.** $a^2 + 2a + 1$

7. $x^2 + 2x + 1$ **8.** $x^2 + x - 2$ **9.** $x^2 + 4x + 3$

10. $x^2 + x - 6$ **11.** $x^2 + 7x + 12$ **12.** $x^2 + x - 12$

13. $x^2 + 3x - 4$ **14.** $x^2 - 7x + 10$ **15.** $x^2 + 2x - 15$

16. $x^2 + 11x + 30$ **17.** $x^2 - x - 30$ **18.** $x^2 + 11x + 24$

19. $2x^2 - 98$ **20.** $3x^2 - 75$ **21.** $x^2 + 3x + xy + 3y$

22. $3y - 21 + xy - 7x$ **23.** $x^2 + 6x - 16$ **24.** $x^2 - 3x - 28$

25. $4x^3 + 20x^2 - 56x$ **26.** $6x^3 - 6x^2 - 120x$ **27.** $12x^2 + 34x + 24$

28. $8a^2 + 6ab - 5b^2$ **29.** $4a^2 - b^2$ **30.** $28 - 13x - 6x^2$

31. $20 - 3x - 2x^2$

32. $x^2 - 2x + 4$

33. $a^2 + a - 3$

34. $6y^2 + y - 15$

35. $4x^2 - x - 5$

36. $x^2y - y^3$

37. $4t^2 + 36$

38. $x^2 + x + xy + y$

39. $ax + 2x + a + 2$

40. $18x^3 - 63x^2 + 9x$

41. $12a^3 - 24a^2 + 4a$

42. $x^2 + 14x - 32$

43. $x^2 - 14x - 48$

44. $16a^2 - 56ab + 49b^2$

45. $25p^2 - 70pq + 49q^2$

46. $7x^2 + 24xy + 9y^2$

47. $125 - 8y^3$

48. $64x^3 + 27$

49. $-x^2 - x + 30$

50. $-x^2 + 6x - 8$

51. $14 + 5x - x^2$

52. $3 - 2x - x^2$

53. $3x^4y + 6x^3y - 72x^2y$

54. $2x^3y + 8x^2y^2 - 10xy^3$

55. $5x^3y^2 - 40x^2y^3 + 35xy^4$

56. $4x^4y - 8x^3y - 60x^2y$

57. $12x^3y + 243xy$

58. $6x^3y^2 + 8xy^2$

59. $4 - x^2$

60. $9 - y^2$

61. $3rs - s + 12r - 4$

62. $x^3 - 2x^2 + 3x - 6$

63. $4x^2 - 8xy - 3x + 6y$

64. $4x^2 - 2xy - 7yz + 14xz$

65. $6x^2 + 18xy + 12y^2$

66. $12x^2 + 46xy - 8y^2$

67. $xy^2 - 4x + 3y^2 - 12$

68. $x^2y^2 - 9x^2 + 3y^2 - 27$

69. $5(x + y) + x(x + y)$

70. $7(x - y) + y(x - y)$

71. $14t^2 - 9t + 1$

72. $3t^2 - 5t + 1$

73. $3x^2 + 2x - 5$

74. $7x^2 + 19x - 6$

75. $x^2 + 9xy - 36y^2$

76. $3x^2 + 10xy - 8y^2$

77. $1 - 8ab - 20a^2b^2$

78. $1 - 7ab - 60a^2b^2$

79. $9 - 10x^2 + x^4$

80. $36 - 13x^2 + x^4$

81. $x^4 - 14x^2 - 32$

82. $x^4 - 22x^2 - 75$

83. $x^2 - 23x + 120$

84. $y^2 + 22y + 96$

85. $6x^3 - 28x^2 + 16x$

86. $6y^3 - 8y^2 - 30y$

87. $27x^3 - 125y^3$

88. $216y^3 - z^3$

89. $x^3y^3 + 8z^3$

90. $27a^3b^3 + 8$

91. $2xy - 72x^3y$

92. $2x^3 - 18x$

93. $x^3 + 6x^2 - 4x - 24$

94. $x^3 - 2x^2 - 36x + 72$

95. $6a^3 + 10a^2$

96. $4n^2 - 6n$

97. $a^2(a + 2) + 2(a + 2)$

98. $a - b + x(a - b)$

99. $x^3 - 28 + 7x^2 - 4x$

100. $a^3 - 45 - 9a + 5a^2$

Concept Extensions

Factor.

101. $(x - y)^2 - z^2$

102. $(x + 2y)^2 - 9$

103. $81 - (5x + 1)^2$

104. $b^2 - (4a + c)$

105. Explain why it makes good sense to factor out the GCF first, before using other methods of factoring.

106. The sum of two squares usually does not factor. Is the sum of two squares $9x^2 + 81y^2$ factorable?

107. Which of the following are equivalent to $(x + 10)(x - 7)$

  a. $(x - 7)(x + 10)$ **b.** $-1(x + 10)(x - 7)$ **c.** $-1(x + 10)(7 - x)$ **d.** $-1(-x - 10)(7 - x)$

6.5   SOLVING QUADRATIC EQUATIONS BY FACTORING

Objectives

1   Define quadratic equation.

2   Solve quadratic equations by factoring.

3   Solve equations with degree greater than 2 by factoring.

①   Linear equations, while versatile, are not versatile enough to model many real-life phenomena. For example, let's suppose an object is dropped from the top of a 256-foot cliff and we want to know how long before the object strikes the ground. The answer to this question is found by solving the equation $-16t^2 + 256 = 0$. (See Example 1 in the next section.) This equation is called a **quadratic equation** because it contains a variable with an exponent of 2, and no other variable in the equation contains an exponent greater than 2. In this section, we solve quadratic equations by factoring.

256 feet

Quadratic Equation

A quadratic equation is one that can be written in the form

$$ax^2 + bx + c = 0,$$

where a, b, and c are real numbers, and $a \neq 0$.

Notice that the degree of the polynomial $ax^2 + bx + c$ is 2. Here are a few more examples of quadratic equations.

Quadratic Equations

$$3x^2 + 5x + 6 = 0 \qquad x^2 = 9 \qquad y^2 + y = 1$$

The form $ax^2 + bx + c = 0$ is called the **standard form** of a quadratic equation. The quadratic equations $3x^2 + 5x + 6 = 0$ and $-16t^2 + 256 = 0$ are in standard form. One side of the equation is 0 and the other side is a polynomial of degree 2 written in descending powers of the variable.

②   Some quadratic equations can be solved by making use of factoring and the **zero factor theorem**.

Zero Factor Theorem

If a and b are real numbers and if $ab = 0$, then $a = 0$ or $b = 0$.

This theorem states that if the product of two numbers is 0 then at least one of the numbers must be 0.

EXAMPLE 1

Solve $(x - 3)(x + 1) = 0$.

Solution   If this equation is to be a true statement, then either the factor $x - 3$ must be 0 or the factor $x + 1$ must be 0. In other words, either

$$x - 3 = 0 \qquad \text{or} \qquad x + 1 = 0$$

If we solve these two linear equations, we have

$$x = 3 \quad \text{or} \quad x = -1$$

Thus, 3 and -1 are both solutions of the equation $(x - 3)(x + 1) = 0$. To check, we replace x with 3 in the original equation. Then we replace x with -1 in the original equation.

Check

$$(x - 3)(x + 1) = 0 \qquad\qquad\qquad (x - 3)(x + 1) = 0$$
$$(3 - 3)(3 + 1) \stackrel{?}{=} 0 \quad \text{Replace } x \text{ with 3.} \quad (-1 - 3)(-1 + 1) \stackrel{?}{=} 0 \quad \text{Replace } x \text{ with } -1.$$
$$0(4) = 0 \quad \text{True} \qquad\qquad\qquad (-4)(0) = 0 \quad \text{True}$$

The solutions are 3 and -1, or we say that the solution set is $\{-1, 3\}$.

> ### Helpful Hint
>
> The zero factor property says that *if a product is 0, then a factor is 0*.
>
> If $a \cdot b = 0$, then $a = 0$ or $b = 0$.
>
> If $x(x + 5) = 0$, then $x = 0$ or $x + 5 = 0$.
>
> If $(x + 7)(2x - 3) = 0$, then $x + 7 = 0$ or $2x - 3 = 0$.
>
> Use this property only when the product is 0. For example, if $a \cdot b = 8$, we do not know the value of a or b. The values may be $a = 2, b = 4$ or $a = 8, b = 1$, or any other two numbers whose product is 8.

EXAMPLE 2

Solve $x^2 - 9x = -20$.

Solution   First, write the equation in standard form; then factor.

$$x^2 - 9x = -20$$
$$x^2 - 9x + 20 = 0 \qquad \text{Write in standard form by adding 20 to both sides.}$$
$$(x - 4)(x - 5) = 0 \qquad \text{Factor.}$$

Next, use the zero factor theorem and set each factor equal to 0.

$$x - 4 = 0 \quad \text{or} \quad x - 5 = 0 \qquad \text{Set each factor equal to 0.}$$
$$x = 4 \quad \text{or} \quad x = 5 \qquad \text{Solve.}$$

Check the solutions by replacing x with each value in the original equation. The solutions are 4 and 5.

The following steps may be used to solve a quadratic equation by factoring.

> ### Solving Quadratic Equations by Factoring
>
> **Step 1.** Write the equation in standard form: $ax^2 + bx + c = 0$.
>
> **Step 2.** Factor the quadratic completely.
>
> **Step 3.** Set each factor containing a variable equal to 0.
>
> **Step 4.** Solve the resulting equations.
>
> **Step 5.** Check each solution in the original equation.

Since it is not always possible to factor a quadratic polynomial, not all quadratic equations can be solved by factoring. Other methods of solving quadratic equations are presented in Chapter 11.

EXAMPLE 3

Solve $x(2x - 7) = 4$.

Solution First, write the equation in standard form; then factor.

$$x(2x - 7) = 4$$
$$2x^2 - 7x = 4 \qquad \text{Multiply.}$$
$$2x^2 - 7x - 4 = 0 \qquad \text{Write in standard form.}$$
$$(2x + 1)(x - 4) = 0 \qquad \text{Factor.}$$

$$2x + 1 = 0 \quad \text{or} \quad x - 4 = 0 \quad \text{Set each factor equal to zero.}$$
$$2x = -1 \quad \text{or} \quad x = 4 \quad \text{Solve.}$$
$$x = -\frac{1}{2}$$

Check both solutions $-\dfrac{1}{2}$ and 4.

> **Helpful Hint**
>
> To solve the equation $x(2x - 7) = 4$, above do **not** set each factor equal to 4. Remember that to apply the zero factor property, one side of the equation *must be* 0 and the other side of the equation must be in factored form.

EXAMPLE 4

Solve $-2x^2 - 4x + 30 = 0$.

Solution The equation is in standard form so we begin by factoring out a common factor of -2.

$$-2x^2 - 4x + 30 = 0$$
$$-2(x^2 + 2x - 15) = 0 \qquad \text{Factor out } -2.$$
$$-2(x + 5)(x - 3) = 0 \qquad \text{Factor the quadratic.}$$

Next, set each factor **containing a variable** equal to 0.

$$x + 5 = 0 \quad \text{or} \quad x - 3 = 0 \quad \text{Set each factor containing a variable equal to 0.}$$
$$x = -5 \quad \text{or} \quad x = 3 \quad \text{Solve.}$$

Note that the factor -2 is a constant term containing no variables and can never equal 0. The solutions are -5 and 3.

3 Some equations involving polynomials of degree higher than 2 may also be solved by factoring and then applying the zero factor theorem.

EXAMPLE 5

Solve $3x^3 - 12x = 0$.

Solution Factor the left side of the equation. Begin by factoring out the common factor of $3x$.

$$3x^3 - 12x = 0$$
$$3x(x^2 - 4) = 0 \qquad \text{Factor out the GCF } 3x.$$
$$3x(x + 2)(x - 2) = 0 \qquad \text{Factor } x^2 - 4, \text{ a difference of squares.}$$

$$3x = 0 \quad \text{or} \quad x + 2 = 0 \quad \text{or} \quad x - 2 = 0 \qquad \textit{Set each factor}$$
$$\textit{equal to 0.}$$

$$x = 0 \quad \text{or} \quad\quad x = -2 \quad \text{or} \quad\quad x = 2 \qquad \textit{Solve.}$$

Thus, the equation $3x^3 - 12x = 0$ has three solutions: 0, -2, and 2. To check, replace x with each solution in the original equation.

Let $x = 0$.	**Let $x = -2$.**	**Let $x = 2$.**
$3(0)^3 - 12(0) \overset{?}{=} 0$	$3(-2)^3 - 12(-2) \overset{?}{=} 0$	$3(2)^3 - 12(2) \overset{?}{=} 0$
$0 = 0$	$3(-8) + 24 \overset{?}{=} 0$	$3(8) - 24 \overset{?}{=} 0$
	$0 = 0$	$0 = 0$

Substituting 0, -2, or 2 into the original equation results each time in a true equation. The solutions are 0, -2, and 2.

EXAMPLE 6

Solve $(5x - 1)(2x^2 + 15x + 18) = 0$.

Solution

$$(5x - 1)(2x^2 + 15x + 18) = 0$$
$$(5x - 1)(2x + 3)(x + 6) = 0 \qquad \textit{Factor the trinomial.}$$

$$5x - 1 = 0 \quad \text{or} \quad 2x + 3 = 0 \quad \text{or} \quad x + 6 = 0 \qquad \textit{Set each factor equal to 0.}$$

$$5x = 1 \quad \text{or} \quad\quad 2x = -3 \quad \text{or} \quad x = -6 \qquad \textit{Solve.}$$

$$x = \frac{1}{5} \quad \text{or} \quad\quad x = -\frac{3}{2}$$

The solutions are $\frac{1}{5}$, $-\frac{3}{2}$, and -6. Check by replacing x with each solution in the original equation. The solutions are -6, $-\frac{3}{2}$, and $\frac{1}{5}$.

EXAMPLE 7

Solve $2x^3 - 4x^2 - 30x = 0$.

Solution   Begin by factoring out the GCF $2x$.

$$2x^3 - 4x^2 - 30x = 0$$
$$2x(x^2 - 2x - 15) = 0 \qquad \textit{Factor out the GCF 2x.}$$
$$2x(x - 5)(x + 3) = 0 \qquad \textit{Factor the quadratic.}$$

$$2x = 0 \quad \text{or} \quad x - 5 = 0 \quad \text{or} \quad x + 3 = 0 \qquad \textit{Set each factor containing a variable equal to 0.}$$

$$x = 0 \quad \text{or} \quad\quad x = 5 \quad \text{or} \quad\quad x = -3 \qquad \textit{Solve.}$$

Check by replacing x with each solution in the cubic equation. The solutions are -3, 0, and 5.

In Chapter 3, we graphed linear equations in two variables, such as $y = 5x - 6$. Recall that to find the x-intercept of the graph of a linear equation, let $y = 0$ and solve for x. This is also how to find the x-intercepts of the graph of a **quadratic equation in two variables**, such as $y = x^2 - 5x + 4$.

EXAMPLE 8

Find the x-intercepts of the graph of $y = x^2 - 5x + 4$.

Solution Let $y = 0$ and solve for x.

$$y = x^2 - 5x + 4$$
$$0 = x^2 - 5x + 4 \qquad \text{Let } y = 0.$$
$$0 = (x - 1)(x - 4) \qquad \text{Factor.}$$
$$x - 1 = 0 \quad \text{or} \quad x - 4 = 0 \qquad \text{Set each factor equal to 0.}$$
$$x = 1 \quad \text{or} \quad x = 4 \qquad \text{Solve.}$$

The x-intercepts of the graph of $y = x^2 - 5x + 4$ are $(1, 0)$ and $(4, 0)$.

The graph of $y = x^2 - 5x + 4$ is shown in the margin.

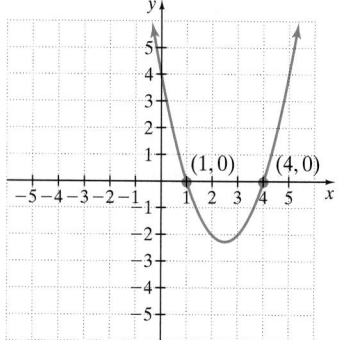

In general, a quadratic equation in two variables is one that can be written in the form $y = ax^2 + bx + c$ where $a \neq 0$. The graph of such an equation is called a **parabola** and will open up or down depending on the value of a.

Notice that the x-intercepts of the graph of $y = ax^2 + bx + c$ are the real number solutions of $0 = ax^2 + bx + c$. Also, the real number solutions of $0 = ax^2 + bx + c$ are the x-intercepts of the graph of $y = ax^2 + bx + c$. We study more about graphs of quadratic equations in two variables in Chapter 11.

Graph of $y = ax^2 + bx + c$
x-intercepts are solutions of $0 = ax^2 + bx + c$

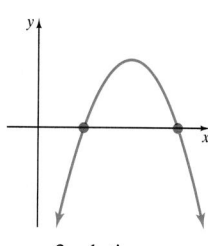

no solution   1 solution   2 solutions   2 solutions

Graphing Calculator Explorations

$y = x^2 + 4x - 3$

$x = .63829787 \quad y = -.0393843$

A grapher may be used to find solutions of a quadratic equation whether the related quadratic polynomial is factorable or not. For example, let's use a grapher to approximate the solutions of $0 = x^2 + 4x - 3$. To do so, graph $y_1 = x^2 + 4x - 3$. Recall that the x-intercepts of this graph are the solutions of $0 = x^2 + 4x - 3$.

Notice that the graph appears to have an x-intercept between -5 and -4 and one between 0 and 1. Many graphers contain a TRACE feature. This feature activates a graph cursor that can be used to *trace* along a graph while the corresponding x- and y-coordinates are shown on the screen. Use the TRACE feature to confirm that x-intercepts lie between -5 and -4 and also 0 and 1. To approximate the x-intercepts to the nearest tenth, use a ROOT or a ZOOM feature on your grapher or redefine the viewing window. (A ROOT feature calculates the x-intercept. A ZOOM feature magnifies the viewing window around a specific location such as the graph cursor.) If we redefine the window to $[0, 1]$ on the x-axis and $[-1, 1]$ on the y-axis, the following graph is generated.

By using the TRACE feature, we can conclude that one x-intercept is approximately 0.6 to the nearest tenth. By repeating these steps for the other x-intercept, we find that it is approximately -4.6.

(continued)

Use a grapher to approximate the real number solutions to the nearest tenth. If an equation has no real number solution, state so.

1. $3x^2 - 4x - 6 = 0$

2. $x^2 - x - 9 = 0$

3. $2x^2 + x + 2 = 0$

4. $-4x^2 - 5x - 4 = 0$

5. $-x^2 + x + 5 = 0$

6. $10x^2 + 6x - 3 = 0$

STUDY SKILLS REMINDER

Continue your outline started in Section 2.3. Write how to recognize and how to solve quadratic and higher degree equations by factoring in your own words. For example:

Solving Equations and Inequalities

I. Equations

    A. Linear equations (Sec. 2.3)

    B. Quadratic and Higher Degree Equations–Recognize: *highest power on variable in equation is at least 2 when written in standard form (one side 0)*–Solve: write in standard form and try to factor the polynomial on one side of the equation. If it factors, set each factor equal to 0.

II. Inequalities

    A. Linear inequalities (Sec. 2.7)

See Appendix A for summary exercises.

MENTAL MATH

Solve each equation by inspection.

1. $(a - 3)(a - 7) = 0$

2. $(a - 5)(a - 2) = 0$

3. $(x + 8)(x + 6) = 0$

4. $(x + 2)(x + 3) = 0$

5. $(x + 1)(x - 3) = 0$

6. $(x - 1)(x + 2) = 0$

EXERCISE SET 6.5

STUDY GUIDE/SSM   CD/VIDEO   PH MATH TUTOR CENTER   MathXL®Tutorials ON CD   MathXL®   MyMathLab®

Solve each equation. See Example 1.

1. $(x - 2)(x + 1) = 0$

2. $(x + 3)(x + 2) = 0$

3. $x(x + 6) = 0$

4. $2x(x - 7) = 0$

5. $(2x + 3)(4x - 5) = 0$

6. $(3x - 2)(5x + 1) = 0$

7. $(2x - 7)(7x + 2) = 0$

8. $(9x + 1)(4x - 3) = 0$

9. Write a quadratic equation that has two solutions, 6 and -1. Leave the polynomial in the equation in factored form.

10. Write a quadratic equation that has two solutions, 0 and -2. Leave the polynomial in the equation in factored form.

Solve each equation. See Examples 2 through 4.

11. $x^2 - 13x + 36 = 0$

12. $x^2 + 2x - 63 = 0$

13. $x^2 + 2x - 8 = 0$

14. $x^2 - 5x + 6 = 0$

15. $x^2 - 4x = 32$

16. $x^2 - 5x = 24$

17. $x(3x - 1) = 14$

18. $x(4x - 11) = 3$

19. $3x^2 + 19x - 72 = 0$

20. $36x^2 + x - 21 = 0$

21. Write a quadratic equation in standard form that has two solutions, 5 and 7.

22. Write an equation that has three solutions, 0, 1, and 2.

Solve each equation. See Examples 5 through 7.

23. $x^3 - 12x^2 + 32x = 0$

24. $x^3 - 14x^2 + 49x = 0$

25. $(4x - 3)(16x^2 - 24x + 9) = 0$

26. $(2x + 5)(4x^2 - 10x + 25) = 0$

27. $4x^3 - x = 0$

28. $4y^3 - 36y = 0$

29. $32x^3 - 4x^2 - 6x = 0$

30. $15x^3 + 24x^2 - 63x = 0$

MIXED PRACTICE

Solve each equation. Be careful. Some of the equations are quadratic and higher degree and some are linear.

31. $x(x + 7) = 0$

32. $y(6 - y) = 0$

33. $(x + 5)(x - 4) = 0$

34. $(x - 8)(x - 1) = 0$

35. $x^2 - x = 30$

36. $x^2 + 13x = -36$

37. $6y^2 - 22y - 40 = 0$

38. $3x^2 - 6x - 9 = 0$

39. $(2x + 3)(2x^2 - 5x - 3) = 0$

40. $(2x - 9)(x^2 + 5x - 36) = 0$

41. $x^2 - 15 = -2x$

42. $x^2 - 26 = -11x$

43. $x^2 - 16x = 0$

44. $x^2 + 5x = 0$

45. $-18y^2 - 33y + 216 = 0$

46. $-20y^2 + 145y - 35 = 0$

47. $12x^2 - 59x + 55 = 0$

48. $30x^2 - 97x + 60 = 0$

49. $18x^2 + 9x - 2 = 0$

50. $28x^2 - 27x - 10 = 0$

51. $x(6x + 7) = 5$

52. $4x(8x + 9) = 5$

53. $4(x - 7) = 6$

54. $5(3 - 4x) = 9$

55. $5x^2 - 6x - 8 = 0$

56. $9x^2 + 6x + 2 = 0$

57. $(y - 2)(y + 3) = 6$

58. $(y - 5)(y - 2) = 28$

59. $4y^2 - 1 = 0$

60. $4y^2 - 81 = 0$

61. $t^2 + 13t + 22 = 0$

62. $x^2 - 9x + 18 = 0$

63. $5t - 3 = 12$

64. $9 - t = -1$

65. $x^2 + 6x - 17 = -26$

66. $x^2 - 8x - 4 = -20$

67. $12x^2 + 7x - 12 = 0$

68. $30x^2 - 11x - 30 = 0$

69. $10t^3 - 25t - 15t^2 = 0$

70. $36t^3 - 48t - 12t^2 = 0$

Find the x-intercepts of the graph of each equation. See Example 8.

71. $y = (3x + 4)(x - 1)$

72. $y = (5x - 3)(x - 4)$

73. $y = x^2 - 3x - 10$

74. $y = x^2 + 7x + 6$

75. $y = 2x^2 + 11x - 6$

76. $y = 4x^2 + 11x + 6$

For Exercises 77 through 82, match each equation with its graph. See Example 8.

a.

b.

c.

d.

e.

f.

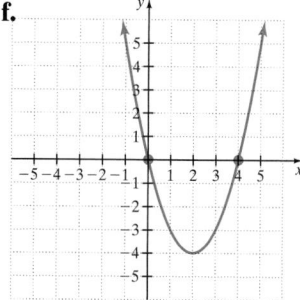

77. $y = (x + 2)(x - 1)$

78. $y = (x - 5)(x + 2)$

79. $y = x(x + 3)$

80. $y = x(x - 4)$

81. $y = 2x^2 - 8$

82. $y = 2x^2 - 2$

REVIEW AND PREVIEW

Perform the following operations. Write all results in lowest terms. See Section 1.3.

83. $\dfrac{3}{5} + \dfrac{4}{9}$

84. $\dfrac{2}{3} + \dfrac{3}{7}$

85. $\dfrac{7}{10} - \dfrac{5}{12}$

86. $\dfrac{5}{9} - \dfrac{5}{12}$

87. $\dfrac{7}{8} \div \dfrac{7}{15}$ **88.** $\dfrac{5}{12} - \dfrac{3}{10}$ **89.** $\dfrac{4}{5} \cdot \dfrac{7}{8}$ **90.** $\dfrac{3}{7} \cdot \dfrac{12}{17}$

Concept Extensions

91. A compass is accidentally thrown upward and out of an air balloon at a height of 300 feet. The height, y, of the compass at time x in seconds is given by the equation

$$y = -16x^2 + 20x + 300$$

300 feet

a. Find the height of the compass at the given times by filling in the table below.

time x	0	1	2	3	4	5	6
height y							

b. Use the table to determine when the compass strikes the ground.

c. Use the table to approximate the maximum height of the compass.

d. Plot the points (x, y) on a rectangular coordinate system and connect them with a smooth curve. Explain your results.

92. A rocket is fired upward from the ground with an initial velocity of 100 feet per second. The height, y, of the rocket at any time x is given by the equation

$$y = -16x^2 + 100x$$

a. Find the height of the rocket at the given times by filling in the table below.

time x	0	1	2	3	4	5	6	7
height y								

b. Use the table to approximate when the rocket strikes the ground to the nearest second.

c. Use the table to approximate the maximum height of the rocket.

d. Plot the points (x, y) on a rectangular coordinate system and connect them with a smooth curve. Explain your results.

Solve each equation. First, multiply the binomial.

To solve $(x - 6)(2x - 3) = (x + 2)(x + 9)$, see below.

$$(x - 6)(2x - 3) = (x + 2)(x + 9)$$
$$2x^2 - 15x + 18 = x^2 + 11x + 18$$
$$x^2 - 26x = 0$$
$$x(x - 26) = 0$$
$$x = 0 \quad \text{or} \quad x - 26 = 0$$
$$x = 26$$

93. $(x - 3)(3x + 4) = (x + 2)(x - 6)$

94. $(2x - 3)(x + 6) = (x - 9)(x + 2)$

95. $(2x - 3)(x + 8) = (x - 6)(x + 4)$

96. $(x + 6)(x - 6) = (2x - 9)(x + 4)$

6.6 QUADRATIC EQUATIONS AND PROBLEM SOLVING

Objective

1 Solve problems that can be modeled by quadratic equations.

1 Some problems may be modeled by quadratic equations. To solve these problems, we use the same problem-solving steps that were introduced in Section 2.4. When solving these problems, keep in mind that a solution of an equation that models a problem may not be a solution to the problem. For example, a person's age or the length of a rectangle is always a positive number. Discard solutions that do not make sense as solutions of the problem.

EXAMPLE 1

FINDING THE LENGTH OF TIME

For a TV commercial, a piece of luggage is dropped from a cliff 256 feet above the ground to show the durability of the luggage. Neglecting air resistance, the height h in feet of the luggage above the ground after t seconds is given by the quadratic equation

$$h = -16t^2 + 256$$

Find how long it takes for the luggage to hit the ground.

Solution **1.** UNDERSTAND. Read and reread the problem. Then draw a picture of the problem.

256 feet

The equation $h = -16t^2 + 256$ models the height of the falling luggage at time t. Familiarize yourself with this equation by finding the height of the luggage at $t = 1$ second and $t = 2$ seconds.

When $t = 1$ second, the height of the suitcase is

$$h = -16(1)^2 + 256 = 240 \text{ feet.}$$

When $t = 2$ seconds, the height of the suitcase is

$$h = -16(2)^2 + 256 = 192 \text{ feet.}$$

2. TRANSLATE. To find how long it takes the luggage to hit the ground, we want to know the value of t for which the height $h = 0$.

$$0 = -16t^2 + 256$$

3. SOLVE. We solve the quadratic equation by factoring.

$$0 = -16t^2 + 256$$
$$0 = -16(t^2 - 16)$$
$$0 = -16(t - 4)(t + 4)$$
$$t - 4 = 0 \quad \text{or} \quad t + 4 = 0$$
$$t = 4 \quad\quad\quad t = -4$$

4. INTERPRET. Since the time t cannot be negative, the proposed solution is 4 seconds.

Check: Verify that the height of the luggage when t is 4 seconds is 0.
When $t = 4$ seconds, $h = -16(4)^2 + 256 = -256 + 256 = 0$ feet.

State: The solution checks and the luggage hits the ground 4 seconds after it is dropped.

EXAMPLE 2

FINDING A NUMBER

The square of a number plus three times the number is 70. Find the number.

Solution

1. UNDERSTAND. Read and reread the problem. Suppose that the number is 5. The square of 5 is 5^2 or 25. Three times 5 is 15. Then $25 + 15 = 40$, not 70, so the number must be greater than 5. Remember, the purpose of proposing a number, such as 5, is to better understand the problem. Now that we do, we will let $x =$ the number.

2. TRANSLATE.

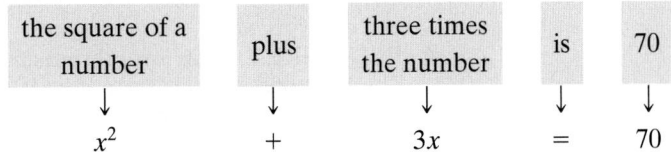

the square of a number	plus	three times the number	is	70
↓	↓	↓	↓	↓
x^2	$+$	$3x$	$=$	70

3. SOLVE.
$$x^2 + 3x = 70$$
$$x^2 + 3x - 70 = 0 \qquad \text{Subtract 70 from both sides.}$$
$$(x + 10)(x - 7) = 0 \qquad \text{Factor.}$$
$$x + 10 = 0 \quad \text{or} \quad x - 7 = 0 \qquad \text{Set each factor equal to 0.}$$
$$x = -10 \qquad\qquad x = 7 \qquad \text{Solve.}$$

4. INTERPRET.

Check: The square of -10 is $(-10)^2$, or 100. Three times -10 is $3(-10)$ or -30. Then $100 + (-30) = 70$, the correct sum, so -10 checks.
The square of 7 is 7^2 or 49. Three times 7 is $3(7)$, or 21. Then $49 + 21 = 70$, the correct sum, so 7 checks.
State: There are two numbers. They are -10 and 7.

EXAMPLE 3

FINDING THE BASE AND HEIGHT OF A SAIL

The height of a triangular sail is 2 meters less than twice the length of the base. If the sail has an area of 30 square meters, find the length of its base and the height.

Solution

1. UNDERSTAND. Read and reread the problem. Since we are finding the length of the base and the height, we let

$$x = \text{the length of the base}$$

and since the height is 2 meters less than twice the base,

$$2x - 2 = \text{the height}$$

An illustration is shown on the next page.

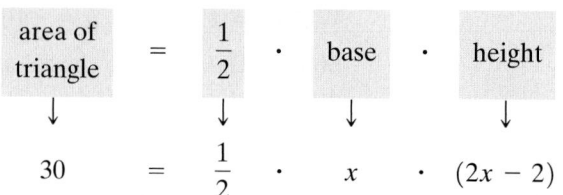

2. TRANSLATE. We are given that the area of the triangle is 30 square meters, so we use the formula for area of a triangle.

area of triangle	=	$\frac{1}{2}$	·	base	·	height

$$30 = \frac{1}{2} \cdot x \cdot (2x - 2)$$

3. SOLVE. Now we solve the quadratic equation.

$$30 = \frac{1}{2}x(2x - 2)$$

$30 = x^2 - x$ Multiply.

$x^2 - x - 30 = 0$ Write in standard form.

$(x - 6)(x + 5) = 0$ Factor.

$x - 6 = 0$ or $x + 5 = 0$ Set each factor equal to 0.

$x = 6$ $x = -5$

4. INTERPRET. Since x represents the length of the base, we discard the solution -5. The base of a triangle cannot be negative. The base is then 6 meters and the height is $2(6) - 2 = 10$ meters.

Check: To check this problem, we recall that $\frac{1}{2}$ base · height = area, or

$$\frac{1}{2}(6)(10) = 30$$ The required area

State: The base of the triangular sail is 6 meters and the height is 10 meters.

The next example makes use of the **Pythagorean theorem** and consecutive integers. Before we review this theorem, recall that a **right triangle** is a triangle that contains a 90° or right angle. The **hypotenuse** of a right triangle is the side opposite the right angle and is the longest side of the triangle. The **legs** of a right triangle are the other sides of the triangle.

Pythagorean Theorem

In a right triangle, the sum of the squares of the lengths of the two legs is equal to the square of the length of the hypotenuse.

$$(\text{leg})^2 + (\text{leg})^2 = (\text{hypotenuse})^2 \quad \text{or} \quad a^2 + b^2 = c^2$$

Study the following diagrams for a review of consecutive integers.

Consecutive integers:

If x is the first integer: x, $x + 1$, $x + 2$

Consecutive even integers:

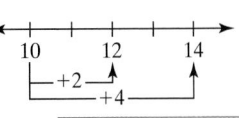

> **Helpful Hint**
>
> This 2 means that even numbers are 2 units between each other.

If x is the first even integer: x, $x + 2$, $x + 4$

Consecutive odd integers:

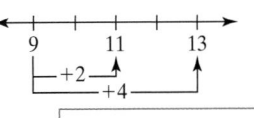

> **Helpful Hint**
>
> This 2 means that odd numbers are 2 units between each other.

If x is the first odd integer: x, $x + 2$, $x + 4$

EXAMPLE 4

FINDING THE DIMENSIONS OF A TRIANGLE

Find the lengths of the sides of a right triangle if the lengths can be expressed as three consecutive even integers.

Solution

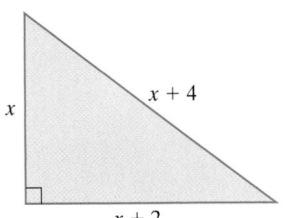

1. UNDERSTAND.   Read and reread the problem. Let's suppose that the length of one leg of the right triangle is 4 units. Then the other leg is the next even integer, or 6 units, and the hypotenuse of the triangle is the next even integer, or 8 units. Remember that the hypotenuse is the longest side. Let's see if a triangle with sides of these lengths forms a right triangle. To do this, we check to see whether the Pythagorean theorem holds true.

$$4^2 + 6^2 \stackrel{?}{=} 8^2$$
$$16 + 36 \stackrel{?}{=} 64$$
$$52 = 64 \qquad \text{False}$$

Our proposed numbers do not check, but we now have a better understanding of the problem.

We let x, $x + 2$, and $x + 4$ be three consecutive even integers. Since these integers represent lengths of the sides of a right triangle, we have

$$x = \text{one leg}$$
$$x + 2 = \text{other leg}$$
$$x + 4 = \text{hypotenuse (longest side)}$$

2. TRANSLATE. By the Pythagorean theorem, we have that

$$(\text{hypotenuse})^2 = (\text{leg})^2 + (\text{leg})^2$$
$$(x + 4)^2 = (x)^2 + (x + 2)^2$$

3. SOLVE. Now we solve the equation.

$$(x + 4)^2 = x^2 + (x + 2)^2$$
$$x^2 + 8x + 16 = x^2 + x^2 + 4x + 4 \qquad \text{Multiply.}$$
$$x^2 + 8x + 16 = 2x^2 + 4x + 4 \qquad \text{Combine like terms.}$$
$$x^2 - 4x - 12 = 0 \qquad \text{Write in standard form.}$$
$$(x - 6)(x + 2) = 0 \qquad \text{Factor.}$$
$$x - 6 = 0 \qquad \text{or} \qquad x + 2 = 0 \qquad \text{Set each factor equal to 0.}$$
$$x = 6 \qquad\qquad\qquad x = -2$$

4. INTERPRET. We discard $x = -2$ since length cannot be negative. If $x = 6$, then $x + 2 = 8$ and $x + 4 = 10$.

Check: Verify that $(\text{hypotenuse})^2 = (\text{leg})^2 + (\text{leg})^2$, or $10^2 = 6^2 + 8^2$, or $100 = 36 + 64$.

State: The sides of the right triangle have lengths 6 units, 8 units, and 10 units.

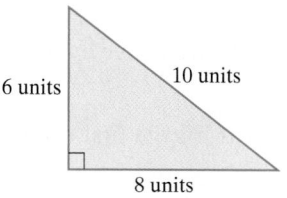

Spotlight on

DECISION ♂ MAKING

Suppose you are a landscaper. You are landscaping a public park and have just put in a flower bed measuring 8 feet by 12 feet. You would also like to surround the bed with a decorative floral border consisting of low-growing, spreading plants. Each plant will cover approximately 1 square foot when mature, and you have 224 plants to use. How wide of a strip of ground should you prepare around the flower bed for the border? Explain.

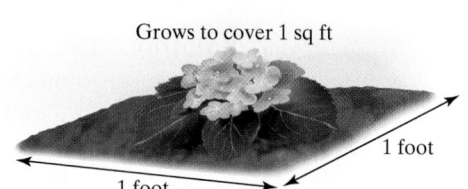

Grows to cover 1 sq ft

1 foot

1 foot

EXERCISE SET 6.6

STUDY GUIDE/SSM   CD/ VIDEO   PH MATH TUTOR CENTER   MathXL®Tutorials ON CD   MathXL®   MyMathLab®

MIXED PRACTICE

See Examples 1 through 4 for all exercises. Represent each given condition using a single variable, x.

△ **1.** The length and width of a rectangle whose length is 4 centimeters more than its width

△ **2.** The length and width of a rectangle whose length is twice its width

3. Two consecutive odd integers

4. Two consecutive even integers

△ **5.** The base and height of a triangle whose height is one more than four times its base

△ **6.** The base and height of a trapezoid whose base is three less than five times its height

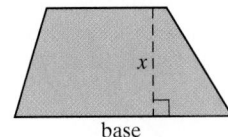

base

Use the information given to find the dimensions of each figure.

△ **7.** The *area* of the square is 121 square units. Find the length of its sides.

△ **8.** The *area* of the rectangle is 84 square inches. Find its length and width.

△ **9.** The *perimeter* of the quadrilateral is 120 centimeters. Find the lengths of the sides.

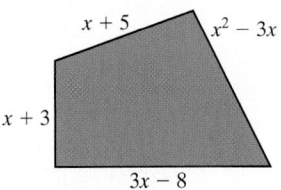

△ **10.** The *perimeter* of the triangle is 85 feet. Find the lengths of its sides.

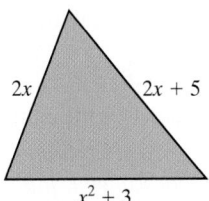

△ **11.** The *area* of the parallelogram is 96 square miles. Find its base and height.

△ **12.** The *area* of the circle is 25π square kilometers. Find its radius.

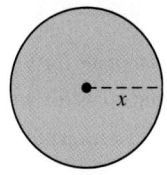

Solve.

13. An object is thrown upward from the top of an 80-foot building with an initial velocity of 64 feet per second. The height h of the object after t seconds is given by the quadratic equation $h = -16t^2 + 64t + 80$. When will the object hit the ground?

14. A hang glider pilot accidentally drops her compass from the top of a 400-foot cliff. The height h of the compass after t seconds is given by the quadratic equation $h = -16t^2 + 400$. When will the compass hit the ground?

△ **15.** The length of a rectangle is 7 centimeters less than twice its width. Its area is 30 square centimeters. Find the dimensions of the rectangle.

△ **16.** The length of a rectangle is 9 inches more than its width. Its area is 112 square inches. Find the dimensions of the rectangle.

The equation $D = \frac{1}{2}n(n - 3)$ gives the number of diagonals D for a polygon with n sides. For example, a polygon with 6 sides has $D = \frac{1}{2} \cdot 6(6 - 3)$ or $D = 9$ diagonals. (See if you can count all 9 diagonals. Some are shown in the figure.) Use this equation, $D = \frac{1}{2}n(n - 3)$, for Exercises 17 through 20.

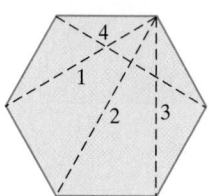

△ **17.** Find the number of diagonals for a polygon that has 12 sides.

△ **18.** Find the number of diagonals for a polygon that has 15 sides.

△ **19.** Find the number of sides n for a polygon that has 35 diagonals.

△ **20.** Find the number of sides n for a polygon that has 14 diagonals.

Solve.

21. The sum of a number and its square is 132. Find the number(s).

22. The sum of a number and its square is 182. Find the number(s).

△ **23.** Two boats travel at a right angle to each other after leaving the same dock at the same time. One hour later the boats are 17 miles apart. If one boat travels 7 miles per hour faster than the other boat, find the rate of each boat.

△ **24.** The side of a square equals the width of a rectangle. The length of the rectangle is 6 meters longer than its width. The sum of the areas of the square and the rectangle is 176 square meters. Find the side of the square.

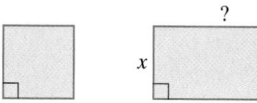

25. The sum of two numbers is 20, and the sum of their squares is 218. Find the numbers.

26. The sum of two numbers is 25, and the sum of their squares is 325. Find the numbers.

△ **27.** If the sides of a square are increased by 3 inches, the area becomes 64 square inches. Find the length of the sides of the original square.

△ **28.** If the sides of a square are increased by 5 meters, the area becomes 100 square meters. Find the length of the sides of the original square.

🔒 **29.** One leg of a right triangle is 4 millimeters longer than the smaller leg and the hypotenuse is 8 millimeters longer than the smaller leg. Find the lengths of the sides of the triangle.

△ **30.** One leg of a right triangle is 9 centimeters longer than the other leg and the hypotenuse is 45 centimeters. Find the lengths of the legs of the triangle.

△ **31.** The length of the base of a triangle is twice its height. If the area of the triangle is 100 square kilometers, find the height.

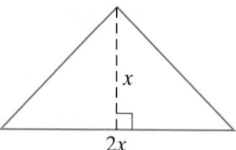

△ **32.** The height of a triangle is 2 millimeters less than the base. If the area is 60 square millimeters, find the base.

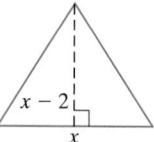

△ **33.** Find the length of the shorter leg of a right triangle if the longer leg is 12 feet more than the shorter leg and the hypotenuse is 12 feet less than twice the shorter leg.

△ **34.** Find the length of the shorter leg of a right triangle if the longer leg is 10 miles more than the shorter leg and the hypotenuse is 10 miles less than twice the shorter leg.

35. An object is dropped from the top of the 625-foot-tall Waldorf-Astoria Hotel on Park Avenue in New York City. (*Source: World Almanac* research) The height h of the object after t seconds is given by the equation $h = -16t^2 + 625$. Find how many seconds pass before the object reaches the ground.

625 feet

36. A 6-foot-tall person drops an object from the top of the Westin Peachtree Plaza in Atlanta, Georgia. The Westin building is 723 feet tall. (*Source: World Almanac* research) The height h of the object after t seconds is given by the equation $h = -16t^2 + 729$. Find how many seconds pass before the object reaches the ground.

723 feet

37. At the end of 2 years, P dollars invested at an interest rate r compounded annually increases to an amount, A dollars, given by

$$A = P(1 + r)^2$$

Find the interest rate if $100 increased to $144 in 2 years.

38. At the end of 2 years, P dollars invested at an interest rate r compounded annually increases to an amount, A dollars, given by

$$A = P(1 + r)^2$$

Find the interest rate if $2000 increased to $2420 in 2 years.

△ **39.** Find the dimensions of a rectangle whose width is 7 miles less than its length and whose area is 120 square miles.

△ **40.** Find the dimensions of a rectangle whose width is 2 inches less than half its length and whose area is 160 square inches.

41. If the cost, C, for manufacturing x units of a certain product is given by $C = x^2 - 15x + 50$, find the number of units manufactured at a cost of $9500.

42. If a switchboard handles n telephones, the number C of telephone connections it can make simultaneously is given by the equation $C = \dfrac{n(n-1)}{2}$. Find how many telephones are handled by a switchboard making 120 telephone connections simultaneously.

REVIEW AND PREVIEW

The following double line graph shows a comparison of the number of farms in the United States and the size of the average farm. Use this graph to answer Exercises 43 through 49. See Section 3.1.

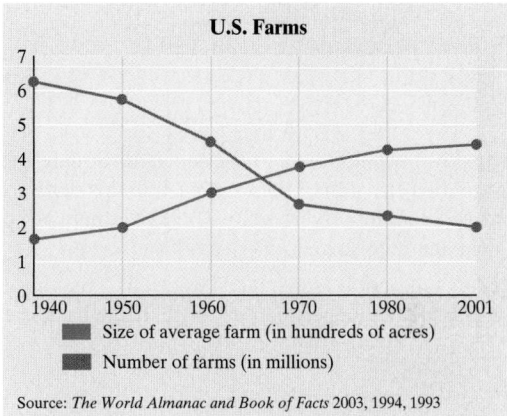

U.S. Farms

■ Size of average farm (in hundreds of acres)
■ Number of farms (in millions)

Source: *The World Almanac and Book of Facts* 2003, 1994, 1993

△ **43.** Approximate the size of the average farm in 1940.

△ **44.** Approximate the size of the average farm in 2001.

45. Approximate the number of farms in 1940.

46. Approximate the number of farms in 2001.

47. Approximate the year that the colored lines in this graph intersect.

48. In your own words, explain the meaning of the point of intersection in the graph.

49. Describe the trends shown in this graph and speculate as to why these trends have occurred.

Write each fraction in simplest form. See Section 1.3.

50. $\dfrac{20}{35}$ **51.** $\dfrac{24}{32}$

52. $\dfrac{27}{18}$ **53.** $\dfrac{15}{27}$

54. $\dfrac{14}{42}$ **55.** $\dfrac{45}{50}$

Concept Extensions

△ **56.** According to the International America's Cup Class (IACC) rule, a sailboat competing in the America's Cup match must have a 110-foot-tall mast and a combined mainsail and jib sail area of 3000 square feet. (*Source:* America's Cup Organizing Committee) A design for an IACC-class sailboat calls for the mainsail to be 60% of the combined sail area. If the height of the triangular mainsail is 28 feet more than twice the length of the boom, find the length of the boom and the height of the mainsail.

110 feet

mainsail | jib sail

boom

57. Write down two numbers whose sum is 10. Square each number and find the sum of the squares. Use this work to write a word problem like Exercise 21. Then give the word problem to a classmate to solve.

58. A rectangular pool is surrounded by a walk 4 meters wide. The pool is 6 meters longer than its width. If the total area is 576

square meters more than the area of the pool, find the dimensions of the pool.

4

4

x

$x + 6$

59. A rectangular garden is surrounded by a walk of uniform width. The area of the garden is 180 square yards. If the dimensions of the garden plus the walk are 16 yards by 24 yards, find the width of the walk.

x

180 sq yards

16 yd

x

24 yd

CHAPTER 6 PROJECT

Choosing Among Building Options

Whether putting in a new floor, hanging new wallpaper, or retiling a bathroom, it may be necessary to choose among several different materials with different pricing schemes. If a fixed amount of money is available for projects like these, it can be helpful to compare the choices by calculating how much area can be covered by a fixed dollar-value of material.

In this project, you will have the opportunity to choose among three different choices of materials for building a patio around a swimming pool. This project may be completed by working in groups or individually.

x | 10

15

$2x + 10$

x

$2x + 15$

Situation: Suppose you have just had a 10-foot-by-15-foot in-ground swimming pool installed in your backyard. You have $3000 left from the building project that you would like to spend on surrounding the pool with a patio, equally wide on all sides (see figure). You have talked to several local suppliers about options for building this patio and must choose among the following.

Option	Material	Price
A	Poured cement	$5 per square foot
B	Brick	$7.50 per square foot plus a $30 flat fee for delivering the bricks
C	Outdoor carpeting	$4.50 per square foot plus $10.86 per foot of the pool's perimeter to install edging

1. Find the area of the swimming pool.

2. Write an algebraic expression for the total area of the region containing both the pool and the patio.

△ **3.** Use subtraction to find an algebraic expression for the area of just the patio (not including the area of the pool).

△ **4.** Find the perimeter of the swimming pool alone.

5. For each patio material option, write an algebraic expression for the total cost of installing the patio based on its area and the given price information.

6. If you plan to spend the entire $3000 on the patio, how wide would the patio in option A be?

7. If you plan to spend the entire $3000 on the patio, how wide would the patio in option B be?

8. If you plan to spend the entire $3000 on the patio, how wide would the patio in option C be?

9. Which option would you choose? Why? Discuss the pros and cons of each option.

6 CHAPTER VOCABULARY CHECK

Fill in each blank with one of the words or phrases listed below.

factoring   quadratic equation   perfect square trinomial   greatest common factor

1. An equation that can be written in the form $ax^2 + bx + c = 0$ (with a not 0) is called a _____.

2. _____ is the process of writing an expression as a product.

3. The _____ of a list of terms is the product of all common factors.

4. A trinomial that is the square of some binomial is called a _____.

CHAPTER 6 HIGHLIGHTS

Definitions and Concepts	Examples

Section 6.1   The Greatest Common Factor and Factoring by Grouping

Factoring is the process of writing an expression as a product.	Factor: $6 = 2 \cdot 3$ $x^2 + 5x + 6 = (x + 2)(x + 3)$
To Find the GCF of a List of Integers	Find the GCF of 12, 36, and 48.
Step 1. Write each number as a product of primes.	$12 = 2 \cdot 2 \cdot 3$
Step 2. Identify the common prime factors.	$36 = 2 \cdot 2 \cdot 3 \cdot 3$
Step 3. The product of all common factors is the greatest common factor. If there are no common prime factors, the GCF is 1.	$48 = 2 \cdot 2 \cdot 2 \cdot 2 \cdot 3$ $GCF = 2 \cdot 2 \cdot 3 = 12$
The GCF of a list of common variables raised to powers is the variable raised to the smallest exponent in the list.	The GCF of z^5, z^3, and z^{10} is z^3.
The GCF of a list of terms is the product of all common factors.	Find the GCF of $8x^2y$, $10x^3y^2$, and $26x^2y^3$. The GCF of 8, 10, and 26 is 2. The GCF of x^2, x^3, and x^2 is x^2. The GCF of y, y^2, and y^3 is y. The GCF of the terms is $2x^2y$.

(continued)

Definitions and Concepts	Examples

Section 6.1 The Greatest Common Factor and Factoring by Grouping

To Factor by Grouping

Step 1. Arrange the terms so that the first two terms have a common factor and the last two have a common factor.

Step 2. For each pair of terms, factor out the pair's GCF.

Step 3. If there is now a common binomial factor, factor it out.

Step 4. If there is no common binomial factor, begin again, rearranging the terms differently. If no rearrangement leads to a common binomial factor, the polynomial cannot be factored.

Factor $10ax + 15a - 6xy - 9y$.

Step 1. $10ax + 15a - 6xy - 9y$

Step 2. $5a(2x + 3) - 3y(2x + 3)$

Step 3. $(2x + 3)(5a - 3y)$

Section 6.2 Factoring Trinomials of the Form $x^2 + bx + c$

To factor a trinomial of the form $x^2 + bx + c$, look for two numbers whose product is c and whose sum is b. The factored form is

$$(x + \text{one number})(x + \text{other number})$$

Factor: $x^2 + 7x + 12$

$3 + 4 = 7 \qquad 3 \cdot 4 = 12$

$(x + 3)(x + 4)$

Section 6.3 Factoring Trinomials of the Form $ax^2 + bx + c$

Method 1: To factor $ax^2 + bx + c$, try various combinations of factors of ax^2 and c until a middle term of bx is obtained when checking.

Factor: $3x^2 + 14x - 5$

Factors of $3x^2$: $3x, x$

Factors of -5: $-1, 5$ and $1, -5$.

$$(3x - \underline{1})(x + 5)$$
$$\underbrace{}_{\substack{-1x \\ 15x}}$$
$$14x \qquad \text{\textit{Correct middle term}}$$

Method 2: Factor $ax^2 + bx + c$ by grouping.

Step 1. Find two numbers whose product is $a \cdot c$ and and whose sum is b.

Step 2. Rewrite bx, using the factors found in Step 1.

Step 3. Factor by grouping.

Factor: $3x^2 + 14x - 5$

Step 1. Find two numbers whose product is $3 \cdot (-5)$ or -15 whose sum is 14. They are 15 and -1.

Step 2. $3x^2 + 14x - 5$
$$= 3x^2 + 15x - 1x - 5$$

Step 3. $= 3x(x + 5) - 1(x + 5)$
$$= (x + 5)(3x - 1)$$

A **perfect square trinomial** is a trinomial that is the square of some binomial.

Perfect square trinomial = square of binomial
$$x^2 + 4x + 4 = (x + 2)^2$$
$$25x^2 - 10x + 1 = (5x - 1)^2 \qquad \textit{(continued)}$$

Definitions and Concepts	**Examples**

Section 6.3 Factoring Trinomials of the Form $ax^2 + bx + c$

Factoring Perfect Square Trinomials: $$a^2 + 2ab + b^2 = (a + b)^2$$ $$a^2 - 2ab + b^2 = (a - b)^2$$	Factor: $$x^2 + 6x + 9 = x^2 + 2 \cdot x \cdot 3 + 3^2 = (x + 3)^2$$ $$4x^2 - 12x + 9 = (2x)^2 - 2 \cdot 2x \cdot 3 + 3^2 = (2x - 3)^2$$

Section 6.4 Factoring Binomials

Difference of Squares $$a^2 - b^2 = (a + b)(a - b)$$ *Sum or Difference of Cubes* $$a^3 + b^3 = (a + b)(a^2 - ab + b^2)$$ $$a^3 - b^3 = (a - b)(a^2 + ab + b^2)$$	Factor: $$x^2 - 9 = x^2 - 3^2 = (x + 3)(x - 3)$$ $$y^3 + 8 = y^3 + 2^3 = (y + 2)(y^2 - 2y + 4)$$ $$125z^3 - 1 = (5z)^3 - 1^3 = (5z - 1)(25z^2 + 5z + 1)$$

Integrated Review—Choosing a Factoring Strategy

To Factor a Polynomial,	Factor: $2x^4 - 6x^2 - 8$
Step 1. Factor out the GCF.	**Step 1.** $2x^4 - 6x^2 - 8 = 2(x^4 - 3x^2 - 4)$
Step 2. **a.** If two terms,	
i. $a^2 - b^2 = (a + b)(a - b)$	
ii. $a^3 - b^3 = (a - b)(a^2 + ab + b^2)$	
iii. $a^3 + b^3 = (a + b)(a^2 - ab + b^2)$	
b. If three terms,	**Step 2. b. ii.** $= 2(x^2 + 1)(x^2 - 4)$
i. $a^2 + 2ab + b^2 = (a + b)^2$	
ii. Methods in Sections 6.2 and 6.3	
c. If four or more terms, try factoring by grouping.	
Step 3. See if any factors can be factored further.	**Step 3.** $= 2(x^2 + 1)(x + 2)(x - 2)$
Step 4. Check by multiplying.	**Step 4.** Check by multiplying. $$2(x^2 + 1)(x + 2)(x - 2) = 2(x^2 + 1)(x^2 - 4)$$ $$= 2(x^4 - 3x^2 - 4)$$ $$= 2x^4 - 6x^2 - 8$$

Section 6.5 Solving Quadratic Equations by Factoring

A **quadratic equation** is an equation that can be written in the form $ax^2 + bx + c = 0$ with a not 0. The form $ax^2 + bx + c = 0$ is called the **standard form** of a quadratic equation.	*Quadratic Equation* *Standard Form* $$x^2 = 16 \qquad\qquad x^2 - 16 = 0$$ $$y = -2y^2 + 5 \qquad 2y^2 + y - 5 = 0$$

(continued)

Definitions and Concepts	**Examples**

Section 6.5 Solving Quadratic Equations by Factoring

Zero Factor Theorem

If a and b are real numbers and if $ab = 0$, then $a = 0$ or $b = 0$.

To solve quadratic equations by factoring,

Step 1. Write the equation in standard form: $ax^2 + bx + c = 0$.

Step 2. Factor the quadratic.

Step 3. Set each factor containing a variable equal to 0.

Step 4. Solve the equations.

Step 5. Check in the original equation.

If $(x + 3)(x - 1) = 0$, then $x + 3 = 0$ or $x - 1 = 0$

Solve: $3x^2 = 13x - 4$

Step 1. $3x^2 - 13x + 4 = 0$

Step 2. $(3x - 1)(x - 4) = 0$

Step 3. $3x - 1 = 0$ or $x - 4 = 0$

Step 4. $3x = 1$ or $x = 4$

$$x = \frac{1}{3}$$

Step 5. Check both $\frac{1}{3}$ and 4 in the original equation.

Section 6.6 Quadratic Equations and Problem Solving

Problem-Solving Steps

1. UNDERSTAND the problem.

A garden is in the shape of a rectangle whose length is two feet more than its width. If the area of the garden is 35 square feet, find its dimensions.

1. Read and reread the problem. Guess a solution and check your guess.
   Let x be the width of the rectangular garden. Then $x + 2$ is the length.

x

$x + 2$

2. TRANSLATE.

2. In words:

length	·	width	=	area

Translate: $(x + 2)$ · x = 35

3. SOLVE.

3.
$$(x + 2)x = 35$$
$$x^2 + 2x - 35 = 0$$
$$(x - 5)(x + 7) = 0$$
$$x - 5 = 0 \quad \text{or} \quad x + 7 = 0$$
$$x = 5 \quad \text{or} \quad x = -7$$

4. INTERPRET.

4. Discard the solution of -7 since x represents width.
   Check: If x is 5 feet then $x + 2 = 5 + 2 = 7$ feet. The area of a rectangle whose width is 5 feet and whose length is 7 feet is (5 feet)(7 feet) or 35 square feet.
   State: The garden is 5 feet by 7 feet.

6 CHAPTER REVIEW

(6.1) *Complete the factoring.*

1. $6x^2 - 15x = 3x(\quad)$

2. $2x^3y - 6x^2y^2 - 8xy^3 = 2xy(\quad)$

Factor the GCF from each polynomial.

3. $20x^2 + 12x$

4. $6x^2y^2 - 3xy^3$

5. $-8x^3y + 6x^2y^2$

6. $3x(2x + 3) - 5(2x + 3)$

7. $5x(x + 1) - (x + 1)$

Factor.

8. $3x^2 - 3x + 2x - 2$

9. $6x^2 + 10x - 3x - 5$

10. $3a^2 + 9ab + 3b^2 + ab$

(6.2) *Factor each trinomial.*

11. $x^2 + 6x + 8$

12. $x^2 - 11x + 24$

13. $x^2 + x + 2$

14. $x^2 - 5x - 6$

15. $x^2 + 2x - 8$

16. $x^2 + 4xy - 12y^2$

17. $x^2 + 8xy + 15y^2$

18. $3x^2y + 6xy^2 + 3y^3$

19. $72 - 18x - 2x^2$

20. $32 + 12x - 4x^2$

(6.3) *Factor each trinomial.*

21. $2x^2 + 11x - 6$

22. $4x^2 - 7x + 4$

23. $4x^2 + 4x - 3$

24. $6x^2 + 5xy - 4y^2$

25. $6x^2 - 25xy + 4y^2$

26. $18x^2 - 60x + 50$

27. $2x^2 - 23xy - 39y^2$

28. $4x^2 - 28xy + 49y^2$

29. $18x^2 - 9xy - 20y^2$

30. $36x^3y + 24x^2y^2 - 45xy^3$

(6.4) *Factor each binomial.*

31. $4x^2 - 9$

32. $9t^2 - 25s^2$

33. $16x^2 + y^2$

34. $x^3 - 8y^3$

35. $8x^3 + 27$

36. $2x^3 + 8x$

37. $54 - 2x^3y^3$

38. $9x^2 - 4y^2$

39. $16x^4 - 1$

40. $x^4 + 16$

REVIEW AND PREVIEW

Factor.

41. $2x^2 + 5x - 12$

42. $3x^2 - 12$

43. $x(x - 1) + 3(x - 1)$

44. $x^2 + xy - 3x - 3y$

45. $4x^2y - 6xy^2$

46. $8x^2 - 15x - x^3$

47. $125x^3 + 27$

48. $24x^2 - 3x - 18$

49. $(x + 7)^2 - y^2$

50. $x^2(x + 3) - 4(x + 3)$

(6.5) *Solve the following equations.*

51. $(x + 6)(x - 2) = 0$

52. $3x(x + 1)(7x - 2) = 0$

53. $4(5x + 1)(x + 3) = 0$

54. $x^2 + 8x + 7 = 0$

55. $x^2 - 2x - 24 = 0$

56. $x^2 + 10x = -25$

57. $x(x - 10) = -16$

58. $(3x - 1)(9x^2 + 3x + 1) = 0$

59. $56x^2 - 5x - 6 = 0$

60. $20x^2 - 7x - 6 = 0$

61. $5(3x + 2) = 4$

62. $6x^2 - 3x + 8 = 0$

63. $12 - 5t = -3$

64. $5x^3 + 20x^2 + 20x = 0$

65. $4t^3 - 5t^2 - 21t = 0$

(6.6) *Solve the following problems.*

△ **66.** A flag for a local organization is in the shape of a rectangle whose length is 15 inches less than twice its width. If the area of the flag is 500 square inches, find its dimensions.

△ **67.** The base of a triangular sail is four times its height. If the area of the triangle is 162 square yards, find the base.

68. Find two consecutive positive integers whose product is 380.

69. A rocket is fired from the ground with an initial velocity of 440 feet per second. Its height h after t seconds is given by the equation

$$h = -16t^2 + 440t$$

a. Find how many seconds pass before the rocket reaches a height of 2800 feet. Explain why two answers are obtained.

b. Find how many seconds pass before the rocket reaches the ground again.

△ **70.** An architect's squaring instrument is in the shape of a right triangle. Find the length of the long leg of the right triangle if the hypotenuse is 8 centimeters longer than the long leg and the short leg is 8 centimeters shorter than the long leg.

STUDY SKILLS REMINDER

Are You Prepared for a Test on Chapter 6?

Below is a list of some *common trouble areas* for topics covered in Chapter 6. After studying for your test—but before taking your test—read these.

▶ Don't forget that the first step to factor any polynomial is to first factor out any common factors.

$$9x^2 - 36 = 9(x^2 - 4) = 9(x + 2)(x - 2)$$

▶ Can you completely factor $x^4 - 24x^2 - 25$?

$$x^4 - 24x^2 - 25 = (x^2 - 25)(x^2 + 1)$$
$$= (x + 5)(x - 5)(x^2 + 1)$$

▶ Remember that to use the zero factor property to solve a quadratic equation, one side of the equation must be 0 and the other side must be a factored polynomial.

$$x(x - 2) = 3 \quad \textit{Cannot use zero factor property.}$$
$$x^2 - 2x - 3 = 0$$
$$(x - 3)(x + 1) = 0 \quad \textit{Now we can use zero factor property.}$$

$$x - 3 = 0 \quad \text{or} \quad x + 1 = 0$$
$$x = 3 \quad \text{or} \quad x = -1$$

Remember: This is simply a sampling of selected topics given to check your understanding. For a review of Chapter 6 in your text, see the material at the end of this chapter.

CHAPTER 6 TEST

Remember to use your Chapter Test Prep Video CD to help you study and view solutions to the test questions you need help with.

Factor each polynomial completely. If a polynomial cannot be factored, write "prime."

1. $y^2 - 8y - 48$

2. $x^2 + x - 10$

3. $9x^3 + 39x^2 + 12x$

4. $3a^2 + 3ab - 7a - 7b$

5. $3x^2 - 5x + 2$

6. $x^2 + 14xy + 24y^2$

7. $180 - 5x^2$

8. $6t^2 - t - 5$

9. $xy^2 - 7y^2 - 4x + 28$

10. $x - x^5$

11. $-xy^3 - x^3y$

12. $64x^3 - 1$

13. $8y^3 - 64$

Solve each equation.

14. $x^2 + 5x = 14$ **15.** $x(x + 6) = 7$

16. $3x(2x - 3)(3x + 4) = 0$ **17.** $5t^3 - 45t = 0$

18. $t^2 - 2t - 15 = 0$ **19.** $6x^2 = 15x$

Solve each problem.

△ **20.** A deck for a home is in the shape of a triangle. The length of the base of the triangle is 9 feet longer than its altitude. If the area of the triangle is 68 square feet, find the length of the base.

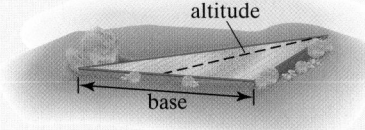

altitude

base

21. The sum of two numbers is 17 and the sum of their squares is 145. Find the numbers.

22. An object is dropped from the top of the Woolworth Building on Broadway in New York City. The height h of the object after t seconds is given by the equation

$$h = -16t^2 + 784$$

Find how many seconds pass before the object reaches the ground.

6 CHAPTER CUMULATIVE REVIEW

1. Translate each sentence into a mathematical statement.
  a. Nine is less than or equal to eleven.
  b. Eight is greater than one.
  c. Three is not equal to four.

2. Insert $<$ or $>$ in the space to make each statement true.
  a. $|-5|$ $|-3|$
  b. $|0|$ $|-2|$

3. Write each fraction in lowest terms.
  a. $\dfrac{42}{49}$ **b.** $\dfrac{11}{27}$ **c.** $\dfrac{88}{20}$

4. Evaluate $\dfrac{x}{y} + 5x$ if $x = 20$ and $y = 10$.

5. Simplify: $\dfrac{8 + 2 \cdot 3}{2^2 - 1}$

6. Evaluate $\dfrac{x}{y} + 5x$ if $x = -20$ and $y = 10$.

7. Add.
  a. $3 + (-7) + (-8)$
  b. $[7 + (-10)] + [-2 + |-4|]$

8. Evaluate $\dfrac{x}{y} + 5x$ if $x = -20$ and $y = -10$.

9. Multiply.
  a. $(-6)(4)$ **b.** $2(-1)$
  c. $(-5)(-10)$

10. Simplify: $5 - 2(3x - 7)$

11. Simplify each expression by combining like terms.
  a. $7x - 3x$ **b.** $10y^2 + y^2$
  c. $8x^2 + 2x - 3x$

12. Solve: $0.8y + 0.2(y - 1) = 1.8$

Solve.

13. $\dfrac{y}{7} = 20$ **14.** $\dfrac{x}{-7} = -4$

15. $-3x = 33$ **16.** $-\dfrac{2}{3}x = -22$

17. $8(2 - t) = -5t$ **18.** $-z = \dfrac{7z + 3}{5}$

19. A 10-foot board is to be cut into two pieces so that the longer piece is 4 times the shorter. Find the length of each piece.

20. Solve $3x + 9 \le 5(x - 1)$. Write the solution set using interval notation.

21. Graph the linear equation $y = -\dfrac{1}{3}x$.

22. Is the ordered pair $(-1, 2)$ a solution of $-7x - 8y = -9$?

23. Is the line passing through the points $(-6, 0)$ and $(-2, 3)$ parallel to the line passing through the points $(5, 4)$ and $(9, 7)$?

24. Find the slope of the line through $(5, -6)$ and $(5, 2)$.

25. Evaluate each expression for the given value of x.
  a. $2x^3$; x is 5   **b.** $\dfrac{9}{x^2}$; x is -3

26. Find the slope and y-intercept of the line whose equation is $7x - 3y = 2$.

27. Find the degree of each term.

   a. $3x^2$ **b.** $5x^3yz$ **c.** y

   d. $12x^2yz^3$ **e.** 5

28. Find an equation of the vertical line through $(0, 7)$.

29. Add $11x^3 - 12x^2 + x - 3$ and $x^3 - 10x + 5$.

30. Find an equation of the line with slope 4 and y-intercept $\left(0, \dfrac{1}{2}\right)$. Write the equation in standard form.

31. Multiply $(3x + 2)(2x - 5)$.

32. Write an equation of the line through $(-4, 0)$ and $(6, -1)$. Write the equation in standard form.

33. Multiply $(3y + 1)^2$.

34. Solve the system: $\begin{cases} -x + 3y = 18 \\ -3x + 2y = 19 \end{cases}$

35. Simplify by writing each expression with positive exponents only.

   a. 3^{-2} **b.** $2x^{-3}$

   c. $2^{-1} + 4^{-1}$ **d.** $(-2)^{-4}$

36. Simplify: $\dfrac{(5a^7)^2}{a^5}$

37. Write each number in scientific notation.

   a. 367,000,000

   b. 0.000003

   c. 20,520,000,000

   d. 0.00085

38. Multiply: $(3x - 7y)^2$

39. Divide $x^2 + 7x + 12$ by $x + 3$ using long division.

40. Simplify: $\dfrac{(xy)^{-3}}{(x^5y^6)^3}$

41. Find the GCF of each list of terms.

   a. $x^3, x^7,$ and x^5 **b.** $y, y^4,$ and y^7

Factor.

42. $z^3 + 7z + z^2 + 7$

43. $x^2 + 7x + 12$

44. $2x^3 + 2x^2 - 84x$

45. $8x^2 - 22x + 5$

46. $-4x^2 - 23x + 6$

47. $25a^2 - 9b^2$

48. $9xy^2 - 16x$

49. Solve $(x - 3)(x + 1) = 0$.

50. Solve $x^2 - 13x = -36$.

Rational Expressions

In this chapter, we expand our knowledge of algebraic expressions to include another category called rational expressions, such as $\frac{x+1}{x}$. We explore the operations of addition, subtraction, multiplication, and division for these algebraic fractions, using principles similar to the principles for number fractions. Thus, the material in this chapter will make full use of your knowledge of number fractions.

A WIDELY KNOWN ASPECT OF FORENSIC SCIENCE

Forensic science is the application of science and technology to the resolution of criminal and civil issues. One of the most widely known aspects of forensic science is criminalistics—the analysis of crime scenes and the examination of physical evidence. Criminalistic investigations are often carried out in what is popularly known as a crime laboratory.

Crime lab workers might be expected to analyze and identify chemical compounds, human hair and tissues, or materials such as metals, glass, or wood. They might also make measurements on items of physical evidence and compare footprints, firearms, or bullets. Anyone wishing to work in criminalistics, or the forensic sciences in general, should have excellent problem-solving and communication skills and also have a good mathematical background.

In the Spotlight on Decision Making feature on page 415, you will have the opportunity, as a forensic lab technician, to decide what material a piece of physical evidence is made of.

7.1 RATIONAL FUNCTIONS AND SIMPLIFYING RATIONAL EXPRESSIONS

Objectives

1 Define a rational expression and a rational fuction and find the domain of a rational function.

2 Simplify rational expressions.

3 Use rational functions in applications.

Recall that a *rational number*, or *fraction*, is a number that can be written as the quotient $\frac{p}{q}$ of two integers p and q as long as q is not 0. A **rational expression** is an expression that can be written as the quotient $\frac{P}{Q}$ of two polynomials P and Q as long as Q is not 0.

Examples of Rational Expressions

$$\frac{3x + 7}{2} \qquad \frac{5x^2 - 3}{x - 1} \qquad \frac{7x - 2}{2x^2 + 7x + 6}$$

Rational expressions are sometimes used to describe functions. For example, we call the function $f(x) = \frac{x^2 + 2}{x - 3}$ a **rational function** since $\frac{x^2 + 2}{x - 3}$ is a rational expression.

1 As with fractions, a rational expression is **undefined** if the denominator is 0. If a variable in a rational expression is replaced with a number that makes the denominator 0, we say that the rational expression is **undefined** for this value of the variable. For example, the rational expression $\frac{x^2 + 2}{x - 3}$ is undefined when x is 3, because replacing x with 3 results in a denominator of 0. For this reason, we must exclude 3 from the domain of the function $f(x) = \frac{x^2 + 2}{x - 3}$.

The domain of f is then

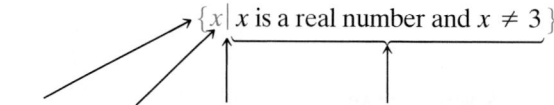

$$\{x \mid x \text{ is a real number and } x \neq 3\}$$

"The set of all x such that x is a real number and x is not equal to 3."
In this section, we will use this set builder notations to write domains.

Unless told otherwise, we assume that the domain of a function described by an equation is the set of all real numbers for which the equation is defined.

EXAMPLE 1

Find the domain of each rational function.

a. $f(x) = \dfrac{8x^3 + 7x^2 + 20}{2}$ **b.** $g(x) = \dfrac{5x^2 - 3}{x - 1}$ **c.** $f(x) = \dfrac{7x - 2}{x^2 - 2x - 15}$

Solution   The domain of each function will contain all real numbers except those values that make the denominator 0.

a. No matter what the value of x, the denominator of $f(x) = \dfrac{8x^3 + 7x^2 + 20}{2}$ is never 0, so the domain of f is $\{x \mid x \text{ is a real number}\}$.

b. To find the values of x that make the denominator of $g(x)$ equal to 0, we solve the equation "denominator $= 0$":

$$x - 1 = 0, \quad \text{or} \quad x = 1$$

The domain must exclude 1 since the rational expression is undefined when x is 1. The domain of g is $\{x \mid x \text{ is a real number and } x \neq 1\}$.

c. We find the domain by setting the denominator equal to 0.

$$x^2 - 2x - 15 = 0 \quad \text{Set the denominator equal to 0 and solve.}$$
$$(x - 5)(x + 3) = 0$$
$$x - 5 = 0 \quad \text{or} \quad x + 3 = 0$$
$$x = 5 \quad \text{or} \quad x = -3$$

If x is replaced with 5 or with -3, the rational expression is undefined.

The domain of f is $\{x \mid x \text{ is a real number and } x \neq 5, x \neq -3\}$.

✔ **CONCEPT CHECK**

For which of these values (if any) is the rational expression $\dfrac{x - 3}{x^2 + 2}$ undefined?

a. 2

b. 3

c. -2

d. 0

e. None of these

 Recall that a fraction is in lowest terms or simplest form if the numerator and denominator have no common factors other than 1 (or -1). For example, $\dfrac{3}{13}$ is in lowest terms since 3 and 13 have no common factors other than 1 (or -1).

To **simplify** a rational expression, or to write it in lowest terms, we use the fundamental principle of rational expressions.

Fundamental Principle of Rational Expressions

For any rational expression $\dfrac{P}{Q}$ and any polynomial R, where $R \neq 0$,

$$\frac{PR}{QR} = \frac{P}{Q}$$

Thus, the fundamental principle says that multiplying or dividing the numerator and denominator of a rational expression by the same nonzero polynomial yields an equivalent rational expression.

To simplify a rational expression such as $\dfrac{(x+2)^2}{x^2-4}$, factor the numerator and the denominator and then use the fundamental principle of rational expressions to divide out common factors.

$$\frac{(x+2)^2}{x^2-4} = \frac{(x+2)(x+2)}{(x+2)(x-2)} = \frac{x+2}{x-2}$$

This means that the rational expression $\dfrac{(x+2)^2}{x^2-4}$ has the same value as the rational expression $\dfrac{x+2}{x-2}$ for all values of x except 2 and -2. (Remember that when x is 2, the denominators of both rational expressions are 0 and that when x is -2, the original rational expression has a denominator of 0.)

As we simplify rational expressions, we will assume that the simplified rational expression is equivalent to the original rational expression for all real numbers except those for which either denominator is 0.

In general, the following steps may be used to simplify rational expressions or to write a rational expression in lowest terms.

> ## Simplifying or Writing a Rational Expression in Lowest Terms
>
> **Step 1:** Completely factor the numerator and denominator of the rational expression.
>
> **Step 2:** Apply the fundamental principle of rational expressions to divide out factors common to both the numerator and denominator.

For now, we assume that variables in a rational expression do not represent values that make the denominator 0.

EXAMPLE 2

Simplify each rational expression.

a. $\dfrac{24x^6y^5}{8x^7y}$ **b.** $\dfrac{2x^2}{10x^3-2x^2}$

Solution **a.** The GCF of the numerator and denominator is $8x^6 + y$.

$$\frac{24x^6y^5}{8x^7y} = \frac{(8x^6y)\,3y^4}{(8x^6y)\,x} = \frac{3y^4}{x}$$

b. Factor out $2x^2$ from the denominator. Then divide numerator and denominator by their GCF, $2x^3$.

$$\frac{2x^2}{10x^3-2x^2} = \frac{2x^2}{2x^2\,(5x-1)} = \frac{1}{5x-1}$$

When the terms in the numerator of a rational expression differ by sign from the terms of the denominator, the polynomials are opposites of each other and the expression simplifies to -1. To see this, study Example 3b on the next page.

EXAMPLE 3

Simplify each rational expression.

a. $\dfrac{2 + x}{x + 2}$ **b.** $\dfrac{2 - x}{x - 2}$

Solution   **a.** $\dfrac{2 + x}{x + 2} = \dfrac{x + 2}{x + 2} = 1$ *By the commutative property of addition, $2 + x = x + 2$.*

b. $\dfrac{2 - x}{x - 2}$

The terms in the numerator of $\dfrac{2 - x}{x - 2}$ differ by sign from the terms of the denominator, so the polynomials are opposites of each other and the expression simplifies to -1. To see this, we factor out -1 from the numerator or the denominator. If -1 is factored from the numerator, then

Helpful Hint

When the numerator and the denominator of a rational expression are opposites of each other, the expression simplifies to -1.

$$\dfrac{2 - x}{x - 2} = \dfrac{-1(-2 + x)}{x - 2} = \dfrac{-1(x - 2)}{x - 2} = \dfrac{-1}{1} = -1$$

If -1 is factored from the denominator, the result is the same.

$$\dfrac{2 - x}{x - 2} = \dfrac{2 - x}{-1(-x + 2)} = \dfrac{2 - x}{-1(2 - x)} = \dfrac{1}{-1} = -1$$

EXAMPLE 4

Simplify $\dfrac{18 - 2x^2}{x^2 - 2x - 3}$.

Solution

$$\dfrac{18 - 2x^2}{x^2 - 2x - 3} = \dfrac{2(9 - x^2)}{(x + 1)(x - 3)} \quad \text{Factor.}$$

$$= \dfrac{2(3 + x)(3 - x)}{(x + 1)(x - 3)} \quad \text{Factor completely.}$$

$$= \dfrac{2(3 + x) \cdot -1(x - 3)}{(x + 1)(x - 3)} \quad \begin{array}{l}\text{Notice the opposites } 3 - x \\ \text{and } x - 3. \text{ Write } 3 - x \\ \text{as } -1(x - 3) \text{ and simplify.}\end{array}$$

$$= -\dfrac{2(3 + x)}{x + 1}$$

Helpful Hint

Recall that for a fraction $\dfrac{a}{b}$,

$$\dfrac{a}{-b} = \dfrac{-a}{b} = -\dfrac{a}{b}$$

For example

$$\dfrac{-(x + 1)}{(x + 2)} = \dfrac{(x + 1)}{-(x + 2)} = -\dfrac{x + 1}{x + 2}$$

✔ **CONCEPT CHECK**

Which of the following expressions are equivalent to $\dfrac{x}{8-x}$?

    a. $\dfrac{-x}{x-8}$ **b.** $\dfrac{-x}{8-x}$ **c.** $\dfrac{x}{x-8}$ **d.** $\dfrac{-x}{-8+x}$

EXAMPLE 5

Simplify each rational expression.

   a. $\dfrac{x^3+8}{2+x}$ **b.** $\dfrac{2y^2+2}{y^3-5y^2+y-5}$

Solution   **a.** $\dfrac{x^3+8}{2+x} = \dfrac{(x+2)(x^2-2x+4)}{x+2}$ Factor the sum of the two cubes.

               $= x^2 - 2x + 4$ Divide out common factors.

   b. $\dfrac{2y^2+2}{y^3-5y^2+y-5} = \dfrac{2(y^2+1)}{(y^3-5y^2)+(y-5)}$ Factor the numerator.

               $= \dfrac{2(y^2+1)}{y^2(y-5)+1(y-5)}$ Factor the denominator by grouping.

               $= \dfrac{2(y^2+1)}{(y-5)(y^2+1)}$

               $= \dfrac{2}{y-5}$ Divide out common factors.

✔ **CONCEPT CHECK**

Does $\dfrac{n}{n+2}$ simplify to $\dfrac{1}{2}$? Why or why not?

    Rational functions occur often in real-life situations.

EXAMPLE 6

COST FOR PRESSING COMPACT DISCS

For the ICL Production Company, the rational function $C(x) = \dfrac{2.6x+10{,}000}{x}$ describes the company's cost per disc of pressing x compact discs. Find the cost per disc for pressing

   a. 100 compact discs   **b.** 1000 compact discs

Solution   **a.** $C(100) = \dfrac{2.6(100)+10{,}000}{100} = \dfrac{10{,}260}{100} = 102.6$

The cost per disc for pressing 100 compact discs is $102.60.

Concept Check Answer:
a and d

   b. $C(1000) = \dfrac{2.6(1000)+10{,}000}{1000} = \dfrac{12{,}600}{1000} = 12.6$

Concept Check Answer:
no; answers may vary

The cost per disc for pressing 1000 compact discs is $12.60. Notice that as more compact discs are produced, the cost per disc decreases.

Spotlight on DECISION MAKING

Suppose you are a forensic lab technician. You have been asked to try to identify a piece of metal found at a crime scene. You know that one way to analyze the piece of metal is to find its density (mass per unit volume), using the formula $density = \dfrac{mass}{volume}$. After weighing the metal, you find its mass as 36.2 grams. You have also determined the volume of the metal to be 4.5 milliliters. Use this information to decide which type of metal this piece is most likely made of and explain your reasoning. What other chatacteristics might help the identification?

Densities	
Metal	**Density(g/ml)**
Aluminum	2.7
Iron	7.8
Lead	11.5
Silver	10.5

Graphing Calculator Explorations

Recall that since the rational expression $\dfrac{7x - 2}{(x - 2)(x + 5)}$ is not defined when $x = 2$ or when $x = -5$, we say that the domain of the rational function $f(x) = \dfrac{7x - 2}{(x - 2)(x + 5)}$ is all real numbers except 2 and -5. This domain can be written as $\{x \mid x$ is a real number and $x \neq 2$, $x \neq -5\}$. This means that the graph of $f(x)$ should not cross the vertical lines $x = 2$ and $x = -5$. The graph of $f(x)$ in *connected* mode is shown at left. In connected mode the graphing calculator tries to connect all dots of the graph so that the result is a smooth curve. This is what has happened in the graph. Notice that the graph appears to contain vertical lines at $x = 2$ and at $x = -5$. We know that this cannot happen because the function is not defined at $x = 2$ and at $x = -5$. We also know that this cannot happen because the graph of this function would not pass the vertical line test.

If we graph $f(x)$ in *dot* mode, the graph appears as follows on the next page. In dot mode the graphing calculator will not connect dots with a smooth curve. Notice that the vertical lines have disappeared, and we have a better picture of the graph. The graph, however, actually appears more like the hand-drawn graph on the next page. By using a Table feature, a Calculate Value feature, or by tracing, we can see that the function is not defined at $x = 2$ and at $x = -5$.

(continued)

Find the domain of each rational function. Then graph each rational function and use the graph to confirm the domain.

1. $f(x) = \dfrac{x+1}{x^2 - 4}$

2. $g(x) = \dfrac{5x}{x^2 - 9}$

3. $h(x) = \dfrac{x^2}{2x^2 + 7x - 4}$

4. $f(x) = \dfrac{3x+2}{4x^2 - 19x - 5}$

EXERCISE SET 7.1

    Math XP

STUDY   CD/   PH MATH   MathXL®Tutorials   MathXL®   MyMathLab®
GUIDE/SSM   VIDEO   TUTOR CENTER   ON CD

Find each function value. See Example 6.

 1. $f(x) = \dfrac{x+8}{2x-1}; f(2), f(0), f(-1)$

2. $f(y) = \dfrac{y-2}{-5+y}; f(-5), f(0), f(10)$

3. $g(x) = \dfrac{x^2+8}{x^3 - 25x}; g(3), g(-2), g(1)$

4. $s(t) = \dfrac{t^3+1}{t^2+1}; s(-1), s(1), s(2)$

Find the domain of each rational function. See Example 2.

5. $f(x) = \dfrac{5x-7}{4}$

6. $g(x) = \dfrac{4-3x}{2}$

7. $s(t) = \dfrac{t^2+1}{2t}$

8. $v(t) = -\dfrac{5t+t^2}{3t}$

9. $f(x) = \dfrac{3x}{7-x}$

10. $f(x) = \dfrac{-4x}{-2+x}$

11. $C(x) = \dfrac{x+3}{x^2-4}$

12. $R(x) = \dfrac{5}{x^2-7x}$

13. $g(x) = \dfrac{5x}{3x^2 - 14x - 5}$ **14.** $h(x) = \dfrac{-x}{4x^2 + 11x - 3}$

15. $R(x) = \dfrac{3 + 2x}{x^3 + x^2 - 2x}$ **16.** $h(x) = \dfrac{5 - 3x}{2x^2 - 14x + 20}$

MIXED PRACTICE

Simplify each rational expression. See Examples 3 through 5.

17. $\dfrac{10x^3}{18x}$ **18.** $\dfrac{48a^7}{16a^{10}}$

19. $\dfrac{9x^6 y^3}{18x^2 y^5}$ **20.** $\dfrac{10ab^5}{15a^3 b^5}$

🔒 **21.** $\dfrac{x + 5}{5 + x}$ 🔒 **22.** $\dfrac{x - 5}{5 - x}$

23. $\dfrac{x - 1}{1 - x^2}$ **24.** $\dfrac{10 + 5x}{x^2 + 2x}$

25. $\dfrac{8x - 4}{6x - 3}$ **26.** $\dfrac{6x - 12}{15x - 30}$

🔒 **27.** $\dfrac{2x - 14}{7 - x}$ **28.** $\dfrac{9 - x}{5x - 45}$

29. $\dfrac{x^2 - 2x - 3}{x^2 - 6x + 9}$ **30.** $\dfrac{x^2 + 10x + 25}{x^2 + 8x + 15}$

31. $\dfrac{2x^2 + 12x + 18}{5x^2 + 17x + 6}$ **32.** $\dfrac{x^2 - 4}{2x^2 + 8x + 8}$

33. $\dfrac{30x + 6}{5x^2 + x}$ **34.** $\dfrac{3x + 4}{9x^2 + 4}$

35. $\dfrac{2x^2 - x - 3}{2x^3 - 3x^2 + 2x - 3}$ **36.** $\dfrac{3x^2 - 5x - 2}{6x^3 + 2x^2 + 3x + 1}$

37. $\dfrac{8q^2}{16q^3 - 16q^2}$ **38.** $\dfrac{3y}{6y^2 - 30y}$

39. $\dfrac{x^2 + 6x - 40}{10 + x}$ **40.** $\dfrac{x^2 - 8x + 16}{4 - x}$

41. $\dfrac{x^3 - 125}{5 - x}$ **42.** $\dfrac{4x + 4}{2x^3 + 2}$

🔒 **43.** $\dfrac{8x^3 - 27}{4x - 6}$ **44.** $\dfrac{9x^2 - 15x + 25}{27x^3 + 125}$

Solve. See Example 6.

45. The dose of medicine prescribed for a child depends on the child's age A in years and the adult dose D for the medication. Young's Rule is a formula used by pediatricians that gives a child's dose C as

$$C = \dfrac{DA}{A + 12}$$

Suppose that an 8-year-old child needs medication, and the normal adult dose is 1000 mg. What size dose should the child receive?

46. During a storm, water treatment engineers monitor how quickly rain is falling. If too much rain comes too fast, there is a danger of sewers backing up. A formula that gives the rainfall intensity i in millimeters per hour for a certain strength storm in eastern Virginia is

$$i = \dfrac{5840}{t + 29}$$

where t is the duration of the storm in minutes. What rainfall intensity should engineers expect for a storm of this strength in eastern Virginia that lasts for 80 minutes? Round your answer to one decimal place.

47. Calculating body-mass index is a way to gauge whether a person should lose weight. Doctors recommend that body-mass index values fall between 19 and 25. The formula for body-mass index B is

$$B = \dfrac{705w}{h^2}$$

where w is weight in pounds and h is height in inches. Should a 148-pound person who is 5 feet 6 inches tall lose weight?

48. Anthropologists and forensic scientists use a measure called the cephalic index to help classify skulls. The cephalic index of a skull with width W and length L from front to back is given by the formula

$$C = \dfrac{100W}{L}$$

A long skull has an index value less than 75, a medium skull has an index value between 75 and 85, and a broad skull has an index value over 85. Find the cephalic index of a skull that is 5 inches wide and 6.4 inches long. Classify the skull.

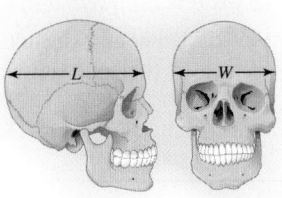

49. The total revenue $R(x)$ from the sale of a popular music compact disc is approximately given by the function

$$R(x) = \frac{150x^2}{x^2 + 3}$$

where x is the number of years since the CD has been released and revenue $R(x)$ is in millions of dollars.

a. Find the total revenue generated by the end of the first year.

b. Find the total revenue generated by the end of the second year.

c. Find the total revenue generated in the second year only.

50. For a certain model fax machine, the manufacturing cost $C(x)$ per machine is given by the function

$$C(x) = \frac{250x + 10,000}{x}$$

where x is the number of fax machines manufactured and cost $C(x)$ is in dollars per machine.

a. Find the cost per fax machine when manufacturing 100 fax machines.

b. Find the cost per fax machine when manufacturing 1000 fax machines.

c. Does the cost per machine decrease or increase when more machines are manufactured? Explain why this is so.

REVIEW AND PREVIEW

Perform the indicated operations. See Section 1.3.

51. $\dfrac{1}{3} \cdot \dfrac{9}{11}$

52. $\dfrac{5}{27} \cdot \dfrac{2}{5}$

53. $\dfrac{1}{3} \div \dfrac{1}{4}$

54. $\dfrac{7}{8} \div \dfrac{1}{2}$

55. $\dfrac{5}{6} \cdot \dfrac{10}{11} \cdot \dfrac{2}{3}$

56. $\dfrac{4}{3} \cdot \dfrac{1}{7} \cdot \dfrac{10}{13}$

57. $\dfrac{13}{20} \div \dfrac{2}{9}$

58. $\dfrac{8}{15} \div \dfrac{5}{8}$

Concept Extensions

59. In your own words, explain how to find the domain of a rational function.

60. In your own words, explain how to simplify a rational expression or to write it in lowest terms.

61. Recall that the fundamental principle applies to common *factors* only. Which of the following are *not* true? Explain why.

a. $\dfrac{3-1}{3+5} = -\dfrac{1}{5}$

b. $\dfrac{2x + 10}{2} = x + 5$

c. $\dfrac{37}{72} = \dfrac{3}{2}$

d. $\dfrac{2x + 6}{2} = x + 3$

62. Graph a portion of the function $f(x) = \dfrac{20x}{100 - x}$. To do so, complete the given table, plot the points, and then connect the plotted points with a smooth curve.

x	0	10	30	50	70	90	95	99
y or $f(x)$								

63. The domain of the function $f(x) = \dfrac{1}{x}$ is all real numbers except 0. This means that the graph of this function will be in two pieces: one piece corresponding to x values less than 0 and one piece corresponding to x values greater than 0. Graph the function by completing the following tables, separately plotting the points, and connecting each set of plotted points with a smooth curve.

x	$\frac{1}{4}$	$\frac{1}{2}$	1	2	4
y or $f(x)$					

x	-4	-2	-1	$-\frac{1}{2}$	$-\frac{1}{4}$
y or $f(x)$					

64. The function $f(x) = \dfrac{100,000x}{100 - x}$ models the cost in dollars for removing x percent of the pollutants from a bayou in which a nearby company dumped creosote.

a. What is the domain of $f(x)$?

b. Find the cost of removing 30% of the pollutants from the bayou. (*Hint:* Find $f(30)$.)

c. Find the cost of removing 60% of the pollutants and then 80% of the pollutants.

d. Find $f(90)$, then $f(95)$, and then $f(99)$. What happens to the cost as x approaches 100%?

65. The total revenue from the sale of a popular book is approximated by the rational function $R(x) = \dfrac{1000x^2}{x^2 + 4}$ where x is the number of years since publication and $R(x)$ is the total revenue in millions of dollars.

a. Find the total revenue at the end of the first year.

b. Find the total revenue at the end of the second year.

c. Find the revenue during the second year only.

How does the graph of $y = \frac{x^2 - 9}{x - 3}$ *compare to the graph of* $y = x + 3$? *Recall that* $\frac{x^2 - 9}{x - 3} = \frac{(x + 3)(x - 3)}{x - 3} = x + 3$ *as long as x is not 3. This means that the graph of* $y = \frac{x^2 - 9}{x - 3}$ *is the same as the graph of* $y = x + 3$ *with* $x \neq 3$. *To graph* $y = \frac{x^2 - 9}{x - 3}$, *then, graph the linear equation* $y = x + 3$ *and place an open dot on the graph at 3. This open dot or interruption of the line at 3 means* $x \neq 3$.

66. Graph $y = \frac{x^2 - 25}{x + 5}$.

67. Graph $y = \frac{x^2 - 16}{x - 4}$.

68. Graph $y = \frac{x^2 + x - 12}{x + 4}$.

69. Graph $y = \frac{x^2 - 6x + 8}{x - 2}$.

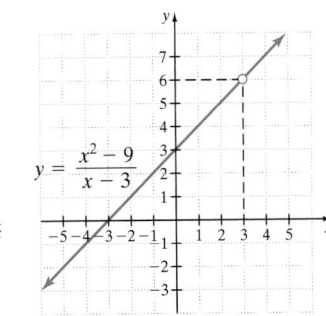

7.2   *MULTIPLYING AND DIVIDING RATIONAL EXPRESSIONS*

Objectives

1 Multiply rational expressions.

2 Divide rational expressions.

1   Just as simplifying rational expressions is similar to simplifying number fractions, multiplying and dividing rational expressions is similar to multiplying and dividing number fractions. To find the product of fractions and rational expressions, multiply the numerators and multiply the denominators.

$$\frac{3}{5} \cdot \frac{1}{4} = \frac{3 \cdot 1}{5 \cdot 4} = \frac{3}{20} \quad \text{and} \quad \frac{x}{y + 1} \cdot \frac{x + 3}{y - 1} = \frac{x(x + 3)}{(y + 1)(y - 1)}$$

Multiplying Rational Expressions

Let $P, Q, R,$ and S be polynomials. Then

$$\frac{P}{Q} \cdot \frac{R}{S} = \frac{PR}{QS}$$

as long as $Q \neq 0$ and $S \neq 0$.

EXAMPLE 1

Multiply.

a. $\frac{25x}{2} \cdot \frac{1}{y^3}$

b. $\frac{-7x^2}{5y} \cdot \frac{3y^5}{14x^2}$

Solution To multiply rational expressions, multiply the numerators and then multiply the denominators of both expressions. Then simplify if possible.

a. $\dfrac{25x}{2} \cdot \dfrac{1}{y^3} = \dfrac{25x \cdot 1}{2 \cdot y^3} = \dfrac{25x}{2y^3}$

The expression $\dfrac{25x}{2y^3}$ is in simplest form.

b. $\dfrac{-7x^2}{5y} \cdot \dfrac{3y^5}{14x^2} = \dfrac{-7x^2 \cdot 3y^5}{5y \cdot 14x^2}$ Multiply.

The expression $\dfrac{-7x^2 \cdot 3y^5}{5y \cdot 14x^2}$ is not in simplest form, so we factor the numerator and the denominator and apply the fundamental principle.

$$= \dfrac{-1 \cdot 7 \cdot 3 \cdot x^2 \cdot y \cdot y^4}{5 \cdot 2 \cdot 7 \cdot x^2 \cdot y}$$

$$= -\dfrac{3y^4}{10}$$

When multiplying rational expressions, it is usually best to factor each numerator and denominator. This will help us when we apply the fundamental principle to write the product in lowest terms.

EXAMPLE 2

Multiply: $\dfrac{x^2 + x}{3x} \cdot \dfrac{6}{5x + 5}$

Solution $\dfrac{x^2 + x}{3x} \cdot \dfrac{6}{5x + 5} = \dfrac{x(x + 1)}{3x} \cdot \dfrac{2 \cdot 3}{5(x + 1)}$ Factor numerators and denominators.

$$= \dfrac{x(x + 1) \cdot 2 \cdot 3}{3x \cdot 5(x + 1)}$$ Multiply.

$$= \dfrac{2}{5}$$ Simplify.

The following steps may be used to multiply rational expressions.

Multiplying Rational Expressions

Step 1. Completely factor numerators and denominators.

Step 2. Multiply numerators and multiply denominators.

Step 3. Simplify or write the product in lowest terms by applying the fundamental principle to all common factors.

✔ **CONCEPT CHECK**

Which of the following is a true statement?

a. $\dfrac{1}{3} \cdot \dfrac{1}{2} = \dfrac{1}{5}$ **b.** $\dfrac{2}{x} \cdot \dfrac{5}{x} = \dfrac{10}{x}$ **c.** $\dfrac{3}{x} \cdot \dfrac{1}{2} = \dfrac{3}{2x}$ **d.** $\dfrac{x}{7} \cdot \dfrac{x + 5}{4} = \dfrac{2x + 5}{28}$

EXAMPLE 3

Multiply: $\dfrac{3x + 3}{5x - 5x^2} \cdot \dfrac{2x^2 + x - 3}{4x^2 - 9}$

Solution

$\dfrac{3x + 3}{5x - 5x^2} \cdot \dfrac{2x^2 + x - 3}{4x^2 - 9} = \dfrac{3(x + 1)}{5x(1 - x)} \cdot \dfrac{(2x + 3)(x - 1)}{(2x - 3)(2x + 3)}$ Factor.

$= \dfrac{3(x + 1)(2x + 3)(x - 1)}{5x(1 - x)(2x - 3)(2x + 3)}$ Multiply.

$= \dfrac{3(x + 1)(x - 1)}{5x(1 - x)(2x - 3)}$ Apply the fundamental principle.

Next, recall that $x - 1$ and $1 - x$ are opposites so that $x - 1 = -1(1 - x)$.

$= \dfrac{3(x + 1)(-1)(1 - x)}{5x(1 - x)(2x - 3)}$ Write $x - 1$ as $-1(1 - x)$.

$= \dfrac{-3(x + 1)}{5x(2x - 3)}$ or $-\dfrac{3(x + 1)}{5x(2x - 3)}$ Apply the fundamental principle.

2 We can divide by a rational expression in the same way we divide by a fraction. To divide by a fraction, multiply by its reciprocal.

> **Helpful Hint**
> Don't forget how to find reciprocals. The reciprocal of $\dfrac{a}{b}$ is $\dfrac{b}{a}$, $a \neq 0$, $b \neq 0$.

For example, to divide $\frac{3}{2}$ by $\frac{7}{8}$, multiply $\frac{3}{2}$ by $\frac{8}{7}$.

$$\frac{3}{2} \div \frac{7}{8} = \frac{3}{2} \cdot \frac{8}{7} = \frac{3 \cdot 4 \cdot 2}{2 \cdot 7} = \frac{12}{7}$$

Dividing Rational Expressions

Let P, Q, R, and S be polynomials. Then,

$$\frac{P}{Q} \div \frac{R}{S} = \frac{P}{Q} \cdot \frac{S}{R} = \frac{PS}{QR}$$

as long as $Q \neq 0$, $S \neq 0$, and $R \neq 0$.

EXAMPLE 4

Divide: $\dfrac{3x^3y^7}{40} \div \dfrac{4x^3}{y^2}$

Solution

$$\dfrac{3x^3y^7}{40} \div \dfrac{4x^3}{y^2} = \dfrac{3x^3y^7}{40} \cdot \dfrac{y^2}{4x^3} \qquad \text{Multiply by the reciprocal of } \dfrac{4x^3}{y^2}.$$

$$= \dfrac{3x^3y^9}{160x^3}$$

$$= \dfrac{3y^9}{160} \qquad \text{Simplify.}$$

EXAMPLE 5

Divide $\dfrac{(x-1)(x+2)}{10}$ by $\dfrac{2x+4}{5}$

Solution

$$\dfrac{(x-1)(x+2)}{10} \div \dfrac{2x+4}{5} = \dfrac{(x-1)(x+2)}{10} \cdot \dfrac{5}{2x+4} \qquad \begin{array}{l}\text{Multiply by the reciprocal of}\\ \dfrac{2x+4}{5}.\end{array}$$

$$= \dfrac{(x-1)(x+2) \cdot 5}{5 \cdot 2 \cdot 2 \cdot (x+2)} \qquad \text{Factor and multiply.}$$

$$= \dfrac{x-1}{4} \qquad \text{Simplify.}$$

The following may be used to divide by a rational expression.

Dividing by a Rational Expression

Multiply by its reciprocal.

EXAMPLE 6

Divide: $\dfrac{6x+2}{x^2-1} \div \dfrac{3x^2+x}{x-1}$

Solution

$$\dfrac{6x+2}{x^2-1} \div \dfrac{3x^2+x}{x-1} = \dfrac{6x+2}{x^2-1} \cdot \dfrac{x-1}{3x^2+x} \qquad \text{Multiply by the reciprocal.}$$

$$= \dfrac{2(3x+1)(x-1)}{(x+1)(x-1) \cdot x(3x+1)} \qquad \text{Factor and multiply.}$$

$$= \dfrac{2}{x(x+1)} \qquad \text{Simplify.}$$

EXAMPLE 7

Divide: $\dfrac{2x^2-11x+5}{5x-25} \div \dfrac{4x-2}{10}$

Solution

$$\dfrac{2x^2-11x+5}{5x-25} \div \dfrac{4x-2}{10} = \dfrac{2x^2-11x+5}{5x-25} \cdot \dfrac{10}{4x-2} \qquad \text{Multiply by the reciprocal.}$$

$$= \dfrac{(2x-1)(x-5) \cdot 2 \cdot 5}{5(x-5) \cdot 2(2x-1)} \qquad \text{Factor and multiply.}$$

$$= \dfrac{1}{1} \quad \text{or} \quad 1 \qquad \text{Simplify.}$$

Now that we know how to multiply fractions and rational expressions, we can use this knowledge to help us convert between units of measure. To do so, we will use **unit fractions**. A unit fraction is a fraction that equals 1. For example, since 12 in. = 1 ft, we have the unit fractions

$$\frac{12 \text{ in.}}{1 \text{ ft}} = 1 \quad \text{and} \quad \frac{1 \text{ ft}}{12 \text{ in.}} = 1$$

EXAMPLE 8

CONVERTING FROM SQUARE YARDS TO SQUARE FEET

The largest casino in the world is the Foxwoods Resort Casino in Ledyard, CT. The gaming area for this casino is 21,444 *square yards*. Find the size of the gaming area in *square feet*. (*Source: The Guinness Book of Records*)

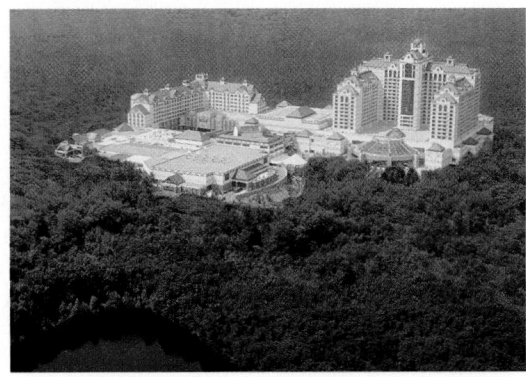

Solution There are 9 square feet in 1 square yard.

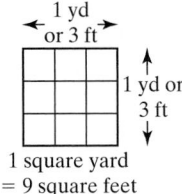

1 yd or 3 ft

1 yd or 3 ft

1 square yard = 9 square feet

Unit fraction

$$21{,}444 \text{ square yards} = 21{,}444 \text{ sq. yd} \cdot \frac{9 \text{ sq. ft}}{1 \text{ sq. yd}}$$

$$= 192{,}996 \text{ square feet}$$

> **Helpful Hint**
>
> When converting a unit of measurement, if possible, write the unit fraction so that **the numerator contains the units we are converting to** and **the denominator contains the original units.** For example, suppose we want to convert 48 inches to feet.
>
> Unit fraction
>
> $$48 \text{ in.} = \frac{48 \text{ in.}}{1} \cdot \frac{1 \text{ ft}}{12 \text{ in.}} \quad \begin{array}{l}\leftarrow \text{Units converting to} \\ \leftarrow \text{Original units}\end{array}$$
>
> $$= \frac{48}{12} \text{ ft} = 4 \text{ ft}$$

MENTAL MATH

Find the following products. See Example 1.

1. $\dfrac{2}{y} \cdot \dfrac{x}{3}$ **2.** $\dfrac{3x}{4} \cdot \dfrac{1}{y}$ **3.** $\dfrac{5}{7} \cdot \dfrac{y^2}{x^2}$ **4.** $\dfrac{x^5}{11} \cdot \dfrac{4}{z^3}$ **5.** $\dfrac{9}{x} \cdot \dfrac{x}{5}$ **6.** $\dfrac{y}{7} \cdot \dfrac{3}{y}$

EXERCISE SET 7.2

STUDY GUIDE/SSM | CD/VIDEO | PH MATH TUTOR CENTER | MathXL®Tutorials ON CD | MathXL® | MyMathLab®

Multiply. Simplify if possible. See Examples 1 through 3.

 1. $\dfrac{8x}{2} \cdot \dfrac{x^5}{4x^2}$

2. $\dfrac{6x^2}{10x^3} \cdot \dfrac{5x}{12}$

3. $-\dfrac{5a^2b}{30a^2b^2} \cdot b^3$

4. $-\dfrac{9x^3y^2}{18xy^5} \cdot y^3$

5. $\dfrac{4}{x+2} \cdot \dfrac{x}{7}$

6. $\dfrac{9}{y-6} \cdot \dfrac{y}{5}$

7. $\dfrac{x}{2x-14} \cdot \dfrac{x^2-7x}{5}$

8. $\dfrac{4x-24}{20x} \cdot \dfrac{5}{x-6}$

9. $\dfrac{6x+6}{5} \cdot \dfrac{10}{36x+36}$

10. $\dfrac{x^2+x}{8} \cdot \dfrac{16}{x+1}$

11. $\dfrac{m^2-n^2}{m+n} \cdot \dfrac{m}{m^2-mn}$

12. $\dfrac{(m-n)^2}{m+n} \cdot \dfrac{m}{m^2-mn}$

13. $\dfrac{x^2-25}{x^2-3x-10} \cdot \dfrac{x+2}{x}$

14. $\dfrac{a^2+6a+9}{a^2-4} \cdot \dfrac{a+3}{a-2}$

△ **15.** Find the area of the following rectangle.

$\dfrac{2x}{x^2-25}$ feet

$\dfrac{x+5}{9x^3}$ feet

△ **16.** Find the area of the following square.

$\dfrac{2x}{5x^2+3x}$ meters

Divide. Simplify if possible. See Examples 4 through 7.

17. $\dfrac{5x^7}{2x^5} \div \dfrac{10x}{4x^3}$

18. $\dfrac{9y^4}{6y} \div \dfrac{y^2}{3}$

19. $\dfrac{8x^2}{y^3} \div \dfrac{4x^2y^3}{6}$

20. $\dfrac{7a^2b}{3ab^2} \div \dfrac{21a^2b^2}{14ab}$

21. $\dfrac{(x-6)(x+4)}{4x} \div \dfrac{2x-12}{8x^2}$

22. $\dfrac{(x+3)^2}{5} \div \dfrac{5x+15}{25}$

23. $\dfrac{3x^2}{x^2-1} \div \dfrac{x^5}{(x+1)^2}$

24. $\dfrac{(x+1)}{(x+1)(2x+3)} \div \dfrac{20}{2x+3}$

25. $\dfrac{m^2-n^2}{m+n} \div \dfrac{m}{m^2+nm}$

26. $\dfrac{(m-n)^2}{m+n} \div \dfrac{m^2-mn}{m}$

 27. $\dfrac{x+2}{7-x} \div \dfrac{x^2-5x+6}{x^2-9x+14}$

28. $(x-3) \div \dfrac{x^2+3x-18}{x}$

29. $\dfrac{x^2+7x+10}{1-x} \div \dfrac{x^2+2x-15}{x-1}$

30. $\dfrac{a^2-b^2}{9} \div \dfrac{3b-3a}{27x^2}$

✎ **31.** Explain how to multiply rational expressions.

✎ **32.** Explain how to divide rational expressions.

MIXED PRACTICE

Perform the indicated operations.

33. $\dfrac{5a^2b}{30a^2b^2} \cdot \dfrac{1}{b^3}$

34. $\dfrac{9x^2y^2}{42xy^5} \cdot \dfrac{6}{x^5}$

35. $\dfrac{12x^3y}{8xy^7} \div \dfrac{7x^5y}{6x}$

36. $\dfrac{4y^2z}{3y^7z^7} \div \dfrac{12y}{6z}$

37. $\dfrac{5x-10}{12} \div \dfrac{4x-8}{8}$

38. $\dfrac{6x+6}{5} \div \dfrac{3x+3}{10}$

39. $\dfrac{x^2+5x}{8} \cdot \dfrac{9}{3x+15}$

40. $\dfrac{3x^2+12x}{6} \cdot \dfrac{9}{2x+8}$

41. $\dfrac{7}{6p^2+q} \div \dfrac{14}{18p^2+3q}$

42. $\dfrac{5x-10}{12} \div \dfrac{4x-8}{8}$

 43. $\dfrac{3x+4y}{x^2+4xy+4y^2} \cdot \dfrac{x+2y}{2}$

44. $\dfrac{2a+2b}{3} \div \dfrac{a^2-b^2}{a-b}$

45. $\dfrac{x^2-9}{x^2+8} \div \dfrac{3-x}{2x^2+16}$

46. $\dfrac{x^2-y^2}{3x^2+3xy} \cdot \dfrac{3x^2+6x}{3x^2-2xy-y^2}$

47. $\dfrac{(x+2)^2}{x-2} \div \dfrac{x^2-4}{2x-4}$

48. $\dfrac{x^2-4}{2y} \div \dfrac{2-x}{6xy}$

49. $\dfrac{a^2+7a+12}{a^2+5a+6} \cdot \dfrac{a^2+8a+15}{a^2+5a+4}$

50. $\dfrac{b^2+2b-3}{b^2+b-2} \cdot \dfrac{b^2-4}{b^2+6b+8}$

51. $\dfrac{1}{-x-4} \div \dfrac{x^2-7x}{x^2-3x-28}$

52. $\dfrac{x^2-10x+21}{7-x} \div (x+3)$

53. $\dfrac{x^2-5x-24}{2x^2-2x-24} \cdot \dfrac{4x^2+4x-24}{x^2-10x+16}$

54. $\dfrac{a^2-b^2}{a} \cdot \dfrac{a+b}{a^2+ab}$

55. $(x-5) \div \dfrac{5-x}{x^2+2}$

56. $\dfrac{2x^2 + 3xy + y^2}{x^2 - y^2} \div \dfrac{1}{2x + 2y}$

57. $\dfrac{x^2 - y^2}{x^2 - 2xy + y^2} \cdot \dfrac{y - x}{x + y}$ **58.** $\dfrac{x + 3}{x^2 - 9} \cdot \dfrac{x^2 - 8x + 15}{5x}$

59. Find the quotient of $\dfrac{x^2 - 9}{2x}$ and $\dfrac{x + 3}{8x^4}$.

60. Find the quotient of $\dfrac{4x^2 + 4x + 1}{4x + 2}$ and $\dfrac{4x + 2}{16}$.

Convert as indicated. See Example 8.

61. 10 square feet = _____ square inches.

62. 1008 square inches = ____ square feet.

63. The Pentagon, headquarters for the Department of Defense, contains 3,707,745 square feet of office space. Convert this to square yards. (Round to the nearest square yard.) (*Source: World Almanac*, 2003)

64. The Empire State building in Manhattan contains approximately 137,300 square yards of office space. Convert this to square feet.

65. 50 miles per hour = ____ feet per second (round to the nearest whole).

66. 10 feet per second = ____ miles per hour (round to the nearest tenth).

67. The speed of sound is 5023 feet per second in ocean water whose temperature is 77°F. Convert this speed of sound to miles per hour. Round to the nearest tenth. (*Source: CRC Handbook of Chemistry and Physics*, 65th edition)

68. On October 15, 1997, Andy Green broke the world land speed record. He reached a speed of approximately 763 mph in his car Thrust SSC in the Black Rock Desert, Nevada. Find this speed in feet per second. Round to the nearest whole. (*Source: The World Almanac and Book of Facts*, 2003)

REVIEW AND PREVIEW

Perform each operation. See Section 1.3.

69. $\dfrac{1}{5} + \dfrac{4}{5}$ **70.** $\dfrac{3}{15} + \dfrac{6}{15}$

71. $\dfrac{9}{9} - \dfrac{19}{9}$ **72.** $\dfrac{4}{3} - \dfrac{8}{3}$

73. $\dfrac{6}{5} + \left(\dfrac{1}{5} - \dfrac{8}{5}\right)$ **74.** $-\dfrac{3}{2} + \left(\dfrac{1}{2} - \dfrac{3}{2}\right)$

See Section 3.2.

75. Graph the linear equation $x - 2y = 6$.

76. Graph the linear equation $5x + y = 10$.

Concept Extensions

Simplify. These expressions contain 4-term polynomials and sums and differences of cubes.

77. $\dfrac{a^2 + ac + ba + bc}{a - b} \div \dfrac{a + c}{a + b}$

78. $\dfrac{x^2 + 2x - xy - 2y}{x^2 - y^2} \div \dfrac{2x + 4}{x + y}$

79. $\dfrac{3x^2 + 8x + 5}{x^2 + 8x + 7} \cdot \dfrac{x + 7}{x^2 + 4}$

80. $\dfrac{16x^2 + 2x}{16x^2 + 10x + 1} \cdot \dfrac{1}{4x^2 + 2x}$

81. $\dfrac{x^3 + 8}{x^2 - 2x + 4} \cdot \dfrac{4}{x^2 - 4}$

82. $\dfrac{9y}{3y - 3} \cdot \dfrac{y^3 - 1}{y^3 + y^2 + y}$ **83.** $\dfrac{a^2 - ab}{6a^2 + 6ab} \div \dfrac{a^3 - b^3}{a^2 - b^2}$

84. $\dfrac{x^3 + 27y^3}{6x} \div \dfrac{x^2 - 9y^2}{x^2 - 3xy}$

Perform the following operations.

85. $\left(\dfrac{x^2 - y^2}{x^2 + y^2} \div \dfrac{x^2 - y^2}{3x}\right) \cdot \dfrac{x^2 + y^2}{6}$

86. $\left(\dfrac{x^2 - 9}{x^2 - 1} \cdot \dfrac{x^2 + 2x + 1}{2x^2 + 9x + 9}\right) \div \dfrac{2x + 3}{1 - x}$

87. $\left(\dfrac{2a + b}{b^2} \cdot \dfrac{3a^2 - 2ab}{ab + 2b^2}\right) \div \dfrac{a^2 - 3ab + 2b^2}{5ab - 10b^2}$

88. $\left(\dfrac{x^2y^2 - xy}{4x - 4y} \div \dfrac{3y - 3x}{8x - 8y}\right) \cdot \dfrac{y - x}{8}$

89. On a recent day, 1 euro was equivalent to 1.09 U.S. dollars. If you had wanted to exchange $2000 U.S. to euros on that day for a European vacation, how much would you have received? Round to the nearest hundredth. (*Source: International Monetary Fund*)

90. An environmental technician finds that warm water from an industrial process is being discharged into a nearby pond at a rate of 30 gallons per minute. Plant regulations state that the flow rate should be no more than 0.1 cubic feet per second. Is the flow rate of 30 gallons per minute in violation of the plant regulations? (*Hint:* 1 cubic foot is equivalent to 7.48 gallons.)

7.3

ADDING AND SUBTRACTING RATIONAL EXPRESSIONS WITH COMMON DENOMINATORS AND LEAST COMMON DENOMINATOR

Objectives

1 Add and subtract rational expressions with the same denominator.

2 Find the least common denominator of a list of rational expressions.

3 Write a rational expression as an equivalent expression whose denominator is given.

1 Like multiplication and division, addition and subtraction of rational expressions is similar to addition and subtraction of rational numbers. In this section, we add and subtract rational expressions with a common (or the same) denominator.

Add: $\dfrac{6}{5} + \dfrac{2}{5}$ | Add: $\dfrac{9}{x+2} + \dfrac{3}{x+2}$

Add the numerators and place the sum over the common denominator.

$$\frac{6}{5} + \frac{2}{5} = \frac{6+2}{5}$$
$$= \frac{8}{5} \quad \text{Simplify.}$$

$$\frac{9}{x+2} + \frac{3}{x+2} = \frac{9+3}{x+2}$$
$$= \frac{12}{x+2} \quad \text{Simplify.}$$

Adding and Subtracting Rational Expressions with Common Denominators

If $\dfrac{P}{R}$ and $\dfrac{Q}{R}$ are rational expressions, then

$$\frac{P}{R} + \frac{Q}{R} = \frac{P+Q}{R} \quad \text{and} \quad \frac{P}{R} - \frac{Q}{R} = \frac{P-Q}{R}$$

To add or subtract rational expressions, add or subtract numerators and place the sum or difference over the common denominator.

EXAMPLE 1

Add: $\dfrac{5m}{2n} + \dfrac{m}{2n}$

Solution

$$\frac{5m}{2n} + \frac{m}{2n} = \frac{5m+m}{2n} \qquad \text{Add the numerators.}$$

$$= \frac{6m}{2n} \qquad \text{Simplify the numerator by combining like terms.}$$

$$= \frac{3m}{n} \qquad \text{Simplify by applying the fundamental principle.}$$

EXAMPLE 2

Subtract: $\dfrac{2y}{2y-7} - \dfrac{7}{2y-7}$

Solution $\dfrac{2y}{2y-7} - \dfrac{7}{2y-7} = \dfrac{2y-7}{2y-7}$ Subtract the numerators.

$\phantom{\dfrac{2y}{2y-7} - \dfrac{7}{2y-7}} = \dfrac{1}{1}$ or 1 Simplify.

EXAMPLE 3

Subtract: $\dfrac{3x^2+2x}{x-1} - \dfrac{10x-5}{x-1}$

Solution $\dfrac{3x^2+2x}{x-1} - \dfrac{10x-5}{x-1} = \dfrac{(3x^2+2x)-(10x-5)}{x-1}$ Subtract the numerators. Notice the parentheses.

> **Helpful Hint**
> Parentheses are inserted so that the entire numerator, $10x-5$, is subtracted.

$ = \dfrac{3x^2+2x-10x+5}{x-1}$ Use the distributive property.

$ = \dfrac{3x^2-8x+5}{x-1}$ Combine like terms.

$ = \dfrac{(x-1)(3x-5)}{x-1}$ Factor.

$ = 3x-5$ Simplify.

> **Helpful Hint**
> Notice how the numerator $10x-5$ has been subtracted in Example 3.
>
> This $-$ sign applies to the entire numerator of $10x-5$.   So parentheses are inserted here to indicate this.
>
> $$\dfrac{3x^2+2x}{x-1} - \dfrac{10x-5}{x-1} = \dfrac{3x^2+2x-(10x-5)}{x-1}$$

2   To add and subtract fractions with **unlike** denominators, first find a least common denominator (LCD), and then write all fractions as equivalent fractions with the LCD.

For example, suppose we add $\frac{8}{3}$ and $\frac{2}{5}$. The LCD of denominators 3 and 5 is 15, since 15 is the least common multiple (LCM) of 3 and 5. That is, 15 is the smallest number that both 3 and 5 divide into evenly. Rewrite each fraction so that its denominator is 15. (Notice how we apply the fundamental principle of rational expressions.)

$$\frac{8}{3} + \frac{2}{5} = \frac{8(5)}{3(5)} + \frac{2(3)}{5(3)} = \frac{40}{15} + \frac{6}{15} = \frac{40+6}{15} = \frac{46}{15}$$

We are multiplying by **1**.

To add or subtract rational expressions with unlike denominators, we also first find an LCD and then write all rational expressions as equivalent expressions with the LCD. The **least common denominator (LCD) of a list of rational expressions** is a polynomial of least degree whose factors include all the factors of the denominators in the list.

Finding the Least Common Denominator (LCD)

Step 1. Factor each denominator completely.

Step 2. The least common denominator (LCD) is the product of all unique factors found in Step 1, each raised to a power equal to the greatest number of times that the factor appears in any one factored denominator.

EXAMPLE 4

Find the LCD for each pair.

a. $\dfrac{1}{8}, \dfrac{3}{22}$ **b.** $\dfrac{7}{5x}, \dfrac{6}{15x^2}$

Solution **a.** Start by finding the prime factorization of each denominator.

$$8 = 2 \cdot 2 \cdot 2 = 2^3 \quad \text{and} \quad 22 = 2 \cdot 11$$

Next, write the product of all the unique factors, each raised to a power equal to the greatest number of times that the factor appears in any denominator.

The greatest number of times that the factor   appears is .
The greatest number of times that the factor 11 appears is 1.

$$\text{LCD} = 2^3 \cdot 11^1 = 8 \cdot 11 = 88$$

b. Factor each denominator.

$$5x = 5 \cdot x \quad \text{and} \quad 15x^2 = 3 \cdot 5 \cdot x^2$$

The greatest number of times that the factor 5 appears is 1.
The greatest number of times that the factor 3 appears is 1.
The greatest number of times that the factor x appears is 2.

$$\text{LCD} = 3^1 \cdot 5^1 \cdot x^2 = 15x^2$$

EXAMPLE 5

Find the LCD of

a. $\dfrac{7x}{x+2}$ and $\dfrac{5x^2}{x-2}$ **b.** $\dfrac{3}{x}$ and $\dfrac{6}{x+4}$

Solution **a.** The denominators $x + 2$ and $x - 2$ are completely factored already. The factor $x + 2$ appears once and the factor $x - 2$ appears once.

$$\text{LCD} = (x + 2)(x - 2)$$

b. The denominators x and $x + 4$ cannot be factored further. The factor x appeared once and the factor $x + 4$ appears once.

$$\text{LCD} = x(x + 4)$$

EXAMPLE 6

Find the LCD of $\dfrac{6m^2}{3m + 15}$ and $\dfrac{2}{(m + 5)^2}$.

Solution   Factor each denominator.

$$3m + 15 = 3(m + 5)$$
$$(m + 5)^2 \text{ is already factored.}$$

The greatest number of times that the factor 3 appears is 1.
The greatest number of times that the factor $m + 5$ appears *in any one
denominator* is 2.

$$\text{LCD} = 3(m + 5)^2$$

✔ **CONCEPT CHECK**

Choose the correct LCD of $\dfrac{x}{(x + 1)^2}$ and $\dfrac{5}{x + 1}$.

a. $x + 1$ **b.** $(x + 1)^2$ **c.** $(x + 1)^3$ **d.** $5x(x + 1)^2$

EXAMPLE 7

Find the LCD of $\dfrac{t - 10}{t^2 - t - 6}$ and $\dfrac{t + 5}{t^2 + 3t + 2}$.

Solution   Start by factoring each denominator.

$$t^2 - t - 6 = (t - 3)(t + 2)$$
$$t^2 + 3t + 2 = (t + 1)(t + 2)$$
$$\text{LCD} = (t - 3)(t + 2)(t + 1)$$

EXAMPLE 8

Find the LCD of $\dfrac{2}{x - 2}$ and $\dfrac{10}{2 - x}$.

Solution   The denominators $x - 2$ and $2 - x$ are opposites. That is, $2 - x = -1(x - 2)$. Use
$x - 2$ or $2 - x$ as the LCD.

$$\text{LCD} = x - 2 \qquad \text{or} \qquad \text{LCD} = 2 - x$$

Concept Check Answer:
b

3 Next we practice writing a rational expression as an equivalent rational expression with a given denominator. To do this, we apply the fundamental principle, which says that $\frac{PR}{QR} = \frac{P}{Q}$, or equivalently that $\frac{P}{Q} = \frac{PR}{QR}$. This can be seen by recalling that multiplying an expression by 1 produces an equivalent expression. In other words,

$$\frac{P}{Q} = \frac{P}{Q} \cdot 1 = \frac{P}{Q} \cdot \frac{R}{R} = \frac{PR}{QR}.$$

EXAMPLE 9

Write $\dfrac{4b}{9a}$ as an equivalent fraction with the given denominator.

$$\frac{4b}{9a} = \frac{}{27a^2b}$$

Solution Ask yourself: "What do we multiply $9a$ by to get $27a^2b$?" The answer is $3ab$, since $9a(3ab) = 27a^2b$. Multiply the numerator and denominator by $3ab$.

$$\frac{4b}{9a} = \frac{4b(3ab)}{9a(3ab)} = \frac{12ab^2}{27a^2b}$$

EXAMPLE 10

Write the rational expression as an equivalent rational expression with the given denominator.

$$\frac{5}{x^2 - 4} = \frac{}{(x - 2)(x + 2)(x - 4)}$$

Solution First, factor the denominator $x^2 - 4$ as $(x - 2)(x + 2)$.

If we multiply the original denominator $(x - 2)(x + 2)$ by $x - 4$, the result is the new denominator $(x + 2)(x - 2)(x - 4)$. Thus, multiply the numerator and the denominator by $x - 4$.

$$\frac{5}{x^2 - 4} = \underbrace{\frac{5}{(x - 2)(x + 2)}}_{\substack{\text{Factored} \\ \text{denominator}}} = \frac{5(x - 4)}{(x - 2)(x + 2)(x - 4)}$$

$$= \frac{5x - 20}{(x - 2)(x + 2)(x - 4)}$$

MENTAL MATH

Perform the indicated operations.

1. $\dfrac{2}{3} + \dfrac{1}{3}$

2. $\dfrac{5}{11} + \dfrac{1}{11}$

3. $\dfrac{3x}{9} + \dfrac{4x}{9}$

4. $\dfrac{3y}{8} + \dfrac{2y}{8}$

5. $\dfrac{8}{9} - \dfrac{7}{9}$

6. $-\dfrac{4}{12} - \dfrac{3}{12}$

7. $\dfrac{7}{5} - \dfrac{10y}{5}$

8. $\dfrac{12x}{7} - \dfrac{4x}{7}$

EXERCISE SET 7.3

STUDY GUIDE/SSM   CD/VIDEO   PH MATH TUTOR CENTER   MathXL®Tutorials ON CD   MathXL®   MyMathLab®

Add or subtract as indicated. Simplify the result if possible. See Examples 1 through 3.

1. $\dfrac{a}{13} + \dfrac{9}{13}$

2. $\dfrac{x+1}{7} + \dfrac{6}{7}$

3. $\dfrac{9}{3+y} + \dfrac{y+1}{3+y}$

4. $\dfrac{9}{y+9} + \dfrac{y}{y+9}$

5. $\dfrac{4m}{3n} + \dfrac{5m}{3n}$

6. $\dfrac{3p}{2} + \dfrac{11p}{2}$

7. $\dfrac{2x+1}{x-3} + \dfrac{3x+6}{x-3}$

8. $\dfrac{4p-3}{2p+7} + \dfrac{3p+8}{2p+7}$

9. $\dfrac{7}{8} - \dfrac{3}{8}$

10. $\dfrac{4}{5} - \dfrac{13}{5}$

11. $\dfrac{4m}{m-6} - \dfrac{24}{m-6}$

12. $\dfrac{8y}{y-2} - \dfrac{16}{y-2}$

13. $\dfrac{2x^2}{x-5} - \dfrac{25+x^2}{x-5}$

14. $\dfrac{6x^2}{2x-5} - \dfrac{25+2x^2}{2x-5}$

15. $\dfrac{-3x^2-4}{x-4} - \dfrac{12-4x^2}{x-4}$

16. $\dfrac{7x^2-9}{2x-5} - \dfrac{16+3x^2}{2x-5}$

17. $\dfrac{2x+3}{x+1} - \dfrac{x+2}{x+1}$

18. $\dfrac{1}{x^2-2x-15} - \dfrac{4-x}{x^2-2x-15}$

19. $\dfrac{3}{x^3} + \dfrac{9}{x^3}$

20. $\dfrac{5}{xy} + \dfrac{8}{xy}$

21. $\dfrac{5}{x+4} - \dfrac{10}{x+4}$

22. $\dfrac{4}{2x+1} - \dfrac{8}{2x+1}$

23. $\dfrac{x}{x+y} - \dfrac{2}{x+y}$

24. $\dfrac{y+1}{y+2} - \dfrac{3}{y+2}$

25. $\dfrac{8x}{2x+5} + \dfrac{20}{2x+5}$

26. $\dfrac{12y-5}{3y-1} + \dfrac{1}{3y-1}$

27. $\dfrac{5x+4}{x-1} - \dfrac{2x+7}{x-1}$

28. $\dfrac{x^2+9x}{x+7} - \dfrac{4x+14}{x+7}$

29. $\dfrac{a}{a^2+2a-15} - \dfrac{3}{a^2+2a-15}$

30. $\dfrac{3y}{y^2+3y-10} - \dfrac{6}{y^2+3y-10}$

31. $\dfrac{2x+3}{x^2-x-30} - \dfrac{x-2}{x^2-x-30}$

32. $\dfrac{3x-1}{x^2+5x-6} - \dfrac{2x-7}{x^2+5x-6}$

△ **33.** A square-shaped pasture has a side of length $\dfrac{5}{x-2}$ meters. Express its perimeter as a rational expression.

$\dfrac{5}{x-2}$ meters

△ **34.** The following trapezoid has sides of indicated length. Find its perimeter.

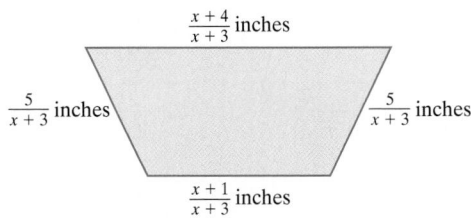

$\dfrac{x+4}{x+3}$ inches

$\dfrac{5}{x+3}$ inches

$\dfrac{5}{x+3}$ inches

$\dfrac{x+1}{x+3}$ inches

35. Describe the process for adding and subtracting two rational expressions with the same denominators.

36. Explain the similarities between subtracting $\frac{3}{8}$ from $\frac{7}{8}$ and subtracting $\frac{6}{x+3}$ from $\frac{9}{x+3}$.

Find the LCD for the following lists of rational expressions. See Examples 4 through 8.

37. $\frac{2}{3}, \frac{4}{33}$

38. $\frac{8}{20}, \frac{4}{15}$

39. $\frac{19}{2x}, \frac{5}{4x^3}$

40. $\frac{17x}{4y^5}, \frac{2}{8y}$

41. $\frac{9}{8x}, \frac{3}{2x+4}$

42. $\frac{1}{6y}, \frac{3x}{4y+12}$

43. $\frac{1}{3x+3}, \frac{8}{2x^2+4x+2}$

44. $\frac{19x+5}{4x-12}, \frac{3}{2x^2-12x+18}$

45. $\frac{5}{x-8}, \frac{3}{8-x}$

46. $\frac{2x+5}{3x-7}, \frac{5}{7-3x}$

47. $\frac{4+x}{8x^2(x-1)^2}, \frac{17}{10x^3(x-1)}$

48. $\frac{2x+3}{9x(x+2)}, \frac{9x+5}{12(x+2)^2}$

49. $\frac{9x+1}{2x+1}, \frac{3x-5}{2x-1}$

50. $\frac{5}{4x-2}, \frac{7}{4x+2}$

51. $\frac{5x+1}{2x^2+7x-4}, \frac{3x}{2x^2+5x-3}$

52. $\frac{4}{x^2+4x+3}, \frac{4x-2}{x^2+10x+21}$

53. Write some instructions to help a friend who is having difficulty finding the LCD of two rational expressions.

54. Explain why the LCD of rational expressions $\frac{7}{x+1}$ and $\frac{9x}{(x+1)^2}$ is $(x+1)^2$ and not $(x+1)^3$.

Rewrite each rational expression as an equivalent rational expression whose denominator is the given polynomial. See Examples 9 and 10.

55. $\frac{3}{2x} = \frac{}{4x^2}$

56. $\frac{3}{9y^5} = \frac{}{72y^9}$

57. $\frac{6}{3a} = \frac{}{12ab^2}$

58. $\frac{17a}{4y^2x} = \frac{}{32y^3x^2z}$

59. $\frac{9}{x+3} = \frac{}{2(x+3)}$

60. $\frac{4x+1}{3x+6} = \frac{}{3y(x+2)}$

61. $\frac{9a+2}{5a+10} = \frac{}{5b(a+2)}$

62. $\frac{5+y}{2x^2+10} = \frac{}{4(x^2+5)}$

63. $\frac{x}{x^2+6x+8} = \frac{}{(x+4)(x+2)(x+1)}$

64. $\frac{5x}{x^2+2x-3} = \frac{}{(x-1)(x-5)(x+3)}$

65. $\frac{9y-1}{15x^2-30} = \frac{}{30x^2-60}$

66. $\frac{8x+3}{7y^2-21} = \frac{}{21y^2-63}$

67. $\frac{5}{2x^2-9x-5} = \frac{}{3x(2x+1)(x-7)(x-5)}$

68. $\frac{x-9}{3x^2+10x+3} = \frac{}{x(x+3)(x+5)(3x+1)}$

REVIEW AND PREVIEW

Perform each operation. See Section 1.3.

69. $\frac{2}{3} + \frac{5}{7}$

70. $\frac{9}{10} - \frac{3}{5}$

71. $\frac{2}{6} - \frac{3}{4}$

72. $\frac{11}{15} + \frac{5}{9}$

Solve the following quadratic equations by factoring. See Section 6.5.

73. $x(x-3) = 0$

74. $2x(x+5) = 0$

75. $x^2 + 6x + 5 = 0$

76. $x^2 - 6x + 5 = 0$

Concept Extensions

Multiple choice. Select the correct result.

77. $\frac{3}{x} + \frac{y}{x} =$

  a. $\frac{3+y}{x^2}$ **b.** $\frac{3+y}{2x}$ **c.** $\frac{3+y}{x}$

78. $\frac{3}{x} - \frac{y}{x} =$

  a. $\frac{3-y}{x^2}$ **b.** $\frac{3-y}{2x}$ **c.** $\frac{3-y}{x}$

79. $\frac{3}{x} \cdot \frac{y}{x} =$

  a. $\frac{3y}{x}$ **b.** $\frac{3y}{x^2}$ **c.** $3y$

80. $\frac{3}{x} \div \frac{y}{x} =$

  a. $\frac{3}{y}$ **b.** $\frac{y}{3}$ **c.** $\frac{3}{x^2y}$

Write each rational expression as an equivalent expression with a denominator of $x - 2$.

81. $\dfrac{5}{2 - x}$

82. $\dfrac{8y}{2 - x}$

83. $-\dfrac{7 + x}{2 - x}$

84. $\dfrac{x - 3}{-(x - 2)}$

85. The planet Mercury revolves around the sun in 88 Earth days. It takes Jupiter 4332 Earth days to make one revolution around the sun. (*Source:* National Space Science Data Center) If the two planets are aligned as shown in the figure, how long will it take for them to align again?

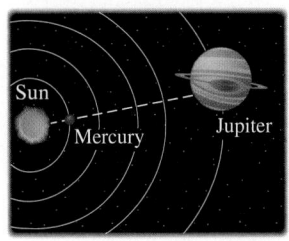

86. You are throwing a barbecue and you want to make sure that you purchase the same number of hot dogs as hot dog buns. Hot dogs come 8 to a package and hot dog buns come 12 to a package. What is the least number of each type of package you should buy?

87. An algebra student approaches you with a problem. He's tried to subtract two rational expressions, but his result does not match the book's. Check to see if the student has made an error. If so, correct his work shown below.

$$\frac{2x - 6}{x - 5} - \frac{x + 4}{x - 5}$$
$$= \frac{2x - 6 - x + 4}{x - 5}$$
$$= \frac{x - 2}{x - 5}$$

Objective

1   Add and subtract rational expressions with unlike denominators.

1   In the previous section, we practiced all the skills we need to add and subtract rational expressions with unlike or different denominators. The steps are as follows:

Adding or Subtracting Rational Expressions with Unlike Denominators

Step 1. Find the LCD of the rational expressions.

Step 2. Rewrite each rational expression as an equivalent expression whose denominator is the LCD found in Step 1.

Step 3. Add or subtract numerators and write the sum or difference over the common denominator.

Step 4. Simplify or write the rational expression in simplest form.

EXAMPLE 1

Perform each indicated operation.

a. $\dfrac{a}{4} - \dfrac{2a}{8}$ **b.** $\dfrac{3}{10x^2} + \dfrac{7}{25x}$

Solution

a. First, we must find the LCD. Since $4 = 2^2$ and $8 = 2^3$, the LCD $= 2^3 = 8$. Next we write each fraction as an equivalent fraction with the denominator 8, then we subtract.

$$\frac{a}{4} - \frac{2a}{8} = \frac{a(2)}{4(2)} - \frac{2a}{8} = \frac{2a}{8} - \frac{2a}{8} = \frac{2a - 2a}{8} = \frac{0}{8} = 0$$

b. Since $10x^2 = 2 \cdot 5 \cdot x \cdot x$ and $25x = 5 \cdot 5 \cdot x$, the LCD $= 2 \cdot 5^2 \cdot x^2 = 50x^2$. We write each fraction as an equivalent fraction with a denominator of $50x^2$.

$$\frac{3}{10x^2} + \frac{7}{25x} = \frac{3(5)}{10x^2(5)} + \frac{7(2x)}{25x(2x)}$$

$$= \frac{15}{50x^2} + \frac{14x}{50x^2}$$

$$= \frac{15 + 14x}{50x^2} \qquad \text{Add numerators. Write the sum}$$
$$\text{over the common denominator.}$$

EXAMPLE 2

Subtract: $\dfrac{6x}{x^2 - 4} - \dfrac{3}{x + 2}$

Solution

Since $x^2 - 4 = (x + 2)(x - 2)$, the LCD $= (x - 2)(x + 2)$. We write equivalent expressions with the LCD as denominators.

$$\frac{6x}{x^2 - 4} - \frac{3}{x + 2} = \frac{6x}{(x - 2)(x + 2)} - \frac{3(x - 2)}{(x + 2)(x - 2)}$$

$$= \frac{6x - 3(x - 2)}{(x + 2)(x - 2)} \qquad \text{Subtract numerators. Write the difference}$$
$$\text{over the common denominator.}$$

$$= \frac{6x - 3x + 6}{(x + 2)(x - 2)} \qquad \text{Apply the distributive property in the}$$
$$\text{numerator.}$$

$$= \frac{3x + 6}{(x + 2)(x - 2)} \qquad \text{Combine like terms in the numerator.}$$

Next we factor the numerator to see if this rational expression can be simplified.

$$= \frac{3(x + 2)}{(x + 2)(x - 2)} \qquad \text{Factor.}$$

$$= \frac{3}{x - 2} \qquad \text{Apply the fundamental principle to simplify.}$$

EXAMPLE 3

Add: $\dfrac{2}{3t} + \dfrac{5}{t+1}$

Solution   The LCD is $3t(t+1)$. We write each rational expression as an equivalent rational expression with a denominator of $3t(t+1)$.

$$\frac{2}{3t} + \frac{5}{t+1} = \frac{2(t+1)}{3t(t+1)} + \frac{5(3t)}{(t+1)(3t)}$$

$$= \frac{2(t+1) + 5(3t)}{3t(t+1)} \qquad \text{Add numerators. Write the sum over the common denominator.}$$

$$= \frac{2t + 2 + 15t}{3t(t+1)} \qquad \text{Apply the distributive property in the numerator.}$$

$$= \frac{17t + 2}{3t(t+1)} \qquad \text{Combine like terms in the numerator.}$$

EXAMPLE 4

Subtract: $\dfrac{7}{x-3} - \dfrac{9}{3-x}$

Solution   To find a common denominator, we notice that $x-3$ and $3-x$ are opposites. That is, $3-x = -(x-3)$. We write the denominator $3-x$ as $-(x-3)$ and simplify.

$$\frac{7}{x-3} - \frac{9}{3-x} = \frac{7}{x-3} - \frac{9}{-(x-3)}$$

$$= \frac{7}{x-3} - \frac{-9}{x-3} \qquad \text{Apply } \frac{a}{-b} = \frac{-a}{b}.$$

$$= \frac{7 - (-9)}{x-3} \qquad \text{Subtract numerators. Write the difference over the common denominator.}$$

$$= \frac{16}{x-3}$$

EXAMPLE 5

Add: $1 + \dfrac{m}{m+1}$

Solution   Recall that 1 is the same as $\dfrac{1}{1}$. The LCD of and $\dfrac{1}{1}$ and $\dfrac{m}{m+1}$ is $m+1$.

$$1 + \frac{m}{m+1} = \frac{1}{1} + \frac{m}{m+1} \qquad \text{Write } \mathbf{1} \text{ as } \tfrac{1}{1}.$$

$$= \frac{1(m+1)}{1(m+1)} + \frac{m}{m+1} \qquad \text{Multiply both the numerator and the denominator of } \tfrac{1}{1} \text{ by } m+1.$$

$$= \frac{m + 1 + m}{m + 1}$$

Add numerators. Write the sum over the common denominator.

$$= \frac{2m + 1}{m + 1}$$

Combine like terms in the numerator.

EXAMPLE 6

Subtract: $\dfrac{3}{2x^2 + x} - \dfrac{2x}{6x + 3}$

Solution First, we factor the denominators.

$$\frac{3}{2x^2 + x} - \frac{2x}{6x + 3} = \frac{3}{x(2x + 1)} - \frac{2x}{3(2x + 1)}$$

The LCD is $3x(2x + 1)$. We write equivalent expressions with denominators of $3x(2x + 1)$.

$$= \frac{3(3)}{x(2x + 1)(3)} - \frac{2x(x)}{3(2x + 1)(x)}$$

$$= \frac{9 - 2x^2}{3x(2x + 1)}$$

Subtract numerators. Write the difference over the common denominator.

EXAMPLE 7

Add: $\dfrac{2x}{x^2 + 2x + 1} + \dfrac{x}{x^2 - 1}$

Solution First we factor the denominators.

$$\frac{2x}{x^2 + 2x + 1} + \frac{x}{x^2 - 1} = \frac{2x}{(x + 1)(x + 1)} + \frac{x}{(x + 1)(x - 1)}$$

Now we write the rational expressions as equivalent expressions with denominators of $(x + 1)(x + 1)(x - 1)$, the LCD.

$$= \frac{2x(x - 1)}{(x + 1)(x + 1)(x - 1)} + \frac{x(x + 1)}{(x + 1)(x - 1)(x + 1)}$$

$$= \frac{2x(x - 1) + x(x + 1)}{(x + 1)^2(x - 1)}$$

Add numerators. Write the sum over the common denominator.

$$= \frac{2x^2 - 2x + x^2 + x}{(x + 1)^2(x - 1)}$$

Apply the distributive property in the numerator.

$$= \frac{3x^2 - x}{(x + 1)^2(x - 1)} \quad \text{or} \quad \frac{x(3x - 1)}{(x + 1)^2(x - 1)}$$

The numerator was factored as a last step to see if the rational expression could be simplified further. Since there are no factors common to the numerator and the denominator, we can't simplify further.

MENTAL MATH

Match each exercise with the *first* step needed to perform the operation. Do not actually perform the operation.

1. $\dfrac{3}{4} - \dfrac{y}{4}$ **2.** $\dfrac{2}{a} \cdot \dfrac{3}{(a+6)}$ **3.** $\dfrac{x+1}{x} \div \dfrac{x-1}{x}$ **4.** $\dfrac{9}{x-2} - \dfrac{x}{x+2}$

A. Multiply the first rational expression by the reciprocal of the second rational expression.
B. Find the LCD. Write each expression as an equivalent expression with the LCD as denominator.
C. Multiply numerators, then multiply denominators.
D. Subtract numerators. Place the difference over a common denominator.

EXERCISE SET 7.4

STUDY   CD/   PH MATH   MathXL®Tutorials   MathXL®   MyMathLab®
GUIDE/SSM   VIDEO   TUTOR CENTER   ON CD

Perform the indicated operations. See Example 1.

1. $\dfrac{4}{2x} + \dfrac{9}{3x}$

2. $\dfrac{15}{7a} + \dfrac{8}{6a}$

3. $\dfrac{15a}{b} + \dfrac{6b}{5}$

4. $\dfrac{4c}{d} + \dfrac{8x}{5}$

5. $\dfrac{3}{x} + \dfrac{5}{2x^2}$

6. $\dfrac{14}{3x^2} + \dfrac{6}{x}$

Perform the indicated operations. See Examples 2 and 3.

7. $\dfrac{6}{x+1} + \dfrac{9}{2x+2}$

8. $\dfrac{8}{x+4} - \dfrac{3}{3x+12}$

9. $\dfrac{15}{2x-4} + \dfrac{x}{x^2-4}$

10. $\dfrac{3}{x+2} - \dfrac{1}{x^2-4}$

11. $\dfrac{3}{4x} + \dfrac{8}{x-2}$

12. $\dfrac{x}{x+1} + \dfrac{3}{x-1}$

13. $\dfrac{5}{y^2} - \dfrac{y}{2y+1}$

14. $\dfrac{x}{4x-3} - \dfrac{3}{8x-6}$

15. In your own words, explain how to add two rational expressions with unlike denominators.

16. In your own words, explain how to subtract two rational expressions with unlike denominators.

Add or subtract as indicated. See Example 4.

17. $\dfrac{6}{x-3} + \dfrac{8}{3-x}$

18. $\dfrac{9}{x-3} + \dfrac{9}{3-x}$

19. $\dfrac{-8}{x^2-1} - \dfrac{7}{1-x^2}$

20. $\dfrac{-9}{25x^2-1} + \dfrac{7}{1-25x^2}$

21. $\dfrac{x}{x^2-4} - \dfrac{2}{4-x^2}$

22. $\dfrac{5}{2x-6} - \dfrac{3}{6-2x}$

Add or subtract as indicated. See Example 5.

23. $\dfrac{5}{x} + 2$

24. $\dfrac{7}{x^2} - 5x$

25. $\dfrac{5}{x-2} + 6$

26. $\dfrac{6y}{y+5} + 1$

27. $\dfrac{y+2}{y+3} - 2$

28. $\dfrac{7}{2x-3} - 3$

29. Two angles are said to be complementary if their sum is 90°. If one angle measures $\dfrac{40}{x}$ degrees, find the measure of its complement.

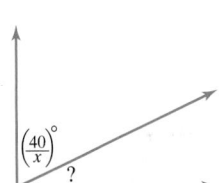

30. Two angles are said to be supplementary if their sum is 180°. If one angle measures $\dfrac{x+2}{x}$ degrees, find the measure of its supplement.

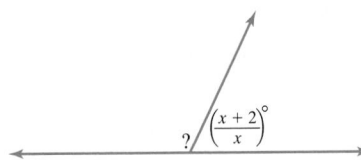

MIXED PRACTICE

Perform the indicated operations. See Examples 1 through 7.

31. $\dfrac{5x}{x+2} - \dfrac{3x-4}{x+2}$

32. $\dfrac{7x}{x-3} - \dfrac{4x+9}{x-3}$

33. $\dfrac{3x^4}{x} - \dfrac{4x^2}{x^2}$

34. $\dfrac{5x}{6} + \dfrac{15x^2}{2}$

35. $\dfrac{1}{x+3} - \dfrac{1}{(x+3)^2}$

36. $\dfrac{5x}{(x-2)^2} - \dfrac{3}{x-2}$

37. $\dfrac{4}{5b} + \dfrac{1}{b - 1}$

38. $\dfrac{1}{y + 5} + \dfrac{2}{3y}$

39. $\dfrac{2}{m} + 1$

40. $\dfrac{6}{x} - 1$

41. $\dfrac{6}{1 - 2x} - \dfrac{4}{2x - 1}$

42. $\dfrac{10}{3n - 4} - \dfrac{5}{4 - 3n}$

43. $\dfrac{7}{(x + 1)(x - 1)} + \dfrac{8}{(x + 1)^2}$

44. $\dfrac{5x + 2}{(x + 1)(x + 5)} - \dfrac{2}{x + 5}$

45. $\dfrac{x}{x^2 - 1} - \dfrac{2}{x^2 - 2x + 1}$

46. $\dfrac{x}{x^2 - 4} - \dfrac{5}{x^2 - 4x + 4}$

 47. $\dfrac{3a}{2a + 6} - \dfrac{a - 1}{a + 3}$

48. $\dfrac{1}{x + y} - \dfrac{y}{x^2 - y^2}$

49. $\dfrac{5}{2 - x} + \dfrac{x}{2x - 4}$

50. $\dfrac{-1}{a - 2} + \dfrac{4}{4 - 2a}$

51. $\dfrac{-7}{y^2 - 3y + 2} - \dfrac{2}{y - 1}$

52. $\dfrac{2}{x^2 + 4x + 4} + \dfrac{1}{x + 2}$

53. $\dfrac{13}{x^2 - 5x + 6} - \dfrac{5}{x - 3}$

54. $\dfrac{27}{y^2 - 81} + \dfrac{3}{2(y + 9)}$

55. $\dfrac{8}{(x + 2)(x - 2)} + \dfrac{4}{(x + 2)(x - 3)}$

56. $\dfrac{5}{6x^2(x + 2)} + \dfrac{4x}{x(x + 2)^2}$

57. $\dfrac{5}{9x^2 - 4} + \dfrac{2}{3x - 2}$

58. $\dfrac{4}{x^2 - x - 6} + \dfrac{x}{x^2 + 5x + 6}$

59. $\dfrac{x + 8}{x^2 - 5x - 6} + \dfrac{x + 1}{x^2 - 4x - 5}$

 60. $\dfrac{x}{x^2 + 12x + 20} - \dfrac{1}{x^2 + 8x - 20}$

REVIEW AND PREVIEW

Solve the following linear and quadratic equations. See Sections 2.3 and 6.5.

61. $3x + 5 = 7$

62. $5x - 1 = 8$

63. $2x^2 - x - 1 = 0$

64. $4x^2 - 9 = 0$

Simplify the following rational expressions. See Section 7.1.

65. $\dfrac{2 + x}{x + 2}$

66. $\dfrac{y^2 + y}{y + y^2}$

67. $\dfrac{2 - x}{x - 2}$

68. $\dfrac{z - 4}{4 - z}$

Concept Extensions

△ **69.** The length of the rectangle is $\dfrac{3}{y - 5}$ feet, while its width is $\dfrac{2}{y}$ feet. Find its perimeter and then find its area.

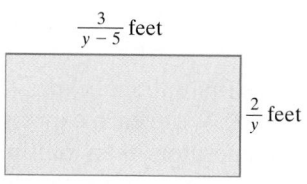

$\dfrac{3}{y - 5}$ feet

$\dfrac{2}{y}$ feet

70. A board of length $\dfrac{3}{x + 4}$ inches was cut into two pieces. If one piece is $\dfrac{1}{x - 4}$ inches, express the length of the other board as a rational expression.

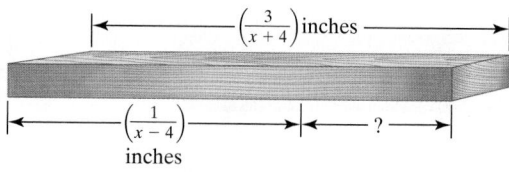

$\left(\dfrac{3}{x + 4}\right)$ inches

$\left(\dfrac{1}{x - 4}\right)$ inches   ?

Perform the indicated operations. Addition, subtraction, multiplication, and division of rational expressions are included here.

71. $\dfrac{15x}{x + 8} \cdot \dfrac{2x + 16}{3x}$

72. $\dfrac{9z + 5}{15} \cdot \dfrac{5z}{81z^2 - 25}$

73. $\dfrac{8x + 7}{3x + 5} - \dfrac{2x - 3}{3x + 5}$

74. $\dfrac{2z^2}{4z - 1} - \dfrac{z - 2z^2}{4z - 1}$

75. $\dfrac{5a + 10}{18} \div \dfrac{a^2 - 4}{10a}$

76. $\dfrac{9}{x^2 - 1} \div \dfrac{12}{3x + 3}$

77. $\dfrac{5}{x^2 - 3x + 2} + \dfrac{1}{x - 2}$

78. $\dfrac{4}{2x^2 + 5x - 3} + \dfrac{2}{x + 3}$

79. Explain when the LCD is the product of the denominators.

80. Explain when the LCD is the same as one of the denominators of a rational expression to be added or subtracted.

Add or subtract as indicated.

81. $\dfrac{5}{x^2 - 4} + \dfrac{2}{x^2 - 4x + 4} - \dfrac{3}{x^2 - x - 6}$

82. $\dfrac{8}{x^2 + 6x + 5} - \dfrac{3x}{x^2 + 4x - 5} + \dfrac{2}{x^2 - 1}$

83. $\dfrac{5 + x}{x^3 - 27} + \dfrac{x}{x^3 + 3x^2 + 9x}$

84. $\dfrac{x + 5}{x^3 + 1} - \dfrac{3}{2x^2 - 2x + 2}$

85. The dose of medicine prescribed for a child depends on the child's age A in years and the adult dose D for the medication. Two expressions that give a child's dose are Young's

Rule, $\dfrac{DA}{A + 12}$, and Cowling's Rule, $\dfrac{D(A + 1)}{24}$. Find an expression for the difference in the doses given by these expressions.

86. In ice hockey, penalty killing percentage is a statistic calculated as $1 - \dfrac{G}{P}$, where G = opponent's power play goals and P = opponent's power play opportunities. Subtract these terms.

7.5   SOLVING EQUATIONS CONTAINING RATIONAL EXPRESSIONS

Objectives

1 Solve equations containing rational expressions.

2 Solve equations containing rational expressions for a specified variable.

1   In Chapter 2, we solved equations containing fractions. In this section, we continue the work we began in Chapter 2 by solving equations containing rational expressions.

Examples of Equations Containing Rational Expressions

$$\frac{x}{5} + \frac{x + 2}{9} = 8 \quad \text{and} \quad \frac{x + 1}{9x - 5} = \frac{2}{3x}$$

To solve equations such as these, use the multiplication property of equality to clear the equation of fractions by multiplying both sides of the equation by the LCD.

EXAMPLE 1

Solve: $\dfrac{x}{2} + \dfrac{8}{3} = \dfrac{1}{6}$

Solution   The LCD of denominators 2, 3, and 6 is 6, so we multiply both sides of the equation by 6.

$$6\left(\frac{x}{2} + \frac{8}{3}\right) = 6\left(\frac{1}{6}\right)$$

$$6\left(\frac{x}{2}\right) + 6\left(\frac{8}{3}\right) = 6\left(\frac{1}{6}\right) \qquad \text{Use the distributive property.}$$

> **Helpful Hint**
>
> Make sure that *each* term is multiplied by the LCD, 6.

$$3 \cdot x + 16 = 1 \qquad \text{Multiply and simplify.}$$
$$3x = -15 \qquad \text{Subtract 16 from both sides.}$$
$$x = -5 \qquad \text{Divide both sides by 3.}$$

Check   To check, we replace x with -5 in the original equation.

$$\frac{x}{2} + \frac{8}{3} = \frac{1}{6}$$

$$\frac{-5}{2} + \frac{8}{3} \stackrel{?}{=} \frac{1}{6} \qquad \text{Replace } x \text{ with } -5.$$

$$\frac{1}{6} = \frac{1}{6} \qquad \text{True}$$

This number checks, so the solution is -5.

EXAMPLE 2

Solve: $\dfrac{t-4}{2} - \dfrac{t-3}{9} = \dfrac{5}{18}$

Solution The LCD of denominators 2, 9, and 18 is 18, so we multiply both sides of the equation by 18.

$$18\left(\dfrac{t-4}{2} - \dfrac{t-3}{9}\right) = 18\left(\dfrac{5}{18}\right)$$

$$18\left(\dfrac{t-4}{2}\right) - 18\left(\dfrac{t-3}{9}\right) = 18\left(\dfrac{5}{18}\right) \qquad \text{Use the distributive property.}$$

$$9(t-4) - 2(t-3) = 5 \qquad \text{Simplify.}$$

$$9t - 36 - 2t + 6 = 5 \qquad \text{Use the distributive property.}$$

$$7t - 30 = 5 \qquad \text{Combine like terms.}$$

$$7t = 35$$

$$t = 5 \qquad \text{Solve for } t.$$

> **Helpful Hint**
> Multiply *each* term by 18.

Check

$$\dfrac{t-4}{2} - \dfrac{t-3}{9} = \dfrac{5}{18}$$

$$\dfrac{5-4}{2} - \dfrac{5-3}{9} \overset{?}{=} \dfrac{5}{18} \qquad \text{Replace } t \text{ with } 5.$$

$$\dfrac{1}{2} - \dfrac{2}{9} \overset{?}{=} \dfrac{5}{18} \qquad \text{Simplify.}$$

$$\dfrac{5}{18} = \dfrac{5}{18} \qquad \text{True}$$

The solution is 5.

Recall from Section 7.1 that a rational expression is defined for all real numbers except those that make the denominator of the expression 0. This means that if an equation contains *rational expressions with variables in the denominator*, we must be certain that the proposed solution does not make the denominator 0. If replacing the variable with the proposed solution makes the denominator 0, the rational expression is undefined and this proposed solution must be rejected.

EXAMPLE 3

Solve: $3 - \dfrac{6}{x} = x + 8$

Solution In this equation, 0 cannot be a solution because if x is 0, the rational expression $\dfrac{6}{x}$ is undefined. The LCD is x, so we multiply both sides of the equation by x.

$$x\left(3 - \dfrac{6}{x}\right) = x(x + 8)$$

$$x(3) - x\left(\dfrac{6}{x}\right) = x \cdot x + x \cdot 8 \qquad \text{Use the distributive property.}$$

$$3x - 6 = x^2 + 8x \qquad \text{Simplify.}$$

> **Helpful Hint**
> Multiply *each* term by x.

Now we write the quadratic equation in standard form and solve for x.

$$0 = x^2 + 5x + 6$$
$$0 = (x + 3)(x + 2) \qquad \text{Factor.}$$
$$x + 3 = 0 \quad \text{or} \quad x + 2 = 0 \qquad \text{Set each factor equal to 0 and solve.}$$
$$x = -3 \qquad\qquad x = -2$$

Notice that neither -3 nor -2 makes the denominator in the original equation equal to 0.

Check To check these solutions, we replace x in the original equation by -3, and then by -2.

If $x = -3$:

$$3 - \frac{6}{x} = x + 8$$
$$3 - \frac{6}{-3} \stackrel{?}{=} -3 + 8$$
$$3 - (-2) \stackrel{?}{=} 5$$
$$5 = 5 \qquad \text{True}$$

If $x = -2$:

$$3 - \frac{6}{x} = x + 8$$
$$3 - \frac{6}{-2} \stackrel{?}{=} -2 + 8$$
$$3 - (-3) \stackrel{?}{=} 6$$
$$6 = 6 \qquad \text{True}$$

Both -3 and -2 are solutions.

The following steps may be used to solve an equation containing rational expressions.

Solving an Equation Containing Rational Expressions

Step 1. Multiply both sides of the equation by the LCD of all rational expressions in the equation.

Step 2. Remove any grouping symbols and solve the resulting equation.

Step 3. Check the solution in the original equation.

EXAMPLE 4

Solve: $\dfrac{4x}{x^2 - 25} + \dfrac{2}{x - 5} = \dfrac{1}{x + 5}$

Solution The denominator $x^2 - 25$ factors as $(x + 5)(x - 5)$. The LCD is then $(x + 5)(x - 5)$, so we multiply both sides of the equation by this LCD.

$$(x + 5)(x - 5)\left(\frac{4x}{(x + 5)(x - 5)} + \frac{2}{x - 5}\right) \qquad \text{Multiply by the LCD. Notice that } -5 \text{ and } 5 \text{ cannot be solutions.}$$
$$= (x + 5)(x - 5)\left(\frac{1}{x + 5}\right)$$
$$(x + 5)(x - 5) \cdot \frac{4x}{x^2 - 25} + (x + 5)(x - 5) \cdot \frac{2}{x - 5} \qquad \text{Use the distributive property.}$$
$$= (x + 5)(x - 5) \cdot \frac{1}{x + 5}$$

$$4x + 2(x + 5) = x - 5 \qquad \text{Simplify.}$$
$$4x + 2x + 10 = x - 5 \qquad \text{Use the distributive property.}$$
$$6x + 10 = x - 5 \qquad \text{Combine like terms.}$$
$$5x = -15$$
$$x = -3 \qquad \text{Divide both sides by 5.}$$

Check   Check by replacing x with -3 in the original equation. The solution is -3.

EXAMPLE 5

Solve: $\dfrac{2x}{x - 4} = \dfrac{8}{x - 4} + 1$

Solution   Multiply both sides by the LCD, $x - 4$.

$$(x - 4)\left(\frac{2x}{x - 4}\right) = (x - 4)\left(\frac{8}{x - 4} + 1\right) \qquad \text{Multiply by the LCD.}$$
$$\qquad\qquad\qquad\qquad\qquad\qquad\qquad \text{Notice that 4 cannot be a solution.}$$
$$(x - 4)\cdot\frac{2x}{x - 4} = (x - 4)\cdot\frac{8}{x - 4} + (x - 4)\cdot 1 \qquad \text{Use the distributive property.}$$
$$2x = 8 + (x - 4) \qquad \text{Simplify.}$$
$$2x = 4 + x$$
$$x = 4$$

Notice that 4 makes the denominator 0 in the original equation. Therefore, 4 is *not* a solution and this equation has *no solution*.

> **Helpful Hint**
>
> As we can see from Example 5, it is important to check the proposed solution(s) in the *original* equation.

✔ **CONCEPT CHECK**

When can we clear fractions by multiplying through by the LCD?

a. When adding or subtracting rational expressions
b. When solving an equation containing rational expressions
c. Both of these   **d.** Neither of these

EXAMPLE 6

Solve: $x + \dfrac{14}{x - 2} = \dfrac{7x}{x - 2} + 1$

Solution   Notice the denominators in this equation. We can see that 2 can't be a solution. The LCD is $x - 2$, so we multiply both sides of the equation by $x - 2$.

$$(x - 2)\left(x + \frac{14}{x - 2}\right) = (x - 2)\left(\frac{7x}{x - 2} + 1\right)$$
$$(x - 2)(x) + (x - 2)\left(\frac{14}{x - 2}\right) = (x - 2)\left(\frac{7x}{x - 2}\right) + (x - 2)(1)$$

Concept Check Answer:
b

$$x^2 - 2x + 14 = 7x + x - 2 \quad \text{Simplify.}$$
$$x^2 - 2x + 14 = 8x - 2 \qquad \text{Combine like terms.}$$
$$x^2 - 10x + 16 = 0 \qquad \text{Write the quadratic}$$
$$\text{equation in standard form.}$$
$$(x - 8)(x - 2) = 0 \qquad \text{Factor.}$$
$$x - 8 = 0 \quad \text{or} \quad x - 2 = 0 \qquad \text{Set each factor equal to 0.}$$
$$x = 8 \qquad\qquad x = 2 \qquad \text{Solve.}$$

As we have already noted, 2 can't be a solution of the original equation. So we need only replace x with 8 in the original equation. We find that 8 is a solution; the only solution is 8.

2   The last example in this section is an equation containing several variables. We are directed to solve for one of them. The steps used in the preceeding examples can be applied to solve equations for a specified variable as well.

EXAMPLE 7

Solve $\dfrac{1}{a} + \dfrac{1}{b} = \dfrac{1}{x}$ for x.

Solution   (This type of equation often models a work problem, as we shall see in Section 7.6.) The LCD is abx, so we multiply both sides by abx.

$$abx\left(\frac{1}{a} + \frac{1}{b}\right) = abx\left(\frac{1}{x}\right)$$
$$abx\left(\frac{1}{a}\right) + abx\left(\frac{1}{b}\right) = abx \cdot \frac{1}{x}$$
$$bx + ax = ab \qquad \text{Simplify.}$$
$$x(b + a) = ab \qquad \text{Factor out } x \text{ from each term on the left side.}$$
$$\frac{x(b + a)}{b + a} = \frac{ab}{b + a} \qquad \text{Divide both sides by } b + a.$$
$$x = \frac{ab}{b + a} \qquad \text{Simplify.}$$

This equation is now solved for x.

Graphing Calculator Explorations

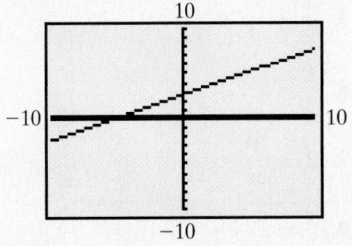

A grapher may be used to check solutions of equations containing rational expressions. For example, to check the solution of Example 1, $\frac{x}{2} + \frac{8}{3} = \frac{1}{6}$, graph $y_1 = x/2 + 8/3$ and $y_2 = 1/6$.

Use TRACE and ZOOM, or use INTERSECT, to find the point of intersection. The point of intersection has an x-value of -5, so the solution of the equation is -5.

Use a grapher to check the examples of this section.

1. Example 2   **2.** Example 3   **3.** Example 4

4. Example 5   **5.** Example 6

STUDY SKILLS REMINDER

Continue your outline started in Section 2.3. Write how to recognize and how to solve equations containing rational expressions in your own words. For example:

Solving Equations and Inequalities

I. Equations

    A. Linear equations (Sec. 2.3)

    B. Quadratic and Higher Degree Equations (Sec. 6.5)

    C. Equations with Rational Expressions—Recognize: *equation with rational expressions* — Solve: multiply both sides of the equation by the LCD—then simplify to decide whether the equation is linear or quadratic

II. Inequalities

    A. Linear inequalities (Sec. 2.7)

See Appendix A for exercises.

MENTAL MATH

Solve each equation for the variable.

1. $\dfrac{x}{5} = 2$ **2.** $\dfrac{x}{8} = 4$ **3.** $\dfrac{z}{6} = 6$ **4.** $\dfrac{y}{7} = 8$

EXERCISE SET 7.5

STUDY   CD/   PH MATH   MathXL®Tutorials   MathXL®   MyMathLab®
GUIDE/SSM   VIDEO   TUTOR CENTER   ON CD

Solve each equation. See Examples 1 through 3.

1. $\dfrac{x}{5} + 3 = 9$

2. $\dfrac{x}{5} - 2 = 9$

3. $\dfrac{x}{2} + \dfrac{5x}{4} = \dfrac{x}{12}$

4. $\dfrac{x}{6} + \dfrac{4x}{3} = \dfrac{x}{18}$

5. $\dfrac{y}{7} - \dfrac{3y}{2} = 1$

6. $\dfrac{a}{9} - \dfrac{4a}{3} = 2$

7. $2 + \dfrac{10}{x} = x + 5$

8. $6 + \dfrac{5}{y} = y - \dfrac{2}{y}$

9. $\dfrac{a}{5} = \dfrac{a - 3}{2}$

10. $\dfrac{2b}{5} = \dfrac{b + 2}{6}$

11. $\dfrac{x - 3}{5} + \dfrac{x - 2}{2} = \dfrac{1}{2}$

12. $\dfrac{a + 5}{4} + \dfrac{a + 5}{2} = \dfrac{a}{8}$

13. $\dfrac{x + 1}{3} - \dfrac{x - 2}{4} = \dfrac{5}{6}$

14. $\dfrac{x + 2}{15} - \dfrac{x - 1}{3} = \dfrac{7}{15}$

Solve each equation. See Examples 4 through 6.

15. $\dfrac{9}{2a - 5} = -2$

16. $\dfrac{6}{4 - 3x} = 3$

17. $\dfrac{y}{y + 4} + \dfrac{4}{y + 4} = 3$

18. $\dfrac{5y}{y + 1} - \dfrac{3}{y + 1} = 4$

19. $\dfrac{4y}{y - 3} - 3 = \dfrac{3y - 1}{y + 3}$

20. $\dfrac{2x}{x + 2} - 2 = \dfrac{x - 8}{x - 2}$

21. $\dfrac{4y}{y - 4} + 5 = \dfrac{5y}{y - 4}$

22. $\dfrac{2a}{a + 2} - 5 = \dfrac{7a}{a + 2}$

23. $\dfrac{7}{x - 2} + 1 = \dfrac{x}{x + 2}$

24. $1 + \dfrac{3}{x + 1} = \dfrac{x}{x - 1}$

25. $\dfrac{x + 1}{x + 3} = \dfrac{2x^2 - 15x}{x^2 + x - 6} - \dfrac{x - 3}{x - 2}$

26. $\dfrac{3}{x + 3} = \dfrac{12x + 19}{x^2 + 7x + 12} - \dfrac{5}{x + 4}$

27. $\dfrac{y}{2y + 2} + \dfrac{2y - 16}{4y + 4} = \dfrac{2y - 3}{y + 1}$

28. $\dfrac{1}{x + 2} = \dfrac{4}{x^2 - 4} - \dfrac{1}{x - 2}$

MIXED PRACTICE

Solve each equation.

29. $\dfrac{2x}{7} - 5x = 9$

30. $\dfrac{4x}{8} - 5x = 10$

31. $\dfrac{2}{y} + \dfrac{1}{2} = \dfrac{5}{2y}$

32. $\dfrac{6}{3y} + \dfrac{3}{y} = 1$

33. $\dfrac{4x + 10}{7} = \dfrac{8}{2}$

34. $\dfrac{1}{2} = \dfrac{x + 1}{8}$

35. $2 + \dfrac{3}{a - 3} = \dfrac{a}{a - 3}$

36. $\dfrac{2y}{y - 2} - \dfrac{4}{y - 2} = 4$

37. $\dfrac{5}{x} + \dfrac{2}{3} = \dfrac{7}{2x}$

38. $\dfrac{5}{3} - \dfrac{3}{2x} = \dfrac{5}{4}$

39. $\dfrac{2a}{a + 4} = \dfrac{3}{a - 1}$

40. $\dfrac{5}{3x - 8} = \dfrac{x}{x - 2}$

41. $\dfrac{x + 1}{3} - \dfrac{x - 1}{6} = \dfrac{1}{6}$

42. $\dfrac{3x}{5} - \dfrac{x - 6}{3} = \dfrac{1}{5}$

43. $\dfrac{4r - 1}{r^2 + 5r - 14} + \dfrac{2}{r + 7} = \dfrac{1}{r - 2}$

44. $\dfrac{2t + 3}{t - 1} - \dfrac{2}{t + 3} = \dfrac{5 - 6t}{t^2 + 2t - 3}$

45. $\dfrac{t}{t - 4} = \dfrac{t + 4}{6}$

46. $\dfrac{15}{x + 4} = \dfrac{x - 4}{x}$

47. $\dfrac{x}{2x + 6} + \dfrac{x + 1}{3x + 9} = \dfrac{2}{4x + 12}$

48. $\dfrac{a}{5a - 5} - \dfrac{a - 2}{2a - 2} = \dfrac{5}{4a - 4}$

Solve each equation for the indicated variable. See Example 7.

49. $\dfrac{D}{R} = T$; for R

50. $\dfrac{A}{W} = L$; for W

51. $\dfrac{3}{x} = \dfrac{5y}{x + 2}$; for y

52. $\dfrac{7x - 1}{2x} = \dfrac{5}{y}$; for y

53. $\dfrac{3a + 2}{3b - 2} = -\dfrac{4}{2a}$; for b

54. $\dfrac{6x + y}{7x} = \dfrac{3x}{h}$; for h

55. $\dfrac{A}{BH} = \dfrac{1}{2}$; for B

56. $\dfrac{V}{\pi r^2 h} = 1$; for h

57. $\dfrac{C}{\pi r} = 2$; for r

58. $\dfrac{3V}{A} = H$; for V

59. $\dfrac{1}{a} = \dfrac{1}{b} + \dfrac{1}{c}$; for a

60. $\dfrac{1}{2} - \dfrac{1}{x} = \dfrac{1}{y}$; for x

61. $\dfrac{m^2}{6} - \dfrac{n}{3} = \dfrac{p}{2}$; for n

62. $\dfrac{x^2}{r} + \dfrac{y^2}{t} = 1$; for r

REVIEW AND PREVIEW

Identify the x- and y-intercepts. See Section 3.3.

63.

64.

65.

66.

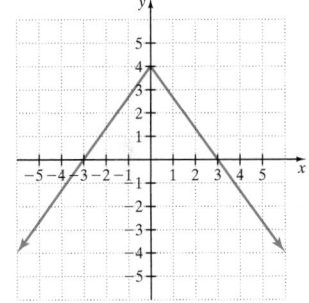

Concept Extensions

67. Explain the difference between solving an equation such as $\dfrac{x}{2} + \dfrac{3}{4} = \dfrac{x}{4}$ for x and performing an operation such as adding $\dfrac{x}{2} + \dfrac{3}{4}$.

68. When solving an equation such as $\dfrac{y}{4} = \dfrac{y}{2} - \dfrac{1}{4}$, we may multiply all terms by 4. When subtracting two rational expressions such as $\dfrac{y}{2} - \dfrac{1}{4}$, we may not. Explain why.

Determine whether each of the following is an equation or an expression. If it is an equation, then solve it for its variable. If it is an expression, perform the indicated operation.

69. $\dfrac{1}{x} + \dfrac{5}{9}$

70. $\dfrac{1}{x} + \dfrac{5}{9} = \dfrac{2}{3}$

71. $\dfrac{5}{x - 1} - \dfrac{2}{x} = \dfrac{5}{x(x - 1)}$

72. $\dfrac{5}{x - 1} - \dfrac{2}{x}$

Recall that two angles are supplementary if the sum of their measures is 180°. Find the measures of the following supplementary angles.

73.

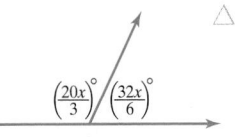

74.

Recall that two angles are complementary if the sum of their measures is 90°. Find the measures of the following complementary angles.

△ **75.**

△ **76.**

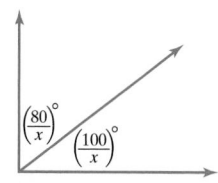

Solve each equation.

77. $\dfrac{5}{a^2 + 4a + 3} + \dfrac{2}{a^2 + a - 6} - \dfrac{3}{a^2 - a - 2} = 0$

78. $-\dfrac{2}{a^2 + 2a - 8} + \dfrac{1}{a^2 + 9a + 20} = \dfrac{-4}{a^2 + 3a - 10}$

INTEGRATED REVIEW   SUMMARY ON RATIONAL EXPRESSIONS

It is important to know the difference between performing operations with rational expressions and solving an equation containing rational expressions. Study the examples below.

Performing Operations with Rational Expressions

Adding:

$$\frac{1}{x} + \frac{1}{x + 5} = \frac{1 \cdot (x + 5)}{x(x + 5)} + \frac{1 \cdot x}{x(x + 5)} = \frac{x + 5 + x}{x(x + 5)} = \frac{2x + 5}{x(x + 5)}$$

Subtracting:

$$\frac{3}{x} - \frac{5}{x^2 y} = \frac{3 \cdot xy}{x \cdot xy} - \frac{5}{x^2 y} = \frac{3xy - 5}{x^2 y}$$

Multiplying:

$$\frac{2}{x} \cdot \frac{5}{x - 1} = \frac{2 \cdot 5}{x(x - 1)} = \frac{10}{x(x - 1)}$$

Dividing:

$$\frac{4}{2x + 1} \div \frac{x - 3}{x} = \frac{4}{2x + 1} \cdot \frac{x}{x - 3} = \frac{4x}{(2x + 1)(x - 3)}$$

Solving an Equation Containing Rational Expressions

To solve an equation containing rational expressions, we clear the equation of fractions by multiplying both sides by the LCD.

$$\frac{3}{x} - \frac{5}{x - 1} = \frac{1}{x(x - 1)}$$ Note that x can't be 0 or 1.

$$x(x - 1)\left(\frac{3}{x}\right) - x(x - 1)\left(\frac{5}{x - 1}\right) = x(x - 1) \cdot \frac{1}{x(x - 1)}$$ Multiply both sides by the LCD.

$$3(x - 1) - 5x = 1$$ Simplify.

$$3x - 3 - 5x = 1$$ Use the distributive property.

$$-2x - 3 = 1$$ Combine like terms.

$$-2x = 4$$ Add 3 to both sides.

$$x = -2$$ Divide both sides by -2.

Determine whether each of the following is an equation or an expression. If it is an equation, solve it for its variable. If it is an expression, perform the indicated operation.

1. $\dfrac{1}{x} + \dfrac{2}{3}$

2. $\dfrac{3}{a} + \dfrac{5}{6}$

3. $\dfrac{1}{x} + \dfrac{2}{3} = \dfrac{3}{x}$

4. $\dfrac{3}{a} + \dfrac{5}{6} = 1$

5. $\dfrac{2}{x - 1} - \dfrac{1}{x}$

6. $\dfrac{4}{x - 3} - \dfrac{1}{x}$

7. $\dfrac{2}{x + 1} - \dfrac{1}{x} = 1$

8. $\dfrac{4}{x - 3} - \dfrac{1}{x} = \dfrac{6}{x(x - 3)}$

9. $\dfrac{15x}{x + 8} \cdot \dfrac{2x + 16}{3x}$

10. $\dfrac{9z + 5}{15} \cdot \dfrac{5z}{81z^2 - 25}$

11. $\dfrac{2x + 1}{x - 3} + \dfrac{3x + 6}{x - 3}$

12. $\dfrac{4p - 3}{2p + 7} + \dfrac{3p + 8}{2p + 7}$

13. $\dfrac{x + 5}{7} = \dfrac{8}{2}$

14. $\dfrac{1}{2} = \dfrac{x - 1}{8}$

15. $\dfrac{5a + 10}{18} \div \dfrac{a^2 - 4}{10a}$

16. $\dfrac{9}{x^2 - 1} + \dfrac{12}{3x + 3}$

17. $\dfrac{x + 2}{3x - 1} + \dfrac{5}{(3x - 1)^2}$

18. $\dfrac{4}{(2x - 5)^2} + \dfrac{x + 1}{2x - 5}$

19. $\dfrac{x - 7}{x} - \dfrac{x + 2}{5x}$

20. $\dfrac{9}{x^2 - 4} + \dfrac{2}{x + 2} = \dfrac{-1}{x - 2}$

21. $\dfrac{3}{x + 3} = \dfrac{5}{x^2 - 9} - \dfrac{2}{x - 3}$

22. $\dfrac{10x - 9}{x} - \dfrac{x - 4}{3x}$

7.6   PROPORTION AND PROBLEM SOLVING WITH RATIONAL EQUATIONS

Objectives

1 Use proportions to solve problems.

2 Solve problems about numbers.

3 Solve problems about work.

4 Solve problems about distance.

1 A **ratio** is the quotient of two numbers or two quantities. For example, the ratio of 2 to 5 can be written as $\frac{2}{5}$, the quotient of 2 and 5.

If two ratios are equal, we say the ratios are **in proportion** to each other. A **proportion** is a mathematical statement that two ratios are equal.

For example, the equation $\frac{1}{2} = \frac{4}{8}$ is a proportion, as is $\frac{x}{5} = \frac{8}{10}$, because both sides of the equations are ratios. When we want to emphasize the equation as a proportion, we

read the proportion $\dfrac{1}{2} = \dfrac{4}{8}$ as "one is to two as four is to eight"

In a proportion, cross products are equal. To understand cross products, let's start with the proportion

$$\frac{a}{b} = \frac{c}{d}$$

and multiply both sides by the LCD, bd.

$$bd\left(\frac{a}{b}\right) = bd\left(\frac{c}{d}\right) \qquad \text{Multiply both sides by the LCD, } bd.$$
$$ad = bc \qquad \text{Simplify.}$$

Cross product   Cross product

Notice why ad and bc are called cross products.

$$\frac{a}{b} = \frac{c}{d}$$

Cross Products

If $\dfrac{a}{b} = \dfrac{c}{d}$, then $ad = bc$.

For example, if

$$\frac{1}{2} = \frac{4}{8}, \quad \text{then} \quad 1 \cdot 8 = 2 \cdot 4 \quad \text{or}$$
$$8 = 8$$

EXAMPLE 1

Solve for x: $\dfrac{45}{x} = \dfrac{5}{7}$

Solution   This is an equation with rational expressions, and also a proportion. Below are two ways to solve.

Since this is a rational equation, we can use the methods of the previous section.

$$\frac{45}{x} = \frac{5}{7}$$

$7x \cdot \dfrac{45}{x} = 7x \cdot \dfrac{5}{7}$ Multiply both sides by LCD **7x**.

$7 \cdot 45 = x \cdot 5$ Divide out common factors.

$315 = 5x$ Multiply.

$\dfrac{315}{5} = \dfrac{5x}{5}$ Divide both sides by **5**.

$63 = x$ Simplify.

Since this is also a proportion, we may set cross products equal.

$$\frac{45}{x} = \frac{5}{7}$$

$45 \cdot 7 = x \cdot 5$ Set cross products equal.

$315 = 5x$ Multiply.

$\dfrac{315}{5} = \dfrac{5x}{5}$ Divide both sides by **5**.

$63 = x$ Simplify.

Check   Both methods give us a solution of 63. To check, substitute 63 for x in the original proportion. The solution is 63.

In this section, if the rational equation is a proportion, we will use cross products to solve.

Proportions can be used to model and solve many real-life problems. When using proportions in this way, it is important to judge whether the solution is reasonable. Doing so helps us to decide if the proportion has been formed correctly. We use the same problem-solving steps that were introduced in Section 2.4.

EXAMPLE 2

CALCULATING THE COST OF RECORDABLE COMPACT DISCS

Three boxes of CD-Rs (recordable compact discs) cost $37.47. How much should 5 boxes cost?

Solution   **1. UNDERSTAND.**   Read and reread the problem. We know that the cost of 5 boxes is more than the cost of 3 boxes, or $37.47, and less than the cost of 6 boxes, which is double the cost of 3 boxes, or 2($37.47) = $74.94. Let's suppose that 5 boxes cost $60.00. To check, we see if 3 boxes is to 5 boxes as the *price* of 3 boxes is to the *price*

of 5 boxes. In other words, we see if

$$\frac{3 \text{ boxes}}{5 \text{ boxes}} = \frac{\text{price of 3 boxes}}{\text{price of 5 boxes}}$$

or

$$\frac{3}{5} = \frac{37.47}{60.00}$$

$$3(60.00) = 5(37.47) \qquad \textit{Set cross products equal.}$$

or

$$180.00 = 187.35 \qquad \textit{Not a true statement.}$$

Thus, $60 is not correct, but we now have a better understanding of the problem.

Let x = price of 5 boxes of CD-Rs.

2. TRANSLATE.

$$\frac{3 \text{ boxes}}{5 \text{ boxes}} = \frac{\text{price of 3 boxes}}{\text{price of 5 boxes}}$$

$$\frac{3}{5} = \frac{37.47}{x}$$

3. SOLVE.

$$\frac{3}{5} = \frac{37.47}{x}$$

$$3x = 5(37.47) \qquad \textit{Set cross products equal.}$$

$$3x = 187.35$$

$$x = 62.45 \qquad \textit{Divide both sides by 3.}$$

4. INTERPRET.

Check: Verify that 3 boxes is to 5 boxes as $37.47 is to $62.45. Also, notice that our solution is a reasonable one as discussed in Step 1.

State: Five boxes of CD-Rs cost $62.45.

> **Helpful Hint**
>
> The proportion $\dfrac{5 \text{ boxes}}{3 \text{ boxes}} = \dfrac{\text{price of 5 boxes}}{\text{price of 3 boxes}}$ could also have been used to solve the problem above. Notice that the cross products are the same.

Similar triangles have the same shape but not necessarily the same size. In similar triangles, the measures of corresponding angles are equal, and corresponding sides are in proportion.

If triangle ABC and triangle XYZ shown are similar, then we know that the measure of angle A = the measure of angle X, the measure of angle B = the measure of angle Y, and the measure of angle C = the measure of angle Z. We also know that corresponding sides are in proportion: $\frac{a}{x} = \frac{b}{y} = \frac{c}{z}$.

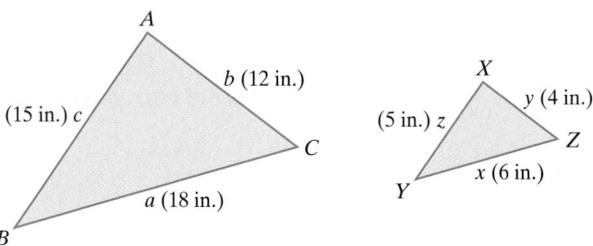

In this section, we will position similar triangles so that they have the same orientation.

To show that corresponding sides are in proportion for the triangles above, we write the ratios of the corresponding sides.

$$\frac{a}{x} = \frac{18}{6} = 3 \qquad \frac{b}{y} = \frac{12}{4} = 3 \qquad \frac{c}{z} = \frac{15}{5} = 3$$

△ **EXAMPLE 3**

FINDING THE LENGTH OF A SIDE OF A TRIANGLE

If the following two triangles are similar, find the missing length x.

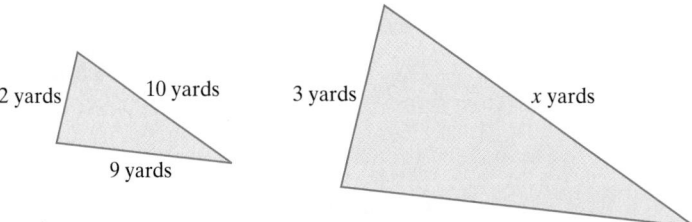

Solution
1. UNDERSTAND. Read the problem and study the figure.
2. TRANSLATE. Since the triangles are similar, their corresponding sides are in proportion and we have

$$\frac{2}{3} = \frac{10}{x}$$

3. SOLVE. To solve, we multiply both sides by the LCD, $3x$, or cross multiply.

$$2x = 30$$
$$x = 15 \qquad \text{Divide both sides by 2.}$$

4. INTERPRET.
   Check: To check, replace x with 15 in the original proportion and see that a true statement results.
   State: The missing length is 15 yards.

Suppose you coach your company's softball team. Two new employees are interested in playing on the company team, and you have only one open position. Employee A reports that last season he had 32 hits in 122 times at bat. Employee B reports that last season she had 19 hits in 56 times at bat. Which would you try to recruit first? Why? What other factors would you want to consider?

2   Let's continue to solve problems. The remaining problems are all modeled by rational equations.

EXAMPLE 4

FINDING AN UNKNOWN NUMBER

The quotient of a number and 6, minus $\frac{5}{3}$, is the quotient of the number and 2. Find the number.

Solution   **1.** UNDERSTAND.   Read and reread the problem. Suppose that the unknown number is 2, then we see if the quotient of 2 and 6, or $\frac{2}{6}$, minus $\frac{5}{3}$ is equal to the quotient of 2 and 2, or $\frac{2}{2}$.

$$\frac{2}{6} - \frac{5}{3} = \frac{1}{3} - \frac{5}{3} = -\frac{4}{3}, \text{ not } \frac{2}{2}$$

Don't forget that the purpose of a proposed solution is to better understand the problem.

Let x = the unknown number.

2. TRANSLATE.

In words:	the quotient of x and 6	minus	$\frac{5}{3}$	is	the quotient of x and 2
	↓	↓	↓	↓	↓
Translate:	$\frac{x}{6}$	$-$	$\frac{5}{3}$	$=$	$\frac{x}{2}$

3. SOLVE.   Here, we solve the equation $\frac{x}{6} - \frac{5}{3} = \frac{x}{2}$. We begin by multiplying both sides of the equation by the LCD, 6.

$$6\left(\frac{x}{6} - \frac{5}{3}\right) = 6\left(\frac{x}{2}\right)$$

$$6\left(\frac{x}{6}\right) - 6\left(\frac{5}{3}\right) = 6\left(\frac{x}{2}\right) \qquad \text{Apply the distributive property.}$$

$$x - 10 = 3x \qquad \text{Simplify.}$$

$$-10 = 2x \qquad \text{Subtract } x \text{ from both sides.}$$

$$-\frac{10}{2} = \frac{2x}{2} \qquad \text{Divide both sides by 2.}$$

$$-5 = x \qquad \text{Simplify.}$$

4. INTERPRET.

Check: To check, we verify that "the quotient of -5 and 6 minus $\frac{5}{3}$ is the quotient of -5 and 2," or $-\frac{5}{6} - \frac{5}{3} = -\frac{5}{2}$.

State: The unknown number is -5.

3 The next example is often called a work problem. Work problems usually involve people or machines doing a certain task.

EXAMPLE 5

FINDING WORK RATES

Sam Waterton and Frank Schaffer work in a plant that manufactures automobiles. Sam can complete a quality control tour of the plant in 3 hours while his assistant, Frank, needs 7 hours to complete the same job. The regional manager is coming to inspect the plant facilities, so both Sam and Frank are directed to complete a quality control tour together. How long will this take?

Solution

1. UNDERSTAND. Read and reread the problem. The key idea here is the relationship between the **time** (hours) it takes to complete the job and the **part of the job** completed in 1 unit of time (hour). For example, if the **time** it takes Sam to complete the job is 3 hours, the **part of the job** he can complete in 1 hour is $\frac{1}{3}$. Similarly, Frank can complete $\frac{1}{7}$ of the job in 1 hour.

Let $x =$ the **time** in hours it takes Sam and Frank to complete the job together.

Then $\frac{1}{x} =$ the **part of the job** they complete in 1 hour.

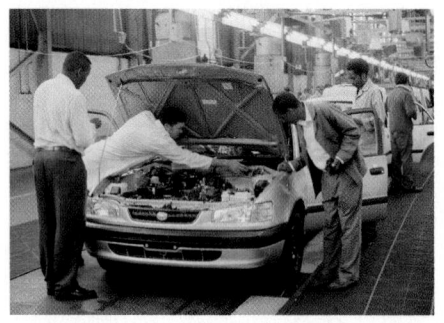

	Hours to Complete Total Job	Part of Job Completed in 1 Hour
Sam	3	$\frac{1}{3}$
Frank	7	$\frac{1}{7}$
Together	x	$\frac{1}{x}$

2. TRANSLATE.

In words:

part of job Sam completed in 1 hour	added to	part of job Frank completed in 1 hour	is equal to	part of job they completed together in 1 hour
↓	↓	↓	↓	↓

Translate: $\frac{1}{3}$ $+$ $\frac{1}{7}$ $=$ $\frac{1}{x}$

3. SOLVE. Here, we solve the equation $\frac{1}{3} + \frac{1}{7} = \frac{1}{x}$. We begin by multiplying both sides of the equation by the LCD, $21x$.

$$21x\left(\frac{1}{3}\right) + 21x\left(\frac{1}{7}\right) = 21x\left(\frac{1}{x}\right)$$

$$7x + 3x = 21 \qquad \text{Simplify.}$$

$$10x = 21$$

$$x = \frac{21}{10} \quad \text{or} \quad 2\frac{1}{10} \text{hours}$$

4. INTERPRET.

Check: Our proposed solution is $2\frac{1}{10}$ hours. This proposed solution is reasonable since $2\frac{1}{10}$ hours is more than half of Sam's time and less than half of Frank's time. Check this solution in the originally *stated* problem.

State: Sam and Frank can complete the quality control tour in $2\frac{1}{10}$ hours.

4   Next we look at a problem solved by the distance formula.

EXAMPLE 6

FINDING SPEEDS OF VEHICLES

A car travels 180 miles in the same time that a truck travels 120 miles. If the car's speed is 20 miles per hour faster than the truck's, find the car's speed and the truck's speed.

Solution   **1. UNDERSTAND.**   Read and reread the problem. Suppose that the truck's speed is 45 miles per hour. Then the car's speed is 20 miles per hour more, or 65 miles per hour.

We are given that the car travels 180 miles in the same time that the truck travels 120 miles. To find the time it takes the car to travel 180 miles, remember that since $d = rt$, we know that $\frac{d}{r} = t$.

Car's Time	**Truck's Time**
$t = \dfrac{d}{r} = \dfrac{180}{65} = 2\dfrac{50}{65} = 2\dfrac{10}{13}$ hours	$t = \dfrac{d}{r} = \dfrac{120}{45} = 2\dfrac{30}{45} = 2\dfrac{2}{3}$ hours

Since the times are not the same, our proposed solution is not correct. But we have a better understanding of the problem.

Let $x =$ the speed of the truck.

Since the car's speed is 20 miles per hour faster than the truck's, then

$$x + 20 = \text{the speed of the car}$$

Use the formula $d = r \cdot t$ or **d**istance $=$ **r**ate $\cdot$ **t**ime. Prepare a chart to organize the information in the problem.

Helpful Hint

If $d = r \cdot t$,

then $t = \dfrac{d}{r}$

or $time = \dfrac{distance}{rate}$.

	Distance	=	*Rate*	$\cdot$	*Time*
Truck	120		x		$\dfrac{120}{x}$ $\leftarrow$ distance $\leftarrow$ rate
Car	180		$x + 20$		$\dfrac{180}{x + 20}$ $\leftarrow$ distance $\leftarrow$ rate

2. TRANSLATE. Since the car and the truck traveled the same amount of time, we have that

In words:

car's time	=	truck's time
↓		↓

Translate: $\dfrac{180}{x + 20} = \dfrac{120}{x}$

3. SOLVE. We begin by multiplying both sides of the equation by the LCD, $x(x + 20)$, or cross multiplying.

$$\frac{180}{x + 20} = \frac{120}{x}$$
$$180x = 120(x + 20)$$
$$180x = 120x + 2400 \quad \text{Use the distributive property.}$$
$$60x = 2400 \quad \text{Subtract } 120x \text{ from both sides.}$$
$$x = 40 \quad \text{Divide both sides by 60.}$$

4. INTERPRET. The speed of the truck is 40 miles per hour. The speed of the car must then be $x + 20$ or 60 miles per hour.

Check: Find the time it takes the car to travel 180 miles and the time it takes the truck to travel 120 miles.

Car's Time
$$t = \frac{d}{r} = \frac{180}{60} = 3 \text{ hours}$$

Truck's Time
$$t = \frac{d}{r} = \frac{120}{40} = 3 \text{ hours}$$

Since both travel the same amount of time, the proposed solution is correct.

State: The car's speed is 60 miles per hour and the truck's speed is 40 miles per hour.

Spotlight on
DECISION
MAKING

Suppose you must select a child care center for your two-year-old daughter. You have compiled information on two possible choices in the table shown at the right. Which child care center would you choose? Why?

	Center A	Center B
Weekly cost	$130	$110
Number of 2-year-olds	7	13
Number of adults for 2-year-olds	2	3
Distance from home	6 miles	11 miles

EXERCISE SET 7.6

STUDY GUIDE/SSM CD/VIDEO PH MATH TUTOR CENTER MathXL®Tutorials ON CD MathXL® MyMathLab®

Solve each proportion. See Example 1.

1. $\dfrac{2}{3} = \dfrac{x}{6}$

2. $\dfrac{x}{2} = \dfrac{16}{6}$

5. $\dfrac{x + 1}{2x + 3} = \dfrac{2}{3}$

6. $\dfrac{x + 1}{x + 2} = \dfrac{5}{3}$

3. $\dfrac{x}{10} = \dfrac{5}{9}$

4. $\dfrac{9}{4x} = \dfrac{6}{2}$

7. $\dfrac{9}{5} = \dfrac{12}{3x + 2}$

8. $\dfrac{6}{11} = \dfrac{27}{3x - 2}$

9. For which of the following equations can we immediately use cross products to solve for x?

a. $\dfrac{2 - x}{5} = \dfrac{1 + x}{3}$ **b.** $\dfrac{2}{5} - x = \dfrac{1 + x}{3}$

10. For what value of x is $\dfrac{x}{x - 1}$ in proportion to $\dfrac{x + 1}{x}$? Explain your result.

Solve. See Example 2.

11. The ratio of the weight of an object on Earth to the weight of the same object on Pluto is 100 to 3. If an elephant weighs 4100 pounds on Earth, find the elephant's weight on Pluto.

12. If a 170-pound person weighs approximately 65 pounds on Mars, how much does a 9000-pound satellite weigh?

13. There are 110 calories per 28.4 grams of Crispy Rice cereal. Find how many calories are in 42.6 grams of this cereal.

14. On an architect's blueprint, 1 inch corresponds to 4 feet. Find the length of a wall represented by a line that is $3\frac{7}{8}$ inches long on the blueprint.

Find the unknown length x or y in the following pairs of similar triangles. See Example 3.

15.

16.

17.

18.

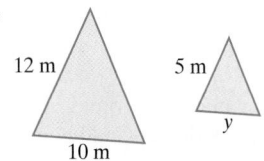

Solve the following. See Example 4.

19. Three times the reciprocal of a number equals 9 times the reciprocal of 6. Find the number.

20. Twelve divided by the sum of x and 2 equals the quotient of 4 and the difference of x and 2. Find x.

21. If twice a number added to 3 is divided by the number plus 1, the result is three halves. Find the number.

22. A number added to the product of 6 and the reciprocal of the number equals -5. Find the number.

See Example 5.

23. Smith Engineering found that an experienced surveyor surveys a roadbed in 4 hours. An apprentice surveyor needs 5 hours to survey the same stretch of road. If the two work together, find how long it takes them to complete the job.

24. An experienced bricklayer constructs a small wall in 3 hours. The apprentice completes the job in 6 hours. Find how long it takes if they work together.

25. In 2 minutes, a conveyor belt moves 300 pounds of recyclable aluminum from the delivery truck to a storage area. A smaller belt moves the same quantity of cans the same distance in 6 minutes. If both belts are used, find how long it takes to move the cans to the storage area.

26. Find how long it takes the conveyor belts described in Exercise 25 to move 1200 pounds of cans. (*Hint:* Think of 1200 pounds as four 300-pound jobs.)

See Example 6.

27. A jogger begins her workout by jogging to the park, a distance of 12 miles. She then jogs home at the same speed but along a different route. This return trip is 18 miles and her time is one hour longer. Find her jogging speed. Complete the accompanying chart and use it to find her jogging speed.

	distance	=	rate	·	time
Trip to park	12				
Return trip	18				

28. A boat can travel 9 miles upstream in the same amount of time it takes to travel 11 miles downstream. If the current of the river is 3 miles per hour, complete the chart below and use it to find the speed of the boat in still water.

	distance	=	rate	·	time
Upstream	9		$r - 3$		
Downstream	11		$r + 3$		

29. A cyclist rode the first 20-mile portion of his workout at a constant speed. For the 16-mile cooldown portion of his workout, he reduced his speed by 2 miles per hour. Each portion of the workout took the same time. Find the cyclist's speed during the first portion and find his speed during the cooldown portion.

30. A semi-truck travels 300 miles through the flatland in the same amount of time that it travels 180 miles through mountains.

The rate of the truck is 20 miles per hour slower in the mountains than in the flatland. Find both the flatland rate and mountain rate.

31. A human factors expert recommends that there be at least 9 square feet of floor space in a college classroom for every student in the class. Find the minimum floor space that 40 students need.

32. Due to space problems at a local university, a 20-foot by 12-foot conference room is converted into a classroom. Find the maximum number of students the room can accommodate. (See Exercise 31.)

MIXED PRACTICE

Solve the following.

33. One-fourth equals the quotient of a number and 8. Find the number.

34. Four times a number added to 5 is divided by 6. The result is $\frac{7}{2}$. Find the number.

35. Marcus and Tony work for Lombardo's Pipe and Concrete. Mr. Lombardo is preparing an estimate for a customer. He knows that Marcus lays a slab of concrete in 6 hours. Tony lays the same size slab in 4 hours. If both work on the job and the cost of labor is $45.00 per hour, decide what the labor estimate should be.

36. Mr. Dodson can paint his house by himself in 4 days. His son needs an additional day to complete the job if he works by himself. If they work together, find how long it takes to paint the house.

37. While road testing a new make of car, the editor of a consumer magazine finds that he can go 10 miles into a 3-mile-per-hour wind in the same amount of time he can go 11 miles with a 3-mile-per-hour wind behind him. Find the speed of the car in still air.

38. A fisherman on Pearl River rows 9 miles downstream in the same amount of time he rows 3 miles upstream. If the current is 6 miles per hour, find how long it takes him to cover the 12 miles.

△ **39.** Find the unknown length y.

△ **40.**

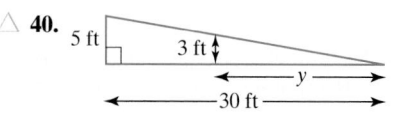

41. Ken Hall, a tailback, holds the high school sports record for total yards rushed in a season. In 1953, he rushed for 4045 total yards in 12 games. Find his average rushing yards per game.

42. To estimate the number of people in Jackson, population 50,000, who have no health insurance, 250 people were polled. Of those polled, 39 had no insurance. How many people in the city might we expect to be uninsured?

43. Two divided by the difference of a number and 3 minus 4 divided by a number plus 3, equals 8 times the reciprocal of the difference of the number squared and 9. What is the number?

44. If 15 times the reciprocal of a number is added to the ratio of 9 times a number minus 7 and the number plus 2, the result is 9. What is the number?

45. A pilot flies 630 miles with a tail wind of 35 miles per hour. Against the wind, he flies only 455 miles in the same amount of time. Find the rate of the plane in still air.

46. A marketing manager travels 1080 miles in a corporate jet and then an additional 240 miles by car. If the car ride takes one hour longer than the jet ride takes, and if the rate of the jet is 6 times the rate of the car, find the time the manager travels by jet and find the time the manager travels by car.

47. To mix weed killer with water correctly, it is necessary to mix 8 teaspoons of weed killer with 2 gallons of water. Find how many gallons of water are needed to mix with the entire box if it contains 36 teaspoons of weed killer.

48. A recent headline read, "Women Earn Bigger Checks in 1 of Every 6 Couples." If there are 23,000 couples in a nearby metropolitan area, how many women would you expect to earn bigger paychecks?

49. A cyclist rides 16 miles per hour on level ground on a still day. He finds that he rides 48 miles with the wind behind him in the same amount of time that he rides 16 miles into the wind. Find the rate of the wind.

50. The current on a portion of the Mississippi River is 3 miles per hour. A barge can go 6 miles upstream in the same amount of time it takes to go 10 miles downstream. Find the speed of the boat in still water.

51. One custodian cleans a suite of offices in 3 hours. When a second worker is asked to join the regular custodian, the job takes only $1\frac{1}{2}$ hours. How long does it take the second worker to do the same job alone?

52. One person proofreads a copy for a small newspaper in 4 hours. If a second proofreader is also employed, the job can be done in $2\frac{1}{2}$ hours. How long does it take for the second proofreader to do the same job alone?

△ **53.** An architect is completing the plans for a triangular deck. Use the diagram below to find the missing dimension.

△ **54.** A student wishes to make a small model of a triangular mainsail in order to study the effects of wind on the sail. The smaller model will be the same shape as a regular-size sailboat's mainsail. Use the following diagram to find the missing dimensions.

55. The manufacturers of cans of salted mixed nuts state that the ratio of peanuts to other nuts is 3 to 2. If 324 peanuts are in a can, find how many other nuts should also be in the can.

56. There are 1280 calories in a 14-ounce portion of Eagle Brand Milk. Find how many calories are in 2 ounces of Eagle Brand Milk.

57. One pipe fills a storage pool in 20 hours. A second pipe fills the same pool in 15 hours. When a third pipe is added and all three are used to fill the pool, it takes only 6 hours. Find how long it takes the third pipe to do the job.

58. One pump fills a tank 2 times as fast as another pump. If the pumps work together, they fill the tank in 18 minutes. How long does it take for each pump to fill the tank?

59. Andrew and Timothy Larson volunteer at a local recycling plant. Andrew can sort a batch of recyclables in 2 hours alone while his brother Timothy needs 3 hours to complete the same job. If they work together, how long will it take them to sort one batch?

60. A car travels 280 miles in the same time that a motorcycle travels 240 miles. If the car's speed is 10 miles per hour more than the motorcycle's, find the speed of the car and the speed of the motorcycle.

61. In 6 hours, an experienced cook prepares enough pies to supply a local restaurant's daily order. Another cook prepares the same number of pies in 7 hours. Together with a third cook, they prepare the pies in 2 hours. Find the work rate of the third cook.

62. Mrs. Smith balances the company books in 8 hours. It takes her assistant half again as long to do the same job. If they work together, find how long it takes them to balance the books.

63. One pump fills a tank 3 times as fast as another pump. If the pumps work together, they fill the tank in 21 minutes. How long does it take for each pump to fill the tank?

REVIEW AND PREVIEW

Find the slope of the line through each pair of points. Use the slope to determine whether the line is vertical, horizontal, or moves upward or downward from left to right. See Section 3.4.

64. $(-2, 5), (4, -3)$

65. $(0, 4), (2, 10)$

Simplify. Follow the circled steps in the order shown.

66. $\dfrac{\frac{3}{4} + \frac{1}{4}}{\frac{3}{8} + \frac{13}{8}}$ $\left.\begin{array}{l} \\ \\ \end{array}\right\}$ ① Add. ← ③ Divide. ② Add.

67. $\dfrac{\frac{9}{5} + \frac{6}{5}}{\frac{17}{6} + \frac{7}{6}}$ $\left.\begin{array}{l} \\ \\ \end{array}\right\}$ ① Add. ← ③ Divide. ② Add.

68. $\dfrac{\frac{2}{5} + \frac{1}{5}}{\frac{7}{10} + \frac{7}{10}}$ $\left.\begin{array}{l} \\ \\ \end{array}\right\}$ ① Add. ← ③ Divide. ② Add.

69. $\dfrac{\frac{1}{4} + \frac{5}{4}}{\frac{3}{8} + \frac{7}{8}}$ $\left.\begin{array}{l} \\ \\ \end{array}\right\}$ ① Add. ← ③ Divide. ② Add.

Concept Extensions

The following bar graph shows the capacity of the United States to generate electricity from the wind in the years shown. Use this graph for Exercises 70 and 71.

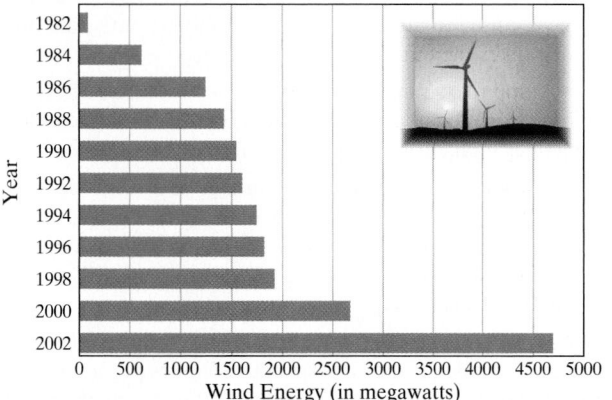

70. Find the approximate increase in megawatt capacity during the 2-year period from 1982 to 1984.

71. Find the approximate increase in megawatt capacity during the 2-year period from 2000 to 2002.

In general, 1000 megawatts will serve the average electricity needs of 560,000 people. Use this fact and the preceding graph to answer Exercises 72 and 73.

72. In 2000, the number of megawatts that were generated from wind would serve the electricity needs of how many people?

73. How many megawatts of electricity are needed to serve the city or town in which you live?

74. If x is 10, is $\frac{2}{x}$ in proportion to $\frac{x}{50}$? Explain why or why not.

One of the great algebraists of ancient times was a man named Diophantus. Little is known of his life other than that he lived and worked in Alexandria. Some historians believe he lived during the first century of the Christian era, about the time of Nero. The only clue to his personal life is the following epigram found in a collection called the Palatine Anthology.

God granted him youth for a sixth of his life and added a twelfth part to this. He clothed his cheeks in down. He lit him the light of wedlock after a seventh part and five years after his marriage, He granted him a son. Alas, lateborn wretched child. After attaining the measure of half his father's life, cruel fate overtook him, thus leaving Diophantus during the last four years of his life only such consolation as the science of numbers. How old was Diophantus at his death?*

We are looking for Diophantus' age when he died, so let x represent that age. If we sum the parts of his life, we should get the total age.

Parts of his life
$$\begin{cases} \frac{1}{6}x + \frac{1}{12}x \text{ is the time of his youth.} \\ \frac{1}{7}x \text{ is the time between his youth and when he married.} \\ 5 \text{ years is the time between his marriage and the birth of his son.} \\ \frac{1}{2}x \text{ is the time Diophantus had with his son.} \\ 4 \text{ years is the time between his son's death and his own.} \end{cases}$$

The sum of these parts should equal Diophantus' age when he died.

$$\frac{1}{6}\cdot x + \frac{1}{12}\cdot x + \frac{1}{7}\cdot x + 5 + \frac{1}{2}\cdot x + 4 = x$$

75. Solve the epigram.

76. How old was Diophantus when his son was born? How old was the son when he died?

77. Solve the following epigram:

I was four when my mother packed my lunch and sent me off to school. Half my life was spent in school and another sixth was spent on a farm. Alas, hard times befell me. My crops and cattle fared poorly and my land was sold. I returned to school for 3 years and have spent one tenth of my life teaching. How old am I?

78. Write an epigram describing your life. Be sure that none of the time periods in your epigram overlap.

79. A hyena spots a giraffe 0.5 mile away and begins running toward it. The giraffe starts running away from the hyena just as the hyena begins running toward it. A hyena can run at a speed of 40 mph and a giraffe can run at 32 mph. How long will it take for the hyena to overtake the giraffe? (*Source: World Almanac and Book of Facts*)

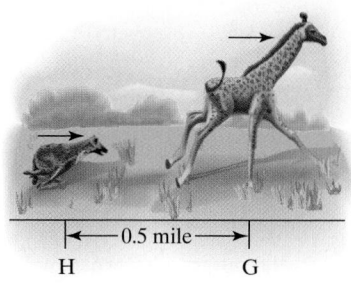

*From *The Nature and Growth of Modern Mathematics*, Edna Kramer, 1970, Fawcett Premier Books, Vol. 1, pages 107–108.

7.7   *SIMPLIFYING COMPLEX FRACTIONS*

Objectives

1   Simplify complex fractions by simplifying the numerator and denominator and then dividing.

2   Simplify complex fractions by multiplying by a common denominator.

3   Simplify expressions with negative exponents.

1   A rational expression whose numerator, denominator, or both contain one or more rational expressions is called a **complex rational expression** or a **complex fraction**.

Complex Fractions

$$\dfrac{\dfrac{1}{a}}{\dfrac{b}{2}} \qquad \dfrac{\dfrac{x}{2y^2}}{\dfrac{6x-2}{9y}} \qquad \dfrac{x+\dfrac{1}{y}}{y+1}$$

The parts of a complex fraction are

$$\dfrac{\left.\dfrac{x}{y+2}\right\}}{\left.7+\dfrac{1}{y}\right\}}$$

← Numerator of complex fraction

← Main fraction bar

← Denominator of complex fraction

Our goal in this section is to simplify complex fractions. A complex fraction is simplified when it is in the form $\dfrac{P}{Q}$, where P and Q are polynomials that have no common factors. Two methods of simplifying complex fractions are introduced. The first method evolves from the definition of a fraction as a quotient.

Simplifying A Complex Fraction: Method I

Step 1: Simplify the numerator and the denominator of the complex fraction so that each is a single fraction.

Step 2: Perform the indicated division by multiplying the numerator of the complex fraction by the reciprocal of the denominator of the complex fraction.

Step 3: Simplify if possible.

EXAMPLE 1

Simplify each complex fraction.

a. $\dfrac{\dfrac{2x}{27y^2}}{\dfrac{6x^2}{9}}$ **b.** $\dfrac{\dfrac{5x}{x+2}}{\dfrac{10}{x-2}}$ **c.** $\dfrac{\dfrac{x}{y^2}+\dfrac{1}{y}}{\dfrac{y}{x^2}+\dfrac{1}{x}}$

Solution   **a.** The numerator of the complex fraction is already a single fraction, and so is the denominator. Perform the indicated division by multiplying the numerator, $\dfrac{2x}{27y^2}$, by the reciprocal of the denominator, $\dfrac{6x^2}{9}$. Then simplify.

$$\dfrac{\dfrac{2x}{27y^2}}{\dfrac{6x^2}{9}} = \dfrac{2x}{27y^2} \div \dfrac{6x^2}{9}$$

$$= \dfrac{2x}{27y^2} \cdot \dfrac{9}{6x^2} \qquad \text{Multiply by the reciprocal of } \dfrac{6x^2}{9}.$$

$$= \dfrac{2x \cdot 9}{27y^2 \cdot 6x^2}$$

$$= \dfrac{1}{9xy^2}$$

b. $\dfrac{\left\{\dfrac{5x}{x+2}\right.}{\left\{\dfrac{10}{x-2}\right.} = \dfrac{5x}{x+2} \div \dfrac{10}{x-2} = \dfrac{5x}{x+2} \cdot \dfrac{x-2}{10} \qquad \text{Multiply by the reciprocal of } \dfrac{10}{x-2}.$

> **Helpful Hint**
> Both the numerator and denominator are single fractions, so we perform the indicated division.

$$= \dfrac{5x(x-2)}{2 \cdot 5(x+2)}$$

$$= \dfrac{x(x-2)}{2(x+2)} \qquad \text{Simplify.}$$

c. First simplify the numerator and the denominator of the complex fraction separately so that each is a single fraction. Then perform the indicated division.

$$\dfrac{\dfrac{x}{y^2} + \dfrac{1}{y}}{\dfrac{y}{x^2} + \dfrac{1}{x}} = \dfrac{\dfrac{x}{y^2} + \dfrac{1 \cdot y}{y \cdot y}}{\dfrac{y}{x^2} + \dfrac{1 \cdot x}{x \cdot x}} \qquad \text{Simplify the numerator. The LCD is } y^2.$$

$$= \dfrac{\dfrac{x+y}{y^2}}{\dfrac{y+x}{x^2}} \qquad \text{Simplify the denominator. The LCD is } x^2.$$

$$= \dfrac{x+y}{y^2} \div \dfrac{y+x}{x^2} \qquad \text{Add.}$$

$$= \dfrac{x+y}{y^2} \cdot \dfrac{x^2}{y+x} \qquad \text{Add.}$$

$$= \dfrac{x^2(x+y)}{y^2(y+x)} \qquad \text{Multiply by the reciprocal of } \dfrac{y+x}{x^2}.$$

$$= \dfrac{x^2}{y^2} \qquad \text{Simplify.}$$

✔ **CONCEPT CHECK**

Which of the following are equivalent to $\dfrac{\dfrac{1}{x}}{\dfrac{3}{y}}$?

a. $\dfrac{1}{x} \div \dfrac{3}{y}$ **b.** $\dfrac{1}{x} \cdot \dfrac{y}{3}$ **c.** $\dfrac{1}{x} \div \dfrac{y}{3}$

2   Next we look at another method of simplifying complex fractions. With this method we multiply the numerator and the denominator of the complex fraction by the LCD of all fractions in the complex fraction.

Simplifying A Complex Fraction: Method II

Step 1: Multiply the numerator and the denominator of the complex fraction by the LCD of the fractions in both the numerator and the denominator.

Step 2: Simplify.

EXAMPLE 2

Simplify each complex fraction.

a. $\dfrac{\dfrac{5x}{x+2}}{\dfrac{10}{x-2}}$ **b.** $\dfrac{\dfrac{x}{y^2}+\dfrac{1}{y}}{\dfrac{y}{x^2}+\dfrac{1}{x}}$

Solution   **a.** The least common denominator of $\dfrac{5x}{x+2}$ and $\dfrac{10}{x-2}$ is $(x+2)(x-2)$. Multiply both the numerator, $\dfrac{5x}{x+2}$, and the denominator, $\dfrac{10}{x-2}$, by the LCD.

$$\dfrac{\dfrac{5x}{x+2}}{\dfrac{10}{x-2}} = \dfrac{\left(\dfrac{5x}{x+2}\right)\cdot(x+2)(x-2)}{\left(\dfrac{10}{x-2}\right)\cdot(x+2)(x-2)}$$ Multiply numerator and denominator by the LCD.

$$= \dfrac{5x\cdot(x-2)}{2\cdot5\cdot(x+2)}$$ Simplify.

$$= \dfrac{x(x-2)}{2(x+2)}$$ Simplify.

Concept Check Answer:
a and b

b. The least common denominator of $\dfrac{x}{y^2}, \dfrac{1}{y}, \dfrac{y}{x^2}$, and $\dfrac{1}{x}$ is x^2y^2.

$$\dfrac{\dfrac{x}{y^2}+\dfrac{1}{y}}{\dfrac{y}{x^2}+\dfrac{1}{x}} = \dfrac{\left(\dfrac{x}{y^2}+\dfrac{1}{y}\right)\cdot x^2y^2}{\left(\dfrac{y}{x^2}+\dfrac{1}{x}\right)\cdot x^2y^2}$$ *Multiply the numerator and denominator by the LCD.*

$$= \dfrac{\dfrac{x}{y^2}\cdot x^2y^2+\dfrac{1}{y}\cdot x^2y^2}{\dfrac{y}{x^2}\cdot x^2y^2+\dfrac{1}{x}\cdot x^2y^2}$$ *Use the distributive property.*

$$= \dfrac{x^3+x^2y}{y^3+xy^2}$$ *Simplify.*

$$= \dfrac{x^2(x+y)}{y^2(y+x)}$$ *Factor.*

$$= \dfrac{x^2}{y^2}$$ *Simplify.*   ⬭

3   If an expression contains negative exponents, write the expression as an equivalent expression with positive exponents.

EXAMPLE 3

Simplify: $\dfrac{x^{-1}+2xy^{-1}}{x^{-2}-x^{-2}y^{-1}}$

Solution   This fraction does not appear to be a complex fraction. If we write it by using only positive exponents, however, we see that it is a complex fraction.

$$\dfrac{x^{-1}+2xy^{-1}}{x^{-2}-x^{-2}y^{-1}} = \dfrac{\dfrac{1}{x}+\dfrac{2x}{y}}{\dfrac{1}{x^2}-\dfrac{1}{x^2y}}$$

The LCD of $\dfrac{1}{x}, \dfrac{2x}{y}, \dfrac{1}{x^2}$, and $\dfrac{1}{x^2y}$ is x^2y. Multiply both the numerator and denominator by x^2y.

$$= \dfrac{\left(\dfrac{1}{x}+\dfrac{2x}{y}\right)\cdot x^2y}{\left(\dfrac{1}{x^2}-\dfrac{1}{x^2y}\right)\cdot x^2y}$$

$$= \dfrac{\dfrac{1}{x}\cdot x^2y+\dfrac{2x}{y}\cdot x^2y}{\dfrac{1}{x^2}\cdot x^2y-\dfrac{1}{x^2y}\cdot x^2y}$$ *Apply the distributive property.*

$$= \dfrac{xy+2x^3}{y-1} \quad \text{or} \quad \dfrac{x(y+2x^2)}{y-1}$$ *Simplify.*   ⬭

EXAMPLE 4

Simplify: $\dfrac{(2x)^{-1} + 1}{2x^{-1} - 1}$

Solution

$\dfrac{(2x)^{-1} + 1}{2x^{-1} - 1} = \dfrac{\dfrac{1}{2x} + 1}{\dfrac{2}{x} - 1}$

Write using positive exponents.

> **Helpful Hint**
> Don't forget that
> $(2x)^{-1} = \dfrac{1}{2x}$, but
> $2x^{-1} = 2 \cdot \dfrac{1}{x} = \dfrac{2}{x}$.

$= \dfrac{\left(\dfrac{1}{2x} + 1\right) \cdot 2x}{\left(\dfrac{2}{x} - 1\right) \cdot 2x}$

The LCD of $\dfrac{1}{2x}$ and $\dfrac{2}{x}$ is $2x$.

$= \dfrac{\dfrac{1}{2x} \cdot 2x + 1 \cdot 2x}{\dfrac{2}{x} \cdot 2x - 1 \cdot 2x}$

Use distributive property.

$= \dfrac{1 + 2x}{4 - 2x}$ or $\dfrac{1 + 2x}{2(2 - x)}$

Simplify.

STUDY SKILLS REMINDER

Are You Satisfied with Your Performance in this Course thus Far?

If not, ask yourself the following questions:

▶ Am I attending all class periods and arriving on time?

▶ Am I working and checking my homework assignments?

▶ Am I getting help when I need it?

▶ In addition to my instructor, am I using the supplements to this text that could help me? For example, the tutorial video lessons? The tutorial software?

▶ Am I satisfied with my performance on quizzes and tests?

If you answered no to *any* of these questions, read or reread Section 1.1 for suggestions in these areas. Also, you may want to contact your instructor for additional feedback.

EXERCISE SET 7.7

STUDY GUIDE/SSM   CD/VIDEO   PH MATH TUTOR CENTER   MathXL®Tutorials ON CD   MathXL®   MyMathLab®

Simplify each complex fraction. See Examples 1 and 2.

 1. $\dfrac{\dfrac{10}{3x}}{\dfrac{5}{6x}}$

2. $\dfrac{\dfrac{15}{2x}}{\dfrac{5}{6x}}$

3. $\dfrac{1 + \dfrac{2}{5}}{2 + \dfrac{3}{5}}$

4. $\dfrac{2 + \dfrac{1}{7}}{3 - \dfrac{4}{7}}$

5. $\dfrac{\dfrac{4}{x-1}}{\dfrac{x}{x-1}}$

6. $\dfrac{\dfrac{x}{x+2}}{\dfrac{2}{x+2}}$

7. $\dfrac{1 - \dfrac{2}{x}}{x + \dfrac{4}{9x}}$

8. $\dfrac{5 - \dfrac{3}{x}}{x + \dfrac{2}{3x}}$

9. $\dfrac{\dfrac{4x^2 - y^2}{xy}}{\dfrac{2}{y} - \dfrac{1}{x}}$

10. $\dfrac{\dfrac{x^2 - 9y^2}{xy}}{\dfrac{1}{y} - \dfrac{3}{x}}$

11. $\dfrac{\dfrac{x+1}{3}}{\dfrac{2x-1}{6}}$

12. $\dfrac{\dfrac{x+3}{12}}{\dfrac{4x-5}{15}}$

13. $\dfrac{\dfrac{2}{x} + \dfrac{3}{x^2}}{\dfrac{4}{x^2} - \dfrac{9}{x}}$

14. $\dfrac{\dfrac{2}{x^2} + \dfrac{1}{x}}{\dfrac{4}{x^2} - \dfrac{1}{x}}$

15. $\dfrac{\dfrac{1}{x} + \dfrac{2}{x^2}}{x + \dfrac{8}{x^2}}$

16. $\dfrac{\dfrac{1}{y} + \dfrac{3}{y^2}}{y + \dfrac{27}{y^2}}$

17. $\dfrac{\dfrac{4}{5-x} + \dfrac{5}{x-5}}{\dfrac{2}{x} + \dfrac{3}{x-5}}$

18. $\dfrac{\dfrac{3}{x-4} - \dfrac{2}{4-x}}{\dfrac{2}{x-4} - \dfrac{2}{x}}$

 19. $\dfrac{\dfrac{x+2}{x} - \dfrac{2}{x-1}}{\dfrac{x+1}{x} + \dfrac{x+1}{x-1}}$

20. $\dfrac{\dfrac{5}{a+2} - \dfrac{1}{a-2}}{\dfrac{3}{2+a} + \dfrac{6}{2-a}}$

21. $\dfrac{\dfrac{2}{x} + 3}{\dfrac{4}{x^2} - 9}$

22. $\dfrac{2 + \dfrac{1}{x}}{4x - \dfrac{1}{x}}$

23. $\dfrac{1 - \dfrac{x}{y}}{\dfrac{x^2}{y^2} - 1}$

24. $\dfrac{1 - \dfrac{2}{x}}{x - \dfrac{4}{x}}$

25. $\dfrac{\dfrac{-2x}{x-y}}{\dfrac{y}{x^2}}$

26. $\dfrac{\dfrac{7y}{x^2 + xy}}{\dfrac{y^2}{x^2}}$

27. $\dfrac{\dfrac{2}{x} + \dfrac{1}{x^2}}{\dfrac{y}{x^2}}$

28. $\dfrac{\dfrac{5}{x^2} - \dfrac{2}{x}}{\dfrac{1}{x} + 2}$

29. $\dfrac{\dfrac{x}{9} - \dfrac{1}{x}}{1 + \dfrac{3}{x}}$

30. $\dfrac{\dfrac{x}{4} - \dfrac{4}{x}}{1 - \dfrac{4}{x}}$

31. $\dfrac{\dfrac{x-1}{x^2-4}}{1 + \dfrac{1}{x-2}}$

32. $\dfrac{\dfrac{2}{x+5} + \dfrac{4}{x+3}}{\dfrac{3x+13}{x^2+8x+15}}$

Simplify. See Examples 3 and 4.

33. $\dfrac{x^{-1}}{x^{-2} + y^{-2}}$

34. $\dfrac{a^{-3} + b^{-1}}{a^{-2}}$

 35. $\dfrac{2a^{-1} + 3b^{-2}}{a^{-1} - b^{-1}}$

36. $\dfrac{x^{-1} + y^{-1}}{3x^{-2} + 5y^{-2}}$

37. $\dfrac{1}{x - x^{-1}}$

38. $\dfrac{x^{-2}}{x + 3x^{-1}}$

39. $\dfrac{a^{-1} + 1}{a^{-1} - 1}$

40. $\dfrac{a^{-1} - 4}{4 + a^{-1}}$

41. $\dfrac{3x^{-1} + (2y)^{-1}}{x^{-2}}$

42. $\dfrac{5x^{-2} - 3y^{-1}}{x^{-1} + y^{-1}}$

43. $\dfrac{2a^{-1} + (2a)^{-1}}{a^{-1} + 2a^{-2}}$

44. $\dfrac{a^{-1} + 2a^{-2}}{2a^{-1} + (2a)^{-1}}$

45. $\dfrac{5x^{-1} + 2y^{-1}}{x^{-2}y^{-2}}$

46. $\dfrac{x^{-2}y^{-2}}{5x^{-1} + 2y^{-1}}$

47. $\dfrac{5x^{-1} - 2y^{-1}}{25x^{-2} - 4y^{-2}}$

48. $\dfrac{3x^{-1} + 3y^{-1}}{4x^{-2} - 9y^{-2}}$

REVIEW AND PREVIEW

Simplify. See Section 5.1.

49. $\dfrac{3x^3y^2}{12x}$

50. $\dfrac{-36xb^3}{9xb^2}$

51. $\dfrac{144x^5y^5}{-16x^2y}$

52. $\dfrac{48x^3y^2}{-4xy}$

Solve the following. See Section 3.7.

53. If $P(x) = -x^2$, find $P(-3)$.

54. If $f(x) = x^2 - 6$, find $f(-1)$.

Concept Extensions

55. When the source of a sound is traveling toward a listener, the pitch that the listener hears due to the Doppler effect is given by the complex rational compression $\dfrac{a}{1 - \dfrac{s}{770}}$, where a is the actual pitch of the sound and s is the speed of the sound source. Simplify this expression.

56. In baseball, the earned run average (ERA) statistic gives the average number of earned runs scored on a pitcher per game. It is computed with the following expression: $\dfrac{E}{\dfrac{I}{9}}$, where E is the number of earned runs scored on a pitcher and I is the total number of innings pitched by the pitcher. Simplify this expression.

57. Which of the following are equivalent to $\dfrac{\dfrac{1}{x}}{\dfrac{3}{y}}$?

a. $\dfrac{1}{x} \div \dfrac{3}{y}$ **b.** $\dfrac{1}{x} \cdot \dfrac{y}{3}$ **c.** $\dfrac{1}{x} \div \dfrac{y}{3}$

58. In your own words, explain one method for simplifying a complex fraction.

Simplify.

59. $\dfrac{1}{1 + (1 + x)^{-1}}$

60. $\dfrac{(x + 2)^{-1} + (x - 2)^{-1}}{(x^2 - 4)^{-1}}$

61. $\dfrac{x}{1 - \dfrac{1}{1 + \dfrac{1}{x}}}$

62. $\dfrac{x}{1 - \dfrac{1}{1 - \dfrac{1}{x}}}$

63. $\dfrac{\dfrac{2}{y^2} - \dfrac{5}{xy} - \dfrac{3}{x^2}}{\dfrac{2}{y^2} + \dfrac{7}{xy} + \dfrac{3}{x^2}}$

64. $\dfrac{\dfrac{2}{x^2} - \dfrac{1}{xy} - \dfrac{1}{y^2}}{\dfrac{1}{x^2} - \dfrac{3}{xy} + \dfrac{2}{y^2}}$

65. $\dfrac{3(a + 1)^{-1} + 4a^{-2}}{(a^3 + a^2)^{-1}}$

66. $\dfrac{9x^{-1} - 5(x - y)^{-1}}{4(x - y)^{-1}}$

*In the study of calculus, the difference quotient $\dfrac{f(a + h) - f(a)}{h}$ is often found and simplified. Find and simplify this quotient for each function $f(x)$ by following steps **a** through **d**.*

a. *Find $(a + h)$.*

b. *Find $f(a)$.*

c. *Use steps **a** and **b** to find $\dfrac{f(a + h) - f(a)}{h}$*

d. *Simplify the result of step **c**.*

67. $f(x) = \dfrac{1}{x}$

68. $f(x) = \dfrac{5}{x}$

69. $\dfrac{3}{x + 1}$

70. $\dfrac{2}{x^2}$

CHAPTER 7 PROJECT

Comparing Dosage Formulas

In this project, you will have the opportunity to investigate two well-known formulas for predicting the correct doses of medication for children. This project may be completed by working in groups or individually.

Young's Rule and Cowling's Rule are dose formulas for prescribing medicines to children. Unlike formulas for, say area or distance, these dose formulas describe only an approximate relationship. The formulas relate a child's age A in years and an adult dose D of medication to the proper child's dose C. The formulas are most accurate when applied to children between the ages of 2 and 13.

$$\text{Young's Rule:} \quad C = \frac{DA}{A + 12}$$

$$\text{Cowling's Rule:} \quad C = \frac{D(A + 1)}{24}$$

Age A	Young's Rule	Cowling's Rule	Difference
2			
3			
4			
5			
6			
7			
8			
9			
10			
11			
12			
13			

1. Let the adult dose $D = 1000$ mg. Complete the Young's Rule and Cowling's Rule columns of the following table comparing the doses predicted by both formulas for ages 2 through 13.

2. Use the data from the table in Question 1 to form sets of ordered pairs of the form (age, child's dose) for each formula. Graph the ordered pairs for each formula on the same graph. Describe the shapes of the graphed data.

3. Use your table, graph, or both, to decide whether either formula will consistently predict a larger dose than the other. If so, which one? If not, is there an age at which the doses predicted by one becomes greater than the doses predicted by the other? If so, estimate that age.

4. Use your graph to estimate for what age the difference in the two predicted doses is greatest.

5. Return to the table in Question 1 and complete the last column, titled "Difference," by finding the absolute value of the difference between the Young's dose and the Cowling's dose for each age. Use this column in the table to verify your graphical estimate found in Question 4.

6. Does Cowling's Rule ever predict exactly the adult dose? If so, at what age? Explain. Does Young's Rule ever predict exactly the adult dose? If so, at what age? Explain.

7. Many doctors prefer to use formulas that relate doses to factors other than a child's age. Why is age not necessarily the most important factor when predicting a child's dose? What other factors might be used?

7 CHAPTER VOCABULARY CHECK

Fill in each blank with one of the words or phrases listed below.

rational expression	complex fraction	ratio	proportion
cross products	direct variation	inverse variation	

1. A _____ is the quotient of two numbers.

2. $\dfrac{x}{2} = \dfrac{7}{16}$ is an example of a _____ .

3. If $\dfrac{a}{b} = \dfrac{c}{d}$, then ad and bc are called _____ .

4. A _____ is an expression that can be written in the form $\dfrac{P}{Q}$, where P and Q are polynomials and Q is not 0.

5. In a _____ , the numerator or denominator or both may contain fractions.

6. The equation $y = \dfrac{k}{x}$ is an example of _____ .

7. The equation $y = kx$ is an example of _____ .

CHAPTER 7 HIGHLIGHTS

Definitions and Concepts	**Examples**

Section 7.1 Rational Functions and Simplifying Rational Expressions

Definitions and Concepts	Examples
A **rational expression** is an expression that can be written in the form $\frac{P}{Q}$, where P and Q are polynomials and Q does not equal 0.	Rational Expressions $$\frac{7y^3}{4}, \frac{x^2 + 6x + 1}{x - 3}, \frac{-5}{s^3 + 8}$$
To find values for which a rational expression is undefined, find values for which the denominator is 0.	Find any values for which the expression $\dfrac{5y}{y^2 - 4y + 3}$ is undefined. $$\begin{aligned} y^2 - 4y + 3 &= 0 && \text{Set denominator equal to 0.} \\ (y - 3)(y - 1) &= 0 && \text{Factor.} \\ y - 3 = 0 \text{ or } y - 1 &= 0 && \text{Set each factor equal to 0.} \\ y = 3 \text{ or } y &= 1 && \text{Solve.} \end{aligned}$$ The expression is undefined when y is 3 and when y is 1.
Fundamental Principle of Rational Expressions If P and Q are polynomials, and Q and R are not 0, then $$\frac{PR}{QR} = \frac{P}{Q}$$	By the fundamental principle, $$\frac{(x - 3)(x + 1)}{x(x + 1)} = \frac{x - 3}{x}$$ as long as $x \neq 0$ and $x \neq -1$.
To simplify a rational expression, **Step 1.** Factor the numerator and denominator. **Step 2.** Apply the fundamental principle to divide out common factors.	Simplify: $\dfrac{4x + 20}{x^2 - 25}$ $$\frac{4x + 20}{x^2 - 25} = \frac{4(x + 5)}{(x + 5)(x - 5)} = \frac{4}{x - 5}$$

Definitions and Concepts	**Examples**

Section 7.2 Multiplying and Dividing Rational Expressions

To multiply rational expressions,

Step 1. Factor numerators and denominators.

Step 2. Multiply numerators and multiply denominators.

Step 3. Write the product in simplest form.

$$\frac{P}{Q} \cdot \frac{R}{S} = \frac{PR}{QS}$$

Multiply: $\dfrac{4x + 4}{2x - 3} \cdot \dfrac{2x^2 + x - 6}{x^2 - 1}$

$$\frac{4x + 4}{2x - 3} \cdot \frac{2x^2 + x - 6}{x^2 - 1} = \frac{4(x + 1)}{2x - 3} \cdot \frac{(2x - 3)(x + 2)}{(x + 1)(x - 1)}$$

$$= \frac{4(x + 1)(2x - 3)(x + 2)}{(2x - 3)(x + 1)(x - 1)}$$

$$= \frac{4(x + 2)}{x - 1}$$

To divide by a rational expression, multiply by the reciprocal.

$$\frac{P}{Q} \div \frac{R}{S} = \frac{P}{Q} \cdot \frac{S}{R} = \frac{PS}{QR}$$

Divide: $\dfrac{15x + 5}{3x^2 - 14x - 5} \div \dfrac{15}{3x - 12}$

$$\frac{15x + 5}{3x^2 - 14x - 5} \div \frac{15}{3x - 12} = \frac{5(3x + 1)}{(3x + 1)(x - 5)} \cdot \frac{3(x - 4)}{3 \cdot 5}$$

$$= \frac{x - 4}{x - 5}$$

Section 7.3 Adding and Subtracting Rational Expressions with Common Denominators and Least Common Denominator

To add or subtract rational expressions with the same denominator, add or subtract numerators, and place the sum or difference over a common denominator.

$$\frac{P}{R} + \frac{Q}{R} = \frac{P + Q}{R}$$

$$\frac{P}{R} - \frac{Q}{R} = \frac{P - Q}{R}$$

Perform indicated operations.

$$\frac{5}{x + 1} + \frac{x}{x + 1} = \frac{5 + x}{x + 1}$$

$$\frac{2y + 7}{y^2 - 9} - \frac{y + 4}{y^2 - 9} = \frac{(2y + 7) - (y + 4)}{y^2 - 9}$$

$$= \frac{2y + 7 - y - 4}{y^2 - 9}$$

$$= \frac{y + 3}{(y + 3)(y - 3)}$$

$$= \frac{1}{y - 3}$$

To find the least common denominator (LCD),

Step 1. Factor the denominators.

Step 2. The LCD is the product of all unique factors, each raised to a power equal to the greatest number of times that it appears in any one factored denominator.

Find the LCD for

$$\frac{7x}{x^2 + 10x + 25} \quad \text{and} \quad \frac{11}{3x^2 + 15x}$$

$$x^2 + 10x + 25 = (x + 5)(x + 5)$$

$$3x^2 + 15x = 3x(x + 5)$$

LCD is $3x(x + 5)(x + 5)$ or $3x(x + 5)^2$

Definitions and Concepts	**Examples**

Section 7.4 Adding and Subtracting Rational Expressions with Unlike Denominators

To add or subtract rational expressions with unlike denominators,

Step 1. Find the LCD.

Step 2. Rewrite each rational expression as an equivalent expression whose denominator is the LCD.

Step 3. Add or subtract numerators and place the sum or difference over the common denominator.

Step 4. Write the result in simplest form.

Perform the indicated operation.

$$\frac{9x + 3}{x^2 - 9} - \frac{5}{x - 3}$$

$$= \frac{9x + 3}{(x + 3)(x - 3)} - \frac{5}{x - 3}$$

LCD is $(x + 3)(x - 3)$.

$$= \frac{9x + 3}{(x + 3)(x - 3)} - \frac{5(x + 3)}{(x - 3)(x + 3)}$$

$$= \frac{9x + 3 - 5(x + 3)}{(x + 3)(x - 3)}$$

$$= \frac{9x + 3 - 5x - 15}{(x + 3)(x - 3)}$$

$$= \frac{4x - 12}{(x + 3)(x - 3)}$$

$$= \frac{4(x - 3)}{(x + 3)(x - 3)} = \frac{4}{x + 3}$$

Section 7.5 Solving Equations Containing Rational Expressions

To solve an equation containing rational expressions,

Step 1. Multiply both sides of the equation by the LCD of all rational expressions in the equation.

Step 2. Remove any grouping symbols and solve the resulting equation.

Step 3. Check the solution in the original equation.

Solve: $\dfrac{5x}{x + 2} + 3 = \dfrac{4x - 6}{x + 2}$

$$(x + 2)\left(\frac{5x}{x + 2} + 3\right) = (x + 2)\left(\frac{4x - 6}{x + 2}\right)$$

$$(x + 2)\left(\frac{5x}{x + 2}\right) + (x + 2)(3) = (x + 2)\left(\frac{4x - 6}{x + 2}\right)$$

$$5x + 3x + 6 = 4x - 6$$

$$4x = -12$$

$$x = -3$$

The solution checks and the solution is -3.

Section 7.6 Proportion and Problem Solving with Rational Equations

A **ratio** is the quotient of two numbers or two quantities. A **proportion** is a mathematical statement that two ratios are equal.

Proportions

$$\frac{2}{3} = \frac{8}{12} \qquad \frac{x}{7} = \frac{15}{35}$$

Cross products:

If $\dfrac{a}{b} = \dfrac{c}{d}$, then $ad = bc$.

Cross Products

$$\frac{2}{3} = \frac{8}{12} \quad \begin{array}{l} 3 \cdot 8 \text{ or } 24 \\ 2 \cdot 12 \text{ or } 24 \end{array}$$

(continued)

Definitions and Concepts	Examples

Section 7.6 Proportion and Problem Solving with Rational Equations

Solve: $\dfrac{3}{4} = \dfrac{x}{x-1}$

$\dfrac{3}{4} = \dfrac{x}{x-1}$

$3(x-1) = 4x$ — Set cross products equal.

$3x - 3 = 4x$

$-3 = x$

Problem-Solving Steps

1. UNDERSTAND. Read and reread the problem.

A small plane and a car leave Kansas City, Missouri, and head for Minneapolis, Minnesota, a distance of 450 miles. The speed of the plane is 3 times the speed of the car, and the plane arrives 6 hours ahead of the car. Find the speed of the car.

Let x = the speed of the car.

Then $3x$ = the speed of the plane.

	Distance	=	Rate	·	Time
Car	450		x		$\dfrac{450}{x}$ $\left(\dfrac{\text{distance}}{\text{rate}}\right)$
Plane	450		$3x$		$\dfrac{450}{3x}$ $\left(\dfrac{\text{distance}}{\text{rate}}\right)$

In words:

plane's time	+	6 hours	=	car's time

2. TRANSLATE.

Translate: $\dfrac{450}{3x}$ + 6 = $\dfrac{450}{x}$

$\dfrac{450}{3x} + 6 = \dfrac{450}{x}$

3. SOLVE.

$3x\left(\dfrac{450}{3x}\right) + 3x(6) = 3x\left(\dfrac{450}{x}\right)$

$450 + 18x = 1350$

$18x = 900$

$x = 50$

4. INTERPRET.

Check the solution by replacing x with 50 in the original equation. *State* the conclusion: The speed of the car is 50 miles per hour.

Definitions and Concepts	Examples

Section 7.7 Simplifying Complex Fractions

Method 1: Simplify the numerator and the denominator so that each is a single fraction. Then perform the indicated division and simplify if possible.

Simplify $\dfrac{\dfrac{x+2}{x}}{x - \dfrac{4}{x}}$.

Method 1: $\dfrac{\dfrac{x+2}{x}}{\dfrac{x \cdot x}{1 \cdot x} - \dfrac{4}{x}} = \dfrac{\dfrac{x+2}{x}}{\dfrac{x^2-4}{x}}$

$= \dfrac{x+2}{x} \cdot \dfrac{x}{(x+2)(x-2)} = \dfrac{1}{x-2}$

Method 2: Multiply the numerator and the denominator of the complex fraction by the LCD of the fractions in both the numerator and the denominator. Then simplify if possible.

Method 2: $\dfrac{\left(\dfrac{x+2}{x}\right) \cdot x}{\left(x - \dfrac{4}{x}\right) \cdot x} = \dfrac{x+2}{x \cdot x - \dfrac{4}{x} \cdot x}$

$= \dfrac{x+2}{x^2-4} = \dfrac{x+2}{(x+2)(x-2)} = \dfrac{1}{x-2}$

CHAPTER REVIEW

(7.1) Find the domain for each rational function.

1. $F(x) = \dfrac{-3x^2}{x-5}$

2. $h(x) = \dfrac{4x}{3x-12}$

3. $f(x) = \dfrac{x^3+2}{x^2+8x}$

4. $G(x) = \dfrac{20}{3x^2-48}$

Write each rational expression in lowest terms.

5. $\dfrac{15x^4}{45x^2}$

6. $\dfrac{x+2}{2+x}$

7. $\dfrac{18m^6p^2}{10m^4p}$

8. $\dfrac{x-12}{12-x}$

9. $\dfrac{5x-15}{25x-75}$

10. $\dfrac{22x+8}{11x+4}$

11. $\dfrac{2x}{2x^2-2x}$

12. $\dfrac{x+7}{x^2-49}$

13. $\dfrac{2x^2+4x-30}{x^2+x-20}$

14. $\dfrac{xy-3x+2y-6}{x^2+4x+4}$

15. The average cost of manufacturing x bookcases is given by the rational function

$$C(x) = \dfrac{35x+4200}{x}$$

a. Find the average cost per bookcase of manufacturing 50 bookcases.

b. Find the average cost per bookcase of manufacturing 100 bookcases.

c. As the number of bookcase increases, does the average cost per bookcase increase or decrease? (See parts **a** and **b**.)

(7.2) Perform the indicated operations and simplify.

16. $\dfrac{15x^3y^2}{z} \cdot \dfrac{z}{5xy^3}$

17. $\dfrac{-y^3}{8} \cdot \dfrac{9x^2}{y^3}$

18. $\dfrac{x^2-9}{x^2-4} \cdot \dfrac{x-2}{x+3}$

19. $\dfrac{2x+5}{x-6} \cdot \dfrac{2x}{-x+6}$

20. $\dfrac{x^2-5x-24}{x^2-x-12} \div \dfrac{x^2-10x+16}{x^2+x-6}$

21. $\dfrac{4x + 4y}{xy^2} \div \dfrac{3x + 3y}{x^2y}$

22. $\dfrac{x^2 + x - 42}{x - 3} \cdot \dfrac{(x - 3)^2}{x + 7}$

23. $\dfrac{2a + 2b}{3} \cdot \dfrac{a - b}{a^2 - b^2}$

24. $\dfrac{x^2 - 9x + 14}{x^2 - 5x + 6} \cdot \dfrac{x + 2}{x^2 - 5x - 14}$

25. $(x - 3) \cdot \dfrac{x}{x^2 + 3x - 18}$

26. $\dfrac{2x^2 - 9x + 9}{8x - 12} \div \dfrac{x^2 - 3x}{2x}$

27. $\dfrac{x^2 - y^2}{x^2 + xy} \div \dfrac{3x^2 - 2xy - y^2}{3x^2 + 6x}$

28. $\dfrac{x^2 - y^2}{8x^2 - 16xy + 8y^2} \div \dfrac{x + y}{4x - y}$

29. $\dfrac{x - y}{4} \div \dfrac{y^2 - 2y - xy + 2x}{16x + 24}$

30. $\dfrac{y - 3}{4x + 3} \div \dfrac{9 - y^2}{4x^2 - x - 3}$

(7.3) Perform the indicated operations and simplify.

31. $\dfrac{5x - 4}{3x - 1} + \dfrac{6}{3x - 1}$

32. $\dfrac{4x - 5}{3x^2} - \dfrac{2x + 5}{3x^2}$

33. $\dfrac{9x + 7}{6x^2} - \dfrac{3x + 4}{6x^2}$

Find the LCD of each pair of rational expressions.

34. $\dfrac{x + 4}{2x}, \dfrac{3}{7x}$

35. $\dfrac{x - 2}{x^2 - 5x - 24}, \dfrac{3}{x^2 + 11x + 24}$

Rewrite the following rational expressions as equivalent expressions whose denominator is the given polynomial.

36. $\dfrac{x + 2}{x^2 + 11x + 18}, (x + 2)(x - 5)(x + 9)$

37. $\dfrac{3x - 5}{x^2 + 4x + 4}, (x + 2)^2(x + 3)$

(7.4) Perform the indicated operations and simplify.

38. $\dfrac{4}{5x^2} - \dfrac{6}{y}$

39. $\dfrac{2}{x - 3} - \dfrac{4}{x - 1}$

40. $\dfrac{x + 7}{x + 3} - \dfrac{x - 3}{x + 7}$

41. $\dfrac{4}{x + 3} - 2$

42. $\dfrac{3}{x^2 + 2x - 8} + \dfrac{2}{x^2 - 3x + 2}$

43. $\dfrac{2x - 5}{6x + 9} - \dfrac{4}{2x^2 + 3x}$

44. $\dfrac{x - 1}{x^2 - 2x + 1} - \dfrac{x + 1}{x - 1}$

45. $\dfrac{x - 1}{x^2 + 4x + 4} + \dfrac{x - 1}{x + 2}$

Find the perimeter and the area of each figure.

△ **46.**

△ **47.**

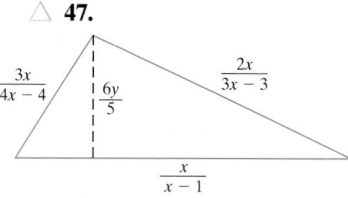

(7.5) Solve each equation for the variable or perform the indicated operation.

48. $\dfrac{x + 4}{9} = \dfrac{5}{9}$

49. $\dfrac{n}{10} = 9 - \dfrac{n}{5}$

50. $\dfrac{5y - 3}{7} = \dfrac{15y - 2}{28}$

51. $\dfrac{2}{x + 1} - \dfrac{1}{x - 2} = -\dfrac{1}{2}$

52. $\dfrac{1}{a + 3} + \dfrac{1}{a - 3} = -\dfrac{5}{a^2 - 9}$

53. $\dfrac{y}{2y + 2} + \dfrac{2y - 16}{4y + 4} = \dfrac{y - 3}{y + 1}$

54. $\dfrac{4}{x + 3} + \dfrac{8}{x^2 - 9} = 0$

55. $\dfrac{2}{x - 3} - \dfrac{4}{x + 3} = \dfrac{8}{x^2 - 9}$

56. $\dfrac{x - 3}{x + 1} - \dfrac{x - 6}{x + 5} = 0$

57. $x + 5 = \dfrac{6}{x}$

Solve the equation for the indicated variable.

58. $\dfrac{4A}{5b} = x^2$, for b

59. $\dfrac{x}{7} + \dfrac{y}{8} = 10$, for y

(7.6) Solve each proportion.

60. $\dfrac{x}{2} = \dfrac{12}{4}$

61. $\dfrac{20}{1} = \dfrac{x}{25}$

62. $\dfrac{2}{x - 1} = \dfrac{3}{x + 3}$

63. $\dfrac{4}{y - 3} = \dfrac{2}{y - 3}$

Solve.

64. A machine can process 300 parts in 20 minutes. Find how many parts can be processed in 45 minutes.

65. As his consulting fee, Mr. Visconti charges $90.00 per day. Find how much he charges for 3 hours of consulting. Assume an 8-hour work day.

66. One fundraiser can address 100 letters in 35 minutes. Find how many he can address in 55 minutes.

67. Five times the reciprocal of a number equals the sum of $\frac{3}{2}$ times the reciprocal of the number and $\frac{7}{6}$. What is the number?

68. The reciprocal of a number equals the reciprocal of the difference of 4 and the number. Find the number.

69. A car travels 90 miles in the same time that a car traveling 10 miles per hour slower travels 60 miles. Find the speed of each car.

70. The speed of a bayou near Lafayette, Louisiana, is 4 miles per hour. A paddleboat travels 48 miles upstream in the same amount of time it takes to travel 72 miles downstream. Find the speed of the boat in still water.

71. When Mark and Maria manicure Mr. Stergeon's lawn, it takes them 5 hours. If Mark works alone, it takes 7 hours. Find how long it takes Maria alone.

72. It takes pipe A 20 days to fill a fish pond. Pipe B takes 15 days. Find how long it takes both pipes together to fill the pond.

Given that the pairs of triangles are similar, find each missing length x.

 73.

 74.

75.

△ **76.**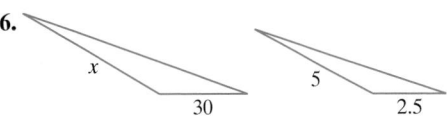

(7.7) Simplify each complex fraction.

77. $\dfrac{\dfrac{5x}{27}}{-\dfrac{10xy}{21}}$

78. $\dfrac{\dfrac{8x}{x^2 - 9}}{\dfrac{4}{x + 3}}$

79. $\dfrac{\dfrac{3}{5} + \dfrac{2}{7}}{\dfrac{1}{5} + \dfrac{5}{6}}$

80. $\dfrac{\dfrac{2}{a} + \dfrac{1}{2a}}{a + \dfrac{a}{2}}$

81. $\dfrac{\dfrac{1}{a} + \dfrac{1}{b}}{\dfrac{1}{ab}}$

82. $\dfrac{\dfrac{6}{x + 2} + 4}{\dfrac{8}{x + 2} - 4}$

83. $\dfrac{\dfrac{x - 3}{x + 3} + \dfrac{x + 3}{x - 3}}{\dfrac{x - 3}{x + 3} - \dfrac{x + 3}{x - 3}}$

84. $\dfrac{\dfrac{3}{x - 1} - \dfrac{2}{1 - x}}{\dfrac{2}{x - 1} - \dfrac{2}{x}}$

85. $\dfrac{x + y^{-1}}{\dfrac{x}{y}}$

86. $\dfrac{x - xy^{-1}}{\dfrac{1 + x}{y}}$

87. $\dfrac{y^{-2}}{1 - y^{-2}}$

88. $\dfrac{4 + x^{-1}}{3 + x^{-1}}$

STUDY SKILLS REMINDER

Are You Preparing for a Test on Chapter 7?

Below I have listed *a common trouble* area for topics covered in Chapter 7. After studying for your test, but before taking your test, read this.

Do you know the differences between how to perform operations such as $\frac{4}{x} + \frac{2}{3}$ or $\frac{4}{x} \div \frac{2}{x}$ and how to solve an equation such as $\frac{4}{x} + \frac{2}{3} = 1$?

$$\frac{4}{x} + \frac{2}{3} = \frac{4 \cdot 3}{x \cdot 3} + \frac{2 \cdot x}{3 \cdot x} = \frac{12}{3x} + \frac{2x}{3x} = \frac{12 + 2x}{3x} \text{ or } \frac{2(6 + x)}{3x}, \text{ the sum.}$$

Addition—write each expression as an equivalent expression with the same LCD denominator.

$$\frac{4}{x} \div \frac{2}{x} = \frac{4}{x} \cdot \frac{x}{2} = \frac{4 \cdot x}{x \cdot 2} = \frac{4}{2} = 2, \text{ the quotient.}$$

Division—multiply the first rational expression by the reciprocal of the second.

Solve for x: $\dfrac{4}{x} + \dfrac{2}{3} = 1$ — Equation to be solved.

$3x\left(\dfrac{4}{x} + \dfrac{2}{3}\right) = 3x \cdot 1$ — Multiply both sides of the equation by the LCD $3x$.

$3x\left(\dfrac{4}{x}\right) + 3x\left(\dfrac{2}{3}\right) = 3x \cdot 1$ — Use the distributive property.

$12 + 2x = 3x$ — Multiply and simplify.

$12 = x$ — Subtract $2x$ from both sides.

The solution is 12. Check the solution in the original equation.

Remember: This is simply a sampling of selected topics given to check your understanding. For a review of Chapter 7 in your text, see the material at the end of this chapter.

CHAPTER 7 TEST

Remember to use your Chapter Test Prep Video CD to help you study and view solutions to the test questions you need help with.

Find the domain of each rational function.

1. $f(x) = \dfrac{5x^2}{1 - x}$

2. $g(x) = \dfrac{9x^2 - 9}{x^2 + 4x + 3}$

Simplify each rational expression.

3. $\dfrac{3x - 6}{5x - 10}$

4. $\dfrac{x + 6}{x^2 + 12x + 36}$

5. $\dfrac{x + 3}{x^3 + 27}$

6. $\dfrac{2m^3 - 2m^2 - 12m}{m^2 - 5m + 6}$

7. $\dfrac{ay + 3a + 2y + 6}{ay + 3a + 5y + 15}$

8. $\dfrac{y - x}{x^2 - y^2}$

Perform the indicated operation and simplify if possible.

9. $\dfrac{3}{x - 1} \cdot (5x - 5)$

10. $\dfrac{y^2 - 5y + 6}{2y + 4} \cdot \dfrac{y + 2}{2y - 6}$

11. $\dfrac{5}{2x + 5} - \dfrac{6}{2x + 5}$

12. $\dfrac{5a}{a^2 - a - 6} - \dfrac{2}{a - 3}$

13. $\dfrac{6}{x^2 - 1} + \dfrac{3}{x + 1}$

14. $\dfrac{x^2 - 9}{x^2 - 3x} \div \dfrac{xy + 5x + 3y + 15}{2x + 10}$

15. $\dfrac{x + 2}{x^2 + 11x + 18} + \dfrac{5}{x^2 - 3x - 10}$

Solve each equation.

16. $\dfrac{4}{y} - \dfrac{5}{3} = \dfrac{-1}{5}$

17. $\dfrac{5}{y + 1} = \dfrac{4}{y + 2}$

18. $\dfrac{a}{a - 3} = \dfrac{3}{a - 3} - \dfrac{3}{2}$

19. $x - \dfrac{14}{x - 1} = 4 - \dfrac{2x}{x - 1}$

Simplify each complex fraction.

20. $\dfrac{\dfrac{5x^2}{yz^2}}{\dfrac{10x}{z^3}}$

21. $\dfrac{5 - \dfrac{1}{y^2}}{\dfrac{1}{y} + \dfrac{2}{y^2}}$

22. In a sample of 85 fluorescent bulbs, 3 were found to be defective. At this rate, how many defective bulbs should be found in 510 bulbs?

23. One number plus five times its reciprocal is equal to six. Find the number.

24. A pleasure boat traveling down the Red River takes the same time to go 14 miles upstream as it takes to go 16 miles downstream. If the current of the river is 2 miles per hour, find the speed of the boat in still water.

25. An inlet pipe can fill a tank in 12 hours. A second pipe can fill the tank in 15 hours. If both pipes are used, find how long it takes to fill the tank.

△ **26.** Given that the two triangles are similar, find x.

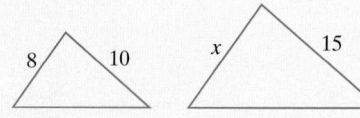

27. For a certain computer desk, the average cost C (in dollars) per desk manufactured is

$$C = \frac{100x + 3000}{x}$$

where x is the number of desks manufactured.

a. Find the average cost per desk when manufacturing 200 computer desks.

b. Find the average cost per desk when manufacturing 1000 computer desks.

7 CHAPTER CUMULATIVE REVIEW

1. Write each sentence as an equation. Let x represent the unknown number.

  a. The quotient of 15 and a number is 4.

  b. Three subtracted from 12 is a number.

  c. Four times a number, added to 17, is 21.

2. Write each sentence as an equation. Let x represent the unknown number.

  a. The difference of 12 and a number is -45.

  b. The product of 12 and a number is -45.

  c. A number less 10 is twice the number.

3. Rajiv Puri invested part of his \$20,000 inheritance in a mutual funds account that pays 7% simple interest yearly and the rest in a certificate of deposit that pays 9% simple interest yearly. At the end of one year, Rajiv's investments earned \$1550. Find the amount he invested at each rate.

4. The number of non-business bankruptcies has increased over the years. In 2002, the number of non-business bankruptcies was 80,000 less than twice the number in 1994. If the total of non-business bankruptcies for these two years is 2,290,000 find the number of non-business bankruptcies for each year. (*Source:* American Bankruptcy Institute)

5. Graph $x - 3y = 6$ by finding and plotting intercepts.

6. Find the slope of the line whose equation is $7x + 2y = 9$.

7. Use the product rule to simplify each expressing.

  a. $4^2 \cdot 4^5$ **b.** $x^4 \cdot x^6$ **c.** $y^3 \cdot y$

  d. $y^3 \cdot y^2 \cdot y^7$ **e.** $(-5)^7 \cdot (-5)^8$ **f.** $a^2 \cdot b^2$

8. Simplify.

  a. $\dfrac{x^9}{x^7}$ **b.** $\dfrac{x^{19}y^5}{xy}$ **c.** $(x^5y^2)^3$ **d.** $(-3a^2b)(5a^3b)$

9. Subtract $(12z^5 - 12z^3 + z) - (-3z^4 + z^3 + 12z)$.

10. Subtract $(9x^2 - 6x + 2)$ from $(x + 1)$.

11. Multiply $(3a + b)^3$.

12. Multiply $(2x + 1)(5x^2 - x + 2)$.

13. Use a special product to square each binomial.

  a. $(t + 2)^2$ **b.** $(p - q)^2$

  c. $(2x + 5)^2$ **d.** $(x^2 - 7y)^2$

14. Multiply.

  a. $(x + 9)^2$ **b.** $(2x + 1)(2x - 1)$

  c. $8x(x^2 + 1)(x^2 - 1)$

15. Simplify each expression. Write results using positive exponents only.

  a. $\dfrac{1}{x^{-3}}$ **b.** $\dfrac{1}{3^{-4}}$

  c. $\dfrac{p^{-4}}{q^{-9}}$ **d.** $\dfrac{5^{-3}}{2^{-5}}$

16. Simplify. Write results with positive exponents.

    a. 5^{-3} **b.** $\dfrac{9}{x^{-7}}$ **c.** $\dfrac{11^{-1}}{7^{-2}}$

17. Divide $\dfrac{4x^2 + 7 + 8x^3}{2x + 3}$.

18. Divide $(4x^3 - 9x + 2)$ by $(x - 4)$.

19. Find the GCF of each list of numbers.

    a. 28 and 40   **b.** 55 and 21

    c. 15, 18, and 66

20. Find the GCF of $9x^2$, $6x^3$, and $21x^5$.

Factor.

21. $-9a^5 + 18a^2 - 3a$

22. $7x^6 - 7x^5 + 7x^4$

23. $3m^2 - 24m - 60$

24. $-2a^2 + 10a + 12$

25. $3x^3 + 11x + 6$

26. $10m^2 - 7m + 1$

27. $x^2 + 12x + 36$

28. $4x^2 + 12x + 3$

29. $x^2 + 4$

30. $x^2 - 4$

31. $x^3 + 8$

32. $27y^3 - 1$

33. $2x^3 + 3x^2 - 2x - 3$

34. $3x^3 + 5x^2 - 12x - 20$

35. $12m^2 - 3n^2$

36. $x^5 - x$

37. Solve: $x(2x - 7) = 4$

38. Solve: $3x^2 + 5x = 2$

39. Find the x-intercepts of the graph of $y = x^2 - 5x + 4$.

40. Find the x-intercepts of the graph of $y = x^2 - x - 6$

41. The height of a triangular sail is 2 meters less than twice the length of the base. If the sail has an area of 30 square meters, find the length of its base and the height.

42. The height of a parallelogram is 5 feet more than three times its base. If the area of the parallelogram is 182 square feet, find the length of its base and height.

43. Simplify: $\dfrac{18 - 2x^2}{x^2 - 2x - 3}$

44. Simplify: $\dfrac{2x^2 - 50}{4x^4 - 20x^3}$

45. Divide: $\dfrac{6x + 2}{x^2 - 1} \div \dfrac{3x^2 + x}{x - 1}$

46. Multiply: $\dfrac{6x^2 - 18x}{3x^2 - 2x} \cdot \dfrac{15x - 10}{x^2 - 9}$

47. Simplify: $\dfrac{(2x)^{-1} + 1}{2x^{-1} - 1}$

48. Simplify: $\dfrac{\dfrac{m}{3} + \dfrac{n}{6}}{\dfrac{m + n}{12}}$

Functions and Graphs

In Section 3.7, we introduced the notion of relation and to the notion of function, perhaps the single most important and useful concept in all of mathematics. In this chapter, we explore this concept further.

The Centers for Disease Control and Prevention (CDC), located in Atlanta, Georgia, was founded by the United States government as the primary federal agency devoted to public health. Public health officials there work constantly to provide information about current health issues. Prevention of diseases, injury, and disability is another focus of the public health specialists at the CDC.

The CDC employs approximately 8500 people in Atlanta, Washington, D.C., and at its many field sites nationwide. Positions at the CDC include medical officers, public health advisors, epidemiologists, behavioral scientists, statisticians, and information and technology specialists.

In the Spotlight on Decision Making feature on page 484, you will have the opportunity as a public health official to make a decision about a public awareness campaign concerning the eradication of mumps in the United States.

Link: www.cdc.gov *(Centers for Disease Control and Prevention)*
Source of text: (same)

8.1   *GRAPHING AND WRITING LINEAR FUNCTIONS*

Objectives

1 Graph linear functions.

2 Write an equation of a line using function notation.

3 Find equations of parallel and perpendicular lines.

1   In this section, we identify and graph **linear functions.** Let's decide what the phrase *linear function* means and later in this section a formal definition will be given. By the vertical line test, we know that all linear equations except those whose graphs are vertical lines are functions. Thus, all linear equations except those of the form $x = c$ (vertical lines) are linear functions. For example, we know from Section 3.1 that $y = 2x$ is a linear equation in two variables. Its graph is shown.

x	$y = 2x$
1	2
0	0
-1	-2

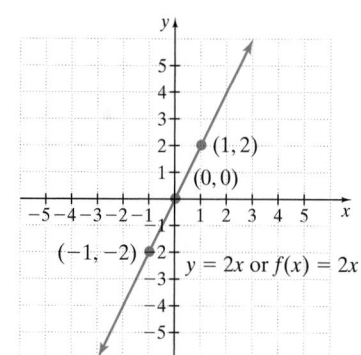

Because this graph passes the vertical line test, we know that $y = 2x$ is a function. If we want to emphasize that this equation describes a function, we may write $y = 2x$ as $f(x) = 2x$.

EXAMPLE 1

Graph $g(x) = 2x + 1$. Compare this graph with the graph of $f(x) = 2x$.

Solution   To graph $g(x) = 2x + 1$, find three ordered pair solutions. We will use the same x-values as above.

$$g(0) = 2 \cdot 0 + 1 = 1$$

$$g(-1) = 2(-1) + 1 = -1$$

$$g(1) = 2 \cdot 1 + 1 = 3$$

The graph of $g(x) = 2x + 1$ is the same as the graph of $y = 2x + 1$ since y is a function of x, or $y = g(x)$.

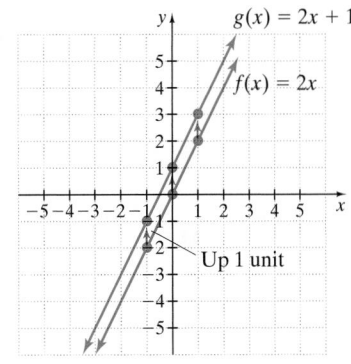

x	$f(x) = 2x$	$g(x) = 2x + 1$
0	0	1
−1	−2	−1
1	2	3

add **1**

Notice that y-values for the graph of $g(x) = 2x + 1$ are obtained by adding 1 to each y-value of each corresponding point of the graph of $f(x) = 2x$. The graph of $g(x) = 2x + 1$ is the same as the graph of $f(x) = 2x$ shifted upward 1 unit.

If a linear function is solved for y, we can easily use function notation to describe it by replacing y with $f(x)$. Recall the slope-intercept form of a linear equation, $y = mx + b$. Since this form is solved for y, we use it to define a linear function.

In general, a **linear function** is a function that can be written in the form $f(x) = mx + b$. For example, $g(x) = 2x + 1$ is in this form, with $m = 2$ and $b = 1$. Just as for $y = mx + b$, m is the slope of the line and $(0, b)$ is the y-intercept.

EXAMPLE 2

Graph the linear functions $f(x) = -3x$ and $g(x) = -3x - 6$ on the same set of axes.

Solution   To graph $f(x)$ and $g(x)$, find ordered pair solutions.

subtract **6**

x	$f(x) = -3x$	$g(x) = -3x - 6$
0	0	−6
1	−3	−9
−1	3	−3
−2	6	0

subtract **6**

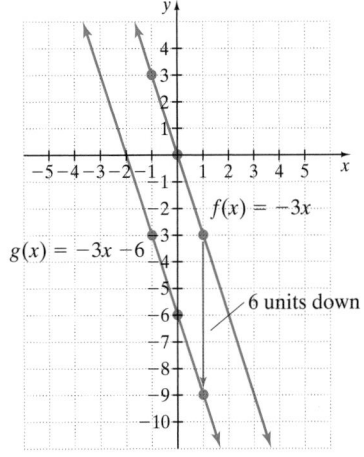

Each y-value for the graph of $g(x) = -3x - 6$ is obtained by subtracting 6 from the y-value of the corresponding point of the graph of $f(x) = -3x$. The graph of $g(x) = -3x - 6$ is the same as the graph of $f(x) = -3x$ shifted downward 6 units.

Notice that the y-intercept of the graph of $g(x) = -3x - 6$ in the preceding figure is $(0, -6)$. Once again, if *a linear function is written in the form* $f(x) = mx + b$ *or* $y = mx + b$, *the y-intercept is* $(0, b)$. This is because if x is 0, then $f(x) = mx + b$ becomes $f(0) = m \cdot 0 + b = b$, and we have the ordered pair solution $(0, b)$.

EXAMPLE 3

Find the slope and y-intercept of the graph of each linear function.

a. $f(x) = \dfrac{1}{2}x + \dfrac{3}{7}$
  b. $y = -2.5x - 3.2$

Solution   **a.** The slope of $f(x) = \dfrac{1}{2}x + \dfrac{3}{7}$ is $\dfrac{1}{2}$ and the y-intercept is $\left(0, \dfrac{3}{7}\right)$.

b. The slope of $y = -2.5x - 3.2$ is -2.5 and the y-intercept is $(0, -3.2)$.

EXAMPLE 4

Graph $x - 3y = 6$.

Solution   This equation does describe a linear function since it is a linear equation whose graph is *not* a vertical line. Let's graph by finding intercepts.

Let $y = 0$ to find the x-intercept and $x = 0$ to find the y-intercept.

$$\begin{array}{ll} \text{If } y = 0 \quad \text{then} & \text{If } x = 0 \quad \text{then} \\ x - 3(0) = 6 & 0 - 3y = 6 \\ x - 0 = 6 & -3y = 6 \\ x = 6 & y = -2 \end{array}$$

The x-intercept is $(6, 0)$ and the y-intercept is $(0, -2)$. We find a third ordered pair solution to check our work. If we let $y = -1$, then $x = 3$. Plot the points $(6, 0)$, $(0, -2)$, and $(3, -1)$. The graph of $x - 3y = 6$ is the line drawn through these points, as shown. We can see that this graph passes the vertical line test, as expected since the equation describes a function.

x	y
6	0
0	-2
3	-1

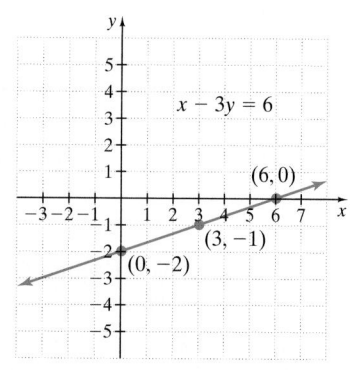

2   Once again, notice that the equation $x - 3y = 6$ describes a linear function—"linear" because it's a linear equation whose graph is a line and "function" because the graph passes the vertical line test.

If we want to emphasize that the equation $x - 3y = 6$ from Example 4 describes a function, first solve the equation for y.

$$x - 3y = 6$$

$$-3y = -x + 6 \qquad \text{Subtract } x \text{ from both sides.}$$

$$\frac{-3y}{-3} = \frac{-x}{-3} + \frac{6}{-3} \qquad \text{Divide both sides by } -3.$$

$$y = \frac{1}{3}x - 2 \qquad \text{Simplify.}$$

Next, let
$$y = f(x)$$

$$f(x) = \frac{1}{3}x - 2$$

> ### Helpful Hint
> Any linear equation that describes a function can be written using function notation. To do so,
>
> **1.** Solve the equation for y and then
>
> **2.** Replace y with $f(x)$, as we did above.

Let's review writing an equation of a line through two points. In the next example, we write the equation using function notation.

EXAMPLE 5

Find an equation of the line through points $(4, 0)$ and $(-4, -5)$. Write the equation using function notation.

Solution   First, find the slope of the line.

$$m = \frac{-5 - 0}{-4 - 4} = \frac{-5}{-8} = \frac{5}{8}$$

Next, make use of the point–slope form. Replace (x_1, y_1) by either $(4, 0)$ or $(-4, -5)$ in the point–slope equation. We will choose the point $(4, 0)$. The line through $(4, 0)$ with slope $\frac{5}{8}$ is

$$y - y_1 = m(x - x_1) \qquad \text{Point–slope form.}$$

$$y - 0 = \frac{5}{8}(x - 4) \qquad \text{Let } m = \frac{5}{8} \text{ and } (x_1, y_1) = (4, 0).$$

$$8y = 5(x - 4) \qquad \text{Multiply both sides by 8.}$$

$$8y = 5x - 20 \qquad \text{Apply the distributive property.}$$

To write the equation using function notation, we solve for y, then replace y with $f(x)$.

$$8y = 5x - 20$$

$$y = \frac{5}{8}x - \frac{20}{8} \qquad \text{Divide both sides by 8.}$$

$$f(x) = \frac{5}{8}x - \frac{5}{2} \qquad \text{Write using function notation.}$$

3   Next, we write equations of parallel and perpendicular lines.

△ **EXAMPLE 6**

Find an equation of the line containing the point (4, 4) and parallel to the line $2x + 3y = -6$. Write the equation in standard form and then using function notation.

Solution   Because the line we want to find is *parallel* to the line $2x + 3y = -6$, the two lines must have equal slopes. Find the slope of $2x + 3y = -6$ by writing it in the form $y = mx + b$. In other words, solve the equation for y.

$$2x + 3y = -6$$

$$3y = -2x - 6 \qquad \text{Subtract } 2x \text{ from both sides.}$$

$$y = \frac{-2x}{3} - \frac{6}{3} \qquad \text{Divide by 3.}$$

$$y = -\frac{2}{3}x - 2 \qquad \text{Write in slope-intercept form.}$$

The slope of this line is $-\frac{2}{3}$. Thus, a line parallel to this line will also have a slope of $-\frac{2}{3}$.

The equation we are asked to find describes a line containing the point (4, 4) with a slope of $-\frac{2}{3}$. We use the point-slope form.

$$y - y_1 = m(x - x_1)$$

$$y - 4 = -\frac{2}{3}(x - 4) \qquad \text{Let } m = -\frac{2}{3}, \, x_1 = 4, \text{ and } y_1 = 4.$$

$$3(y - 4) = -2(x - 4) \qquad \text{Multiply both sides by 3.}$$

$$3y - 12 = -2x + 8 \qquad \text{Apply the distributive property.}$$

$$2x + 3y = 20 \qquad \text{Write in standard form.}$$

To write the equation using function notation, we solve for y.

$$3y = -2x + 20 \qquad \text{Subtract } 2x \text{ from both sides.}$$

$$y = -\frac{2}{3}x + \frac{20}{3} \qquad \text{Divide both sides by 3.}$$

$$f(x) = -\frac{2}{3}x + \frac{20}{3} \qquad \text{Write using function notation.}$$

> **Helpful Hint**
>
> From the previous example, multiply both sides of the equation $2x + 3y = 20$ by -1 and it becomes $-2x - 3y = -20$. Both equations are in standard form, and their graphs are the same as $f(x) = -\frac{2}{3}x + \frac{20}{3}$.

EXAMPLE 7

Write a function that describes the line containing the point $(4, 4)$ and is perpendicular to the line $2x + 3y = -6$.

Solution   In the previous example, we found that the slope of the line $2x + 3y = -6$ is $-\dfrac{2}{3}$. A line perpendicular to this line will have a slope that is the negative reciprocal of $-\dfrac{2}{3}$, or $\dfrac{3}{2}$.

From the point-slope equation, we have

$$y - y_1 = m(x - x_1)$$

$$y - 4 = \frac{3}{2}(x - 4) \qquad \text{Let } x_1 = 4, y_1 = 4 \text{ and } m = \frac{3}{2}.$$

$$2(y - 4) = 3(x - 4) \qquad \text{Multiply both sides by 2.}$$

$$2y - 8 = 3x - 12 \qquad \text{Apply the distributive property.}$$

$$2y = 3x - 4 \qquad \text{Add 8 to both sides.}$$

$$y = \frac{3}{2}x - 2 \qquad \text{Divide both sides by 2.}$$

$$f(x) = \frac{3}{2}x - 2 \qquad \text{Write using function notation.}$$

Forms of Linear Equations

$Ax + By = C$	**Standard form** of a linear equation A and B are not both 0.
$y = mx + b$	**Slope–intercept form** of a linear equation The slope is m, and the y-intercept is $(0, b)$.
$y - y_1 = m(x - x_1)$	**Point–slope form** of a linear equation The slope is m, and (x_1, y_1) is a point on the line.
$y = c$	**Horizontal line** The slope is 0, and the y-intercept is $(0, c)$.
$x = c$	**Vertical line** The slope is undefined and the x-intercept is $(c, 0)$.
$f(x) = mx + b$	**Linear function form** The slope is m and the y-intercept is $(0, b)$.

Parallel and Perpendicular Lines

Nonvertical parallel lines have the same slope. The product of the slopes of two nonvertical perpendicular lines is -1.

Spotlight on

DECISION
of MAKING

Suppose you are a public health official. In 1993, the International Task Force for Disease Eradication (ITFDE) identified mumps as one of six infectious diseases that could probably be eradicated worldwide with current technology. The ITFDE defined "eradication" as reducing the incidence of a disease to zero. Does the graph of reported mumps cases in the United States support the possibility of U.S. mumps eradication? Explain.

Suppose U.S. officials would like to see mumps eradicated by 2010. If this goal does not currently seem possible, your department will increase eradication efforts with the launch of a new public awareness campaign. Will the new public awareness campaign be necessary? (*Hint:* Use the data for the years 2000 and 2001 to help you decide.)

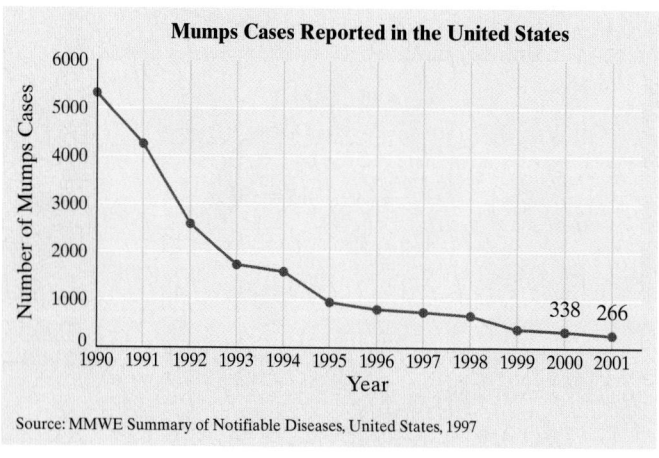

Mumps Cases Reported in the United States

Source: MMWE Summary of Notifiable Diseases, United States, 1997

MENTAL MATH

State the slope and the *y*-intercept of the graph of each function. See Example 3.

1. $f(x) = -4x + 12$

2. $f(x) = \frac{2}{3}x - \frac{7}{2}$

3. $f(x) = 5x$

4. $f(x) = -x$

5. $f(x) = \frac{1}{2}x + 6$

6. $f(x) = -\frac{2}{3}x + 5$

Decide whether the lines, when graphed, are parallel, perpendicular, or neither.

7. $f(x) = 12x + 6$
$g(x) = 12x - 2$

8. $f(x) = -5x + 8$
$g(x) = -5x - 8$

9. $f(x) = -9x + 3$
$g(x) = \frac{3}{2}x - 7$

10. $f(x) = 2x - 12$
$g(x) = \frac{1}{2}x - 6$

EXERCISE SET 8.1

| STUDY | CD/ | PH MATH | MathXL®Tutorials | MathXL® | MyMathLab® |
| GUIDE/SSM | VIDEO | TUTOR CENTER | ON CD | | |

Graph each linear function. See Examples 1 and 2.

1. $f(x) = -2x$

2. $f(x) = 2x$

3. $f(x) = -2x + 3$

4. $f(x) = 2x + 6$

5. $f(x) = \frac{1}{2}x$

6. $f(x) = \frac{1}{3}x$

7. $f(x) = \frac{1}{2}x - 4$

8. $f(x) = \frac{1}{3}x - 2$

The graph of $f(x) = 5x$ follows. Use this graph to match each linear function with its graph. See Examples 1 through 3.

$y = 5x$

A

B

C

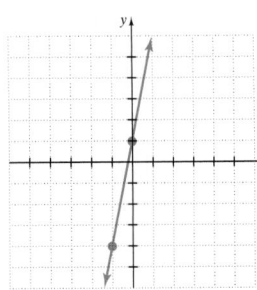

D

9. $f(x) = 5x - 3$

10. $f(x) = 5x - 2$

11. $f(x) = 5x + 1$

12. $f(x) = 5x + 3$

Graph each linear function by finding x-and y-intercepts. Then write each equation using function notation. See Example 4.

13. $x - y = 3$

14. $x - y = -4$

15. $x = 5y$

16. $2x = y$

17. $-x + 2y = 6$

18. $x - 2y = -8$

19. $2x - 4y = 8$

20. $2x + 3y = 6$

21. In your own words, explain how to find x- and y-intercepts.

22. Explain why it is a good idea to use three points to graph a linear equation.

Find an equation of the line passing through the given points. Use function notation to write the equation. See Example 5.

23. $(2, 0), (4, 6)$

24. $(3, 0), (7, 8)$

25. $(-2, 5), (-6, 13)$

26. $(7, -4), (2, 6)$

27. $(-3, -8), (-6, -9)$

28. $(8, -3), (4, -8)$

Find an equation of each line. Write the equation using function notation. See Examples 6 and 7.

29. Through $(3, 8)$; parallel to $f(x) = 4x - 2$

30. Through $(1, 5)$; parallel to $f(x) = 3x - 4$

31. Through $(2, -5)$; perpendicular to $3y = x - 6$

32. Through $(-4, 8)$; perpendicular to $2x - 3y = 1$

33. Through $(-2, -3)$; parallel to $3x + 2y = 5$

34. Through $(-2, -3)$; perpendicular to $3x + 2y = 5$

MIXED PRACTICE

Graph each linear equation.

35. $x + 2y = 8$

36. $x - 3y = 3$

37. $3x + 5y = 7$

38. $3x - 2y = 5$

39. $5 = 6x - y$

40. $4 = x - 3y$

41. $-x + 10y = 11$

42. $-x + 9 = -y$

43. $x + 3 = 0$

44. $f(x) = \dfrac{3}{2}$

45. $f(x) = \dfrac{3}{4}x + 2$

46. $f(x) = \dfrac{4}{3}x + 2$

47. $f(x) = x$

48. $f(x) = -x$

49. $f(x) = \dfrac{1}{2}x$

50. $f(x) = -2x$

51. $f(x) = 4x - \dfrac{1}{3}$

52. $f(x) = -3x + \dfrac{3}{4}$

Find the equation of each line. Write the equation in standard form and using function notation.

53. Slope 2; through $(-2, 3)$

54. Slope 3; through $(-4, 2)$

55. Through $(-7, -4)$ and $(0, -6)$

56. Through $(2, -8)$ and $(-4, -3)$

57. Through $(6, -2)$; parallel to the line $2x + 4y = 9$

58. Through $(8, -3)$; parallel to the line $6x + 2y = 5$

59. Through $(6, 1)$; parallel to the line $8x - y = 9$

60. Through $(3, 5)$; perpendicular to the line $2x - y = 8$

REVIEW AND PREVIEW

If $f(x) = x^2$, $g(x) = |x|$, and $h(x) = 3x - 4$, find each function value. See Section 3.7.

61. $f(3)$

62. $f(-3)$

63. $g(-2)$

64. $g(2)$

65. $h(1)$

66. $h(-1)$

Concept Extensions

Solve.

67. Broyhill Furniture found that it takes 2 hours to manufacture each table for one of its special dining room sets. Each chair takes 3 hours to manufacture. A total of 1500 hours is available to produce tables and chairs of this style. The linear equation that models this situation is $2x + 3y = 1500$, where x represents the number of tables produced and y the number of chairs produced.

   a. Complete the ordered pair solution (0,) of this equation. Describe the manufacturing situation this solution corresponds to.

   b. Complete the ordered pair solution (, 0) for this equation. Describe the manufacturing situation this solution corresponds to.

   c. If 50 tables are produced, find the greatest number of chairs the company can make.

68. While manufacturing two different camera models, Kodak found that the basic model costs \$55 to produce, whereas the deluxe model costs \$75. The weekly budget for these two models is limited to \$33,000 in production costs. The linear equation that models this situation is $55x + 75y = 33,000$, where x represents the number of basic models and y the number of deluxe models.

   a. Complete the ordered pair solution (0,) of this equation. Describe the manufacturing situation this solution corresponds to.

   b. Complete the ordered pair solution (, 0) of this equation. Describe the manufacturing situation this solution corresponds to.

   c. If 350 deluxe models are produced, find the greatest number of basic models that can be made in one week.

69. The cost of renting a car for a day is given by the linear function $C(x) = 0.2x + 24$, where $C(x)$ is in dollars and x is the number of miles driven.

   a. Find the cost of driving the car 200 miles.

   b. Graph $C(x) = 0.2x + 24$.

   c. How can you tell from the graph of $C(x)$ that as the number of miles driven increases, the total cost increases also?

70. The cost of renting a piece of machinery is given by the linear function $C(x) = 4x + 10$, where $C(x)$ is in dollars and x is given in hours.

   a. Find the cost of renting the piece of machinery for 8 hours.

   b. Graph $C(x) = 4x + 10$.

   c. How can you tell from the graph of $C(x)$ that as the number of hours increases, the total cost increases also?

71. The yearly cost of tuition (in state) and required fees for attending a public two-year college full time can be estimated by the linear function $f(x) = 53.6x + 849.88$, where x is the number of years after 1990 and $f(x)$ is the total cost. (*Source:* U.S. National Center for Education Statistics)

   a. Use this function to approximate the yearly cost of attending a two-year college in the year 2010. [*Hint:* Find $f(20)$.]

   b. Use the given function to predict in what year the yearly cost of tuition and required fees will exceed \$2000. [*Hint:* Let $f(x) = 2000$ and solve for x.]

   c. Use this function to approximate the yearly cost of attending a two-year college in the present year. If you attend a two-year college, is this amount greater than or less than the amount that is currently charged by the college that you attend?

72. The yearly cost of tuition (in state) and required fees for attending a public four-year college can be estimated by the linear function $f(x) = 174.4x + 2074.38$, where x is the number of years after 1990 and $f(x)$ is the total cost in dollars. (*Source:* U.S. National Center for Education Statistics)

   a. Use this function to approximate the yearly cost of attending a four-year college in the year 2010. [*Hint:* Find $f(20)$.]

   b. Use the given function to predict in what year the yearly cost of tuition and required fees will exceed \$5000. [*Hint:* Let $f(x) = 5000$ and solve for x.]

   c. Use this function to approximate the yearly cost of attending a four-year college in the present year. If you attend a four-year college, is this amount greater than or less than the amount that is currently charged by the college that you attend?

The dosage in milligrams D of Ivermectin, a heartworm preventive, for a dog who weighs x pounds is given by

$$D(x) = \frac{136}{25}x;$$

73. Find the proper dosage for a dog that weighs 30 pounds.

74. Find the proper dosage for a dog that weighs 50 pounds.

75. The per capita consumption (in pounds) of all poultry in the United States is approximated by the function $C(x) = 1.69x + 87.54$, where x is the number of years since 1995. (*Source*: Based on actual and estimated data from the Economic Research Service, U.S. Department of Agriculture, 1995–2002)

   a. Find and interpret $C(5)$.

   b. Estimate the per capita consumption of all poultry in the United States in 2002.

76. The average length of U.S. hospital stays has been decreasing, following the equation $y = -0.09x + 8.02$ where x is the number of years since 1970. (*Source*: National Center for Health Statistics)

   a. What was the length of the average hospital stay in 1995?

   b. If this trend continues, what will the average length be in 2007?

77. In 2002, the median price of an existing home in the United States was approximately $147,802. In 1999, the median price of an existing home was $133,300. Let y be the median price of an existing home in the year x, where $x = 0$ represents 1999. (*Source*: National Association of REALTORS®)

   a. Write a linear function that models the median existing home price in terms of the year x. [*Hint*: The line must pass through the points $(0, 133{,}300)$ and $(3, 147{,}802)$]

   b. Use this equation to predict the median existing home price for the year 2008.

   c. Interpret the slope of the equation found in part **a**.

78. The number of births (in thousands) in the United States in 2000 was 5380. The number of births (in thousands) in the United States in 1997 was 3895. Let y be the number of births (in thousands) in the year x, where $x = 0$ represents 1997. (*Source*: National Center for Health Statistics)

   a. Write a linear function that models the number of births (in thousands) in terms of the year x. (See hint for Exercise 77a.)

   b. Use this function to predict the number of births in the United States for the year 2010.

   c. Interpret the slope of the equation in part a.

79. The number of people employed in the United States as medical assistants was 757 thousand in 2000. By the year 2010, this number is expected to rise to 1052 thousand. Let y be the number of medical assistants (in thousands) employed in the United States in the year x, where $x = 0$ represents 2000. (*Source*: Bureau of Labor Statistics)

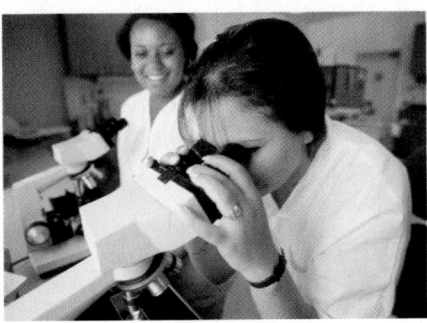

   a. Write a linear function that models the number of people (in thousands) employed as medical assistants in the year x. (See hint for Exercise 77a.)

   b. Use this function to estimate the number of people who will be employed as medical assistants in the year 2004.

80. The number of people employed in the United States as systems analysts was 431 thousand in 2000. By the year 2010, this number is expected to rise to 689 thousand. Let y be the number of systems analysts (in thousands) employed in the United States in the year x, where $x = 0$ represents 2000. (*Source*: Bureau of Labor Statistics)

   a. Write a linear function that models the number of people (in thousands) employed as systems analysts in the year x. (See hint for Exercise 77a.)

   b. Use this function to estimate the number of people who will be employed as systems analysts in the year 2008.

Example:

Find an equation of the perpendicular bisector of the line segment whose endpoints are $(2, 6)$ and $(0, -2)$.

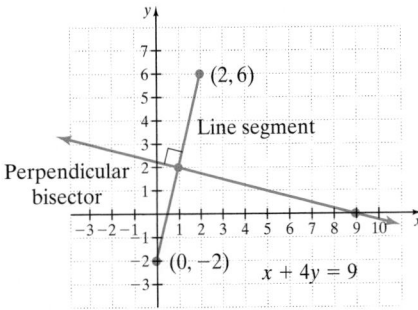

Solution:

A perpendicular bisector is a line that contains the midpoint of the given segment and is perpendicular to the segment.

Step 1: The midpoint of the segment with endpoints $(2, 6)$ and $(0, -2)$ is $(1, 2)$.

Step 2: The slope of the segment containing points $(2, 6)$ and $(0, -2)$ is 4.

Step 3: A line perpendicular to this line segment will have slope of $-\frac{1}{4}$.

Step 4: The equation of the line through the midpoint $(1, 2)$ with a slope of $-\frac{1}{4}$ will be the equation of the perpendicular bisector. This equation in standard form is $x + 4y = 9$.

Find an equation of the perpendicular bisector of the line segment whose endpoints are given. See the previous example.

△ **81.** $(3, -1); (-5, 1)$ **82.** $(-6, -3); (-8, -1)$

△ **83.** $(-2, 6); (-22, -4)$ **84.** $(5, 8); (7, 2)$

△ **85.** $(2, 3); (-4, 7)$ **86.** $(-6, 8); (-4, -2)$

📱 *Use a graphing calculator to verify the results of each exercise.*

87. Exercise 9   **88.** Exercise 10

89. Exercise 17   **90.** Exercise 18

91. The graph of $f(x)$ or $y = -4x$ is given below.

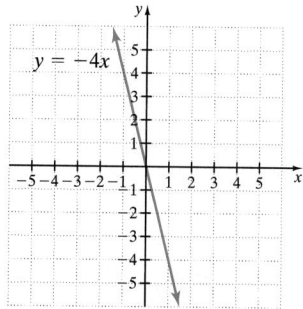

Without actually graphing, describe the shape and location of

a. $y = -4x + 2$ **b.** $y = -4x - 5$

8.2 GRAPHING NONLINEAR FUNCTIONS

Objectives

1. Review function notation.
2. Graph nonlinear functions.
3. Shift graphs of functions.

In the previous section, we studied linear equations that described functions. Not all equations in two variables are linear equations, and not all graphs of equations in two variables are lines. In Chapter 3, we saw graphs of nonlinear equations, some of which were functions since they passed the vertical line test. In this section, we study the functions whose graphs are not lines. First, let's review the function notation.

1 Suppose we have a function f such that $f(2) = -1$. Recall this means that when $x = 2, f(x)$, or $y = -1$. Thus, the graph of f passes through $(2, -1)$.

> **Helpful Hint**
>
> Remember that $f(x)$ is a special symbol in mathematics used to denote a function. The symbol $f(x)$ is read "f of x." It does *not* mean $f \cdot x$ (f times x).

✔ **CONCEPT CHECK**

Suppose $y = f(x)$ and we are told that $f(3) = 9$. Which is not true?

a. When $x = 3$, $y = 9$.
b. A possible function is $f(x) = x^2$.
c. A point on the graph of the function is $(3, 9)$.
d. A possible function is $f(x) = 2x + 4$.

If it helps, think of a function, f, as a machine that has been programmed with a certain correspondence or rule. An input value (a member of the domain) is then fed into the machine, the machine does the correspondence or rule and the result is the output (a member of the range).

EXAMPLE 1

Given the graphs of the functions f and g, find each function value by inspecting the graphs.

 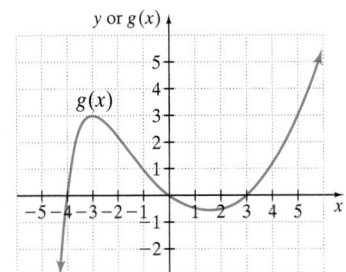

a. $f(4)$ **b.** $f(-2)$ **c.** $g(5)$ **d.** $g(0)$
e. Find all x-values such that $f(x) = 1$.
f. Find all x-values such that $g(x) = 0$.

Solution

a. To find $f(4)$, find the y-value when $x = 4$. We see from the graph that when $x = 4$, y or $f(x) = 2$. Thus, $f(4) = 2$.

b. $f(-2) = 1$ from the ordered pair $(-2, 1)$.

c. $g(5) = 3$ from the ordered pair $(5, 3)$.

d. $g(0) = 0$ from the ordered pair $(0, 0)$.

e. To find x-values such that $f(x) = 1$, I'm looking for any ordered pairs on the graph of f whose $f(x)$ or y-value is 1. They are $(2, 1)$ and $(-2, 1)$. Thus $f(2) = 1$ and $f(-2) = 1$. The x-values are 2 and -2.

f. Find ordered pairs on the graph of g whose $g(x)$ or y-value is 0. They are $(3, 0)$ $(0, 0)$ and $(-4, 0)$. Thus $g(3) = 0$, $g(0) = 0$, and $g(-4) = 0$. The x-values are 3, 0, and -4.

Concept Check Answer:
d

2 Now let's practice graphing nonlinear functions.

EXAMPLE 2

Graph $f(x) = x^2$.

Solution This equation is not linear because the x^2 term does not allow us to write it in the form $Ax + By = C$. Its graph is not a line. We begin by finding ordered pair solutions. Because $f(x) = y$, feel free to think of this equation as $f(x) = x^2$ or $y = x^2$. This graph is solved for y, we choose x-values and find corresponding y-values.

If $x = -3$, then $f(-3) = (-3)^2$, or 9.

If $x = -2$, then $f(-2) = (-2)^2$, or 4.

If $x = -1$, then $f(-1) = (-1)^2$, or 1.

If $x = 0$, then $f(0) = 0^2$, or 0.

If $x = 1$, then $f(1) = 1^2$, or 1.

If $x = 2$, then $f(2) = 2^2$, or 4.

If $x = 3$, then $f(3) = 3^2$, or 9.

x	y or $f(x)$
-3	9
-2	4
-1	1
0	0
1	1
2	4
3	9

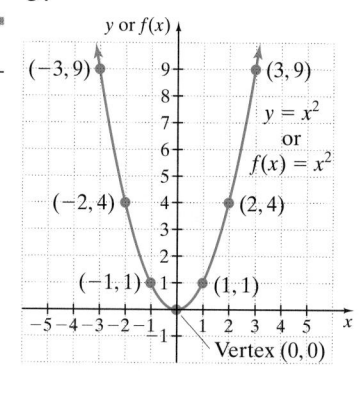

Study the table a moment and look for patterns. Notice that the ordered pair solution $(0, 0)$ contains the smallest y-value because any other x-value squared will give a positive result. This means that the point $(0, 0)$ will be the lowest point on the graph. Also notice that all other y or $f(x)$-values correspond to two different x-values. For example, $3^2 = 9$ and also $(-3)^2 = 9$. This means that the graph will be a mirror image of itself across the y-axis. Connect the plotted points with a smooth curve to sketch its graph.

This curve is given a special name, a parabola. We will study more about parabolas in later chapters.

EXAMPLE 3

Graph $f(x) = |x|$.

Solution This is not a linear equation since it cannot be written in the form $Ax + By = C$. Its graph is not a line. Because we do not know the shape of this graph, we find many ordered pair solutions. We will choose x-values and substitute to find corresponding y-values.

If $x = -3$, then $f(-3) = |-3|$, or 3.

If $x = -2$, then $f(-2) = |-2|$, or 2.

If $x = -1$, then $f(-1) = |-1|$, or 1.

If $x = 0$, then $f(0) = |0|$, or 0.

If $x = 1$, then $f(1) = |1|$, or 1.

If $x = 2$, then $f(2) = |2|$, or 2.

If $x = 3$, then $f(3) = |3|$, or 3.

x	y or $f(x)$
-3	3
-2	2
-1	1
0	0
1	1
2	2
3	3

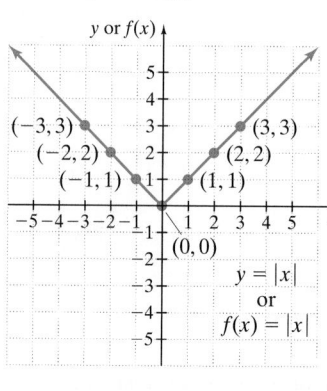

Again, study the table of values for a moment and notice any patterns.

From the plotted ordered pairs, we see that the graph of this absolute value equation is V-shaped.

3   Recall from Section 8.1 that the graph of $g(x) = 2x + 1$, for example, is the same as the graph of $f(x) = 2x$ shifted upward 1 unit. Let's see if this same shifting occurs for nonlinear functions.

EXAMPLE 4

Graph $g(x) = |x| + 2$.

Solution   Recall that the graph of $f(x) = |x|$ is V-shaped with the point of the V at $(0,0)$. Let's see if the graph of $g(x) = |x| + 2$ is the same as the graph of $f(x) = |x|$ moved upward 2 units.

x	y or $f(x)$
-3	5
-2	4
-1	3
0	2
1	3
2	4
3	5

If $x = -3, f(-3) = |-3| + 2 = 5$

If $x = -2, f(-2) = |-2| + 2 = 4$

If $x = -1, f(-1) = |-1| + 2 = 3$

If $x = 0, f(0) = |0| + 2 = 2$

If $x = 1, f(1) = |1| + 2 = 3$

If $x = 2, f(2) = |2| + 2 = 4$

If $x = 3, f(3) = |3| + 2 = 5$

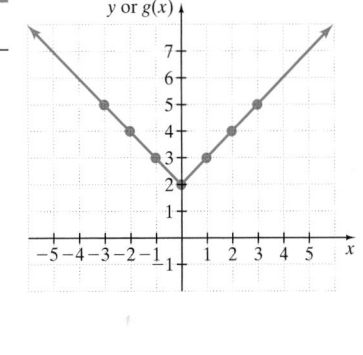

Indeed, notice that the graph of $g(x) = |x| + 2$ is the same as the graph of $f(x) = |x|$ moved upward 2 units. This is true in general.

For any function $f(x)$, the graph of $f(x) + k$ is the same as the graph of $f(x)$ shifted k units upward if k is positive and $|k|$ units downward if k is negative.

In Section 11.5, we study shifting and reflecting of $f(x) = x^2$ further.

Graphing Calculator Explorations

It is possible to use a graphing calculator to sketch the graph of more than one equation on the same set of axes. For example, graph the functions $f(x) = x^2$ and $g(x) = x^2 + 4$ on the same set of axes.

To graph on the same set of axes, press the $\boxed{Y=}$ key and enter the equations on the first two lines.

$$Y_1 = x^2$$
$$Y_2 = x^2 + 4$$

Then press the $\boxed{\text{GRAPH}}$ key as usual. The screen should look like this.

Notice that the graph of y or $g(x) = x^2 + 4$ is the graph of $y = x^2$ moved 4 units upward.

Graph each pair of functions on the same set of axes. Describe the similarities and differences in their graphs.

1. $f(x) = |x|$
   $g(x) = |x| + 1$

2. $f(x) = x^2$
   $h(x) = x^2 - 5$

3. $f(x) = x$
   $H(x) = x - 6$

4. $f(x) = |x|$
   $G(x) = |x| + 3$

5. $f(x) = -x^2$
   $F(x) = -x^2 + 7$

6. $f(x) = x$
   $F(x) = x + 2$

EXERCISE SET 8.2

STUDY GUIDE/SSM · CD/VIDEO · PH MATH TUTOR CENTER · MathXL®Tutorials ON CD · MathXL® · MyMathLab®

Use the graph of the following function f(x) to find each value. See Example 1.

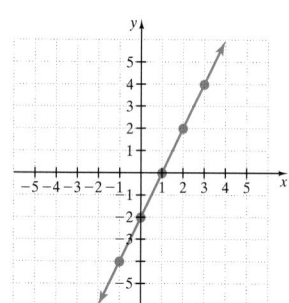

1. $f(1)$
2. $f(0)$
3. $f(-1)$
4. $f(2)$
5. Find x such that $f(x) = 4$.
6. Find x such that $f(x) = -6$.

Use the graph of the functions below to answer Exercises 7 through 18. See Example 1.

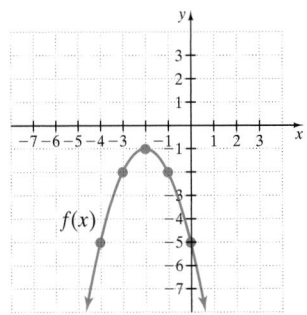

7. If $f(1) = -10$ write the corresponding ordered pair.

8. If $f(-5) = -10$, write the corresponding ordered pair.

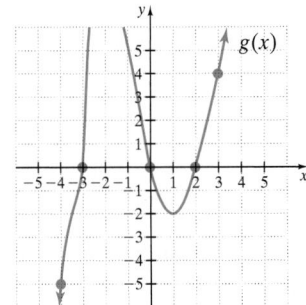

9. If $g(4) = 56$, write the corresponding ordered pair.

10. If $g(-2) = 8$, write the corresponding ordered pair.

11. Find $f(-1)$.

12. Find $f(-2)$.

13. Find $g(2)$.

14. Find $g(-4)$.

15. Find all values of x such that $f(x) = -5$.

16. Find all values of x such that $f(x) = -2$.

17. Find all values of x such that $g(x) = 4$.

18. Find all values of x such that $g(x) = 0$.

MIXED PRACTICE

Graph each function. See Examples 2 through 4.

19. $f(x) = x^2 + 3$

20. $g(x) = -2x^2$

21. $h(x) = |x| - 2$

22. $f(x) = |x - 2|$

23. $g(x) = 2x^2$

24. $h(x) = 5|x|$

25. $f(x) = 5x - 1$

26. $g(x) = -3x + 2$

27. $h(x) = |x + 1| + 3$

28. $f(x) = 3x^2 + x$

29. $g(x) = -|x|$

30. $h(x) = -4|x|$

31. $h(x) = x^3$

32. $f(x) = x^3 - 2$

Recall that the graph of $y = |x|$ is

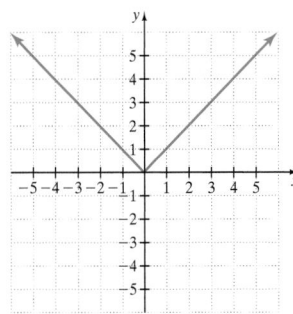

Without actually graphing, match each equation with its graph.

a. $f(x) = |x| - 1$

b. $f(x) = |x| + 1$

c. $f(x) = |x| - 3$

d. $f(x) = |x| + 3$

33.

34.

35.

36.

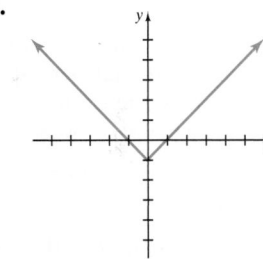

37. What is the greatest number of x-intercepts that a function may have? Explain your answer.

38. What is the greatest number of y-intercepts that a function may have? Explain your answer.

The nonlinear function $A(r) = \pi r^2$ may be used to find the area of a circle if we are given its radius.

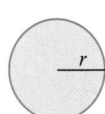

39. Find the area of a circle whose radius is 5 centimeters. (Do not approximate π.)

40. Find the area of a circular garden whose radius is 8 feet. (Do not approximate π.)

The nonlinear function $V(x) = x^3$ may be used to find the volume of a cube if we are given the length x of a side.

△ **41.** Find the volume of a cube whose side is 14 inches.

△ **42.** Find the volume of a die whose side is 1.7 centimeters.

REVIEW AND PREVIEW

Perform each operation. See Sections 5.2 and 5.3

43. $(5x - 7) - (9x - 1)$

44. $(5x - 7)(9x - 1)$

45. $(5x - 7) + (9x - 1)$

46. $(x^2 - 2x + 1) - (3x^2 + 10)$

Concept Extensions

For Exercises 47 through 50, match each description with the graph that best illustrates it.

47. Moe worked 40 hours per week until the fall semester started. He quit and didn't work again until he worked 60 hours a week during the holiday season starting mid-December break.

48. Kawana worked 40 hours a week for her father during the summer. She slowly cut back her hours to not working at all during the fall semester. During the holiday season in December, she started working again and increased her hours to 60 hours per week.

49. Wendy worked from July through February, never quitting. She worked between 10 and 30 hours per week.

50. Bartholomew worked from July through February. The rest of the time, he worked between 10 and 40 hours per week. During the holiday season between mid-November and the beginning of January, he worked 40 hours per week.

a.

b.

c.

d.

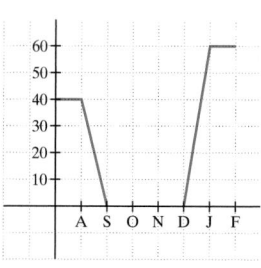

This graph of a nonlinear function shows hourly minimum wages and the years it increased. Use this graph for Exercises 51 through 54.

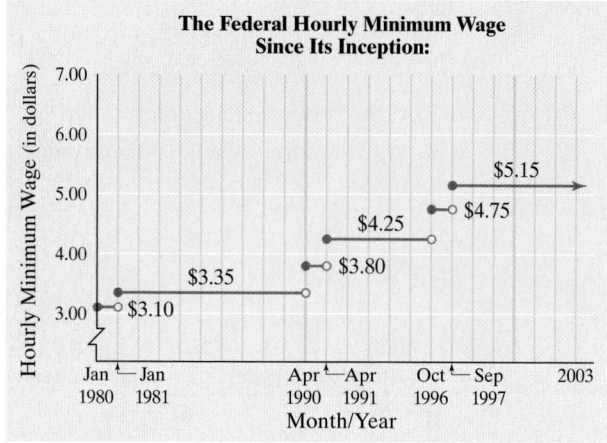

51. What was the first year that the minimum hourly wage rose above $4.00?

52. What was the first year that the minimum hourly wage rose above $5.00?

53. Why do you think that this graph is shaped the way it is?

54. The federal hourly minimum wage started in 1938 at $0.25. How much has it increased by in 2003?

55. Graph $f(x) = x^2 - 4x + 7$. Let $x = 0, 1, 2, 3, 4$ to generate ordered pair solutions.

56. Graph $f(x) = x^2 + 2x + 3$. Let $x = -3, -2, -1, 0, 1$ to generate ordered pair solutions.

57. The function $f(x) = [x]$ is called the greatest integer function and its graph is similar to the graph above. The value of $[x]$ is the greatest integer less than or equal to x. For example, $f(1.5) = [1.5] = 1$ since the greatest integer ≤ 1.5 is 1. Find the graph of $f(x) = [x]$.

SUMMARY ON FUNCTIONS

Find the slope and y-intercept of the graph of each function.

1. $f(x) = 3x - 5$

2. $f(x) = \frac{5}{2}x - \frac{7}{2}$

Determine whether each pair of lines is parallel, perpendicular, or neither.

3. $f(x) = 8x - 6$ **4.** $f(x) = \frac{2}{3}x + 1$

   $g(x) = 8x + 6$ $2y + 3x = 1$

Find the equation of each line. Write the equation in the form $f(x) = mx + b$.

5. Through $(1, 6)$ and $(5, 2)$

6. Through $(2, -8)$ and $(-6, -5)$

7. Through $(-1, -5)$; parallel to $3x - y = 5$

8. Through $(0, 4)$; perpendicular to

   $4x - 5y = 10$

9. Through $(2, -3)$; perpendicular to

   $4x + y = \frac{2}{3}$

10. Through $(-1, 0)$; parallel to $5x + 2y = 2$

Determine whether each function is linear or not. Then graph the function.

11. $f(x) = 4x - 2$

12. $f(x) = 6x - 5$

13. $g(x) = |x| + 3$

14. $h(x) = |x| + 2$

15. $f(x) = 2x^2$

16. $F(x) = 3x^2$

17. $h(x) = x^2 - 3$

18. $G(x) = x^2 + 3$

19. $F(x) = -2x$

20. $H(x) = -3x$

21. $G(x) = |x + 2|$

22. $g(x) = |x - 1|$

23. $f(x) = \frac{1}{3}x - 1$

24. $f(x) = \frac{1}{2}x - 3$

25. $g(x) = -\frac{3}{2}x + 1$

26. $G(x) = -\frac{2}{3}x + 1$

8.3   *THE ALGEBRA OF FUNCTIONS; COMPOSITE FUNCTIONS*

Objectives

1 Add, subtract, multiply, and divide functions.

2 Construct composite functions.

1   As we have seen in earlier chapters, it is possible to add, subtract, multiply, and divide functions. Although we have not stated it as such, the sums, differences, products, and quotients of functions are themselves functions. For example, if $f(x) = 3x$ and $g(x) = x + 1$, their product, $f(x) \cdot g(x) = 3x(x + 1) = 3x^2 + 3x$, is a new function. We can use the notation $(f \cdot g)(x)$ to denote this new function. Finding the sum, difference, product, and quotient of functions to generate new functions is called the **algebra of functions.**

Algebra of Functions

Let f and g be functions. New functions from f and g are defined as follows.

Sum	$(f + g)(x) = f(x) + g(x)$
Difference	$(f - g)(x) = f(x) - g(x)$
Product	$(f \cdot g)(x) = f(x) \cdot g(x)$
Quotient	$\left(\dfrac{f}{g}\right)(x) = \dfrac{f(x)}{g(x)}, g(x) \neq 0$

EXAMPLE 1

If $f(x) = x - 1$ and $g(x) = 2x - 3$, find

a. $(f + g)(x)$ **b.** $(f - g)(x)$ **c.** $(f \cdot g)(x)$ **d.** $\left(\dfrac{f}{g}\right)(x)$

Solution Use the algebra of functions and replace $f(x)$ by $x - 1$ and $g(x)$ by $2x - 3$. Then we simplify.

a. $(f + g)(x) = f(x) + g(x)$
$$= (x - 1) + (2x - 3)$$
$$= 3x - 4$$

b. $(f - g)(x) = f(x) - g(x)$
$$= (x - 1) - (2x - 3)$$
$$= x - 1 - 2x + 3$$
$$= -x + 2$$

c. $(f \cdot g)(x) = f(x) \cdot g(x)$
$$= (x - 1)(2x - 3)$$
$$= 2x^2 - 5x + 3$$

d. $\left(\dfrac{f}{g}\right)(x) = \dfrac{f(x)}{g(x)} = \dfrac{x - 1}{2x - 3}$, where $x \neq \dfrac{3}{2}$

There is an interesting but not surprising relationship between the graphs of functions and the graphs of their sum, difference, product, and quotient. For example, the graph of $(f + g)(x)$ can be found by adding the graph of $f(x)$ to the graph of $g(x)$. We add two graphs by adding y-values of corresponding x-values.

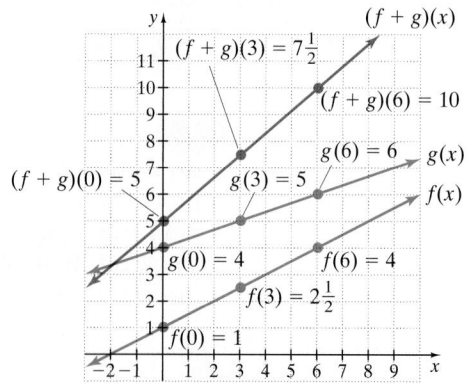

2   Another way to combine functions is called **function composition.** To understand this new way of combining functions, study the diagrams below. The left diagram shows degrees Celsius $f(x)$ as a function of degrees Fahrenheit x. The right diagram shows Kelvins $g(x)$ as a function of degrees Celsius x. (The Kelvin scale is a temperature scale devised by Lord Kelvin in 1848.) The function represented by the first diagram we will call f, and the second function we will call g.

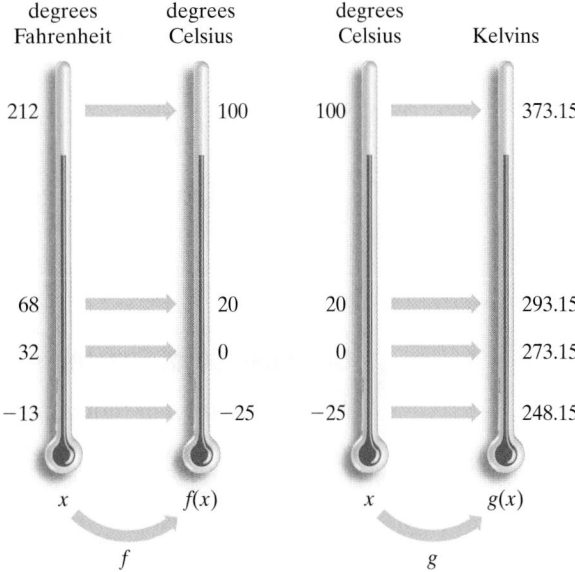

Suppose that we want a function that shows a direct conversion from degrees Fahrenheit to Kelvins. In other words, suppose that a function is needed that shows Kelvins as a function of degrees Fahrenheit. This can easily be done because the output of the first function $f(x)$ is the same as the input of the second function. If we use $f(x)$ to represent this, then we get the following diagram.

For example $g(f(-13)) = 248.15$, and so on.

Since the output of the first function is used as the input of the second function, we write the new function as $g(f(x))$. The new function is formed from the composition of the other two functions. The mathematical symbol for this composition is $(g \circ f)(x)$. Thus, $(g \circ f)(x) = g(f(x))$.

It is possible to find an equation for the composition of the two functions f and g. In other words, we can find a function that converts degrees Fahrenheit directly to Kelvins. The function $f(x) = \frac{5}{9}(x - 32)$ converts degrees Fahrenheit to degrees Celsius, and the function $g(x) = x + 273.15$ converts degrees Celsius to Kelvins. Thus,

$$(g \circ f)(x) = g(f(x)) = g\left(\frac{5}{9}(x - 32)\right) = \frac{5}{9}(x - 32) + 273.15$$

In general, the notation **$g(f(x))$** means "g composed with f" and can be written as $(g \circ f)(x)$. Also $f(g(x))$, or $(f \circ g)(x)$, means "f composed with g."

Composition of Functions

The composition of functions f and g is
$$(f \circ g)(x) = f(g(x))$$

Helpful Hint

$(f \circ g)(x)$ does not mean the same as $(f \cdot g)(x)$.

$$(f \circ g)(x) = f(g(x)) \text{ while } (f \cdot g)(x) = f(x) \cdot g(x)$$

Composition of functions   Multiplication of functions

EXAMPLE 2

If $f(x) = x^2$ and $g(x) = x + 3$, find each composition.

a. $(f \circ g)(2)$ and $(g \circ f)(2)$

b. $(f \circ g)(x)$ and $(g \circ f)(x)$

Solution **a.** $(f \circ g)(2) = f(g(2))$

$= f(5)$ Replace $g(2)$ with 5. [Since $g(x) = x + 3$, then $g(2) = 2 + 3 = 5$.]

$= 5^2 = 25$

$(g \circ f)(2) = g(f(2))$

$= g(4)$ Since $f(x) = x^2$, then $f(2) = 2^2 = 4$.

$= 4 + 3 = 7$

b.

$$(f \circ g)(x) = f(g(x))$$

$$= f(x + 3) \qquad \text{Replace } g(x) \text{ with } x + 3.$$

$$= (x + 3)^2 \qquad f(x + 3) = (x + 3)^2$$

$$= x^2 + 6x + 9 \qquad \text{Square } (x + 3).$$

$$(g \circ f)(x) = g(f(x))$$

$$= g(x^2) \qquad \text{Replace } f(x) \text{ with } x^2.$$

$$= x^2 + 3 \qquad g(x^2) = x^2 + 3$$

EXAMPLE 3

If $f(x) = |x|$ and $g(x) = x - 2$, find each composition.

a. $(f \circ g)(x)$

Solution **b.** $(g \circ f)(x)$

a. $(f \circ g)(x) = f(g(x)) = f(x - 2) = |x - 2|$

b. $(g \circ f)(x) = g(f(x)) = g(|x|) = |x| - 2$

> **Helpful Hint**
> In Examples 2 and 3, notice that $(g \circ f)(x) \neq (f \circ g)(x)$. In general, $(g \circ f)(x)$ *may* or *may not* equal $(f \circ g)(x)$.

EXAMPLE 4

If $f(x) = 5x, g(x) = x - 2$, and $h(x) = \sqrt{x}$, write each function as a composition using two of the given functions.

a. $F(x) = \sqrt{x - 2}$

b. $G(x) = 5x - 2$

Solution **a.** Notice the order in which the function F operates on an input value x. First, 2 is subtracted from x. This is the function $g(x) = x - 2$. Then the square root *of that result* is taken. The square root function is $h(x) = \sqrt{x}$. This means that $F = h \circ g$. To check, we find $h \circ g$.

$$F(x) = (h \circ g)(x) = h(g(x)) = h(x - 2) = \sqrt{x - 2}$$

b. Notice the order in which the function G operates on an input value x. First, x is multiplied by 5, and then 2 is subtracted from the result. This means that $G = g \circ f$. To check, we find $g \circ f$.

$$G(x) = (g \circ f)(x) = g(f(x)) = g(5x) = 5x - 2$$

Graphing Calculator Explorations

If $f(x) = \frac{1}{2}x + 2$ and $g(x) = \frac{1}{3}x^2 + 4$, then

$$(f + g)(x) = f(x) + g(x)$$
$$= \left(\frac{1}{2}x + 2\right) + \left(\frac{1}{3}x^2 + 4\right)$$
$$= \frac{1}{3}x^2 + \frac{1}{2}x + 6.$$

To visualize this addition of functions with a graphing calculator, graph

$$Y_1 = \frac{1}{2}x + 2, \qquad Y_2 = \frac{1}{3}x^2 + 4, \qquad Y_3 = \frac{1}{3}x^2 + \frac{1}{2}x + 6$$

Use a TABLE feature to verify that for a given x value, $Y_1 + Y_2 = Y_3$. For example, verify that when $x = 0$, $Y_1 = 2$, $Y_2 = 4$, and $Y_3 = 2 + 4 = 6$.

The graph on the left shows $y = \frac{1}{3}x^2 + \frac{1}{2}x + 6$, $y = \frac{1}{3}x^2 + 4$, and $y = \frac{1}{2}x + 2$ plotted on axes from -10 to 10.

MENTAL MATH

Match each function with its definition.

1. $(f \circ g)(x)$ **4.** $(g \circ f)(x)$ **A.** $g[f(x)]$ **D.** $\dfrac{f(x)}{g(x)}$

2. $(f \cdot g)(x)$ **5.** $\left(\dfrac{f}{g}\right)(x)$ **B.** $f(x) + g(x)$ **E.** $f(x) \cdot g(x)$

3. $(f - g)(x)$ **6.** $(f + g)(x)$ **C.** $f[g(x)]$ **F.** $f(x) - g(x)$

EXERCISE SET 8.3

STUDY GUIDE/SSM   CD/VIDEO   PH MATH TUTOR CENTER   MathXL®Tutorials ON CD   MathXL®   MyMathLab®

For the functions f and g, find a. $(f + g)(x)$, b. $(f - g)(x)$, c. $(f \cdot g)(x)$, and d. $\left(\dfrac{f}{g}\right)(x)$. See Example 1.

1. $f(x) = x - 7, g(x) = 2x + 1$

2. $f(x) = x + 4, g(x) = 5x - 2$

3. $f(x) = x^2 + 1, g(x) = 5x$

4. $f(x) = x^2 - 2, g(x) = 3x$

5. $f(x) = |x|, g(x) = x + 5$

6. $f(x) = x^3, g(x) = x - 3$

7. $f(x) = -3x, g(x) = 5x^2$

8. $f(x) = 4x^3, g(x) = -6x$

If $f(x) = x^2 - 6x + 2$, $g(x) = -2x$, and $h(x) = |x|$, find each composition. See Example 2.

9. $(f \circ g)(2)$ **10.** $(h \circ f)(-2)$

11. $(g \circ f)(-1)$ **12.** $(f \circ h)(1)$

13. $(g \circ h)(0)$ **14.** $(h \circ g)(0)$

Find $(f \circ g)(x)$ and $(g \circ f)(x)$. See Examples 2 and 3.

15. $f(x) = x^2 + 1, g(x) = 5x$

16. $f(x) = x - 3, g(x) = x^2$

17. $f(x) = 2x - 3, g(x) = x + 7$

18. $f(x) = x + 10, g(x) = 3x + 1$

19. $f(x) = x^3 + x - 2, g(x) = -2x$

20. $f(x) = -4x, g(x) = x^3 + x^2 - 6$

21. $f(x) = |x|, g(x) = -5x + 2$

22. $f(x) = 7x - 1, g(x) = 8x + 2$

If $f(x) = 3x$, $g(x) = |x|$, and $h(x) = x^2 + 2$, write each function as a composition using two of the given functions. See Example 4.

23. $H(x) = |x^2 + 2|$

24. $G(x) = |3x|$

25. $F(x) = 9x^2 + 2$

26. $H(x) = 3x^2 + 6$

27. $G(x) = 3|x|$

28. $F(x) = x^2 + 2$

Find $f(x)$ and $g(x)$ so that the given function $h(x) = (f \circ g)(x)$.

29. $h(x) = (x + 2)^2$

30. $h(x) = |x - 1|$

31. $h(x) = |x + 5| + 2$

32. $h(x) = (3x + 4)^2 + 3$

33. $h(x) = \dfrac{1}{2x - 3}$

34. $h(x) = \dfrac{1}{x + 10}$

REVIEW AND PREVIEW

Solve each equation for y. See Section 2.5.

35. $x = y + 2$

36. $x = y - 5$

37. $x = 3y$

38. $x = -6y$

39. $x = -2y - 7$

40. $x = 4y + 7$

Concept Extensions

Given that $f(-1) = 4 \quad g(-1) = -4$
$f(0) = 5 \quad g(0) = -3$
$f(2) = 7 \quad g(2) = -1$
$f(7) = 1 \quad g(7) = 4$

Find each function value.

41. $(f + g)(2)$

42. $(f - g)(2)$

43. $(f \circ g)(2)$

44. $(g \circ f)(2)$

45. $(f \cdot g)(7)$

46. $(f \cdot g)(0)$

47. $\left(\dfrac{f}{g}\right)(-1)$

48. $\left(\dfrac{g}{f}\right)(-1)$

49. If you are given $f(x)$ and $g(x)$, explain in your own words how to find $(f \circ g)(x)$, and then how to find $(g \circ f)(x)$.

50. Given $f(x)$ and $g(x)$, describe in your own words the difference between $(f \circ g)(x)$ and $(f \cdot g)(x)$.

Solve.

51. Business people are concerned with cost functions, revenue functions, and profit functions. Recall that the profit $P(x)$ obtained from x units of a product is equal to the revenue $R(x)$ from selling the x units minus the cost $C(x)$ of manufacturing the x units. Write an equation expressing this relationship among $C(x)$, $R(x)$, and $P(x)$.

52. Suppose the revenue $R(x)$ for x units of a product can be described by $R(x) = 25x$, and the cost $C(x)$ can be described by $C(x) = 50 + x^2 + 4x$. Find the profit $P(x)$ for x units.

STUDY SKILLS REMINDER

How Are You Doing?

If you haven't done so yet, take a few moments and think about how you are doing in this course. Are you working toward your goal of successfully completing this course? Is your performance on homework, quizzes, and tests satisfactory? If not, you might want to see your instructor to see if he/she has any suggestions on how you can improve your performance. Let me once again remind you that, in addition to your instructor, there are many places to get help with your mathematics course. A few suggestions are below.

▶ This text has an accompanying video lesson for every section in this text.

▶ The back of this book contains answers to odd-numbered exercises and selected solutions.

▶ A tutorial software program is available with this text. It has lessons corresponding to each section in the text.

▶ There is a student solutions manual available that contains worked-out solutions to odd-numbered exercises as well as solutions to every exercise in the Integrated Reviews, Chapter Reviews, Chapter Tests, and Cumulative Reviews.

▶ Don't forget to check with your instructor for other local resources available to you, such as a tutor center.

8.4   VARIATION AND PROBLEM SOLVING

Objectives

1. Solve problems involving direct variation.
2. Solve problems involving inverse variation.
3. Solve problems involving joint variation.
4. Solve problems involving combined variation.

We conclude this chapter with a special family of functions, called variation.

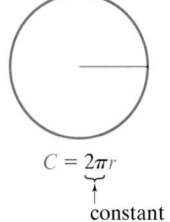

$C = 2\pi r$
constant

1   A very familiar example of direct variation is the relationship of the circumference C of a circle to its radius r. The formula $C = 2\pi r$ expresses that the circumference is always 2π times the radius. In other words, C is always a constant multiple (2π) of r. Because it is, we say that **C varies directly as r**, that **C varies directly with r**, or that **C is directly proportional to r**.

Direct Variation

y **varies directly as** *x*, or *y* **is directly proportional to** *x*, if there is a nonzero constant *k* such that

$$y = kx$$

The number *k* is called the **constant of variation** or the **constant of proportionality**.

In the above definition, the relationship described between *x* and *y* is a linear one. In other words, the graph of $y = kx$ is a line. The slope of the line is *k*, and the line passes through the origin. Notice that the direct variation equation describes a linear function since it is in the form $y = mx + b$ where *b* is 0.

For example, the graph of the direct variation equation $C = 2\pi r$ is shown. The horizontal axis represents the radius *r*, and the vertical axis is the circumference *C*. From the graph we can read that when the radius is 6 units, the circumference is approximately 38 units. Also, when the circumference is 45 units, the radius is between 7 and 8 units. Notice that as the radius increases, the circumference increases.

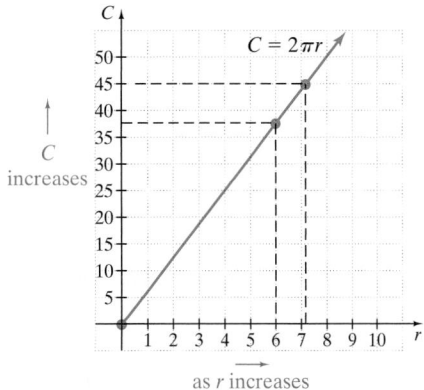

As expected, the graph above does pass the vertical line test. It is the graph of a function.

EXAMPLE 1

Suppose that *y* varies directly as *x*. If *y* is 5 when *x* is 30, find the constant of variation and the direct variation equation.

Solution   Since *y* varies directly as *x*, we write $y = kx$. If $y = 5$ when $x = 30$, we have that

$$y = kx$$
$$5 = k(30) \quad \text{Replace } y \text{ with } 5 \text{ and } x \text{ with } 30.$$
$$\frac{1}{6} = k \quad \text{Solve for } k.$$

The constant of variation is $\frac{1}{6}$.

After finding the constant of variation *k*, the direct variation equation can be written as $y = \frac{1}{6}x$.

EXAMPLE 2

USING DIRECT VARIATION AND HOOKE'S LAW

Hooke's law states that the distance a spring stretches is directly proportional to the weight attached to the spring. If a 40-pound weight attached to the spring stretches the spring 5 inches, find the distance that a 65-pound weight attached to the spring stretches the spring.

Solution   1. UNDERSTAND.   Read and reread the problem. Notice that we are given that the distance a spring stretches is **directly proportional** to the weight attached. We let

$$d = \text{the distance stretched}$$
$$w = \text{the weight attached}$$

The constant of variation is represented by k.

65 lbs

2. TRANSLATE.   Because d is directly proportional to w, we write

$$d = kw$$

3. SOLVE.   When a weight of 40 pounds is attached, the spring stretches 5 inches. That is, when $w = 40$, $d = 5$.

$$d = kw$$
$$5 = k(40) \quad \text{Replace } d \text{ with 5 and } w \text{ with 40.}$$
$$\frac{1}{8} = k \qquad \text{Solve for } k.$$

Now when we replace k with $\frac{1}{8}$ in the equation

$$d = kw, \text{ we have}$$
$$d = \frac{1}{8}w$$

To find the stretch when a weight of 65 pounds is attached, we replace w with 65 to find d.

$$d = \frac{1}{8}(65)$$
$$= \frac{65}{8} = 8\frac{1}{8} \quad \text{or} \quad 8.125$$

4. INTERPRET.

Check: Check the proposed solution of 8.125 inches in the original problem.

State: The spring stetches 8.125 inches when a 65-pound weight is attached.

2   When y is proportional to the **reciprocal** of another variable x, we say that **y varies inversely as x,** or that **y is inversely proportional to x.** An example of the inverse variation relationship is the relationship between the pressure that a gas exerts and the volume of its container. As the volume of a container decreases, the pressure of the gas it contains increases.

Inverse Variation

y varies inversely as x, or **y is inversely proportional to x,** if there is a nonzero constant k such that

$$y = \frac{k}{x}$$

The number k is called the **constant of variation** or the **constant of proportionality.**

Notice that $y = \dfrac{k}{x}$ is a rational equation. Its graph for $k > 0$ and $x > 0$ is shown. From the graph, we can see that as x increases, y decreases.

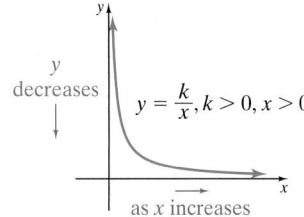

Once again, the graph above passes the vertical line test. It is the graph of a function.

EXAMPLE 3

Suppose that u varies inversely as w. If u is 3 when w is 5, find the constant of variation and the inverse variation equation.

Solution   Since u varies inversely as w, we have $u = \dfrac{k}{w}$. We let $u = 3$ and $w = 5$, and we solve for k.

$$u = \frac{k}{w}$$

$$3 = \frac{k}{5} \quad \text{Let } u = 3 \text{ and } w = 5.$$

$$15 = k \quad \text{Multiply both sides by 5.}$$

The constant of variation k is 15. This gives the inverse variation equation

$$u = \frac{15}{w}$$

EXAMPLE 4

USING INVERSE VARIATION AND BOYLE'S LAW

Boyle's law says that if the temperature stays the same, the pressure P of a gas is inversely proportional to the volume V. If a cylinder in a steam engine has a pressure of 960 kilopascals when the volume is 1.4 cubic meters, find the pressure when the volume increases to 2.5 cubic meters.

Solution   1. UNDERSTAND.   Read and reread the problem. Notice that we are given that the pressure of a gas is *inversely proportional* to the volume. We will let P = the pressure and V = the volume. The constant of variation is represented by k.

2. TRANSLATE.   Because P is inversely proportional to V, we write

$$P = \frac{k}{V}$$

When $P = 960$ kilopascals, the volume $V = 1.4$ cubic meters. We use this information to find k.

$$960 = \frac{k}{1.4} \quad \text{Let } P = 960 \text{ and } V = 1.4.$$

$$1344 = k \quad \text{Multiply both sides by 1.4.}$$

Thus, the value of k is 1344. Replacing k with 1344 in the variation equation, we have

$$P = \frac{1344}{V}$$

Next we find P when V is 2.5 cubic meters.

3. SOLVE.

$$P = \frac{1344}{2.5} \quad \text{Let } V = 2.5.$$

$$= 537.6$$

4. INTERPRET.

Check: Check the proposed solution in the original problem.
State: When the volume is 2.5 cubic meters, the pressure is 537.6 kilopascals.

3   Sometimes the ratio of a variable to the product of many other variables is constant. For example, the ratio of distance traveled to the product of speed and time traveled is always 1.

$$\frac{d}{rt} = 1 \quad \text{or} \quad d = rt$$

Such a relationship is called **joint variation.**

Joint Variation

If the ratio of a variable y to the product of two or more variables is constant, then y **varies jointly as,** or **is jointly proportional to,** the other variables. If

$$y = kxz$$

then the number k is the **constant of variation** or the **constant of proportionality.**

 CONCEPT CHECK

Which type of variation is represented by the equation $xy = 8$? Explain.

    a. Direct variation   **b.** Inverse variation   **c.** Joint variation

 EXAMPLE 5

EXPRESSING SURFACE AREA

The lateral surface area of a cylinder varies jointly as its radius and height. Express this surface area S in terms of radius r and height h.

Solution   Because the surface area varies jointly as the radius r and the height h, we equate S to a constant multiple of r and h.

$$S = krh$$

In the equation, $S = krh$, it can be determined that the constant k is 2π, and we then have the formula $S = 2\pi rh$. (The lateral surface area formula does not include the areas of the two circular bases.)

4   Some examples of variation involve combinations of direct, inverse, and joint variation. We will call these variations **combined variation.**

Concept Check Answer:
b; answers may vary

△ EXAMPLE 6

FINDING COLUMN WEIGHT

The maximum weight that a circular column can support is directly proportional to the fourth power of its diameter and is inversely proportional to the square of its height. A 2-meter-diameter column that is 8 meters in height can support 1 ton. Find the weight that a 1-meter-diameter column that is 4 meters in height can support.

Solution

1. UNDERSTAND.   Read and reread the problem. Let w = weight, d = diameter, h = height, and k = the constant of variation.

2. TRANSLATE.   Since w is directly proportional to d^4 and inversely proportional to h^2, we have

$$w = \frac{kd^4}{h^2}$$

3. SOLVE.   To find k, we are given that a 2-meter-diameter column that is 8 meters in height can support 1 ton. That is, $w = 1$ when $d = 2$ and $h = 8$, or

$$1 = \frac{k \cdot 2^4}{8^2} \qquad \text{Let } w = 1, d = 2, \text{ and } h = 8.$$

$$1 = \frac{k \cdot 16}{64}$$

$$4 = k \qquad \text{Solve for } k.$$

Now replace k with 4 in the equation $w = \dfrac{kd^4}{h^2}$ and we have

$$w = \frac{4d^4}{h^2}$$

To find weight w for a 1-meter-diameter column that is 4 meters in height, let $d = 1$ and $h = 4$.

$$w = \frac{4 \cdot 1^4}{4^2}$$

$$w = \frac{4}{16} = \frac{1}{4}$$

4. INTERPRET.

Check: Check the proposed solution in the original problem.

State: The 1-meter-diameter column that is 4 meters in height can hold $\dfrac{1}{4}$ ton of weight.

Spotlight on DECISION MAKING

Suppose you are a painting contractor. You have been hired to paint the ceilings of a one-story home, whose layout is shown in the figure. The amount of paint you need is directly proportional to the area that is to be painted. You know that 450 square feet can be painted with 4 quarts of paint. Quarts of paint cost $5.95 each and gallons of paint cost $21.50 each. You have estimated the cost of the paint to be about $50. Is your estimate correct? If so, explain why. If not, how much more should the estimate be?

MENTAL MATH

State whether each equation represents direct, inverse, or joint variation.

1. $y = 5x$

2. $y = \dfrac{700}{x}$

3. $y = 5xz$

4. $y = \dfrac{1}{2}abc$

5. $y = \dfrac{9.1}{x}$

6. $y = 2.3x$

7. $y = \dfrac{2}{3}x$

8. $y = 3.1\,st$

EXERCISE SET 8.4

STUDY GUIDE/SSM   CD/ VIDEO   PH MATH TUTOR CENTER   MathXL®Tutorials ON CD   MathXL®   MyMathLab®

MIXED PRACTICE

Write an equation to describe each variation. Use k for the constant of proportionality. See Examples 1 through 6.

1. y varies directly as x

2. p varies directly as q

3. a varies inversely as b

4. y varies inversely as x

5. y varies jointly as x and z

6. y varies jointly as q, r, and t

7. y varies inversely as x^3

8. y varies inversely as a^4

9. y varies directly as x and inversely as p^2

10. y varies directly as a^5 and inversely as b

If y varies directly as x, find the constant of variation k and the direct variation equation for each situation. See Example 1.

 11. $y = 4$ when $x = 20$

12. $y = 5$ when $x = 30$

13. $y = 6$ when $x = 4$

14. $y = 12$ when $x = 8$

15. $y = 7$ when $x = \dfrac{1}{2}$

16. $y = 11$ when $x = \dfrac{1}{3}$

17. $y = 0.2$ when $x = 0.8$

18. $y = 0.4$ when $x = 2.5$

Solve. See Example 2.

19. The weight of a synthetic ball varies directly with the cube of its radius. A ball with a radius of 2 inches weighs 1.20 pounds. Find the weight of a ball of the same material with a 3-inch radius.

20. At sea, the distance to the horizon is directly proportional to the square root of the elevation of the observer. If a person who is 36 feet above the water can see 7.4 miles, find how far a person 64 feet above the water can see. Round answer to one decimal place.

21. The amount P of pollution varies directly with the population N of people. Kansas City has a population of 450,000 and produces 260,000 tons of pollutants. Find how many tons of pollution we should expect St. Louis to produce, if we know that its population is 980,000. Round answer to the nearest whole ton.

22. Charles' law states that if the pressure P stays the same, the volume V of a gas is directly proportional to its temperature T. If a balloon is filled with 20 cubic meters of a gas at a temperature of 300 K, find the new volume if the temperature rises to 360 K while the pressure stays the same.

If y varies inversely as x, find the constant of variation k and the inverse variation equation for each situation. See Example 3.

 23. $y = 6$ when $x = 5$ **24.** $y = 20$ when $x = 9$

25. $y = 100$ when $x = 7$ **26.** $y = 63$ when $x = 3$

27. $y = \dfrac{1}{8}$ when $x = 16$ **28.** $y = \dfrac{1}{10}$ when $x = 40$

29. $y = 0.2$ when $x = 0.7$ **30.** $y = 0.6$ when $x = 0.3$

Solve. See Example 4.

31. Pairs of markings a set distance apart are made on highways so that police can detect drivers exceeding the speed limit. Over a fixed distance, the speed R varies inversely with the time T. In one particular pair of markings, R is 45 mph when T is 6 seconds. Find the speed of a car that travels the given distance in 5 seconds.

32. The weight of an object on or above the surface of Earth varies inversely as the square of the distance between the object and Earth's center. If a person weighs 160 pounds on Earth's surface, find the individual's weight if he moves 200 miles above Earth. Round answer to the nearest pound. (Assume that Earth's radius is 4000 miles.)

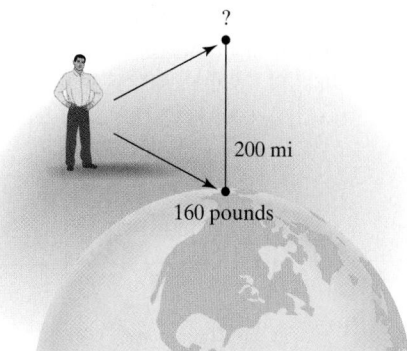

33. If the voltage V in an electric circuit is held constant, the current I is inversely proportional to the resistance R. If the current is 40 amperes when the resistance is 270 ohms, find the current when the resistance is 150 ohms.

34. Because it is more efficient to produce larger numbers of items, the cost of producing recordable disks is inversely proportional to the number produced. If 4000 can be produced at a cost of \$0.45 each, find the cost per disk when 6000 are produced.

35. The intensity I of light varies inversely as the square of the distance d from the light source. If the distance from the light source is doubled, determine what happens to the intensity of light at the new location.

△ **36.** The maximum weight that a circular column can hold is inversely proportional to the square of its height. If an 8-foot column can hold 2 tons, find how much weight a 10-foot column can hold.

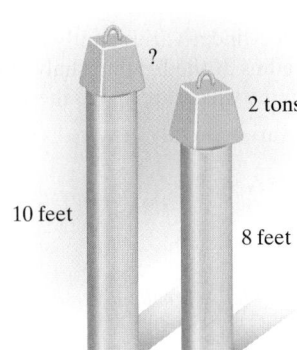

Write each statement as an equation. See Example 5.

37. x varies jointly as y and z.

38. P varies jointly as R and the square of S.

39. r varies jointly as s and the cube of t.

40. a varies jointly as b and c.

For each statement, find the constant of variation and the variation equation. See Examples 5 and 6.

41. y varies directly as the cube of x; $y = 9$ when $x = 3$

42. y varies directly as the cube of x; $y = 32$ when $x = 4$

43. y varies directly as the square root of x; $y = 0.4$ when $x = 4$

44. y varies directly as the square root of x; $y = 2.1$ when $x = 9$

45. y varies inversely as the square of x; $y = 0.052$ when $x = 5$

46. y varies inversely as the square of x; $y = 0.011$ when $x = 10$

47. y varies jointly as x and the cube of z; $y = 120$ when $x = 5$ and $z = 2$

48. y varies jointly as x and the square of z; $y = 360$ when $x = 4$ and $z = 3$

Solve. See Examples 5 and 6.

🔒 **49.** The maximum weight that a rectangular beam can support varies jointly as its width and the square of its height and inversely as its length. If a beam $\frac{1}{2}$ foot wide, $\frac{1}{3}$ foot high, and 10 feet long can support 12 tons, find how much a similar beam can support if the beam is $\frac{2}{3}$ foot wide, $\frac{1}{2}$ foot high, and 16 feet long.

50. The number of cars manufactured on an assembly line at a General Motors plant varies jointly as the number of workers and the time they work. If 200 workers can produce 60 cars in 2 hours, find how many cars 240 workers should be able to make in 3 hours.

△ **51.** The volume of a cone varies jointly as the square of its radius and its height. If the volume of a cone is 32π cubic inches when the radius is 4 inches and the height is 6 inches, find the volume of a cone when the radius is 3 inches and the height is 5 inches.

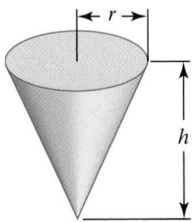

52. When a wind blows perpendicularly against a flat surface, its force is jointly proportional to the surface area and the speed of the wind. A sail whose surface area is 12 square feet experiences a 20-pound force when the wind speed is 10 miles per hour. Find the force on an 8-square-foot sail if the wind speed is 12 miles per hour.

53. The horsepower that can be safely transmitted to a shaft varies jointly as the shaft's angular speed of rotation (in revolutions per minute) and the cube of its diameter. A 2-inch shaft making 120 revolutions per minute safely transmits 40 horsepower. Find how much horsepower can be safely transmitted by a 3-inch shaft making 80 revolutions per minute.

△ **54.** The maximum weight that a rectangular beam can support varies jointly as its width and the square of its height and inversely as its length. If a beam $\frac{1}{3}$ foot wide, 1 foot high, and 10 feet long can support 3 tons, find how much weight a similar beam can support if it is 1 foot wide, $\frac{1}{3}$ foot high, and 9 feet long.

55. The atmospheric pressure y (in millibars) is inversely proportional to the altitude x (in kilometers). If the atmospheric pressure is 400 millibars at an altitude of 8 kilometers, find the atmospheric pressure at an altitude of 4 kilometers.

REVIEW AND PREVIEW

Find the exact circumference and area of each circle. See the inside cover for a list of geometric formulas.

△ **56.**

4 in.

△ **57.**

6 cm

△ **58.**

9 cm

△ **59.**

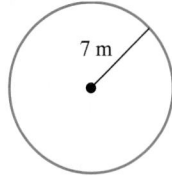

7 m

Simplify. See Sections 1.5 through 1.7.

60. $|-1.2|$

61. $|-3|$

62. $-|7|$

63. $|0|$

64. $-\left|-\dfrac{1}{2}\right|$

65. $-\left|\dfrac{1}{5}\right|$

66. $\left(\dfrac{2}{3}\right)^3$

67. $\left(\dfrac{5}{11}\right)^2$

Concept Extensions

68. The horsepower to drive a boat varies directly as the cube of the speed of the boat. If the speed of the boat is to double, determine the corresponding increase in horsepower required.

69. The volume of a cylinder varies jointly as the height and the square of the radius. If the height is halved and the radius is doubled, determine what happens to the volume.

70. Suppose that y varies directly as x. If x is doubled, what is the effect on y?

71. Suppose that y varies directly as x^2. If x is doubled, what is the effect on y?

Complete the following table for the inverse variation $y = \dfrac{k}{x}$ over each given value of k. Plot the points on a rectangular coordinate system.

x	$\dfrac{1}{4}$	$\dfrac{1}{2}$	1	2	4
$y = \dfrac{k}{x}$					

72. $k = 1$

73. $k = 3$

74. $k = 5$

75. $k = \dfrac{1}{2}$

CHAPTER 8 PROJECT

Modeling Electricity Production

According to the U.S. Department of Energy, energy produced by renewable sources (including hydroelectric, geothermal, wind, and solar powers) accounted for nearly 8% of the United States' total production of energy in 2004. Wind energy can be harnessed by windmills to produce electricity, but it is the least utilized of these renewable energy sources. However, progressive communities are experimenting with fields of windmills for communal electricity needs, examining exactly how wind speed affects the amount of electricity produced.

A community in California is experimenting with electricity generated by windmills. City engineers are analyzing data they have gathered about their field of windmills. The engineers are familiar with other research demonstrating that the amount of electricity that a windmill generates hourly (in watt-hours) is

directly proportional to the cube of the wind speed (in miles per hour). In this project, you will use this fact to help the engineers analyze the amount of electricity generated by the windmill field. This project may be completed by working in groups or individually.

1. The city engineers have documented that when the wind speed is exactly 10 miles per hour, a windmill generates electricity at a rate of 15 watt-hours. Find a formula that models the relationship between the wind speed and the amount of electricity generated hourly by a windmill.

2. Complete the following table for the given wind speeds. Then use the table to estimate the wind speed required to obtain: **a.** 100 watt-hours, and **b.** 400 watt-hours.

3. Plot the ordered pairs from the table. Describe the trend shown by the graph.

4. The engineers' data show that for several days the wind speed was more or less steady at 20 miles per hour, and the windmills generated the expected 120 watt-hours. According to the weather forecast for the coming few days, wind speed will fluctuate wildly, but will still average 20 miles per hour. Should the engineers still expect the windmills to generate 120 watt-hours? Demonstrate your reasoning with a numerical example.

5. During one three-day period, each windmill generated 150 watt-hours. The forecast predicts that the wind speed for the coming few days will drop by half. How many watt-hours should the engineers now expect each windmill to generate? In general, if the wind speed yields c watt-hours, how many watt-hours does half the wind speed yield?

Wind Speed (miles per hour)	Electricity Generated (watt-hours)
15	
17	
19	
21	
23	
25	
27	
29	
31	
33	
35	

VOCABULARY CHECK

Fill in each blank with one of the words or phrases listed below.

linear function   composition   jointly   directly   inversely

1. The _____ of functions f and g is $(f \circ g)(x) = f(g(x))$.

2. A _____ is a function that can be written in the form $f(x) = mx + b$.

3. In the equation $y = kx$, y varies _____ as x.

4. In the equation $y = \dfrac{k}{x}$, y varies _____ as x.

5. In the equation $y = kxz$, y varies _____ as x and z.

CHAPTER 8 HIGHLIGHTS

Definitions and Concepts	**Examples**

Section 8.1   Graphing and Writing Linear Functions

A **linear function** is a function that can be written in the form $f(x) = mx + b$.

To graph a linear function, find three ordered pair solutions. Graph the solutions and draw a line through the plotted points.

Linear Functions

$$f(x) = -3, g(x) = 5x, h(x) = -\frac{1}{3}x - 7$$

Graph $f(x) = -2x$.

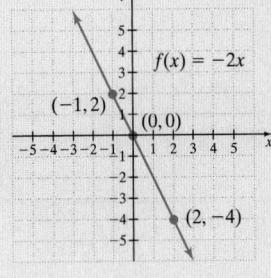

x	y or $f(x)$
-1	2
0	0
2	-4

(continued)

Definitions and Concepts	**Examples**

Section 8.1 Graphing and Writing Linear Functions

The point–slope form of the equation of a line is $y - y_1 = m(x - x_1)$, where m is the slope of the line and (x_1, y_1) is a point on the line.

Find an equation of the line parallel to $g(x) = 2x - 1$ and containing the point $(1, -4)$. Write the equation using function notation. Since we want a parallel line, use the same slope of $g(x)$, which is 2.

$$y - y_1 = m(x - x_1)$$
$$y - (-4) = 2(x - 1)$$
$$y + 4 = 2x - 2$$
$$y = 2x - 6 \qquad \text{Solve for } y.$$
$$f(x) = 2x - 6 \qquad \text{Let } y = f(x)$$

Section 8.2 Graphing Nonlinear Functions

The graph of $y = mx + b$ is the same as the graph of $y = mx$, but shifted $|b|$ units up if b is positive and $|b|$ units down if b is negative.

Graph $g(x) = -2x + 3$.

This is the same as the graph of $f(x) = -2x$ shifted 3 units up.

To graph a function that is not linear, find a sufficient number of ordered pair solutions so that a pattern may be discovered.

Graph $f(x) = x^3 + 2$.

x	y or $f(x)$
-2	-6
-1	1
0	2
1	3
2	10

Definitions and Concepts	Examples

Section 8.3 The Algebra of Functions; Composite Functions

Algebra of Functions

Sum $(f + g)(x) = f(x) + g(x)$

Difference $(f - g)(x) = f(x) - g(x)$

Product $(f \cdot g)(x) = f(x) \cdot g(x)$

Quotient $\left(\dfrac{f}{g}\right)(x) = \dfrac{f(x)}{g(x)}, g(x) \neq 0$

Composite Functions

The notation $(f \circ g)(x)$ means "f composed with g."

$$(f \circ g)(x) = f(g(x))$$

$$(g \circ f)(x) = g(f(x))$$

If $f(x) = 7x$ and $g(x) = x^2 + 1$,

$$(f + g)(x) = f(x) + g(x) = 7x + x^2 + 1$$

$$(f - g)(x) = f(x) - g(x) = 7x - (x^2 + 1)$$

$$= 7x - x^2 - 1$$

$$(f \cdot g)(x) = f(x) \cdot g(x) = 7x(x^2 + 1)$$

$$= 7x^3 + 7x$$

$$\left(\frac{f}{g}\right)(x) = \frac{f(x)}{g(x)} = \frac{7x}{x^2 + 1}$$

If $f(x) = x^2 + 1$ and $g(x) = x - 5$, find $(f \circ g)(x)$.

$$(f \circ g)(x) = f(g(x))$$

$$= f(x - 5)$$

$$= (x - 5)^2 + 1$$

$$= x^2 - 10x + 26$$

Section 8.4 Variation and Problem Solving

y **varies directly as** *x*, or *y* is **directly proportional to** *x*, if there is a nonzero constant *k* such that

$$y = kx$$

y **varies inversely as** *x*, or *y* is **inversely proportional to** x, if there is a nonzero constant *k* such that

$$y = \frac{k}{x}$$

y **varies jointly as** *x* and *z* or *y* is **jointly proportional to** *x* and *z* if there is a nonzero constant *k* such that

$$y = kxz$$

The circumference of a circle C varies directly as its radius r.

$$C = 2\underset{k}{\underbrace{\pi}}r$$

Pressure P varies inversely with volume V.

$$P = \frac{k}{V}$$

The lateral surface area S of a cylinder varies jointly as its radius r and height h.

$$S = 2\underset{k}{\underbrace{\pi}}rh$$

8 CHAPTER REVIEW

(8.1) *Graph each linear function.*

1. $f(x) = x$

2. $f(x) = -\dfrac{1}{3}x$

3. $g(x) = 4x - 1$

4. $F(x) = -\dfrac{2}{3}x + 2$

The graph of $f(x) = 3x$ is sketched below. Use this graph to match each linear function with its graph.

A

B

C

D
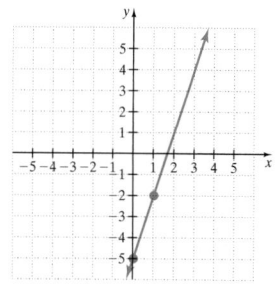

5. $f(x) = 3x + 1$

6. $f(x) = 3x - 2$

7. $f(x) = 3x + 2$

8. $f(x) = 3x - 5$

Find the slope and y-intercept of each function.

9. $f(x) = \dfrac{2}{5}x - \dfrac{4}{3}$

10. $f(x) = -\dfrac{2}{7}x + \dfrac{3}{2}$

Find the equation of each line satisfying the given conditions. Write each equation using function notation.

11. Slope $-\dfrac{2}{3}$; y-intercept $(0, 4)$

12. Slope -1; y-intercept $(0, -2)$

△ **13.** Through $(2, -6)$; parallel to $6x + 3y = 5$

△ **14.** Through $(-4, -2)$; parallel to $3x + 2y = 8$

△ **15.** Through $(-6, -1)$; perpendicular to $4x + 3y = 5$

△ **16.** Through $(-4, 5)$; perpendicular to $2x - 3y = 6$

17. In 2002, the percent of U.S. drivers wearing seat belts was 81%. The number of drivers wearing seat belts in 1990 was 65%. Let y be the number of drivers wearing seat belts in the year x, where $x = 0$ represents 1990. (*Source:* Strategis Group for Personal Communications Asso.)

    a. Write a linear function that models the percent of U.S. drivers wearing seat belts in terms of the year x. [*Hint:* Write 2 ordered pairs of the form (years past 1990, percent of drivers).]

    b. Use this function to predict the number of U.S. drivers wearing seat belts in the year 2009. (Round to the nearest percent.)

18. In 1998, the number of people (in millions) reporting arthritis was 43. The number of people (in millions) predicted to be reporting arthritis in 2020 is 60. Let y be the number of people (in millions) reporting arthritis in the year x, where $x = 0$ represents 1998. (*Source:* Arthritis Foundation)

    a. Write a linear equation that models the number of people (in millions) reporting arthritis in terms of the year x (See the hint for Exercise 17.)

    b. Use this equation to predict the number of people reporting arthritis in 2010. (Round to the nearest

Decide whether the lines are parallel, perpendicular, or neither.

△ **19.** $f(x) = -2x + 6$
$g(x) = 2x - 1$

△ **20.** $-x + 3y = 2$
$6x - 18y = 3$

The cost C, in dollars, of renting a minivan for a day is given by the linear function $C(x) = 0.3x + 42$, where x is number of miles driven.

21. Find the cost of renting the minivan for a day and driving it 150 miles.

22. Graph $C(x) = 0.3x + 42$.

(8.2) Use the graph of the function below to answer Exercises 23 through 26.

23. Find $f(-1)$.

24. Find $f(1)$.

25. Find all values of x such that $f(x) = 1$.

26. Find all values of x such that $f(x) = -1$.

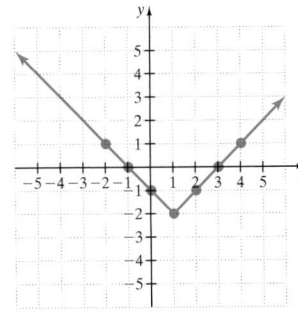

Determine whether each function is linear or not. Then graph the function.

27. $f(x) = 3x$

28. $f(x) = 5x$

29. $g(x) = |x| + 4$

30. $h(x) = x^2 + 4$

31. $F(x) = -\dfrac{1}{2}x + 2$

32. $G(x) = -x + 5$

33. $y = -1.36x$

34. $y = 2.1x + 5.9$

35. $H(x) = (x - 2)^2$

36. $f(x) = -|x - 3|$

(8.3) If $f(x) = x - 5$ and $g(x) = 2x + 1$, find

37. $(f + g)(x)$

38. $(f - g)(x)$

39. $(f \cdot g)(x)$

40. $\left(\dfrac{g}{f}\right)(x)$

If $f(x) = x^2 - 2$, $g(x) = x + 1$, and $h(x) = x^3 - x^2$, find each composition.

41. $(f \circ g)(x)$

42. $(g \circ f)(x)$

43. $(h \circ g)(2)$

44. $(f \circ f)(x)$

45. $(f \circ g)(-1)$

46. $(h \circ h)(2)$

(8.4) Solve each variation problem.

47. A is directly proportional to B. If $A = 6$ when $B = 14$, find A when $B = 21$.

48. C is inversely proportional to D. If $C = 12$ when $D = 8$, find C when $D = 24$.

48. According to Boyle's law, the pressure exerted by a gas is inversely proportional to the volume, as long as the temperature stays the same. If a gas exerts a pressure of 1250 pounds per square inch when the volume is 2 cubic feet, find the volume when the pressure is 800 pounds per square inch.

△ **50.** The surface area of a sphere varies directly as the square of its radius. If the surface area is 36π square inches when the radius is 3 inches, find the surface area when the radius is 4 inches.

CHAPTER 8 TEST

Remember to use your Chapter Test Prep Video CD to help you study and view solutions to the test questions you need help with.

Use the graph of the function f to find each value.

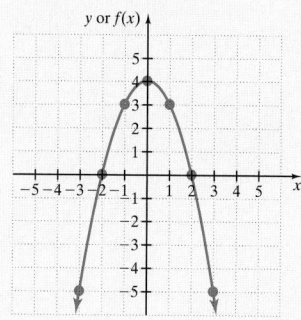

1. Find $f(1)$.

2. Find $f(-3)$.

3. Find all values of x such that $f(x) = 0$.

4. Find all values of x such that $f(x) = 4$.

Graph each line.

5. $2x - 3y = -6$

6. $f(x) = \dfrac{2}{3}x$

Graph each nonlinear function. Suggested x-values have been given for ordered pair solutions.

7. $f(x) = (x - 1)^2$ Let $x = -2, -1, 0, 1, 2, 3, 4$

8. $g(x) = |x| + 2$ Let $x = -3, -2, -1, 0, 1, 2, 3$

Find an equatin of each line satisfying the conditions given. Write the equations using function notation.

△ **9.** Through $(4, -2)$ and $(6, -3)$

△ **10.** Through $(-1, 2)$; perpendicular to $3x - y = 4$

11. Parallel to $2y + x = 3$; through $(3, -2)$

12. The average yearly earnings for high school graduates age 18 and older is given by the linear function

$$f(x) = 732x + 21{,}428$$

where x is the number of years since 1996 that a person graduated. (*Source:* U.S. Census Bureau)

a. Find the average earnings in 1998 for high school graduates.

b. Predict the average earnings for high school graduates in the year 2005.

c. Predict the first year that the average earnings for high school graduates will be greater than $30,000.

d. Find and interpret the slope of this equation.

e. Find and interpret the y-intercept of this equation.

If $f(x) = x^2 + 5$ and $g(x) = x - 2$, find the following.

13. $(f - g)(x)$

14. $(f \cdot g)(x)$

If $f(x) = x$, $g(x) = x - 7$, and $h(x) = x^2 - 6x + 5$, find the following.

15. $(f \circ h)(0)$

16. $(g \circ f)(x)$

17. $(g \circ h)(x)$

18. Suppose that W is inversely proportional to V. If $W = 20$ when $V = 12$, find W when $V = 15$.

19. Suppose that Q is jointly proportional to R and the square of S. If $Q = 24$ when $R = 3$ and $S = 4$, find Q when $R = 2$ and $S = 3$.

20. When an anvil is dropped into a gorge, the speed with which it strikes the ground is directly proportional to the square root of the distance it falls. An anvil that falls 400 feet hits the ground at a speed of 160 feet per second. Find the height of a cliff over the gorge if a dropped anvil hits the ground at a speed of 128 feet per second.

CHAPTER CUMULATIVE REVIEW

1. Simplify $3[4 + 2(10 - 1)]$.
2. Simplify $5[3 + 6(8 - 5)]$.

Find the value of each expression when $x = 2$ and $y = -5$.

3. **a.** $\dfrac{x - y}{12 + x}$ **b.** $x^2 - y$

4. **a.** $\dfrac{x + y}{3y}$ **b.** $y^2 - x$

Solve.

5. $-3x = 33$

6. $\dfrac{2}{3}y = 7$

7. $8(2 - t) = -5t$

8. $5x - 9 = 5x - 29$

9. Solve $y = mx + b$ for x.

10. Solve $y = 7x - 2$ for x.

11. $-4x + 7 \geq -9$ Write the solution using interval notation rotation.

12. $-5x - 6 < 3x + 1$ Write the solution using interval notation.

13. Find the slope of a line perpendicular to the line passing through the points $(-1, 7)$ and $(2, 2)$.

14. Find the slope of a line parallel to the line passing through the points $(0, 7)$ and $(-1, 0)$.

15. If $g(x) = x^2 - 3$, find
    a. $g(2)$ **b.** $g(-2)$
    c. $g(0)$.

16. If $f(x) = 3 - x^2$, find
    a. $f(2)$ **b.** $f(-2)$
    c. $f(0)$

17. Solve the system: $\begin{cases} 2x + y = 10 \\ x = y + 2 \end{cases}$

18. Solve the system: $\begin{cases} 3y = x + 10 \\ 2x + 5y = 24 \end{cases}$

19. Solve the system: $\begin{cases} -x - \dfrac{y}{2} = \dfrac{5}{2} \\ -\dfrac{x}{2} + \dfrac{y}{4} = 0 \end{cases}$

20. Solve the system: $\begin{cases} \dfrac{x}{2} + y = \dfrac{5}{6} \\ 2x - y = \dfrac{5}{6} \end{cases}$

21. Divide $x^2 + 7x + 12$ by $x + 3$ using long division.

22. Divide: $\dfrac{5x^2y - 6xy + 2}{6xy}$

Factor completely.

23. **a.** $6t + 18$
    b. $y^5 - y^7$

24. **a.** $5y - 20$
    b. $z^{10} - z^3$

25. $x^2 + 4x - 12$

26. $x^2 - 10x + 21$

27. $10x^2 - 13xy - 3y^2$

28. $12a^2 + 5ab - 2b^2$

29. $x^3 + 8$

30. $y^3 - 27$

Solve.

31. $x^2 - 9x = -20$

32. $y^2 - 5y = -6$

Simplify.

33. **a.** $\dfrac{24x^6y^5}{8x^7y}$

     b. $\dfrac{2x^2}{10x^3 - 2x^2}$

34. **a.** $\dfrac{33x^4y^2}{3xy}$

     b. $\dfrac{9y}{90y^2 + 9y}$

Perform indicated operations.

35. $\dfrac{3x + 3}{5x - 5x^2} \cdot \dfrac{2x^2 + x - 3}{4x^2 - 9}$

36. $\dfrac{2x}{x - 6} - \dfrac{x + 6}{x - 6}$

37. $\dfrac{3x^2 + 2x}{x - 1} - \dfrac{10x - 5}{x - 1}$

38. $\dfrac{9}{y^2} - 4y$

Solve.

39. $3 - \dfrac{6}{x} = x + 8$

40. $\dfrac{x}{2} + \dfrac{x}{5} = \dfrac{x - 7}{20}$

Find an equation of the line through the given points. Write the equation using function notation.

41. $(4, 0)$ and $(-4, -5)$

42. $(-1, 3)$ and $(-2, 7)$

Inequalities and Absolute Value

Our ability to solve problems using mathematics depends in part on our ability to solve equations and inequalities. In this chapter, we solve absolute value equations and inequalities in one variable and graph their solutions on number lines. We conclude this chapter by graphing linear inequalities in two variables.

Since the beginning of civilization, education has been necessary for humanity to survive. The first public schools in the United States were established in Massachusetts in 1647. These schools were open to all children and were supported by public funds. After the Revolutionary War, attempts to unify public education resulted in the establishment of state public school systems.

Today, a large team of educators, principals, counselors, librarians, school board members, and a school superintendent administer today's complex system of public schools. Administrators establish policy and procedure as well as manage the budget and other business aspects involved in the maintenance of a large organization. Most jobs in school administration require a master's or doctoral degree in an education-related field of study.

In the Spotlight on Decision Making feature on page 526, you will have the opportunity to make a decision about long-term planning related to school enrollment as a school superintendent.

Link: www.aasa.org *(American Association of School Administrators)*
Sources of text: www.bls.gov/oco/ocos007.htm *(Bureau of Labor Statistics), World Book Millennium 2000 (encyclopedia on CD) "Education"*

9.1   COMPOUND INEQUALITIES

Objectives

1 Find the intersection of two sets.

2 Solve compound inequalities containing **and**.

3 Find the union of two sets.

4 Solve compound inequalities containing **or**.

Two inequalities joined by the words **and** or **or** are called **compound inequalities.**

Compound Inequalities

$$x + 3 < 8 \text{ and } x > 2$$

$$\frac{2x}{3} \geq 5 \text{ or } -x + 10 < 7$$

1   The solution set of a compound inequality formed by the word **and** is the **intersection** of the solution sets of the two inequalities. We use the symbol ∩ to represent "intersection."

> ### Intersection of Two Sets
>
> The intersection of two sets, A and B, is the set of all elements common to both sets. A intersect B is denoted by
>
> $$A \cap B$$
>

EXAMPLE 1

If $A = \{x \mid x \text{ is an even number greater than 0 and less than 10}\}$ and $B = \{3, 4, 5, 6\}$, find $A \cap B$.

Solution   Let's list the elements in set A.

$$A = \{2, 4, 6, 8\}$$

The numbers 4 and 6 are in both sets. The intersection is $\{4, 6\}$.

2   A value is a solution of a compound inequality formed by the word **and** if it is a solution of *both* inequalities. For example, the solution set of the compound inequality $x \leq 5$ and $x \geq 3$ contains all values of x that make the inequality $x \leq 5$ a true statement **and** the inequality $x \geq 3$ a true statement. The first graph shown below is the graph of $x \leq 5$, the second graph is the graph of $x \geq 3$, and the third graph shows the intersection of the two graphs. The third graph is the graph of $x \leq 5$ **and** $x \geq 3$.

$\{x \mid x \leq 5\}$ $(-\infty, 5]$

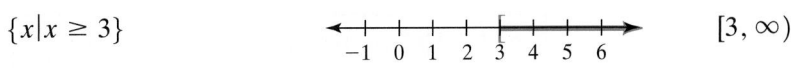

$\{x | x \geq 3\}$ $[3, \infty)$

$\{x | x \leq 5 \text{ and } x \geq 3\}$ $[3, 5]$

Since $x \geq 3$ is the same as $3 \leq x$, the compound inequality $3 \leq x$ and $x \leq 5$ can be written in a more compact form as $3 \leq x \leq 5$. The solution set $\{x | 3 \leq x \leq 5\}$ includes all numbers that are greater than or equal to 3 and at the same time less than or equal to 5. In interval notation, the solution set is $[3, 5]$.

EXAMPLE 2

Solve $x - 7 < 2$ and $2x + 1 < 9$.

Soution   First we solve each inequality separately.

$$
\begin{array}{rcl}
x - 7 < 2 & \text{and} & 2x + 1 < 9 \\
x < 9 & \text{and} & 2x < 8 \\
x < 9 & \text{and} & x < 4
\end{array}
$$

Now we can graph the two intervals on two number lines and find their intersection. Their intersection is shown on the third number line below.

$\{x | x < 9\}$ $(-\infty, 9)$

$\{x | x < 4\}$ $(-\infty, 4)$

$\{x | x < 9 \text{ and } x < 4\} = \{x | x < 4\}$ $(-\infty, 4)$

The solution set is $(-\infty, 4)$.

EXAMPLE 3

Solve $2x \geq 0$ and $4x - 1 \leq -9$.

Solution   First we solve each inequality separately.

$$
\begin{array}{rcl}
2x \geq 0 & \text{and} & 4x - 1 \leq -9 \\
x \geq 0 & \text{and} & 4x \leq -8 \\
x \geq 0 & \text{and} & x \leq -2
\end{array}
$$

Now we can graph the two intervals and find their intersection.

$\{x | x \geq 0\}$ $[0, \infty)$

$\{x | x \leq -2\}$ $(-\infty, -2]$

$\{x | x \geq 0 \text{ and } x \leq -2\} = \varnothing$ $\varnothing$

There is no number that is greater than or equal to 0 *and* less than or equal to -2. The solution set is $\varnothing$.

> ▶ **Helpful Hint**
>
> Example 3 shows that some compound inequalities have no solution. Also, some have all real numbers as solutions.

To solve a compound inequality written in a compact form, such as $2 < 4 - x < 7$, we get x alone in the "middle part." Since a compound inequality is really two inequalities in one statement, we must perform the same operations on all three parts of the inequality.

EXAMPLE 4

Solve $2 < 4 - x < 7$.

Solution To get x alone, we first subtract 4 from all three parts.

$$2 < 4 - x < 7$$
$$2 - 4 < 4 - x - 4 < 7 - 4 \qquad \text{Subtract 4 from all three parts.}$$
$$-2 < -x < 3 \qquad \text{Simplify.}$$
$$\frac{-2}{-1} > \frac{-x}{-1} > \frac{3}{-1} \qquad \begin{array}{l}\text{Divide all three parts by } -1 \text{ and} \\ \text{reverse the inequality symbols.}\end{array}$$

> ▶ **Helpful Hint**
> Don't forget to reverse both inequality symbols.

$$2 > x > -3$$

This is equivalent to $-3 < x < 2$.

The solution set in interval notation is $(-3, 2)$, and its graph is shown.

EXAMPLE 5

Solve $-1 \le \dfrac{2x}{3} + 5 \le 2$.

Solution First, clear the inequality of fractions by multiplying all three parts by the LCD of 3.

$$-1 \le \frac{2x}{3} + 5 \le 2$$
$$3(-1) \le 3\left(\frac{2x}{3} + 5\right) \le 3(2) \qquad \text{Multiply all three parts by the LCD of 3.}$$
$$-3 \le 2x + 15 \le 6 \qquad \text{Use the distributive property and multiply.}$$
$$-3 - 15 \le 2x + 15 - 15 \le 6 - 15 \qquad \text{Subtract 15 from all three parts.}$$
$$-18 \le 2x \le -9 \qquad \text{Simplify.}$$
$$\frac{-18}{2} \le \frac{2x}{2} \le \frac{-9}{2} \qquad \text{Divide all three parts by 2.}$$
$$-9 \le x \le -\frac{9}{2} \qquad \text{Simplify.}$$

The graph of the solution is shown.

$$-\frac{9}{2}$$

$-10\ -9\ -8\ -7\ -6\ -5\ -4\ -3$

The solution set in interval notation is $\left[-9, -\dfrac{9}{2}\right]$.

3 The solution set of a compound inequality formed by the word **or** is the **union** of the solution sets of the two inequalities. We use the symbol $\cup$ to denote "union."

Union of Two Sets

The union of two sets, A and B, is the set of elements that belong to *either* of the sets. A union B is denoted by

$$A \cup B \qquad A \quad B$$

$A \cup B$

EXAMPLE 6

If $A = \{x \mid x \text{ is an even number greater than 0 and less than 10}\}$ and $B = \{3, 4, 5, 6\}$ Find $A \cup B$.

Solution Recall from Example 1 that $A = \{2, 4, 6, 8\}$. The numbers that are in either set or both sets are $\{2, 3, 4, 5, 6, 8\}$. This set is the union.

4 A value is a solution of a compound inequality formed by the word **or** if it is a solution of **either** inequality. For example, the solution set of the compound inequality $x \le 1$ **or** $x \ge 3$ contains all numbers that make the inequality $x \le 1$ a true statement **or** the inequality $x \ge 3$ a true statement.

$\{x \mid x \le 1\}$ $\begin{array}{c} \\ -1\ 0\ 1\ 2\ 3\ 4\ 5\ 6 \end{array}$ $(-\infty, 1]$

$\{x \mid x \ge 3\}$ $\begin{array}{c} \\ -1\ 0\ 1\ 2\ 3\ 4\ 5\ 6 \end{array}$ $[3, \infty)$

$\{x \mid x \le 1 \text{ or } x \ge 3\}$ $\begin{array}{c} \\ -1\ 0\ 1\ 2\ 3\ 4\ 5\ 6 \end{array}$ $(-\infty, 1] \cup [3, \infty)$

In interval notation, the set $\{x \mid x \le 1 \text{ or } x \ge 3\}$ is written as $(-\infty, 1] \cup [3, \infty)$.

EXAMPLE 7

Solve $5x - 3 \le 10$ or $x + 1 \ge 5$.

Solution First we solve each inequality separately.

$$
\begin{aligned}
5x - 3 &\le 10 \quad && \text{or} & x + 1 &\ge 5 \\
5x &\le 13 \quad && \text{or} & x &\ge 4 \\
x &\le \frac{13}{5} \quad && \text{or} & x &\ge 4
\end{aligned}
$$

Now we can graph each interval and find their union.

$$\left\{x \,\middle|\, x \leq \frac{13}{5}\right\}$$

$$\left(-\infty, \frac{13}{5}\right]$$

$$\{x \,|\, x \geq 4\}$$

$$[4, \infty)$$

$$\left\{x \,\middle|\, x \leq \frac{13}{5} \text{ or } x \geq 4\right\}$$

$$\left(-\infty, \frac{13}{5}\right] \cup [4, \infty)$$

The solution set is $\left(-\infty, \dfrac{13}{5}\right] \cup [4, \infty)$.

EXAMPLE 8

Solve $-2x - 5 < -3$ or $6x < 0$.

Solution First we solve each inequality separately.

$$
\begin{array}{rclcrcl}
-2x - 5 & < & -3 & \text{or} & 6x & < & 0 \\
-2x & < & 2 & \text{or} & x & < & 0 \\
x & > & -1 & \text{or} & x & < & 0
\end{array}
$$

Now we can graph each interval and find their union.

$$\{x \,|\, x > -1\}$$

$$(-1, \infty)$$

$$\{x \,|\, x < 0\}$$

$$(-\infty, 0)$$

$$\{x \,|\, x > -1 \text{ or } x < 0\}$$
= all real numbers

$$(-\infty, \infty)$$

The solution set is $(-\infty, \infty)$.

Spotlight on
DECISION
✗MAKING

Suppose you are the superintendent of Copley Public Schools. You are aware that the general population of Copley is increasing and that enrollment at the schools is steadily rising. The high school can house a maximum of 1200 students. Once this maximum has been exceeded, temporary classrooms must be erected to handle the overflow.

You have been studying the changes in population and conclude that the equation $y = 30x + 1025$ models the high school enrollment x years from now. As you prepare a long-term planning report, you must decide whether temporary classrooms will be needed in the next 10 years. If so, when is the latest that funding for temporary classrooms could be added to the annual budget?

STUDY SKILLS REMINDER

Continue your outline started in Section 2.3. Write how to recognize and how to solve compound inequalities in your own words. For example:

Solving Equations and Inequalities

I. Equations
    A. Linear equations—(Section 2.3)
    B. Quadratic equations—(Section 6.5)
    C. Equations with rational expressions—(Section 7.5)

II. Inequalities
    A. Linear inequalities—(Section 2.7)
    B. **Compound Inequalities**—Recognize: *2 inequality signs or 2 inequalities separated by "and" or "or"*—Solve: if 2 inequality signs, then solve for *x* in the middle, remember that "and" means intersection and "or" means union.

See Appendix A for summary exercises.

EXERCISE SET 9.1

STUDY   CD/   PH MATH   MathXL®Tutorials   MathXL®   MyMathLab®
GUIDE/SSM   VIDEO   TUTOR CENTER   ON CD

If $A = \{x \mid x \text{ is an even integer}\}$, $B = \{x \mid x \text{ is an odd integer}\}$, $C = \{2, 3, 4, 5\}$, and $D = \{4, 5, 6, 7\}$, list the elements of each set. See Examples 1 and 6.

1. $C \cup D$ **2.** $C \cap D$

3. $A \cap D$ **4.** $A \cup D$

5. $A \cup B$ **6.** $A \cap B$

7. $B \cap D$ **8.** $B \cup D$

9. $B \cup C$ **10.** $B \cap C$

11. $A \cap C$ **12.** $A \cup C$

Solve each compound inequality. Graph the solution set and write it in interval notation. See Examples 2 and 3.

13. $x < 5$ and $x > -2$

14. $x \le 7$ and $x \le 1$

15. $x + 1 \ge 7$ and $3x - 1 \ge 5$

16. $-2x < -8$ and $x - 5 < 5$

17. $4x + 2 \le -10$ and $2x \le 0$ **18.** $x + 4 > 0$ and $4x > 0$

Solve each compound inequality. Graph the solution set and write it in interval notation. See Examples 4 and 5.

19. $5 < x - 6 < 11$ **20.** $-2 \le x + 3 \le 0$

21. $-2 \le 3x - 5 \le 7$ **22.** $1 < 4 + 2x < 7$

23. $1 \le \dfrac{2}{3}x + 3 \le 4$ **24.** $-2 < \dfrac{1}{2}x - 5 < 1$

25. $-5 \le \dfrac{x + 1}{4} \le -2$ **26.** $-4 \le \dfrac{2x + 5}{3} \le 1$

Solve each compound inequality. Graph the solution set and write it in interval notation. See Examples 7 and 8.

27. $x < -1$ or $x > 0$

28. $x \le 1$ or $x \le -3$

29. $-2x \le -4$ or $5x - 20 \ge 5$

30. $x + 4 < 0$ or $6x > -12$

31. $3(x - 1) < 12$ or $x + 7 > 10$

32. $5(x - 1) \geq -5$ or $5 - x \leq 11$

33. Explain how solving an and–compound inequality is similar to finding the intersection of two sets.

34. Explain how solving an or–compound inequality is similar to finding the union of two sets.

MIXED PRACTICE

Solve each compound inequality. Graph the solution set and write it in interval notation.

35. $x < 2$ and $x > -1$

36. $x < 5$ and $x < 1$

37. $x < 2$ or $x > -1$

38. $x < 5$ or $x < 1$

39. $x \geq -5$ and $x \geq -1$

40. $x \leq 0$ or $x \geq -3$

41. $x \geq -5$ or $x \geq -1$

42. $x \leq 0$ and $x \geq -3$

43. $0 \leq 2x - 3 \leq 9$

44. $3 < 5x + 1 < 11$

45. $\dfrac{1}{2} < x - \dfrac{3}{4} < 2$

46. $\dfrac{2}{3} < x + \dfrac{1}{2} < 4$

47. $x + 3 \geq 3$ and $x + 3 \leq 2$

48. $2x - 1 \geq 3$ and $-x > 2$

49. $3x \geq 5$ or $-x - 6 < 1$

50. $\dfrac{3}{8}x + 1 \leq 0$ or $-2x < -4$

51. $0 < \dfrac{5 - 2x}{3} < 5$

52. $-2 < \dfrac{-2x - 1}{3} < 2$

53. $-6 < 3(x - 2) \leq 8$

54. $-5 < 2(x + 4) < 8$

55. $-x + 5 > 6$ and $1 + 2x \leq -5$

56. $5x \leq 0$ and $-x + 5 < 8$

57. $3x + 2 \leq 5$ or $7x > 29$

58. $-x < 7$ or $3x + 1 < -20$

59. $5 - x > 7$ and $2x + 3 \geq 13$

60. $-2x < -6$ or $1 - x > -2$

61. $-\dfrac{1}{2} \leq \dfrac{4x - 1}{6} < \dfrac{5}{6}$

62. $-\dfrac{1}{2} \leq \dfrac{3x - 1}{10} < \dfrac{1}{2}$

63. $\dfrac{1}{15} < \dfrac{8 - 3x}{15} < \dfrac{4}{5}$

64. $-\dfrac{1}{4} < \dfrac{6 - x}{12} < -\dfrac{1}{6}$

65. $0.3 < 0.2x - 0.9 < 1.5$

66. $-0.7 \leq 0.4x + 0.8 < 0.5$

REVIEW AND PREVIEW

Evaluate the following. See Sections 1.5 and 1.6.

67. $|-7| - |19|$

68. $|-7 - 19|$

69. $-(-6) - |-10|$

70. $|-4| - (-4) + |-20|$

Find by inspection all values for x that make each equation true.

71. $|x| = 7$

72. $|x| = 5$

73. $|x| = 0$

74. $|x| = -2$

Concept Extensions

The formula for converting Fahrenheit temperatures to Celsius temperatures is $C = \frac{5}{9}(F - 32)$. Use this formula for Exercises 75 and 76.

75. During a recent year, the temperatures in Chicago ranged from $-29°$ to $35°C$. Use a compound inequality to convert these temperatures to Fahrenheit temperatures.

76. In Oslo, the average temperature ranges from $-10°$ to $18°$ Celsius. Use a compound inequality to convert these temperatures to the Fahrenheit scale.

Solve.

77. Christian D'Angelo has scores of 68, 65, 75, and 78 on his algebra tests. Use a compound inequality to find the scores he can make on his final exam to receive a C in the course. The final exam counts as two tests, and a C is received if the final course average is from 70 to 79.

78. Wendy Wood has scores of 80, 90, 82, and 75 on her chemistry tests. Use a compound inequality to find the range of scores she can make on her final exam to receive a B in the course. The final exam counts as two tests, and a B is received if the final course average is from 80 to 89.

Use the graph to answer Exercises 79 and 80.

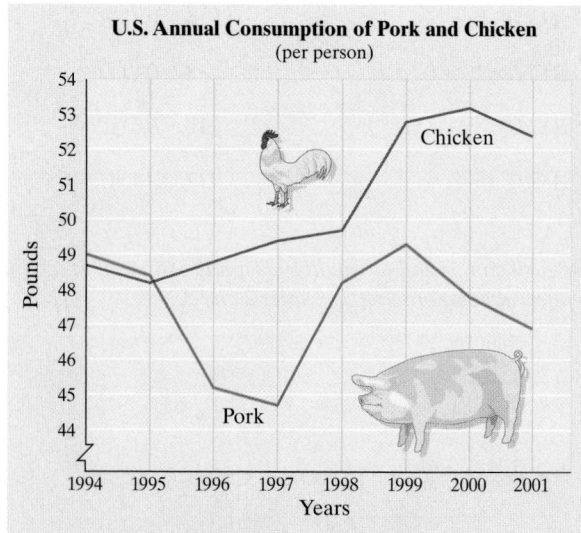

U.S. Annual Consumption of Pork and Chicken
(per person)

Source: Based on data from Economic Research Service, U.S. Department of Agriculture, *Agricultural Outlook*, June/July, 1999.

79. For what years was the consumption of pork greater than 48 pounds per person *and* the consumption of chicken greater than 48 pounds per person?

80. For what years was the consumption of pork less than 48 pounds per person *or* the consumption of chicken greater than 49 pounds per person?

For Exercises 81 through 86, study the following example.

Example

*To solve $x - 6 < 3x < 2x + 5$, notice that this inequality contains a variable not only in the middle, but also on the left and the right. When this occurs, we solve by rewriting the inequality using the word **and**.*

$$x - 6 < 3x \quad \text{and} \quad 3x < 2x + 5$$

$$-6 < 2x \quad \text{and} \quad x < 5$$

$$-3 < x$$

$$x > -3 \quad \text{and} \quad x < 5$$

$x > -3$

$x < 5$

$-3 < x < 5$, or $(-3, 5)$

In Exercises 81 through 86 use the above example to solve the compound inequality for x.

81. $2x - 3 < 3x + 1 < 4x - 5$

82. $x + 3 < 2x + 1 < 4x + 6$

83. $-3(x - 2) \leq 3 - 2x \leq 10 - 3x$

84. $7x - 1 \leq 7 + 5x \leq 3(1 + 2x)$

85. $5x - 8 < 2(2 + x) < -2(1 + 2x)$

86. $1 + 2x < 3(2 + x) < 1 + 4x$

9.2   ABSOLUTE VALUE EQUATIONS

Objective

1 Solve absolute value equations.

1 In Chapter 1, we defined the absolute value of a number as its distance from 0 on a number line.

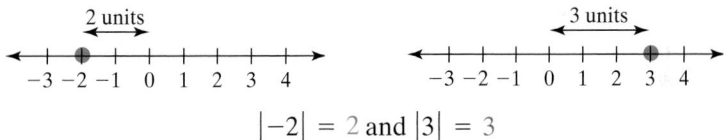

$|-2| = 2$ and $|3| = 3$

In this section, we concentrate on solving equations containing the absolute value of a variable or a variable expression. Examples of absolute value equations are

$$|x| = 3 \qquad -5 = |2y + 7| \qquad |z - 6.7| = |3z + 1.2|$$

Since distance and absolute value are so closely related, absolute value equations and inequalities (see Section 9.3) are extremely useful in solving distance-type problems, such as calculating the possible error in a measurement. For the absolute value equation $|x| = 3$, its solution set will contain all numbers whose distance from 0 is 3 units. Two numbers are 3 units away from 0 on the number line: 3 and -3.

Thus, the solution set of the equation $|x| = 3$ is $\{3, -3\}$. This suggests the following:

Solving Equations of the Form $|X| = a$

If a is a positive number, then $|X| = a$ is equivalent to $X = a$ or $X = -a$.

EXAMPLE 1

Solve $|p| = 2$.

Solution Since 2 is positive, $|p| = 2$ is equivalent to $p = 2$ or $p = -2$.

To check, let $p = 2$ and then $p = -2$ in the original equation.

$|p| = 2$	*Original equation*		$|p| = 2$	*Original equation*
$|2| = 2$	*Let $p = 2$.*		$|-2| = 2$	*Let $p = -2$.*
$2 = 2$	*True*		$2 = 2$	*True*

The solutions are 2 and -2 or the solution set is $\{2, -2\}$.

If the expression inside the absolute value bars is more complicated than a single variable x, we can still apply the absolute value property.

> **Helpful Hint**
>
> For the equation $|X| = a$ in the box on the previous page, X can be a single variable or a variable expression.

EXAMPLE 2

Solve $|5w + 3| = 7$.

Solution Here the expression inside the absolute value bars is $5w + 3$. If we think of the expression $5w + 3$ as x in the absolute value property, we see that $|x| = 7$ is equivalent to

$$x = 7 \quad \text{or} \quad x = -7$$

Then substitute $5w + 3$ for x, and we have

$$5w + 3 = 7 \quad \text{or} \quad 5w + 3 = -7$$

Solve these two equations for w.

$$
\begin{aligned}
5w + 3 &= 7 &\text{or} \quad 5w + 3 &= -7 \\
5w &= 4 &\text{or} \quad 5w &= -10 \\
w &= \frac{4}{5} &\text{or} \quad w &= -2
\end{aligned}
$$

Check To check, let $w = -2$ and then $w = \frac{4}{5}$ in the original equation.

Let $w = -2$

$$
\begin{aligned}
|5(-2) + 3| &= 7 \\
|-10 + 3| &= 7 \\
|-7| &= 7 \\
7 &= 7 \quad \text{True}
\end{aligned}
$$

Let $w = \frac{4}{5}$

$$
\begin{aligned}
\left|5\left(\frac{4}{5}\right) + 3\right| &= 7 \\
|4 + 3| &= 7 \\
|7| &= 7 \\
7 &= 7 \quad \text{True}
\end{aligned}
$$

Both solutions check, and the solutions are -2 and $\frac{4}{5}$ or the solution set is $\left\{-2, \frac{4}{5}\right\}$.

EXAMPLE 3

Solve $\left|\dfrac{x}{2} - 1\right| = 11$.

Solution $\left|\dfrac{x}{2} - 1\right| = 11$ is equivalent to

$$\dfrac{x}{2} - 1 = 11 \quad \text{or} \quad \dfrac{x}{2} - 1 = -11$$

$$2\left(\dfrac{x}{2} - 1\right) = 2(11) \quad \text{or} \quad 2\left(\dfrac{x}{2} - 1\right) = 2(-11) \qquad \text{Clear fractions.}$$

$$x - 2 = 22 \quad \text{or} \quad x - 2 = -22 \qquad \text{Apply the distributive property.}$$

$$x = 24 \quad \text{or} \quad x = -20$$

The solutions are 24 and -20.

> **Helpful Hint**
>
> If the equation has a single absolute value expression containing variables, isolate the absolute value expression first.

To apply the absolute value rule, first make sure that the absolute value expression is isolated.

EXAMPLE 4

Solve $|2x| + 5 = 7$.

Solution   We want the absolute value expression alone on one side of the equation, so begin by subtracting 5 from both sides. Then apply the absolute value property.

$$|2x| + 5 = 7$$

$$|2x| = 2 \qquad \text{Subtract 5 from both sides.}$$

$$2x = 2 \quad \text{or} \quad 2x = -2$$

$$x = 1 \quad \text{or} \quad x = -1$$

The solutions are -1 and 1.

EXAMPLE 5

Solve $|y| = 0$.

Solution   We are looking for all numbers whose distance from 0 is zero units. The only number is 0. The solution is 0.

The next two examples illustrate a special case for absolute value equations. This special case occurs when an isolated absolute value is equal to a negative number.

EXAMPLE 6

Solve $2|x| + 25 = 23$.

Solution   First, isolate the absolute value.

$$2|x| + 25 = 23$$

$$2|x| = -2 \qquad \text{Subtract 25 from both sides.}$$

$$|x| = -1 \qquad \text{Divide both sides by 2.}$$

The absolute value of a number is never negative, so this equation has no solution. The solution set is $\{\ \}$ or $\varnothing$.

EXAMPLE 7

Solve $\left|\dfrac{3x + 1}{2}\right| = -2$.

Solution Again, the absolute value of any expression is never negative, so no solution exists. The solution set is $\{\ \ \}$ or $\varnothing$.

Given two absolute value expressions, we might ask, when are the absolute values of two expressions equal? To see the answer, notice that

$$|2| = |2|, \quad |-2| = |-2|, \quad |-2| = |2|, \quad \text{and} \quad |2| = |-2|$$

same   same   opposites   opposites

Two absolute value expressions are equal when the expressions inside the absolute value bars are equal to or are opposites of each other.

EXAMPLE 8

Solve $|3x + 2| = |5x - 8|$.

Solution This equation is true if the expressions inside the absolute value bars are equal to or are opposites of each other.

$$3x + 2 = 5x - 8 \quad \text{or} \quad 3x + 2 = -(5x - 8)$$

Next, solve each equation.

$$
\begin{aligned}
3x + 2 &= 5x - 8 &\text{or}\quad 3x + 2 &= -5x + 8 \\
-2x + 2 &= -8 &\text{or}\quad 8x + 2 &= 8 \\
-2x &= -10 &\text{or}\quad 8x &= 6 \\
x &= 5 &\text{or}\quad x &= \frac{3}{4}
\end{aligned}
$$

The solutions are $\dfrac{3}{4}$ and 5.

EXAMPLE 9

Solve $|x - 3| = |5 - x|$.

Solution

$$
\begin{aligned}
x - 3 &= 5 - x &\text{or}\quad x - 3 &= -(5 - x) \\
2x - 3 &= 5 &\text{or}\quad x - 3 &= -5 + x \\
2x &= 8 &\text{or}\quad x - 3 - x &= -5 + x - x \\
x &= 4 &\text{or}\quad -3 &= -5 \qquad \text{False}
\end{aligned}
$$

Recall from Section 2.3 that when an equation simplifies to a false statement, the equation has no solution. Thus, the only solution for the original absolute value equation is 4.

CONCEPT CHECK

True or false? Absolute value equations always have two solutions. Explain your answer.

The following box summarizes the methods shown for solving absolute value equations.

Absolute Value Equations

$|X| = a$ $\begin{cases} \text{If } a \text{ is positive, then solve } X = a \text{ or } X = -a. \\ \text{If } a \text{ is } 0, \text{ solve } X = 0. \\ \text{If } a \text{ is negative, the equation } |X| = a \text{ has no solution.} \end{cases}$

$|X| = |Y|$ Solve $X = Y$ or $X = -Y$.

STUDY SKILLS REMINDER

Continue your outline started in Section 2.3. Write how to recognize and how to solve absolute value equations in your own words. For example:

Solving Equations and Inequalities

I. Equations
   A. Linear equations—(Section 2.3)
   B. Quadratic equations—(Section 6.5)
   C. Equations with Rational Expressions—(Section 7.5)
   D. **Absolute value equations**—Recognize: *linear equation with absolute value bars*—Solve: if possible, first isolate absolute value: if $|x| = a$ (positive), then solve $x = a$ or $x = -a$, if $|x| = 0$, then solve $x = 0$, if $|x| = -a$, then there is no solution, if $|x| = |y|$, then solve $x = y$ or $x = -y$.

II. Inequalities
   A. Linear inequalities—(Section 2.7)
   B. Compound inequalities—(Section 9.1)

See Appendix A for summary exercises.

Concept Check Answer:
false; answers may vary

MENTAL MATH

Simplify each expression.

1. $|-7|$

2. $|-8|$

3. $-|5|$

4. $-|10|$

5. $-|-6|$

6. $-|-3|$

7. $|-3| + |-2| + |-7|$

8. $|-1| + |-6| + |-8|$

EXERCISE SET 9.2

STUDY
GUIDE/SSM

CD/
VIDEO

PH MATH
TUTOR CENTER

MathXL®Tutorials
ON CD

MathXL®

MyMathLab®

Solve each absolute value equation. See Examples 1 through 7.

 1. $|x| = 7$

2. $|y| = 15$

3. $|3x| = 12.6$

4. $|6n| = 12.6$

5. $|2x - 5| = 9$

6. $|6 + 2n| = 4$

7. $\left|\dfrac{x}{2} - 3\right| = 1$

8. $\left|\dfrac{n}{3} + 2\right| = 4$

9. $|z| + 4 = 9$

10. $|x| + 1 = 3$

11. $|3x| + 5 = 14$

12. $|2x| - 6 = 4$

13. $|2x| = 0$

14. $|7z| = 0$

15. $|4n + 1| + 10 = 4$

16. $|3z - 2| + 8 = 1$

17. $|5x - 1| = 0$

18. $|3y + 2| = 0$

19. Write an absolute value equation representing all numbers x whose distance from 0 is 5 units.

20. Write an absolute value equation representing all numbers x whose distance from 0 is 2 units.

Solve. See Examples 8 and 9.

21. $|5x - 7| = |3x + 11|$

22. $|9y + 1| = |6y + 4|$

23. $|z + 8| = |z - 3|$

24. $|2x - 5| = |2x + 5|$

25. Describe how solving an absolute value equation such as $|2x - 1| = 3$ is similar to solving an absolute value equation such as $|2x - 1| = |x - 5|$.

26. Describe how solving an absolute value equation such as $|2x - 1| = 3$ is different from solving an absolute value equation such as $|2x - 1| = |x - 5|$.

MIXED PRACTICE

Solve each absolute value equation.

27. $|x| = 4$

28. $|x| = 1$

29. $|y| = 0$

30. $|y| = 8$

31. $|z| = -2$

32. $|y| = -9$

33. $|7 - 3x| = 7$

34. $|4m + 5| = 5$

35. $|6x| - 1 = 11$

36. $|7z| + 1 = 22$

37. $|4p| = -8$

38. $|5m| = -10$

39. $|x - 3| + 3 = 7$

40. $|x + 4| - 4 = 1$

41. $\left|\dfrac{z}{4} + 5\right| = -7$

42. $\left|\dfrac{c}{5} - 1\right| = -2$

43. $|9v - 3| = -8$

44. $|1 - 3b| = -7$

45. $|8n + 1| = 0$

46. $|5x - 2| = 0$

47. $|1 + 6c| - 7 = -3$

48. $|2 + 3m| - 9 = -7$

49. $|5x + 1| = 11$

50. $|8 - 6c| = 1$

51. $|4x - 2| = |-10|$

52. $|3x + 5| = |-4|$

53. $|5x + 1| = |4x - 7|$

54. $|3 + 6n| = |4n + 11|$

55. $|6 + 2x| = -|-7|$

56. $|4 - 5y| = -|-3|$

57. $|2x - 6| = |10 - 2x|$

58. $|4n + 5| = |4n + 3|$

59. $\left|\dfrac{2x - 5}{3}\right| = 7$

60. $\left|\dfrac{1 + 3n}{4}\right| = 4$

61. $2 + |5n| = 17$

62. $8 + |4m| = 24$

63. $\left|\dfrac{2x - 1}{3}\right| = |-5|$

64. $\left|\dfrac{5x + 2}{2}\right| = |-6|$

65. $|2y - 3| = |9 - 4y|$

66. $|5z - 1| = |7 - z|$

67. $\left|\dfrac{3n + 2}{8}\right| = |-1|$

68. $\left|\dfrac{2r - 6}{5}\right| = |-2|$

69. $|x + 4| = |7 - x|$

70. $|8 - y| = |y + 2|$

71. $\left|\dfrac{8c - 7}{3}\right| = -|-5|$

72. $\left|\dfrac{5d + 1}{6}\right| = -|-9|$

73. Explain why some absolute value equations have two solutions.

74. Explain why some absolute value equations have one solution.

REVIEW AND PREVIEW

The circle graph shows the U.S. cheese consumption for 2001. Use this graph to answer Exercises 75 through 77. See Sections 1.3 and 3.1.

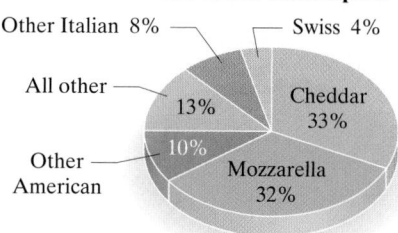

U.S. Cheese Consumption

Other Italian 8% — Swiss 4%

All other

Cheddar 33%

13%

Other American

10%

Mozzarella 32%

75. What percent of cheese consumption came from cheddar cheese?

76. A circle contains 360°. Find the number of degrees in the 4% sector for swiss cheese.

77. If a family consumed 120 pounds of cheese in 2001, find the amount of mozzarella we might expect they consumed.

List five integer solutions of each inequality.

78. $|x| \leq 3$

79. $|x| \geq -2$

80. $|y| > -10$

81. $|y| < 0$

Concept Extensions

82. Write an absolute value equation representing all numbers x whose distance from 0 is 5 units.

83. Write an absolute value equation representing all numbers x whose distance from 0 is 2 units.

Write each as an equivalent absolute value.

84. $x = 6$ or $x = -6$

85. $2x - 1 = 4$ or $2x - 1 = -4$

86. $x - 2 = 3x - 4$ or $x - 2 = -(3x - 4)$

For what value(s) of c will an absolute value equation of the form $|ax + b| = c$ have

a. one solution?

b. no solution?

c. two solutions?

9.3 ABSOLUTE VALUE INEQUALITIES

Objectives

1 Solve absolute value inequalities of the form $|x| < a$.

2 Solve absolute value inequalities of the form $|x| > a$.

1   The solution set of an absolute value inequality such as $|x| < 2$ contains all numbers whose distance from 0 is less than 2 units, as shown below.

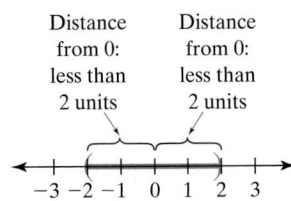

The solution set is $\{x | -2 < x < 2\}$, or $(-2, 2)$ in interval notation.

EXAMPLE 1

Solve $|x| \leq 3$.

Solution   The solution set of this inequality contains all numbers whose distance from 0 is less than or equal to 3. Thus 3, -3, and all numbers between 3 and -3 are in the solution set.

The solution set is $[-3, 3]$.

In general, we have the following.

> ## Solving Absolute Value Inequalities of the Form $|X| < a$
>
> If a is a positive number, then $|X| < a$ is equivalent to $-a < X < a$.

This property also holds true for the inequality symbol $\leq$.

EXAMPLE 2

Solve $|m - 6| < 2$.

Solution Replace X with $m - 6$ and a with 2 in the preceding property, and we see that

$$|m - 6| < 2 \quad \text{is equivalent to} \quad -2 < m - 6 < 2$$

Solve this compound inequality for m by adding 6 to all three parts.

$$-2 < m - 6 < 2$$
$$-2 + 6 < m - 6 + 6 < 2 + 6 \qquad \text{Add 6 to all three parts.}$$
$$4 < m < 8 \qquad \text{Simplify.}$$

The solution set is $(4, 8)$, and its graph is shown.

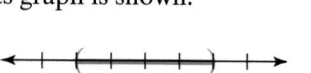

> ### ▶ Helpful Hint
> Before using an absolute value inequality property, isolate the absolute value expression on one side of the inequality.

EXAMPLE 3

Solve $|5x + 1| + 1 \leq 10$.

Solution First, isolate the absolute value expression by subtracting 1 from both sides.

$$|5x + 1| + 1 \leq 10$$
$$|5x + 1| \leq 10 - 1 \qquad \text{Subtract 1 from both sides.}$$
$$|5x + 1| \leq 9 \qquad \text{Simplify.}$$

Since 9 is positive, we apply the absolute value property for $|X| \leq a$.

$$-9 \leq 5x + 1 \leq 9$$
$$-9 - 1 \leq 5x + 1 - 1 \leq 9 - 1 \qquad \text{Subtract 1 from all three parts.}$$
$$-10 \leq 5x \leq 8 \qquad \text{Simplify.}$$
$$-2 \leq x \leq \frac{8}{5} \qquad \text{Divide all three parts by 5.}$$

The solution set is $\left[-2, \frac{8}{5} \right]$, and the graph is shown above.

EXAMPLE 4

Solve $\left| 2x - \dfrac{1}{10} \right| < -13$.

Solution The absolute value of a number is always nonnegative and can never be less than -13. Thus this absolute value inequality has no solution. The solution set is $\{\ \ \}$ or $\varnothing$.

2 Let us now solve an absolute value inequality of the form $|X| > a$, such as $|x| \geq 3$. The solution set contains all numbers whose distance from 0 is 3 or more units. Thus the graph of the solution set contains 3 and all points to the right of 3 on the number line or -3 and all points to the left of -3 on the number line.

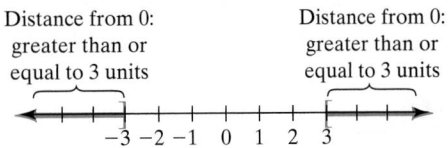

This solution set is written as $\{x \mid x \leq -3 \text{ or } x \geq 3\}$. In interval notation, the solution is $(-\infty, -3] \cup [3, \infty)$, since "or" means "union." In general, we have the following.

Solving Absolute Value Inequalities of the Form $|X| > a$

If a is a positive number, then $|X| > a$ is equivalent to $X < -a$ or $X > a$.

This property also holds true for the inequality symbol $\geq$.

EXAMPLE 5

Solve $|y - 3| > 7$.

Solution Since 7 is positive, we apply the property for $|X| > a$.

$$|y - 3| > 7 \text{ is equivalent to } y - 3 < -7 \text{ or } y - 3 > 7$$

Next, solve the compound inequality.

$$
\begin{array}{lll}
y - 3 < -7 & \text{or} & y - 3 > 7 \\
y - 3 + 3 < -7 + 3 & \text{or} & y - 3 + 3 > 7 + 3 \quad \text{Add 3 to both sides.} \\
y < -4 & \text{or} & y > 10 \quad \text{Simplify.}
\end{array}
$$

The solution set is $(-\infty, -4) \cup (10, \infty)$, and its graph is shown.

Examples 6 and 8 illustrate special cases of absolute value inequalities. These special cases occur when an isolated absolute value expression is less than, less than or equal to, greater than, or greater than or equal to a negative number or 0.

EXAMPLE 6

Solve $|2x + 9| + 5 > 3$.

Solution First isolate the absolute value expression by subtracting 5 from both sides.

$$
\begin{array}{ll}
|2x + 9| + 5 > 3 & \\
|2x + 9| + 5 - 5 > 3 - 5 & \text{Subtract 5 from both sides.} \\
|2x + 9| > -2 & \text{Simplify.}
\end{array}
$$

The absolute value of any number is always nonnegative and thus is always greater than -2. This inequality and the original inequality are true for all values of x. The solution set is $\{x \mid x \text{ is a real number}\}$ or $(-\infty, \infty)$ and its graph is shown.

✔ **CONCEPT CHECK**

Without taking any solution steps, how do you know that the absolute value inequality $|3x - 2| > -9$ has a solution? What is its solution?

EXAMPLE 7

Solve $\left| \dfrac{x}{3} - 1 \right| - 7 \geq -5$.

Solution First, isolate the absolute value expression by adding 7 to both sides.

$$\left| \frac{x}{3} - 1 \right| - 7 \geq -5$$

$$\left| \frac{x}{3} - 1 \right| - 7 + 7 \geq -5 + 7 \qquad \text{Add 7 to both sides.}$$

$$\left| \frac{x}{3} - 1 \right| \geq 2 \qquad \text{Simplify.}$$

Next, write the absolute value inequality as an equivalent compound inequality and solve.

$$\frac{x}{3} - 1 \leq -2 \qquad \text{or} \qquad \frac{x}{3} - 1 \geq 2$$

$$3\left(\frac{x}{3} - 1\right) \leq 3(-2) \qquad \text{or} \qquad 3\left(\frac{x}{3} - 1\right) \geq 3(2) \qquad \text{Clear the inequalities of fractions.}$$

$$x - 3 \leq -6 \qquad \text{or} \qquad x - 3 \geq 6 \qquad \text{Apply the distributive property.}$$

$$x \leq -3 \qquad \text{or} \qquad x \geq 9 \qquad \text{Add 3 to both sides.}$$

The solution set is $(-\infty, -3] \cup [9, \infty)$, and its graph is shown.

EXAMPLE 8

Solve $\left| \dfrac{2(x + 1)}{3} \right| \leq 0$.

Solution Recall that "$\leq$" means "less than or equal to." The absolute value of any expression will never be less than 0, but it may be equal to 0. Thus, to solve $\left| \dfrac{2(x + 1)}{3} \right| \leq 0$ we solve $\left| \dfrac{2(x + 1)}{3} \right| = 0$

Concept Check Answer:
$(-\infty, \infty)$ since the absolute value is always nonnegative

$$\frac{2(x + 1)}{3} = 0$$

$$3\left[\frac{2(x + 1)}{3}\right] = 3(0) \qquad \text{Clear the equation of fractions.}$$

$$2x + 2 = 0 \qquad \text{Apply the distributive property.}$$

$$2x = -2 \qquad \text{Subtract 2 from both sides.}$$

$$x = -1 \qquad \text{Divide both sides by 2.}$$

The solution set is $\{-1\}$.

The following box summarizes the types of absolute value equations and inequalities.

Solving Absolute Value Equations and Inequalities With $a > 0$

Algebraic Solution	Solution Graph
$|X| = a$ is equivalent to $X = a$ or $X = -a$.	
$|X| < a$ is equivalent to $-a < X < a$.	
$|X| > a$ is equivalent to $X < -a$ or $X > a$.	

STUDY SKILLS REMINDER

Continue your outline started in Section 2.3. Write how to recognize and how to solve absolute value inequalities in your own words. For example:

Solving Equations and Inequalities

I. Equations
    A. Linear equations (Section 2.3)
    B. Quadratic equations—(Section 6.5)
    C. Equations with rational expressions—(Section 7.5)
    D. Absolute value equations (Section 9.2)

II. Inequalities
    A. Linear inequalities (Section 2.7)
    B. Compound inequalities (Section 9.1)
    C. **Absolute value inequalities**—Recognize: *inequality with absolute value bars*—Solve: if possible, first isolate absolute value: if $|x| > a$ (positive), solve $x < -a$ or $x > a$, if $|x| < a$ (positive) then solve $-a < x < a$.

See Appendix A for summary exercises.

MENTAL MATH

Match each absolute value inequality with an equivalent statement.

1. $|2x + 1| = 3$ **A.** $2x + 1 > 3$ or $2x + 1 < -3$

2. $|2x + 1| \leq 3$ **B.** $2x + 1 \geq 3$ or $2x + 1 \leq -3$

3. $|2x + 1| < 3$ **C.** $-3 < 2x + 1 < 3$

4. $|2x + 1| \geq 3$ **D.** $2x + 1 = 3$ or $2x + 1 = -3$

5. $|2x + 1| > 3$ **E.** $-3 \leq 2x + 1 \leq 3$

EXERCISE SET 9.3

STUDY GUIDE/SSM   CD/VIDEO   PH MATH TUTOR CENTER   MathXL®Tutorials ON CD   MathXL®   MyMathLab®

Solve each inequality. Then graph the solution set. See Examples 1 through 4.

1. $|x| \leq 4$ 2. $|x| < 6$

3. $|x - 3| < 2$ 4. $|y| \leq 5$

5. $|x + 3| < 2$ 6. $|x + 4| < 6$

7. $|2x + 7| \leq 13$ 8. $|5x - 3| \leq 18$

9. $|x| + 7 \leq 12$ 10. $|x| + 6 \leq 7$

 11. $|3x - 1| < -5$ 12. $|8x - 3| < -2$

13. $|x - 6| - 7 \leq -1$ 14. $|z + 2| - 7 < -3$

Solve each inequality. Graph the solution set. See Examples 5 through 7.

15. $|x| > 3$ 16. $|y| \geq 4$

17. $|x + 10| \geq 14$ 18. $|x - 9| \geq 2$

19. $|x| + 2 > 6$ 20. $|x| - 1 > 3$

21. $|5x| > -4$ 22. $|4x - 11| > -1$

23. $|6x - 8| + 3 > 7$ 24. $|10 + 3x| + 1 > 2$

Solve each inequality. Graph the solution set. See Example 8.

25. $|x| \leq 0$

26. $|x| \geq 0$

27. $|8x + 3| > 0$

28. $|5x - 6| < 0$

MIXED PRACTICE

Solve each inequality. Graph the solution set.

29. $|x| \leq 2$ 30. $|z| < 6$

31. $|y| > 1$ 32. $|x| \geq 10$

33. $|x - 3| < 8$

34. $|-3 + x| \leq 10$

35. $|0.6x - 3| > 0.6$

36. $|1 + 0.3x| \geq 0.1$

37. $5 + |x| \leq 2$

38. $8 + |x| < 1$

39. $|x| > -4$

40. $|x| \leq -7$

41. $|2x - 7| \leq 11$

42. $|5x + 2| < 8$

43. $|x + 5| + 2 \geq 8$

44. $|-1 + x| - 6 > 2$

45. $|x| > 0$

46. $|x| < 0$

47. $9 + |x| > 7$

48. $5 + |x| \geq 4$

49. $6 + |4x - 1| \leq 9$

50. $-3 + |5x - 2| \leq 4$

51. $\left|\dfrac{2}{3}x + 1\right| > 1$

52. $|5x - 1| \geq 2$

53. $|5x + 3| < -6$

54. $|4 + 9x| \geq -6$

55. $|8x + 3| \geq 0$

56. $|5x - 6| \leq 0$

57. $|1 + 3x| + 4 < 5$

58. $|7x - 3| - 1 \leq 10$

59. $\left|\dfrac{x + 6}{3}\right| > 2$

60. $\left|\dfrac{7 + x}{2}\right| \geq 4$

61. $-15 + |2x - 7| \leq -6$

62. $-9 + |3 + 4x| < -4$

63. $\left|2x + \dfrac{3}{4}\right| - 7 \leq -2$

64. $\left|\dfrac{3}{5} + 4x\right| - 6 < -1$

Solve each equation or inequality for x.

65. $|2x - 3| < 7$

66. $|2x - 3| > 7$

67. $|2x - 3| = 7$

68. $|5 - 6x| = 29$

69. $|x - 5| \geq 12$

70. $|x + 4| \geq 20$

71. $|9 + 4x| = 0$

72. $|9 + 4x| \geq 0$

73. $|2x + 1| + 4 < 7$

74. $8 + |5x - 3| \geq 11$

75. $|3x - 5| + 4 = 5$

76. $|8x| = -5$

77. $|x + 11| = -1$

78. $|4x - 4| = -3$

79. $\left|\dfrac{2x - 1}{3}\right| = 6$

80. $\left|\dfrac{6 - x}{4}\right| = 5$

81. $\left|\dfrac{3x - 5}{6}\right| > 5$

82. $\left|\dfrac{4x - 7}{5}\right| < 2$

REVIEW AND PREVIEW

Recall the formula:

$$\text{Probability of an event} = \dfrac{\text{number of ways that the event can occur}}{\text{number of possible outcomes}}$$

Find the probability of rolling each number on a single toss of a die. (Recall that a die is a cube with each of its six sides containing 1, 2, 3, 4, 5, and 6 black dots, respectively.) See Section 2.3.

83. $P(\text{rolling a } 2)$

84. $P(\text{rolling a } 5)$

85. $P(\text{rolling a } 7)$

86. $P(\text{rolling a } 0)$

87. $P(\text{rolling a } 1 \text{ or } 3)$

88. $P(\text{rolling a } 1, 2, 3, 4, 5, \text{ or } 6)$

Consider the equation $3x - 4y = 12$. For each value of x or y given, find the corresponding value of the other variable that makes the statement true. See Section 3.2.

89. If $x = 2$, find y

90. If $y = -1$, find x

91. If $y = -3$, find x

92. If $x = 4$, find y

Concept Extensions

93. Write an absolute value inequality representing all numbers x whose distance from 0 is less than 7 units.

94. Write an absolute value inequality representing all numbers x whose distance from 0 is greater than 4 units.

95. Write $-5 \leq x \leq 5$ as an equivalent inequality containing an absolute value.

96. Write $x > 1$ or $x < -1$ as an equivalent inequality containing an absolute value.

97. Describe how solving $|x - 3| = 5$ is different from solving $|x - 3| < 5$.

98. Describe how solving $|x + 4| = 0$ is similar to solving $|x + 4| \leq 0$.

The expression $|x_T - x|$ is defined to be the absolute error in x, where x_T is the true value of a quantity and x is the measured value or value as stored in a computer.

99. If the true value of a quantity is 3.5 and the absolute error must be less than 0.05, find the acceptable measured values.

100. If the true value of a quantity is 0.2 and the approximate value stored in a computer is $\dfrac{51}{256}$, find the absolute error.

INTEGRATED REVIEW SOLVING EQUATIONS AND INEQUALITIES

Solve each equation or inequality. Write inequality solution sets in interval rotation. For inequalities containing "and" or "or" also graph the solution set and write it in interval rotation.

1. $x < 7$ and $x > -5$

2. $x < 7$ or $x > -5$

3. $|4x - 3| = 1$

4. $|2x + 1| < 5$

5. $|6x| - 9 \geq -3$

6. $|x - 7| = |2x + 11|$

7. $-5 \leq \dfrac{3x - 8}{2} \leq 2$

8. $|9x - 1| = -3$

9. $3x + 2 \leq 5$ or $-3x \geq 0$

10. $3x + 2 \leq 5$ and $-3x \geq 0$

11. $|3 - x| - 5 \leq -2$

12. $\left|\dfrac{4x + 1}{5}\right| = |-1|$

Match each equation or inequality on the left with an equivalent statement on the right.

13. $|2x + 1| = 5$

A. $2x + 1 > 5$ or $2x + 1 < -5$

14. $|2x + 1| < 5$

B. $2x + 1 = 5$ or $2x + 1 = -5$

15. $|2x + 1| > 5$

C. $x < 5$

16. $x < 3$ or $x < 5$

D. $x < 3$

17. $x < 3$ and $x < 5$

E. $-5 < 2x + 1 < 5$

9.4 GRAPHING LINEAR INEQUALITIES IN TWO VARIABLES AND SYSTEMS OF LINEAR INEQUALITIES

Objective

1 Graph a linear inequality in two variables.

2 Solve a system of linear inequalities.

1 In this section, we continue our work with systems in Chapter 4 by solving systems of linear inequalities. Before that we first need to learn to graph a single linear inequality in two variables.

Recall that a linear equation in two variables is an equation that can be written in the form $Ax + By = C$ where A, B, and C are real numbers and A and B are not both 0. The definition of a linear inequality is the same except that the equal sign is replaced with an inequality sign.

A **linear inequality in two variables** is an inequality that can be written in one of the forms:

$$Ax + By < C \qquad Ax + By \leq C$$
$$Ax + By > C \qquad Ax + By \geq C$$

where A, B, and C are real numbers and A and B are not both 0. Just as for linear equations in x and y, an ordered pair is a **solution** of an inequality in x and y if replacing the variables by coordinates of the ordered pair results in a true statement.

To graph a linear inequality in two variables, we will begin by graphing a related equation. For example, to graph $x - y \leq 1$, we begin by graphing $x - y = 1$.

The linear equation $x - y = 1$ is graphed next. Recall that all points on the line correspond to ordered pairs that satisfy the equation $x - y = 1$. It can be shown that all the points above the line $x - y = 1$ have coordinates that satisfy the inequality $x - y < 1$. Similarly, all points below the line have coordinates that satisfy the inequality $x - y > 1$. To see this, a few points have been selected on one side of the line. These points all satisfy $x - y < 1$ as shown in the table below.

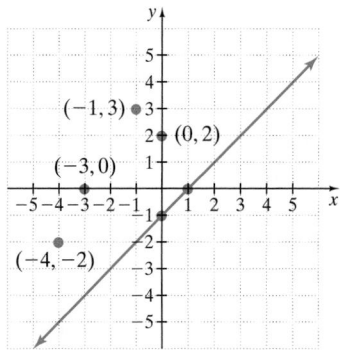

Point	$x - y < 1$	
Check $(0, 2)$	$0 - 2 < 1$	
	$-2 < 1$	True
Check $(-3, 0)$	$-3 - 0 < 1$	
	$-3 < 1$	True
Check $(-4, -2)$	$-4 - (-2) < 1$	
	$-2 < 1$	True
Check $(-1, 3)$	$-1 - 3 < 1$	
	$-4 < 1$	True

The graph of $x - y < 1$ is in blue, and the graph of $x - y > 1$ is in red.

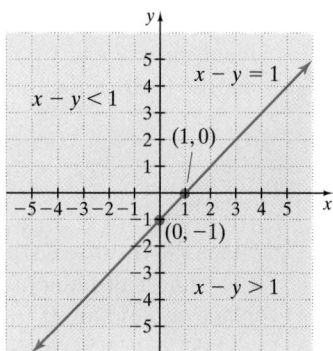

The region above the line and the region below the line are called **half-planes**. Every line divides the plane (similar to a sheet of paper extending indefinitely in all directions) into two half-planes; the line is called the **boundary**.

How do we graph $x - y \leq 1$? Recall that the inequality $x - y \leq 1$ means

$$x - y = 1 \quad \text{or} \quad x - y < 1$$

Thus, the graph of $x - y \leq 1$ is the half-plane $x - y < 1$ along with the boundary line $x - y = 1$.

> ## Graphing a Linear Inequality in Two Variables
>
> **Step 1.** Graph the boundary line found by replacing the inequality sign with an equal sign. If the inequality sign is $>$ or $<$, graph a dashed boundary line (indicating that the points on the line are not solutions of the inequality). If the inequality sign is $\geq$ or $\leq$, graph a solid boundary line (indicating that the points on the line are solutions of the inequality).
>
> **Step 2.** Choose a point, not on the boundary line, as a test point. Substitute the coordinates of this test point into the original inequality.
>
> **Step 3.** If a true statement is obtained in Step 2, shade the half-plane that contains the test point. If a false statement is obtained, shade the half-plane that does not contain the test point.

EXAMPLE 1

Graph: $x + y < 7$

Solution First we graph the boundary line by graphing the equation $x + y = 7$. We graph this boundary as a dashed line because the inequality sign is $<$, and thus the points on the line are not solutions of the inequality $x + y < 7$.

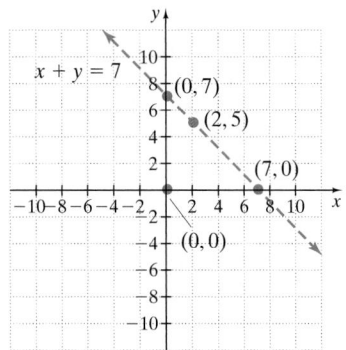

Next, choose a test point, being careful not to choose a point on the boundary line. We choose $(0, 0)$. Substitute the coordinates of $(0, 0)$ into $x + y < 7$.

$$x + y < 7 \qquad \textit{Original inequality}$$
$$0 + 0 \overset{?}{<} 7 \qquad \textit{Replace x with 0 and y with 0.}$$
$$0 < 7 \qquad \textit{True}$$

Since the result is a true statement, $(0, 0)$ is a solution of $x + y < 7$, and every point in the same half-plane as $(0, 0)$ is also a solution. To indicate this, shade the entire half-plane containing $(0, 0)$, as shown.

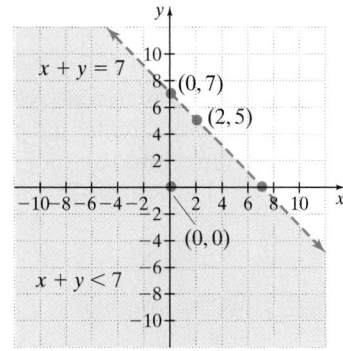

EXAMPLE 2

Graph: $2x - y \geq 3$

Solution   Graph the boundary line by graphing $2x - y = 3$. Draw this line as a solid line since the inequality sign is $\geq$, and thus the points on the line are solutions of $2x - y \geq 3$. Once again, $(0, 0)$ is a convenient test point since it is not on the boundary line.

Substitute 0 for x and 0 for y into the **original inequality**.

$$2x - y \geq 3$$
$$2(0) - 0 \overset{?}{\geq} 3 \qquad \text{Let } x = 0 \text{ and } y = 0.$$
$$0 \geq 3 \qquad \text{False}$$

Since the statement is false, no point in the half-plane containing $(0, 0)$ is a solution. Shade the half-plane that does not contain $(0, 0)$. Every point in the shaded half-plane and every point on the boundary line satisfies $2x - y \geq 3$.

Helpful Hint

When graphing an inequality, make sure the test point is substituted into the **original inequality**. For Example 2, we substituted the test point $(0, 0)$ into the **original inequality** $2x - y \geq 3$, *not* $2x - y = 3$.

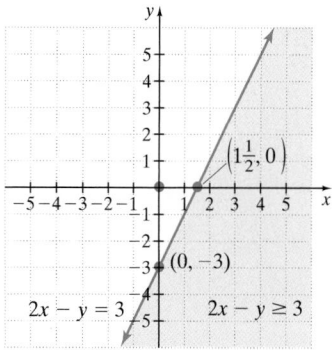

EXAMPLE 3

Graph: $x < 2y$

Solution   We find the boundary line by graphing $x = 2y$. The boundary line is a dashed line since the inequality symbol is $<$. We cannot use $(0, 0)$ as a test point because it is a point on the boundary line. We choose instead $(0, 2)$.

$$x < 2y$$
$$0 \overset{?}{<} 2(2) \qquad \text{Let } x = 0 \text{ and } y = 2.$$
$$0 < 4 \qquad \text{True}$$

Since the statement is true, we shade the half-plane that contains the test point $(0, 2)$, as shown below.

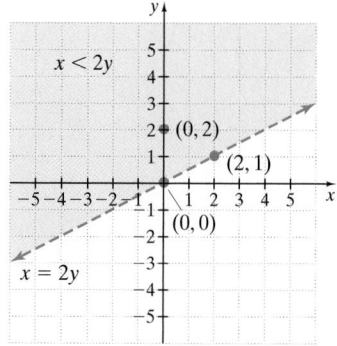

EXAMPLE 4

Graph: $5x + 4y \leq 20$

Solution We graph the solid boundary line $5x + 4y = 20$ and choose $(0, 0)$ as the test point.

$$5x + 4y \leq 20$$
$$5(0) + 4(0) \overset{?}{\leq} 20 \qquad \text{Let } x = 0 \text{ and } y = 0.$$
$$0 \leq 20 \qquad \text{True}$$

We shade the half-plane that contains $(0, 0)$, as shown.

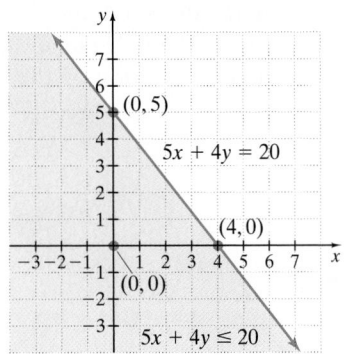

EXAMPLE 5

Graph: $y > 3$

Solution We graph the dashed boundary line $y = 3$ and choose $(0, 0)$ as the test point. (Recall that the graph of $y = 3$ is a horizontal line with y-intercept 3.)

$$y > 3$$
$$0 \overset{?}{>} 3 \qquad \text{Let } y = 0.$$
$$0 > 3 \qquad \text{False}$$

We shade the half-plane that does not contain $(0, 0)$, as shown below.

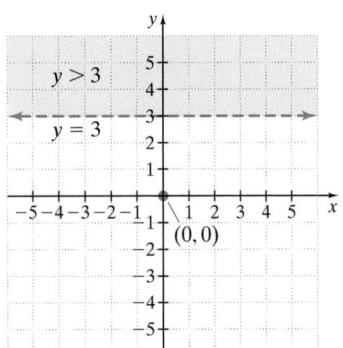

2   Just as two linear equations make a system of linear equations, two linear inequalities make a **system of linear inequalities**. Systems of inequalities are very important in a process called linear programming. Many businesses use linear programming to find the most profitable way to use limited resources such as employees, machines, or buildings.

A **solution of a system of linear inequalities** is an ordered pair that satisfies each inequality in the system. The set of all such ordered pairs is the solution set of the system. Graphing this set gives us a picture of the solution set. We can graph a system of inequalities by graphing each inequality in the system and identifying the region of overlap.

EXAMPLE 6

Graph the solution of the system: $\begin{cases} 3x \geq y \\ x + 2y \leq 8 \end{cases}$

Solution   We begin by graphing each inequality on the same set of axes. The graph of the solution of the system is the region contained in the graphs of both inequalities. It is their intersection.

First, graph $3x \geq y$. The boundary line is the graph of $3x = y$. Sketch a solid boundary line since the inequality $3x \geq y$ means $3x > y$ or $3x = y$. The test point $(1, 0)$ satisfies the inequality, so shade the half-plane that includes $(1, 0)$.

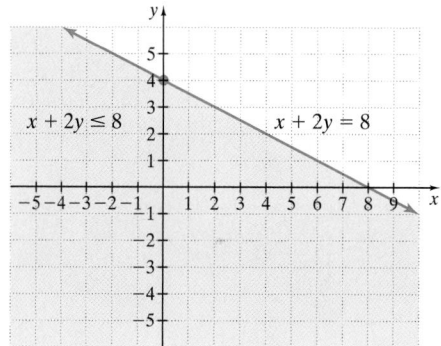

Next, sketch a solid boundary line $x + 2y = 8$ on the same set of axes. The test point $(0, 0)$ satisfies the inequality $x + 2y \leq 8$, so shade the half-plane that includes $(0, 0)$. (For clarity, the graph of $x + 2y \leq 8$ is shown on a separate set of axes.)

An ordered pair solution of the system must satisfy both inequalities. These solutions are points that lie in both shaded regions. The solution of the system is the purple shaded region as seen below. This solution includes parts of both boundary lines.

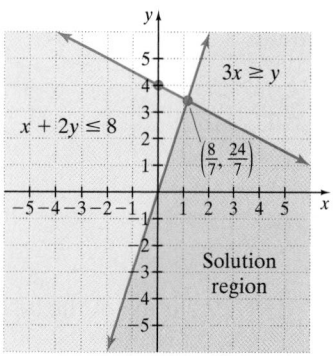

In linear programming, it is sometimes necessary to find the coordinates of the **corner point**: the point at which the two boundary lines intersect. To find the point of intersection, solve the related linear system

$$\begin{cases} 3x = y \\ x + 2y = 8 \end{cases}$$

by the substitution method or the addition method. The lines intersect at $\left(\frac{8}{7}, \frac{24}{7}\right)$, the corner point of the graph.

Graphing the Solution of a System of Linear Inequalities

Step 1. Graph each inequality in the system on the same set of axes.

Step 2. The solutions of the system are the points common to the graphs of all the inequalities in the system.

EXAMPLE 7

Graph the solution of the system: $\begin{cases} x - y < 2 \\ x + 2y > -1 \end{cases}$

Solution Graph both inequalities on the same set of axes. Both boundary lines are dashed lines since the inequality symbols are $<$ and $>$. The solution of the system is the region shown by the purple shading. In this example, the boundary lines are not a part of the solution.

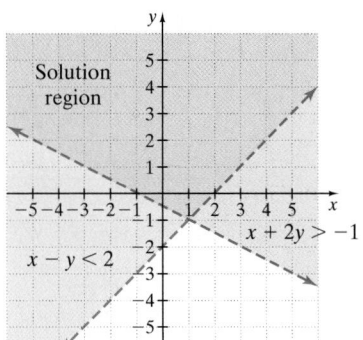

EXAMPLE 8

Graph the solution of the system: $\begin{cases} -3x + 4y < 12 \\ x \geq 2 \end{cases}$

Solution Graph both inequalities on the same set of axes.

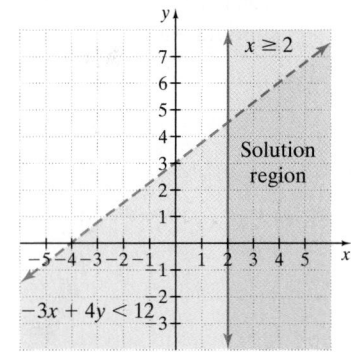

The solution of the system is the purple shaded region, including a portion of the line $x = 2$.

MENTAL MATH

State whether the graph of each inequality includes its corresponding boundary line.

1. $y \geq x + 4$ **2.** $x - y > -7$ **3.** $y \geq x$ **4.** $x > 0$

Decide whether $(0, 0)$ is a solution of each given inequality.

5. $x + y > -5$ **6.** $2x + 3y < 10$ **7.** $x - y \leq -1$ **8.** $\dfrac{2}{3}x + \dfrac{5}{6}y > 4$

EXERCISE SET 9.4

STUDY GUIDE/SSM   CD/ VIDEO   PH MATH TUTOR CENTER   MathXL®Tutorials ON CD   MathXL®   MyMathLab®

Determine which ordered pairs given are solutions of the linear inequality in two variables. See Example 1.

1. $x - y > 3;\ (2, -1), (5, 1)$

2. $y - x < -2;\ (2, 1), (5, -1)$

3. $3x - 5y \leq -4;\ (-1, -1), (4, 0)$

4. $2x + y \geq 10;\ (-1, -4), (5, 0)$

 5. $x < -y;\ (0, 2), (-5, 1)$

6. $y > 3x;\ (0, 0), (-1, -4)$

MIXED PRACTICE

Graph each inequality. See Examples 2 through 5.

7. $x + y \leq 1$ **8.** $x + y \geq -2$ **9.** $2x + y > -4$

10. $x + 3y \leq 3$ **11.** $x + 6y \leq -6$ **12.** $7x + y > -14$

13. $2x + 5y > -10$ **14.** $5x + 2y \leq 10$ **15.** $x + 2y \leq 3$

16. $2x + 3y > -5$ **17.** $2x + 7y > 5$ **18.** $3x + 5y \leq -2$

19. $x - 2y \geq 3$ **20.** $4x + y \leq 2$ **21.** $5x + y < 3$

22. $x + 2y > -7$ **23.** $4x + y < 8$ **24.** $9x + 2y \geq -9$

25. $y \geq 2x$ **26.** $x < 5y$ **27.** $x \geq 0$

28. $y \leq 0$ **29.** $y \leq -3$ **30.** $x > -\dfrac{2}{3}$

31. $2x - 7y > 0$ **32.** $5x + 2y \leq 0$ **33.** $3x - 7y \geq 0$

34. $-2x - 9y > 0$ **35.** $x > y$ **36.** $x \leq -y$

37. $x - y \leq 6$ **38.** $x - y > 10$ **39.** $-\dfrac{1}{4}y + \dfrac{1}{3}x > 1$

40. $\dfrac{1}{2}x - \dfrac{1}{3}y \leq -1$ **41.** $-x < 0.4y$ **42.** $0.3x \geq 0.1y$

In Exercises 43–48, match each inequality with its graph.

  a. $x > 2$

  b. $y < 2$

  c. $y < 2x$

  d. $y \leq -3x$

  e. $2x + 3y < 6$

  f. $3x + 2y > 6$

43.

44.

45.

46.

47.

48.

Graph the solution of each system of linear inequalities. See Examples 6 through 8.

49. $\begin{cases} y \geq x + 1 \\ y \geq 3 - x \end{cases}$

50. $\begin{cases} y \geq x - 3 \\ y \geq -1 - x \end{cases}$

51. $\begin{cases} y < 3x - 4 \\ y \leq x + 2 \end{cases}$

52. $\begin{cases} y \leq 2x + 1 \\ y > x + 2 \end{cases}$

53. $\begin{cases} y \leq -2x - 2 \\ y \geq x + 4 \end{cases}$

54. $\begin{cases} y \leq 2x + 4 \\ y \geq -x - 5 \end{cases}$

55. $\begin{cases} y \geq -x + 2 \\ y \leq 2x + 5 \end{cases}$

56. $\begin{cases} y \geq x - 5 \\ y \leq -3x + 3 \end{cases}$

57. $\begin{cases} x \geq 3y \\ x + 3y \leq 6 \end{cases}$

58. $\begin{cases} -2x < y \\ x + 2y < 3 \end{cases}$

59. $\begin{cases} y + 2x \geq 0 \\ 5x - 3y \leq 12 \end{cases}$

60. $\begin{cases} y + 2x \leq 0 \\ 5x + 3y \geq -2 \end{cases}$

61. $\begin{cases} 3x - 4y \geq -6 \\ 2x + y \leq 7 \end{cases}$

62. $\begin{cases} 4x - y \geq -2 \\ 2x + 3y \leq -8 \end{cases}$

63. $\begin{cases} x \leq 2 \\ y \geq -3 \end{cases}$

64. $\begin{cases} x \geq -3 \\ y \geq -2 \end{cases}$

65. $\begin{cases} y \geq 1 \\ x < -3 \end{cases}$

66. $\begin{cases} y > 2 \\ x \geq -1 \end{cases}$

67. $\begin{cases} 2x + 3y < -8 \\ x \geq -4 \end{cases}$

68. $\begin{cases} 3x + 2y \leq 6 \\ x < 2 \end{cases}$

69. $\begin{cases} 2x - 5y \leq 9 \\ y \leq -3 \end{cases}$

70. $\begin{cases} 2x + 5y \leq -10 \\ y \geq 1 \end{cases}$

71. $\begin{cases} y \geq \dfrac{1}{2}x + 2 \\ y \leq \dfrac{1}{2}x - 3 \end{cases}$

72. $\begin{cases} y \geq \dfrac{-3}{2}x + 3 \\ y < \dfrac{-3}{2}x + 6 \end{cases}$

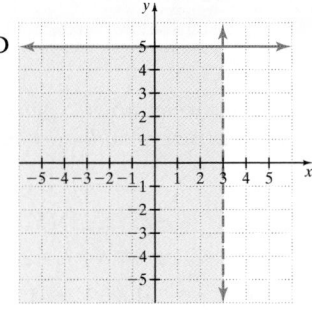

REVIEW AND PREVIEW

Find the square of each expression. For example, the square of 7 is 7^2 or 49. The square of 5x is $(5x)^2$ or $25x^2$. See Section 1.4 and 5.1.

73. 4

74. 3

75. $6x$

76. $11y$

77. $10y^3$

78. $8x^5$

Concept Extensions

For each system of inequalities, choose the corresponding graph.

79. $\begin{cases} y < 5 \\ x > 3 \end{cases}$

80. $\begin{cases} y > 5 \\ x < 3 \end{cases}$

81. $\begin{cases} y \leq 5 \\ x < 3 \end{cases}$

82. $\begin{cases} y > 5 \\ x \geq 3 \end{cases}$

83. Explain how to decide which region to shade to show the solution region of the following system.

$$\begin{cases} x \geq 3 \\ y \geq -2 \end{cases}$$

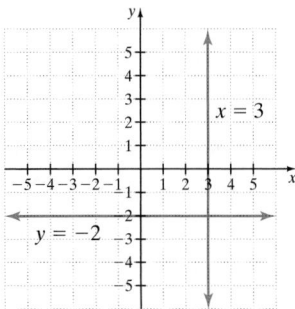

84. Describe the location of the solution region of the system

$$\begin{cases} x > 0 \\ y > 0. \end{cases}$$

85. Write an inequality whose solutions are all pairs of numbers x and y whose sum is at least 13. Graph the inequality.

86. Write an inequality whose solutions are all the pairs of numbers x and y whose sum is at most -4. Graph the inequality.

87. Explain why a point on the boundary line should not be chosen as the test point.

	Department/Program	Actual Expenditure
I.	**Board of Health**	
	Immunization Programs	$14,800
	Inspections	$41,900
II.	**Fire Department**	
	Equipment	$375,000
	Salaries	$268,500
III.	**Libraries**	
	Book/Periodical Purchases	$107,300
	Equipment	$29,000
	Salaries	$118,400
IV.	**Parks and Recreation**	
	Maintenance	$82,500
	Playground Equipment	$45,000
	Salaries	$118,000
	Summer Programs	$96,200
V.	**Police Department**	
	Equipment	$328,000
	Salaries	$405,000
VI.	**Public Works**	
	Recycling	$48,100
	Sewage	$92,500
	Snow Removal & Road Salt	$268,300
	Street Maintenance	$284,000
	Water Treatment	$94,100
	TOTAL	$2,816,600

THE TOWN CRIER
Annual Budget Set at Town Meeting
ANYTOWN, USA (MG)—This year's annual budget is as follows:

	Amount Budgeted
BOARD OF HEALTH	
Immunization Programs	$15,000
Inspections	$50,000
FIRE DEPARTMENT	
Equipment	$450,000
Salaries	$275,000
LIBRARIES	
Book/Periodical Purchases	$90,000
Equipment	$30,000
Salaries	$120,000
PARKS AND RECREATION	
Maintenance	$70,000
Playground Equipment	$50,000
Salaries	$140,000
Summer Programs	$80,000
POLICE DEPARTMENT	
Equipment	$300,000
Salaries	$400,000
PUBLIC WORKS	
Recycling	$50,000
Sewage	$100,000
Snow Removal & Road Salt	$200,000
Street Maintenance	$250,000
Water Treatment	$100,000
TOTAL	**$2,770,000**

BUDGET WORKSHEET

Budget category	Budgeted amount	Minimum allowed	Actual expense	Maximum allowed	Within budget?	Amt over/ under budget
Immunization Programs	$15,000	$13,200	$14,800	$16,800	Yes	Under $200

STUDY SKILLS REMINDER

Are You Preparing for a Test on Chapter 9?

Below I have listed some common trouble areas for topics covered in Chapter 9. After studying for your test—but before taking your test—read these.

▶ Remember to reverse the direction of the inequality symbol when multiplying or dividing both sides of an inequality by a negative number.

$$-11x < 33 \qquad \text{Direction of arrow is reversed.}$$
$$\frac{-11x}{-11} > \frac{33}{-11}$$
$$x > -3$$

▶ Remember the differences when solving absolute value equations and inequalities.

$$|x + 1| = 3$$
$$x + 1 = 3 \quad \text{or} \quad x + 1 = -3$$
$$x = 2 \quad \text{or} \quad x = -4$$
$$\{2, -4\}$$

$$|x + 1| < 3$$
$$-3 < x + 1 < 3$$
$$-3 - 1 < x < 3 - 1$$
$$-4 < x < 2$$
$$(-4, 2)$$

$$|x + 1| > 3$$
$$x + 1 < -3 \quad \text{or} \quad x + 1 > 3$$
$$x < -4 \quad \text{or} \quad x > 2$$
$$(-\infty, -4) \cup (2, \infty)$$

Remember: This is simply a checklist of common trouble areas. For a review of Chapter 9, see the Highlights and Chapter Review at the end of this chapter.

CHAPTER VOCABULARY CHECK

Fill in each blank with one of the words or phrases listed below.

compound inequality   solution   absolute value   union   intersection   system of linear inequalities

1. The statement "$x < 5$ or $x > 7$" is called a(n) _____.
2. The _____ of two sets is the set of all elements common to both sets.
3. The _____ of two sets is the set of all elements that belong to either of the sets.
4. A number's distance from 0 is called its _____.
5. When a variable in an equation is replaced by a number and the resulting equation is true, then that number is called a(n) _____ of the equation.
6. Two or more linear inequalities are called a _____.

CHAPTER 9   HIGHLIGHTS

Definitions and Concepts	Examples

Section 9.1 Compound Inequalities

Two inequalities joined by the words **and** or **or** are called **compound inequalities.**

Compound inequalities:

$$x - 7 \leq 4 \quad \text{and} \quad x \geq -21$$
$$2x + 7 > x - 3 \quad \text{or} \quad 5x + 2 > -3$$

The solution set of a compound inequality formed by the word **and** is the **intersection** $\cap$ of the solution sets of the two inequalities.

Solve for x:

$$x < 5 \text{ and } x < 3$$

$\{x | x < 5\}$ ◄—+—+—+—+—+—+—)—+—► $(-\infty, 5)$
　　　　　　　　$-1 \ 0 \ 1 \ 2 \ 3 \ 4 \ 5 \ 6$

$\{x | x < 3\}$ ◄—+—+—+—+—)—+—+—+—► $(-\infty, 3)$
　　　　　　　　$-1 \ 0 \ 1 \ 2 \ 3 \ 4 \ 5 \ 6$

$\{x | x < 3$ ◄—+—+—+—+—)—+—+—+—► $(-\infty, 3)$
and $x < 5\}$　$-1 \ 0 \ 1 \ 2 \ 3 \ 4 \ 5 \ 6$

The solution set of a compound inequality formed by the word **or** is the **union**, $\cup$, of the solution sets of the two inequalities.

Solve for x:

$$x - 2 \geq -3 \quad \text{or} \quad 2x \leq -4$$
$$x \geq -1 \quad \text{or} \quad x \leq -2$$

$\{x | x \geq -1\}$ ◄—+—+—+—[—+—+—+—+—► $[-1, \infty)$
　　　　　　　$-4 \ -3 \ -2 \ -1 \ 0 \ 1 \ 2 \ 3$

$\{x | x \leq -2\}$ ◄—+—+—]—+—+—+—+—+—► $(-\infty, -2]$
　　　　　　　$-4 \ -3 \ -2 \ -1 \ 0 \ 1 \ 2 \ 3$

$\{x | x \leq -2$ ◄—+—+—]—+—[—+—+—+—► $(-\infty, -2]$
or $x \geq -1\}$　$-4 \ -3 \ -2 \ -1 \ 0 \ 1 \ 2 \ 3$　$\cup [-1, \infty)$

Section 9.2 Absolute Value Equations

If a is a positive number, then $|x| = a$ is equivalent to $x = a$ or $x = -a$.

Solve for y:

$$|5y - 1| - 7 = 4$$

$|5y - 1| = 11$

$5y - 1 = 11 \quad \text{or} \quad 5y - 1 = -11 \qquad$ Add 7.

$5y = 12 \quad \text{or} \qquad 5y = -10 \qquad$ Add 1.

$y = \dfrac{12}{5} \qquad \text{or} \quad y = -2 \qquad$ Divide by 5.

The solutions are -2 and $\frac{12}{5}$.

If a is negative, then $|x| = a$ has no solution.

Solve for x:

$$\left| \frac{x}{2} - 7 \right| = -1$$

The solution set is $\{ \ \}$ or $\emptyset$.

(continued)

Definitions and Concepts	Examples

Section 9.2 Absolute Value Equations

If an absolute value equation is of the form $|x| = |y|$, solve $x = y$ or $x = -y$.

Solve for x:
$$|x - 7| = |2x + 1|$$

$x - 7 = 2x + 1$ or $x - 7 = -(2x + 1)$

$\quad\quad x = 2x + 8$ $x - 7 = -2x - 1$

$\quad\quad -x = 8$ $x = -2x + 6$

$\quad\quad x = -8$ or $3x = 6$

$\quad\quad\quad\quad\quad\quad\quad\quad\quad\quad x = 2$

The solutions are -8 and 2.

Section 9.3 Absolute Value Inequalities

If a is a positive number, then $|x| < a$ is equivalent to $-a < x < a$.

Solve for y:
$$|y - 5| \le 3$$
$$-3 \le y - 5 \le 3$$
$$-3 + 5 \le y - 5 + 5 \le 3 + 5 \quad \text{Add } 5.$$
$$2 \le y \le 8$$

The solution set is $[2, 8]$.

If a is a positive number, then $|x| > a$ is equivalent to $x < -a$ or $x > a$.

Solve for x:
$$\left|\frac{x}{2} - 3\right| > 7$$

$\dfrac{x}{2} - 3 < -7$ or $\dfrac{x}{2} - 3 > 7$

$x - 6 < -14$ or $x - 6 > 14$ Multiply by 2.

$\quad\quad x < -8$ or $x > 20$ Add 6.

The solution set is $(-\infty, -8) \cup (20, \infty)$.

Section 9.4 Graphing Linear Inequalities in Two Variables and Systems of Linear Inequalities

A **linear inequality in two variables** is an inequality that can be written in one of the forms:

$$Ax + By < C \quad\quad Ax + By \le C$$
$$Ax + By > C \quad\quad Ax + By \ge C$$

To graph a linear inequality

1. Graph the boundary line by graphing the related equation. Draw the line solid if the inequality symbol is $\le$ or $\ge$. Draw the line dashed if the inequality symbol is $<$ or $>$.

2. Choose a test point not on the line. Substitute its coordinates into the original inequality.

Linear Inequalities

$$2x - 5y < 6 \quad\quad x \ge -5$$
$$y > -8x \quad\quad y \le 2$$

Graph $2x - y \le 4$.

1. Graph $2x - y = 4$. Draw a solid line because the inequality symbol is $\le$.

2. Check the test point $(0, 0)$ in the inequality.
$2x - y \le 4$.

$$2 \cdot 0 - 0 \le 4 \quad \text{Let } x = 0 \text{ and } y = 0.$$

$$0 \le 4 \quad \text{True}$$

(continued)

Definitions and Concepts	Examples

Section 9.4 Graphing Linear Inequalities in Two Variables and Systems of Linear Inequalities

3. If the resulting inequality is true, shade the half-plane that contains the test point. If the inequality is not true, shade the half-plane that does not contain the test point.

3. The inequality is true so we shade the half-plane containing $(0, 0)$.

$2x - y \leq 4$

$2x - y = 4$

A system of linear inequalities consists of two or more linear inequalities.

To graph a system of inequalities, graph each inequality in the system. The overlapping region is the solution of the system.

System of Linear Inequalities

$$\begin{cases} x - y \geq 3 \\ y \leq -2x \end{cases}$$

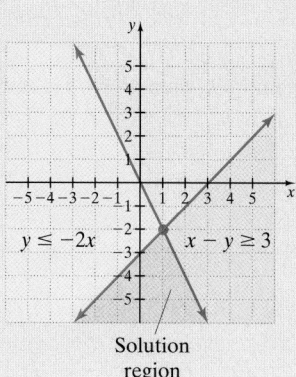

$y \leq -2x$ $x - y \geq 3$

Solution region

CHAPTER REVIEW

(9.1) *Solve each inequality.*

1. $1 \leq 4x - 7 \leq 3$

2. $-2 \leq 8 + 5x < -1$

3. $-3 < 4(2x - 1) < 12$

4. $-6 < x - (3 - 4x) < -3$

5. $\dfrac{1}{6} < \dfrac{4x - 3}{3} \leq \dfrac{4}{5}$

6. $0 \leq \dfrac{2(3x + 4)}{5} \leq 3$

7. $x \leq 2$ and $x > -5$ **8.** $x \leq 2$ or $x > -5$

9. $3x - 5 > 6$ or $-x < -5$

10. $-2x \leq 6$ and $-2x + 3 < -7$

(9.2) Solve each absolute value equation.

11. $|x - 7| = 9$

12. $|8 - x| = 3$

13. $|2x + 9| = 9$

14. $|-3x + 4| = 7$

15. $|3x - 2| + 6 = 10$

16. $5 + |6x + 1| = 5$

17. $-5 = |4x - 3|$

18. $|5 - 6x| + 8 = 3$

19. $|7x| - 26 = -5$

20. $-8 = |x - 3| - 10$

21. $\left|\dfrac{3x - 7}{4}\right| = 2$

22. $\left|\dfrac{9 - 2x}{5}\right| = -3$

23. $|6x + 1| = |15 + 4x|$

24. $|x - 3| = |7 + 2x|$

(9.3) Solve each absolute value inequality. Graph the solution set and write in interval notation.

25. $|5x - 1| < 9$

26. $|6 + 4x| \geq 10$

27. $|3x| - 8 > 1$

28. $9 + |5x| < 24$

29. $|6x - 5| \leq -1$

30. $|6x - 5| \geq -1$

31. $\left|3x + \dfrac{2}{5}\right| \geq 4$

32. $\left|\dfrac{4x - 3}{5}\right| < 1$

33. $\left|\dfrac{x}{3} + 6\right| - 8 > -5$

34. $\left|\dfrac{4(x - 1)}{7}\right| + 10 < 2$

(9.4) Graph the following inequalities.

35. $3x - 4y \leq 0$

36. $3x - 4y \geq 0$

37. $x + 6y < 6$

38. $x + y > -2$

39. $y \geq -7$

40. $y \leq -4$

41. $-x \leq y$

42. $x \geq -y$

Graph the solutions of the following systems of linear inequalities.

43. $\begin{cases} y \geq 2x - 3 \\ y \leq -2x + 1 \end{cases}$

44. $\begin{cases} y \leq -3x - 3 \\ y \leq 2x + 7 \end{cases}$

45. $\begin{cases} x + 2y > 0 \\ x - y \leq 6 \end{cases}$

46. $\begin{cases} x - 2y \geq 7 \\ x + y \leq -5 \end{cases}$

47. $\begin{cases} 3x - 2y \leq 4 \\ 2x + y \geq 5 \end{cases}$

48. $\begin{cases} 4x - y \leq 0 \\ 3x - 2y \geq -5 \end{cases}$

49. $\begin{cases} -3x + 2y > -1 \\ y < -2 \end{cases}$

50. $\begin{cases} -2x + 3y > -7 \\ x \geq -2 \end{cases}$

CHAPTER 9 TEST

Remember to use your Chapter Test Prep Video CD to help you study and view solutions to the test questions you need help with.

Solve each equation or inequality.

1. $|6x - 5| - 3 = -2$

2. $|8 - 2t| = -6$

3 $|x - 5| = |x + 2|$

4. $-3 < 2(x - 3) \leq 4$

5. $|3x + 1| > 5$

6. $|x - 5| - 4 < -2$

7. $x \leq -2 \text{ and } x \leq -5$

8. $x \leq -2 \text{ or } x \leq -5$

9. $-x > 1 \text{ and } 3x + 3 \geq x - 3$

10. $6x + 1 > 5x + 4 \text{ or } 1 - x > -4$

11. $\left|\dfrac{5x - 7}{2}\right| = 4$

12. $|17x - \frac{1}{5}| > -2$

Graph each linear equality.

13. $y \geq -4x$

14. $2x - 3y > -6$

Graph the solutions of the following systems of linear inequalities.

15. $\begin{cases} y + 2x \leq 4 \\ y \geq 2 \end{cases}$

16. $\begin{cases} 2y - x \geq 1 \\ x + y \geq -4 \end{cases}$

CHAPTER CUMULATIVE REVIEW

1. Find the value of each expression when $x = 2$ and $y = -5$.

    a. $\dfrac{x - y}{12 + x}$ **b.** $x^2 - y$

2. Find the value of each expression when $x = -4$ and $y = 7$.

    a. $\dfrac{x - y}{7 - x}$ **b.** $x^2 + 2y$

3. Simplify each expression.

    a. $\dfrac{(-12)(-3) + 3}{-7 - (-2)}$ **b.** $\dfrac{2(-3)^2 - 20}{-5 + 4}$

4. Simplify each expression.

    a. $\dfrac{4(-3) - (-6)}{-8 + 4}$ **b.** $\dfrac{3 + (-3)(-2)^3}{-1 - (-4)}$

5. Simplify each expression by combining like terms.

    a. $2x + 3x + 5 + 2$

    b. $-5a - 3 + a + 2$

    c. $4y - 3y^2$

    d. $2.3x + 5x - 6$

    e. $-\dfrac{1}{2}b + b$

6. Simplify each expression by combining like terms.

    a. $4x - 3 + 7 - 5x$

    b. $-6y + 3y - 8 + 8y$

    c. $2 + 8.1a + a - 6$

    d. $2x^2 - 2x$

7. Solve $2x + 3x - 5 + 7 = 10x + 3 - 6x - 4$.

8. Solve $6y - 11 + 4 + 2y = 8 + 15y - 8y$.

9. Complete the table for the equation $y = 3x$.

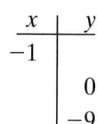

x	y
-1	
	0
	-9

10. Complete the table for the equation $2x + y = 6$.

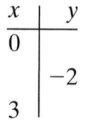

x	y
0	
	-2
3	

11. Identify the x- and y-intercepts.

  a.

  b.

  c.

  d.

  e.

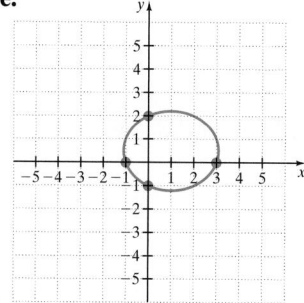

12. Identify the x- and y-intercepts.

  a.

  b.

c.

d.
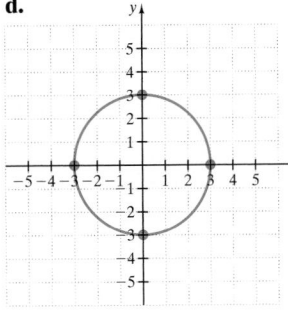

13. Determine whether the graphs of $y = -\dfrac{1}{5}x + 1$ and $2x + 10y = 30$ are parallel lines, perpendicular lines, or neither.

14. Determine whether the graphs of $y = 3x + 7$ and $x + 3y = -15$ are parallel lines, perpendicular lines, or neither.

15. Find an equation of the line with y-intercept $(0, -3)$ and slope of $\frac{1}{4}$.

16. Find an equation of a line with y-intercept $(0, 4)$ and slope of -2.

17. Find an equation of the line parallel to the line $y = 5$ and passing through $(-2, -3)$.

18. Find an equation of the line perpendicular to $y = 2x + 4$ and passing through $(1, 5)$.

19. Which of the following linear equations are functions?
a. $y = x$
b. $y = 2x + 1$
c. $y = 5$
d. $x = -1$

20. Which of the following linear equations are functions?
a. $2x + 3 = y$
b. $x + 4 = 0$
c. $\dfrac{1}{2}y = 2x$
d. $y = 0$

21. Which of the following ordered pairs is a solution of the given system?
$$\begin{cases} 2x - 3y = 6 \\ x = 2y \end{cases}$$
a. $(12, 6)$
b. $(0, -2)$

22. Which of the following ordered pairs is a solution of the given system?
$$\begin{cases} 2x + y = 4 \\ x + y = 2 \end{cases}$$
a. $(1, 1)$
b. $(2, 0)$

23. Add $11x^3 - 12x^2 + x - 3$ and $x^3 - 10x + 5$.

24. Combine like terms to simplify.
$4a^2 + 3a - 2a^2 + 7a - 5$.

25. Factor $x^2 + 5yx + 6y^2$.

26. Factor $3x^2 + 15x + 18$.

27. Divide $\dfrac{3x^3y^7}{40} \div \dfrac{4x^3}{y^2}$.

28. Divide $\dfrac{12x^2y^3}{5} \div \dfrac{3y^2}{x}$.

29. Subtract $\dfrac{2y}{2y - 7} - \dfrac{7}{2y - 7}$.

30. Subtract $\dfrac{-4x^2}{x + 1} - \dfrac{4x}{x + 1}$.

31. Add $\dfrac{2x}{x^2 + 2x + 1} + \dfrac{x}{x^2 - 1}$.

32. Add $\dfrac{3x}{x^2 + 5x + 6} + \dfrac{1}{x^2 + 2x - 3}$.

33. Solve $\dfrac{x}{2} + \dfrac{8}{3} = \dfrac{1}{6}$.

34. Solve $\dfrac{1}{21} + \dfrac{x}{7} = \dfrac{5}{3}$.

35. Solve the following system of equations by graphing.
$$\begin{cases} 2x + y = 7 \\ 2y = -4x \end{cases}$$

36. Solve the following system by graphing.
$$\begin{cases} y = x + 2 \\ 2x + y = 5 \end{cases}$$

37. Solve the system.
$$\begin{cases} 7x - 3y = -14 \\ -3x + y = 6 \end{cases}$$

38. Solve the system.
$$\begin{cases} 5x + y = 3 \\ y = -5x \end{cases}$$

39. Solve the system.
$$\begin{cases} 3x - 2y = 2 \\ -9x + 6y = -6 \end{cases}$$

40. Solve the system.
$$\begin{cases} -2x + y = 7 \\ 6x - 3y = -21 \end{cases}$$

41. Graph the solution of the system
$$\begin{cases} -3x + 4y < 12 \\ x \geq 2 \end{cases}$$

42. Graph the solution of the system.

$$\begin{cases} 2x - y \le 6 \\ \quad\; y \ge 2 \end{cases}$$

43. Simplify the following.

a. $\left(\dfrac{-5x^2}{y^3} \right)^2$

b. $\dfrac{\left(x^3\right)^4 x}{x^7}$

c. $\dfrac{\left(2x\right)^5}{x^3}$

d. $\dfrac{\left(a^2 b\right)^3}{a^3 b^2}$

44. Simplify.

a. $\left(\dfrac{-6x}{y^3} \right)^3$

b. $\dfrac{a^2 b^7}{\left(2b^2\right)^5}$

c. $\dfrac{\left(3y\right)^2}{y^2}$

d. $\dfrac{\left(x^2 y^4\right)^2}{x y^3}$

45. Solve $(5x - 1)(2x^2 + 15x + 18) = 0$.

46. Solve $(x + 1)(2x^2 - 3x - 5) = 0$.

47. Solve $\dfrac{45}{x} = \dfrac{5}{7}$.

48. Solve $\dfrac{2x + 7}{3} = \dfrac{x - 6}{2}$.

Rational Exponents, Radicals, and Complex Numbers

In this chapter, radical notation is reviewed, and then rational exponents are introduced. As the name implies, rational exponents are exponents that are rational numbers. We present an interpretation of rational exponents that is consistent with the meaning and rules already established for integer exponents, and we present two forms of notation for roots: radical and exponent. We conclude this chapter with complex numbers, a natural extension of the real number system.

The National Aeronautics and Space Administration (NASA) has been making advances in science and technology since it was established in 1958. NASA employees proudly strive to accomplish their mission to explore space, learn about our own planet, and encourage future exploration of the universe. Future projects include the Kepler mission, which will search for habitable planets in deep space, and the Space Interferometry Mission, which will make continuous observations of space over 95 million kilometers from Earth.

In addition to astronauts, NASA employs many scientists of different disciplines, among them engineers, mathematicians, computer programmers, microbiologists, geologists, and medical doctors. NASA also offers summer internship and student employment programs, both paid and unpaid.

In the Spotlight on Decision Making feature on page 568, you will have the opportunity to explore a new moon orbiting the planet Uranus as a NASA scientist.

Source: `www.nasa.gov` *(National Aeronautics and Space Administration)*

10.1 RADICALS AND RADICAL FUNCTIONS

Objectives

1 Find square roots.
2 Approximate roots using a calculator.
3 Find cube roots.
4 Find nth roots.
5 Find $\sqrt[n]{a^n}$ where a is a real number.
6 Graph square and cube root functions.

1 To find a **square root** of a number a, we find a number that was squared to get a. Thus, because

$$5^2 = 25 \quad \text{and} \quad (-5)^2 = 25, \text{ then}$$

both 5 and -5 are square roots of 25.

We denote the **nonnegative**, or **principal, square root** with the **radical sign**.

$$\sqrt{25} = 5$$

We denote the **negative square root** with the **negative radical sign**.

$$-\sqrt{25} = -5$$

An expression containing a radical sign is called a **radical expression**. An expression within, or "under," a radical sign is called a **radicand**.

radical expression : $\sqrt{a}$ ↙ radical sign   ↖ radicand

Principal and Negative Square Roots

If a is a nonnegative number, then

$\sqrt{a}$ is the **principal, or nonnegative square root** of a
$-\sqrt{a}$ is the **negative square root** of a

 EXAMPLE 1

Simplify. Assume that all variables represent positive numbers.

a. $\sqrt{36}$ **b.** $\sqrt{0}$ **c.** $\sqrt{\dfrac{4}{49}}$ **d.** $\sqrt{0.25}$

e. $\sqrt{x^6}$ **f.** $\sqrt{9x^{12}}$ **g.** $-\sqrt{81}$

Solution   **a.** $\sqrt{36} = 6$ because $6^2 = 36$ and 6 is not negative.

b. $\sqrt{0} = 0$ because $0^2 = 0$ and 0 is not negative.

c. $\sqrt{\dfrac{4}{49}} = \dfrac{2}{7}$ because $\left(\dfrac{2}{7}\right)^2 = \dfrac{4}{49}$ and $\dfrac{2}{7}$ is not negative.

d. $\sqrt{0.25} = 0.5$ because $(0.5)^2 = 0.25$.

e. $\sqrt{x^6} = x^3$ because $(x^3)^2 = x^6$.

f. $\sqrt{9x^{12}} = 3x^6$ because $(3x^6)^2 = 9x^{12}$.

g. $-\sqrt{81} = -9$. The negative in front of the radical indicates the negative square root of 81.

Can we find the square root of a negative number, say $\sqrt{-4}$? That is, can we find a real number whose square is -4? No, there is no real number whose square is -4, and we say that $\sqrt{-4}$ is not a real number. In general:

The square root of a negative number is not a real number.

> **Helpful Hint**
>
> - Remember: $\sqrt{0} = 0$
> - Don't forget, the square root of a negative number, such as $\sqrt{-9}$, is not a real number. In Section 10.7, we will see what kind of a number $\sqrt{-9}$ is.

2   Recall that numbers such as 1, 4, 9, and 25 are called **perfect squares**, since $1 = 1^2, 4 = 2^2, 9 = 3^2$, and $25 = 5^2$. Square roots of perfect square radicands simplify to rational numbers. What happens when we try to simplify a root such as $\sqrt{3}$? Since there is no rational number whose square is 3, then $\sqrt{3}$ is not a rational number. It is called an **irrational number**, and we can find a decimal **approximation** of it. To find decimal approximations, use a calculator. For example, an approximation for $\sqrt{3}$ is

$$\sqrt{3} \approx 1.732$$
$$\uparrow$$

approximation symbol

To see if the approximation is reasonable, notice that since

$$1 < 3 < 4, \text{ then}$$
$$\sqrt{1} < \sqrt{3} < \sqrt{4}, \text{ or}$$
$$1 < \sqrt{3} < 2.$$

We found $\sqrt{3} \approx 1.732$, a number between 1 and 2, so our result is reasonable.

EXAMPLE 2

Use a calculator to approximate $\sqrt{20}$. Round the approximation to 3 decimal places and check to see that your approximation is reasonable.

$$\sqrt{20} \approx 4.472$$

Solution   Is this reasonable? Since $16 < 20 < 25$, then $\sqrt{16} < \sqrt{20} < \sqrt{25}$, or $4 < \sqrt{20} < 5$. The approximation is between 4 and 5 and thus is reasonable.

3 Finding roots can be extended to other roots such as cube roots. For example, since $2^3 = 8$, we call 2 the **cube root** of 8. In symbols, we write

$$\sqrt[3]{8} = 2$$

Cube Root

The **cube root** of a real number a is written as $\sqrt[3]{a}$, and

$$\sqrt[3]{a} = b \text{ only if } b^3 = a$$

From this definition, we have

$$\sqrt[3]{64} = 4 \text{ since } 4^3 = 64$$
$$\sqrt[3]{-27} = -3 \text{ since } (-3)^3 = -27$$
$$\sqrt[3]{x^3} = x \text{ since } x^3 = x^3$$

Notice that, unlike with square roots, *it is possible to have a negative radicand when finding a cube root.* This is so because the *cube* of a negative number is a negative number. Therefore, the *cube root* of a negative number is a negative number.

EXAMPLE 3

Find the cube roots.

a. $\sqrt[3]{1}$ **b.** $\sqrt[3]{-64}$ **c.** $\sqrt[3]{\dfrac{8}{125}}$

d. $\sqrt[3]{x^6}$ **e.** $\sqrt[3]{-27x^9}$

Solution   **a.** $\sqrt[3]{1} = 1$ because $1^3 = 1$.

b. $\sqrt[3]{-64} = -4$ because $(-4)^3 = -64$.

c. $\sqrt[3]{\dfrac{8}{125}} = \dfrac{2}{5}$ because $\left(\dfrac{2}{5}\right)^3 = \dfrac{8}{125}$.

d. $\sqrt[3]{x^6} = x^2$ because $(x^2)^3 = x^6$.

e. $\sqrt[3]{-27x^9} = -3x^3$ because $(-3x^3)^3 = -27x^9$.

4 Just as we can raise a real number to powers other than 2 or 3, we can find roots other than square roots and cube roots. In fact, we can find the **nth root** of a number, where n is any natural number. In symbols, the nth root of a is written as $\sqrt[n]{a}$, where n is called the **index**. The index 2 is usually omitted for square roots.

> **Helpful Hint**
>
> If the index is even, such as $\sqrt{}$, $\sqrt[4]{}$, $\sqrt[6]{}$, and so on, the radicand must be non-negative for the root to be a real number. For example,
>
> $$\sqrt[4]{16} = 2, \text{ but } \sqrt[4]{-16} \text{ is not a real number.}$$
> $$\sqrt[6]{64} = 2, \text{ but } \sqrt[6]{-64} \text{ is not a real number.}$$
>
> If the index is odd, such as $\sqrt[3]{}$, $\sqrt[5]{}$, and so on, the radicand may be any real number. For example,
>
> $$\sqrt[3]{64} = 4 \text{ and } \sqrt[3]{-64} = -4$$
> $$\sqrt[5]{32} = 2 \text{ and } \sqrt[5]{-32} = -2$$

✔ **CONCEPT CHECK**

Which one is not a real number?

a. $\sqrt[3]{-15}$ b. $\sqrt[4]{-15}$ c. $\sqrt[5]{-15}$ d. $\sqrt{(-15)^2}$

EXAMPLE 4

Simplify the following expressions.

a. $\sqrt[4]{81}$ b. $\sqrt[5]{-243}$ c. $-\sqrt{25}$

d. $\sqrt[4]{-81}$ e. $\sqrt[3]{64x^3}$

Solution a. $\sqrt[4]{81} = 3$ because $3^4 = 81$ and 3 is positive.

b. $\sqrt[5]{-243} = -3$ because $(-3)^5 = -243$.

c. $-\sqrt{25} = -5$ because -5 is the opposite of $\sqrt{25}$.

d. $\sqrt[4]{-81}$ is not a real number. There is no real number that, when raised to the fourth power, is -81.

e. $\sqrt[3]{64x^3} = 4x$ because $(4x)^3 = 64x^3$.

⑤   Recall that the notation $\sqrt{a^2}$ indicates the positive square root of a^2 only. For example,

$$\sqrt{(-5)^2} = \sqrt{25} = 5$$

When variables are present in the radicand and it is unclear whether the variable represents a positive number or a negative number, absolute value bars are sometimes needed to ensure that the result is a positive number. For example,

$$\sqrt{x^2} = |x|$$

This ensures that the result is positive. This same situation may occur when the index is any *even* positive integer. When the index is any *odd* positive integer, absolute value bars are not necessary.

Concept Check Answer:
b

> **Finding $\sqrt[n]{a^n}$**
>
> If n is an *even* positive integer, then $\sqrt[n]{a^n} = |a|$.
>
> If n is an *odd* positive integer, then $\sqrt[n]{a^n} = a$.

EXAMPLE 5

Simplify.

a. $\sqrt{(-3)^2}$ **b.** $\sqrt{x^2}$ **c.** $\sqrt[4]{(x-2)^4}$ **d.** $\sqrt[3]{(-5)^3}$

e. $\sqrt[5]{(2x-7)^5}$ **f.** $\sqrt{25x^2}$ **g.** $\sqrt{x^2 + 2x + 1}$

Solution

a. $\sqrt{(-3)^2} = |-3| = 3$ When the index is even, the absolute value bars ensure us that our result is not negative.

b. $\sqrt{x^2} = |x|$

c. $\sqrt[4]{(x-2)^4} = |x-2|$

d. $\sqrt[3]{(-5)^3} = -5$

e. $\sqrt[5]{(2x-7)^5} = 2x - 7$ Absolute value bars are not needed when the index is odd.

f. $\sqrt{25x^2} = 5|x|$

g. $\sqrt{x^2 + 2x + 1} = \sqrt{(x+1)^2} = |x+1|$

6   Recall that an equation in x and y describes a function if each x-value is paired with exactly one y-value. With this in mind, does the equation

$$y = \sqrt{x}$$

describe a function? First, notice that replacement values for x must be nonnegative real numbers, since $\sqrt{x}$ is not a real number if $x < 0$. The notation $\sqrt{x}$ denotes the principal square root of x, so for every nonnegative number x, there is exactly one number, $\sqrt{x}$. Therefore, $y = \sqrt{x}$ describes a function, and we may write it as

$$f(x) = \sqrt{x}$$

In general, radical functions are functions of the form

$$f(x) = \sqrt[n]{x}.$$

Recall that the domain of a function in x is the set of all possible replacement values of x. This means that if n is even, the domain is the set of all nonnegative numbers, or $\{x | x \geq 0\}$. If n is odd, the domain is the set of all real numbers. Keep this in mind as we find function values.

EXAMPLE 6

If $f(x) = \sqrt{x-4}$ and $g(x) = \sqrt[3]{x+2}$, find each function value.

a. $f(8)$ **b.** $f(6)$

c. $g(-1)$ **d.** $g(1)$

Solution **a.** $f(8) = \sqrt{8 - 4} = \sqrt{4} = 2$
b. $f(6) = \sqrt{6 - 4} = \sqrt{2}$
c. $g(-1) = \sqrt[3]{-1 + 2} = \sqrt[3]{1} = 1$
d. $g(1) = \sqrt[3]{1 + 2} = \sqrt[3]{3}$

> **Helpful Hint**
>
> Notice that for the function $f(x) = \sqrt{x - 4}$, the domain includes all real numbers that make the radicand ≥ 0. To see what numbers these are, solve $x - 4 \geq 0$ and find that $x \geq 4$. The domain is $\{x \mid x \geq 4\}$.
>
> The domain of the cube root function $g(x) = \sqrt[3]{x + 2}$ is the set of real numbers.

EXAMPLE 7

Graph the square root function $f(x) = \sqrt{x}$.

Solution To graph, we identify the domain, evaluate the function for several values of x, plot the resulting points, and connect the points with a smooth curve. Since $\sqrt{x}$ represents the nonnegative square root of x, the domain of this function is the set of all nonnegative numbers, $\{x \mid x \geq 0\}$, or $[0, \infty)$. We have approximated $\sqrt{3}$ below to help us locate the point corresponding to $\left(3, \sqrt{3}\right)$.

x	$f(x) = \sqrt{x}$
0	0
1	1
3	$\sqrt{3} \approx 1.7$
4	2
9	3

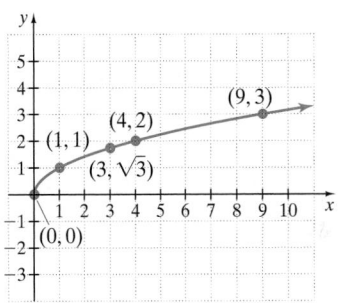

Notice that the graph of this function passes the vertical line test, as expected.

The equation $f(x) = \sqrt[3]{x}$ also describes a function. Here x may be any real number, so the domain of this function is the set of all real numbers, or $(-\infty, \infty)$. A few function values are given next.

$$f(0) = \sqrt[3]{0} = 0$$
$$f(1) = \sqrt[3]{1} = 1$$
$$f(-1) = \sqrt[3]{-1} = -1$$

$$\left.\begin{array}{l} f(6) = \sqrt[3]{6} \\ f(-6) = \sqrt[3]{-6} \end{array}\right\}$$ Here, there is no rational number whose cube is 6. Thus, the radicals do not simplify to rational numbers.

$$f(8) = \sqrt[3]{8} = 2$$
$$f(-8) = \sqrt[3]{-8} = -2$$

EXAMPLE 8

Graph the function $f(x) = \sqrt[3]{x}$.

Solution   To graph, we identify the domain, plot points, and connect the points with a smooth curve. The domain of this function is the set of all real numbers. The table comes from the function values obtained earlier. We have approximated $\sqrt[3]{6}$ and $\sqrt[3]{-6}$ for graphing purposes.

x	$f(x) = \sqrt[3]{x}$
0	0
1	1
-1	-1
6	$\sqrt[3]{6} \approx 1.8$
-6	$\sqrt[3]{-6} \approx -1.8$
8	2
-8	-2

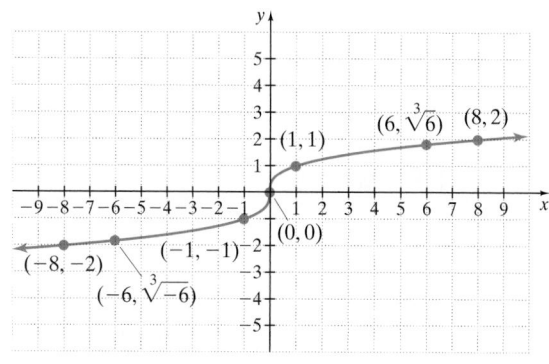

The graph of this function passes the vertical line test, as expected.

Spotlight on DECISION MAKING

Suppose you are a scientist working for NASA. A new moon, S/2001U1, has been discovered orbiting the planet Uranus in our outer solar system. You have been asked to check whether it is possible for this moon to have an oxygen atmosphere. You can do so by comparing the average speed of an oxygen molecule (480 meters per second) to the moon's **escape velocity**, the speed an object must travel to permanently leave the moon's gravitational pull. If the moon's escape velocity is greater than the average speed of oxygen molecules, then it is possible for the moon to retain oxygen in its atmosphere—that is, if oxygen exists on the moon at all.

Data about the new moon are listed in the table. Use that along with the escape velocity formula given to the right to decide whether it is possible for S/2001U1 to have an oxygen atmosphere.

S/2001U1 Parameters	
Mass	9.07×10^{20} kg
Radius	620,000 m
Visual geometric albedo	0.07
Orbital period	7.2 days

$$v = \sqrt{\frac{2GM}{r}}, \text{ where}$$

v is the escape velocity (in meters per second, m/s),

M is the mass of the moon (in kilograms, kg),

r is the radius of the moon (in meters, m), and

G is the universal constant of gravitation where

$$\left(G = 6.67 \times 10^{-11} \frac{\text{m}^3}{\text{kg} \cdot \text{s}^2}\right).$$

MENTAL MATH

Choose the correct letter. No pencil is needed, just think your way through these.

1. Which radical is not a real number?

   a. $\sqrt{3}$ **b.** $-\sqrt{11}$ **c.** $\sqrt[3]{-10}$ **d.** $\sqrt{-10}$

2. Which radical(s) simplify to 3?

   a. $\sqrt{9}$ **b.** $\sqrt{-9}$ **c.** $\sqrt[3]{27}$ **d.** $\sqrt[3]{-27}$

3. Which radical(s) simplify to −3?

   a. $\sqrt{9}$ **b.** $\sqrt{-9}$ **c.** $\sqrt[3]{27}$ **d.** $\sqrt[3]{-27}$

4. Which radical does not simplify to a whole number?

   a. $\sqrt{64}$ **b.** $\sqrt[3]{64}$ **c.** $\sqrt{8}$ **d.** $\sqrt[3]{8}$

EXERCISE SET 10.1

STUDY GUIDE/SSM   CD/VIDEO   PH MATH TUTOR CENTER   MathXL®Tutorials ON CD   MathXL®   MyMathLab®

Simplify. Assume that variables represent positive real numbers. See Example 1.

1. $\sqrt{100}$ 2. $\sqrt{400}$

3. $\sqrt{\dfrac{1}{4}}$ 4. $\sqrt{\dfrac{9}{25}}$

5. $\sqrt{0.0001}$ 6. $\sqrt{0.04}$

7. $-\sqrt{36}$ 8. $-\sqrt{9}$

9. $\sqrt{x^{10}}$ 10. $\sqrt{x^{16}}$

11. $\sqrt{16y^6}$ 12. $\sqrt{64y^{20}}$

Use a calculator to approximate each square root to 3 decimal places. Check to see that each approximation is reasonable. See Example 2.

13. $\sqrt{7}$ 14. $\sqrt{11}$

15. $\sqrt{38}$ 16. $\sqrt{56}$

17. $\sqrt{200}$ 18. $\sqrt{300}$

Find each cube root. See Example 3.

19. $\sqrt[3]{64}$ 20. $\sqrt[3]{27}$

21. $\sqrt[3]{\dfrac{1}{8}}$ 22. $\sqrt[3]{\dfrac{27}{64}}$

23. $\sqrt[3]{-1}$ 24. $\sqrt[3]{-125}$

25. $\sqrt[3]{x^{12}}$ 26. $\sqrt[3]{x^{15}}$

27. $\sqrt[3]{-27x^9}$ 28. $\sqrt[3]{-64x^6}$

Find each root. Assume that all variables represent nonnegative real numbers. See Example 4.

29. $-\sqrt[4]{16}$ 30. $\sqrt[5]{-243}$

31. $\sqrt[4]{-16}$ 32. $\sqrt{-16}$

33. $\sqrt[5]{-32}$ 34. $\sqrt[5]{-1}$

35. $\sqrt[5]{x^{20}}$ 36. $\sqrt[4]{x^{20}}$

37. $\sqrt[6]{64x^{12}}$ 38. $\sqrt[5]{-32x^{15}}$

39. $\sqrt{81x^4}$ 40. $\sqrt[4]{81x^4}$

41. $\sqrt[4]{256x^8}$ 42. $\sqrt{256x^8}$

Simplify. Assume that the variables represent any real number. See Example 5.

43. $\sqrt{(-8)^2}$ 44. $\sqrt{(-7)^2}$

45. $\sqrt[3]{(-8)^3}$ 46. $\sqrt[5]{(-7)^5}$

47. $\sqrt{4x^2}$ 48. $\sqrt[4]{16x^4}$

49. $\sqrt[3]{x^3}$ 50. $\sqrt[5]{x^5}$

51. $\sqrt{(x-5)^2}$ 52. $\sqrt{(y-6)^2}$

53. $\sqrt{x^2 + 4x + 4}$

   (*Hint:* Factor the polynomial first.)

54. $\sqrt{x^2 - 8x + 16}$

   (*Hint:* Factor the polynomial first.)

MIXED PRACTICE

Simplify each radical. Assume that all variables represent positive real numbers.

55. $-\sqrt{121}$

56. $-\sqrt[3]{125}$

57. $\sqrt[3]{8x^3}$

58. $\sqrt{16x^8}$

59. $\sqrt{y^{12}}$

60. $\sqrt[3]{y^{12}}$

61. $\sqrt{25a^2b^{20}}$

62. $\sqrt{9x^4y^6}$

63. $\sqrt[3]{-27x^{12}y^9}$

64. $\sqrt[3]{-8a^{21}b^6}$

65. $\sqrt[4]{a^{16}b^4}$

66. $\sqrt[4]{x^8y^{12}}$

67. $\sqrt[5]{-32x^{10}y^5}$

68. $\sqrt[5]{-243z^{15}}$

69. $\sqrt{\dfrac{25}{49}}$

70. $\sqrt{\dfrac{4}{81}}$

71. $\sqrt{\dfrac{x^2}{4y^2}}$

72. $\sqrt{\dfrac{y^{10}}{9x^6}}$

73. $-\sqrt[3]{\dfrac{z^{21}}{27x^3}}$

74. $-\sqrt[3]{\dfrac{64a^3}{b^9}}$

75. $\sqrt[4]{\dfrac{x^4}{16}}$

76. $\sqrt[4]{\dfrac{y^4}{81x^4}}$

If $f(x) = \sqrt{2x + 3}$ and $g(x) = \sqrt[3]{x - 8}$, find the following function values. See Example 6.

77. $f(0)$

78. $g(0)$

79. $g(7)$

80. $f(-1)$

81. $g(-19)$

82. $f(3)$

83. $f(2)$

84. $g(1)$

Identify the domain and then graph each function. See Example 7.

85. $f(x) = \sqrt{x} + 2$

86. $f(x) = \sqrt{x} - 2$

87. $f(x) = \sqrt{x - 3}$; use the following table.

x	$f(x)$
3	
4	
7	
12	

88. $f(x) = \sqrt{x + 1}$; use the following table.

x	$f(x)$
-1	
0	
3	
8	

Identify the domain and then graph each function. See Example 8.

89. $f(x) = \sqrt[3]{x} + 1$

90. $f(x) = \sqrt[3]{x} - 2$

91. $g(x) = \sqrt[3]{x - 1}$; use the following table.

x	$g(x)$
1	
2	
0	
9	
-7	

92. $g(x) = \sqrt[3]{x + 1}$; use the following table.

x	$g(x)$
-1	
0	
-2	
7	
-9	

REVIEW AND PREVIEW

Simplify each exponential expression. See Sections 5.1., 5.3, and 5.5.

93. $(-2x^3y^2)^5$

94. $(4y^6z^7)^3$

95. $(-3x^2y^3z^5)(20x^5y^7)$

96. $(-14a^5bc^2)(2abc^4)$

97. $\dfrac{7x^{-1}y}{14(x^5y^2)^{-2}}$

98. $\dfrac{(2a^{-1}b^2)^3}{(8a^2b)^{-2}}$

Concept Extensions

99. Explain why $\sqrt{-64}$ is not a real number.

100. Explain why $\sqrt[3]{-64}$ is a real number.

For Exercises 101 through 104, do not use a calculator.

101. $\sqrt{160}$ is closest to

    a. 10   **b.** 13   **c.** 20   **d.** 40

102. $\sqrt{1000}$ is closest to

    a. 10   **b.** 30   **c.** 100   **d.** 500

103. The perimeter of the triangle is closest to

    a. 12   **b.** 18   **c.** 66   **d.** 132

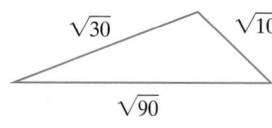

104. The length of the bent wire is closest to

   a. 5   **b.** $\sqrt{28}$ **c.** 7   **d.** 14

The Mosteller formula for calculating adult body surface area is

$B = \sqrt{\dfrac{hw}{3131}}$, *where B is an individual's body surface area in square meters, h is the individual's height in inches, and w is the individual's weight in pounds. Use this information to answer Exercises 105 and 106. Round answers to 2 decimal places.*

105. Find the body surface area of an individual who is 66 inches tall and who weighs 135 pounds.

106. Find the body surface area of an individual who is 74 inches tall and who weighs 225 pounds.

107. Suppose that a friend tells you that $\sqrt{13} \approx 5.7$. Without a calculator, how can you convince your friend that he or she must have made an error?

108. Escape velocity is the minimum speed that an object must reach to escape a planet's pull of gravity. Escape velocity v is given by the equation $v = \sqrt{\dfrac{2Gm}{r}}$, where m is the mass of the planet, r is its radius, and G is the universal gravitational constant, which has a value of $G = 6.67 \times 10^{-11}$ m³/ kg·s². The mass of Earth is 5.97×10^{24} kg and its radius is 6.37×10^6 m. Use this information to find the escape velocity for Earth. Round to the nearest whole number. (*Source:* National Space Science Data Center)

Use a graphing calculator to verify the domain of each function and its graph.

109. Exercise 85

110. Exercise 86

111. Exercise 89

112. Exercise 90

10.2 RATIONAL EXPONENTS

Objectives

1 Understand the meaning of $a^{1/n}$.

2 Understand the meaning of $a^{m/n}$.

3 Understand the meaning of $a^{-m/n}$.

4 Use rules for exponents to simplify expressions that contain rational exponents.

5 Use rational exponents to simplify radical expressions.

1 So far in this text, we have not defined expressions with rational exponents such as $3^{1/2}$, $x^{2/3}$, and $-9^{-1/4}$. We will define these expressions so that the rules for exponents will apply to these rational exponents as well.

Suppose that $x = 5^{1/3}$. Then

using rules for
exponents

$$x^3 = (5^{1/3})^3 = 5^{1/3 \cdot 3} = 5^1 \text{ or } 5$$

Since $x^3 = 5$, then x is the number whose cube is 5, or $= \sqrt[3]{5}$. Notice that we also know that $x = 5^{1/3}$. This means

$$5^{1/3} = \sqrt[3]{5}$$

Definition of $a^{1/n}$

If n is a positive integer greater than 1 and $\sqrt[n]{a}$ is a real number, then

$$a^{1/n} = \sqrt[n]{a}$$

Notice that the denominator of the rational exponent corresponds to the index of the radical.

EXAMPLE 1

Use radical notation to write the following. Simplify if possible.

a. $4^{1/2}$ **b.** $64^{1/3}$

c. $x^{1/4}$ **d.** $0^{1/6}$

e. $-9^{1/2}$ **f.** $(81x^8)^{1/4}$

g. $(5y)^{1/3}$

Solution
a. $4^{1/2} = \sqrt{4} = 2$ **b.** $64^{1/3} = \sqrt[3]{64} = 4$

c. $x^{1/4} = \sqrt[4]{x}$ **d.** $0^{1/6} = \sqrt[6]{0} = 0$

e. $-9^{1/2} = -\sqrt{9} = -3$ **f.** $(81x^8)^{1/4} = \sqrt[4]{81x^8} = 3x^2$

g. $(5y)^{1/3} = \sqrt[3]{5y}$

2   As we expand our use of exponents to include $\dfrac{m}{n}$, we define their meaning so that rules for exponents still hold true. For example, by properties of exponents,

$$8^{2/3} = (8^{1/3})^2 = \left(\sqrt[3]{8}\right)^2 \qquad \text{or}$$
$$8^{2/3} = (8^2)^{1/3} = \sqrt[3]{8^2}$$

Definition of $a^{m/n}$

If m and n are positive integers greater than 1 with $\dfrac{m}{n}$ in lowest terms, then

$$a^{m/n} = \sqrt[n]{a^m} = \left(\sqrt[n]{a}\right)^m$$

as long as $\sqrt[n]{a}$ is a real number.

Notice that the denominator n of the rational exponent corresponds to the index of the radical. The numerator m of the rational exponent indicates that the base is to be raised to the mth power. This means

$$8^{2/3} = \sqrt[3]{8^2} = \sqrt[3]{64} = 4 \qquad \text{or}$$
$$8^{2/3} = \left(\sqrt[3]{8}\right)^2 = 2^2 = 4$$

From simplifying $8^{2/3}$, can you see that it doesn't matter whether you raise to a power first and then take the nth root or you take the nth root first and then raise to a power?

> **Helpful Hint**
> Most of the time, $\left(\sqrt[n]{a}\right)^m$ will be easier to calculate than $\sqrt[n]{a^m}$.

EXAMPLE 2

Use radical notation to write the following. Then simplify if possible.

a. $4^{3/2}$ **b.** $-16^{3/4}$ **c.** $(-27)^{2/3}$

d. $\left(\dfrac{1}{9}\right)^{3/2}$ **e.** $(4x-1)^{3/5}$

Solution

a. $4^{3/2} = \left(\sqrt{4}\right)^3 = 2^3 = 8$

b. $-16^{3/4} = -\left(\sqrt[4]{16}\right)^3 = -(2)^3 = -8$

c. $(-27)^{2/3} = \left(\sqrt[3]{-27}\right)^2 = (-3)^2 = 9$

d. $\left(\dfrac{1}{9}\right)^{3/2} = \left(\sqrt{\dfrac{1}{9}}\right)^3 = \left(\dfrac{1}{3}\right)^3 = \dfrac{1}{27}$

e. $(4x-1)^{3/5} = \sqrt[5]{(4x-1)^3}$

> **Helpful Hint**
> The *denominator* of a rational exponent is the index of the corresponding radical. For example, $x^{1/5} = \sqrt[5]{x}$ and $z^{2/3} = \sqrt[3]{z^2}$, or $z^{2/3} = \left(\sqrt[3]{z}\right)^2$.

3   The rational exponents we have given meaning to exclude negative rational numbers. To complete the set of definitions, we define $a^{-m/n}$.

Definition of $a^{-m/n}$

$$a^{-m/n} = \dfrac{1}{a^{m/n}}$$

as long as $a^{m/n}$ is a nonzero real number.

EXAMPLE 3

Write each expression with a positive exponent, and then simplify.

a. $16^{-3/4}$ **b.** $(-27)^{-2/3}$

Solution **a.** $16^{-3/4} = \dfrac{1}{16^{3/4}} = \dfrac{1}{\left(\sqrt[4]{16}\right)^3} = \dfrac{1}{2^3} = \dfrac{1}{8}$

b. $(-27)^{-2/3} = \dfrac{1}{(-27)^{2/3}} = \dfrac{1}{\left(\sqrt[3]{-27}\right)^2} = \dfrac{1}{(-3)^2} = \dfrac{1}{9}$

> ### Helpful Hint
>
> If an expression contains a negative rational exponent, such as $9^{-3/2}$, you may want to first write the expression with a positive exponent and then interpret the rational exponent. Notice that the sign of the base is not affected by the sign of its exponent. For example,
>
> $$9^{-3/2} = \frac{1}{9^{3/2}} = \frac{1}{\left(\sqrt{9}\right)^3} = \frac{1}{27}$$
>
> Also,
>
> $$(-27)^{-1/3} = \frac{1}{(-27)^{1/3}} = -\frac{1}{3}$$

✔ **CONCEPT CHECK**

Which one is correct?

a. $-8^{2/3} = \dfrac{1}{4}$ **b.** $8^{-2/3} = -\dfrac{1}{4}$ **c.** $8^{-2/3} = -4$ **d.** $-8^{-2/3} = -\dfrac{1}{4}$

4   It can be shown that the properties of integer exponents hold for rational exponents. By using these properties and definitions, we can now simplify expressions that contain rational exponents.

These rules are repeated here for review.

Note: For the remainder of this chapter, we will assume that variables represent positive real numbers. Since this is so, we need not insert absolute value bars when we simplify even roots.

Concept Check Answer:
d

Summary of Exponent Rules

If m and n are rational numbers, and a, b, and c are numbers for which the expressions below exist, then

Product rule for exponents: $\qquad a^m \cdot a^n = a^{m+n}$

Power rule for exponents: $\qquad (a^m)^n = a^{m \cdot n}$

Power rules for products and quotients: $\qquad (ab)^n = a^n b^n \qquad$ and

$$\left(\frac{a}{c}\right)^n = \frac{a^n}{c^n}, c \neq 0$$

Quotient rule for exponents: $\qquad \dfrac{a^m}{a^n} = a^{m-n}, a \neq 0$

Zero exponent: $\qquad a^0 = 1, a \neq 0$

Negative exponent: $\qquad a^{-n} = \dfrac{1}{a^n}, a \neq 0$

EXAMPLE 4

Use properties of exponents to simplify. Write results with only positive exponents.

a. $b^{1/3} \cdot b^{5/3}$ $\qquad$ **b.** $x^{1/2}x^{1/3}$ $\qquad$ **c.** $\dfrac{7^{1/3}}{7^{4/3}}$

d. $y^{-4/7} \cdot y^{6/7}$ $\qquad$ **e.** $\dfrac{(2x^{2/5}y^{-1/3})^5}{x^2 y}$

Solution **a.** $b^{1/3} \cdot b^{5/3} = b^{(1/3+5/3)} = b^{6/3} = b^2$

b. $x^{1/2}x^{1/3} = x^{(1/2+1/3)} = x^{3/6+2/6} = x^{5/6}$ $\qquad$ Use the product rule.

c. $\dfrac{7^{1/3}}{7^{4/3}} = 7^{1/3-4/3} = 7^{-3/3} = 7^{-1} = \dfrac{1}{7}$ $\qquad$ Use the quotient rule.

d. $y^{-4/7} \cdot y^{6/7} = y^{-4/7+6/7} = y^{2/7}$ $\qquad$ Use the product rule.

e. We begin by using the power rule $(ab)^m = a^m b^m$ to simplify the numerator.

$$\frac{(2x^{2/5}y^{-1/3})^5}{x^2 y} = \frac{2^5(x^{2/5})^5(y^{-1/3})^5}{x^2 y} = \frac{32x^2 y^{-5/3}}{x^2 y} \qquad \text{Use the power rule and simplify}$$

$$= 32x^{2-2}y^{-5/3-3/3} \qquad \text{Apply the quotient rule.}$$
$$= 32x^0 y^{-8/3}$$
$$= \frac{32}{y^{8/3}}$$

EXAMPLE 5

Multiply.

a. $z^{2/3}(z^{1/3} - z^5)$ **b.** $(x^{1/3} - 5)(x^{1/3} + 2)$

Solution

a. $z^{2/3}(z^{1/3} - z^5) = z^{2/3}z^{1/3} - z^{2/3}z^5$ Apply the distributive property.

$= z^{(2/3+1/3)} - z^{(2/3+5)}$ Use the product rule.

$= z^{3/3} - z^{(2/3+15/3)}$

$= z - z^{17/3}$

b. $(x^{1/3} - 5)(x^{1/3} + 2) = x^{2/3} + 2x^{1/3} - 5x^{1/3} - 10$ Think of $(x^{1/3} - 5)$ and $(x^{1/3} + 2)$

as 2 binomials, and FOIL.

$= x^{2/3} - 3x^{1/3} - 10$

EXAMPLE 6

Factor $x^{-1/2}$ from the expression $3x^{-1/2} - 7x^{5/2}$. Assume that all variables represent positive numbers.

Solution

$$3x^{-1/2} - 7x^{5/2} = (x^{-1/2})(3) - (x^{-1/2})(7x^{6/2})$$
$$= x^{-1/2}(3 - 7x^3)$$

To check, multiply $x^{-1/2}(3 - 7x^3)$ to see that the product is $3x^{-1/2} - 7x^{5/2}$.

5   Some radical expressions are easier to simplify when we first write them with rational exponents. We can simplify some radical expressions by first writing the expression with rational exponents. Use properties of exponents to simplify, and then convert back to radical notation.

EXAMPLE 7

Use rational exponents to simplify. Assume that variables represent positive numbers.

a. $\sqrt[8]{x^4}$ **b.** $\sqrt[6]{25}$ **c.** $\sqrt[4]{r^2s^6}$

Solution

a. $\sqrt[8]{x^4} = x^{4/8} = x^{1/2} = \sqrt{x}$

b. $\sqrt[6]{25} = 25^{1/6} = (5^2)^{1/6} = 5^{2/6} = 5^{1/3} = \sqrt[3]{5}$

c. $\sqrt[4]{r^2s^6} = (r^2s^6)^{1/4} = r^{2/4}s^{6/4} = r^{1/2}s^{3/2} = (rs^3)^{1/2} = \sqrt{rs^3}$

EXAMPLE 8

Use rational exponents to write as a single radical.

a. $\sqrt{x} \cdot \sqrt[4]{x}$ **b.** $\dfrac{\sqrt{x}}{\sqrt[3]{x}}$ **c.** $\sqrt[3]{3} \cdot \sqrt{2}$

Solution   **a.** $\sqrt{x} \cdot \sqrt[4]{x} = x^{1/2} \cdot x^{1/4} = x^{1/2 + 1/4}$
$$= x^{3/4} = \sqrt[4]{x^3}$$

b. $\dfrac{\sqrt{x}}{\sqrt[3]{x}} = \dfrac{x^{1/2}}{x^{1/3}} = x^{1/2 - 1/3} = x^{3/6 - 2/6}$
$$= x^{1/6} = \sqrt[6]{x}$$

c. $\sqrt[3]{3} \cdot \sqrt{2} = 3^{1/3} \cdot 2^{1/2}$ Write with rational exponents.

$= 3^{2/6} \cdot 2^{3/6}$ Write the exponents so that they have the same denominator.

$= (3^2 \cdot 2^3)^{1/6}$ Use $a^n b^n = (ab)^n$

$= \sqrt[6]{3^2 \cdot 2^3}$ Write with radical notation.

$= \sqrt[6]{72}$ Multiply $3^2 \cdot 2^3$.

Spotlight on **DECISION MAKING**

Suppose you are a telecommunications industry analyst. A colleague has just formulated a mathematical model for the number of cellular telephone subscriptions in the United States from 1985 to 2002. The model is $y = 0.162x^{22/5}$ where y is the number of cellular telephone subscriptions x years after 1980. The actual data from 1985 to 2002 are listed in the table.

Your colleague has asked your help in evaluating whether this model represents the actual data well. By comparing the numbers of subscriptions given by the model to the actual data given in the table, decide whether this mathematical model is acceptable. Explain your reasoning.

U.S. Cellular Telephone Subscriptions, 1985–2002

Year	Subscriptions (in thousands)	Year	Subscriptions (in thousands)
1985	204	1994	19,283
1986	500	1995	28,154
1987	884	1996	38,195
1988	1607	1997	48,706
1989	2692	1998	60,831
1990	4367	1999	76,285
1991	6380	2000	97,036
1992	8893	2001	118,398
1993	13,067	2002	134,561

(*Source:* The CTIA Semi-Annual Wireless Survey)

MENTAL MATH

Choose the correct letter for each exercise. Letters will be used more than once. No pencil is needed. Just think about the meaning of each expression.

A = 2, B = −2, C = not a real number

1. $4^{1/2}$ **2.** $-4^{1/2}$ **3.** $(-4)^{1/2}$ **4.** $8^{1/3}$ **5.** $-8^{1/3}$

6. $(-8)^{1/3}$ **7.** $(-32)^{1/5}$ **8.** $(-16)^{1/4}$ **9.** $-16^{1/4}$ **10.** $-32^{1/5}$

EXERCISE SET 10.2

STUDY GUIDE/SSM   CD/ VIDEO   PH MATH TUTOR CENTER   MathXL®Tutorials ON CD   MathXL®   MyMathLab®

Use radical notation to write each expression. Simplify if possible. See Example 1.

1. $49^{1/2}$ **2.** $64^{1/3}$

3. $27^{1/3}$ **4.** $8^{1/3}$

5. $\left(\dfrac{1}{16}\right)^{1/4}$ **6.** $\left(\dfrac{1}{64}\right)^{1/2}$

7. $169^{1/2}$ **8.** $81^{1/4}$

9. $2m^{1/3}$ **10.** $(2m)^{1/3}$

11. $(9x^4)^{1/2}$ **12.** $(16x^8)^{1/2}$

13. $(-27)^{1/3}$ **14.** $-64^{1/2}$

15. $-16^{1/4}$ **16.** $(-32)^{1/5}$

Use radical notation to write each expression. Simplify if possible. See Example 2.

17. $16^{3/4}$ **18.** $4^{5/2}$

19. $(-64)^{2/3}$ **20.** $(-8)^{4/3}$

21. $(-16)^{3/4}$ **22.** $(-9)^{3/2}$

23. $(2x)^{3/5}$ **24.** $2x^{3/5}$

25. $(7x + 2)^{2/3}$ **26.** $(x - 4)^{3/4}$

27. $\left(\dfrac{16}{9}\right)^{3/2}$ **28.** $\left(\dfrac{49}{25}\right)^{3/2}$

Write with positive exponents. Simplify if possible. See Example 3.

29. $8^{-4/3}$ **30.** $64^{-2/3}$

31. $(-64)^{-2/3}$ **32.** $(-8)^{-4/3}$

33. $(-4)^{-3/2}$ **34.** $(-16)^{-5/4}$

35. $x^{-1/4}$ **36.** $y^{-1/6}$

37. $\dfrac{1}{a^{-2/3}}$ **38.** $\dfrac{1}{n^{-8/9}}$

39. $\dfrac{5}{7x^{-3/4}}$ **40.** $\dfrac{2}{3y^{-5/7}}$

41. Explain how writing x^{-7} with positive exponents is similar to writing $x^{-1/4}$ with positive exponents.

42. Explain how writing $2x^{-5}$ with positive exponents is similar to writing $2x^{-3/4}$ with positive exponents.

Use the properties of exponents to simplify each expression. Write with positive exponents. See Example 4.

43. $a^{2/3}a^{5/3}$ **44.** $b^{9/5}b^{8/5}$

45. $x^{-2/5} \cdot x^{7/5}$ **46.** $y^{4/3} \cdot y^{-1/3}$

47. $3^{1/4} \cdot 3^{3/8}$ **48.** $5^{1/2} \cdot 5^{1/6}$

49. $\dfrac{y^{1/3}}{y^{1/6}}$ **50.** $\dfrac{x^{3/4}}{x^{1/8}}$

51. $(4u^2)^{3/2}$ **52.** $(32^{1/5}x^{2/3})^3$

53. $\dfrac{b^{1/2}b^{3/4}}{-b^{1/4}}$ **54.** $\dfrac{a^{1/4}a^{-1/2}}{a^{2/3}}$

55. $\dfrac{(3x^{1/4})^3}{x^{1/12}}$ **56.** $\dfrac{(2x^{1/5})^4}{x^{3/10}}$

Multiply. See Example 5.

57. $y^{1/2}(y^{1/2} - y^{2/3})$ **58.** $x^{1/2}(x^{1/2} + x^{3/2})$

59. $x^{2/3}(2x - 2)$ **60.** $3x^{1/2}(x + y)$

61. $(2x^{1/3} + 3)(2x^{1/3} - 3)$ **62.** $(y^{1/2} + 5)(y^{1/2} + 5)$

Factor the common factor from the given expression. See Example 6.

63. $x^{8/3}; x^{8/3} + x^{10/3}$ **64.** $x^{3/2}; x^{5/2} - x^{3/2}$

65. $x^{1/5}; x^{2/5} - 3x^{1/5}$ **66.** $x^{2/7}; x^{3/7} - 2x^{2/7}$

67. $x^{-1/3}; 5x^{-1/3} + x^{2/3}$ **68.** $x^{-3/4}; x^{-3/4} + 3x^{1/4}$

Use rational exponents to simplify each radical. Assume that all variables represent positive numbers. See Example 7.

69. $\sqrt[6]{x^3}$

70. $\sqrt[9]{a^3}$

71. $\sqrt[6]{4}$

72. $\sqrt[4]{36}$

73. $\sqrt[4]{16x^2}$

74. $\sqrt[8]{4y^2}$

75. $\sqrt[8]{x^4y^4}$

76. $\sqrt[9]{y^6z^3}$

Use rational expressions to write as a single radical expression. See Example 8.

77. $\sqrt[3]{y} \cdot \sqrt[5]{y^2}$

78. $\sqrt[3]{y^2} \cdot \sqrt[6]{y}$

79. $\dfrac{\sqrt[3]{b^2}}{\sqrt[4]{b}}$

80. $\dfrac{\sqrt[4]{a}}{\sqrt[5]{a}}$

81. $\dfrac{\sqrt[3]{a^2}}{\sqrt[6]{a}}$

82. $\dfrac{\sqrt[5]{b^2}}{\sqrt[10]{b^3}}$

83. $\sqrt{3} \cdot \sqrt[3]{4}$

84. $\sqrt[3]{5} \cdot \sqrt{2}$

85. $\sqrt[5]{7} \cdot \sqrt[3]{y}$

86. $\sqrt[4]{5} \cdot \sqrt[3]{x}$

REVIEW AND PREVIEW

Write each integer as a product of two integers such that one of the factors is a perfect square. For example, write 18 as $9 \cdot 2$, because 9 is a perfect square.

87. 75

88. 20

89. 48

90. 45

Write each integer as a product of two integers such that one of the factors is a perfect cube. For example, write 24 as $8 \cdot 3$, because 8 is a perfect cube.

91. 16

92. 56

93. 54

94. 80

Concept Extensions

Basal metabolic rate (BMR) is the number of calories per day a person needs to maintain life. A person's basal metabolic rate $B(w)$ in calories per day can be estimated with the function $B(w) = 70w^{3/4}$, where w is the person's weight in kilograms. Use this information to answer Exercises 95 and 96.

95. Estimate the BMR for a person who weighs 60 kilograms. Round to the nearest calorie. (*Note:* 60 kilograms is approximately 132 pounds.)

96. Estimate the BMR for a person who weighs 90 kilograms. Round to the nearest calorie. (*Note:* 90 kilograms is approximately 198 pounds.)

The number of cellular telephone subscriptions in the United States from 1994 through 2002 can be modeled by the function $f(x) = 1.54x^{9/5}$, where y is the number of cellular telephone subscriptions in millions, x years after 1990. (Source: Based on data from the Cellular Telecommunications & Internet Association, 1994–2000) Use this information to answer Exercises 97 and 98.

97. Use this model to estimate the number of cellular telephone subscriptions in the United States in 2000. Round to the nearest tenth of a million.

98. Predict the number of cellular telephone subscriptions in the United States in 2007. Round to the nearest tenth of a million.

Fill in each box with the correct expression.

99. $\boxed{} \cdot a^{2/3} = a^{3/3}$, or a

100. $\boxed{} \cdot x^{1/8} = x^{4/8}$, or $x^{1/2}$

101. $\dfrac{\boxed{}}{x^{-2/5}} = x^{3/5}$

102. $\dfrac{\boxed{}}{y^{-3/4}} = y^{4/4}$, or y

Use a calculator to write a four-decimal-place approximation of each number.

103. $8^{1/4}$

104. $20^{1/5}$

105. $18^{3/5}$

106. $76^{5/7}$

107. In physics, the speed of a wave traveling over a stretched string with tension t and density u is given by the expression $\dfrac{\sqrt{t}}{\sqrt{u}}$. Write this expression with rational exponents.

108. In electronics, the angular frequency of oscillations in a certain type of circuit is given by the expression $(LC)^{-1/2}$. Use radical notation to write this expression.

10.3   SIMPLIFYING RADICAL EXPRESSIONS

Objectives

1   Use the product rule for radicals.

2   Use the quotient rule for radicals.

3   Simplify radicals.

1   It is possible to simplify some radicals that do not evaluate to rational numbers. To do so, we use a product rule and a quotient rule for radicals. To discover the product rule, notice the following pattern.

$$\sqrt{9} \cdot \sqrt{4} = 3 \cdot 2 = 6$$
$$\sqrt{9 \cdot 4} = \sqrt{36} = 6$$

Since both expressions simplify to 6, it is true that

$$\sqrt{9} \cdot \sqrt{4} = \sqrt{9 \cdot 4}$$

This pattern suggests the following product rule for radicals.

Product Rule for Radicals

If $\sqrt[n]{a}$ and $\sqrt[n]{b}$ are real numbers, then

$$\sqrt[n]{a} \cdot \sqrt[n]{b} = \sqrt[n]{ab}$$

Notice that the product rule is the relationship $a^{1/n} \cdot b^{1/n} = (ab)^{1/n}$ stated in radical notation.

EXAMPLE 1

Multiply.

a. $\sqrt{3} \cdot \sqrt{5}$ **b.** $\sqrt{21} \cdot \sqrt{x}$ **c.** $\sqrt[3]{4} \cdot \sqrt[3]{2}$

d. $\sqrt[4]{5y^2} \cdot \sqrt[4]{2x^3}$ **e.** $\sqrt{\dfrac{2}{a}} \cdot \sqrt{\dfrac{b}{3}}$

Solution   **a.** $\sqrt{3} \cdot \sqrt{5} = \sqrt{3 \cdot 5} = \sqrt{15}$

b. $\sqrt{21} \cdot \sqrt{x} = \sqrt{21x}$

c. $\sqrt[3]{4} \cdot \sqrt[3]{2} = \sqrt[3]{4 \cdot 2} = \sqrt[3]{8} = 2$

d. $\sqrt[4]{5y^2} \cdot \sqrt[4]{2x^3} = \sqrt[4]{5y^2 \cdot 2x^3} = \sqrt[4]{10y^2x^3}$

e. $\sqrt{\dfrac{2}{a}} \cdot \sqrt{\dfrac{b}{3}} = \sqrt{\dfrac{2}{a} \cdot \dfrac{b}{3}} = \sqrt{\dfrac{2b}{3a}}$

2   To discover a quotient rule for radicals, notice the following pattern.

$$\sqrt{\frac{4}{9}} = \frac{2}{3}$$

$$\frac{\sqrt{4}}{\sqrt{9}} = \frac{2}{3}$$

Since both expressions simplify to $\frac{2}{3}$, it is true that

$$\sqrt{\frac{4}{9}} = \frac{\sqrt{4}}{\sqrt{9}}$$

This pattern suggests the following quotient rule for radicals.

Quotient Rule for Radicals

If $\sqrt[n]{a}$ and $\sqrt[n]{b}$ are real numbers and $\sqrt[n]{b}$ is not zero, then

$$\sqrt[n]{\frac{a}{b}} = \frac{\sqrt[n]{a}}{\sqrt[n]{b}}$$

Notice that the quotient rule is the relationship $\left(\frac{a}{b}\right)^{1/n} = \frac{a^{1/n}}{b^{1/n}}$ stated in radical notation. We can use the quotient rule to simplify radical expressions by reading the rule from left to right, or to divide radicals by reading the rule from right to left.

For example,

$$\sqrt{\frac{x}{16}} = \frac{\sqrt{x}}{\sqrt{16}} = \frac{\sqrt{x}}{4} \qquad \text{Using } \sqrt[n]{\frac{a}{b}} = \frac{\sqrt[n]{a}}{\sqrt[n]{b}}$$

$$\frac{\sqrt{75}}{\sqrt{3}} = \sqrt{\frac{75}{3}} = \sqrt{25} = 5 \qquad \text{Using } \frac{\sqrt[n]{a}}{\sqrt[n]{b}} = \sqrt[n]{\frac{a}{b}}$$

Note: *Recall that from Section 10.2 on, we assume that variables represent positive real numbers. Since this is so, we need not insert absolute value bars when we simplify even roots.*

EXAMPLE 2

Use the quotient rule to simplify.

a. $\sqrt{\frac{25}{49}}$

b. $\sqrt{\frac{x}{9}}$

c. $\sqrt[3]{\frac{8}{27}}$

d. $\sqrt[4]{\frac{3}{16y^4}}$

Solution

a. $\sqrt{\dfrac{25}{49}} = \dfrac{\sqrt{25}}{\sqrt{49}} = \dfrac{5}{7}$

b. $\sqrt{\dfrac{x}{9}} = \dfrac{\sqrt{x}}{\sqrt{9}} = \dfrac{\sqrt{x}}{3}$

c. $\sqrt[3]{\dfrac{8}{27}} = \dfrac{\sqrt[3]{8}}{\sqrt[3]{27}} = \dfrac{2}{3}$

d. $\sqrt[4]{\dfrac{3}{16y^4}} = \dfrac{\sqrt[4]{3}}{\sqrt[4]{16y^4}} = \dfrac{\sqrt[4]{3}}{2y}$

3   Both the product and quotient rules can be used to simplify a radical. If the product rule is read from right to left, we have that $\sqrt[n]{ab} = \sqrt[n]{a} \cdot \sqrt[n]{b}$. This is used to simplify the following radicals.

EXAMPLE 3

Simplify the following.

a. $\sqrt{50}$ b. $\sqrt[3]{24}$ c. $\sqrt{26}$ d. $\sqrt[4]{32}$

Solution

a. Factor 50 such that one factor is the largest perfect square that divides 50. The largest perfect square factor of 50 is 25, so we write 50 as $25 \cdot 2$ and use the product rule for radicals to simplify.

> **Helpful Hint**
> Don't forget that, for example, $5\sqrt{2}$ means $5 \cdot \sqrt{2}$.

$$\sqrt{50} = \sqrt{25 \cdot 2} = \sqrt{25} \cdot \sqrt{2} = 5\sqrt{2}$$

       The largest perfect square factor of 50.

b. $\sqrt[3]{24} = \sqrt[3]{8 \cdot 3} = \sqrt[3]{8} \cdot \sqrt[3]{3} = 2\sqrt[3]{3}$

       The largest perfect cube factor of 24.

c. $\sqrt{26}$ The largest perfect square factor of 26 is 1, so $\sqrt{26}$ cannot be simplified further.

d. $\sqrt[4]{32} = \sqrt[4]{16 \cdot 2} = \sqrt[4]{16} \cdot \sqrt[4]{2} = 2\sqrt[4]{2}$

       The largest fourth power factor of 32.

After simplifying a radical such as a square root, always check the radicand to see that it contains no other perfect square factors. It may, if the largest perfect square factor of the radicand was not originally recognized. For example,

$$\sqrt{200} = \sqrt{4 \cdot 50} = \sqrt{4} \cdot \sqrt{50} = 2\sqrt{50}$$

Notice that the radicand 50 still contains the perfect square factor 25. This is because 4 is not the largest perfect square factor of 200. We continue as follows.

$$2\sqrt{50} = 2\sqrt{25 \cdot 2} = 2 \cdot \sqrt{25} \cdot \sqrt{2} = 2 \cdot 5 \cdot \sqrt{2} = 10\sqrt{2}$$

The radical is now simplified since 2 contains no perfect square factors (other than 1).

In general, we say that a radicand of the form $\sqrt[n]{a}$ is simplified when the radicand a contains no factors that are perfect nth powers (other than 1 or -1).

Helpful Hint

To help you recognize largest perfect power factors of a radicand, it will help if you are familiar with some perfect powers. A few are listed below.

Perfect Squares	1,	4,	9,	16,	25,	36,	49,	64,	81,	100,	121,	144
	1^2	2^2	3^2	4^2	5^2	6^2	7^2	8^2	9^2	10^2	11^2	12^2

Perfect Cubes	1,	8,	27,	64,	125
	1^3	2^3	3^3	4^3	5^3

Perfect Fourth Powers	1,	16,	81,	256
	1^4	2^4	3^4	4^4

EXAMPLE 4

Use the product rule to simplify.

a. $\sqrt{25x^3}$ **b.** $\sqrt[3]{54x^6y^8}$ **c.** $\sqrt[4]{81z^{11}}$

Solution   **a.** $\sqrt{25x^3} = \sqrt{25x^2 \cdot x}$ Find the largest perfect square factor.

$\qquad = \sqrt{25x^2} \cdot \sqrt{x}$ Apply the product rule.

$\qquad = 5x\sqrt{x}$ Simplify.

b. $\sqrt[3]{54x^6y^8} = \sqrt[3]{27 \cdot 2 \cdot x^6 \cdot y^6 \cdot y^2}$ Factor the radicand and identify perfect cube factors.

$\qquad = \sqrt[3]{27x^6y^6 \cdot 2y^2}$

$\qquad = \sqrt[3]{27x^6y^6} \cdot \sqrt[3]{2y^2}$ Apply the product rule.

$\qquad = 3x^2y^2\sqrt[3]{2y^2}$ Simplify.

c. $\sqrt[4]{81z^{11}} = \sqrt[4]{81 \cdot z^8 \cdot z^3}$ Factor the radicand and identify perfect fourth power factors.

$\qquad = \sqrt[4]{81z^8} \cdot \sqrt[4]{z^3}$ Apply the product rule.

$\qquad = 3z^2\sqrt[4]{z^3}$ Simplify.

EXAMPLE 5

Use the quotient rule to divide, and simplify if possible.

a. $\dfrac{\sqrt{20}}{\sqrt{5}}$

b. $\dfrac{\sqrt{50x}}{2\sqrt{2}}$

c. $\dfrac{7\sqrt[3]{48x^4y^8}}{\sqrt[3]{6y^2}}$

d. $\dfrac{2\sqrt[4]{32a^8b^6}}{\sqrt[4]{a^{-1}b^2}}$

Solution

a. $\dfrac{\sqrt{20}}{\sqrt{5}} = \sqrt{\dfrac{20}{5}}$ Apply the quotient rule.

$\qquad\quad = \sqrt{4}$ Simplify.

$\qquad\quad = 2$ Simplify.

b. $\dfrac{\sqrt{50x}}{2\sqrt{2}} = \dfrac{1}{2}\cdot\sqrt{\dfrac{50x}{2}}$ Apply the quotient rule.

$\qquad\quad = \dfrac{1}{2}\cdot\sqrt{25x}$ Simplify.

$\qquad\quad = \dfrac{1}{2}\cdot\sqrt{25}\cdot\sqrt{x}$ Factor 25x.

$\qquad\quad = \dfrac{1}{2}\cdot 5\cdot\sqrt{x}$ Simplify.

$\qquad\quad = \dfrac{5}{2}\sqrt{x}$

c. $\dfrac{7\sqrt[3]{48x^4y^8}}{\sqrt[3]{6y^2}} = 7\cdot\sqrt[3]{\dfrac{48x^4y^8}{6y^2}}$ Apply the quotient rule.

$\qquad\quad = 7\cdot\sqrt[3]{8x^4y^6}$ Simplify.

$\qquad\quad = 7\sqrt[3]{8x^3y^6\cdot x}$ Factor.

$\qquad\quad = 7\cdot\sqrt[3]{8x^3y^6}\cdot\sqrt[3]{x}$ Apply the product rule.

$\qquad\quad = 7\cdot 2xy^2\cdot\sqrt[3]{x}$ Simplify.

$\qquad\quad = 14xy^2\sqrt[3]{x}$

d. $\dfrac{2\sqrt[4]{32a^8b^6}}{\sqrt[4]{a^{-1}b^2}} = 2\sqrt[4]{\dfrac{32a^8b^6}{a^{-1}b^2}} = 2\sqrt[4]{32a^9b^4} = 2\sqrt[4]{16\cdot a^8\cdot b^4\cdot 2\cdot a}$

$\qquad\quad = 2\sqrt[4]{16a^8b^4}\cdot\sqrt[4]{2a} = 2\cdot 2a^2b\cdot\sqrt[4]{2a} = 4a^2b\sqrt[4]{2a}$

✔ **CONCEPT CHECK**

Find and correct the error:

$$\dfrac{\sqrt[3]{27}}{\sqrt{9}} = \sqrt[3]{\dfrac{27}{9}} = \sqrt[3]{3}$$

Concept Check Answer:

$$\dfrac{\sqrt[3]{27}}{\sqrt{9}} = \dfrac{3}{3} = 1$$

EXERCISE SET 10.3

STUDY GUIDE/SSM   CD/ VIDEO   PH MATH TUTOR CENTER   MathXL®Tutorials ON CD   MathXL®   MyMathLab®

Use the product rule to multiply. See Example 1.

1. $\sqrt{7} \cdot \sqrt{2}$

2. $\sqrt{11} \cdot \sqrt{10}$

3. $\sqrt[4]{8} \cdot \sqrt[4]{2}$

4. $\sqrt[4]{27} \cdot \sqrt[4]{3}$

5. $\sqrt[3]{4} \cdot \sqrt[3]{9}$

6. $\sqrt[3]{10} \cdot \sqrt[3]{5}$

7. $\sqrt{2} \cdot \sqrt{3x}$

8. $\sqrt{3y} \cdot \sqrt{5x}$

9. $\sqrt{\dfrac{7}{x}} \cdot \sqrt{\dfrac{2}{y}}$

10. $\sqrt{\dfrac{6}{m}} \cdot \sqrt{\dfrac{n}{5}}$

11. $\sqrt[4]{4x^3} \cdot \sqrt[4]{5}$

12. $\sqrt[4]{ab^2} \cdot \sqrt[4]{27ab}$

Use the quotient rule to simplify. See Examples 2 and 3.

13. $\sqrt{\dfrac{6}{49}}$

14. $\sqrt{\dfrac{8}{81}}$

15. $\sqrt{\dfrac{2}{49}}$

16. $\sqrt{\dfrac{5}{121}}$

17. $\sqrt[4]{\dfrac{x^3}{16}}$

18. $\sqrt[4]{\dfrac{y}{81x^4}}$

19. $\sqrt[3]{\dfrac{4}{27}}$

20. $\sqrt[3]{\dfrac{3}{64}}$

21. $\sqrt[4]{\dfrac{8}{x^8}}$

22. $\sqrt[4]{\dfrac{a^3}{81}}$

23. $\sqrt[3]{\dfrac{2x}{81y^{12}}}$

24. $\sqrt[3]{\dfrac{3}{8x^6}}$

25. $\sqrt{\dfrac{x^2y}{100}}$

26. $\sqrt{\dfrac{y^2z}{36}}$

27. $\sqrt{\dfrac{5x^2}{4y^2}}$

28. $\sqrt{\dfrac{y^{10}}{9x^6}}$

29. $-\sqrt[3]{\dfrac{z^7}{27x^3}}$

30. $-\sqrt[3]{\dfrac{64a}{b^9}}$

Simplify. See Examples 3 and 4.

31. $\sqrt{32}$

32. $\sqrt{27}$

33. $\sqrt[3]{192}$

34. $\sqrt[3]{108}$

35. $5\sqrt{75}$

36. $3\sqrt{8}$

37. $\sqrt{24}$

38. $\sqrt{20}$

39. $\sqrt{100x^5}$

40. $\sqrt{64y^9}$

41. $\sqrt[3]{16y^7}$

42. $\sqrt[3]{64y^9}$

43. $\sqrt[4]{a^8b^7}$

44. $\sqrt[5]{32z^{12}}$

45. $\sqrt{y^5}$

46. $\sqrt[3]{y^5}$

47. $\sqrt{25a^2b^3}$

48. $\sqrt{9x^5y^7}$

49. $\sqrt[5]{-32x^{10}y}$

50. $\sqrt[5]{-243z^9}$

51. $\sqrt[3]{50x^{14}}$

52. $\sqrt[3]{40y^{10}}$

53. $-\sqrt{32a^8b^7}$

54. $-\sqrt{20ab^6}$

55. $\sqrt{9x^7y^9}$

56. $\sqrt{12r^9s^{12}}$

57. $\sqrt[3]{125r^9s^{12}}$

58. $\sqrt[3]{8a^6b^9}$

Use the quotient rule to divide. Then simplify if possible. See Example 5.

59. $\dfrac{\sqrt{14}}{\sqrt{7}}$

60. $\dfrac{\sqrt{45}}{\sqrt{9}}$

61. $\dfrac{\sqrt[3]{24}}{\sqrt[3]{3}}$

62. $\dfrac{\sqrt[3]{10}}{\sqrt[3]{2}}$

63. $\dfrac{5\sqrt[4]{48}}{\sqrt[4]{3}}$

64. $\dfrac{7\sqrt[4]{162}}{\sqrt[4]{2}}$

65. $\dfrac{\sqrt{x^5y^3}}{\sqrt{xy}}$

66. $\dfrac{\sqrt{a^7b^6}}{\sqrt{a^3b^2}}$

67. $\dfrac{8\sqrt[3]{54m^7}}{\sqrt[3]{2m}}$

68. $\dfrac{\sqrt[3]{128x^3}}{-3\sqrt[3]{2x}}$

69. $\dfrac{3\sqrt{100x^2}}{2\sqrt{2x^{-1}}}$

70. $\dfrac{\sqrt{270y^2}}{5\sqrt{3y^{-4}}}$

71. $\dfrac{\sqrt[4]{96a^{10}b^3}}{\sqrt[4]{3a^2b^3}}$

72. $\dfrac{\sqrt[5]{64x^{10}y^3}}{\sqrt[5]{2x^3y^{-7}}}$

REVIEW AND PREVIEW

Perform each indicated operation. See Sections 2.1 and 5.3.

73. $6x + 8x$

74. $(6x)(8x)$

75. $(2x + 3)(x - 5)$

76. $(2x + 3) + (x - 5)$

77. $9y^2 - 8y^2$

78. $(9y^2)(-8y^2)$

79. $-3(x + 5)$

80. $-3 + x + 5$

81. $(x - 4)^2$

82. $(2x + 1)^2$

Concept Extensions

△ **83.** The formula for the surface area A of a cone with height h and radius r is given by

$$A = \pi r \sqrt{r^2 + h^2}$$

a. Find the surface area of a cone whose height is 3 centimeters and whose radius is 4 centimeters.

b. Approximate to two decimal places the surface area of a cone whose height is 7.2 feet and whose radius is 6.8 feet.

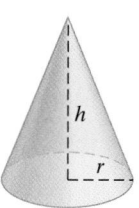

△ **84.** Before Mount Vesuvius, a volcano in Italy, erupted violently in 79 A.D., its height was 4190 feet. Vesuvius was roughly cone-shaped, and its base had a radius of approximately 25,200 feet. Use the formula for the surface area of a cone, given in Exercise 83, to approximate the surface area this volcano had before it erupted. (*Source:* Global Volcanism Network)

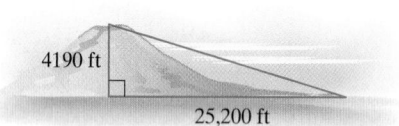

4190 ft

25,200 ft

85. The owner of Knightime Video has determined that the demand equation for renting older releases is given by the equation $F(x) = 0.6\sqrt{49 - x^2}$, where x is the price in dollars per two-day rental and $F(x)$ is the number of times the video is demanded per week.

a. Approximate to one decimal place the demand per week of an older release if the rental price is $3 per two-day rental.

b. Approximate to one decimal place the demand per week of an older release if the rental price is $5 per two-day rental.

c. Explain how the owner of the video store can use this equation to predict the number of copies of each tape that should be in stock.

10.4 ADDING, SUBTRACTING, AND MULTIPLYING RADICAL EXPRESSIONS

Objectives

1 Add or subtract radical expressions.

2 Multiply radical expressions.

1 We have learned that sums or differences of like terms can be simplified. To simplify these sums or differences, we use the distributive property. For example,

$$2x + 3x = (2 + 3)x = 5x \quad \text{and} \quad 7x^2y - 4x^2y = (7 - 4)x^2y = 3x^2y$$

The distributive property can also be used to add **like radicals**.

> ### Like Radicals
> Radicals with the same index and the same radicand are like radicals.

For example, $2\sqrt{7} + 3\sqrt{7} = (2 + 3)\sqrt{7} = 5\sqrt{7}$. Also,

$$5\sqrt{3x} - 7\sqrt{3x} = (5 - 7)\sqrt{3x} = -2\sqrt{3x}$$

The expression $2\sqrt{7} + 2\sqrt[3]{7}$ cannot be simplified further since $2\sqrt{7}$ and $2\sqrt[3]{7}$ are not like radicals.

EXAMPLE 1

Add or subtract as indicated.

a. $4\sqrt{11} + 8\sqrt{11}$ **b.** $5\sqrt[3]{3x} - 7\sqrt[3]{3x}$ **c.** $2\sqrt{7} + 2\sqrt[3]{7}$

Solution   **a.** $4\sqrt{11} + 8\sqrt{11} = (4 + 8)\sqrt{11} = 12\sqrt{11}$

b. $5\sqrt[3]{3x} - 7\sqrt[3]{3x} = (5 - 7)\sqrt[3]{3x} = -2\sqrt[3]{3x}$

c. $2\sqrt{7} + 2\sqrt[3]{7}$
This expression cannot be simplified since $2\sqrt{7}$ and $2\sqrt[3]{7}$ do not contain like radicals.

When adding or subtracting radicals, always check first to see whether any radicals can be simplified.

 CONCEPT CHECK

True or false:

$$\sqrt{a} + \sqrt{b} = \sqrt{a + b}?$$

Explain.

EXAMPLE 2

Add or subtract. Assume that variables represent positive real numbers.

a. $\sqrt{20} + 2\sqrt{45}$ **b.** $\sqrt[3]{54} - 5\sqrt[3]{16} + \sqrt[3]{2}$ **c.** $\sqrt{27x} - 2\sqrt{9x} + \sqrt{72x}$

d. $\sqrt[3]{98} + \sqrt{98}$ **e.** $\sqrt[3]{48y^4} + \sqrt[3]{6y^4}$

Solution   First, simplify each radical. Then add or subtract any like radicals.

a. $\sqrt{20} + 2\sqrt{45} = \sqrt{4 \cdot 5} + 2\sqrt{9 \cdot 5}$ Factor **20** and **45**.

$\qquad\qquad\qquad = \sqrt{4} \cdot \sqrt{5} + 2 \cdot \sqrt{9} \cdot \sqrt{5}$ Use the product rule.

$\qquad\qquad\qquad = 2 \cdot \sqrt{5} + 2 \cdot 3 \cdot \sqrt{5}$ Simplify $\sqrt{4}$ and $\sqrt{9}$.

$\qquad\qquad\qquad = 2\sqrt{5} + 6\sqrt{5}$ Add like radicals.

$\qquad\qquad\qquad = 8\sqrt{5}$

b. $\sqrt[3]{54} - 5\sqrt[3]{16} + \sqrt[3]{2}$

$\qquad = \sqrt[3]{27} \cdot \sqrt[3]{2} - 5 \cdot \sqrt[3]{8} \cdot \sqrt[3]{2} + \sqrt[3]{2}$ Factor and use the product rule.

$\qquad = 3 \cdot \sqrt[3]{2} - 5 \cdot 2 \cdot \sqrt[3]{2} + \sqrt[3]{2}$ Simplify $\sqrt[3]{27}$ and $\sqrt[3]{8}$.

$\qquad = 3\sqrt[3]{2} - 10\sqrt[3]{2} + \sqrt[3]{2}$ Write $5 \cdot 2$ as **10**.

$\qquad = -6\sqrt[3]{2}$ Combine like radicals.

c. $\sqrt{27x} - 2\sqrt{9x} + \sqrt{72x}$

$= \sqrt{9} \cdot \sqrt{3x} - 2 \cdot \sqrt{9} \cdot \sqrt{x} + \sqrt{36} \cdot \sqrt{2x}$ Factor and use the product rule.

$= 3 \cdot \sqrt{3x} - 2 \cdot 3 \cdot \sqrt{x} + 6 \cdot \sqrt{2x}$ Simplify $\sqrt{9}$ and $\sqrt{36}$.

$= 3\sqrt{3x} - 6\sqrt{x} + 6\sqrt{2x}$ Write $2 \cdot 3$ as 6.

Helpful Hint

None of these terms contain like radicals. We can simplify no further.

d. $\sqrt[3]{98} + \sqrt{98} = \sqrt[3]{98} + \sqrt{49} \cdot \sqrt{2}$ Factor and use the product rule.

$= \sqrt[3]{98} + 7\sqrt{2}$ No further simplification is possible.

e. $\sqrt[3]{48y^4} + \sqrt[3]{6y^4} = \sqrt[3]{8y^3} \cdot \sqrt[3]{6y} + \sqrt[3]{y^3} \cdot \sqrt[3]{6y}$ Factor and use the product rule.

$= 2y\sqrt[3]{6y} + y\sqrt[3]{6y}$ Simplify $\sqrt[3]{8y^3}$ and $\sqrt[3]{y^3}$.

$= 3y\sqrt[3]{6y}$ Combine like radicals.

EXAMPLE 3

Add or subtract as indicated.

a. $\dfrac{\sqrt{45}}{4} - \dfrac{\sqrt{5}}{3}$ **b.** $\sqrt[3]{\dfrac{7x}{8}} + 2\sqrt[3]{7x}$

Solution   **a.** $\dfrac{\sqrt{45}}{4} - \dfrac{\sqrt{5}}{3} = \dfrac{3\sqrt{5}}{4} - \dfrac{\sqrt{5}}{3}$ To subtract, notice that the LCD is 12.

$= \dfrac{3\sqrt{5} \cdot 3}{4 \cdot 3} - \dfrac{\sqrt{5} \cdot 4}{3 \cdot 4}$ Write each expression as an equivalent expression with a denominator of 12.

$= \dfrac{9\sqrt{5}}{12} - \dfrac{4\sqrt{5}}{12}$ Multiply factors in the numerator and the denominator.

$= \dfrac{5\sqrt{5}}{12}$ Subtract.

b. $\sqrt[3]{\dfrac{7x}{8}} + 2\sqrt[3]{7x} = \dfrac{\sqrt[3]{7x}}{\sqrt[3]{8}} + 2\sqrt[3]{7x}$ Apply the quotient rule for radicals.

$= \dfrac{\sqrt[3]{7x}}{2} + 2\sqrt[3]{7x}$ Simplify.

$= \dfrac{\sqrt[3]{7x}}{2} + \dfrac{2\sqrt[3]{7x} \cdot 2}{2}$ Write each expression as an equivalent expression with a denominator of 2.

$= \dfrac{\sqrt[3]{7x}}{2} + \dfrac{4\sqrt[3]{7x}}{2}$

$= \dfrac{5\sqrt[3]{7x}}{2}$ Add.

2   We can multiply radical expressions by using many of the same properties used to multiply polynomial expressions. For instance, to multiply $\sqrt{2}\left(\sqrt{6} - 3\sqrt{2}\right)$, we use the distributive property and multiply $\sqrt{2}$ by each term inside the parentheses.

$$\sqrt{2}\left(\sqrt{6} - 3\sqrt{2}\right) = \sqrt{2}\left(\sqrt{6}\right) - \sqrt{2}\left(3\sqrt{2}\right) \quad \text{Use the distributive property.}$$
$$= \sqrt{2 \cdot 6} - 3\sqrt{2 \cdot 2}$$
$$= \sqrt{2 \cdot 2 \cdot 3} - 3 \cdot 2 \quad \text{Use the product rule for radicals.}$$
$$= 2\sqrt{3} - 6$$

EXAMPLE 4

Multiply.

a. $\sqrt{3}\left(5 + \sqrt{30}\right)$ **b.** $\left(\sqrt{5} - \sqrt{6}\right)\left(\sqrt{7} + 1\right)$ **c.** $\left(7\sqrt{x} + 5\right)\left(3\sqrt{x} - \sqrt{5}\right)$

d. $\left(4\sqrt{3} - 1\right)^2$ **e.** $\left(\sqrt{2x} - 5\right)\left(\sqrt{2x} + 5\right)$ **f.** $\left(\sqrt{x - 3} + 5\right)^2$

Solution   **a.** $\sqrt{3}\left(5 + \sqrt{30}\right) = \sqrt{3}(5) + \sqrt{3}\left(\sqrt{30}\right)$
$$= 5\sqrt{3} + \sqrt{3 \cdot 30}$$
$$= 5\sqrt{3} + \sqrt{3 \cdot 3 \cdot 10}$$
$$= 5\sqrt{3} + 3\sqrt{10}$$

b. To multiply, we can use the FOIL method.

$$\left(\sqrt{5} - \sqrt{6}\right)\left(\sqrt{7} + 1\right) = \overset{\text{First}}{\sqrt{5} \cdot \sqrt{7}} + \overset{\text{Outer}}{\sqrt{5} \cdot 1} - \overset{\text{Inner}}{\sqrt{6} \cdot \sqrt{7}} - \overset{\text{Last}}{\sqrt{6} \cdot 1}$$
$$= \sqrt{35} + \sqrt{5} - \sqrt{42} - \sqrt{6}$$

c. $\left(7\sqrt{x} + 5\right)\left(3\sqrt{x} - \sqrt{5}\right) = 7\sqrt{x}\left(3\sqrt{x}\right) - 7\sqrt{x}\left(\sqrt{5}\right) + 5\left(3\sqrt{x}\right) - 5\left(\sqrt{5}\right)$
$$= 21x - 7\sqrt{5x} + 15\sqrt{x} - 5\sqrt{5}$$

d. $\left(4\sqrt{3} - 1\right)^2 = \left(4\sqrt{3} - 1\right)\left(4\sqrt{3} - 1\right)$
$$= 4\sqrt{3}\left(4\sqrt{3}\right) - 4\sqrt{3}(1) - 1\left(4\sqrt{3}\right) - 1(-1)$$
$$= 16 \cdot 3 - 4\sqrt{3} - 4\sqrt{3} + 1$$
$$= 48 - 8\sqrt{3} + 1$$
$$= 49 - 8\sqrt{3}$$

e. $\left(\sqrt{2x} - 5\right)\left(\sqrt{2x} + 5\right) = \sqrt{2x} \cdot \sqrt{2x} + 5\sqrt{2x} - 5\sqrt{2x} - 5 \cdot 5$
$$= 2x - 25$$

f. $\underbrace{\left(\sqrt{x - 3}\right.}_{a} + \underbrace{5}_{b})^2 = \underbrace{\left(\sqrt{x - 3}\right)^2}_{a^2} + \underbrace{2 \cdot}_{+ 2 \cdot} \underbrace{\sqrt{x - 3}}_{a} \cdot \underbrace{5}_{\cdot b} + \underbrace{5^2}_{+ b^2}$

$$= x - 3 + 10\sqrt{x - 3} + 25 \quad \text{Simplify.}$$
$$= x + 22 + 10\sqrt{x - 3} \quad \text{Combine like terms.}$$

MENTAL MATH

Simplify. Assume that all variables represent positive real numbers.

1. $2\sqrt{3} + 4\sqrt{3}$ **2.** $5\sqrt{7} + 3\sqrt{7}$ **3.** $8\sqrt{x} - 5\sqrt{x}$

4. $3\sqrt{y} + 10\sqrt{y}$ **5.** $7\sqrt[3]{x} + 5\sqrt[3]{x}$ **6.** $8\sqrt[3]{z} - 2\sqrt[3]{z}$

Add or Subtract if possible.

7. $\sqrt{11} + \sqrt{11}$ **8.** $\sqrt{11} + \sqrt[3]{11}$ **9.** $9\sqrt{13} - \sqrt{13}$

10. $9\sqrt{13} - \sqrt[4]{13}$ **11.** $8\sqrt[3]{2x} + 3\sqrt[3]{2x} - \sqrt[3]{2x}$ **12.** $8\sqrt[3]{2x} + 3\sqrt[3]{2x^2} - \sqrt[3]{2x}$

EXERCISE SET 10.4

STUDY GUIDE/SSM   CD/VIDEO   PH MATH TUTOR CENTER   MathXL®Tutorials ON CD   MathXL®   MyMathLab®

Add or subtract. See Examples 1 through 3.

1. $\sqrt{8} - \sqrt{32}$ **2.** $\sqrt{27} - \sqrt{75}$

3. $2\sqrt{2x^3} + 4x\sqrt{8x}$ **4.** $3\sqrt{45x^3} + x\sqrt{5x}$

 5. $2\sqrt{50} - 3\sqrt{125} + \sqrt{98}$ **6.** $4\sqrt{32} - \sqrt{18} + 2\sqrt{128}$

7. $\sqrt[3]{16x} - \sqrt[3]{54x}$ **8.** $2\sqrt[3]{3a^4} - 3a\sqrt[3]{81a}$

9. $\sqrt{9b^3} - \sqrt{25b^3} + \sqrt{49b^3}$

10. $\sqrt{4x^7} + 9x^2\sqrt{x^3} - 5x\sqrt{x^5}$

11. $\dfrac{5\sqrt{2}}{3} + \dfrac{2\sqrt{2}}{5}$ **12.** $\dfrac{\sqrt{3}}{2} + \dfrac{4\sqrt{3}}{3}$

13. $\sqrt[3]{\dfrac{11}{8}} - \dfrac{\sqrt[3]{11}}{6}$ **14.** $\dfrac{2\sqrt[3]{4}}{7} - \dfrac{\sqrt[3]{4}}{14}$

15. $\dfrac{\sqrt{20x}}{9} + \sqrt{\dfrac{5x}{9}}$ **16.** $\dfrac{3x\sqrt{7}}{5} + \sqrt{\dfrac{7x^2}{100}}$

17. $7\sqrt{9} - 7 + \sqrt{3}$ **18.** $\sqrt{16} - 5\sqrt{10} + 7$

19. $2 + 3\sqrt{y^2} - 6\sqrt{y^2} + 5$

20. $3\sqrt{7} - \sqrt[3]{x} + 4\sqrt{7} - 3\sqrt[3]{x}$

21. $3\sqrt{108} - 2\sqrt{18} - 3\sqrt{48}$

22. $-\sqrt{75} + \sqrt{12} - 3\sqrt{3}$

23. $-5\sqrt[3]{625} + \sqrt[3]{40}$ **24.** $-2\sqrt[3]{108} - \sqrt[3]{32}$

25. $\sqrt{9b^3} - \sqrt{25b^3} + \sqrt{16b^3}$

26. $\sqrt{4x^7y^5} + 9x^2\sqrt{x^3y^5} - 5xy\sqrt{x^5y^3}$

27. $5y\sqrt{8y} + 2\sqrt{50y^3}$ **28.** $3\sqrt{8x^2y^3} - 2x\sqrt{32y^3}$

29. $\sqrt[3]{54xy^3} - 5\sqrt[3]{2xy^3} + y\sqrt[3]{128x}$

30. $2\sqrt[3]{24x^3y^4} + 4x\sqrt[3]{81y^4}$

31. $6\sqrt[3]{11} + 8\sqrt{11} - 12\sqrt{11}$ **32.** $3\sqrt[3]{5} + 4\sqrt{5}$

33. $-2\sqrt[4]{x^7} + 3\sqrt[4]{16x^7}$

34. $6\sqrt[3]{24x^3} - 2\sqrt[3]{81x^3} - x\sqrt[3]{3}$

35. $\dfrac{4\sqrt{3}}{3} - \dfrac{\sqrt{12}}{3}$ **36.** $\dfrac{\sqrt{45}}{10} + \dfrac{7\sqrt{5}}{10}$

37. $\dfrac{\sqrt[3]{8x^4}}{7} + \dfrac{3x\sqrt[3]{x}}{7}$ **38.** $\dfrac{\sqrt[4]{48}}{5x} - \dfrac{2\sqrt[4]{3}}{10x}$

39. $\sqrt{\dfrac{28}{x^2}} + \sqrt{\dfrac{7}{4x^2}}$ **40.** $\dfrac{\sqrt{99}}{5x} - \sqrt{\dfrac{44}{x^2}}$

41. $\sqrt[3]{\dfrac{16}{27}} - \dfrac{\sqrt[3]{54}}{6}$ **42.** $\dfrac{\sqrt[3]{3}}{10} + \sqrt[3]{\dfrac{24}{125}}$

43. $-\dfrac{\sqrt[3]{2x^4}}{9} + \sqrt[3]{\dfrac{250x^4}{27}}$ **44.** $\dfrac{\sqrt[4]{y^5}}{8} + \dfrac{5y\sqrt[4]{y^2}}{4}$

△ **45.** Find the perimeter of the trapezoid.

2√12 in.   3√3 in.   √12 in.   2√27 in.

△ **46.** Find the perimeter of the triangle.

√8 m   √32 m   √45 m

Multiply, and then simplify if possible. See Example 4.

47. $\sqrt{7}(\sqrt{5} + \sqrt{3})$

48. $\sqrt{5}(\sqrt{15} - \sqrt{35})$

49. $(\sqrt{5} - \sqrt{2})^2$

50. $(3x - \sqrt{2})(3x - \sqrt{2})$

51. $\sqrt{3x}(\sqrt{3} - \sqrt{x})$

52. $\sqrt{5y}(\sqrt{y} + \sqrt{5})$

53. $(2\sqrt{x} - 5)(3\sqrt{x} + 1)$

54. $(8\sqrt{y} + z)(4\sqrt{y} - 1)$

55. $(\sqrt[3]{a} - 4)(\sqrt[3]{a} + 5)$

56. $(\sqrt[3]{a} + 2)(\sqrt[3]{a} + 7)$

57. $6(\sqrt{2} - 2)$

58. $\sqrt{5}(6 - \sqrt{5})$

59. $\sqrt{2}(\sqrt{2} + x\sqrt{6})$

60. $\sqrt{3}(\sqrt{3} - 2\sqrt{5x})$

61. $(2\sqrt{7} + 3\sqrt{5})(\sqrt{7} - 2\sqrt{5})$

62. $(\sqrt{6} - 4\sqrt{2})(3\sqrt{6} + 1)$

63. $(\sqrt{x} - y)(\sqrt{x} + y)$

64. $(3\sqrt{x} + 2)(\sqrt{3x} - 2)$

65. $(\sqrt{3} + x)^2$

66. $(\sqrt{y} - 3x)^2$

67. $(\sqrt{5x} - 3\sqrt{2})(\sqrt{5x} - 3\sqrt{3})$

68. $(5\sqrt{3x} - \sqrt{y})(4\sqrt{x} + 1)$

69. $(\sqrt[3]{4} + 2)(\sqrt[3]{2} - 1)$

70. $(\sqrt[3]{3} + \sqrt[3]{2})(\sqrt[3]{9} - \sqrt[3]{4})$

71. $(\sqrt[3]{x} + 1)(\sqrt[3]{x} - 4\sqrt{x} + 7)$

72. $(\sqrt[3]{3x} + 3)(\sqrt[3]{2x} - 3x - 1)$

73. $(\sqrt{x - 1} + 5)^2$

74. $(\sqrt{3x + 1} + 2)^2$

75. $(\sqrt{2x + 5} - 1)^2$

76. $(\sqrt{x - 6} - 7)^2$

REVIEW AND PREVIEW

Factor each numerator and denominator. Then simplify if possible. See Section 7.1.

77. $\dfrac{2x - 14}{2}$

78. $\dfrac{8x - 24y}{4}$

79. $\dfrac{7x - 7y}{x^2 - y^2}$

80. $\dfrac{x^3 - 8}{4x - 8}$

81. $\dfrac{6a^2b - 9ab}{3ab}$

82. $\dfrac{14r - 28r^2s^2}{7rs}$

83. $\dfrac{-4 + 2\sqrt{3}}{6}$

84. $\dfrac{-5 + 10\sqrt{7}}{5}$

Concept Extensions

85. Find the perimeter and area of the rectangle.

$\sqrt{125}$ ft

$3\sqrt{20}$ ft

86. Find the area and perimeter of the trapezoid. (*Hint:* The area of a trapezoid is the product of half the height $6\sqrt{3}$ meters and the sum of the bases $2\sqrt{63}$ and $7\sqrt{7}$ meters.)

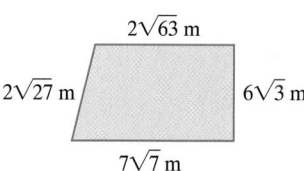

$2\sqrt{63}$ m

$2\sqrt{27}$ m

$6\sqrt{3}$ m

$7\sqrt{7}$ m

87. a. Add: $\sqrt{3} + \sqrt{3}$

b. Multiply: $\sqrt{3} \cdot \sqrt{3}$

c. Describe the differences in parts **a** and **b**.

88. Multiply: $(\sqrt{2} + \sqrt{3} - 1)^2$

89. Explain how simplifying $2x + 3x$ is similar to simplifying $2\sqrt{x} + 3\sqrt{x}$.

90. Explain how multiplying $(x - 2)(x + 3)$ is similar to multiplying $(\sqrt{x} - \sqrt{2})(\sqrt{x} + 3)$.

10.5 RATIONALIZING DENOMINATORS AND NUMERATORS OF RADICAL EXPRESSIONS

Objectives

1 Rationalize denominators.

2 Rationalize denominators having two terms.

3 Rationalize numerators.

1   Often in mathematics, it is helpful to write a radical expression such as $\dfrac{\sqrt{3}}{\sqrt{2}}$ either without a radical in the denominator or without a radical in the numerator. The process of writing this expression as an equivalent expression but without a radical in the denominator is called **rationalizing the denominator**. To rationalize the denominator of $\dfrac{\sqrt{3}}{\sqrt{2}}$, we use the fundamental principle of fractions and multiply the numerator and the denominator by $\sqrt{2}$. Recall that this is the same as multiplying by $\dfrac{\sqrt{2}}{\sqrt{2}}$, which simplifies to 1.

$$\frac{\sqrt{3}}{\sqrt{2}} = \frac{\sqrt{3}\cdot\sqrt{2}}{\sqrt{2}\cdot\sqrt{2}} = \frac{\sqrt{6}}{\sqrt{4}} = \frac{\sqrt{6}}{2}$$

EXAMPLE 1

Rationalize the denominator of each expression.

a. $\dfrac{2}{\sqrt{5}}$ **b.** $\dfrac{2\sqrt{16}}{\sqrt{9x}}$ **c.** $\sqrt[3]{\dfrac{1}{2}}$

Solution   **a.** To rationalize the denominator, we multiply the numerator and denominator by a factor that makes the radicand in the denominator a perfect square.

$$\frac{2}{\sqrt{5}} = \frac{2\cdot\sqrt{5}}{\sqrt{5}\cdot\sqrt{5}} = \frac{2\sqrt{5}}{5} \qquad \text{\textit{The denominator is now rationalized.}}$$

b. First, we simplify the radicals and then rationalize the denominator.

$$\frac{2\sqrt{16}}{\sqrt{9x}} = \frac{2(4)}{3\sqrt{x}} = \frac{8}{3\sqrt{x}}$$

To rationalize the denominator, multiply the numerator and denominator by $\sqrt{x}$. Then

$$\frac{8}{3\sqrt{x}} = \frac{8\cdot\sqrt{x}}{3\sqrt{x}\cdot\sqrt{x}} = \frac{8\sqrt{x}}{3x}$$

c. $\sqrt[3]{\dfrac{1}{2}} = \dfrac{\sqrt[3]{1}}{\sqrt[3]{2}} = \dfrac{1}{\sqrt[3]{2}}$. Now we rationalize the denominator. Since $\sqrt[3]{2}$ is a cube root, we want to multiply by a value that will make the radicand 2 a perfect cube. If we multiply $\sqrt[3]{2}$ by $\sqrt[3]{2^2}$, we get $\sqrt[3]{2^3} = \sqrt[3]{8} = 2$.

$$\frac{1\cdot\sqrt[3]{2^2}}{\sqrt[3]{2}\cdot\sqrt[3]{2^2}} = \frac{\sqrt[3]{4}}{\sqrt[3]{2^3}} = \frac{\sqrt[3]{4}}{2} \qquad \text{\textit{Multiply the numerator and denominator by } } \sqrt[3]{2^2} \text{\textit{ and then simplify.}}$$

✔ **CONCEPT CHECK**

Determine by which number both the numerator and denominator can be multiplied to rationalize the denominator of the radical expression.

a. $\dfrac{1}{\sqrt[3]{7}}$ **b.** $\dfrac{1}{\sqrt[4]{8}}$

EXAMPLE 2

Rationalize the denominator of $\sqrt{\dfrac{7x}{3y}}$.

Solution

$$\sqrt{\frac{7x}{3y}} = \frac{\sqrt{7x}}{\sqrt{3y}}$$

Use the quotient rule. No radical may be simplified further.

$$= \frac{\sqrt{7x} \cdot \sqrt{3y}}{\sqrt{3y} \cdot \sqrt{3y}}$$

Multiply numerator and denominator by $\sqrt{3y}$ so that the radicand in the denominator is a perfect square.

$$= \frac{\sqrt{21xy}}{3y}$$

Use the product rule in the numerator and denominator. Remember that $\sqrt{3y} \cdot \sqrt{3y} = 3y$.

EXAMPLE 3

Rationalize the denominator of $\dfrac{\sqrt[4]{x}}{\sqrt[4]{81y^5}}$.

Solution   First, simplify each radical if possible.

$$\frac{\sqrt[4]{x}}{\sqrt[4]{81y^5}} = \frac{\sqrt[4]{x}}{\sqrt[4]{81y^4} \cdot \sqrt[4]{y}}$$

Use the product rule in the denominator.

$$= \frac{\sqrt[4]{x}}{3y\sqrt[4]{y}}$$

Write $\sqrt[4]{81y^4}$ as $3y$.

$$= \frac{\sqrt[4]{x} \cdot \sqrt[4]{y^3}}{3y\sqrt[4]{y} \cdot \sqrt[4]{y^3}}$$

Multiply numerator and denominator by $\sqrt[4]{y^3}$ so that the radicand in the denominator is a perfect fourth power.

$$= \frac{\sqrt[4]{xy^3}}{3y\sqrt[4]{y^4}}$$

Use the product rule in the numerator and denominator.

$$= \frac{\sqrt[4]{xy^3}}{3y^2}$$

In the denominator, $\sqrt[4]{y^4} = y$ and $3y \cdot y = 3y^2$.

2   Remember the product of the sum and difference of two terms?

$$(a + b)(a - b) = a^2 - b^2$$

Concept Check Answer:
a. $\sqrt[3]{7^2}$ or $\sqrt[3]{49}$ **b.** $\sqrt[4]{2}$

These two expressions are called conjugates of each other.

To rationalize a numerator or denominator that is a sum or difference of two terms, we use conjugates. To see how and why this works, let's rationalize the denominator of the expression $\dfrac{5}{\sqrt{3} - 2}$. To do so, we multiply both the numerator and the denominator by $\sqrt{3} + 2$, the **conjugate** of the denominator $\sqrt{3} - 2$, and see what happens.

$$\frac{5}{\sqrt{3} - 2} = \frac{5(\sqrt{3} + 2)}{(\sqrt{3} - 2)(\sqrt{3} + 2)}$$

$$= \frac{5(\sqrt{3} + 2)}{(\sqrt{3})^2 - 2^2} \quad \text{Multiply the sum and difference of two terms: } (a + b)(a - b) = a^2 - b^2.$$

$$= \frac{5(\sqrt{3} + 2)}{3 - 4}$$

$$= \frac{5(\sqrt{3} + 2)}{-1}$$

$$= -5(\sqrt{3} + 2) \quad \text{or} \quad -5\sqrt{3} - 10$$

Notice in the denominator that the product of $(\sqrt{3} - 2)$ and its conjugate, $(\sqrt{3} + 2)$, is -1. In general, the product of an expression and its conjugate will contain no radical terms. This is why, when rationalizing a denominator or a numerator containing two terms, we multiply by its conjugate. Examples of conjugates are

$$\sqrt{a} - \sqrt{b} \quad \text{and} \quad \sqrt{a} + \sqrt{b}$$
$$x + \sqrt{y} \quad \text{and} \quad x - \sqrt{y}$$

EXAMPLE 4

Rationalize each denominator.

a. $\dfrac{2}{3\sqrt{2} + 4}$ **b.** $\dfrac{\sqrt{6} + 2}{\sqrt{5} - \sqrt{3}}$ **c.** $\dfrac{2\sqrt{m}}{3\sqrt{x} + \sqrt{m}}$

Solution   **a.** Multiply the numerator and denominator by the conjugate of the denominator, $3\sqrt{2} + 4$.

$$\frac{2}{3\sqrt{2} + 4} = \frac{2(3\sqrt{2} - 4)}{(3\sqrt{2} + 4)(3\sqrt{2} - 4)}$$

$$= \frac{2(3\sqrt{2} - 4)}{(3\sqrt{2})^2 - 4^2}$$

$$= \frac{2(3\sqrt{2} - 4)}{18 - 16}$$

$$= \frac{2(3\sqrt{2} - 4)}{2}, \quad \text{or} \quad 3\sqrt{2} - 4$$

It is often helpful to leave a numerator in factored form to help determine whether the expression can be simplified.

b. Multiply the numerator and denominator by the conjugate of $\sqrt{5} - \sqrt{3}$.

$$\frac{\sqrt{6} + 2}{\sqrt{5} - \sqrt{3}} = \frac{(\sqrt{6} + 2)(\sqrt{5} + \sqrt{3})}{(\sqrt{5} - \sqrt{3})(\sqrt{5} + \sqrt{3})}$$

$$= \frac{\sqrt{6}\sqrt{5} + \sqrt{6}\sqrt{3} + 2\sqrt{5} + 2\sqrt{3}}{(\sqrt{5})^2 - (\sqrt{3})^2}$$

$$= \frac{\sqrt{30} + \sqrt{18} + 2\sqrt{5} + 2\sqrt{3}}{5 - 3}$$

$$= \frac{\sqrt{30} + 3\sqrt{2} + 2\sqrt{5} + 2\sqrt{3}}{2}$$

c. Multiply by the conjugate of $3\sqrt{x} + \sqrt{m}$ to eliminate the radicals from the denominator.

$$\frac{2\sqrt{m}}{3\sqrt{x} + \sqrt{m}} = \frac{2\sqrt{m}(3\sqrt{x} - \sqrt{m})}{(3\sqrt{x} + \sqrt{m})(3\sqrt{x} - \sqrt{m})} = \frac{6\sqrt{mx} - 2m}{(3\sqrt{x})^2 - (\sqrt{m})^2}$$

$$= \frac{6\sqrt{mx} - 2m}{9x - m}$$

---

3   As mentioned earlier, it is also often helpful to write an expression such as $\dfrac{\sqrt{3}}{\sqrt{2}}$ as an equivalent expression without a radical in the numerator. This process is called **rationalizing the numerator**. To rationalize the numerator of $\dfrac{\sqrt{3}}{\sqrt{2}}$, we multiply the numerator and the denominator by $\sqrt{3}$.

$$\frac{\sqrt{3}}{\sqrt{2}} = \frac{\sqrt{3} \cdot \sqrt{3}}{\sqrt{2} \cdot \sqrt{3}} = \frac{\sqrt{9}}{\sqrt{6}} = \frac{3}{\sqrt{6}}$$

EXAMPLE 5

Rationalize the numerator of $\dfrac{\sqrt{7}}{\sqrt{45}}$.

Solution   First we simplify $\sqrt{45}$.

$$\frac{\sqrt{7}}{\sqrt{45}} = \frac{\sqrt{7}}{\sqrt{9 \cdot 5}} = \frac{\sqrt{7}}{3\sqrt{5}}$$

Next we rationalize the numerator by multiplying the numerator and the denominator by $\sqrt{7}$.

$$\frac{\sqrt{7}}{3\sqrt{5}} = \frac{\sqrt{7} \cdot \sqrt{7}}{3\sqrt{5} \cdot \sqrt{7}} = \frac{7}{3\sqrt{5 \cdot 7}} = \frac{7}{3\sqrt{35}}$$

EXAMPLE 6

Rationalize the numerator of $\dfrac{\sqrt[3]{2x^2}}{\sqrt[3]{5y}}$.

Solution The numerator and the denominator of this expression are already simplified. To rationalize the numerator, $\sqrt[3]{2x^2}$, we multiply the numerator and denominator by a factor that will make the radicand a perfect cube. If we multiply $\sqrt[3]{2x^2}$ by $\sqrt[3]{4x}$, we get $\sqrt[3]{8x^3} = 2x$.

$$\frac{\sqrt[3]{2x^2}}{\sqrt[3]{5y}} = \frac{\sqrt[3]{2x^2} \cdot \sqrt[3]{4x}}{\sqrt[3]{5y} \cdot \sqrt[3]{4x}} = \frac{\sqrt[3]{8x^3}}{\sqrt[3]{20xy}} = \frac{2x}{\sqrt[3]{20xy}}$$

EXAMPLE 7

Rationalize the numerator of $\dfrac{\sqrt{x} + 2}{5}$.

Solution We multiply the numerator and the denominator by the conjugate of the numerator, $\sqrt{x} + 2$.

$$\frac{\sqrt{x} + 2}{5} = \frac{\left(\sqrt{x} + 2\right)\left(\sqrt{x} - 2\right)}{5\left(\sqrt{x} - 2\right)} \quad \text{Multiply by } \sqrt{x} - 2, \text{ the conjugate of } \sqrt{x} + 2.$$

$$= \frac{\left(\sqrt{x}\right)^2 - 2^2}{5\left(\sqrt{x} - 2\right)} \quad (a + b)(a - b) = a^2 - b^2.$$

$$= \frac{x - 4}{5\left(\sqrt{x} - 2\right)}$$

STUDY SKILLS REMINDER

How Are Your Homework Assignments Going?

By now, you should have good homework habits. If not, it's never too late to begin. Why is it so important in mathematics to keep up with homework? You probably now know the answer to that question. You have probably realized by now that many concepts in mathematics build on each other. Your understanding of one chapter in mathematics usually depends on your understanding of the previous chapter's material.

Don't forget that completing your homework assignment involves a lot more than attempting a few of the problems assigned.

To complete a homework assignment, remember these four things:

1. Attempt all of it.
2. Check it.
3. Correct it.
4. If needed, ask questions about it.

MENTAL MATH

Find the conjugate of each expression.

1. $\sqrt{2} + x$

2. $\sqrt{3} + y$

3. $5 - \sqrt{a}$

4. $6 - \sqrt{b}$

5. $7\sqrt{5} + 8\sqrt{x}$

6. $9\sqrt{2} - 6\sqrt{y}$

EXERCISE SET 10.5

STUDY GUIDE/SSM | CD/ VIDEO | PH MATH TUTOR CENTER | MathXL®Tutorials ON CD | MathXL® | MyMathLab®

Rationalize each denominator. See Examples 1 through 3.

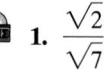 **1.** $\dfrac{\sqrt{2}}{\sqrt{7}}$

2. $\dfrac{\sqrt{3}}{\sqrt{2}}$

3. $\sqrt{\dfrac{1}{5}}$

4. $\sqrt{\dfrac{1}{2}}$

5. $\sqrt[3]{\dfrac{3}{4}}$

6. $\sqrt[3]{\dfrac{2}{9}}$

7. $\dfrac{4}{\sqrt[3]{3}}$

8. $\dfrac{6}{\sqrt[3]{9}}$

9. $\dfrac{3}{\sqrt{8x}}$

10. $\dfrac{5}{\sqrt{27a}}$

11. $\dfrac{3}{\sqrt[3]{4x^2}}$

12. $\dfrac{5}{\sqrt[3]{3y}}$

13. $\sqrt{\dfrac{4}{x}}$

14. $\sqrt{\dfrac{25}{y}}$

15. $\dfrac{9}{\sqrt{3a}}$

16. $\dfrac{x}{\sqrt{5}}$

17. $\dfrac{3}{\sqrt[3]{2}}$

18. $\dfrac{5}{\sqrt[3]{9}}$

19. $\dfrac{2\sqrt{3}}{\sqrt{7}}$

20. $\dfrac{-5\sqrt{2}}{\sqrt{11}}$

21. $\sqrt{\dfrac{2x}{5y}}$

22. $\sqrt{\dfrac{13a}{2b}}$

23. $\sqrt[4]{\dfrac{81}{8}}$

24. $\sqrt[4]{\dfrac{1}{9}}$

25. $\sqrt[4]{\dfrac{16}{9x^7}}$

26. $\sqrt[5]{\dfrac{32}{m^6 n^{13}}}$

27. $\dfrac{5a}{\sqrt[5]{8a^9 b^{11}}}$

28. $\dfrac{9y}{\sqrt[4]{4y^9}}$

Rationalize each denominator. See Example 4.

29. $\dfrac{6}{2 - \sqrt{7}}$

30. $\dfrac{3}{\sqrt{7} - 4}$

31. $\dfrac{-7}{\sqrt{x} - 3}$

32. $\dfrac{-8}{\sqrt{y} + 4}$

33. $\dfrac{\sqrt{2} - \sqrt{3}}{\sqrt{2} + \sqrt{3}}$

34. $\dfrac{\sqrt{3} + \sqrt{4}}{\sqrt{2} + \sqrt{3}}$

35. $\dfrac{\sqrt{a} + 1}{2\sqrt{a} - \sqrt{b}}$

36. $\dfrac{2\sqrt{a} - 3}{2\sqrt{a} - \sqrt{b}}$

37. $\dfrac{8}{1 + \sqrt{10}}$

38. $\dfrac{-3}{\sqrt{6} - 2}$

39. $\dfrac{\sqrt{x}}{\sqrt{x} + \sqrt{y}}$

40. $\dfrac{2\sqrt{a}}{2\sqrt{x} - \sqrt{y}}$

41. $\dfrac{2\sqrt{3} + \sqrt{6}}{4\sqrt{3} - \sqrt{6}}$

42. $\dfrac{4\sqrt{5} + \sqrt{2}}{2\sqrt{5} - \sqrt{2}}$

Rationalize each numerator. See Examples 5 and 6.

43. $\sqrt{\dfrac{5}{3}}$

44. $\sqrt{\dfrac{3}{2}}$

45. $\sqrt{\dfrac{18}{5}}$

46. $\sqrt{\dfrac{12}{7}}$

47. $\dfrac{\sqrt{4x}}{7}$

48. $\dfrac{\sqrt{3x^5}}{6}$

49. $\dfrac{\sqrt[3]{5y^2}}{\sqrt[3]{4x}}$

50. $\dfrac{\sqrt[3]{4x}}{\sqrt[3]{z^4}}$

51. $\sqrt{\dfrac{2}{5}}$

52. $\sqrt{\dfrac{3}{7}}$

53. $\dfrac{\sqrt{2x}}{11}$

54. $\dfrac{\sqrt{y}}{7}$

55. $\sqrt[3]{\dfrac{7}{8}}$

56. $\sqrt[3]{\dfrac{25}{2}}$

57. $\dfrac{\sqrt[3]{3x^5}}{10}$

58. $\sqrt[3]{\dfrac{9y}{7}}$

59. $\sqrt{\dfrac{18x^4y^6}{3z}}$

60. $\sqrt{\dfrac{8x^5y}{2z}}$

61. When rationalizing the denominator of $\dfrac{\sqrt{5}}{\sqrt{7}}$, explain why both the numerator and the denominator must be multiplied by $\sqrt{7}$.

62. When rationalizing the numerator of $\dfrac{\sqrt{5}}{\sqrt{7}}$, explain why both the numerator and the denominator must be multiplied by $\sqrt{5}$.

Rationalize each numerator. See Example 7.

63. $\dfrac{2 - \sqrt{11}}{6}$

64. $\dfrac{\sqrt{15} + 1}{2}$

65. $\dfrac{2 - \sqrt{7}}{-5}$

66. $\dfrac{\sqrt{5} + 2}{\sqrt{2}}$

67. $\dfrac{\sqrt{x} + 3}{\sqrt{x}}$

68. $\dfrac{5 + \sqrt{2}}{\sqrt{2x}}$

69. $\dfrac{\sqrt{2} - 1}{\sqrt{2} + 1}$

70. $\dfrac{\sqrt{8} - \sqrt{3}}{\sqrt{2} + \sqrt{3}}$

71. $\dfrac{\sqrt{x} + 1}{\sqrt{x} - 1}$

72. $\dfrac{\sqrt{x} + \sqrt{y}}{\sqrt{x} - \sqrt{y}}$

REVIEW AND PREVIEW

Solve each equation. See Sections 2.3 and 6.5.

73. $2x - 7 = 3(x - 4)$

74. $9x - 4 = 7(x - 2)$

75. $(x - 6)(2x + 1) = 0$

76. $(y + 2)(5y + 4) = 0$

77. $x^2 - 8x = -12$

78. $x^3 = x$

Concept Extensions

79. The formula of the radius r of a sphere with surface area A is

$$r = \sqrt{\dfrac{A}{4\pi}}$$

Rationalize the denominator of the radical expression in this formula.

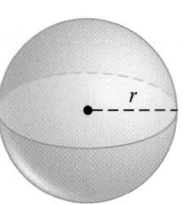

80. The formula for the radius r of a cone with height 7 centimeters and volume V is

$$r = \sqrt{\dfrac{3V}{7\pi}}$$

Rationalize the numerator of the radical expression in this formula.

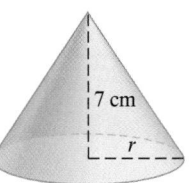

7 cm

r

81. Explain why rationalizing the denominator does not change the value of the original expression.

82. Explain why rationalizing the numerator does not change the value of the original expression.

RADICALS AND RATIONAL EXPONENTS

INTEGRATED REVIEW

Throughout this review, assume that all variables represent positive real numbers. Find each root.

1. $\sqrt{81}$

2. $\sqrt[3]{-8}$

3. $\sqrt[4]{\dfrac{1}{16}}$

4. $\sqrt{x^6}$

5. $\sqrt[3]{y^9}$

6. $\sqrt{4y^{10}}$

7. $\sqrt[5]{-32y^5}$

8. $\sqrt[4]{81b^{12}}$

Use radical notation to rewrite each expression. Simplify if possible.

9. $36^{1/2}$

10. $(3y)^{1/4}$

11. $64^{-2/3}$

12. $(x + 1)^{3/5}$

Use the properties of exponents to simplify each expression. Write with positive exponents.

13. $y^{-1/6} \cdot y^{7/6}$

14. $\dfrac{(2x^{1/3})^4}{x^{5/6}}$

15. $\dfrac{x^{1/4}x^{3/4}}{x^{-1/4}}$

16. $4^{1/3} \cdot 4^{2/5}$

Use rational exponents to simplify each radical.

17. $\sqrt[3]{8x^6}$

18. $\sqrt[12]{a^9b^6}$

Use rational exponents to write each as a single radical expression.

19. $\sqrt[4]{x} \cdot \sqrt{x}$

20. $\sqrt{5} \cdot \sqrt[3]{2}$

Simplify.

21. $\sqrt{40}$

22. $\sqrt[4]{16x^7y^{10}}$

23. $\sqrt[3]{54x^4}$

24. $\sqrt[5]{-64b^{10}}$

Multiply or divide. Then simplify if possible.

25. $\sqrt{5} \cdot \sqrt{x}$

26. $\sqrt[3]{8x} \cdot \sqrt[3]{8x^2}$

27. $\dfrac{\sqrt{98y^6}}{\sqrt{2y}}$

28. $\dfrac{\sqrt[4]{48a^9b^3}}{\sqrt[4]{ab^3}}$

Perform each indicated operation.

29. $\sqrt{20} - \sqrt{75} + 5\sqrt{7}$

30. $\sqrt[3]{54y^4} - y\sqrt[3]{16y}$

31. $\sqrt{3}\left(\sqrt{5} - \sqrt{2}\right)$

32. $\left(\sqrt{7} + \sqrt{3}\right)^2$

33. $\left(2x - \sqrt{5}\right)\left(2x + \sqrt{5}\right)$

34. $\left(\sqrt{x + 1} - 1\right)^2$

Rationalize each denominator.

35. $\sqrt{\dfrac{7}{3}}$

36. $\dfrac{5}{\sqrt[3]{2x^2}}$

37. $\dfrac{\sqrt{3} - \sqrt{7}}{2\sqrt{3} + \sqrt{7}}$

Rationalize each numerator.

38. $\sqrt{\dfrac{7}{3}}$

39. $\sqrt[3]{\dfrac{9y}{11}}$

40. $\dfrac{\sqrt{x} - 2}{\sqrt{x}}$

10.6   RADICAL EQUATIONS AND PROBLEM SOLVING

Objectives

1   Solve equations that contain radical expressions.

2   Use the Pythagorean theorem to model problems.

1   In this section, we present techniques to solve equations containing radical expressions such as

$$\sqrt{2x - 3} = 9$$

We use the power rule to help us solve these radical equations.

> **Power Rule**
>
> If both sides of an equation are raised to the same power, **all** solutions of the original equation are **among** the solutions of the new equation.

This property *does not* say that raising both sides of an equation to a power yields an equivalent equation. A solution of the new equation *may or may not* be a solution of the original equation. For example, $(-2)^2 = 2^2$, but $-2 \neq 2$. Thus, *each solution of the new equation must be checked* to make sure it is a solution of the original equation. Recall that a proposed solution that is not a solution of the original equation is called an **extraneous solution**.

EXAMPLE 1

Solve $\sqrt{2x - 3} = 9$.

Solution　We use the power rule to square both sides of the equation to eliminate the radical.

$$\sqrt{2x - 3} = 9$$
$$\left(\sqrt{2x - 3}\right)^2 = 9^2$$
$$2x - 3 = 81$$
$$2x = 84$$
$$x = 42$$

Now we, check the solution in the original equation.

Check:
$$\sqrt{2x - 3} = 9$$
$$\sqrt{2(42) - 3} \stackrel{?}{=} 9 \qquad \text{Let } x = 42.$$
$$\sqrt{84 - 3} \stackrel{?}{=} 9$$
$$\sqrt{81} \stackrel{?}{=} 9$$
$$9 = 9 \qquad \text{True}$$

The solution checks, so we conclude that the solution is 42 or the solution set is $\{42\}$.

To solve a radical equation, first isolate a radical on one side of the equation.

EXAMPLE 2

Solve $\sqrt{-10x - 1} + 3x = 0$.

Solution　First, isolate the radical on one side of the equation. To do this, we subtract $3x$ from both sides.

$$\sqrt{-10x - 1} + 3x = 0$$
$$\sqrt{-10x - 1} + 3x - 3x = 0 - 3x$$
$$\sqrt{-10x - 1} = -3x$$

Next we use the power rule to eliminate the radical.

$$\left(\sqrt{-10x - 1}\right)^2 = (-3x)^2$$
$$-10x - 1 = 9x^2$$

Since this is a quadratic equation, we can set the equation equal to 0 and try to solve by factoring.

$$9x^2 + 10x + 1 = 0$$
$$(9x + 1)(x + 1) = 0 \qquad \text{Factor.}$$
$$9x + 1 = 0 \quad \text{or} \quad x + 1 = 0 \qquad \text{Set each factor equal to 0.}$$
$$x = -\frac{1}{9} \quad \text{or} \quad x = -1$$

Check: Let $x = -\dfrac{1}{9}$.

$$\sqrt{-10x - 1} + 3x = 0$$
$$\sqrt{-10\left(-\frac{1}{9}\right) - 1} + 3\left(-\frac{1}{9}\right) \stackrel{?}{=} 0$$
$$\sqrt{\frac{10}{9} - \frac{9}{9}} - \frac{3}{9} \stackrel{?}{=} 0$$
$$\sqrt{\frac{1}{9}} - \frac{1}{3} \stackrel{?}{=} 0$$
$$\frac{1}{3} - \frac{1}{3} = 0 \quad \text{True}$$

Let $x = -1$.

$$\sqrt{-10x - 1} + 3x = 0$$
$$\sqrt{-10(-1) - 1} + 3(-1) \stackrel{?}{=} 0$$
$$\sqrt{10 - 1} - 3 \stackrel{?}{=} 0$$
$$\sqrt{9} - 3 \stackrel{?}{=} 0$$
$$3 - 3 = 0 \quad \text{True}$$

Both solutions check. The solutions are $-\dfrac{1}{9}$ and -1 or the solution set is $\left\{ -\dfrac{1}{9}, -1 \right\}$.

The following steps may be used to solve a radical equation.

Solving a Radical Equation

Step 1: Isolate one radical on one side of the equation.

Step 2: Raise each side of the equation to a power equal to the index of the radical and simplify.

Step 3: If the equation still contains a radical term, repeat Steps 1 and 2. If not, solve the equation.

Step 4: Check all proposed solutions in the original equation.

EXAMPLE 3

Solve $\sqrt[3]{x + 1} + 5 = 3$.

Solution   First we isolate the radical by subtracting 5 from both sides of the equation.

$$\sqrt[3]{x + 1} + 5 = 3$$
$$\sqrt[3]{x + 1} = -2$$

Next we raise both sides of the equation to the third power to eliminate the radical.

$$\left(\sqrt[3]{x + 1}\right)^3 = (-2)^3$$
$$x + 1 = -8$$
$$x = -9$$

The solution checks in the original equation, so the solution is -9.

EXAMPLE 4

Solve $\sqrt{4 - x} = x - 2$.

Solution

$$\sqrt{4 - x} = x - 2$$
$$\left(\sqrt{4 - x}\right)^2 = (x - 2)^2$$
$$4 - x = x^2 - 4x + 4$$
$$x^2 - 3x = 0 \qquad \text{Write the quadratic equation in standard form.}$$
$$x(x - 3) = 0 \qquad \text{Factor.}$$
$$x = 0 \quad \text{or} \quad x - 3 = 0 \qquad \text{Set each factor equal to } 0.$$
$$x = 3$$

Check:

$$\sqrt{4 - x} = x - 2 \qquad\qquad \sqrt{4 - x} = x - 2$$
$$\sqrt{4 - 0} \stackrel{?}{=} 0 - 2 \quad \text{Let } x = 0. \qquad \sqrt{4 - 3} \stackrel{?}{=} 3 - 2 \quad \text{Let } x = 3.$$
$$2 = -2 \quad \text{False} \qquad\qquad\qquad 1 = 1 \quad \text{True}$$

The proposed solution 3 checks, but 0 does not. Since 0 is an extraneous solution, the only solution is 3.

> **Helpful Hint**
>
> In Example 4, notice that $(x - 2)^2 = x^2 - 4x + 4$. Make sure binomials are squared correctly.

✔ **CONCEPT CHECK**

How can you immediately tell that the equation $\sqrt{2y + 3} = -4$ has no real solution?

EXAMPLE 5

Solve $\sqrt{2x + 5} + \sqrt{2x} = 3$.

Solution   We get one radical alone by subtracting $\sqrt{2x}$ from both sides.

$$\sqrt{2x + 5} + \sqrt{2x} = 3$$
$$\sqrt{2x + 5} = 3 - \sqrt{2x}$$

Now we use the power rule to begin eliminating the radicals. First we square both sides.

$$\left(\sqrt{2x + 5}\right)^2 = \left(3 - \sqrt{2x}\right)^2$$
$$2x + 5 = 9 - 6\sqrt{2x} + 2x \quad \text{Multiply } \left(3 - \sqrt{2x}\right)\left(3 - \sqrt{2x}\right).$$

Concept Check Answer:
answers may vary

There is still a radical in the equation, so we get a radical alone again. Then we square both sides.

$$2x + 5 = 9 - 6\sqrt{2x} + 2x \quad \text{Get the radical alone.}$$

$$6\sqrt{2x} = 4$$

$$36(2x) = 16 \quad \text{Square both sides of the equation to eliminate the radical.}$$

$$72x = 16 \quad \text{Multiply.}$$

$$x = \frac{16}{72} \quad \text{Solve.}$$

$$x = \frac{2}{9} \quad \text{Simplify.}$$

The proposed solution, $\frac{2}{9}$, checks in the original equation. The solution is $\frac{2}{9}$.

Helpful Hint

Make sure expressions are squared correctly. In Example 5, we squared $\left(3 - \sqrt{2x}\right)$ as

$$\left(3 - \sqrt{2x}\right)^2 = \left(3 - \sqrt{2x}\right)\left(3 - \sqrt{2x}\right)$$
$$= 3 \cdot 3 - 3\sqrt{2x} - 3\sqrt{2x} + \sqrt{2x} \cdot \sqrt{2x}$$
$$= 9 - 6\sqrt{2x} + 2x$$

✔ **CONCEPT CHECK**

What is wrong with the following solution?

$$\sqrt{2x + 5} + \sqrt{4 - x} = 8$$
$$\left(\sqrt{2x + 5} + \sqrt{4 - x}\right)^2 = 8^2$$
$$(2x + 5) + (4 - x) = 64$$
$$x + 9 = 64$$
$$x = 55$$

---

2   Recall that the Pythagorean theorem states that in a right triangle, the length of the hypotenuse squared equals the sum of the lengths of each of the legs squared.

Pythagorean Theorem

If a and b are the lengths of the legs of a right triangle and c is the length of the hypotenuse, then $a^2 + b^2 = c^2$.

Concept Check Answer:
$\left(\sqrt{2x + 5} + \sqrt{4 - x}\right)^2$ is not
$(2x + 5) + (4 - x)$.

⚠ (**EXAMPLE 6**)

Find the length of the unknown leg of the right triangle.

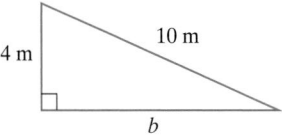

Solution In the formula $a^2 + b^2 = c^2$, c is the hypotenuse. Here, $c = 10$, the length of the hypotenuse, and $a = 4$. We solve for b. Then $a^2 + b^2 = c^2$ becomes

$$4^2 + b^2 = 10^2$$
$$16 + b^2 = 100$$
$$b^2 = 84 \qquad \text{Subtract } \mathbf{16} \text{ from both sides.}$$
$$b = \pm\sqrt{84} = \pm\sqrt{4 \cdot 21} = \pm 2\sqrt{21}$$

Since b is a length and thus is positive, we will use the positive value only.

The unknown leg of the triangle is $2\sqrt{21}$ meters long.

⚠ (**EXAMPLE 7**)

CALCULATING PLACEMENT OF A WIRE

A 50-foot supporting wire is to be attached to a 75-foot antenna. Because of surrounding buildings, sidewalks, and roadways, the wire must be anchored exactly 20 feet from the base of the antenna.

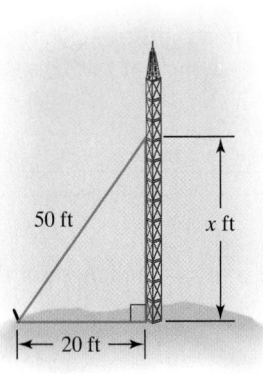

a. How high from the base of the antenna is the wire attached?

b. Local regulations require that a supporting wire be attached at a height no less than $\frac{3}{5}$ of the total height of the antenna. From part **a**, have local regulations been met?

Solution **1.** UNDERSTAND.   Read and reread the problem. From the diagram we notice that a right triangle is formed with hypotenuse 50 feet and one leg 20 feet. Let x be the height from the base of the antenna to the attached wire.

2. TRANSLATE.   Use the Pythagorean theorem.

$$a^2 + b^2 = c^2$$
$$20^2 + x^2 = 50^2 \qquad a = 20, c = 50$$

3. SOLVE. $20^2 + x^2 = 50^2$

$$400 + x^2 = 2500$$
$$x^2 = 2100 \qquad \text{Subtract 400 from both sides.}$$
$$x = \pm\sqrt{2100}$$
$$= \pm 10\sqrt{21}$$

4. INTERPRET.   *Check* the work and *state* the solution.

Check: We will use only the positive value, $x = 10\sqrt{21}$ because x represents length. The wire is attached exactly $10\sqrt{21}$ feet from the base of the pole, or approximately 45.8 feet.

State: The supporting wire must be attached at a height no less than $\frac{3}{5}$ of the total height of the antenna. This height is $\frac{3}{5}$ (75 feet), or 45 feet. Since we know from part **a** that the wire is to be attached at a height of approximately 45.8 feet, local regulations have been met.

Graphing Calculator Explorations

We can use a graphing calculator to solve radical equations. For example, to use a graphing calculator to approximate the solutions of the equation solved in Example 4, we graph the following.

$$Y_1 = \sqrt{4 - x} \qquad \text{and} \qquad Y_2 = x - 2$$

The x-value of the point of intersection is the solution. Use the Intersect feature or the Zoom and Trace features of your graphing calculator to see that the solution is 3.

Use a graphing calculator to solve each radical equation. Round all solutions to the nearest hundredth.

1. $\sqrt{x + 7} = x$

2. $\sqrt{3x + 5} = 2x$

3. $\sqrt{2x + 1} = \sqrt{2x + 2}$

4. $\sqrt{10x - 1} = \sqrt{-10x + 10} - 1$

5. $1.2x = \sqrt{3.1x + 5}$

6. $\sqrt{1.9x^2 - 2.2} = -0.8x + 3$

STUDY SKILLS REMINDER

Continue your outline started in Section 2.3. Write how to recognize and how to solve equations containing radical terms in your own words. For example:

Solving Equations and Inequalities

1. I. Equations
    - **A.** Linear equations (Sec. 2.3)
    - **B.** Quadratic Equations (Sec. 6.5)
    - **C.** Equations with Rational Expressions (Sec. 7.5)
    - **D.** Absolute value equations (Sec. 9.2)
    - **E. Equations with Radicals** — Recognize: *equation with a radical term* — Solve: isolate a radical and then raise each side of the equation to a power equal to the index of the radical. You may have to repeat this. Remember to check possible solutions in the original equation.

2. II. Inequalities
    - **A.** Linear inequalities (Sec. 2.7)
    - **B.** Compound inequalities (Sec. 9.1)
    - **C.** Absolute value inequalities (Sec. 9.3)

Spotlight on
DECISION *&* MAKING

Suppose you are a psychologist studying a person's ability to recognize patterns. You theorize that IQ is linked to pattern-recognition ability and design a research study using human subjects to test your theory. As part of your study, you have decided to form three groups of test subjects based roughly on IQ. Group *A* will consist of subjects having IQ's under 90, Group *B* will consist of subjects having IQ's from 90 to 105, and Group *C* will consist of subjects having IQ's over 105.

While preparing for your research study, you came across the findings of another psychologist suggesting that the number S of nonsense syllables that a person can repeat consecutively depends on his or her IQ score I according to the equation $S = 2\sqrt{I} - 9$. Because administering IQ tests can be time-consuming and because your groupings by IQ need only be approximate, you decide to use this equation as a quick way to assign test subjects to Groups *A*, *B*, and *C*. Each subject is individually screened by listening to a string of 20 random nonsense syllables and then repeating as many as possible. The results for the first 5 test subjects are listed in the table. For each subject, decide to which group—*A, B,* or *C*—the subject should be assigned.

Test Subject Screening

Subject	S—the number of Nonsense Syllables Successfully Repeated
1	11
2	13
3	9
4	12
5	10

EXERCISE SET 10.6

STUDY GUIDE/SSM   CD/ VIDEO   PH MATH TUTOR CENTER   MathXL®Tutorials ON CD   MathXL®   MyMathLab®

Solve. See Examples 1 and 2.

1. $\sqrt{2x} = 4$

2. $\sqrt{3x} = 3$

3. $\sqrt{x-3} = 2$

4. $\sqrt{x+1} = 5$

5. $\sqrt{2x} = -4$

6. $\sqrt{5x} = -5$

7. $\sqrt{4x-3} - 5 = 0$

8. $\sqrt{x-3} - 1 = 0$

9. $\sqrt{2x-3} - 2 = 1$

10. $\sqrt{3x+3} - 4 = 8$

Solve. See Example 3.

11. $\sqrt[3]{6x} = -3$

12. $\sqrt[3]{4x} = -2$

13. $\sqrt[3]{x-2} - 3 = 0$

14. $\sqrt[3]{2x-6} - 4 = 0$

Solve. See Examples 4 and 5.

15. $\sqrt{13-x} = x - 1$

16. $\sqrt{2x-3} = 3 - x$

17. $x - \sqrt{4-3x} = -8$

18. $2x + \sqrt{x+1} = 8$

19. $\sqrt{y+5} = 2 - \sqrt{y-4}$

20. $\sqrt{x+3} + \sqrt{x-5} = 3$

21. $\sqrt{x-3} + \sqrt{x+2} = 5$

22. $\sqrt{2x-4} - \sqrt{3x+4} = -2$

MIXED PRACTICE

Solve. See Examples 1 through 5.

23. $\sqrt{3x-2} = 5$

24. $\sqrt{5x-4} = 9$

25. $-\sqrt{2x} + 4 = -6$

26. $-\sqrt{3x+9} = -12$

27. $\sqrt{3x+1} + 2 = 0$

28. $\sqrt{3x+1} - 2 = 0$

29. $\sqrt[4]{4x+1} - 2 = 0$

30. $\sqrt[4]{2x-9} - 3 = 0$

31. $\sqrt{4x-3} = 7$

32. $\sqrt{3x+9} = 6$

33. $\sqrt[3]{6x-3} - 3 = 0$

34. $\sqrt[3]{3x+4} = 7$

35. $\sqrt[3]{2x-3} - 2 = -5$

36. $\sqrt[3]{x-4} - 5 = -7$

37. $\sqrt{x+4} = \sqrt{2x-5}$

38. $\sqrt{3y+6} = \sqrt{7y-6}$

39. $x - \sqrt{1-x} = -5$

40. $x - \sqrt{x-2} = 4$

41. $\sqrt[3]{-6x-1} = \sqrt[3]{-2x-5}$

42. $x + \sqrt{x+5} = 7$

43. $\sqrt{5x-1} - \sqrt{x+2} = 3$

44. $\sqrt{2x-1} - 4 = -\sqrt{x-4}$

45. $\sqrt{2x-1} = \sqrt{1-2x}$

46. $\sqrt{7x-4} = \sqrt{4-7x}$

47. $\sqrt{3x+4} - 1 = \sqrt{2x+1}$

48. $\sqrt{x-2} + 3 = \sqrt{4x+1}$

49. $\sqrt{y+3} - \sqrt{y-3} = 1$

50. $\sqrt{x+1} - \sqrt{x-1} = 2$

Find the length of the unknown side of each triangle. See Example 6.

51.

6 ft

3 ft

52.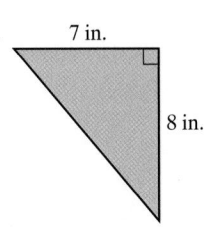

7 in.

8 in.

53.

3 m

7 m

54.

4 cm

7 cm

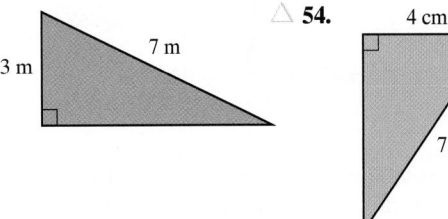

Find the length of the unknown side of each triangle. Give the exact length and a one-decimal-place approximation. See Example 6.

55.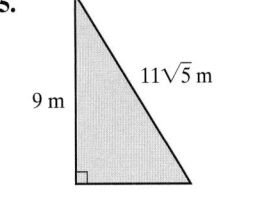

9 m

$11\sqrt{5}$ m

56.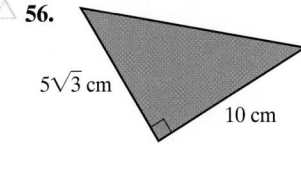

$5\sqrt{3}$ cm

10 cm

57.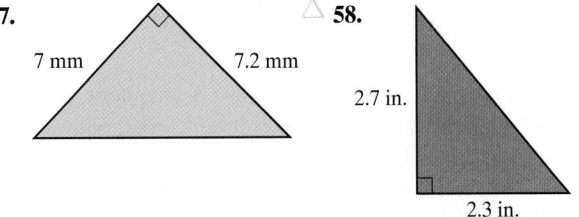

7 mm

7.2 mm

58.

2.7 in.

2.3 in.

Solve. See Example 7. Give exact answers and two-decimal-place approximations where appropriate.

59. A wire is needed to support a vertical pole 15 feet high. The cable will be anchored to a stake 8 feet from the base of the pole. How much cable is needed?

15 ft

8 ft

△ **60.** The tallest structure in the United States is a TV tower in Blanchard, North Dakota. Its height is 2063 feet. A 2382-foot length of wire is to be used as a guy wire attached to the top of the tower. Approximate to the nearest foot how far from the base of the tower the guy wire must be anchored. (*Source:* U.S. Geological Survey)

2382 ft   2063 ft

?

△ **61.** A spotlight is mounted on the eaves of a house 12 feet above the ground. A flower bed runs between the house and the sidewalk, so the closest the ladder can be placed to the house is 5 feet. How long a ladder is needed so that an electrician can reach the place where the light is mounted?

12 ft

5 ft

△ **62.** A wire is to be attached to support a telephone pole. Because of surrounding buildings, sidewalks, and roadways, the wire must be anchored exactly 15 feet from the base of the pole. Telephone company workers have only 30 feet of cable, and 2 feet of that must be used to attach the cable to the pole and to the stake on the ground. How high from the base of the pole can the wire be attached?

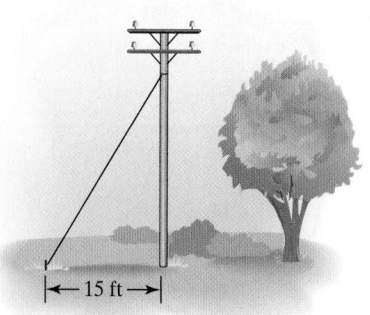

|← 15 ft →|

△ **63.** The radius of the Moon is 1080 miles. Use the formula for the radius r of a sphere given its surface area A,

$$r = \sqrt{\frac{A}{4\pi}}$$

to find the surface area of the Moon. Round to the nearest square mile. (*Source:* National Space Science Data Center)

64. Police departments find it very useful to be able to approximate the speed of a car when they are given the distance that the car skidded before it came to a stop. If the road surface is wet concrete, the function $S(x) = \sqrt{10.5x}$ is used, where $S(x)$ is the speed of the car in miles per hour and x is the distance skidded in feet. Find how fast a car was moving if it skidded 280 feet on wet concrete.

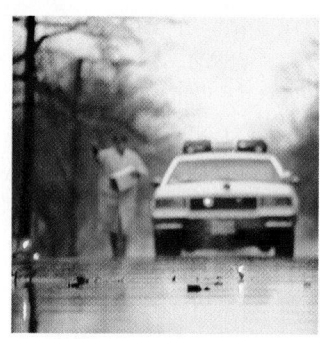

65. The formula $v = \sqrt{2gh}$ gives the velocity v, in feet per second, of an object when it falls h feet accelerated by gravity g, in feet per second squared. If g is approximately 32 feet per second squared, find how far an object has fallen if its velocity is 80 feet per second.

66. Two tractors are pulling a tree stump from a field. If two forces A and B pull at right angles (90°) to each other, the size of the resulting force R is given by the formula $R = \sqrt{A^2 + B^2}$. If tractor A is exerting 600 pounds of force and the resulting force is 850 pounds, find how much force tractor B is exerting.

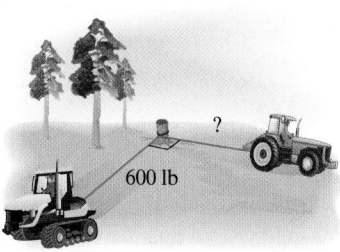

?

600 lb

In psychology, it has been suggested that the number S of nonsense syllables that a person can repeat consecutively depends on his or her IQ score I according to the equation $S = 2\sqrt{I} - 9$.

67. Use this relationship to estimate the IQ of a person who can repeat 11 nonsense syllables consecutively.

68. Use this relationship to estimate the IQ of a person who can repeat 15 nonsense syllables consecutively.

*The **period** of a pendulum is the time it takes for the pendulum to make one full back-and-forth swing. The period of a pendulum depends on the length of the pendulum. The formula for the period*

P, *in seconds, is* $P = 2\pi\sqrt{\dfrac{l}{32}}$ *where l is the length of the pendulum in feet. Use this formula for Exercises 69 through 74.*

69. Find the period of a pendulum whose length is 2 feet. Give an exact answer and a two-decimal-place approximation.

70. Klockit sells a 43-inch lyre pendulum. Find the period of this pendulum. Round your answer to 2 decimal places. (Hint: First convert inches to feet.)

71. Find the length of a pendulum whose period is 4 seconds. Round your answer to 2 decimal places.

72. Find the length of a pendulum whose period is 3 seconds. Round your answer to 2 decimal places.

73. Study the relationship between period and pendulum length in Exercises 69 through 72 and make a conjecture about this relationship.

74. Galileo experimented with pendulums. He supposedly made conjectures about pendulums of equal length with different bob weights. Try this experiment. Make two pendulums 3 feet long. Attach a heavy weight (lead) to one and a light

weight (a cork) to the other. Pull both pendulums back the same angle measure and release. Make a conjecture from your observations. (There is more about pendulums in the Chapter 10 Project.)

If the three lengths of the sides of a triangle are known, Heron's formula can be used to find its area. If a, b, and c are the three lengths of the sides, Heron's formula for area is:

$$A = \sqrt{s(s - a)(s - b)(s - c)}$$

where s is half the perimeter of the triangle, or $s = \dfrac{1}{2}(a + b + c)$.

Use this formula to find the area of each triangle. Give an exact answer and then a two-decimal place approximation.

△ **75.**

6 mi   10 mi

14 mi

△ **76.**

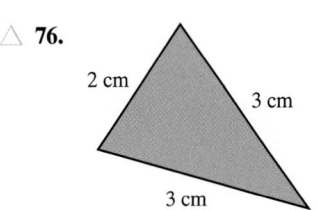

2 cm   3 cm

3 cm

77. Describe when Heron's formula might be useful.

78. In your own words, explain why you think s in *Heron's formula* is called the *semiperimeter*.

The maximum distance D(h) in kilometers that a person can see from a height h kilometers above the ground is given by the function $D(h) = 111.7\sqrt{h}$. Use this function for Exercises 79 and 80. Round your answers to two decimal places.

79. Find the height that would allow a person to see 80 kilometers.

80. Find the height that would allow a person to see 40 kilometers.

REVIEW AND PREVIEW

Use the vertical line test to determine whether each graph represents the graph of a function. See Section 3.7.

81.   **82.**

83.

84.

85.

86.

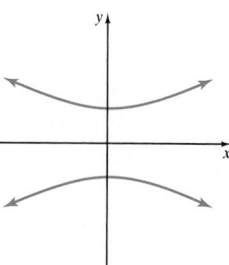

Simplify. See Section 7.7.

87. $\dfrac{\dfrac{x}{6}}{\dfrac{2x}{3} + \dfrac{1}{2}}$

88. $\dfrac{\dfrac{1}{y} + \dfrac{4}{5}}{\dfrac{-3}{20}}$

89. $\dfrac{\dfrac{z}{5} + \dfrac{1}{10}}{\dfrac{z}{20} - \dfrac{z}{5}}$

90. $\dfrac{\dfrac{1}{y} + \dfrac{1}{x}}{\dfrac{1}{y} - \dfrac{1}{x}}$

Concept Extensions

91. Solve: $\sqrt{\sqrt{x+3} + \sqrt{x}} = \sqrt{3}$

92. Explain why proposed solutions of radical equations must be checked.

93. The cost $C(x)$ in dollars per day to operate a small delivery service is given by $C(x) = 80\sqrt[3]{x} + 500$, where x is the number of deliveries per day. In July, the manager decides that it is necessary to keep delivery costs below $1620.00. Find the greatest number of deliveries this company can make per day and still keep overhead below $1620.00.

94. Consider the equations $\sqrt{2x} = 4$ and $\sqrt[3]{2x} = 4$.
   a. Explain the difference in solving these equations.
   b. Explain the similarity in solving these equations.

Example

For Exercises 95 through 98, see the example below.
Solve $(t^2 - 3t) - 2\sqrt{t^2 - 3t} = 0$.

Solution

Substitution can be used to make this problem somewhat simpler. Since $t^2 - 3t$ occurs more than once, let $x = t^2 - 3t$.

$$(t^2 - 3t) - 2\sqrt{t^2 - 3t} = 0$$
$$x - 2\sqrt{x} = 0 \qquad \text{Let } x = t^2 - 3t.$$
$$x = 2\sqrt{x}$$
$$x^2 = \left(2\sqrt{x}\right)^2$$
$$x^2 = 4x$$
$$x^2 - 4x = 0$$
$$x(x - 4) = 0$$
$$x = 0 \quad \text{or} \quad x - 4 = 0$$
$$x = 4$$

Now we "undo" the substitution.
$x = 0$ Replace x with $t^2 - 3t$.

$$t^2 - 3t = 0$$
$$t(t - 3) = 0$$
$$t = 0 \quad \text{or} \quad t - 3 = 0$$
$$t = 3$$

$x = 4$ Replace x with $t^2 - 3t$.

$$t^2 - 3t = 4$$
$$t^2 - 3t - 4 = 0$$
$$(t - 4)(t + 1) = 0$$
$$t - 4 = 0 \quad \text{or} \quad t + 1 = 0$$
$$t = 4 \qquad t = -1$$

In this problem, we have four possible solutions: $0, 3, 4,$ and -1. All four solutions check in the original equation, so the solutions are $-1, 0, 3, 4$.

Solve. See the preceding example.

95. $3\sqrt{x^2 - 8x} = x^2 - 8x$

96. $\sqrt{(x^2 - x) + 7} = 2(x^2 - x) - 1$

97. $7 - (x^2 - 3x) = \sqrt{(x^2 - 3x)} + 5$

98. $x^2 + 6x = 4\sqrt{x^2 + 6x}$

10.7   COMPLEX NUMBERS

Objectives

1. Define imaginary and complex numbers.
2. Add or subtract complex numbers.
3. Multiply complex numbers.
4. Divide complex numbers.
5. Raise i to powers.

1 Our work with radical expressions has excluded expressions such as $\sqrt{-16}$ because $\sqrt{-16}$ is not a real number; there is no real number whose square is -16. In this section, we discuss a number system that includes roots of negative numbers. This number system is the **complex number system**, and it includes the set of real numbers as a subset. The complex number system allows us to solve equations such as $x^2 + 1 = 0$ that have no real number solutions. The set of complex numbers includes the **imaginary unit**.

> **Imaginary Unit**
>
> The imaginary unit, written i, is the number whose square is -1. That is,
>
> $$i^2 = -1 \quad \text{and} \quad i = \sqrt{-1}$$

To write the square root of a negative number in terms of i, use the property that if a is a positive number, then

$$\sqrt{-a} = \sqrt{-1} \cdot \sqrt{a}$$
$$= i \cdot \sqrt{a}$$

Using i, we can write $\sqrt{-16}$ as

$$\sqrt{-16} = \sqrt{-1 \cdot 16} = \sqrt{-1} \cdot \sqrt{16} = i \cdot 4, \text{ or } 4i$$

EXAMPLE 1

Write with i notation.

a. $\sqrt{-36}$ **b.** $\sqrt{-5}$ **c.** $-\sqrt{-20}$

> **Helpful Hint**
> Since $\sqrt{5}i$ can easily be confused with $\sqrt{5i}$, we write $\sqrt{5}i$ as $i\sqrt{5}$.

Solution   **a.** $\sqrt{-36} = \sqrt{-1 \cdot 36} = \sqrt{-1} \cdot \sqrt{36} = i \cdot 6, \text{ or } 6i$
b. $\sqrt{-5} = \sqrt{-1(5)} = \sqrt{-1} \cdot \sqrt{5} = i\sqrt{5}$.
c. $-\sqrt{-20} = -\sqrt{-1 \cdot 20} = -\sqrt{-1} \cdot \sqrt{4 \cdot 5} = -i \cdot 2\sqrt{5} = -2i\sqrt{5}$

The product rule for radicals does not necessarily hold true for imaginary numbers. *To multiply square roots of negative numbers, first we write each number in terms of the imaginary unit i.* For example, to multiply $\sqrt{-4}$ and $\sqrt{-9}$, we first write each number in the form bi.

$$\sqrt{-4}\sqrt{-9} = 2i(3i) = 6i^2 = 6(-1) = -6$$

We will also use this method to simplify quotients of square roots of negative numbers.

EXAMPLE 2

Multiply or divide as indicated.

a. $\sqrt{-3} \cdot \sqrt{-5}$ **b.** $\sqrt{-36} \cdot \sqrt{-1}$

c. $\sqrt{8} \cdot \sqrt{-2}$ **d.** $\dfrac{\sqrt{-125}}{\sqrt{5}}$

Solution

a. $\sqrt{-3} \cdot \sqrt{-5} = i\sqrt{3}\left(i\sqrt{5}\right) = i^2\sqrt{15} = -1\sqrt{15} = -\sqrt{15}$

b. $\sqrt{-36} \cdot \sqrt{-1} = 6i(i) = 6i^2 = 6(-1) = -6$

c. $\sqrt{8} \cdot \sqrt{-2} = 2\sqrt{2}\left(i\sqrt{2}\right) = 2i\left(\sqrt{2}\sqrt{2}\right) = 2i(2) = 4i$

d. $\dfrac{\sqrt{-125}}{\sqrt{5}} = \dfrac{i\sqrt{125}}{\sqrt{5}} = i\sqrt{25} = 5i$

Now that we have practiced working with the imaginary unit, we define complex numbers.

Complex Numbers

A **complex number** is a number that can be written in the form $a + bi$, where a and b are real numbers.

Notice that the set of real numbers is a subset of the complex numbers since any real number can be written in the form of a complex number. For example,

$$16 = 16 + 0i$$

In general, a complex number $a + bi$ is a real number if $b = 0$. Also, a complex number is called a **pure imaginary number** if $a = 0$. For example,

$$3i = 0 + 3i \qquad \text{and} \qquad i\sqrt{7} = 0 + i\sqrt{7}$$

are pure imaginary numbers.

The following diagram shows the relationship between complex numbers and their subsets.

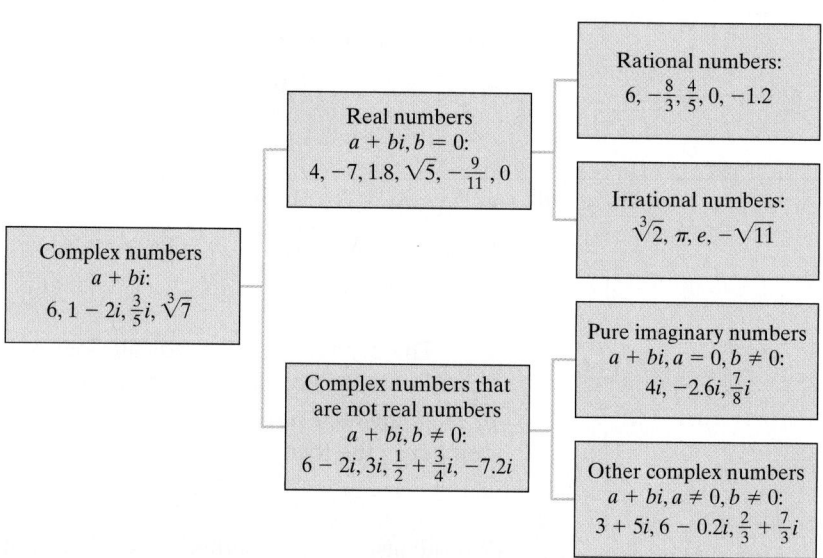

✔ **CONCEPT CHECK**

True or false? Every complex number is also a real number.

2   Two complex numbers $a + bi$ and $c + di$ are equal if and only if $a = c$ and $b = d$. Complex numbers can be added or subtracted by adding or subtracting their real parts and then adding or subtracting their imaginary parts.

Sum or Difference of Complex Numbers

If $a + bi$ and $c + di$ are complex numbers, then their sum is

$$(a + bi) + (c + di) = (a + c) + (b + d)i$$

Their difference is

$$(a + bi) - (c + di) = a + bi - c - di = (a - c) + (b - d)i$$

EXAMPLE 3

Add or subtract the complex numbers. Write the sum or difference in the form $a + bi$.

a. $(2 + 3i) + (-3 + 2i)$ **b.** $5i - (1 - i)$ **c.** $(-3 - 7i) - (-6)$

Solution   **a.** $(2 + 3i) + (-3 + 2i) = (2 - 3) + (3 + 2)i = -1 + 5i$

b. $5i - (1 - i) = 5i - 1 + i$

$$= -1 + (5 + 1)i$$

$$= -1 + 6i$$

c. $(-3 - 7i) - (-6) = -3 - 7i + 6$

$$= (-3 + 6) - 7i$$

$$= 3 - 7i$$

3   To multiply two complex numbers of the form $a + bi$, we multiply as though they are binomials. Then we use the relationship $i^2 = -1$ to simplify.

EXAMPLE 4

Multiply the complex numbers. Write the product in the form $a + bi$.

a. $-7i \cdot 3i$ **b.** $3i(2 - i)$ **c.** $(2 - 5i)(4 + i)$

d. $(2 - i)^2$ **e.** $(7 + 3i)(7 - 3i)$

Concept Check Answer:
false

Solution **a.** $-7i \cdot 3i = -21i^2$

$= -21(-1)$ Replace i^2 with -1.

$= 21$

b. $3i(2 - i) = 3i \cdot 2 - 3i \cdot i$ Use the distributive property.

$= 6i - 3i^2$ Multiply.

$= 6i - 3(-1)$ Replace i^2 with -1.

$= 6i + 3$

$= 3 + 6i$

Use the FOIL order below. (First, Outer, Inner, Last)

c. $(2 - 5i)(4 + i) = 2(4) + 2(i) - 5i(4) - 5i(i)$

 F O I L

$= 8 + 2i - 20i - 5i^2$

$= 8 - 18i - 5(-1)$ $i^2 = -1$.

$= 8 - 18i + 5$

$= 13 - 18i$

d. $(2 - i)^2 = (2 - i)(2 - i)$

$= 2(2) - 2(i) - 2(i) + i^2$

$= 4 - 4i + (-1)$ $i^2 = -1$.

$= 3 - 4i$

e. $(7 + 3i)(7 - 3i) = 7(7) - 7(3i) + 3i(7) - 3i(3i)$

$= 49 - 21i + 21i - 9i^2$

$= 49 - 9(-1)$ $i^2 = -1$.

$= 49 + 9$

$= 58$

Notice that if you add, subtract, or multiply two complex numbers, just like real numbers, the result is a complex number.

4   From Example 4e, notice that the product of $7 + 3i$ and $7 - 3i$ is a real number. These two complex numbers are called **complex conjugates** of one another. In general, we have the following definition.

Complex Conjugates

The complex numbers $(a + bi)$ and $(a - bi)$ are called **complex conjugates** of each other, and $(a + bi)(a - bi) = a^2 + b^2$.

To see that the product of a complex number $a + bi$ and its conjugate $a - bi$ is the real number $a^2 + b^2$ we multiply.

$$(a + bi)(a - bi) = a^2 - abi + abi - b^2 i^2$$
$$= a^2 - b^2(-1)$$
$$= a^2 + b^2$$

We use complex conjugates to divide by a complex number.

EXAMPLE 5

Divide. Write in the form $a + bi$.

a. $\dfrac{2 + i}{1 - i}$

b. $\dfrac{7}{3i}$

Solution   **a.** Multiply the numerator and denominator by the complex conjugate of $1 - i$ to eliminate the imaginary number in the denominator.

$$\frac{2 + i}{1 - i} = \frac{(2 + i)(1 + i)}{(1 - i)(1 + i)}$$
$$= \frac{2(1) + 2(i) + 1(i) + i^2}{1^2 - i^2}$$
$$= \frac{2 + 3i - 1}{1 + 1} \qquad \text{Here, } i^2 = -1.$$
$$= \frac{1 + 3i}{2} \quad \text{or} \quad \frac{1}{2} + \frac{3}{2}i$$

> **Helpful Hint**
>
> Recall that division can be checked by multiplication.
>
> To check that $\dfrac{2 + i}{1 - i} = \dfrac{1}{2} + \dfrac{3}{2}i$, in Example 5a, multiply $\left(\dfrac{1}{2} + \dfrac{3}{2}i\right)(1 - i)$ to verify that the product is $2 + i$.

b. Multiply the numerator and denominator by the conjugate of $3i$. Note that $3i = 0 + 3i$, so its conjugate is $0 - 3i$ or $-3i$.

$$\frac{7}{3i} = \frac{7(-3i)}{(3i)(-3i)} = \frac{-21i}{-9i^2} = \frac{-21i}{-9(-1)} = \frac{-21i}{9} = \frac{-7i}{3} \quad \text{or} \quad 0 - \frac{7}{3}i$$

5 We can use the fact that $i^2 = -1$ to find higher powers of i. To find i^3, we rewrite it as the product of i^2 and i.

$$i^3 = i^2 \cdot i = (-1)i = -i$$
$$i^4 = i^2 \cdot i^2 = (-1) \cdot (-1) = 1$$

We continue this process and use the fact that $i^4 = 1$ and $i^2 = -1$ to simplify i^5 and i^6.

$$i^5 = i^4 \cdot i = 1 \cdot i = i$$
$$i^6 = i^4 \cdot i^2 = 1 \cdot (-1) = -1$$

If we continue finding powers of i, we generate the following pattern. Notice that the values $i, -1, -i$, and 1 repeat as i is raised to higher and higher powers.

$i^1 = i$	$i^5 = i$	$i^9 = i$
$i^2 = -1$	$i^6 = -1$	$i^{10} = -1$
$i^3 = -i$	$i^7 = -i$	$i^{11} = -i$
$i^4 = 1$	$i^8 = 1$	$i^{12} = 1$

This pattern allows us to find other powers of i. To do so, we will use the fact that $i^4 = 1$ and rewrite a power of i in terms of i^4.

For example, $i^{22} = i^{20} \cdot i^2 = (i^4)^5 \cdot i^2 = 1^5 \cdot (-1) = 1 \cdot (-1) = -1$.

EXAMPLE 6

Find the following powers of i.

a. i^7 **b.** i^{20} **c.** i^{46} **d.** i^{-12}

Solution

a. $i^7 = i^4 \cdot i^3 = 1(-i) = -i$

b. $i^{20} = (i^4)^5 = 1^5 = 1$

c. $i^{46} = i^{44} \cdot i^2 = (i^4)^{11} \cdot i^2 = 1^{11}(-1) = -1$

d. $i^{-12} = \dfrac{1}{i^{12}} = \dfrac{1}{(i^4)^3} = \dfrac{1}{(1)^3} = \dfrac{1}{1} = 1$

MENTAL MATH

Simplify. See Example 1.

 1. $\sqrt{-81}$ **2.** $\sqrt{-49}$ **3.** $\sqrt{-7}$ **4.** $\sqrt{-3}$

5. $-\sqrt{16}$ **6.** $-\sqrt{4}$ **7.** $\sqrt{-64}$ **8.** $\sqrt{-100}$

EXERCISE SET 10.7

STUDY GUIDE/SSM   CD/VIDEO   PH MATH TUTOR CENTER   MathXL®Tutorials ON CD   MathXL®   MyMathLab®

Write in terms of i. See Example 1.

1. $\sqrt{-24}$
2. $\sqrt{-32}$
3. $-\sqrt{-36}$
4. $-\sqrt{-121}$
5. $8\sqrt{-63}$
6. $4\sqrt{-20}$
7. $-\sqrt{54}$
8. $\sqrt{-63}$

Multiply or divide. See Example 2.

9. $\sqrt{-2}\cdot\sqrt{-7}$
10. $\sqrt{-11}\cdot\sqrt{-3}$
11. $\sqrt{-5}\cdot\sqrt{-10}$
12. $\sqrt{-2}\cdot\sqrt{-6}$
13. $\sqrt{16}\cdot\sqrt{-1}$
14. $\sqrt{3}\cdot\sqrt{-27}$
15. $\dfrac{\sqrt{-9}}{\sqrt{3}}$
16. $\dfrac{\sqrt{49}}{\sqrt{-10}}$
17. $\dfrac{\sqrt{-80}}{\sqrt{-10}}$
18. $\dfrac{\sqrt{-40}}{\sqrt{-8}}$

Add or subtract. Write the sum or difference in the form $a + bi$. See Example 3.

19. $(4 - 7i) + (2 + 3i)$
20. $(2 - 4i) - (2 - i)$
21. $(6 + 5i) - (8 - i)$
22. $(8 - 3i) + (-8 + 3i)$
23. $6 - (8 + 4i)$
24. $(9 - 4i) - 9$

Multiply. Write the product in the form $a + bi$. See Example 4.

25. $6i(2 - 3i)$
26. $5i(4 - 7i)$
27. $\left(\sqrt{3} + 2i\right)\left(\sqrt{3} - 2i\right)$
28. $\left(\sqrt{5} - 5i\right)\left(\sqrt{5} + 5i\right)$
29. $(4 - 2i)^2$
30. $(6 - 3i)^2$

Write each quotient in the form $a + bi$. See Example 5.

31. $\dfrac{4}{i}$
32. $\dfrac{5}{6i}$
33. $\dfrac{7}{4 + 3i}$
34. $\dfrac{9}{1 - 2i}$
35. $\dfrac{3 + 5i}{1 + i}$
36. $\dfrac{6 + 2i}{4 - 3i}$
37. $\dfrac{5 - i}{3 - 2i}$
38. $\dfrac{6 - i}{2 + i}$

MIXED PRACTICE

Perform the indicated operation. Write the result in the form $a + bi$.

39. $(7i)(-9i)$
40. $(-6i)(-4i)$
41. $(6 - 3i) - (4 - 2i)$
42. $(-2 - 4i) - (6 - 8i)$

43. $(6 - 2i)(3 + i)$
44. $(2 - 4i)(2 - i)$
45. $(8 - 3i) + (2 + 3i)$
46. $(7 + 4i) + (4 - 4i)$
47. $(1 - i)(1 + i)$
48. $(6 + 2i)(6 - 2i)$
49. $\dfrac{16 + 15i}{-3i}$
50. $\dfrac{2 - 3i}{-7i}$
51. $(9 + 8i)^2$
52. $(4 - 7i)^2$
53. $\dfrac{2}{3 + i}$
54. $\dfrac{5}{3 - 2i}$
55. $(5 - 6i) - 4i$
56. $(6 - 2i) + 7i$
57. $\dfrac{2 - 3i}{2 + i}$
58. $\dfrac{6 + 5i}{6 - 5i}$
59. $(2 + 4i) + (6 - 5i)$
60. $(5 - 3i) + (7 - 8i)$

Find each power of i. See Example 6.

61. i^8
62. i^{10}
63. i^{21}
64. i^{15}
65. i^{11}
66. i^{40}
67. i^{-6}
68. i^{-9}
69. $(2i)^6$
70. $(5i)^4$
71. $(-3i)^5$
72. $(-2i)^7$

REVIEW AND PREVIEW

Recall that the sum of the measures of the angles of a triangle is $180°$. Find the unknown angle in each triangle.

73.

74.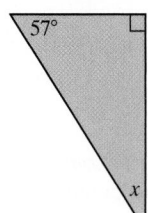

Use synthetic division to divide the following. See Section 5.7.

75. $(x^3 - 6x^2 + 3x - 4) \div (x - 1)$

76. $(5x^4 - 3x^2 + 2) \div (x + 2)$

Thirty people were recently polled about their average monthly balance in their checking accounts. The results of this poll are shown in the following histogram. Use this graph to answer Exercises 77 through 82. See Section 1.2.

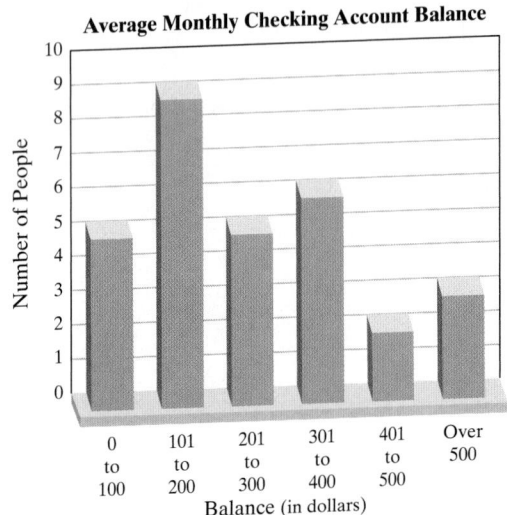

Average Monthly Checking Account Balance

77. How many people polled reported an average checking balance of $201 to $300?

78. How many people polled reported an average checking balance of $0 to $100?

79. How many people polled reported an average checking balance of $200 or less?

80. How many people polled reported an average checking balance of $301 or more?

81. What percent of people polled reported an average checking balance of $201 to $300?

82. What percent of people polled reported an average checking balance of $0 to $100?

Concept Extensions

Write in the form a + bi.

83. $i^3 + i^4$

84. $i^8 - i^7$

85. $i^6 + i^8$

86. $i^4 + i^{12}$

87. $2 + \sqrt{-9}$

88. $5 - \sqrt{-16}$

89. $\dfrac{6 + \sqrt{-18}}{3}$

90. $\dfrac{4 - \sqrt{-8}}{2}$

91. $\dfrac{5 - \sqrt{-75}}{10}$

92. Describe how to find the conjugate of a complex number.

93. Explain why the product of a complex number and its complex conjugate is a real number.

Simplify.

94. $\left(8 - \sqrt{-3}\right) - \left(2 + \sqrt{-12}\right)$

95. $\left(8 - \sqrt{-4}\right) - \left(2 + \sqrt{-16}\right)$

96. Determine whether $2i$ is a solution of $x^2 + 4 = 0$.

97. Determine whether $-1 + i$ is a solution of $x^2 + 2x = -2$.

CHAPTER 10 PROJECT

Calculating the Length and Period of a Pendulum

Materials:

- string (at least 1 meter long)
- weight
- meter stick
- stopwatch
- calculator

This activity may be completed by working in groups or individually. Make a simple pendulum by securely tying the string to a weight.

The formula relating a pendulum's period T (in seconds) to its length l (in centimeters) is

$$T = 2\pi\sqrt{\frac{l}{980}}$$

The **period** of a pendulum is defined as the time it takes the pendulum to complete one full back-and-forth swing. In this activity, you will be measuring your simple pendulum's period with a stopwatch. Because the periods will be only a few seconds long, it will be more accurate for you to time a total of five complete swings and then find the average time of one complete swing.

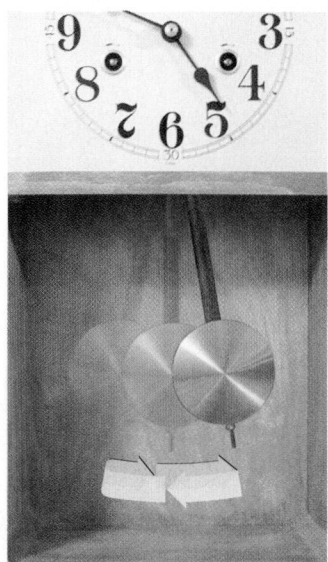

1. For each of the pendulum (string) lengths given in Table 1, measure the time required for 5 complete swings and record it in the appropriate column. Next, divide this value by 5 to find the measured period of the pendulum for the given length and record it in the Measured Period T_m column in the table. Use the given formula to calculate the theoretical period T for the same pendulum length and record it in the appropriate column. (Round to two decimal places.) Find and record in the last column the difference between the measured period and the theoretical period.

2. For each of the periods T given in Table 2, use the given formula and calculate the theoretical pendulum length l required to yield the given period. Record l in the appropriate column; round to one decimal place. Next, use this length l and measure and record the time for 5 complete swings. Divide this value by 5 to find the measured period T_m, and record it. Then find and record in the last column the difference between the theoretical period and the measured period.

3. Use the general trends you find in the tables to describe the relationship between a pendulum's period and its length.

4. Discuss the differences you found between the values of the theoretical period and the measured period. What factors contributed to these differences?

Table 1

Length l (centimeters)	Time for 5 Swings (seconds)	Measured Period T_m (seconds)	Theoretical Period T (seconds)	Difference $\lvert T - T_m \rvert$
30				
55				
70				

Table 2

Period T (seconds)	Theoretical Length l (centimeters)	Time for 5 Swings (seconds)	Measured Period T_m (seconds)	Difference $\lvert T - T_m \rvert$
1				
1.25				
2				

STUDY SKILLS REMINDER

Are You Preparing for a Test on Chapter 10?

Below I have listed some common trouble areas for topics covered in Chapter 10. After studying for your test, but before taking your test, read these.

▶ Remember how to convert an expression with rational expressions to one with radicals and one with radicals to one with rational expressions.

$$7^{2/3} = \sqrt[3]{7^2} \text{ or } \left(\sqrt[3]{7}\right)^2$$
$$\sqrt[5]{4^3} = 4^{3/5}$$

▶ Remember the difference between $\sqrt{x} + \sqrt{x}$ and $\sqrt{x} \cdot \sqrt{x}, x > 0$.

$$\sqrt{x} + \sqrt{x} = 2\sqrt{x}$$
$$\sqrt{x} \cdot \sqrt{x} = x$$

▶ Don't forget the difference between rationalizing the denominator of $\sqrt{\dfrac{2}{x}}$ and rationalizing the denominator of $\dfrac{\sqrt{2}}{\sqrt{x} + 1}, x > 0$.

$$\sqrt{\frac{2}{x}} = \frac{\sqrt{2}}{\sqrt{x}} = \frac{\sqrt{2} \cdot \sqrt{x}}{\sqrt{x} \cdot \sqrt{x}} = \frac{\sqrt{2x}}{x}$$
$$\frac{\sqrt{2}}{\sqrt{x} + 1} = \frac{\sqrt{2}\left(\sqrt{x} - 1\right)}{\left(\sqrt{x} + 1\right)\left(\sqrt{x} - 1\right)} = \frac{\sqrt{2}\left(\sqrt{x} - 1\right)}{x - 1}$$

Remember: This is simply a checklist of common trouble areas. For a review of Chapter 10, see the Highlights and Chapter Review at the end of this chapter.

10 CHAPTER VOCABULARY CHECK

Fill in each blank with one of the words or phrases listed below.

| index | rationalizing | conjugate | principal square root | cube root |
| complex number | like radicals | radicand | imaginary unit | |

1. The _____ of $\sqrt{3} + 2$ is $\sqrt{3} - 2$.

2. The _____ of a nonnegative number a is written as $\sqrt{a}$.

3. The process of writing a radical expression as an equivalent expression but without a radical in the denominator is called _____ the denominator.

4. The _____, written i, is the number whose square is -1.

5. The _____ of a number is written as $\sqrt[3]{a}$.

6. In the notation $\sqrt[n]{a}$, n is called the _____ and a is called the _____.

7. Radicals with the same index and the same radicand are called _____.

8. A _____ is a number that can be written in the form $a + bi$ where a and b are real numbers.

CHAPTER 10 HIGHLIGHTS

Definitions and Concepts	Examples

Section 10.1 Radicals and Radical Functions

The **positive**, or **principal**, **square root** of a nonnegative number a is written as $\sqrt{a}$.

$$\sqrt{a} = b \text{ only if } b^2 = a \text{ and } b \geq 0$$

The **negative square root of** a is written as $-\sqrt{a}$.

The **cube root** of a real number a is written as $\sqrt[3]{a}$.

$$\sqrt[3]{a} = b \text{ only if } b^3 = a$$

If n is an even positive integer, then $\sqrt[n]{a^n} = |a|$.

If n is an odd positive integer, then $\sqrt[n]{a^n} = a$.

A **radical function** in x is a function defined by an expression containing a root of x.

$$\sqrt{36} = 6 \qquad \sqrt{\frac{9}{100}} = \frac{3}{10}$$

$$-\sqrt{36} = -6 \qquad \sqrt{0.04} = 0.2$$

$$\sqrt[3]{27} = 3 \qquad \sqrt[9]{-\frac{1}{8}} = -\frac{1}{2}$$

$$\sqrt[3]{y^6} = y^2 \qquad \sqrt[3]{64x^9} = 4x^3$$

$$\sqrt{(-3)^2} = |-3| = 3$$

$$\sqrt[3]{(-7)^3} = -7$$

If $\quad f(x) = \sqrt{x} + 2,$

$$f(1) = \sqrt{(1)} + 2 = 1 + 2 = 3$$

$$f(3) = \sqrt{(3)} + 2 \approx 3.73$$

Section 10.2 Rational Exponents

$a^{1/n} = \sqrt[n]{a}$ if $\sqrt[n]{a}$ is a real number.

If m and n are positive integers greater than 1 with $\frac{m}{n}$ in lowest terms and $\sqrt[n]{a}$ is a real number, then

$$a^{m/n} = (a^{1/n})^m = \left(\sqrt[n]{a}\right)^m$$

$a^{-m/n} = \dfrac{1}{a^{m/n}}$ as long as $a^{m/n}$ is a nonzero number.

Exponent rules are true for rational exponents.

$$81^{1/2} = \sqrt{81} = 9$$

$$(-8x^3)^{1/3} = \sqrt[3]{-8x^3} = -2x$$

$$4^{5/2} = \left(\sqrt{4}\right)^5 = 2^5 = 32$$

$$27^{2/3} = \left(\sqrt[3]{27}\right)^2 = 3^2 = 9$$

$$16^{-3/4} = \frac{1}{16^{3/4}} = \frac{1}{\left(\sqrt[4]{16}\right)^3} = \frac{1}{2^3} = \frac{1}{8}$$

$$x^{2/3} \cdot x^{-5/6} = x^{2/3 - 5/6} = x^{-1/6} = \frac{1}{x^{1/6}}$$

$$(8^4)^{1/2} = 8^2 = 64$$

$$\frac{a^{4/5}}{a^{-2/5}} = a^{4/5 - (-2/5)} = a^{6/5}$$

(continued)

Definitions and Concepts	Examples

Section 10.3 Simplifying Radical Expressions

Product and Quotient Rules

If $\sqrt[n]{a}$ and $\sqrt[n]{b}$ are real numbers,

$$\sqrt[n]{a} \cdot \sqrt[n]{b} = \sqrt[n]{a \cdot b}$$

$$\frac{\sqrt[n]{a}}{\sqrt[n]{b}} = \sqrt[n]{\frac{a}{b}}, \text{ provided } \sqrt[n]{b} \neq 0$$

A radical of the form $\sqrt[n]{a}$ is **simplified** when a contains no factors that are perfect nth powers.

Multiply or divide as indicated:

$$\sqrt{11} \cdot \sqrt{3} = \sqrt{33}$$

$$\frac{\sqrt[3]{40x}}{\sqrt[3]{5x}} = \sqrt[3]{8} = 2$$

$$\sqrt{40} = \sqrt{4 \cdot 10} = 2\sqrt{10}$$

$$\sqrt{36x^5} = \sqrt{36x^4 \cdot x} = 6x^2\sqrt{x}$$

$$\sqrt[3]{24x^7y^3} = \sqrt[3]{8x^6y^3 \cdot 3x} = 2x^2y\sqrt[3]{3x}$$

$$\sqrt{36x^4 \cdot x} = 6x^2\sqrt{x}$$

Section 10.4 Adding, Subtracting, and Multiplying Radical Expressions

Radicals with the same index and the same radicand are **like radicals**.

The distributive property can be used to add like radicals.

Radical expressions are multiplied by using many of the same properties used to multiply polynomials.

$$5\sqrt{6} + 2\sqrt{6} = (5+2)\sqrt{6} = 7\sqrt{6}$$
$$= \sqrt[3]{3x} - 10\sqrt[3]{3x} + 3\sqrt[3]{10x}$$
$$= (-1-10)\sqrt[3]{3x} + 3\sqrt[3]{10x}$$
$$= -11\sqrt[3]{3x} + 3\sqrt[3]{10x}$$

Multiply.

$$\left(\sqrt{5} - \sqrt{2x}\right)\left(\sqrt{2} + \sqrt{2x}\right)$$
$$= \sqrt{10} + \sqrt{10x} - \sqrt{4x} - 2x$$
$$= \sqrt{10} + \sqrt{10x} - 2\sqrt{x} - 2x$$
$$\left(2\sqrt{3} - \sqrt{8x}\right)\left(2\sqrt{3} + \sqrt{8x}\right)$$
$$= 4(3) - 8x = 12 - 8x$$

Section 10.5 Rationalizing Denominators and Numerators of Radical Expressions

The **conjugate** of $a + b$ is $a - b$.

The process of writing the denominator of a radical expression without a radical is called **rationalizing the denominator**.

The conjugate of $\sqrt{7} + \sqrt{3}$ is $\sqrt{7} - \sqrt{3}$

Rationalize each denominator.

$$\frac{\sqrt{5}}{\sqrt{3}} = \frac{\sqrt{5} \cdot \sqrt{3}}{\sqrt{3} \cdot \sqrt{3}} = \frac{\sqrt{15}}{3}$$

$$\frac{6}{\sqrt{7} + \sqrt{3}} = \frac{6\left(\sqrt{7} - \sqrt{3}\right)}{\left(\sqrt{7} + \sqrt{3}\right)\left(\sqrt{7} - \sqrt{3}\right)}$$
$$= \frac{6\left(\sqrt{7} - \sqrt{3}\right)}{7 - 3}$$
$$= \frac{6\left(\sqrt{7} - \sqrt{3}\right)}{4} = \frac{3\left(\sqrt{7} - \sqrt{3}\right)}{2}$$

(continued)

Definitions and Concepts	**Examples**

The process of writing the numerator of a radical expression without a radical is called **rationalizing the numerator.**

Rationalize each numerator.

$$\frac{\sqrt[3]{9}}{\sqrt[3]{5}} = \frac{\sqrt[3]{9} \cdot \sqrt[3]{3}}{\sqrt[3]{5} \cdot \sqrt[3]{3}} = \frac{\sqrt[3]{27}}{\sqrt[3]{15}} = \frac{3}{\sqrt[3]{15}}$$

$$\frac{\sqrt{9} + \sqrt{3x}}{12} = \frac{(\sqrt{9} + \sqrt{3x})(\sqrt{9} - \sqrt{3x})}{12(\sqrt{9} - \sqrt{3x})}$$

$$= \frac{9 - 3x}{12(\sqrt{9} - \sqrt{3x})}$$

$$= \frac{3(3 - x)}{3 \cdot 4(3 - \sqrt{3x})} = \frac{3 - x}{4(3 - \sqrt{3x})}$$

To Solve a Radical Equation

Step 1: Write the equation so that one radical is by itself on one side of the equation.

Step 2: Raise each side of the equation to a power equal to the index of the radical and simplify.

Step 3: If the equation still contains a radical, repeat Steps 1 and 2. If not, solve the equation.

Step 4: Check all proposed solutions in the original equation.

Solve $x = \sqrt{4x + 9} + 3$.

1. $x - 3 = \sqrt{4x + 9}$

2. $\quad (x - 3)^2 = (\sqrt{4x + 9})^2$
$x^2 - 6x + 9 = 4x + 9$

3. $\quad x^2 - 10x = 0$
$\quad x(x - 10) = 0$
$\qquad\qquad x = 0 \quad\text{or}\quad x = 10$

4. The proposed solution 10 checks, but 0 does not. The solution is 10.

$$i^2 = -1 \text{ and } i = \sqrt{-1}$$

A **complex number** is a number that can be written in the form $a + bi$, where a and b are real numbers.

Simplify $\sqrt{-9}$.

$$\sqrt{-9} = \sqrt{-1 \cdot 9} = \sqrt{-1} \cdot \sqrt{9} = i \cdot 3 \text{ or } 3i$$

Complex Numbers	*Written in form $a + bi$*
12	$12 + 0i$
$-5i$	$0 + (-5)i$
$-2 - 3i$	$-2 + (-3)i$

Multiply,

$$\sqrt{-3} \cdot \sqrt{-7} = i\sqrt{3} \cdot i\sqrt{7}$$

$$= i^2\sqrt{21}$$

$$= -\sqrt{21}$$

(continued)

Definitions and Concepts	**Examples**
Section 10.7 Complex Numbers	

To add or subtract complex numbers, add or subtract their real parts and then add or subtract their imaginary parts.	Perform each indicated operation. $$(-3 + 2i) - (7 - 4i) = -3 + 2i - 7 + 4i$$ $$= -10 + 6i$$
To multiply complex numbers, multiply as though they are binomials.	$$(-7 - 2i)(6 + i) = -42 - 7i - 12i - 2i^2$$ $$= -42 - 19i - 2(-1)$$ $$= -42 - 19i + 2$$ $$= -40 - 19i$$
The complex numbers $(a + bi)$ and $(a - bi)$ are called **complex conjugates.**	The complex conjugate of $$(3 + 6i) \text{ is } (3 - 6i).$$ Their product is a real number. $$(3 - 6i)(3 + 6i) = 9 - 36i^2$$ $$= 9 - 36(-1) = 9 + 36 = 45$$
To divide complex numbers, multiply the numerator and the denominator by the conjugate of the denominator.	Divide. $$\frac{4}{2 - i} = \frac{4(2 + i)}{(2 - i)(2 + i)}$$ $$= \frac{4(2 + i)}{4 - i^2}$$ $$= \frac{4(2 + i)}{5}$$ $$= \frac{8 + 4i}{5} = \frac{8}{5} + \frac{4}{5}i$$

10 CHAPTER REVIEW

(10.1) Find the root. Assume that all variables represent positive numbers.

1. $\sqrt{81}$

2. $\sqrt[4]{81}$

3. $\sqrt[3]{-8}$

4. $\sqrt[4]{-16}$

5. $-\sqrt{\dfrac{1}{49}}$

6. $\sqrt{x^{64}}$

7. $-\sqrt{36}$

8. $\sqrt[3]{64}$

9. $\sqrt[3]{-a^6 b^9}$

10. $\sqrt{16a^4 b^{12}}$

11. $\sqrt[5]{32a^5 b^{10}}$

12. $\sqrt[5]{-32x^{15} y^{20}}$

13. $\sqrt{\dfrac{x^{12}}{36y^2}}$

14. $\sqrt[3]{\dfrac{27y^3}{z^{12}}}$

Simplify. Use absolute value bars when necessary.

15. $\sqrt{(-x)^2}$

16. $\sqrt[4]{(x^2 - 4)^4}$

17. $\sqrt[3]{(-27)^3}$

18. $\sqrt[5]{(-5)^5}$

19. $-\sqrt[5]{x^5}$

20. $\sqrt[4]{16(2y + z)^{12}}$

21. $\sqrt{25(x - y)^{10}}$

22. $\sqrt[5]{-y^5}$

23. $\sqrt[9]{-x^9}$

Identify the domain and then graph each function.

24. $f(x) = \sqrt{x} + 3$

25. $g(x) = \sqrt[3]{x} - 3$; use the accompanying table.

x	-5	2	3	4	11
$g(x)$					

(10.2) *Evaluate the following.*

26. $\left(\dfrac{1}{81}\right)^{1/4}$

27. $\left(-\dfrac{1}{27}\right)^{1/3}$

28. $(-27)^{-1/3}$

29. $(-64)^{-1/3}$

30. $-9^{3/2}$

31. $64^{-1/3}$

32. $(-25)^{5/2}$

33. $\left(\dfrac{25}{49}\right)^{-3/2}$

34. $\left(\dfrac{8}{27}\right)^{-2/3}$

35. $\left(-\dfrac{1}{36}\right)^{-1/4}$

Write with rational exponents.

36. $\sqrt[3]{x^2}$

37. $\sqrt[5]{5x^2y^3}$

Write with radical notation.

38. $y^{4/5}$

39. $5(xy^2z^5)^{1/3}$

40. $(x + 2y)^{-1/2}$

Simplify each expression. Assume that all variables represent positive numbers. Write with only positive exponents.

41. $a^{1/3}a^{4/3}a^{1/2}$

42. $\dfrac{b^{1/3}}{b^{4/3}}$

43. $(a^{1/2}a^{-2})^3$

44. $(x^{-3}y^6)^{1/3}$

45. $\left(\dfrac{b^{3/4}}{a^{-1/2}}\right)^8$

46. $\dfrac{x^{1/4}x^{-1/2}}{x^{2/3}}$

47. $\left(\dfrac{49c^{5/3}}{a^{-1/4}b^{5/6}}\right)^{-1}$

48. $a^{-1/4}(a^{5/4} - a^{9/4})$

Use a calculator and write a three-decimal-place approximation.

49. $\sqrt{20}$

50. $\sqrt[3]{-39}$

51. $\sqrt[4]{726}$

52. $56^{1/3}$

53. $-78^{3/4}$

54. $105^{-2/3}$

Use rational exponents to write each radical with the same index. Then multiply.

55. $\sqrt[3]{2} \cdot \sqrt{7}$

56. $\sqrt[3]{3} \cdot \sqrt[4]{x}$

(10.3) *Perform the indicated operations and then simplify if possible. For the remainder of this review, assume that variables represent positive numbers only.*

57. $\sqrt{3} \cdot \sqrt{8}$

58. $\sqrt[3]{7y} \cdot \sqrt[3]{x^2z}$

59. $\dfrac{\sqrt{44x^3}}{\sqrt{11x}}$

60. $\dfrac{\sqrt[4]{a^6b^{13}}}{\sqrt[4]{a^2b}}$

Simplify.

61. $\sqrt{60}$

62. $-\sqrt{75}$

63. $\sqrt[3]{162}$

64. $\sqrt[3]{-32}$

65. $\sqrt{36x^7}$

66. $\sqrt[3]{24a^5b^7}$

67. $\sqrt{\dfrac{p^{17}}{121}}$

68. $\sqrt[3]{\dfrac{y^5}{27x^6}}$

69. $\sqrt[4]{\dfrac{xy^6}{81}}$

70. $\sqrt{\dfrac{2x^3}{49y^4}}$

△ **71.** The formula for the radius r of a circle of area A is

$$r = \sqrt{\dfrac{A}{\pi}}$$

    a. Find the exact radius of a circle whose area is 25 square meters.

    b. Approximate to two decimal places the radius of a circle whose area is 104 square inches.

(10.4) *Perform the indicated operation.*

72. $x\sqrt{75xy} - \sqrt{27x^3y}$

73. $2\sqrt{32x^2y^3} - xy\sqrt{98y}$

74. $\sqrt[3]{128} + \sqrt[3]{250}$

75. $3\sqrt[4]{32a^5} - a\sqrt[4]{162a}$

76. $\dfrac{5}{\sqrt{4}} + \dfrac{\sqrt{3}}{3}$

77. $\sqrt{\dfrac{8}{x^2}} - \sqrt{\dfrac{50}{16x^2}}$

78. $2\sqrt{50} - 3\sqrt{125} + \sqrt{98}$

79. $2a\sqrt[4]{32b^5} - 3b\sqrt[4]{162a^4b} + \sqrt[4]{2a^4b^5}$

Multiply and then simplify if possible.

80. $\sqrt{3}(\sqrt{27} - \sqrt{3})$

81. $(\sqrt{x} - 3)^2$

82. $(\sqrt{5} - 5)(2\sqrt{5} + 2)$

83. $(2\sqrt{x} - 3\sqrt{y})(2\sqrt{x} + 3\sqrt{y})$

84. $(\sqrt{a} + 3)(\sqrt{a} - 3)$

85. $(\sqrt[3]{a} + 2)^2$

86. $(\sqrt[3]{5x} + 9)(\sqrt[3]{5x} - 9)$

87. $(\sqrt[3]{a} + 4)(\sqrt[3]{a^2} - 4\sqrt[3]{a} + 16)$

(10.5) *Rationalize each denominator.*

88. $\dfrac{3}{\sqrt{7}}$

89. $\sqrt{\dfrac{x}{12}}$

90. $\dfrac{5}{\sqrt[3]{4}}$

91. $\sqrt{\dfrac{24x^5}{3y^2}}$

92. $\sqrt[3]{\dfrac{15x^6y^7}{z^2}}$

93. $\dfrac{5}{2 - \sqrt{7}}$

94. $\dfrac{3}{\sqrt{y} - 2}$

95. $\dfrac{\sqrt{2} - \sqrt{3}}{\sqrt{2} + \sqrt{3}}$

Rationalize each numerator.

96. $\dfrac{\sqrt{11}}{3}$

97. $\sqrt{\dfrac{18}{y}}$

98. $\dfrac{\sqrt[3]{9}}{7}$

99. $\sqrt{\dfrac{24x^5}{3y^2}}$

100. $\sqrt[3]{\dfrac{xy^2}{10z}}$

101. $\dfrac{\sqrt{x}+5}{-3}$

(10.6) *Solve each equation for the variable.*

102. $\sqrt{y-7}=5$

103. $\sqrt{2x+10}=4$

104. $\sqrt[3]{2x-6}=4$

105. $\sqrt{x+6}=\sqrt{x+2}$

106. $2x-5\sqrt{x}=3$

107. $\sqrt{x+9}=2+\sqrt{x-7}$

Find each unknown length.

 108.

3 cm

3 cm

 109.

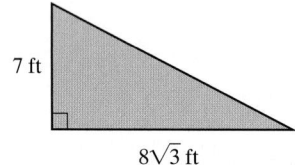

7 ft

$8\sqrt{3}$ ft

110. Beverly Hillis wants to determine the distance x across a pond on her property. She is able to measure the distances shown on the following diagram. Find how wide the lake is at the crossing point, indicated by the triangle, to the nearest tenth of a foot.

65 ft

40 ft

△ **111.** A pipe fitter needs to connect two underground pipelines that are offset by 3 feet, as pictured in the diagram. Neglecting the joints needed to join the pipes, find the length of the shortest possible connecting pipe rounded to the nearest hundredth of a foot.

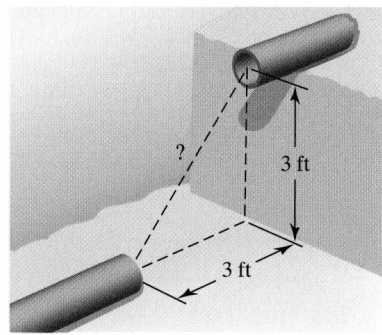

?

3 ft

3 ft

(10.7) *Perform the indicated operation and simplify. Write the result in the form $a + bi$.*

112. $\sqrt{-8}$

113. $-\sqrt{-6}$

114. $\sqrt{-4}+\sqrt{-16}$

115. $\sqrt{-2}\cdot\sqrt{-5}$

116. $(12-6i)+(3+2i)$

117. $(-8-7i)-(5-4i)$

118. $\left(\sqrt{3}+\sqrt{2}\right)+\left(3\sqrt{2}-\sqrt{-8}\right)$

119. $2i(2-5i)$

120. $-3i(6-4i)$

121. $(3+2i)(1+i)$

122. $(2-3i)^2$

123. $\left(\sqrt{6}-9i\right)\left(\sqrt{6}+9i\right)$

124. $\dfrac{2+3i}{2i}$

125. $\dfrac{1+i}{-3i}$

CHAPTER 10 TEST

Remember to use your Chapter Test Prep Video CD to help you study and view solutions to the test questions you need help with.

Raise to the power or find the root. Assume that all variables represent positive numbers. Write with only positive exponents.

1. $\sqrt{216}$

2. $-\sqrt[4]{x^{64}}$

3. $\left(\dfrac{1}{125}\right)^{1/3}$

4. $\left(\dfrac{1}{125}\right)^{-1/3}$

5. $\left(\dfrac{8x^3}{27}\right)^{2/3}$

6. $\sqrt[3]{-a^{18}b^9}$

7. $\left(\dfrac{64c^{4/3}}{a^{-2/3}b^{5/6}}\right)^{1/2}$

8. $a^{-2/3}(a^{5/4} - a^3)$

Find the root. Use absolute value bars when necessary.

9. $\sqrt[4]{(4xy)^4}$

10. $\sqrt[3]{(-27)^3}$

Rationalize the denominator. Assume that all variables represent positive numbers.

11. $\sqrt{\dfrac{9}{y}}$

12. $\dfrac{4 - \sqrt{x}}{4 + 2\sqrt{x}}$

13. $\dfrac{\sqrt[3]{ab}}{\sqrt[3]{ab^2}}$

14. Rationalize the numerator of $\dfrac{\sqrt{6} + x}{8}$ and simplify.

Perform the indicated operations. Assume that all variables represent positive numbers.

15. $\sqrt{125x^3} - 3\sqrt{20x^3}$

16. $\sqrt{3}(\sqrt{16} - \sqrt{2})$

17. $(\sqrt{x} + 1)^2$

18. $(\sqrt{2} - 4)(\sqrt{3} + 1)$

19. $(\sqrt{5} + 5)(\sqrt{5} - 5)$

Use a calculator to approximate each to three decimal places.

20. $\sqrt{561}$

21. $386^{-2/3}$

Solve.

22. $x = \sqrt{x - 2} + 2$

23. $\sqrt{x^2 - 7} + 3 = 0$

24. $\sqrt[3]{x + 5} = \sqrt[3]{2x - 1}$

Perform the indicated operation and simplify. Write the result in the form $a + bi$.

25. $\sqrt{-2}$

26. $-\sqrt{-8}$

27. $(12 - 6i) - (12 - 3i)$

28. $(6 - 2i)(6 + 2i)$

29. $(4 + 3i)^2$

30. $\dfrac{1 + 4i}{1 - i}$

△ **31.** Find x.

32. Identify the domain of $g(x)$. Then complete the accompanying table and graph $g(x)$.
$$g(x) = \sqrt{x} + 2$$

x	−2	−1	2	7
$g(x)$				

Solve.

33. The function $V(r) = \sqrt{2.5r}$ can be used to estimate the maximum safe velocity V in miles per hour at which a car can travel if it is driven along a curved road with a *radius of curvature* r in feet. To the nearest whole number, find the maximum safe speed if a cloverleaf exit on an expressway has a radius of curvature of 300 feet.

34. Use the formula from Exercise 33 to find the radius of curvature if the safe velocity is 30 mph.

CHAPTER CUMULATIVE REVIEW

1. Simplify each expression.

a. $\dfrac{(-12)(-3) + 3}{-7 - (-2)}$

b. $\dfrac{2(-3)^2 - 20}{-5 + 4}$

2. Simplify each expression.

a. $2(x - 3) + (5x + 3)$

b. $4(3x + 2) - 3(5x - 1)$

c. $7x + 2(x - 7) - 3x$

3. Solve: $\dfrac{x}{2} - 1 = \dfrac{2}{3}x - 3$

4. Solve: $\dfrac{a - 1}{2} + a = 2 - \dfrac{2a + 7}{8}$

5. A 10-foot board is to be cut into two pieces so that the longer piece is 4 times the shorter. Find the length of each piece.

6. The Smith family owns a lake house 121.5 miles from home. If it takes them $4\frac{1}{2}$ hours round-trip to drive from their house to their lake house, find their average speed.

7. Determine the number of solutions of the system.
$$\begin{cases} 3x - y = 4 \\ x + 2y = 8 \end{cases}$$

8. Solve: $|3x - 2| + 5 = 5$

9. Solve the system:
$$\begin{cases} x + 2y = 7 \\ 2x + 2y = 13 \end{cases}$$

10. Solve: $\left|\dfrac{x}{2} - 1\right| \le 0$.

11. Solve the system:
$$\begin{cases} 2x - y = 7 \\ 8x - 4y = 1 \end{cases}$$

12. Graph $y = |x - 2|$.

13. Eric Daly, a chemistry teaching assistant, needs 10 liters of 20% saline solution (salt water) for his 2 p.m. laboratory class. Unfortunately, the only mixtures on hand are a 5% saline solution and a 25% saline solution. How much of each solution should he mix to produce the 20% solution?

14. Find the domain and the range of each relation. Use the vertical line test to determine whether each graph is the graph of a function.

a.

b.

c.

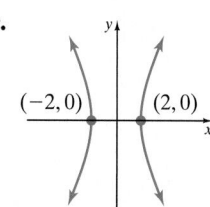

15. If $P(x) = 3x^2 - 2x - 5$, find the following.

a. $P(1)$

b. $P(-2)$

16. Graph $f(x) = -2$.

17. Divide $10x^3 - 5x^2 + 20x$ by $5x$.

18. Find the slope of $y = -3$.

19. Use synthetic division to divide $2x^3 - x^2 - 13x$ by $x - 3$.

20. Use the substitution method to solve the system.
$$\begin{cases} \dfrac{x}{6} - \dfrac{y}{2} = 1 \\ \dfrac{x}{3} - \dfrac{y}{4} = 2 \end{cases}$$

21. Factor $x^2 + 5yx + 6y^2$.

22. At a seasonal clearance sale, Nana Long spent $33.75. She paid $3.50 for tee-shirts and $4.25 for shorts. If she bought 9 items, how many of each item did she buy?

23. Simplify each rational expression.

a. $\dfrac{x^3 + 8}{2 + x}$

b. $\dfrac{2y^2 + 2}{y^3 - 5y^2 + y - 5}$

24. Use scientific notation to simplify and write the answer in scientific notation. $\dfrac{0.0000035 \times 4000}{0.28}$

25. Divide: $\dfrac{3x^3 y^7}{40} \div \dfrac{4x^3}{y^2}$.

26. Subtract $(2x - 5)$ from the sum of $(5x^2 - 3x + 6)$ and $(4x^2 + 5x - 3)$.

27. Subtract: $\dfrac{2y}{2y - 7} - \dfrac{7}{2y - 7}$

28. Multiply and simplify the product if possible.

a. $(y - 2)(3y + 4)$

b. $(3y - 1)(2y^2 + 3y - 1)$

29. Add: $\dfrac{2x}{x^2 + 2x + 1} + \dfrac{x}{x^2 - 1}$

30. Factor $x^3 - x^2 + 4x - 4$

31. Simply each complex fraction.

a. $\dfrac{\dfrac{5x}{x + 2}}{\dfrac{10}{x - 2}}$

b. $\dfrac{\dfrac{x}{y^2} + \dfrac{1}{y}}{\dfrac{y}{x^2} + \dfrac{1}{x}}$

32. Simplify each rational expression.

a. $\dfrac{a^3 - 8}{2 - a}$

b. $\dfrac{3a^2 - 3}{a^3 + 5a^2 - a - 5}$

33. Solve: $\dfrac{x}{2} + \dfrac{8}{3} = \dfrac{1}{6}$

34. Perform the indicated operations.

a. $\dfrac{3}{xy^2} - \dfrac{2}{3x^2 y}$

b. $\dfrac{5x}{x + 3} - \dfrac{2x}{x - 3}$

c. $\dfrac{x}{x - 2} - \dfrac{5}{2 - x}$

35. If the following two triangles are similar, find the missing length x.

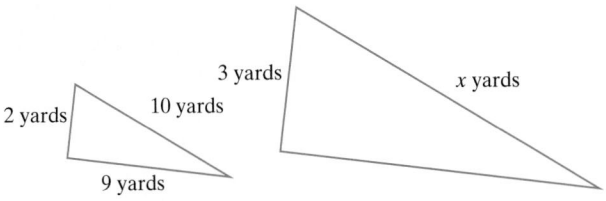

36. Simplify each complex fraction.

a. $\dfrac{\dfrac{y - 2}{16}}{\dfrac{2y + 3}{12}}$

b. $\dfrac{\dfrac{x}{16} - \dfrac{1}{x}}{1 - \dfrac{4}{x}}$

37. Find the cube roots.

a. $\sqrt[3]{1}$

b. $\sqrt[3]{-64}$

c. $\sqrt[3]{\dfrac{8}{125}}$

d. $\sqrt[3]{x^6}$

e. $\sqrt[3]{-27x^9}$

38. Divide $x^3 - 2x^2 + 3x - 6$ by $x - 2$.

39. Write each expression with a positive exponent, and then simplify.

a. $16^{-3/4}$

b. $(-27)^{-2/3}$

40. Use synthetic division to divide $4y^3 - 12y^2 - y + 12$ by $y - 3$.

41. Rational the numerator of $\dfrac{\sqrt{x} + 2}{5}$

42. Solve: $\dfrac{28}{9 - a^2} = \dfrac{2a}{a - 3} + \dfrac{6}{a + 3}$

43. Suppose that u varies inversely as w. If u is 3 when w is 5, find the constant of variation and the inverse variation equation.

44. Suppose that y varies directly as x. If $y = 0.51$ when $x = 3$, find the constant of variation and the direct variation equation.

Quadratic Equations and Functions

An important part of the study of algebra is learning to model and solve problems. Often, the model of a problem is a quadratic equation or a function containing a second-degree polynomial. In this chapter, we continue the work begun in Chapter 6, when we solved polynomial equations in one variable by factoring. Two additional methods of solving quadratic equations are analyzed, as well as methods of solving nonlinear inequalities in one variable.

Community theater groups began to formulate in the United States in the early part of the twentieth century. While many major cities have large theater districts, over 5000 theater companies nationwide entertain smaller communities. Community theater groups provide a creative outlet for aspiring actors, musicians, technicians, producers, and directors.

Producers are responsible for the business end of the show: obtaining financing, setting a budget, securing the performance location, advertising, setting ticket prices, and hiring the director. Directors are in charge of the artistic decisions in the show. They choose the script, audition actors, select costumes and sets, conduct rehearsals, and oversee the cast and crew.

In the Spotlight on Decision Making feature on page 688, you will have the opportunity to decide on optimal ticket prices as a community theater member.

Link: www.communitytheater.org *(Community Theater Green Room)*

Source of text: World Book Millenium 2000 (encyclopedia on CD) "Theater," http://www.bls.gov/oco/ocos093.htm *(Bureau of Labor Statistics)*

11.1 SOLVING QUADRATIC EQUATIONS BY COMPLETING THE SQUARE

Objectives

1. Use the square root property to solve quadratic equations.
2. Solve quadratic equations by completing the square.
3. Use quadratic equations to solve problems.

1 In Chapter 6, we solved quadratic equations by factoring. Recall that a **quadratic,** or **second-degree, equation** is an equation that can be written in the form $ax^2 + bx + c = 0$, where a, b, and c are real numbers and a is not 0. To solve a quadratic equation such as $x^2 = 9$ by factoring, we use the zero-factor theorem. To use the zero-factor theorem, the equation must first be written in standard form, $ax^2 + bx + c = 0$.

$$x^2 = 9$$

$$x^2 - 9 = 0 \qquad \text{Subtract 9 from both sides.}$$

$$(x + 3)(x - 3) = 0 \qquad \text{Factor.}$$

$$x + 3 = 0 \quad \text{or} \quad x - 3 = 0 \qquad \text{Set each factor equal to 0.}$$

$$x = -3 \qquad\quad x = 3 \qquad \text{Solve.}$$

The solution set is $\{-3, 3\}$, the positive and negative square roots of 9. Not all quadratic equations can be solved by factoring, so we need to explore other methods. Notice that the solutions of the equation $x^2 = 9$ are two numbers whose square is 9.

$$3^2 = 9 \qquad \text{and} \qquad (-3)^2 = 9$$

Thus, we can solve the equation $x^2 = 9$ by taking the square root of both sides. Be sure to include both $\sqrt{9}$ and $-\sqrt{9}$ as solutions since both $\sqrt{9}$ and $-\sqrt{9}$ are numbers whose square is 9.

$$x^2 = 9$$

$$\sqrt{x^2} = \pm\sqrt{9} \qquad \begin{array}{l}\text{The notation } \pm\sqrt{9} \text{ (read as "plus or minus}\\ \sqrt{9}\text{") indicates the pair of numbers } +\sqrt{9}\\ \text{and } -\sqrt{9}.\end{array}$$

$$x = \pm 3$$

This illustrates the square root property.

> **Helpful Hint**
>
> The notation ± 3, for example, is read as "plus or minus 3." It is a shorthand notation for the pair of numbers $+3$ and -3.

Square Root Property

If b is a real number and if $a^2 = b$, then $a = \pm\sqrt{b}$.

EXAMPLE 1

Use the square root property to solve $x^2 = 50$.

Solution

$$x^2 = 50$$
$$x = \pm\sqrt{50} \quad \text{Use the square root property.}$$
$$x = \pm 5\sqrt{2} \quad \text{Simplify the radical.}$$

Check

Let $x = 5\sqrt{2}$.
$$x^2 = 50$$
$$\left(5\sqrt{2}\right)^2 \stackrel{?}{=} 50$$
$$25 \cdot 2 \stackrel{?}{=} 50$$
$$50 = 50 \quad \text{True}$$

Let $x = -5\sqrt{2}$.
$$x^2 = 50$$
$$\left(-5\sqrt{2}\right)^2 \stackrel{?}{=} 50$$
$$25 \cdot 2 \stackrel{?}{=} 50$$
$$50 = 50 \quad \text{True}$$

The solutions are $5\sqrt{2}$ and $-5\sqrt{2}$, or the solution set is $\{-5\sqrt{2}, 5\sqrt{2}\}$.

EXAMPLE 2

Use the square root property to solve $2x^2 = 14$.

Solution First we get the squared variable alone on one side of the equation.

$$2x^2 = 14$$
$$x^2 = 7 \quad \text{Divide both sides by 2.}$$
$$x = \pm\sqrt{7} \quad \text{Use the square root property.}$$

Check to see that the solutions are $\sqrt{7}$ and $-\sqrt{7}$, or the solution set is $\{-\sqrt{7}, \sqrt{7}\}$.

EXAMPLE 3

Use the square root property to solve $(x + 1)^2 = 12$.

Solution

$$(x + 1)^2 = 12$$
$$x + 1 = \pm\sqrt{12} \quad \text{Use the square root property.}$$
$$x + 1 = \pm 2\sqrt{3} \quad \text{Simplify the radical.}$$
$$x = -1 \pm 2\sqrt{3} \quad \text{Subtract 1 from both sides.}$$

Check Below is a check for $-1 + 2\sqrt{3}$. The check for $-1 - 2\sqrt{3}$ is almost the same and is left for you to do on your own.

$$(x + 1)^2 = 12$$
$$\left(-1 + 2\sqrt{3} + 1\right)^2 \stackrel{?}{=} 12$$
$$\left(2\sqrt{3}\right)^2 \stackrel{?}{=} 12$$
$$4 \cdot 3 \stackrel{?}{=} 12$$
$$12 = 12 \quad \text{True}$$

The solutions are $-1 + 2\sqrt{3}$ and $-1 - 2\sqrt{3}$.

EXAMPLE 4

Use the square root property to solve $(2x - 5)^2 = -16$.

Solution

$$(2x - 5)^2 = -16$$

$2x - 5 = \pm\sqrt{-16}$ Use the square root property.

$2x - 5 = \pm 4i$ Simplify the radical.

$2x = 5 \pm 4i$ Add **5** to both sides.

$x = \dfrac{5 \pm 4i}{2}$ Divide both sides by **2**.

The solutions are $\dfrac{5 + 4i}{2}$ and $\dfrac{5 - 4i}{2}$.

✔ CONCEPT CHECK

How do you know just by looking that $(x - 2)^2 = -4$ has complex, but not real solutions?

2   Notice from Examples 3 and 4 that, if we write a quadratic equation so that one side is the square of a binomial, we can solve by using the square root property. To write the square of a binomial, we write perfect square trinomials. Recall that a perfect square trinomial is a trinomial that can be factored into two identical binomial factors.

Perfect Square Trinomials	*Factored Form*
$x^2 + 8x + 16$	$(x + 4)^2$
$x^2 - 6x + 9$	$(x - 3)^2$
$x^2 + 3x + \dfrac{9}{4}$	$\left(x + \dfrac{3}{2}\right)^2$

Notice that for each perfect square trinomial, **the constant term of the trinomial is the square of half the coefficient of the x-term.** For example,

$$x^2 + 8x + 16 \qquad x^2 - 6x + 9$$

$$\frac{1}{2}(8) = 4 \text{ and } 4^2 = 16 \qquad \frac{1}{2}(-6) = -3 \text{ and } (-3)^2 = 9$$

The process of writing a quadratic equation so that one side is a perfect square trinomial is called **completing the square.**

EXAMPLE 5

Solve $p^2 + 2p = 4$ by completing the square.

Solution   First, add the square of half the coefficient of p to both sides so that the resulting trinomial will be a perfect square trinomial. The coefficient of p is 2.

$$\frac{1}{2}(2) = 1 \quad \text{and} \quad 1^2 = 1$$

Add 1 to both sides of the original equation.

$$p^2 + 2p = 4$$

$$p^2 + 2p + 1 = 4 + 1 \qquad \text{Add } \mathbf{1} \text{ to both sides.}$$

$$(p + 1)^2 = 5 \qquad \text{Factor the trinomial; simplify the right side.}$$

We may now use the square root property and solve for p.

$$p + 1 = \pm\sqrt{5} \qquad \text{Use the square root property.}$$

$$p = -1 \pm \sqrt{5} \qquad \text{Subtract } \mathbf{1} \text{ from both sides.}$$

Notice that there are two solutions: $-1 + \sqrt{5}$ and $-1 - \sqrt{5}$.

EXAMPLE 6

Solve $m^2 - 7m - 1 = 0$ for m by completing the square.

Solution   First, add 1 to both sides of the equation so that the left side has no constant term.

$$m^2 - 7m - 1 = 0$$

$$m^2 - 7m = 1$$

Now find the constant term that makes the left side a perfect square trinomial by squaring half the coefficient of m. Add this constant to both sides of the equation.

$$\frac{1}{2}(-7) = -\frac{7}{2} \quad \text{and} \quad \left(-\frac{7}{2}\right)^2 = \frac{49}{4}$$

$$m^2 - 7m + \frac{49}{4} = 1 + \frac{49}{4} \qquad \text{Add } \frac{\mathbf{49}}{\mathbf{4}} \text{ to both sides of the equation.}$$

$$\left(m - \frac{7}{2}\right)^2 = \frac{53}{4} \qquad \begin{array}{l}\text{Factor the perfect square trinomial}\\ \text{and simplify the right side.}\end{array}$$

$$m - \frac{7}{2} = \pm\sqrt{\frac{53}{4}} \qquad \text{Apply the square root property.}$$

$$m = \frac{7}{2} \pm \frac{\sqrt{53}}{2} \qquad \begin{array}{l}\text{Add } \frac{\mathbf{7}}{\mathbf{2}} \text{ to both sides and}\\ \text{simplify } \sqrt{\frac{53}{4}}.\end{array}$$

$$m = \frac{7 \pm \sqrt{53}}{2} \qquad \text{Simplify.}$$

The solutions are $\dfrac{7 + \sqrt{53}}{2}$ and $\dfrac{7 - \sqrt{53}}{2}$.

EXAMPLE 7

Solve $2x^2 - 8x + 3 = 0$.

Solution Our procedure for finding the constant term to complete the square works only if the coefficient of the squared variable term is 1. Therefore, to solve this equation, the first step is to divide both sides by 2, the coefficient of x^2.

$$2x^2 - 8x + 3 = 0$$

$$x^2 - 4x + \frac{3}{2} = 0 \qquad \text{Divide both sides by 2.}$$

$$x^2 - 4x = -\frac{3}{2} \qquad \text{Subtract } \frac{3}{2} \text{ from both sides.}$$

Next find the square of half of -4.

$$\frac{1}{2}(-4) = -2 \quad \text{and} \quad (-2)^2 = 4$$

Add 4 to both sides of the equation to complete the square.

$$x^2 - 4x + 4 = -\frac{3}{2} + 4$$

$$(x - 2)^2 = \frac{5}{2} \qquad \text{Factor the perfect square and simplify the right side.}$$

$$x - 2 = \pm\sqrt{\frac{5}{2}} \qquad \text{Apply the square root property.}$$

$$x - 2 = \pm\frac{\sqrt{10}}{2} \qquad \text{Rationalize the denominator.}$$

$$x = 2 \pm \frac{\sqrt{10}}{2} \qquad \text{Add 2 to both sides.}$$

$$= \frac{4}{2} \pm \frac{\sqrt{10}}{2} \qquad \text{Find the common denominator.}$$

$$= \frac{4 \pm \sqrt{10}}{2} \qquad \text{Simplify.}$$

The solutions are $\frac{4 + \sqrt{10}}{2}$ and $\frac{4 - \sqrt{10}}{2}$.

The following steps may be used to solve a quadratic equation such as $ax^2 + bx + c = 0$ by completing the square. This method may be used whether or not the polynomial $ax^2 + bx + c$ is factorable.

Solving a Quadratic Equation in x By Completing the Square

Step 1: If the coefficient of x^2 is 1, go to Step 2. Otherwise, divide both sides of the equation by the coefficient of x^2.

Step 2: Isolate all variable terms on one side of the equation.

Step 3: Complete the square for the resulting binomial by adding the square of half of the coefficient of x to both sides of the equation.

Step 4: Factor the resulting perfect square trinomial and write it as the square of a binomial.

Step 5: Use the square root property to solve for x.

EXAMPLE 8

Solve $3x^2 - 9x + 8 = 0$ by completing the square.

Solution $3x^2 - 9x + 8 = 0$

Step 1: $x^2 - 3x + \dfrac{8}{3} = 0$ Divide both sides of the equation by **3**.

Step 2: $x^2 - 3x = -\dfrac{8}{3}$ Subtract $\dfrac{8}{3}$ from both sides.

Since $\dfrac{1}{2}(-3) = -\dfrac{3}{2}$ and $\left(-\dfrac{3}{2}\right)^2 = \dfrac{9}{4}$, we add $\dfrac{9}{4}$ to both sides of the equation.

Step 3: $x^2 - 3x + \dfrac{9}{4} = -\dfrac{8}{3} + \dfrac{9}{4}$

Step 4: $\left(x - \dfrac{3}{2}\right)^2 = -\dfrac{5}{12}$ Factor the perfect square trinomial.

Step 5: $x - \dfrac{3}{2} = \pm\sqrt{-\dfrac{5}{12}}$ Apply the square root property.

$x - \dfrac{3}{2} = \pm\dfrac{i\sqrt{5}}{2\sqrt{3}}$ Simplify the radical.

$x - \dfrac{3}{2} = \pm\dfrac{i\sqrt{15}}{6}$ Rationalize the denominator.

$x = \dfrac{3}{2} \pm \dfrac{i\sqrt{15}}{6}$ Add $\dfrac{3}{2}$ to both sides.

$= \dfrac{9}{6} \pm \dfrac{i\sqrt{15}}{6}$ Find a common denominator.

$= \dfrac{9 \pm i\sqrt{15}}{6}$ Simplify.

The solutions are $\dfrac{9 + i\sqrt{15}}{6}$ and $\dfrac{9 - i\sqrt{15}}{6}$.

3 Recall the **simple interest** formula $I = Prt$, where I is the interest earned, P is the principal, r is the rate of interest, and t is time. If \$100 is invested at a simple interest rate of 5% annually, at the end of 3 years the total interest I earned is

$$I = P \cdot r \cdot t$$

or

$$I = 100 \cdot 0.05 \cdot 3 = \$15$$

and the new principal is

$$\$100 + \$15 = \$115$$

Most of the time, the interest computed on money borrowed or money deposited is **compound interest.** Compound interest, unlike simple interest, is computed on original principal *and* on interest already earned. To see the difference between simple interest and compound interest, suppose that \$100 is invested at a rate of 5% compounded annually. To find the total amount of money at the end of 3 years, we calculate as follows.

$$I = P \cdot r \cdot t$$

First year: Interest $= \$100 \cdot 0.05 \cdot 1 = \$5.00$
              New principal $= \$100.00 + \$5.00 = \$105.00$
Second year: Interest $= \$105.00 \cdot 0.05 \cdot 1 = \$5.25$
              New principal $= \$105.00 + \$5.25 = \$110.25$
Third year: Interest $= \$110.25 \cdot 0.05 \cdot 1 \approx \$5.51$
              New principal $= \$110.25 + \$5.51 = \$115.76$

At the end of the third year, the total compound interest earned is \$15.76, whereas the total simple interest earned is \$15.

It is tedious to calculate compound interest as we did above, so we use a compound interest formula. The formula for calculating the total amount of money when interest is compounded annually is

$$A = P(1 + r)^t$$

where P is the original investment, r is the interest rate per compounding period, and t is the number of periods. For example, the amount of money A at the end of 3 years if \$100 is invested at 5% compounded annually is

$$A = \$100(1 + 0.05)^3 \approx \$100(1.1576) = \$115.76$$

as we previously calculated.

EXAMPLE 9

FINDING INTEREST RATES

Find the interest rate r if \$2000 compounded annually grows to \$2420 in 2 years.

Solution

1. UNDERSTAND the problem. Since the \$2000 is compounded annually, we use the compound interest formula. For this example, make sure that you understand the formula for compounding interest annually.

2. TRANSLATE. We substitute the given values into the formula.

$$A = P(1 + r)^t$$

$$2420 = 2000(1 + r)^2 \quad \text{Let } A = 2420, P = 2000, \text{ and } t = 2.$$

3. SOLVE. Solve the equation for r.

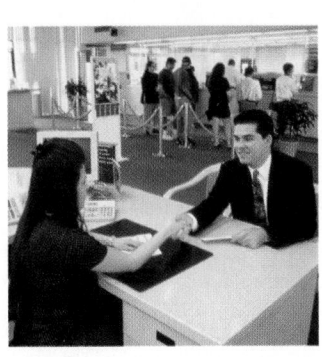

$$2420 = 2000(1 + r)^2$$

$$\frac{2420}{2000} = (1 + r)^2 \qquad \text{Divide both sides by \textbf{2000}.}$$

$$\frac{121}{100} = (1 + r)^2 \qquad \text{Simplify the fraction.}$$

$$\pm\sqrt{\frac{121}{100}} = 1 + r \qquad \text{Use the square root property.}$$

$$\pm\frac{11}{10} = 1 + r \qquad \text{Simplify.}$$

$$-1 \pm \frac{11}{10} = r$$

$$-\frac{10}{10} \pm \frac{11}{10} = r$$

$$\frac{1}{10} = r \quad \text{or} \quad -\frac{21}{10} = r$$

4. INTERPRET. The rate cannot be negative, so we reject $-\dfrac{21}{10}$.

Check: $\dfrac{1}{10} = 0.10 = 10\%$ per year. If we invest \$2000 at 10% compounded annually, in 2 years the amount in the account would be $2000(1 + 0.10)^2 = 2420$ dollars, the desired amount.

State: The interest rate is 10% compounded annually.

Graphing Calculator Explorations

In Section 6.5, we showed how we can use a grapher to approximate real number solutions of a quadratic equation written in standard form. We can also use a grapher to solve a quadratic equation when it is not written in standard form. For example, to solve $(x + 1)^2 = 12$, the quadratic equation in Example 3, we graph the following on the same set of axes. Use Xmin $= -10$, Xmax $= 10$, Ymin $= -13$, and Ymax $= 13$.

$$Y_1 = (x + 1)^2 \quad \text{and} \quad Y_2 = 12$$

Use the Intersect feature or the Zoom and Trace features to locate the points of intersection of the graphs. (See your manuals for specific instructions.) The x-values of these points are the solutions of $(x + 1)^2 = 12$. The solutions, rounded to two decimal places, are 2.46 and -4.46.

Check to see that these numbers are approximations of the exact solutions $-1 \pm 2\sqrt{3}$.

(continued)

Use a graphing calculator to solve each quadratic equation. Round all solutions to the nearest hundredth.

1. $x(x - 5) = 8$

2. $x(x + 2) = 5$

3. $x^2 + 0.5x = 0.3x + 1$

4. $x^2 - 2.6x = -2.2x + 3$

5. Use a grapher and solve $(2x - 5)^2 = -16$, Example 4 in this section, using the window

$$\text{Xmin} = -20$$

$$\text{Xmax} = 20$$

$$\text{Xscl} = 1$$

$$\text{Ymin} = -20$$

$$\text{Ymax} = 20$$

$$\text{Yscl} = 1$$

Explain the results. Compare your results with the solution found in Example 4.

6. What are the advantages and disadvantages of using a grapher to solve quadratic equations?

EXERCISE SET 11.1

STUDY GUIDE/SSM · CD/VIDEO · PH MATH TUTOR CENTER · MathXL®Tutorials ON CD · MathXL® · MyMathLab®

Use the square root property to solve each equation. These equations have real-number solutions. See Examples 1 through 3.

 1. $x^2 = 16$

2. $x^2 = 49$

3. $x^2 - 7 = 0$

4. $x^2 - 11 = 0$

5. $x^2 = 18$

6. $y^2 = 20$

7. $3z^2 - 30 = 0$

8. $2x^2 = 4$

9. $(x + 5)^2 = 9$

10. $(y - 3)^2 = 4$

11. $(z - 6)^2 = 18$

12. $(y + 4)^2 = 27$

13. $(2x - 3)^2 = 8$

14. $(4x + 9)^2 = 6$

Use the square root property to solve each equation. See Examples 1 through 4.

15. $x^2 + 9 = 0$

16. $x^2 + 4 = 0$

17. $x^2 - 6 = 0$

18. $y^2 - 10 = 0$

19. $2z^2 + 16 = 0$

20. $3p^2 + 36 = 0$

21. $(x - 1)^2 = -16$

22. $(y + 2)^2 = -25$

23. $(z + 7)^2 = 5$

24. $(x + 10)^2 = 11$

25. $(x + 3)^2 = -8$

26. $(y - 4)^2 = -18$

Add the proper constant to each binomial so that the resulting trinomial is a perfect square trinomial. Then factor the trinomial.

27. $x^2 + 16x +$ _____

28. $y^2 + 2y +$ _____

29. $z^2 - 12z +$ _____

30. $x^2 - 8x +$ _____

31. $p^2 + 9p +$ _____

32. $n^2 + 5n +$ _____

33. $x^2 + x +$ _____

34. $y^2 - y +$ _____

MIXED PRACTICE

Solve each equation by completing the square. These equations have real number solutions. See Examples 5 through 7.

35. $x^2 + 8x = -15$

36. $y^2 + 6y = -8$

37. $x^2 + 6x + 2 = 0$

38. $x^2 - 2x - 2 = 0$

39. $x^2 + x - 1 = 0$

40. $x^2 + 3x - 2 = 0$

41. $x^2 + 2x - 5 = 0$

42. $y^2 + y - 7 = 0$

43. $3p^2 - 12p + 2 = 0$

44. $2x^2 + 14x - 1 = 0$

45. $4y^2 - 12y - 2 = 0$

46. $6x^2 - 3 = 6x$

47. $2x^2 + 7x = 4$

48. $3x^2 - 4x = 4$

49. $x^2 - 4x - 5 = 0$

50. $y^2 + 6y - 8 = 0$

51. $x^2 + 8x + 1 = 0$

52. $x^2 - 10x + 2 = 0$

53. $3y^2 + 6y - 4 = 0$

54. $2y^2 + 12y + 3 = 0$

55. $2x^2 - 3x - 5 = 0$

56. $5x^2 + 3x - 2 = 0$

Solve each equation by completing the square. See Examples 5 through 8.

57. $y^2 + 2y + 2 = 0$

58. $x^2 + 4x + 6 = 0$

59. $x^2 - 6x + 3 = 0$

60. $x^2 - 7x - 1 = 0$

61. $2a^2 + 8a = -12$

62. $3x^2 + 12x = -14$

63. $5x^2 + 15x - 1 = 0$

64. $16y^2 + 16y - 1 = 0$

65. $2x^2 - x + 6 = 0$

66. $4x^2 - 2x + 5 = 0$

67. $x^2 + 10x + 28 = 0$

68. $y^2 + 8y + 18 = 0$

69. $z^2 + 3z - 4 = 0$

70. $y^2 + y - 2 = 0$

71. $2x^2 - 4x + 3 = 0$

72. $9x^2 - 36x = -40$

73. $3x^2 + 3x = 5$

74. $5y^2 - 15y = 1$

Use the formula $A = P(1 + r)^t$ to solve Exercises 75 through 78. See Example 9.

75. Find the rate r at which \$3000 grows to \$4320 in 2 years.

76. Find the rate r at which \$800 grows to \$882 in 2 years.

77. Find the rate at which \$810 grows to \$1000 in 2 years.

78. Find the rate at which \$2000 grows to \$2880 in 2 years.

79. In your own words, what is the difference between simple interest and compound interest?

80. If you are depositing money in an account that pays 4%, would you prefer the interest to be simple or compound? Explain why.

81. If you are borrowing money at a rate of 10%, would you prefer the interest to be simple or compound? Explain why.

REVIEW AND PREVIEW

Simplify each expression. See Section 10.1.

82. $\dfrac{3}{4} - \sqrt{\dfrac{25}{16}}$

83. $\dfrac{3}{5} + \sqrt{\dfrac{16}{25}}$

84. $\dfrac{1}{2} - \sqrt{\dfrac{9}{4}}$

85. $\dfrac{9}{10} - \sqrt{\dfrac{49}{100}}$

Simplify each expression. See Section 10.5.

86. $\dfrac{6 + 4\sqrt{5}}{2}$

87. $\dfrac{10 - 20\sqrt{3}}{2}$

88. $\dfrac{3 - 9\sqrt{5}}{6}$

89. $\dfrac{12 - 8\sqrt{7}}{16}$

Evaluate $\sqrt{b^2 - 4ac}$ for each set of values. See Section 10.3.

90. $a = 2, b = 4, c = -1$

91. $a = 1, b = 6, c = 2$

92. $a = 3, b = -1, c = -2$

93. $a = 1, b = -3, c = -1$

Concept Extensions

Find two possible missing terms so that each is a perfect square trinomial.

94. $x^2 +$ ___ $+ 16$

95. $y^2 +$ ___ $+ 9$

96. $z^2 +$ ___ $+ \dfrac{25}{4}$

97. $x^2 +$ ___ $+ \dfrac{1}{4}$

Neglecting air resistance, the distance $s(t)$ in feet traveled by a freely falling object is given by the function $s(t) = 16t^2$, where t is time in seconds. Use this formula to solve Exercises 98 through 101. Round answers to two decimal places.

98. The Petronas Towers in Kuala Lumpur, built in 1997, are the tallest buildings in Malaysia. Each tower is 1483 feet tall. How long would it take an object to fall to the ground from the top of one of the towers? (*Source:* Council on Tall Buildings and Urban Habitat, Lehigh University)

99. The height of the Chicago Beach Tower Hotel, built in 1998 in Dubai, United Arab Emirates, is 1053 feet. How long would it take an object to fall to the ground from the top of the building? (*Source:* Council on Tall Buildings and Urban Habitat, Lehigh University)

100. The height of the Nurek Dam in Tajikistan (part of the former USSR that borders Afghanistan) is 984 feet. How long would it take an object to fall from the top to the base of the dam? (*Source:* U.S. Committee on Large Dams of the International Commission on Large Dams)

101. The Hoover Dam, located on the Colorado River on the border of Nevada and Arizona near Las Vegas, is 725 feet tall. How long would it take an object to fall from the top to the base of the dam? (*Source:* U.S. Committee on Large Dams of the International Commission on Large Dams)

Solve.

△ **102.** The area of a square room is 225 square feet. Find the dimensions of the room.

△ **103.** The area of a circle is 36π square inches. Find the radius of the circle.

△ **104.** An isosceles right triangle has legs of equal length. If the hypotenuse is 20 centimeters long, find the length of each leg.

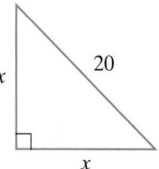

△ **105.** A 27-inch TV is advertised in the *Daily Sentry* newspaper. If 27 inches is the measure of the diagonal of the picture tube, find the measure of each side of the picture tube.

A common equation used in business is a demand equation. It expresses the relationship between the unit price of some commodity and the quantity demanded. For Exercises 106 and 107, p represents the unit price and x represents the quantity demanded in thousands.

106. A manufacturing company has found that the demand equation for a certain type of scissors is given by the equation $p = -x^2 + 47$. Find the demand for the scissors if the price is $11 per pair.

107. Acme, Inc., sells desk lamps and has found that the demand equation for a certain style of desk lamp is given by the equation $p = -x^2 + 15$. Find the demand for the desk lamp if the price is $7 per lamp.

11.2 SOLVING QUADRATIC EQUATIONS BY THE QUADRATIC FORMULA

Objectives

1 Solve quadratic equations by using the quadratic formula.

2 Determine the number and type of solutions of a quadratic equation by using the discriminant.

3 Solve geometric problems modeled by quadratic equations.

1 Any quadratic equation can be solved by completing the square. Since the same sequence of steps is repeated each time we complete the square, let's complete the square for a general quadratic equation, $ax^2 + bx + c = 0, a \neq 0$. By doing so, we find a pattern for the solutions of a quadratic equation known as the **quadratic formula.**

Recall that to complete the square for an equation such as $ax^2 + bx + c = 0$, we first divide both sides by the coefficient of x^2.

$$ax^2 + bx + c = 0$$

$$x^2 + \frac{b}{a}x + \frac{c}{a} = 0 \qquad \text{Divide both sides by } a, \text{ the coefficient of } x^2.$$

$$x^2 + \frac{b}{a}x = -\frac{c}{a} \qquad \text{Subtract the constant } \frac{c}{a} \text{ from both sides.}$$

Next, find the square of half $\frac{b}{a}$, the coefficient of x.

$$\frac{1}{2}\left(\frac{b}{a}\right) = \frac{b}{2a} \quad \text{and} \quad \left(\frac{b}{2a}\right)^2 = \frac{b^2}{4a^2}$$

Add this result to both sides of the equation.

$$x^2 + \frac{b}{a}x + \frac{b^2}{4a^2} = -\frac{c}{a} + \frac{b^2}{4a^2} \qquad \text{Add } \frac{b^2}{4a^2} \text{ to both sides.}$$

$$x^2 + \frac{b}{a}x + \frac{b^2}{4a^2} = \frac{-c \cdot 4a}{a \cdot 4a} + \frac{b^2}{4a^2} \qquad \text{Find a common denominator on the right side.}$$

$$x^2 + \frac{b}{a}x + \frac{b^2}{4a^2} = \frac{b^2 - 4ac}{4a^2} \qquad \text{Simplify the right side.}$$

$$\left(x + \frac{b}{2a}\right)^2 = \frac{b^2 - 4ac}{4a^2} \qquad \text{Factor the perfect square trinomial on the left side.}$$

$$x + \frac{b}{2a} = \pm\sqrt{\frac{b^2 - 4ac}{4a^2}} \qquad \text{Apply the square root property.}$$

$$x + \frac{b}{2a} = \pm\frac{\sqrt{b^2 - 4ac}}{2a} \qquad \text{Simplify the radical.}$$

$$x = -\frac{b}{2a} \pm \frac{\sqrt{b^2 - 4ac}}{2a} \qquad \text{Subtract } \frac{b}{2a} \text{ from both sides.}$$

$$x = \frac{-b \pm \sqrt{b^2 - 4ac}}{2a} \qquad \text{Simplify.}$$

This equation identifies the solutions of the general quadratic equation in standard form and is called the quadratic formula. It can be used to solve any equation written in standard form $ax^2 + bx + c = 0$ as long as a is not 0.

Quadratic Formula

A quadratic equation written in the form $ax^2 + bx + c = 0$ has the solutions

$$x = \frac{-b \pm \sqrt{b^2 - 4ac}}{2a}$$

EXAMPLE 1

Solve $3x^2 + 16x + 5 = 0$ for x.

Solution This equation is in standard form, so $a = 3$, $b = 16$, and $c = 5$. Substitute these values into the quadratic formula.

$$x = \frac{-b \pm \sqrt{b^2 - 4ac}}{2a} \qquad \text{Quadratic formula.}$$

$$= \frac{-16 \pm \sqrt{16^2 - 4(3)(5)}}{2 \cdot 3} \qquad \text{Use } a = 3, b = 16, \text{ and } c = 5.$$

$$= \frac{-16 \pm \sqrt{256 - 60}}{6}$$

$$= \frac{-16 \pm \sqrt{196}}{6} = \frac{-16 \pm 14}{6}$$

$$x = \frac{-16 + 14}{6} = -\frac{1}{3} \quad \text{or} \quad x = \frac{-16 - 14}{6} = -\frac{30}{6} = -5$$

The solutions are $-\dfrac{1}{3}$ and -5, or the solution set is $\left\{-\frac{1}{3}, -5\right\}$.

EXAMPLE 2

Solve $2x^2 - 4x = 3$.

Solution First write the equation in standard form by subtracting 3 from both sides.

$$2x^2 - 4x - 3 = 0$$

Helpful Hint

To replace a, b, and c correctly in the quadratic formula, write the quadratic equation in standard form $ax^2 + bx + c = 0$.

Now $a = 2$, $b = -4$, and $c = -3$. Substitute these values into the quadratic formula.

$$x = \frac{-b \pm \sqrt{b^2 - 4ac}}{2a}$$

$$= \frac{-(-4) \pm \sqrt{(-4)^2 - 4(2)(-3)}}{2 \cdot 2}$$

$$= \frac{4 \pm \sqrt{16 + 24}}{4}$$

$$= \frac{4 \pm \sqrt{40}}{4} = \frac{4 \pm 2\sqrt{10}}{4}$$

$$= \frac{2(2 \pm \sqrt{10})}{2 \cdot 2} = \frac{2 \pm \sqrt{10}}{2}$$

The solutions are $\dfrac{2 + \sqrt{10}}{2}$ and $\dfrac{2 - \sqrt{10}}{2}$, or the solution set is $\left\{ \dfrac{2 - \sqrt{10}}{2}, \dfrac{2 + \sqrt{10}}{2} \right\}$.

> ### Helpful Hint
>
> To simplify the expression $\dfrac{4 \pm 2\sqrt{10}}{4}$ in the preceding example, note that 2 is factored out of both terms of the numerator *before* simplifying.
>
> $$\frac{4 \pm 2\sqrt{10}}{4} = \frac{2(2 \pm \sqrt{10})}{2 \cdot 2} = \frac{2 \pm \sqrt{10}}{2}$$

✔ **CONCEPT CHECK**

For the quadratic equation $x^2 = 7$, which substitution is correct?

a. $a = 1, b = 0,$ and $c = -7$

b. $a = 1, b = 0,$ and $c = 7$

c. $a = 0, b = 0,$ and $c = 7$

d. $a = 1, b = 1,$ and $c = -7$

EXAMPLE 3

Solve $\dfrac{1}{4}m^2 - m + \dfrac{1}{2} = 0$.

Solution   We could use the quadratic formula with $a = \dfrac{1}{4}, b = -1,$ and $c = \dfrac{1}{2}$. Instead, we find a simpler, equivalent standard form equation whose coefficients are not fractions.

Multiply both sides of the equation by 4 to clear fractions.

$$4\left(\frac{1}{4}m^2 - m + \frac{1}{2} \right) = 4 \cdot 0$$

$$m^2 - 4m + 2 = 0 \qquad \text{Simplify.}$$

Substitute $a = 1, b = -4,$ and $c = 2$ into the quadratic formula and simplify.

$$m = \frac{-(-4) \pm \sqrt{(-4)^2 - 4(1)(2)}}{2 \cdot 1} = \frac{4 \pm \sqrt{16 - 8}}{2}$$

$$= \frac{4 \pm \sqrt{8}}{2} = \frac{4 \pm 2\sqrt{2}}{2} = \frac{2(2 \pm \sqrt{2})}{2}$$

$$= 2 \pm \sqrt{2}$$

Concept Check Answer:
a

The solutions are $2 + \sqrt{2}$ and $2 - \sqrt{2}$.

EXAMPLE 4

Solve $x = -3x^2 - 3$.

Solution The equation in standard form is $3x^2 + x + 3 = 0$. Thus, let $a = 3$, $b = 1$, and $c = 3$ in the quadratic formula.

$$x = \frac{-1 \pm \sqrt{1^2 - 4(3)(3)}}{2 \cdot 3} = \frac{-1 \pm \sqrt{1 - 36}}{6} = \frac{-1 \pm \sqrt{-35}}{6} = \frac{-1 \pm i\sqrt{35}}{6}$$

The solutions are $\dfrac{-1 + i\sqrt{35}}{6}$ and $\dfrac{-1 - i\sqrt{35}}{6}$.

✔ **CONCEPT CHECK**

What is the first step in solving $-3x^2 = 5x - 4$ using the quadratic formula?

In Example 1, the equation $3x^2 + 16x + 5 = 0$ had 2 real roots, $-\dfrac{1}{3}$ and -5. In Example 4, the equation $3x^2 + x + 3 = 0$ (written in standard form) had no real roots. How do their related graphs compare? Recall that the x-intercepts of $f(x) = 3x^2 + 16x + 5$ occur where $f(x) = 0$ or where $3x^2 + 16x + 5 = 0$. Since this equation has 2 real roots, the graph has 2 x-intercepts. Similarly, since the equation $3x^2 + x + 3 = 0$ has no real roots, the graph of $f(x) = 3x^2 + x + 3$ has no x-intercepts.

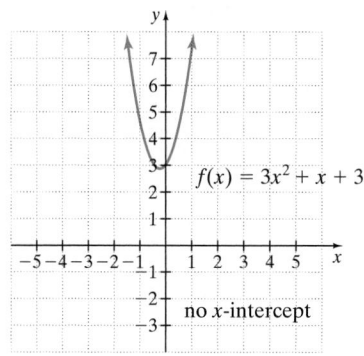

2 In the quadratic formula, $x = \dfrac{-b \pm \sqrt{b^2 - 4ac}}{2a}$, the radicand $b^2 - 4ac$ is called the **discriminant** because, by knowing its value, we can **discriminate** among the possible number and type of solutions of a quadratic equation. Possible values of the discriminant and their meanings are summarized next.

Discriminant

The following table corresponds the discriminant $b^2 - 4ac$ of a quadratic equation of the form $ax^2 + bx + c = 0$ with the number and type of solutions of the equation.

$b^2 - 4ac$	Number and Type of Solutions
Positive	Two real solutions
Zero	One real solution
Negative	Two complex but not real solutions

EXAMPLE 5

Use the discriminant to determine the number and type of solutions of each quadratic equation.

a. $x^2 + 2x + 1 = 0$ **b.** $3x^2 + 2 = 0$ **c.** $2x^2 - 7x - 4 = 0$

Solution

a. In $x^2 + 2x + 1 = 0$, $a = 1$, $b = 2$, and $c = 1$. Thus,

$$b^2 - 4ac = 2^2 - 4(1)(1) = 0$$

Since $b^2 - 4ac = 0$, this quadratic equation has one real solution.

b. In this equation, $a = 3$, $b = 0$, $c = 2$. Then $b^2 - 4ac = 0 - 4(3)(2) = -24$. Since $b^2 - 4ac$ is negative, the quadratic equation has two complex but not real solutions.

c. In this equation, $a = 2$, $b = -7$, and $c = -4$. Then

$$b^2 - 4ac = (-7)^2 - 4(2)(-4) = 81$$

Since $b^2 - 4ac$ is positive, the quadratic equation has two real solutions.

The discriminant helps us determine the number and type of solutions of a quadratic equation, $ax^2 + bx + c = 0$. Recall that the solutions of this equation are the same as the x-intercepts of its related graph $f(x) = ax^2 + bx + c$. This means that the discriminant of $ax^2 + bx + c = 0$ also tells us the number of x-intercepts for the graph of $f(x) = ax^2 + bx + c$, or equivalently $y = ax^2 + bx + c$.

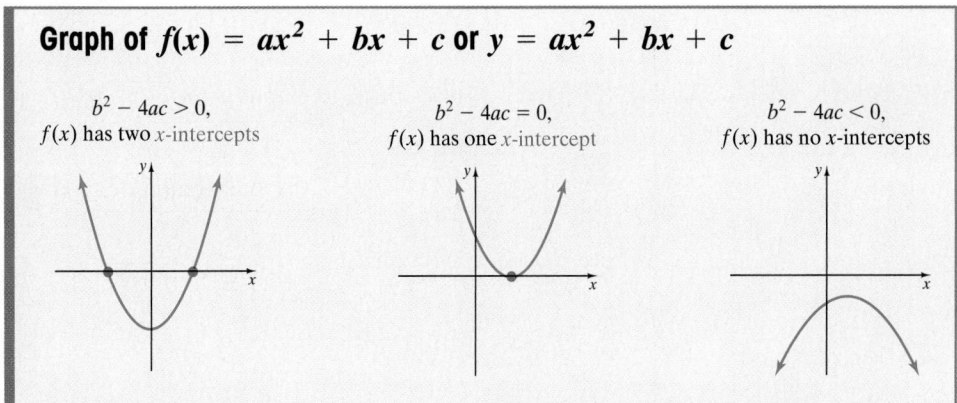

Graph of $f(x) = ax^2 + bx + c$ or $y = ax^2 + bx + c$

| $b^2 - 4ac > 0$, $f(x)$ has two x-intercepts | $b^2 - 4ac = 0$, $f(x)$ has one x-intercept | $b^2 - 4ac < 0$, $f(x)$ has no x-intercepts |

3   The quadratic formula is useful in solving problems that are modeled by quadratic equations.

⚠️ **EXAMPLE 6**

CALCULATING DISTANCE SAVED

At a local university, students often leave the sidewalk and cut across the lawn to save walking distance. Given the diagram below of a favorite place to cut across the lawn, approximate how many feet of walking distance a student saves by cutting across the lawn instead of walking on the sidewalk.

Solution

1. UNDERSTAND. Read and reread the problem. In the diagram, notice that a triangle is formed. Since the corner of the block forms a right angle, we use the Pythagorean theorem for right triangles. You may want to review this theorem.

2. TRANSLATE. By the Pythagorean theorem, we have

   In words: $(\text{leg})^2 + (\text{leg})^2 = (\text{hypotenuse})^2$

   Translate: $x^2 + (x + 20)^2 = 50^2$

3. SOLVE. Use the quadratic formula to solve.

   $$x^2 + x^2 + 40x + 400 = 2500 \quad \text{Square } (x + 20) \text{ and } 50.$$

   $$2x^2 + 40x - 2100 = 0 \quad \text{Set the equation equal to 0.}$$

   $$x^2 + 20x - 1050 = 0 \quad \text{Divide by 2.}$$

   Here, $a = 1, b = 20, c = -1050$. By the quadratic formula,

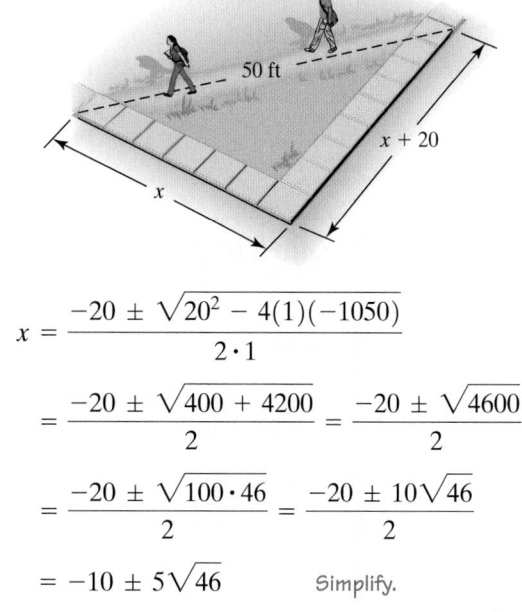
50 ft

$x + 20$

x

   $$x = \frac{-20 \pm \sqrt{20^2 - 4(1)(-1050)}}{2 \cdot 1}$$

   $$= \frac{-20 \pm \sqrt{400 + 4200}}{2} = \frac{-20 \pm \sqrt{4600}}{2}$$

   $$= \frac{-20 \pm \sqrt{100 \cdot 46}}{2} = \frac{-20 \pm 10\sqrt{46}}{2}$$

   $$= -10 \pm 5\sqrt{46} \quad \text{Simplify.}$$

4. INTERPRET.

   Check: Your calculations in the quadratic formula. The length of a side of a triangle can't be negative, so we reject $-10 - 5\sqrt{46}$. Since $-10 + 5\sqrt{46} \approx 24$ feet, the walking distance along the sidewalk is

   $$x + (x + 20) \approx 24 + (24 + 20) = 68 \text{ feet.}$$

   State: A student saves about $68 - 50$ or 18 feet of walking distance by cutting across the lawn.

EXAMPLE 7

CALCULATING LANDING TIME

An object is thrown upward from the top of a 200-foot cliff with a velocity of 12 feet per second. The height h in feet of the object after t seconds is

$$h = -16t^2 + 12t + 200$$

How long after the object is thrown will it strike the ground? Round to the nearest tenth of a second.

200 ft

Solution   1. UNDERSTAND. Read and reread the problem.

2. TRANSLATE. Since we want to know when the object strikes the ground, we want to know when the height $h = 0$, or

$$0 = -16t^2 + 12t + 200$$

3. SOLVE. First we divide both sides of the equation by -4.

$$0 = 4t^2 - 3t - 50 \qquad \text{Divide both sides by } -4.$$

Here, $a = 4$, $b = -3$, and $c = -50$. By the quadratic formula,

$$t = \frac{-(-3) \pm \sqrt{(-3)^2 - 4(4)(-50)}}{2 \cdot 4}$$

$$= \frac{3 \pm \sqrt{9 + 800}}{8}$$

$$= \frac{3 \pm \sqrt{809}}{8}$$

4. INTERPRET.

Check: We check our calculations from the quadratic formula. Since the time won't be negative, we reject the proposed solution

$$\frac{3 - \sqrt{809}}{8}.$$

State: The time it takes for the object to strike the ground is exactly

$$\frac{3 + \sqrt{809}}{8} \text{ seconds} \approx 3.9 \text{ seconds}.$$

STUDY SKILLS REMINDER

Continue your outline started in Section 2.3. For this section, you only need to expand on your earlier section – **I.C.**– on solving quadratic equations. For example:

Solving Equations and Inequalities

I. Equations

   A. Linear equations (Sec. 2.3)

   B. Quadratic and Higher Degree Equations—

      1. If the polynomial is of the form $(ax + b)^2 = c$, use the square root property and solve. If not, got to Step 2.

      2. Write the equation in standard form (one side is 0.)

      3. If the polynomial on one side factors, solve by factoring.

      4. If the polynomial does not factor, solve by the quadratic formula.

   C. Equations with rational expressions (Sec. 7.5)

   D. Absolute value equations (Sec. 9.2)

   E. Equations with radicals (Sec. 10.6)

II. Inequalities

   A. Linear inequalities (Sec. 2.7)

   B. Compound inequalities (Sec. 9.1)

   C. Absolute value inequalities (Sec. 9.3)

Spotlight on DECISION MAKING

Suppose you are a registered dietician. Recently, you read an article in a nutrition journal that described a relationship between weight and the Recommended Dietary Allowance (RDA) for vitamin A in children up to age 10. The relationship is $y = 0.149x^2 - 4.475x + 406.478$, where y is the RDA for vitamin A in micrograms for a child whose weight is x pounds. (*Source:* Food and Nutrition Board, National Academy of Sciences—Institute of Medicine, 1989)

    You are working with a 4-year-old patient who weighs 40 pounds. After analyzing her diet, you are able to determine that she is currently getting an average of 400 micrograms of vitamin A daily. Decide whether her current vitamin A intake is adequate. If not, how much more is needed each day? In either case, determine how much weight she will need to gain before a daily intake of 500 micrograms of vitamin A is appropriate.

MENTAL MATH

Identify the values of $a, b,$ and c in each quadratic equation.

1. $x^2 + 3x + 1 = 0$ **2.** $2x^2 - 5x - 7 = 0$ **3.** $7x^2 - 4 = 0$

4. $x^2 + 9 = 0$ **5.** $6x^2 - x = 0$ **6.** $5x^2 + 3x = 0$

EXERCISE SET 11.2

STUDY GUIDE/SSM   CD/VIDEO   PH MATH TUTOR CENTER   MathXL®Tutorials ON CD   MathXL®   MyMathLab®

Use the quadratic formula to solve each equation. These equations have real number solutions. See Examples 1 through 3.

1. $m^2 + 5m - 6 = 0$

2. $p^2 + 11p - 12 = 0$

3. $2y = 5y^2 - 3$

4. $5x^2 - 3 = 14x$

5. $x^2 - 6x + 9 = 0$

6. $y^2 + 10y + 25 = 0$

7. $x^2 + 7x + 4 = 0$

8. $y^2 + 5y + 3 = 0$

9. $8m^2 - 2m = 7$

10. $11n^2 - 9n = 1$

11. $3m^2 - 7m = 3$

12. $x^2 - 13 = 5x$

13. $\frac{1}{2}x^2 - x - 1 = 0$

14. $\frac{1}{6}x^2 + x + \frac{1}{3} = 0$

15. $\frac{2}{5}y^2 + \frac{1}{5}y = \frac{3}{5}$

16. $\frac{1}{8}x^2 + x = \frac{5}{2}$

17. $\frac{1}{3}y^2 - y - \frac{1}{6} = 0$

18. $\frac{1}{2}y^2 = y + \frac{1}{2}$

19. Solve Exercise 1 by factoring. Explain the result.

20. Solve Exercise 2 by factoring. Explain the result.

Use the quadratic formula to solve each equation. See Example 4.

21. $6 = -4x^2 + 3x$

22. $9x^2 + x + 2 = 0$

23. $(x + 5)(x - 1) = 2$

24. $x(x + 6) = 2$

25. $10y^2 + 10y + 3 = 0$

26. $3y^2 + 6y + 5 = 0$

The solutions of the quadratic equation $ax^2 + bx + c = 0$ are

$$\frac{-b + \sqrt{b^2 - 4ac}}{2a} \quad and \quad \frac{-b - \sqrt{b^2 - 4ac}}{2a}.$$

Use the discriminant to determine the number and type of solutions of each equation. See Example 5.

27. $9x - 2x^2 + 5 = 0$

28. $5 - 4x + 12x^2 = 0$

29. $4x^2 + 12x = -9$

30. $9x^2 + 1 = 6x$

31. $3x = -2x^2 + 7$

32. $3x^2 = 5 - 7x$

33. $6 = 4x - 5x^2$

34. $8x = 3 - 9x^2$

MIXED PRACTICE

Use the quadratic formula to solve each equation. These equations have real number solutions.

35. $x^2 + 5x = -2$

36. $y^2 - 8 = 4y$

37. $(m + 2)(2m - 6) = 5(m - 1) - 12$

38. $7p(p - 2) + 2(p + 4) = 3$

39. $\frac{x^2}{3} - x = \frac{5}{3}$

40. $\frac{x^2}{2} - 3 = -\frac{9}{2}x$

41. $x(6x + 2) - 3 = 0$

42. $x(7x + 1) = 2$

Use the quadratic formula to solve each equation.

43. $x^2 + 6x + 13 = 0$

44. $x^2 + 2x + 2 = 0$

 45. $\frac{2}{5}y^2 + \frac{1}{5}y + \frac{3}{5} = 0$

46. $\frac{1}{8}x^2 + x + \frac{5}{2} = 0$

47. $\frac{1}{2}y^2 = y - \frac{1}{2}$

48. $\frac{2}{3}x^2 - \frac{20}{3}x = -\frac{100}{6}$

49. $(n - 2)^2 = 15n$

50. $\left(p - \frac{1}{2}\right)^2 = \frac{p}{2}$

Solve. See Examples 6 and 7.

51. Nancy, Thelma, and John Varner live on a corner lot. Often, neighborhood children cut across their lot to save walking distance. Given the diagram below, approximate to the nearest foot how many feet of walking distance is saved by cutting across their property instead of walking around the lot.

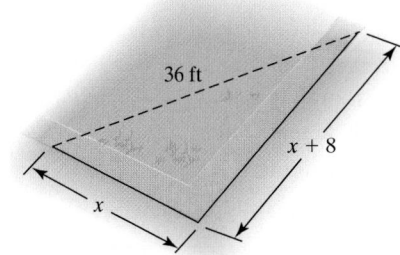

36 ft

$x + 8$

x

△ **52.** Given the diagram below, approximate to the nearest foot how many feet of walking distance a person saves by cutting across the lawn instead of walking on the sidewalk.

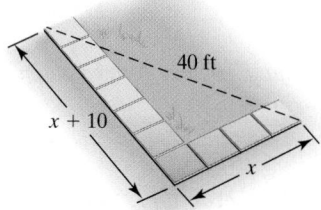

△ **53.** The hypotenuse of an isosceles right triangle is 2 centimeters longer than either of its legs. Find the exact length of each side. (*Hint:* An isosceles right triangle is a right triangle whose legs are the same length.)

△ **54.** The hypotenuse of an isosceles right triangle is one meter longer than either of its legs. Find the length of each side.

△ **55.** Uri Chechov's rectangular dog pen for his Irish setter must have an area of 400 square feet. Also, the length must be 10 feet longer than the width. Find the dimensions of the pen.

△ **56.** An entry in the Peach Festival Poster Contest must be rectangular and have an area of 1200 square inches. Furthermore, its length must be 20 inches longer than its width. Find the dimensions each entry must have.

△ **57.** A holding pen for cattle must be square and have a diagonal length of 100 meters.
   a. Find the length of a side of the pen.
   b. Find the area of the pen.

△ **58.** A rectangle is three times longer than it is wide. It has a diagonal of length 50 centimeters.
   a. Find the dimensions of the rectangle.
   b. Find the perimeter of the rectangle.

△ **59.** The heaviest reported door in the world is the 708.6 ton radiation shield door in the National Institute for Fusion Science at Toki, Japan. If the height of the door is 1.1 feet longer than its width, and its front area (neglecting depth) is 1439.9 square feet,

find its width and height [Interesting note: the door is 6.6 feet thick.] (Source: *Guiness World Records*.)

△ **60.** Christi and Robbie Wegmann are constructing a rectangular stained glass window whose length is 7.3 inches longer than its width. If the area of the window is 569.9 square inches, find its width and length.

61. If a point *B* divides a line segment such that the smaller portion is to the larger portion as the larger is to the whole, the whole is the length of the *golden ratio*.

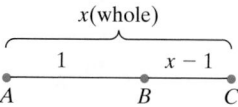

The golden ratio was thought by the Greeks to be the most pleasing to the eye, and many of their buildings contained numerous examples of the golden ratio. The value of the golden ratio is the positive solution of

$$\begin{matrix} \text{(smaller)} \\ \text{(larger)} \end{matrix} \quad \frac{x-1}{1} = \frac{1}{x} \quad \begin{matrix} \text{(larger)} \\ \text{(whole)} \end{matrix}$$

Find this value.

△ **62.** The base of a triangle is four more than twice its height. If the area of the triangle is 42 square centimeters, find its base and height.

The Wollomombi Falls in Australia have a height of 1100 feet. A pebble is thrown upward from the top of the falls with an initial velocity of 20 feet per second. The height of the pebble h after t seconds is given by the equation $h = -16t^2 + 20t + 1100$. Use this equation for Exercises 63 and 64.

63. How long after the pebble is thrown will it hit the ground? Round to the nearest tenth of a second.

64. How long after the pebble is thrown will it be 550 feet from the ground? Round to the nearest tenth of a second.

A ball is thrown downward from the top of a 180-foot building with an initial velocity of 20 feet per second. The height of the ball h after t seconds is given by the equation $h = -16t^2 - 20t + 180$. Use this equation to answer Exercises 65 and 66.

180 ft

50 ft

65. How long after the ball is thrown will it strike the ground? Round the result to the nearest tenth of a second.

66. How long after the ball is thrown will it be 50 feet from the ground? Round the result to the nearest tenth of a second.

REVIEW AND PREVIEW

Solve each equation. See Sections 7.5 and 10.6.

67. $\sqrt{5x - 2} = 3$

68. $\sqrt{y + 2} + 7 = 12$

69. $\dfrac{1}{x} + \dfrac{2}{5} = \dfrac{7}{x}$

70. $\dfrac{10}{z} = \dfrac{5}{z} - \dfrac{1}{3}$

Factor. See Section 11.3.

71. $x^4 + x^2 - 20$

72. $2y^4 + 11y^2 - 6$

73. $z^4 - 13z^2 + 36$

74. $x^4 - 1$

Concept Extensions

Use the quadratic formula and a calculator to approximate each solution to the nearest tenth.

75. $2x^2 - 6x + 3 = 0$

76. $3.6x^2 + 1.8x - 4.3 = 0$

The accompanying graph shows the daily low temperatures for one week in New Orleans, Louisiana.

77. Which day of the week shows the greatest decrease in low temperature?

78. Which day of the week shows the greatest increase in low temperature?

79. Which day of the week had the lowest low temperature?

80. Use the graph to estimate the low temperature on Thursday.

Notice that the shape of the temperature graph is similar to the curve drawn. In fact, this graph can be modeled by the quadratic function $f(x) = 3x^2 - 18x + 56$, where f(x) is the temperature in degrees Fahrenheit and x is the number of days from Sunday. (This graph is shown in red.) Use this function to answer Exercises 81 and 82.

81. Use the quadratic function given to approximate the temperature on Thursday. Does your answer agree with the graph above?

82. Use the function given and the quadratic formula to find when the temperature was 35° F. [*Hint:* Let $f(x) = 35$ and solve for *x*.] Round your answer to one decimal place and interpret your result. Does your answer agree with the graph above?

83. Wal-Mart Stores' net income can be modeled by the quadratic function $f(x) = 112.5x^2 + 498.7x + 5454$, where f(x) is net income in millions of dollars and x is the number of years after 2000. (*Source:* Based on data from Wal-Mart Stores, Inc.)

  a. Find Wal-Mart's net income in 2002.

  b. If the trend described by the model continues, predict the year after 2000 in which Wal-Mart's net income will be $15,000 million. Round to the nearest whole year.

84. The number of inmates in custody in U.S. prisons and jails can be modeled by the quadratic function $p(x) = -3892.4x^2 + 91{,}152x + 1{,}576{,}254$ where p(x) is the number of inmates and x is the number of years after 1995. (*Source:* Based on data from the Bureau of Justice Statistics, U.S. Department of Justice, 1995–2002) Round **a** and **b** to the nearest ten thousand.

  a. Find the number of prison inmates in the United States in 2000.

  b. Find the number of prison inmates in the United States in 2002.

The solutions of the quadratic equation $ax^2 + bx + c = 0$ are $\dfrac{-b + \sqrt{b^2 - 4ac}}{2a}$ and $\dfrac{-b - \sqrt{b^2 - 4ac}}{2a}$.

85. Show that the sum of these solutions is $\dfrac{-b}{a}$.

86. Show that the product of these solutions is $\dfrac{c}{a}$.

Use the quadratic formula to solve each quadratic equation.

87. $3x^2 - \sqrt{12}x + 1 = 0$,

  (*Hint:* $a = 3, b = -\sqrt{12}, c = 1$)

88. $5x^2 + \sqrt{20}x + 1 = 0$

89. $x^2 + \sqrt{2}x + 1 = 0$

90. $x^2 - \sqrt{2}x + 1 = 0$

91. $2x^2 - \sqrt{3}x - 1 = 0$

92. $7x^2 + \sqrt{7}x - 2 = 0$

93. Use a graphing calculator to solve Exercises 63 and 65.

94. Use a graphing calculator to solve Exercises 64 and 66.

Recall that the discriminant also tells us the number of x-intercepts of the related function.

95. Check the results of Exercise 27 by graphing $y = 9x - 2x^2 + 5$.

96. Check the results of Exercise 28 by graphing $y = 5 - 4x + 12x^2$.

11.3   SOLVING EQUATIONS BY USING QUADRATIC METHODS

Objectives

1 Solve various equations that are quadratic in form.

2 Solve problems that lead to quadratic equations.

1   In this section, we discuss various types of equations that can be solved in part by using the methods for solving quadratic equations.

Once each equation is simplified, you may want to use these steps when deciding what method to use to solve the quadratic equation.

Solving a Quadratic Equation

Step 1: If the equation is in the form $(ax + b)^2 = c$, use the square root property and solve. If not, go to Step 2.

Step 2: Write the equation in standard form: $ax^2 + bx + c = 0$.

Step 3: Try to solve the equation by the factoring method. If not possible, go to Step 4.

Step 4: Solve the equation by the quadratic formula.

The first example is a radical equation that becomes a quadratic equation once we square both sides.

EXAMPLE 1

Solve $x - \sqrt{x} - 6 = 0$.

Solution

Recall that to solve a radical equation, first get the radical alone on one side of the equation. Then square both sides.

$$x - 6 = \sqrt{x} \quad \text{Add } \sqrt{x} \text{ to both sides.}$$
$$(x - 6)^2 = \left(\sqrt{x}\right)^2 \quad \text{Square both sides.}$$
$$x^2 - 12x + 36 = x$$

$$x^2 - 13x + 36 = 0 \qquad \textit{Set the equation equal to 0.}$$

$$(x - 9)(x - 4) = 0$$

$$x - 9 = 0 \quad \text{or} \quad x - 4 = 0$$

$$x = 9 \qquad\qquad x = 4$$

Check

$$\text{Let } x = 9 \qquad\qquad\qquad \text{Let } x = 4$$

$$x - \sqrt{x} - 6 = 0 \qquad\qquad x - \sqrt{x} - 6 = 0$$

$$9 - \sqrt{9} - 6 \overset{?}{=} 0 \qquad\qquad 4 - \sqrt{4} - 6 \overset{?}{=} 0$$

$$9 - 3 - 6 \overset{?}{=} 0 \qquad\qquad 4 - 2 - 6 \overset{?}{=} 0$$

$$0 = 0 \quad \textit{True} \qquad\qquad -4 = 0 \quad \textit{False}$$

The solution is 9 or the solution set is $\{9\}$.

EXAMPLE 2

Solve $\dfrac{3x}{x - 2} - \dfrac{x + 1}{x} = \dfrac{6}{x(x - 2)}$.

Solution   In this equation, x cannot be either 2 or 0, because these values cause denominators to equal zero. To solve for x, we first multiply both sides of the equation by $x(x - 2)$ to clear the fractions. By the distributive property, this means that we multiply each term by $x(x - 2)$.

$$x(x - 2)\left(\frac{3x}{x - 2}\right) - x(x - 2)\left(\frac{x + 1}{x}\right) = x(x - 2)\left[\frac{6}{x(x - 2)}\right]$$

$$3x^2 - (x - 2)(x + 1) = 6 \qquad \textit{Simplify.}$$

$$3x^2 - (x^2 - x - 2) = 6 \qquad \textit{Multiply.}$$

$$3x^2 - x^2 + x + 2 = 6$$

$$2x^2 + x - 4 = 0 \qquad \textit{Simplify.}$$

This equation cannot be factored using integers, so we solve by the quadratic formula.

$$x = \frac{-1 \pm \sqrt{1^2 - 4(2)(-4)}}{2 \cdot 2} \qquad \textit{Use } a = 2, b = 1, \text{ and } c = -4 \text{ in the quadratic formula.}$$

$$= \frac{-1 \pm \sqrt{1 + 32}}{4} \qquad \textit{Simplify.}$$

$$= \frac{-1 \pm \sqrt{33}}{4}$$

Neither proposed solution will make the denominators 0.

The solutions are $\dfrac{-1 + \sqrt{33}}{4}$ and $\dfrac{-1 - \sqrt{33}}{4}$ or the solution set is

$$\left\{\frac{-1 + \sqrt{33}}{4}, \frac{-1 - \sqrt{33}}{4}\right\}.$$

EXAMPLE 3

Solve $p^4 - 3p^2 - 4 = 0$.

Solution First we factor the trinomial.

$$p^4 - 3p^2 - 4 = 0$$

$$(p^2 - 4)(p^2 + 1) = 0 \qquad \text{Factor.}$$

$$(p - 2)(p + 2)(p^2 + 1) = 0 \qquad \text{Factor further.}$$

$$p - 2 = 0 \quad \text{or} \quad p + 2 = 0 \quad \text{or} \quad p^2 + 1 = 0 \quad \text{Set each factor equal}$$
$$\qquad\qquad\qquad\qquad\qquad\qquad\qquad\qquad\text{to 0 and solve.}$$
$$p = 2 \qquad\qquad p = -2 \qquad\qquad p^2 = -1$$

$$p = \pm\sqrt{-1} = \pm i$$

The solutions are $2, -2, i$ and $-i$.

> **Helpful Hint**
>
> Example 3 can be solved using substitution also. Think of $p^4 - 3p^2 - 4 = 0$ as
>
> $(p^2)^2 - 3p^2 - 4 = 0$ Then let $x = p^2$, and solve and substitute back. The
>   ↓   ↙   solutions will be the same.
> $x^2 - 3x - 4 = 0$

✔ **CONCEPT CHECK**

a. True or false? The maximum number of solutions that a quadratic equation can have is 2.

b. True or false? The maximum number of solutions that an equation in quadratic form can have is 2.

EXAMPLE 4

Solve $(x - 3)^2 - 3(x - 3) - 4 = 0$.

Solution Notice that the quantity $(x - 3)$ is repeated in this equation. Sometimes it is helpful to substitute a variable (in this case other than x) for the repeated quantity. We will let $y = x - 3$. Then

$$(x - 3)^2 - 3(x - 3) - 4 = 0$$

becomes

$$y^2 - 3y - 4 = 0 \quad \text{Let } x - 3 = y.$$

$$(y - 4)(y + 1) = 0 \quad \text{Factor.}$$

To solve, we use the zero factor property.

$$y - 4 = 0 \quad \text{or} \quad y + 1 = 0 \quad \text{Set each factor equal to 0.}$$

$$y = 4 \qquad\qquad y = -1 \quad \text{Solve.}$$

Concept Check Answer:

a. true   **b.** false

Helpful Hint

When using substitution, don't forget to substitute back to the original variable.

To find values of x, we substitute back. That is, we substitute $x - 3$ for y.

$$x - 3 = 4 \quad \text{or} \quad x - 3 = -1$$

$$x = 7 \qquad\qquad x = 2$$

Both 2 and 7 check. The solutions are 2 and 7.

EXAMPLE 5

Solve $x^{2/3} - 5x^{1/3} + 6 = 0$.

Solution   The key to solving this equation is recognizing that $x^{2/3} = (x^{1/3})^2$. We replace $x^{1/3}$ with m so that

$$(x^{1/3})^2 - 5x^{1/3} + 6 = 0$$

becomes

$$m^2 - 5m + 6 = 0$$

Now we solve by factoring.

$$m^2 - 5m + 6 = 0$$

$$(m - 3)(m - 2) = 0 \qquad\qquad \text{Factor.}$$

$$m - 3 = 0 \quad \text{or} \quad m - 2 = 0 \quad \text{Set each factor equal to 0.}$$

$$m = 3 \qquad\qquad m = 2$$

Since $m = x^{1/3}$, we have

$$x^{1/3} = 3 \qquad \text{or} \quad x^{1/3} = 2$$

$$x = 3^3 = 27 \quad \text{or} \qquad x = 2^3 = 8$$

Both 8 and 27 check. The solutions are 8 and 27.

2   The next example is a work problem. This problem is modeled by a rational equation that simplifies to a quadratic equation.

EXAMPLE 6

FINDING WORK TIME

Together, an experienced word processor and an apprentice word processor can create a word document in 6 hours. Alone, the experienced word processor can create the document 2 hours faster than the apprentice word processor can. Find the time in which each person can create the word document alone.

Solution   1. UNDERSTAND. Read and reread the problem. The key idea here is the relationship between the *time* (hours) it takes to complete the job and the *part of the job* completed in one unit of time (hour). For example, because they can complete the job together in 6 hours, the *part of the job* they can complete in 1 hour is $\frac{1}{6}$.

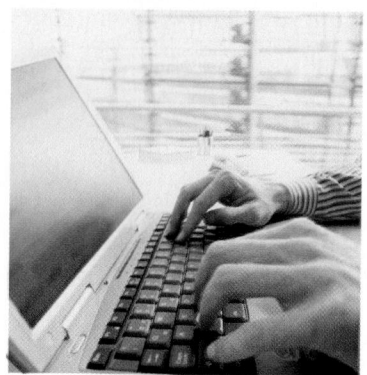

Let

x = the *time* in hours it takes the apprentice word processor to complete the job alone

$x - 2$ = the *time* in hours it takes the experienced word processor to complete the job alone

We can summarize in a chart the information discussed

	Total Hours to Complete Job	*Part of Job Completed in 1 Hour*
Apprentice word processor	x	$\dfrac{1}{x}$
Experienced word processor	$x - 2$	$\dfrac{1}{x - 2}$
Together	6	$\dfrac{1}{6}$

2. TRANSLATE.

In words:

part of job completed by apprentice word processor in 1 hour	added to	part of job completed by experienced word processor in 1 hour	is equal to	part of job completed together in 1 hour
↓	↓	↓	↓	↓

Translate: $\quad \dfrac{1}{x} \qquad + \qquad \dfrac{1}{x - 2} \qquad = \qquad \dfrac{1}{6}$

3. SOLVE.

$$\frac{1}{x} + \frac{1}{x - 2} = \frac{1}{6}$$

$$6x(x - 2)\left(\frac{1}{x} + \frac{1}{x - 2}\right) = 6x(x - 2) \cdot \frac{1}{6} \qquad \text{Multiply both sides by the LCD } 6x(x - 2).$$

$$6x(x - 2) \cdot \frac{1}{x} + 6x(x - 2) \cdot \frac{1}{x - 2} = 6x(x - 2) \cdot \frac{1}{6} \qquad \text{Use the distributive property.}$$

$$6(x - 2) + 6x = x(x - 2)$$

$$6x - 12 + 6x = x^2 - 2x$$

$$0 = x^2 - 14x + 12$$

Now we can substitute $a = 1, b = -14$, and $c = 12$ into the quadratic formula and simplify.

$$x = \frac{-(-14) \pm \sqrt{(-14)^2 - 4(1)(12)}}{2 \cdot 1} = \frac{14 \pm \sqrt{148}}{2}$$

Using a calculator or a square root table, we see that $\sqrt{148} \approx 12.2$ rounded to one decimal place. Thus,

$$x \approx \frac{14 \pm 12.2}{2}$$

$$x \approx \frac{14 + 12.2}{2} = 13.1 \quad \text{or} \quad x \approx \frac{14 - 12.2}{2} = 0.9$$

4. INTERPRET.

Check: If the apprentice word processor completes the job alone in 0.9 hours, the experienced word processor completes the job alone in $x - 2 = 0.9 - 2 = -1.1$ hours. Since this is not possible, we reject the solution of 0.9. The approximate solution thus is 13.1 hours.

State: The apprentice word processor can complete the job alone in approximately 13.1 hours, and the experienced word processor can complete the job alone in approximately

$$x - 2 = 13.1 - 2 = 11.1 \text{ hours}.$$

EXAMPLE 7

Beach x mph   Fargo

400 mi
$x + 10$ mph

400 mi

FINDING DRIVING SPEEDS

Beach and Fargo are about 400 miles apart. A salesperson travels from Fargo to Beach one day at a certain speed. She returns to Fargo the next day and drives 10 mph faster. Her total travel time was $14\frac{2}{3}$ hours. Find her speed to Beach and the return speed to Fargo.

Solution   **1.** UNDERSTAND. Read and reread the problem. Let

$$x = \text{the speed to Beach, so}$$

$$x + 10 = \text{the return speed to Fargo.}$$

Then organize the given information in a table.

> **Helpful Hint**
>
> Since $d = rt$, then $t = \dfrac{d}{r}$.
>
> The time column was completed using $\dfrac{d}{r}$.

	distance	=	rate	·	time	
To Beach	400		x		$\dfrac{400}{x}$	← distance ← rate
Return to Fargo	400		$x + 10$		$\dfrac{400}{x + 10}$	← distance ← rate

2. TRANSLATE.

In words:

time to Beach	+	return time to Fargo	=	$14\frac{2}{3}$ hours

Translate:

$$\frac{400}{x} + \frac{400}{x + 10} = \frac{44}{3}$$

3. SOLVE.

$$\frac{400}{x} + \frac{400}{x+10} = \frac{44}{3}$$

$$\frac{100}{x} + \frac{100}{x+10} = \frac{11}{3}$$ Divide both sides by **4.**

$$3x(x+10)\left(\frac{100}{x} + \frac{100}{x+10}\right) = 3x(x+10) \cdot \frac{11}{3}$$ Multiply both sides by the LCD $3x(x+10)$.

$$3x(x+10) \cdot \frac{100}{x} + 3x(x+10) \cdot \frac{100}{x+10} = 3x(x+10) \cdot \frac{11}{3}$$ Use the distributive property.

$$3(x+10) \cdot 100 + 3x \cdot 100 = x(x+10) \cdot 11$$

$$300x + 3000 + 300x = 11x^2 + 110x$$

$$0 = 11x^2 - 490x - 3000$$ Set equation equal to 0.

$$0 = (11x+60)(x-50)$$ Factor.

$$11x + 60 = 0 \quad \text{or} \quad x - 50 = 0$$ Set each factor equal to 0.

$$x = -\frac{60}{11} \text{ or } -5\frac{5}{11}; \quad x = 50$$

4. INTERPRET.

Check: The speed is not negative, so it's not $-5\frac{5}{11}$. The number 50 does check.

State: The speed to Beach was 50 mph and her return speed to Fargo was 60 mph.

EXERCISE SET 11.3

STUDY GUIDE/SSM   CD/ VIDEO   PH MATH TUTOR CENTER   MathXL®Tutorials ON CD   MathXL®   MyMathLab®

Solve. See Example 1.

1. $2x = \sqrt{10 + 3x}$

2. $3x = \sqrt{8x + 1}$

3. $x - 2\sqrt{x} = 8$

4. $x - \sqrt{2x} = 4$

5. $\sqrt{9x} = x + 2$

6. $\sqrt{16x} = x + 3$

Solve. See Example 2.

7. $\frac{2}{x} + \frac{3}{x-1} = 1$

8. $\frac{6}{x^2} = \frac{3}{x+1}$

9. $\frac{3}{x} + \frac{4}{x+2} = 2$

10. $\frac{5}{x-2} + \frac{4}{x+2} = 1$

11. $\frac{7}{x^2 - 5x + 6} = \frac{2x}{x-3} - \frac{x}{x-2}$

12. $\frac{11}{2x^2 + x - 15} = \frac{5}{2x-5} - \frac{x}{x+3}$

Solve. See Example 3.

13. $p^4 - 16 = 0$

14. $x^4 + 2x^2 - 3 = 0$

15. $4x^4 + 11x^2 = 3$

16. $z^4 = 81$

17. $z^4 - 13z^2 + 36 = 0$

18. $9x^4 + 5x^2 - 4 = 0$

Solve. See Examples 4 and 5.

19. $x^{2/3} - 3x^{1/3} - 10 = 0$ **20.** $x^{2/3} + 2x^{1/3} + 1 = 0$

21. $(5n + 1)^2 + 2(5n + 1) - 3 = 0$

22. $(m - 6)^2 + 5(m - 6) + 4 = 0$

23. $2x^{2/3} - 5x^{1/3} = 3$ **24.** $3x^{2/3} + 11x^{1/3} = 4$

25. $1 + \dfrac{2}{3t - 2} = \dfrac{8}{(3t - 2)^2}$ **26.** $2 - \dfrac{7}{x + 6} = \dfrac{15}{(x + 6)^2}$

27. $20x^{2/3} - 6x^{1/3} - 2 = 0$ **28.** $4x^{2/3} + 16x^{1/3} = -15$

MIXED PRACTICE

Solve. See Examples 1 through 5.

29. $a^4 - 5a^2 + 6 = 0$

30. $x^4 - 12x^2 + 11 = 0$

31. $\dfrac{2x}{x - 2} + \dfrac{x}{x + 3} = -\dfrac{5}{x + 3}$

32. $\dfrac{5}{x - 3} + \dfrac{x}{x + 3} = \dfrac{19}{x^2 - 9}$

33. $(p + 2)^2 = 9(p + 2) - 20$

34. $2(4m - 3)^2 - 9(4m - 3) = 5$

35. $2x = \sqrt{11x + 3}$ **36.** $4x = \sqrt{2x + 3}$

37. $x^{2/3} - 8x^{1/3} + 15 = 0$

38. $x^{2/3} - 2x^{1/3} - 8 = 0$

39. $y^3 + 9y - y^2 - 9 = 0$

40. $x^3 + x - 3x^2 - 3 = 0$

41. $2x^{2/3} + 3x^{1/3} - 2 = 0$

42. $6x^{2/3} - 25x^{1/3} - 25 = 0$

43. $x^{-2} - x^{-1} - 6 = 0$

44. $y^{-2} - 8y^{-1} + 7 = 0$

45. $x - \sqrt{x} = 2$ **46.** $x - \sqrt{3x} = 6$

47. $\dfrac{x}{x - 1} + \dfrac{1}{x + 1} = \dfrac{2}{x^2 - 1}$

48. $\dfrac{x}{x - 5} + \dfrac{5}{x + 5} = -\dfrac{1}{x^2 - 25}$

49. $p^4 - p^2 - 20 = 0$

50. $x^4 - 10x^2 + 9 = 0$

51. $2x^3 = -54$

52. $y^3 - 216 = 0$

53. $1 = \dfrac{4}{x - 7} + \dfrac{5}{(x - 7)^2}$

54. $3 + \dfrac{1}{2p + 4} = \dfrac{10}{(2p + 4)^2}$

55. $27y^4 + 15y^2 = 2$

56. $8z^4 + 14z^2 = -5$

Solve. See Examples 6 and 7.

57. A jogger ran 3 miles, decreased her speed by 1 mile per hour, and then ran another 4 miles. If her total time jogging was $1\dfrac{3}{5}$ hours, find her speed for each part of her run.

58. Mark Keaton's workout consists of jogging for 3 miles, and then riding his bike for 5 miles at a speed 4 miles per hour faster than he jogs. If his total workout time is 1 hour, find his jogging speed and his biking speed.

59. A Chinese restaurant in Mandeville, Louisiana, has a large goldfish pond around the restaurant. Suppose that an inlet pipe and a hose together can fill the pond in 8 hours. The inlet pipe alone can complete the job in one hour less time than the hose alone. Find the time that the hose can complete the job alone and the time that the inlet pipe can complete the job alone. Round each to the nearest tenth of an hour.

60. A water tank on a farm in Flatonia, Texas, can be filled with a large inlet pipe and a small inlet pipe in 3 hours. The large inlet pipe alone can fill the tank in 2 hours less time than the small inlet pipe alone. Find the time to the nearest tenth of an hour each pipe can fill the tank alone.

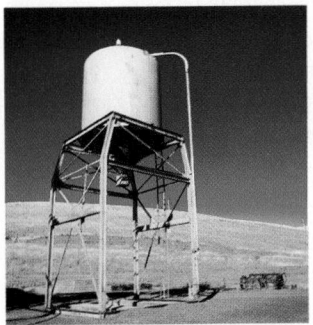

61. Roma Sherry drove 330 miles from her hometown to Tucson. During her return trip, she was able to increase her speed by 11 mph. If her return trip took 1 hour less time, find her original speed and her speed returning home.

62. A salesperson drove to Portland, a distance of 300 miles. During the last 80 miles of his trip, heavy rainfall forced him to decrease his speed by 15 mph. If his total driving time was 6 hours, find his original speed and his speed during the rainfall.

63. Bill Shaughnessy and his son Billy can clean the house together in 4 hours. When the son works alone, it takes him an hour longer to clean than it takes his dad alone. Find how long to the nearest tenth of an hour it takes the son to clean alone.

64. Together, Noodles and Freckles eat a 50-pound bag of dog food in 30 days. Noodles by himself eats a 50-pound bag in 2 weeks less time than Freckles does by himself. How many days to the nearest whole day would a 50-pound bag of dog food last Freckles?

65. The product of a number and 4 less than the number is 96. Find the number.

66. A whole number increased by its square is two more than twice itself. Find the number.

△ **67.** Suppose that an open box is to be made from a square sheet of cardboard by cutting out squares from each corner as shown and then folding along the dotted lines. If the box is to have a volume of 300 cubic centimeters, find the original dimensions of the sheet of cardboard.

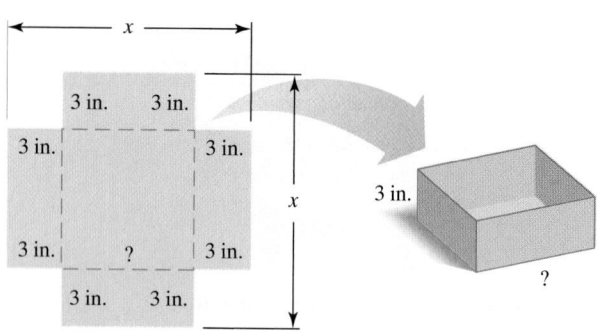

a. The ? in the drawing above will be the length (and also the width) of the box as shown in the drawing above. Represent this length in terms of x.

b. Use the formula for volume of a box, $V = l \cdot w \cdot h$, to write an equation in x.

c. Solve the equation for x and give the dimensions of the sheet of cardboard. Check your solution.

△ **68.** Suppose that an open box is to be made from a square sheet of cardboard by cutting out squares from each corner as shown and then folding along the dotted lines. If the box is to have a volume of 128 cubic inches, find the original dimensions of the sheet of cardboard. (*Hint:* Use Exercise 67 Parts **a, b,** and **c** to help you.)

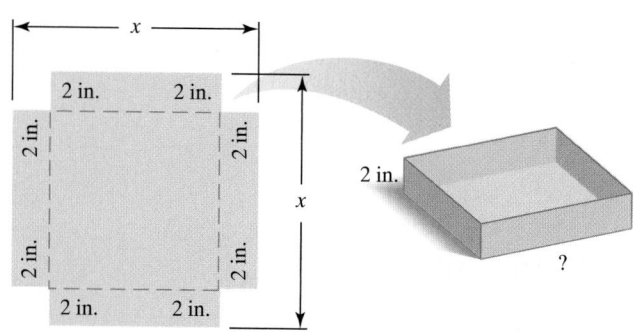

△ **69.** A sprinkler that sprays water in a circular motion is to be used to water a square garden. If the area of the garden is 920 square feet, find the smallest whole number *radius* that the sprinkler can be adjusted to so that the entire garden is watered.

△ **70.** Suppose that a square field has an area of 6270 square feet. See Exercise 69 and find a new sprinkler radius.

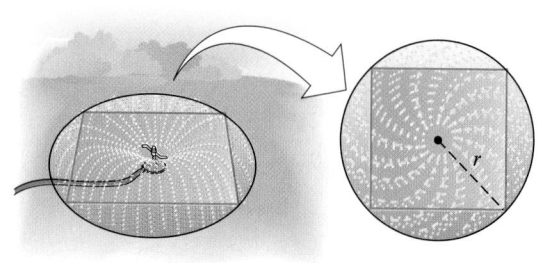

REVIEW AND PREVIEW

Solve each inequality. See Section 2.7.

71. $\dfrac{5x}{3} + 2 \le 7$

72. $\dfrac{2x}{3} + \dfrac{1}{6} \ge 2$

73. $\dfrac{y - 1}{15} > -\dfrac{2}{5}$

74. $\dfrac{z - 2}{12} < \dfrac{1}{4}$

Find the domain and range of each graphed relation. Decide which relations are also functions. See Section 3.7.

75.

76.

77.

78.

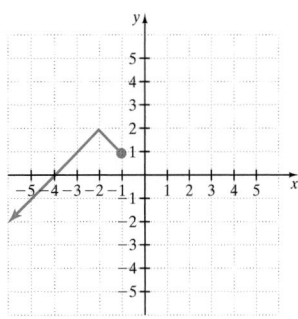

Concept Extensions

79. Write a polynomial equation that has three solutions: 2, 5, and −7.

80. Write a polynomial equation that has three solutions: 0, 2i, and −2i.

81. During the 2003 Grand Prix of Miami auto race, Adrian Fernandez posted the fastest lap speed but Mario Dominguez won the race. The track is 7920 feet (1.5 miles) long. Fernandez's fastest lap speed was 0.88 feet per second faster than Dominguez's fastest lap speed. Traveling at these fastest speeds, Mario Dominguez would have taken 0.38 seconds longer than Fernandez to complete a lap. (*Source:* Championship Auto Racing Teams, Inc.)

  a. Find Mario Dominguez's fastest lap speed during the race. Round to two decimal places.

  b. Find Adrian Fernandez's fastest lap speed during the race. Round to two decimal places.

  c. Convert each speed to miles per hour. Round to one decimal place.

82. Use a graphing calculator to solve Exercise 29. Compare the solution with the solution from Exercise 29. Explain any differences.

SUMMARY ON SOLVING QUADRATIC EQUATIONS

INTEGRATED REVIEW

Use the square root property to solve each equation.

1. $x^2 - 10 = 0$

2. $x^2 - 14 = 0$

3. $(x - 1)^2 = 8$

4. $(x + 5)^2 = 12$

Solve each equation by completing the square.

5. $x^2 + 2x - 12 = 0$

6. $x^2 - 12x + 11 = 0$

7. $3x^2 + 3x = 5$

8. $16y^2 + 16y = 1$

Use the quadratic formula to solve each equation

9. $2x^2 - 4x + 1 = 0$

10. $\frac{1}{2}x^2 + 3x + 2 = 0$

11. $x^2 + 4x = -7$

12. $x^2 + x = -3$

Solve each equation. Use a method of your choice.

13. $x^2 + 3x + 6 = 0$

14. $2x^2 + 18 = 0$

15. $x^2 + 17x = 0$

16. $4x^2 - 2x - 3 = 0$

17. $(x - 2)^2 = 27$

18. $\frac{1}{2}x^2 - 2x + \frac{1}{2} = 0$

19. $3x^2 + 2x = 8$

20. $2x^2 = -5x - 1$

21. $x(x - 2) = 5$

22. $x^2 - 31 = 0$

23. $5x^2 - 55 = 0$

24. $5x^2 + 55 = 0$

25. $x(x + 5) = 66$

26. $5x^2 + 6x - 2 = 0$

27. $2x^2 + 3x = 1$

△ **28.** The diagonal of a square room measures 20 feet. Find the exact length of a side of the room. Then approximate the length to the nearest tenth of a foot.

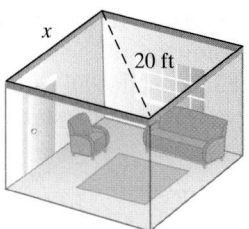

29. Jack and Lucy Hoag together can prepare a crawfish boil for a large party in 4 hours. Lucy alone can complete the job in 2 hours less time than Jack alone. Find the time that each person can prepare the crawfish boil alone. Round each time to the nearest tenth of an hour.

30. Diane Gray exercises at Total Body Gym. On the treadmill, she runs 5 miles, then increases her speed by 1 mile per hour and runs an additional 2 miles. If her total time on the treadmill is $1\frac{1}{3}$ hours, find her speed during each part of her run.

11.4   NONLINEAR INEQUALITIES IN ONE VARIABLE

Objectives

1 Solve polynomial inequalities of degree 2 or greater.

2 Solve inequalities that contain rational expressions with variables in the denominator.

1   Just as we can solve linear inequalities in one variable, so can we also solve quadratic inequalities in one variable. A **quadratic inequality** is an inequality that can be written so that one side is a quadratic expression and the other side is 0. Here are examples of quadratic inequalities in one variable. Each is written in **standard form.**

$$x^2 - 10x + 7 \leq 0 \quad 3x^2 + 2x - 6 > 0$$
$$2x^2 + 9x - 2 < 0 \quad x^2 - 3x + 11 \geq 0$$

A solution of a quadratic inequality in one variable is a value of the variable that makes the inequality a true statement.

The value of an expression such as $x^2 - 3x - 10$ will sometimes be positive, sometimes negative, and sometimes 0, depending on the value substituted for x. To solve the inequality $x^2 - 3x - 10 < 0$, we are looking for all values of x that make the expression $x^2 - 3x - 10$ **less than 0,** or **negative.** To understand how we find these values, we'll study the graph of the quadratic function $y = x^2 - 3x - 10$.

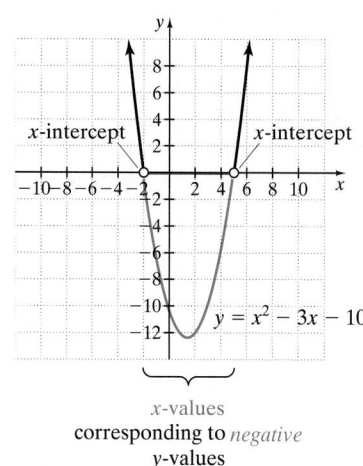

x-values
corresponding to *negative*
y-values

Notice that the x-values for which y is positive are separated from the x values for which y is negative by the x-intercepts. (Recall that the x-intercepts correspond to values of x for which $y = 0$.) Thus, the solution set of $x^2 - 3x - 10 < 0$ consists of all real numbers from -2 to 5, or in interval notation, $(-2, 5)$.

It is not necessary to graph $y = x^2 - 3x - 10$ to solve the related inequality $x^2 - 3x - 10 < 0$. Instead, we can draw a number line representing the x-axis and keep the following in mind: *A region on the number line for which the value of $x^2 - 3x - 10$ is positive is separated from a region on the number line for which the value of $x^2 - 3x - 10$ is negative by a value for which the expression is 0.*

Let's find these values for which the expression is 0 by solving the related equation:

$$x^2 - 3x - 10 = 0$$

$$(x - 5)(x + 2) = 0 \qquad \text{Factor.}$$

$$x - 5 = 0 \quad \text{or} \quad x + 2 = 0 \qquad \text{Set each factor equal to 0.}$$

$$x = 5 \quad x = -2 \qquad \text{Solve.}$$

These two numbers -2 and 5, divide the number line into three regions. We will call the regions A, B, and C. These regions are important because, if the value of $x^2 - 3x - 10$ is negative when a number from a region is substituted for x, then $x^2 - 3x - 10$ is negative when any number in that region is substituted for x. The same is true if the value of $x^2 - 3x - 10$ is positive for a particular value of x in a region.

To see whether the inequality $x^2 - 3x - 10 < 0$ is true or false in each region, we choose a test point from each region and substitute its value for x in the inequality

$x^2 - 3x - 10 < 0$. If the resulting inequality is true, the region containing the test point is a solution region.

Region	Test Point Value	Factored Form $(x - 5)(x + 2) < 0$	Result
A	-3	$(-8)(-1) < 0$	
B	0	$(-5)(2) < 0$	True
C	6	$(1)(8) < 0$	

The values in region B satisfy the inequality. The numbers -2 and 5 are not included in the solution set since the inequality symbol is $<$. The solution set is $(-2, 5)$, and its graph is shown.

EXAMPLE 1

Solve $(x + 3)(x - 3) > 0$.

Solution First we solve the related equation $(x + 3)(x - 3) = 0$.

$$(x + 3)(x - 3) = 0$$

$$x + 3 = 0 \quad \text{or} \quad x - 3 = 0$$

$$x = -3 \qquad\qquad x = 3$$

The two numbers -3 and 3 separate the number line into three regions, A, B, and C.

Now we substitute the value of a test point from each region. If the test value satisfies the inequality, every value in the region containing the test value is a solution.

Region	Test Point Value	$(x + 3)(x - 3) > 0$	Result
A	-4	$(-1)(-7) > 0$	
B	0	$(3)(-3) > 0$	False
C	4	$(7)(1) > 0$	

The points in regions A and C satisfy the inequality. The numbers -3 and 3 are not included in the solution since the inequality symbol is $>$. The solution set is $(-\infty, -3) \cup (3, \infty)$, and its graph is shown.

The following steps may be used to solve a polynomial inequality.

> ### Solving a Polynomial Inequality
>
> **Step 1:** Write the inequality in standard form and then solve the related equation.
> **Step 2:** Separate the number line into regions with the solutions from Step 1.
> **Step 3:** For each region, choose a test point and determine whether its value satisfies the *original inequality*.
> **Step 4:** The solution set includes the regions whose test point value is a solution. If the inequality symbol is $\leq$ or $\geq$, the values from Step 1 are solutions; if $<$ or $>$, they are not.

✔ **CONCEPT CHECK**

When choosing a test point in Step 4, why would the solutions from Step 2 not make good choices for test points?

EXAMPLE 2

Solve $x^2 - 4x \leq 0$.

Solution   First we solve the related equation $x^2 - 4x = 0$.

$$x^2 - 4x = 0$$

$$x(x - 4) = 0$$

$$x = 0 \quad \text{or} \quad x = 4$$

The numbers 0 and 4 separate the number line into three regions, A, B, and C.

Check a test value in each region in the original inequality. Values in region B satisfy the inequality. The numbers 0 and 4 are included in the solution since the inequality symbol is $\leq$. The solution set is $[0, 4]$, and its graph is shown.

EXAMPLE 3

Solve $(x + 2)(x - 1)(x - 5) \leq 0$.

Solution   First we solve $(x + 2)(x - 1)(x - 5) = 0$. By inspection, we see that the solutions are $-2, 1$, and 5. They separate the number line into four regions, A, B, C, and D. Next we check test points from each region.

Concept Check Answer:
The solutions found in Step 2 have a value of 0 in the original inequality.

	Region	Test Point Value	$(x + 2)(x - 1)$ $(x - 5) \leq 0$	Result
	A	-3	$(-1)(-4)(-8) \leq 0$	True
	B	0	$(2)(-1)(-5) \leq 0$	False
	C	2	$(4)(1)(-3) \leq 0$	True
	D	6	$(8)(5)(1) \leq 0$	False

The solution set is $(-\infty, -2] \cup [1, 5]$, and its graph is shown. We include the numbers $-2, 1$, and 5 because the inequality symbol is $\leq$.

2   Inequalities containing rational expressions with variables in the denominator are solved by using a similar procedure.

EXAMPLE 4

Solve $\dfrac{x + 2}{x - 3} \leq 0$.

Solution   First we find all values that make the denominator equal to 0. To do this, we solve $x - 3 = 0$ and find that $x = 3$.

Next, we solve the related equation $\dfrac{x + 2}{x - 3} = 0$.

$$\frac{x + 2}{x - 3} = 0$$

$$x + 2 = 0 \quad \text{Multiply both sides by the LCD, } x - 3.$$

$$x = -2$$

Now we place these numbers on a number line and proceed as before, checking test point values in the original inequality.

Choose -3 from region A.

$$\frac{x + 2}{x - 3} \leq 0$$

$$\frac{-3 + 2}{-3 - 3} \leq 0$$

$$\frac{-1}{-6} \leq 0$$

$$\frac{1}{6} \leq 0 \quad \text{False}$$

Choose 0 from region B.

$$\frac{x + 2}{x - 3} \leq 0$$

$$\frac{0 + 2}{0 - 3} \leq 0$$

$$-\frac{2}{3} \leq 0 \quad \text{True}$$

Choose 4 from region C.

$$\frac{x + 2}{x - 3} \leq 0$$

$$\frac{4 + 2}{4 - 3} \leq 0$$

$$6 \leq 0 \quad \text{False}$$

The solution set is $[-2, 3)$. This interval includes -2 because -2 satisfies the original inequality. This interval does not include 3, because 3 would make the denominator 0.

The following steps may be used to solve a rational inequality with variables in the denominator.

Solving a Rational Inequality

Step 1: Solve for values that make all denominators 0.

Step 2: Solve the related equation.

Step 3: Separate the number line into regions with the solutions from Steps 1 and 2.

Step 4: For each region, choose a test point and determine whether its value satisfies the *original inequality.*

Step 5: The solution set includes the regions whose test point value is a solution. Check whether to include values from Step 2. Be sure *not* to include values that make any denominator 0.

EXAMPLE 5

Solve $\dfrac{5}{x + 1} < -2$.

Solution First we find values for x that make the denominator equal to 0.

$$x + 1 = 0$$

$$x = -1$$

Next we solve $\dfrac{5}{x + 1} = -2$.

$$(x + 1) \cdot \frac{5}{x + 1} = (x + 1) \cdot -2 \qquad \text{Multiply both sides by the LCD, } x + 1.$$

$$5 = -2x - 2 \qquad \text{Simplify.}$$

$$7 = -2x$$

$$-\frac{7}{2} = x$$

We use these two solutions to divide a number line into three regions and choose test points. Only a test point value from region B satisfies the *original inequality.* The solution set is $\left(-\dfrac{7}{2}, -1 \right)$, and its graph is shown.

STUDY SKILLS REMINDER

Continue your outline started in Section 2.3. Write how to recognize and how to solve nonlinear inequalities in your own words. For example:

Solving Equations and Inequalities

I. Equations
   A. Linear equations (Sec. 2.3)
   B. Absolute value equations (Sec. 9.2)
   C. Quadratic and higher degree equations (Sec. 6.5 and Chapter 11)
   D. Equations with rational expressions (Sec. 7.5)
   E. Equations with radicals (Sec. 10.6)

II. Inequalities
   A. Linear inequalities (Sec. 2.7)
   B. Compound inequalities (Sec. 9.1)
   C. Absolute value inequalities (Sec. 9.3)
   D. Nonlinear inequalities—
      1. Polynomial inequalities — Recognize: *inequalities that are quadratic or higher degree* — Solve: write in standard form (equal to 0). Solve the related equation and use solutions to divide the number line into regions. Check regions with test values.
      2. Rational inequalities — Recognize: *inequalities that contain a variable in the denominator* — Solve: find values that make all denominators 0 and find solutions to the related equation. Use these numbers to divide the number line into regions. Check regions with test values in the original inequality.

EXERCISE SET 11.4

    Math XL

STUDY   CD/   PH MATH   MathXL®Tutorials   MathXL®   MyMathLab®
GUIDE/SSM   VIDEO   TUTOR CENTER   ON CD

Solve each quadratic inequality. Graph the solution set and write the solution set in interval notation. See Examples 1 through 3.

1. $(x + 1)(x + 5) > 0$

2. $(x + 1)(x + 5) \leq 0$

3. $(x - 3)(x + 4) \leq 0$

4. $(x + 4)(x - 1) > 0$

5. $x^2 - 7x + 10 \leq 0$

6. $x^2 + 8x + 15 \geq 0$

7. $3x^2 + 16x < -5$

8. $2x^2 - 5x < 7$

9. $(x - 6)(x - 4)(x - 2) > 0$

10. $(x - 6)(x - 4)(x - 2) \leq 0$

11. $x(x - 1)(x + 4) \leq 0$

12. $x(x - 6)(x + 2) > 0$

13. $(x^2 - 9)(x^2 - 4) > 0$

14. $(x^2 - 16)(x^2 - 1) \leq 0$

Solve each inequality. Graph the solution set and write the solution set in interval notation. See Example 4.

15. $\dfrac{x+7}{x-2} < 0$ **16.** $\dfrac{x-5}{x-6} > 0$ **17.** $\dfrac{5}{x+1} > 0$

18. $\dfrac{3}{y-5} < 0$ **19.** $\dfrac{x+1}{x-4} \geq 0$ **20.** $\dfrac{x+1}{x-4} \leq 0$

Solve each inequality. Graph the solution set and write the solution set in interval notation. See Example 5.

21. $\dfrac{3}{x-2} < 4$ **22.** $\dfrac{-2}{y+3} > 2$ **23.** $\dfrac{x^2+6}{5x} \geq 1$

24. $\dfrac{y^2+15}{8y} \leq 1$

MIXED PRACTICE

Solve each inequality. Graph the solution set and write the solution set in interval notation.

25. $(x-8)(x+7) > 0$ **26.** $(x-5)(x+1) < 0$

27. $(2x-3)(4x+5) \leq 0$ **28.** $(6x+7)(7x-12) > 0$

 29. $x^2 > x$ **30.** $x^2 < 25$

31. $(2x-8)(x+4)(x-6) \leq 0$

32. $(3x-12)(x+5)(2x-3) \geq 0$

33. $6x^2 - 5x \geq 6$

34. $12x^2 + 11x \leq 15$

35. $4x^3 + 16x^2 - 9x - 36 > 0$

36. $x^3 + 2x^2 - 4x - 8 < 0$

37. $x^4 - 26x^2 + 25 \geq 0$

38. $16x^4 - 40x^2 + 9 \leq 0$

39. $(2x-7)(3x+5) > 0$

40. $(4x-9)(2x+5) < 0$

 41. $\dfrac{x}{x-10} < 0$ **42.** $\dfrac{x+10}{x-10} > 0$

43. $\dfrac{x-5}{x+4} \geq 0$ **44.** $\dfrac{x-3}{x+2} \leq 0$

45. $\dfrac{x(x+6)}{(x-7)(x+1)} \geq 0$

46. $\dfrac{(x-2)(x+2)}{(x+1)(x-4)} \leq 0$

47. $\dfrac{-1}{x-1} > -1$ **48.** $\dfrac{4}{y+2} < -2$

49. $\dfrac{x}{x+4} \leq 2$ **50.** $\dfrac{4x}{x-3} \geq 5$

51. $\dfrac{z}{z-5} \geq 2z$ **52.** $\dfrac{p}{p+4} \leq 3p$

53. $\dfrac{(x+1)^2}{5x} > 0$ **54.** $\dfrac{(2x-3)^2}{x} < 0$

REVIEW AND PREVIEW

Recall that the graph of $f(x) + K$ is the same as the graph of $f(x)$ shifted K units upward if $K > 0$ and $|K|$ units downward if $K < 0$. Use the graph of $f(x) = |x|$ below to sketch the graph of each function. (See Sections 8.1 and 8.2.)

55. $g(x) = |x| + 2$ **56.** $H(x) = |x| - 2$

57. $F(x) = |x| - 1$ **58.** $h(x) = |x| + 5$

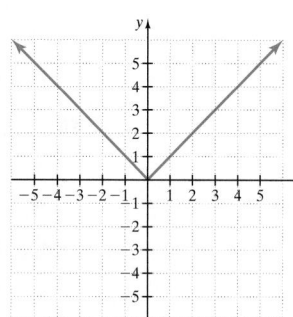

Use the graph of $f(x) = x^2$ below to sketch the graph of each function.

59. $F(x) = x^2 - 3$ **60.** $h(x) = x^2 - 4$

61. $H(x) = x^2 + 1$ **62.** $g(x) = x^2 + 3$

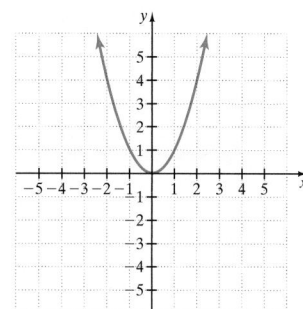

Concept Extensions

Find all numbers that satisfy each of the following.

63. Explain why $\dfrac{x+2}{x-3} > 0$ and $(x+2)(x-3) > 0$ have the same solutions.

64. Explain why $\dfrac{x+2}{x-3} \geq 0$ and $(x+2)(x-3) \geq 0$ do not have the same solutions.

65. A number minus its reciprocal is less than zero. Find the numbers.

66. Twice a number added to its reciprocal is nonnegative. Find the numbers.

67. The total profit function $P(x)$ for a company producing x thousand units is given by
$$P(x) = -2x^2 + 26x - 44$$
Find the values of x for which the company makes a profit.
[*Hint:* The company makes a profit when $P(x) > 0$.]

68. A projectile is fired straight up from the ground with an initial velocity of 80 feet per second. Its height $s(t)$ in feet at any time t is given by the function
$$s(t) = -16t^2 + 80t$$
Find the interval of time for which the height of the projectile is greater than 96 feet.

Use a graphing calculator to check each exercise.

69. Exercise 25.

70. Exercise 26.

71. Exercise 37.

72. Exercise 38.

11.5   QUADRATIC FUNCTIONS AND THEIR GRAPHS

Objectives

1 Graph quadratic functions of the form $f(x) = x^2 + k$.

2 Graph quadratic functions of the form $f(x) = (x - h)^2$.

3 Graph quadratic functions of the form $f(x) = (x - h)^2 + k$.

4 Graph quadratic functions of the form $f(x) = ax^2$.

5 Graph quadratic functions of the form $f(x) = a(x - h)^2 + k$.

1 We first graphed the quadratic equation $y = x^2$ in Section 3.1. In Section 3.2, we learned that this graph defines a function, and we wrote $y = x^2$ as $f(x) = x^2$. In these sections, we discovered that the graph of a quadratic function is a parabola opening upward or downward. In this section, we continue our study of quadratic functions and their graphs.

First, let's recall the definition of a quadratic function.

> **Quadratic Function**
>
> A quadratic function is a function that can be written in the form $f(x) = ax^2 + bx + c$, where a, b, and c are real numbers and $a \neq 0$.

Notice that equations of the form $y = ax^2 + bx + c$, where $a \neq 0$, define quadratic functions, since y is a function of x or $y = f(x)$.

Recall that if $a > 0$, the parabola opens upward and if $a < 0$, the parabola opens downward. Also, the vertex of a parabola is the lowest point if the parabola opens upward and the highest point if the parabola opens downward. The axis of symmetry is the vertical line that passes through the vertex.

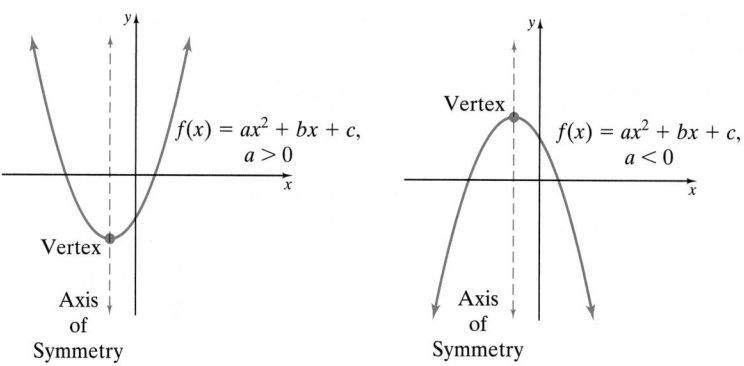

EXAMPLE 1

Graph $f(x) = x^2$ and $g(x) = x^2 + 3$ on the same set of axes.

Solution First we construct a table of values for $f(x)$ and plot the points. Notice that for each x-value, the corresponding value of $g(x)$ must be 3 more than the corresponding value of $f(x)$ since $f(x) = x^2$ and $g(x) = x^2 + 3$. In other words, the graph of $g(x) = x^2 + 3$ is the same as the graph of $f(x) = x^2$ shifted upward 3 units. The axis of symmetry for both graphs is the y-axis.

x	$f(x) = x^2$	$g(x) = x^2 + 3$
-2	4	7
-1	1	4
0	0	3
1	1	4
2	4	7

Each y-value is increased by 3.

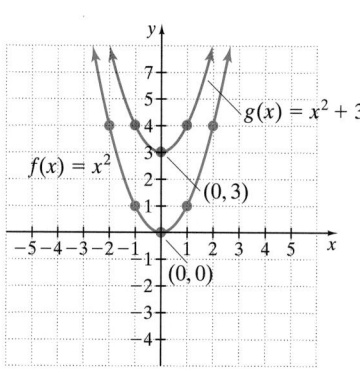

In general, we have the following properties.

Graphing the Parabola Defined by $f(x) = x^2 + k$

If k is positive, the graph of $f(x) = x^2 + k$ is the graph of $y = x^2$ shifted upward k units.
If k is negative, the graph of $f(x) = x^2 + k$ is the graph of $y = x^2$ shifted downward $|k|$ units.
The vertex is $(0, k)$, and the axis of symmetry is the y-axis.

EXAMPLE 2

Graph each function.

a. $F(x) = x^2 + 2$ **b.** $g(x) = x^2 - 3$

Solution   **a.** $F(x) = x^2 + 2$

The graph of $F(x) = x^2 + 2$ is obtained by shifting the graph of $y = x^2$ upward 2 units.

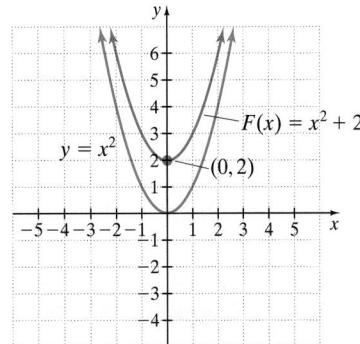

b. $g(x) = x^2 - 3$

The graph of $g(x) = x^2 - 3$ is obtained by shifting the graph of $y = x^2$ downward 3 units.

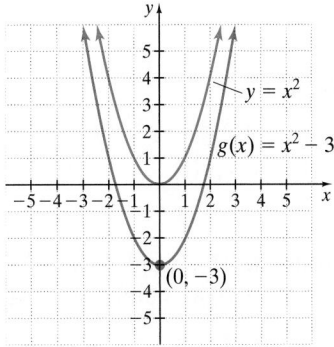

2   Now we will graph functions of the form $f(x) = (x - h)^2$.

EXAMPLE 3

Graph $f(x) = x^2$ and $g(x) = (x - 2)^2$ on the same set of axes.

Solution   By plotting points, we see that for each x-value, the corresponding value of $g(x)$ is the same as the value of $f(x)$ when the x-value is increased by 2. Thus, the graph of $g(x) = (x - 2)^2$ is the graph of $f(x) = x^2$ shifted to the right 2 units. The axis of symmetry for the graph of $g(x) = (x - 2)^2$ is also shifted 2 units to the right and is the line $x = 2$.

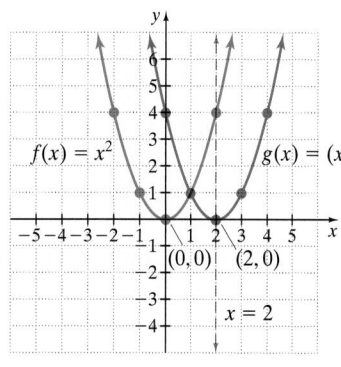

x	$f(x) = x^2$	x	$g(x) = (x - 2)^2$
-2	4	0	4
-1	1	1	1
0	0	2	0
1	1	3	1
2	4	4	4

Each x-value
Increased by 2
corresponds to
same y-value.

In general, we have the following properties.

Graphing the Parabola Defined by $f(x) = (x - h)^2$

If h is positive, the graph of $f(x) = (x - h)^2$ is the graph of $y = x^2$ shifted to the right h units.

If h is negative, the graph of $f(x) = (x - h)^2$ is the graph of $y = x^2$ shifted to the left $|h|$ units.

The vertex is $(h, 0)$, and the axis of symmetry is the vertical line $x = h$.

EXAMPLE 4

Graph each function.

a. $G(x) = (x - 3)^2$ **b.** $F(x) = (x + 1)^2$

Solution   **a.** The graph of $G(x) = (x - 3)^2$ is obtained by shifting the graph of $y = x^2$ to the right 3 units. The graph of $G(x)$ is below on the left.

b. The equation $F(x) = (x + 1)^2$ can be written as $F(x) = [x - (-1)]^2$. The graph of $F(x) = [x - (-1)]^2$ is obtained by shifting the graph of $y = x^2$ to the left 1 unit. The graph of $F(x)$ is below on the right.

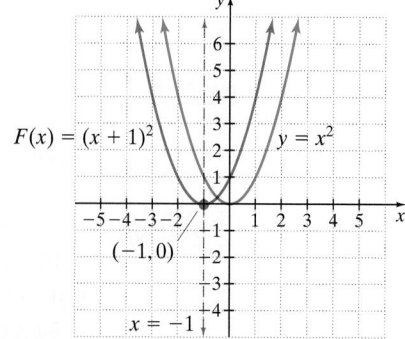

3   As we will see in graphing functions of the form $f(x) = (x - h)^2 + k$, it is possible to combine vertical and horizontal shifts.

Graphing the Parabola Defined by $f(x) = (x - h)^2 + k$

The parabola has the same shape as $y = x^2$.
The vertex is (h, k), and the axis of symmetry is the vertical line $x = h$.

EXAMPLE 5

Graph $F(x) = (x - 3)^2 + 1$.

Solution   The graph of $F(x) = (x - 3)^2 + 1$ is the graph of $y = x^2$ shifted 3 units to the right and 1 unit up. The vertex is then $(3, 1)$, and the axis of symmetry is $x = 3$. A few ordered pair solutions are plotted to aid in graphing.

x	$F(x) = (x - 3)^2 + 1$
1	5
2	2
4	2
5	5

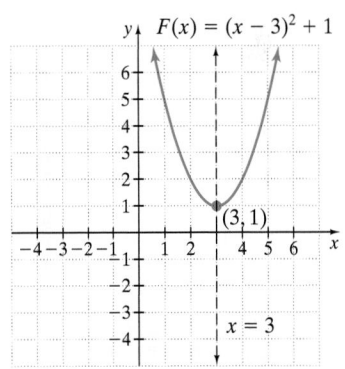

4   Next, we discover the change in the shape of the graph when the coefficient of x^2 is not 1.

EXAMPLE 6

Graph $f(x) = x^2$, $g(x) = 3x^2$, and $h(x) = \dfrac{1}{2}x^2$ on the same set of axes.

Solution   Comparing the tables of values, we see that for each x-value, the corresponding value of $g(x)$ is triple the corresponding value of $f(x)$. Similarly, the value of $h(x)$ is half the value of $f(x)$.

x	$f(x) = x^2$
-2	4
-1	1
0	0
1	1
2	4

x	$g(x) = 3x^2$
-2	12
-1	3
0	0
1	3
2	12

x	$h(x) = \dfrac{1}{2}x^2$
-2	2
-1	$\dfrac{1}{2}$
0	0
1	$\dfrac{1}{2}$
2	2

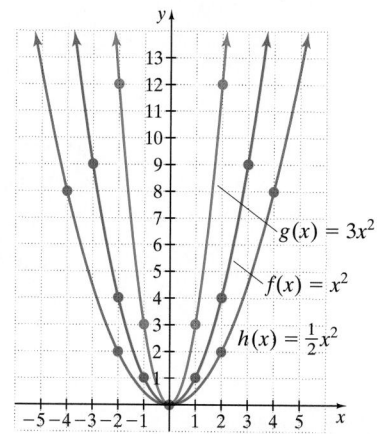

The result is that the graph of $g(x) = 3x^2$ is narrower than the graph of $f(x) = x^2$ and the graph of $h(x) = \dfrac{1}{2}x^2$ is wider. The vertex for each graph is $(0, 0)$, and the axis of symmetry is the y-axis.

Graphing the Parabola Defined by $f(x) = ax^2$

If a is positive, the parabola opens upward, and if a is negative, the parabola opens downward.
If $|a| > 1$, the graph of the parabola is narrower than the graph of $y = x^2$.
If $|a| < 1$, the graph of the parabola is wider than the graph of $y = x^2$.

EXAMPLE 7

Graph $f(x) = -2x^2$.

Solution Because $a = -2$, a negative value, this parabola opens downward. Since $|-2| = 2$ and $2 > 1$, the parabola is narrower than the graph of $y = x^2$. The vertex is $(0, 0)$, and the axis of symmetry is the y-axis. We verify this by plotting a few points.

x	$f(x) = -2x^2$
-2	-8
-1	-2
0	0
1	-2
2	-8

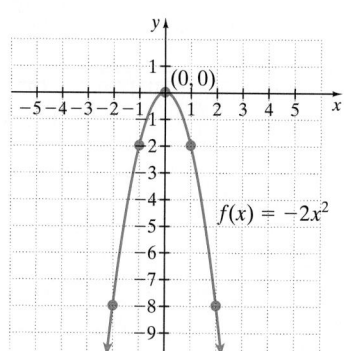

5 Now we will see the shape of the graph of a quadratic function of the form $f(x) = a(x - h)^2 + k$.

EXAMPLE 8

Graph $g(x) = \dfrac{1}{2}(x + 2)^2 + 5$. Find the vertex and the axis of symmetry.

Solution The function $g(x) = \dfrac{1}{2}(x + 2)^2 + 5$ may be written as $g(x) = \dfrac{1}{2}[x - (-2)]^2 + 5$.

Thus, this graph is the same as the graph of $y = x^2$ shifted 2 units to the left and 5 units up, and it is wider because a is $\dfrac{1}{2}$. The vertex is $(-2, 5)$, and the axis of symmetry is $x = -2$. We plot a few points to verify.

x	$g(x) = \dfrac{1}{2}(x + 2)^2 + 5$
-4	7
-3	$5\dfrac{1}{2}$
-2	5
-1	$5\dfrac{1}{2}$
0	7

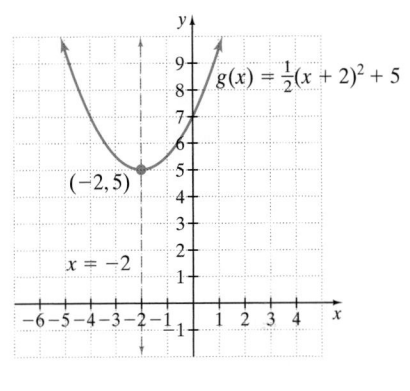

In general, the following holds.

Graph of a Quadratic Function

The graph of a quadratic function written in the form $f(x) = a(x - h)^2 + k$ is a parabola with vertex (h, k). If $a > 0$, the parabola opens upward, and if $a < 0$, the parabola opens downward. The axis of symmetry is the line whose equation is $x = h$.

✔ **CONCEPT CHECK**

Which description of the graph of $f(x) = -0.35(x + 3)^2 - 4$ is correct?

a. The graph opens downward and has its vertex at $(-3, 4)$.

b. The graph opens upward and has its vertex at $(-3, 4)$.

c. The graph opens downward and has its vertex at $(-3, -4)$.

d. The graph is narrower than the graph of $y = x^2$.

Concept Check Answer:

c

Graphing Calculator Explorations

Use a graphing calculator to graph the first function of each pair that follows. Then use its graph to predict the graph of the second function. Check your prediction by graphing both on the same set of axes.

1. $F(x) = \sqrt{x}; G(x) = \sqrt{x} + 1$

2. $g(x) = x^3; H(x) = x^3 - 2$

3. $H(x) = |x|; f(x) = |x - 5|$

4. $h(x) = x^3 + 2; g(x) = (x - 3)^3 + 2$

5. $f(x) = |x + 4|; F(x) = |x + 4| + 3$

6. $G(x) = \sqrt{x} - 2; g(x) = \sqrt{x - 4} - 2$

MENTAL MATH

State the vertex of the graph of each quadratic function.

1. $f(x) = x^2$ **2.** $f(x) = -5x^2$ **3.** $g(x) = (x - 2)^2$ **4.** $g(x) = (x + 5)^2$

5. $f(x) = 2x^2 + 3$ **6.** $h(x) = x^2 - 1$ **7.** $g(x) = (x + 1)^2 + 5$ **8.** $h(x) = (x - 10)^2 - 7$

EXERCISE SET 11.5

STUDY GUIDE/SSM   CD/VIDEO   PH MATH TUTOR CENTER   MathXL®Tutorials ON CD   MathXL®   MyMathLab®

Sketch the graph of each quadratic function. Label the vertex, and sketch and label the axis of symmetry. See Examples 1 through 5.

1. $f(x) = x^2 - 1$ **2.** $g(x) = x^2 + 3$

3. $h(x) = x^2 + 5$ **4.** $h(x) = x^2 - 4$

5. $g(x) = x^2 + 7$ **6.** $f(x) = x^2 - 2$

7. $f(x) = (x - 5)^2$ **8.** $g(x) = (x + 5)^2$

9. $h(x) = (x + 2)^2$ **10.** $H(x) = (x - 1)^2$

11. $G(x) = (x + 3)^2$ **12.** $f(x) = (x - 6)^2$

13. $f(x) = (x - 2)^2 + 5$ **14.** $g(x) = (x - 6)^2 + 1$

15. $h(x) = (x + 1)^2 + 4$ **16.** $G(x) = (x + 3)^2 + 3$

17. $g(x) = (x + 2)^2 - 5$ **18.** $h(x) = (x + 4)^2 - 6$

Sketch the graph of each quadratic function. Label the vertex, and sketch and label the axis of symmetry. See Examples 6 and 7.

19. $g(x) = -x^2$ **20.** $f(x) = 5x^2$

21. $h(x) = \frac{1}{3}x^2$ **22.** $g(x) = -3x^2$

23. $H(x) = 2x^2$ **24.** $f(x) = -\frac{1}{4}x^2$

Sketch the graph of each quadratic function. Label the vertex, and sketch and label the axis of symmetry. See Example 8.

25. $f(x) = 2(x - 1)^2 + 3$ **26.** $g(x) = 4(x - 4)^2 + 2$

27. $h(x) = -3(x + 3)^2 + 1$ **28.** $f(x) = -(x - 2)^2 - 6$

29. $H(x) = \frac{1}{2}(x - 6)^2 - 3$ **30.** $G(x) = \frac{1}{5}(x + 4)^2 + 3$

MIXED PRACTICE

Sketch the graph of each quadratic function. Label the vertex, and sketch and label the axis of symmetry.

31. $f(x) = -(x - 2)^2$ **32.** $g(x) = -(x + 6)^2$

33. $F(x) = -x^2 + 4$ **34.** $H(x) = -x^2 + 10$

35. $F(x) = 2x^2 - 5$ **36.** $g(x) = \dfrac{1}{2}x^2 - 2$

37. $h(x) = (x - 6)^2 + 4$ **38.** $f(x) = (x - 5)^2 + 2$

39. $F(x) = \left(x + \dfrac{1}{2}\right)^2 - 2$ **40.** $H(x) = \left(x + \dfrac{1}{2}\right)^2 - 3$

41. $F(x) = \dfrac{3}{2}(x + 7)^2 + 1$ **42.** $g(x) = -\dfrac{3}{2}(x - 1)^2 - 5$

43. $f(x) = \dfrac{1}{4}x^2 - 9$ **44.** $H(x) = \dfrac{3}{4}x^2 - 2$

45. $G(x) = 5\left(x + \dfrac{1}{2}\right)^2$ **46.** $F(x) = 3\left(x - \dfrac{3}{2}\right)^2$

47. $h(x) = -(x - 1)^2 - 1$ **48.** $f(x) = -3(x + 2)^2 + 2$

49. $g(x) = \sqrt{3}(x + 5)^2 + \dfrac{3}{4}$

50. $G(x) = \sqrt{5}(x - 7)^2 - \dfrac{1}{2}$

51. $h(x) = 10(x + 4)^2 - 6$

52. $h(x) = 8(x + 1)^2 + 9$

53. $f(x) = -2(x - 4)^2 + 5$

54. $G(x) = -4(x + 9)^2 - 1$

REVIEW AND PREVIEW

Add the proper constant to each binomial so that the resulting trinomial is a perfect square trinomial. See Section 11.1.

55. $x^2 + 8x$ **56.** $y^2 + 4y$

57. $z^2 - 16z$ **58.** $x^2 - 10x$

59. $y^2 + y$ **60.** $z^2 - 3z$

Solve by completing the square. See Section 11.1.

61. $x^2 + 4x = 12$ **62.** $y^2 + 6y = -5$

63. $z^2 + 10z - 1 = 0$ **64.** $x^2 + 14x + 20 = 0$

65. $z^2 - 8z = 2$ **66.** $y^2 - 10y = 3$

Concept Extensions

Write the equation of the parabola that has the same shape as $f(x) = 5x^2$ but with the following vertex.

67. $(2, 3)$ **68.** $(1, 6)$

69. $(-3, 6)$ **70.** $(4, -1)$

The shifting properties covered in this section apply to the graphs of all functions. Given the accompanying graph of $y = f(x)$, sketch the graph of each of the following.

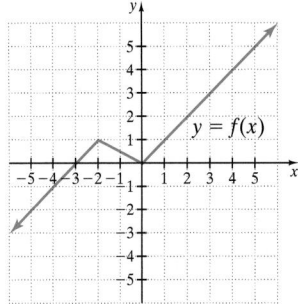

71. $y = f(x) + 1$ **72.** $y = f(x) - 2$

73. $y = f(x - 3)$ **74.** $y = f(x + 3)$

75. $y = f(x + 2) + 2$ **76.** $y = f(x - 1) + 1$

11.6   *FURTHER GRAPHING OF QUADRATIC FUNCTIONS*

Objectives

1 Write quadratic functions in the form $y = a(x - h)^2 + k$.

2 Derive a formula for finding the vertex of a parabola.

3 Find the minimum or maximum value of a quadratic function.

1   We know that the graph of a quadratic function is a parabola. If a quadratic function is written in the form

$$f(x) = a(x - h)^2 + k$$

we can easily find the vertex (h, k) and graph the parabola. To write a quadratic function in this form, complete the square. (See Section 11.1 for a review of completing the square.)

EXAMPLE 1

Graph $f(x) = x^2 - 4x - 12$. Find the vertex and any intercepts.

Solution   The graph of this quadratic function is a parabola. To find the vertex of the parabola, we will write the function in the form $y = (x - h)^2 + k$. To do this, we complete the square on the binomial $x^2 - 4x$. To simplify our work, we let $f(x) = y$.

$$y = x^2 - 4x - 12 \qquad \text{Let } f(x) = y.$$

$$y + 12 = x^2 - 4x \qquad \text{Add } \mathbf{12} \text{ to both sides to get}$$
$$\text{the } x\text{-variable terms alone.}$$

Now we add the square of half of -4 to both sides.

$$\frac{1}{2}(-4) = -2 \quad \text{and} \quad (-2)^2 = 4$$

$$y + 12 + 4 = x^2 - 4x + 4 \qquad \text{Add } \mathbf{4} \text{ to both sides.}$$

$$y + 16 = (x - 2)^2 \qquad \text{Factor the trinomial.}$$

$$y = (x - 2)^2 - 16 \qquad \text{Subtract } \mathbf{16} \text{ from both sides.}$$

$$f(x) = (x - 2)^2 - 16 \qquad \text{Replace } y \text{ with } f(x).$$

From this equation, we can see that the vertex of the parabola is $(2, -16)$, a point in quadrant IV, and the axis of symmetry is the line $x = 2$.

Notice that $a = 1$. Since $a > 0$, the parabola opens upward. This parabola opening upward with vertex $(2, -16)$ will have two x-intercepts and one y-intercept. (See the Helpful Hint after this example.)

x-intercepts: let y or $f(x) = 0$ y-intercept: let $x = 0$

$$f(x) = x^2 - 4x - 12 \qquad\qquad f(x) = x^2 - 4x - 12$$

$$0 = x^2 - 4x - 12 \qquad\qquad f(0) = 0^2 - 4 \cdot 0 - 12$$

$$0 = (x - 6)(x + 2) \qquad\qquad = -12$$

$$0 = x - 6 \quad \text{or} \quad 0 = x + 2$$

$$6 = x \qquad\qquad -2 = x$$

The two x-intercepts are $(6, 0)$ and $(-2, 0)$. The y-intercept is $(0, -12)$. The sketch of $f(x) = x^2 - 4x - 12$ is shown.

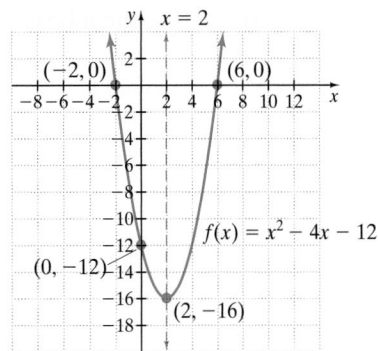

Notice that the axis of symmetry is always halfway between the x-intercepts. For the example above, halfway between -2 and 6 is $\dfrac{-2 + 6}{2} = 2$, and the axis of symmetry is $x = 2$.

Helpful Hint

Parabola Opens Upward
Vertex in I or II: no x-intercept
Vertex in III or IV: 2 x-intercepts

Parabola Opens Downward
Vertex in I or II: 2 x-intercepts
Vertex in III or IV: no x-intercept.

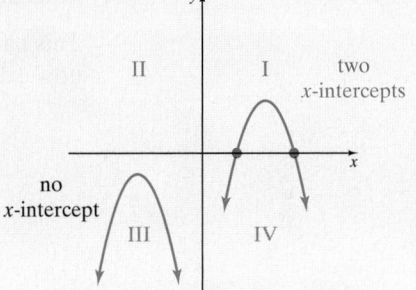

EXAMPLE 2

Graph $f(x) = 3x^2 + 3x + 1$. Find the vertex and any intercepts.

Solution   Replace $f(x)$ with y and complete the square on x to write the equation in the form $y = a(x - h)^2 + k$.

$$y = 3x^2 + 3x + 1 \qquad \text{Replace } f(x) \text{ with } y.$$

$$y - 1 = 3x^2 + 3x \qquad \text{Isolate } x\text{-variable terms.}$$

Factor 3 from the terms $3x^2 + 3x$ so that the coefficient of x^2 is 1.

$$y - 1 = 3(x^2 + x) \qquad \text{Factor out 3.}$$

The coefficient of x in the parentheses above is 1. Then $\dfrac{1}{2}(1) = \dfrac{1}{2}$ and $\left(\dfrac{1}{2}\right)^2 = \dfrac{1}{4}$.

Since we are adding $\dfrac{1}{4}$ inside the parentheses, we are really adding $3\left(\dfrac{1}{4}\right)$, so we *must* add $3\left(\dfrac{1}{4}\right)$ to the left side.

$$y - 1 + 3\left(\frac{1}{4}\right) = 3\left(x^2 + x + \frac{1}{4}\right)$$

$$y - \frac{1}{4} = 3\left(x + \frac{1}{2}\right)^2 \qquad \text{Simplify the left side and factor the right side.}$$

$$y = 3\left(x + \frac{1}{2}\right)^2 + \frac{1}{4} \qquad \text{Add } \frac{1}{4} \text{ to both sides.}$$

$$f(x) = 3\left(x + \frac{1}{2}\right)^2 + \frac{1}{4} \qquad \text{Replace } y \text{ with } f(x).$$

Then $a = 3$, $h = -\dfrac{1}{2}$, and $k = \dfrac{1}{4}$. This means that the parabola opens upward with vertex $\left(-\dfrac{1}{2}, \dfrac{1}{4}\right)$ and that the axis of symmetry is the line $x = -\dfrac{1}{2}$.

To find the y-intercept, let $x = 0$. Then

$$f(0) = 3(0)^2 + 3(0) + 1 = 1$$

Thus the y-intercept is $(0, 1)$.

This parabola has no x-intercepts since the vertex is in the second quadrant and opens upward. Use the vertex, axis of symmetry, and y-intercept to sketch the parabola.

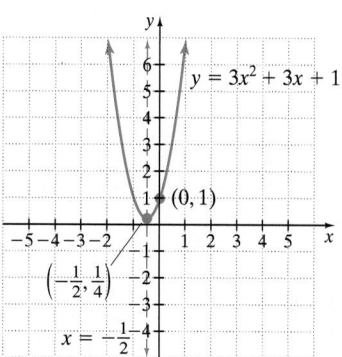

EXAMPLE 3

Graph $f(x) = -x^2 - 2x + 3$. Find the vertex and any intercepts.

Solution We write $f(x)$ in the form $a(x - h)^2 + k$ by completing the square. First we replace $f(x)$ with y.

$$f(x) = -x^2 - 2x + 3$$

$$y = -x^2 - 2x + 3$$

$$y - 3 = -x^2 - 2x$$

Subtract 3 from both sides to get the
x-variable terms alone.

$$y - 3 = -1(x^2 + 2x)$$

Factor -1 from the terms $-x^2 - 2x$.

The coefficient of *x* is 2. Then $\frac{1}{2}(2) = 1$ and $1^2 = 1$. We add 1 to the right side inside the parentheses and add $-1(1)$ to the left side.

$$y - 3 - 1(1) = -1(x^2 + 2x + 1)$$

$$y - 4 = -1(x + 1)^2$$

Simplify the left side and
factor the right side.

$$y = -1(x + 1)^2 + 4$$

Add 4 to both sides.

$$\underline{f(x) = -1(x + 1)^2 + 4}$$

Replace *y* with *f*(*x*).

> **Helpful Hint**
>
> This can be written as
> $f(x) = -1[x - (-1)]^2 + 4$.
> Notice that the vertex is
> $(-1, 4)$.

Since $a = -1$, the parabola opens downward with vertex $(-1, 4)$ and axis of symmetry $x = -1$.

To find the *y*-intercept, we let $x = 0$ and solve for *y*. Then

$$f(0) = -0^2 - 2(0) + 3 = 3$$

Thus, $(0, 3)$ is the *y*-intercept.

To find the *x*-intercepts, we let *y* or $f(x) = 0$ and solve for *x*.

$$f(x) = -x^2 - 2x + 3$$

$$0 = -x^2 - 2x + 3$$

Let $f(x) = 0$.

Now we divide both sides by -1 so that the coefficient of x^2 is 1.

$$\frac{0}{-1} = \frac{-x^2}{-1} - \frac{2x}{-1} + \frac{3}{-1}$$

Divide both sides by -1.

$$0 = x^2 + 2x - 3$$

Simplify.

$$0 = (x + 3)(x - 1)$$

Factor.

$$x + 3 = 0 \quad \text{or} \quad x - 1 = 0$$

Set each factor equal to 0.

$$x = -3 \qquad x = 1$$

Solve.

The *x*-intercepts are $(-3, 0)$ and $(1, 0)$. Use these points to sketch the parabola.

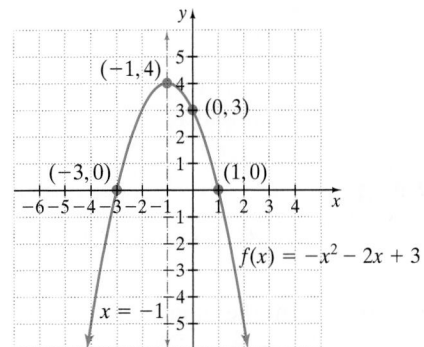

2 There is also a formula that may be used to find the vertex of a parabola. Now that we have practiced completing the square, we will show that the x-coordinate of the vertex of the graph of $f(x)$ or $y = ax^2 + bx + c$ can be found by the formula $x = \dfrac{-b}{2a}$. To do so, we complete the square on x and write the equation in the form $y = a(x - h)^2 + k$.

First, isolate the x-variable terms by subtracting c from both sides.

$$y = ax^2 + bx + c$$

$$y - c = ax^2 + bx$$

Next, factor a from the terms $ax^2 + bx$.

$$y - c = a\left(x^2 + \frac{b}{a}x\right)$$

Next, add the square of half of $\dfrac{b}{a}$, or $\left(\dfrac{b}{2a}\right)^2 = \dfrac{b^2}{4a^2}$, to the right side inside the parentheses.

Because of the factor a, what we really added was $a\left(\dfrac{b^2}{4a^2}\right)$ and this must be added to the left side.

$$y - c + \left(\right) = \left(x^2 + \frac{b}{a}x + \right)$$

$$y - c + \frac{b^2}{4a} = a\left(x + \frac{b}{2a}\right)^2 \qquad \text{Simplify the left side and factor the right side.}$$

$$y = a\left(x + \frac{b}{2a}\right)^2 + c - \frac{b^2}{4a} \qquad \begin{array}{l}\text{Add } c \text{ to both sides and sub-}\\ \text{tract } \dfrac{b^2}{4a} \text{ from both sides.}\end{array}$$

Compare this form with $f(x)$ or $y = a(x - h)^2 + k$ and see that h is $\dfrac{-b}{2a}$, which means that the x-coordinate of the vertex of the graph of $f(x) = ax^2 + bx + c$ is $\dfrac{-b}{2a}$.

Vertex Formula

The graph of $f(x) = ax^2 + bx + c$, when $a \neq 0$, is a parabola with vertex

$$\left(\frac{-b}{2a}, f\left(\frac{-b}{2a}\right)\right)$$

Let's use this formula to find the vertex of the parabola we graphed in Example 1.

EXAMPLE 4

Find the vertex of the graph of $f(x) = x^2 - 4x - 12$.

Solution In the quadratic function $f(x) = x^2 - 4x - 12$, notice that $a = 1, b = -4$, and $c = -12$. Then

$$\frac{-b}{2a} = \frac{-(-4)}{2(1)} = 2$$

The x-value of the vertex is 2. To find the corresponding $f(x)$ or y-value, find $f(2)$. Then

$$f(2) = 2^2 - 4(2) - 12 = 4 - 8 - 12 = -16$$

The vertex is $(2, -16)$. These results agree with our findings in Example 1.

3   The vertex of a parabola gives us some important information about its corresponding quadratic function. The quadratic function whose graph is a parabola that opens upward has a minimum value, and the quadratic function whose graph is a parabola that opens downward has a maximum value. The $f(x)$ or y-value of the vertex is the minimum or maximum value of the function.

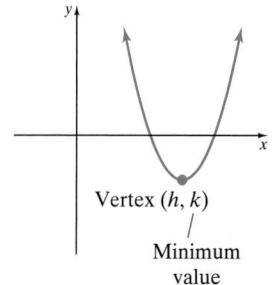

✔ **CONCEPT CHECK**

Without making any calculations, tell whether the graph of $f(x) = 7 - x - 0.3x^2$ has a maximum value or a minimum value. Explain your reasoning.

EXAMPLE 5

FINDING MAXIMUM HEIGHT

A rock is thrown upward from the ground. Its height in feet above ground after t seconds is given by the function $f(t) = -16t^2 + 20t$. Find the maximum height of the rock and the number of seconds it took for the rock to reach its maximum height.

Solution   **1.** UNDERSTAND. The maximum height of the rock is the largest value of $f(t)$. Since the function $f(t) = -16t^2 + 20t$ is a quadratic function, its graph is a parabola. It opens downward since $-16 < 0$. Thus, the maximum value of $f(t)$ is the $f(t)$ or y-value of the vertex of its graph.

2. TRANSLATE. To find the vertex (h, k), notice that for $f(t) = -16t^2 + 20t$, $a = -16, b = 20$, and $c = 0$. We will use these values and the vertex formula

$$\left(\frac{-b}{2a}, f\left(\frac{-b}{2a} \right) \right)$$

3. SOLVE.

$$h = \frac{-b}{2a} = \frac{-20}{-32} = \frac{5}{8}$$

$$f\left(\frac{5}{8} \right) = -16\left(\frac{5}{8} \right)^2 + 20\left(\frac{5}{8} \right)$$

Concept Check Answer:
$f(x)$ has a maximum value since it opens downward.

$$= -16\left(\frac{25}{64}\right) + \frac{25}{2}$$

$$= -\frac{25}{4} + \frac{50}{4} = \frac{25}{4}$$

4. INTERPRET. The graph of $f(t)$ is a parabola opening downward with vertex $\left(\frac{5}{8}, \frac{25}{4}\right)$.

This means that the rock's maximum height is $\frac{25}{4}$ feet, or $6\frac{1}{4}$ feet, which was reached in $\frac{5}{8}$ second.

Spotlight on

DECISION
✈MAKING

Suppose you are a member of a community theater group, the Slidell Players. For an upcoming performance of *Grease*, your group must decide on a ticket price. The graph shows the relationship between ticket price and box office receipts for past Slidell Players performances.

What ticket price would you suggest that the Slidell Players charge for its performance of *Grease*? Explain your reasoning.

EXERCISE SET 11.6

STUDY GUIDE/SSM CD/VIDEO PH MATH TUTOR CENTER MathXL®Tutorials ON CD MathXL® MyMathLab®

Find the vertex of the graph of each quadratic function. See Examples 1 through 4.

1. $f(x) = x^2 + 8x + 7$ **2.** $f(x) = x^2 + 6x + 5$

3. $f(x) = -x^2 + 10x + 5$ **4.** $f(x) = -x^2 - 8x + 2$

5. $f(x) = 5x^2 - 10x + 3$ **6.** $f(x) = -3x^2 + 6x + 4$

7. $f(x) = -x^2 + x + 1$ **8.** $f(x) = x^2 - 9x + 8$

Match each function with its graph. See Examples 1 through 4.

A

$(-1, -4)$

B

$(1, -4)$

C

$(-2, -1)$

D
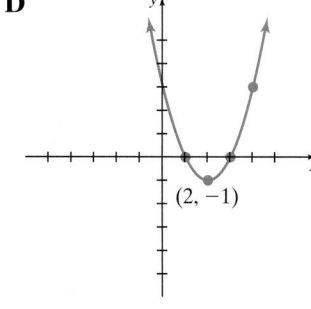
$(2, -1)$

9. $f(x) = x^2 - 4x + 3$ **10.** $f(x) = x^2 + 2x - 3$

11. $f(x) = x^2 - 2x - 3$ **12.** $f(x) = x^2 + 4x + 3$

MIXED PRACTICE

Find the vertex of the graph of each quadratic function. Determine whether the graph opens upward or downward, find any intercepts, and sketch the graph. See Examples 1 through 3.

13. $f(x) = x^2 + 4x - 5$ **14.** $f(x) = x^2 + 2x - 3$

15. $f(x) = -x^2 + 2x - 1$ **16.** $f(x) = -x^2 + 4x - 4$

17. $f(x) = x^2 - 4$ **18.** $f(x) = x^2 - 1$

19. $f(x) = 4x^2 + 4x - 3$ **20.** $f(x) = 2x^2 - x - 3$

21. $f(x) = x^2 + 8x + 15$ **22.** $f(x) = x^2 + 10x + 9$

23. $f(x) = x^2 - 6x + 5$ **24.** $f(x) = x^2 - 4x + 3$

25. $f(x) = x^2 - 4x + 5$ **26.** $f(x) = x^2 - 6x + 11$

27. $f(x) = 2x^2 + 4x + 5$ **28.** $f(x) = 3x^2 + 12x + 16$

29. $f(x) = -2x^2 + 12x$ **30.** $f(x) = -4x^2 + 8x$

31. $f(x) = x^2 + 1$ **32.** $f(x) = x^2 + 4$

33. $f(x) = x^2 - 2x - 15$ **34.** $f(x) = x^2 - 4x + 3$

35. $f(x) = -5x^2 + 5x$ **36.** $f(x) = 3x^2 - 12x$

37. $f(x) = -x^2 + 2x - 12$ **38.** $f(x) = -x^2 + 8x - 17$

39. $f(x) = 3x^2 - 12x + 15$ **40.** $f(x) = 2x^2 - 8x + 11$

41. $f(x) = x^2 + x - 6$ **42.** $f(x) = x^2 + 3x - 18$

43. $f(x) = -2x^2 - 3x + 35$ **44.** $f(x) = 3x^2 - 13x - 10$

Solve. See Example 5.

45. If a projectile is fired straight upward from the ground with an initial speed of 96 feet per second, then its height h in feet after t seconds is given by the equation

$$h(t) = -16t^2 + 96t$$

Find the maximum height of the projectile.

46. The cost C in dollars of manufacturing x bicycles at Holladay's Production Plant is given by the function $C(x) = 2x^2 - 800x + 92{,}000$.

  a. Find the number of bicycles that must be manufactured to minimize the cost.

  b. Find the minimum cost.

47. If Rheam Gaspar throws a ball upward with an initial speed of 32 feet per second, then its height h in feet after t seconds is given by the equation

$$h(t) = -16t^2 + 32t$$

Find the maximum height of the ball.

48. The Utah Ski Club sells calendars to raise money. The profit P, in cents, from selling x calendars is given by the equation $P(x) = 360x - x^2$.

  a. Find how many calendars must be sold to maximize profit.

  b. Find the maximum profit.

49. Find two numbers whose sum is 60 and whose product is as large as possible. [*Hint:* Let x and $60 - x$ be the two positive numbers. Their product can be described by the function $f(x) = x(60 - x)$.]

50. Find two numbers whose sum is 11 and whose product is as large as possible. (Use the hint for Exercise 49.)

51. Find two numbers whose difference is 10 and whose product is as small as possible. (Use the hint for Exercise 49.)

52. Find two numbers whose difference is 8 and whose product is as small as possible.

53. The length and width of a rectangle must have a sum of 40. Find the dimensions of the rectangle that will have the maximum area. (Use the hint for Exercise 49.)

54. The length and width of a rectangle must have a sum of 50. Find the dimensions of the rectangle that will have maximum area.

REVIEW AND PREVIEW

Sketch the graph of each function. See Section 11.5.

55. $f(x) = x^2 + 2$ **56.** $f(x) = (x - 3)^2$

57. $g(x) = x + 2$ **58.** $h(x) = x - 3$

59. $f(x) = (x + 5)^2 + 2$ **60.** $f(x) = 2(x - 3)^2 + 2$

61. $f(x) = 3(x - 4)^2 + 1$ **62.** $f(x) = (x + 1)^2 + 4$

63. $f(x) = -(x - 4)^2 + \dfrac{3}{2}$ **64.** $f(x) = -2(x + 7)^2 + \dfrac{1}{2}$

Concept Extensions

Find the vertex of the graph of each quadratic function. Determine whether the graph opens upward or downward, find the y-intercept, approximate the x-intercepts to one decimal place, and sketch the graph.

65. $f(x) = x^2 + 10x + 15$ **66.** $f(x) = x^2 - 6x + 4$

67. $f(x) = 3x^2 - 6x + 7$ **68.** $f(x) = 2x^2 + 4x - 1$

Find the maximum or minimum value of each function. Approximate to two decimal places.

69. $f(x) = 2.3x^2 - 6.1x + 3.2$

70. $f(x) = 7.6x^2 + 9.8x - 2.1$

71. $f(x) = -1.9x^2 + 5.6x - 2.7$

72. $f(x) = -5.2x^2 - 3.8x + 5.1$

73. The number of inmates in custody in U.S. prisons and jails can be modeled by the quadratic function

$$p(x) = -x^2 + 93x + 1128$$

where $p(x)$ is the number of inmates in thousands and x is the number of years after 1990. (*Source*: Based on data from the Bureau of Justice Statistics, U.S. Department of Justice, 1990–2000)

  a. Will this function have a maximum or a minimum? How can you tell?

  b. According to this model, in what year will the number of prison and jail inmates in custody in the United States be at its maximum/minimum?

  c. What is the maximum/minimum number of inmates predicted?

74. Methane is a gas produced by landfills, natural gas systems, and coal mining that contributes to the greenhouse effect and global warming. Projected methane emissions in the United States can be modeled by the quadratic function

$$f(x) = -0.072x^2 + 1.93x + 173.9$$

where $f(x)$ is the amount of methane produced in million metric tons and x is the number of years after 2000. (*Source*: Based on data from the U.S. Environmental Protection Agency, 2000–2020)

  a. According to this model, what will U.S. emissions of methane be in 2009?

  b. Will this function have a maximum or a minimum? How can you tell?

  c. In what year will methane emissions in the United States be at their maximum/minimum? Round to the nearest whole year.

  d. What is the level of methane emissions for that year? (Use your rounded answer from part **c**.)

Use a graphing calculator to check each exercise.

75. Exercise 27   **76.** Exercise 28

77. Exercise 37   **78.** Exercise 38

CHAPTER 11 PROJECT

Fitting a Quadratic Model to Data

Throughout the twentieth century, the eating habits of Americans changed noticeably. Americans started consuming less whole milk and butter, and started consuming more skim and low-fat milk and margarine. We also started eating more poultry and fish. In this project, you will have the opportunity to investigate trends in per capita consumption of poultry during the twentieth century. This project may be completed by working in groups or individually.

    We will start by finding a quadratic model, $y = ax^2 + bx + c$, that has ordered pair solutions that correspond to the data for U.S. per capita consumption of poultry given in the table. To do so, substitute each data pair into the equation. Each time, the result is an equation in three unknowns: a, b, and c. Because there are three pairs of data, we can form a system of three linear equations in three

unknowns. Solving for the values of a, b, and c gives a quadratic model that represents the given data.

U.S. PER CAPITA CONSUMPTION OF POULTRY (IN POUNDS)

Year	x	Poultry Consumption, y (in pounds)
1909	9	11
1969	69	33
2001	101	66

(*Source*: Economic research service, U.S. Department of Agriculture)

1. Write the system of equations that must be solved to find the values of *a*, *b*, and *c* needed for a quadratic model of the given data.

2. Solve the system of equations for *a*, *b*, and *c*. Recall the various methods of solving linear systems used in Chapter 4. You might consider using matrices, Cramer's rule, or a graphing calculator to do so. Round to the nearest thousandth.

3. Write the quadratic model for the data. Note that the variable *x* represents the number of years after 1900.

4. In 1939, the actual U.S. per capita consumption of poultry was 12 pounds per person. Based on this information, how accurate do you think this model is for years other than those given in the table?

5. Use your model to estimate the per capita consumption of poultry in 1950.

6. According to the model, in what year was per capita consumption of poultry 50 pounds per person?

7. In what year was the per capita consumption of poultry at its lowest level? What was that level?

8. Who might be interested in a model like this and how would it be helpful?

STUDY SKILLS REMINDER

Are You Preparing for a Test on Chapter 11?

Below I have listed some common trouble areas for topics covered in Chapter 11. After studying for your test—but before taking your test—read these.

▶ Don't forget that to solve a quadratic equation such as $x^2 + 6x = 1$, by completing the square, add the square of half of   to both sides.

$$x^2 + 6x = 1$$
$$x^2 + 6x + 9 = 1 + 9 \qquad \text{Add \textbf{9} to both sides,}$$
$$(x + 3)^2 = 10 \qquad \left(\frac{1}{2}(6) = 3 \text{ and } 3^2 = 9 \right)$$
$$x + 3 = \pm \sqrt{10}$$
$$x = -3 \pm \sqrt{10}$$

▶ Remember to write a quadratic equation in standard form $(ax^2 + bx + c = 0)$ before using the quadratic formula to solve.

$$x(4x - 1) = 1$$
$$4x^2 - x - 1 = 0 \qquad \qquad \text{Write in standard form.}$$
$$x = \frac{-(-1) \pm \sqrt{(-1)^2 - 4(4)(-1)}}{2 \cdot 4} \qquad \begin{array}{l} \text{Use the quadratic} \\ \text{formula with} \\ a = 4, b = -1, \\ \text{and } c = -1. \end{array}$$

$$x = \frac{1 \pm \sqrt{17}}{8} \qquad \text{Simplify.}$$

▶ Review the steps for solving a quadratic equation in general on page 654.

▶ Don't forget how to graph a quadratic function in the form $f(x) = a(x - h)^2 + k$. The graph of $f(x) = -2(x - 3)^2 - 1$

opens
downward $\overbrace{\text{shift 3}}$ $\overbrace{\text{shift 1}}$
            units right   unit down

narrower

(continued)

Remember: This is simply a checklist of common trouble areas. For a review of Chapter 11, see the Highlights and Chapter Review at the end of this chapter.

CHAPTER VOCABULARY CHECK

Fill in each blank with one of the words or phrases listed below.

quadratic formula	quadratic	discriminant	$\pm\sqrt{b}$
completing the square	quadratic inequality	(h, k)	$(0, k)$
$(h, 0)$	$\dfrac{-b}{2a}$		

1. The _____ helps us know find the number and type of solutions of a quadratic equation.

2. If $a^2 = b$, then $a = $ _____

3. The graph of $f(x) = ax^2 + bx + c$ where a is not 0 is a parabola whose vertex has x-value of ___.

4. A(n) _____ is an inequality that can be written so that one side is a quadratic expression and the other side is 0.

5. The process of writing a quadratic equation so that one side is a perfect square trinomial is called _____.

6. The graph of $f(x) = x^2 + k$ has vertex _____.

7. The graph of $f(x) = (x - h)^2$ has vertex _____.

8. The graph of $f(x) = (x - h)^2 + k$ has vertex _____.

9. The formula $x = \dfrac{-b \pm \sqrt{b^2 - 4ac}}{2a}$ is called the _____.

10. A _____ equation is one that can be written in the form $ax^2 + bx + c = 0$ where a, b, and c are real numbers and a is not 0.

CHAPTER 11 HIGHLIGHTS

Definitions and Concepts	**Examples**

Section 11.1   Solving Quadratic Equations by Completing the Square

Square root property

If b is a real number and if $a^2 = b$, then $a = \pm\sqrt{b}$.

To solve a quadratic equation in x by completing the square

Step 1: If the coefficient of x^2 is not 1, divide both sides of the equation by the coefficient of x^2.

Step 2: Isolate the variable terms.

Step 3: Complete the square by adding the square of half of the coefficient of x to both sides.

Step 4: Write the resulting trinomial as the square of a binomial.

Step 5: Apply the square root property and solve for x.

Solve $(x + 3)^2 = 14$.
$$x + 3 = \pm\sqrt{14}$$
$$x = -3 \pm \sqrt{14}$$

Solve $3x^2 - 12x - 18 = 0$.

1. $x^2 - 4x - 6 = 0$

2. $x^2 - 4x = 6$

3. $\frac{1}{2}(-4) = -2$ and $(-2)^2 = 4$
$$x^2 - 4x + 4 = 6 + 4$$

4. $\quad (x - 2)^2 = 10$

5. $\quad x - 2 = \pm\sqrt{10}$
$$x = 2 \pm \sqrt{10}$$

Section 11.2   Solving Quadratic Equations by the Quadratic Formula

A quadratic equation written in the form $ax^2 + bx + c = 0$ has solutions

$$x = \frac{-b \pm \sqrt{b^2 - 4ac}}{2a}$$

Solve $x^2 - x - 3 = 0$.
$$a = 1, b = -1, c = -3$$
$$x = \frac{-(-1) \pm \sqrt{(-1)^2 - 4(1)(-3)}}{2 \cdot 1}$$
$$x = \frac{1 \pm \sqrt{13}}{2}$$

Section 11.3   Solving Equations by Using Quadratic Methods

Substitution is often helpful in solving an equation that contains a repeated variable expression.

Solve $(2x + 1)^2 - 5(2x + 1) + 6 = 0$.
Let $m = 2x + 1$. Then

$$m^2 - 5m + 6 = 0 \qquad \text{Let } m = 2x + 1.$$
$$(m - 3)(m - 2) = 0$$
$$m = 3 \quad \text{or} \quad m = 2$$
$$2x + 1 = 3 \quad \text{or} \quad 2x + 1 = 2 \quad \text{Substitute back.}$$
$$x = 1 \quad \text{or} \quad x = \frac{1}{2}$$

Definitions and Concepts	Examples

Section 11.4 Nonlinear Inequalities in One Variable

To solve a polynomial inequality

Step 1: Write the inequality in standard form.

Step 2: Solve the related equation.

Step 3: Use solutions from Step 2 to separate the number line into regions.

Step 4: Use test points to determine whether values in each region satisfy the original inequality.

Step 5: Write the solution set as the union of regions whose test point value is a solution.

Solve $x^2 \geq 6x$.

1. $x^2 - 6x \geq 0$

2. $x^2 - 6x = 0$

$x(x - 6) = 0$

$x = 0 \quad \text{or} \quad x = 6$

3.

4.

Region	Test Point Value	$x^2 \geq 6x$	Result
A	-2	$(-2)^2 \geq 6(-2)$	True
B	1	$1^2 \geq 6(1)$	False
C	7	$7^2 \geq 6(7)$	True

5.

The solution set is $(-\infty, 0] \cup [6, \infty)$.

To solve a rational inequality

Step 1: Solve for values that make all denominators 0.

Step 2: Solve the related equation.

Step 3: Use solutions from Steps 1 and 2 to separate the number line into regions.

Step 4: Use test points to determine whether values in each region satisfy the original inequality.

Step 5: Write the solution set as the union of regions whose test point value is a solution.

Solve $\dfrac{6}{x - 1} < -2$.

1. $x - 1 = 0$ Set denominator equal to 0.

$x = 1$

2. $\dfrac{6}{x - 1} = -2$

$6 = -2(x - 1)$ Multiply by $(x - 1)$.

$6 = -2x + 2$

$4 = -2x$

$-2 = x$

3.

4. Only a test value from region B satisfies the original inequality.

5.

The solution set is $(-2, 1)$.

Definitions and Concepts	**Examples**

Section 11.5 Quadratic Functions and Their Graphs

Graph of a quadratic function

The graph of a quadratic function written in the form $f(x) = a(x - h)^2 + k$ is a parabola with vertex (h, k). If $a > 0$, the parabola opens upward; if $a < 0$, the parabola opens downward. The axis of symmetry is the line whose equation is $x = h$.

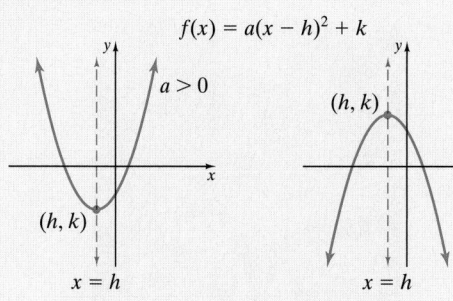

Graph $g(x) = 3(x - 1)^2 + 4$.

The graph is a parabola with vertex $(\ ,4)$ and axis of symmetry $x = 1$. Since $a = 3$ is positive, the graph opens upward.

Section 11.6 Further Graphing of Quadratic Functions

The graph of $f(x) = ax^2 + bx + c$, where $a \neq 0$, is a parabola with vertex

$$\left(\frac{-b}{2a}, f\left(\frac{-b}{2a} \right) \right)$$

Graph $f(x) = x^2 - 2x - 8$. Find the vertex and x- and y-intercepts.

$$\frac{-b}{2a} = \frac{-(-2)}{2 \cdot 1} = 1$$

$$f(1) = 1^2 - 2(1) - 8 = -9$$

The vertex is $(1, -9)$.

$$0 = x^2 - 2x - 8$$
$$0 = (x - 4)(x + 2)$$
$$x = 4 \quad \text{or} \quad x = -2$$

The x-intercepts are $(4, 0)$ and $(-2, 0)$.

$$f(0) = 0^2 - 2 \cdot 0 - 8 = -8$$

The y-intercept is $(0, -8)$.

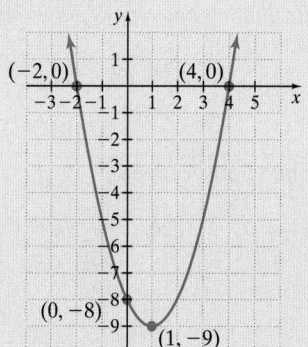

CHAPTER REVIEW

(11.1) *Solve by factoring.*

1. $x^2 - 15x + 14 = 0$

2. $x^2 - x - 30 = 0$

3. $10x^2 = 3x + 4$

4. $7a^2 = 29a + 30$

Solve by using the square root property.

5. $4m^2 = 196$

6. $9y^2 = 36$

7. $(9n + 1)^2 = 9$

8. $(5x - 2)^2 = 2$

Solve by completing the square.

9. $z^2 + 3z + 1 = 0$

10. $x^2 + x + 7 = 0$

11. $(2x + 1)^2 = x$

12. $(3x - 4)^2 = 10x$

13. If P dollars are originally invested, the formula $A = P(1 + r)^2$ gives the amount A in an account paying interest rate r compounded annually after 2 years. Find the interest rate r such that $2500 increases to $2717 in 2 years. Round the result to the nearest hundredth of a percent.

△ **14.** Two ships leave a port at the same time and travel at the same speed. One ship is traveling due north and the other due east. In a few hours, the ships are 150 miles apart. How many miles has each ship traveled? Give an exact answer and a one-decimal-place approximation.

(11.2) *If the discriminant of a quadratic equation has the given value, determine the number and type of solutions of the equation.*

15. -8

16. 48

17. 100

18. 0

Solve by using the quadratic formula.

19. $x^2 - 16x + 64 = 0$

20. $x^2 + 5x = 0$

21. $x^2 + 11 = 0$

22. $2x^2 + 3x = 5$

23. $6x^2 + 7 = 5x$

24. $9a^2 + 4 = 2a$

25. $(5a - 2)^2 - a = 0$

26. $(2x - 3)^2 = x$

27. Cadets graduating from military school usually toss their hats high into the air at the end of the ceremony. One cadet threw his hat so that its distance $d(t)$ in feet above the ground t seconds after it was thrown was $d(t) = -16t^2 + 30t + 6$.

a. Find the distance above the ground of the hat 1 second after it was thrown.

b. Find the time it takes the hat to hit the ground. Give an exact time and a one-decimal-place approximation.

△ **28.** The hypotenuse of an isosceles right triangle is 6 centimeters longer than either of the legs. Find the length of the legs.

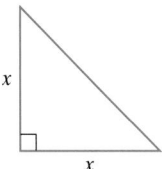

(11.3) *Solve each equation for the variable.*

29. $x^3 = 27$

30. $y^3 = -64$

31. $\dfrac{5}{x} + \dfrac{6}{x - 2} = 3$

32. $\dfrac{7}{8} = \dfrac{8}{x^2}$

33. $x^4 - 21x^2 - 100 = 0$

34. $5(x + 3)^2 - 19(x + 3) = 4$

35. $x^{2/3} - 6x^{1/3} + 5 = 0$

36. $x^{2/3} - 6x^{1/3} = -8$

37. $a^6 - a^2 = a^4 - 1$

38. $y^{-2} + y^{-1} = 20$

39. Two postal workers, Jerome Grant and Tim Bozik, can sort a stack of mail in 5 hours. Working alone, Tim can sort the mail in 1 hour less time than Jerome can. Find the time that each postal worker can sort the mail alone. Round the result to one decimal place.

40. A negative number decreased by its reciprocal is $-\dfrac{24}{5}$. Find the number.

(11.4) Solve each inequality for x. Graph the solution set and write each solution set in interval notation.

41. $2x^2 - 50 \le 0$

42. $\dfrac{1}{4}x^2 < \dfrac{1}{16}$

43. $(2x - 3)(4x + 5) \ge 0$

44. $(x^2 - 16)(x^2 - 1) > 0$

45. $\dfrac{x - 5}{x - 6} < 0$

46. $\dfrac{x(x + 5)}{4x - 3} \ge 0$

47. $\dfrac{(4x + 3)(x - 5)}{x(x + 6)} > 0$

48. $(x + 5)(x - 6)(x + 2) \le 0$

49. $x^3 + 3x^2 - 25x - 75 > 0$

50. $\dfrac{x^2 + 4}{3x} \le 1$

51. $\dfrac{(5x + 6)(x - 3)}{x(6x - 5)} < 0$

52. $\dfrac{3}{x - 2} > 2$

(11.5) Sketch the graph of each function. Label the vertex and the axis of symmetry.

53. $f(x) = x^2 - 4$

54. $g(x) = x^2 + 7$

55. $H(x) = 2x^2$

56. $h(x) = -\dfrac{1}{3}x^2$

57. $F(x) = (x - 1)^2$

58. $G(x) = (x + 5)^2$

59. $f(x) = (x - 4)^2 - 2$

60. $f(x) = -3(x - 1)^2 + 1$

(11.6) Sketch the graph of each function. Find the vertex and the intercepts.

61. $f(x) = x^2 + 10x + 25$

62. $f(x) = -x^2 + 6x - 9$

63. $f(x) = 4x^2 - 1$

64. $f(x) = -5x^2 + 5$

65. Find the vertex of the graph of $f(x) = -3x^2 - 5x + 4$. Determine whether the graph opens upward or downward, find the y-intercept, approximate the x-intercepts to one decimal place, and sketch the graph.

66. The function $h(t) = -16t^2 + 120t + 300$ gives the height in feet of a projectile fired from the top of a building in t seconds.

   a. When will the object reach a height of 350 feet? Round your answer to one decimal place.

   b. Explain why part **a** has two answers.

67. Find two numbers whose product is as large as possible, given that their sum is 420.

68. Write an equation of a quadratic function whose graph is a parabola that has vertex $(-3, 7)$ and that passes through the origin.

CHAPTER 11 TEST

Remember to use your Chapter Test Prep Video CD to help you study and view solutions to the test questions you need help with.

Solve each equation for the variable.

1. $5x^2 - 2x = 7$

2. $(x + 1)^2 = 10$

3. $m^2 - m + 8 = 0$

4. $y^2 - 3y = 5$

5. $\dfrac{4}{x + 2} + \dfrac{2x}{x - 2} = \dfrac{6}{x^2 - 4}$

6. $x^5 + 3x^4 = x + 3$

7. $(x + 1)^2 - 15(x + 1) + 56 = 0$

Solve the equation for the variable by completing the square.

8. $x^2 - 6x = -2$

9. $2a^2 + 5 = 4a$

Solve each inequality for x. Graph the solution set and then write the solution set in interval notation.

10. $2x^2 - 7x > 15$

11. $(x^2 - 16)(x^2 - 25) \ge 0$

12. $\dfrac{5}{x + 3} < 1$

13. $\dfrac{7x - 14}{x^2 - 9} \le 0$

Graph each function. Label the vertex.

14. $f(x) = 3x^2$

15. $G(x) = -2(x - 1)^2 + 5$

Graph each function. Find and label the vertex, y-intercept, and x-intercepts (if any).

16. $h(x) = x^2 - 4x + 4$ **17.** $F(x) = 2x^2 - 8x + 9$

18. Dave and Sandy Hartranft can paint a room together in 4 hours. Working alone, Dave can paint the room in 2 hours less time than Sandy can. Find how long it takes Sandy to paint the room alone.

19. A stone is thrown upward from a bridge. The stone's height in feet, $s(t)$, above the water t seconds after the stone is thrown is a function given by the equation
$s(t) = -16t^2 + 32t + 256$.

  a. Find the maximum height of the stone.

b. Find the time it takes the stone to hit the water. Round the answer to two decimal places.

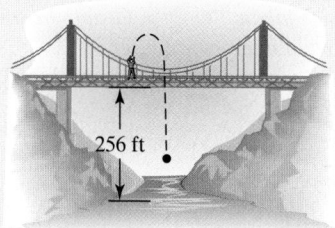

256 ft

△ **20.** Given the diagram shown, approximate to the nearest foot how many feet of walking distance a person saves by cutting across the lawn instead of walking on the sidewalk.

20 ft

$x + 8$

x

STUDY SKILLS REMINDER

Are You Prepared for Your Final Exam?

To prepare for your final exam, try the following study techniques.

▶ Review the material that you will be responsible for on your exam. Also check your notebook for any lecture notes that you highlighted.

▶ Review any formulas that you may need to memorize.

▶ Check to see if your instructor or math department will be conducting a final exam review.

▶ Check with your instructor to see whether there are final exams from previous semesters/quarters that are available to students for study.

▶ Use your previously taken tests as a practice final exam. To do so, rewrite the test questions in mixed order on blank sheets of paper. This will help you prepare for exam conditions.

▶ If you are unsure of a few topics, see your instructor or visit a learning lab for further assistance. Also, viewing the video segment of a troublesome section will help.

▶ If you need further exercises to work, try the chapter tests at the end of appropriate chapters.

  Good luck! I hope you have enjoyed this textbook and your algebra course.

CHAPTER CUMULATIVE REVIEW

1. Find the value of each expression when $x = 2$ and $y = -5$.

   a. $\dfrac{x - y}{12 + x}$ **b.** $x^2 - y$.

2. Solve $|3x - 2| = -5$.

3. Simplify each expression by combining like terms.

   a. $2x + 3x + 5 + 2$ **b.** $-5a - 3 + a + 2$

   c. $4y - 3y^2$ **d.** $2.3x + 5x - 6$

   e. $-\dfrac{1}{2}b + b$

4. Use the addition method to solve the system.

$$\begin{cases} -6x + y = 5 \\ 4x - 2y = 6 \end{cases}$$

5. Solve the following system of equations by graphing.

$$\begin{cases} 2x + y = 7 \\ 2y = -4x \end{cases}$$

6. Simplify. Use positive exponents to write each answer.

   a. $(a^{-2}bc^3)^{-3}$

   b. $\left(\dfrac{a^{-4}b^2}{c^3}\right)^{-2}$

   c. $\left(\dfrac{3a^8b^2}{12a^5b^5}\right)^{-2}$

7. Solve the system:

$$\begin{cases} 7x - 3y = -14 \\ -3x + y = 6 \end{cases}$$

8. Multiply.

   a. $(4a - 3)(7a - 2)$

   b. $(2a + b)(3a - 5b)$

9. Simplify the following.

   a. $\left(\dfrac{-5x^2}{y^3}\right)^2$ **b.** $\dfrac{(x^3)^4 x}{x^7}$

   c. $\dfrac{(2x)^5}{x^3}$ **d.** $\dfrac{(a^2b)^3}{a^3b^2}$

10. Factor.

   a. $9x^3 + 27x^2 - 15x$

   b. $2x(3y - 2) - 5(3y - 2)$

   c. $2xy + 6x - y - 3$

11. If $P(x) = 2x^3 - 4x^2 + 5$

   a. Find $P(2)$ by substitution.

   b. Use synthetic division to find the remainder when $P(x)$ is divided by $x - 2$.

12. Factor $x^2 - 2x - 48$.

13. Solve $(5x - 1)(2x^2 + 15x + 18) = 0$.

14. Factor. $2ax^2 - 12axy + 18ay^2$

15. Write the rational expression in lowest terms.

$$\dfrac{2x^2}{10x^3 - 2x^2}$$

16. Solve $2(a^2 + 2) - 8 = -2a(a - 2) - 5$.

17. Simplify.

$$\dfrac{x^{-1} + 2xy^{-1}}{x^{-2} - x^{-2}y^{-1}}.$$

18. Find the vertex and any intercepts of $f(x) = x^2 + x - 12$.

19. Factor $4m^2 - 4m + 1$.

20. Simplify $\dfrac{x^2 - 4x + 4}{2 - x}$.

21. The square of a number plus three times the number is 70. Find the number.

22. Subtract. $\dfrac{a + 1}{a^2 - 6a + 8} - \dfrac{3}{16 - a^2}$

23. Use the product rule to simplify.

   a. $\sqrt{25x^3}$

   b. $\sqrt[3]{54x^6y^8}$

   c. $\sqrt[4]{81z^{11}}$

24. Simplify $\dfrac{(2a)^{-1} + b^{-1}}{a^{-1} + (2b)^{-1}}$.

25. Rationalize the denominator of each expression.

a. $\dfrac{2}{\sqrt{5}}$

b. $\dfrac{2\sqrt{16}}{\sqrt{9x}}$

c. $\sqrt[3]{\dfrac{1}{2}}$

26. Divide $x^3 - 3x^2 - 10x + 24$ by $x + 3$.

27. Solve $\sqrt{2x + 5} + \sqrt{2x} = 3$.

28. If $P(x) = 4x^3 - 2x^2 + 3$,

  a. Find $P(-2)$ by substitution.

  b. Use synthetic division to find the remainder when $P(x)$ is divided by $x + 2$.

29. Solve $\dfrac{x}{2} + \dfrac{8}{3} = \dfrac{1}{6}$.

30. Solve $\dfrac{x + 3}{x^2 + 5x + 6} = \dfrac{3}{2x + 4} - \dfrac{1}{x + 3}$.

31. The quotient of a number and 6, minus $\dfrac{5}{3}$, is the quotient of the number and 2. Find the number.

32. Mr. Briley can roof his house in 24 hours. His son can roof the same house in 40 hours. If they work together, how long will it take to roof the house?

33. Suppose that y varies directly as x. If y is 5 when x is 30, find the constant of variation and the direct variation equation.

34. Suppose that y varies inversely as x. If y is 8 when x is 24, find the constant of variation and the inverse variation equation.

35. Simplify.

  a. $\sqrt{(-3)^2}$

  b. $\sqrt{x^2}$

  c. $\sqrt[4]{(x - 2)^4}$

  d. $\sqrt[3]{(-5)^3}$

  e. $\sqrt[5]{(2x - 7)^5}$

  f. $\sqrt{25x^2}$

  g. $\sqrt{x^2 + 2x + 1}$

36. Simplify. Assume that the variables represent any real number.

  a. $\sqrt{(-2)^2}$

  b. $\sqrt{y^2}$

  c. $\sqrt[4]{(a - 3)^4}$

  d. $\sqrt[3]{(-6)^3}$

  e. $\sqrt[5]{(3x - 1)^5}$

37. Use rational exponents to simplify. Assume that variables represent positive numbers.

  a. $\sqrt[8]{x^4}$

  b. $\sqrt[6]{25}$

  c. $\sqrt[4]{r^2 s^6}$

38. Use rational exponents to simplify. Assume that variables represent positive numbers.

  a. $\sqrt[4]{5^2}$

  b. $\sqrt[12]{x^3}$

  c. $\sqrt[6]{x^2 y^4}$

39. Divide. Write in the form $a + bi$.

  a. $\dfrac{2 + i}{1 - i}$

  b. $\dfrac{7}{3i}$

40. Write each product in the form of $a + bi$.

  a. $3i(5 - 2i)$

  b. $(6 - 5i)^2$

  c. $\left(\sqrt{3} + 2i\right)\left(\sqrt{3} - 2i\right)$

41. Use the square root property to solve $(x + 1)^2 = 12$.

42. Use the square root property to solve $(y - 1)^2 = 24$.

43. Solve $x - \sqrt{x} - 6 = 0$.

44. Use the quadratic formula to solve. $m^2 = 4m + 8$

12

Exponential and Logarithmic Functions

In this chapter, we discuss two closely related functions: exponential and logarithmic functions. These functions are vital to applications in economics, finance, engineering, the sciences, education, and other fields. Models of tumor growth and learning curves are two examples of the uses of exponential and logarithmic functions.

High quality of products or goods is essential to the success of any manufacturer. A quality assurance inspector is a person whose job is to determine whether a product meets industry standards for quality and safety. Inspectors test and retest samples at every step in the production process.

A quality assurance inspector is a person with a good eye for detail. Inspectors should be familiar with all aspects of the production process so that they can spot problems quickly. They should be able to work as members of a team, and they should possess the writing skills necessary to communicate problems in reports. Quality assurance inspectors also use technical tools such as problem solving, statistical quality control, and computer skills.

In the Spotlight on Decision Making feature on page 732, you will have the opportunity to make a decision about the reliability of a DVD player as a quality assurance inspector for an electronics manufacturer.

Link: www.asq.org *(American Society for Quality)*

Source of text: www.workforcedevelopment.com/engineering/qualitycontrol.html

12.1   *INVERSE FUNCTIONS*

Objectives

① Determine whether a function is a one-to-one function.

② Use the horizontal line test to decide whether a function is a one-to-one function.

③ Find the inverse of a function.

④ Find the equation of the inverse of a function.

⑤ Graph functions and their inverses.

⑥ Determine whether two functions are inverses of each other.

① In the next section, we begin a study of two new functions: exponential and logarithmic functions. As we learn more about these functions, we will discover that they share a special relation to each other: They are inverses of each other.

Before we study these functions, we need to learn about inverses. We begin by defining one-to-one functions.

Study the following diagram.

Recall from Section 8.3 that since each degrees Fahrenheit (input) corresponds to exactly one degrees Celsius (output), this pairing of inputs and outputs does describe a function. Also notice that each output corresponds to exactly one input. This type of function is given a special name—a one-to-one function.

Does the set $f = \{(0, 1), (2, 2), (-3, 5), (7, 6)\}$ describe a one-to-one function? It is a function since each x-value corresponds to a unique y-value. For this particular function f, each y-value also corresponds to a unique x-value. Thus, this function is also a **one-to-one function.**

One-to-one Function

For a **one-to-one function,** each x-value (input) corresponds to only one y-value (output), and each y-value (output) corresponds to only one x-value (input).

EXAMPLE 1

Determine whether each function described is one-to-one.

a. $f = \{(6,2),(5,4),(-1,0),(7,3)\}$

b. $g = \{(3,9),(-4,2),(-3,9),(0,0)\}$

c. $h = \{(1,1),(2,2),(10,10),(-5,-5)\}$

d.

Mineral (Input)	Talc	Gypsum	Diamond	Topaz	Stibnite
Hardness on the Mohs Scale (Output)	1	2	10	8	2

e.

f.

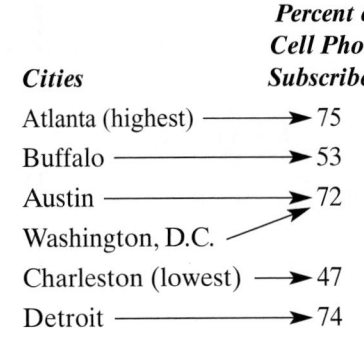

Cities	Percent of Cell Phone Subscribers
Atlanta (highest)	75
Buffalo	53
Austin	72
Washington, D.C.	
Charleston (lowest)	47
Detroit	74

Solution

a. f is one-to-one since each y-value corresponds to only one x-value.

b. g is not one-to-one because the y-value 9 in $(3,9)$ and $(-3,9)$ corresponds to two different x-values.

c. h is a one-to-one function since each y-value corresponds to only one x-value.

d. This table does not describe a one-to-one function since the output 2 corresponds to two different inputs, gypsum and stibnite.

e. This graph does not describe a one-to-one function since the y-value -1 corresponds to three different x-values, -2, -1, and 3.

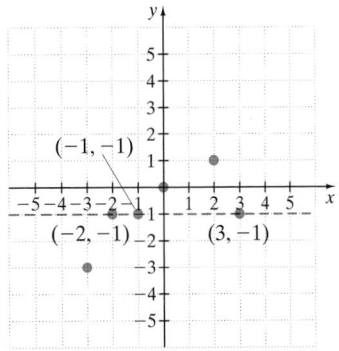

f. The mapping is not one-to-one since 72% corresponds to Austin and Washington, D.C.

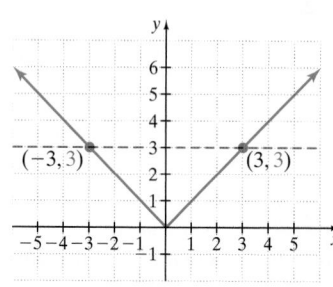

2 Recall that we recognize the graph of a function when it passes the vertical line test. Since every x-value of the function corresponds to exactly one y-value, each vertical line intersects the function's graph at most once. The graph shown (left), for instance, is the graph of a function.

Is this function a *one-to-one* function? The answer is no. To see why not, notice that the y-value of the ordered pair $(-3, 3)$, for example, is the same as the y-value of the ordered pair $(3, 3)$. In other words, the y-value 3 corresponds to two x-values, -3 and 3. This function is therefore not one-to-one.

To test whether a graph is the graph of a one-to-one function, apply the vertical line test to see if it is a function, and then apply a similar **horizontal line test** to see if it is a one-to-one function.

Horizontal Line Test

If every horizontal line intersects the graph of a function at most once, then the function is a one-to-one function.

EXAMPLE 2

Determine whether each graph is the graph of a one-to-one function.

a.

b.

c.

d.

e.

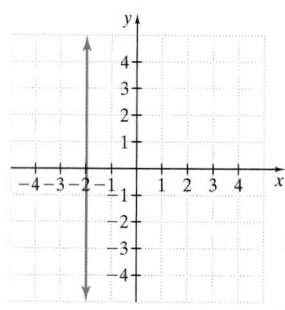

Solution   Graphs **a, b, c**, and **d** all pass the vertical line test, so only these graphs are graphs of functions. But, of these, only **b** and **c** pass the horizontal line test, so only **b** and **c** are graphs of one-to-one functions.

> ### Helpful Hint
> All linear equations are one-to-one functions except those whose graphs are horizontal or vertical lines. A vertical line does not pass the vertical line test and hence is not the graph of a function. A horizontal line is the graph of a function but does not pass the horizontal line test and hence is not the graph of a one-to-one function.
>
>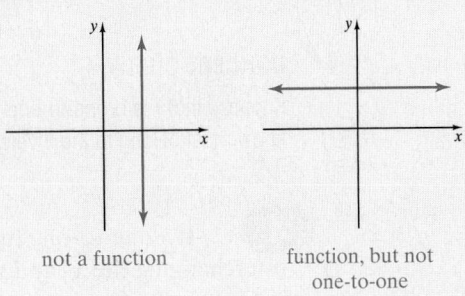
>
> not a function   function, but not one-to-one

f

Input: degrees Fahrenheit

Output: degrees Celsius

212 → 100

68 → 20

32 → 0

−13 → −25

f^{-1}

3   One-to-one functions are special in that their graphs pass both the vertical and horizontal line tests. They are special, too, in another sense: For each one-to-one function, we can find its **inverse function** by switching the coordinates of the ordered pairs of the function, or the inputs and the outputs. For example, the inverse of the one-to-one function $f = \{(2, -3), (5, 10), (9, 1)\}$ is $\{(-3, 2), (10, 5), (1, 9)\}$. For a function f, we use the notation f^{-1}, read "f inverse," to denote its inverse function. Notice that since the coordinates of each ordered pair have been switched, the domain (set of inputs) of f is the range (set of outputs) of f^{-1}, and the range of f is the domain of f^{-1}.

The diagram to the left shows the inverse of the one-to-one function f with ordered pairs of the form (degrees Fahrenheit, degrees Celsuis) is the function f^{-1} with ordered pairs of the form (degrees Celsuis, degrees Fahrenheit). Notice that the ordered pair $(-13, -25)$ of the function, for example, becomes the ordered pair $(-25, -13)$ of its inverse.

> ## Inverse Function
> The inverse of a one-to-one function f is the one-to-one function f^{-1} that consists of the set of all ordered pairs (y, x) where (x, y) belongs to f.

> ### Helpful Hint
> If a function is not one-to-one, it does not have an inverse function.

EXAMPLE 3

Find the inverse of the one-to-one function.

$$f = \{(0, 1), (-2, 7), (3, -6), (4, 4)\}$$

Solution $f^{-1} = \{(1, 0), (7, -2), (-6, 3), (4, 4)\}$

Switch coordinates
of each ordered pair.

> **Helpful Hint**
>
> The symbol f^{-1} is the single symbol used to denote the inverse of the function f.
> It is read as "f inverse." This symbol *does not mean* $\dfrac{1}{f}$.

✔ **CONCEPT CHECK**

Suppose that f is a one-to-one function. If the ordered pair (1, 5) belongs to f, name one point that we know must belong to the inverse function, f^{-1}.

4 If a one-to-one function f is defined as a set of ordered pairs, we can find f^{-1} by interchanging the x- and y-coordinates of the ordered pairs. If a one-to-one function f is given in the form of an equation, we can find f^{-1} by using a similar procedure.

Finding the Inverse of a One-to-One Function $f(x)$

Step 1: Replace $f(x)$ with y.
Step 2: Interchange x and y.
Step 3: Solve the equation for y.
Step 4: Replace y with the notation $f^{-1}(x)$.

EXAMPLE 4

Find an equation of the inverse of $f(x) = x + 3$.

Solution $f(x) = x + 3$

Step 1: $y = x + 3$ Replace $f(x)$ with y.

Step 2: $x = y + 3$ Interchange x and y.

Step 3: $x - 3 = y$ Solve for y.

Step 4: $f^{-1}(x) = x - 3$ Replace y with $f^{-1}(x)$.

The inverse of $f(x) = x + 3$ is $f^{-1}(x) = x - 3$. Notice that, for example,

$$f(1) = 1 + 3 = 4 \quad \text{and} \quad f^{-1}(4) = 4 - 3 = 1$$

Ordered pair: $(1, 4)$ Ordered pair: $(4, 1)$

Concept Check Answer:
(5, 1)

The coordinates are
switched, as expected.

EXAMPLE 5

Find the equation of the inverse of $f(x) = 3x - 5$. Graph f and f^{-1} on the same set of axes.

Solution $f(x) = 3x - 5$

Step 1: $y = 3x - 5$ Replace $f(x)$ with y.

Step 2: $x = 3y - 5$ Interchange x and y.

Step 3: $3y = x + 5$ Solve for y.

$$y = \frac{x + 5}{3}$$

Step 4: $f^{-1}(x) = \dfrac{x + 5}{3}$ Replace y with $f^{-1}(x)$.

Now we graph $f(x)$ and $f^{-1}(x)$ on the same set of axes. Both $f(x) = 3x - 5$ and $f^{-1}(x) = \dfrac{x + 5}{3}$ are linear functions, so each graph is a line.

$f(x) = 3x - 5$

x	$y = f(x)$
1	-2
0	-5
$\dfrac{5}{3}$	0

$f^{-1}(x) = \dfrac{x + 5}{3}$

x	$y = f^{-1}(x)$
-2	1
-5	0
0	$\dfrac{5}{3}$

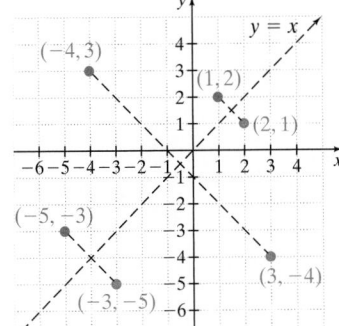

5   Notice that the graphs of f and f^{-1} in Example 5 are mirror images of each other, and the "mirror" is the dashed line $y = x$. This is true for every function and its inverse. For this reason, we say that *the graphs of f and f^{-1} are symmetric about the line $y = x$.*

To see why this happens, study the graph of a few ordered pairs and their switched coordinates in the diagram to the left.

EXAMPLE 6

Graph the inverse of each function.

Solution The function is graphed in blue and the inverse is graphed in red.

a.

b.

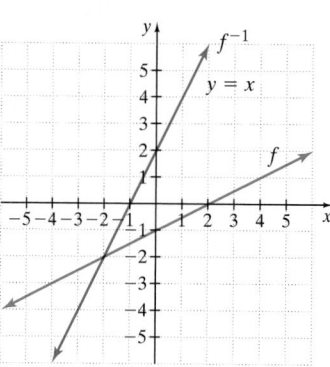

6 Notice in the table of values in Example 5 that $f(0) = -5$ and $f^{-1}(-5) = 0$, as expected. Also, for example, $f(1) = -2$ and $f^{-1}(-2) = 1$. In words, we say that for some input x, the function f^{-1} takes the output of x, called $f(x)$, back to x.

$$x \rightarrow f(x) \quad \text{and} \quad f^{-1}(f(x)) \rightarrow x$$
$$\downarrow \quad \downarrow \qquad\qquad\qquad \downarrow \quad \downarrow$$
$$f(0) = -5 \quad \text{and} \quad f^{-1}(-5) = 0$$
$$f(1) = -2 \quad \text{and} \quad f^{-1}(-2) = 1$$

In general,

> If f is a one-to-one function, then the inverse of f is the function f^{-1} such that
> $$(f^{-1} \circ f)(x) = x \quad \text{and} \quad (f \circ f^{-1})(x) = x$$

EXAMPLE 7

Show that if $f(x) = 3x + 2$, then $f^{-1}(x) = \dfrac{x - 2}{3}$.

Solution See that $(f^{-1} \circ f)(x) = x$ and $(f \circ f^{-1})(x) = x$.

$$(f^{-1} \circ f)(x) = f^{-1}(f(x))$$

$$= f^{-1}(3x + 2) \qquad \text{Replace } f(x) \text{ with } 3x + 2.$$

$$= \frac{3x + 2 - 2}{3}$$

$$= \frac{3x}{3}$$

$$= x$$

$$(f \circ f^{-1})(x) = f(f^{-1}(x))$$

$$= f\left(\frac{x-2}{3}\right) \qquad \text{Replace } f^{-1}(x) \text{ with } \frac{x-2}{3}.$$

$$= 3\left(\frac{x-2}{3}\right) + 2$$

$$= x - 2 + 2$$

$$= x$$

Graphing Calculator Explorations

A graphing calculator can be used to visualize the results of Example 7. Recall that the graph of a function f and its inverse f^{-1} are mirror images of each other across the line $y = x$. To see this for the function from Example 7, use a square window and graph

the given function: $Y_1 = 3x + 2$

its inverse: $\quad Y_2 = \dfrac{x-2}{3}$

and the line: $\quad Y_3 = x$

Exercises will follow in Exercise Set 12.1.

EXERCISE SET 12.1

| STUDY GUIDE/SSM | CD/ VIDEO | PH MATH TUTOR CENTER | MathXL®Tutorials ON CD | MathXL® | MyMathLab® |

Determine whether each function is a one-to-one function. If it is one-to-one, list the inverse function by switching coordinates, or inputs and outputs. See Examples 1 and 3.

1. $f = \{(-1, -1), (1, 1), (0, 2), (2, 0)\}$

2. $g = \{(8, 6), (9, 6), (3, 4), (-4, 4)\}$

3. $h = \{(10, 10)\}$

4. $r = \{(1, 2), (3, 4), (5, 6), (6, 7)\}$

5. $f = \{(11, 12), (4, 3), (3, 4), (6, 6)\}$

6. $g = \{(0, 3), (3, 7), (6, 7), (-2, -2)\}$

7.

Month of 2003 (Input)	January	March	May	July	September
Unemployment rate in percent	5.7	5.8	6.1	6.2	6.1

(*Source:* U.S. Department of Housing and Urban Development)

8.

State (Input)	Wisconson	Ohio	Georgia	Colorado	California	Arizona
Electoral Votes (Output)	10	20	15	9	55	10

9.

State (Input)	California	Vermont	Virginia	Texas	South Dakota
Rank in Population (Output)	1	49	12	2	46

(*Source:* U.S. Bureau of the Census)

△ **10.**

Shape (Input)	Triangle	Pentagon	Quadrilateral	Hexagon	Decagon
Number of Sides (Output)	3	5	4	6	10

(*Source:* U.S. Bureau of the Census)

Given the one-to-one function $f(x) = x^3 + 2$, *find the following.*
[*Hint: You do not need to find the equation for* $f^{-1}(x)$.]

11. a. $f(1)$
   b. $f^{-1}(3)$

12. a. $f(0)$
   b. $f^{-1}(2)$

13. a. $f(-1)$
   b. $f^{-1}(1)$

14. a. $f(-2)$
   b. $f^{-1}(-6)$

Determine whether the graph of each function is the graph of a one-to-one function. See Example 2.

15.

16.

 17.

18.

19.

20.

21.

22.

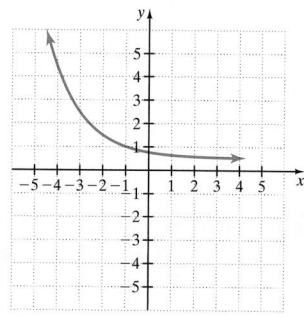

Each of the following functions is one-to-one. Find the inverse of each function and graph the function and its inverse on the same set of axes. See Examples 4 and 5.

23. $f(x) = x + 4$

24. $f(x) = x - 5$

 25. $f(x) = 2x - 3$

26. $f(x) = 4x + 9$

27. $f(x) = \dfrac{1}{2}x - 1$

28. $f(x) = -\dfrac{1}{2}x + 2$

29. $f(x) = x^3$

30. $f(x) = x^3 - 1$

Find the inverse of each one-to-one function. See Examples 4 and 5.

31. $f(x) = 5x + 2$

32. $f(x) = 6x - 1$

33. $f(x) = \dfrac{x - 2}{5}$

34. $f(x) = \dfrac{4x - 3}{2}$

35. $f(x) = \sqrt[3]{x}$

36. $f(x) = \sqrt[3]{x + 1}$

37. $f(x) = \dfrac{5}{3x + 1}$

38. $f(x) = \dfrac{7}{2x + 4}$

39. $f(x) = (x + 2)^3$

40. $f(x) = (x - 5)^3$

Graph the inverse of each function on the same set of axes. See Example 6.

41.

42.

43.

44.

45.

46.

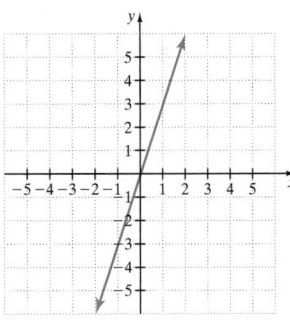

Solve. See Example 7.

47. If $f(x) = 2x + 1$, show that $f^{-1}(x) = \dfrac{x - 1}{2}$.

48. If $f(x) = 3x - 10$, show that $f^{-1}(x) = \dfrac{x + 10}{3}$.

49. If $f(x) = x^3 + 6$, show that $f^{-1}(x) = \sqrt[3]{x - 6}$.

50. If $f(x) = x^3 - 5$, show that $f^{-1}(x) = \sqrt[3]{x + 5}$.

REVIEW AND PREVIEW

Evaluate each of the following. See Section 10.2.

51. $25^{1/2}$

52. $49^{1/2}$

53. $16^{3/4}$

54. $27^{2/3}$

55. $9^{-3/2}$

56. $81^{-3/4}$

If $f(x) = 3^x$, find the following. In Exercises 59 and 60, give an exact answer and a two-decimal-place approximation. See Sections 3.7 and 5.1.

57. $f(2)$

58. $f(0)$

59. $f\left(\frac{1}{2}\right)$

60. $f\left(\frac{2}{3}\right)$

Concept Extensions

For Exercises 61 and 62,

    a. Write the ordered pairs for $f(x)$ whose points are highlighted. (Include the points whose coordinates are given.)

    b. Write the corresponding ordered pairs for the inverse of f, f^{-1}.

    c. Graph the ordered pairs for f^{-1} found in part **b.**

    d. Graph $f^{-1}(x)$ by drawing a smooth curve through the plotted points.

61.

62.

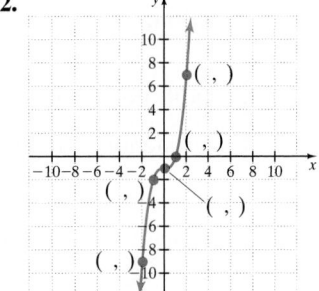

63. If you are given the graph of a function, describe how you can tell from the graph whether a function has an inverse.

64. Describe the appearance of the graphs of a function and its inverse.

Find the inverse of each given one-to-one function. Then graph the function and its inverse on a square window.

65. $f(x) = 3x + 1$

66. $f(x) = -2x - 6$

67. $f(x) = \sqrt[3]{x + 1}$

68. $f(x) = x^3 - 3$

12.2 EXPONENTIAL FUNCTIONS

Objectives

1. Graph exponential functions.
2. Solve equations of the form $b^x = b^y$.
3. Solve problems modeled by exponential equations.

1   In earlier chapters, we gave meaning to exponential expressions such as 2^x, where x is a rational number. For example,

$$2^3 = 2 \cdot 2 \cdot 2 \qquad \text{Three factors, each factor is } 2$$

$$2^{3/2} = (2^{1/2})^3 = \sqrt{2} \cdot \sqrt{2} \cdot \sqrt{2} \quad \text{Three factors, each factor is } \sqrt{2}$$

When x is an irrational number (for example, $\sqrt{3}$), what meaning can we give to $2^{\sqrt{3}}$?

It is beyond the scope of this book to give precise meaning to 2^x if x is irrational. We can confirm your intuition and say that $2^{\sqrt{3}}$ is a real number, and since $1 < \sqrt{3} < 2$, then $2^1 < 2^{\sqrt{3}} < 2^2$. We can also use a calculator and approximate $2^{\sqrt{3}}$: $2^{\sqrt{3}} \approx 3.321997$. In fact, as long as the base b is positive, b^x is a real number for all real numbers x. Finally, the rules of exponents apply whether x is rational or irrational, as long as b is positive. In this section, we are interested in functions of the form $f(x) = b^x$, where $b > 0$. A function of this form is called an **exponential function.**

Exponential Function

A function of the form

$$f(x) = b^x$$

is called an **exponential function** if $b > 0$, b is not 1, and x is a real number.

Next, we practice graphing exponential functions.

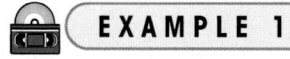 **EXAMPLE 1**

Graph the exponential functions defined by $f(x) = 2^x$ and $g(x) = 3^x$ on the same set of axes.

Solution   Graph each function by plotting points. Set up a table of values for each of the two functions.

$f(x) = 2^x$

x	0	1	2	3	-1	-2
$f(x)$	1	2	4	8	$\dfrac{1}{2}$	$\dfrac{1}{4}$

$g(x) = 3^x$

x	0	1	2	3	-1	-2
$g(x)$	1	3	9	27	$\dfrac{1}{3}$	$\dfrac{1}{9}$

If each set of points is plotted and connected with a smooth curve, the following graphs result.

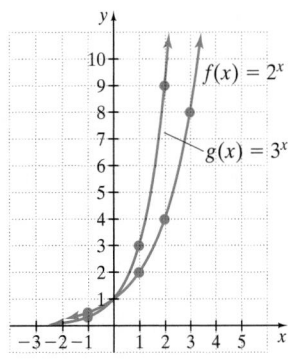

A number of things should be noted about the two graphs of exponential functions in Example 1. First, the graphs show that $f(x) = 2^x$ and $g(x) = 3^x$ are one-to-one functions since each graph passes the vertical and horizontal line tests. The y-intercept of each graph is $(0, 1)$, but neither graph has an x-intercept. From the graph, we can also see that the domain of each function is all real numbers and that the range is $(0, \infty)$. We can also see that as x-values are increasing, y-values are increasing also.

EXAMPLE 2

Graph the exponential functions $y = \left(\dfrac{1}{2}\right)^x$ and $y = \left(\dfrac{1}{3}\right)^x$ on the same set of axes.

Solution   As before, plot points and connect them with a smooth curve.

$y = \left(\dfrac{1}{2}\right)^x$

x	0	1	2	3	-1	-2
y	1	$\dfrac{1}{2}$	$\dfrac{1}{4}$	$\dfrac{1}{8}$	2	4

$y = \left(\dfrac{1}{3}\right)^x$

x	0	1	2	3	-1	-2
y	1	$\dfrac{1}{3}$	$\dfrac{1}{9}$	$\dfrac{1}{27}$	3	9

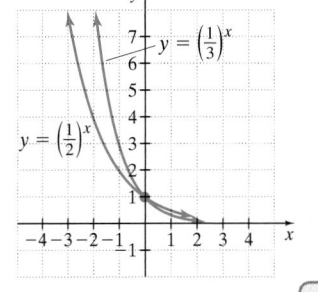

Each function in Example 2 again is a one-to-one function. The y-intercept of both is $(0, 1)$. The domain is the set of all real numbers, and the range is $(0, \infty)$.

Notice the difference between the graphs of Example 1 and the graphs of Example 2. An exponential function is always increasing if the base is greater than 1.

When the base is between 0 and 1, the graph is always decreasing. The following figures summarize these characteristics of exponential functions.

$$f(x) = b^x, \quad b > 0, \quad b \neq 1$$

- one-to-one function
- y-intercept $(0, 1)$
- no x-intercept

- domain: $(-\infty, \infty)$
- range: $(0, \infty)$

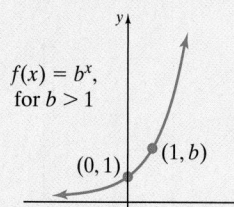

$f(x) = b^x$, for $b > 1$

$(0, 1)$ $(1, b)$

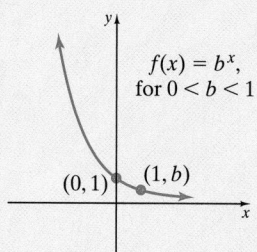

$f(x) = b^x$, for $0 < b < 1$

$(0, 1)$ $(1, b)$

EXAMPLE 3

Graph the exponential function $f(x) = 3^{x+2}$.

Solution As before, we find and plot a few ordered pair solutions. Then we connect the points with a smooth curve.

$y = 3^{x+2}$

x	0	-1	-2	-3	-4
y	9	3	1	$\dfrac{1}{3}$	$\dfrac{1}{9}$

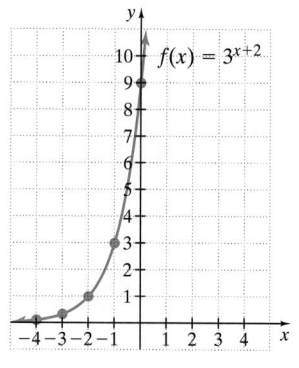

$f(x) = 3^{x+2}$

✔ **CONCEPT CHECK**

Which functions are exponential functions?

a. $f(x) = x^3$ **b.** $g(x) = \left(\dfrac{2}{3}\right)^x$ **c.** $h(x) = 5^{x-2}$ **d.** $w(x) = (2x)^2$

---

2 We have seen that an exponential function $y = b^x$ is a one-to-one function. Another way of stating this fact is a property that we can use to solve exponential equations.

Uniqueness of b^x

Let $b > 0$ and $b \neq 1$. Then $b^x = b^y$ is equivalent to $x = y$.

Concept Check Answer:
b, c

EXAMPLE 4

Solve each equation for x.

a. $2^x = 16$ **b.** $9^x = 27$ **c.** $4^{x+3} = 8^x$

Solution   **a.** We write 16 as a power of 2 and then use the uniqueness of b^x to solve.

$$2^x = 16$$
$$2^x = 2^4$$

Since the bases are the same and are nonnegative, by the uniqueness of b^x, we then have that the exponents are equal. Thus,

$$x = 4$$

The solution is 4, or the solution set is $\{4\}$.

b. Notice that both 9 and 27 are powers of 3.

$$9^x = 27$$
$$(3^2)^x = 3^3 \qquad \text{Write 9 and 27 as powers of 3.}$$
$$3^{2x} = 3^3$$
$$2x = 3 \qquad \text{Apply the uniqueness of } b^x.$$
$$x = \frac{3}{2} \qquad \text{Divide by 2.}$$

To check, replace x with $\frac{3}{2}$ in the original expression, $9^x = 27$. The solution is $\frac{3}{2}$.

c. Write both 4 and 8 as powers of 2.

$$4^{x+3} = 8^x$$
$$(2^2)^{x+3} = (2^3)^x$$
$$2^{2x+6} = 2^{3x}$$
$$2x + 6 = 3x \qquad \text{Apply the uniqueness of } b^x.$$
$$6 = x \qquad \text{Subtract } 2x \text{ from both sides.}$$

The solution is 6.

There is one major problem with the preceding technique. Often the two sides of an equation cannot easily be written as powers of a common base. We explore how to solve an equation such as $4 = 3^x$ with the help of **logarithms** later.

3   The bar graph on the next page shows the increase in the number of cellular phone users. Notice that the graph of the exponential function $y = 34(1.254)^x$ approximates the heights of the bars.

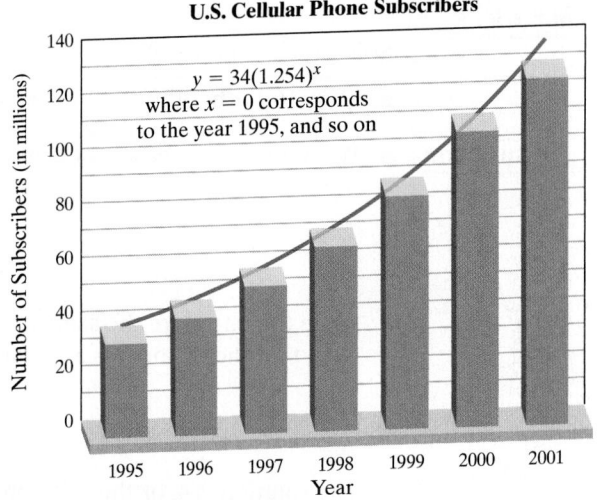

U.S. Cellular Phone Subscribers

$y = 34(1.254)^x$ where $x = 0$ corresponds to the year 1995, and so on

Source: The CTIA Semi-Annual Wireless Survey

The graph above shows just one example of how the world abounds with patterns that can be modeled by exponential functions. To make these applications realistic, we use numbers that warrant a calculator. Another application of an exponential function has to do with interest rates on loans.

The exponential function defined by $A = P\left(1 + \dfrac{r}{n}\right)^{nt}$ models the dollars A accrued (or owed) after P dollars are invested (or loaned) at an annual rate of interest r compounded n times each year for t years. This function is known as the compound interest formula.

EXAMPLE 5

USING THE COMPOUND INTEREST FORMULA

Find the amount owed at the end of 5 years if $1600 is loaned at a rate of 9% compounded monthly.

Solution We use the formula $A = P\left(1 + \dfrac{r}{n}\right)^{nt}$, with the following values.

$P = \$1600$ (the amount of the loan)

$r = 9\% = 0.09$ (the annual rate of interest)

$n = 12$ (the number of times interest is compounded each year)

$t = 5$ (the duration of the loan, in years)

$A = P\left(1 + \dfrac{r}{n}\right)^{nt}$ *Compound interest formula*

$= 1600\left(1 + \dfrac{0.09}{12}\right)^{12(5)}$ *Substitute known values.*

$= 1600(1.0075)^{60}$

To approximate A, use the $\boxed{y^x}$ or $\boxed{\wedge}$ key on your calculator.

$\boxed{2505.0896}$

Thus, the amount A owed is approximately $2505.09.

EXAMPLE 6

ESTIMATING PERCENT OF RADIOACTIVE MATERIAL

As a result of the Chernobyl nuclear accident, radioactive debris was carried through the atmosphere. One immediate concern was the impact that the debris had on the milk supply. The percent y of radioactive material in raw milk after t days is estimated by $y = 100(2.7)^{-0.1t}$. Estimate the expected percent of radioactive material in the milk after 30 days.

Solution

Replace t with 30 in the given equation.

$$y = 100(2.7)^{-0.1t}$$

$$= 100(2.7)^{-0.1(30)} \qquad \text{Let } t = 30.$$

$$= 100(2.7)^{-3}$$

To approximate the percent y, the following keystrokes may be used on a scientific calculator.

$$\boxed{2.7}\ \boxed{y^x}\ \boxed{3}\ \boxed{+/-}\ \boxed{=}\ \boxed{\times}\ \boxed{100}\ \boxed{=}$$

The display should read

$$\boxed{5.0805263}$$

Thus, approximately 5% of the radioactive material still remained in the milk supply after 30 days.

Graphing Calculator Explorations

We can use a graphing calculator and its TRACE feature to solve Example 6 graphically.

To estimate the expected percent of radioactive material in the milk after 30 days, enter $Y_1 = 100(2.7)^{-0.1x}$. (The variable t in Example 6 is changed to x here to better accomodate our work on the graphing calculator.) The graph does not appear on a standard viewing window, so we need to determine an appropriate viewing window. Because it doesn't make sense to look at radioactivity *before* the Chernobyl nuclear accident, we use Xmin = 0. We are interested in finding the percent of radioactive material in the milk when $x = 30$, so we choose Xmax = 35 to leave enough space to see the graph at $x = 30$. Because the values of y are percents, it seems appropriate that $0 \le y \le 100$. (We also use Xscl = 1 and Yscl = 10.) Now we graph the function.

We can use the TRACE feature to obtain an approximation of the expected percent of radioactive material in the milk when $x = 30$. (A TABLE feature may also be used to approximate the percent.) To obtain a better approximation, let's use the ZOOM feature several times to zoom in near $x = 30$.

(continued)

The percent of radioactive material in the milk 30 days after the Chernobyl accident was 5.08%, accurate to two decimal places.

Use a graphing calculator to find each percent. Approximate your solutions so that they are accurate to two decimal places.

1. Estimate the expected percent of radioactive material in the milk 2 days after the Chernobyl nuclear accident.

2. Estimate the expected percent of radioactive material in the milk 10 days after the Chernobyl nuclear accident.

3. Estimate the expected percent of radioactive material in the milk 15 days after the Chernobyl nuclear accident.

4. Estimate the expected percent of radioactive material in the milk 25 days after the Chernobyl nuclear accident.

EXERCISE SET 12.2

STUDY GUIDE/SSM   CD/VIDEO   PH MATH TUTOR CENTER   MathXL®Tutorials ON CD   MathXL®   MyMathLab®

Graph each exponential function. See Examples 1 through 3.

1. $y = 4^x$

2. $y = 5^x$

3. $y = 2^x + 1$

4. $y = 3^x - 1$

5. $y = \left(\dfrac{1}{4}\right)^x$

6. $y = \left(\dfrac{1}{5}\right)^x$

7. $y = \left(\dfrac{1}{2}\right)^x - 2$

8. $y = \left(\dfrac{1}{3}\right)^x + 2$

9. $y = -2^x$

10. $y = -3^x$

11. $y = -\left(\dfrac{1}{4}\right)^x$

12. $y = -\left(\dfrac{1}{5}\right)^x$

13. $f(x) = 2^{x+1}$

14. $f(x) = 3^{x-1}$

15. $f(x) = 4^{x-2}$

16. $f(x) = 2^{x+3}$

C

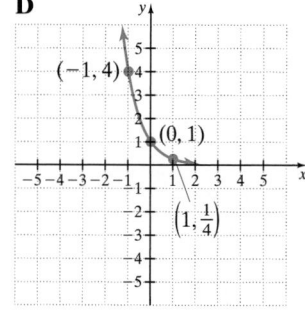

Graph C: points $(-1, 2)$, $(0, 1)$, $\left(1, \dfrac{1}{2}\right)$

D

Graph D: points $(-1, 4)$, $(0, 1)$, $\left(1, \dfrac{1}{4}\right)$

Match each exponential equation with its graph below or in the next column.

17. $f(x) = \left(\dfrac{1}{2}\right)^x$

18. $f(x) = 2^x$

19. $f(x) = \left(\dfrac{1}{4}\right)^x$

20. $f(x) = 3^x$

A

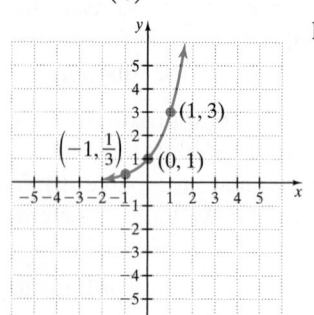

Graph A: points $\left(-1, \dfrac{1}{3}\right)$, $(0, 1)$, $(1, 3)$

B

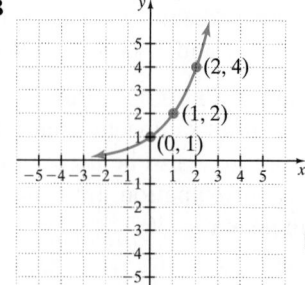

Graph B: points $(0, 1)$, $(1, 2)$, $(2, 4)$

Solve each equation for x. See Example 4.

21. $3^x = 27$

22. $6^x = 36$

 23. $16^x = 8$

24. $64^x = 16$

25. $32^{2x-3} = 2$

26. $9^{2x+1} = 81$

27. $\dfrac{1}{4} = 2^{3x}$

28. $\dfrac{1}{27} = 3^{2x}$

29. $5^x = 625$

30. $2^x = 64$

31. $4^x = 8$

32. $32^x = 4$

 33. $27^{x+1} = 9$

34. $125^{x-2} = 25$

35. $81^{x-1} = 27^{2x}$

36. $4^{3x-7} = 32^{2x}$

Solve. Unless otherwise indicated, round results to one decimal place. See Example 6.

37. One type of uranium has a daily radioactive decay rate of 0.4%. If 30 pounds of this uranium is available today, find how much will still remain after 50 days. Use $y = 30(2.7)^{-0.004t}$, and let t be 50.

38. The nuclear waste from an atomic energy plant decays at a rate of 3% each century. If 150 pounds of nuclear waste is disposed of, find how much of it will still remain after 10 centuries. Use $y = 150(2.7)^{-0.03t}$, and let t be 10.

39. National Park Service personnel are trying to increase the size of the bison population of Theodore Roosevelt National Park. If 260 bison currently live in the park, and if the population's rate of growth is 2.5% annually, find how many bison (rounded to the nearest whole) there should be in 10 years. Use $y = 260(2.7)^{0.025t}$.

40. The size of the rat population of a wharf area grows at a rate of 8% monthly. If there are 200 rats in January, find how many rats (rounded to the nearest whole) should be expected by next January. Use $y = 200(2.7)^{0.08t}$.

41. A rare isotope of a nuclear material is very unstable, decaying at a rate of 15% each second. Find how much isotope remains 10 seconds after 5 grams of the isotope is created. Use $y = 5(2.7)^{-0.15t}$.

42. An accidental spill of 75 grams of radioactive material in a local stream has led to the presence of radioactive debris decaying at a rate of 4% each day. Find how much debris still remains after 14 days. Use $y = 75(2.7)^{-0.04t}$.

43. Retail revenue from shopping on the Internet is currently growing at a rate of 56% per year. In 2000, a total of $42.1 billion in revenue was collected through Internet retail sales. Answer the following questions using $y = 42.1(1.56)^t$ where y is Internet revenue in billions of dollars and t is the number of years after 2000. Round answers to the nearest tenth of a billion dollars. (*Source:* Based on data from eMarketer)

   a. According to the model, what level of retail revenues from Internet shopping was expected in 2001?

   b. Predict the level of Internet shopping revenues in 2009.

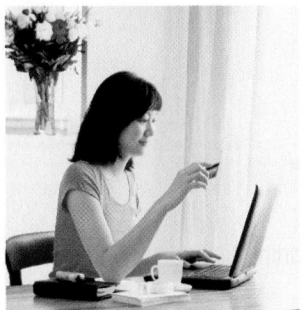

44. An unusually wet spring has caused the size of the Cape Cod mosquito population to increase by 8% each day. If an estimated 200,000 mosquitoes are on Cape Cod on May 12, find how many mosquitoes will inhabit the Cape on May 25. Use $y = 200,000(2.7)^{0.08t}$. Round to the nearest thousand.

45. The equation $y = 120.882(1.012)^x$ models the population of the United States from 1930 through 2000. In the equation, y is the population in millions and x represents the number of years after 1930. Round answers to the nearest tenth of a million.(*Source:* Based on data from the U.S. Bureau of the Census)

   a. Estimate the population of the United States in 1970.

   b. Assuming this equation continues to be valid in the future, use the equation to predict the population of the United States in 2020.

46. Carbon dioxide (CO_2) is a greenhouse gas that contributes to global warming. Partially due to the combustion of fossil fuels, the amount of CO_2 in Earth's atmosphere has been increasing by 0.4% annually over the past century. In 2000, the concentration of CO_2 in the atmosphere was 369.4 parts per million by volume. To make the following predictions, use $y = 369.4(1.004)^t$ where y is the concentration of CO_2 in parts per million and t is the number of years after 2000. Round answers to the nearest tenth. (*Sources:* Based on data from the United Nations Environment Programme and the Carbon Dioxide Information Analysis Center)

   a. Predict the concentration of CO_2 in the atmosphere in the year 2006.

   b. Predict the concentration of CO_2 in the atmosphere in the year 2030.

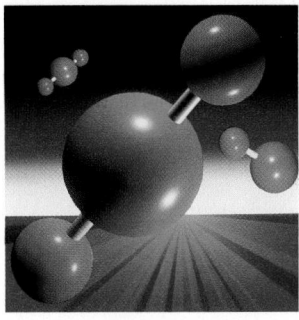

Solve. Use $A = P\left(1 + \dfrac{r}{n}\right)^{nt}$. Round answers to two decimal places. See Example 5.

47. Find the amount Erica owes at the end of 3 years if $6000 is loaned to her at a rate of 8% compounded monthly.

48. Find the amount owed at the end of 5 years if $3000 is loaned at a rate of 10% compounded quarterly.

49. Find the total amount Janina has in a college savings account if $2000 was invested and earned 6% compounded semiannually for 12 years.

50. Find the amount accrued if $500 is invested and earns 7% compounded monthly for 4 years.

The formula $y = 34(1.254)^x$ gives the number of cellular phone users y (in millions) in the United States for the years 1995 through 2001. In this formula, $x = 0$ corresponds to 1995, $x = 1$ corresponds to 1996, and so on. Use this formula to solve Exercises 51 and 52. Round results to the nearest whole million.

51. Use this model to predict the number of cellular phone users in the year 2008.

52. Use this model to predict the number of cellular phone users in the year 2010.

REVIEW AND PREVIEW

Solve each equation. See Sections 2.3 and 6.5.

53. $5x - 2 = 18$

54. $3x - 7 = 11$

55. $3x - 4 = 3(x + 1)$

56. $2 - 6x = 6(1 - x)$

57. $x^2 + 6 = 5x$

58. $18 = 11x - x^2$

By inspection, find the value for x that makes each statement true.

59. $2^x = 8$

60. $3^x = 9$

61. $5^x = \dfrac{1}{5}$

62. $4^x = 1$

Concept Extensions

63. Explain why the graph of an exponential function $y = b^x$ contains the point $(1, b)$.

64. Explain why an exponential function $y = b^x$ has a y-intercept of $(0, 1)$.

Graph.

65. $y = |3^x|$

66. $y = \left|\left(\dfrac{1}{3}\right)^x\right|$

67. $y = 3^{|x|}$

68. $y = \left(\dfrac{1}{3}\right)^{|x|}$

69. Graph $y = 2^x$ and $y = \left(\dfrac{1}{2}\right)^{-x}$ on the same set of axes. Describe what you see and why.

70. Graph $y = 2^x$ and $x = 2^y$ on the same set of axes. Describe what you see.

Use a graphing calculator to solve. Estimate each result to two decimal places.

71. Verify the results of Exercise 37.

72. From Exercise 37, estimate the number of pounds of uranium that will be available after 100 days.

73. From Exercise 37, estimate the number of pounds of uranium that will be available after 120 days.

74. Verify the results of Exercise 42.

75. From Exercise 42, estimate the amount of debris that remains after 10 days.

76. From Exercise 42, estimate the amount of debris that remains after 20 days.

12.3 LOGARITHMIC FUNCTIONS

Objectives

1 Write exponential equations with logarithmic notation and write logarithmic equations with exponential notation.

2 Solve logarithmic equations by using exponential notation.

3 Identify and graph logarithmic functions.

1   Since the exponential function $f(x) = 2^x$ is a one-to-one function, it has an inverse.

We can create a table of values for f^{-1} by switching the coordinates in the accompanying table of values for $f(x) = 2^x$.

x	$y = f(x)$	x	$y = f^{-1}(x)$
-3	$\dfrac{1}{8}$	$\dfrac{1}{8}$	-3
-2	$\dfrac{1}{4}$	$\dfrac{1}{4}$	-2
-1	$\dfrac{1}{2}$	$\dfrac{1}{2}$	-1
0	1	1	0
1	2	2	1
2	4	4	2
3	8	8	3

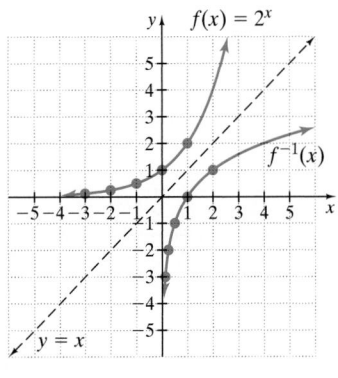

The graphs of $f(x)$ and its inverse are shown above. Notice that the graphs of f and f^{-1} are symmetric about the line $y = x$, as expected.

Now we would like to be able to write an equation for f^{-1}. To do so, we follow the steps for finding an inverse.

$$f(x) = 2^x$$

Step 1: Replace $f(x)$ by y. $y = 2^x$

Step 2: Interchange x and y. $x = 2^y$

Step 3: Solve for y.

At this point, we are stuck. To solve this equation for y, a new notation, the **logarithmic notation,** is needed. The symbol $\log_b x$ means "the power to which b is raised in order to produce a result of x."

$$\log_b x = y \quad \text{means} \quad b^y = x$$

We say that $\log_b x$ is "the logarithm of x to the base b" or "the log of x to the base b."

Logarithmic Definition

If $b > 0$ and $b \neq 1$, then

$$y = \log_b x \text{ means } x = b^y$$

for every $x > 0$ and every real number y.

Before returning to the function $x = 2^y$ and solving it for y in terms of x, let's practice using the new notation $\log_b x$.

> **Helpful Hint**
>
> Notice that a *logarithm* is an *exponent*. In other words, $\log_3 9$ is the *power* that we raise 3 to in order to get 9.

It is important to be able to write exponential equations from logarithmic notation, and vice versa. The following table shows examples of both forms.

Logarithmic Equation	*Corresponding Exponential Equation*
$\log_3 9 = 2$	$3^2 = 9$
$\log_6 1 = 0$	$6^0 = 1$
$\log_2 8 = 3$	$2^3 = 8$
$\log_4 \dfrac{1}{16} = -2$	$4^{-2} = \dfrac{1}{16}$
$\log_8 2 = \dfrac{1}{3}$	$8^{1/3} = 2$

EXAMPLE 1

Write as an exponential equation.

a. $\log_5 25 = 2$ **b.** $\log_6 \dfrac{1}{6} = -1$ **c.** $\log_2 \sqrt{2} = \dfrac{1}{2}$

Solution **a.** $\log_5 25 = 2$ means $5^2 = 25$

b. $\log_6 \dfrac{1}{6} = -1$ means $6^{-1} = \dfrac{1}{6}$

c. $\log_2 \sqrt{2} = \dfrac{1}{2}$ means $2^{1/2} = \sqrt{2}$

EXAMPLE 2

Write as a logarithmic equation.

a. $9^3 = 729$ **b.** $6^{-2} = \dfrac{1}{36}$ **c.** $5^{1/3} = \sqrt[3]{5}$

Solution **a.** $9^3 = 729$ means $\log_9 729 = 3$

b. $6^{-2} = \dfrac{1}{36}$ means $\log_6 \dfrac{1}{36} = -2$

c. $5^{1/3} = \sqrt[3]{5}$ means $\log_5 \sqrt[3]{5} = \dfrac{1}{3}$

EXAMPLE 3

Find the value of each logarithmic expression.

a. $\log_4 16$ **b.** $\log_{10} \dfrac{1}{10}$ **c.** $\log_9 3$

Solution **a.** $\log_4 16 = 2$ because $4^2 = 16$

b. $\log_{10} \dfrac{1}{10} = -1$ because $10^{-1} = \dfrac{1}{10}$

c. $\log_9 3 = \dfrac{1}{2}$ because $9^{1/2} = \sqrt{9} = 3$

> **Helpful Hint**
> Another method for evaluating logarithms such as those in Example 3 is to set the expression equal to x and then write them in exponential form to find x. For example:
>
> **a.** $\log_4 16 = x$ means $4^x = 16$. Since $4^2 = 16$, $x = 2$ or $\log_4 16 = 2$.
>
> **b.** $\log_{10} \dfrac{1}{10} = x$ means $10^x = \dfrac{1}{10}$. Since $10^{-1} = \dfrac{1}{10}$, $x = -1$ or $\log_{10} \dfrac{1}{10} = -1$.
>
> **c.** $\log_9 3 = x$ means $9^x = 3$. Since $9^{1/2} = 3$, $x = \dfrac{1}{2}$ or $\log_9 3 = \dfrac{1}{2}$.

2   The ability to interchange the logarithmic and exponential forms of a statement is often the key to solving logarithmic equations.

EXAMPLE 4

Solve each equation for x.

a. $\log_4 \dfrac{1}{4} = x$ **b.** $\log_5 x = 3$ **c.** $\log_x 25 = 2$

d. $\log_3 1 = x$ **e.** $\log_b 1 = x$

Solution   **a.** $\log_4 \dfrac{1}{4} = x$ means $4^x = \dfrac{1}{4}$. Solve $4^x = \dfrac{1}{4}$ for x.

$$4^x = \frac{1}{4}$$
$$4^x = 4^{-1}$$

Since the bases are the same, by the uniqueness of b^x, we have that

$$x = -1$$

The solution is -1. To check, see that $\log_4 \dfrac{1}{4} = -1$, since $4^{-1} = \dfrac{1}{4}$.

b. $\log_5 x = 3$ means $5^3 = x$ or

$$x = 125$$

The solution is 125.

c. $\log_x 25 = 2$ means $x^2 = 25$ and $x > 0$ and $x \neq 1$. Thus,

$$x = 5$$

Even though $(-5)^2 = 25$, the base b of a logarithm must be positive. The solution is 5.

d. $\log_3 1 = x$ means $3^x = 1$. Either solve this equation by inspection or solve by writing 1 as 3^0 as shown.

$$3^x = 3^0 \qquad \text{Write \textbf{1} as } 3^0.$$

$$x = 0 \quad \text{Apply the uniqueness of } b^x.$$

The solution is 0.

e. $\log_b 1 = x$ means $b^x = 1$ and $b > 0$ and $b \neq 1$.

$$b^x = b^0 \quad \text{Write 1 as } b^0.$$

$$x = 0 \quad \text{Apply the uniqueness of } b^x.$$

The solution is 0.

In Example **4e** we proved an important property of logarithms. That is, $\log_b 1$ is always 0. This property as well as two important others are given next.

Properties of Logarithms

If b is a real number, $b > 0$, and $b \neq 1$, then

1. $\log_b 1 = 0$
2. $\log_b b^x = x$
3. $b^{\log_b x} = x$

To see that **2.** $\log_b b^x = x$, change the logarithmic form to exponential form. Then, $\log_b b^x = x$ means $b^x = b^x$. In exponential form, the statement is true, so in logarithmic form, the statement is also true. To understand **3.** $b^{\log_b x} = x$, write this exponential equation as an equivalent logarithm.

EXAMPLE 5

Simplify.

a. $\log_3 3^2$ **b.** $\log_7 7^{-1}$ **c.** $5^{\log_5 3}$ **d.** $2^{\log_2 6}$

Solution   **a.** From Property 2, $\log_3 3^2 = 2$.
b. From Property 2, $\log_7 7^{-1} = -1$.
c. From Property 3, $5^{\log_5 3} = 3$.
d. From Property 3, $2^{\log_2 6} = 6$.

3   Let us now return to the function $f(x) = 2^x$ and write an equation for its inverse, $f^{-1}(x)$. Recall our earlier work.

$$f(x) = 2^x$$

Step 1: Replace $f(x)$ by y. $y = 2^x$

Step 2: Interchange x and y. $x = 2^y$

Having gained proficiency with the notation $\log_b x$, we can now complete the steps for writing the inverse equation by writing $x = 2^y$ as an equivalent logarithm.

Step 1: Solve for y. $y = \log_2 x$

Step 2: Replace y with $f^{-1}(x)$. $f^{-1}(x) = \log_2 x$

Thus, $f^{-1}(x) = \log_2 x$ defines a function that is the inverse function of the function $f(x) = 2^x$. The function $f^{-1}(x)$ or $y = \log_2 x$ is called a **logarithmic function**.

> ## Logarithmic Function
>
> If x is a positive real number, b is a constant positive real number, and b is not 1, then a **logarithmic function** is a function that can be defined by
>
> $$f(x) = \log_b x$$
>
> The domain of f is the set of positive real numbers, and the range of f is the set of real numbers.

✔ **CONCEPT CHECK**

Let $f(x) = \log_3 x$ and $g(x) = 3^x$. These two functions are inverses of each other. Since $(2, 9)$ is an ordered pair solution of $g(x)$ or $g(2) = 9$, what ordered pair do we know to be a solution of $f(x)$? Also, find $f(9)$. Explain why.

We can explore logarithmic functions by graphing them.

EXAMPLE 6

Graph the logarithmic function $y = \log_2 x$.

Solution First we write the equation with exponential notation as $2^y = x$. Then we find some ordered pair solutions that satisfy this equation. Finally, we plot the points and connect them with a smooth curve. The domain of this function is $(0, \infty)$, and the range is all real numbers.

Since $x = 2^y$ is solved for x, we choose y-values and compute corresponding x-values.

If $y = 0, x = 2^0 = 1$

If $y = 1, x = 2^1 = 2$

If $y = 2, x = 2^2 = 4$

If $y = -1, x = 2^{-1} = \dfrac{1}{2}$

$x = 2^y$	y
1	0
2	1
4	2
$\dfrac{1}{2}$	-1

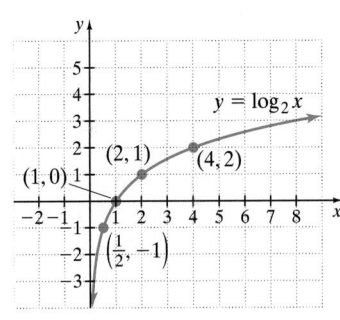

EXAMPLE 7

Graph the logarithmic function $f(x) = \log_{1/3} x$.

Solution Replace $f(x)$ with y, and write the result with exponential notation.

$$f(x) = \log_{1/3} x$$

$$y = \log_{1/3} x \qquad \text{Replace } f(x) \text{ with } y.$$

$$\left(\frac{1}{3}\right)^y = x \qquad \text{Write in exponential form.}$$

Concept Check Answer:
$(9, 2)$; $f(9) = 2$; answers may vary

Now we can find ordered pair solutions that satisfy $\left(\dfrac{1}{3}\right)^y = x$, plot these points, and connect them with a smooth curve.

If $y = 0, x = \left(\dfrac{1}{3}\right)^0 = 1$

If $y = 1, x = \left(\dfrac{1}{3}\right)^1 = \dfrac{1}{3}$

If $y = -1, x = \left(\dfrac{1}{3}\right)^{-1} = 3$

If $y = -2, x = \left(\dfrac{1}{3}\right)^{-2} = 9$

$x = \left(\dfrac{1}{3}\right)^y$	y
1	0
$\dfrac{1}{3}$	1
3	-1
9	-2

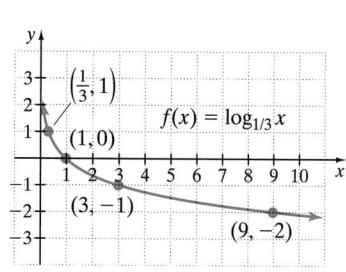

The domain of this function is $(0, \infty)$, and the range is the set of all real numbers.

The following figures summarize characteristics of logarithmic functions.

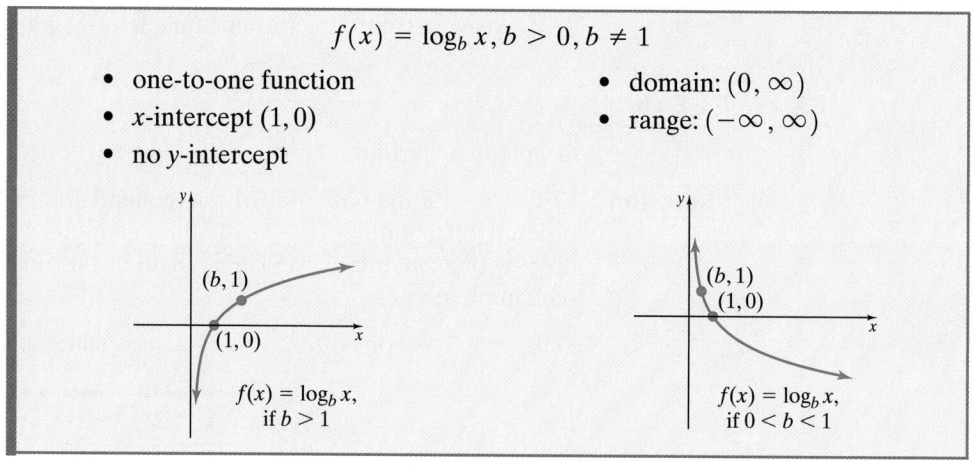

$$f(x) = \log_b x, b > 0, b \neq 1$$

- one-to-one function
- x-intercept $(1, 0)$
- no y-intercept

- domain: $(0, \infty)$
- range: $(-\infty, \infty)$

$f(x) = \log_b x,$ if $b > 1$

$f(x) = \log_b x,$ if $0 < b < 1$

Spotlight on DECISION MAKING

Suppose you are the Webmaster for a small but growing company. One of your duties is to ensure that your company's newly established Web site can adequately handle the number of visitors to it. You decide to find a mathematical model for recent Web site usage statistics to help predict future numbers of visitors. Ultimately, you would like to use this model to predict when your Web site's server capacity must be expanded.

The first step in finding a model for the usage statistics is to decide what type of mathematical model to use: linear, quadratic, exponential, or logarithmic. The graph shows the number of visitors to your company's Web site in each of the first five months since it was established. Use the graph to decide which type of mathematical model to use. Explain your reasoning.

Month	1	2	3	4	5
Visitors	166,511	1,320,978	1,996,298	2,475,445	2,847,100

EXERCISE SET 12.3

STUDY GUIDE/SSM   CD/VIDEO   PH MATH TUTOR CENTER   MathXL®Tutorials ON CD   MathXL®   MyMathLab®

Write each as an exponential equation. See Example 1.

1. $\log_6 36 = 2$ **2.** $\log_2 32 = 5$

3. $\log_3 \frac{1}{27} = -3$ **4.** $\log_5 \frac{1}{25} = -2$

5. $\log_{10} 1000 = 3$ **6.** $\log_{10} 10 = 1$

7. $\log_e x = 4$ **8.** $\log_e \frac{1}{e} = -1$

9. $\log_e \frac{1}{e^2} = -2$ **10.** $\log_e y = 7$

11. $\log_7 \sqrt{7} = \frac{1}{2}$ **12.** $\log_{11} \sqrt[4]{11} = \frac{1}{4}$

Write each as a logarithmic equation. See Example 2.

13. $2^4 = 16$ **14.** $5^3 = 125$

15. $10^2 = 100$ **16.** $10^4 = 10,000$

17. $e^3 = x$ **18.** $e^5 = y$

19. $10^{-1} = \frac{1}{10}$ **20.** $10^{-2} = \frac{1}{100}$

21. $4^{-2} = \frac{1}{16}$ **22.** $3^{-4} = \frac{1}{81}$

23. $5^{1/2} = \sqrt{5}$ **24.** $4^{1/3} = \sqrt[3]{4}$

Find the value of each logarithmic expression. See Examples 3 and 5.

 25. $\log_2 8$ **26.** $\log_3 9$

27. $\log_3 \frac{1}{9}$ **28.** $\log_2 \frac{1}{32}$

29. $\log_{25} 5$ **30.** $\log_8 \frac{1}{2}$

31. $\log_{1/2} 2$ **32.** $\log_{2/3} \frac{4}{9}$

33. $\log_7 1$ **34.** $\log_9 9$

35. $\log_2 2^4$ **36.** $\log_6 6^{-2}$

37. $\log_{10} 100$ **38.** $\log_{10} \frac{1}{10}$

39. $3^{\log_3 5}$ **40.** $5^{\log_5 7}$

41. $\log_3 81$ **42.** $\log_2 16$

43. $\log_4 \frac{1}{64}$ **44.** $\log_3 \frac{1}{9}$

45. Explain why negative numbers are not included as logarithmic bases.

46. Explain why 1 is not included as a logarithmic base.

Solve each equation for x. See Example 4.

47. $\log_3 9 = x$ **48.** $\log_2 8 = x$

49. $\log_3 x = 4$ **50.** $\log_2 x = 3$

51. $\log_x 49 = 2$ **52.** $\log_x 8 = 3$

53. $\log_2 \frac{1}{8} = x$ **54.** $\log_3 \frac{1}{81} = x$

55. $\log_3 \frac{1}{27} = x$ **56.** $\log_5 \frac{1}{125} = x$

57. $\log_8 x = \frac{1}{3}$ **58.** $\log_9 x = \frac{1}{2}$

59. $\log_4 16 = x$ **60.** $\log_2 16 = x$

61. $\log_{3/4} x = 3$ **62.** $\log_{2/3} x = 2$

63. $\log_x 100 = 2$ **64.** $\log_x 27 = 3$

Simplify. See Example 5.

65. $\log_5 5^3$ **66.** $\log_6 6^2$

67. $2^{\log_2 3}$ **68.** $7^{\log_7 4}$

69. $\log_9 9$ **70.** $\log_8 (8)^{-1}$

Graph each logarithmic function. Label any intercepts. See Examples 6 and 7.

71. $y = \log_3 x$ **72.** $y = \log_2 x$

73. $f(x) = \log_{1/4} x$ **74.** $f(x) = \log_{1/2} x$

75. $f(x) = \log_5 x$ **76.** $f(x) = \log_6 x$

77. $f(x) = \log_{1/6} x$ **78.** $f(x) = \log_{1/5} x$

REVIEW AND PREVIEW

Simplify each rational expression. See Section 7.1.

79. $\dfrac{x + 3}{3 + x}$ **80.** $\dfrac{x - 5}{5 - x}$

81. $\dfrac{x^2 - 8x + 16}{2x - 8}$ **82.** $\dfrac{x^2 - 3x - 10}{2 + x}$

Add or subtract as indicated. See Section 7.4.

83. $\dfrac{2}{x} + \dfrac{3}{x^2}$ **84.** $\dfrac{3x}{x + 3} + \dfrac{9}{x + 3}$

85. $\dfrac{m^2}{m + 1} - \dfrac{1}{m + 1}$ **86.** $\dfrac{5}{y + 1} - \dfrac{4}{y - 1}$

Solve by first writing as an exponential.

87. $\log_7(5x - 2) = 1$ **88.** $\log_3(2x + 4) = 2$

89. Simplify: $\log_3(\log_5 125)$ **90.** Simplify: $\log(\log_4(\log_2 16))$

Concept Extensions

Graph each function and its inverse function on the same set of axes. Label any intercepts.

91. $y = 4^x$; $y = \log_4 x$ **92.** $y = 3^x$; $y = \log_3 x$

93. $y = \left(\dfrac{1}{3}\right)^x$; $y = \log_{1/3} x$ **94.** $y = \left(\dfrac{1}{2}\right)^x$; $y = \log_{1/2} x$

95. The formula $\log_{10}(1 - k) = \dfrac{-0.3}{H}$ models the relationship between the half-life H of a radioactive material and its rate of decay k. Find the rate of decay of the iodine isotope I-131 if its half-life is 8 days. Round to 4 decimal places.

96. Explain why the graph of the function $y = \log_b x$ contains the point $(1, 0)$ no matter what b is.

97. $\text{Log}_3 10$ is between which two integers? Explain your answer.

12.4 PROPERTIES OF LOGARITHMS

Objectives

1 Use the product property of logarithms.

2 Use the quotient property of logarithms.

3 Use the power property of logarithms.

4 Use the properties of logarithms together.

In the previous section we explored some basic properties of logarithms. We now introduce and explore additional properties. Because a logarithm is an exponent, logarithmic properties are just restatements of exponential properties.

1 The first of these properties is called the **product property of logarithms,** because it deals with the logarithm of a product.

Product Property of Logarithms

If x, y, and b are positive real numbers and $b \neq 1$, then

$$\log_b xy = \log_b x + \log_b y$$

To prove this, let $\log_b x = M$ and $\log_b y = N$. Now write each logarithm with exponential notation.

$$\log_b x = M \quad \text{is equivalent to} \quad b^M = x$$

$$\log_b y = N \quad \text{is equivalent to} \quad b^N = y$$

Multiply the left sides and the right sides of the exponential equations, and we have that

$$xy = (b^M)(b^N) = b^{M+N}$$

If we write the equation $xy = b^{M+N}$ in equivalent logarithmic form, we have

$$\log_b xy = M + N$$

But since $M = \log_b x$ and $N = \log_b y$, we can write

$$\log_b xy = \log_b x + \log_b y \quad \text{Let } M = \log_b x \text{ and } N = \log_b y.$$

In other words, the logarithm of a product is the sum of the logarithms of the factors. This property is sometimes used to simplify logarithmic expressions.

In the examples that follow, assume that variables represent positive numbers.

EXAMPLE 1

Write each sum as a single logarithm.

a. $\log_{11} 10 + \log_{11} 3$ **b.** $\log_3 \dfrac{1}{2} + \log_3 12$ **c.** $\log_2(x + 2) + \log_2 x$

Solution In each case, both terms have a common logarithmic base.

a. $\log_{11}10 + \log_{11}3 = \log_{11}(10 \cdot 3)$ *Apply the product property.*

$= \log_{11}30$

b. $\log_3\dfrac{1}{2} + \log_3 12 = \log_3\left(\dfrac{1}{2} \cdot 12\right) = \log_3 6$

c. $\log_2(x + 2) + \log_2 x = \log_2[(x + 2) \cdot x] = \log_2(x^2 + 2x)$

> **Helpful Hint**
> Check your logarithm properties. Make sure you understand that $\log_2(x + 2)$ *is not* $\log_2 x + \log_2 2$.

2 The second property is the **quotient property of logarithms.**

> **Quotient Property of Logarithms**
> If x, y, and b are positive real numbers and $b \neq 1$, then
> $$\log_b\frac{x}{y} = \log_b x - \log_b y$$

The proof of the quotient property of logarithms is similar to the proof of the product property. Notice that the quotient property says that the logarithm of a quotient is the difference of the logarithms of the dividend and divisor

✔ **CONCEPT CHECK**

Which of the following is the correct way to rewrite $\log_5\dfrac{7}{2}$?

a. $\log_5 7 - \log_5 2$ **b.** $\log_5(7 - 2)$ **c.** $\dfrac{\log_5 7}{\log_5 2}$ **d.** $\log_5 14$

EXAMPLE 2

Write each difference as a single logarithm.

a. $\log_{10} 27 - \log_{10} 3$ **b.** $\log_5 8 - \log_5 x$ **c.** $\log_3(x^2 + 5) - \log_3(x^2 + 1)$

Solution All terms have a common logarithmic base.

a. $\log_{10} 27 - \log_{10} 3 = \log_{10}\dfrac{27}{3} = \log_{10} 9$

b. $\log_5 8 - \log_5 x = \log_5\dfrac{8}{x}$

c. $\log_3(x^2 + 5) - \log_3(x^2 + 1) = \log_3\dfrac{x^2 + 5}{x^2 + 1}$ *Apply the quotient property.*

Concept Check Answer:
a

3 The third and final property we introduce is the **power property of logarithms.**

> ### Power Property of Logarithms
> If x and b are positive real numbers, $b \neq 1$, and r is a real number, then
> $$\log_b x^r = r \log_b x$$

EXAMPLE 3

Use the power property to rewrite each expression.

a. $\log_5 x^3$ **b.** $\log_4 \sqrt{2}$

Solution **a.** $\log_5 x^3 = 3 \log_5 x$

b. $\log_4 \sqrt{2} = \log_4 2^{1/2} = \dfrac{1}{2} \log_4 2$

4 Many times we must use more than one property of logarithms to simplify a logarithmic expression.

EXAMPLE 4

Write as a single logarithm.

a. $2 \log_5 3 + 3 \log_5 2$ **b.** $3 \log_9 x - \log_9(x + 1)$ **c.** $\log_4 25 + \log_4 3 - \log_4 5$

Solution In each case, all terms have a common logarithmic base.

a. $2 \log_5 3 + 3 \log_5 2 = \log_5 3^2 + \log_5 2^3$ Apply the power property.

$\qquad\qquad\qquad = \log_5 9 + \log_5 8$

$\qquad\qquad\qquad = \log_5(9 \cdot 8)$ Apply the product property.

$\qquad\qquad\qquad = \log_5 72$

b. $3 \log_9 x - \log_9(x + 1) = \log_9 x^3 - \log_9(x + 1)$ Apply the power property.

$\qquad\qquad\qquad = \log_9 \dfrac{x^3}{x + 1}$ Apply the quotient property.

c. Use both the product and quotient properties.

$\log_4 25 + \log_4 3 - \log_4 5 = \log_4(25 \cdot 3) - \log_4 5$ Apply the product property.

$\qquad\qquad\qquad = \log_4 75 - \log_4 5$ Simplify.

$\qquad\qquad\qquad = \log_4 \dfrac{75}{5}$ Apply the quotient property.

$\qquad\qquad\qquad = \log_4 15$ Simplify.

EXAMPLE 5

Write each expression as sums or differences of multiples of logarithms.

a. $\log_3 \dfrac{5 \cdot 7}{4}$ **b.** $\log_2 \dfrac{x^5}{y^2}$

Solution   **a.** $\log_3 \dfrac{5 \cdot 7}{4} = \log_3(5 \cdot 7) - \log_3 4$ *Apply the quotient property.*

$\qquad\qquad\qquad = \log_3 5 + \log_3 7 - \log_3 4$ *Apply the product property.*

b. $\log_2 \dfrac{x^5}{y^2} = \log_2(x^5) - \log_2(y^2)$ *Apply the quotient property.*

$\qquad\qquad\qquad = 5 \log_2 x - 2 \log_2 y$ *Apply the power property.*

> **Helpful Hint**
>
> Notice that we are not able to simplify further a logarithmic expression such as $\log_5(2x - 1)$. None of the basic properties gives a way to write the logarithm of a difference in some equivalent form.

✔ **CONCEPT CHECK**

What is **wrong** with the following?

$$\log_{10}(x^2 + 5) = \log_{10} x^2 + \log_{10} 5$$
$$= 2 \log_{10} x + \log_{10} 5$$

Use a numerical example to demonstrate that the result is incorrect.

EXAMPLE 6

If $\log_b 2 = 0.43$ and $\log_b 3 = 0.68$, use the properties of logarithms to evaluate.

a. $\log_b 6$ **b.** $\log_b 9$ **c.** $\log_b \sqrt{2}$

Solution   **a.** $\log_b 6 = \log_b(2 \cdot 3)$ *Write 6 as $2 \cdot 3$.*

$\qquad\qquad = \log_b 2 + \log_b 3$ *Apply the product property.*

$\qquad\qquad = 0.43 + 0.68$ *Substitute given values.*

$\qquad\qquad = 1.11$ *Simplify.*

b. $\log_b 9 = \log_b 3^2$ *Write 9 as 3^2.*

$\qquad\qquad = 2 \log_b 3$

$\qquad\qquad = 2(0.68)$ *Substitute 0.68 for $\log_b 3$.*

$\qquad\qquad = 1.36$ *Simplify.*

c. First, recall that $\sqrt{2} = 2^{1/2}$. Then

$\log_b \sqrt{2} = \log_b 2^{1/2}$ *Write $\sqrt{2}$ as $2^{1/2}$.*

$\qquad\qquad = \dfrac{1}{2} \log_b 2$ *Apply the power property.*

$\qquad\qquad = \dfrac{1}{2}(0.43)$ *Substitute the given value.*

$\qquad\qquad = 0.215$ *Simplify.*

Concept Check Answer:
The properties do not give any way to simplify the logarithm of a sum; answers may vary.

A summary of the basic properties of logarithms that we have developed so far is given next.

Properties of Logarithms

If x, y, and b are positive real numbers, $b \neq 1$, and r is a real number, then

1. $\log_b 1 = 0$

2. $\log_b b^x = x$

3. $b^{\log_b x} = x$

4. $\log_b xy = \log_b x + \log_b y$ Product property

5. $\log_b \dfrac{x}{y} = \log_b x - \log_b y$ Quotient property

6. $\log_b x^r = r \log_b x$ Power property

Spotlight on DECISION MAKING

Suppose you are a quality assurance inspector for an electronics manufacturer. Your department has conducted reliability studies of a new model of DVD player. Your studies show that the DVD player's reliability can be described by the exponential function $R(t) = 2.7^{-(1/3)t}$, where the reliability R is the probability that the DVD player is still working t years after it is manufactured.

The marketing department asks for your input in choosing a warranty period for the DVD player. Popular warranty periods for similar competing DVD players are 1 year, 2 years, and 3 years. Using the graph of the reliability for this DVD player, which warranty period would you recommend? Explain your reasoning. What other factors would you want to consider?

MENTAL MATH

Select the correct choice.

1. $\log 12 + \log 3 = \log$ _____

   a. 36   **b.** 15   **c.** 4   **d.** 9

2. $\log 12 - \log 3 = \log$ _____

   a. 36   **b.** 15   **c.** 4   **d.** 9

3. $7 \log 2 =$ _____

   a. $\log 14$ **b.** $\log 2^7$ **c.** $\log 7^2$ **d.** $(\log 2)^7$

4. The base of $\log 7$ is _____

   a. e **b.** 7   **c.** 10   **d.** no base

5. The base of $\ln 7$ is _____

   a. e **b.** 7   **c.** 10   **d.** no base

6. $\log_5 5^2 =$ _____

   a. 25   **b.** 2   **c.** 5^{5^2} **d.** 32

EXERCISE SET 12.4

STUDY GUIDE/SSM | CD/VIDEO | PH MATH TUTOR CENTER | MathXL®Tutorials ON CD | MathXL® | MyMathLab®

Write each sum as the logarithm of a single expression. Assume that variables represent positive numbers. See Example 1.

 1. $\log_5 2 + \log_5 7$

2. $\log_3 8 + \log_3 4$

3. $\log_4 9 + \log_4 x$

4. $\log_2 x + \log_2 y$

5. $\log_{10} 5 + \log_{10} 2 + \log_{10}(x^2 + 2)$

6. $\log_6 3 + \log_6(x + 4) + \log_6 5$

Write each as the logarithm of a single expression. Assume that variables represent positive numbers. See Examples 2 and 4.

7. $\log_5 12 - \log_5 4$

8. $\log_7 20 - \log_7 4$

9. $\log_2 x - \log_2 y$

10. $\log_3 12 - \log_3 z$

11. $\log_4 2 + \log_4 10 - \log_4 5$

12. $\log_6 18 + \log_6 2 - \log_6 9$

Use the power property to rewrite each expression. See Example 3.

13. $\log_3 x^2$

14. $\log_2 x^5$

15. $\log_4 5^{-1}$

16. $\log_6 7^{-2}$

17. $\log_5 \sqrt{y}$

18. $\log_5 \sqrt[3]{x}$

MIXED PRACTICE

Write each as a single logarithm. Assume that variables represent positive numbers. See Example 4.

19. $2 \log_2 5$

20. $3 \log_5 2$

 21. $3 \log_5 x + 6 \log_5 z$

22. $2 \log_7 y + 6 \log_7 z$

23. $\log_{10} x - \log_{10}(x + 1) + \log_{10}(x^2 - 2)$

24. $\log_9(4x) - \log_9(x - 3) + \log_9(x^3 + 1)$

25. $\log_4 5 + \log_4 7$

26. $\log_3 2 + \log_3 5$

 27. $\log_3 8 - \log_3 2$

28. $\log_5 12 - \log_5 3$

29. $\log_7 6 + \log_7 3 - \log_7 4$

30. $\log_8 5 + \log_8 15 - \log_8 20$

31. $3 \log_4 2 + \log_4 6$

32. $2 \log_3 5 + \log_3 2$

33. $3 \log_2 x + \dfrac{1}{2}\log_2 x - 2 \log_2(x + 1)$

34. $2 \log_5 x + \dfrac{1}{3}\log_5 x - 3 \log_5(x + 5)$

35. $2 \log_8 x - \dfrac{2}{3}\log_8 x + 4 \log_8 x$

36. $5 \log_6 x - \dfrac{3}{4}\log_6 x + 3 \log_6 x$

Write each expression as a sum or difference of multiples of logarithms. Assume that variables represent positive numbers. See Example 5.

37. $\log_2 \dfrac{7 \cdot 11}{3}$

38. $\log_5 \dfrac{2 \cdot 9}{13}$

39. $\log_3 \dfrac{4y}{5}$

40. $\log_4 \dfrac{2}{9z}$

41. $\log_2 \dfrac{x^3}{y}$

42. $\log_5 \dfrac{x}{y^4}$

43. $\log_b \sqrt{7x}$

44. $\log_b \sqrt{\dfrac{3}{y}}$

45. $\log_7 \dfrac{5x}{4}$

46. $\log_9 \dfrac{7}{y}$

47. $\log_5 x^3(x + 1)$

48. $\log_2 y^3 z$

49. $\log_6 \dfrac{x^2}{x + 3}$

50. $\log_3 \dfrac{(x + 5)^2}{x}$

If $\log_b 3 = 0.5$ and $\log_b 5 = 0.7$, evaluate the following. See Example 6. If necessary, round to three decimal places.

51. $\log_b \dfrac{5}{3}$

52. $\log_b 25$

53. $\log_b 15$

54. $\log_b \dfrac{3}{5}$

55. $\log_b \sqrt[3]{5}$

56. $\log_b \sqrt[4]{3}$

If $\log_b 2 = 0.43$ and $\log_b 3 = 0.68$, evaluate the following.

57. $\log_b 8$

58. $\log_b 81$

59. $\log_b \dfrac{3}{9}$

60. $\log_b \dfrac{4}{32}$

61. $\log_b \sqrt{\dfrac{2}{3}}$

62. $\log_b \sqrt{\dfrac{3}{2}}$

REVIEW AND PREVIEW

63. Graph the functions $y = 10^x$ and $y = \log_{10} x$ on the same set of axes. See Section 12.4.

Evaluate each expression. See Section 12.4.

64. $\log_{10} 100$

65. $\log_{10} \dfrac{1}{10}$

66. $\log_7 7^2$

67. $\log_7 \sqrt{7}$

Concept Extensions

Answer the following true or false. Study your logarithm properties carefully before answering.

68. $\log_2 x^3 = 3 \log_2 x$

69. $\log_3(x + y) = \log_3 x + \log_3 y$

70. $\dfrac{\log_7 10}{\log_7 5} = \log_7 2$

71. $\log_7 \dfrac{14}{8} = \log_7 14 - \log_7 8$

72. $\dfrac{\log_7 x}{\log_7 y} = (\log_7 x) - (\log_7 y)$

73. $(\log_3 6) \cdot (\log_3 4) = \log_3 24$

74. It is true that $\log 8 = \log(8 \cdot 1) = \log 8 + \log 1$. Explain how $\log 8$ can equal $\log 8 + \log 1$.

INTEGRATED REVIEW   FUNCTIONS AND PROPERTIES OF LOGARITHMS

Determine whether each is a one-to-one function. If it is, find its inverse.

1. $f = \{(-2, 6), (4, 8), (2, -6), (3, 3)\}$ **2.** $g = \{(4, 2), (-1, 3), (5, 3), (7, 1)\}$

Determine whether the graph of each function is one-to-one.

3.

4.

5.
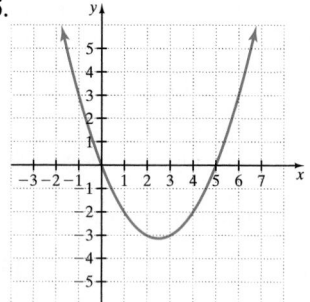

Each function listed is one-to-one. Find the inverse of each function.

6. $f(x) = 3x$ **7.** $f(x) = x + 4$

8. $f(x) = 5x - 1$ **9.** $f(x) = 3x + 2$

Graph each function.

10. $y = \left(\dfrac{1}{2}\right)^x$ **11.** $y = 2^x + 1$

12. $y = \log_3 x$ **13.** $y = \log_{1/3} x$

Solve.

14. $2^x = 8$ **15.** $9 = 3^{x-5}$ **16.** $4^{x-1} = 8^{x+2}$ **17.** $25^x = 125^{x-1}$

18. $\log_4 16 = x$ **19.** $\log_{49} 7 = x$ **20.** $\log_2 x = 5$

21. $\log_x 64 = 3$ **22.** $\log_x \dfrac{1}{125} = -3$ **23.** $\log_3 x = -2$

Write each as a single logarithm.

24. $5 \log_2 x$ **25.** $x \log_2 5$

26. $3 \log_5 x - 5 \log_5 y$ **27.** $9 \log_5 x + 3 \log_5 y$

28. $\log_2 x + \log_2(x - 3) - \log_2(x^2 + 4)$ **29.** $\log_3 y - \log_3(y + 2) + \log_3(y^3 + 11)$

Write each expression as a sum or difference of multiples of logarithms.

30. $\log_7 \dfrac{9x^2}{y}$ **31.** $\log_6 \dfrac{5y}{z^2}$

12.5   COMMON LOGARITHMS, NATURAL LOGARITHMS, AND CHANGE OF BASE

Objectives

1. Identify common logarithms and approximate them by calculator.
2. Evaluate common logarithms of powers of 10.
3. Identify natural logarithms and approximate them by calculator.
4. Evaluate natural logarithms of powers of e.
5. Use the change of base formula.

     In this section we look closely at two particular logarithmic bases. These two logarithmic bases are used so frequently that logarithms to their bases are given special names. **Common logarithms** are logarithms to base 10. **Natural logarithms** are logarithms to base e, which we introduce in this section. The work in this section is based on the use of the calculator, which has both the "common log" $\boxed{\text{LOG}}$ and the "natural log" $\boxed{\text{LN}}$ keys.

1 Logarithms to base 10, common logarithms, are used frequently because our number system is a base 10 decimal system. The notation $\log x$ means the same as $\log_{10} x$.

Common Logarithms

$$\log x \text{ means } \log_{10} x$$

EXAMPLE 1

Use a calculator to approximate $\log 7$ to four decimal places.

Solution Press the following sequence of keys.

$$\boxed{7} \;\; \boxed{\text{LOG}} \quad \text{or} \quad \boxed{\text{LOG}} \;\; \boxed{7} \;\; \boxed{\text{ENTER}}$$

To four decimal places,

$$\log 7 \approx 0.8451$$

2 To evaluate the common log of a power of 10, a calculator is not needed. According to the property of logarithms,

$$\log_b b^x = x$$

It follows that if b is replaced with 10, we have

$$\log 10^x = x$$

> **Helpful Hint**
> Remember that $\log 10^x$ means $\log_{10} 10^x = x$.

EXAMPLE 2

Find the exact value of each logarithm.

a. $\log 10$ **b.** $\log 1000$ **c.** $\log \dfrac{1}{10}$ **d.** $\log \sqrt{10}$

Solution   **a.** $\log 10 = \log 10^1 = 1$ **b.** $\log 1000 = \log 10^3 = 3$

       c. $\log \dfrac{1}{10} = \log 10^{-1} = -1$ **d.** $\log \sqrt{10} = \log 10^{1/2} = \dfrac{1}{2}$

As we will soon see, equations containing common logs are useful models of many natural phenomena.

EXAMPLE 3

Solve $\log x = 1.2$ for x. Give an exact solution, and then approximate the solution to four decimal places.

Solution   Remember that the base of a common log is understood to be 10.

> **Helpful Hint**
> The understood base is 10.

$$\log x = 1.2$$
$$10^{1.2} = x \qquad \text{Write with exponential notation.}$$

The exact solution is $10^{1.2}$. To four decimal places, $x \approx 15.8489$.

The Richter scale measures the intensity, or magnitude, of an earthquake. The formula for the magnitude R of an earthquake is $R = \log\left(\dfrac{a}{T}\right) + B$, where a is the amplitude in micrometers of the vertical motion of the ground at the recording station, T is the number of seconds between successive seismic waves, and B is an adjustment factor that takes into account the weakening of the seismic wave as the distance increases from the epicenter of the earthquake.

(**EXAMPLE 4**)

FINDING THE MAGNITUDE OF AN EARTHQUAKE

Find an earthquake's magnitude on the Richter scale if a recording station measures an amplitude of 300 micrometers and 2.5 seconds between waves. Assume that B is 4.2. Approximate the solution to the nearest tenth.

Solution   Substitute the known values into the formula for earthquake intensity.

$$R = \log\left(\frac{a}{T}\right) + B \qquad \text{Richter scale formula}$$

$$= \log\left(\frac{300}{2.5}\right) + 4.2 \qquad \text{Let } a = 300, T = 2.5, \text{ and } B = 4.2.$$

$$= \log(120) + 4.2$$

$$\approx 2.1 + 4.2 \qquad \text{Approximate log 120 by 2.1.}$$

$$= 6.3$$

This earthquake had a magnitude of 6.3 on the Richter scale.

3   **Natural logarithms** are also frequently used, especially to describe natural events; hence the label "natural logarithm." Natural logarithms are logarithms to the base e, which is a constant approximately equal to 2.7183. The number e is an irrational number, as is π. The notation $\log_e x$ is usually abbreviated to $\ln x$. (The abbreviation ln is read "el en.")

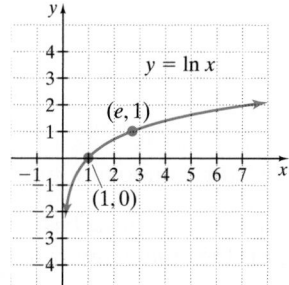

> **Natural Logarithms**
>
> $\ln x$ means $\log_e x$

The graph of $y = \ln x$ is shown to the left.

EXAMPLE 5

Use a calculator to approximate ln 8 to four decimal places.

Solution Press the following sequence of keys.

$$\boxed{8}\ \boxed{\text{LN}}\quad\text{or}\quad\boxed{\text{LN}}\ \boxed{8}\ \boxed{\text{ENTER}}$$

To four decimal places,

$$\ln 8 \approx 2.0794$$

4 As a result of the property $\log_b b^x = x$, we know that $\log_e e^x = x$, or $\ln e^x = x$.

Since $\ln e^x = x$, then $\ln e^5 = 5$, $\ln e^{22} = 22$ and so on. Also,

$$\ln e^1 = 1 \text{ or simply } \ln e = 1.$$

That is why the graph of $y = \ln x$ shown on the previous page passes through $(e, 1)$. If $x = e$, then $y = \ln e = 1$, thus the ordered pair $(e, 1)$.

EXAMPLE 6

Find the exact value of each natural logarithm.

a. $\ln e^3$ **b.** $\ln \sqrt[5]{e}$

Solution **a.** $\ln e^3 = 3$ **b.** $\ln \sqrt[5]{e} = \ln e^{1/5} = \dfrac{1}{5}$

EXAMPLE 7

Solve $\ln 3x = 5$. Give an exact solution, and then approximate the solution to four decimal places.

Solution Remember that the base of a natural logarithm is understood to be e.

$$\ln 3x = 5$$

> **Helpful Hint**
> The understood base is e.

$$e^5 = 3x \quad \text{Write with exponential notation.}$$

$$\frac{e^5}{3} = x \quad \text{Solve for } x.$$

The exact solution is $\dfrac{e^5}{3}$. To four decimal places,

$$x \approx 49.4711.$$

Recall from Section 12.2 the formula $A = P\left(1 + \dfrac{r}{n}\right)^{nt}$ for compound interest, where n represents the number of compoundings per year. When interest is compounded continuously, the formula $A = Pe^{rt}$ is used, where r is the annual interest rate and interest is compounded continuously for t years.

EXAMPLE 8

FINDING FINAL LOAN PAYMENT

Find the amount owed at the end of 5 years if $1600 is loaned at a rate of 9% compounded continuously.

Solution Use the formula $A = Pe^{rt}$, where

$$P = \$1600 \text{ (the size of the loan)}$$

$$r = 9\% = 0.09 \text{ (the rate of interest)}$$

$$t = 5 \text{ (the 5-year duration of the loan)}$$

$$A = Pe^{rt}$$

$$= 1600e^{0.09(5)} \qquad \text{Substitute in known values.}$$

$$= 1600e^{0.45}$$

Now we can use a calculator to approximate the solution.

$$A \approx 2509.30$$

The total amount of money owed is $2509.30.

5   Calculators are handy tools for approximating natural and common logarithms. Unfortunately, some calculators cannot be used to approximate logarithms to bases other than e or 10—at least not directly. In such cases, we use the change of base formula.

Change of Base

If a, b, and c are positive real numbers and neither b nor c is 1, then

$$\log_b a = \frac{\log_c a}{\log_c b}$$

EXAMPLE 9

Approximate $\log_5 3$ to four decimal places.

Solution Use the change of base property to write $\log_5 3$ as a quotient of logarithms to base 10.

$$\log_5 3 = \frac{\log 3}{\log 5} \qquad \text{Use the change of base property. In the change of base property, we let } a = 3, b = 5, \text{ and } c = 10.$$

$$\approx \frac{0.4771213}{0.69897} \qquad \text{Approximate logarithms by calculator.}$$

$$\approx 0.6826062 \qquad \text{Simplify by calculator.}$$

To four decimal places, $\log_5 3 \approx 0.6826$.

Concept Check Answer:

$f(x) = \dfrac{\log x}{\log 5}$

✔ **CONCEPT CHECK**

If a graphing calculator cannot directly evaluate logarithms to base 5, describe how you could use the graphing calculator to graph the function $f(x) = \log_5 x$.

STUDY SKILLS REMINDER

Do You Remember What to Do the Day of an Exam?

On the day of an exam, don't forget to try the following:

▶ Allow yourself plenty of time to arrive.

▶ Read the directions on the test carefully.

▶ Read each problem carefully as you take your test. Make sure that you answer the question asked.

▶ Watch your time and pace yourself so that you may attempt each problem on your test.

▶ If you have time, check your work and answers.

▶ Do not turn your test in early. If you have extra time, spend it double-checking your work.

Good luck!

EXERCISE SET 12.5

STUDY GUIDE/SSM · CD/VIDEO · PH MATH TUTOR CENTER · MathXL®Tutorials ON CD · MathXL® · MyMathLab®

Use a calculator to approximate each logarithm to four decimal places. See Examples 1 and 5.

1. $\log 8$

2. $\log 6$

3. $\log 2.31$

4. $\log 4.86$

5. $\ln 2$

6. $\ln 3$

7. $\ln 0.0716$

8. $\ln 0.0032$

9. $\log 12.6$

10. $\log 25.9$

11. $\ln 5$

12. $\ln 7$

13. $\log 41.5$

14. $\ln 41.5$

15. Use a calculator and try to approximate $\log 0$. Describe what happens and explain why.

16. Use a calculator and try to approximate $\ln 0$. Describe what happens and explain why.

Find the exact value. See Examples 2 and 6.

 17. $\log 100$

18. $\log 10{,}000$

19. $\log\left(\dfrac{1}{1000}\right)$

20. $\log\left(\dfrac{1}{100}\right)$

21. $\ln e^2$

22. $\ln e^4$

23. $\ln \sqrt[4]{e}$

24. $\ln \sqrt[5]{e}$

25. $\log 10^3$

26. $\ln e^5$

27. $\ln e^2$

28. $\log 10^7$

29. $\log 0.0001$

30. $\log 0.001$

31. $\ln \sqrt{e}$

32. $\log \sqrt{10}$

33. Without using a calculator, explain which of $\log 50$ or $\ln 50$ must be larger.

34. Without using a calculator, explain which of $\log 50^{-1}$ or $\ln 50^{-1}$ must be larger.

Solve each equation for x. Give an exact solution and a four-decimal-place approximation. See Examples 3 and 7.

35. $\log x = 1.3$

36. $\log x = 2.1$

37. $\log 2x = 1.1$

38. $\log 3x = 1.3$

39. $\ln x = 1.4$

40. $\ln x = 2.1$

41. $\ln(3x - 4) = 2.3$

42. $\ln(2x + 5) = 3.4$

43. $\log x = 2.3$

44. $\log x = 3.1$

45. $\ln x = -2.3$

46. $\ln x = -3.7$

47. $\log (2x + 1) = -0.5$

48. $\log(3x - 2) = -0.8$

49. $\ln 4x = 0.18$

50. $\ln 3x = 0.76$

Approximate each logarithm to four decimal places. See Example 9.

51. $\log_2 3$

52. $\log_3 2$

53. $\log_{1/2} 5$

54. $\log_{1/3} 2$

55. $\log_4 9$

56. $\log_9 4$

57. $\log_3 \dfrac{1}{6}$

58. $\log_6 \dfrac{2}{3}$

59. $\log_8 6$

60. $\log_6 8$

Use the formula $R = \log\left(\dfrac{a}{T}\right) + B$ *to find the intensity* R *on the Richter scale of the earthquakes that fit the descriptions given. Round answers to one decimal place. See Example 4.*

61. Amplitude a is 200 micrometers, time T between waves is 1.6 seconds, and B is 2.1.

62. Amplitude a is 150 micrometers, time T between waves is 3.6 seconds, and B is 1.9.

63. Amplitude a is 400 micrometers, time T between waves is 2.6 seconds, and B is 3.1.

64. Amplitude a is 450 micrometers, time T between waves is 4.2 seconds, and B is 2.7.

Use the formula $A = Pe^{rt}$ *to solve. See Example 8.*

65. Find how much money Dana Jones has after 12 years if $1400 is invested at 8% interest compounded continuously.

66. Determine the size of an account in which $3500 earns 6% interest compounded continuously for 1 year.

67. Find the amount of money Barbara Mack owes at the end of 4 years if 6% interest is compounded continuously on her $2000 debt.

68. Find the amount of money for which a $2500 certificate of deposit is redeemable if it has been paying 10% interest compounded continuously for 3 years.

REVIEW AND PREVIEW

Solve each equation for x. *See Sections 2.3 and 6.5.*

69. $6x - 3(2 - 5x) = 6$

70. $2x + 3 = 5 - 2(3x - 1)$

71. $2x + 3y = 6x$

72. $4x - 8y = 10x$

73. $x^2 + 7x = -6$

74. $x^2 + 4x = 12$

Solve each system of equations. See Sections 4.1 through 4.3.

75. $\begin{cases} x + 2y = -4 \\ 3x - y = 9 \end{cases}$

76. $\begin{cases} 5x + y = 5 \\ -3x - 2y = -10 \end{cases}$

Concept Extensions

77. Without using a calculator, explain which of log 50 or ln 50 must be larger.

78. Without using a calculator, explain which of $\log 50^{-1}$ or $\ln 50^{-1}$ must be larger.

Graph each function by finding ordered pair solutions, plotting the solutions, and then drawing a smooth curve through the plotted points.

79. $f(x) = e^x$

80. $f(x) = e^{2x}$

81. $f(x) = e^{-3x}$

82. $f(x) = e^{-x}$

83. $f(x) = e^x + 2$

84. $f(x) = e^x - 3$

85. $f(x) = e^{x-1}$

86. $f(x) = e^{x+4}$

87. $f(x) = 3e^x$

88. $f(x) = -2e^x$

89. $f(x) = \ln x$

90. $f(x) = \log x$

91. $f(x) = -2 \log x$

92. $f(x) = 3 \ln x$

93. $f(x) = \log(x + 2)$

94. $f(x) = \log(x - 2)$

95. $f(x) = \ln x - 3$

96. $f(x) = \ln x + 3$

97. Graph $f(x) = e^x$ (Exercise 79), $f(x) = e^x + 2$ (Exercise 83), and $f(x) = e^x - 3$ (Exercise 84) on the same screen. Discuss any trends shown on the graphs.

98. Graph $f(x) = \ln x$ (Exercise 89), $f(x) = \ln x - 3$ (Exercise 95), and $f(x) = \ln x + 3$ (Exercise 96). Discuss any trends shown on the graphs.

12.6   EXPONENTIAL AND LOGARITHMIC EQUATIONS AND APPLICATIONS

Objectives

1   Solve exponential equations.

2   Solve logarithmic equations.

3   Solve problems that can be modeled by exponential and logarithmic equations.

1   In Section 12.2 we solved exponential equations such as $2^x = 16$ by writing 16 as a power of 2 and applying the uniqueness of b^x.

$$2^x = 16$$

$$2^x = 2^4 \qquad \text{Write } \mathbf{16} \text{ as } \mathbf{2^4}.$$

$$x = 4 \qquad \text{Use the uniqueness of } b^x.$$

Solving the equation in this manner is possible since 16 is a power of 2. If solving an equation such as $2^x = $ *a number,* where the number is not a power of 2, we use logarithms. For example, to solve an equation such as $3^x = 7$, we use the fact that $f(x) = \log_b x$ is a one-to-one function. Another way of stating this fact is as a property of equality.

Logarithm Property of Equality

Let a, b, and c be real numbers such that $\log_b a$ and $\log_b c$ are real numbers and b is not 1. Then

$$\log_b a = \log_b c \text{ is equivalent to } a = c$$

EXAMPLE 1

Solve $3^x = 7$.

Solution To solve, we use the logarithm property of equality and take the logarithm of both sides. For this example, we use the common logarithm.

$$3^x = 7$$

$$\log 3^x = \log 7 \qquad \text{Take the common log of both sides.}$$

$$x \log 3 = \log 7 \qquad \text{Apply the power property of logarithms.}$$

$$x = \frac{\log 7}{\log 3} \qquad \text{Divide both sides by } \log 3.$$

The exact solution is $\dfrac{\log 7}{\log 3}$. If a decimal approximation is preferred,

$$\frac{\log 7}{\log 3} \approx \frac{0.845098}{0.4771213} \approx 1.7712 \text{ to four decimal places}.$$

The solution is $\dfrac{\log 7}{\log 3}$, or *approximately* 1.7712.

2   By applying the appropriate properties of logarithms, we can solve a broad variety of logarithmic equations.

EXAMPLE 2

Solve $\log_4(x - 2) = 2$.

Solution Notice that $x - 2$ must be positive, so x must be greater than 2. With this in mind, we first write the equation with exponential notation.

$$\log_4(x - 2) = 2$$

$$4^2 = x - 2$$

$$16 = x - 2$$

$$18 = x \qquad \text{Add 2 to both sides.}$$

Check   To check, we replace x with 18 in the original equation.

$$\log_4(x - 2) = 2$$

$$\log_4(18 - 2) \stackrel{?}{=} 2 \qquad \text{Let } x = \mathbf{18}.$$

$$\log_4 16 \stackrel{?}{=} 2$$

$$4^2 = 16 \qquad \text{True.}$$

The solution is 18.

EXAMPLE 3

Solve $\log_2 x + \log_2(x - 1) = 1$.

Solution   Notice that $x - 1$ must be positive, so x must be greater than 1. We use the product property on the left side of the equation.

$$\log_2 x + \log_2(x - 1) = 1$$

$$\log_2 x(x - 1) = 1 \qquad \text{Apply the product property.}$$

$$\log_2(x^2 - x) = 1$$

Next we write the equation with exponential notation and solve for x.

$$2^1 = x^2 - x$$

$$0 = x^2 - x - 2 \qquad \qquad \text{Subtract 2 from both sides.}$$

$$0 = (x - 2)(x + 1) \qquad \qquad \text{Factor.}$$

$$0 = x - 2 \quad \text{or} \quad 0 = x + 1 \quad \text{Set each factor equal to 0.}$$

$$2 = x \qquad \qquad -1 = x$$

Recall that -1 cannot be a solution because x must be greater than 1. If we forgot this, we would still reject -1 after checking. To see this, we replace x with -1 in the original equation.

$$\log_2 x + \log_2(x - 1) = 1$$

$$\log_2(-1) + \log_2(-1 - 1) \stackrel{?}{=} 1 \quad \text{Let } x = \mathbf{-1}.$$

Because the logarithm of a negative number is undefined, -1 is rejected. Check to see that the solution is 2.

EXAMPLE 4

Solve $\log (x + 2) - \log x = 2$.

We use the quotient property of logarithms on the left side of the equation.

Solution
$$\log (x + 2) - \log x = 2$$

$$\log \frac{x + 2}{x} = 2 \qquad \text{Apply the quotient property.}$$

$$10^2 = \frac{x + 2}{x} \qquad \text{Write using exponential notation.}$$

$$100 = \frac{x + 2}{x} \qquad \text{Simplify.}$$

$$100x = x + 2 \qquad \text{Multiply both sides by } x.$$

$$99x = 2 \qquad \text{Subtract } x \text{ from both sides.}$$

$$x = \frac{2}{99} \qquad \text{Divide both sides by } 99.$$

Verify that the solution is $\frac{2}{99}$.

3   Logarithmic and exponential functions are used in a variety of scientific, technical, and business settings. A few examples follow.

EXAMPLE 5

ESTIMATING POPULATION SIZE

The population size y of a community of lemmings varies according to the relationship $y = y_0 e^{0.15t}$. In this formula, t is time in months, and y_0 is the initial population at time 0. Estimate the population after 6 months if there were originally 5000 lemmings.

Solution We substitute 5000 for y_0 and 6 for t.

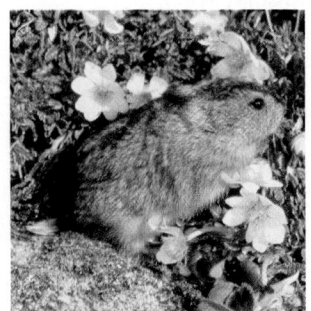

$$y = y_0 e^{0.15t}$$

$$= 5000 e^{0.15(6)} \qquad \text{Let } t = 6 \text{ and } y_0 = 5000.$$

$$= 5000 e^{0.9} \qquad \text{Multiply.}$$

Using a calculator, we find that $y \approx 12,298.016$. In 6 months the population will be approximately 12,300 lemmings.

EXAMPLE 6

DOUBLING AN INVESTMENT

How long does it take an investment of $2000 to double if it is invested at 5% interest compounded quarterly? The necessary formula is $A = P\left(1 + \frac{r}{n}\right)^{nt}$, where A is the accrued (or owed) amount, P is the principal invested, r is the annual rate of interest, n is the number of compounding periods per year, and t is the number of years.

Solution   We are given that $P = \$2000$ and $r = 5\% = 0.05$. Compounding quarterly means 4 times a year, so $n = 4$. The investment is to double, so A must be $\$4000$. Substitute these values and solve for t.

$$A = P\left(1 + \frac{r}{n}\right)^{nt}$$

$$4000 = 2000\left(1 + \frac{0.05}{4}\right)^{4t} \qquad \text{Substitute in known values.}$$

$$4000 = 2000(1.0125)^{4t} \qquad \text{Simplify } 1 + \frac{0.05}{4}.$$

$$2 = (1.0125)^{4t} \qquad \text{Divide both sides by } \mathbf{2000}.$$

$$\log 2 = \log 1.0125^{4t} \qquad \text{Take the logarithm of both sides.}$$

$$\log 2 = 4t(\log 1.0125) \qquad \text{Apply the power property.}$$

$$\frac{\log 2}{4 \log 1.0125} = t \qquad \text{Divide both sides by } \mathbf{4 \log 1.0125}.$$

$$13.949408 \approx t \qquad \text{Approximate by calculator.}$$

Thus, it takes nearly 14 years for the money to double in value.

Graphing Calculator Explorations

Use a graphing calculator to find how long it takes an investment of $\$1500$ to triple if it is invested at 8% interest compounded monthly.

First, let $P = \$1500$, $r = 0.08$, and $n = 12$ (for 12 months) in the formula

$$A = P\left(1 + \frac{r}{n}\right)^{nt}$$

Notice that when the investment has tripled, the accrued amount A is $\$4500$. Thus,

$$4500 = 1500\left(1 + \frac{0.08}{12}\right)^{12t}$$

Determine an appropriate viewing window and enter and graph the equations

$$Y_1 = 1500\left(1 + \frac{0.08}{12}\right)^{12x}$$

and

$$Y_2 = 4500$$

The point of intersection of the two curves is the solution. The x-coordinate tells how long it takes for the investment to triple.

Use a TRACE feature or an INTERSECT feature to approximate the coordinates of the point of intersection of the two curves. It takes approximately 13.78 years, or 13 years and 9 months, for the investment to triple in value to $\$4500$.

(continued)

Use this graphical solution method to solve each problem. Round each answer to the nearest hundredth.

1. Find how long it takes an investment of $5000 to grow to $6000 if it is invested at 5% interest compounded quarterly.

2. Find how long it takes an investment of $1000 to double if it is invested at 4.5% interest compounded daily. (Use 365 days in a year.)

3. Find how long it takes an investment of $10,000 to quadruple if it is invested at 6% interest compounded monthly.

4. Find how long it takes $500 to grow to $800 if it is invested at 4% interest compounded semiannually.

STUDY SKILLS REMINDER

Continue your outline started in Section 2.3. Write how to recognize and how to solve exponential and logarithmic equations in your own words. For example:

Solving Equations and Inequalities

I. Equations
  A. Linear equations (Section 2.3)
  B. Quadratic and higher degree equations (Section 6.5 and Chapter 11)
  C. Equations with rational expressions (Section 7.5)
  D. Absolute value equations (Section 9.2)
  E. Equations with radicals (Section 10.6)
  F. **Exponential equations**—Recognize: *equations where the variable is in the exponent*—Solve:
    1. If we can write the bases the same, then set the exponents equal to each other and solve
    2. If we can't write the bases the same, then solve using logarithms
  G. **Logarithmic equations**—Recognize: *equations with logarithms*—Solve: Write the equation so that there is a single logarithm on one side and a constant on the other side. Then use the definition of a logarithm to write the logarithm as an equivalent exponential and solve.

II. Inequalities
  A. Linear inequalities (Section 2.7)
  B. Compound inequalities (Section 9.1)
  C. Absolute value inequalities (Section 9.3)
  D. Nonlinear inequalities (Section 11.4)
    1. Polynomial inequalities
    2. Rational inequalities

EXERCISE SET 12.6

STUDY   CD/   PH MATH   MathXL®Tutorials   MathXL®   MyMathLab®
GUIDE/SSM   VIDEO   TUTOR CENTER   ON CD

Solve each equation. Give an exact solution, and also approximate the solution to four decimal places. See Example 1.

 1. $3^x = 6$

2. $4^x = 7$

3. $3^{2x} = 3.8$

4. $5^{3x} = 5.6$

5. $2^{x-3} = 5$

6. $8^{x-2} = 12$

7. $9^x = 5$

8. $3^x = 11$

9. $4^{x+7} = 3$

10. $6^{x+3} = 2$

11. $7^{3x-4} = 11$

12. $5^{2x-6} = 12$

13. $e^{6x} = 5$

14. $e^{2x} = 8$

MIXED PRACTICE

Solve each equation. See Examples 1 through 4.

15. $\log_2(x + 5) = 4$

16. $\log_6(x^2 - x) = 1$

17. $\log_3 x^2 = 4$

18. $\log_2 x^2 = 6$

19. $\log_4 2 + \log_4 x = 0$

20. $\log_3 5 + \log_3 x = 1$

21. $\log_2 6 - \log_2 x = 3$

22. $\log_4 10 - \log_4 x = 2$

23. $\log_4 x + \log_4(x + 6) = 2$ **24.** $\log_3 x + \log_3(x + 6) = 3$

25. $\log_5(x + 3) - \log_5 x = 2$ **26.** $\log_6(x + 2) - \log_6 x = 2$

27. $\log_3(x - 2) = 2$

28. $\log_2(x - 5) = 3$

29. $\log_4(x^2 - 3x) = 1$

30. $\log_8(x^2 - 2x) = 1$

31. $\ln 5 + \ln x = 0$

32. $\ln 3 + \ln (x - 1) = 0$

33. $3 \log x - \log x^2 = 2$

34. $2 \log x - \log x = 3$

35. $\log_2 x + \log_2(x + 5) = 1$ **36.** $\log_4 x + \log_4(x + 7) = 1$

37. $\log_4 x - \log_4(2x - 3) = 3$

38. $\log_2 x - \log_2(3x + 5) = 4$

39. $\log_2 x + \log_2(3x + 1) = 1$

40. $\log_3 x + \log_3(x - 8) = 2$

Solve. See Example 5.

41. The size of the wolf population at Isle Royale National Park increases at a rate of 4.3% per year. If the size of the current population is 83 wolves, find how many there should be in 5 years. Use $y = y_0 e^{0.043t}$ and round to the nearest whole.

42. The number of victims of a flu epidemic is increasing at a rate of 7.5% per week. If 20,000 persons are currently infected, find in how many days we can expect 45,000 to have the flu. Use $y = y_0 e^{0.075t}$ and round to the nearest whole. (Hint: Don't forget to convert your answer to days.)

43. The size of the population of Senegal is increasing at a rate of 2.6% per year. If 10,589,571 people lived in Senegal in 2002, find how many inhabitants there will be by 2008. Round to the nearest ten-thousand. Use $y = y_0 e^{0.026t}$.

44. In 2002, 1046 million people were citizens of India. Find how long it will take India's population to reach a size of 1500 million (that is, 1.5 billion) if the population size is growing at a rate of 1.7% per year. Use $y = y_0 e^{0.017t}$ and round to the nearest tenth. (*Source: U.S. Bureau of the Census, International Data Base*)

45. In 2002, Russia had a population of 144,979 thousand. At that time, Russia's population was declining at a rate of 1.8% per year. If this continues, how long will it take for Russia's population to reach 120,000 thousand? Use $y = y_0 e^{-0.018t}$ and round to the nearest tenth. (*Source: U.S. Bureau of the Census, International Data Base*)

46. The population of Italy has been decreasing at a rate of 0.1% per year. If there were 57,715,625 people living in Italy in 2002, how many inhabitants will there be by 2020? Use $y = y_0 e^{-0.001t}$ and round to the nearest hundred. (*Source: U.S. Bureau of the Census, International Data Base*)

Use the formula $A = P\left(1 + \dfrac{r}{n}\right)^{nt}$ to solve these compound interest problems. Round to the nearest tenth. See Example 6.

47. Find how long it takes $600 to double if it is invested at 7% interest compounded monthly.

48. Find how long it takes $600 to double if it is invested at 12% interest compounded monthly.

49. Find how long it takes a $1200 investment to earn $200 interest if it is invested at 9% interest compounded quarterly.

50. Find how long it takes a $1500 investment to earn $200 interest if it is invested at 10% compounded semiannually.

51. Find how long it takes $1000 to double if it is invested at 8% interest compounded semiannually.

52. Find how long it takes $1000 to double if it is invested at 8% interest compounded monthly.

The formula $w = 0.00185h^{2.67}$ is used to estimate the normal weight w of a boy h inches tall. Use this formula to solve the height-weight problems. Round to the nearest tenth.

53. Find the expected weight of a boy who is 35 inches tall.

54. Find the expected weight of a boy who is 43 inches tall.

55. Find the expected height of a boy who weighs 85 pounds.

56. Find the expected height of a boy who weighs 140 pounds.

The formula $P = 14.7e^{-0.21x}$ gives the average atmospheric pressure P, in pounds per square inch, at an altitude x, in miles above sea level. Use this formula to solve these pressure problems. Round answers to the nearest tenth.

57. Find the average atmospheric pressure of Denver, which is 1 mile above sea level.

58. Find the average atmospheric pressure of Pikes Peak, which is 2.7 miles above sea level.

59. Find the elevation of a Delta jet if the atmospheric pressure outside the jet is 7.5 lb/in.^{2}.

60. Find the elevation of a remote Himalayan peak if the atmospheric pressure atop the peak is 6.5 lb/in.^{2}.

Psychologists call the graph of the formula $t = \dfrac{1}{c}\ln\left(\dfrac{A}{A - N}\right)$ the learning curve, since the formula relates time t passed, in weeks, to a measure N of learning achieved, to a measure A of maximum learning possible, and to a measure c of an individual's learning style. Round to the nearest week.

61. Norman is learning to type. If he wants to type at a rate of 50 words per minute (N is 50) and his expected maximum rate is 75 words per minute (A is 75), find how many weeks it should take him to achieve his goal. Assume that c is 0.09.

62. An experiment with teaching chimpanzees sign language shows that a typical chimp can master a maximum of 65 signs. Find how many weeks it should take a chimpanzee to master 30 signs if c is 0.03.

63. Janine is working on her dictation skills. She wants to take dictation at a rate of 150 words per minute and believes that the maximum rate she can hope for is 210 words per minute. Find how many weeks it should take her to achieve the 150 words per minute level if c is 0.07.

64. A psychologist is measuring human capability to memorize nonsense syllables. Find how many weeks it should take a subject to learn 15 nonsense syllables if the maximum possible to learn is 24 syllables and c is 0.17.

REVIEW AND PREVIEW

If $x = -2$, $y = 0$, and $z = 3$, find the value of each expression. See Section 1.3.

65. $\dfrac{x^2 - y + 2z}{3x}$

66. $\dfrac{x^3 - 2y + z}{2z}$

67. $\dfrac{3z - 4x + y}{x + 2z}$

68. $\dfrac{4y - 3x + z}{2x + y}$

Find the inverse function of each one-to-one function. See Section 12.1.

69. $f(x) = 5x + 2$

70. $f(x) = \dfrac{x - 3}{4}$

Concept Extensions

The formula $y = y_0e^{kt}$ gives the population size y of a population that experiences an annual rate of population growth k (given as a decimal). In this formula, t is time in years and y_0 is the initial population at time 0. Use this formula to solve Exercises 71 and 72.

71. In 1990, the population of Arizona was 3,665,228. By 2000, the population had grown to 5,130,632. Find the annual rate of population growth over this period. Round your answer to the nearest tenth of a percent. (*Source:* U.S. Bureau of the Census)

72. In 1990, the population of Nevada was 1,201,833. By 2000, the population had grown to 1,998,257. Find the annual rate of population growth over this period. Round your answer to the nearest tenth of a percent. (*Source:* U.S. Bureau of the Census)

73. When solving a logarithmic equation, explain why you must check possible solutions in the original equation.

74. Solve $5^x = 9$ by taking the common logarithm of both sides of the equation. Next, solve this equation by taking the natural logarithm of both sides. Compare your solutions. Are they the same? Why or why not?

Use a graphing calculator to solve each equation. For example, to solve Exercise 75, let $Y_1 = e^{0.3x}$ and $Y_2 = 8$, and graph the equations. The x-value of the point of intersection is the solution. Round all solutions to two decimal places.

75. $e^{0.3x} = 8$

76. $10^{0.5x} = 7$

77. $2\log(-5.6x + 1.3) = -x - 1$

78. $\ln(1.3x - 2.1) = -3.5x + 5$

79. Check Exercise 11.

80. Check Exercise 12.

81. Check Exercise 31.

82. Check Exercise 32.

Modeling Temperature

When a cold object is placed in a warm room, the object's temperature gradually rises until it becomes, or nearly becomes, room temperature. Similarly, if a hot object is placed in a cooler room, the object's temperature gradually falls to room temperature. The way in which a cold or hot object warms up or cools off is modeled by an exact mathematical relationship, known as Newton's law of cooling. This law relates the temperature of an object to the time elapsed since its warming or cooling began. In this project, you will have the opportunity to investigate this model of cooling and warming. This project may be completed by working in groups or individually.

To investigate Newton's law of cooling in this project, you will collect experimental data in one of two methods: Method 1, using a stopwatch and thermometer, or Method 2, using Texas Instruments' Calculator-Based Laboratory (CBL™) or Second Generation Calculator-Based Laboratory (CBL 2™).

Method 1 Materials

- Container of either cold or hot liquid
- Thermometer
- Stopwatch
- Graphing calculator with regression capabilities

Method 2 Materials

- Container of either cold or hot liquid
- A TI-82, TI-83, or TI-85 graphing calculator with unit-to-unit link cable
- CBL™ or CBL 2™ unit with temperature probe

DATA TABLE

Time, *t*	Temperature, *T*
0	

Steps for Collecting Data with Method 1:

a. Insert the thermometer into the liquid and allow a thermometer reading to register. Take a temperature reading T as you start the stopwatch (at $t = 0$) and record it in the accompanying data table.

b. Continue taking temperature readings at uniform intervals anywhere between 5 and 10 minutes long. At each reading use the stopwatch to measure the length of time that has elapsed since the temperature readings started with your first reading at $t = 0$. Record your time t and liquid temperature T in the data table. Gather data for six to twelve readings.

c. Plot the data from the data table. Plot t on the horizontal axis and T on the vertical axis.

Steps for Collecting Data with Method 2:

a. Enter the HEAT program appropriate for your calculator.

b. Prepare the CBL or CBL 2 and the graphing calculator. Insert the temperature probe into the liquid.

c. Start the HEAT program on the graphing calculator and follow its instructions to begin collecting data. The program will collect 36 temperature readings in degrees Celsius and plot them in real time with t on the horizontal axis and T on the vertical axis.

1. Which of the following mathematical models best fits the data you collected? Explain your reasoning. (Assume $a > 0$.)

    a. $T = ab^t + c$

    b. $T = ab^{-t} + c$

    c. $T = -ab^{-t} + c$

    d. $T = \ln(-ax + b) + c$

    e. $T = -\ln(-ax + b) + c$

2. What does the constant c represent in the model you chose? What is the value of c in this activity?

3. (Optional) Subtract the value of c from each of your observations of T. Enter the new ordered pairs $(t, T - c)$ into a graphing calculator. Use the exponential or logarithmic regression feature to find a model for your experimental data. Graph the ordered pairs $(t, T - c)$ with the model you found. How well does the model fit the data? How does the model compare with your selection from Question 1?

Graphing Calculator Programs

TI-83 Program

```
PROGRAM:HEAT82
:PlotsOff
:Func
:FnOff
:AxesOn
:ClrDraw
:ClrList L3, L4
:-10 → Ymin
:90 → Ymax
:10 → Yscl
:ClrHome
:{1, 0} → L1
:Send (L1)
:{1, 1, 1} → L1
:Send (L1)
:36 → dim L3
:36 → dim L4
:Disp "HOW MUCH TIME"
:Disp "BETWEEN POINTS"
:Disp "IN SECONDS?"
:Input T
:-2*T → Xmin
:36*T → Xmax
:T → Xscl
:seq(K, K, T, 36*T, T) → L3
:ClrHome
:Disp "PRESS ENTER"
:Disp "TO START"
:Pause
:ClrHome
:{3, T, -1, 0} → L1
:Send(L1)
:For (K, 1, 36, 1)
:Get (L4 (K))
:Pt-On (L3 (K), L4 (K))
:End
:ClrHome
:Plot1 (Scatter, L3, L4, ·)
:DispGraph
:Stop
```

TI-85 Program

```
PROGRAM:HEAT85
:Func
:FnOff
:AxesOn
:ClDrw
:1 → dimL L3: 1 → dimL L4
:-10 → yMin
:90 → yMax
:10 → yScl
:ClLCD
:{1, 0} → L1
:Outpt ("CBLSEND", L1)
:{1, 1, 1} → L1
:Outpt ("CBLSEND", L1)
:36 → dimL L3
:36 → dimL L4
:Disp "HOW MUCH TIME"
:Disp "BETWEEN POINTS"
:Disp "IN SECONDS"
:Input T
:-2*T → xMin
:36*T → xMax
:T → xScl
:seq(K, K, T, 36*T, T) → L3
:ClLCD
:Disp "PRESS ENTER"
:Disp "TO START"
:Pause
:ClLCD
:{3, T, -1, 0} → L1
:Outpt ("CBLSEND", L1)
:For (K, 1, 36, 1)
:Input "CBLGET", L4 (K)
:PtOn (L3(K), L4(K))
:End
:ClLCD
:Scatter L3, L4
:DispG
:Stop
```

STUDY SKILLS REMINDER

Are You Preparing for a Test on Chapter 12?

Below I have listed some common trouble areas for topics covered in Chapter 12. After studying for your test—but before taking your test—read these.

▶ Don't forget that f^{-1} is a special notation used to denote the inverse of a function.

Let's find the inverse of the invertible function $f(x) = 3x - 5$.

$$f(x) = 3x - 5$$

$$y = 3x - 5 \quad \text{Replace } f(x) \text{ with } y.$$

$$x = 3y - 5 \quad \text{Switch variables.}$$

$$x + 5 = 3y$$

$$\frac{x + 5}{3} = y \quad \text{Solve for } y.$$

$$f^{-1}(x) = \frac{x + 5}{3} \quad \text{Replace } y \text{ with } f^{-1}(x).$$

▶ Don't forget that $y = \log_b x$ means $b^y = x$.

Thus, $3 = \log_5 125$ means $5^3 = 125$.

▶ Remember rules for logarithms.

$$\log_b 3x = \log_b 3 + \log_b x$$

$\log_b(3 + x)$ cannot be simplified in the same manner.

Remember: This is simply a checklist of common trouble areas. For a review of Chapter 12, see the Highlights and Chapter Review at the end of this chapter.

12 CHAPTER VOCABULARY CHECK

Fill in each blank with one of the words or phrases listed below.

inverse	common	natural	horizontal
vertical	logarithmic	symmetric	exponential

1. For each one-to-one function, we can find its _____ function by switching the coordinates of the ordered pairs of the function.

2. A function of the form $f(x) = b^x$ is called an _____ function if $b > 0$, b is not 1, and x is a real number.

3. The graphs of f and f^{-1} are _____ about the line $y = x$.

4. _____ logarithms are logarithms to base e.

5. _____ logarithms are logarithms to base 10.

6. To see whether a graph is the graph of a one-to-one function, apply the _____ line test to see if it is a function, and then apply the _____ line test to see if it is a one-to-one function.

7. A _____ function is a function that can be defined by $f(x) = \log_b x$ where x is a positive real number, b is a constant positive real number, and b is not 1.

CHAPTER 12 HIGHLIGHTS

Definitions and Concepts	Examples

Section 12.1   Inverse Functions

If f is a function, then f is a **one-to-one function** only if each y-value (output) corresponds to only one x-value (input).

Horizontal Line Test

If every horizontal line intersects the graph of a function at most once, then the function is a one-to-one function.

Determine whether each graph is a one-to-one function.

Graphs **A** and **C** pass the vertical line test, so only these are graphs of functions. Of graphs **A** and **C**, only graph **A** passes the horizontal line test, so only graph **A** is the graph of a one-to-one function.

The **inverse** of a one-to-one function f is the one-to-one function f^{-1} that is the set of all ordered pairs (b, a) such that (a, b) belongs to f.

To Find the Inverse of a One-to-One Function f(x)

Step 1: Replace $f(x)$ with y.

Step 2: Interchange x and y.

Step 3: Solve for y.

Step 4: Replace y with $f^{-1}(x)$.

Find the inverse of $f(x) = 2x + 7$.

$$y = 2x + 7 \quad \text{Replace } f(x) \text{ with } y.$$

$$x = 2y + 7 \quad \text{Interchange } x \text{ and } y.$$

$$2y = x - 7 \quad \text{Solve for } y.$$

$$y = \frac{x - 7}{2}$$

$$f^{-1}(x) = \frac{x - 7}{2} \quad \text{Replace } y \text{ with } f^{-1}(x).$$

The inverse of $f(x) = 2x + 7$ is $f^{-1}(x) = \dfrac{x - 7}{2}$.

| **Definitions and Concepts** | **Examples** |

Section 12.2 Exponential Functions

A function of the form $f(x) = b^x$ is an **exponential function,** where $b > 0$, $b \neq 1$, and x is a real number.

Graph the exponential function $y = 4^x$.

x	y
-2	$\frac{1}{16}$
-1	$\frac{1}{4}$
0	1
1	4
2	16

Uniqueness of b^x

If $b > 0$ and $b \neq 1$, then $b^x = b^y$ is equivalent to $x = y$.

Solve $2^{x+5} = 8$.

$$2^{x+5} = 2^3 \qquad \text{Write 8 as } 2^3.$$
$$x + 5 = 3 \qquad \text{Use the uniqueness of } b^x.$$
$$x = -2 \qquad \text{Subtract 5 from both sides.}$$

Section 12.3 Logarithmic Functions

Logarithmic Definition

If $b > 0$ and $b \neq 1$, then

$$y = \log_b x \quad \text{means} \quad x = b^y$$

for any positive number x and real number y.

Logarithmic Form	**Corresponding Exponential Statement**
$\log_5 25 = 2$	$5^2 = 25$
$\log_9 3 = \dfrac{1}{2}$	$9^{1/2} = 3$

Properties of Logarithms

If b is a real number, $b > 0$ and $b \neq 1$, then

$$\log_b 1 = 0, \quad \log_b b^x = x, \quad b^{\log_b x} = x$$

$$\log_5 1 = 0, \quad \log_7 7^2 = 2, \quad 3^{\log_3 6} = 6$$

Logarithmic Function

If $b > 0$ and $b \neq 1$, then a **logarithmic function** is a function that can be defined as

$$f(x) = \log_b x$$

The domain of f is the set of positive real numbers, and the range of f is the set of real numbers.

Graph $y = \log_3 x$.

Write $y = \log_3 x$ as $3^y = x$. Plot the ordered pair solutions listed in the table, and connect them with a smooth curve.

x	y
3	1
1	0
$\frac{1}{3}$	-1
$\frac{1}{9}$	-2

Definitions and Concepts	**Examples**

Section 12.4 Properties of Logarithms

Let x, y, and b be positive numbers and $b \neq 1$.

Product Property

$$\log_b xy = \log_b x + \log_b y$$

Quotient Property

$$\log_b \frac{x}{y} = \log_b x - \log_b y$$

Power Property

$$\log_b x^r = r \log_b x$$

Write as a single logarithm.

$$2 \log_5 6 + \log_5 x - \log_5(y + 2)$$

$$= \log_5 6^2 + \log_5 x - \log_5(y + 2) \quad \text{Power property}$$

$$= \log_5 36 \cdot x - \log_5(y + 2) \quad \text{Product property}$$

$$= \log_5 \frac{36x}{y + 2} \quad \text{Quotient property}$$

Section 12.5 Common Logarithms, Natural Logarithms, and Change of Base

Common Logarithms

$$\log x \quad \text{means} \quad \log_{10} x$$

Natural Logarithms

$$\ln x \quad \text{means} \quad \log_e x$$

Continuously Compounded Interest Formula

$$A = Pe^{rt}$$

where r is the annual interest rate for P dollars invested for t years.

$$\log 5 = \log_{10} 5 \approx 0.69897$$

$$\ln 7 = \log_e 7 \approx 1.94591$$

Find the amount in an account at the end of 3 years if $1000 is invested at an interest rate of 4% compounded continuously.

Here, $t = 3$ years, $P = \$1000$, and $r = 0.04$.

$$A = Pe^{rt}$$

$$= 1000e^{0.04(3)}$$

$$\approx \$1127.50$$

Section 12.6 Exponential and Logarithmic Equations and Applications

Logarithm Property of Equality

Let $\log_b a$ and $\log_b c$ be real numbers and $b \neq 1$. Then

$$\log_b a = \log_b c \text{ is equivalent to } a = c$$

Solve $2^x = 5$.

$$\log 2^x = \log 5 \quad \text{Log property of equality}$$

$$x \log 2 = \log 5 \quad \text{Power property}$$

$$x = \frac{\log 5}{\log 2} \quad \text{Divide both sides by } \log 2.$$

$$x \approx 2.3219 \quad \text{Use a calculator.}$$

12 CHAPTER REVIEW

(12.1) *Determine whether each function is a one-to-one function. If it is one-to-one, list the elements of its inverse.*

1. $h = \{(-9, 14), (6, 8), (-11, 12), (15, 15)\}$

2. $f = \{(-5, 5), (0, 4), (13, 5), (11, -6)\}$

3.

U.S. Region (Input)	West	Midwest	South	Northeast
Rank in Automobile Thefts (Output)	2	4	1	3

 4.

Shape (Input)	Square	Triangle	Parallelogram	Rectangle
Number of Sides (Output)	4	3	4	4

Given that $f(x) = \sqrt{x + 2}$ *is a one-to-one function, find the following.*

5. a. $f(7)$
   b. $f^{-1}(3)$

6. a. $f(-1)$
   b. $f^{-1}(1)$

Determine whether each function is a one-to-one function.

7.

8.

9.

10.

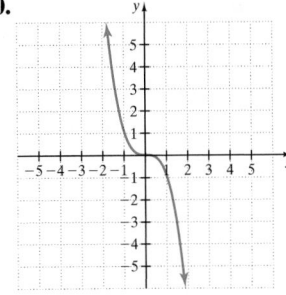

Find an equation defining the inverse function of the given one-to-one function.

11. $f(x) = x - 9$

12. $f(x) = x + 8$

13. $f(x) = 6x + 11$

14. $f(x) = 12x$

15. $f(x) = x^3 - 5$

16. $f(x) = \sqrt[3]{x + 2}$

17. $g(x) = \dfrac{12x - 7}{6}$

18. $r(x) = \dfrac{13}{2}x - 4$

On the same set of axes, graph the given one-to-one function and its inverse.

19. $g(x) = \sqrt{x}$

20. $h(x) = 5x - 5$

21. Find the inverse of the one-to-one function $f(x) = 2x - 3$. Then graph both $f(x)$ and $f^{-1}(x)$ on a graphing calculator using a square window.

(12.2) *Solve each equation for x.*

22. $4^x = 64$

23. $3^x = \dfrac{1}{9}$

24. $2^{3x} = \dfrac{1}{16}$

25. $5^{2x} = 125$

26. $9^{x+1} = 243$

27. $8^{3x-2} = 4$

Graph each exponential function.

28. $y = 3^x$

29. $y = \left(\dfrac{1}{3}\right)^x$

30. $y = 4 \cdot 2^x$

31. $y = 2^x + 4$

Use the formula $A = P\left(1 + \dfrac{r}{n}\right)^{nt}$ to solve the interest problems. In this formula,

A = amount accrued (or owed)

P = principal invested (or loaned)

r = rate of interest

n = number of compounding periods per year

t = time in years

32. Find the amount accrued if $1600 is invested at 9% interest compounded semiannually for 7 years.

33. A total of $800 is invested in a 7% certificate of deposit for which interest is compounded quarterly. Find the value that this certificate will have at the end of 5 years.

34. Use a graphing calculator to verify the results of Exercise 32.

(12.3) *Write each equation with logarithmic notation.*

35. $49 = 7^2$

46. $2^{-4} = \dfrac{1}{16}$

Write each logarithmic equation with exponential notation.

37. $\log_{1/2} 16 = -4$

38. $\log_{0.4} 0.064 = 3$

Solve for x.

39. $\log_4 x = -3$

40. $\log_3 x = 2$

41. $\log_3 1 = x$

42. $\log_4 64 = x$

43. $\log_x 64 = 2$

44. $\log_x 81 = 4$

45. $\log_4 4^5 = x$

46. $\log_7 7^{-2} = x$

47. $5^{\log_5 4} = x$

48. $2^{\log_2 9} = x$

49. $\log_2(3x - 1) = 4$

50. $\log_3(2x + 5) = 2$

51. $\log_4(x^2 - 3x) = 1$

52. $\log_8(x^2 + 7x) = 1$

Graph each pair of equations on the same coordinate system.

53. $y = 2^x$ and $y = \log_2 x$

54. $y = \left(\dfrac{1}{2}\right)^x$ and $y = \log_{1/2} x$

(12.4) *Write each of the following as single logarithms.*

55. $\log_3 8 + \log_3 4$

56. $\log_2 6 + \log_2 3$

57. $\log_7 15 - \log_7 20$

58. $\log 18 - \log 12$

59. $\log_{11} 8 + \log_{11} 3 - \log_{11} 6$

60. $\log_5 14 + \log_5 3 - \log_5 21$

61. $2 \log_5 x - 2 \log_5(x + 1) + \log_5 x$

62. $4 \log_3 x - \log_3 x + \log_3(x + 2)$

Use properties of logarithms to write each expression as a sum or difference of multiples of logarithms.

63. $\log_3 \dfrac{x^3}{x + 2}$

64. $\log_4 \dfrac{x + 5}{x^2}$

65. $\log_2 \dfrac{3x^2 y}{z}$

66. $\log_7 \dfrac{yz^3}{x}$

If $\log_b 2 = 0.36$ and $\log_b 5 = 0.83$, find the following.

67. $\log_b 50$

68. $\log_b \dfrac{4}{5}$

(12.5) *Use a calculator to approximate the logarithm to four decimal places.*

69. $\log 3.6$

70. $\log 0.15$

71. $\ln 1.25$

72. $\ln 4.63$

Find the exact value.

73. $\log 1000$

74. $\log \dfrac{1}{10}$

75. $\ln \dfrac{1}{e}$

76. $\ln e^4$

Solve each equation for x.

77. $\ln(2x) = 2$

78. $\ln(3x) = 1.6$

79. $\ln(2x - 3) = -1$

80. $\ln(3x + 1) = 2$

Use the formula $\ln \dfrac{I}{I_0} = -kx$ to solve radiation problems. In this formula,

x = depth in millimeters

I = intensity of radiation

I_0 = initial intensity

k = a constant measure dependent on the material

Round answers to two decimal places.

81. Find the depth at which the intensity of the radiation passing through a lead shield is reduced to 3% of the original intensity if the value of k is 2.1.

82. If k is 3.2, find the depth at which 2% of the original radiation will penetrate.

Approximate the logarithm to four decimal places.

83. $\log_5 1.6$

84. $\log_3 4$

Use the formula $A = Pe^{rt}$ to solve the interest problems in which interest is compounded continuously. In this formula,

A = amount accrued (or owed)

P = principal invested (or loaned)

r = rate of interest

t = time in years

85. Bank of New York offers a 5-year, 6% continuously compounded investment option. Find the amount accrued if $1450 is invested.

86. Find the amount to which a $940 investment grows if it is invested at 11% compounded continuously for 3 years.

(12.6) *Solve each exponential equation for x. Give an exact solution and also approximate the solution to four decimal places.*

87. $3^{2x} = 7$

88. $6^{3x} = 5$

89. $3^{2x+1} = 6$

90. $4^{3x+2} = 9$

91. $5^{3x-5} = 4$

92. $8^{4x-2} = 3$

93. $2 \cdot 5^{x-1} = 1$

94. $3 \cdot 4^{x+5} = 2$

Solve the equation for x.

95. $\log_5 2 + \log_5 x = 2$

96. $\log_3 x + \log_3 10 = 2$

97. $\log (5x) - \log (x + 1) = 4$

98. $\ln (3x) - \ln (x - 3) = 2$

99. $\log_2 x + \log_2 2x - 3 = 1$

100. $-\log_6(4x + 7) + \log_6 x = 1$

Use the formula $y = y_0 e^{kt}$ to solve the population growth problems. In this formula,

y = size of population

y_0 = initial count of population

k = rate of growth written as a decimal

t = time

Round each answer to the nearest whole.

101. The population of mallard ducks in Nova Scotia is expected to grow at a rate of 6% per week during the spring migration. If 155,000 ducks are already in Nova Scotia, find how many are expected by the end of 4 weeks.

102. The population of Indonesia is growing at a rate of 1.5% per year. If the population in 2002 was 232,073,071, find the expected population by the year 2010. (*Source:* U.S. Bureau of the Census, International Data Base)

103. Japan is experiencing an annual growth rate of 0.1%. In 2002, the population of Japan was 126,975,000. How long will it take for the population to be 130,000,000? (*Source:* U.S. Bureau of the Census, International Data Base)

104. In 2002, Canada had a population of 31,902,268. How long will it take Canada to double in population if its growth rate is 0.8% annually? (*Source:* U.S. Bureau of the Census, International Data Base)

105. Egypt's population is increasing at a rate of 1.6% per year. How long will it take for its 2002 population of 70,712,345 to double in size? (*Source:* U.S. Bureau of the Census, International Data Base)

Use the compound interest equation $A = P\left(1 + \dfrac{r}{n}\right)^{nt}$ to solve the following. (See the directions for Exercises 32 and 33 for an explanation of this formula. Round answers to the nearest tenth.)

106. Find how long it will take a $5000 investment to grow to $10,000 if it is invested at 8% interest compounded quarterly.

107. An investment of $6000 has grown to $10,000 while the money was invested at 6% interest compounded monthly. Find how long it was invested.

Use a graphing calculator to solve each equation. Round all solutions to two decimal places.

108. $e^x = 2$

109. $10^{0.3x} = 7$

CHAPTER 12 TEST Remember to use your Chapter Test Prep Video CD to help you study and view solutions to the test questions you need help with.

On the same set of axes, graph the given one-to-one function and its inverse.

1. $f(x) = 7x - 14$

Determine whether the given graph is the graph of a one-to-one function.

2.

3.

Determine whether each function is one-to-one. If it is one-to-one, find an equation or a set of ordered pairs that defines the inverse function of the given function.

4. $y = 6 - 2x$

5. $f = \{(0,0), (2,3), (-1,5)\}$

6.

Word (Input)	Dog	Cat	House	Desk	Circle
First Letter of Word (Output)	d	c	h	d	c

Use the properties of logarithms to write each expression as a single logarithm.

7. $\log_3 6 + \log_3 4$

8. $\log_5 x + 3 \log_5 x - \log_5(x + 1)$

9. Write the expression $\log_6 \dfrac{2x}{y^3}$ as the sum or difference of multiples of logarithms.

10. If $\log_b 3 = 0.79$ and $\log_b 5 = 1.16$, find the value of $\log_b \dfrac{3}{25}$.

11. Approximate $\log_7 8$ to four decimal places.

12. Solve $8^{x-1} = \dfrac{1}{64}$ for x. Give an exact solution.

13. Solve $3^{2x+5} = 4$ for x. Give an exact solution, and also approximate the solution to four decimal places.

Solve each logarithmic equation for x. Give an exact solution.

14. $\log_3 x = -2$

15. $\ln \sqrt{e} = x$

16. $\log_8(3x - 2) = 2$

17. $\log_5 x + \log_5 3 = 2$

18. $\log_4(x + 1) - \log_4(x - 2) = 3$

19. Solve $\ln (3x + 7) = 1.31$ accurate to four decimal places.

20. Graph $y = \left(\dfrac{1}{2}\right)^x + 1$.

21. Graph the functions $y = 3^x$ and $y = \log_3 x$ on the same coordinate system.

Use the formula $A = P\left(1 + \dfrac{r}{n}\right)^{nt}$ to solve Exercises 22 and 23.

22. Find the amount in the account if $4000 is invested for 3 years at 9% interest compounded monthly.

23. Find how long it will take $2000 to grow to $3000 if the money is invested at 7% interest compounded semiannually. Round to the nearest whole.

Use the population growth formula $y = y_0 e^{kt}$ to solve Exercises 24 and 25.

24. The prairie dog population of the Grand Rapids area now stands at 57,000 animals. If the population is growing at a rate of 2.6% annually, find how many prairie dogs there will be in that area 5 years from now.

25. In an attempt to save an endangered species of wood duck, naturalists would like to increase the wood duck population from 400 to 1000 ducks. If the annual population growth rate is 6.2%, find how long it will take the naturalists to reach their goal. Round to the nearest whole year.

26. The formula $\log (1 + k) = \dfrac{0.3}{D}$ relates the doubling time D, in days, and the growth rate k for a population of mice. Find the rate at which the population is increasing if the doubling time is 56 days. Round to the nearest tenth of a percent.

CHAPTER CUMULATIVE REVIEW

1. Divide. Write all quotients in lowest terms.

  a. $\dfrac{4}{5} \div \dfrac{5}{16}$ **b.** $\dfrac{7}{10} \div 14$ **c.** $\dfrac{3}{8} \div \dfrac{3}{10}$

2. Solve $\dfrac{1}{3}(x - 2) = \dfrac{1}{4}(x + 1)$.

3. Graph $f(x) = x^2$.

4. Find the equation of a line through $(-2, 6)$ and perpendicular to $f(x) = -3x + 4$. Write the equation using function notation.

5. Solve the system.

$$\begin{cases} x - 5y - 2z = 6 \\ -2x + 10y + 4z = -12 \\ \dfrac{1}{2}x - \dfrac{5}{2}y - z = 3 \end{cases}$$

6. Line l and line m are parallel lines cut by transversal t. Find the values of x and y.

7. Simplify each expression.

  a. $-3 + [(-2 - 5) - 2]$ **b.** $2^3 - |10| + [-6 - (-5)]$

8. Use the power rules to simplify the following. Use positive exponents to write all results.

  a. $(4a^3)^2$ **b.** $\left(-\dfrac{2}{3}\right)^3$

  c. $\left(\dfrac{4a^5}{b^3}\right)^3$ **d.** $\left(\dfrac{3^{-2}}{x}\right)^{-3}$

  e. $(a^{-2}b^3c^{-4})^{-2}$

9. For the ICL Production Company, the rational function $C(x) = \dfrac{2.6x + 10,000}{x}$ describes the company's cost per disc of pressing x compact discs. Find the cost per disc for pressing:

  a. 100 compact discs   **b.** 1000 compact discs

10. Multiply.

  a. $(3x - 1)^2$

  b. $\left(\dfrac{1}{2}x + 3\right)\left(\dfrac{1}{2}x - 3\right)$

  c. $(2x - 5)(6x + 7)$

11. Solve $12a - 8a = 10 + 2a - 13 - 7$.

12. Perform the indicated operation and simplify if possible.

$$\dfrac{5}{x - 2} + \dfrac{3}{x^2 + 4x + 4} - \dfrac{6}{x + 2}$$

13. Divide $\dfrac{8x^2y^2 - 16xy + 2x}{4xy}$.

14. Simplify each complex fraction.

  a. $\dfrac{\dfrac{a}{5}}{a - 1}$ **b.** $\dfrac{\dfrac{3}{2 + a} + \dfrac{6}{2 - a}}{\dfrac{5}{a + 2} - \dfrac{1}{a - 2}}$ **c.** $\dfrac{x^{-1} + y^{-1}}{xy}$

15. Factor $3m^2 - 24m - 60$.

16. Factor $5x^2 - 85x + 350$.

17. Subtract $\dfrac{3x^2 + 2x}{x - 1} - \dfrac{10x - 5}{x - 1}$.

18. Use synthetic division to divide $(8x^2 - 12x - 7) \div (x - 2)$.

19. Simplify the following expressions.

  a. $\sqrt[4]{81}$ **b.** $\sqrt[5]{-243}$ **c.** $-\sqrt{25}$

  d. $\sqrt[4]{-81}$ **e.** $\sqrt[3]{64x^3}$

20. Solve $\dfrac{1}{a + 5} = \dfrac{1}{3a + 6} - \dfrac{a + 2}{a^2 + 7x + 10}$

21. Use rational exponents to write as a single radical.

  a. $\sqrt{x} \cdot \sqrt[4]{x}$ **b.** $\dfrac{\sqrt{x}}{\sqrt[3]{x}}$ **c.** $\sqrt[3]{3} \cdot \sqrt{2}$

22. Suppose that y varies directly as x. If $y = \dfrac{1}{2}$ when $x = 12$, find the constant of variation and the direct variation equation.

23. Multiply.

   a. $\sqrt{3}\left(5 + \sqrt{30}\right)$ **b.** $\left(\sqrt{5} - \sqrt{6}\right)\left(\sqrt{7} + 1\right)$

   c. $\left(7\sqrt{x} + 5\right)\left(3\sqrt{x} - \sqrt{5}\right)$ **d.** $\left(4\sqrt{3} - 1\right)^2$

   e. $\left(\sqrt{2x} - 5\right)\left(\sqrt{2x} + 5\right)$ **f.** $\left(\sqrt{x-3} + 5\right)^2$

24. Find each root. Assume that all variables represent nonnegative real numbers.

   a. $\sqrt[4]{81}$ **b.** $\sqrt[3]{-27}$ **c.** $\sqrt{\dfrac{9}{64}}$

   d. $\sqrt[4]{x^{12}}$ **e.** $\sqrt[3]{-125y^6}$

25. Rationalize the denominator of $\dfrac{\sqrt[4]{x}}{\sqrt[4]{81y^5}}$.

26. Multiply.

   a. $a^{1/4}\left(a^{3/4} - a^8\right)$ **b.** $\left(x^{1/2} - 3\right)\left(x^{1/2} + 5\right)$

27. Solve $\sqrt{4 - x} = x - 2$.

28. Use the quotient rule to divide and simplify if possible.

   a. $\dfrac{\sqrt{54}}{\sqrt{6}}$ **b.** $\dfrac{\sqrt{108a^2}}{3\sqrt{3}}$ **c.** $\dfrac{3\sqrt[3]{81a^5b^{10}}}{\sqrt[3]{3b^4}}$

29. Solve $3x^2 - 9x + 8 = 0$ by completing the square.

30. Add or subtract as indicated.

   a. $\dfrac{\sqrt{20}}{3} + \dfrac{\sqrt{5}}{4}$ **b.** $\sqrt[3]{\dfrac{24x}{27}} - \dfrac{\sqrt[3]{3x}}{2}$

31. Solve $\dfrac{3x}{x-2} - \dfrac{x+1}{x} = \dfrac{6}{x(x-2)}$.

32. Rationalize the denominator. $\sqrt[3]{\dfrac{27}{m^4n^8}}$

33. Solve $x^2 - 4x \leq 0$.

34. Find the length of the unknown side of the triangle.

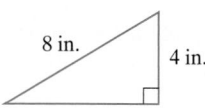
8 in. 4 in.

35. Graph $F(x) = (x - 3)^2 + 1$.

36. Find the followings powers of i.

   a. i^8 **b.** i^{21}

   c. i^{42} **d.** i^{-13}

37. Solve $\dfrac{45}{x} = \dfrac{5}{7}$.

38. Solve $4x^2 + 8x - 1 = 0$ by completing the square.

39. Find an equation of the inverse of $f(x) = x + 3$.

40. Solve by using the quadratic formula.

$$\left(x - \dfrac{1}{2}\right)^2 = \dfrac{x}{2}$$

41. Find the value of each logarithmic expression.

   a. $\log_4 16$ **b.** $\log_{10} \dfrac{1}{10}$ **c.** $\log_9 3$

42. Graph $f(x) = -(x + 1)^2 + 1$. Find the vertex and the axis of symmetry.

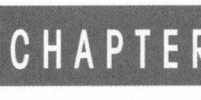

CHAPTER

13

Conic Sections

In Chapter 11, we analyzed some of the important connections between a parabola and its equation. Parabolas are interesting in their own right but are more interesting still because they are part of a collection of curves known as conic sections. This chapter is devoted to quadratic equations in two variables and their conic section graphs: the parabola, circle, ellipse, and hyperbola.

stronomers are scientists who study objects in space. About 55% of professional astronomers are employed as faculty members at colleges or universities. Observational astronomers work in laboratories, observatories, or governmental agencies. They spend many hours looking through telescopes or observing and analyzing data from satellites or spacecraft. A smaller portion of astronomers work in the business sector, conducting research or working in areas closely related to astronomy.

Astronomers with undergraduate degrees may work in planetariums teaching the public about astronomy, or as science teachers in high schools. Students pursuing a degree in astronomy will concentrate their course work in physics, mathematics, geology, chemistry, and computer science.

In the Spotlight on Decision Making feature on page 778, you will have the opportunity to describe the shape of a comet's orbit as an astronomer.

Link: www.aas.org *(American Astronomical Society)*
Source of text: www.aas.org/education/career.html *(AAS Careers in Astronomy brochure online)*

13.1 THE PARABOLA AND THE CIRCLE

Objectives

1 Graph parabolas of the form $x = a(y - k)^2 + h$ and $y = a(x - h)^2 + k$.

2 Use the distance formula and the midpoint formula.

3 Graph circles of the form $(x - h)^2 + (y - k)^2 = r^2$.

4 Write the equation of a circle, given its center and radius.

5 Find the center and the radius of a circle, given its equation.

Conic sections derive their name because each conic section is the intersection of a right circular cone and a plane. The circle, parabola, ellipse, and hyperbola are the conic sections.

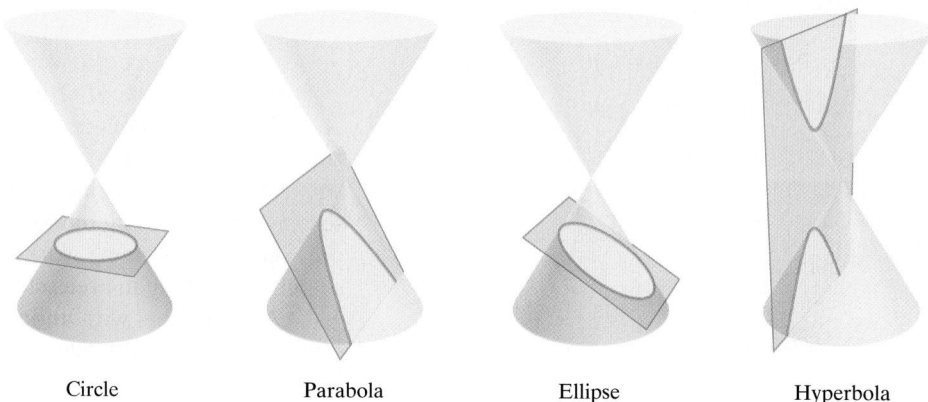

Circle   Parabola   Ellipse   Hyperbola

1   Thus far, we have seen that *f(x)* or $y = a(x - h)^2 + k$ is the equation of a parabola that opens upward if $a > 0$ or downward if $a < 0$. Parabolas can also open left or right, or even on a slant. Equations of these parabolas are not functions of x, of course, since a parabola opening any way other than upward or downward fails the vertical line test. In this section, we introduce parabolas that open to the left and to the right. Parabolas opening on a slant will not be developed in this book.

Just as $y = a(x - h)^2 + k$ is the equation of a parabola that opens upward or downward, $x = a(y - k)^2 + h$ is the equation of a parabola that opens to the right or to the left. The parabola opens to the right if $a > 0$ and to the left if $a < 0$. The parabola has vertex (h, k), and its axis of symmetry is the line $y = k$.

Parabolas

$$y = a(x - h)^2 + k$$

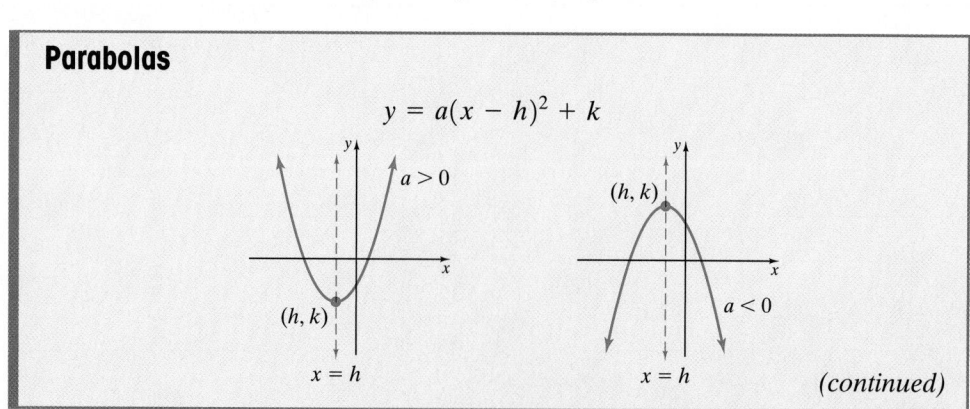

(continued)

$$x = a(y - k)^2 + h$$

(h, k) $y = k$ $a > 0$

$a < 0$ $y = k$ (h, k)

The equations $y = a(x - h)^2 + k$ and $x = a(y - k)^2 + h$ are called **standard forms.**

✔ **CONCEPT CHECK**

Does the graph of the parabola given by the equation $x = -3y^2$ open to the left, to the right, upward, or downward?

EXAMPLE 1

Graph the parabola $x = 2y^2$.

Solution Written in standard form, the equation $x = 2y^2$ is $x = 2(y - 0)^2 + 0$ with $a = 2, h = 0$, and $k = 0$. Its graph is a parabola with vertex $(0, 0)$, and its axis of symmetry is the line $y = 0$. Since $a > 0$, this parabola opens to the right. The table shows a few more ordered pair solutions of $x = 2y^2$. Its graph is also shown.

x	y
8	-2
2	-1
0	0
2	1
8	2

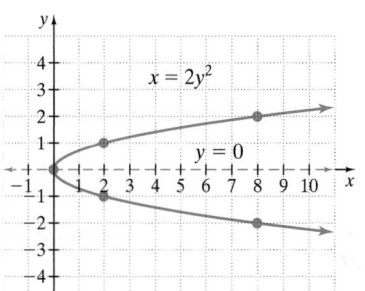

$x = 2y^2$

$y = 0$

EXAMPLE 2

Graph the parabola $x = -3(y - 1)^2 + 2$.

Solution The equation $x = -3(y - 1)^2 + 2$ is in the form $x = a(y - k)^2 + h$ with $a = -3$, $k = 1$, and $h = 2$. Since $a < 0$, the parabola opens to the left. The vertex (h, k) is $(2, 1)$, and the axis of symmetry is the line $y = 1$. When $y = 0, x = -1$, so the x-intercept is $(-1, 0)$. Again, we obtain a few ordered pair solutions and then graph the parabola.

x	y
2	1
-1	0
-1	2
-10	3
-10	-1

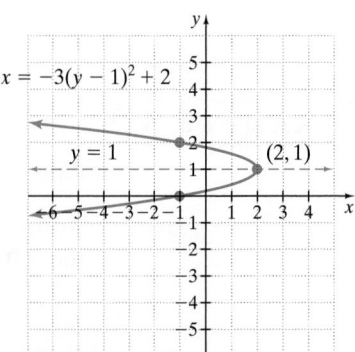

$x = -3(y - 1)^2 + 2$

$y = 1$ $(2, 1)$

Concept Check Answer:
to the left

EXAMPLE 3

Graph $y = -x^2 - 2x + 15$.

Solution Complete the square on x to write the equation in standard form.

$$y - 15 = -x^2 - 2x \qquad \text{Subtract 15 from both sides.}$$
$$y - 15 = -1(x^2 + 2x) \qquad \text{Factor} -1 \text{ from the terms } -x^2 - 2x.$$

The coefficient of x is 2. Find the square of half of 2.

$$\frac{1}{2}(2) = 1 \quad \text{and} \quad 1^2 = 1$$

$$y - 15 - 1(1) = -1(x^2 + 2x + 1) \qquad \text{Add} -1(1) \text{ to both sides.}$$

$$y - 16 = -1(x + 1)^2 \qquad \begin{array}{l}\text{Simplify the left side and}\\\text{factor the right side.}\end{array}$$

$$y = -(x + 1)^2 + 16 \qquad \text{Add 16 to both sides.}$$

The equation is now in standard form $y = a(x - h)^2 + k$ with $a = -1, h = -1$, and $k = 16$.

The vertex is then (h, k), or $(-1, 16)$.

A second method for finding the vertex is by using the formula $\dfrac{-b}{2a}$.

$$x = \frac{-(-2)}{2(-1)} = \frac{2}{-2} = -1$$
$$y = -(-1)^2 - 2(-1) + 15 = -1 + 2 + 15 = 16$$

Again, we see that the vertex is $(-1, 16)$, and the axis of symmetry is the vertical line $x = -1$. The y-intercept is $(0, 15)$. Now we can use a few more ordered pair solutions to graph the parabola.

x	y
-1	16
0	15
-2	15
1	12
-3	12
3	0
-5	0

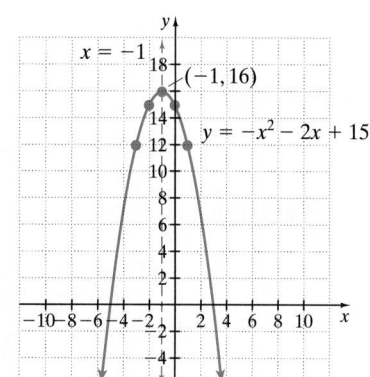

EXAMPLE 4

Graph $x = 2y^2 + 4y + 5$.

Solution Notice that this equation is quadratic in y, so its graph is a parabola that opens to the left or the right. We can complete the square on y or we can use the formula $\dfrac{-b}{2a}$ to find the vertex.

Since the equation is quadratic in y, the formula gives us the y-value of the vertex.

$$y = \frac{-4}{2 \cdot 2} = \frac{-4}{4} = -1$$

$$x = 2(-1)^2 + 4(-1) + 5 = 2 \cdot 1 - 4 + 5 = 3$$

The vertex is $(3, -1)$, and the axis of symmetry is the line $y = -1$. The parabola opens to the right since $a > 0$. The x-intercept is $(5, 0)$.

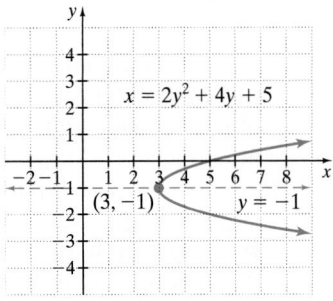

2 Another conic section is the circle. Before we review the circle, we need a formula to calculate the distance between points of the Cartesian coordinate system. To find the distance between two points, we use the distance formula, which is derived from the Pythagorean theorem.

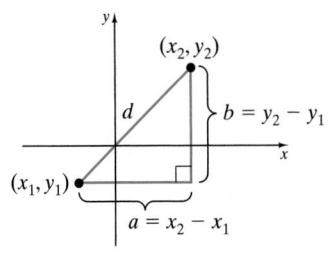

To find the distance d between two points (x_1, y_1) and (x_2, y_2) as shown to the left, notice that the length of leg a is $x_2 - x_1$ and that the length of leg b is $y_2 - y_1$.

Thus, the Pythagorean theorem tells us that

$$d^2 = a^2 + b^2$$

or

$$d^2 = (x_2 - x_1)^2 + (y_2 - y_1)^2$$

or

$$d = \sqrt{(x_2 - x_1)^2 + (y_2 - y_1)^2}$$

This formula gives us the distance between any two points on the real plane.

Distance Formula

The distance d between two points (x_1, y_1) and (x_2, y_2) is given by

$$d = \sqrt{(x_2 - x_1)^2 + (y_2 - y_1)^2}$$

EXAMPLE 5

Find the distance between $(2, -5)$ and $(1, -4)$. Give an exact distance and a three-decimal-place approximation.

Solution To use the distance formula, it makes no difference which point we call (x_1, y_1) and which point we call (x_2, y_2). We will let $(x_1, y_1) = (2, -5)$ and $(x_2, y_2) = (1, -4)$.

$$d = \sqrt{(x_2 - x_1)^2 + (y_2 - y_1)^2}$$
$$= \sqrt{(1 - 2)^2 + [-4 - (-5)]^2}$$
$$= \sqrt{(-1)^2 + (1)^2}$$
$$= \sqrt{1 + 1}$$
$$= \sqrt{2} \approx 1.414$$

The distance between the two points is exactly $\sqrt{2}$ units, or approximately 1.414 units.

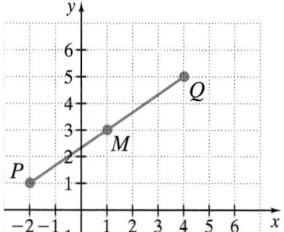

The **midpoint** of a line segment is the **point** located exactly halfway between the two endpoints of the line segment. On the graph to the left, the point M is the midpoint of line segment PQ. Thus, the distance between M and P equals the distance between M and Q.

The x-coordinate of M is at half the distance between the x-coordinates of P and Q, and the y-coordinate of M is at half the distance between the y-coordinates of P and Q. That is, the x-coordinate of M is the average of the x-coordinates of P and Q; the y-coordinate of M is the average of the y-coordinates of P and Q.

Midpoint Formula

The midpoint of the line segment whose endpoints are (x_1, y_1) and (x_2, y_2) is the point with coordinates

$$\left(\frac{x_1 + x_2}{2}, \frac{y_1 + y_2}{2} \right)$$

EXAMPLE 6

Find the midpoint of the line segment that joins points $P(-3, 3)$ and $Q(1, 0)$.

Solution   Use the midpoint formula. It makes no difference which point we call (x_1, y_1) or which point we call (x_2, y_2). Let $(x_1, y_1) = (-3, 3)$ and $(x_2, y_2) = (1, 0)$.

$$\text{midpoint} = \left(\frac{x_1 + x_2}{2}, \frac{y_1 + y_2}{2} \right)$$
$$= \left(\frac{-3 + 1}{2}, \frac{3 + 0}{2} \right)$$
$$= \left(\frac{-2}{2}, \frac{3}{2} \right)$$
$$= \left(-1, \frac{3}{2} \right)$$

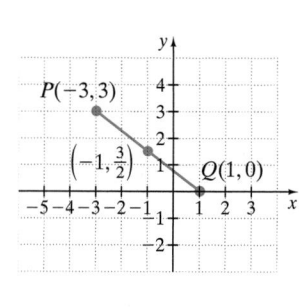

The midpoint of the segment is $\left(-1, \frac{3}{2} \right)$.

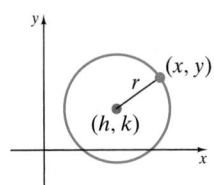

3   Another conic section is the **circle.** A circle is the set of all points in a plane that are the same distance from a fixed point called the **center.** The distance is called the **radius** of the circle. To find a standard equation for a circle, let (h, k) represent the center of the circle, and let (x, y) represent any point on the circle. The distance between (h, k) and (x, y) is defined to be the circle's radius, r units. We can find this distance r by using the distance formula.

$$r = \sqrt{(x - h)^2 + (y - k)^2}$$

$$r^2 = (x - h)^2 + (y - k)^2 \qquad \text{Square both sides.}$$

Circle

The graph of $(x - h)^2 + (y - k)^2 = r^2$ is a circle with center (h, k) and radius r.

The equation $(x - h)^2 + (y - k)^2 = r^2$ is called **standard form.**
    If an equation can be written in the standard form

$$(x - h)^2 + (y - k)^2 = r^2$$

then its graph is a circle, which we can draw by graphing the center (h, k) and using the radius r.

EXAMPLE 7

Graph $x^2 + y^2 = 4$.

Solution   The equation can be written in standard form as

$$(x - 0)^2 + (y - 0)^2 = 2^2$$

The center of the circle is $(0, 0)$, and the radius is 2. Its graph is shown.

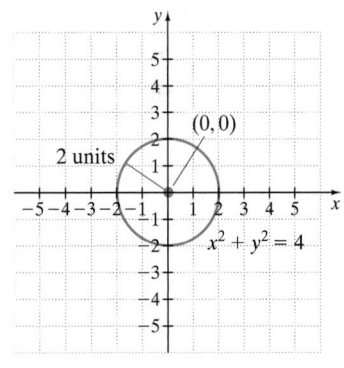

> ### Helpful Hint
> Notice the difference between the equation of a circle and the equation of a parabola. The equation of a circle contains both x^2 and y^2 terms on the same side of the equation with equal coefficients. The equation of a parabola has either an x^2 term or a y^2 term but not both.

EXAMPLE 8

Graph $(x + 1)^2 + y^2 = 8$.

Solution The equation can be written as $(x + 1)^2 + (y - 0)^2 = 8$ with $h = -1, k = 0$, and $r = \sqrt{8}$. The center is $(-1, 0)$, and the radius is $\sqrt{8} = 2\sqrt{2} \approx 2.8$.

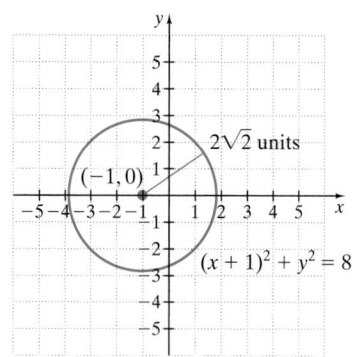

✔ **CONCEPT CHECK**

In the graph of the equation $(x - 3)^2 + (y - 2)^2 = 5$, what is the distance between the center of the circle and any point on the circle?

4 Since a circle is determined entirely by its center and radius, this information is all we need to write the equation of a circle.

EXAMPLE 9

Find an equation of the circle with center $(-7, 3)$ and radius 10.

Solution Using the given values $h = -7$, $k = 3$, and $r = 10$, we write the equation

$$(x - h)^2 + (y - k)^2 = r^2$$

or

$$[x - (-7)]^2 + (y - 3)^2 = 10^2 \qquad \text{Substitute the given values.}$$

or

$$(x + 7)^2 + (y - 3)^2 = 100$$

5 To find the center and the radius of a circle from its equation, write the equation in standard form. To write the equation of a circle in standard form, we complete the square on both x and y.

EXAMPLE 10

Graph $x^2 + y^2 + 4x - 8y = 16$.

Solution Since this equation contains x^2 and y^2 terms on the same side of the equation with equal coefficients, its graph is a circle. To write the equation in standard form, group the terms involving x and the terms involving y, and then complete the square on each variable.

Concept Check Answer:
$\sqrt{5}$ units

$$(x^2 + 4x) + (y^2 - 8y) = 16$$

Thus, $\frac{1}{2}(4) = 2$ and $2^2 = 4$. Also, $\frac{1}{2}(-8) = -4$ and $(-4)^2 = 16$. Add 4 and then 16 to both sides.

$$(x^2 + 4x + 4) + (y^2 - 8y + 16) = 16 + 4 + 16$$

$$(x + 2)^2 + (y - 4)^2 = 36 \qquad \text{Factor.}$$

This circle has the center $(-2, 4)$ and radius 6, as shown.

Graphing Calculator Explorations

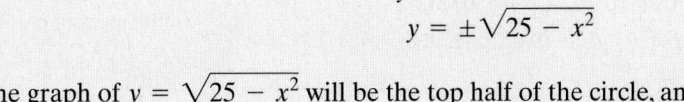

To graph an equation such as $x^2 + y^2 = 25$ with a graphing calculator, we first solve the equation for y.

$$x^2 + y^2 = 25$$
$$y^2 = 25 - x^2$$
$$y = \pm\sqrt{25 - x^2}$$

The graph of $y = \sqrt{25 - x^2}$ will be the top half of the circle, and the graph of $y = -\sqrt{25 - x^2}$ will be the bottom half of the circle.

To graph, press $\boxed{Y=}$ and enter $Y_1 = \sqrt{25 - x^2}$ and $Y_2 = -\sqrt{25 - x^2}$. Insert parentheses around $25 - x^2$ so that $\sqrt{25 - x^2}$ and not $\sqrt{25} - x^2$ is graphed.

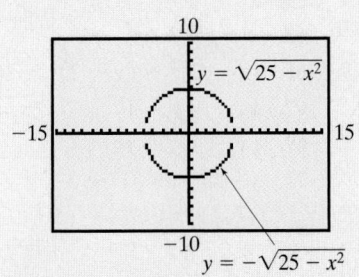

The top graph to the left does not appear to be a circle because we are currently using a standard window and the screen is rectangular. This causes the tick marks on the x-axis to be farther apart than the tick marks on the y-axis and, thus, creates the distorted circle. If we want the graph to appear circular, we must define a square window by using a feature of the graphing calculator or by redefining the window to show the x-axis from -15 to 15 and the y-axis from -10 to 10. Using a square window, the graph appears as shown on the bottom to the left.

Use a graphing calculator to graph each circle.

1. $x^2 + y^2 = 55$

2. $x^2 + y^2 = 20$

3. $5x^2 + 5y^2 = 50$

4. $6x^2 + 6y^2 = 105$

5. $2x^2 + 2y^2 - 34 = 0$

6. $4x^2 + 4y^2 - 48 = 0$

7. $7x^2 + 7y^2 - 89 = 0$

8. $3x^2 + 3y^2 - 35 = 0$

MENTAL MATH

The graph of each equation is a parabola. Determine whether the parabola opens upward, downward, to the left, or to the right.

1. $y = x^2 - 7x + 5$

2. $y = -x^2 + 16$

3. $x = -y^2 - y + 2$

4. $x = 3y^2 + 2y - 5$

5. $y = -x^2 + 2x + 1$

6. $x = -y^2 + 2y - 6$

EXERCISE SET 13.1

STUDY GUIDE/SSM   CD/VIDEO   PH MATH TUTOR CENTER   MathXL®Tutorials ON CD   MathXL®   MyMathLab®

The graph of each equation is a parabola. Find the vertex of the parabola and sketch its graph. See Examples 1 through 4.

1. $x = 3y^2$

2. $x = -2y^2$

 3. $x = (y - 2)^2 + 3$

4. $x = (y - 4)^2 - 1$

5. $y = 3(x - 1)^2 + 5$

6. $x = -4(y - 2)^2 + 2$

7. $x = y^2 + 6y + 8$

8. $x = y^2 - 6y + 6$

9. $y = x^2 + 10x + 20$

10. $y = x^2 + 4x - 5$

11. $x = -2y^2 + 4y + 6$

12. $x = 3y^2 + 6y + 7$

Find the distance between each pair of points. Approximate the distance in Exercises 21 and 22 to two decimal places. See Example 5.

13. $(5, 1)$ and $(8, 5)$

14. $(2, 3)$ and $(14, 8)$

15. $(-3, 2)$ and $(1, -3)$

16. $(3, -2)$ and $(-4, 1)$

17. $(-9, 4)$ and $(-8, 1)$

18. $(-5, -2)$ and $(-6, -6)$

19. $\left(0, -\sqrt{2}\right)$ and $\left(\sqrt{3}, 0\right)$

20. $\left(-\sqrt{5}, 0\right)$ and $\left(0, \sqrt{7}\right)$

21. $(1.7, -3.6)$ and $(-8.6, 5.7)$

22. $(9.6, 2.5)$ and $(-1.9, -3.7)$

23. $\left(2\sqrt{3}, \sqrt{6}\right)$ and $\left(-\sqrt{3}, 4\sqrt{6}\right)$

24. $\left(5\sqrt{2}, -4\right)$ and $\left(-3\sqrt{2}, -8\right)$

Find the midpoint of the line segment whose endpoints are given. See Example 6.

25. $(6, -8), (2, 4)$

26. $(3, 9), (7, 11)$

27. $(-2, -1), (-8, 6)$

28. $(-3, -4), (6, -8)$

29. $(7, 3), (-1, -3)$

30. $(-2, 5), (-1, 6)$

31. $\left(\frac{1}{2}, \frac{3}{8}\right), \left(-\frac{3}{2}, \frac{5}{8}\right)$

32. $\left(-\frac{2}{5}, \frac{7}{15}\right), \left(-\frac{2}{5}, -\frac{4}{15}\right)$

33. $\left(\sqrt{2}, 3\sqrt{5}\right), \left(\sqrt{2}, -2\sqrt{5}\right)$

34. $\left(\sqrt{8}, -\sqrt{12}\right), \left(3\sqrt{2}, 7\sqrt{3}\right)$

35. $(4.6, -3.5), (7.8, -9.8)$

36. $(-4.6, 2.1), (-6.7, 1.9)$

The graph of each equation is a circle. Find the center and the radius, and then sketch. See Examples 7, 8, and 10.

37. $x^2 + y^2 = 9$

38. $x^2 + y^2 = 25$

39. $x^2 + (y - 2)^2 = 1$

40. $(x - 3)^2 + y^2 = 9$

41. $(x - 5)^2 + (y + 2)^2 = 1$

42. $(x + 3)^2 + (y + 3)^2 = 4$

43. $x^2 + y^2 + 6y = 0$

44. $x^2 + 10x + y^2 = 0$

45. $x^2 + y^2 + 2x - 4y = 4$

46. $x^2 + 6x - 4y + y^2 = 3$

47. $x^2 + y^2 - 4x - 8y - 2 = 0$

48. $x^2 + y^2 - 2x - 6y - 5 = 0$

Write an equation of the circle with the given center and radius. See Example 9.

49. $(2, 3); 6$

50. $(-7, 6); 2$

51. $(0, 0); \sqrt{3}$

52. $(0, -6); \sqrt{2}$

53. $(-5, 4); 3\sqrt{5}$

54. the origin; $4\sqrt{7}$

55. If you are given a list of equations of circles and parabolas and none are in standard form, explain how you would determine which is an equation of a circle and which is an equation of a parabola. Explain also how you would distinguish the upward or downward parabolas from the left-opening or right-opening parabolas.

MIXED PRACTICE

Sketch the graph of each equation. If the graph is a parabola, find its vertex. If the graph is a circle, find its center and radius.

56. $x = y^2 + 2$

57. $x = y^2 - 3$

58. $y = (x + 3)^2 + 3$

59. $y = (x - 2)^2 - 2$

60. $x^2 + y^2 = 49$

61. $x^2 + y^2 = 1$

62. $x = (y - 1)^2 + 4$

63. $x = (y + 3)^2 - 1$

64. $(x + 3)^2 + (y - 1)^2 = 9$

65. $(x - 2)^2 + (y - 2)^2 = 16$

66. $x = -2(y + 5)^2$

67. $x = -(y - 1)^2$

68. $x^2 + (y + 5)^2 = 5$

69. $(x - 4)^2 + y^2 = 7$

70. $y = 3(x - 4)^2 + 2$

71. $y = 5(x + 5)^2 + 3$

72. $2x^2 + 2y^2 = \frac{1}{2}$

73. $\frac{x^2}{8} + \frac{y^2}{8} = 2$

74. $y = x^2 - 2x - 15$

75. $y = x^2 + 7x + 6$

76. $x^2 + y^2 + 6x + 10y - 2 = 0$

77. $x^2 + y^2 + 2x + 12y - 12 = 0$

78. $x = y^2 + 6y + 2$

79. $x = y^2 + 8y - 4$

80. $x^2 + y^2 - 8y + 5 = 0$

81. $x^2 - 10y + y^2 + 4 = 0$

82. $x = -2y^2 - 4y$

83. $x = -3y^2 + 30y$

84. $\dfrac{x^2}{3} + \dfrac{y^2}{3} = 2$

85. $5x^2 + 5y^2 = 25$

86. $y = 4x^2 - 40x + 105$

87. $y = 5x^2 - 20x + 16$

REVIEW AND PREVIEW

Graph each equation. See Section 3.3.

88. $y = 2x + 5$

89. $y = -3x + 3$

90. $y = 3$

91. $x = -2$

Rationalize each denominator and simplify if possible. See Section 10.5.

92. $\dfrac{1}{\sqrt{3}}$

93. $\dfrac{\sqrt{5}}{\sqrt{8}}$

94. $\dfrac{4\sqrt{7}}{\sqrt{6}}$

95. $\dfrac{10}{\sqrt{5}}$

Concept Extensions

△ **96.** Determine whether the triangle with vertices $(2, 6)$, $(0, -2)$, and $(5, 1)$ is an isosceles triangle.

△ **97.** In 1893, Pittsburgh bridge builder George Ferris designed and built a gigantic revolving steel wheel whose height was 264 feet and diameter was 250 feet. This Ferris wheel opened at the 1893 exposition in Chicago. It had 36 wooden cars, each capable of holding 60 passengers. (*Source:* The Handy Science Answer Book)

   a. What was the radius of this Ferris wheel?
   b. How close is the wheel to the ground?
   c. How high is the center of the wheel from the ground?
   d. Using the axes in the drawing, what are the coordinates of the center of the wheel?
   e. Use parts **a** and **d** to write the equation of the wheel.

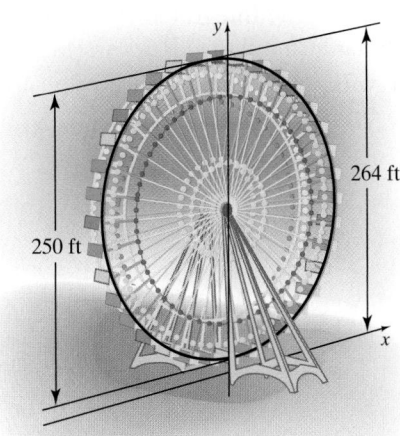

△ **98.** The world's largest-diameter Ferris wheel currently operating is the Cosmoclock 21 at Yokohama City, Japan. It has a

60-armed wheel, its diameter is 100 meters and it has a height of 105 meters. (*Source:* The Handy Science Answer Book)

   a. What is the radius of this Ferris wheel?
   b. How close is the wheel to the ground?
   c. How high is the center of the wheel from the ground?
   d. Using the axes in the drawing, what are the coordinates of the center of the wheel?
   e. Use parts **a** and **d** to write the equation of the wheel.

△ **99.** Two surveyors need to find the distance across a lake. They place a reference pole at point A in the diagram. Point B is 3 meters east and 1 meter north of the reference point A. Point C is 19 meters east and 13 meters north of point A. Find the distance across the lake, from B to C.

△ **100.** Cindy Brown, an architect, is drawing plans on grid paper for a circular pool with a fountain in the middle. The paper is marked off in centimeters, and each centimeter represents 1 foot. On the paper, the diameter of the "pool" is 20 centimeters, and "fountain" is the point $(0, 0)$.

   a. Sketch the architect's drawing. Be sure to label the axes.
   b. Write an equation that describes the circular pool.
   c. Cindy plans to place a circle of lights around the fountain such that each light is 5 feet from the fountain. Write an equation for the circle of lights and sketch the circle on your drawing.

101. A bridge constructed over a bayou has a supporting arch in the shape of a parabola. Find an equation of the parabolic arch if the length of the road over the arch is 100 meters and the maximum height of the arch is 40 meters.

Use a graphing calculator to verify each exercise. Use a square viewing window.

102. Exercise 84.

103. Exercise 85.

104. Exercise 86.

105. Exercise 87.

STUDY SKILLS REMINDER

How Are You Doing?

If you haven't done so yet, take a few moments and think about how you are doing in this course. Are you working toward your goal of successfully completing this course? Is your performance on homework, quizzes, and tests satisfactory? If not, you might want to see your instructor to see if he/she has any suggestions on how you can improve your performance. Let me once again remind you that, in addition to your instructor, there are many places to get help with your mathematics course. A few suggestions are below.

▶ This text has an accompanying video lesson for every section in this text.

▶ The back of this book contains answers to odd-numbered exercises.

▶ A tutorial software program is available with this text. It contains lessons corresponding to each section in the text.

▶ There is a student solutions manual available that contains worked-out solutions to odd-numbered exercises as well as solutions to every exercise in the Integrated Reviews, Chapter Reviews, Chapter Tests, and Cumulative Reviews.

▶ Don't forget to check with your instructor for other local resources available to you, such as a tutor center.

13.2 THE ELLIPSE AND THE HYPERBOLA

Objectives

1 Define and graph an ellipse.

2 Define and graph a hyperbola.

1 An **ellipse** can be thought of as the set of points in a plane such that the sum of the distances of those points from two fixed points is constant. Each of the two fixed

points is called a **focus.** (The plural of focus is **foci.**) The point midway between the foci is called the **center.**

An ellipse may be drawn by hand by using two thumbtacks, a piece of string, and a pencil. Secure the two thumbtacks in a piece of cardboard, for example, and tie each end of the string to a tack. Use your pencil to pull the string tight and draw the ellipse. The two thumbtacks are the foci of the drawn ellipse.

  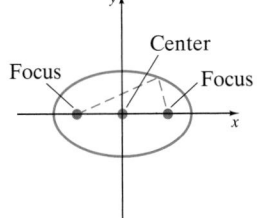

Ellipse with Center $(0, 0)$

The graph of an equation of the form $\dfrac{x^2}{a^2} + \dfrac{y^2}{b^2} = 1$ is an ellipse with center $(0, 0)$.

The x-intercepts are $(a, 0)$ and $(-a, 0)$, and the y-intercepts are $(0, b)$, and $(0, -b)$.

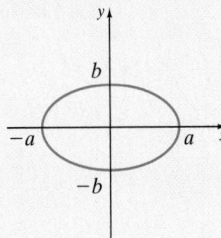

The **standard form** of an ellipse with center $(0, 0)$ is $\dfrac{x^2}{a^2} + \dfrac{y^2}{b^2} = 1$.

EXAMPLE 1

Graph $\dfrac{x^2}{9} + \dfrac{y^2}{16} = 1$.

Solution   The equation is of the form $\dfrac{x^2}{a^2} + \dfrac{y^2}{b^2} = 1$, with $a = 3$ and $b = 4$, so its graph is an ellipse with center $(0, 0)$, x-intercepts $(3, 0)$ and $(-3, 0)$, and y-intercepts $(0, 4)$ and $(0, -4)$.

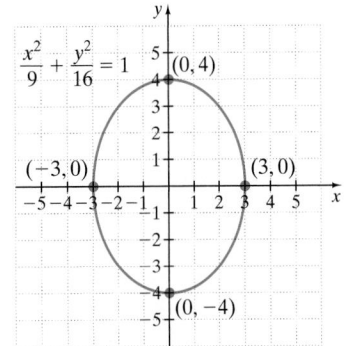

EXAMPLE 2

Graph $4x^2 + 16y^2 = 64$.

Solution Although this equation contains a sum of squared terms in x and y on the same side of an equation, this is not the equation of a circle since the coefficients of x^2 and y^2 are not the same. The graph of this equation is an ellipse. Since the standard form of the equation of an ellipse has 1 on one side, divide both sides of this equation by 64.

$$4x^2 + 16y^2 = 64$$

$$\frac{4x^2}{64} + \frac{16y^2}{64} = \frac{64}{64} \qquad \text{Divide both sides by 64.}$$

$$\frac{x^2}{16} + \frac{y^2}{4} = 1 \qquad \text{Simplify.}$$

We now recognize the equation of an ellipse with $a = 4$ and $b = 2$. This ellipse has center $(0, 0)$, x-intercepts $(4, 0)$ and $(-4, 0)$, and y-intercepts $(0, 2)$ and $(0, -2)$.

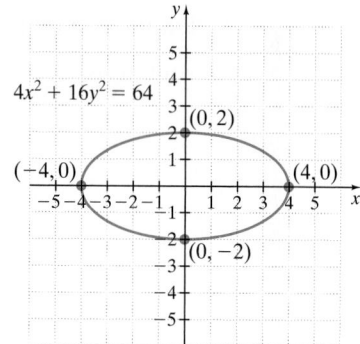

The center of an ellipse is not always $(0, 0)$, as shown in the next example.

EXAMPLE 3

Graph $\dfrac{(x + 3)^2}{25} + \dfrac{(y - 2)^2}{36} = 1$.

Solution The center of this ellipse is found in a way that is similar to finding the center of a circle. This ellipse has center $(-3, 2)$. Notice that $a = 5$ and $b = 6$. To find four points on the

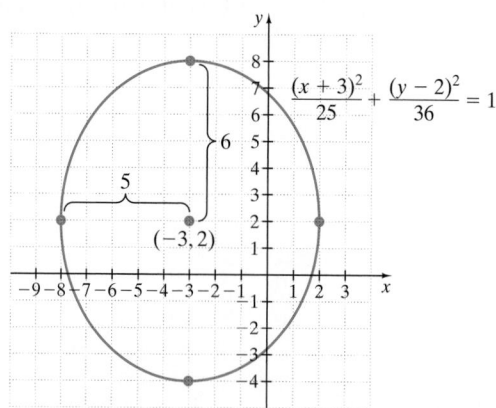

graph of the ellipse, first graph the center, $(-3, 2)$. Since $a = 5$, count 5 units right and then 5 units left of the point with coordinates $(-3, 2)$. Next, since $b = 6$, start at $(-3, 2)$ and count 6 units up and then 6 units down to find two more points on the ellipse.

✔ CONCEPT CHECK

In the graph of the equation $\dfrac{x^2}{64} + \dfrac{y^2}{36} = 1$, which distance is longer: the distance between the x-intercepts or the distance between the y-intercepts? How much longer? Explain.

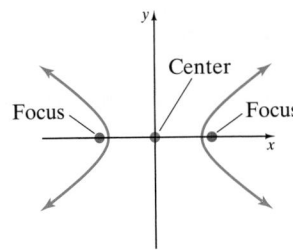

2 The final conic section is the **hyperbola.** A hyperbola is the set of points in a plane such that the absolute value of the difference of the distances from two fixed points is constant. Each of the two fixed points is called a **focus.** The point midway between the foci is called the **center.**

Using the distance formula, we can show that the graph of $\dfrac{x^2}{a^2} - \dfrac{y^2}{b^2} = 1$ is a hyperbola with center $(0, 0)$ and x-intercepts $(a, 0)$ and $(-a, 0)$. Also, the graph of $\dfrac{y^2}{b^2} - \dfrac{x^2}{a^2} = 1$ is a hyperbola with center $(0, 0)$ and y-intercepts $(0, b)$ and $(0, -b)$.

Hyperbola with Center $(0, 0)$

The graph of an equation of the form $\dfrac{x^2}{a^2} - \dfrac{y^2}{b^2} = 1$ is a hyperbola with center $(0, 0)$ and x-intercepts $(a, 0)$ and $(-a, 0)$.

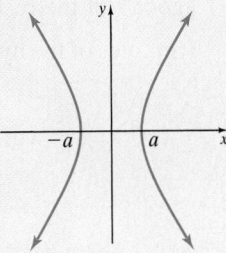

The graph of an equation of the form $\dfrac{y^2}{b^2} - \dfrac{x^2}{a^2} = 1$ is a hyperbola with center $(0, 0)$ and y-intercepts $(0, b)$ and $(0, -b)$.

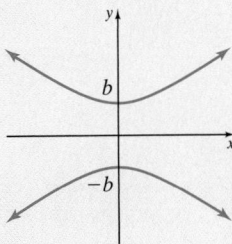

The equations $\dfrac{x^2}{a^2} - \dfrac{y^2}{b^2} = 1$ and $\dfrac{y^2}{b^2} - \dfrac{x^2}{a^2} = 1$ are the **standard forms** for the equation of a hyperbola.

Concept Check Answer:
x-intercepts, by 4 units

> ### Helpful Hint
> Notice the difference between the equation of an ellipse and a hyperbola. The equation of the ellipse contains x^2 and y^2 terms on the same side of the equation with same-sign coefficients. For a hyperbola, the coefficients on the same side of the equation have different signs.

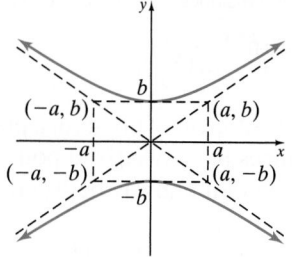

Graphing a hyperbola such as $\dfrac{y^2}{b^2} - \dfrac{x^2}{a^2} = 1$ is made easier by recognizing one of its important characteristics. Examining the figure to the left, notice how the sides of the branches of the hyperbola extend indefinitely and seem to approach the dashed lines in the figure. These dashed lines are called the **asymptotes** of the hyperbola.

To sketch these lines, or asymptotes, draw a rectangle with vertices (a, b), $(-a, b)$, $(a, -b)$, and $(-a, -b)$. The asymptotes of the hyperbola are the extended diagonals of this rectangle.

EXAMPLE 4

Graph $\dfrac{x^2}{16} - \dfrac{y^2}{25} = 1$.

Solution This equation has the form $\dfrac{x^2}{a^2} - \dfrac{y^2}{b^2} = 1$, with $a = 4$ and $b = 5$. Thus, its graph is a hyperbola that opens to the left and right. It has center $(0, 0)$ and x-intercepts $(4, 0)$ and $(-4, 0)$. To aid in graphing the hyperbola, we first sketch its asymptotes. The extended diagonals of the rectangle with corners $(4, 5)$, $(4, -5)$, $(-4, 5)$, and $(-4, -5)$ are the asymptotes of the hyperbola. Then we use the asymptotes to aid in sketching the hyperbola.

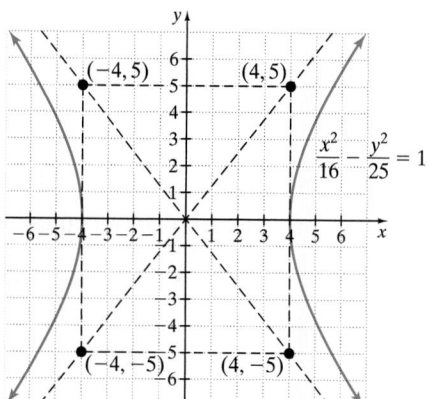

EXAMPLE 5

Graph $4y^2 - 9x^2 = 36$.

Solution Since this is a difference of squared terms in x and y on the same side of the equation, its graph is a hyperbola, as opposed to an ellipse or a circle. The standard form of the equation of a hyperbola has a 1 on one side, so divide both sides of the equation by 36.

$$4y^2 - 9x^2 = 36$$

$$\frac{4y^2}{36} - \frac{9x^2}{36} = \frac{36}{36} \qquad \text{Divide both sides by 36.}$$

$$\frac{y^2}{9} - \frac{x^2}{4} = 1 \qquad \text{Simplify.}$$

The equation is of the form $\dfrac{y^2}{b^2} - \dfrac{x^2}{a^2} = 1$, with $a = 2$ and $b = 3$, so the hyperbola is centered at $(0, 0)$ with y-intercepts $(0, 3)$ and $(0, -3)$. The sketch of the hyperbola is shown.

Graphing Calculator Explorations

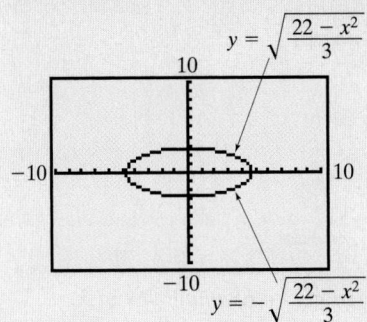

To find the graph of an ellipse by using a graphing calculator, use the same procedure as for graphing a circle. For example, to graph $x^2 + 3y^2 = 22$, first solve for y.

$$3y^2 = 22 - x^2$$

$$y^2 = \frac{22 - x^2}{3}$$

$$y = \pm\sqrt{\frac{22 - x^2}{3}}$$

Next press the key and enter $Y_1 = \sqrt{\dfrac{22 - x^2}{3}}$ and $Y_2 = -\sqrt{\dfrac{22 - x^2}{3}}$. (Insert

two sets of parentheses in the radicand as $\sqrt{((22 - x^2)/3)}$ so that the desired graph is obtained.) The graph appears to the left.

Use a graphing calculator to graph each ellipse.

1. $10x^2 + y^2 = 32$ **2.** $x^2 + 6y^2 = 35$

3. $20x^2 + 5y^2 = 100$ **4.** $4y^2 + 12x^2 = 48$

5. $7.3x^2 + 15.5y^2 = 95.2$ **6.** $18.8x^2 + 36.1y^2 = 205.8$

Spotlight on

DECISION ☀ MAKING

Suppose you are an astronomer. You know that the orbits of stars, planets, comets, asteroids, and satellites all have the shape of one of the conic sections. *Eccentricity* is a measure used to describe the shape and elongation of an orbital path. The table shows ranges of eccentricities for the different types of conic sections.

For each of the following comets known to pass through our solar system, decide what type of orbit the comet has based on its eccentricity *e*. Describe how the shape of the comet's orbit affects how often it passes through our solar system. (For more exercises on eccentricity, see Exercise Set 13.2.)

a. Spacewatch (1997 P2), $e = 1.02851919$

b. Whipple, $e = 0.25871336$

c. Lee (1999 H1), $e = 0.99973749$

d. Giacobini-Zinner, $e = 0.70647162$

e. Tabur (1997 N1), $e = 1.00004712$

Conic Section	Eccentricity e
Circle	$e = 0$
Ellipse	$0 < e < 1$
Parabola	$e = 1$
Hyperbola	$e > 1$

MENTAL MATH

Identify the graph of each equation as an ellipse or a hyperbola.

1. $\dfrac{x^2}{16} + \dfrac{y^2}{4} = 1$

2. $\dfrac{x^2}{16} - \dfrac{y^2}{4} = 1$

3. $x^2 - 5y^2 = 3$

4. $-x^2 + 5y^2 = 3$

5. $-\dfrac{y^2}{25} + \dfrac{x^2}{36} = 1$

6. $\dfrac{y^2}{25} + \dfrac{x^2}{36} = 1$

EXERCISE SET 13.2

STUDY GUIDE/SSM · CD/VIDEO · PH MATH TUTOR CENTER · MathXL®Tutorials ON CD · MathXL® · MyMathLab®

Sketch the graph of each equation. See Examples 1 and 2.

1. $\dfrac{x^2}{4} + \dfrac{y^2}{25} = 1$

2. $\dfrac{x^2}{9} + y^2 = 1$

3. $\dfrac{x^2}{16} + \dfrac{y^2}{9} = 1$

4. $x^2 + \dfrac{y^2}{4} = 1$

 5. $9x^2 + 4y^2 = 36$

6. $x^2 + 4y^2 = 16$

7. $4x^2 + 25y^2 = 100$

8. $36x^2 + y^2 = 36$

Sketch the graph of each equation. See Example 3.

9. $\dfrac{(x + 1)^2}{36} + \dfrac{(y - 2)^2}{49} = 1$

10. $\dfrac{(x - 3)^2}{9} + \dfrac{(y + 3)^2}{16} = 1$

 11. $\dfrac{(x - 1)^2}{4} + \dfrac{(y - 1)^2}{25} = 1$

12. $\dfrac{(x + 3)^2}{16} + \dfrac{(y + 2)^2}{4} = 1$

Sketch the graph of each equation. See Examples 4 and 5.

 13. $\dfrac{x^2}{4} - \dfrac{y^2}{9} = 1$

14. $\dfrac{x^2}{36} - \dfrac{y^2}{36} = 1$

15. $\dfrac{y^2}{25} - \dfrac{x^2}{16} = 1$

16. $\dfrac{y^2}{25} - \dfrac{x^2}{49} = 1$

17. $x^2 - 4y^2 = 16$

18. $4x^2 - y^2 = 36$

19. $16y^2 - x^2 = 16$

20. $4y^2 - 25x^2 = 100$

21. If you are given a list of equations of circles, parabolas, ellipses, and hyperbolas, explain how you could distinguish the different conic sections from their equations.

MIXED PRACTICE

Identify whether each equation, when graphed, will be a parabola, circle, ellipse, or hyperbola. Sketch the graph of each equation.

22. $(x - 7)^2 + (y - 2)^2 = 4$ **23.** $y = x^2 + 4$

24. $y = x^2 + 12x + 36$ **25.** $\dfrac{x^2}{4} + \dfrac{y^2}{9} = 1$

26. $\dfrac{y^2}{9} - \dfrac{x^2}{9} = 1$ **27.** $\dfrac{x^2}{16} - \dfrac{y^2}{4} = 1$

28. $\dfrac{x^2}{16} + \dfrac{y^2}{4} = 1$ **29.** $x^2 + y^2 = 16$

30. $x = y^2 + 4y - 1$ **31.** $x = -y^2 + 6y$

32. $9x^2 - 4y^2 = 36$ **33.** $9x^2 + 4y^2 = 36$

34. $\dfrac{(x - 1)^2}{49} + \dfrac{(y + 2)^2}{25} = 1$

35. $y^2 = x^2 + 16$

36. $\left(x + \dfrac{1}{2}\right)^2 + \left(y - \dfrac{1}{2}\right)^2 = 1$

37. $y = -2x^2 + 4x - 3$

REVIEW AND PREVIEW

Solve each inequality. See Section 2.7.

38. $x < 5$ and $x < 1$

39. $x < 5$ or $x < 1$

40. $2x - 1 \geq 7$ or $-3x \leq -6$

41. $2x - 1 \geq 7$ and $-3x \leq -6$

Perform the indicated operations. See Sections 5.1 and 5.2.

42. $(2x^3)(-4x^2)$ **43.** $2x^3 - 4x^3$
44. $-5x^2 + x^2$ **45.** $(-5x^2)(x^2)$

Concept Extensions

46. We know that $x^2 + y^2 = 25$ is the equation of a circle. Rewrite the equation so that the right side is equal to 1. Which type of conic section does this equation form resemble? In fact, the circle is a special case of this type of conic section. Describe the conditions under which this type of conic section is a circle.

The orbits of stars, planets, comets, asteroids, and satellites all have the shape of one of the conic sections. Astronomers use a measure called eccentricity to describe the shape and elongation of an orbital path. For the circle and ellipse, eccentricity e is calculated with the formula $e = \dfrac{c}{d}$, where $c^2 = |a^2 - b^2|$ and d is the larger value of a or b. For a hyperbola, eccentricity e is calculated with the formula $e = \dfrac{c}{d}$, where $c^2 = a^2 + b^2$ and the value of d is equal to a if the hyperbola has x-intercepts or equal to b if the hyperbola has y-intercepts. Use equations A–H to answer Exercises 47–56.

A $\dfrac{x^2}{36} - \dfrac{y^2}{13} = 1$ **B** $\dfrac{x^2}{4} + \dfrac{y^2}{4} = 1$ **C** $\dfrac{x^2}{25} + \dfrac{y^2}{16} = 1$

D $\dfrac{y^2}{25} - \dfrac{x^2}{39} = 1$ **E** $\dfrac{x^2}{17} + \dfrac{y^2}{81} = 1$ **F** $\dfrac{x^2}{36} + \dfrac{y^2}{36} = 1$

G $\dfrac{x^2}{16} - \dfrac{y^2}{65} = 1$ **H** $\dfrac{x^2}{144} + \dfrac{y^2}{140} = 1$

47. Identify the type of conic section represented by each of the equations A–H.

48. For each of the equations A–H, identify the values of a^2 and b^2.

49. For each of the equations A–H, calculate the value of c^2 and c.

50. For each of the equations A–H, find the value of d.

51. For each of the equations A–H, calculate the eccentricity e.

52. What do you notice about the values of e for the equations you identified as ellipses?

53. What do you notice about the values of e for the equations you identified as circles?

54. What do you notice about the values of e for the equations you identified as hyperbolas?

55. The eccentricity of a parabola is exactly 1. Use this information and the observations you made in Exercises 31, 32, and 33 to describe a way that could be used to identify the type of conic section based on its eccentricity value.

56. Graph each of the conic sections given in equations A–H. What do you notice about the shape of the ellipses for increasing values of eccentricity? Which is the most elliptical? Which is the least elliptical, that is, the most circular?

57. A planet's orbit about the Sun can be described as an ellipse. Consider the Sun as the origin of a rectangular coordinate system. Suppose that the x-intercepts of the elliptical path of the planet are $\pm130,000,000$ and that the y-intercepts are $\pm125,000,000$. Write the equation of the elliptical path of the planet.

58. Comets orbit the Sun in elongated ellipses. Consider the Sun as the origin of a rectangular coordinate system. Suppose that the equation of the path of the comet is

$$\frac{(x - 1,782,000,000)^2}{3.42 \cdot 10^{23}} + \frac{(y - 356,400,000)^2}{1.368 \cdot 10^{22}} = 1$$

Find the center of the path of the comet.

59. Use a graphing calculator to verify Exercise 5.

60. Use a graphing calculator to verify Exercise 6.

For Exercises 61 through 66, see the example below.

Example

Sketch the graph of $\dfrac{(x-2)^2}{25} - \dfrac{(y-1)^2}{9} = 1$.

Solution

This hyperbola has center $(2, 1)$. Notice that $a = 5$ and $b = 3$.

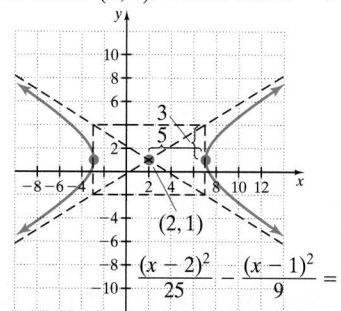

$$\dfrac{(x-2)^2}{25} - \dfrac{(x-1)^2}{9} = 1$$

Sketch the graph of each equation.

61. $\dfrac{(x-1)^2}{4} - \dfrac{(y+1)^2}{25} = 1$ **62.** $\dfrac{(x+2)^2}{9} - \dfrac{(y-1)^2}{4} = 1$

63. $\dfrac{y^2}{16} - \dfrac{(x+3)^2}{9} = 1$ **64.** $\dfrac{(y+4)^2}{4} - \dfrac{x^2}{25} = 1$

65. $\dfrac{(x+5)^2}{16} - \dfrac{(y+2)^2}{25} = 1$ **66.** $\dfrac{(x-3)^2}{9} - \dfrac{(y-2)^2}{4} = 1$

INTEGRATED REVIEW   GRAPHING CONIC SECTIONS

Following is a summary of conic sections.

Conic Sections

	Standard Form	**Graph**
Parabola	$y = a(x-h)^2 + k$	
Parabola	$x = a(y-k)^2 + h$	
Circle	$(x-h)^2 + (y-k)^2 = r^2$	
Ellipse	$\dfrac{x^2}{a^2} + \dfrac{y^2}{b^2} = 1$	
Hyperbola	$\dfrac{x^2}{a^2} - \dfrac{y^2}{b^2} = 1$	

(continued)

Hyperbola	$\dfrac{y^2}{b^2} - \dfrac{x^2}{a^2} = 1$	

Identify whether each equation, when graphed, will be a parabola, circle, ellipse, or hyperbola. Then graph each equation.

1. $(x - 7)^2 + (y - 2)^2 = 4$ **2.** $y = x^2 + 4$ **3.** $y = x^2 + 12x + 36$

4. $\dfrac{x^2}{4} + \dfrac{y^2}{9} = 1$ **5.** $\dfrac{y^2}{9} - \dfrac{x^2}{9} = 1$ **6.** $\dfrac{x^2}{16} - \dfrac{y^2}{4} = 1$

7. $\dfrac{x^2}{16} + \dfrac{y^2}{4} = 1$ **8.** $x^2 + y^2 = 16$ **9.** $x = y^2 + 4y - 1$

10. $x = -y^2 + 6y$ **11.** $9x^2 - 4y^2 = 36$ **12.** $9x^2 + 4y^2 = 36$

13. $\dfrac{(x - 1)^2}{49} + \dfrac{(y + 2)^2}{25} = 1$ **14.** $y^2 = x^2 + 16$ **15.** $\left(x + \dfrac{1}{2}\right)^2 + \left(y - \dfrac{1}{2}\right)^2 = 1$

13.3   SOLVING NONLINEAR SYSTEMS OF EQUATIONS

Objectives

1 Solve a nonlinear system by substitution.

2 Solve a nonlinear system by elimination.

In Chapter 4, we used graphing, substitution, and elimination methods to find solutions of systems of linear equations in two variables. We now apply these same methods to nonlinear systems of equations in two variables. A **nonlinear system of equations** is a system of equations at least one of which is not linear. Since we will be graphing the equations in each system, we are interested in real number solutions only.

1 First, nonlinear systems are solved by the substitution method.

EXAMPLE 1

Solve the system.

$$\begin{cases} x^2 - 3y = 1 \\ x - y = 1 \end{cases}$$

Solution   We can solve this system by substitution if we solve one equation for one of the variables. Solving the first equation for x is not the best choice since doing so introduces a radical. Also, solving for y in the first equation introduces a fraction. We solve the second equation for y.

$x - y = 1$ *Second equation*

$x - 1 = y$ *Solve for y.*

Replace y with $x - 1$ in the first equation, and then solve for x.

$$x^2 - 3y = 1 \qquad \text{First equation}$$

$$x^2 - 3\overbrace{(x - 1)}^{} = 1 \qquad \text{Replace } y \text{ with } x - 1.$$

$$x^2 - 3x + 3 = 1$$

$$x^2 - 3x + 2 = 0$$

$$(x - 2)(x - 1) = 0$$

$$x = 2 \quad \text{or} \quad x = 1$$

Let $x = 2$ and then let $x = 1$ in the equation $y = x - 1$ to find corresponding y-values.

$$\begin{array}{ll}
\text{Let } x = 2. & \text{Let } x = 1. \\
y = x - 1 & y = x - 1 \\
y = 2 - 1 = 1 & y = 1 - 1 = 0
\end{array}$$

The solutions are $(2, 1)$ and $(1, 0)$ or the solution set is $\{(2, 1), (1, 0)\}$. Check both solutions in both equations. Both solutions satisfy both equations, so both are solutions of the system. The graph of each equation in the system is shown next. Intersections of the graphs are at $(2, 1)$ and $(1, 0)$.

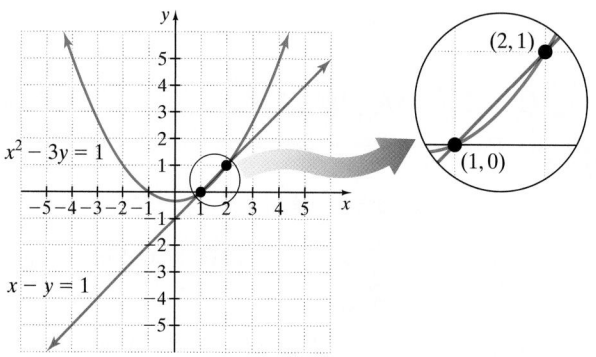

EXAMPLE 2

Solve the system.

$$\begin{cases} y = \sqrt{x} \\ x^2 + y^2 = 6 \end{cases}$$

Solution This system is ideal for substitution since y is expressed in terms of x in the first equation. Notice that if $y = \sqrt{x}$, then both x and y must be nonnegative if they are real numbers. Substitute $\sqrt{x}$ for y in the second equation, and solve for x.

$$x^2 + y^2 = 6$$

$$x^2 + \left(\sqrt{x}\right)^2 = 6 \qquad \text{Let } y = \sqrt{x}.$$

$$x^2 + x = 6$$

$$x^2 + x - 6 = 0$$

$$(x + 3)(x - 2) = 0$$

$$x = -3 \quad \text{or} \quad x = 2$$

The solution -3 is discarded because we have noted that x must be nonnegative. To see this, let $x = -3$ in the first equation. Then let $x = 2$ in the first equation to find a corresponding y-value.

Let $x = -3$.   Let $x = 2$.

$y = \sqrt{x}$ $y = \sqrt{x}$

$y = \sqrt{-3}$ *Not a real number* $y = \sqrt{2}$

Since we are interested only in real number solutions, the only solution is $\left(2, \sqrt{2}\right)$. Check to see that this solution satisfies both equations. The graph of each equation in the system is shown next.

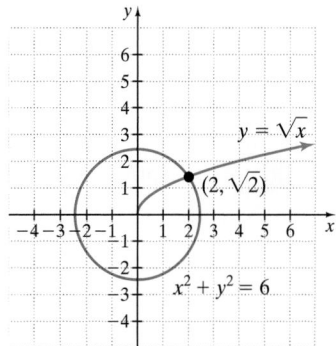

EXAMPLE 3

Solve the system.

$$\begin{cases} x^2 + y^2 = 4 \\ x + y = 3 \end{cases}$$

Solution   We use the substitution method and solve the second equation for x.

$$x + y = 3 \qquad \textit{Second equation}$$
$$x = 3 - y$$

Now we let $x = 3 - y$ in the first equation.

$$x^2 + y^2 = 4 \qquad \textit{First equation}$$
$$(3 - y)^2 + y^2 = 4 \qquad \textit{Let } x = 3 - y.$$
$$9 - 6y + y^2 + y^2 = 4$$
$$2y^2 - 6y + 5 = 0$$

By the quadratic formula, where $a = 2, b = -6$, and $c = 5$, we have

$$y = \frac{6 \pm \sqrt{(-6)^2 - 4 \cdot 2 \cdot 5}}{2 \cdot 2} = \frac{6 \pm \sqrt{-4}}{4}$$

Since $\sqrt{-4}$ is not a real number, there is no real solution, or $\varnothing$. Graphically, the circle and the line do not intersect, as shown on the following page.

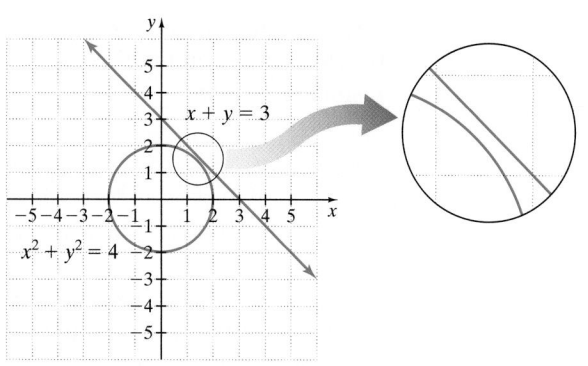

✔ **CONCEPT CHECK**

Without solving, how can you tell that $x^2 + y^2 = 9$ and $x^2 + y^2 = 16$ do not have any points of intersection?

2   Some nonlinear systems may be solved by the elimination method.

EXAMPLE 4

Solve the system.

$$\begin{cases} x^2 + 2y^2 = 10 \\ x^2 - y^2 = 1 \end{cases}$$

Solution We will use the elimination, or addition, method to solve this system. To eliminate x^2 when we add the two equations, multiply both sides of the second equation by -1. Then

$$\begin{cases} x^2 + 2y^2 = 10 \\ (-1)(x^2 - y^2) = -1 \cdot 1 \end{cases}$$ *is equivalent to* $$\begin{cases} x^2 + 2y^2 = 10 \\ \underline{-x^2 + y^2 = -1} \end{cases}$$ *Add.*

$$3y^2 = 9$$ *Divide both*

$$y^2 = 3$$ *sides by 3.*

$$y = \pm\sqrt{3}$$

To find the corresponding x-values, we let $y = \sqrt{3}$ and $y = -\sqrt{3}$ in either original equation. We choose the second equation.

Let $y = \sqrt{3}$.

$$x^2 - y^2 = 1$$
$$x^2 - \left(\sqrt{3}\right)^2 = 1$$
$$x^2 - 3 = 1$$
$$x^2 = 4$$
$$x = \pm\sqrt{4} = \pm 2$$

Let $y = -\sqrt{3}$.

$$x^2 - y^2 = 1$$
$$x^2 - \left(-\sqrt{3}\right)^2 = 1$$
$$x^2 - 3 = 1$$
$$x^2 = 4$$
$$x = \pm\sqrt{4} = \pm 2$$

Concept Check Answer:
$x^2 + y^2 = 9$ is a circle inside the circle $x^2 + y^2 = 16$, therefore they do not have any points of intersection.

The solutions are $\left(2, \sqrt{3}\right), \left(-2, \sqrt{3}\right), \left(2, -\sqrt{3}\right),$ and $\left(-2, -\sqrt{3}\right)$. Check all four ordered pairs in both equations of the system. The graph of each equation in this system is shown on the following page.

$x^2 - y^2 = 1$

$(-2, \sqrt{3})$ $(2, \sqrt{3})$

$x^2 + 2y^2 = 10$

$(-2, -\sqrt{3})$ $(2, -\sqrt{3})$

EXERCISE SET 13.3

STUDY GUIDE/SSM   CD/ VIDEO   PH MATH TUTOR CENTER   MathXL®Tutorials ON CD   MathXL®   MyMathLab®

MIXED PRACTICE

Solve each nonlinear system of equations for real solutions. See Examples 1 through 4.

1. $\begin{cases} x^2 + y^2 = 25 \\ 4x + 3y = 0 \end{cases}$

2. $\begin{cases} x^2 + y^2 = 25 \\ 3x + 4y = 0 \end{cases}$

3. $\begin{cases} x^2 + 4y^2 = 10 \\ y = x \end{cases}$

4. $\begin{cases} 4x^2 + y^2 = 10 \\ y = x \end{cases}$

5. $\begin{cases} y^2 = 4 - x \\ x - 2y = 4 \end{cases}$

6. $\begin{cases} x^2 + y^2 = 4 \\ x + y = -2 \end{cases}$

7. $\begin{cases} x^2 + y^2 = 9 \\ 16x^2 - 4y^2 = 64 \end{cases}$

8. $\begin{cases} 4x^2 + 3y^2 = 35 \\ 5x^2 + 2y^2 = 42 \end{cases}$

9. $\begin{cases} x^2 + 2y^2 = 2 \\ x - y = 2 \end{cases}$

10. $\begin{cases} x^2 + 2y^2 = 2 \\ x^2 - 2y^2 = 6 \end{cases}$

11. $\begin{cases} y = x^2 - 3 \\ 4x - y = 6 \end{cases}$

12. $\begin{cases} y = x + 1 \\ x^2 - y^2 = 1 \end{cases}$

13. $\begin{cases} y = x^2 \\ 3x + y = 10 \end{cases}$

14. $\begin{cases} 6x - y = 5 \\ xy = 1 \end{cases}$

15. $\begin{cases} y = 2x^2 + 1 \\ x + y = -1 \end{cases}$

16. $\begin{cases} x^2 + y^2 = 9 \\ x + y = 5 \end{cases}$

17. $\begin{cases} y = x^2 - 4 \\ y = x^2 - 4x \end{cases}$

18. $\begin{cases} x = y^2 - 3 \\ x = y^2 - 3y \end{cases}$

19. $\begin{cases} 2x^2 + 3y^2 = 14 \\ -x^2 + y^2 = 3 \end{cases}$

20. $\begin{cases} 4x^2 - 2y^2 = 2 \\ -x^2 + y^2 = 2 \end{cases}$

21. $\begin{cases} x^2 + y^2 = 1 \\ x^2 + (y + 3)^2 = 4 \end{cases}$

22. $\begin{cases} x^2 + 2y^2 = 4 \\ x^2 - y^2 = 4 \end{cases}$

23. $\begin{cases} y = x^2 + 2 \\ y = -x^2 + 4 \end{cases}$

24. $\begin{cases} x = -y^2 - 3 \\ x = y^2 - 5 \end{cases}$

25. $\begin{cases} 3x^2 + y^2 = 9 \\ 3x^2 - y^2 = 9 \end{cases}$

26. $\begin{cases} x^2 + y^2 = 25 \\ x = y^2 - 5 \end{cases}$

27. $\begin{cases} x^2 + 3y^2 = 6 \\ x^2 - 3y^2 = 10 \end{cases}$

28. $\begin{cases} x^2 + y^2 = 1 \\ y = x^2 - 9 \end{cases}$

29. $\begin{cases} x^2 + y^2 = 36 \\ y = \dfrac{1}{6}x^2 - 6 \end{cases}$

30. $\begin{cases} x^2 + y^2 = 16 \\ y = -\dfrac{1}{4}x^2 + 4 \end{cases}$

REVIEW AND PREVIEW

Graph each inequality in two variables. See Section 9.4.

31. $x > -3$

32. $y \le 1$

33. $y < 2x - 1$

34. $3x - y \le 4$

Find the perimeter of each geometric figure. See Section 5.2.

△ 35.

x inches $(2x - 5)$ inches

$(5x - 20)$ inches

△ 36.

$(3x + 2)$ centimeters

△ 37. $(x^2 + 3x + 1)$ meters

x^2 meters

△ 38. $2x^2$ feet

$4x$ feet $(3x^2 + 1)$ feet

$(3x^2 + 7)$ feet

Concept Extensions

39. How many real solutions are possible for a system of equations whose graphs are a circle and a parabola? Draw diagrams to illustrate each possibility.

40. How many real solutions are possible for a system of equations whose graphs are an ellipse and a line? Draw diagrams to illustrate each possibility.

41. The sum of the squares of two numbers is 130. The difference of the squares of the two numbers is 32. Find the two numbers.

42. The sum of the squares of two numbers is 20. Their product is 8. Find the two numbers.

△ **43.** During the development stage of a new rectangular keypad for a security system, it was decided that the area of the rectangle should be 285 square centimeters and the perimeter should be 68 centimeters. Find the dimensions of the keypad.

△ **44.** A rectangular holding pen for cattle is to be designed so that its perimeter is 92 feet and its area is 525 feet. Find the dimensions of the holding pen.

Recall that in business, a demand function expresses the quantity of a commodity demanded as a function of the commodity's unit price. A supply function expresses the quantity of a commodity supplied as a function of the commodity's unit price. When the quantity produced and supplied is equal to the quantity demanded, then we have what is called **market equilibrium**.

45. The demand function for a certain compact disc is given by the function
$$p = -0.01x^2 - 0.2x + 9$$
and the corresponding supply function is given by
$$p = 0.01x^2 - 0.1x + 3$$
where p is in dollars and x is in thousands of units. Find the equilibrium quantity and the corresponding price by solving the system consisting of the two given equations.

46. The demand function for a certain style of picture frame is given by the function
$$p = -2x^2 + 90$$
and the corresponding supply function is given by
$$p = 9x + 34$$
where p is in dollars and x is in thousands of units. Find the equilibrium quantity and the corresponding price by solving the system consisting of the two given equations.

 Use a graphing calculator to verify the results of each exercise.

47. Exercise 3.   **48.** Exercise 4.
49. Exercise 23.   **50.** Exercise 24.

STUDY SKILLS REMINDER

Are You Satisfied with Your Performance on a Particular Quiz or Exam?

If not, don't forget to analyze your quiz or exam and look for common errors.

Were most of your errors a result of

▶ *Carelessness?* If your errors were careless, did you turn in your work before the allotted time expired? If so, resolve next time to use the entire time allotted. Any extra time can be spent checking your work.

▶ *Running out of time?* If so, make a point to better manage your time on your next exam. A few suggestions are to work any questions that you are unsure of last and to check your work after all questions have been answered.

▶ *Not understanding a concept?* If so, review that concept and correct your work. Remember next time to make sure that all concepts on a quiz or exam are understood before the exam.

13.4   NONLINEAR INEQUALITIES AND SYSTEMS OF INEQUALITIES

Objectives

1. Graph a nonlinear inequality.
2. Graph a system of nonlinear inequalities.

1   We can graph a nonlinear inequality in two variables such as $\dfrac{x^2}{9} + \dfrac{y^2}{16} \leq 1$ in a way similar to the way we graphed a linear inequality in two variables in Section 9.4. First, we graph the related equation $\dfrac{x^2}{9} + \dfrac{y^2}{16} = 1$. The graph of the equation is our boundary. Then, using test points, we determine and shade the region whose points satisfy the inequality.

EXAMPLE 1

Graph $\dfrac{x^2}{9} + \dfrac{y^2}{16} \leq 1$.

Solution   First, graph the equation $\dfrac{x^2}{9} + \dfrac{y^2}{16} = 1$. Sketch a solid curve since the graph of $\dfrac{x^2}{9} + \dfrac{y^2}{16} \leq 1$ includes the graph of $\dfrac{x^2}{9} + \dfrac{y^2}{16} = 1$. The graph is an ellipse, and it divides the plane into two regions, the "inside" and the "outside" of the ellipse. To determine which region contains the solutions, select a test point in either region and determine whether the coordinates of the point satisfy the inequality. We choose $(0,0)$ as the test point.

$$\frac{x^2}{9} + \frac{y^2}{16} \leq 1$$

$$\frac{0^2}{9} + \frac{0^2}{16} \leq 1 \qquad \text{Let } x = 0 \text{ and } y = 0.$$

$$0 \leq 1 \qquad \text{True}$$

Since this statement is true, the solution set is the region containing $(0,0)$. The graph of the solution set includes the points on and inside the ellipse, as shaded in the figure.

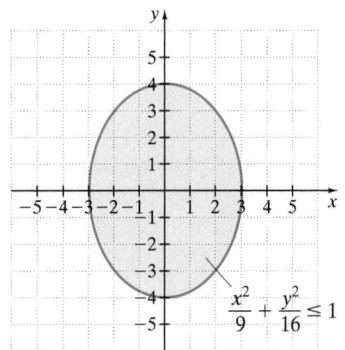

EXAMPLE 2

Graph $4y^2 > x^2 + 16$.

Solution The related equation is $4y^2 = x^2 + 16$. Subtract x^2 from both sides and divide both sides by 16, and we have $\dfrac{y^2}{4} - \dfrac{x^2}{16} = 1$, which is a hyperbola. Graph the hyperbola as a dashed curve since the graph of $4y^2 > x^2 + 16$ does *not* include the graph of $4y^2 = x^2 + 16$. The hyperbola divides the plane into three regions. Select a test point in each region—not on a boundary line—to determine whether that region contains solutions of the inequality.

Test region A with $(0, 4)$	*Test region B with* $(0, 0)$	*Test region C with* $(0, -4)$
$4y^2 > x^2 + 16$	$4y^2 > x^2 + 16$	$4y^2 > x^2 + 16$
$4(4)^2 > 0^2 + 16$	$4(0)^2 > 0^2 + 16$	$4(-4)^2 > 0^2 + 16$
$64 > 16$ True	$0 > 16$ False	$64 > 16$ True

The graph of the solution set includes the shaded regions A and C only, not the boundary.

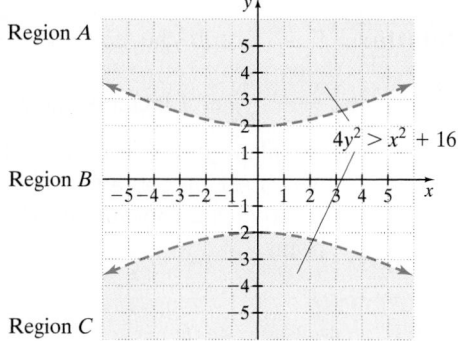

2 In Section 9.4 we graphed systems of linear inequalities. Recall that the graph of a system of inequalities is the intersection of the graphs of the inequalities.

EXAMPLE 3

Graph the system.

$$\begin{cases} x \le 1 - 2y \\ y \le x^2 \end{cases}$$

Solution We graph each inequality on the same set of axes. The intersection is shown in the third graph on the following page. It is the purple shaded region along with its boundary lines. The coordinates of the points of intersection can be found by solving the related system.

$$\begin{cases} x = 1 - 2y \\ y = x^2 \end{cases}$$

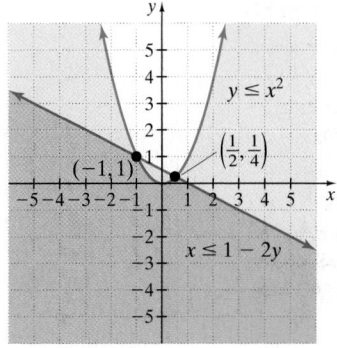

EXAMPLE 4

Graph the system.

$$\begin{cases} x^2 + y^2 < 25 \\ \dfrac{x^2}{9} - \dfrac{y^2}{25} < 1 \\ \qquad y < x + 3 \end{cases}$$

Solution   We graph each inequality. The graph of $x^2 + y^2 < 25$ contains points "inside" the circle that has center $(0, 0)$ and radius 5. The graph of $\dfrac{x^2}{9} - \dfrac{y^2}{25} < 1$ is the region between the two branches of the hyperbola with x-intercepts -3 and 3 and center $(0, 0)$. The graph of $y < x + 3$ is the region "below" the line with slope 1 and y-intercept $(0, 3)$. The graph of the solution set of the system is the intersection of all the graphs, the darkest shaded region shown. The boundary of this region is not part of the solution.

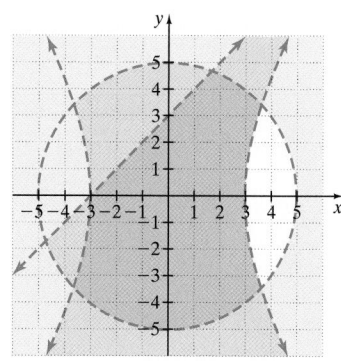

Spotlight on
DECISION
❦ MAKING

Suppose you are an architect. You have just designed a bridge over a two-lane road. Spanning the road is an arch in the shape of a half-ellipse that is 40 feet wide at the base of the arch and is 15 feet tall at the center of the arch. A colleague has just pointed out that the bridge must have a 13-foot clearance for vehicles on the road. The road is 22 feet wide, and its center line falls directly beneath the highest point of the arch. Decide whether your current bridge design will allow 13-foot-tall vehicles to pass on the road beneath it or if your bridge must be redesigned. Explain your reasoning. (*Hint:* Envision a 13-foot-tall semi truck passing under the bridge in the right-hand lane of the road. Try checking whether points along the top of the truck would fall within the ellipse that defines the arch of the bridge. Remember that the truck can drive within any portion of the right-hand lane.)

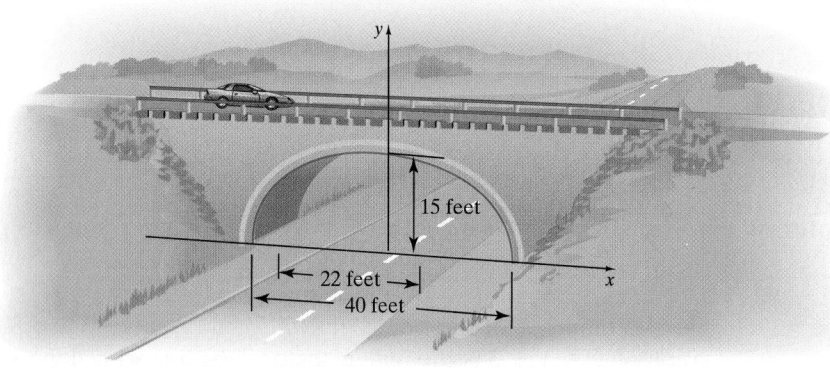

15 feet

← 22 feet →
←——— 40 feet ———→

EXERCISE SET 13.4

STUDY GUIDE/SSM   CD/VIDEO   PH MATH TUTOR CENTER   MathXL®Tutorials ON CD   MathXL®   MyMathLab®

Graph each inequality. See Examples 1 and 2.

1. $y < x^2$

2. $y < -x^2$

3. $x^2 + y^2 \geq 16$

4. $x^2 + y^2 < 36$

5. $\dfrac{x^2}{4} - y^2 < 1$

6. $x^2 - \dfrac{y^2}{9} \geq 1$

7. $y > (x - 1)^2 - 3$

8. $y > (x + 3)^2 + 2$

9. $x^2 + y^2 \leq 9$

10. $x^2 + y^2 > 4$

11. $y > -x^2 + 5$

12. $y < -x^2 + 5$

13. $\dfrac{x^2}{4} + \dfrac{y^2}{9} \leq 1$

14. $\dfrac{x^2}{25} + \dfrac{y^2}{4} \geq 1$

15. $\dfrac{y^2}{4} - x^2 \leq 1$

16. $\dfrac{y^2}{16} - \dfrac{x^2}{9} > 1$

17. $y < (x - 2)^2 + 1$ **18.** $y > (x - 2)^2 + 1$

19. $y \le x^2 + x - 2$ **20.** $y > x^2 + x - 2$

Graph each system. See Examples 3 and 4.

21. $\begin{cases} 2x - y < 2 \\ \quad\quad y \le -x^2 \end{cases}$ **22.** $\begin{cases} x - 2y > 4 \\ \quad\quad y > -x^2 \end{cases}$

23. $\begin{cases} 4x + 3y \ge 12 \\ x^2 + y^2 < 16 \end{cases}$ **24.** $\begin{cases} 3x - 4y \le 12 \\ x^2 + y^2 < 16 \end{cases}$

25. $\begin{cases} x^2 + y^2 \le 9 \\ x^2 + y^2 \ge 1 \end{cases}$ **26.** $\begin{cases} x^2 + y^2 \ge 9 \\ x^2 + y^2 \ge 16 \end{cases}$

27. $\begin{cases} y > x^2 \\ y \ge 2x + 1 \end{cases}$ **28.** $\begin{cases} y \le -x^2 + 3 \\ y \le 2x - 1 \end{cases}$

29. $\begin{cases} x > y^2 \\ y > 0 \end{cases}$ **30.** $\begin{cases} \quad x < (y + 1)^2 + 2 \\ x + y \ge 3 \end{cases}$

🔒 **31.** $\begin{cases} x^2 + y^2 > 9 \\ \quad\quad y > x^2 \end{cases}$ **32.** $\begin{cases} x^2 + y^2 \le 9 \\ \quad\quad y < x^2 \end{cases}$

33. $\begin{cases} \dfrac{x^2}{4} + \dfrac{y^2}{9} \ge 1 \\ x^2 + y^2 \ge 4 \end{cases}$ **34.** $\begin{cases} x^2 + (y - 2)^2 \ge 9 \\ \dfrac{x^2}{4} + \dfrac{y^2}{25} < 1 \end{cases}$

35. $\begin{cases} x^2 - y^2 \ge 1 \\ \quad\quad y \ge 0 \end{cases}$ **36.** $\begin{cases} x^2 - y^2 \ge 1 \\ \quad\quad x \ge 0 \end{cases}$

37. $\begin{cases} x + y \ge 1 \\ 2x + 3y < 1 \\ \quad\quad x > -3 \end{cases}$ **38.** $\begin{cases} x - y < -1 \\ 4x - 3y > 0 \\ \quad\quad y > 0 \end{cases}$

39. $\begin{cases} x^2 - y^2 < 1 \\ \dfrac{x^2}{16} + y^2 \le 1 \\ \quad\quad x \ge -2 \end{cases}$ **40.** $\begin{cases} x^2 - y^2 \ge 1 \\ \dfrac{x^2}{16} + \dfrac{y^2}{4} \le 1 \\ \quad\quad y \ge 1 \end{cases}$

REVIEW AND PREVIEW

Determine which graph is the graph of a function. See Section 3.7.

41.   **42.**

43.   **44.**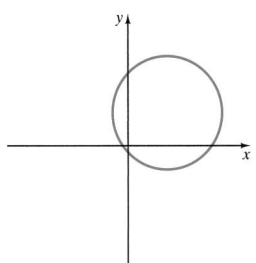

Find each function value if $f(x) = 3x^2 - 2$. See Section 3.7.

45. $f(-1)$ **46.** $f(-3)$

47. $f(a)$ **48.** $f(b)$

Concept Extensions

49. Discuss how graphing a linear inequality such as $x + y < 9$ is similar to graphing a nonlinear inequality such as $x^2 + y^2 < 9$.

50. Discuss how graphing a linear inequality such as $x + y < 9$ is different from graphing a nonlinear inequality such as $x^2 + y^2 < 9$.

51. Graph the system: $\begin{cases} y \le x^2 \\ y \ge x + 2 \\ x \ge 0 \\ y \ge 0 \end{cases}$

CHAPTER 13 PROJECT

Modeling Conic Sections

In this project, you will have the opportunity to construct and investigate a model of an ellipse. You will need two thumbtacks or nails, graph paper, cardboard, tape, string, a pencil, and a ruler. This project may be completed by working in groups or individually.

Follow these steps, answering any questions as you go.

1. Draw an *x*-axis and a *y*-axis on the graph paper as shown in Figure 1.

2. Place the graph paper on the cardboard and attach it with tape.

3. Locate two points on the *x*-axis, each about $1\frac{1}{2}$ inches from the origin and on opposite sides of the origin (see Figure 1). Insert thumbtacks (or nails) at each of these locations.

4. Fasten a 9-inch piece of string to the thumbtacks as shown in Figure 2. Use your pencil to draw and keep the string taut while you carefully move the pencil in a path all around the thumbtacks.

5. Using the grid of the graph paper as a guide, find an approximate equation of the ellipse you drew.

6. Experiment by moving the tacks closer together or farther apart and drawing new ellipses. What do you observe?

7. Write a paragraph explaining why the figure drawn by the pencil is an ellipse. How might you use the same materials to draw a circle?

8. (Optional) Choose one of the ellipses you drew with the string and pencil. Use a ruler to draw any six tangent lines to the ellipse. (A line is tangent to the ellipse if it intersects, or just touches, the ellipse at only one point. See Figure 3.) Extend the tangent lines to yield six points of intersection among the tangents. Use a straightedge to draw a line connecting each pair of opposite points of intersection. What do you observe? Repeat with a different ellipse. Can you make a conjecture about the relationship among the lines that connect opposite points of intersection?

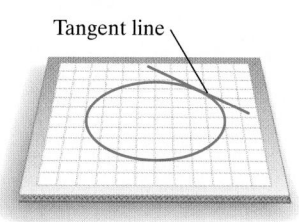

Figure 1   Figure 2   Figure 3

STUDY SKILLS REMINDER

Are You Preparing for a Test on Chapter 13?

Below I have listed some common trouble areas for topics covered in Chapter 13. After studying for your test—but before taking your test—read these.

▶ Don't forget to review all the standard forms for the conic sections.

▶ Remember that the midpoint of a segment is a *point*. The x-coordinate is the average of the x-coordinates of the endpoints of the segment and the y-coordinate is the average of the y-coordinates of the endpoints of the segment.

The midpoint of the segment joining $(-1, 5)$ and $(3, 4)$ is $\left(\dfrac{-1+3}{2}, \dfrac{5+4}{2}\right)$ or $\left(1, \dfrac{9}{2}\right)$.

▶ Remember that the distance formula gives the *distance* between two points.

The distance between $(-1, 5)$ and $(3, 4)$ is

$$\sqrt{(3-(-1))^2 + (4-5)^2} = \sqrt{4^2 + (-1)^2}$$
$$= \sqrt{16+1} = \sqrt{17} \text{ units}$$

▶ Don't forget that both methods, substitution and elimination, are available for solving nonlinear systems of equations.

$$\begin{cases} x^2 + y^2 = 7 \\ 2x^2 - 3y^2 = 4 \end{cases} \text{ is equivalent to}$$

$$\begin{array}{l} \begin{cases} 3x^2 + 3y^2 = 21 \\ 2x^2 - 3y^2 = 4 \end{cases} \\ \overline{5x^2 = 25} \\ x^2 = 5 \\ x = \pm\sqrt{5} \end{array}$$

(continued)

Let $x = \pm\sqrt{5}$ in either original equation, and $y = \pm\sqrt{2}$, the solution set is $\left\{\left(\sqrt{5}, \sqrt{2}\right), \left(-\sqrt{5}, \sqrt{2}\right), \left(\sqrt{5}, -\sqrt{2}\right), \left(-\sqrt{5}, -\sqrt{2}\right)\right\}$.

Remember: This is simply a checklist of common trouble areas. For a review of Chapter 13, see the Highlights and Chapter Review at the end of this chapter.

13 CHAPTER VOCABULARY CHECK

Fill in each blank with one of the words or phrases listed below.

circle   midpoint   radius   distance
center   ellipse   hyperbola   nonlinear system of equations

1. The _____ formula is $d = \sqrt{(x_2 - x_1)^2 + (y_2 - y_1)^2}$.

2. A(n) _____ is the set of all points in a plane that are the same distance from a fixed point, called the _____.

3. A _____ is a system of equations at least one of which is not linear.

4. A(n) _____ is the set of points on a plane such that the sum of the distances of those points from two fixed points is a constant.

5. In a circle, the distance from the center to a point of the circle is called its _____.

6. A(n) _____ is the set of points in a plane such that the absolute value of the difference of the distance from two fixed points is constant.

7. The _____ formula is $\left(\dfrac{x_1 + x_2}{2}, \dfrac{y_1 + y_2}{2} \right)$.

CHAPTER 13 HIGHLIGHTS

Definitions and Concepts	**Examples**

Section 13.1   The Parabola and the Circle

Parabolas

$$y = a(x - h)^2 + k$$

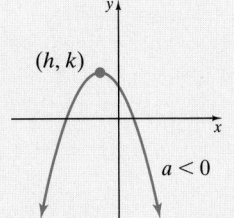

Graph

$$x = 3y^2 - 12y + 13.$$
$$x - 13 = 3y^2 - 12y$$
$$x - 13 + 3(4) = 3(y^2 - 4y + 4) \qquad \text{Add } 3(4) \text{ to}$$
$$x = 3(y - 2)^2 + 1 \qquad \text{both sides.}$$

Since $a = 3$, this parabola opens to the right with vertex $(1, 2)$. Its axis of symmetry is $y = 2$. The x-intercept is $(13, 0)$.

(continued)

Definitions and Concepts	**Examples**

Section 13.1 The Parabola and the Circle

$$x = a(y - k)^2 + h$$

Distance formula

The distance d between two points (x_1, y_1) and (x_2, y_2) is given by

$$d = \sqrt{(x_2 - x_1)^2 + (y_2 - y_1)^2}$$

Find the distance between points $(-1, 6)$ and $(-2, -4)$. Let $(x_1, y_1) = (-1, 6)$ and $(x_2, y_2) = (-2, -4)$.

$$\begin{aligned} d &= \sqrt{(x_2 - x_1)^2 + (y_2 - y_1)^2} \\ &= \sqrt{(-2 - (-1))^2 + (-4 - 6)^2} \\ &= \sqrt{1 + 100} = \sqrt{101} \end{aligned}$$

Midpoint formula

The midpoint of the line segment whose endpoints are (x_1, y_1) and (x_2, y_2) is the point with coordinates

$$\left(\frac{x_1 + x_2}{2}, \frac{y_1 + y_2}{2} \right)$$

Find the midpoint of the line segment whose endpoints are $(-1, 6)$ and $(-2, -4)$.

$$\left(\frac{-1 + (-2)}{2}, \frac{6 + (-4)}{2} \right)$$

The midpoint is

$$\left(-\frac{3}{2}, 1 \right)$$

Circle

The graph of $(x - h)^2 + (y - k)^2 = r^2$ is a circle with center (h, k) and radius r.

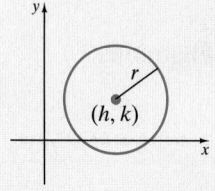

Graph $x^2 + (y + 3)^2 = 5$.

This equation can be written as

$$(x - 0)^2 + (y + 3)^2 = 5 \text{ with } h = 0,$$
$$k = -3, \text{ and } r = \sqrt{5}.$$

The center of this circle is $(0, -3)$, and the radius is $\sqrt{5}$.

Definitions and Concepts	**Examples**

Ellipse with center $(0, 0)$

The graph of an equation of the form $\dfrac{x^2}{a^2} + \dfrac{y^2}{b^2} = 1$ is an ellipse with center $(0, 0)$. The x-intercepts are $(a, 0)$ and $(-a, 0)$, and the y-intercepts are $(0, b)$ and $(0, -b)$.

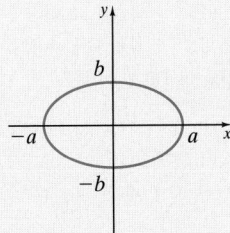

Hyperbola with center $(0, 0)$

The graph of an equation of the form $\dfrac{x^2}{a^2} - \dfrac{y^2}{b^2} = 1$ is a hyperbola with center $(0, 0)$ and x-intercepts $(a, 0)$ and $(-a, 0)$.

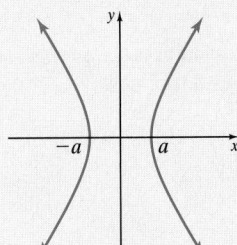

The graph of an equation of the form $\dfrac{y^2}{b^2} - \dfrac{x^2}{a^2} = 1$ is a hyperbola with center $(0, 0)$ and y-intercepts $(0, b)$ and $(0, -b)$.

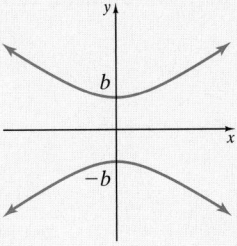

Graph $4x^2 + 9y^2 = 36$.

$$\frac{x^2}{9} + \frac{y^2}{4} = 1 \qquad \text{Divide by } \mathbf{36}.$$

$$\frac{x^2}{3^2} + \frac{y^2}{2^2} = 1$$

The ellipse has center $(0, 0)$, x-intercepts $(3, 0)$ and $(-3, 0)$, and y-intercepts $(0, 2)$ and $(0, -2)$.

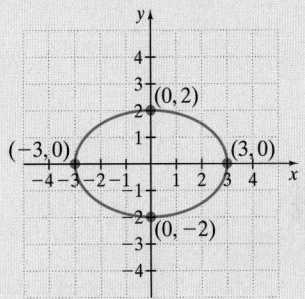

Graph $\dfrac{x^2}{9} - \dfrac{y^2}{4} = 1$. Here $a = 3$ and $b = 2$.

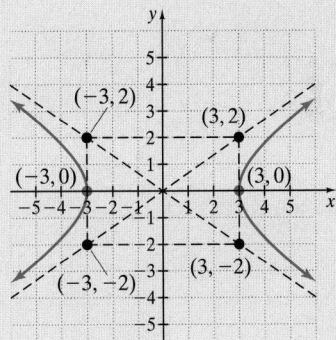

Definitions and Concepts	**Examples**

Section 13.3 Solving Nonlinear Systems of Equations

A **nonlinear system of equations** is a system of equations at least one of which is not linear. Both the substitution method and the elimination method may be used to solve a nonlinear system of equations.

Solve the nonlinear system $\begin{cases} y = x + 2 \\ 2x^2 + y^2 = 3 \end{cases}$

Substitute $x + 2$ for y in the second equation.

$$2x^2 + y^2 = 3$$
$$2x^2 + (x + 2)^2 = 3$$
$$2x^2 + x^2 + 4x + 4 = 3$$
$$3x^2 + 4x + 1 = 0$$
$$(3x + 1)(x + 1) = 0$$
$$x = -\frac{1}{3}, x = -1$$

If $x = -\frac{1}{3}$, $y = x + 2 = -\frac{1}{3} + 2 = \frac{5}{3}$.

If $x = -1$, $y = x + 2 = -1 + 2 = 1$.

The solutions are $\left(-\frac{1}{3}, \frac{5}{3}\right)$ and $(-1, 1)$.

Section 13.4 Nonlinear Inequalities and Systems of Inequalities

The graph of a system of inequalities is the intersection of the graphs of the inequalities.

Graph the system $\begin{cases} x \geq y^2 \\ x + y \leq 4 \end{cases}$

The graph of the system is the pink shaded region along with its boundary lines.

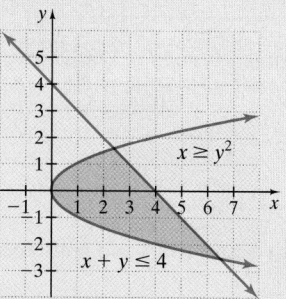

CHAPTER REVIEW

(13.1) *Find the distance between each pair of points. For Exercises 7 and 8, round the distance to two decimal places.*

1. $(-6, 3)$ and $(8, 4)$

2. $(3, 5)$ and $(8, 9)$

3. $(-4, -6)$ and $(-1, 5)$

4. $(-1, 5)$ and $(2, -3)$

5. $\left(-\sqrt{2}, 0\right)$ and $\left(0, -4\sqrt{6}\right)$

6. $\left(-\sqrt{5}, -\sqrt{11}\right)$ and $\left(-\sqrt{5}, -3\sqrt{11}\right)$

7. $(7.4, -8.6)$ and $(-1.2, 5.6)$

8. $(2.3, 1.8)$ and $(10.7, -9.2)$

Find the midpoint of the line segment whose endpoints are given.

9. $(2, 6)$ and $(-12, 4)$ **10.** $(-3, 8)$ and $(11, 24)$

11. $(-6, -5)$ and $(-9, 7)$ **12.** $(4, -6)$ and $(-15, 2)$

13. $\left(0, -\dfrac{3}{8}\right)$ and $\left(\dfrac{1}{10}, 0\right)$ **14.** $\left(\dfrac{3}{4}, -\dfrac{1}{7}\right)$ and $\left(-\dfrac{1}{4}, -\dfrac{3}{7}\right)$

15. $\left(\sqrt{3}, -2\sqrt{6}\right)$ and $\left(\sqrt{3}, -4\sqrt{6}\right)$

16. $\left(-5\sqrt{3}, 2\sqrt{7}\right)$ and $\left(-3\sqrt{3}, 10\sqrt{7}\right)$

Write an equation of the circle with the given center and radius.

17. center $(-4, 4)$ radius 3

18. center $(5, 0)$, radius 5

19. center $(-7, -9)$, radius $\sqrt{11}$

20. center $(0, 0)$, radius $\dfrac{7}{2}$

Sketch the graph of the equation. If the graph is a circle, find its center. If the graph is a parabola, find its vertex.

21. $x^2 + y^2 = 7$ **22.** $x = 2(y - 5)^2 + 4$

23. $x = -(y + 2)^2 + 3$ **24.** $(x - 1)^2 + (y - 2)^2 = 4$

25. $y = -x^2 + 4x + 10$ **26.** $x = -y^2 - 4y + 6$

27. $x = \dfrac{1}{2}y^2 + 2y + 1$ **28.** $y = -3x^2 + \dfrac{1}{2}x + 4$

29. $x^2 + y^2 + 2x + y = \dfrac{3}{4}$ **30.** $x^2 + y^2 + 3y = \dfrac{7}{4}$

31. $4x^2 + 4y^2 + 16x + 8y = 1$ **32.** $3x^2 + 6x + 3y^2 = 9$

33. $y = x^2 + 6x + 9$ **34.** $x = y^2 + 6y + 9$

35. Write an equation of the circle centered at $(5.6, -2.4)$ with diameter 6.2.

(13.2) Sketch the graph of each equation.

36. $x^2 + \dfrac{y^2}{4} = 1$ **37.** $x^2 - \dfrac{y^2}{4} = 1$

38. $\dfrac{y^2}{4} - \dfrac{x^2}{16} = 1$ **39.** $\dfrac{y^2}{4} + \dfrac{x^2}{16} = 1$

40. $\dfrac{x^2}{5} + \dfrac{y^2}{5} = 1$ **41.** $\dfrac{x^2}{5} - \dfrac{y^2}{5} = 1$

42. $-5x^2 + 25y^2 = 125$ **43.** $4y^2 + 9x^2 = 36$

44. $\dfrac{(x - 2)^2}{4} + (y - 1)^2 = 1$ **45.** $\dfrac{(x + 3)^2}{9} + \dfrac{(y - 4)^2}{25} = 1$

46. $x^2 - y^2 = 1$ **47.** $36y^2 - 49x^2 = 1764$

48. $y^2 = x^2 + 9$ **49.** $x^2 = 4y^2 - 16$

50. $100 - 25x^2 = 4y^2$

Sketch the graph of each equation. Identify whether each equation, when graphed, will be a parabola, circle, ellipse, or hyperbola.

51. $y = x^2 + 4x + 6$ **52.** $y^2 = x^2 + 6$

53. $y^2 + x^2 = 4x + 6$ **54.** $y^2 + 2x^2 = 4x + 6$

55. $x^2 + y^2 - 8y = 0$ **56.** $x - 4y = y^2$

57. $x^2 - 4 = y^2$ **58.** $x^2 = 4 - y^2$

59. $6(x - 2)^2 + 9(y + 5)^2 = 36$

60. $36y^2 = 576 + 16x^2$ **61.** $\dfrac{x^2}{16} - \dfrac{y^2}{25} = 1$

62. $3(x - 7)^2 + 3(y + 4)^2 = 1$

Use a graphing calculator to verify the results of each exercise.

63. Exercise 39.   **64.** Exercise 40.

65. Exercise 51.   **66.** Exercise 58.

(13.3) Solve each system of equations.

67. $\begin{cases} y = 2x - 4 \\ y^2 = 4x \end{cases}$ **68.** $\begin{cases} x^2 + y^2 = 4 \\ x - y = 4 \end{cases}$

69. $\begin{cases} y = x + 2 \\ y = x^2 \end{cases}$ **70.** $\begin{cases} y = x^2 - 5x + 1 \\ y = -x + 6 \end{cases}$

71. $\begin{cases} 4x - y^2 = 0 \\ 2x^2 + y^2 = 16 \end{cases}$ **72.** $\begin{cases} x^2 + 4y^2 = 16 \\ x^2 + y^2 = 4 \end{cases}$

73. $\begin{cases} x^2 + y^2 = 10 \\ 9x^2 + y^2 = 18 \end{cases}$ **74.** $\begin{cases} x^2 + 2y = 9 \\ 5x - 2y = 5 \end{cases}$

75. $\begin{cases} y = 3x^2 + 5x - 4 \\ y = 3x^2 - x + 2 \end{cases}$ **76.** $\begin{cases} x^2 - 3y^2 = 1 \\ 4x^2 + 5y^2 = 21 \end{cases}$

△ **77.** Find the length and the width of a room whose area is 150 square feet and whose perimeter is 50 feet.

78. What is the greatest number of real solutions possible for a system of two equations whose graphs are an ellipse and a hyperbola?

(13.4) Graph the inequality or system of inequalities.

79. $y \le -x^2 + 3$ **80.** $x^2 + y^2 < 9$

81. $x^2 - y^2 < 1$ **82.** $\dfrac{x^2}{4} + \dfrac{y^2}{9} \ge 1$

83. $\begin{cases} 2x \le 4 \\ x + y \ge 1 \end{cases}$ **84.** $\begin{cases} 3x + 4y \le 12 \\ x - 2y > 6 \end{cases}$

85. $\begin{cases} y > x^2 \\ x + y \ge 3 \end{cases}$ **86.** $\begin{cases} x^2 + y^2 \le 16 \\ x^2 + y^2 \ge 4 \end{cases}$

87. $\begin{cases} x^2 + y^2 < 4 \\ x^2 - y^2 \le 1 \end{cases}$ **88.** $\begin{cases} x^2 + y^2 < 4 \\ y \ge x^2 - 1 \\ x \ge 0 \end{cases}$

CHAPTER 13 TEST

Remember to use the Chapter Test Prep Video CD to help you study and view solutions to the test questions you need help with.

1. Find the distance between the points $(-6, 3)$ and $(-8, -7)$.

2. Find the distance between the points $(-2\sqrt{5}, \sqrt{10})$ and $(-\sqrt{5}, 4\sqrt{10})$.

3. Find the midpoint of the line segment whose endpoints are $(-2, -5)$ and $(-6, 12)$.

Sketch the graph of each equation.

4. $x^2 + y^2 = 36$

5. $x^2 - y^2 = 36$

6. $16x^2 + 9y^2 = 144$

7. $y = x^2 - 8x + 16$

8. $x^2 + y^2 + 6x = 16$

9. $x = y^2 + 8y - 3$

10. $\dfrac{(x-4)^2}{16} + \dfrac{(y-3)^2}{9} = 1$

11. $y^2 - x^2 = 1$

Solve each system.

12. $\begin{cases} x^2 + y^2 = 26 \\ x^2 - 2y^2 = 23 \end{cases}$

13. $\begin{cases} y = x^2 - 5x + 6 \\ y = 2x \end{cases}$

Graph the solution of each system.

14. $\begin{cases} 2x + 5y \geq 10 \\ y \geq x^2 + 1 \end{cases}$

15. $\begin{cases} \dfrac{x^2}{4} + y^2 \leq 1 \\ x + y > 1 \end{cases}$

16. $\begin{cases} x^2 + y^2 \geq 4 \\ x^2 + y^2 < 16 \\ y \geq 0 \end{cases}$

17. A bridge has an arch in the shape of a half-ellipse. If the equation of the ellipse, measured in feet, is $100x^2 + 225y^2 = 22{,}500$, find the height of the arch from the road and the width of the arch.

CHAPTER CUMULATIVE REVIEW

1. Solve $2x \geq 0$ and $4x - 1 \leq -9$.

2. Solve $3x + 4 > 1$ and $2x - 5 \leq 9$. Write the solution in interval notation.

3. Solve $5x - 3 \leq 10$ or $x + 1 \geq 5$.

4. Find the slope of the line that goes through $(3, 2)$ and $(1, -4)$.

Solve.

5. $|5w + 3| = 7$

6. Two planes leave Greensboro, one traveling north and the other south. After 2 hours they are 650 miles apart. If one plane is flying 25 mph faster than the other, what is the speed of each?

7. $\left|\dfrac{x}{2} - 1\right| = 11$

8. Use the quotient rule to simplify.

   a. $\dfrac{4^8}{4^3}$

   b. $\dfrac{y^{11}}{y^5}$

   c. $\dfrac{32x^7}{4x^6}$

   d. $\dfrac{18a^{12}b^6}{12a^8b^6}$

9. Solve $|3x + 2| = |5x - 8|$.

10. Factor.

   a. $3y^2 + 14y + 15$

   b. $20a^5 + 54a^4 + 10a^3$

   c. $(y - 3)^2 - 2(y - 3) - 8$

11. Solve $|m - 6| < 2$.

12. Perform the indicated operation and simplify if possible.

$$\dfrac{2}{3a - 15} - \dfrac{a}{25 - a^2}$$

13. Simplify $\dfrac{x^{-1} + 2xy^{-1}}{x^{-2} - x^{-2}y^{-1}}$.

14. Simplify each complex fraction.

   a. $(a^{-1} - b^{-1})^{-1}$

   b. $\dfrac{2 - \dfrac{1}{x}}{4x - \dfrac{1}{x}}$

15. Solve $|2x + 9| + 5 > 3$.

16. Solve $\dfrac{2}{x + 3} = \dfrac{1}{x^2 - 9} - \dfrac{1}{x - 3}$

17. Use the remainder theorem and synthetic division to find $P(4)$ if
$P(x) = 4x^6 - 25x^5 + 35x^4 + 17x^2$.

18. Suppose that y varies inversely as x. If $y = 3$ when $x = \frac{2}{3}$, find the constant of variation and the direct variation equation.

19. Find the cube roots.

  a. $\sqrt[3]{1}$ **b.** $\sqrt[3]{-64}$

  c. $\sqrt[3]{\dfrac{8}{125}}$ **d.** $\sqrt[3]{x^6}$

  e. $\sqrt[3]{-27x^9}$

20. Multiply and simplify if possible.

  a. $\sqrt{5}\left(2 + \sqrt{15}\right)$

  b. $\left(\sqrt{3} - \sqrt{5}\right)\left(\sqrt{7} - 1\right)$

  c. $\left(2\sqrt{5} - 1\right)^2$

  d. $\left(3\sqrt{2} + 5\right)\left(3\sqrt{2} - 5\right)$

21. Multiply.

  a. $z^{2/3}\left(z^{1/3} - z^5\right)$ **b.** $\left(x^{1/3} - 5\right)\left(x^{1/3} + 2\right)$

22. Rationalize the denominator $\dfrac{-2}{\sqrt{3} + 3}$.

23. Use the quotient rule to divide, and simplify if possible.

  a. $\dfrac{\sqrt{20}}{\sqrt{5}}$ **b.** $\dfrac{\sqrt{50x}}{2\sqrt{2}}$

  c. $\dfrac{7\sqrt[3]{48x^4y^8}}{\sqrt[3]{6y^2}}$ **d.** $\dfrac{2\sqrt[4]{32a^8b^6}}{\sqrt[4]{a^{-1}b^2}}$

24. Solve $\sqrt{2x - 3} = x - 3$.

25. Add or subtract as indicated.

  a. $\dfrac{\sqrt{45}}{4} - \dfrac{\sqrt{5}}{3}$ **b.** $\sqrt[3]{\dfrac{7x}{8}} + 2\sqrt[3]{7x}$

26. Use the discriminant to determine the number and type of solutions for $9x^2 - 6x = -4$.

27. Rationalize the denominator of $\sqrt{\dfrac{7x}{3y}}$.

28. Solve $\dfrac{4}{x - 2} - \dfrac{x}{x + 2} = \dfrac{16}{x^2 - 4}$

29. Solve $\sqrt{2x - 3} = 9$.

30. Solve $x^3 + 2x^2 - 4x \geq 8$

31. Find the following powers of i.

  a. i^7 **b.** i^{20}

  c. i^{46} **d.** i^{-12}

32. Graph $f(x) = (x + 2)^2 - 1$

33. Solve $p^2 + 2p = 4$ by completing the square.

34. Find the maximum value of $f(x) = -x^2 - 6x + 4$.

35. Solve $\dfrac{1}{4}m^2 - m + \dfrac{1}{2} = 0$.

36. Find the inverse of $f(x) = \dfrac{x + 1}{2}$.

37. Solve $p^4 - 3p^2 - 4 = 0$.

38. Use the quotient rule to simplify.

  a. $\dfrac{\sqrt{32}}{\sqrt{4}}$ **b.** $\dfrac{\sqrt[3]{240y^2}}{5\sqrt[3]{3y^{-4}}}$

  c. $\dfrac{\sqrt[5]{64x^9y^2}}{\sqrt[5]{2x^2y^{-8}}}$

39. Solve $\dfrac{x + 2}{x - 3} \leq 0$.

40. Graph $4x^2 + 9y^2 = 36$.

41. Graph $g(x) = \dfrac{1}{2}(x + 2)^2 + 5$. Find the vertex and the axis of symmetry.

42. Solve each equation for x.

  a. $64^x = 4$ **b.** $125^{x-3} = 25$

  c. $\dfrac{1}{81} = 3^{2x}$

43. Find the vertex of the graph of $f(x) = x^2 - 4x - 12$.

44. Graph the system: $\begin{cases} x + 2y < 8 \\ \qquad y \geq x^2 \end{cases}$

45. Find the distance between $(2, -5)$ and $(1, -4)$. Give an exact distance and a three-decimal-place approximation.

46. Solve the system $\begin{cases} x^2 + y^2 = 36 \\ \qquad y = x + 6 \end{cases}$

CHAPTER 14

Sequences, Series, and the Binomial Theorem

Having explored in some depth the concept of function, we turn now in this final chapter to *sequences*. In one sense, a sequence is simply an ordered list of numbers. In another sense, a sequence is itself a function. Phenomena modeled by such functions are everywhere around us. The starting place for all mathematics is the sequence of natural numbers: 1, 2, 3, 4, and so on.

Sequences lead us to *series*, which are a sum of ordered numbers. Through series we gain new insight, for example about the expansion of a binomial $(a + b)^n$, the concluding topic of this book.

As Baby Boomers approach retirement age and life spans lengthen overall, more and more people will seek professional assistance with financial planners.

Financial planners work for investment firms, accounting firms, insurance companies, banks, credit counseling organizations, law firms, or in private practice. They give advice on retirement planning, insurance needs, investment options, estate planning, tax strategies, and employee benefits. Certified financial planners use math and problem-solving skills in such tasks as analyzing clients' current cash flow, estimating cash needs for future goals, and calculating investment returns.

In the Spotlight on Decision Making feature on page 823, you will have the opportunity to make a decision about reaching a client's retirement goals as a certified financial planner.

Source: Certified Financial Planner Board of Standards website

14.1 SEQUENCES

Objectives

1   Write the terms of a sequence given its general term.
2   Find the general term of a sequence.
3   Solve applications that involve sequences.

Suppose that a town's present population of 100,000 is growing by 5% each year. After the first year, the town's population will be

$$100{,}000 + 0.05(100{,}000) = 105{,}000$$

After the second year, the town's population will be

$$105{,}000 + 0.05(105{,}000) = 110{,}250$$

After the third year, the town's population will be

$$110{,}250 + 0.05(110{,}250) \approx 115{,}763$$

If we continue to calculate, the town's yearly population can be written as the **infinite sequence** of numbers

$$105{,}000, 110{,}250, 115{,}763, \ldots$$

If we decide to stop calculating after a certain year (say, the fourth year), we obtain the **finite sequence**

$$105{,}000, 110{,}250, 115{,}763, 121{,}551$$

Sequences

An infinite sequence is a function whose domain is the set of natural numbers $\{1, 2, 3, 4, \ldots\}$.
    A finite sequence is a function whose domain is the set of natural numbers $\{1, 2, 3, 4, \ldots, n\}$, where n is some natural number.

---

1   Given the sequence $2, 4, 8, 16, \ldots$, we say that each number is a **term** of the sequence. Because a sequence is a function, we could describe it by writing $f(n) = 2^n$, where n is a natural number. Instead, we use the notation

$$a_n = 2^n$$

Some function values are

$$a_1 = 2^1 = 2 \qquad \text{First term of the sequence}$$
$$a_2 = 2^2 = 4 \qquad \text{Second term}$$
$$a_3 = 2^3 = 8 \qquad \text{Third term}$$
$$a_4 = 2^4 = 16 \qquad \text{Fourth term}$$
$$a_{10} = 2^{10} = 1024 \qquad \text{Tenth term}$$

The nth term of the sequence a_n is called the **general term.**

> ▶ **Helpful Hint**
>
> If it helps, think of a sequence as simply a list of values in which a position is assigned. For the first sequence in this section,
>
> Value: 2, 4, 8, 16, ... , 1024
>   ↑   ↑   ↑ ↑   ↑
> Position   1st 2nd 3rd 4th   10th

EXAMPLE 1

Write the first five terms of the sequence whose general term is given by

$$a_n = n^2 - 1$$

Solution   Evaluate a_n, where n is 1, 2, 3, 4, and 5.

$$a_n = n^2 - 1$$
$$a_1 = 1^2 - 1 = 0 \qquad \text{Replace } n \text{ with } \mathbf{1.}$$
$$a_2 = 2^2 - 1 = 3 \qquad \text{Replace } n \text{ with } \mathbf{2.}$$
$$a_3 = 3^2 - 1 = 8 \qquad \text{Replace } n \text{ with } \mathbf{3.}$$
$$a_4 = 4^2 - 1 = 15 \qquad \text{Replace } n \text{ with } \mathbf{4.}$$
$$a_5 = 5^2 - 1 = 24 \qquad \text{Replace } n \text{ with } \mathbf{5.}$$

Thus, the first five terms of the sequence $a_n = n^2 - 1$ are 0, 3, 8, 15, and 24.

EXAMPLE 2

If the general term of a sequence is given by $a_n = \dfrac{(-1)^n}{3n}$, find

a. the first term of the sequence

b. a_8

c. the one-hundredth term of the sequence

d. a_{15}

Solution   **a.** $a_1 = \dfrac{(-1)^1}{3(1)} = -\dfrac{1}{3} \qquad$ Replace n with **1.**

b. $a_8 = \dfrac{(-1)^8}{3(8)} = \dfrac{1}{24} \qquad$ Replace n with **8.**

c. $a_{100} = \dfrac{(-1)^{100}}{3(100)} = \dfrac{1}{300} \qquad$ Replace n with **100.**

d. $a_{15} = \dfrac{(-1)^{15}}{3(15)} = -\dfrac{1}{45} \qquad$ Replace n with **15.**

2   Suppose we know the first few terms of a sequence and want to find a general term that fits the pattern of the first few terms.

EXAMPLE 3

Find a general term a_n of the sequence whose first few terms are given.

a. $1, 4, 9, 16, \ldots$

b. $\dfrac{1}{1}, \dfrac{1}{2}, \dfrac{1}{3}, \dfrac{1}{4}, \dfrac{1}{5}, \ldots$

c. $-3, -6, -9, -12, \ldots$

d. $\dfrac{1}{2}, \dfrac{1}{4}, \dfrac{1}{8}, \dfrac{1}{16}, \ldots$

Solution   **a.** These numbers are the squares of the first four natural numbers, so a general term might be $a_n = n^2$.

b. These numbers are the reciprocals of the first five natural numbers, so a general term might be $a_n = \dfrac{1}{n}$.

c. These numbers are the product of -3 and the first four natural numbers, so a general term might be $a_n = -3n$.

d. Notice that the denominators double each time.

$$\frac{1}{2}, \quad \frac{1}{2 \cdot 2}, \quad \frac{1}{2(2 \cdot 2)}, \quad \frac{1}{2(2 \cdot 2 \cdot 2)}$$

or

$$\frac{1}{2^1}, \quad \frac{1}{2^2}, \quad \frac{1}{2^3}, \quad \frac{1}{2^4}$$

We might then suppose that the general term is $a_n = \dfrac{1}{2^n}$.

3   Sequences model many phenomena of the physical world, as illustrated by the following example.

EXAMPLE 4

FINDING A PUPPY'S WEIGHT GAIN

The amount of weight, in pounds, a puppy gains in each month of its first year is modeled by a sequence whose general term is $a_n = n + 4$, where n is the number of the month. Write the first five terms of the sequence, and find how much weight the puppy should gain in its fifth month.

Solution   Evaluate $a_n = n + 4$ when n is $1, 2, 3, 4,$ and 5.

$$a_1 = 1 + 4 = 5$$
$$a_2 = 2 + 4 = 6$$
$$a_3 = 3 + 4 = 7$$
$$a_4 = 4 + 4 = 8$$
$$a_5 = 5 + 4 = 9$$

The puppy should gain 9 pounds in its fifth month.

Spotlight on

DECISION

✐ MAKING

Suppose you are considering two job offers. The first job offer pays $11.50 per hour and guarantees a $0.65-per-hour raise each year. The second job offer pays $10.75 per hour and guarantees a $1.10-per-hour raise each year. If one of your goals is to be earning at least $15 per hour in 5 years, which job offer would you accept? Explain. What other factors would you want to consider?

EXERCISE SET 14.1

| STUDY GUIDE/SSM | CD/ VIDEO | PH MATH TUTOR CENTER | MathXL®Tutorials ON CD | MathXL® | MyMathLab® |

Write the first five terms of each sequence whose general term is given. See Example 1.

1. $a_n = n + 4$

2. $a_n = 5 - n$

3. $a_n = (-1)^n$

4. $a_n = (-2)^n$

5. $a_n = \dfrac{1}{n + 3}$

6. $a_n = \dfrac{1}{7 - n}$

7. $a_n = 2n$

8. $a_n = -6n$

9. $a_n = -n^2$

10. $a_n = n^2 + 2$

11. $a_n = 2^n$

12. $a_n = 3^{n-2}$

13. $a_n = 2n + 5$

14. $a_n = 1 - 3n$

15. $a_n = (-1)^n n^2$

16. $a_n = (-1)^{n+1}(n - 1)$

Find the indicated term for each sequence whose general term is given. See Example 2.

17. $a_n = 3n^2; a_5$

18. $a_n = -n^2; a_{15}$

19. $a_n = 6n - 2; a_{20}$

20. $a_n = 100 - 7n; a_{50}$

21. $a_n = \dfrac{n + 3}{n}; a_{15}$

22. $a_n = \dfrac{n}{n + 4}; a_{24}$

23. $a_n = (-3)^n; a_6$

24. $a_n = 5^{n+1}; a_3$

25. $a_n = \dfrac{n - 2}{n + 1}; a_6$

26. $a_n = \dfrac{n + 3}{n + 4}; a_8$

27. $a_n = \dfrac{(-1)^n}{n}; a_8$

28. $a_n = \dfrac{(-1)^n}{2n}; a_{100}$

29. $a_n = -n^2 + 5; a_{10}$

30. $a_n = 8 - n^2; a_{20}$

31. $a_n = \dfrac{(-1)^n}{n + 6}; a_{19}$

32. $a_n = \dfrac{n - 4}{(-2)^n}; a_6$

Find a general term a_n for each sequence whose first four terms are given. See Example 3.

33. $3, 7, 11, 15$

34. $2, 7, 12, 17$

35. $-2, -4, -8, -16$

36. $-4, 16, -64, 256$

37. $\dfrac{1}{3}, \dfrac{1}{9}, \dfrac{1}{27}, \dfrac{1}{81}$

38. $\dfrac{2}{5}, \dfrac{2}{25}, \dfrac{2}{125}, \dfrac{2}{625}$

Solve. See Example 4.

39. The distance, in feet, that a Thermos dropped from a cliff falls in each consecutive second is modeled by a sequence whose general term is $a_n = 32n - 16$, where n is the number of seconds. Find the distance the Thermos falls in the second, third, and fourth seconds.

40. The population size of a culture of bacteria triples every hour such that its size is modeled by the sequence $a_n = 50(3)^{n-1}$, where n is the number of the hour just beginning. Find the size of the culture at the beginning of the fourth hour and the size of the culture at the beginning of the first hour.

41. Mrs. Laser agrees to give her son Mark an allowance of $0.10 on the first day of his 14-day vacation, $0.20 on the second day, $0.40 on the third day, and so on. Write an equation of a sequence whose terms correspond to Mark's allowance. Find the allowance Mark will receive on the last day of his vacation.

42. A small theater has 10 rows with 12 seats in the first row, 15 seats in the second row, 18 seats in the third row, and so on. Write an equation of a sequence whose terms correspond to the seats in each row. Find the number of seats in the eighth row.

43. The number of cases of a new infectious disease is doubling every year such that the number of cases is modeled by a sequence whose general term is $a_n = 75(2)^{n-1}$, where n is the number of the year just beginning. Find how many cases there will be at the beginning of the sixth year. Find how many cases there were at the beginning of the first year.

44. A new college had an initial enrollment of 2700 students in 2000, and each year the enrollment increases by 150 students. Find the enrollment for each of 5 years, beginning with 2000.

45. An endangered species of sparrow had an estimated population of 800 in 2000, and scientists predict that its population will decrease by half each year. Estimate the population in 2004. Estimate the year the sparrow will be extinct.

46. A **Fibonacci sequence** is a special type of sequence in which the first two terms are 1 and each term thereafter is the sum of the two previous terms: $1, 1, 2, 3, 5, 8, \ldots$. Many plants and animals seem to grow according to a Fibonacci sequence, including pine cones, pineapple scales, nautilus shells, and certain flowers. Write the first 15 terms of the Fibonacci sequence.

REVIEW AND PREVIEW

Sketch the graph of each quadratic function. See Section 11.5.

47. $f(x) = (x - 1)^2 + 3$

48. $f(x) = (x - 2)^2 + 1$

49. $f(x) = 2(x + 4)^2 + 2$

50. $f(x) = 3(x - 3)^2 + 4$

Find the distance between each pair of points. See Section 13.1.

51. $(-4, -1)$ and $(-7, -3)$

52. $(-2, -1)$ and $(-1, 5)$

53. $(2, -7)$ and $(-3, -3)$

54. $(10, -14)$ and $(5, -11)$

Concept Extensions

Find the first five terms of each sequence. Round each term after the first to four decimal places.

55. $a_n = \dfrac{1}{\sqrt{n}}$

56. $\dfrac{\sqrt{n}}{\sqrt{n} + 1}$

57. $a_n = \left(1 + \dfrac{1}{n}\right)^n$

58. $a_n = \left(1 + \dfrac{0.05}{n}\right)^n$

14.2 ARITHMETIC AND GEOMETRIC SEQUENCES

Objectives

1 Identify arithmetic sequences and their common differences.

2 Identify geometric sequences and their common ratios.

1 Find the first four terms of the sequence whose general term is $a_n = 5 + (n - 1)3$.

$$a_1 = 5 + (1 - 1)3 = 5 \qquad \text{Replace } n \text{ with } 1.$$
$$a_2 = 5 + (2 - 1)3 = 8 \qquad \text{Replace } n \text{ with } 2.$$
$$a_3 = 5 + (3 - 1)3 = 11 \qquad \text{Replace } n \text{ with } 3.$$
$$a_4 = 5 + (4 - 1)3 = 14 \qquad \text{Replace } n \text{ with } 4.$$

The first four terms are $5, 8, 11,$ and 14. Notice that the difference of any two successive terms is 3.

$$8 - 5 = 3$$
$$11 - 8 = 3$$
$$14 - 11 = 3$$
$$\vdots$$
$$a_n - a_{n-1} = 3$$

nth term   previous term

Because the difference of any two successive terms is a constant, we call the sequence an **arithmetic sequence,** or an **arithmetic progression.** The constant difference d in successive terms is called the **common difference.** In this example, d is 3.

> ## Arithmetic Sequence and Common Difference
>
> An **arithmetic sequence** is a sequence in which each term (after the first) differs from the preceding term by a constant amount d. The constant d is called the **common difference** of the sequence.

The sequence $2, 6, 10, 14, 18, \ldots$ is an arithmetic sequence. Its common difference is 4. Given the first term a_1 and the common difference d of an arithmetic sequence, we can find any term of the sequence.

EXAMPLE 1

Write the first five terms of the arithmetic sequence whose first term is 7 and whose common difference is 2.

Solution

$$a_1 = 7$$
$$a_2 = 7 + 2 = 9$$
$$a_3 = 9 + 2 = 11$$
$$a_4 = 11 + 2 = 13$$
$$a_5 = 13 + 2 = 15$$

The first five terms are $7, 9, 11, 13, 15$.

Notice the general pattern of the terms in Example 1.

$$a_1 = 7$$
$$a_2 = 7 + 2 = 9 \qquad \text{or} \quad a_2 = a_1 + d$$
$$a_3 = 9 + 2 = 11 \qquad \text{or} \quad a_3 = a_2 + d = (a_1 + d) + d = a_1 + 2d$$
$$a_4 = 11 + 2 = 13 \quad \text{or} \quad a_4 = a_3 + d = (a_1 + 2d) + d = a_1 + 3d$$
$$a_5 = 13 + 2 = 15 \quad \text{or} \quad a_5 = a_4 + d = (a_1 + 3d) + d = a_1 + 4d$$

$\longrightarrow$ (subscript $-$ **1**) is multiplier $\longrightarrow$

The pattern on the right suggests that the general term a_n of an arithmetic sequence is given by

$$a_n = a_1 + (n - 1)d$$

> ## General Term of an Arithmetic Sequence
>
> The general term a_n of an arithmetic sequence is given by
> $$a_n = a_1 + (n - 1)d$$
> where a_1 is the first term and d is the common difference.

EXAMPLE 2

Consider the arithmetic sequence whose first term is 3 and common difference is -5.

a. Write an expression for the general term a_n.

b. Find the twentieth term of this sequence.

Solution   **a.** Since this is an arithmetic sequence, the general term a_n is given by $a_n = a_1 + (n-1)d$. Here, $a_1 = 3$ and $d = -5$, so

$$a_n = 3 + (n-1)(-5) \qquad \text{Let } a_1 = 3 \text{ and } d = -5.$$
$$= 3 - 5n + 5 \qquad \text{Multiply.}$$
$$= 8 - 5n \qquad \text{Simplify.}$$

b. $a_n = 8 - 5n$
$$a_{20} = 8 - 5 \cdot 20 \qquad \text{Let } n = 20.$$
$$= 8 - 100 = -92$$

EXAMPLE 3

Find the eleventh term of the arithmetic sequence whose first three terms are $2, 9$, and 16.

Solution   Since the sequence is arithmetic, the eleventh term is

$$a_{11} = a_1 + (11 - 1)d = a_1 + 10d$$

We know a_1 is the first term of the sequence, so $a_1 = 2$. Also, d is the constant difference of terms, so $d = a_2 - a_1 = 9 - 2 = 7$. Thus,

$$a_{11} = a_1 + 10d$$
$$= 2 + 10 \cdot 7 \qquad \text{Let } a_1 = 2 \text{ and } d = 7.$$
$$= 72$$

EXAMPLE 4

If the third term of an arithmetic sequence is 12 and the eighth term is 27, find the fifth term.

Solution   We need to find a_1 and d to write the general term, which then enables us to find a_5, the fifth term. The given facts about terms a_3 and a_8 lead to a system of linear equations.

$$\begin{cases} a_3 = a_1 + (3-1)d \\ a_8 = a_1 + (8-1)d \end{cases} \quad \text{or} \quad \begin{cases} 12 = a_1 + 2d \\ 27 = a_1 + 7d \end{cases}$$

Next, we solve the system $\begin{cases} 12 = a_1 + 2d \\ 27 = a_1 + 7d \end{cases}$ by elimination. Multiply both sides of the second equation by -1 so that

$$\begin{cases} 12 = a_1 + 2d \\ -1(27) = -1(a_1 + 7d) \end{cases} \quad \begin{array}{c} \text{simplifies} \\ \text{to} \end{array} \quad \begin{cases} 12 = a_1 + 2d \\ \underline{-27 = -a_1 - 7d} \end{cases}$$
$$-15 = -5d \qquad \text{Add the equations.}$$
$$3 = d \qquad \text{Divide both sides by } -5.$$

To find a_1, let $d = 3$ in $12 = a_1 + 2d$. Then

$$12 = a_1 + 2(3)$$
$$12 = a_1 + 6$$
$$6 = a_1$$

Thus, $a_1 = 6$ and $d = 3$, so

$$a_n = 6 + (n-1)(3)$$
$$= 6 + 3n - 3$$
$$= 3 + 3n$$

and

$$a_5 = 3 + 3 \cdot 5 = 18$$

EXAMPLE 5

FINDING SALARY

Donna Theime has an offer for a job starting at $40,000 per year and guaranteeing her a raise of $1600 per year for the next 5 years. Write the general term for the arithmetic sequence that models Donna's potential annual salaries, and find her salary for the fourth year.

Solution The first term, a_1, is 40,000, and d is 1600. So

$$a_n = 40,000 + (n-1)(1600) = 38,400 + 1600n$$
$$a_4 = 38,400 + 1600 \cdot 4 = 44,800$$

Her salary for the fourth year will be $44,800.

2 We now investigate a **geometric sequence,** also called a **geometric progression.** In the sequence $5, 15, 45, 135, \ldots$, each term after the first is the *product* of 3 and the preceding term. This pattern of multiplying by a constant to get the next term defines a geometric sequence. The constant is called the **common ratio** because it is the ratio of any term (after the first) to its preceding term.

$$\frac{15}{5} = 3$$

$$\frac{45}{15} = 3$$

$$\frac{135}{45} = 3$$

$$\vdots$$

$$\begin{array}{l} n\text{th term} \longrightarrow \\ \text{previous term} \longrightarrow \end{array} \frac{a_n}{a_{n-1}} = 3$$

Geometric Sequence and Common Ratio

A **geometric sequence** is a sequence in which each term (after the first) is obtained by multiplying the preceding term by a constant r. The constant r is called the **common ratio** of the sequence.

The sequence $12, 6, 3, \dfrac{3}{2}, \ldots$ is geometric since each term after the first is the product of the previous term and $\dfrac{1}{2}$.

EXAMPLE 6

Write the first five terms of a geometric sequence whose first term is 7 and whose common ratio is 2.

Solution
$$a_1 = 7$$
$$a_2 = 7(2) = 14$$
$$a_3 = 14(2) = 28$$
$$a_4 = 28(2) = 56$$
$$a_5 = 56(2) = 112$$

The first five terms are $7, 14, 28, 56,$ and 112.

Notice the general pattern of the terms in Example 6.

$$a_1 = 7$$

$$a_2 = 7(2) = 14 \quad \text{or} \quad a_2 = a_1(r)$$

$$a_3 = 14(2) = 28 \quad \text{or} \quad a_3 = a_2(r) = (a_1 \cdot r) \cdot r = a_1 r^2$$

$$a_4 = 28(2) = 56 \quad \text{or} \quad a_4 = a_3(r) = (a_1 \cdot r^2) \cdot r = a_1 r^3$$

$$a_5 = 56(2) = 112 \quad \text{or} \quad a_5 = a_4(r) = (a_1 \cdot r^3) \cdot r = a_1 r^4$$

$$\text{(subscript} -1) \text{ is power}$$

The pattern on the right above suggests that the general term of a geometric sequence is given by $a_n = a_1 r^{n-1}$.

General Term of a Geometric Sequence

The general term a_n of a geometric sequence is given by

$$a_n = a_1 r^{n-1}$$

where a_1 is the first term and r is the common ratio.

EXAMPLE 7

Find the eighth term of the geometric sequence whose first term is 12 and whose common ratio is $\dfrac{1}{2}$.

Solution   Since this is a geometric sequence, the general term a_n is given by

$$a_n = a_1 r^{n-1}$$

Here $a_1 = 12$ and $r = \dfrac{1}{2}$, so $a_n = 12\left(\dfrac{1}{2}\right)^{n-1}$. Evaluate a_n for $n = 8$.

$$a_8 = 12\left(\frac{1}{2}\right)^{8-1} = 12\left(\frac{1}{2}\right)^7 = 12\left(\frac{1}{128}\right) = \frac{3}{32}$$

EXAMPLE 8

Find the fifth term of the geometric sequence whose first three terms are $2, -6$, and 18.

Solution   Since the sequence is geometric and $a_1 = 2$, the fifth term must be $a_1 r^{5-1}$, or $2r^4$. We know that r is the common ratio of terms, so r must be $\dfrac{-6}{2}$, or -3. Thus,

$$a_5 = 2r^4$$

$$a_5 = 2(-3)^4 = 162$$

EXAMPLE 9

If the second term of a geometric sequence is $\frac{5}{4}$ and the third term is $\frac{5}{16}$, find the first term and the common ratio.

Solution Notice that $\frac{5}{16} \div \frac{5}{4} = \frac{1}{4}$, so $r = \frac{1}{4}$. Then

$$a_2 = a_1\left(\frac{1}{4}\right)^{2-1}$$

$$\frac{5}{4} = a_1\left(\frac{1}{4}\right)^1, \text{ or } a_1 = 5 \qquad \text{Replace } a_2 \text{ with } \frac{5}{4}.$$

The first term is 5.

EXAMPLE 10

PREDICTING POPULATION OF A BACTERIAL CULTURE

The population size of a bacterial culture growing under controlled conditions is doubling each day. Predict how large the culture will be at the beginning of day 7 if it measures 10 units at the beginning of day 1.

Solution Since the culture doubles in size each day, the population sizes are modeled by a geometric sequence. Here $a_1 = 10$ and $r = 2$. Thus,

$$a_n = a_1 r^{n-1} = 10(2)^{n-1} \quad \text{and} \quad a_7 = 10(2)^{7-1} = 640$$

The bacterial culture should measure 640 units at the beginning of day 7.

Spotlight on **DECISION** *✻* **MAKING**

Suppose you are a research biologist studying a particular strain of bacteria that grows at a rate of 1.5 times per hour. For a particular experiment, you will start with a culture of 200 units of bacteria and will allow the culture to grow for 7 hours. Decide whether a culture dish that holds 5000 units will be large enough for this experiment. If not, would a dish that holds 10,000 units be a better choice?

EXERCISE SET 14.2

| STUDY GUIDE/SSM | CD/ VIDEO | PH MATH TUTOR CENTER | MathXL®Tutorials ON CD | MathXL® | MyMathLab® |

Write the first five terms of the arithmetic or geometric sequence whose first term, a_1, and common difference, d, or common ratio, r, are given. See Examples 1 and 6.

1. $a_1 = 4; d = 2$

2. $a_1 = 3; d = 10$

 3. $a_1 = 6; d = -2$

4. $a_1 = -20; d = 3$

5. $a_1 = 1; r = 3$

6. $a_1 = -2; r = 2$

7. $a_1 = 48; r = \frac{1}{2}$

8. $a_1 = 1; r = \frac{1}{3}$

Find the indicated term of each sequence. See Examples 2 and 7.

9. The eighth term of the arithmetic sequence whose first term is 12 and whose common difference is 3

10. The twelfth term of the arithmetic sequence whose first term is 32 and whose common difference is -4

11. The fourth term of the geometric sequence whose first term is 7 and whose common ratio is -5

12. The fifth term of the geometric sequence whose first term is 3 and whose common ratio is 3

13. The fifteenth term of the arithmetic sequence whose first term is -4 and whose common difference is -4

14. The sixth term of the geometric sequence whose first term is 5 and whose common ratio is -4

Find the indicated term of each sequence. See Examples 3 and 8.

15. The ninth term of the arithmetic sequence $0, 12, 24, \ldots$

16. The thirteenth term of the arithmetic sequence $-3, 0, 3, \ldots$

17. The twenty-fifth term of the arithmetic sequence $20, 18, 16, \ldots$

18. The ninth term of the geometric sequence $5, 10, 20, \ldots$

19. The fifth term of the geometric sequence $2, -10, 50, \ldots$

20. The sixth term of the geometric sequence $\dfrac{1}{2}, \dfrac{3}{2}, \dfrac{9}{2}, \ldots$

Find the indicated term of each sequence. See Examples 4 and 9.

21. The eighth term of the arithmetic sequence whose fourth term is 19 and whose fifteenth term is 52

22. If the second term of an arithmetic sequence is 6 and the tenth term is 30, find the twenty-fifth term.

23. If the second term of an arithmetic sequence is -1 and the fourth term is 5, find the ninth term.

24. If the second term of a geometric sequence is 15 and the third term is 3, find a_1 and r.

25. If the second term of a geometric sequence is $-\dfrac{4}{3}$ and the third term is $\dfrac{8}{3}$, find a_1 and r.

26. If the third term of a geometric sequence is 4 and the fourth term is -12, find a_1 and r.

27. Explain why 14, 10, and 6 may be the first three terms of an arithmetic sequence when it appears we are subtracting instead of adding to get the next term.

28. Explain why 80, 20, and 5 may be the first three terms of a geometric sequence when it appears we are dividing instead of multiplying to get the next term.

MIXED PRACTICE

Given are the first three terms of a sequence that is either arithmetic or geometric. If the sequence is arithmetic, find a_1 and d. If a sequence is geometric, find a_1 and r.

29. $2, 4, 6$

30. $8, 16, 24$

31. $5, 10, 20$

32. $2, 6, 18$

33. $\dfrac{1}{2}, \dfrac{1}{10}, \dfrac{1}{50}$

34. $\dfrac{2}{3}, \dfrac{4}{3}, 2$

35. $x, 5x, 25x$

36. $y, -3y, 9y$

37. $p, p + 4, p + 8$

38. $t, t - 1, t - 2$

Find the indicated term of each sequence.

39. The twenty-first term of the arithmetic sequence whose first term is 14 and whose common difference is $\dfrac{1}{4}$

40. The fifth term of the geometric sequence whose first term is 8 and whose common ratio is -3

41. The fourth term of the geometric sequence whose first term is 3 and whose common ratio is $-\dfrac{2}{3}$

42. The fourth term of the arithmetic sequence whose first term is 9 and whose common difference is 5

43. The fifteenth term of the arithmetic sequence $\dfrac{3}{2}, 2, \dfrac{5}{2}, \ldots$

44. The eleventh term of the arithmetic sequence $2, \dfrac{5}{3}, \dfrac{4}{3}, \ldots$

45. The sixth term of the geometric sequence $24, 8, \dfrac{8}{3}, \ldots$

46. The eighteenth term of the arithmetic sequence $5, 2, -1, \ldots$

47. If the third term of an arithmetic sequence is 2 and the seventeenth term is -40, find the tenth term.

48. If the third term of a geometric sequence is -28 and the fourth term is -56, find a_1 and r.

Solve. See Examples 5 and 10.

49. An auditorium has 54 seats in the first row, 58 seats in the second row, 62 seats in the third row, and so on. Find the general term of this arithmetic sequence and the number of seats in the twentieth row.

50. A triangular display of cans in a grocery store has 20 cans in the first row, 17 cans in the next row, and so on, in an arithmetic sequence. Find the general term and the number of cans in the fifth row. Find how many rows there are in the display and how many cans are in the top row.

51. The initial size of a virus culture is 6 units, and it triples its size every day. Find the general term of the geometric sequence that models the culture's size.

52. A real estate investment broker predicts that a certain property will increase in value 15% each year. Thus, the yearly property values can be modeled by a geometric sequence whose common ratio r is 1.15. If the initial property value was $500,000, write the first four terms of the sequence and predict the value at the end of the third year.

53. A rubber ball is dropped from a height of 486 feet, and it continues to bounce one-third the height from which it last fell. Write out the first five terms of this geometric sequence and find the general term. Find how many bounces it takes for the ball to rebound less than 1 foot.

54. On the first swing, the length of the arc through which a pendulum swings is 50 inches. The length of each successive swing is 80% of the preceding swing. Determine whether this sequence is arithmetic or geometric. Find the length of the fourth swing.

55. Jose takes a job that offers a monthly starting salary of $4000 and guarantees him a monthly raise of $125 during his first year of training. Find the general term of this arithmetic sequence and his monthly salary at the end of his training.

56. At the beginning of Claudia Schaffer's exercise program, she rides 15 minutes on the Lifecycle. Each week she increases her riding time by 5 minutes. Write the general term of this arithmetic sequence, and find her riding time after 7 weeks. Find how many weeks it takes her to reach a riding time of 1 hour.

57. If a radioactive element has a half-life of 3 hours, then x grams of the element dwindles to $\dfrac{x}{2}$ grams after 3 hours. If a nuclear reactor has 400 grams of that radioactive element, find the amount of radioactive material after 12 hours.

REVIEW AND PREVIEW

Evaluate. See Section 1.7.

58. $5(1) + 5(2) + 5(3) + 5(4)$

59. $\dfrac{1}{3(1)} + \dfrac{1}{3(2)} + \dfrac{1}{3(3)}$

60. $2(2 - 4) + 3(3 - 4) + 4(4 - 4)$

61. $3^0 + 3^1 + 3^2 + 3^3$

62. $\dfrac{1}{4(1)} + \dfrac{1}{4(2)} + \dfrac{1}{4(3)}$

63. $\dfrac{8 - 1}{8 + 1} + \dfrac{8 - 2}{8 + 2} + \dfrac{8 - 3}{8 + 3}$

Concept Extensions

Write the first four terms of the arithmetic or geometric sequence whose first term, a_1, and common difference, d, or common ratio, r, are given.

64. $a_1 = \$3720, d = -\$268.50$

65. $a_1 = \$11{,}782.40, r = 0.5$

66. $a_1 = 26.8, r = 2.5$

67. $a_1 = 19.652; d = -0.034$

68. Describe a situation in your life that can be modeled by a geometric sequence. Write an equation for the sequence.

69. Describe a situation in your life that can be modeled by an arithmetic sequence. Write an equation for the sequence.

14.3 SERIES

Objectives

1 Identify finite and infinite series and use summation notation.

2 Find partial sums.

1 A person who conscientiously saves money by saving first $100 and then saving $10 more each month than he saved the preceding month is saving money according to the arithmetic sequence

$$a_n = 100 + 10(n - 1)$$

Following this sequence, he can predict how much money he should save for any particular month. But if he also wants to know how much money *in total* he has saved, say, by the fifth month, he must find the *sum* of the first five terms of the sequence

$$\underbrace{100}_{a_1} + \underbrace{100 + 10}_{a_2} + \underbrace{100 + 20}_{a_3} + \underbrace{100 + 30}_{a_4} + \underbrace{100 + 40}_{a_5}$$

A sum of the terms of a sequence is called a **series** (the plural is also "series"). As our example here suggests, series are frequently used to model financial and natural phenomena.

A series is a **finite series** if it is the sum of a finite number of terms. A series is an **infinite series** if it is the sum of all the terms of the sequence. For example,

Sequence	Series	
$5, 9, 13$	$5 + 9 + 13$	Finite; sum of **3** terms
$5, 9, 13, \ldots$	$5 + 9 + 13 + \cdots$	Infinite
$4, -2, 1, -\dfrac{1}{2}, \dfrac{1}{4}$	$4 + (-2) + 1 + \left(-\dfrac{1}{2}\right) + \left(\dfrac{1}{4}\right)$	Finite; sum of **5** terms
$4, -2, 1, \ldots$	$4 + (-2) + 1 + \cdots$	Infinite
$3, 6, \ldots, 99$	$3 + 6 + \cdots + 99$	Finite; sum of **33** terms

A shorthand notation for denoting a series when the general term of the sequence is known is called **summation notation.** The Greek uppercase letter **sigma**, Σ, is used to mean "sum." The expression $\sum\limits_{n=1}^{5}(3n + 1)$ is read "the sum of $3n + 1$ as n goes from 1 to 5"; this expression means the sum of the first five terms of the sequence whose general term is $a_n = 3n + 1$. Often, the variable i is used instead of n in summation notation: $\sum\limits_{i=1}^{5}(3i + 1)$. Whether we use n, i, k, or some other variable, the variable is called the **index of summation.** The notation $i = 1$ below the symbol Σ indicates the beginning value of i, and the number 5 above the symbol Σ indicates the ending value of i. Thus, the terms of the sequence are found by successively replacing i with the natural numbers $1, 2, 3, 4, 5$. To find the sum, we write out the terms and then add.

$$\sum_{i=1}^{5}(3i + 1) = (3 \cdot 1 + 1) + (3 \cdot 2 + 1) + (3 \cdot 3 + 1)$$
$$+ (3 \cdot 4 + 1) + (3 \cdot 5 + 1)$$
$$= 4 + 7 + 10 + 13 + 16 = 50$$

EXAMPLE 1

Evaluate.

a. $\sum\limits_{i=0}^{6} \dfrac{i - 2}{2}$ **b.** $\sum\limits_{i=3}^{5} 2^i$

Solution **a.** $\sum\limits_{i=0}^{6} \dfrac{i - 2}{2} = \dfrac{0 - 2}{2} + \dfrac{1 - 2}{2} + \dfrac{2 - 2}{2} + \dfrac{3 - 2}{2} + \dfrac{4 - 2}{2} + \dfrac{5 - 2}{2} + \dfrac{6 - 2}{2}$

$$= (-1) + \left(-\dfrac{1}{2}\right) + 0 + \dfrac{1}{2} + 1 + \dfrac{3}{2} + 2$$

$$= \dfrac{7}{2}, \text{ or } 3\dfrac{1}{2}$$

b. $\displaystyle\sum_{i=3}^{5} 2^i = 2^3 + 2^4 + 2^5$

$$= 8 + 16 + 32$$
$$= 56$$

EXAMPLE 2

Write each series with summation notation.

a. $3 + 6 + 9 + 12 + 15$

b. $\dfrac{1}{2} + \dfrac{1}{4} + \dfrac{1}{8} + \dfrac{1}{16}$

Solution   **a.** Since the *difference* of each term and the preceding term is 3, the terms correspond to the first five terms of the arithmetic sequence $a_n = a_1 + (n-1)d$ with $a_1 = 3$ and $d = 3$. So $a_n = 3 + (n-1)3 = 3n$, when simplified. Thus, in summation notation,

$$3 + 6 + 9 + 12 + 15 = \sum_{i=1}^{5} 3i.$$

b. Since each term is the *product* of the preceding term and $\dfrac{1}{2}$, these terms correspond to the first four terms of the geometric sequence $a_n = a_1 r^{n-1}$. Here $a_1 = \dfrac{1}{2}$ and $r = \dfrac{1}{2}$, so $a_n = \left(\dfrac{1}{2}\right)\left(\dfrac{1}{2}\right)^{n-1} = \left(\dfrac{1}{2}\right)^{1+(n-1)} = \left(\dfrac{1}{2}\right)^n$. In summation notation,

$$\frac{1}{2} + \frac{1}{4} + \frac{1}{8} + \frac{1}{16} = \sum_{i=1}^{4}\left(\frac{1}{2}\right)^i$$

2   The sum of the first n terms of a sequence is a finite series known as a **partial sum**, S_n. Thus, for the sequence $a_1, a_2, \ldots, a_n$, the first three partial sums are

$$S_1 = a_1$$
$$S_2 = a_1 + a_2$$
$$S_3 = a_1 + a_2 + a_3$$

In general, S_n is the sum of the first n terms of a sequence.

$$S_n = \sum_{i=1}^{n} a_n$$

EXAMPLE 3

Find the sum of the first three terms of the sequence whose general term is $a_n = \dfrac{n+3}{2n}$.

Solution
$$S_3 = \sum_{i=1}^{3} \frac{i+3}{2i} = \frac{1+3}{2 \cdot 1} + \frac{2+3}{2 \cdot 2} + \frac{3+3}{2 \cdot 3}$$

$$= 2 + \frac{5}{4} + 1 = 4\frac{1}{4}$$

The next example illustrates how these sums model real-life phenomena.

> **EXAMPLE 4**

NUMBER OF BABY GORILLAS BORN

The number of baby gorillas born at the San Diego Zoo is a sequence defined by $a_n = n(n - 1)$, where n is the number of years the zoo has owned gorillas. Find the *total* number of baby gorillas born in the *first 4 years*.

Solution   To solve, find the sum

$$S_4 = \sum_{i=1}^{4} i(i - 1)$$

$$= 1(1 - 1) + 2(2 - 1) + 3(3 - 1) + 4(4 - 1)$$

$$= 0 + 2 + 6 + 12 = 20$$

There were 20 gorillas born in the first 4 years.

EXERCISE SET 14.3

STUDY GUIDE/SSM   CD/VIDEO   PH MATH TUTOR CENTER   MathXL®Tutorials ON CD   MathXL®   MyMathLab®

Evaluate. See Example 1.

1. $\displaystyle\sum_{i=1}^{4} (i - 3)$

2. $\displaystyle\sum_{i=1}^{5} (i + 6)$

 3. $\displaystyle\sum_{i=4}^{7} (2i + 4)$

4. $\displaystyle\sum_{i=2}^{3} (5i - 1)$

5. $\displaystyle\sum_{i=2}^{4} (i^2 - 3)$

6. $\displaystyle\sum_{i=3}^{5} i^3$

7. $\displaystyle\sum_{i=1}^{3} \left(\frac{1}{i + 5}\right)$

8. $\displaystyle\sum_{i=2}^{4} \left(\frac{2}{i + 3}\right)$

 9. $\displaystyle\sum_{i=1}^{3} \frac{1}{6i}$

10. $\displaystyle\sum_{i=1}^{3} \frac{1}{3i}$

11. $\displaystyle\sum_{i=2}^{6} 3i$

12. $\displaystyle\sum_{i=3}^{6} -4i$

13. $\displaystyle\sum_{i=3}^{5} i(i + 2)$

14. $\displaystyle\sum_{i=2}^{4} i(i - 3)$

15. $\displaystyle\sum_{i=1}^{5} 2^i$

16. $\displaystyle\sum_{i=1}^{4} 3^{i-1}$

17. $\displaystyle\sum_{i=1}^{4} \frac{4i}{i + 3}$

18. $\displaystyle\sum_{i=2}^{5} \frac{6 - i}{6 + i}$

Write each series with summation notation. See Example 2.

19. $1 + 3 + 5 + 7 + 9$

20. $4 + 7 + 10 + 13$

21. $4 + 12 + 36 + 108$

22. $5 + 10 + 20 + 40 + 80 + 160$

23. $12 + 9 + 6 + 3 + 0 + (-3)$

24. $5 + 1 + (-3) + (-7)$

25. $12 + 4 + \dfrac{4}{3} + \dfrac{4}{9}$

26. $80 + 20 + 5 + \dfrac{5}{4} + \dfrac{5}{16}$

27. $1 + 4 + 9 + 16 + 25 + 36 + 49$

28. $1 + (-4) + 9 + (-16)$

Find each partial sum. See Example 3.

29. Find the sum of the first two terms of the sequence whose general term is $a_n = (n + 2)(n - 5)$.

30. Find the sum of the first six terms of the sequence whose general term is $a_n = (-1)^n$.

31. Find the sum of the first two terms of the sequence whose general term is $a_n = n(n - 6)$.

32. Find the sum of the first seven terms of the sequence whose general term is $a_n = (-1)^{n-1}$.

33. Find the sum of the first four terms of the sequence whose general term is $a_n = (n + 3)(n + 1)$.

34. Find the sum of the first five terms of the sequence whose general term is $a_n = \dfrac{(-1)^n}{2n}$.

35. Find the sum of the first four terms of the sequence whose general term is $a_n = -2n$.

36. Find the sum of the first five terms of the sequence whose general term is $a_n = (n - 1)^2$.

37. Find the sum of the first three terms of the sequence whose general term is $a_n = -\dfrac{n}{3}$.

38. Find the sum of the first three terms of the sequence whose general term is $a_n = (n + 4)^2$.

Solve. See Example 4.

39. A gardener is making a triangular planting with 1 tree in the first row, 2 trees in the second row, 3 trees in the third row, and so on for 10 rows. Write the sequence that describes the number of trees in each row. Find the total number of trees planted.

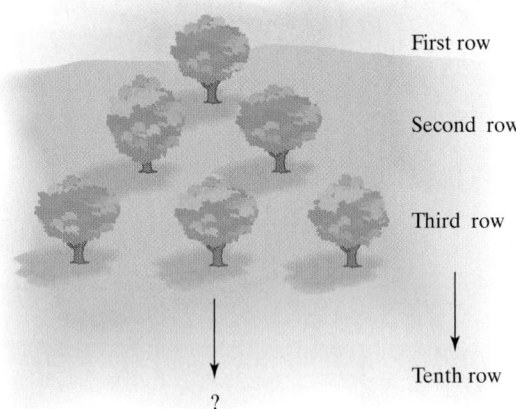

First row

Second row

Third row

Tenth row

?

40. Some surfers at the beach form a human pyramid with 2 surfers in the top row, 3 surfers in the second row, 4 surfers in the third row, and so on. If there are 6 rows in the pyramid, write the sequence that describes the number of surfers in each row of the pyramid. Find the total number of surfers.

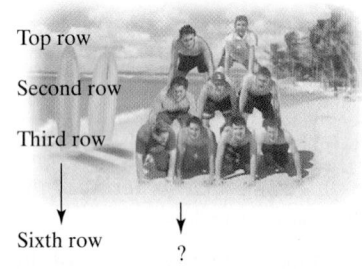

Top row

Second row

Third row

Sixth row

?

41. A culture of fungus starts with 6 units and doubles every day. Write the general term of the sequence that describes the growth of this fungus. Find the number of fungus units there will be at the beginning of the fifth day.

42. A bacterial colony begins with 100 bacteria and doubles every 6 hours. Write the general term of the sequence describing the growth of the bacteria. Find the number of bacteria there will be after 24 hours.

43. A bacterial colony begins with 50 bacteria and doubles every 12 hours. Write the sequence that describes the growth of the bacteria. Find the number of bacteria there will be after 48 hours.

44. The number of otters born each year in a new aquarium forms a sequence whose general term is $a_n = (n - 1)(n + 3)$. Find the number of otters born in the third year, and find the total number of otters born in the first three years.

45. The number of opossums killed each month on a new highway forms the sequence whose general term is $a_n = (n + 1)(n + 2)$, where n is the number of the months. Find the number of opossums killed in the fourth month, and find the total number killed in the first four months.

46. In 2003 the population of an endangered fish was estimated by environmentalists to be decreasing each year. The size of the population in a given year is $24 - 4n$ thousand fish fewer than the previous year. Find the decrease in population in 2005, if year 1 is 2003. Find the total decrease in the fish population for the years 2003 through 2005.

47. The amount of decay in pounds of a radioactive isotope each year is given by the sequence whose general term is $a_n = 100(0.5)^n$, where n is the number of the year. Find the amount of decay in the fourth year, and find the total amount of decay in the first four years.

48. Susan has a choice between two job offers. Job A has an annual starting salary of $20,000 with guaranteed annual raises of $1200 for the next four years, whereas job B has an annual starting salary of $18,000 with guaranteed annual raises of $2500 for the next four years. Compare the fifth partial sums for each sequence to determine which job would pay Susan more money over the next 5 years.

49. A pendulum swings a length of 40 inches on its first swing. Each successive swing is $\dfrac{4}{5}$ of the preceding swing. Find the length of the fifth swing and the total length swung during the first five swings. (Round to the nearest tenth of an inch.)

50. Explain the difference between a sequence and a series.

REVIEW AND PREVIEW

Evaluate. See Section 1.7.

51. $\dfrac{5}{1 - \dfrac{1}{2}}$

52. $\dfrac{-3}{1 - \dfrac{1}{7}}$

53. $\dfrac{\dfrac{1}{3}}{1 - \dfrac{1}{10}}$

54. $\dfrac{\dfrac{6}{11}}{1 - \dfrac{1}{10}}$

55. $\dfrac{3(1 - 2^4)}{1 - 2}$

56. $\dfrac{2(1 - 5^3)}{1 - 5}$

57. $\dfrac{10}{2}(3 + 15)$

58. $\dfrac{12}{2}(2 + 19)$

Concept Extensions

59. a. Write the sum $\sum\limits_{i=1}^{7}(i + i^2)$ without summation notation.

  b. Write the sum $\sum\limits_{i=1}^{7}i + \sum\limits_{i=1}^{7}i^2$ without summation notation.

  c. Compare the results of parts **a** and **b.**

  d. Do you think the following is true or false? Explain your answer.

$$\sum_{i=1}^{n}(a_n + b_n) = \sum_{i=1}^{n}a_n + \sum_{i=1}^{n}b_n$$

60. a. Write the sum $\sum\limits_{i=1}^{6}5i^3$ without summation notation.

  b. Write the expression $5 \cdot \sum\limits_{i=1}^{6}i^3$ without summation notation.

  c. Compare the results of parts **a** and **b.**

  d. Do you think the following is true or false? Explain your answer.

$$\sum_{i=1}^{n}c \cdot a_n = c \cdot \sum_{i=1}^{n}a_n \text{ where } c \text{ is a constant}$$

INTEGRATED REVIEW — SEQUENCES AND SERIES

Write the first five terms of each sequence whose general term is given.

1. $a_n = n - 3$

2. $a_n = \dfrac{7}{1 + n}$

3. $a_n = 3^{n-1}$

4. $a_n = n^2 - 5$

Find the indicated term for each sequence.

5. $(-2)^n$; a_6

6. $-n^2 + 2$; a_4

7. $\dfrac{(-1)^n}{n}$; a_{40}

8. $\dfrac{(-1)^n}{2n}$; a_{41}

Write the first five terms of the arithmetic or geometric sequence whose first term is a_1, and common difference, d, or common ratio, r, are given.

9. $a_1 = 7$; $d = -3$

10. $a_1 = -3$; $r = 5$

11. $a_1 = 45$; $r = \dfrac{1}{3}$

12. $a_1 = -12$; $d = 10$

Find the indicated term of each sequence.

13. The tenth term of the arithmetic sequence whose first term is 20 and whose common difference is 9.

14. The sixth term of the geometric sequence whose first term is 64 and whose common ratio is $\dfrac{3}{4}$.

15. The seventh term of the geometric sequence $6, -12, 24, \ldots$

16. The twentieth term of the arithmetic sequence $-100, -85, -70, \ldots$

17. The fifth term of the arithmetic sequence whose fourth term is -5 and whose tenth term is -35.

18. The fifth term of the geometric sequence whose fourth term is 1 and whose seventh term is $\dfrac{1}{125}$.

Evaluate.

19. $\displaystyle\sum_{i=1}^{4} 5i$

20. $\displaystyle\sum_{i=1}^{7} (3i + 2)$

21. $\displaystyle\sum_{i=3}^{7} 2^{i-4}$

22. $\displaystyle\sum_{i=2}^{5} \frac{i}{i + 1}$

Find each partial sum.

23. Find the sum of the first three terms of the sequence whose general term is $a_n = n(n - 4)$.

24. Find the sum of the first ten terms of the sequence whose general term is $a_n = (-1)^n(n + 1)$.

14.4   PARTIAL SUMS OF ARITHMETIC AND GEOMETRIC SEQUENCES

Objectives

1. Find the partial sum of an arithmetic sequence.
2. Find the partial sum of a geometric sequence.
3. Find the sum of the terms of an infinite geometric sequence.

1   Partial sums S_n are relatively easy to find when n is small—that is, when the number of terms to add is small. But when n is large, finding S_n can be tedious. For a large n, S_n is still relatively easy to find if the addends are terms of an arithmetic sequence or a geometric sequence.

For an arithmetic sequence, $a_n = a_1 + (n - 1)d$ for some first term a_1 and some common difference d. So S_n, the sum of the first n terms, is

$$S_n = a_1 + (a_1 + d) + (a_1 + 2d) + \cdots + (a_1 + (n - 1)d)$$

We might also find S_n by "working backward" from the nth term a_n, finding the preceding term a_{n-1}, by subtracting d each time.

$$S_n = a_n + (a_n - d) + (a_n - 2d) + \cdots + (a_n - (n - 1)d)$$

Now add the left sides of these two equations and add the right sides.

$$2S_n = (a_1 + a_n) + (a_1 + a_n) + (a_1 + a_n) + \cdots + (a_1 + a_n)$$

The d terms subtract out, leaving n sums of the first term, a_1, and last term, a_n. Thus, we write

$$2S_n = n(a_1 + a_n)$$

or

$$S_n = \frac{n}{2}(a_1 + a_n)$$

> ### Partial Sum S_n of an Arithmetic Sequence
>
> The partial sum S_n of the first n terms of an arithmetic sequence is given by
>
> $$S_n = \frac{n}{2}(a_1 + a_n)$$
>
> where a_1 is the first term of the sequence and a_n is the nth term.

EXAMPLE 1

Use the partial sum formula to find the sum of the first six terms of the arithmetic sequence $2, 5, 8, 11, 14, 17, \ldots$.

Solution Use the formula for S_n of an arithmetic sequence, replacing n with 6, a_1 with 2, and a_n with 17.

$$S_n = \frac{n}{2}(a_1 + a_n) = \frac{6}{2}(2 + 17) = 3(19) = 57$$

EXAMPLE 2

Find the sum of the first 30 positive integers.

Solution Because $1, 2, 3, \ldots, 30$ is an arithmetic sequence, use the formula for S_n with $n = 30$, $a_1 = 1$, and $a_n = 30$. Thus,

$$S_n = \frac{n}{2}(a_1 + a_n) = \frac{30}{2}(1 + 30) = 15(31) = 465$$

EXAMPLE 3

STACKING ROLLS OF CARPET

Rolls of carpet are stacked in 20 rows with 3 rolls in the top row, 4 rolls in the next row, and so on, forming an arithmetic sequence. Find the total number of carpet rolls if there are 22 rolls in the bottom row.

3 rolls

4 rolls

5 rolls

Solution The list $3, 4, 5, \ldots, 22$ is the first 20 terms of an arithmetic sequence. Use the formula for S_n with $a_1 = 3$, $a_n = 22$, and $n = 20$ terms. Thus,

$$S_{20} = \frac{20}{2}(3 + 22) = 10(25) = 250$$

There are a total of 250 rolls of carpet.

2 We can also derive a formula for the partial sum S_n of the first n terms of a geometric series. If $a_n = a_1 r^{n-1}$, then

$$S_n = a_1 + a_1 r + a_1 r^2 + \cdots + a_1 r^{n-1}$$

$$\uparrow \quad \uparrow \quad \uparrow \qquad\qquad \uparrow$$

1st 2nd 3rd nth
term term term   term

Multiply each side of the equation by $-r$.

$$-rS_n = -a_1 r - a_1 r^2 - a_1 r^3 - \cdots - a_1 r^n$$

Add the two equations.

$$S_n - rS_n = a_1 + (a_1 r - a_1 r) + (a_1 r^2 - a_1 r^2) + (a_1 r^3 - a_1 r^3) + \cdots - a_1 r^n$$
$$S_n - rS_n = a_1 - a_1 r^n$$

Now factor each side.

$$S_n(1 - r) = a_1(1 - r^n)$$

Solve for S_n by dividing both sides by $1 - r$. Thus,

$$S_n = \frac{a_1(1 - r^n)}{1 - r}$$

as long as r is not 1.

Partial Sum S_n of a Geometric Sequence

The partial sum S_n of the first n terms of a geometric sequence is given by

$$S_n = \frac{a_1(1 - r^n)}{1 - r}$$

where a_1 is the first term of the sequence, r is the common ratio, and $r \neq 1$.

EXAMPLE 4

Find the sum of the first six terms of the geometric sequence 5, 10, 20, 40, 80, 160.

Solution Use the formula for the partial sum S_n of the terms of a geometric sequence. Here, $n = 6$, the first term $a_1 = 5$, and the common ratio $r = 2$.

$$S_n = \frac{a_1(1 - r^n)}{1 - r}$$

$$S_6 = \frac{5(1 - 2^6)}{1 - 2} = \frac{5(-63)}{-1} = 315$$

EXAMPLE 5

FINDING AMOUNT OF DONATION

A grant from an alumnus to a university specified that the university was to receive $800,000 during the first year and 75% of the preceding year's donation during each of the following 5 years. Find the total amount donated during the 6 years.

Solution   The donations are modeled by the first six terms of a geometric sequence. Evaluate S_n when $n = 6$, $a_1 = 800,000$, and $r = 0.75$.

$$S_6 = \frac{800,000[1 - (0.75)^6]}{1 - 0.75}$$

$$= \$2,630,468.75$$

The total amount donated during the 6 years is $2,630,468.75.

3   Is it possible to find the sum of all the terms of an infinite sequence? Examine the partial sums of the geometric sequence $\dfrac{1}{2}, \dfrac{1}{4}, \dfrac{1}{8}, \ldots$.

$$S_1 = \frac{1}{2}$$

$$S_2 = \frac{1}{2} + \frac{1}{4} = \frac{3}{4}$$

$$S_3 = \frac{1}{2} + \frac{1}{4} + \frac{1}{8} = \frac{7}{8}$$

$$S_4 = \frac{1}{2} + \frac{1}{4} + \frac{1}{8} + \frac{1}{16} = \frac{15}{16}$$

$$S_5 = \frac{1}{2} + \frac{1}{4} + \frac{1}{8} + \frac{1}{16} + \frac{1}{32} = \frac{31}{32}$$

$$\vdots$$

$$S_{10} = \frac{1}{2} + \frac{1}{4} + \frac{1}{8} + \cdots + \frac{1}{2^{10}} = \frac{1023}{1024}$$

Even though each partial sum is larger than the preceding partial sum, we see that each partial sum is closer to 1 than the preceding partial sum. If n gets larger and larger, then S_n gets closer and closer to 1. We say that 1 is the **limit** of S_n and also that 1 is the sum of the terms of this infinite sequence. In general, if $|r| < 1$, the following formula gives the sum of the terms of an infinite geometric sequence.

Sum of the Terms of an Infinite Geometric Sequence

The sum S_∞ of the terms of an infinite geometric sequence is given by

$$S_\infty = \frac{a_1}{1 - r}$$

where a_1 is the first term of the sequence, r is the common ratio, and $|r| < 1$. If $|r| \geq 1$, S_∞ does not exist.

What happens for other values of r? For example, in the following geometric sequence, $r = 3$.

$$6, 18, 54, 162, \ldots$$

Here, as n increases, the sum S_n increases also. This time, though, S_n does not get closer and closer to a fixed number but instead increases without bound.

EXAMPLE 6

Find the sum of the terms of the geometric sequence $2, \dfrac{2}{3}, \dfrac{2}{9}, \dfrac{2}{27}, \ldots$.

Solution For this geometric sequence, $r = \dfrac{1}{3}$. Since $|r| < 1$, we may use the formula for S_∞ of a geometric sequence with $a_1 = 2$ and $r = \dfrac{1}{3}$.

$$S_\infty = \frac{a_1}{1 - r} = \frac{2}{1 - \dfrac{1}{3}} = \frac{2}{\dfrac{2}{3}} = 3$$

The formula for the sum of the terms of an infinite geometric sequence can be used to write a repeating decimal as a fraction. For example,

$$0.33\overline{3} = \frac{3}{10} + \frac{3}{100} + \frac{3}{1000} + \cdots$$

This sum is the sum of the terms of an infinite geometric sequence whose first term a_1 is $\dfrac{3}{10}$ and whose common ratio r is $\dfrac{1}{10}$. Using the formula for S_∞,

$$S_\infty = \frac{a_1}{1 - r} = \frac{\dfrac{3}{10}}{1 - \dfrac{1}{10}} = \frac{1}{3}$$

So, $0.33\overline{3} = \dfrac{1}{3}$.

EXAMPLE 7

DISTANCE TRAVELED BY A PENDULUM

On its first pass, a pendulum swings through an arc whose length is 24 inches. On each pass thereafter, the arc length is 75% of the arc length on the preceding pass. Find the total distance the pendulum travels before it comes to rest.

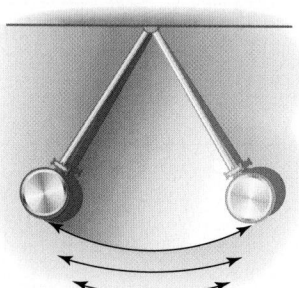

Solution We must find the sum of the terms of an infinite geometric sequence whose first term, a_1, is 24 and whose common ratio, r, is 0.75. Since $|r| < 1$, we may use the formula for S_∞.

$$S_\infty = \frac{a_1}{1 - r} = \frac{24}{1 - 0.75} = \frac{24}{0.25} = 96$$

The pendulum travels a total distance of 96 inches before it comes to rest.

Spotlight on

DECISION

⸙ MAKING

Suppose you are a certified financial planner. You are working with a 30-year-old client whose goal is to retire at age 65 with a sum of $500,000 to live on. This year she just started making the maximum annual contribution of $2000 to a Roth IRA (Individual Retirement Account) that pays 8% interest compounded annually.

You know that if she continues to make a $2000 contribution at the beginning of each year, by the end of the nth year, her account increases in value by the nth term of the geometric sequence $a_n = 2000(1.08)^n$, which considers both the annual $2000 contribution and her earned interest. Decide whether your client will be able to reach her retirement goal by making only the maximum Roth IRA contribution each year until she retires, or if you should suggest an additional investment to help her reach her goal. (Note: Use a partial sum to find the value of the Roth IRA at the end of 35 years. To find a_1 of the geometric sequence, be sure to evaluate the equation for the nth term at $n = 1$.)

EXERCISE SET 14.4

STUDY   CD/   PH MATH   MathXL®Tutorials   MathXL®   MyMathLab®
GUIDE/SSM   VIDEO   TUTOR CENTER   ON CD

Use the partial sum formula to find the partial sum of the given arithmetic or geometric sequence. See Examples 1 and 4.

1. Find the sum of the first six terms of the arithmetic sequence $1, 3, 5, 7, \ldots$.

2. Find the sum of the first seven terms of the arithmetic sequence $-7, -11, -15, \ldots$.

3. Find the sum of the first five terms of the geometric sequence $4, 12, 36, \ldots$.

4. Find the sum of the first eight terms of the geometric sequence $-1, 2, -4, \ldots$.

5. Find the sum of the first six terms of the arithmetic sequence $3, 6, 9, \ldots$.

6. Find the sum of the first four terms of the arithmetic sequence $-4, -8, -12, \ldots$.

7. Find the sum of the first four terms of the geometric sequence $2, \dfrac{2}{5}, \dfrac{2}{25}, \ldots$.

8. Find the sum of the first five terms of the geometric sequence $\dfrac{1}{3}, -\dfrac{2}{3}, \dfrac{4}{3}, \ldots$.

Solve. See Example 2.

9. Find the sum of the first ten positive integers.

10. Find the sum of the first eight negative integers.

11. Find the sum of the first four positive odd integers.

12. Find the sum of the first five negative odd integers.

Find the sum of the terms of each infinite geometric sequence. See Example 6.

13. $12, 6, 3, \ldots$

14. $45, 15, 5, \ldots$

15. $\dfrac{1}{10}, \dfrac{1}{100}, \dfrac{1}{1000}, \ldots$

16. $\dfrac{3}{5}, \dfrac{3}{20}, \dfrac{3}{80}, \ldots$

17. $-10, -5, -\dfrac{5}{2}, \ldots$

18. $-16, -4, -1, \ldots$

19. $2, -\dfrac{1}{4}, \dfrac{1}{32}, \ldots$

20. $-3, \dfrac{3}{5}, -\dfrac{3}{25}, \ldots$

21. $\dfrac{2}{3}, -\dfrac{1}{3}, \dfrac{1}{6}, \ldots$

22. $6, -4, \dfrac{8}{3}, \ldots$

MIXED PRACTICE

Solve.

23. Find the sum of the first ten terms of the sequence $-4, 1, 6, \ldots, 41$ where 41 is the tenth term.

24. Find the sum of the first twelve terms of the sequence $-3, -13, -23, \ldots, -113$ where -113 is the twelfth term.

25. Find the sum of the first seven terms of the sequence $3, \dfrac{3}{2}, \dfrac{3}{4}, \ldots$.

26. Find the sum of the first five terms of the sequence $-2, -6, -18, \ldots$.

27. Find the sum of the first five terms of the sequence $-12, 6, -3, \ldots$.

28. Find the sum of the first four terms of the sequence $-\dfrac{1}{4}, -\dfrac{3}{4}, -\dfrac{9}{4}, \ldots$.

29. Find the sum of the first twenty terms of the sequence $\dfrac{1}{2}, \dfrac{1}{4}, 0, \ldots, -\dfrac{17}{4}$ where $-\dfrac{17}{4}$ is the twentieth term.

30. Find the sum of the first fifteen terms of the sequence $-5, -9, -13, \ldots, -61$ where -61 is the fifteenth term.

31. If a_1 is 8 and r is $-\dfrac{2}{3}$, find S_3.

32. If a_1 is 10, a_{18} is $\dfrac{3}{2}$, and d is $-\dfrac{1}{2}$, find S_{18}.

Solve. See Example 3.

33. Modern Car Company has come out with a new car model. Market analysts predict that 4000 cars will be sold in the first month and that sales will drop by 50 cars per month after that during the first year. Write out the first five terms of the sequence, and find the number of sold cars predicted for the twelfth month. Find the total predicted number of sold cars for the first year.

34. A company that sends faxes charges $3 for the first page sent and $0.10 less than the preceding page for each additional page sent. The cost per page forms an arithmetic sequence. Write the first five terms of this sequence, and use a partial sum to find the cost of sending a nine-page document.

35. Sal has two job offers: Firm A starts at $22,000 per year and guarantees raises of $1000 per year, whereas Firm B starts at $20,000 and guarantees raises of $1200 per year. Over a 10-year period, determine the more profitable offer.

36. The game of pool uses 15 balls numbered 1 to 15. In the variety called rotation, a player who sinks a ball receives as many points as the number on the ball. Use an arithmetic series to find the score of a player who sinks all 15 balls.

Solve. See Example 5.

37. A woman made $30,000 during the first year she owned her business and made an additional 10% over the previous year in each subsequent year. Find how much she made during her fourth year of business. Find her total earnings during the first four years.

38. In free fall, a parachutist falls 16 feet during the first second, 48 feet during the second second, 80 feet during the third second, and so on. Find how far she falls during the eighth second. Find the total distance she falls during the first 8 seconds.

39. A trainee in a computer company takes 0.9 times as long to assemble each computer as he took to assemble the preceding computer. If it took him 30 minutes to assemble the first computer, find how long it takes him to assemble the fifth computer. Find the total time he takes to assemble the first five computers. (Round to the nearest minute.)

40. On a gambling trip to Reno, Carol doubled her bet each time she lost. If her first losing bet was $5 and she lost six consecutive bets, find how much she lost on the sixth bet. Find the total amount lost on these six bets.

Solve. See Example 7.

41. A ball is dropped from a height of 20 feet and repeatedly rebounds to a height that is $\dfrac{4}{5}$ of its previous height. Find the total distance the ball covers before it comes to rest.

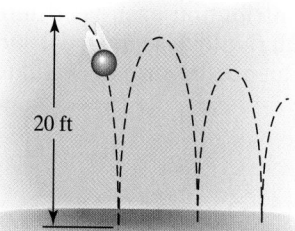

20 ft

42. A rotating flywheel coming to rest makes 300 revolutions in the first minute and in each minute thereafter makes $\dfrac{2}{5}$ as many revolutions as in the preceding minute. Find how many revolutions the wheel makes before it comes to rest.

Solve.

43. In the pool game of rotation, player A sinks balls numbered 1 to 9, and player B sinks the rest of the balls. Use arithmetic series to find each player's score (see Exercise 36).

44. A godfather deposited $250 in a savings account on the day his godchild was born. On each subsequent birthday he deposited $50 more than he deposited the previous year. Find how much money he deposited on his godchild's twenty-first birthday. Find the total amount deposited over the 21 years.

45. During the holiday rush a business can rent a computer system for $200 the first day, with the rental fee decreasing $5 for each additional day. Find the fee paid for 20 days during the holiday rush.

46. The spraying of a field with insecticide killed 6400 weevils the first day, 1600 the second day, 400 the third day, and so on. Find the total number of weevils killed during the first 5 days.

47. A college student humorously asks his parents to charge him room and board according to this geometric sequence: $0.01 for the first day of the month, $0.02 for the second day, $0.04 for the third day, and so on. Find the total room and board he would pay for 30 days.

48. Following its television advertising campaign, a bank attracted 80 new customers the first day, 120 the second day, 160 the third day, and so on, in an arithmetic sequence. Find how many new customers were attracted during the first 5 days following its television campaign.

REVIEW AND PREVIEW

Evaluate. See Section 1.7.

49. $6 \cdot 5 \cdot 4 \cdot 3 \cdot 2 \cdot 1$

50. $8 \cdot 7 \cdot 6 \cdot 5 \cdot 4 \cdot 3 \cdot 2 \cdot 1$

51. $\dfrac{3 \cdot 2 \cdot 1}{2 \cdot 1}$

52. $\dfrac{5 \cdot 4 \cdot 3 \cdot 2 \cdot 1}{3 \cdot 2 \cdot 1}$

Multiply. See Section 5.4.

53. $(x + 5)^2$

54. $(x - 2)^2$

55. $(2x - 1)^3$

56. $(3x + 2)^3$

Concept Extensions

57. Write $0.88\overline{8}$ as an infinite geometric series and use the formula for S_∞ to write it as a rational number.

58. Write $0.54\overline{54}$ as an infinite geometric series and use the formula S_∞ to write it as a rational number.

59. Explain whether the sequence $5, 5, 5, \ldots$ is arithmetic, geometric, neither, or both.

60. Describe a situation in everyday life that can be modeled by an infinite geometric series.

14.5   THE BINOMIAL THEOREM

Objectives

1   Use Pascal's triangle to expand binomials.

2   Evaluate factorials.

3   Use the binomial theorem to expand binomials.

4   Find the nth term in the expansion of a binomial raised to a positive power.

In this section, we learn how to **expand** binomials of the form $(a + b)^n$ easily. Expanding a binomial such as $(a + b)^n$ means to write the factored form as a sum. First, we review the patterns in the expansions of $(a + b)^n$.

$(a + b)^0 = 1$	1 term
$(a + b)^1 = a + b$	2 terms
$(a + b)^2 = a^2 + 2ab + b^2$	3 terms
$(a + b)^3 = a^3 + 3a^2b + 3ab^2 + b^3$	4 terms
$(a + b)^4 = a^4 + 4a^3b + 6a^2b^2 + 4ab^3 + b^4$	5 terms
$(a + b)^5 = a^5 + 5a^4b + 10a^3b^2 + 10a^2b^3 + 5ab^4 + b^5$	6 terms

Notice the following patterns.

1. The expansion of $(a + b)^n$ contains $n + 1$ terms. For example, for $(a + b)^3$, $n = 3$, and the expansion contains $3 + 1$ terms, or 4 terms.

2. The first term of the expansion of $(a + b)^n$ is a^n, and the last term is b^n.

3. The powers of a decrease by 1 for each term, whereas the powers of b increase by 1 for each term.

4. For each term of the expansion of $(a + b)^n$, the sum of the exponents of a and b is n. (For example, the sum of the exponents of $5a^4b$ is $4 + 1$, or 5, and the sum of the exponents of $10a^3b^2$ is $3 + 2$, or 5.)

1 There are patterns in the coefficients of the terms as well. Written in a triangular array, the coefficients are called **Pascal's triangle.**

$(a + b)^0$:						1						$n = 0$
$(a + b)^1$:					1		1					$n = 1$
$(a + b)^2$:				1		2		1				$n = 2$
$(a + b)^3$:			1		3		3		1			$n = 3$
$(a + b)^4$:		1		4		6		4		1		$n = 4$
$(a + b)^5$:	1		5		10		10		5		1	$n = 5$

Each row in Pascal's triangle begins and ends with 1. Any other number in a row is the sum of the two closest numbers above it. Using this pattern, we can write the next row, for $n = 6$, by first writing the number 1. Then we can add the consecutive numbers in the row for $n = 5$ and write each sum "between and below" the pair. We complete the row by writing a 1.

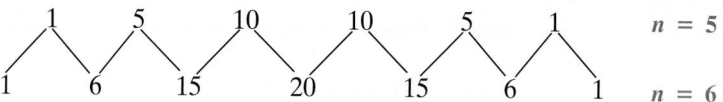

We can use Pascal's triangle and the patterns noted to expand $(a + b)^n$ without actually multiplying any terms.

EXAMPLE 1

Expand $(a + b)^6$.

Solution Using the $n = 6$ row of Pascal's triangle as the coefficients and following the patterns noted, $(a + b)^6$ can be expanded as

$$a^6 + 6a^5b + 15a^4b^2 + 20a^3b^3 + 15a^2b^4 + 6ab^5 + b^6$$

2 For a large n, the use of Pascal's triangle to find coefficients for $(a + b)^n$ can be tedious. An alternative method for determining these coefficients is based on the concept of a **factorial**.

The **factorial of n**, written $n!$ (read "n factorial"), is the product of the first n consecutive natural numbers.

> ### Factorial of n: $n!$
>
> If n is a natural number, then $n! = n(n - 1)(n - 2)(n - 3) \cdots \cdot 3 \cdot 2 \cdot 1$. The factorial of 0, written 0!, is defined to be 1.

For example, $3! = 3 \cdot 2 \cdot 1 = 6$, $5! = 5 \cdot 4 \cdot 3 \cdot 2 \cdot 1 = 120$, and $0! = 1$.

EXAMPLE 2

Evaluate each expression.

a. $\dfrac{5!}{6!}$

b. $\dfrac{10!}{7!3!}$

c. $\dfrac{3!}{2!1!}$

d. $\dfrac{7!}{7!0!}$

Solution **a.** $\dfrac{5!}{6!} = \dfrac{5\cdot4\cdot3\cdot2\cdot1}{6\cdot5\cdot4\cdot3\cdot2\cdot1} = \dfrac{1}{6}$

b. $\dfrac{10!}{7!3!} = \dfrac{10\cdot9\cdot8\cdot7!}{7!\cdot3\cdot2\cdot1} = \dfrac{10\cdot9\cdot8}{3\cdot2\cdot1} = 10\cdot3\cdot4 = 120$

c. $\dfrac{3!}{2!1!} = \dfrac{3\cdot2\cdot1}{2\cdot1\cdot1} = 3$

d. $\dfrac{7!}{7!0!} = \dfrac{7!}{7!\cdot1} = 1$

> **Helpful Hint**
>
> We can use a calculator with a factorial key to evaluate a factorial. A calculator uses scientific notation for large results.

3 It can be proved, although we won't do so here, that the coefficients of terms in the expansion of $(a + b)^n$ can be expressed in terms of factorials. Following patterns 1 through 4 given earlier and using the factorial expressions of the coefficients, we have what is known as the **binomial theorem.**

> **Binomial Theorem**
>
> If n is a positive integer, then
> $$(a + b)^n = a^n + \frac{n}{1!}a^{n-1}b^1 + \frac{n(n-1)}{2!}a^{n-2}b^2$$
> $$+ \frac{n(n-1)(n-2)}{3!}a^{n-3}b^3 + \cdots + b^n$$

We call the formula for $(a + b)^n$ given by the binomial theorem the **binomial formula.**

EXAMPLE 3

Use the binomial theorem to expand $(x + y)^{10}$.

Solution Let $a = x, b = y$, and $n = 10$ in the binomial formula.

$$(x + y)^{10} = x^{10} + \frac{10}{1!}x^9y + \frac{10\cdot9}{2!}x^8y^2 + \frac{10\cdot9\cdot8}{3!}x^7y^3 + \frac{10\cdot9\cdot8\cdot7}{4!}x^6y^4$$
$$+ \frac{10\cdot9\cdot8\cdot7\cdot6}{5!}x^5y^5 + \frac{10\cdot9\cdot8\cdot7\cdot6\cdot5}{6!}x^4y^6$$
$$+ \frac{10\cdot9\cdot8\cdot7\cdot6\cdot5\cdot4}{7!}x^3y^7$$
$$+ \frac{10\cdot9\cdot8\cdot7\cdot6\cdot5\cdot4\cdot3}{8!}x^2y^8$$
$$+ \frac{10\cdot9\cdot8\cdot7\cdot6\cdot5\cdot4\cdot3\cdot2}{9!}xy^9 + y^{10}$$
$$= x^{10} + 10x^9y + 45x^8y^2 + 120x^7y^3 + 210x^6y^4 + 252x^5y^5 + 210x^4y^6$$
$$+ 120x^3y^7 + 45x^2y^8 + 10xy^9 + y^{10}$$

EXAMPLE 4

Use the binomial theorem to expand $(x + 2y)^5$.

Solution   Let $a = x$ and $b = 2y$ in the binomial formula.

$$(x + 2y)^5 = x^5 + \frac{5}{1!}x^4(2y) + \frac{5 \cdot 4}{2!}x^3(2y)^2 + \frac{5 \cdot 4 \cdot 3}{3!}x^2(2y)^3$$

$$+ \frac{5 \cdot 4 \cdot 3 \cdot 2}{4!}x(2y)^4 + (2y)^5$$

$$= x^5 + 10x^4y + 40x^3y^2 + 80x^2y^3 + 80xy^4 + 32y^5$$

EXAMPLE 5

Use the binomial theorem to expand $(3m - n)^4$.

Solution   Let $a = 3m$ and $b = -n$ in the binomial formula.

$$(3m - n)^4 = (3m)^4 + \frac{4}{1!}(3m)^3(-n) + \frac{4 \cdot 3}{2!}(3m)^2(-n)^2$$

$$+ \frac{4 \cdot 3 \cdot 2}{3!}(3m)(-n)^3 + (-n)^4$$

$$= 81m^4 - 108m^3n + 54m^2n^2 - 12mn^3 + n^4$$

4   Sometimes it is convenient to find a specific term of a binomial expansion without writing out the entire expansion. By studying the expansion of binomials, a pattern forms for each term. This pattern is most easily stated for the $(r + 1)$st term.

$(r + 1)st$ Term in a Binomial Expansion

The $(r + 1)$st term of the expansion of $(a + b)^n$ is $\frac{n!}{r!(n - r)!}a^{n-r}b^r$.

EXAMPLE 6

Find the eighth term in the expansion of $(2x - y)^{10}$.

Solution   Use the formula, with $n = 10, a = 2x, b = -y$, and $r + 1 = 8$. Notice that, since $r + 1 = 8, r = 7$.

$$\frac{n!}{r!(n - r)!}a^{n-r}b^r = \frac{10!}{7!3!}(2x)^3(-y)^7$$

$$= 120(8x^3)(-y^7)$$

$$= -960x^3y^7$$

EXERCISE SET 14.5

STUDY GUIDE/SSM   CD/VIDEO   PH MATH TUTOR CENTER   MathXL®Tutorials ON CD   MathXL®   MyMathLab®

Use Pascal's triangle to expand the binomial. See Example 1.

1. $(m + n)^3$

2. $(x + y)^4$

3. $(c + d)^5$

4. $(a + b)^6$

5. $(y - x)^5$

6. $(q - r)^7$

7. Explain how to generate a row of Pascal's triangle.

8. Write the $n = 8$ row of Pascal's triangle.

Evaluate each expression. See Example 2.

9. $\dfrac{8!}{7!}$

10. $\dfrac{6!}{0!}$

 11. $\dfrac{7!}{5!}$

12. $\dfrac{8!}{5!}$

13. $\dfrac{10!}{7!2!}$

14. $\dfrac{9!}{5!3!}$

15. $\dfrac{8!}{6!0!}$

16. $\dfrac{10!}{4!6!}$

Use the binomial formula to expand each binomial. See Examples 3 through 5.

17. $(a + b)^7$

18. $(x + y)^8$

19. $(a + 2b)^5$

20. $(x + 3y)^6$

21. $(q + r)^9$

22. $(b + c)^6$

23. $(4a + b)^5$

24. $(3m + n)^4$

25. $(5a - 2b)^4$

26. $(m - 4)^6$

27. $(2a + 3b)^3$

28. $(4 - 3x)^5$

29. $(x + 2)^5$

30. $(3 + 2a)^4$

Find the indicated term. See Example 6.

31. The fifth term of the expansion of $(c - d)^5$

32. The fourth term of the expansion of $(x - y)^6$

33. The eighth term of the expansion of $(2c + d)^7$

34. The tenth term of the expansion of $(5x - y)^9$

35. The fourth term of the expansion of $(2r - s)^5$

36. The first term of the expansion of $(3q - 7r)^6$

37. The third term of the expansion of $(x + y)^4$

38. The fourth term of the expansion of $(a + b)^8$

39. The second term of the expansion of $(a + 3b)^{10}$

40. The third term of the expansion of $(m + 5n)^7$

REVIEW AND PREVIEW

Sketch the graph of each function. Decide whether each function is one-to-one. See Sections 8.2, 11.5, and 12.1.

41. $f(x) = |x|$

42. $g(x) = 3(x - 1)^2$

43. $H(x) = 2x + 3$

44. $F(x) = -2$

45. $f(x) = x^2 + 3$

46. $h(x) = -(x + 1)^2 - 4$

Concept Extensions

47. Expand the expression $\left(\sqrt{x} + \sqrt{3}\right)^5$

48. Find the term containing x^2 in the expansion of $\left(\sqrt{x} - \sqrt{5}\right)^6$

Evaluate the following.

The notation $\dbinom{n}{r}$ means $\dfrac{n!}{r!(n - r)!}$. For example,

$$\binom{5}{3} = \frac{5!}{3!(5 - 3)!} = \frac{5!}{3!2!} = \frac{5 \cdot 4 \cdot 3 \cdot 2 \cdot 1}{(3 \cdot 2 \cdot 1) \cdot (2 \cdot 1)} = 10.$$

49. $\dbinom{9}{5}$

50. $\dbinom{4}{3}$

51. $\dbinom{8}{2}$

52. $\dbinom{12}{11}$

53. Show that $\dbinom{n}{n} = 1$ for any whole number n.

STUDY SKILLS REMINDER

Are You Prepared for Your Final Exam?

To prepare for your final exam, try the following study techniques.

▶ Review the material that you will be responsible for on your exam. Also check your notebook for any lecture notes that you highlighted.

▶ Review any formulas that you may need to memorize.

▶ Check to see if your instructor or math department will be conducting a final exam review.

▶ Check with your instructor to see whether there are final exams from previous semesters/quarters that are available to students for study.

▶ Use your previously taken tests as a practice final exam. To do so, rewrite the test questions in mixed order on blank sheets of paper. This will help you prepare for exam conditions.

▶ If you are unsure of a few topics, see your instructor or visit a learning lab for further assistance. Also, viewing the video segment of a troublesome section will help.

▶ If you need further exercises to work, try the chapter tests at the end of appropriate chapters.

Good luck! I hope you have enjoyed this textbook and your algebra course.

CHAPTER 14 PROJECT

Modeling College Tuition

Annual college tuition has steadily increased since 1970. According to the College Board, by the 2000–2001 academic year, the average annual tuition at a public four-year university had increased to $3506. Similarly, average annual tuition at private four-year universities had grown to $15,531.

Over the past few years, annual tuition at four-year public universities has been increasing at an average rate of 4.5% per year. Over the same time period, annual tuition at four-year private universities has been increasing at an average rate of $804.30 per year. In this project, you will have the opportunity to model and investigate the trend in increasing tuition at public and private universities. This project may be completed by working in groups or individually.

1. Using the information given in the introductory paragraphs, decide whether the sequence of public university tuitions is arithmetic or geometric.

2. Using the information given in the introductory paragraphs, decide whether the sequence of private university tuitions is arithmetic or geometric.

3. Find the general term of the sequence that describes the pattern of average annual tuition for four-year public universities. Let $n = 1$ represent the 2000–2001 academic year.

4. Find the general term of the sequence that describes the pattern of average annual tuition for four-year private universities. Let $n = 1$ represent the 2000–2001 academic year.

5. Assuming that the rate of tuition increase remains the same, use the general term equation from Question 3 to find the average annual tuition at a four-year public university for the 2003–2004 academic year.

6. Assuming that the rate of tuition increase remains the same, use the general term equation from Question 4 to find the

average annual tuition at a four-year private university for the 2004–2005 academic year.

7. Use partial sums to find the average cost of a four-year college education at a public university for a student who started college in the 2000–2001 academic year.

8. Use partial sums to find the average cost of a four-year college education at a private university for a student who starts col-

lege in the 2003–2004 academic year. (Hint: One way to do this is to find S_7 and subtract S_3 from it. If you use this method, explain why this gives the desired sum.)

9. (Optional) Use newspapers or news magazines to find a situation that can be modeled by a sequence. Briefly describe the situation, and decide whether it is an arithmetic or a geometric sequence. Find an equation of the general term of the sequence.

Academic Year	n
2000–2001	$n = 1$
2001–2002	$n = 2$
2002–2003	$n = 3$
2003–2004	$n = 4$
2004–2005	$n = 5$
2005–2006	$n = 6$
2006–2007	$n = 7$

CHAPTER VOCABULARY CHECK

Fill in each blank with one of the words or phrases listed below.

general term	common difference	finite sequence	common ratio
Pascal's triangle	infinite sequence	factorial of n	arithmetic sequence
geometric sequence	series		

1. A(n) _____ is a function whose domain is the set of natural numbers $\{1, 2, 3, \ldots, n\}$, where n is some natural number.

2. The _____ , written $n!$, is the product of the first n consecutive natural numbers.

3. A(n) _____ is a function whose domain is the set of natural numbers.

4. A(n) _____ is a sequence in which each term (after the first) is obtained by multiplying the preceding term by a constant amount r. The constant r is called the _____ of the sequence.

5. A sum of the terms of a sequence is called a _____

6. The nth term of the sequence a_n is called the _____ .

7. A(n) _____ is a sequence in which each term (after the first) differs from the preceding term by a constant amount d. The constant d is called the _____ of the sequence.

8. A(n) triangle array of the coefficients of the terms of the expansions of $(a + b)^n$ is called _____ .

CHAPTER 14 HIGHLIGHTS

Definitions and Concepts	**Examples**

Section 14.1 Sequences

An **infinite sequence** is a function whose domain is the set of natural numbers $\{1, 2, 3, 4, \dots\}$.

Infinite Sequence

$$2, 4, 6, 8, 10, \dots$$

A **finite sequence** is a function whose domain is the set of natural numbers $\{1, 2, 3, 4, \dots, n\}$, where n is some natural number.

Finite Sequence

$$1, -2, 3, -4, 5, -6$$

The notation a_n, where n is a natural number, is used to denote a sequence.

Write the first four terms of the sequence whose general term is $a_n = n^2 + 1$.

$$a_1 = 1^2 + 1 = 2$$
$$a_2 = 2^2 + 1 = 5$$
$$a_3 = 3^2 + 1 = 10$$
$$a_4 = 4^2 + 1 = 17$$

Section 14.2 Arithmetic and Geometric Sequences

An **arithmetic sequence** is a sequence in which each term differs from the preceding term by a constant amount d, called the **common difference.**

Arithmetic Sequence

$$5, 8, 11, 14, 17, 20, \dots$$

Here, $a_1 = 5$ and $d = 3$.

The general term is

The **general term** a_n of an arithmetic sequence is given by

$$a_n = a_1 + (n - 1)d$$

where a_1 is the first term and d is the common difference.

$$a_n = a_1 + (n - 1)d \text{ or}$$
$$a_n = 5 + (n - 1)3$$

A **geometric sequence** is a sequence in which each term is obtained by multiplying the preceding term by a constant r, called the **common ratio.**

Geometric Sequence

$$12, -6, 3, -\frac{3}{2}, \dots$$

Here $a_1 = 12$ and $r = -\frac{1}{2}$.

The general term is

The **general term** a_n of a geometric sequence is given by

$$a_n = a_1 r^{n-1}$$

where a_1 is the first term and r is the common ratio.

$$a_n = a_1 r^{n-1} \text{ or}$$
$$a_n = 12\left(-\frac{1}{2}\right)^{n-1}$$

Definitions and Concepts	**Examples**

Section 14.3 Series

A sum of the terms of a sequence is called a **series.**

A shorthand notation for denoting a series is called **summation notation:**

index of summation $\rightarrow$ $\displaystyle\sum_{i=1}^{4}$ $\rightarrow$ Greek letter sigma used to mean sum

Sequence	Series	
$3, 7, 11, 15$	$3 + 7 + 11 + 15$	finite
$3, 7, 11, 15, \ldots$	$3 + 7 + 11 + 15 + \cdots$	infinite

$$\sum_{i=1}^{4} 3^i = 3^1 + 3^2 + 3^3 + 3^4$$
$$= 3 + 9 + 27 + 81$$
$$= 120$$

Section 14.4 Partial Sums of Arithmetic and Geometric Sequences

Partial sum, S_n, of the first n terms of an arithmetic sequence:

$$S_n = \frac{n}{2}(a_1 + a_n)$$

where a_1 is the first term and a_n is the nth term.

Partial sum, S_n, of the first n terms of a geometric sequence:

$$S_n = \frac{a_1(1 - r^n)}{1 - r}$$

where a_1 is the first term, r is the common ratio, and $r \neq 1$.

Sum of the terms of an infinite geometric sequence:

$$S_\infty = \frac{a_1}{1 - r}$$

where a_1 is the first term, r is the common ratio, and $|r| < 1$. (If $|r| \geq 1$, S_∞ does not exist.)

The sum of the first five terms of the arithmetic sequence
$$12, 24, 36, 48, 60, \ldots \text{ is}$$
$$S_n = \frac{5}{2}(12 + 60) = 180$$

The sum of the first five terms of the geometric sequence
$$15, 30, 60, 120, 240, \ldots \text{ is}$$
$$S_5 = \frac{15(1 - 2^5)}{1 - 2} = 465$$

The sum of the terms of the infinite geometric sequence
$$1, \frac{1}{3}, \frac{1}{9}, \frac{1}{27}, \ldots \text{ is}$$
$$S_\infty = \frac{1}{1 - \frac{1}{3}} = \frac{3}{2}$$

Section 14.5 The Binomial Theorem

The **factorial of n,** written $n!$, is the product of the first n consecutive natural numbers.

Binomial Theorem

If n is a positive integer, then

$$(a + b)^n = a^n + \frac{n}{1!}a^{n-1}b^1 + \frac{n(n - 1)}{2!}a^{n-2}b^2$$
$$+ \frac{n(n - 1)(n - 2)}{3!}a^{n-3}b^3 + \cdots + b^n$$

$$5! = 5 \cdot 4 \cdot 3 \cdot 2 \cdot 1 = 120$$

Expand $(3x + y)^4$.

$$(3x + y)^4 = (3x)^4 + \frac{4}{1!}(3x)^3(y)^1$$
$$+ \frac{4 \cdot 3}{2!}(3x)^2(y)^2 + \frac{4 \cdot 3 \cdot 2}{3!}(3x)^1 y^3 + y^4$$
$$= 81x^4 + 108x^3 y + 54x^2 y^2 + 12xy^3 + y^4$$

CHAPTER REVIEW

(14.1) Find the indicated term(s) of the given sequence.

1. The first five terms of the sequence $a_n = -3n^2$

2. The first five terms of the sequence $a_n = n^2 + 2n$

3. The one-hundredth term of the sequence $a_n = \dfrac{(-1)^n}{100}$

4. The fiftieth term of the sequence $a_n = \dfrac{2n}{(-1)^2}$

5. The general term a_n of the sequence $\dfrac{1}{6}, \dfrac{1}{12}, \dfrac{1}{18}, \dots$.

6. The general term a_n of the sequence $-1, 4, -9, 16, \dots$.

Solve the following applications.

7. The distance in feet that an olive falling from rest in a vacuum will travel during each second is given by an arithmetic sequence whose general term is $a_n = 32n - 16$, where n is the number of the second. Find the distance the olive will fall during the fifth, sixth, and seventh seconds.

8. A culture of yeast doubles every day in a geometric progression whose general term is $a_n = 100(2)^{n-1}$, where n is the number of the day just ending. Find how many days it takes the yeast culture to measure at least 10,000. Find the original measure of the yeast culture.

9. The Centers for Disease Control and Prevention (CDC) reported that a new type of virus infected approximately 450 people during 2003, the year it was first discovered. The CDC predicts that during the next decade the virus will infect three times as many people each year as the year before. Write out the first five terms of this geometric sequence, and predict the number of infected people there will be in 2007.

10. The first row of an amphitheater contains 50 seats, and each row thereafter contains 8 additional seats. Write the first ten terms of this arithmetic progression, and find the number of seats in the tenth row.

(14.2)

11. Find the first five terms of the geometric sequence whose first term is -2 and whose common ratio is $\dfrac{2}{3}$.

12. Find the first five terms of the arithmetic sequence whose first term is 12 and whose common difference is -1.5.

13. Find the thirtieth term of the arithmetic sequence whose first term is -5 and whose common difference is 4.

14. Find the eleventh term of the arithmetic sequence whose first term is 2 and whose common difference is $\dfrac{3}{4}$.

15. Find the twentieth term of the arithmetic sequence whose first three terms are 12, 7, and 2.

16. Find the sixth term of the geometric sequence whose first three terms are 4, 6, and 9.

17. If the fourth term of an arithmetic sequence is 18 and the twentieth term is 98, find the first term and the common difference.

18. If the third term of a geometric sequence is -48 and the fourth term is 192, find the first term and the common ratio.

19. Find the general term of the sequence $\dfrac{3}{10}, \dfrac{3}{100}, \dfrac{3}{1000}, \dots$.

20. Find a general term that satisfies the terms shown for the sequence $50, 58, 66, \dots$.

Determine whether each of the following sequences is arithmetic, geometric, or neither. If a sequence is arithmetic, find a_1 and d. If a sequence is geometric, find a_1 and r.

21. $\dfrac{8}{3}, 4, 6, \dots$

22. $-10.5, -6.1, -1.7$

23. $7x, -14x, 28x$

24. $3x^2, 9x^4, 81x^8, \dots$

Solve the following applications.

25. To test the bounce of a racquetball, the ball is dropped from a height of 8 feet. The ball is judged "good" if it rebounds at least 75% of its previous height with each bounce. Write out the first six terms of this geometric sequence (round to the nearest tenth). Determine if a ball is "good" that rebounds to a height of 2.5 feet after the fifth bounce.

26. A display of oil cans in an auto parts store has 25 cans in the bottom row, 21 cans in the next row, and so on, in an arithmetic progression. Find the general term and the number of cans in the top row.

27. Suppose that you save $1 the first day of a month, $2 the second day, $4 the third day, continuing to double your savings each day. Write the general term of this geometric sequence and find the amount you will save on the tenth day. Estimate the amount you will save on the thirtieth day of the month, and check your estimate with a calculator.

28. On the first swing, the length of an arc through which a pendulum swings is 30 inches. The length of the arc for each successive swing is 70% of the preceding swing. Find the length of the arc for the fifth swing.

29. Rosa takes a job that has a monthly starting salary of $900 and guarantees her a monthly raise of $150 during her 6-month training period. Find the general term of this sequence and her salary at the end of her training.

30. A sheet of paper is $\dfrac{1}{512}$-inch thick. By folding the sheet in half, the total thickness will be $\dfrac{1}{256}$ inch. A second fold produces a total thickness of $\dfrac{1}{128}$ inch. Estimate the thickness of the stack after 15 folds, and then check your estimate with a calculator.

(14.3) *Write out the terms and find the sum for each of the following.*

31. $\displaystyle\sum_{i=1}^{5}(2i - 1)$

32. $\displaystyle\sum_{i=1}^{5}i(i + 2)$

33. $\displaystyle\sum_{i=2}^{4}\dfrac{(-1)^i}{2i}$

34. $\displaystyle\sum_{i=3}^{5}5(-1)^{i-1}$

Find the partial sum of the given sequence.

35. S_4 of the sequence $a_n = (n - 3)(n + 2)$

36. S_6 of the sequence $a_n = n^2$

37. S_5 of the sequence $a_n = -8 + (n - 1)3$

38. S_3 of the sequence $a_n = 5(4)^{n-1}$

Write the sum with Σ notation.

39. $1 + 3 + 9 + 27 + 81 + 243$

40. $6 + 2 + (-2) + (-6) + (-10) + (-14) + (-18)$

41. $\dfrac{1}{4} + \dfrac{1}{16} + \dfrac{1}{64} + \dfrac{1}{256}$

42. $1 + \left(-\dfrac{3}{2}\right) + \dfrac{9}{4}$

Solve.

43. A yeast colony begins with 20 yeast and doubles every 8 hours. Write the sequence that describes the growth of the yeast, and find the total yeast after 48 hours.

44. The number of cranes born each year in a new aviary forms a sequence whose general term is $a_n = n^2 + 2n - 1$. Find the number of cranes born in the fourth year and the total number of cranes born in the first four years.

45. Harold has a choice between two job offers. Job A has an annual starting salary of $39,500 with guaranteed annual raises of $2200 for the next four years, whereas job B has an annual starting salary of $41,000 with guaranteed annual raises of $1400 for the next four years. Compare the salaries for the fifth year under each job offer.

46. A sample of radioactive waste is decaying such that the amount decaying in kilograms during year n is $a_n = 200(0.5)^n$. Find the amount of decay in the third year, and the total amount of decay in the first three years.

(14.4) *Find the partial sum of the given sequence.*

47. The sixth partial sum of the sequence $15, 19, 23, \ldots$.

48. The ninth partial sum of the sequence $5, -10, 20, \ldots$.

49. The sum of the first 30 odd positive integers

50. The sum of the first 20 positive multiples of 7

51. The sum of the first 20 terms of the sequence $8, 5, 2, \ldots$.

52. The sum of the first eight terms of the sequence $\dfrac{3}{4}, \dfrac{9}{4}, \dfrac{27}{4}, \ldots$.

53. S_4 if $a_1 = 6$ and $r = 5$

54. S_{100} if $a_1 = -3$ and $d = -6$

Find the sum of each infinite geometric sequence.

55. $5, \dfrac{5}{2}, \dfrac{5}{4}, \ldots$

56. $18, -2, \dfrac{2}{9}, \ldots$

57. $-20, -4, -\dfrac{4}{5}, \ldots$

58. $0.2, 0.02, 0.002, \ldots$

Solve.

59. A frozen yogurt store owner cleared $20,000 the first year he owned his business and made an additional 15% over the previous year in each subsequent year. Find how much he made during his fourth year of business. Find his total earnings during the first 4 years (round to the nearest dollar).

60. On his first morning in a television assembly factory, a trainee takes 0.8 times as long to assemble each television as he took to assemble the one before. If it took him 40 minutes to assemble the first television, find how long it takes him to assemble the fourth television. Find the total time he takes to assemble the first four televisions. (Round to the nearest minute.)

61. During the harvest season a farmer can rent a combine machine for $100 the first day, with the rental fee decreasing $7 for each additional day. Find how much the farmer pays for the rental on the seventh day. Find how much total rent the farmer pays for 7 days.

62. A rubber ball is dropped from a height of 15 feet and rebounds 80% of its previous height after each bounce. Find the total distance the ball travels before it comes to rest.

63. After a pond was sprayed once with insecticide, 1800 mosquitoes were killed the first day, 600 the second day, 200 the third day, and so on. Find the total number of mosquitoes killed during the first 6 days after the spraying (round to the nearest unit).

64. See Exercise 63. Find the day on which the insecticide is no longer effective, and find the total number of mosquitoes killed (round to the nearest mosquito).

65. Use the formula S_∞ to write $0.55\bar{5}$ as a fraction.

66. A movie theater has 27 seats in the first row, 30 seats in the second row, 33 seats in the third row, and so on. Find the total number of seats in the theater if there are 20 rows.

(14.5) Use Pascal's triangle to expand each binomial.

67. $(x + z)^5$

68. $(y - r)^6$

69. $(2x + y)^4$

70. $(3y - z)^4$

Use the binomial formula to expand the following.

71. $(b + c)^8$

72. $(x - w)^7$

73. $(4m - n)^4$

74. $(p - 2r)^5$

Find the indicated term.

75. The fourth term of the expansion of $(a + b)^7$

76. The eleventh term of the expansion of $(y + 2z)^{10}$

CHAPTER 14 TEST

Remember to use your Chapter Test Prep Video CD to help you study and view solutions to the test questions you need help with.

Find the indicated term(s) of the given sequence.

1. The first five terms of the sequence $a_n = \dfrac{(-1)^n}{n + 4}$

2. The eightieth term of the sequence $a_n = 10 + 3(n - 1)$

3. The general term of the sequence $\dfrac{2}{5}, \dfrac{2}{25}, \dfrac{2}{125}, \ldots$

4. The general term of the sequence $-9, 18, -27, 36, \ldots$

Find the partial sum of the given sequence.

5. S_5 of the sequence $a_n = 5(2)^{n-1}$

6. S_{30} of the sequence $a_n = 18 + (n - 1)(-2)$

7. S_∞ of the sequence $a_1 = 24$ and $r = \dfrac{1}{6}$

8. S_∞ of the sequence $\dfrac{3}{2}, -\dfrac{3}{4}, \dfrac{3}{8}, \ldots$

9. $\displaystyle\sum_{i=1}^{4} i(i - 2)$ **10.** $\displaystyle\sum_{i=2}^{4} 5(2)^i(-1)^{i-1}$

Expand each binomial.

11. $(a - b)^6$ **12.** $(2x + y)^5$

Solve the following applications.

13. The population of a small town is growing yearly according to the sequence defined by $a_n = 250 + 75(n - 1)$, where n is the number of the year just beginning. Predict the population at the beginning of the tenth year. Find the town's initial population.

14. A gardener is making a triangular planting with one shrub in the first row, three shrubs in the second row, five shrubs in the third row, and so on, for eight rows. Write the finite series of this sequence, and find the total number of shrubs planted.

15. A pendulum swings through an arc of length 80 centimeters on its first swing. On each successive swing, the length of the arc is $\dfrac{3}{4}$ the length of the arc on the preceding swing. Find the length of the arc on the fourth swing, and find the total arc length for the first four swings.

16. See Exercise 15. Find the total arc length before the pendulum comes to rest.

17. A parachutist in free-fall falls 16 feet during the first second, 48 feet during the second second, 80 feet during the third second, and so on. Find how far he falls during the tenth second. Find the total distance he falls during the first 10 seconds.

18. Use the formula S_∞ to write $0.42\overline{42}$ as a fraction.

CHAPTER CUMULATIVE REVIEW

1. Evaluate.

    a. $(-2)^3$ **b.** -2^3

    c. $(-3)^2$ **d.** -3^2

2. Simplify each expression.

    a. $3a - (4a + 3)$

    b. $(5x - 3) + (2x + 6)$

    c. $4(2x - 5) - 3(5x + 1)$

3. Subtract $4x - 2$ from $2x - 3$. Simplify, if possible.

4. Sara bought a digital camera for \$344.50 including tax. If the tax rate is 6%, what was the price of the camera before taxes.

5. Write an equation of the line with y-intercept $(0, -3)$ and slope of $\dfrac{1}{4}$.

6. Find an equation of a line through $(3, -2)$ and parallel to $3x - 2y = 6$. Write the equation using function notation.

7. Find an equation of the line through $(2, 5)$ and $(-3, 4)$. Write the equation in standard form.

8. Solve $y^3 + 5y^2 - y = 5$

9. Use synthetic division to divide $(x^4 - 2x^3 - 11x^2 + 5x + 34)$ by $(x + 2)$.

10. Perform the indicated operation and simplify if possible.
$$\frac{5}{3a - 6} - \frac{a}{a - 2} + \frac{3 + 2a}{5a - 10}$$

11. Simplify the following.

    a. $\sqrt{50}$ **b.** $\sqrt[3]{24}$

    c. $\sqrt{26}$ **d.** $\sqrt[4]{32}$

12. Solve $\sqrt{3x + 6} - \sqrt{7x - 6} = 0$

13. Find the interest rate r if \$2000 compounded annually grows to \$2420 in 2 years.

14. Rationalize each denominator.

    a. $\sqrt[3]{\dfrac{4}{3x}}$ **b.** $\dfrac{\sqrt{2} + 1}{\sqrt{2} - 1}$

15. Solve $(x - 3)^2 - 3(x - 3) - 4 = 0$.

16. Solve $\dfrac{10}{(2x + 4)^2} - \dfrac{1}{2x + 4} = 3$

17. Solve $\dfrac{5}{x + 1} < -2$.

18. Graph $f(x) = (x + 2)^2 - 6$. Find the vertex and axis of symmetry.

19. A rock is thrown upward from the ground. Its height in feet above ground after t seconds is given by the function $f(t) = -16t^2 + 20t$. Find the maximum height of the rock and the number of seconds it took for the rock to reach its maximum height.

20. Find the vertex of $f(x) = x^2 + 3x - 18$

21. If $f(x) = x^2$ and $g(x) = x + 3$, find each composition.

    a. $(f \circ g)(2)$ and $(g \circ f)(2)$

    b. $(f \circ g)(x)$ and $(g \circ f)(x)$

22. Find the inverse of $f(x) = -2x + 3$

23. Find the inverse of the one-to-one function.
$f = \{(0, 1), (-2, 7), (3, -6), (4, 4)\}$

24. If $f(x) = x^2 - 2$ and $g(x) = x + 1$, find each composition.

    a. $(f \circ g)(2)$ and $(g \circ f)(2)$

    b. $(f \circ g)(x)$ and $(g \circ f)(x)$

25. Solve each equation for x.

    a. $2^x = 16$ **b.** $9^x = 27$

    c. $4^{x+3} = 8^x$

26. Solve each equation.

    a. $\log_2 32 = x$ **b.** $\log_4 \dfrac{1}{64} = x$

    c. $\log_{\frac{1}{2}} x = 5$

27. Simplify.

    a. $\log_3 3^2$ **b.** $\log_7 7^{-1}$

    c. $5^{\log_5 3}$ **d.** $2^{\log_2 6}$

28. Solve each equation for x.

    a. $4^x = 64$ **b.** $8^x = 32$

    c. $9^{x+4} = 243^x$

29. Write each sum as a single logarithm.

    a. $\log_{11} 10 + \log_{11} 3$ **b.** $\log_3 \dfrac{1}{2} + \log_3 12$

    c. $\log_2(x + 2) + \log_2 x$

30. Find the exact value.

    a. $\log 100{,}000$

    b. $\log 10^{-3}$

    c. $\ln \sqrt[5]{e}$

    d. $\ln e^4$

31. Find the amount owed at the end of 5 years if \$1600 is loaned at a rate of 9% compounded continuously.

32. Write each expression as a single logarithm.

a. $\log_6 5 + \log_6 4$

b. $\log_8 12 - \log_8 4$

c. $2 \log_2 x + 3 \log_2 x - 2 \log_2(x - 1)$

33. Solve $3^x = 7$.

34. Using $A = P\left(1 + \dfrac{r}{n}\right)^{nt}$, find how long it takes $5000 to double if it is invested at 2% interest compounded quarterly. Round to the nearest tenth.

35. Solve $\log_4(x - 2) = 2$.

36. Solve $\log_4 10 - \log_4 x = 2$

37. Graph $\dfrac{x^2}{16} - \dfrac{y^2}{25} = 1$.

38. Find the distance between $(8, 5)$ and $(-2, 4)$.

39. Solve the system: $\begin{cases} y = \sqrt{x} \\ x^2 + y^2 = 6 \end{cases}$

40. Solve the system: $\begin{cases} x^2 + y^2 = 36 \\ x - y = 6 \end{cases}$

41. Graph $\dfrac{x^2}{9} + \dfrac{y^2}{16} \leq 1$.

42. Graph $\begin{cases} y \geq x^2 \\ y \leq 4 \end{cases}$.

43. Write the first five terms of the sequence whose general term is given by $a_n = n^2 - 1$.

44. If the general term of a sequence is $a_n = \dfrac{n}{n + 4}$, find a_8.

45. Find the eleventh term of the arithmetic sequence whose first three terms are $2, 9,$ and 16.

46. Find the sixth term of the geometric sequence $2, 10, 50, \ldots$

47. Evaluate.

a. $\displaystyle\sum_{i=0}^{6} \dfrac{i - 2}{2}$ **b.** $\displaystyle\sum_{i=3}^{5} 2^i$

48. Evaluate.

a. $\displaystyle\sum_{i=0}^{4} i(i + 1)$ **b.** $\displaystyle\sum_{i=0}^{3} 2^i$

49. Find the sum of the first 30 positive integers.

50. Find the third term of the expansion of $(x - y)^6$.

Solving Equations and Inequalities— Study Skills Practice

Each set of exercises corresponds to the Study Skills Reminder in the section noted. After reviewing the outline in that section's Study Skills Reminder, solve each equation or inequality.

Section 2.3

Solve.

1. $3x - 4 = 3(2x - 1) + 7$

2. $\dfrac{7}{5} + \dfrac{y}{10} = 2$

3. $5 + 2x = 5(x + 1)$

4. $5 + 2x = 2(x + 1)$

Section 2.7

Solve. Write inequality solutions in interval notation.

1. $\dfrac{x + 3}{2} > 1$

2. $4(x - 2) + 3x \geq 9(x - 1) - 2$

3. $\dfrac{x - 2}{2} - \dfrac{x - 4}{3} = \dfrac{5}{6}$

4. $6(x + 1) - 2 = 6x + 4$

Section 6.5

Solve. Write inequality solutions in interval notation.

1. $2x^2 - 17x = 9$

2. $14x - 17x + 6 = 6(x + 1)$

3. $x(x + 5) = 36$

4. $\dfrac{3x - 5}{4} > 1$

5. $3(2x - 1) < 9$

6. $\dfrac{2}{3}x \leq \dfrac{5}{6}$

Section 7.5

Solve. Write inequality solutions in interval notation.

1. $\dfrac{x}{10} - \dfrac{1}{2} = \dfrac{7}{5x}$

2. $1 - 5x = 9$

3. $x(2x - 11) = 21$

4. $5(x - 3) + 2x = 7(x + 1) - 22$

5. $-8 + 2x - 4 \le -2$

6. $x^3 = 25x$

Section 9.1

Solve. Write inequality solutions in interval notation.

1. $x - 2 \le 1$ and $3x - 1 \ge -4$

2. $-2 < x - 1 < 5$

3. $-2x + 2.5 = -7.7$

4. $-5x > 20$

5. $x \le -3$ or $x \le -5$

Section 9.2

Solve. Write inequality solutions in interval notation.

1. $|2 + 3x| = 7$

2. $5x < -10$ or $3x - 4 > 2$

3. $\dfrac{5t}{2} - \dfrac{3t}{4} = 7$

4. $|x - 2| = |x + 1|$

5. $5(x - 3) + x + 2 \ge 3(x + 2) + 2x$

6. $|4x + 7| = |-35|$

Section 9.3

Solve. Write inequality solutions in interval notation.

1. $|x - 11| \ge 7$

2. $|x - 11| = 7$

3. $-5 < x - (2x + 3) < 0$

4. $|9x| - 8 = -1$

5. $\dfrac{4x}{5} - 1 = \dfrac{x}{2} + 2$

Section 9.6

Solve. Write inequality solutions in interval notation.

1. $4^x = 8^{x-1}$

2. $2^{x+7} = 5$

3. $x(x - 9) > 0$

4. $\sqrt[3]{6x + 1} - 2 = 0$

5. $\log_4(x^2 - 3x) = 1$

6. $3x^2 + 2x = 3$

7. $\dfrac{6}{x - 2} \ge 3$

8. $\dfrac{x + 3}{5} - \dfrac{7 - x}{3} = \dfrac{3}{10}$

9. $\log_3(2x + 1) - \log_3 x = 1$

10. $|6x - 5| > 1$

Section 10.6

Solve. Write inequality solutions in interval notation.

1. $x(3x + 14) = 5$

2. $\dfrac{2}{x - 2} + \dfrac{3}{x + 2} = \dfrac{7}{x^2 - 4}$

3. $|5x - 4| = |4x + 1|$

4. $\sqrt{6x - 1} = 7$

5. $-2(x - 4) + 3x \le -3(x + 2) - 2$

6. $|-x + 3| > 5$

Section 11.2

Solve. Write inequality solutions in interval notation.

1. $(x - 2)^2 = 17$

2. $x^2 - 5x + 2 = 0$

3. $x^2 - 5x + 6 = 0$

4. $\dfrac{x}{4} - \dfrac{3}{2} > 1$

5. $\sqrt{2x + 30} = x + 3$

6. $|3x + 11| - 12 = -1$

7. $\dfrac{3x^2 - 7}{3x^2 - 8x - 3} = \dfrac{1}{x - 3} + \dfrac{2}{3x + 1}$

8. $\left|\dfrac{x - 4}{3}\right| < 10$

Section 11.4

Solve. Write inequality solutions in interval notation.

1. $x^2 - 3x - 10 = 0$

2. $x^2 - 3x - 10 < 0$

3. $\dfrac{x + 4}{x - 10} = 0$

4. $\dfrac{x + 4}{x - 10} \ge 0$

5. $\sqrt{x - 7} - 12 = -8$

6. $(5x - 2)^2 = 10$

7. $\left|\dfrac{3x + 5}{2}\right| = -9$

8. $2(3x - 6) + 4 = 7(x + 1) - 2x$

9. $-4(x - 3) + 2x < 6x + 4$

10. $x \le 7$ and $2(x + 1) + 3 \le 7$

Operations on Decimals

To **add** or **subtract** decimals, write the numbers vertically with decimal points lined up. Add or subtract as with whole numbers and place the decimal point in the answer directly below the decimal points in the problem.

EXAMPLE 1

Add $5.87 + 23.279 + 0.003$.

Solution

```
   5.87
  23.279
+  0.003
 -------
  29.152
```

EXAMPLE 2

Subtract $32.15 - 11.237$.

Solution

$$
\begin{array}{cccccc}
 & & \overset{1}{} & \overset{11}{} & \overset{4}{} & \overset{10}{} \\
 & 3 & \cancel{2} . & \cancel{1} & \cancel{5} & \cancel{0} \\
- & 1 & 1 . & 2 & 3 & 7 \\
\hline
 & 2 & 0 . & 9 & 1 & 3 \\
\end{array}
$$

To **multiply** decimals, multiply the numbers as if they were whole numbers. The decimal point in the product is placed so that the number of decimal places in the product is the same as the sum of the number of decimal places in the factors.

EXAMPLE 3

Multiply 0.072×3.5.

Solution

```
   0.072    3 decimal places
×    3.5    1 decimal place
 ------
    360
   216
 ------
  0.2520    4 decimal places
```

To **divide** decimals, move the decimal point in the divisor to the right of the last digit. Move the decimal point in the dividend the same number of places that the decimal point in the divisor was moved. The decimal point in the quotient lies directly above the decimal point in the dividend.

EXAMPLE 4

Divide $9.46 \div 0.04$.

Solution
$$
\begin{array}{r}
236.5 \\
04.\overline{)946.0} \\
-8 \\
\hline
14 \\
-12 \\
\hline
26 \\
-24 \\
\hline
20 \\
-20 \\
\hline
\end{array}
$$

APPENDIX B EXERCISE SET

Perform the indicated operations.

1. $9.076 + 8.004$

2. $\begin{array}{r} 6.3 \\ \times\ 0.05 \end{array}$

3. $\begin{array}{r} 27.004 \\ -14.2 \end{array}$

4. $\begin{array}{r} 0.0036 \\ 7.12 \\ 32.502 \\ +0.05 \end{array}$

5. $\begin{array}{r} 107.92 \\ +3.04 \end{array}$

6. $7.2 \div 4$

7. $10 - 7.6$

8. $40 \div 0.25$

9. $126.32 - 97.89$

10. $\begin{array}{r} 3.62 \\ 7.11 \\ 12.36 \\ 4.15 \\ +2.29 \end{array}$

11. $\begin{array}{r} 3.25 \\ \times\ 70 \end{array}$

12. $\begin{array}{r} 26.014 \\ -\ 7.8 \end{array}$

13. $8.1 \div 3$

14. $\begin{array}{r} 1.2366 \\ 0.005 \\ 15.17 \\ +0.97 \end{array}$

15. $55.405 - 6.1711$

16. $8.09 + 0.22$

17. $60 \div 0.75$

18. $20 - 12.29$

19. $7.612 \div 100$

20. $\begin{array}{r} 8.72 \\ 1.12 \\ 14.86 \\ 3.98 \\ +\ 1.99 \end{array}$

21. $12.312 \div 2.7$

22. $0.443 \div 100$

23. $\begin{array}{r} 569.2 \\ 71.25 \\ +\ 8.01 \end{array}$

24. $3.706 - 2.91$

25. $768 - 0.17$

26. $63 \div 0.28$

27. $12 + 0.062$

28. $0.42 + 18$

29. $76 - 14.52$

30. $1.1092 \div 0.47$

31. $3.311 \div 0.43$

32. $7.61 + 0.0004$

33. $\begin{array}{r} 762.12 \\ 89.7 \\ +\ 11.55 \end{array}$

34. $444 \div 0.6$

35. $23.4 - 0.821$

36. $3.7 + 5.6$

37. $476.12 - 112.97$

38. $19.872 \div 0.54$

39. $0.007 + 7$

40. $\begin{array}{r} 51.77 \\ +\ 3.6 \end{array}$

Review of Algebra Topics

Recall that equations model many real-life problems. For example, we can use a linear equation to calculate the increase in households (in millions) with digital cameras.

With the help of your computer, digital cameras allow you to see your pictures and make copies immediately, send them in e-mail or use them on a Web page. Current projected number of households with these cameras are shown in the graph below.

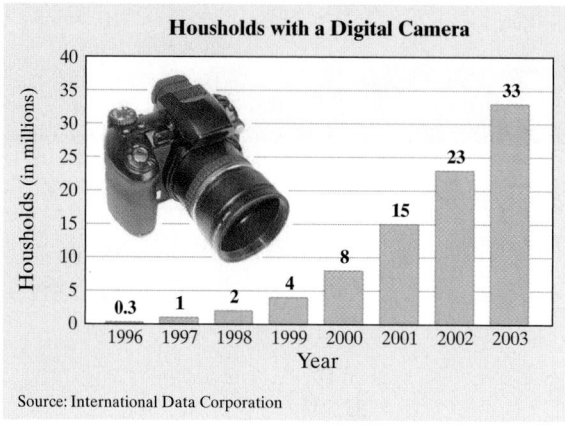

Housholds with a Digital Camera

Source: International Data Corporation

To find the increase in households from 2002 to 2003, for example, we can use the equation below.

In words:	Increase in households	is	households in 2003	minus	households in 2002
Translate:	x	$=$	31.5	$-$	22.8

Since our variable x (increase in households) is by itself on one side of the equation, we can find the value of x by simplifying the right side.

$$x = 8.7$$

The increase in households with digital cameras from 2002 to 2003 is $8.7 million.

The equation, $x = 31.5 - 22.8$ is a linear equation in one variable. In this section, we review solving linear equations and quadratic equations that can be solved by factoring. We will learn other methods for solving quadratic equations in Chapter 10.

C-1   EQUATIONS

EXAMPLE 1

Solve: $2(x - 3) = 5x - 9$.

Solution   First, use the distributive property.

$$2(x - 3) = 5x - 9$$
$$2x - 6 = 5x - 9 \quad \text{Use the distributive property.}$$

This is a linear equation so next, get variable terms on the same side of the equation by subtracting $5x$ from both sides.

$$2x - 6 = 5x - 9 \quad \text{Subtract 5x from both sides.}$$
$$-3x - 6 = -9 \quad \text{Simplify.}$$
$$-3x - 6 = -9 \quad \text{Add 6 to both sides.}$$
$$-3x = -3 \quad \text{Simplify.}$$
$$\frac{-3x}{} = \frac{-3}{} \quad \text{Divide both sides by } -3.$$
$$x = 1$$

Let $x = 1$ in the original equation to see that 1 is the solution.

EXAMPLE 2

Solve: $y\dfrac{y}{3} - \dfrac{y}{4} = \dfrac{1}{6}$.

Solution   First, clear the equation of fractions by multiplying both sides of the equation by 12, the LCD of denominators 3, 4, and 6.

$$\frac{y}{3} - \frac{y}{4} = \frac{1}{6}$$

$$\left(\frac{y}{3} - \frac{y}{4}\right) = \left(\frac{1}{6}\right) \quad \text{Multiply both sides by the LCD 12.}$$

$$\left(\frac{y}{3}\right) - \left(\frac{y}{4}\right) = 2 \quad \text{Apply the distributive property.}$$

$$4y - 3y = 2 \quad \text{Simplify.}$$

This is a linear equation, so we simply combine like terms on the left side to solve.

$$y = 2 \quad \text{Simplify.}$$

Check: To check, let $y = 2$ in the original equation.

$$\frac{}{3} - \frac{}{4} = \frac{1}{6} \quad \text{Original equation}$$

$$\frac{}{3} - \frac{}{4} \stackrel{?}{=} \frac{1}{6} \quad \text{Let } y = 2.$$

$$\frac{8}{12} - \frac{6}{12} \stackrel{?}{=} \frac{1}{6} \quad \text{Write fractions with the LCD.}$$

$$\frac{2}{12} \stackrel{?}{=} \frac{1}{6} \qquad \text{Subtract.}$$

$$\frac{1}{6} = \frac{1}{6} \qquad \text{Simplify.}$$

This is a true statement, so the solution is 2.

EXAMPLE 3

Solve $3(x^2 + 4) + 5 = -6(x^2 + 2x) + 13$.

Solution First, use the distributive property.

$$3(x^2 + 4) + 5 = -6(x^2 + 2x) + 13$$

$$3x^2 + 12 + 5 = -6x^2 - 12x + 13 \qquad \text{Apply the distributive property.}$$

This is a quadratic equation so we write it in standard form. In other words, we write the equation so that one side is 0.

$$9x^2 + 12x + 4 = 0$$

$$(3x + 2)(3x + 2) = 0 \qquad \text{Rewrite the equation so that one side is 0.}$$

$$3x + 2 = 0 \quad \text{or} \quad 3x + 2 = 0 \qquad \text{Factor}$$

$$3x = -2 \quad \text{or} \quad 3x = -2 \qquad \text{Set each factor equal to 0.}$$

Solve each equation.

$$x = -\frac{2}{3} \quad \text{or} \quad x = -\frac{2}{3}$$

The solution is $-\frac{2}{3}$. Check by substituting $-\frac{2}{3}$ into the original equation.

EXAMPLE 4

Solve for $x \dfrac{x + 5}{2} + \dfrac{1}{2} = 2x - \dfrac{x - 3}{8}$.

Solution Multiply both sides of the equation by 8, the LCD of 2 and 8.

$$\left(\frac{x + 5}{2} + \frac{1}{2}\right) = \left(2x - \frac{x - 3}{8}\right) \qquad \text{Multiply both sides by 8.}$$

$$\left(\frac{x + 5}{2}\right) + \quad \cdot\frac{1}{2} = \quad \cdot 2x - \left(\frac{x - 3}{8}\right) \qquad \text{Apply the distributive property.}$$

$$4(x + 5) + 4 = 16x - (x - 3) \qquad \text{Simplify.}$$

$$4x + \quad = 16x - x + 3$$

$$4x + \quad = 15x + 3 \qquad \text{Combine like terms.}$$

This equation is linear, so we get variable terms on one side and numbers on the other side.

$$-11x + 24 = 3 \qquad \text{Subtract 15x from both sides.}$$

$$-11x = -21 \qquad \text{Subtract 24 from both sides.}$$

$$\frac{-11x}{} = \frac{-21}{} \qquad \text{Divide both sides by } -11.$$

$$x = \frac{21}{11} \qquad \text{Simplify.}$$

To check, verify that replacing x with $\dfrac{21}{11}$ makes the original equation true. The solution is $\dfrac{21}{11}$.

EXAMPLE 5

Solve $2x^2 = \dfrac{17}{3}x + 1$.

Solution

$$2x^2 = \frac{17}{3}x + 1$$

$$3(2x^2) = 3\left(\frac{17}{3}x + 1\right) \quad \text{Clear the equation of fractions.}$$

$$6x^2 = 17x + 3 \quad \text{Apply the distributive property.}$$

This equation is quadratic, so we write the equation so that one side is 0.

$$6x^2 - 17x - 3 = 0 \qquad \text{Rewrite the equation in standard form.}$$

$$(6x + 1)(x - 3) = 0 \qquad \text{Factor.}$$

$$6x + 1 = 0 \quad \text{or} \quad x - 3 = 0 \qquad \text{Set each factor equal to zero.}$$

$$6x = -1$$

$$x = -\frac{1}{6} \quad \text{or} \quad x = 3 \qquad \text{Solve each equation.}$$

The solutions are $-\dfrac{1}{6}$ and 3.

APPENDIX C-1 EXERCISE SET

Solve each equation. See Examples 1 through 5.

1. $x^2 + 11x + 24 = 0$

2. $y^2 - 10y + 24 = 0$

3. $3x - 4 - 5x = x + 4 + x$

4. $13x - 15x + 8 = 4x + 2 - 24$

5. $12x^2 + 5x - 2 = 0$

6. $3y^2 - y - 14 = 0$

7. $z^2 + 9 = 10z$

8. $n^2 + n = 72$

9. $5(y + 4) = 4(y + 5)$

10. $6(y - 4) = 3(y - 8)$

11. $0.6x - 10 = 1.4x - 14$

12. $0.3x + 2.4 = 0.1x + 4$

13. $x(5x + 2) = 3$

14. $n(2n - 3) = 2$

15. $6x - 2(x - 3) = 4(x + 1) + 4$

16. $10x - 2(x + 4) = 8(x - 2) + 6$

17. $\dfrac{3}{8} + \dfrac{b}{3} = \dfrac{5}{12}$

18. $\dfrac{a}{2} + \dfrac{7}{4} = 5$

19. $x^2 - 6x = x(8 + x)$

20. $n(3 + n) = n^2 + 4n$

21. $\dfrac{z^2}{6} - \dfrac{z}{2} - 3 = 0$

22. $\dfrac{c^2}{20} - \dfrac{c}{4} + \dfrac{1}{5} = 0$

23. $z + 3(2 + 4z) = 6(z + 1) + 5z$

24. $4(m - 6) - m = 8(m - 3) - 5m$

25. $\dfrac{x^2}{2} + \dfrac{x}{20} = \dfrac{1}{10}$

26. $\dfrac{y^2}{30} = \dfrac{y}{15} + \dfrac{1}{2}$

27. $\dfrac{4t^2}{5} = \dfrac{t}{5} + \dfrac{3}{10}$

28. $\dfrac{5x^2}{6} - \dfrac{7x}{2} + \dfrac{2}{3} = 0$

29. $\dfrac{3t + 1}{8} = \dfrac{5 + 2t}{7} + 2$

30. $4 - \dfrac{2z + 7}{9} = \dfrac{7 - z}{12}$

31. $\dfrac{m - 4}{3} - \dfrac{3m - 1}{5} = 1$

32. $\dfrac{n + 1}{8} - \dfrac{2 - n}{3} = \dfrac{5}{6}$

33. $3x^2 = -x$

34. $y^2 = -5y$

35. $x(x - 3) = x^2 + 5x + 7$

36. $z^2 - 4z + 10 = z(z - 5)$

37. $3(t - 8) + 2t = 7 + t$

38. $7c - 2(3c + 1) = 5(4 - 2c)$

39. $-3(x - 4) + x = 5(3 - x)$

40. $-4(a + 1) - 3a = -7(2a - 3)$

41. $(x - 1)(x + 4) = 24$

42. $(2x - 1)(x + 2) = -3$

43. $\dfrac{x^2}{4} - \dfrac{5}{2}x + 6 = 0$

44. $\dfrac{x^2}{18} + \dfrac{x}{2} + 1 = 0$

45. $y^2 + \dfrac{1}{4} = -y$

46. $\dfrac{x^2}{10} + \dfrac{5}{2} = x$

47. Which solution strategies are incorrect? Why?

  a. Solve $(y - 2)(y + 2) = 4$ by setting each factor equal to 4.

  b. Solve $(x + 1)(x + 3) = 0$ by setting each factor equal to 0.

c. Solve $z^2 + 5z + 6 = 0$ by factoring $z^2 + 5z + 6$ and setting each factor equal to 0.

d. Solve $x^2 + 6x + 8 = 10$ by factoring $x^2 + 6x + 8$ and setting each factor equal to 0.

48. Describe two ways a linear equation differs from a quadratic equation.

Find the value of K such that the equations are equivalent.

49. $3.2x + 4 = 5.4x - 7$
$3.2x = 5.4x + K$

50. $-7.6y - 10 = -1.1y + 12$
$-7.6y = -1.1y + K$

51. $\dfrac{x}{6} + 4 = \dfrac{x}{3}$
$x + K = 2x$

52. $\dfrac{5x}{4} + \dfrac{1}{2} = \dfrac{x}{2}$
$5x + K = 2x$

Solve and check.

53. $2.569x = -12.48534$

54. $-9.112y = -47.537304$

55. $2.86z - 8.1258 = -3.75$

56. $1.25x - 20.175 = -8.15$

REVIEW AND PREVIEW

Translate each phrase into an expression. Use the variable x to represent each unknown number. See Section 1.4.

57. the quotient of 8 and a number

58. the sum of 8 and a number

59. the product of 8 and a number 8x

60. the difference of 8 and a number

61. 2 more than three times a number

62. 5 subtracted from twice a number

C-2   PROBLEM SOLVING

Objectives

1   Write algebraic expressions that can be simplified.

2   Apply the steps for problem solving.

1   In order to prepare for problem solving, we practice writing algebraic expressions that can be simplified.

Our first example involves consecutive integers and perimeter. Recall that *consecutive integers* are integers that follow one another in order. Study the examples of consecutive, even, and odd integers and their representations.

Consecutive Integers: **Consecutive Even Integers:**   **Consecutive Odd Integers:**

                    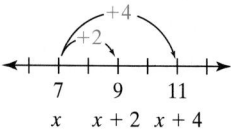

EXAMPLE 1

Write the following as algebraic expressions. Then simplify.

a. The sum of three consecutive integers, if x is the first consecutive integer.

△   **b.** The perimeter of the triangle with sides of length x, $5x$, and $6x - 3$.

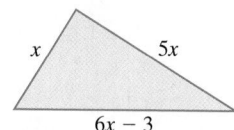

Solution **a.** Recall that if x is the first integer, then the next consecutive integer is 1 more, or $x + 1$ and the next consecutive integer is 1 more than $x + 1$, or $x + 2$.

In words:

first integer	plus	next consecutive integer	plus	next consecutive integer

Translate: x + $(x + 1)$ + $(x + 2)$

Then $x + (x + 1) + (x + 2) = x + x + 1 + x + 2$

$$= 3x + 3 \quad \text{Simplify by combining like terms.}$$

b. The perimeter of a triangle is the sum of the lengths of the sides.

In words:

side	+	side	+	side

Translate: x + $5x$ + $(6x - 3)$

Then $x + 5x + (6x - 3) = x + 5x + 6x - 3$

$$= 12x - 3 \quad \text{Simplify.}$$

EXAMPLE 2

The three busiest airports in the United States are in the cities of Chicago, Atlanta, and Los Angeles. The airport in Atlanta has 15.3 million more arrivals and departures than the Los Angeles airport. The Chicago airport has 5.8 million more arrivals and departures than the Los Angeles airport. Write the sum of the arrivals and departures from these three cities as a simplified algebraic expression. Let x be the number of arrivals and departures at the Los Angeles airport. (*Source: The World Almanac*, 2003)

Solution If $x =$ millions of arrivals and departures at the Los Angeles airport, then

$x + 15.3 =$ millions of arrivals and departures at the Atlanta airport and

$x + 5.8 =$ millions of arrivals and departures at the Chicago airport

Since we want their sum, we have

In words:

arrivals and departures at Los Angeles	+	arrivals and departures at Atlanta	+	arrivals and departures at Chicago.

Translate: x + $(x + 15.3)$ + $(x + 5.8)$

Then $x + (x + 15.3) + (x + 5.8) = x + x + 15.3 + x + 5.8$

$$= 3x + 21.1 \quad \text{Combine like terms.}$$

In Exercise 27, we will find the actual number of arrivals and departures at these airports.

2   Our main purpose for studying algebra is to solve problems. The following problem-solving strategy will be used throughout this text and may also be used to solve real-life problems that occur outside the mathematics classroom.

> ### Helpful Hint
> You may want to begin this section by studying key words and phrases and their translations in Sections 1.2 Objective 5 and 1.4 Objective 4.

> **General Strategy for Problem Solving**
>
> **1.** UNDERSTAND the problem. During this step, become comfortable with the problem. Some ways of doing this are:
>
> Read and reread the problem.
>
> Choose a variable to represent the unknown.
>
> Construct a drawing.
>
> Propose a solution and check. Pay careful attention to how you check your proposed solution. This will help when writing an equation to model the problem.
>
> **2.** TRANSLATE the problem into an equation.
>
> **3.** SOLVE the equation.
>
> **4.** INTERPRET the results: *Check* the proposed solution in the stated problem and *state* your conclusion.

Let's review this strategy by solving a problem involving unknown numbers.

EXAMPLE 3

FINDING UNKNOWN NUMBERS

Find two numbers such that the second number is 3 more than twice the first number and the sum of the two numbers is 72.

Solution   **1.** UNDERSTAND the problem. First let's read and reread the problem and then propose a solution. For example, if the first number is 25, then the second number is 3 more than twice 25, or 53. The sum of 25 and 53 is 78, not the required sum, but we have gained some valuable information about the problem. First, we know that the first number is less than 25 since our guess led to a sum greater than the required sum. Also, we have gained some information as to how to model the problem with an equation.

> **Helpful Hint**
>
> The purpose of proposing a solution is not to guess correctly but to gain confidence and to help understand the problem and how to model it.

Next let's assign a variable and use this variable to represent any other unknown quantities. If we let

the first number $= x$, then

the second number $= 2x + 3$

$\underbrace{\qquad}\quad\uparrow$
$\uparrow\quad$ 3 more than
twice the second number

2. TRANSLATE the problem into an equation. To do so, we use the fact that the sum of the numbers is 72. First let's write this relationship in words and then translate to an equation.

In words:	first number	added to	second number	is	72
	↓	↓	↓	↓	↓
Translate:	x	$+$	$(2x + 3)$	$=$	72

3. SOLVE the equation.

$$x + (2x + 3) = 72$$

$x + 2x + 3 = 72$	Remove parentheses.
$3x + 3 = 72$	Combine like terms.
$3x = 69$	Subtract 3 from both sides.
$x = 23$	Divide both sides by 3.

4. INTERPRET. Here, *we check* our work and *state* the solution. Recall that if the first number $x = 23$, then the second number $2x + 3 = 2 \cdot 23 + 3 = 49$.

Check: Is the second number 3 more than twice the first number? Yes, since 3 more than twice 23 is $46 + 3$, or 49. Also, their sum, $23 + 49 = 72$, is the required sum.

State: The two numbers are 23 and 49.

Many of today's rates and statistics are given as percents. Interest rates, tax rates, nutrition labeling, and percent of households in a given category are just a few examples. Before we practice solving problems containing percents, let's briefly take a moment and review the meaning of percent and how to find a percent of a number.

The word *percent* means "per hundred," and the symbol % is used to denote percent. This means that 23% is 23 per hundred, or $\frac{23}{100}$. Also,

$$41\% = \frac{41}{100} = 0.41$$

To find a percent of a number, we multiply.

EXAMPLE 4

Find 16% of 25.

Solution To find 16% of 25, we find the product of 16% (written as a decimal) and 25.

$$16\% \cdot 25 = 0.16 \cdot 25$$
$$= 4$$

Thus, 16% of 25 is 4.

Next, we solve a problem containing percent.

✔ CONCEPT CHECK

Suppose you are finding 112% of a number x. Which of the following is a correct description of the result? Explain.

a. The result is less than x.
b. The result is equal to x.
c. The result is greater than x.

EXAMPLE 5

FINDING THE ORIGINAL PRICE OF A COMPUTER

Suppose that a computer store just announced an 8% decrease in the price of a particular computer model. If this computer sells for $2162 after the decrease, find the original price of this computer.

Concept Check Answer:
c; the result is greater than x

Solution

1. **UNDERSTAND.** Read and reread the problem. Recall that a percent decrease means a percent of the original price. Let's guess that the original price of the computer is $2500. The amount of decrease is then 8% of $2500, or $(0.08)($2500) = 200. This means that the new price of the computer is the original price minus the decrease, or $2500 - $200 = 2300. Our guess is incorrect, but we now have an idea of how to model this problem. In our model, we will let x = the original price of the computer.

2. **TRANSLATE.**

In words:	original price of computer	minus	8% of original price	is	new price
	↓	↓	↓	↓	↓
Translate:	x	$-$	$0.08x$	$=$	2162

3. **SOLVE** the equation.

$$x - 0.08x = 2162$$
$$0.92x = 2162 \qquad \text{Combine like terms.}$$
$$x = \frac{2162}{0.92} = 2350 \qquad \text{Divide both sides by } 0.92.$$

4. **INTERPRET.**

   Check: If the original price of the computer was $2350, the new price is

   $$\$2350 - (0.08)(\$2350) = \$2350 - \$188$$
   $$= \$2162 \qquad \text{The given new price.}$$

   State: The original price of the computer was $2350.

APPENDIX C-2 EXERCISE SET

STUDY GUIDE/SSM	CD/ VIDEO	PH MATH TUTOR CENTER	MathXL®Tutorials ON CD	MathXL®	MyMathLab®

Write the following as algebraic expressions. Then simplify. See Examples 1 and 2.

△ **1.** The perimeter of the square with side length y.

△ **2.** The perimeter of the rectangle with length x and width $x - 5$.

3. The sum of three consecutive integers if the first integer is z.

4. The sum of three consecutive odd integers if the first integer is x.

5. The total amount of money (in cents) in x nickels and $(x + 3)$ dimes. (*Hint:* the value of a nickel is 5 cents and the value of a dime is 10 cents.)

6. The total amount of money (in cents) in y quarters and $(2y - 1)$ nickels. (Use the hint for Exercise 5.)

△ **7.** A piece of land along Bayou Liberty is to be fenced and subdivided as shown so that each rectangle has the same dimensions. Express the total amount of fencing needed as an algebraic expression in x.

△ **8.** Write the perimeter of the floor plan shown as an algebraic expression in *x*.

Solve. See Example 3.

9. Four times the difference of a number and 2 is the same as 6 times the number, increased by 2. Find the number.

10. Twice the sum of a number and 3 is the same as 1 subtracted from the number. Find the number.

 11. One number is 5 times another number. If the sum of the two numbers is 270, find the numbers.

12. One number is 6 less than another number. If the sum of the two numbers is 150, find the numbers.

Solve. See Example 4.

13. Find 30% of 260.

14. Find 70% of 180.

15. Find 12% of 16.

16. Find 22% of 12.

Solve. See Examples 4 and 5.

17. The United States consists of 2271 million acres of land. Approximately 29% of this land is federally owned. Find the number of acres that are not federally owned. (*Source:* U.S. General Services Administration)

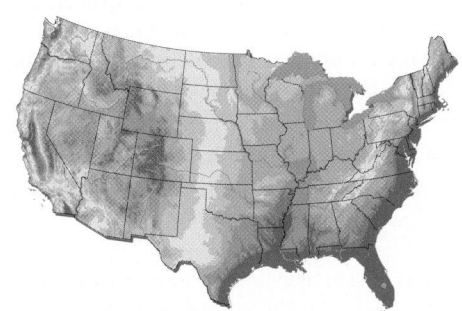

18. The state of Nevada contains the most federally owned acres of land in the United States. If 90% of the state's 70 million acres of land is federally owned, find the number of acres that are not federally owned. (*Source:* U.S. General Services Administration)

19. In 2000, a total of 2342 earthquakes occurred in the United States. Of these, 85% were minor tremors with magnitudes of 3.9 or less on the Richter scale. How many minor earthquakes occurred in the United States in 2000? Round to the nearest whole. (*Source:* U.S. Geological Survey National Earthquake Information Center)

20. Of the 1071 tornadoes that occurred in the United States during 2000, 22.5% occurred during the month of May. How many tornadoes occurred during May 2000? Round to the nearest whole. (*Source:* Storm Prediction Center)

21. According to a recent survey, $33\frac{1}{3}$% of online shoppers in the United States say that they spend more than they intended when shopping online. In a group of 1290 online shoppers, how many would you expect don't spend more than they intended when shopping online? (*Source:* Cyber Dialogue, 04/11/2001)

22. Only 6% of Americans eat a commercially prepared breakfast five times or more per week. As of 2000, Cincinnati, Ohio, had a population of 331,285 people. How many of these people would you expect do not eat a commercially prepared breakfast five or more times per week? Round to the nearest whole. (*Sources:* National Restaurant Association, U.S. Census Bureau)

The following graph is called a circle graph or a pie chart. The circle represents a whole, or in this case, 100%. This particular graph shows the number of times e-mail users check their e-mail. Use this graph to answer Exercises 23 through 26.

How Often We Check Our E-mail

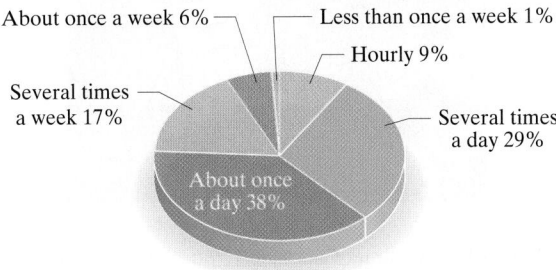

Source: The UCLA Internet Report: "Surveying the Digital Future," UCLA Center for Communication Policy

23. What percent of e-mail users check their e-mail several times per week?

24. Among e-mail users, what is the most popular frequency for checking e-mail?

25. If it is estimated that Fort Wayne, Indiana, has 112,500 e-mail users, how many of these would you expect check their e-mail about once a week?

26. If it is estimated that the city of New Orleans has 265,000 e-mail users, how many of these would you expect check their e-mail either several times a day or hourly?

27. The airports in Chicago, Atlanta, and Los Angeles have a total of 205.9 million annual arrivals and departures. Use this information and Example 2 in this section to find the number from each individual airport.

28. The perimeter of the triangle in Example 1b is 483 feet. Find the length of each side.

29. America West Airlines has 3 models of Boeing aircraft in their fleet. The 737-300 contains 21 more seats than the 737-200. The 757-200 contains 36 less seats than twice the number of seats in the 737-200. Find the number of seats for each aircraft if the total number of seats for the 3 models is 437. (*Source*: America West Airlines)

30. The new governor of California will make $4000 less than the governor of New York, and $89,000 more than the governor of Alaska. If the sum of these 3 salaries is $440,000, find the salary of each governor. (*Source: World Almanac*)

31. A new fax machine was recently purchased for an office in Hopedale for $464.40 including tax. If the tax rate in Hopedale is 8%, find the price of the fax machine before taxes.

32. A premedical student at a local university was complaining that she had just paid $86.11 for her human anatomy book,

including tax. Find the price of the book before taxes if the tax rate at this university is 9%.

33. INVESCO Field at Mile High, home to the Denver Broncos, has 11,675 more seats than Heinz Field, home to the Pittsburgh Steelers. Together, these two stadiums can seat a total of 140,575 NFL fans. How many seats does each stadium have? (*Sources:* Denver Broncos, Pittsburgh Steelers)

34. For the 2001 Major League Baseball season, the opening day payroll for the Minnesota Twins was $46,718,000 less than the opening day payroll for the Colorado Rockies. The total of the opening day payrolls for these two teams was $95,418,000. What was the opening day payroll for each team? (*Source*: Associated Press)

35. In a recent year, the three Internet access companies with the most subscribers were Earthlink, MSN, and AOL. Earthlink had 700,000 more subscribers than MSN and AOL had 3,700,000 more than 5 times the subscribers that MSN had. If the total subscribers for these Internet access companies is 32,400,000, find the number of subscribers for each company. (*Source:* Jupiter Media Metrix)

36. The three tallest hospitals in the world are Guy's Tower in London, Queen Mary Hospital in Hong Kong, and Galter Pavilion in Chicago. These buildings have a total height of 1320 feet. Guy's Tower is 67 feet taller than Galter Pavilion and the Queen Mary Hospital is 47 feet taller than Galter Pavilion. Find the heights of the three hospitals.

37. In 2003, the population of Morocco was 31.2 million. This represented an increase in population of 9.1% from the year 2000. What was the population of Morocco in 2000? Round to the nearest tenth (*Source:* Population Reference Bureau)

38. Palm, Inc. is a worldwide supplier of handheld computing devices. On October 25, 2001, shares of Palm stock closed at $2.44 per share. This represents a 97.4% decrease in stock price from the closing price on March 2, 2000, the day when Palm stock first traded on the NASDAQ exchange. Find the closing price on March 2, 2000. (*Source:* Financial Insight Systems, Inc.)

39. According to government statistics, the number of switchboard operators in the United States is expected to decrease to 185,000 by the year 2008. This represents a decrease of 13.9% from the number of switchboard operators in 1998. (*Source:* U.S. Bureau of Labor Statistics)

   a. Find the number of switchboard operators in 1998. Round to the nearest whole number.

   b. In your own words, explain why you think that the need for switchboard operators is decreasing.

40. The number of deaths caused by tornadoes descreased 59.2% from the 1950s to the 1990s. There were 579 deaths from tornadoes in the 1990s. (*Source:* National Weather Service)

   a. Find the number of deaths caused by tornadoes in the 1950s. Round to the nearest whole number.

b. In your own words, explain why you think that the number of tornado-related deaths has decreased so much since the 1950s.

41. The sum of three consecutive integers is 228. Find the integers.

42. The sum of three consecutive odd integers is 327. Find the integers.

Recall that the sum of the angle measures of a triangle is 180°.

△ **43.** Find the measures of the angles of a triangle if the measure of one angle is twice the measure of a second angle and the third angle measures 3 times the second angle decreased by 12.

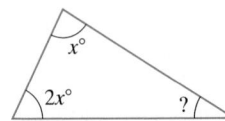

△ **44.** Find the angles of a triangle whose two base angles are equal and whose third angle is 10° less than three times a base angle.

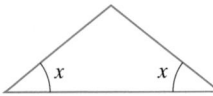

△ **45.** The official manual for traffic signs is the *Manual on Uniform Traffic Control Devices* published by the Government Printing Office. The rectangular sign below has a length 12 inches more than twice its height. If the perimeter of the sign is 312 inches, find its dimensions.

△ **46.** Two frames are needed with the same outside perimeter: one frame in the shape of a square and one in the shape of an equilateral triangle. Each side of the triangle is 6 centimeters longer than each side of the square. Find the dimensions of each frame. (An equilateral triangle has sides that are the same length.)

△ **47.** In a blueprint of a rectangular room, the length is to be 2 centimeters greater than twice its width. Find the dimensions if the perimeter is to be 40 centimeters.

48. A plant food solution contains 5 cups of water for every 1 cup of concentrate. If the solution contains 78 cups of these two ingredients, find the number of cups of concentrate in the solution.

Recall that two angles are complements of each other if their sum is 90°. Two angles are supplements of each other if their sum is 180°. Find the measure of each angle.

△ **49.**

△ **50.**

△ **51.**

△ **52.**

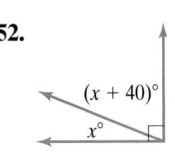

△ **53.** One angle is three times its supplement increased by 20°. Find the measures of the two supplementary angles.

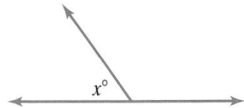

△ **54.** One angle is twice its complement increased by 30°. Find the measures of the two complementary angles.

55. The external tank of a NASA Space Shuttle contains the propellants used for the first 8.5 minutes after launch. Its height is 5 times the sum of its width and 1. If the sum of the height and width is 55.4 meters, find the dimensions of this tank. (*Source:* NASA/Kennedy Space Center)

height

width = *x* meters

56. The blue whale is the largest of whales. Its average weight is 3 times the difference of the average weight of a humpback whale and 5 tons. If the total of the average weights of a blue whale and a humpback whale is 117 tons, find the average weight of each type of whale.

57. Incandescent, fluorescent, and halogen bulbs are lasting longer today than ever before. On average, the number of bulb hours for a fluorescent bulb is 25 times the number of bulb hours for a halogen bulb. The number of bulb hours for an incandescent bulb is 2,500 less than the halogen bulb. If the total number of bulb hours for the three types of bulbs is 105,500, find the number of bulb hours for each type. (*Source:* Popular Science Magazine)

Incandescent bulb	Fluorescent bulb	Halogen bulb
1870	1938	1964

C-3   GRAPHING

Objectives

1 Plot ordered pairs.

2 Graph linear equations.

1   Graphs are widely used today in newspapers, magazines, and all forms of newsletters. A few examples of graphs are shown here.

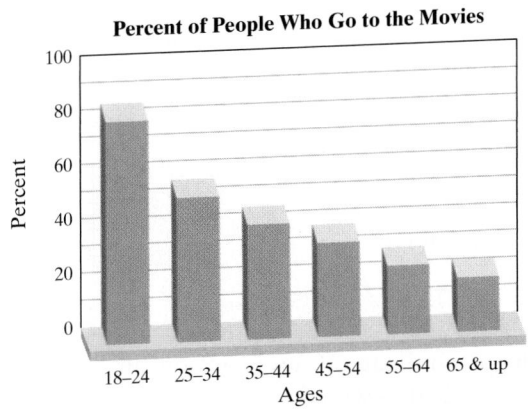

Percent of People Who Go to the Movies

Source: TELENATION/Market Facts, Inc.

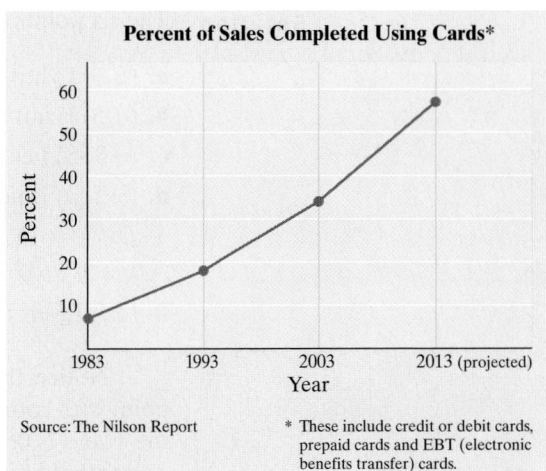

Percent of Sales Completed Using Cards*

Source: The Nilson Report

* These include credit or debit cards, prepaid cards and EBT (electronic benefits transfer) cards.

To review how to read these graphs, we review their origin—the rectangular coordinate system. One way to locate points on a plane is by using a **rectangular coordinate system**, which is also called a **Cartesian coordinate system** after its inventor, René Descartes (1596–1650). The next diagram to the left shows the rectangular coordinate system. For further review of this system, see Section 3.1.

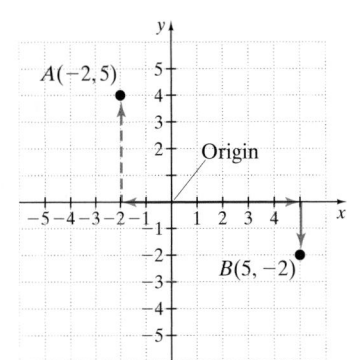

Recall that the location of point A in the figure on the right above is described as 2 units to the left of the origin along the x-axis and 5 units upward parallel to the y-axis. Thus, we identify point A with the ordered pair $(-2, 5)$. Notice that the order of these numbers is critical. The x-value -2 is called the **x-coordinate** and is associated with the x-axis. The y-value 5 is called the **y-coordinate** and is associated with the y-axis. Compare the location of point A with the location of point B, which corresponds to the ordered pair $(5, -2)$.

Keep in mind that **each ordered pair corresponds to exactly one point in the real plane and that each point in the plane corresponds to exactly one ordered pair.** Thus, we may refer to the ordered pair (x, y) as the point (x, y).

EXAMPLE 1

Plot each ordered pair on a Cartesian coordinate system and name the quadrant in which the point is located.

a. $(2, -1)$ **b.** $(0, 5)$ **c.** $(-3, 5)$ **d.** $(-2, 0)$ **e.** $\left(-\dfrac{1}{2}, -4\right)$ **f.** $(1.5, 1.5)$

Solution The six points are graphed as shown.

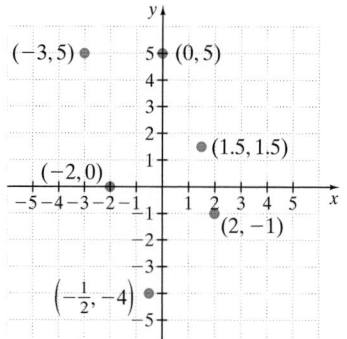

a. $(2, -1)$ lies in quadrant IV.

b. $(0, 5)$ is not in any quadrant.

c. $(-3, 5)$ lies in quadrant II.

d. $(-2, 0)$ is not in any quadrant.

e. $\left(-\dfrac{1}{2}, -4\right)$ is in quadrant III.

f. $(1.5, 1.5)$ is in quadrant I.

Notice that the y-coordinate of any point on the x-axis is 0. For example, the point with coordinates $(-2, 0)$ lies on the x-axis. Also, the x-coordinate of any point on the y-axis is 0. For example, the point with coordinates $(0, 5)$ lies on the y-axis. These points that lie on the axes do not lie in any quadrants.

✔ **CONCEPT CHECK**

Which of the following best describes the location of the point $(3, -6)$ in a rectangular coordinate system?

a. 3 units to the left of the y-axis and 6 units above the x-axis

b. 3 units above the x-axis and 6 units to the left of the y-axis

c. 3 units to the right of the y-axis and 6 units below the x-axis

d. 3 units below the x-axis and 6 units to the right of the y-axis

Recall that an equation such as $3x - y = 12$ is called a linear equation in two variables, and **the graph of every linear equation in two variables is a line.**

Linear Equation in Two Variables

A linear equation in two variables is an equation that can be written in the form

$$Ax + By = C$$

where A and B are not both 0. This form is called **standard form**.

Some examples of equations in standard form:

$$3x - y = 12$$
$$-2.1x + 5.6y = 0$$

▶ **Helpful Hint**

Remember: A linear equation is written in standard form when all of the variable terms are on one side of the equation and the constant is on the other side.

EXAMPLE 2

Graph the equation $y = -2x + 3$.

Solution This is a linear equation. (In standard form it is $2x + y = 3$.) Find three ordered pair solutions, and plot the ordered pairs. The line through the plotted points is the graph. Since the equation is solved for y, let's choose three x-values. Let's let x be 0, 2, and then -1 to find our three ordered pair solutions.

Let $x = 0$

$y = -2x + 3$

$y = -2 \cdot 0 + 3$

$y = 3$ Simplify.

Let $x = 2$

$y = -2x + 3$

$y = -2 \cdot 2 + 3$

$y = -1$ Simplify.

Let $x = -1$

$y = -2x + 3$

$y = -2(-1) + 3$

$y = 5$ Simplify.

The three ordered pairs $(0, 3)$, $(2, -1)$ and $(-1, 5)$ are listed in the table and the graph is shown.

x	y
0	3
2	−1
−1	5

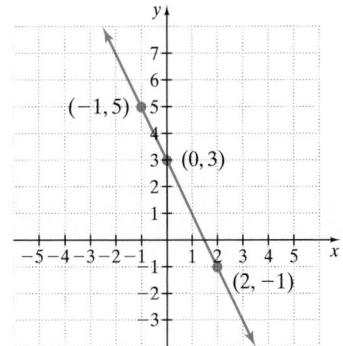

Notice that the graph crosses the y-axis at the point $(0, 3)$. This point is called the **y-intercept**. (You may sometimes see just the number 3 called the y-intercept.) This graph also crosses the x-axis at the point $\left(\frac{3}{2}, 0\right)$. This point is called the **x-intercept**. (You may also see just the number $\frac{3}{2}$ called the x-intercept.)

Since every point on the y-axis has an x-value of 0, we can find the y-intercept of a graph by letting $x = 0$ and solving for y. Also, every point on the x-axis has a y-value of 0. To find the x-intercept, we let $y = 0$ and solve for x.

Finding x- and y-Intercepts

To find an x-intercept, let $y = 0$ and solve for x.
To find a y-intercept, let $x = 0$ and solve for y.

EXAMPLE 3

Graph the linear equation $y = \dfrac{1}{3}x$.

Solution   To graph, we find ordered pair solutions, plot the ordered pairs, and draw a line through the plotted points. We will choose x-values and substitute in the equation. To avoid fractions, we choose x-values that are multiples of 3. To find the y-intercept, we can let $x = 0$.

> **Helpful Hint**
>
> Notice that by using multiples of 3 for x, we avoid fractions.

$$y = \frac{1}{3}x$$

If $x = 0$, then $y = \frac{1}{3}(0)$, or 0.

If $x = 6$, then $y = \frac{1}{3}(6)$, or 2.

If $x = -3$, then $y = \frac{1}{3}(-3)$, or -1.

x	y
6	2
0	0
-3	-1

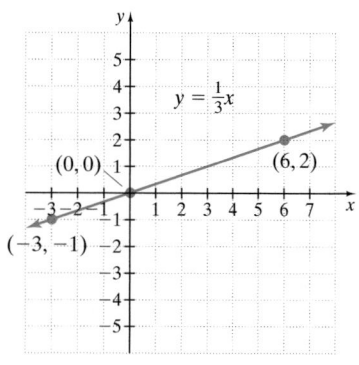

APPENDIX C-3 EXERCISE SET

Determine the coordinates of each point on the graph.

1. Point A

2. Point B

3. Point C

4. Point D

5. Point E

6. Point F

7. Point G

8. Point H

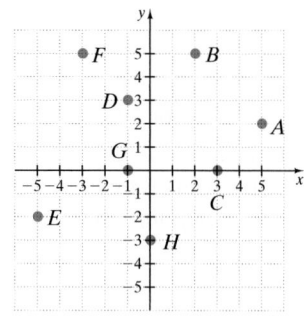

Without graphing, visualize the location of each point. Then give its location by quadrant or x- or y-axis.

9. $(2, 3)$

10. $(0, 5)$

11. $(-2, 7)$

12. $(-3, 0)$

13. $(-1, -4)$

14. $(4, -2)$

15. $(0, -100)$

16. $(10, 30)$

17. $(-10, -30)$

18. $(0, 0)$

19. $(-87, 0)$

20. $(-42, 17)$

Given that x is a positive number and that y is a positive number, determine the quadrant or axis in which each point lies.

21. $(x, -y)$

22. $(-x, y)$

23. $(x, 0)$

24. $(0, -y)$

25. $(-x, -y)$

26. $(0, 0)$

Graph each linear equation. See Examples 2 and 3.

27. $y = -x - 2$

28. $y = -2x + 1$

29. $3x - 4y = 8$

30. $x - 9y = 3$

31. $y = \frac{1}{3}x$

32. $y = \frac{3}{2}x$

33. $y + 4 = 0$

34. $x = -1.5$

Recall that if $f(2) = 7$, for example, this corresponds to the ordered pair (2, 7), for the graph of f. Use this information and the graphs of f and g below to answer Exercises 35 through 42.

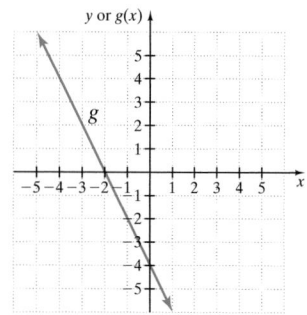

35. $f(4) =$

36. $f(0) =$

37. $g(0) =$

38. $g(-1) =$

39. Find all values for x such that $f(x) = 0$.

40. Find all values for x such that $g(x) = 0$.

41. If $(-1, -2)$ is a point on the graph of g, write this using function notation.

42. If $(-1, 2)$ is a point on the graph of f, write this using function notation.

C-4   POLYNOMIALS AND FACTORING

Objectives

1   Review operations on polynomials.

2   Review factoring polynomials.

FACTORING STRATEGIES

The key to proficiency in factoring polynomials is to practice until you are comfortable with each technique. A strategy for factoring polynomials completely is given next.

Factoring a Polynomial

Step 1: Are there any common factors? If so, factor out the greatest common factor.

Step 2: How many terms are in the polynomial?

    a. If there are *two* terms, decide if one of the following formulas may be applied:

        i. Difference of two squares: $a^2 - b^2 = (a - b)(a + b)$

        ii. Difference of two cubes: $a^3 - b^3 = (a - b)(a^2 + ab + b^2)$

        iii. Sum of two cubes: $a^3 + b^3 = (a + b)(a^2 - ab + b^2)$

    b. If there are *three* terms, try one of the following:

        i. Perfect square trinomial: $a^2 + 2ab + b^2 = (a + b)^2$

                       $a^2 - 2ab + b^2 = (a - b)^2$

        ii. If not a perfect square trinomial, factor by using the methods presented in Section 5.5.

    c. If there are *four* or more terms, try factoring by grouping.

Step 3: See whether any factors in the factored polynomial can be factored further.

A few examples are worked for you below.

EXAMPLE 1

Factor each polynomial completely.

a. $8a^2b - 4ab$ **b.** $36x^2 - 9$ **c.** $2x^2 - 5x - 7$

d. $5p^2 + 5 + qp^2 + q$ **e.** $9x^2 + 24x + 16$ **f.** $y^2 + 25$

Solution   **a. Step 1:** The terms have a common factor of *4ab*, which we factor out.

$$8a^2b - 4ab = 4ab(2a - 1)$$

        Step 2: There are two terms, but the binomial $2a - 1$ is not the difference of two squares or the sum or difference of two cubes.

        Step 3: The factor $2a - 1$ cannot be factored further.

   b. Step 1: Factor out a common factor of 9.

$$36x^2 - 9 = 9(4x^2 - 1)$$

Step 2: The factor $4x^2 - 1$ has two terms, and it is the difference of two squares.
$$9(4x^2 - 1) = 9(2x + 1)(2x - 1)$$

Step 3: No factor with more than one term can be factored further.

c. Step 1: The terms of $2x^2 - 5x - 7$ contain no common factor other than 1 or -1.

Step 2: There are three terms. The trinomial is not a perfect square, so we factor by methods from Section 5.6.
$$2x^2 - 5x - 7 = (2x - 7)(x + 1)$$

Step 3: No factor with more than one term can be factored further.

d. Step 1: There is no common factor of all terms of $5p^2 + 5 + qp^2 + q$.

Step 2: The polynomial has four terms, so try factoring by grouping.
$$5p^2 + 5 + qp^2 + q = (5p^2 + 5) + (qp^2 + q) \quad \text{Group the terms.}$$
$$= 5(p^2 + 1) + q(p^2 + 1)$$
$$= (p^2 + 1)(5 + q)$$

Step 3: No factor can be factored further.

e. Step 1: The terms of $9x^2 + 24x + 16$ contain no common factor other than 1 or -1.

Step 2: The trinomial $9x^2 + 24x + 16$ is a perfect square trinomial, and $9x^2 + 24x + 16 = (3x + 4)^2$.

Step 3: No factor can be factored further.

f. Step 1: There is no common factor of $y^2 + 25$ other than 1.

Step 2: This binomial is the sum of two squares and is prime.

Step 3: The binomial $y^2 + 25$ cannot be factored further.

EXAMPLE 2

Factor each completely.

a. $27a^3 - b^3$ **b.** $3n^2m^4 - 48m^6$ **c.** $2x^2 - 12x + 18 - 2z^2$
d. $8x^4y^2 + 125xy^2$ **e.** $(x - 5)^2 - 49y^2$

Solution   **a.** This binomial is the difference of two cubes.
$$27a^3 - b^3 = (3a)^3 - b^3$$
$$= (3a - b)[(3a)^2 + (3a)(b) + b^2]$$
$$= (3a - b)(9a^2 + 3ab + b^2)$$

b. $3n^2m^4 - 48m^6 = 3m^4(n^2 - 16m^2)$ Factor out the GCF, $3m^4$.
$$= 3m^4(n + 4m)(n - 4m) \quad \text{Factor the difference of squares.}$$

c. $2x^2 - 12x + 18 - 2z^2 = 2(x^2 - 6x + 9 - z^2)$ The GCF is 2.
$$= 2[(x^2 - 6x + 9) - z^2] \quad \text{Group the first three terms together.}$$
$$= 2[(x - 3)^2 - z^2] \quad \text{Factor the perfect square trinomial.}$$
$$= 2[(x - 3) + z][(x - 3) - z] \quad \text{Factor the difference of squares.}$$
$$= 2(x - 3 + z)(x - 3 - z)$$

d. $8x^4y^2 + 125xy^2 = xy^2(8x^3 + 125)$ The GCF is xy^2.

$\qquad\qquad\qquad\quad = xy^2[(2x)^3 + 5^3]$

$\qquad\qquad\qquad\quad = xy^2(2x + 5)[(2x)^2 - (2x)(5) + 5^2]$ Factor the sum of cubes.

$\qquad\qquad\qquad\quad = xy^2(2x + 5)(4x^2 - 10x + 25)$

e. This binomial is the difference of squares.

$$(x - 5)^2 - 49y^2 = (x - 5)^2 - (7y)^2$$
$$= [(x - 5) + 7y][(x - 5) - 7y]$$
$$= (x - 5 + 7y)(x - 5 - 7y)$$

APPENDIX C-4 EXERCISE SET

Operations on Polynomials

Perform each indicated operation.

1. $(-y^2 + 6y - 1) + (3y^2 - 4y - 10)$

2. $(5z^4 - 6z^2 + z + 1) - (7z^4 - 2z + 1)$

3. Subtract $(x - 5)$ from $(x^2 - 6x + 2)$.

4. $(2x^2 + 6x - 5) + (5x^2 - 10x)$

5. $(5x - 3)^2$

6. $(5x^2 - 14x - 3) \div (5x + 1)$

7. $(2x^4 - 3x^2 + 5x - 2) \div (x + 2)$

8. $(4x - 1)(x^2 - 3x - 2)$

Factor completely.

9. $x^2 - 8x + 16 - y^2$

10. $12x^2 - 22x - 20$

11. $x^4 - x$

12. $(2x + 1)^2 - 3(2x + 1) + 2$

13. $14x^2y - 2xy$

14. $24ab^2 - 6ab$

15. $4x^2 - 16$

16. $9x^2 - 81$

17. $3x^2 - 8x - 11$

18. $5x^2 - 2x - 3$

19. $4x^2 + 8x - 12$

20. $6x^2 - 6x - 12$

21. $4x^2 + 36x + 81$

22. $25x^2 + 40x + 16$

23. $8x^3 + 125y^3$

24. $27x^3 - 64y^3$

25. $64x^2y^3 - 8x^2$

26. $27x^5y^4 - 216x^2y$

27. $(x + 5)^3 + y^3$

28. $(y - 1)^3 + 27x^3$

29. $(5a - 3)^2 - 6(5a - 3) + 9$

30. $(4r + 1)^2 + 8(4r + 1) + 16$

31. $7x^2 - 63x$

32. $20x^2 + 23x + 6$

33. $ab - 6a + 7b - 42$

34. $20x^2 - 220x + 600$

35. $x^4 - 1$

36. $15x^2 - 20x$

37. $10x^2 - 7x - 33$

38. $45m^3n^3 - 27m^2n^2$

39. $5a^3b^3 - 50a^3b$

40. $x^4 + x$

41. $16x^2 + 25$

42. $20x^3 + 20y^3$

43. $10x^3 - 210x^2 + 1100x$

44. $9y^2 - 42y + 49$

45. $64a^3b^4 - 27a^3b$

46. $y^4 - 16$

47. $2x^3 - 54$

48. $2sr + 10s - r - 5$

49. $3y^5 - 5y^4 + 6y - 10$

50. $64a^2 + b^2$

51. $100z^3 + 100$

52. $250x^4 - 16x$

53. $4b^2 - 36b + 81$

54. $2a^5 - a^4 + 6a - 3$

55. $(y - 6)^2 + 3(y - 6) + 2$

56. $(c + 2)^2 - 6(c + 2) + 5$

57. Express the area of the shaded region as a polynomial. Factor the polynomial completely.

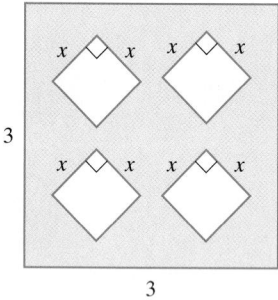

C-5 RATIONAL EXPRESSIONS

Objective

1 Perform operations on rational expressions and solve equations containing rational expressions.

1 It is very important that you remember the difference between performing operations on rational expressions and solving equations containing rational expressions.

Expression to be Added	**Equation to be Solved**
$\dfrac{x}{2} + \dfrac{x}{6}$	$\dfrac{x}{2} + \dfrac{x}{6} = \dfrac{2}{3}$

Write both rational expressions with the LCD, 6, as the denominator.

Multiply both sides by the LCD, 6.

$$\frac{x}{2} + \frac{x}{6} = \frac{x \cdot 3}{2 \cdot 3} + \frac{x}{6}$$

$$6\left(\frac{1}{2} + \frac{x}{6}\right) = 6\left(\frac{2}{3}\right)$$

$$= \frac{3x}{6} + \frac{x}{6}$$

$$3 + x = 4$$

$$= \frac{4x}{6} = \frac{2x}{3}$$

$$x = 1$$

Check to see that the solution is 1.

Remember how to perform operations on rational expressions.

EXAMPLE 1

Multiply $\dfrac{x^3 - 1}{-3x + 3} \cdot \dfrac{15x^2}{x^2 + x + 1}$.

Solution

$$\dfrac{x^3 - 1}{-3x + 3} \cdot \dfrac{15x^2}{x^2 + x + 1} = \dfrac{(x - 1)(x^2 + x + 1)}{-3(x - 1)} \cdot \dfrac{15x^2}{x^2 + x + 1} \qquad \text{Factor.}$$

$$= \dfrac{(x - 1)(x^2 + x + 1) \cdot 3 \cdot 5x^2}{-1 \cdot 3(x - 1)(x^2 + x + 1)} \qquad \text{Factor.}$$

$$= \dfrac{5x^2}{-1} \qquad \qquad \begin{array}{l}\text{Divide out common}\\ \text{factors.}\end{array}$$

$$= -5x^2$$

EXAMPLE 2

Divide $\dfrac{8m^2}{3m^2 - 12} \div \dfrac{40}{2 - m}$.

Solution

$$\dfrac{8m^2}{3m^2 - 12} \div \dfrac{40}{2 - m} = \dfrac{8m^2}{3m^2 - 12} \cdot \dfrac{2 - m}{40} \qquad \text{Multiply by the reciprocal of the divisor.}$$

$$= \dfrac{8m^2(2 - m)}{3(m + 2)(m - 2) \cdot 40} \qquad \text{Factor and multiply.}$$

$$= \dfrac{8m^2 \cdot -1(m - 2)}{3(m + 2)(m - 2) \cdot 8 \cdot 5} \qquad \text{Write } (2 - m) \text{ as } -1(m - 2).$$

$$= -\dfrac{m^2}{15(m + 2)} \qquad \text{Simplify.}$$

EXAMPLE 3

Perform the indicated operation.

$$\dfrac{3x}{x + 2} + \dfrac{2x}{x - 2}$$

Solution   The LCD is the product of the two denominators: $(x + 2)(x - 2)$.

$$\dfrac{3x}{x + 2} + \dfrac{2x}{x - 2} = \dfrac{3x \cdot (x - 2)}{(x + 2) \cdot (x - 2)} + \dfrac{2x \cdot (x + 2)}{(x - 2) \cdot (x + 2)} \qquad \begin{array}{l}\text{Write equivalent rational}\\ \text{expressions.}\end{array}$$

$$= \dfrac{3x(x - 2) + 2x(x + 2)}{(x + 2)(x - 2)} \qquad \text{Add the numerators.}$$

$$= \dfrac{3x^2 - 6x + 2x^2 + 4x}{(x + 2)(x - 2)} \qquad \text{Apply the distributive property.}$$

$$= \dfrac{5x^2 - 2x}{(x + 2)(x - 2)} \qquad \text{Simplify the numerator.}$$

APPENDIX C-5 EXERCISE SET

> **Helpful Hint**
> Remember: Equations can be cleared of fractions; expressions cannot.

Perform each indicated operation and simplify, or solve the equation for the variable.

1. $\dfrac{x}{2} = \dfrac{1}{8} + \dfrac{x}{4}$

2. $\dfrac{x}{4} = \dfrac{3}{2} + \dfrac{x}{10}$

3. $\dfrac{1}{8} + \dfrac{x}{4}$

4. $\dfrac{3}{2} + \dfrac{x}{10}$

5. $\dfrac{4}{x+2} - \dfrac{2}{x-1}$

6. $\dfrac{5}{x-2} - \dfrac{10}{x+4}$

7. $\dfrac{4}{x+2} = \dfrac{2}{x-1}$

8. $\dfrac{5}{x-2} = \dfrac{10}{x+4}$

9. $\dfrac{2}{x^2-4} = \dfrac{1}{x+2} - \dfrac{3}{x-2}$

10. $\dfrac{3}{x^2-25} = \dfrac{1}{x+5} + \dfrac{2}{x-5}$

11. $\dfrac{5}{x^2-3x} + \dfrac{4}{2x-6}$

12. $\dfrac{5}{x^2-3x} \div \dfrac{4}{2x-6}$

13. $\dfrac{x-1}{x+1} + \dfrac{x+7}{x-1} = \dfrac{4}{x^2-1}$

14. $\left(1 - \dfrac{y}{x}\right) \div \left(1 - \dfrac{x}{y}\right)$

15. $\dfrac{a^2-9}{a-6} \cdot \dfrac{a^2-5a-6}{a^2-a-6}$

16. $\dfrac{2}{a-6} + \dfrac{3a}{a^2-5a-6} - \dfrac{a}{5a+5}$

17. $\dfrac{2x+3}{3x-2} = \dfrac{4x+1}{6x+1}$

18. $\dfrac{5x-3}{2x} = \dfrac{10x+3}{4x+1}$

19. $\dfrac{a}{9a^2-1} + \dfrac{2}{6a-2}$

20. $\dfrac{3}{4a-8} - \dfrac{a+2}{a^2-2a}$

21. $-\dfrac{3}{x^2} - \dfrac{1}{x} + 2 = 0$

22. $\dfrac{x}{2x+6} + \dfrac{5}{x^2-9}$

23. $\dfrac{x-8}{x^2-x-2} + \dfrac{2}{x-2}$

24. $\dfrac{x-8}{x^2-x-2} + \dfrac{2}{x-2} = \dfrac{3}{x+1}$

25. $\dfrac{3}{a} - 5 = \dfrac{7}{a} - 1$

26. $\dfrac{7}{3z-9} + \dfrac{5}{z}$

27. a. Which one above is an expression?
b. Describe the first step to simplify this expression.
c. Simplify the expression.

28. a. Which one above is an equation?
b. Describe the first step to solve this equation.
c. Solve the equation.

For each exercise, choose the correct statement.* Each figure represents a real number and no denominators are 0.

29. a. $\dfrac{\triangle + \square}{\triangle} = \square$

b. $\dfrac{\triangle + \square}{\triangle} = 1 + \dfrac{\square}{\triangle}$

c. $\dfrac{\triangle + \square}{\triangle} = \dfrac{\square}{\triangle}$

d. $\dfrac{\triangle + \square}{\triangle} = 1 + \square$

e. $\dfrac{\triangle + \square}{\triangle - \square} = -1$

30. a. $\dfrac{\triangle}{\square} + \dfrac{\square}{\triangle} = \dfrac{\triangle + \square}{\square + \triangle} = 1$

b. $\dfrac{\triangle}{\square} + \dfrac{\square}{\triangle} = \dfrac{\triangle + \square}{\triangle\square}$

c. $\dfrac{\triangle}{\square} + \dfrac{\square}{\triangle} = \triangle\triangle + \square\square$

d. $\dfrac{\triangle}{\square} + \dfrac{\square}{\triangle} = \dfrac{\triangle\triangle + \square\square}{\square\triangle}$

e. $\dfrac{\triangle}{\square} + \dfrac{\square}{\triangle} = \dfrac{\triangle\square}{\square\triangle} = 1$

31. a. $\dfrac{\triangle}{\square} \cdot \dfrac{\bigcirc}{\square} = \dfrac{\triangle\bigcirc}{\square}$

b. $\dfrac{\triangle}{\square} \cdot \dfrac{\bigcirc}{\square} = \triangle\bigcirc$

c. $\dfrac{\triangle}{\square} \cdot \dfrac{\bigcirc}{\square} = \dfrac{\triangle + \bigcirc}{\square + \square}$

d. $\dfrac{\triangle}{\square} \cdot \dfrac{\bigcirc}{\square} = \dfrac{\triangle\bigcirc}{\square\square}$

32. a. $\dfrac{\triangle}{\square} \div \dfrac{\bigcirc}{\triangle} = \dfrac{\triangle\triangle}{\square\bigcirc}$

b. $\dfrac{\triangle}{\square} \div \dfrac{\bigcirc}{\triangle} = \dfrac{\bigcirc\square}{\triangle\triangle}$

c. $\dfrac{\triangle}{\square} \div \dfrac{\bigcirc}{\triangle} = \dfrac{\bigcirc}{\square}$

d. $\dfrac{\triangle}{\square} \div \dfrac{\bigcirc}{\triangle} = \dfrac{\triangle + \triangle}{\square + \bigcirc}$

33. a. $\dfrac{\dfrac{\triangle + \square}{\bigcirc}}{\dfrac{\triangle}{\bigcirc}} = \square$

b. $\dfrac{\dfrac{\triangle + \square}{\bigcirc}}{\dfrac{\triangle}{\bigcirc}} = \dfrac{\triangle\triangle + \triangle\square}{\bigcirc\bigcirc}$

c. $\dfrac{\dfrac{\triangle + \square}{\bigcirc}}{\dfrac{\triangle}{\bigcirc}} = 1 + \square$

d. $\dfrac{\dfrac{\triangle + \square}{\bigcirc}}{\dfrac{\triangle}{\bigcirc}} = \dfrac{\triangle + \square}{\triangle}$

Use $\dfrac{x}{5} - \dfrac{x}{4} = \dfrac{1}{10}$ and $\dfrac{x}{5} - \dfrac{x}{4} + \dfrac{1}{10}$ for Exercises 27 and 28.

*My thanks to Kelly Champagne for permission to use her exercises for 29 through 33.

An Introduction to Using a Graphing Utility

THE VIEWING WINDOW AND INTERPRETING WINDOW SETTINGS

In this appendix, we will use the term **graphing utility** to mean a graphing calculator or a computer software graphing package. All graphing utilities graph equations by plotting points on a screen. While plotting several points can be slow and sometimes tedious for us, a graphing utility can quickly and accurately plot hundreds of points. How does a graphing utility show plotted points? A computer or calculator screen is made up of a grid of small rectangular areas called **pixels**. If a pixel contains a point to be plotted, the pixel is turned "on"; otherwise, the pixel remains "off." The graph of an equation is then a collection of pixels turned "on." The graph of $y = 3x + 1$ from a graphing calculator is shown in Figure A–1. Notice the irregular shape of the line caused by the rectangular pixels.

Figure A-1

The portion of the coordinate plane shown on the screen in Figure A-1 is called the **viewing window** or the **viewing rectangle**. Notice the x-axis and the y-axis on the graph. While tick marks are shown on the axes, they are not labeled. This means that from this screen alone, we do not know how many units each tick mark represents. To see what each tick mark represents and the minimum and maximum values on the axes, check the window setting of the graphing utility. It defines the viewing window. The window of the graph of $y = 3x + 1$ shown in Figure A-1 has the following setting (Figure A-2):

Figure A-2

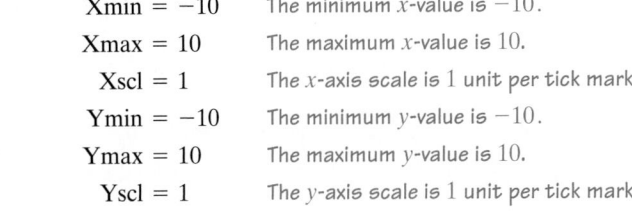

$Xmin = -10$	The minimum x-value is -10.
$Xmax = 10$	The maximum x-value is 10.
$Xscl = 1$	The x-axis scale is 1 unit per tick mark.
$Ymin = -10$	The minimum y-value is -10.
$Ymax = 10$	The maximum y-value is 10.
$Yscl = 1$	The y-axis scale is 1 unit per tick mark.

By knowing the scale, we can find the minimum and the maximum values on the axes simply by counting tick marks. For example, if both the Xscl (x-axis scale) and the Yscl are 1 unit per tick mark on the graph in Figure A-3, we can count the tick marks and find that the minimum x-value is -10 and the maximum x-value is 10. Also, the minimum y-value is -10 and the maximum y-value is 10. If the Xscl (x-axis scale) changes to 2 units per tick mark (shown in Figure A-4), by counting tick marks, we see that the minimum x-value is now -20 and the maximum x-value is now 20.

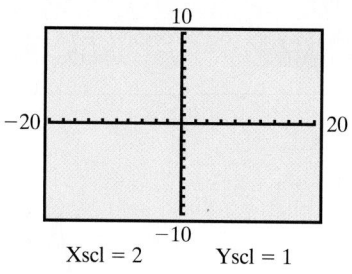

Xscl = 1   Yscl = 1

Figure A-3

Xscl = 2   Yscl = 1

Figure A-4

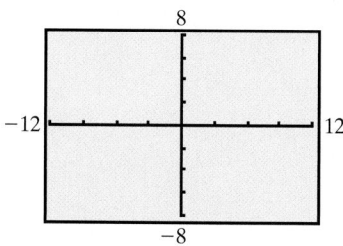

Figure A-5

It is also true that if we know the Xmin and the Xmax values, we can calculate the Xscl by the displayed axes. For example, the Xscl of the graph in Figure A-5 must be 3 units per tick mark for the maximum and minimum x-values to be as shown. Also, the Yscl of that graph must be 2 units per tick mark for the maximum and minimum y-values to be as shown.

We will call the viewing window in Figure A-3 a *standard* viewing window or rectangle. Although a standard viewing window is sufficient for much of this text, special care must be taken to ensure that all key features of a graph are shown. Figures A-6, A-7, and A-8 show the graph of $y = x^2 + 11x - 1$ on three different viewing windows. Note that certain viewing windows for this equation are misleading.

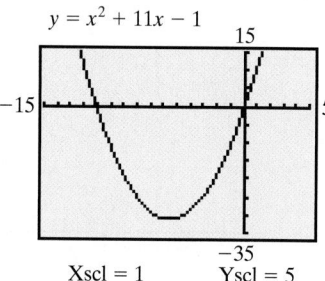

Xscl = 1   Yscl = 5

Figure A-6

Figure A-7

Figure A-8

How do we ensure that all distinguishing features of the graph of an equation are shown? It helps to know about the equation that is being graphed. For example, the equation $y = x^2 + 11x - 1$ is not a linear equation and its graph is not a line. This equation is a quadratic equation and, therefore, its graph is a parabola. By knowing this information, we know that the graph shown in Figure A-6, although correct, is misleading. Of the three viewing rectangles shown, the graph in Figure A-8 is best because it shows more of the distinguishing features of the parabola. Properties of equations needed for graphing will be studied in this text.

VIEWING WINDOW AND INTERPRETING WINDOW SETTINGS EXERCISE SET

In Exercises 1–4, determine whether all ordered pairs listed will lie within a standard viewing rectangle.

1. $(-9, 0), (5, 8), (1, -8)$

2. $(4, 7), (0, 0), (-8, 9)$

3. $(-11, 0), (2, 2), (7, -5)$

4. $(3, 5), (-3, -5), (15, 0)$

In Exercises 5–10, choose an Xmin, Xmax, Ymin, and Ymax so that all ordered pairs listed will lie within the viewing rectangle.

5. $(-90, 0), (55, 80), (0, -80)$

6. $(4, 70), (20, 20), (-18, 90)$

7. $(-11, 0), (2, 2), (7, -5)$

8. $(3, 5), (-3, -5), (15, 0)$

9. $(200, 200), (50, -50), (70, -50)$

10. $(40, 800), (-30, 500), (15, 0)$

Write the window setting for each viewing window shown. Use the following format:

Xmin =	Ymin =
Xmax =	Ymax =
Xscl =	Yscl =

11. **12.** **13.** **14.**

15. **16.**

17. **18.**

 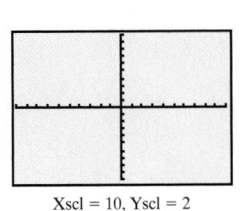

Xscl = 1, Yscl = 3   Xscl = 10, Yscl = 2

19. **20.**

Xscl = 5, Yscl = 10   Xscl = 100, Yscl = 200

GRAPHING EQUATIONS AND SQUARE VIEWING WINDOW

In general, the following steps may be used to graph an equation on a standard viewing window.

> ### Graphing an Equation in X and Y with a Graphing Utility on a Standard Viewing Window
>
> **Step 1:** Solve the equation for y.
> **Step 2:** Using your graphing utility and enter the equation in the form
> $\qquad$ Y = *expression involving x*.
> **Step 3:** Activate the graphing utility.

Special care must be taken when entering the *expression involving x* in Step 2. You must be sure that the graphing utility you are using interprets the expression as you want it to. For example, let's graph $3y = 4x$. To do so,

Step 1: Solve the equation for y.

$$3y = 4x$$
$$\frac{3y}{3} = \frac{4x}{3}$$
$$y = \frac{4}{3}x$$

Step 2: Using your graphing utility, enter the expression $\frac{4}{3}x$ after the Y = prompt. In order for your graphing utility to correctly interpret the expression, you may need to enter $(4/3)x$ or $(4 \div 3)x$.

Step 3: Activate the graphing utility. The graph should appear as in Figure A-9.

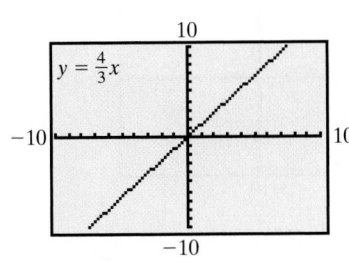

Figure A-9

Distinguishing features of the graph of a line include showing all the intercepts of the line. For example, the window of the graph of the line in Figure A-10 does not show both intercepts of the line, but the window of the graph of the same line in Figure A-11 does show both intercepts. Notice the notation below each graph. This is a shorthand notation of the range setting of the graph. This notation means [Xmin, Xmax] by [Ymin, Ymax].

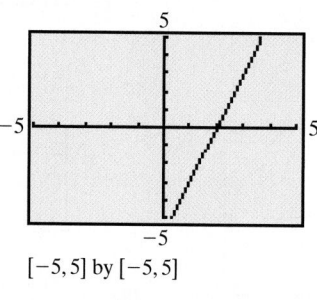

[−5, 5] by [−5, 5]

Figure A-10

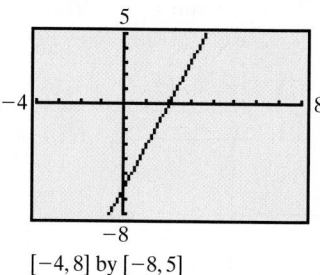

[−4, 8] by [−8, 5]

Figure A-11

On a standard viewing window, the tick marks on the y-axis are closer together than the tick marks on the x-axis. This happens because the viewing window is a rectangle, and so 10 equally spaced tick marks on the positive y-axis will be closer together than 10 equally spaced tick marks on the positive x-axis. This causes the appearance of graphs to be distorted.

For example, notice the different appearances of the same line graphed using different viewing windows. The line in Figure A-12 is distorted because the tick marks along the x-axis are farther apart than the tick marks along the y-axis. The graph of the same line in Figure A-13 is not distorted because the viewing rectangle has been selected so that there is equal spacing between tick marks on both axes.

Figure A-12

Figure A-13

We say that the line in Figure A-13 is graphed on a *square* setting. Some graphing utilities have a built-in program that, if activated, will automatically provide a square setting. A square setting is especially helpful when we are graphing perpendicular lines, circles, or when a true geometric perspective is desired. Some examples of square screens are shown in Figures A-14 and A-15.

Figure A-14

Figure A-15

Other features of a graphing utility such as Trace, Zoom, Intersect, and Table are discussed in appropriate Graphing Calculator Explorations in this text.

GRAPHING EQUATIONS AND SQUARE VIEWING WINDOW EXERCISE SET

Graph each linear equation in two variables, using the two differ-ent range settings given. Determine which setting shows all inter-cepts of a line.

1. $y = 2x + 12$

      Setting A: $[-10, 10]$ by $[-10, 10]$
      Setting B: $[-10, 10]$ by $[-10, 15]$

2. $y = -3x + 25$

      Setting A: $[-5, 5]$ by $[-30, 10]$
      Setting B: $[-10, 10]$ by $[-10, 30]$

3. $y = -x - 41$

      Setting A: $[-50, 10]$ by $[-10, 10]$
      Setting B: $[-50, 10]$ by $[-50, 15]$

4. $y = 6x - 18$

      Setting A: $[-10, 10]$ by $[-20, 10]$
      Setting B: $[-10, 10]$ by $[-10, 10]$

5. $y = \dfrac{1}{2}x - 15$

      Setting A: $[-10, 10]$ by $[-20, 10]$
      Setting B: $[-10, 35]$ by $[-20, 15]$

6. $y = -\dfrac{2}{3}x - \dfrac{29}{3}$

      Setting A: $[-10, 10]$ by $[-10, 10]$
      Setting B: $[-15, 5]$ by $[-15, 5]$

The graph of each equation is a line. Use a graphing utility and a standard viewing window to graph each equation.

7. $3x = 5y$ **8.** $7y = -3x$ **9.** $9x - 5y = 30$

10. $4x + 6y = 20$ **11.** $y = -7$ **12.** $y = 2$

13. $x + 10y = -5$ **14.** $x - 5y = 9$

Graph the following equations using the square setting given. Some keystrokes that may be helpful are given.

15. $y = \sqrt{x}$ $[-12, 12]$ by $[-8, 8]$

      Suggested keystrokes: $\sqrt{}\ x$

16. $y = \sqrt{2x}$ $[-12, 12]$ by $[-8, 8]$

      Suggested keystrokes: $\sqrt{}\ (2x)$

17. $y = x^2 + 2x + 1$ $[-15, 15]$ by $[-10, 10]$

      Suggested keystrokes: $x\verb|^|2 + 2x + 1$

18. $y = x^2 - 5$ $[-15, 15]$ by $[-10, 10]$

      Suggested keystrokes: $x\verb|^|2 - 5$

19. $y = |x|$ $[-9, 9]$ by $[-6, 6]$

      Suggested keystrokes: $ABS\ (x)$

20. $y = |x - 2|$ $[-9, 9]$ by $[-6, 6]$

      Suggested keystrokes: $ABS(x - 2)$

Graph each line. Use a standard viewing window; then, if neces-sary, change the viewing window so that all intercepts of each line show.

21. $x + 2y = 30$ **22.** $1.5x - 3.7y = 40.3$

Solving Systems of Equations by Matrices

Objectives

1 Use matrices to solve a system of two equations.

2 Use matrices to solve a system of three equations.

By now, you may have noticed that the solution of a system of equations depends on the coefficients of the equations in the system and not on the variables. In this section, we introduce solving a system of equations by a **matrix**.

1 A matrix (plural: **matrices**) is a rectangular array of numbers. The following are examples of matrices.

$$\begin{bmatrix} 1 & 0 \\ 0 & 1 \end{bmatrix} \quad \begin{bmatrix} 2 & 1 & 3 & -1 \\ 0 & -1 & 4 & 5 \\ -6 & 2 & 1 & 0 \end{bmatrix} \quad \begin{bmatrix} a & b & c \\ d & e & f \end{bmatrix}$$

The numbers aligned horizontally in a matrix are in the same **row**. The numbers aligned vertically are in the same **column**.

$$\begin{array}{l} \text{row } 1 \rightarrow \\ \text{row } 2 \rightarrow \end{array} \begin{bmatrix} 2 & 1 & 0 \\ -1 & 6 & 2 \end{bmatrix}$$

column 1
column 2
column 3

This matrix has **2** rows and **3** columns. H is called **2 × 3** (read "two by three") matrix.

To see the relationship between systems of equations and matrices, study the example below.

> **Helpful Hint**
>
> Before writing the corresponding matrix associated with a system of equations, make sure that the equations are written in standard form.

System of Equations
(in standard form)

$$\begin{cases} 2x - 3y = 6 & \text{Equation 1} \\ x + y = 0 & \text{Equation 2} \end{cases}$$

Corresponding Matrix

$$\begin{bmatrix} 2 & -3 & 6 \\ 1 & 1 & 0 \end{bmatrix} \begin{array}{l} \text{Row 1} \\ \text{Row 2} \end{array}$$

Notice that the rows of the matrix correspond to the equations in the system. The coefficients of each variable are placed to the left of a vertical dashed line. The constants are placed to the right. Each of these numbers in the matrix is called an **element**.

The method of solving systems by matrices is to write this matrix as an equivalent matrix from which we easily identify the solution. Two matrices are equivalent if they represent systems that have the same solution set. The following **row operations** can be performed on matrices, and the result is an equivalent matrix.

Elementary Row Operations

1. Any two rows in a matrix may be interchanged.
2. The elements of any row may be multiplied (or divided) by the same nonzero number.
3. The elements of any row may be multiplied (or divided) by a nonzero number and added to their corresponding elements in any other row.

> **Helpful Hint**
> Notice that these *row* operations are the same operations that we can perform on *equations* in a system.

To solve a system of two equations in x and y by matrices, write the corresponding matrix associated with the system. Then use elementary row operations to write equivalent matrices until you have a matrix of the form

$$\begin{bmatrix} 1 & a & b \\ 0 & 1 & c \end{bmatrix},$$

where a, b, and c are constants. Why? If a matrix associated with a system of equations is in this form, we can easily solve for x and y. For example,

Matrix		*System of Equations*	
$\begin{bmatrix} 1 & 2 & -3 \\ 0 & 1 & 5 \end{bmatrix}$	corresponds to	$\begin{cases} 1x + 2y = -3 \\ 0x + 1y = 5 \end{cases}$ or	$\begin{cases} x + 2y = -3 \\ y = 5 \end{cases}$

In the second equation, we have $y = 5$. Substituting this in the first equation, we have $x + 2(5) = -3$ or $x = -13$. The solution of the system is the ordered pair $(-13, 5)$.

EXAMPLE 1

Use matrices to solve the system.

$$\begin{cases} x + 3y = 5 \\ 2x - y = -4 \end{cases}$$

Solution   The corresponding matrix is $\begin{bmatrix} 1 & 3 & 5 \\ 2 & -1 & -4 \end{bmatrix}$. We use elementary row operations to write an equivalent matrix that looks like $\begin{bmatrix} 1 & a & b \\ 0 & 1 & c \end{bmatrix}$.

For the matrix given, the element in the first row, first column is already 1, as desired. Next we write an equivalent matrix with a 0 below the 1. To do this, we multiply row 1 by -2 and add to row 2. *We will change only row 2.*

$$\begin{bmatrix} 1 & 3 & 5 \\ -2(1) + 2 & -2(3) + (-1) & -2(5) + (-4) \end{bmatrix} \quad \text{simplifies to} \quad \begin{bmatrix} 1 & 3 & 5 \\ 0 & -7 & -14 \end{bmatrix}$$

$\uparrow \quad \uparrow \qquad \uparrow \qquad \uparrow \qquad \uparrow \qquad \uparrow$

row 1   row 2   row 1   row 2   row 1   row 2

element element   element element   element element

Now we change the -7 to a 1 by use of an elementary row operation. We divide row 2 by -7, then

$$\begin{bmatrix} 1 & 3 & 5 \\ \frac{0}{-7} & \frac{-7}{-7} & \frac{-14}{-7} \end{bmatrix} \quad \text{simplifies to} \quad \begin{bmatrix} 1 & 3 & 5 \\ 0 & 1 & 2 \end{bmatrix}$$

This last matrix corresponds to the system

$$\begin{cases} x + 3y = 5 \\ \quad\quad y = 2 \end{cases}$$

To find x, we let $y = 2$ in the first equation, $x + 3y = 5$.

$$x + 3y = 5 \qquad \text{First equation}$$
$$x + 3(2) = 5 \qquad \text{Let } y = 2.$$
$$x = -1$$

The ordered pair solution is $(-1, 2)$. Check to see that this ordered pair satisfies both equations.

EXAMPLE 2

Use matrices to solve the system.

$$\begin{cases} 2x - y = 3 \\ 4x - 2y = 5 \end{cases}$$

Solution The corresponding matrix is $\begin{bmatrix} 2 & -1 & 3 \\ 4 & -2 & 5 \end{bmatrix}$. To get 1 in the row 1, column 1 position, we divide the elements of row 1 by 2.

$$\begin{bmatrix} \frac{2}{2} & -\frac{1}{2} & \frac{3}{2} \\ 4 & -2 & 5 \end{bmatrix} \quad \text{simplifies to} \quad \begin{bmatrix} 1 & -\frac{1}{2} & \frac{3}{2} \\ 4 & -2 & 5 \end{bmatrix}$$

To get 0 under the 1, we multiply the elements of row 1 by -4 and add the new elements to the elements of row 2.

$$\begin{bmatrix} 1 & -\frac{1}{2} & \frac{3}{2} \\ -4(1) + 4 & -4\left(-\frac{1}{2}\right) - 2 & -4\left(\frac{3}{2}\right) + 5 \end{bmatrix} \quad \text{simplifies to} \quad \begin{bmatrix} 1 & -\frac{1}{2} & \frac{3}{2} \\ 0 & 0 & -1 \end{bmatrix}$$

The corresponding system is $\begin{cases} x - \dfrac{1}{2}y = \dfrac{3}{2} \\ \quad\quad\quad 0 = -1 \end{cases}$. The equation $0 = -1$ is false for all y or x values; hence the system is inconsistent and has no solution.

✔ **CONCEPT CHECK**

Consider the system

$$\begin{cases} 2x - 3y = 8 \\ x + 5y = -3 \end{cases}$$

What is wrong with its corresponding matrix shown below?

$$\begin{bmatrix} 2 & -3 & 8 \\ 0 & 5 & -3 \end{bmatrix}$$

Concept Check Answer:

matrix should be $\begin{bmatrix} 2 & -3 & 8 \\ 1 & 5 & -3 \end{bmatrix}$

2   To solve a system of three equations in three variables using matrices, we will write the corresponding matrix in the form

$$\begin{bmatrix} 1 & a & b & d \\ 0 & 1 & c & e \\ 0 & 0 & 1 & f \end{bmatrix}$$

EXAMPLE 3

Use matrices to solve the system.

$$\begin{cases} x + 2y + z = 2 \\ -2x - y + 2z = 5 \\ x + 3y - 2z = -8 \end{cases}$$

Solution   The corresponding matrix is $\begin{bmatrix} 1 & 2 & 1 & 2 \\ -2 & -1 & 2 & 5 \\ 1 & 3 & -2 & -8 \end{bmatrix}$. Our goal is to write an

equivalent matrix with 1's along the diagonal (see the numbers in red) and 0's below the 1's. The element in row 1, column 1 is already 1. Next we get 0's for each element in the rest of column 1. To do this, first we multiply the elements of row 1 by 2 and add the new elements to row 2. Also, we multiply the elements of row 1 by −1 and add the new elements to the elements of row 3. We *do not change row 1.* Then

$$\begin{bmatrix} 1 & 2 & 1 & 2 \\ 2(1)-2 & 2(2)-1 & 2(1)+2 & 2(2)+5 \\ -1(1)+1 & -1(2)+3 & -1(1)-2 & -1(2)-8 \end{bmatrix} \text{ simplifies to } \begin{bmatrix} 1 & 2 & 1 & 2 \\ 0 & 3 & 4 & 9 \\ 0 & 1 & -3 & -10 \end{bmatrix}$$

We continue down the diagonal and use elementary row operations to get 1 where the element 3 is now. To do this, we interchange rows 2 and 3.

$$\begin{bmatrix} 1 & 2 & 1 & 2 \\ 0 & 3 & 4 & 9 \\ 0 & 1 & -3 & -10 \end{bmatrix} \text{ is equivalent to } \begin{bmatrix} 1 & 2 & 1 & 2 \\ 0 & 1 & -3 & -10 \\ 0 & 3 & 4 & 9 \end{bmatrix}$$

Next we want the new row 3, column 2 element to be 0. We multiply the elements of row 2 by −3 and add the result to the elements of row 3.

$$\begin{bmatrix} 1 & 2 & 1 & 2 \\ 0 & 1 & -3 & -10 \\ -3(0)+0 & -3(1)+3 & -3(-3)+4 & -3(-10)+9 \end{bmatrix} \text{ simplifies to }$$

$$\begin{bmatrix} 1 & 2 & 1 & 2 \\ 0 & 1 & -3 & -10 \\ 0 & 0 & 13 & 39 \end{bmatrix}$$

Finally, we divide the elements of row 3 by 13 so that the final diagonal element is 1.

$$\begin{bmatrix} 1 & 2 & 1 & 2 \\ 0 & 1 & -3 & -10 \\ \frac{0}{13} & \frac{0}{13} & \frac{13}{13} & \frac{39}{13} \end{bmatrix} \text{ simplifies to } \begin{bmatrix} 1 & 2 & 1 & 2 \\ 0 & 1 & -3 & -10 \\ 0 & 0 & 1 & 3 \end{bmatrix}$$

This matrix corresponds to the system

$$\begin{cases} x + 2y + z = 2 \\ y - 3z = -10 \\ z = 3 \end{cases}$$

We identify the z-coordinate of the solution as 3. Next we replace z with 3 in the second equation and solve for y.

$$y - 3z = -10 \quad \text{Second equation}$$
$$y - 3(3) = -10 \quad \text{Let } z = 3.$$
$$y = -1$$

To find x, we let $z = 3$ and $y = -1$ in the first equation.

$$x + 2y + z = 2 \quad \text{First equation}$$
$$x + 2(-1) + 3 = 2 \quad \text{Let } z = 3 \text{ and } y = -1.$$
$$x = 1$$

The ordered triple solution is $(1, -1, 3)$. Check to see that it satisfies all three equations in the original system.

APPENDIX E EXERCISE SET

STUDY GUIDE/SSM   CD/VIDEO   PH MATH TUTOR CENTER   MathXL®Tutorials ON CD   MathXL®   MyMathLab®

Solve each system of linear equations using matrices. See Example 1.

1. $\begin{cases} x + y = 1 \\ x - 2y = 4 \end{cases}$

2. $\begin{cases} 2x - y = 8 \\ x + 3y = 11 \end{cases}$

3. $\begin{cases} x + 3y = 2 \\ x + 2y = 0 \end{cases}$

4. $\begin{cases} 4x - y = 5 \\ 3x - 3y = 0 \end{cases}$

Solve each system of linear equations using matrices. See Example 2.

5. $\begin{cases} x - 2y = 4 \\ 2x - 4y = 4 \end{cases}$

6. $\begin{cases} -x + 3y = 6 \\ 3x - 9y = 9 \end{cases}$

7. $\begin{cases} 3x - 3y = 9 \\ 2x - 2y = 6 \end{cases}$

8. $\begin{cases} 9x - 3y = 6 \\ -18x + 6y = -12 \end{cases}$

Solve each system of linear equations using matrices. See Example 3.

9. $\begin{cases} x + y = 3 \\ 2y = 10 \\ 3x + 2y - 4z = 12 \end{cases}$

10. $\begin{cases} 5x = 5 \\ 2x + y = 4 \\ 3x + y - 5z = -15 \end{cases}$

11. $\begin{cases} 2y - z = -7 \\ x + 4y + z = -4 \\ 5x - y + 2z = 13 \end{cases}$

12. $\begin{cases} 4y + 3z = -2 \\ 5x - 4y = 1 \\ -5x + 4y + z = -3 \end{cases}$

MIXED PRACTICE

Solve each system of linear equations using matrices.

13. $\begin{cases} x - 4 = 0 \\ x + y = 1 \end{cases}$

14. $\begin{cases} 3y = 6 \\ x + y = 7 \end{cases}$

15. $\begin{cases} x + y + z = 2 \\ 2x - z = 5 \\ 3y + z = 2 \end{cases}$

16. $\begin{cases} x + 2y + z = 5 \\ x - y - z = 3 \\ y + z = 2 \end{cases}$

17. $\begin{cases} 5x - 2y = 27 \\ -3x + 5y = 18 \end{cases}$

18. $\begin{cases} 4x - y = 9 \\ 2x + 3y = -27 \end{cases}$

19. $\begin{cases} 4x - 7y = 7 \\ 12x - 21y = 24 \end{cases}$

20. $\begin{cases} 2x - 5y = 12 \\ -4x + 10y = 20 \end{cases}$

21. $\begin{cases} 4x - y + 2z = 5 \\ 2y + z = 4 \\ 4x + y + 3z = 10 \end{cases}$

22. $\begin{cases} 5y - 7z = 14 \\ 2x + y + 4z = 10 \\ 2x + 6y - 3z = 30 \end{cases}$

23. $\begin{cases} 4x + y + z = 3 \\ -x + y - 2z = -11 \\ x + 2y + 2z = -1 \end{cases}$

24. $\begin{cases} x + y + z = 9 \\ 3x - y + z = -1 \\ -2x + 2y - 3z = -2 \end{cases}$

Concept Extensions

25. The percent y of U.S. households that owned a black-and-white television set between the years 1980 and 1993 can be modeled by the linear equation $2.3x + y = 52$, where x represents the number of years after 1980. Similarly, the percent y of U.S. households that owned a microwave oven during this same period can be modeled by the linear equation $-5.4x + y = 14$. (*Source:* Based on data from the Energy Information Administration, U.S. Department of Energy)

   a. The data used to form these two models was incomplete. It is impossible to tell from the data the year in which the percent of households owning black-and-white television sets was the same as the percent of households owning microwave ovens. Use matrix methods to estimate the year in which this occurred.

   b. Did more households own black-and-white television sets or microwave ovens in 1980? In 1993? What trends do these models show? Does this seem to make sense? Why or why not?

   c. According to the models, when will the percent of households owning black-and-white television sets reach 0%?

   d. Do you think your answer to part **c** is accurate? Why or why not?

26. The most popular amusement park in the world (according to annual attendance) is Tokyo Disneyland, whose yearly attendance in thousands can be approximated by the equation $y = 1201x + 16{,}507$ where x is the number of years after 2000. In second place is Walt Disney World's Magic Kingdom, whose yearly attendance, in thousands, can be approximated by $y = -616x + 15{,}400$. Find the last year when attendance in Magic Kingdom was greater than attendance in Tokyo Disneyland. (*Source*: Amusement Business)

27. For the system $\begin{cases} 2x - 3y = 8 \\ x + 5y = -3 \end{cases}$, explain what is wrong with writing the corresponding matrix as $\begin{bmatrix} 2 & 3 & 8 \\ 0 & 5 & -3 \end{bmatrix}$.

Solving Systems of Equations by Determinants

Objectives

1. Define and evaluate a 2×2 determinant.
2. Use Cramer's rule to solve a system of two linear equations in two variables.
3. Define and evaluate a 3×3 determinant.
4. Use Cramer's rule to solve a linear system of three equations in three variables.

1 We have solved systems of two linear equations in two variables in four different ways: graphically, by substitution, by elimination, and by matrices. Now we analyze another method called **Cramer's rule**.

Recall that a matrix is a rectangular array of numbers. If a matrix has the same number of rows and columns, it is called a **square matrix**. Examples of square matrices are

$$\begin{bmatrix} 1 & 6 \\ 5 & 2 \end{bmatrix} \qquad \begin{bmatrix} 2 & 4 & 1 \\ 0 & 5 & 2 \\ 3 & 6 & 9 \end{bmatrix}$$

A **determinant** is a real number associated with a square matrix. The determinant of a square matrix is denoted by placing vertical bars about the array of numbers. Thus,

The determinant of the square matrix $\begin{bmatrix} 1 & 6 \\ 5 & 2 \end{bmatrix}$ is $\begin{vmatrix} 1 & 6 \\ 5 & 2 \end{vmatrix}$.

The determinant of the square matrix $\begin{bmatrix} 2 & 4 & 1 \\ 0 & 5 & 2 \\ 3 & 6 & 9 \end{bmatrix}$ is $\begin{vmatrix} 2 & 4 & 1 \\ 0 & 5 & 2 \\ 3 & 6 & 9 \end{vmatrix}$.

We define the determinant of a 2×2 matrix first. (Recall that 2×2 is read "two by two." It means that the matrix has 2 rows and 2 columns.)

Determinant of a 2×2 Matrix

$$\begin{vmatrix} a & b \\ c & d \end{vmatrix} = ad - bc$$

EXAMPLE 1

Evaluate each determinant.

a. $\begin{vmatrix} -1 & 2 \\ 3 & -4 \end{vmatrix}$

b. $\begin{vmatrix} 2 & 0 \\ 7 & -5 \end{vmatrix}$

Solution   First we identify the values of a, b, c, and d. Then we perform the evaluation.

a. Here $a = -1$, $b = 2$, $c = 3$, and $d = -4$.

$$\begin{vmatrix} -1 & 2 \\ 3 & -4 \end{vmatrix} = ad - bc = (-1)(-4) - (2)(3) = -2$$

b. In this example, $a = 2$, $b = 0$, $c = 7$, and $d = -5$.

$$\begin{vmatrix} 2 & 0 \\ 7 & -5 \end{vmatrix} = ad - bc = 2(-5) - (0)(7) = -10$$

2   To develop Cramer's rule, we solve the system $\begin{cases} ax + by = h \\ cx + dy = k \end{cases}$ using elimination.
First, we eliminate y by multiplying both sides of the first equation by d and both sides of the second equation by $-b$ so that the coefficients of y are opposites. The result is that

$$\begin{cases} d(ax + by) = d \cdot h \\ -b(cx + dy) = -b \cdot k \end{cases} \quad \text{simplifies to} \quad \begin{cases} adx + bdy = hd \\ -bcx - bdy = -kb \end{cases}$$

We now add the two equations and solve for x.

$$\begin{array}{rl} adx + bdy &= hd \\ -bcx - bdy &= -kb \\ \hline adx - bcx &= hd - kb \qquad \text{Add the equations.} \\ (ad - bc)x &= hd - kb \\ x &= \dfrac{hd - kb}{ad - bc} \qquad \text{Solve for } x. \end{array}$$

When we replace x with $\dfrac{hd - kb}{ad - bc}$ in the equation $ax + by = h$ and solve for y, we find that $y = \dfrac{ak - ch}{ad - bc}$.

Notice that the numerator of the value of x is the determinant of

$$\begin{vmatrix} h & b \\ k & d \end{vmatrix} = hd - kb$$

Also, the numerator of the value of y is the determinant of

$$\begin{vmatrix} a & h \\ c & k \end{vmatrix} = ak - hc$$

Finally, the denominators of the values of x and y are the same and are the determinant of

$$\begin{vmatrix} a & b \\ c & d \end{vmatrix} = ad - bc$$

This means that the values of x and y can be written in determinant notation.

$$x = \dfrac{\begin{vmatrix} h & b \\ k & d \end{vmatrix}}{\begin{vmatrix} a & b \\ c & d \end{vmatrix}} \quad \text{and} \quad y = \dfrac{\begin{vmatrix} a & h \\ c & k \end{vmatrix}}{\begin{vmatrix} a & b \\ c & d \end{vmatrix}}$$

For convenience, we label the determinants D, D_x, and D_y.

x-coefficients
y-coefficients

$$\begin{vmatrix} a & b \\ c & d \end{vmatrix} = D \qquad \begin{vmatrix} h & b \\ k & d \end{vmatrix} = D_x \qquad \begin{vmatrix} a & h \\ c & k \end{vmatrix} = D_y$$

x-column replaced   y-column replaced
by constants   by constants

These determinant formulas for the coordinates of the solution of a system are known as **Cramer's rule**.

Cramer's Rule for Two Linear Equations in Two Variables

The solution of the system $\begin{cases} ax + by = h \\ cx + dy = k \end{cases}$ is given by

$$x = \dfrac{\begin{vmatrix} h & b \\ k & d \end{vmatrix}}{\begin{vmatrix} a & b \\ c & d \end{vmatrix}} = \dfrac{D_x}{D} \qquad y = \dfrac{\begin{vmatrix} a & h \\ c & k \end{vmatrix}}{\begin{vmatrix} a & b \\ c & d \end{vmatrix}} = \dfrac{D_y}{D}$$

as long as $D = ad - bc$ is not 0.

When $D = 0$, the system is either inconsistent or the equations are dependent. When this happens, we need to use another method to see which is the case.

EXAMPLE 2

Use Cramer's rule to solve each system.

a. $\begin{cases} 3x + 4y = -7 \\ x - 2y = -9 \end{cases}$ **b.** $\begin{cases} 5x + y = 5 \\ -7x - 2y = -7 \end{cases}$

Solution **a.** First we find D, D_x, and D_y.

$$\begin{array}{ccc} a & b & h \\ \downarrow & \downarrow & \downarrow \end{array}$$
$$\begin{cases} 3x + 4y = -7 \\ x - 2y = -9 \end{cases}$$
$$\begin{array}{ccc} \uparrow & \uparrow & \uparrow \\ c & d & k \end{array}$$

$$D = \begin{vmatrix} a & b \\ c & d \end{vmatrix} = \begin{vmatrix} 3 & 4 \\ 1 & -2 \end{vmatrix} = 3(-2) - 4(1) = -10$$

$$D_x = \begin{vmatrix} h & b \\ k & d \end{vmatrix} = \begin{vmatrix} -7 & 4 \\ -9 & -2 \end{vmatrix} = (-7)(-2) - 4(-9) = 50$$

$$D_y = \begin{vmatrix} a & h \\ c & k \end{vmatrix} = \begin{vmatrix} 3 & -7 \\ 1 & -9 \end{vmatrix} = 3(-9) - (-7)(1) = -20$$

Then $x = \dfrac{D_x}{D} = \dfrac{50}{-10} = -5$ and $y = \dfrac{D_y}{D} = \dfrac{-20}{-10} = 2$. The ordered pair solution is $(-5, 2)$.

As always, check the solution in both original equations.

b. Find D, D_x, and D_y for $\begin{cases} 5x + y = 5 \\ -7x - 2y = -7 \end{cases}$.

$$D = \begin{vmatrix} 5 & 1 \\ -7 & -2 \end{vmatrix} = 5(-2) - (1)(-7) = -3$$

$$D_x = \begin{vmatrix} 5 & 1 \\ -7 & -2 \end{vmatrix} = 5(-2) - (1)(-7) = -3$$

$$D_y = \begin{vmatrix} 5 & 5 \\ -7 & -7 \end{vmatrix} = 5(-7) - 5(-7) = 0$$

Then $x = \dfrac{D_x}{D} = \dfrac{-3}{-3} = 1$ and $y = \dfrac{D_y}{D} = \dfrac{0}{-3} = 0$.

The ordered pair solution is $(1, 0)$. Check this solution in both original equations.

3   A 3×3 determinant can be used to solve a system of three equations in three variables. The determinant of a 3×3 matrix, however, is considerably more complex than a 2×2 one.

> ## Determinant of a 3×3 Matrix
>
> $$\begin{vmatrix} a_1 & b_1 & c_1 \\ a_2 & b_2 & c_2 \\ a_3 & b_3 & c_3 \end{vmatrix} = a_1 \cdot \begin{vmatrix} b_2 & c_2 \\ b_3 & c_3 \end{vmatrix} - a_2 \cdot \begin{vmatrix} b_1 & c_1 \\ b_3 & c_3 \end{vmatrix} + a_3 \cdot \begin{vmatrix} b_1 & c_1 \\ b_2 & c_2 \end{vmatrix}$$

Notice that the determinant of a 3×3 matrix is related to the determinants of three 2×2 matrices. Each determinant of these 2×2 matrices is called a **minor**, and every element of a 3×3 matrix has a minor associated with it. For example, the minor of c_2 is the determinant of the 2×2 matrix found by deleting the row and column containing c_2.

$$\begin{array}{ccc} a_1 & b_1 & c_1 \\ a_2 & b_2 & c_2 \\ a_3 & b_3 & c_3 \end{array} \qquad \text{The minor of } c_2 \text{ is} \qquad \begin{vmatrix} a_1 & b_1 \\ a_3 & b_3 \end{vmatrix}$$

Also, the minor of element a_1 is the determinant of the 2×2 matrix that has no row or column containing a_1.

$$\begin{array}{ccc} a_1 & b_1 & c_1 \\ a_2 & b_2 & c_2 \\ a_3 & b_3 & c_3 \end{array} \qquad \text{The minor of } a_1 \text{ is} \qquad \begin{vmatrix} b_2 & c_2 \\ b_3 & c_3 \end{vmatrix}$$

So the determinant of a 3×3 matrix can be written as

$$a_1 \cdot (\text{minor of } a_1) - a_2 \cdot (\text{minor of } a_2) + a_3 \cdot (\text{minor of } a_3)$$

Finding the determinant by using minors of elements in the first column is called **expanding** by the minors of the first column. *The value of a determinant can be found by expanding by the minors of any row or column.* The following **array of signs** is helpful in determining whether to add or subtract the product of an element and its minor.

$$\begin{array}{ccc} + & - & + \\ - & + & - \\ + & - & + \end{array}$$

If an element is in a position marked $+$, we add. If marked $-$, we subtract.

Concept Check Answer:

$$\begin{array}{cccc} + & - & + & - \\ - & + & - & + \\ + & - & + & - \\ - & + & - & + \end{array}$$

✔ **CONCEPT CHECK**

Suppose you are interested in finding the determinant of a 4×4 matrix. Study the pattern shown in the array of signs for a 3×3 matrix. Use the pattern to expand the array of signs for use with a 4×4 matrix.

EXAMPLE 3

Evaluate by expanding by the minors of the given row or column.

$$\begin{vmatrix} 0 & 5 & 1 \\ 1 & 3 & -1 \\ -2 & 2 & 4 \end{vmatrix}$$

Solution **a.** First column   **b.** Second row

a. The elements of the first column are $0, 1,$ and -2. The first column of the array of signs is $+, -, +$.

$$\begin{vmatrix} 0 & 5 & 1 \\ 1 & 3 & -1 \\ -2 & 2 & 4 \end{vmatrix} = 0 \cdot \begin{vmatrix} 3 & -1 \\ 2 & 4 \end{vmatrix} - 1 \cdot \begin{vmatrix} 5 & 1 \\ 2 & 4 \end{vmatrix} + (-2) \cdot \begin{vmatrix} 5 & 1 \\ 3 & -1 \end{vmatrix}$$

$$= 0(12 - (-2)) - 1(20 - 2) + (-2)(-5 - 3)$$

$$= 0 - 18 + 16 = -2$$

b. The elements of the second row are $1, 3,$ and -1. This time, the signs begin with $-$ and again alternate.

$$\begin{vmatrix} 0 & 5 & 1 \\ 1 & 3 & -1 \\ -2 & 2 & 4 \end{vmatrix} = -1 \cdot \begin{vmatrix} 5 & 1 \\ 2 & 4 \end{vmatrix} + 3 \cdot \begin{vmatrix} 0 & 1 \\ -2 & 4 \end{vmatrix} - (-1) \cdot \begin{vmatrix} 0 & 5 \\ -2 & 2 \end{vmatrix}$$

$$= -1(20 - 2) + 3(0 - (-2)) - (-1)(0 - (-10))$$

$$= -18 + 6 + 10 = -2$$

Notice that the determinant of the 3×3 matrix is the same regardless of the row or column you select to expand by.

✔ **CONCEPT CHECK**

Why would expanding by minors of the second row be a good choice for the determinant
$$\begin{vmatrix} 3 & 4 & -2 \\ 5 & 0 & 0 \\ 6 & -3 & 7 \end{vmatrix}?$$

4   A system of three equations in three variables may be solved with Cramer's rule also. Using the elimination process to solve a system with unknown constants as coefficients leads to the following.

Cramer's Rule for Three Equations in Three Variables

The solution of the system $\begin{cases} a_1x + b_1y + c_1z = k_1 \\ a_2x + b_2y + c_2z = k_2 \\ a_3x + b_3y + c_3z = k_3 \end{cases}$ is given by

$$x = \frac{D_x}{D} \qquad y = \frac{D_y}{D} \qquad \text{and} \qquad z = \frac{D_z}{D}$$

Concept Check Answer:
Two elements of the second row are 0, which makes calculations easier.

where

$$D = \begin{vmatrix} a_1 & b_1 & c_1 \\ a_2 & b_2 & c_2 \\ a_3 & b_3 & c_3 \end{vmatrix} \qquad D_x = \begin{vmatrix} k_1 & b_1 & c_1 \\ k_2 & b_2 & c_2 \\ k_3 & b_3 & c_3 \end{vmatrix}$$

$$D_y = \begin{vmatrix} a_1 & k_1 & c_1 \\ a_2 & k_2 & c_2 \\ a_3 & k_3 & c_3 \end{vmatrix} \qquad D_z = \begin{vmatrix} a_1 & b_1 & k_1 \\ a_2 & b_2 & k_2 \\ a_3 & b_3 & k_3 \end{vmatrix}$$

as long as D is not 0.

EXAMPLE 4

Use Cramer's rule to solve the system.

$$\begin{cases} x - 2y + z = 4 \\ 3x + y - 2z = 3 \\ 5x + 5y + 3z = -8 \end{cases}$$

Solution First we find D, D_x, D_y, and D_z. Beginning with D, we expand by the minors of the first column.

$$D = \begin{vmatrix} 1 & -2 & 1 \\ 3 & 1 & -2 \\ 5 & 5 & 3 \end{vmatrix} = 1 \cdot \begin{vmatrix} 1 & -2 \\ 5 & 3 \end{vmatrix} - 3 \cdot \begin{vmatrix} -2 & 1 \\ 5 & 3 \end{vmatrix} + 5 \cdot \begin{vmatrix} -2 & 1 \\ 1 & -2 \end{vmatrix}$$

$$= 1(3 - (-10)) - 3(-6 - 5) + 5(4 - 1)$$

$$= 13 + 33 + 15 = 61$$

$$D_x = \begin{vmatrix} 4 & -2 & 1 \\ 3 & 1 & -2 \\ -8 & 5 & 3 \end{vmatrix} = 4 \cdot \begin{vmatrix} 1 & -2 \\ 5 & 3 \end{vmatrix} - 3 \cdot \begin{vmatrix} -2 & 1 \\ 5 & 3 \end{vmatrix} + (-8) \cdot \begin{vmatrix} -2 & 1 \\ 1 & -2 \end{vmatrix}$$

$$= 4(3 - (-10)) - 3(-6 - 5) + (-8)(4 - 1)$$

$$= 52 + 33 - 24 = 61$$

$$D_y = \begin{vmatrix} 1 & 4 & 1 \\ 3 & 3 & -2 \\ 5 & -8 & 3 \end{vmatrix} = 1 \cdot \begin{vmatrix} 3 & -2 \\ -8 & 3 \end{vmatrix} - 3 \cdot \begin{vmatrix} 4 & 1 \\ -8 & 3 \end{vmatrix} + 5 \cdot \begin{vmatrix} 4 & 1 \\ 3 & -2 \end{vmatrix}$$

$$= 1(9 - 16) - 3(12 - (-8)) + 5(-8 - 3)$$

$$= -7 - 60 - 55 = -122$$

$$D_z = \begin{vmatrix} 1 & -2 & 4 \\ 3 & 1 & 3 \\ 5 & 5 & -8 \end{vmatrix} = 1 \cdot \begin{vmatrix} 1 & 3 \\ 5 & -8 \end{vmatrix} - 3 \cdot \begin{vmatrix} -2 & 4 \\ 5 & -8 \end{vmatrix} + 5 \cdot \begin{vmatrix} -2 & 4 \\ 1 & 3 \end{vmatrix}$$

$$= 1(-8 - 15) - 3(16 - 20) + 5(-6 - 4)$$

$$= -23 + 12 - 50 = -61$$

From these determinants, we calculate the solution.

$$x = \frac{D_x}{D} = \frac{61}{61} = 1 \qquad y = \frac{D_y}{D} = \frac{-122}{61} = -2 \qquad z = \frac{D_z}{D} = \frac{-61}{61} = -1$$

The ordered triple solution is $(1, -2, -1)$. Check this solution by verifying that it satisfies each equation of the system.

APPENDIX F EXERCISE SET

STUDY GUIDE/SSM   CD/VIDEO   PH MATH TUTOR CENTER   MathXL®Tutorials ON CD   MathXL®   MyMathLab®

Evaluate. See Example 1.

 1. $\begin{vmatrix} 3 & 5 \\ -1 & 7 \end{vmatrix}$ **2.** $\begin{vmatrix} -5 & 1 \\ 0 & -4 \end{vmatrix}$

3. $\begin{vmatrix} 9 & -2 \\ 4 & -3 \end{vmatrix}$ **4.** $\begin{vmatrix} 4 & 0 \\ 9 & 8 \end{vmatrix}$

5. $\begin{vmatrix} -2 & 9 \\ 4 & -18 \end{vmatrix}$ **6.** $\begin{vmatrix} -40 & 8 \\ 70 & -14 \end{vmatrix}$

Use Cramer's rule, if possible, to solve each system of linear equations. See Example 2.

7. $\begin{cases} 2y - 4 = 0 \\ x + 2y = 5 \end{cases}$ **8.** $\begin{cases} 4x - y = 5 \\ 3x - 3 = 0 \end{cases}$

9. $\begin{cases} 3x + y = 1 \\ 2y = 2 - 6x \end{cases}$ **10.** $\begin{cases} y = 2x - 5 \\ 8x - 4y = 20 \end{cases}$

11. $\begin{cases} 5x - 2y = 27 \\ -3x + 5y = 18 \end{cases}$ **12.** $\begin{cases} 4x - y = 9 \\ 2x + 3y = -27 \end{cases}$

Evaluate. See Example 3.

13. $\begin{vmatrix} 2 & 1 & 0 \\ 0 & 5 & -3 \\ 4 & 0 & 2 \end{vmatrix}$ **14.** $\begin{vmatrix} -6 & 4 & 2 \\ 1 & 0 & 5 \\ 0 & 3 & 1 \end{vmatrix}$

15. $\begin{vmatrix} 4 & -6 & 0 \\ -2 & 3 & 0 \\ 4 & -6 & 1 \end{vmatrix}$ **16.** $\begin{vmatrix} 5 & 2 & 1 \\ 3 & -6 & 0 \\ -2 & 8 & 0 \end{vmatrix}$

17. $\begin{vmatrix} 3 & 6 & -3 \\ -1 & -2 & 3 \\ 4 & -1 & 6 \end{vmatrix}$ **18.** $\begin{vmatrix} 2 & -2 & 1 \\ 4 & 1 & 3 \\ 3 & 1 & 2 \end{vmatrix}$

Use Cramer's rule, if possible, to solve each system of linear equations. See Example 4.

19. $\begin{cases} 3x + z = -1 \\ -x - 3y + z = 7 \\ 3y + z = 5 \end{cases}$ **20.** $\begin{cases} 4y - 3z = -2 \\ 8x - 4y = 4 \\ -8x + 4y + z = -2 \end{cases}$

21. $\begin{cases} x + y + z = 8 \\ 2x - y - z = 10 \\ x - 2y + 3z = 22 \end{cases}$ **22.** $\begin{cases} 5x + y + 3z = 1 \\ x - y - 3z = -7 \\ -x + y = 1 \end{cases}$

Evaluate.

23. $\begin{vmatrix} 10 & -1 \\ -4 & 2 \end{vmatrix}$ **24.** $\begin{vmatrix} -6 & 2 \\ 5 & -1 \end{vmatrix}$

25. $\begin{vmatrix} 1 & 0 & 4 \\ 1 & -1 & 2 \\ 3 & 2 & 1 \end{vmatrix}$ **26.** $\begin{vmatrix} 0 & 1 & 2 \\ 3 & -1 & 2 \\ 3 & 2 & -2 \end{vmatrix}$

27. $\begin{vmatrix} \frac{3}{4} & \frac{5}{2} \\ -\frac{1}{6} & \frac{7}{3} \end{vmatrix}$ **28.** $\begin{vmatrix} \frac{5}{7} & \frac{1}{3} \\ \frac{6}{7} & \frac{2}{3} \end{vmatrix}$ **29.** $\begin{vmatrix} 4 & -2 & 2 \\ 6 & -1 & 3 \\ 2 & 1 & 1 \end{vmatrix}$

30. $\begin{vmatrix} 1 & 5 & 0 \\ 7 & 9 & -4 \\ 3 & 2 & -2 \end{vmatrix}$ **31.** $\begin{vmatrix} -2 & 5 & 4 \\ 5 & -1 & 3 \\ 4 & 1 & 2 \end{vmatrix}$ **32.** $\begin{vmatrix} 5 & -2 & 4 \\ -1 & 5 & 3 \\ 1 & 4 & 2 \end{vmatrix}$

MIXED PRACTICE

Use Cramer's rule, if possible, to solve each system of linear equations.

33. $\begin{cases} 2x - 5y = 4 \\ x + 2y = -7 \end{cases}$ **34.** $\begin{cases} 3x - y = 2 \\ -5x + 2y = 0 \end{cases}$

35. $\begin{cases} 4x + 2y = 5 \\ 2x + y = -1 \end{cases}$ **36.** $\begin{cases} 3x + 6y = 15 \\ 2x + 4y = 3 \end{cases}$

37. $\begin{cases} 2x + 2y + z = 1 \\ -x + y + 2z = 3 \\ x + 2y + 4z = 0 \end{cases}$ **38.** $\begin{cases} 2x - 3y + z = 5 \\ x + y + z = 0 \\ 4x + 2y + 4z = 4 \end{cases}$

39. $\begin{cases} \frac{2}{3}x - \frac{3}{4}y = -1 \\ -\frac{1}{6}x + \frac{3}{4}y = \frac{5}{2} \end{cases}$ **40.** $\begin{cases} \frac{1}{2}x - \frac{1}{3}y = -3 \\ \frac{1}{8}x + \frac{1}{6}y = 0 \end{cases}$

41. $\begin{cases} 0.7x - 0.2y = -1.6 \\ 0.2x - y = -1.4 \end{cases}$ **42.** $\begin{cases} -0.7x + 0.6y = 1.3 \\ 0.5x - 0.3y = -0.8 \end{cases}$

43. $\begin{cases} -2x + 4y - 2z = 6 \\ x - 2y + z = -3 \\ 3x - 6y + 3z = -9 \end{cases}$ **44.** $\begin{cases} -x - y + 3z = 2 \\ 4x + 4y - 12z = -8 \\ -3x - 3y + 9z = 6 \end{cases}$

45. $\begin{cases} x - 2y + z = -5 \\ 3y + 2z = 4 \\ 3x - y = -2 \end{cases}$ **46.** $\begin{cases} 4x + 5y = 10 \\ 3y + 2z = -6 \\ x + y + z = 3 \end{cases}$

Concept Extensions

Find the value of x such that each is a true statement.

47. $\begin{vmatrix} 1 & x \\ 2 & 7 \end{vmatrix} = -3$ **48.** $\begin{vmatrix} 6 & 1 \\ -2 & x \end{vmatrix} = 26$

49. If all the elements in a single row of a determinant are zero, to what does the determinant evaluate? Explain your answer.

50. If all the elements in a single column of a determinant are 0, to what does the determinant evaluate? Explain your answer.

51. Suppose you are interested in finding the determinant of a 4×4 matrix. Study the pattern shown in the array of signs for a 3×3 matrix. Use the pattern to expand the array of signs for use with a 4×4 matrix.

52. Why would expanding by minors of the second row be a good choice for the determinant $\begin{vmatrix} 3 & 4 & -2 \\ 5 & 0 & 0 \\ 6 & -3 & 7 \end{vmatrix}$?

Find the value of each determinant. To evaluate a 4×4 determinant, select any row or column and expand by the minors. The array of signs for a 4×4 determinant is the same as for a 3×3 determinant except expanded.

53. $\begin{vmatrix} 5 & 0 & 0 & 0 \\ 0 & 4 & 2 & -1 \\ 1 & 3 & -2 & 0 \\ 0 & -3 & 1 & 2 \end{vmatrix}$ **54.** $\begin{vmatrix} 1 & 7 & 0 & -1 \\ 1 & 3 & -2 & 0 \\ 1 & 0 & -1 & 2 \\ 0 & -6 & 2 & 4 \end{vmatrix}$

55. $\begin{vmatrix} 4 & 0 & 2 & 5 \\ 0 & 3 & -1 & 1 \\ 0 & 0 & 2 & 0 \\ 0 & 0 & 0 & 1 \end{vmatrix}$ **56.** $\begin{vmatrix} 2 & 0 & -1 & 4 \\ 6 & 0 & 4 & 1 \\ 2 & 4 & 3 & -1 \\ 4 & 0 & 5 & -4 \end{vmatrix}$

Mean, Median, and Mode

It is sometimes desirable to be able to describe a set of data, or a set of numbers, by a single "middle" number. Three such **measures of central tendency** are the mean, the median, and the mode.

The most common measure of central tendency is the mean (sometimes called the arithmetic mean or the average). The **mean** of a set of data items, denoted by $\bar{x}$, is the sum of the items divided by the number of items.

EXAMPLE 1

Seven students in a psychology class conducted an experiment on mazes. Each student was given a pencil and asked to successfully complete the same maze. The timed results are below.

Student	Ann	Thanh	Carlos	Jesse	Melinda	Ramzi	Dayni
Time (Seconds)	13.2	11.8	10.7	16.2	15.9	13.8	18.5

a. Who completed the maze in the shortest time? Who completed the maze in the longest time?

b. Find the mean.

c. How many students took longer than the mean time? How many students took shorter than the mean time?

Solution **a.** Carlos completed the maze in 10.7 seconds, the shortest time. Dayni completed the maze in 18.5 seconds, the longest time.

b. To find the mean, $\bar{x}$, find the sum of the data items and divide by 7, the number of items.

$$\bar{x} = \frac{13.2 + 11.8 + 10.7 + 16.2 + 15.9 + 13.8 + 18.5}{7} = \frac{100.1}{7} = 14.3$$

c. Three students, Jesse, Melinda, and Dayni, had times longer than the mean time. Four students, Ann, Thanh, Carlos, and Ramzi, had times shorter than the mean time.

Two other measures of central tendency are the median and the mode.

The **median** of an ordered set of numbers is the middle number. If the number of items is even, the median is the mean of the two middle numbers. The **mode** of a set of numbers is the number that occurs most often. It is possible for a data set to have no mode or more than one mode.

EXAMPLE 2

Find the median and the mode of the following list of numbers. These numbers were high temperatures for fourteen consecutive days in a city in Montana.

$$76, 80, 85, 86, 89, 87, 82, 77, 76, 79, 82, 89, 89, 92$$

Solution   First, write the numbers in order.

$$76, 76, 77, 79, 80, 82, \underbrace{82, 85}_{\substack{two \\ middle\ numbers}}, 86, 87, \underbrace{89, 89, 89}_{mode}, 92$$

Since there are an even number of items, the median is the mean of the two middle numbers.

$$\text{median} = \frac{82 + 85}{2} = 83.5$$

The mode is 89, since 89 occurs most often.

APPENDIX G EXERCISE SET

For each of the following data sets, find the mean, the median, and the mode. If necessary, round the mean to one decimal place.

1. 21, 28, 16, 42, 38

2. 42, 35, 36, 40, 50

3. 7.6, 8.2, 8.2, 9.6, 5.7, 9.1

4. 4.9, 7.1, 6.8, 6.8, 5.3, 4.9

5. 0.2, 0.3, 0.5, 0.6, 0.6, 0.9, 0.2, 0.7, 1.1

6. 0.6, 0.6, 0.8, 0.4, 0.5, 0.3, 0.7, 0.8, 0.1

7. 231, 543, 601, 293, 588, 109, 334, 268

8. 451, 356, 478, 776, 892, 500, 467, 780

The eight tallest buildings in the United States are listed below. Use this table for Exercises 9–12.

Building	Height (feet)
Sears Tower, Chicago, IL	1454
Empire State, New York, NY	1250
Amoco, Chicago, IL	1136
John Hancock Center, Chicago, IL	1127
First Interstate World Center, Los Angeles, CA	1107
Chrysler, New York, NY	1046
NationsBank Tower, Atlanta, GA	1023
Texas Commerce Tower, Houston, TX	1002

9. Find the mean height for the five tallest buildings.

10. Find the median height for the five tallest buildings.

11. Find the median height for the eight tallest buildings.

12. Find the mean height for the eight tallest buildings.

During an experiment, the following times (in seconds) were recorded: 7.8, 6.9, 7.5, 4.7, 6.9, 7.0.

13. Find the mean. Round to the nearest tenth.

14. Find the median.   **15.** Find the mode.

In a mathematics class, the following test scores were recorded for a student: 86, 95, 91, 74, 77, 85.

16. Find the mean. Round to the nearest hundredth.

17. Find the median.   **18.** Find the mode.

The following pulse rates were recorded for a group of fifteen students: 78, 80, 66, 68, 71, 64, 82, 71, 70, 65, 70, 75, 77, 86, 72.

19. Find the mean.   **20.** Find the median.

21. Find the mode.

22. How many rates were higher than the mean?

23. How many rates were lower than the mean?

24. Have each student in your algebra class take his/her pulse rate. Record the data and find the mean, the median, and the mode.

Find the missing numbers in each list of numbers. (These numbers are not necessarily in numerical order)

25. __, __, 16, 18, __

The mode is 21.

The mean is 20.

26. __, __, __, __, 40

The mode is 35.

The median is 37.

The mean is 38.

Review of Angles, Lines, and Special Triangles

The word **geometry** is formed from the Greek words, **geo**, meaning earth, and **metron**, meaning measure. Geometry literally means to measure the earth.

This section contains a review of some basic geometric ideas. It will be assumed that fundamental ideas of geometry such as point, line, ray, and angle are known. In this appendix, the notation $\angle 1$ is read "angle 1" and the notation $m\angle 1$ is read "the measure of angle 1."

We first review types of angles.

Angles

A **right angle** is an angle whose measure is 90°. A right angle can be indicated by a square drawn at the vertex of the angle, as shown below.

An angle whose measure is more than 0° but less than 90° is called an **acute angle**.

An angle whose measure is greater than 90° but less than 180° is called an **obtuse angle**.

An angle whose measure is 180° is called a **straight angle**.

Two angles are said to be **complementary** if the sum of their measures is 90°. Each angle is called the **complement** of the other.

Two angles are said to be **supplementary** if the sum of their measures is 180°. Each angle is called the **supplement** of the other.

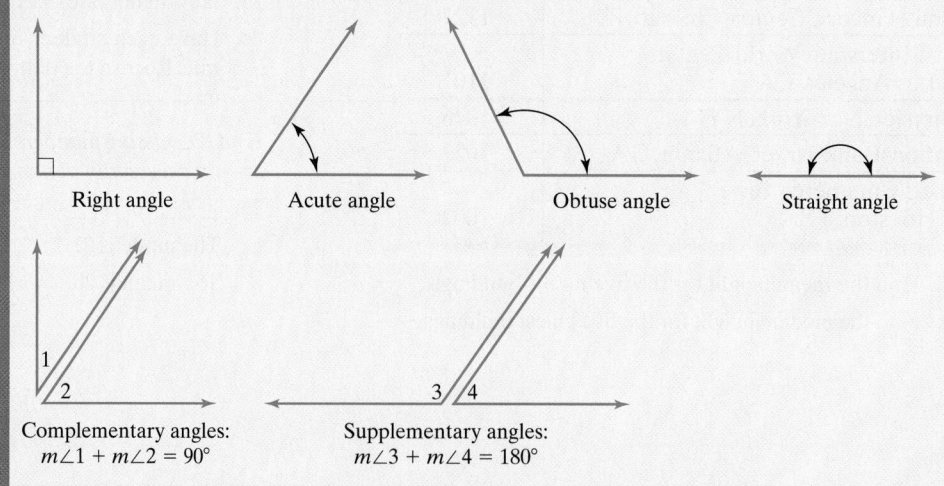

Right angle   Acute angle   Obtuse angle   Straight angle

Complementary angles:
$m\angle 1 + m\angle 2 = 90°$

Supplementary angles:
$m\angle 3 + m\angle 4 = 180°$

EXAMPLE 1

If an angle measures 28°, find its complement.

Solution   Two angles are complementary if the sum of their measures is 90°. The complement of a 28° angle is an angle whose measure is $90° - 28° = 62°$. To check, notice that $28° + 62° = 90°$.

Plane is an undefined term that we will describe. A plane can be thought of as a flat surface with infinite length and width, but no thickness. A plane is two dimensional. The arrows in the following diagram indicate that a plane extends indefinitely and has no boundaries.

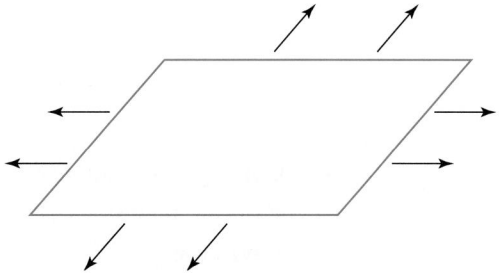

Figures that lie on a plane are called **plane figures**. (See the description of common plane figures in Appendix C.) Lines that lie in the same plane are called **coplanar**.

Lines

Two lines are **parallel** if they lie in the same plane but never meet.

Intersecting lines meet or cross in one point.

Two lines that form right angles when they intersect are said to be **perpendicular**.

| Parallel lines | Intersecting lines | Intersecting lines that are perpendicular |

Two intersecting lines form **vertical angles**. Angles 1 and 3 are vertical angles. Also angles 2 and 4 are vertical angles. It can be shown that **vertical angles have equal measures**.

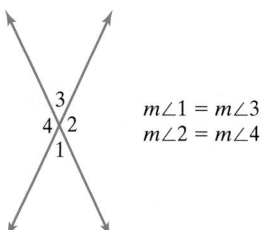

$$m\angle 1 = m\angle 3$$
$$m\angle 2 = m\angle 4$$

Adjacent angles have the same vertex and share a side. Angles 1 and 2 are adjacent angles. Other pairs of adjacent angles are angles 2 and 4, angles 3 and 4, and angles 3 and 1.

A **transversal** is a line that intersects two or more lines in the same plane. Line l is a transversal that intersects lines m and n. The eight angles formed are numbered and certain pairs of these angles are given special names.

Corresponding angles: $\angle 1$ and $\angle 5, \angle 3$ and $\angle 7, \angle 2$ and $\angle 6$, and $\angle 4$ and $\angle 8$.

Exterior angles: $\angle 1, \angle 2, \angle 7$, and $\angle 8$.

Interior angles: $\angle 3, \angle 4, \angle 5$, and $\angle 6$.

Alternate interior angles: $\angle 3$ and $\angle 6, \angle 4$ and $\angle 5$.

These angles and parallel lines are related in the following manner.

Parallel Lines Cut by a Transversal

1. If two parallel lines are cut by a transversal, then

   a. corresponding angles are equal and
   b. alternate interior angles are equal.

2. If corresponding angles formed by two lines and a transversal are equal, then the lines are parallel.

3. If alternate interior angles formed by two lines and a transversal are equal, then the lines are parallel.

EXAMPLE 2

Given that lines m and n are parallel and that the measure of angle 1 is 100°, find the measures of angles 2, 3, and 4.

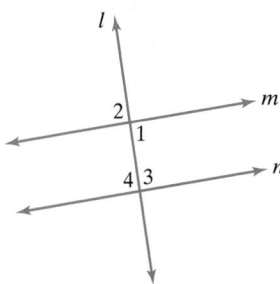

Solution $m\angle 2 = 100°$, since angles 1 and 2 are vertical angles.

$m\angle 4 = 100°$, since angles 1 and 4 are alternate interior angles.

$m\angle 3 = 180° - 100° = 80°$, since angles 4 and 3 are supplementary angles.

A **polygon** is the union of three or more coplanar line segments that intersect each other only at each end point, with each end point shared by exactly two segments.

A **triangle** is a polygon with three sides. The sum of the measures of the three angles of a triangle is 180°. In the following figure, $m\angle 1 + m\angle 2 + m\angle 3 = 180°$.

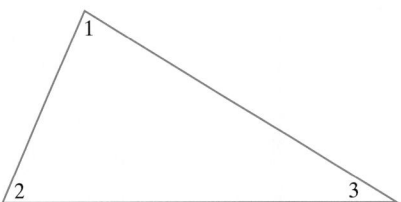

EXAMPLE 3

Find the measure of the third angle of the triangle shown.

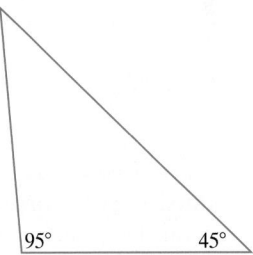

Solution   The sum of the measures of the angles of a triangle is 180°. Since one angle measures 45° and the other angle measures 95°, the third angle measures $180° - 45° - 95° = 40°$.

Two triangles are **congruent** if they have the same size and the same shape. In congruent triangles, the measures of corresponding angles are equal and the lengths of corresponding sides are equal. The following triangles are congruent.

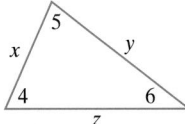

Corresponding angles are equal: $m\angle 1 = m\angle 4$, $m\angle 2 = m\angle 5$, and $m\angle 3 = m\angle 6$. Also, lengths of corresponding sides are equal: $a = x$, $b = y$, and $c = z$.

Any one of the following may be used to determine whether two triangles are congruent.

Congruent Triangles

1. If the measures of two angles of a triangle equal the measures of two angles of another triangle and the lengths of the sides between each pair of angles are equal, the triangles are congruent.

$$m\angle 1 = m\angle 3$$
$$m\angle 2 = m\angle 4$$
and
$$a = x$$

2. If the lengths of the three sides of a triangle equal the lengths of corresponding sides of another triangle, the triangles are congruent.

$$a = x$$
$$b = y$$
and
$$c = z$$

3. If the lengths of two sides of a triangle equal the lengths of corresponding sides of another triangle, and the measures of the angles between each pair of sides are equal, the triangles are congruent.

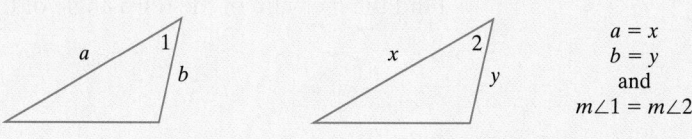

$$a = x$$
$$b = y$$
and
$$m\angle 1 = m\angle 2$$

Two triangles are **similar** if they have the same shape. In similar triangles, the measures of corresponding angles are equal and corresponding sides are in proportion. The following triangles are similar. (All similar triangles drawn in this appendix will be oriented the same.)

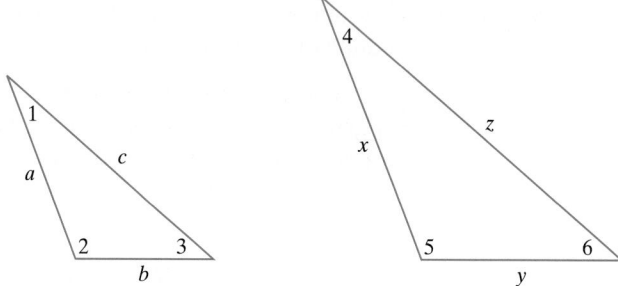

Corresponding angles are equal: $m\angle 1 = m\angle 4$, $m\angle 2 = m\angle 5$, and $m\angle 3 = m\angle 6$.

Also, corresponding sides are proportional: $\dfrac{a}{x} = \dfrac{b}{y} = \dfrac{c}{z}$.

Any one of the following may be used to determine whether two triangles are similar.

Similar Triangles

1. If the measures of two angles of a triangle equal the measures of two angles of another triangle, the triangles are similar.

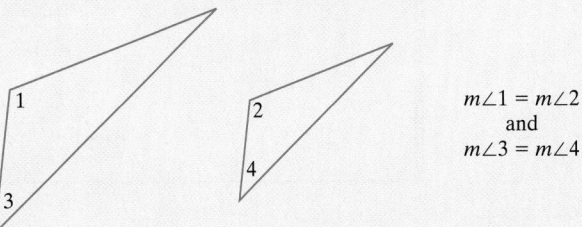

$$m\angle 1 = m\angle 2$$
and
$$m\angle 3 = m\angle 4$$

2. If three sides of one triangle are proportional to three sides of another triangle, the triangles are similar.

$$\frac{a}{x} = \frac{b}{y} = \frac{c}{z}$$

3. If two sides of a triangle are proportional to two sides of another triangle and the measures of the included angles are equal, the triangles are similar.

$$m\angle 1 = m\angle 2$$
and
$$\frac{a}{x} = \frac{b}{y}$$

EXAMPLE 4

Given that the following triangles are similar, find the missing length x.

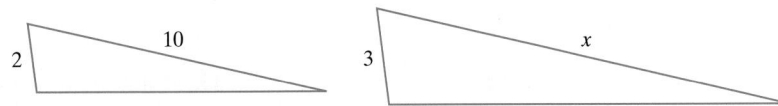

Solution   Since the triangles are similar, corresponding sides are in proportion. Thus, $\dfrac{2}{3} = \dfrac{10}{x}$. To solve this equation for x, we multiply both sides by the LCD, $3x$.

$$3x\left(\frac{2}{3}\right) = 3x\left(\frac{10}{x}\right)$$
$$2x = 30$$
$$x = 15$$

The missing length is 15 units.

A **right triangle** contains a right angle. The side opposite the right angle is called the **hypotenuse**, and the other two sides are called the **legs**. The **Pythagorean theorem** gives a formula that relates the lengths of the three sides of a right triangle.

The Pythagorean Theorem

If a and b are the lengths of the legs of a right triangle, and c is the length of the hypotenuse, then $a^2 + b^2 = c^2$.

EXAMPLE 5

Find the length of the hypotenuse of a right triangle whose legs have lengths of 3 centimeters and 4 centimeters.

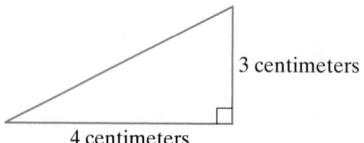

Solution   Because we have a right triangle, we use the Pythagorean theorem. The legs are 3 centimeters and 4 centimeters, so let $a = 3$ and $b = 4$ in the formula.

$$a^2 + b^2 = c^2$$
$$3^2 + 4^2 = c^2$$
$$9 + 16 = c^2$$
$$25 = c^2$$

Since c represents a length, we assume that c is positive. Thus, if c^2 is 25, c must be 5. The hypotenuse has a length of 5 centimeters.

APPENDIX H EXERCISE SET

Find the complement of each angle. See Example 1.

1. $19°$

2. $65°$

3. $70.8°$

4. $45\frac{2}{3}°$

5. $11\frac{1}{4}°$

6. $19.6°$

Find the supplement of each angle.

7. $150°$

8. $90°$

9. $30.2°$

10. $81.9°$

11. $79\frac{1}{2}°$

12. $165\frac{8}{9}°$

13. If lines m and n are parallel, find the measures of angles 1 through 7. See Example 2.

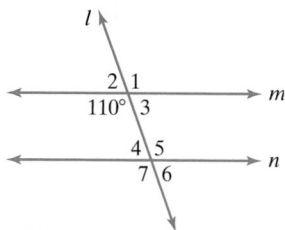

14. If lines *m* and *n* are parallel, find the measures of angles 1 through 5. See Example 2.

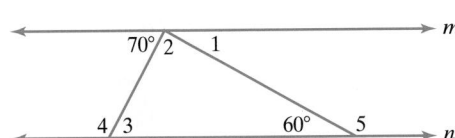

In each of the following, the measures of two angles of a triangle are given. Find the measure of the third angle. See Example 3.

15. $11°, 79°$

16. $8°, 102°$

17. $25°, 65°$

18. $44°, 19°$

19. $30°, 60°$

20. $67°, 23°$

In each of the following, the measure of one angle of a right triangle is given. Find the measures of the other two angles.

21. $45°$

22. $60°$

23. $17°$

24. $30°$

25. $39\frac{3}{4}°$

26. $72.6°$

Given that each of the following pairs of triangles is similar, find the missing lengths. See Example 4.

27.

28.

29.

30.

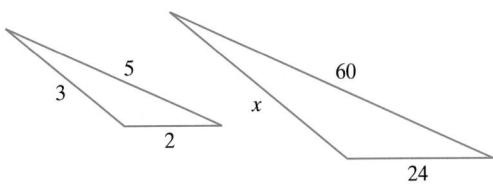

Use the Pythagorean theorem to find the missing lengths in the right triangles. See Example 5.

31.

32.

33.

34.

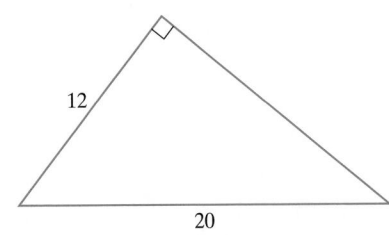

ANSWERS TO SELECTED EXERCISES

CHAPTER 1   REVIEW OF REAL NUMBERS

Exercise Set 1.2 **1.** > **3.** = **5.** < **7.** < **9.** 32 < 212 **11.** 44,300 > 34,611 **13.** true **15.** false **17.** false **19.** true
21. 30 ≤ 45 **23.** 8 < 12 **25.** 5 ≥ 4 **27.** 15 ≠ −2 **29.** 535; −8 **31.** −34,841 **33.** 350; −126 **35.** 1993 **37.** 1993, 1994 **39.** 827 ≥ 818
41. whole, integers, rational, real **43.** integers, rational, real **45.** natural, whole, integers, rational, real **47.** rational, real **49.** irrational, real
51. false **53.** true **55.** true **57.** true **59.** false **61.** > **63.** > **65.** < **67.** < **69.** > **71.** = **73.** < **75.** <
77. −0.04 > −26.7 **79.** Sun **81.** Sun **83.** 20 ≤ 25 **85.** 6 > 0 **87.** −12 < −10 **89.** answers may vary

Mental Math **1.** $\frac{3}{8}$ **3.** $\frac{5}{7}$ **5.** numerator; denominator

Exercise Set 1.3 **1.** $3 \cdot 11$ **3.** $2 \cdot 7 \cdot 7$ **5.** $2 \cdot 2 \cdot 5$ **7.** $3 \cdot 5 \cdot 5$ **9.** $3 \cdot 3 \cdot 5$ **11.** $\frac{1}{2}$ **13.** $\frac{2}{3}$ **15.** $\frac{3}{7}$ **17.** $\frac{3}{5}$ **19.** $\frac{3}{8}$ **21.** $\frac{1}{2}$ **23.** $\frac{6}{7}$ **25.** 15 **27.** $\frac{1}{6}$
29. $\frac{25}{27}$ **31.** $\frac{11}{20}$ sq. mi **33.** $\frac{3}{5}$ **35.** 1 **37.** $\frac{1}{3}$ **39.** $\frac{9}{35}$ **41.** $\frac{21}{30}$ **43.** $\frac{4}{18}$ **45.** $\frac{16}{20}$ **47.** $\frac{23}{21}$ **49.** $1\frac{2}{3}$ **51.** $\frac{5}{66}$ **53.** $\frac{7}{5}$ **55.** $\frac{1}{5}$ **57.** $\frac{3}{8}$ **59.** $\frac{1}{9}$ **61.** $\frac{5}{7}$
63. $\frac{65}{21}$ **65.** $\frac{2}{5}$ **67.** $\frac{9}{7}$ **69.** $\frac{3}{4}$ **71.** $\frac{17}{3}$ **73.** $\frac{7}{26}$ **75.** 1 **77.** $\frac{1}{5}$ **79.** $5\frac{1}{6}$ **81.** $\frac{17}{18}$ **83.** $55\frac{1}{4}$ ft **85.** $6\frac{3}{20}$ m **87.** answers may vary **89.** $3\frac{3}{8}$ mi
91. $\frac{7}{50}$ **93.** $\frac{1}{4}$ **95.** $\frac{111}{613}$ **97.** $\frac{7}{36}$ sq. ft

Calculator Explorations **1.** 625 **3.** 59,049 **5.** 30 **7.** 9857 **9.** 2376

Mental Math **1.** multiply **3.** subtract

Exercise Set 1.4 **1.** 243 **3.** 27 **5.** 1 **7.** 5 **9.** $\frac{1}{125}$ **11.** $\frac{16}{81}$ **13.** 49 **15.** 16 **17.** 1.44 **19.** 17 **21.** 20 **23.** 10 **25.** 21 **27.** 45 **29.** 0 **31.** $\frac{2}{7}$
33. 30 **35.** 2 **37.** $\frac{7}{18}$ **39.** $\frac{27}{10}$ **41.** $\frac{7}{5}$ **43.** no **45. a.** 64 **b.** 43 **c.** 19 **d.** 22 **47.** 9 **49.** 1 **51.** 1 **53.** 11 **55.** 45 **57.** 27 **59.** 132 **61.** $\frac{37}{18}$
63. 16, 64, 144, 256 **65.** yes **67.** no **69.** no **71.** yes **73.** no **75.** $x + 15$ **77.** $x - 5$ **79.** $3x + 22$ **81.** $1 + 2 = 9 \div 3$ **83.** $3 \neq 4 \div 2$
85. $5 + x = 20$ **87.** $13 - 3x = 13$ **89.** $\frac{12}{x} = \frac{1}{2}$ **91.** answers may vary **93.** $(20 - 4) \cdot 4 \div 2$ **95.** 28 m **97.** 12,000 sq. ft **99.** 6.5% **101.** $19.96

Mental Math **1.** negative **3.** 0 **5.** negative

Exercise Set 1.5 **1.** 9 **3.** −14 **5.** 1 **7.** −12 **9.** −5 **11.** −12 **13.** −4 **15.** 7 **17.** −2 **19.** 0 **21.** −19 **23.** 31 **25.** −47 **27.** −2.1 **29.** −8
31. 38 **33.** −13.1 **35.** $\frac{2}{8} = \frac{1}{4}$ **37.** $-\frac{3}{16}$ **39.** $-\frac{13}{10}$ **41.** −8 **43.** −59 **45.** −9 **47.** 5 **49.** 11 **51.** −18 **53.** 19 **55.** −0.7 **57.** −6°
59. −654 ft **61.** −$2127 million **63.** −2 **65.** −6 **67.** 2 **69.** 0 **71.** −6 **73.** answers may vary. **75.** −2 **77.** 0 **79.** $-\frac{2}{3}$
81. answers may vary **83.** yes **85.** no **87.** July **89.** October **91.** 4.7°F **93.** negative **95.** positive

Exercise Set 1.6 **1.** −10 **3.** −5 **5.** 19 **7.** $\frac{1}{6}$ **9.** 2 **11.** −11 **13.** 11 **15.** 5 **17.** 37 **19.** −6.4 **21.** −71 **23.** 0 **25.** 4.1 **27.** $\frac{2}{11}$ **29.** $-\frac{11}{12}$
31. 8.92 **33.** 13 **35.** −5 **37.** −1 **39.** −23 **41.** answers may vary **43.** −26 **45.** −24 **47.** 3 **49.** −45 **51.** −4 **53.** 13 **55.** 6 **57.** 9 **59.** −9
61. −7 **63.** $\frac{7}{5}$ **65.** 21 **67.** $\frac{1}{4}$ **69.** 100° **71.** −23 yd or 23 yd loss **73.** −569 or 569 B.C. **75.** −308 ft **77.** 19,852 ft **79.** 130° **81.** 30° **83.** no
85. no **87.** yes **89.** −4.4°; 2.6°; 12°; 23.5°; 15.3°; 3.9°; −0.3°; −6.3°; −18.2°; −15.7°; −10.3° **91.** October **93.** true **95.** true
97. negative, −2.6466

Calculator Explorations **1.** 38 **3.** −441 **5.** $163.\overline{3}$ or $163\frac{1}{3}$ **7.** 54,499 **9.** 15,625

Mental Math **1.** positive **3.** negative **5.** positive

Exercise Set 1.7 **1.** −24 **3.** −2 **5.** 50 **7.** −12 **9.** 0 **11.** −18 **13.** $\frac{3}{10}$ **15.** $\frac{2}{3}$ **17.** −7 **19.** 0.14 **21.** −800 **23.** −28 **25.** 25 **27.** $-\frac{8}{27}$
29. −121 **31.** $-\frac{1}{4}$ **33.** −30 **35.** 23 **37.** −7 **39.** true **41.** false **43.** 16 **45.** −1 **47.** 25 **49.** −49 **51.** $\frac{1}{9}$ **53.** $\frac{3}{2}$ **55.** $-\frac{1}{14}$ **57.** $-\frac{11}{3}$
59. $\frac{1}{0.2}$ **61.** −6.3 **63.** −9 **65.** 4 **67.** −4 **69.** 0 **71.** −5 **73.** undefined **75.** 3 **77.** −15 **79.** $-\frac{18}{7}$ **81.** $\frac{20}{27}$ **83.** −1 **85.** $-\frac{9}{2}$ **87.** −4
89. 16 **91.** −3 **93.** $-\frac{16}{7}$ **95.** 2 **97.** $\frac{6}{5}$ **99.** −5 **101.** $\frac{3}{2}$ **103.** −21 **105.** 41 **107.** −134 **109.** 3 **111.** 0 **113.** −$5088 million **115.** yes
117. no **119.** yes **121.** answers may vary **123.** 1, −1 **125.** positive **127.** not possible **129.** negative **131.** $-2 + \frac{-15}{3}; -7$ **133.** $2[-5 + (-3)]; -16$

Integrated Review 1. positive 2. positive 3. negative 4. negative 5. positive 6. negative 7. negative 8. positive 9. -35 10. 30
11. 5 12. -5 13. 10 14. -18 15. -2 16. -2 17. $\frac{3}{8}$ 18. $-\frac{11}{42}$ 19. -60 20. 1.9 21. -42 22. -7 23. 2 24. -39 25. 64 26. -81
27. -27 28. 16 29. -1 30. 1 31. -32 32. -32 33. 48 34. -30 35. -6 36. 9 37. -30 38. -44 39. 4 40. -3 41. 2 42. 16
43. 0 44. 19 45. $-\frac{32}{15}$ 46. $\frac{54}{5}$

Exercise Set 1.8 1. $16 + x$ 3. $y \cdot (-4)$ 5. yx 7. $13 + 2x$ 9. $x \cdot (yz)$ 11. $(2 + a) + b$ 13. $(4a) \cdot b$ 15. $a + (b + c)$ 17. $17 + b$
19. $24y$ 21. y 23. $26 + a$ 25. $-72x$ 27. s 29. answers may vary 31. $4x + 4y$ 33. $9x - 54$ 35. $6x + 10$ 37. $28x - 21$
39. $18 + 3x$ 41. $-2y + 2z$ 43. $-21y - 35$ 45. $5x + 20m + 10$ 47. $-4 + 8m - 4n$ 49. $-5x - 2$ 51. $-r + 3 + 7p$ 53. $3x + 4$
55. $-x + 3y$ 57. $6r + 8$ 59. $-36x - 70$ 61. $-16x - 25$ 63. $4(1 + y)$ 65. $11(x + y)$ 67. $-1(5 + x)$ 69. $30(a + b)$
71. commutative property of multiplication 73. associative property of addition 75. distributive property 77. associative property of multiplication
79. identity element of addition 81. distributive property 83. commutative and associative properties of multiplication 85. $-8; \frac{1}{8}$ 87. $-x; \frac{1}{x}$
89. $-2x; \frac{1}{2x}$ 91. no 93. yes 95. answers may vary

Chapter 1 Review 1. $<$ 3. $>$ 5. $<$ 7. $=$ 9. $>$ 11. $4 \geq -3$ 13. $0.03 < 0.3$ 15. a. $1, 3$ b. $0, 1, 3$ c. $-6, 0, 1, 3$ d. $-6, 0, 1, 1\frac{1}{2}, 3, 9.62$
e. π f. $-6, 0, 1, 1\frac{1}{2}, 3, \pi, 9.62$ 17. Friday 19. $2 \cdot 2 \cdot 3 \cdot 3$ 21. $\frac{12}{25}$ 23. $\frac{13}{10}$ 25. $9\frac{3}{8}$ 27. 15 29. $\frac{7}{12}$ 31. $A = \frac{34}{121}$ sq. in.; $P = 2\frac{4}{11}$ in. 33. $2\frac{15}{16}$ lb
35. $11\frac{5}{16}$ lb 37. Odera 39. $3\frac{7}{8}$ lb 41. 16 43. $\frac{4}{49}$ 45. 70 47. 37 49. $\frac{18}{7}$ 51. $20 - 12 = 2 \cdot 4$ 53. 18 55. 5 57. $63°$ 59. no 61. $-\frac{2}{3}$
63. 7 65. -17 67. -5 69. 3.9 71. -14 73. 5 75. -19 77. 15 79. $-\frac{1}{6}$ 81. -48 83. 3 85. undefined 87. undefined 89. -12
91. 9 93. -5 95. commutative property of addition 97. distributive property 99. associative property of addition 101. distributive property
103. multiplicative inverse 105. commutative property of addition

Chapter 1 Test 1. $|-7| > 5$ 2. $9 + 5 \geq 4$ 3. -5 4. -11 5. -3 6. -39 7. 12 8. -2 9. undefined 10. -8 11. $-\frac{1}{3}$ 12. $4\frac{5}{8}$ 13. $-\frac{5}{2}$ or $-2\frac{1}{2}$
14. -32 15. -48 16. 3 17. 0 18. $>$ 19. $>$ 20. $<$ 21. $=$ 22. $2221 < 10{,}993$ 23. a. $1, 7$ b. $0, 1, 7$ c. $-5, -1, 0, 1, 7$
d. $-5, -1, 0, \frac{1}{4}, 1, 7, 11.6$ e. $\sqrt{7}, 3\pi$ f. $-5, -1, 0, \frac{1}{4}, 1, 7, 11.6, \sqrt{7}, 3\pi$ 24. 40 25. 12 26. 22 27. -1 28. associative property of addition
29. commutative property of multiplication 30. distributive property 31. multiplicative inverse 32. 9 33. -3 34. second down 35. yes
36. $17°$ 37. \$650 million 38. \$420

CHAPTER 2 EQUATIONS, INEQUALITIES, AND PROBLEM SOLVING

Mental Math 1. -7 3. 1 5. 17 7. $\frac{1}{8}$ 9. $-\frac{2}{3}$ 11. unlike 13. like 15. unlike

Exercise Set 2.1 1. $15y$ 3. $-15n$ 5. $-1t$ or $-t$ 7. $13w$ 9. $-7b - 9$ 11. $-m - 6$ 13. $5y - 20$ 15. $7d - 11$ 17. $-3x + 2y - 1$ 19. $2x + 14$
21. answers may vary 23. $(6x + 7) + (4x - 10) = 10x - 3$ 25. $(3x - 8) - (7x + 1) = -4x - 9$ 27. $5x^2$ 29. $4x - 3$
31. $8x - 53$ 33. $7.2x - 5.2$ 35. $k - 6$ 37. $0.9m + 1$ 39. $-12y + 16$ 41. $x + 5$ 43. $1.3x + 3.5$ 45. $x + 2$ 47. $-15x + 18$
49. $2k + 10$ 51. $-3x + 5$ 53. $2x - 4$ 55. $\frac{3}{4}x + 12$ 57. $5x + (-2) + 7x = -2 + 12x$ 59. $(m - 9) - (5m - 6) = -4m - 3$
61. $8(x + 6) = 8x + 48$ 63. $2x - (x + 10) = x - 10$ 65. $7\left(\frac{x}{6}\right) = \frac{7x}{6}$ 67. $2 + 3x + (-9) + 4x = 7x - 7$ 69. $(18x - 2)$ ft
71. 2 73. -23 75. -25 77. balanced 79. balanced 81. $(15x + 23)$ in. 83. $5b^2c^3 + b^3c^2$ 85. $5x^2 + 9x$ 87. $-7x^2y$

Mental Math 1. 2 3. 12 5. 17 7. 9 9. 2 11. -5

Exercise Set 2.2 1. 3 3. -2 5. -14 7. 0.5 9. -3 11. -0.7 13. 3 15. 11 17. 0 19. -3 21. 16 23. -4 25. 0
27. 12 29. 10 31. -12 33. 3 35. -2 37. 0 39. answers may vary 41. 10 43. -20 45. 0 47. -5 49. 0 51. $-\frac{3}{2}$ 53. -21
55. $\frac{11}{2}$ 57. 1 59. $-\frac{1}{4}$ 61. 12 63. -30 65. $\frac{9}{10}$ 67. -30 69. 2 71. -2 73. 23 75. $20 - p$ 77. $(10 - x)$ ft 79. $(180 - x)°$ 81. $n + 284$
83. $(m - 60)$ ft 85. $(n + 47{,}628)$ votes 87. $7x$ sq. mi 89. $2x + 2$ 91. $2x + 2$ 93. $7x - 12$ 95. 1 97. $>$ 99. $=$ 101. $(173 - 3x)°$
103. 250 ml 105. answers may vary 107. -6.3 109. -1.23

Calculator Explorations 1. solution 3. not a solution 5. solution

Exercise Set 2.3 1. 1 3. $\frac{9}{2}$ 5. $\frac{3}{2}$ 7. 0 9. 2 11. -5 13. 10 15. 18 17. 1 19. 50 21. 0.2 23. all real numbers 25. no solution
27. no solution 29. answers may vary 31. answers may vary 33. 4 35. -4 37. 3 39. -2 41. 4 43. $\frac{7}{3}$ 45. no solution 47. $\frac{9}{5}$ 49. $\frac{4}{19}$

51. 1 **53.** no solution **55.** $\frac{7}{2}$ **57.** -17 **59.** $\frac{19}{6}$ **61.** all real numbers **63.** 3 **65.** 13 **67.** $2x + 7 = x + 6; -1$ **69.** $3x - 6 = 2x + 8; 14$

71. $\frac{1}{3}x = \frac{5}{6}; \frac{5}{2}$ **73.** $\frac{x}{4} + \frac{1}{2} = \frac{3}{4}; 1$ **75.** $10 - 5x = 3x; \frac{5}{4}$ **77.** $x = 4$ cm, $2x = 8$ cm **79.** -1 **81.** $\frac{1}{5}$ **83.** $(6x - 8)$ m **85.** Fairview

87. 155 **89.** 15.3 **91.** -0.2 **93.** $-\frac{7}{8}$ **95.** no solution

Integrated Review 1. 6 **2.** -17 **3.** 12 **4.** -26 **5.** -3 **6.** -1 **7.** $\frac{27}{2}$ **8.** $\frac{25}{2}$ **9.** 8 **10.** -64 **11.** 2 **12.** -3 **13.** no solution

14. no solution **15.** -2 **16.** -2 **17.** $-\frac{5}{6}$ **18.** $\frac{1}{6}$ **19.** 1 **20.** 6 **21.** 4 **22.** 1 **23.** $\frac{9}{5}$ **24.** $-\frac{6}{5}$ **25.** all real numbers **26.** all real numbers

27. 0 **28.** -1.6 **29.** $\frac{4}{19}$ **30.** $-\frac{5}{19}$ **31.** $\frac{7}{2}$ **32.** $-\frac{1}{4}$

Exercise Set 2.4 1. -25 **3.** $-\frac{3}{4}$ **5.** 234, 235 **7.** Belgium: 32; France: 33; Spain: 34 **9.** 1st piece: 5 in.; 2nd piece: 10 in.; 3rd piece: 25 in.
11. governor of Nebraska: $65,000; governor of Washington: $130,000 **13.** 172 mi **15.** 25 mi **17.** 1st angle: 37.5°; 2nd angle: 37.5°; 3rd angle: 105°
19. A: 60°; B: 120°; C: 120°; D: 60° **21.** 5 ft, 12 ft **23.** 1997: 15.1 million prescriptions; 2001: 20.6 million prescriptions **25.** 45°, 135°
27. 58°, 60°, 62° **29.** 280 mi **31.** Tampa Bay: 48; Oakland: 21 **33.** China: 59 medals; Australia: 58 medals; Germany: 57 medals
35. California: 55; Texas: 34 **37.** Neptune: 8 moons; Uranus: 21 moons; Saturn: 18 moons **39.** -16 **41.** Brown: 66,362; Randall: 53,074 **43.** Illinois
45. Texas: $29.4 million; Florida: $27.2 million **47.** answers may vary **49.** -10 **51.** -9 **53.** -15 **55.** yes: 30°, 60°, 90° **57.** answers may vary
59. answers may vary **61.** c

Exercise Set 2.5 1. $h = 3$ **3.** $h = 3$ **5.** $h = 20$ **7.** $c = 12$ **9.** $r = 2.5$ **11.** $T = 3$ **13.** $h = 15$ **15.** $h = \frac{f}{5g}$ **17.** $w = \frac{v}{lh}$ **19.** $y = 7 - 3x$

21. $R = \frac{A - P}{PT}$ **23.** $A = \frac{3V}{h}$ **25.** $a = P - b - c$ **27.** $h = \frac{S - 2\pi r^2}{2\pi r}$ **29.** 131 ft **31.** $-10°C$ **33.** 6.25 hr **35.** length: 78 ft; width: 52 ft
37. 18 ft, 36 ft, 48 ft **39.** 137.5 mi **41.** 96 piranhas **43.** 2 bags **45.** one 16-in. pizza **47.** 4.65 min **49.** 13 in. **51.** 2.25 hr **53.** 12,090 ft **55.** 50°C

57. 515,509.5 cu. in. **59.** 449 cu. in. **61.** 332.6°F **63.** $\frac{9}{x + 5}$ **65.** $3(x + 4)$ **67.** $3(x - 12)$ **69.** $-109.3°F$ **71.** 500 sec or $8\frac{1}{3}$ min **73.** 608.33 ft

75. 565.5 cu. in. **77.** $25\frac{5}{9}°C$ **79.** It multiplies the area by 4.

Exercise Set 2.6 1. $64 decrease; $192 sale price **3.** $104 **5.** 73% **7.** $666\frac{2}{3}$ mi **9.** 160 mi **11.** 2 gal **13.** $6\frac{2}{3}$ lb **15.** no

17. $11,500 @ 8%; $13,500 @ 9% **19.** $7000 @ 11% profit; $3000 @ 4% loss **21.** $3900 **23.** $30,000 @ 8%; $24,000 @ 10% **25.** $2\frac{2}{9}$ hr **27.** 400 oz

29. 75% increase **31.** $4500 **33.** 230 million **35.** 2.2 mph; 3.3 mph **37.** 1.45% increase **39.** $1\frac{5}{7}$ hr **41.** -4 **43.** $\frac{9}{16}$ **45.** -4 **47.** $>$

49. $=$ **51.** 25 skateboards **53.** 800 books **55.** answers may vary **57.** 7.7% **59.** 19.3% **61.** answers may vary

Mental Math 1. $x > 2$ **3.** $x \geq 8$

Exercise Set 2.7 1. $x \geq 2$ **3.** $x < -5$ **5.** $(-\infty, -1]$ **7.** $\left(-\infty, \frac{1}{2}\right)$

9. $[5, \infty)$ **11.** $(-\infty, -3)$ **13.** $[-5, \infty)$ **15.** $[2, \infty)$
17. $x > -3, (-3, \infty)$ **19.** $x \leq 1, (-\infty, 1]$ **21.** $x > -5, (-5, \infty)$
23. $x \leq -2, (-\infty, -2]$ **25.** $x \leq -8, (-\infty, -8]$ **27.** $x > 4, (4, \infty)$
29. $x \geq 20, [20, \infty)$ **31.** $x > 16, (16, \infty)$ **33.** $x > -3, (-3, \infty)$
35. $x \leq -\frac{2}{3}, \left(-\infty, -\frac{2}{3}\right]$ **37.** $x > \frac{8}{3}, \left(\frac{8}{3}, \infty\right)$ **39.** $x > -13, (-13, \infty)$
41. $x < 0, (-\infty, 0)$ **43.** $x \leq 0, (-\infty, 0]$ **45.** $x > 3, (3, \infty)$
47. $x \leq 0, (-\infty, 0]$ **49.** answers may vary **51.** $(-1, 3)$ **53.** $[2, 3]$
55. $-1 < x < 2, (-1, 2)$ **57.** $4 \leq x \leq 5, [4, 5]$ **59.** $1 < x \leq 5, (1, 5]$
61. $1 < x < 4, (1, 4)$ **63.** $0 < x \leq \frac{14}{3}, \left(0, \frac{14}{3}\right]$ **65.** answers may vary **67.** $x > -10$

69. 86 people **71.** $x \leq 35$ **73.** at least 10% **75.** at least 193 **77.** $x < 200$ recommended; $200 \leq x \leq 240$ borderline; $x > 240$ high

79. $-3 < x < 3$ **81.** $-38.2°F$ to 113°F **83.** 8 **85.** 1 **87.** $\frac{16}{49}$ **89.** $3225 **91.** 1986–1990 **93.** $0.924 \leq d \leq 0.987$ **95.** $(1, \infty)$

97. $\left(-\infty, \frac{5}{8}\right)$

Chapter 2 Review   **1.** $6x$ **3.** $4x - 2$ **5.** $3n - 18$ **7.** $-6x + 7$ **9.** $3x - 7$ **11.** 4   **13.** -6 **15.** -9 **17.** $5; 5$ **19.** $10 - x$ **21.** $(175 - x)°$ **23.** 4   **25.** -1 **27.** -1 **29.** 6   **31.** 2   **33.** no solution   **35.** $\frac{3}{4}$ **37.** 20   **39.** 0   **41.** $\frac{20}{7}$ **43.** $\frac{23}{7}$ **45.** 102   **47.** 3   **49.** 1052 ft   **51.** $307; 955$ **53.** $w = 9$ **55.** $m = \dfrac{y - b}{x}$ **57.** $x = \dfrac{2y - 7}{5}$ **59.** $\pi = \dfrac{C}{D}$ **61.** 15 m   **63.** 1 hr and 20 min   **65.** 80 nickels   **67.** 48 mi   **69.** $(0, \infty)$

71. $[0.5, 1.5)$ **73.** $(-\infty, -4)$ **75.** $(-\infty, 4]$ **77.** $\left(-\dfrac{1}{2}, \dfrac{3}{4}\right)$

79. $\left(-\infty, \dfrac{19}{3}\right]$ **81.** score must be less than 83

Chapter 2 Test   **1.** $y - 10$ **2.** $5.9x + 1.2$ **3.** $-2x + 10$ **4.** $10y + 1$ **5.** -5 **6.** 8   **7.** $\dfrac{7}{10}$ **8.** 0   **9.** 27   **10.** 3   **11.** 0.25   **12.** $\dfrac{25}{7}$ **13.** no solution   **14.** 21   **15.** 7 gal   **16.** \$8500 @ 10%; \$17,000 @ 12%   **17.** $2\dfrac{1}{2}$ hr   **18.** $x = 6$ **19.** $h = \dfrac{V}{\pi r^2}$ **20.** $y = \dfrac{3x - 10}{4}$ **21.** $(-\infty, -2]$

22. $(-\infty, 4)$ **23.** $\left(-1, \dfrac{7}{3}\right)$ **24.** $\left(\dfrac{2}{5}, \infty\right)$

Chapter 2 Cumulative Review   **1. a.** $11, 112$ **b.** $0, 11, 112$ **c.** $-3, -2, 0, 11, 112$ **d.** $-3, -2, 0, \dfrac{1}{4}, 11, 112$ **e.** $\sqrt{2}$ **f.** all numbers in the given set; Sec. 1.2, Ex. 5   **3. a.** 4   **b.** 5   **c.** 0; Sec. 1.2, Ex. 7   **5. a.** $2 \cdot 2 \cdot 2 \cdot 5$ **b.** $3 \cdot 3 \cdot 7$; Sec. 1.3, Ex. 1   **7.** $\dfrac{8}{20}$; Sec. 1.3, Ex. 6

9. 54; Sec. 1.4, Ex. 4   **11.** 2 is a solution; Sec. 1.4, Ex. 7   **13.** -3; Sec. 1.5, Ex. 2   **15.** 2; Sec. 1.5, Ex. 4   **17. a.** 10   **b.** $\dfrac{1}{2}$ **c.** $2x$ **d.** -6; Sec. 1.5, Ex. 10   **19. a.** 9.9   **b.** $-\dfrac{4}{5}$ **c.** $\dfrac{2}{15}$; Sec. 1.6, Ex. 2   **21. a.** $52°$ **b.** $118°$; Sec. 1.6, Ex. 8   **23. a.** -0.06 **b.** $-\dfrac{7}{15}$; Sec. 1.7, Ex. 3   **25. a.** 6   **b.** -12 **c.** $-\dfrac{8}{15}$; Sec. 1.7, Ex. 7   **27. a.** $5 + x$ **b.** $x \cdot 3$; Sec. 1.8, Ex. 1   **29. a.** $8(2 + x)$ **b.** $7(s + t)$; Sec. 1.8, Ex. 5   **31.** $-2x - 1$; Sec. 2.1, Ex. 7   **33.** -1.6; Sec. 2.2, Ex. 1   **35.** 8; Sec. 2.2, Ex. 4   **37.** 140; Sec. 2.2, Ex. 7   **39.** 2; Sec. 2.3, Ex. 1   **41.** 10; Sec. 2.4, Ex. 1   **43.** $\dfrac{V}{wh} = l$; Sec. 2.5, Ex. 5   **45.** $(-\infty, -10]$; Sec. 2.7, Ex. 2

CHAPTER 3   GRAPHING

Mental Math   **1.** answers may vary; Ex. $(5, 5), (7, 3)$ **3.** answers may vary; Ex. $(3, 5), (3, 0)$

Exercise Set 3.1   **1.** France   **3.** France, U.S., Spain   **5.** 39 million   **7.** 20   **9.** 1985   **11.** 1997   **13.** 63 million   **15.** 1900   **17.** 27 million   **19.** answers may vary

21. quadrant I   **23.** no quadrant, x-axis   **25.** quadrant IV   **27.** no quadrant, x-axis   **29.** no quadrant, origin   **31.** no quadrant, y-axis

     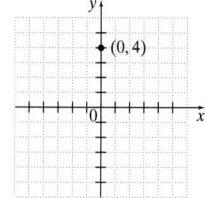

33. $A(0, 0); B\left(3\dfrac{1}{2}, 0\right); C(3, 2); D(-1, 3); E(-2, -2); F(0, -1); G(2, -1)$ **35.** 26 units

37. a. $(2313, 2), (2085, 1), (2711, 21), (2869, 39), (2920, 42), (4038, 99), (1783, 0), (2493, 9)$ **c.** The farther from the equator, the more snowfall.

b.

39. no; yes; yes   **41.** no; yes; no   **43.** no; no   **45.** yes; yes   **47.** $(-4, -2), (4, 0)$ **49.** $(11, -7)$; answers may vary; Ex. $(2, -7)$

51. $(5, 1), \left(1, -\dfrac{1}{7}\right)$ **53.** $\left(0, \dfrac{5}{7}\right), \left(\dfrac{5}{2}, 0\right), (-1, 1)$ **55.** $(3, 0), (3, -0.5), \left(3, \dfrac{1}{4}\right)$ **57.** $(0, 0), (-5, 1), (10, -2)$ **59.** answers may vary

  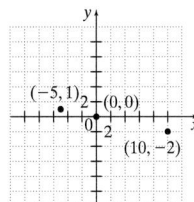

61. a. 13,000; 21,000; 29,000   **b.** 45 desks   **63.** In 1995, there were 670 Target stores.   **65.** year 6: 66 stores; year 7: 60 stores; year 8: 55 stores

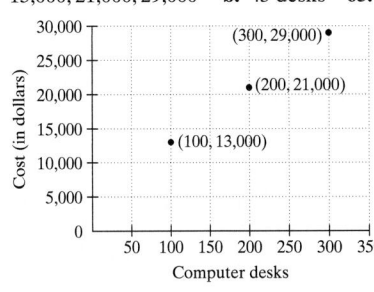

67. $a = b$ **69.** $y = -2x$ **71.** $y = \dfrac{1}{3}x - 2$ **73.** quadrant IV   **75.** quadrants II or III   **77. a.** $(-2, 6)$ **b.** 28 units   **c.** 45 sq. units   **79.** 2001, 2002, 2003

81. 46 million   **83.** 30° north, 90° west   **85.** 40° north, 104° west

Calculator Explorations   1.   **3.**   **5.**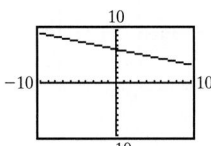

Exercise Set 3.2   1. yes   **3.** yes   **5.** no   **7.** yes

9.   **11.**   **13.**   **15.**   **17.**

19.   **21.**   **23.**   **25.**   **27.**

29.   **31.**   **33.**   **35.**   **37.**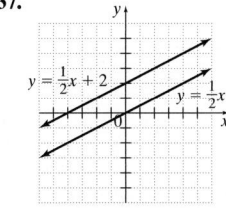

39. c **41.** d **43.** $2790 million dollars **45.**

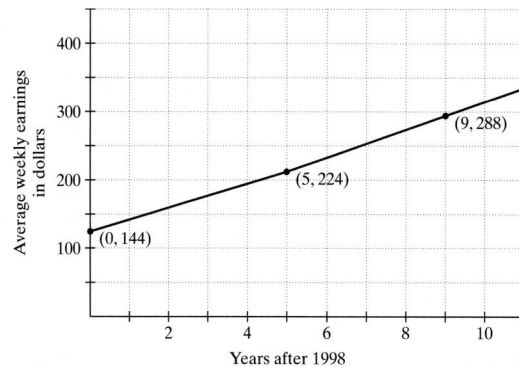

47. $(4, -1)$ **49.** -5 **51.** $-\dfrac{1}{10}$ **53.** $(0, 3), (-3, 0)$ **55.** $(0, 0), (0, 0)$

57. $y = x + 5$ **59.** $2x + 3y = 6$ **61.** answers may vary **63.** answers may vary **65.** $0, 1, 1, 4, 4$

Calculator Explorations **1.**

          3.   **5.**

Mental Math **1.** false **3.** true **5.** false

Exercise Set 3.3 **1.** $(-1, 0); (0, 1)$ **3.** $(-2, 0)$ **5.** $(-1, 0); (1, 0); (0, 1); (0, -2)$ **7.** infinite **9.** 0

11.   **13.**   **15.**   **17.**   **19.**

21.   **23.**   **25.**   **27.**   **29.**

31.   **33.**   **35.**   **37.**   **39.**

41.   **43.**   **45.**   **47.**   **49.**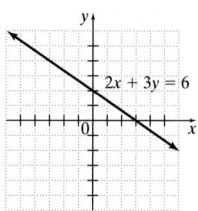

51. C   **53.** E   **55.** B   **57.** $\dfrac{3}{2}$ **59.** 6   **61.** $-\dfrac{6}{5}$ **63. a.** $(0, 569.9)$

b. In 1999, the average price of a digital camera was \$569.90.   **65. a.** $(7.1, 0)$ **b.** 7.1 years after 1995, no music cassettes will be shipped.
c. answers may vary

67. a. $(0, 200)$; answers may vary   **b.** $(400, 0)$; answers may vary

c.   **d.** 300 chairs   **69.** $y = -4$ **71.** answers may vary   **73.** answers may vary

Calculator Explorations   1.   **3.**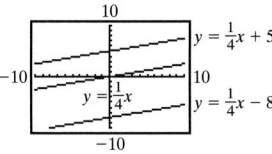

Mental Math   1. upward   **3.** horizontal

Exercise Set 3.4   1. $m = -\dfrac{4}{3}$ **3.** undefined slope   **5.** $m = \dfrac{5}{2}$ **7.** $\dfrac{8}{7}$ **9.** -1 **11.** $-\dfrac{1}{4}$ **13.** undefined   **15.** $-\dfrac{2}{3}$ **17.** undefined   **19.** 0   **21.** line 1
23. line 2   **25.** D   **27.** B   **29.** E   **31.** undefined slope   **33.** $m = 0$ **35.** undefined slope   **37.** $m = 0$ **39. a.** 1   **b.** -1 **41. a.** $\dfrac{9}{11}$ **b.** $-\dfrac{11}{9}$
43. parallel   **45.** perpendicular   **47.** parallel   **49.** neither   **51.** perpendicular   **53.** $-\dfrac{1}{5}$ **55.** $\dfrac{1}{3}$ **57.** 1   **59.** $\dfrac{3}{5}$ **61.** 12.5%   **63.** 40%   **65.** 0.02
67. Every 1 year, there will be 5 million more cell phone users.   **69.** It costs \$0.36 per 1 mile to own and operate a compact car.   **71.** $y = -x + 10$
73. $y = -\dfrac{1}{2}x - 6$ **75.** $y = 5x - 17$ **77.** 28.3   **79.** 1992; 27.6   **81.** from 1992 to 1993   **83.** $x = 6$ **85. a.** $(1994, 782), (2001, 1132)$ **b.** 50

c. For the years 1994 through 2001, the price per acre of U.S. farmland rose \$50 every year.   **87.** The slope through $(-3, 0)$ and $(1, 1)$ is $\dfrac{1}{4}$. The slope
through $(-3, 0)$ and $(-4, 4)$ is -4. The product of the slopes is -1, so the sides are perpendicular.   **89.** -0.25 **91.** 0.875
93. The line becomes steeper.

Integrated Review   1. $m = 2$ **2.** $m = 0$ **3.** $m = -\dfrac{2}{3}$ **4.** undefined

5.   **6.**   **7.**   **8.**

9.   **10.**   **11.**   **12.**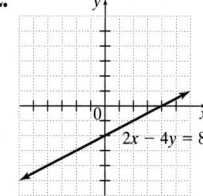

13. parallel   **14.** neither   **15. a.** $(0, 1407)$ **b.** In 2000, there were 1407 million admissions to movie theaters in the U.S.   **c.** 110
d. For the years 2000 through 2002, the number of movie theater admissions increased at a rate of 110 million per year.

Calculator Explorations

1.   **3.**   **5.**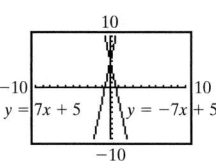

Mental Math   1. $m = 2; (0, -1)$ **3.** $m = 1; \left(0, \dfrac{1}{3}\right)$ **5.** $m = \dfrac{5}{7}; (0, -4)$

Exercise Set 3.5   1. $m = -2; (0, 4)$ **3.** $m = -\dfrac{1}{9}; \left(0, \dfrac{1}{9}\right)$ **5.** $m = \dfrac{4}{3}; (0, -4)$ **7.** $m = -1; (0, 0)$ **9.** $m = 0; (0, -3)$ **11.** $m = \dfrac{1}{5}; (0, 4)$ **13.** B

15. D   **17.** neither   **19.** neither   **21.** perpendicular   **23.** parallel   **25.** answers may vary   **27.** $y = -x + 1$ **29.** $y = 2x + \dfrac{3}{4}$ **31.** $y = \dfrac{2}{7}x$

33.   **35.**   **37.**   **39.**

41.   **43.**   **45.**   **47.**

49. $y = 2x - 14$ **51.** $y = -6x - 11$ **53.** D   **55.** B   **57.** E   **59. a.** $(0, 21), (22, 45)$ **b.** $y = \dfrac{12}{11}x + 21$ **61.** $y = -2x + 5$

63. a. The temperature 100° Celsius is equivalent to 212° Fahrenheit.   **b.** 68°F   **c.** 27°C   **d.** $F = \dfrac{9}{5}C + 32$ **65.** answers may vary

Mental Math   1. $m = 3$; answers may vary; Ex. $(4, 8)$ **3.** $m = -2$; answers may vary; Ex. $(10, -3)$ **5.** $m = \dfrac{2}{5}$; answers may vary; Ex. $(-1, 0)$

Exercise Set 3.6   1. $6x - y = 10$ **3.** $8x + y = -13$ **5.** $x - 2y = 17$ **7.** $2x - y = 4$ **9.** $8x - y = -11$ **11.** $4x - 3y = -1$ **13.** $x = 0$

15. $y = 3$ **17.** $x = -\dfrac{7}{3}$ **19.** $y = 2$ **21.** $y = 5$ **23.** $x = 6$ **25.** $3x + 6y = 10$ **27.** $x - y = -16$ **29.** $x + y = 17$ **31.** $y = 7$

33. $4x + 7y = -18$ **35.** $x + 8y = 0$ **37.** $3x - y = 0$ **39.** $x - y = 0$ **41.** $5x + y = 7$ **43.** $11x + y = -6$ **45.** $x = -\dfrac{3}{4}$ **47.** $y = -3$

49. $7x - y = 4$ **51. a.** $y = 640x + 4760$ **b.** 11,160 vehicles   **53. a.** $s = 32t$ or $y = 32x$ **b.** 128 ft/sec   **55. a.** $y = 0.93x + 70.3$
b. 86.11 persons per sq. mi   **57. a.** $(0, 191), (5, 260)$ **b.** $y = 13.8x + 191$ **c.** \$246.2 million   **59.** $31x - 5y = -5$ **61. a.** $S = -1000p + 13{,}000$
b. 9500 Fun Noodles   **63.** -1 **65.** 5   **67.** no   **69.** yes   **71.** answers may vary   **73. a.** $3x - y = -5$ **b.** $x + 3y = 5$ **75. a.** $3x + 2y = -1$
b. $2x - 3y = 21$

Exercise Set 3.7   1. domain: $\{-7, 0, 2, 10\}$; range: $\{-7, 0, 4, 10\}$ **3.** domain: $\{0, 1, 5\}$; range: $\{-2\}$ **5.** yes   **7.** no   **9.** no   **11.** yes   **13.** yes   **15.** no
17. no   **19.** yes   **21.** yes   **23.** yes   **25.** yes   **27.** no   **29.** no   **31.** 5:20 A.M.   **33.** answers may vary   **35.** \$4.75 per hour   **37.** 1996
39. answers may vary   **41.** $-9, -5, 1$ **43.** $6, 2, 11$ **45.** $-8, 0, 27$ **47.** $2, 0, 3$ **49.** $h(-1) = -5, h(0) = 0, h(4) = 20; (-1, -5), (0, 0), (4, 20)$
51. $h(-1) = 5, h(0) = 3, h(4) = 35; (-1, 5), (0, 3), (4, 35)$ **53.** $h(-1) = 4, h(0) = 3, h(4) = -21; (-1, 4), (0, 3), (4, -21)$
55. $h(-1) = 6, h(0) = 6, h(4) = 6; (-1, 6), (0, 6), (4, 6)$ **57.** $(-\infty, \infty)$ **59.** all real numbers except -5 or $(-\infty, -5) \cup (-5, \infty)$ **61.** $(-\infty, \infty)$
63. domain: $(-\infty, \infty)$; range: $[-4, \infty)$ **65.** domain: $(-\infty, \infty)$; range: $(-\infty, \infty)$ **67.** domain: $(-\infty, \infty)$; range: $\{2\}$ **69.** $(-2, 1)$ **71.** $(-3, -1)$
73. a. 166.38 cm   **b.** 148.25 cm   **75.** answers may vary   **77.** $f(x) = x + 7$ **79. a.** $-3s + 12$ **b.** $-3r + 12$ **81. a.** 132   **b.** $a^2 - 12$

Chapter 3 Review
1. 128 million **3.** 2001 **5.** Cross country skiing; 181 calories **7.** 52 more calories
9. **11.** **13.** **15. a.** $(8.00, 1); (7.50, 10); (6.50, 25); (5.00, 50); (2.00, 100)$
b.

17. no; yes **19.** yes; yes **21.** $(7, 44)$
23. $(-3, 0); (1, 3); (9, 9)$ **25.** $(0, 0); (10, 5); (-10, -5)$ **27.** **29.** **31.**

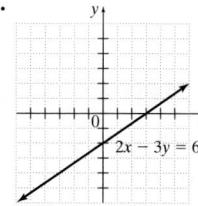

33. \$135 billion **35.** $(0, -3)$ **39.** **41.** **43.**
37. $(-1, 0); (2, 0); (3, 0); (0, -2)$

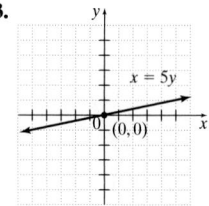

45. $m = \dfrac{1}{5}$ **47.** b **49.** a **51.** $\dfrac{3}{4}$ **53.** 4 **55.** undefined **57.** 0 **59.** Every 1 year, monthly day care costs increase by \$17.75. **61.** $m = -3; (0, 7)$
63. $m = 0; (0, 2)$ **65.** perpendicular **67.** neither **69.** $y = \dfrac{2}{3}x + 6$

71. **73.** **75.** c **77.** b **79.** $3x + y = -5$ **81.** $y = -3$ **83.** $6x + y = 11$ **85.** $x + y = 6$
87. $x = 5$ **89.** $x = 6$ **91. a.** $3x + y = 15$ **b.** $x - 3y = 5$ **93.** yes **95.** yes **97.** yes
99. no **101. a.** 6 **b.** 10 **c.** 5 **103. a.** 45 **b.** -35 **c.** 0 **105.** all real numbers or
$(-\infty, \infty)$ **107.** domain: $[-3, 5]$ range: $[-4, 2]$ **109.** domain: {3} range: $(-\infty, \infty)$

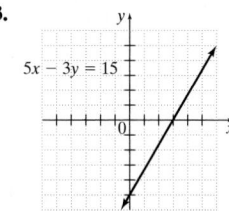

Chapter 3 Test **1. a.** $(1980, 38), (1984, 47), (1988, 51), (1992, 54), (1996, 59), (2000, 55)$
b. **2.** **3.** **4.** **5.**

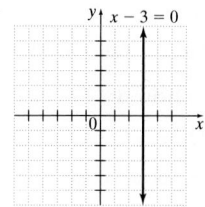

6. $\dfrac{2}{5}$ **7.** 0 **8.** -1 **9.** 3 **10.** undefined **11.** $m = \dfrac{7}{3}; \left(0, -\dfrac{2}{3}\right)$ **12.** neither **13.** $x + 4y = 10$ **14.** $7x + 6y = 0$ **15.** $8x + y = 11$

16. $x = -5$ **17.** $x - 8y = -96$ **18.** yes **19.** no **20. a.** 0 **b.** 0 **c.** 60 **21.** all real numbers except -1 or $(-\infty, -1) \cup (-1, \infty)$
22. domain: $(-\infty, \infty)$; range: $(-\infty, 4]$ **23.** domain: $(-\infty, \infty)$; range: $(-\infty, \infty)$ **24.** 9 P.M. **25.** 4 P.M. **26.** January 1st and December 1st
27. June 1st and end of July **28.** yes; it passes the vertical line test **29.** yes; every location has exactly 1 sunset time per day.

Chapter 3 Cumulative Review 1. a. $<$ **b.** $>$ **c.** $>$; Sec. 1.2, Ex. 1 **3.** $\frac{2}{39}$; Sec. 1.3, Ex. 3 **5.** $\frac{8}{3}$; Sec. 1.4, Ex. 3 **7. a.** -19 **b.** 30 **c.** -0.5

d. $-\frac{4}{5}$ **e.** 6.7 **f.** $\frac{1}{40}$; Sec. 1.5, Ex. 6 **9. a.** -6; **b.** 6.3; Sec. 1.6, Ex. 4 **11. a.** -6 **b.** 0 **c.** $\frac{3}{4}$; Sec. 1.7, Ex. 10 **13. a.** $22 + x$ **b.** $-21x$; Sec. 1.8, Ex. 3

15. a. -3 **b.** 22 **c.** 1 **d.** -1 **e.** $\frac{1}{7}$; Sec. 2.1, Ex. 1 **17.** -1.6; Sec. 2.2, Ex. 2 **19.** 6 Sec. 2.2, Ex. 5 **21.** $3x + 3$; Sec. 2.2, Ex. 8

23. 0; Sec. 2.3, Ex. 4 **25.** 208 Democratic representatives; 223 Republican representatives; Sec. 2.4, Ex. 4 **27.** 40 ft; Sec. 2.5, Ex. 2

29. $\frac{y - b}{m} = x$; Sec. 2.5, Ex. 6 **31.** 40% solution: 8 l; 70% solution: 4 l; Sec. 2.6, Ex. 4 **33.** ◄──────┗━━►; Sec. 2.7, Ex. 1

35. ◄─┗━━━━┛─► $\{x \mid 1 \le x < 4\}$, ; Sec. 2.7, Ex. 10 **37. a.** solution **b.** not a solution **c.** solution; Sec. 3.1, Ex. 5 **39. a.** linear **b.** linear

c. not linear **d.** linear; Sec. 3.2, Ex. 1 **41.** 0; Sec. 3.4, Ex. 3 **43.** $\frac{3}{5}$; Sec. 3.4, Ex. 6

CHAPTER 4   SOLVING SYSTEMS OF LINEAR EQUATIONS AND INEQUALITIES

Calculator Explorations 1. $(0.37, 0.23)$ **3.** $(0.03, -1.89)$

Mental Math 1. 1 solution, $(-1, 3)$ **3.** infinite number of solutions **5.** no solution **7.** 1 solution, $(3, 2)$

Exercise Set 4.1 1. a. no **b.** yes **c.** no **3. a.** no **b.** yes **c.** no **5. a.** yes **b.** yes **c.** yes **7.** answers may vary
9. $(2, 3)$; consistent;   **11.** $(1, -2)$; consistent;   **13.** $(-2, 1)$; consistent;   **15.** $(4, 2)$; consistent;   **17.** no solution; inconsistent;
independent   independent   independent   independent   independent

        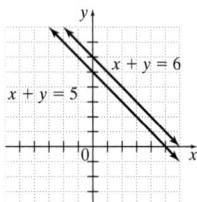

19. infinite number of solutions;   **21.** $(0, -1)$; consistent;   **23.** $(4, -3)$; consistent;   **25.** $(-5, -7)$; consistent;   **27.** $(5, 2)$; consistent;
consistent; dependent   independent   independent   independent   independent

        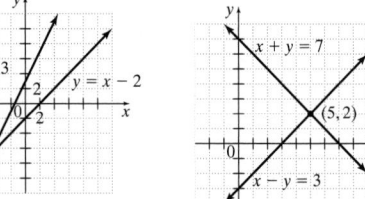

29. answers may vary **31.** intersecting, one solution **33.** parallel, no solution   **61. a.** $(4, 9)$ **b.**
35. identical lines, infinite number of solutions **37.** intersecting, one solution **39.** intersecting, one solution

41. identical lines, infinite number of solutions **43.** parallel, no solution **45.** 2 **47.** $-\frac{2}{5}$ **49.** 2

51. answers may vary **53.** 1984, 1988 **55.** 1996 **57.** answers may vary **59.** answers may vary

Mental Math 1. $(1, 4)$ **3.** infinite number of solutions **5.** $(0, 0)$

Exercise Set 4.2 1. $(2, 1)$ **3.** $(-3, 9)$ **5.** $(4, 2)$ **7.** $(10, 5)$ **9.** $(2, 7)$ **11.** $(-2, 4)$ **13.** $(-2, -1)$ **15.** no solution

17. infinite number of solutions **19.** $(3, -1)$ **21.** $(3, 5)$ **23.** $\left(\frac{2}{3}, -\frac{1}{3}\right)$ **25.** $(-1, -4)$ **27.** $(-6, 2)$ **29.** $(2, 1)$ **31.** no solution **33.** $\left(-\frac{1}{5}, \frac{43}{5}\right)$

35. answers may vary **37.** $(1, -3)$ **39.** $-6x - 4y = -12$ **41.** $-12x + 3y = 9$ **43.** $5n$ **45.** $-15b$ **47. a.** $(12, 480)$ **b.** answers may vary

c.   **49.** $(-2.6, 1.3)$ **51.** $(3.28, 2.11)$

Exercise Set 4.3 **1.** $(1, 2)$ **3.** $(2, -3)$ **5.** $(6, 0)$ **7.** no solution **9.** $\left(2, -\dfrac{1}{2}\right)$ **11.** $(6, -2)$ **13.** infinite number of solutions **15.** $(-2, -5)$
17. $(5, -2)$ **19.** $(-7, 5)$ **21.** $\left(\dfrac{12}{11}, -\dfrac{4}{11}\right)$ **23.** no solution **25.** $\left(\dfrac{3}{2}, 3\right)$ **27.** $(1, 6)$ **29.** infinite number of solutions **31.** $(-2, 0)$
33. answers may vary **35.** $(2, 5)$ **37.** $(-3, 2)$ **39.** $(0, 3)$ **41.** $(5, 7)$ **43.** $\left(\dfrac{1}{3}, 1\right)$ **45.** infinite number of solutions **47.** $(-8.9, 10.6)$
49. $2x + 6 = x - 3$ **51.** $20 - 3x = 2$ **53.** $4(n + 6) = 2n$ **55. a.** 2004
b. skiers: 6.5 million; snow boarders: 6.98 million; rounding in part a caused a difference **57. a.** $b = 15$ **b.** any real number except 15
59. a. answers may vary **b.** answers may vary

Integrated Review **1.** C **2.** D **3.** A **4.** B **5.** $(2, 5)$ **6.** $(4, 2)$ **7.** $(5, -2)$ **8.** $(6, -14)$ **9.** $(-3, 2)$ **10.** $(-4, 3)$ **11.** $(0, 3)$ **12.** $(-2, 4)$
13. $(5, 7)$ **14.** $(-3, -23)$ **15.** $\left(\dfrac{1}{3}, 1\right)$ **16.** $\left(-\dfrac{1}{4}, 2\right)$ **17.** no solution **18.** infinite number of solutions **19.** $(0.5, 3.5)$ **20.** $(-0.75, 1.25)$
21. infinite number of solutions **22.** no solution **23.** answers may vary **24.** answers may vary

Exercise Set 4.4 **1.** a, b, d **3.** Yes; answers may vary. **5.** $(-1, 5, 2)$ **7.** $(-2, 5, 1)$ **9.** $(-2, 3, -1)$ **11.** $\{(x, y, z)|x - 2y + z = -5\}$
13. no solution **15.** $(0, 0, 0)$ **17.** $(-3, -35, -7)$ **19.** $(6, 22, -20)$ **21.** no solution **23.** $(3, 2, 2)$
25. $\{(x, y, z)|x + 2y - 3z = 4\}$ **27.** $(-3, -4, -5)$ **29.** $\left(0, \dfrac{1}{2}, -4\right)$ **31.** $(12, 6, 4)$ **33.** 15 and 30 **35.** 5 **37.** $-\dfrac{5}{3}$
39. answers may vary **41.** answers may vary **43.** $(1, 1, -1)$ **45.** $(1, 1, 0, 2)$ **47.** $(1, -1, 2, 3)$ **49.** answers may vary

Exercise Set 4.5 **1.** 10 and 8 **3. a.** Enterprise class: 1101 ft; Nimitz class: 1092 ft **b.** 3.67 foot ball fields **5.** plane: 520 mph; wind: 40 mph
7. 20 qt of 4%; 40 qt of 1% **9.** United Kingdom; 27,720 students; Spain: 12,292 students **11.** 9 large frames; 13 small frames
13. -10 and -8 **15.** 2005 **17.** tablets: \$0.80; pens: \$0.20 **19.** speed of plane: 630 mph; speed of wind: 90 mph **21. a.** answers may vary
but notice the slope of each function **b.** 2006 **23.** 28 cm; 28 cm; 37 cm **25.** 600 mi **27.** $x = 75; y = 105$ **29.** 625 units
31. 3000 units **33.** 1280 units **35. a.** $R(x) = 450x$ **b.** $C(x) = 200x + 6000$ **c.** 24 desks **37.** 2 units of Mix A; 3 units of Mix B;
1 unit of Mix C **39.** 5 in.; 7 in.; 10 in. **41.** 18, 13, and 9 **43.** 151 free throws; 215 two-point field goals; 39 three-point field goals
45. $x = 60; y = 55; z = 65$ **47.** $5x + 5z = 10$ **49.** $-5y + 2z = 2$ **51.** 1980: 300,000; 2001: 1,400,000 **53.** $a = 3, b = 4, c = -1$
55. $a = 15\dfrac{5}{6}, b = -10\dfrac{5}{6}, c = 1065$; 2250 students in 2009

Chapter 4 Review **1. a.** no **b.** yes **c.** no **3. a.** no **b.** no **c.** yes
5. $(3, -1)$ **7.** $(-3, -2)$ **9.** $\left(\dfrac{1}{2}, \dfrac{1}{2}\right)$

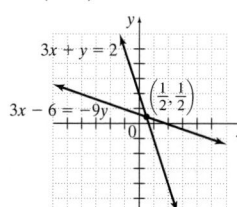

11. intersecting, one solution **13.** identical, infinite number of solutions **15.** $(-1, 4)$ **17.** $(3, -2)$ **19.** no solution; inconsistent **21.** $(3, 1)$
23. $(8, -6)$ **25.** $(-6, 2)$ **27.** $(3, 7)$ **29.** infinite number of solutions; dependent **31.** $\left(2, -2\dfrac{1}{2}\right)$ **33.** $(-6, 15)$ **35.** $(-3, 1)$ **37.** $(2, 0, 2)$
39. $(-\dfrac{1}{2}, \dfrac{3}{4}, 1)$ **41.** no solution **43.** $(1, 1, -2)$ **45.** -6 and 22 **47.** ship: 21.1 mph; current: 3.2 mph **49.** 6% solution: 12.5 cc; 14% solution: 37.5 cc
51. 0.86 hr **53.** 10, 40, 48 **55.** 30 lb of creme-filled: 5 lb of choc-covered nuts: 10 lb of choc-covered raisins

Chapter 4 Test **1.** false **2.** false **3.** true **4.** false **5.** no **6.** yes
7. $(-4, 2)$

8. $(-4, 1)$ **9.** $\left(\dfrac{1}{2}, -2\right)$ **10.** $(4, -2)$ **11.** no solution
12. $(4, -5)$ **13.** $(7, 2)$ **14.** $(5, -2)$ **15.** 120 cc **16.** \$1225 at 5%; \$2775 at 9%
17. Texas: 226 thousand; Missouri: 110 thousand **18.** $(-1, -2, 4)$ **19.** no solution **20.** $23°, 45°, 112°$

Chapter 4 Cumulative Review   1. a. $<$ **b.** $=$ **c.** $>$; Sec. 1.2, Ex. 6   **3. a.** commutative property of multiplication   **b.** associative property of addition   **c.** identity element for addition   **d.** commutative property of multiplication   **e.** multiplicative inverse property   **f.** additive inverse property   **g.** commutative and associative properties of multiplication; Sec. 1.8, Ex. 6   **5.** $-2x - 1$; Sec. 2.1, Ex. 7   **7.** 8; Sec. 2.2, Ex. 4   **9.** 6; Sec. 2.2, Ex. 5

11. 12; Sec. 2.3, Ex. 3   **13.** 10; Sec. 2.4, Ex. 1   **15.** $x = \dfrac{y - b}{m}$; Sec. 2.6, Ex. 6   **17.** $[2, \infty)$; Sec. 2.7, Ex. 3

19.

; Sec. 3.3, Ex. 3   **21.** $-\dfrac{8}{3}$; Sec. 3.4, Ex. 1   **23.** $\dfrac{3}{4}$; Sec. 3.1, Ex. 1   **25.** slope: $\dfrac{3}{4}$; y-intercept: $(0, -1)$; Sec. 3.5, Ex. 3

27. $2x + y = 3$; Sec. 3.6, Ex. 1   **29.** $x = -1$; Sec. 3.6, Ex. 3   **31.** domain: $\{-1, 0, 3\}$; range: $\{-2, 0, 2, 3\}$; Sec. 3.7, Ex. 1

33. a. function   **b.** not a function; Sec. 3.7, Ex. 2   **35.** one solution; Sec. 4.1, Ex. 6   **37.** $\left(6, \dfrac{1}{2}\right)$; Sec. 4.2, Ex. 2

39. $(6, 1)$; Sec. 4.3, Ex. 1   **41.** $(-4, 2, -1)$; Sec. 4.4, Ex. 1   **43.** 7 and 11; Sec. 4.5, Ex. 2

CHAPTER 5   EXPONENTS AND POLYNOMIALS

Mental Math   1. base: 3; exponent: 2   **3.** base: -3; exponent: 6   **5.** base: 4; exponent: 2   **7.** base: 5; exponent: 1; base: 3; exponent: 4   **9.** base: 5; exponent: 1; base: x; exponent: 2

Exercise Set 5.1   1. 49   **3.** -5 **5.** -16 **7.** 16   **9.** 0.00001   **11.** $\dfrac{1}{81}$ **13.** 224   **15.** -250 **17.** answers may vary   **19.** 4   **21.** 135   **23.** 150

25. $\dfrac{32}{5}$ **27.** 343 cu. m   **29.** volume   **31.** x^{10} **33.** $(-3)^{12}$ **35.** $15y^5$ **37.** $-24z^{20}$ **39.** $p^7 q^7$ **41.** $\dfrac{m^9}{n^9}$ **43.** $x^{10}y^{15}$ **45.** $\dfrac{4x^2 z^2}{y^{10}}$ **47.** x^2 **49.** 4

51. $p^6 q^5$ **53.** $\dfrac{y^3}{2}$ **55.** 1   **57.** -2 **59.** 2   **61.** $-\dfrac{27a^6}{b^9}$ **63.** x^{39} **65.** $\dfrac{z^{14}}{625}$ **67.** $7776 m^4 n^3$ **69.** -25 **71.** $\dfrac{1}{64}$ **73.** $81x^2 y^2$ **75.** 1   **77.** 40   **79.** b^6

81. a^9 **83.** $-16x^7$ **85.** $64a^3$ **87.** $36x^2 y^2 z^6$ **89.** $\dfrac{y^{15}}{8x^{12}}$ **91.** x **93.** $2x^2 y$ **95.** $243x^7 y^{24}$ **97.** answers may vary   **99.** $2y - 10$ **101.** $-x - 4$

103. $-x + 5$ **105.** $20x^5$ sq. ft   **107.** $25\pi y^2$ sq. cm   **109.** $27y^{12}$ cu. ft   **111.** x^{9a} **113.** x^{5a} **115.** $x^{5a^2} y^{5ab} z^{5ac}$ **117.** $1045.85

Graphing Calculator Explorations   1. $x^3 - 4x^2 + 7x - 8$ **3.** $-2.1x^2 - 3.2x - 1.7$ **5.** $7.69x^2 - 1.26x + 5.3$

Exercise Set 5.2   1. 0   **3.** 2   **5.** 3   **7.** degree 1; binomial   **9.** degree 2; trinomial   **11.** degree 3; monomial   **13.** degree 3; none of these

15. answers may vary   **17.** 57   **19.** 499   **21.** 1   **23.** $-\dfrac{11}{16}$ **25.** 989 ft   **27.** 477 ft   **29.** $6y$ **31.** $11x - 3$ **33.** $xy + 2x - 1$ **35.** $6x^2 - xy + 16y^2$

37. $18y^2 - 17$ **39.** $3x^2 - 3xy + 6y^2$ **41.** $x^2 - 4x + 8$ **43.** $y^2 + 3$ **45.** $-2x^2 + 5x$ **47.** $-2x^2 - 4x + 15$ **49.** $4x - 13$ **51.** $x^2 + 2$

53. $12x^3 + 8x + 8$ **55.** $7x^3 + 4x^2 + 8x - 10$ **57.** $-18y^2 + 11yx + 14$ **59.** $-x^3 + 8a - 12$ **61.** $5x^2 - 9x - 3$ **63.** $-3x^2 + 3$

65. $8xy^2 + 2x^3 + 3x^2 - 3$ **67.** $7y^2 - 3$ **69.** $5x^2 + 22x + 16$ **71.** $\dfrac{3}{4}x^2 - \dfrac{1}{3}x^2 y - \dfrac{8}{3}x^2 y^2 + \dfrac{3}{2}y^3$ **73.** $-q^4 + q^2 - 3q + 5$ **75.** $15x^2 + 8x - 6$

77. $x^4 - 7x^2 + 5$ **79.** $\dfrac{1}{3}x^2 - x + 1$ **81.** 202 sq. in.   **83. a.** 284 ft   **b.** 536 ft   **c.** 756 ft   **d.** 944 ft   **e.** answers may vary   **f.** 19 sec   **85.** $80,000

87. $40,000   **89.** x^2 **91.** x^4 **93.** $15x - 10$ **95.** $-2x^2 + 10x - 12$ **97.** $3x^{2a} + 2x^a + 0.7$ **99.** $4x^{2y} + 2x^y - 11$ **101.** $(6x^2 + 14y)$ units

103. $4x^2 - 3x + 6$ **105.** $-x^2 - 6x + 10$ **107.** $3x^2 - 12x + 13$ **109.** $15x^2 + 12x - 9$ **111. a.** $2a - 3$ **b.** $-2x - 3$ **c.** $2x + 2h - 3$

113. a. $4a$ **b.** $-4x$ **c.** $4x + 4h$ **115. a.** $4a - 1$ **b.** $-4x - 1$ **c.** $4x + 4h - 1$ **117. a.** 1017 stations   **b.** 3806 stations   **c.** 331 stations   **d.** answers may vary   **119. a.** 3.1 million SUVs   **b.** 5.8 million SUVs   **121. a.** $1622   **b.** $3198   **c.** $6087   **d.** No, $f(x)$ is not linear.

Mental Math   1. $10xy$ **3.** x^7 **5.** $18x^3$ **7.** $-27x^5$ **9.** a^7 **11.** cannot simplify

Exercise Set 5.3   1. $4a^2 - 8a$ **3.** $7x^3 + 14x^2 - 7x$ **5.** $6x^4 - 3x^3$ **7.** $x^2 + 3x$ **9.** $a^2 + 5a - 14$ **11.** $4y^2 - 16y + 16$ **13.** $30x^2 - 79xy + 45y^2$
15. $4x^4 - 20x^2 + 25$ **17.** $x^2 + 5x + 6$ **19.** $x^3 - 5x^2 + 13x - 14$ **21.** $x^4 + 5x^3 - 3x^2 - 11x + 20$ **23.** $10a^3 - 27a^2 + 26a - 12$
25. $x^3 + 6x^2 + 12x + 8$ **27.** $8y^3 - 36y^2 + 54y - 27$ **29.** $2x^3 + 10x^2 + 11x - 3$ **31.** $x^5 - 4x^3 - 7x^2 - 45x + 63$ **33. a.** 25; 13   **b.** 324; 164
c. no; answers may vary   **35.** $2a^2 + 8a$ **37.** $6x^3 - 9x^2 + 12x$ **39.** $15x^2 + 37xy + 18y^2$ **41.** $x^3 + 7x^2 + 16x + 12$ **43.** $49x^2 + 56x + 16$
45. $-6a^4 + 4a^3 - 6a^2$ **47.** $x^3 + 10x^2 + 33x + 36$ **49.** $a^3 + 3a^2 + 3a + 1$ **51.** $x^2 + 2xy + y^2$ **53.** $x^2 - 13x + 42$ **55.** $3a^3 + 6a$
57. $-4y^3 - 12y^2 + 44y$ **59.** $25x^2 - 1$ **61.** $5x^3 - x^2 + 16x + 16$ **63.** $8x^3 - 60x^2 + 150x - 125$ **65.** $32x^3 + 48x^2 - 6x - 20$
67. $49x^2 y^2 - 14xy^2 + y^2$ **69.** $5y^4 - 16y^3 - 4y^2 - 7y - 6$ **71.** $6x^4 - 8x^3 - 7x^2 + 22x - 12$ **73.** $25x^2$ **75.** $9y^6$ **77.** $7000   **79.** $500
81. answers may vary   **83.** $(4x^2 - 25)$ sq. yd   **85.** $(6x^2 - 4x)$ sq. in.   **87.** $(x^2 + 6x + 5)$ sq. units   **89. a.** $a^2 - b^2$ **b.** $4x^2 - 9y^2$ **c.** $16x^2 - 49$
d. answers may vary

Mental Math   1. false   **3.** false

Exercise Set 5.4   **1.** $x^2 + 7x + 12$ **3.** $x^2 + 5x - 50$ **5.** $5x^2 + 4x - 12$ **7.** $4y^2 - 25y + 6$ **9.** $6x^2 + 13x - 5$ **11.** $x^2 - 4x + 4$
13. $4x^2 - 4x + 1$ **15.** $9a^2 - 30a + 25$ **17.** $25x^2 + 90x + 81$ **19.** answers may vary   **21.** $a^2 - 49$ **23.** $9x^2 - 1$ **25.** $9x^2 - \dfrac{1}{4}$ **27.** $81x^2 - y^2$
29. $4x^2 - 0.01$ **31.** $a^2 + 9a + 20$ **33.** $a^2 + 14a + 49$ **35.** $12a^2 - a - 1$ **37.** $x^2 - 4$ **39.** $9a^2 + 6a + 1$ **41.** $4x^3 - x^2y^4 + 4xy - y^5$
43. $x^3 - 3x^2 - 17x + 3$ **45.** $4a^2 - 12a + 9$ **47.** $25x^2 - 36z^2$ **49.** $x^{10} - 8x^5 + 15$ **51.** $x^2 - \dfrac{1}{9}$ **53.** $a^7 - 3a^3 + 11a^4 - 33$
55. $3x^2 - 12x + 12$ **57.** $6b^2 - b - 35$ **59.** $49p^2 - 64$ **61.** $\dfrac{1}{9}a^4 - 49$ **63.** $15x^4 - 5x^3 + 10x^2$ **65.** $4r^2 - 9s^2$ **67.** $9x^2 - 42xy + 49y^2$
69. $16x^2 - 25$ **71.** $64x^2 + 64x + 16$ **73.** $a^2 - \dfrac{1}{4}y^2$ **75.** $\dfrac{1}{25}x^2 - y^2$ **77.** $3a^3 + 2a^2 + 1$ **79.** $\dfrac{5b^5}{7}$ **81.** $-2a^{10}b^5$ **83.** $\dfrac{2y^8}{3}$ **85.** $\dfrac{1}{3}$ **87.** 1
89. $(4x^2 + 4x + 1)$ sq. ft   **91.** $\left(\dfrac{25}{2}a^2 - \dfrac{1}{2}b^2\right)$ sq. units   **93.** $(24x^2 - 32x + 8)$ sq. m   **95.** $x^2 + 10x + 25$ **97.** answers may vary
99. $a^2 + 2ac + c^2 - 25$ **101.** $x^2 - 4x + 4 - y^2$

Integrated Review   **1.** $35x^5$ **2.** $32y^9$ **3.** -16 **4.** 16 **5.** $2x^2 - 9x - 5$ **6.** $3x^2 + 13x - 10$ **7.** $3x - 4$ **8.** $4x + 3$ **9.** $7x^6y^2$ **10.** $\dfrac{10b^6}{7}$
11. $144m^{14}n^{12}$ **12.** $64y^{27}z^{30}$ **13.** $48y^2 - 27$ **14.** $98x^2 - 2$ **15.** $x^{63}y^{45}$ **16.** $27x^{27}$ **17.** $2x^2 - 2x - 6$ **18.** $6x^2 + 13x - 11$ **19.** $2.5y^2 - 6y - 0.2$
20. $8.4x^2 - 6.8x - 5.7$ **21.** $x^2 + 8xy + 16y^2$ **22.** $y^2 - 18yz + 81z^2$ **23.** $2x + 8y$ **24.** $2y - 18z$ **25.** $7x^2 - 10xy + 4y^2$ **26.** $-a^2 - 3ab + 6b^2$
27. $x^3 + 2x^2 - 16x + 3$ **28.** $x^3 - 2x^2 - 5x - 2$ **29.** $6x^5 + 20x^3 - 21x^2 - 70$ **30.** $20x^7 + 25x^3 - 4x^4 - 5$ **31.** $2x^3 - 19x^2 + 44x - 7$
32. $5x^3 + 9x^2 - 17x + 3$ **33.** cannot simplify   **34.** $25x^3y^3$ **35.** $125x^9$ **36.** $\dfrac{x^3}{y^3}$ **37.** $2x$ **38.** x^2

Calculator Explorations   **1.** $5.31\ \text{EE}\ 3$ **3.** $6.6\ \text{EE}\ -9$ **5.** 1.5×10^{13} **7.** 8.15×10^{19}

Mental Math   **1.** $\dfrac{5}{x^2}$ **3.** y^6 **5.** $4y^3$

Exercise Set 5.5   **1.** $\dfrac{1}{64}$ **3.** $\dfrac{1}{16}$ **5.** $\dfrac{7}{x^3}$ **7.** 32 **9.** -64 **11.** $\dfrac{5}{6}$ **13.** p^3 **15.** $\dfrac{q^4}{p^5}$ **17.** $\dfrac{1}{x^3}$ **19.** $\dfrac{4}{3}$ **21.** $-p^4$ **23.** -2 **25.** x^4 **27.** p^4 **29.** m^{11}
31. r^6 **33.** $\dfrac{1}{x^{15}y^9}$ **35.** $\dfrac{1}{x^4}$ **37.** $\dfrac{1}{a^2}$ **39.** $4k^3$ **41.** $3m$ **43.** $-\dfrac{4a^5}{b}$ **45.** $-\dfrac{6x^2}{y^3}$ **47.** $\dfrac{a^{30}}{b^{12}}$ **49.** $\dfrac{1}{x^{10}y^6}$ **51.** $\dfrac{z^2}{4}$ **53.** $\dfrac{1}{32x^5}$ **55.** $\dfrac{49a^4}{b^6}$ **57.** $a^{24}b^8$
59. x^9y^{19} **61.** $-\dfrac{y^8}{8x^2}$ **63.** $-\dfrac{6x}{7y^2}$ **65.** 7.8×10^4 **67.** 1.67×10^{-6} **69.** 6.35×10^{-3} **71.** 1.16×10^6 **73.** 2.0×10^7 **75.** 1.56×10^7
77. 1.36×10^4 **79.** 2.92×10^8 **81.** 0.0000000008673 **83.** 0.033 **85.** $20{,}320$ **87.** $6{,}250{,}000{,}000{,}000{,}000{,}000$ **89.** $9{,}460{,}000{,}000{,}000$
91. 0.000036 **93.** 0.0000000000000000028 **95.** 0.0000005 **97.** $200{,}000$ **99.** $\dfrac{5x^3}{3}$ **101.** $\dfrac{5z^3y^2}{7}$ **103.** $5y - 6 + \dfrac{5}{y}$ **105.** $\dfrac{27}{x^6z^3}$ cu. in.   **107.** -394.5
109. 1.3 sec   **111.** answers may vary   **113.** 2.7×10^9 gal   **115.** answers may vary   **117.** a^m **119.** y^{5a} **121.** $\dfrac{1}{z^{6a+4}}$

Mental Math   **1.** a^2 **3.** a^2 **5.** k^3 **7.** p^5 **9.** k^2

Exercise Set 5.6   **1.** $5p^2 + 6p$ **3.** $-\dfrac{3}{2x} + 3$ **5.** $-3x^2 + x - \dfrac{4}{x^3}$ **7.** $-1 + \dfrac{3}{2x} - \dfrac{7}{4x^4}$ **9.** $5x^3 - 3x + \dfrac{1}{x^2}$ **11.** $(3x^3 + x - 4)$ ft   **13.** $x + 1$
15. $2x + 3$ **17.** $2x + 1 + \dfrac{7}{x - 4}$ **19.** $4x + 9$ **21.** $3a^2 - 3a + 1 + \dfrac{2}{3a + 2}$ **23.** $2b^2 + b + 2 - \dfrac{12}{b + 4}$ **25.** answers may vary   **27.** $(2x + 5)$ m
29. $\dfrac{4}{x} + \dfrac{1}{x^2} + \dfrac{9}{5x^3}$ **31.** $5x - 2 + \dfrac{2}{x + 6}$ **33.** $x^2 - \dfrac{12x}{5} - 1$ **35.** $6x - 1 - \dfrac{1}{x + 3}$ **37.** $4x^2 + 1$ **39.** $2x^2 + 6x - 5 - \dfrac{2}{x - 2}$ **41.** $6x - 1$
43. $-x^3 + 3x^2 - \dfrac{4}{x}$ **45.** $4x + 3 - \dfrac{2}{2x + 1}$ **47.** $2x + 9$ **49.** $2x^2 + 3x - 4$ **51.** $x^2 + 3x + 9$ **53.** $x^2 - x + 1$ **55.** $-3x + 6 - \dfrac{11}{x + 2}$
57. $2b - 1 - \dfrac{6}{2b - 1}$ **59.** $2a^3 + 2a$ **61.** $2x^3 + 14x^2 - 10x$ **63.** $-3x^2y^3 - 21x^3y^2 - 24xy$ **65.** $9a^2b^3c + 36ab^2c - 72ab$
67. The Rolling Stones (1994)   **69.** \$110 million   **71.** $x^3 - x^2 + x$ **73.** $9x^{7a} - 6x^{5a} + 7x^{2a} - 1$

Exercise Set 5.7   **1.** $x + 8$ **3.** $x - 1$ **5.** $x^2 - 5x - 23 - \dfrac{41}{x - 2}$ **7.** $4x + 8 + \dfrac{7}{x - 2}$ **9.** 3 **11.** 73 **13.** -8 **15.** $x^2 + \dfrac{2}{x - 3}$
17. $6x + 7 + \dfrac{1}{x + 1}$ **19.** $2x^3 - 3x^2 + x - 4$ **21.** $3x - 9 + \dfrac{12}{x + 3}$ **23.** $3x^2 - \dfrac{9}{2}x + \dfrac{7}{4} + \dfrac{47}{8(x - \frac{1}{2})}$ **25.** $3x^2 + 3x - 3$
27. $3x^2 + 4x - 8 + \dfrac{20}{x + 1}$ **29.** $x^2 + x + 1$ **31.** $x - 6$ **33.** 1 **35.** -133 **37.** 3 **39.** $-\dfrac{187}{81}$ **41.** $\dfrac{95}{32}$ **43.** answers may vary   **45.** $-\dfrac{5}{6}$ **47.** 54
49. 8 **51.** -32 **53.** 48 **55.** 25 **57.** -2 **59.** $(x + 3)(x^2 + 4) = x^3 + 3x^2 + 4x + 12$ **61.** 0 **63.** $x^3 + 2x^2 + 7x + 28$ **65.** $(x - 1)$ m

Chapter 5 Review   **1.** base: 3; exponent: 2   **3.** base: 5; exponent: 4   **5.** 36 **7.** -65 **9.** 1 **11.** 8 **13.** $-10x^5$ **15.** $\dfrac{b^4}{16}$ **17.** $\dfrac{x^6y^6}{4}$ **19.** $40a^{19}$
21. $\dfrac{3}{64}$ **23.** $\dfrac{1}{x}$ **25.** 5 **27.** 1 **29.** $6a^6b^9$ **31.** 7 **33.** 5 **35. a.** $3, -2, 1, -1, -6; 3, 2, 2, 2, 0$ **b.** 3 **37.** $22; 78; 154.02; 400$ **39.** $-4xy^3 - 3x^3y$

41. $-4x^2 + 10y^2$ **43.** $16xy - 4x^3 + 4x^2 - 9x + 18$ **45.** $x^2 - 6x + 3$ **47.** 290 **49.** 110 **51.** $-x^2 + 2x + 3$

53. 28,770,000 trademark registrations **55.** $-56xz^2$ **57.** $-12x^2a^2y^4$ **59.** $9x - 63$ **61.** $54a - 27$ **63.** $-32y^3 + 48y$

65. $-3a^3b - 3a^2b - 3ab^2$ **67.** $42b^4 - 28b^2 + 14b$ **69.** $6x^2 - 11x - 10$ **71.** $42a^2 + 11a - 3$ **73.** $x^6 + 2x^5 + x^2 + 3x + 2$

75. $x^6 + 8x^4 + 16x^2 - 16$ **77.** $8x^3 - 60x^2 + 150x - 125$ **79.** $15y^3 - 3y^2 + 6y$ **81.** $12a^5 + 28a^3 - 3a^2 - 7$ **83.** $x^6 - 10x^3y + 25y^2$

85. $16x^2 + 16x + 4$ **87.** $x^3 - 3x^2 + 4$ **89.** $25x^2 - 1$ **91.** $a^2 - 4b^2$ **93.** $16a^4 - 4b^2$ **95.** $-\dfrac{1}{49}$ **97.** $\dfrac{1}{16x^4}$ **99.** $\dfrac{9}{4}$ **101.** $\dfrac{1}{42}$ **103.** $-q^3r^3$

105. $\dfrac{r^5}{s^3}$ **107.** $\dfrac{12}{y}$ **109.** $\dfrac{y^6}{3}$ **111.** $\dfrac{x^6y^9}{64z^3}$ **113.** $\dfrac{y^8}{49}$ **115.** $\dfrac{x^3}{y^3}$ **117.** $\dfrac{a^{10}}{b^{10}}$ **119.** $-\dfrac{4}{27}$ **121.** x^{10+3b} **123.** a^{2m+5} **125.** 8.868×10^{-1} **127.** 8.68×10^5

129. 1.5×10^5 **131.** 0.00386 **133.** 893,600 **135.** 0.0000000001 m **137.** 400,000,000,000 **139.** $-a^2 + 3b - 4$ **141.** $4x + \dfrac{7}{x + 5}$

143. $3b^2 - 4b - \dfrac{1}{3b - 2}$ **145.** $-x^2 - 16x - 117 - \dfrac{684}{x - 6}$ **147.** $3x^2 - \dfrac{5}{2}x - \dfrac{1}{4} - \dfrac{5}{8\left(x + \dfrac{3}{2}\right)}$ **149.** $x^2 + 3x + 9 - \dfrac{54}{x - 3}$

151. $3x^3 - 6x^2 + 10x - 20 + \dfrac{50}{x + 2}$ **153.** -9323

Chapter 5 Test **1.** 32 **2.** 81 **3.** -81 **4.** $\dfrac{1}{64}$ **5.** $-15x^{11}$ **6.** y^5 **7.** $\dfrac{1}{r^5}$ **8.** $\dfrac{y^{14}}{x^2}$ **9.** $\dfrac{1}{6xy^8}$ **10.** 5.63×10^5 **11.** 8.63×10^{-5} **12.** 0.0015
13. 62,300 **14.** 0.036 **15. a.** $4, 7, 1, -2; 3, 3, 4, 0$ **b.** 4 **16.** $-2x^2 + 12xy + 11$ **17.** $16x^3 + 7x^2 - 3x - 13$ **18.** $-3x^3 + 5x^2 + 4x + 5$
19. $x^3 + 8x^2 + 3x - 5$ **20.** $3x^3 + 22x^2 + 41x + 14$ **21.** $6x^4 - 9x^3 + 21x^2$ **22.** $3x^2 + 16x - 35$ **23.** $16x^2 - 16x + 4$ **24.** $x^4 - 81b^2$
25. 1001 ft; 985 ft; 857 ft; 601 ft **26.** $\dfrac{x}{2y} + 3 - \dfrac{7}{8y}$ **27.** $x + 2$ **28.** $9x^2 - 6x + 4 - \dfrac{16}{3x + 2}$ **29. a.** 960 ft **b.** 953.44 ft **c.** 11 sec
30. $4x^3 - 15x^2 + 45x - 136 + \dfrac{407}{x + 3}$ **31.** 91

Chapter 5 Cumulative Review **1. a.** true **b.** true **c.** false **d.** true; Sec. 1.2, Ex. 2 **3. a.** $\dfrac{64}{25}$ **b.** $\dfrac{1}{20}$ **c.** $\dfrac{5}{4}$; Sec. 1.3, Ex. 4 **5. a.** 9 **b.** 125 **c.** 16

d. 7 **e.** $\dfrac{9}{49}$; Sec. 1.4, Ex. 1 **7. a.** -10 **b.** -21 **c.** -12; Sec. 1.5, Ex. 3 **9.** -12; Sec. 1.6, Ex. 3 **11. a.** $\dfrac{1}{22}$ **b.** $\dfrac{16}{3}$ **c.** $-\dfrac{1}{10}$ **d.** $-\dfrac{13}{9}$; Sec. 1.7, Ex. 5
13. a. $(5 + 4) + 6$ **b.** $-1 \cdot (2 \cdot 5)$; Sec. 1.8, Ex. 2 **15. a.** $2x + 2y$ **b.** $15 - 10z$ **c.** $5x + 15y - 5z$ **d.** $-2 + y$ **e.** $-3 - x + w$
f. $12x + 28$; Sec. 1.8, Ex. 4 **17. a.** $5x + 10$ **b.** $-2y - 0.6z + 2$ **c.** $-x - y + 2z - 6$; Sec. 2.1, Ex. 5 **19.** 17; Sec. 2.2, Ex. 1
21. 6; Sec. 2.2, Ex. 5 **23.** -10; Sec. 2.3, Ex. 8 **25.** 10; Sec. 2.4, Ex. 1 **27.** width: 4 ft; length: 10 ft; Sec. 2.5, Ex. 4 **29.** $\dfrac{5F - 160}{9} = C$; Sec. 2.5, Ex. 8
31. ◄———┼——┼——►; Sec. 2.7, Ex. 9 **33. a.** $(0, 12)$ **b.** $(2, 6)$ **c.** $(-1, 15)$; Sec. 3.1, Ex. 6

35. ; Sec. 3.2, Ex. 2 **37.** 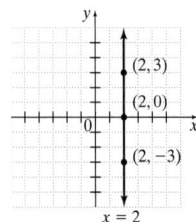 ; Sec. 3.3, Ex. 5 **39.** undefined slope; Sec. 3.4, Ex. 4
41. $12z^5 + 3z^4 - 13z^3 - 11z$; Sec. 5.2, Ex. 9
43. $-6x^7$; Sec. 5.1, Ex. 4 **45. a.** -4 **b.** 11; Sec. 5.2, Ex. 4
47. $4x^2 - 4xy + y^2$; Sec. 5.3, Ex. 6
49. $3m + 1$; Sec. 5.6, Ex. 1

CHAPTER 6 FACTORING POLYNOMIALS

Mental Math **1.** $2 \cdot 7$ **3.** $2 \cdot 5$ **5.** 3 **7.** 3

Exercise Set 6.1 **1.** 4 **3.** 6 **5.** y^2 **7.** xy^2 **9.** 4 **11.** $4y^3$ **13.** $3x^3$ **15.** $9x^2y$ **17.** $15(2x - 1)$ **19.** $6cd(4d^2 - 3c)$ **21.** $-6a^3x(4a - 3)$
23. $4x(3x^2 + 4x - 2)$ **25.** $5xy(x^2 - 3x + 2)$ **27.** answers may vary **29.** $(x + 2)(y + 3)$ **31.** $(y - 3)(x - 4)$ **33.** $(x + y)(2x - 1)$
35. $(x + 3)(5 + y)$ **37.** $(y - 4)(2 + x)$ **39.** $(y - 2)(3x + 8)$ **41.** $(y + 3)(y^2 + 1)$ **43.** $12x^3 - 2x; 2x(6x^2 - 1)$
45. $200x + 25\pi; 25(8x + \pi)$ **47.** $3(x - 2)$ **49.** $2x(16y - 9x)$ **51.** $4(x - 2y + 1)$ **53.** $(x + 2)(8 - y)$ **55.** $-8x^8y^5(5y + 2x)$
57. $-3(x - 4)$ **59.** $6x^3y^2(3y - 2 + x^2)$ **61.** $(x - 2)(y^2 + 1)$ **63.** $(y + 3)(5x + 6)$ **65.** $(x - 2y)(4x - 3)$ **67.** $42yz(3x^3 + 5y^3z^2)$
69. $(3 - x)(5 + y)$ **71.** $2(3x^2 - 1)(2y - 7)$ **73.** $2, 6$ **75.** $-1, -8$ **77.** $-2, 5$ **79.** $-8, 3$ **81.** $x^2 + 7x + 10$ **83.** $a^2 - 15a + 56$
85. answers may vary **87.** factored **89.** not factored **91.** $(n^3 - 6)$ units **93.** $(x^n + 2)(x^n + 3)$ **95.** $(3x^n - 5)(x^n + 7)$
97. a. 44 million **b.** 124 million **c.** $4(2x^2 - 14x + 31)$

Mental Math **1.** $+5$ **3.** -3 **5.** $+2$

Exercise Set 6.2 **1.** $(x + 6)(x + 1)$ **3.** $(x + 5)(x - 4)$ **5.** $(x - 5)(x - 3)$ **7.** $(x - 9)(x - 1)$ **9.** prime **11.** $(x - 6)(x + 3)$ **13.** prime
15. $(x + 5y)(x + 3y)$ **17.** $(x - y)(x - y)$ **19.** $(x - 4y)(x + y)$ **21.** $2(z + 8)(z + 2)$ **23.** $2x(x - 5)(x - 4)$ **25.** $7(x + 3y)(x - y)$
27. product; sum **29.** $(x + 12)(x + 3)$ **31.** $(x - 2)(x + 1)$ **33.** $(r - 12)(r - 4)$ **35.** $(x - 7)(x + 3)$ **37.** $(x + 5y)(x + 2y)$

39. prime **41.** $2(t + 8)(t + 4)$ **43.** $x(x - 6)(x + 4)$ **45.** $(x - 9)(x - 7)$ **47.** $(x + 2y)(x - y)$ **49.** $3(x - 18)(x - 2)$
51. $(x - 24)(x + 6)$ **53.** $6x(x + 4)(x + 5)$ **55.** $2t^3(t - 4)(t - 3)$ **57.** $5xy(x - 8y)(x + 3y)$ **59.** $4y(x^2 + x - 3)$
61. $2b(a - 7b)(a - 3b)$ **63.** $2x^2 + 11x + 5$ **65.** $15y^2 - 17y + 4$ **67.** $9a^2 + 23a - 12$ **69.** $8; 16$ **71.** $6; 26$ **73.** $5; 8; 9$ **75.** $3; 4$
77. $2x^2 + 28x + 66; 2(x + 3)(x + 11)$ **79.** answers may vary **81.** $2y(x + 5)(x + 10)$ **83.** $-12y^3(x^2 + 2x + 3)$ **85.** $(x + 1)(y - 5)(y + 3)$
87. $(x^n + 2)(x^n + 3)$

Mental Math 1. yes **3.** no **5.** yes

Exercise Set 6.3 1. $(2x + 3)(x + 5)$ **3.** $(2x + 1)(x - 5)$ **5.** $(2y + 3)(y - 2)$ **7.** $(4a - 3)^2$ **9.** $(9r - 8)(4r + 3)$ **11.** $(5x + 1)(2x + 3)$
13. $3(7x + 5)(x - 3)$ **15.** $2(2x - 3)(3x + 1)$ **17.** $x(4x + 3)(x - 3)$ **19.** $(x + 11)^2$ **21.** $(x - 8)^2$ **23.** $(4y - 5)^2$ **25.** $(xy - 5)^2$
27. answers may vary **29.** $(2x + 11)(x - 9)$ **31.** $(2x - 7)(2x + 3)$ **33.** $(6x - 7)(5x - 3)$ **35.** $(4x - 9)(6x - 1)$ **37.** $(3x - 4y)^2$
39. $(x - 7y)^2$ **41.** $(2x + 5)(x + 1)$ **43.** prime **45.** $(-2y + 5)(y + 2)$ **47.** $(4x + 3y)^2$ **49.** $2y(4x - 7)(x + 6)$ **51.** $(3x - 2)(x + 1)$
53. $(xy + 2)^2$ **55.** $(7y + 3x)^2$ **57.** $3(x^2 - 14x + 21)$ **59.** $(7a - 6)(6a - 1)$ **61.** $(6x - 7)(3x + 2)$ **63.** $(5p - 7q)^2$ **65.** $(5x + 3)(3x - 5)$
67. $(7t + 1)(t - 4)$ **69.** $a^2 + 2ab + b^2$ **71.** $x^2 - 4$ **73.** $a^3 + 27$ **75.** $y^3 - 125$ **77.** \$75,000 and above **79.** answers may vary **81.** $2; 14$
83. $5; 13$ **85.** 2 **87.** $4; 5$ **89.** $-3xy^2(4x - 5)(x + 1)$ **91.** $-2pq(p - 3q)(15p + q)$ **93.** $(y - 1)^2(4x^2 + 10x + 25)$
95. $(3x^n + 2)(x^n + 5)$

Calculator Explorations

16	14	16
16	14	16
2.89	0.89	2.89
171.61	169.61	171.61
1	-1	1

Mental Math 1. 1^2 **3.** 9^2 **5.** 3^2 **7.** 1^3 **9.** 2^3

Exercise Set 6.4 1. $(x + 2)(x - 2)$ **3.** $(y + 7)(y - 7)$ **5.** $(5y + 3)(5y - 3)$ **7.** $(11 + 10x)(11 - 10x)$ **9.** $3(2x + 3)(2x - 3)$
11. $(13a + 7b)(13a - 7b)$ **13.** $(xy + 1)(xy - 1)$ **15.** $(x^2 + 3)(x^2 - 3)$ **17.** $(7a^2 + 4)(7a^2 - 4)$ **19.** $(x^2 + y^5)(x^2 - y^5)$ **21.** $(x + 6)$
23. $(a + 3)(a^2 - 3a + 9)$ **25.** $(2a + 1)(4a^2 - 2a + 1)$ **27.** $5(k + 2)(k^2 - 2k + 4)$ **29.** $(xy - 4)(x^2y^2 + 4xy + 16)$
31. $(x + 5)(x^2 - 5x + 25)$ **33.** $3x(2x - 3y)(4x^2 + 6xy + 9y^2)$ **35.** $(2x + y)$ **37.** $(x + 11)(x - 11)$ **39.** $(9 + p)(9 - p)$
41. $(2r + 1)(2r - 1)$ **43.** $(3x + 4)(3x - 4)$ **45.** prime **47.** $(3 - t)(9 + 3t + t^2)$ **49.** $8(r - 2)(r^2 + 2r + 4)$ **51.** $(t - 7)(t^2 + 7t + 49)$
53. $(x + 13y)(x - 13y)$ **55.** $(xy + z)(xy - z)$ **57.** $(xy + 1)(x^2y^2 - xy + 1)$ **59.** $(s - 4t)(s^2 + 4st + 16t^2)$ **61.** $2(3r + 2)(3r - 2)$

63. $x(3y + 2)(3y - 2)$ **65.** $25y^2(y + 2)(y - 2)$ **67.** $xy(x + 2y)(x - 2y)$ **69.** $4s^3t^3(2s^3 + 25t^3)$ **71.** $xy^2(27xy - 1)$ **73.** $4x^3 + 2x^2 - 1 + \dfrac{3}{x}$

75. $2x + 1$ **77.** $3x + 4 - \dfrac{2}{x + 3}$ **79.** $(x + 2 + y)(x + 2 - y)$ **81.** $(a + 4)(a - 4)(b - 4)$ **83.** $(x + 3 + 2y)(x + 3 - 2y)$

85. $(x^n + 10)(x^n - 10)$ **87. a.** 777 ft **b.** 441 ft **c.** 7 sec **d.** $(29 + 4t)(29 - 4t)$ **89. a.** 1300 ft **b.** 660 ft **c.** 10 sec **d.** $4(19 + 2t)(19 - 2t)$
91. answers may vary

Integrated Review 1. $(a + b)^2$ **2.** $(a - b)^2$ **3.** $(a - 3)(a + 4)$ **4.** $(a - 5)(a - 2)$ **5.** $(a + 2)(a - 3)$ **6.** $(a + 1)^2$ **7.** $(x + 1)^2$
8. $(x + 2)(x - 1)$ **9.** $(x + 1)(x + 3)$ **10.** $(x + 3)(x - 2)$ **11.** $(x + 3)(x + 4)$ **12.** $(x + 4)(x - 3)$ **13.** $(x + 4)(x - 1)$
14. $(x - 5)(x - 2)$ **15.** $(x + 5)(x - 3)$ **16.** $(x + 6)(x + 5)$ **17.** $(x - 6)(x + 5)$ **18.** $(x + 8)(x + 3)$ **19.** $2(x + 7)(x - 7)$
20. $3(x + 5)(x - 5)$ **21.** $(x + 3)(x + y)$ **22.** $(y - 7)(3 + x)$ **23.** $(x + 8)(x - 2)$ **24.** $(x - 7)(x + 4)$ **25.** $4x(x + 7)(x - 2)$
26. $6x(x - 5)(x + 4)$ **27.** $2(3x + 4)(2x + 3)$ **28.** $(2a - b)(4a + 5b)$ **29.** $(2a + b)(2a - b)$ **30.** $(4 - 3x)(7 + 2x)$ **31.** $(5 - 2x)(4 + x)$
32. prime **33.** prime **34.** $(3y + 5)(2y - 3)$ **35.** $(4x - 5)(x + 1)$ **36.** $y(x + y)(x - y)$ **37.** $4(t^2 + 9)$ **38.** $(x + 1)(x + y)$
39. $(x + 1)(a + 2)$ **40.** $9x(2x^2 - 7x + 1)$ **41.** $4a(3a^2 - 6a + 1)$ **42.** $(x + 16)(x - 2)$ **43.** prime **44.** $(4a - 7b)^2$ **45.** $(5p - 7q)^2$
46. $(7x + 3y)(x + 3y)$ **47.** $(5 - 2y)(25 + 10y + 4y^2)$ **48.** $(4x + 3)(16x^2 - 12x + 9)$ **49.** $-(x - 5)(x + 6)$ **50.** $-(x - 2)(x - 4)$
51. $(7 - x)(2 + x)$ **52.** $(3 + x)(1 - x)$ **53.** $3x^2y(x + 6)(x - 4)$ **54.** $2xy(x + 5y)(x - y)$ **55.** $5xy^2(x - 7y)(x - y)$
56. $4x^2y(x - 5)(x + 3)$ **57.** $3xy(4x^2 + 81)$ **58.** $2xy^2(3x^2 + 4)$ **59.** $(2 + x)(2 - x)$ **60.** $(3 + y)(3 - y)$ **61.** $(s + 4)(3r - 1)$
62. $(x - 2)(x^2 + 3)$ **63.** $(4x - 3)(x - 2y)$ **64.** $(2x - y)(2x + 7z)$ **65.** $6(x + 2y)(x + y)$ **66.** $2(x + 4y)(6x - y)$
67. $(x + 3)(y + 2)(y - 2)$ **68.** $(y + 3)(y - 3)(x^2 + 3)$ **69.** $(5 + x)(x + y)$ **70.** $(x - y)(7 + y)$ **71.** $(7t - 1)(2t - 1)$ **72.** prime
73. $(3x + 5)(x - 1)$ **74.** $(7x - 2)(x + 3)$ **75.** $(x + 12y)(x - 3y)$ **76.** $(3x - 2y)(x + 4y)$ **77.** $(1 - 10ab)(1 + 2ab)$
78. $(1 + 5ab)(1 - 12ab)$ **79.** $(3 + x)(3 - x)(1 + x)(1 - x)$ **80.** $(3 + x)(3 - x)(2 + x)(2 - x)$ **81.** $(x + 4)(x - 4)(x^2 + 2)$
82. $(x + 5)(x - 5)(x^2 + 3)$ **83.** $(x - 15)(x - 8)$ **84.** $(y + 16)(y + 6)$ **85.** $2x(3x - 2)(x - 4)$ **86.** $2y(3y + 5)(y - 3)$
87. $(3x - 5y)(9x^2 + 15xy + 25y^2)$ **88.** $(6y - z)(36y^2 + 6yz + z^2)$ **89.** $(xy + 2z)(x^2y^2 - 2xyz + 4z^2)$ **90.** $(3ab + 2)(9a^2b^2 - 6ab + 4)$
91. $2xy(1 + 6x)(1 - 6x)$ **92.** $2x(x + 3)(x - 3)$ **93.** $(x + 2)(x - 2)(x + 6)$ **94.** $(x - 2)(x + 6)(x - 6)$ **95.** $2a^2(3a + 5)$
96. $2n(2n - 3)$ **97.** $(a^2 + 2)(a + 2)$ **98.** $(a - b)(1 + x)$ **99.** $(x + 2)(x - 2)(x + 7)$ **100.** $(a + 3)(a - 3)(a + 5)$
101. $(x - y + z)(x - y - z)$ **102.** $(x + 2y + 3)(x + 2y - 3)$ **103.** $(9 + 5x + 1)(9 - 5x - 1)$ **104.** $(b + 4a + c)(b - 4a - c)$
105. answers may vary **106.** yes; $9(x^2 + 9y^2)$ **107.** a, c

Calculator Explorations 1. $-0.9, 2.2$ **3.** no real solution **5.** $-1.8, 2.8$

Mental Math 1. $3, 7$ **3.** $-8, -6$ **5.** $-1, 3$

Exercise Set 6.5 **1.** $2, -1$ **3.** $0, -6$ **5.** $-\dfrac{3}{2}, \dfrac{5}{4}$ **7.** $\dfrac{7}{2}, -\dfrac{2}{7}$ **9.** $(x - 6)(x + 1) = 0$ **11.** $9, 4$ **13.** $-4, 2$ **15.** $8, -4$ **17.** $\dfrac{7}{3}, -2$ **19.** $\dfrac{8}{3}, -9$

21. $x^2 - 12x + 35 = 0$ **23.** $0, 8, 4$ **25.** $\dfrac{3}{4}$ **27.** $0, \dfrac{1}{2}, -\dfrac{1}{2}$ **29.** $0, \dfrac{1}{2}, -\dfrac{3}{8}$ **31.** $0, -7$ **33.** $-5, 4$ **35.** $-5, 6$ **37.** $-\dfrac{4}{3}, 5$ **39.** $-\dfrac{3}{2}, -\dfrac{1}{2}, 3$

41. $-5, 3$ **43.** $0, 16$ **45.** $-\dfrac{9}{2}, \dfrac{8}{3}$ **47.** $\dfrac{5}{4}, \dfrac{11}{3}$ **49.** $-\dfrac{2}{3}, \dfrac{1}{6}$ **51.** $-\dfrac{5}{3}, \dfrac{1}{2}$ **53.** $\dfrac{17}{2}$ **55.** $2, -\dfrac{4}{5}$ **57.** $-4, 3$ **59.** $\dfrac{1}{2}, -\dfrac{1}{2}$ **61.** $-2, -11$ **63.** 3

65. -3 **67.** $\dfrac{3}{4}, -\dfrac{4}{3}$ **69.** $0, \dfrac{5}{2}, -1$ **71.** $-\dfrac{4}{3}, 1$ **73.** $-2, 5$ **75.** $-6, \dfrac{1}{2}$ **77.** e **79.** b **81.** c **83.** $\dfrac{47}{45}$ **85.** $\dfrac{17}{60}$ **87.** $\dfrac{15}{8}$ **89.** $\dfrac{7}{10}$

91. a. $300; 304; 276; 216; 124; 0; -156$ **b.** 5 sec **c.** 304 ft **d.**   answers may vary

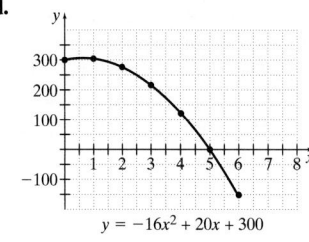

$y = -16x^2 + 20x + 300$

93. $0, \dfrac{1}{2}$ **95.** $0, -15$

Exercise Set 6.6 **1.** width $= x$; length $= x + 4$ **3.** x and $x + 2$ if x is an odd integer **5.** base $= x$; height $= 4x + 1$ **7.** 11 units
9. 15 cm, 13 cm, 70 cm, 22 cm **11.** base $= 16$ mi; height $= 6$ mi **13.** 5 sec **15.** length $= 5$ cm; width $= 6$ cm **17.** 54 diagonals **19.** 10 sides
21. -12 or 11 **23.** slow boat: 8 mph; fast boat: 15 mph **25.** 13 and 7 **27.** 5 in. **29.** 12 mm, 16 mm, 20 mm **31.** 10 km **33.** 36 ft **35.** 6.25 sec
37. 20% **39.** length: 15 mi; width: 8 mi **41.** 105 units **43.** 175 acres **45.** 6.25 million **47.** 1966 **49.** answers may vary **51.** $\dfrac{3}{4}$ **53.** $\dfrac{5}{9}$
55. $\dfrac{9}{10}$ **57.** answers may vary **59.** 3 yd

Chapter 6 Review **1.** $2x - 5$ **3.** $4x(5x + 3)$ **5.** $-2x^2y(4x - 3y)$ **7.** $(x + 1)(5x - 1)$ **9.** $(2x - 1)(3x + 5)$ **11.** $(x + 4)(x + 2)$ **13.** prime
15. $(x + 4)(x - 2)$ **17.** $(x + 5y)(x + 3y)$ **19.** $2(3 - x)(12 + x)$ **21.** $(2x - 1)(x + 6)$ **23.** $(2x + 3)(2x - 1)$ **25.** $(6x - y)(x - 4y)$
27. $(2x + 3y)(x - 13y)$ **29.** $(6x + 5y)(3x - 4y)$ **31.** $(2x + 3)(2x - 3)$ **33.** prime **35.** $(2x + 3)(4x^2 - 6x + 9)$
37. $2(3 - xy)(9 + 3xy + x^2y^2)$ **39.** $(4x^2 + 1)(2x + 1)(2x - 1)$ **41.** $(2x - 3)(x + 4)$ **43.** $(x - 1)(x + 3)$ **45.** $2xy(2x - 3y)$
47. $(5x + 3)(25x^2 - 15x + 9)$ **49.** $(x + 7 + y)(x + 7 - y)$ **51.** $-6, 2$ **53.** $-\dfrac{1}{5}, -3$ **55.** $-4, 6$ **57.** $2, 8$ **59.** $-\dfrac{2}{7}, \dfrac{3}{8}$ **61.** $-\dfrac{2}{5}$ **63.** 3

65. $0, -\dfrac{7}{4}, 3$ **67.** 36 yd **69. a.** 17.5 sec and 10 sec; The rocket reaches a height of 2800 ft on its way up and on its way back down. **b.** 27.5 sec

Chapter 6 Test **1.** $(y - 12)(y + 4)$ **2.** prime **3.** $3x(3x + 1)(x + 4)$ **4.** $(3a - 7)(a + b)$ **5.** $(3x - 2)(x - 1)$ **6.** $(x + 12y)(x + 2y)$
7. $5(6 + x)(6 - x)$ **8.** $(6t + 5)(t - 1)$ **9.** $(y + 2)(y - 2)(x - 7)$ **10.** $x(1 + x^2)(1 + x)(1 - x)$ **11.** $-xy(y^2 + x^2)$
12. $(4x - 1)(16x^2 + 4x + 1)$ **13.** $8(y - 2)(y^2 + 2y + 4)$ **14.** $-7, 2$ **15.** $-7, 1$ **16.** $0, \dfrac{3}{2}, -\dfrac{4}{3}$ **17.** $0, 3, -3$ **18.** $-3, 5$ **19.** $0, \dfrac{5}{2}$ **20.** 17 ft
21. 8 and 9 **22.** 7 sec

Chapter 6 Cumulative Review **1. a.** $9 \le 11$ **b.** $8 > 1$ **c.** $3 \ne 4$; Sec. 1.2, Ex. 3 **3. a.** $\dfrac{6}{7}$ **b.** $\dfrac{11}{27}$ **c.** $\dfrac{22}{5}$; Sec. 1.3, Ex. 2 **5.** $\dfrac{14}{3}$; Sec. 1.4, Ex. 5
7. a. -12 **b.** -1; Sec. 1.5, Ex. 7 **9. a.** -24 **b.** -2 **c.** 50; Sec. 1.7, Ex. 1 **11. a.** $4x$ **b.** $11y^2$ **c.** $8x^2 - x$; Sec. 2.1, Ex. 3
13. 140; Sec. 2.2, Ex. 7 **15.** -11; Sec. 2.2, Ex. 6 **17.** $\dfrac{16}{3}$; Sec. 2.3, Ex. 2 **19.** shorter: 2 ft; longer: 8 ft; Sec. 2.4, Ex. 3
21.

$y = -\dfrac{1}{3}x$

$(-3, 1)$

$(0, 0)$

$(6, -2)$

; Sec. 3.3, Ex. 5 **23.** yes; Sec. 3.4, Ex. 5 **25. a.** 250 **b.** 1; Sec. 5.1, Ex. 2 **27. a.** 2 **b.** 5 **c.** 1 **d.** 6 **e.** 0; Sec. 5.2, Ex. 1
29. $12x^3 - 12x^2 - 9x + 2$; Sec. 5.2, Ex. 8 **31.** $6x^2 - 11x - 10$; Sec. 5.3, Ex. 5 **33.** $9y^2 + 6y + 1$; Sec. 5.4, Ex. 4
35. a. $\dfrac{1}{9}$ **b.** $\dfrac{2}{x^3}$ **c.** $\dfrac{3}{4}$ **d.** $\dfrac{1}{16}$; Sec. 5.5, Ex. 1 **37. a.** 3.67×10^8 **b.** 3.0×10^{-6} **c.** 2.052×10^{10}
d. 8.5×10^{-4}; Sec. 5.5, Ex. 5 **39.** $x + 4$; Sec. 5.6, Ex. 4 **41. a.** x^3 **b.** y; Sec. 6.1, Ex. 2
43. $(x + 3)(x + 4)$; Sec. 6.2, Ex. 1 **45.** $(4x - 1)(2x - 5)$; Sec. 6.3, Ex. 2 **47.** $(5a + 3b)(5a - 3b)$; Sec. 6.4, Ex. 2b
49. $3, -1$; Sec. 6.5, Ex. 1

CHAPTER 7   RATIONAL EXPRESSIONS

Graphing Calculator Explorations   1. $\{x \mid x \text{ is a real number and } x \neq -2, x \neq 2\}$

3. $\left\{x \mid x \text{ is a real number and } x \neq -4, x \neq \dfrac{1}{2}\right\}$

Exercise 7.1   1. $\dfrac{10}{3}, -8, -\dfrac{7}{3}$ **3.** $-\dfrac{17}{18}, \dfrac{2}{7}, -\dfrac{3}{8}$ **5.** $\{x \mid x \text{ is a real number}\}$ **7.** $\{t \mid t \text{ is a real number and } t \neq 0\}$ **9.** $\{x \mid x \text{ is a real number and } x \neq 7\}$

11. $\{x \mid x \text{ is a real number and } x \neq 2, x \neq -2\}$ **13.** $\left\{x \mid x \text{ is a real number and } x \neq 5, x \neq -\dfrac{1}{3}\right\}$ **15.** $\{x \mid x \text{ is a real number and } x \neq -2, x \neq 0, x \neq 1\}$

17. $\dfrac{5x^2}{9}$ **19.** $\dfrac{x^4}{2y^2}$ **21.** 1 **23.** $-\dfrac{1}{1+x}$ **25.** $\dfrac{4}{3}$ **27.** -2 **29.** $\dfrac{x+1}{x-3}$ **31.** $\dfrac{2(x+3)}{5x+2}$ **33.** $\dfrac{6}{x}$ **35.** $\dfrac{x+1}{x^2+1}$ **37.** $\dfrac{1}{2(q-1)}$ **39.** $x-4$

41. $-x^2 - 5x - 25$ **43.** $\dfrac{4x^2+6x+9}{2}$ **45.** 400 mg   **47.** no; $B \approx 24$ **49. a.** \$37.5 million   **b.** \$85.7 million   **c.** \$48.2 million   **51.** $\dfrac{3}{11}$ **53.** $\dfrac{4}{3}$

55. $\dfrac{50}{99}$ **57.** $\dfrac{117}{40}$ **59.** answers may vary   **61.** a, c; answers may vary   **63.** $4, 2, 1, \dfrac{1}{2}, \dfrac{1}{4}; -\dfrac{1}{4}, -\dfrac{1}{2}, -1, -2, -4;$

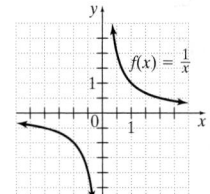

65. a. \$200 million   **b.** \$500 million   **c.** \$300 million

67.   **69.**

Mental Math   1. $\dfrac{2x}{3y}$ **3.** $\dfrac{5y^2}{7x^2}$ **5.** $\dfrac{9}{5}$

Exercise Set 7.2   1. x^4 **3.** $-\dfrac{b^2}{6}$ **5.** $\dfrac{4x}{7(x+2)}$ **7.** $\dfrac{x^2}{10}$ **9.** $\dfrac{1}{3}$ **11.** 1   **13.** $\dfrac{x+5}{x}$ **15.** $\dfrac{2}{9x^2(x-5)}$ sq. ft   **17.** x^4 **19.** $\dfrac{12}{y^6}$ **21.** $x(x+4)$

23. $\dfrac{3(x+1)}{x^3(x-1)}$ **25.** $m^2 - n^2$ **27.** $-\dfrac{x+2}{x-3}$ **29.** $-\dfrac{x+2}{x-3}$ **31.** answers may vary   **33.** $\dfrac{1}{6b^4}$ **35.** $\dfrac{9}{7x^2y^7}$ **37.** $\dfrac{5}{6}$ **39.** $\dfrac{3x}{8}$ **41.** $\dfrac{3}{2}$ **43.** $\dfrac{3x+4y}{2(x+2y)}$

45. $-2(x+3)$ **47.** $\dfrac{2(x+2)}{x-2}$ **49.** $\dfrac{(a+5)(a+3)}{(a+2)(a+1)}$ **51.** $-\dfrac{1}{x}$ **53.** $\dfrac{2(x+3)}{x-4}$ **55.** $-(x^2+2)$ **57.** -1 **59.** $4x^3(x-3)$ **61.** 1440

63. 411,972 sq. yd   **65.** 73   **67.** 3424.8 mph   **69.** 1   **71.** $-\dfrac{10}{9}$ **73.** $-\dfrac{1}{5}$

75.   **77.** $\dfrac{(a+b)^2}{a-b}$ **79.** $\dfrac{3x+5}{x^2+4}$ **81.** $\dfrac{4}{x-2}$ **83.** $\dfrac{a-b}{6(a^2+ab+b^2)}$ **85.** $\dfrac{x}{2}$ **87.** $\dfrac{5a(2a+b)(3a-2b)}{b^2(a-b)(a+2b)}$

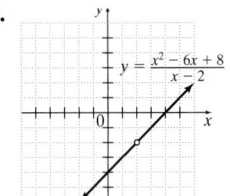

89. 1834.86 euros

Mental Math   1. 1   **3.** $\dfrac{7x}{9}$ **5.** $\dfrac{1}{9}$ **7.** $\dfrac{7-10y}{5}$

Exercise Set 7.3   1. $\dfrac{a+9}{13}$ **3.** $\dfrac{y+10}{3+y}$ **5.** $\dfrac{3m}{n}$ **7.** $\dfrac{5x+7}{x-3}$ **9.** $\dfrac{1}{2}$ **11.** 4   **13.** $x+5$ **15.** $x+4$ **17.** 1   **19.** $\dfrac{12}{x^3}$ **21.** $-\dfrac{5}{x+4}$ **23.** $\dfrac{x-2}{x+y}$

25. 4 **27.** 3 **29.** $\frac{1}{a+5}$ **31.** $\frac{1}{x-6}$ **33.** $\frac{20}{x-2}$m **35.** answers may vary **37.** 33 **39.** $4x^3$ **41.** $8x(x+2)$ **43.** $6(x+1)^2$ **45.** $8-x$ or $x-8$

47. $40x^3(x-1)^2$ **49.** $(2x+1)(2x-1)$ **51.** $(2x-1)(x+4)(x+3)$ **53.** answers may vary **55.** $\frac{6x}{4x^2}$ **57.** $\frac{24b^2}{12ab^2}$ **59.** $\frac{18}{2(x+3)}$ **61.** $\frac{9ab+2b}{5b(a+2)}$

63. $\frac{x^2+x}{(x+4)(x+2)(x+1)}$ **65.** $\frac{18y-2}{30x^2-60}$ **67.** $\frac{15x(x-7)}{3x(2x+1)(x-7)(x-5)}$ **69.** $\frac{29}{21}$ **71.** $-\frac{5}{12}$ **73.** 0,3 **75.** $-5,-1$ **77.** c **79.** b

81. $-\frac{5}{x-2}$ **83.** $\frac{7+x}{x-2}$ **85.** 95,304 Earth days **87.** answers may vary

Mental Math **1.** D **3.** A

Exercise Set 7.4

1. $\frac{5}{x}$ **3.** $\frac{75a+6b^2}{5b}$ **5.** $\frac{6x+5}{2x^2}$ **7.** $\frac{21}{2(x+1)}$ **9.** $\frac{17x+30}{2(x-2)(x+2)}$ **11.** $\frac{35x-6}{4x(x-2)}$ **13.** $\frac{5+10y-y^3}{y^2(2y+1)}$ **15.** answers may vary **17.** $-\frac{2}{x-3}$

19. $-\frac{1}{x^2-1}$ **21.** $\frac{1}{x-2}$ **23.** $\frac{5+2x}{x}$ **25.** $\frac{6x-7}{x-2}$ **27.** $-\frac{y+4}{y+3}$ **29.** $\left(\frac{90x-40}{x}\right)°$ **31.** 2 **33.** $3x^3-4$ **35.** $\frac{x+2}{(x+3)^2}$ **37.** $\frac{9b-4}{5b(b-1)}$

39. $\frac{2+m}{m}$ **41.** $\frac{10}{1-2x}$ **43.** $\frac{15x-1}{(x+1)^2(x-1)}$ **45.** $\frac{x^2-3x-2}{(x-1)^2(x+1)}$ **47.** $\frac{a+2}{2(a+3)}$ **49.** $\frac{x-10}{2(x-2)}$ **51.** $\frac{-3-2y}{(y-1)(y-2)}$ **53.** $\frac{-5x+23}{(x-3)(x-2)}$

55. $\frac{12x-32}{(x+2)(x-2)(x-3)}$ **57.** $\frac{6x+9}{(3x-2)(3x+2)}$ **59.** $\frac{2x^2-2x-46}{(x+1)(x-6)(x-5)}$ **61.** $\frac{2}{3}$ **63.** $-\frac{1}{2},1$ **65.** 1 **67.** -1

69. $\frac{10(y-2)}{y(y-5)}$ft; $\frac{6}{y^2-5y}$sq. ft **71.** 10 **73.** 2 **75.** $\frac{25a}{9(a-2)}$ **77.** $\frac{x+4}{(x-2)(x-1)}$ **79.** answers may vary

81. $\frac{4x^2-15x+6}{(x-2)^2(x+2)(x-3)}$ **83.** $\frac{2(x+1)}{(x-3)(x^2+3x+9)}$ **85.** $\frac{11DA-DA^2-12D}{24(A+12)}$

Calculator Explorations

1.   **3.**   **5.**

Mental Math
1. $x=10$ **3.** $z=36$

Exercise Set 7.5 **1.** 30 **3.** 0 **5.** $-\frac{14}{19}$ **7.** $-5,2$ **9.** 5 **11.** 3 **13.** 0 **15.** $\frac{1}{4}$ **17.** no solution **19.** $-1,12$ **21.** 5 **23.** $-\frac{10}{9}$ **25.** $\frac{11}{14}$

27. no solution **29.** $-\frac{21}{11}$ **31.** 1 **33.** $\frac{9}{2}$ **35.** no solution **37.** $-\frac{9}{4}$ **39.** $-\frac{3}{2},4$ **41.** -2 **43.** $\frac{12}{5}$ **45.** $-2,8$ **47.** $\frac{1}{5}$ **49.** $R=\frac{D}{T}$

51. $y=\frac{3x+6}{5x}$ **53.** $b=-\frac{3a^2+2a-4}{6}$ **55.** $B=\frac{2A}{H}$ **57.** $r=\frac{C}{2\pi}$ **59.** $a=\frac{bc}{c+b}$ **61.** $n=\frac{m^2-3p}{2}$ **63.** $(2,0),(0,-2)$

65. $(-4,0),(-2,0),(3,0),(0,4)$ **67.** answers may vary **69.** $\frac{5x+9}{9x}$ **71.** no solution **73.** $100°,80°$ **75.** $22.5°,67.5°$ **77.** $\frac{17}{4}$

Integrated Review **1.** expression; $\frac{3+2x}{3x}$ **2.** expression; $\frac{18+5a}{6a}$ **3.** equation; 3 **4.** equation; 18 **5.** expression; $\frac{x+1}{x(x-1)}$

6. expression; $\frac{3(x+1)}{x(x-3)}$ **7.** equation; no solution **8.** equation; 1 **9.** expression; 10 **10.** expression; $\frac{z}{3(9z-5)}$ **11.** expression; $\frac{5x+7}{x-3}$

12. expression; $\frac{7p+5}{2p+7}$ **13.** equation; 23 **14.** equation; 5 **15.** expression; $\frac{25a}{9(a-2)}$ **16.** expression; $\frac{4x+5}{(x+1)(x-1)}$

17. expression; $\frac{3x^2+5x+3}{(3x-1)^2}$ **18.** expression; $\frac{2x^2-3x-1}{(2x-5)^2}$ **19.** expression; $\frac{4x-37}{5x}$ **20.** equation; $-\frac{7}{3}$ **21.** equation; $\frac{8}{5}$

22. expression; $\frac{29x-23}{3x}$

Exercise Set 7.6 **1.** 4 **3.** $\frac{50}{9}$ **5.** -3 **7.** $\frac{14}{9}$ **9.** a **11.** 123 lb **13.** 165 cal **15.** $y=21.25$ **17.** $y=5\frac{5}{7}$ft **19.** 2 **21.** -3 **23.** $2\frac{2}{9}$hr

25. $1\frac{1}{2}$min **27.** trip to park rate: r; to park time: $\frac{12}{r}$; return trip rate: r; return time: $\frac{18}{r}-1$; $r=6$ mph **29.** 1st portion: 10 mph; cooldown: 8 mph

31. 360 sq. ft **33.** 2 **35.** \$108.00 **37.** 63 mph **39.** $37\frac{1}{2}$ft **41.** 337 yd/game **43.** 5 **45.** 217 mph **47.** 9 gal **49.** 8 mph **51.** 3 hr **53.** $26\frac{2}{3}$ft

55. 216 other nuts **57.** 20 hr **59.** $1\frac{1}{5}$hr **61.** $5\frac{1}{4}$hr **63.** first pump: 28 min; second pump: 84 min **65.** $m = 3$; upward **67.** $\frac{3}{4}$ **69.** $\frac{6}{5}$
71. 2035 megawatts **73.** answers may vary **75.** 84 yr **77.** 30 yr **79.** 3.75 min

Exercise Set 7.7 **1.** 4 **3.** $\frac{7}{13}$ **5.** $\frac{4}{x}$ **7.** $\frac{9(x-2)}{9x^2+4}$ **9.** $2x + y$ **11.** $\frac{2(x+1)}{2x-1}$ **13.** $\frac{2x+3}{4-9x}$ **15.** $\frac{1}{x^2-2x+4}$ **17.** $\frac{x}{5x-10}$ **19.** $\frac{x-2}{2x-1}$
21. $\frac{x}{2-3x}$ **23.** $-\frac{y}{x+y}$ **25.** $-\frac{2x^3}{y(x-y)}$ **27.** $\frac{2x+1}{y}$ **29.** $\frac{x-3}{9}$ **31.** $\frac{1}{x+2}$ **33.** $\frac{xy^2}{x^2+y^2}$ **35.** $\frac{2b^2+3a}{b(b-a)}$ **37.** $\frac{x}{(x+1)(x-1)}$ **39.** $\frac{1+a}{1-a}$
41. $\frac{x(x+6y)}{2y}$ **43.** $\frac{5a}{2a+4}$ **45.** $5xy^2 + 2x^2y$ **47.** $\frac{xy}{2x+5y}$ **49.** $\frac{x^2y^2}{4}$ **51.** $-9x^3y^4$ **53.** -9 **55.** $\frac{770a}{770-s}$ **57.** a, b **59.** $\frac{1+x}{2+x}$ **61.** $x^2 + x$
63. $\frac{x-3y}{x+3y}$ **65.** $3a^2 + 4a + 4$ **67. a.** $\frac{1}{a+h}$ **b.** $\frac{1}{a}$ **c.** $\frac{\frac{1}{a+h}-\frac{1}{a}}{h}$ **d.** $\frac{-1}{a(a+h)}$ **69. a.** $\frac{3}{a+h+1}$ **b.** $\frac{3}{a+1}$ **c.** $\frac{\frac{3}{a+h+1}-\frac{3}{a+1}}{h}$
d. $\frac{-3}{(a+h+1)(a+1)}$

Chapter 7 Review **1.** $\{x|x \text{ is a real number and } x \neq 5\}$ **3.** $\{x|x \text{ is a real number and } x \neq 0, x \neq -8\}$ **5.** $\frac{x^2}{3}$ **7.** $\frac{9m^2p}{5}$ **9.** $\frac{1}{5}$ **11.** $\frac{1}{x-1}$
13. $\frac{2(x-3)}{x-4}$ **15. a.** \$119 **b.** \$77 **c.** decrease **17.** $-\frac{9x^2}{8}$ **19.** $-\frac{2x(2x+5)}{(x-6)^2}$ **21.** $\frac{4x}{3y}$ **23.** $\frac{2}{3}$ **25.** $\frac{x}{x+6}$ **27.** $\frac{3(x+2)}{3x+y}$ **29.** $-\frac{2(2x+3)}{y-2}$
31. $\frac{5x+2}{3x-1}$ **33.** $\frac{2x+1}{2x^2}$ **35.** $(x-8)(x+8)(x+3)$ **37.** $\frac{3x^2+4x-15}{(x+2)^2(x+3)}$ **39.** $\frac{-2x+10}{(x-3)(x-1)}$ **41.** $\frac{-2x-2}{x+3}$ **43.** $\frac{x-4}{3x}$ **45.** $\frac{x^2+2x-3}{(x+2)^2}$
47. $\frac{29x}{12(x-1)}; \frac{3xy}{5(x-1)}$ **49.** 30 **51.** $-4, 3$ **53.** no solution **55.** 5 **57.** $-6, 1$ **59.** $y = \frac{560-8x}{7}$ **61.** $x = 500$ **63.** no solution **65.** \$33.75
67. 3 **69.** 30 mph; 20 mph **71.** $17\frac{1}{2}$hr **73.** $x = 20$ **75.** $x = 15$ **77.** $-\frac{7}{18y}$ **79.** $\frac{6}{7}$ **81.** $b + a$ **83.** $-\frac{x^2+9}{6x}$ **85.** $\frac{xy+1}{x}$ **87.** $\frac{1}{y^2-1}$

Chapter 7 Test **1.** $\{x|x \text{ is a real number and } x \neq 1\}$ **2.** $\{x|x \text{ is a real number and } x \neq -3, x \neq -1\}$ **3.** $\frac{3}{5}$ **4.** $\frac{1}{x+6}$ **5.** $\frac{1}{x^2-3x+9}$
6. $\frac{2m(m+2)}{m-2}$ **7.** $\frac{a+2}{a+5}$ **8.** $-\frac{1}{x+y}$ **9.** 15 **10.** $\frac{y-2}{4}$ **11.** $-\frac{1}{2x+5}$ **12.** $\frac{3a-4}{(a-3)(a+2)}$ **13.** $\frac{3}{x-1}$ **14.** $\frac{2(x+5)}{x(y+5)}$
15. $\frac{x^2+2x+35}{(x+9)(x+2)(x-5)}$ **16.** $\frac{30}{11}$ **17.** -6 **18.** no solution **19.** $-2, 5$ **20.** $\frac{xz}{2y}$ **21.** $\frac{5y^2-1}{y+2}$ **22.** 18 bulbs **23.** 5 or 1
24. 30 mph **25.** $6\frac{2}{3}$hr **26.** $x = 12$ **27. a.** \$115 **b.** \$103

Chapter 7 Cumulative Review **1. a.** $\frac{15}{x} = 4$ **b.** $12 - 3 = x$ **c.** $4x + 17 = 21$; Sec. 1.4, Ex. 9 **3.** amount at 7%: \$12,500; amount at 9%: \$7500;
Sec. 2.6, Ex. 5 **5.** ; Sec. 3.3, Ex. 2 **7. a.** 4^7 **b.** x^{10} **c.** y^4 **d.** y^{12} **e.** $(-5)^{15}$; **f.** a^2b^2 Sec. 5.1, Ex. 3
9. $12z^5 + 3z^4 - 13z^3 - 11z$; Sec. 5.2, Ex. 9 **11.** $27a^3 + 27a^2b + 9ab^2 + b^3$; Sec. 5.3, Ex. 8
13. a. $t^2 + 4t + 4$ **b.** $p^2 - 2qp + q^2$ **c.** $4x^2 + 20x + 25$
d. $x^4 - 14x^2y + 49y^2$; Sec. 5.4, Ex. 5 **15. a.** x^3 **b.** 81 **c.** $\frac{q^9}{p^4}$ **d.** $\frac{32}{125}$; Sec. 5.5, Ex. 2
17. $4x^2 - 4x + 6 + \frac{-11}{2x+3}$; Sec. 5.6, Ex. 6 **19. a.** 4 **b.** 1 **c.** 3; Sec. 6.1, Ex. 1
21. $-3a(3a^4 - 6a + 1)$; Sec. 6.1, Ex. 5 **23.** $3(m+2)(m-10)$; Sec. 6.2, Ex. 7 **25.** $(3x+2)(x+3)$; Sec. 6.3, Ex. 1 **27.** $(x+6)^2$; Sec. 6.3, Ex. 7
29. prime polynomial; Sec. 6.4, Ex. 5 **31.** $(x+2)(x^2-2x+4)$; Sec. 6.4, Ex. 6 **33.** $(2x+3)(x+1)(x-1)$; Ch. 6 Int. Rev., Ex. 2
35. $3(2m+n)(2m-n)$; Integrated Review, Ex. 3 **37.** $-\frac{1}{2}, 4$; Sec. 6.5, Ex. 3 **39.** $(1, 0), (4, 0)$; Sec. 6.5, Ex. 8
41. base: 6 m; height: 10 m; Sec. 6.6, Ex. 3 **43.** $-\frac{2(3+x)}{x+1}$; Sec. 7.1, Ex. 4 **45.** $\frac{2}{x(x+1)}$; Sec. 7.2, Ex. 6 **47.** $\frac{1-2x}{4-2x}$; Sec. 7.7, Ex. 4

CHAPTER 8 FUNCTIONS AND GRAPHS

Mental Math   1. $m = -4$, y-inter. $(0, 12)$ **3.** $m = 5$, y-inter. $(0, 0)$ **5.** $m = \dfrac{1}{2}$, y-inter. $(0, 6)$ **7.** parallel   **9.** neither

Exercise Set 8.1   1.   **3.**   **5.**   **7.**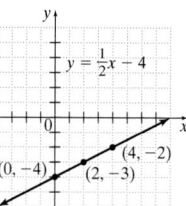

9. C   **11.** D   **13.** $f(x) = x - 3$ **15.** $f(x) = \dfrac{x}{5}$ **17.** $f(x) = \dfrac{1}{2}x + 3$ **19.** $f(x) = \dfrac{1}{2}x - 2$

         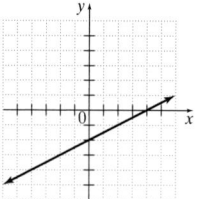

21. answers may vary   **23.** $f(x) = 3x - 6$ **25.** $f(x) = -2x + 1$ **27.** $f(x) = \dfrac{1}{3}x - 7$ **29.** $f(x) = 4x - 4$ **31.** $f(x) = -3x + 1$

33. $f(x) = -\dfrac{3}{2}x - 6$ **35.**   **37.**   **39.**   **41.**

43.   **45.**   **47.**   **49.**   **51.**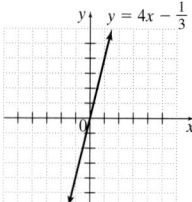

53. $2x - y = -7$; $f(x) = 2x + 7$ **55.** $2x + 7y = -42$; $f(x) = -\dfrac{2}{7}x - 6$ **57.** $x + 2y = 2$; $f(x) = -\dfrac{1}{2}x + 1$

59. $8x - y = 47$; $f(x) = 8x - 47$ **61.** 9   **63.** 2   **65.** -1 **67. a.** $(0, 500)$; if no tables are produced, 500 chairs can be produced
b. $(750, 0)$; if no chairs are produced, 750 tables can be produced   **c.** 466 chairs

69. a. \$64   **b.**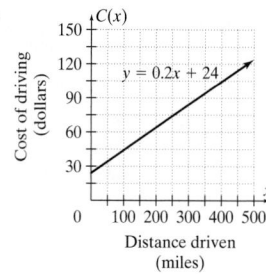

c. The line moves upward from left to right.   **71. a.** \$1921.88 **b.** 2012   **c.** answers may vary
73. 163.2 mg   **75. a.** per capita consumption of poultry was 95.99 lb in 2000   **b.** 99.37 lb
77. a. $f(x) = 4834x + 133{,}300$ **b.** \$176,806   **c.** every year, the median price of a home
increases by \$4834.   **79. a.** $f(x) = 29.5x + 757$ **b.** 875 thousand people   **81.** $-4x + y = 4$
83. $2x + y = -23$ **85.** $3x - 2y = -13$ **87.**

89.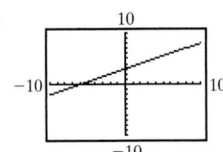

91. a. a line parallel to $y = -4x$ but with x-intercept $(0, 2)$

b. a line parallel to $y = -4x$ but with x-intercept $(0, -5)$

Graphing Calculator Explorations 1.   **3.**   **5.**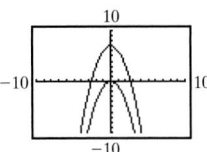

Exercise Set 8.2 1. 0 **3.** -4 **5.** 3 **7.** $(1, -10)$ **9.** $(4, 56)$ **11.** $f(-1) = -2$ **13.** $g(2) = 0$ **15.** $-4, 0$ **17.** 3

19.   **21.**   **23.**   **25.**

27.   **29.**   **31.**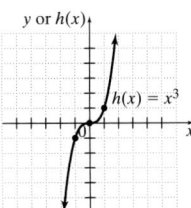

33. d **35.** c **37.** infinite number **39.** 25π sq. cm
41. 2744 cu. in. **43.** $-4x - 6$ **45.** $14x - 8$ **47.** b **49.** c
51. 1991 **53.** answers may vary

55.   **57.**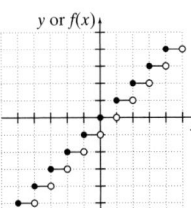

Integrated Review 1. $m = 3; (0, -5)$ **2.** $m = \dfrac{5}{2}; \left(0, -\dfrac{7}{2}\right)$ **3.** parallel **4.** perpendicular **5.** $f(x) = -x + 7$ **6.** $y = -\dfrac{3}{8}x - \dfrac{29}{4}$

7. $f(x) = 3x - 2$ **8.** $f(x) = -\dfrac{5}{4}x + 4$ **9.** $f(x) = \dfrac{1}{4}x - \dfrac{7}{2}$ **10.** $f(x) = -\dfrac{5}{2}x - \dfrac{5}{2}$

11.   **12.**   **13.**   **14.**   **15.**

16.   **17.**   **18.**   **19.**   **20.**

21. **22.** **23.** **24.** **25.**

26.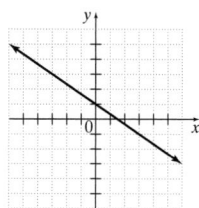

Mental Math   1. C   **3.** F   **5.** D

Exercise Set 8.3   1. a. $3x - 6$ **b.** $-x - 8$ **c.** $2x^2 - 13x - 7$ **d.** $\dfrac{x - 7}{2x + 1}$, where $x \neq -\dfrac{1}{2}$ **3. a.** $x^2 + 5x + 1$ **b.** $x^2 - 5x + 1$ **c.** $5x^3 + 5x$

d. $\dfrac{x^2 + 1}{5x}$, where $x \neq 0$ **5. a.** $|x| + x + 5$ **b.** $|x| - x - 5$ **c.** $x|x| + 5|x|$ **d.** $\dfrac{|x|}{x + 5}$, where $x \neq -5$ **7. a.** $5x^2 - 3x$ **b.** $-5x^2 - 3x$

c. $-15x^3$ **d.** $-\dfrac{3}{5x}$, where $x \neq 0$ **9.** 42   **11.** -18 **13.** 0   **15.** $(f \circ g)(x) = 25x^2 + 1$; $(g \circ f)(x) = 5x^2 + 5$

17. $(f \circ g)(x) = 2x + 11$; $(g \circ f)(x) = 2x + 4$ **19.** $(f \circ g)(x) = -8x^3 - 2x - 2$; $(g \circ f)(x) = -2x^3 - 2x + 4$

21. $(f \circ g)(x) = |-5x + 2|$; $(g \circ f)(x) = -5|x| + 2$ **23.** $H(x) = (g \circ h)(x)$ **25.** $F(x) = (h \circ f)(x)$ **27.** $G(x) = (f \circ g)(x)$

29. answers may vary; for example $g(x) = x + 2$ and $f(x) = x^2$ **31.** answers may vary; for example, $g(x) = x + 5$ and $f(x) = |x| + 2$

33. answers may vary; for example, $g(x) = 2x - 3$ and $f(x) = \dfrac{1}{x}$ **35.** $y = x - 2$ **37.** $y = \dfrac{x}{3}$ **39.** $y = -\dfrac{x + 7}{2}$ **41.** 6   **43.** 4   **45.** 4

47. -1 **49.** answers may vary   **51.** $P(x) = R(x) - C(x)$

Mental Math   1. direct   **3.** joint   **5.** inverse   **7.** direct

Exercise Set 8.4   1. $y = kx$ **3.** $a = \dfrac{k}{b}$ **5.** $y = kxz$ **7.** $y = \dfrac{k}{x^3}$ **9.** $y = \dfrac{kx}{p^2}$ **11.** $k = \dfrac{1}{5}$; $y = \dfrac{1}{5}x$ **13.** $k = \dfrac{3}{2}$; $y = \dfrac{3}{2}x$ **15.** $k = 14$; $y = 14x$

17. $k = 0.25$; $y = 0.25x$ **19.** 4.05 lb   **21.** $P = 566{,}222$ tons   **23.** $k = 30$; $y = \dfrac{30}{x}$ **25.** $k = 700$; $y = \dfrac{700}{x}$ **27.** $k = 2$; $y = \dfrac{2}{x}$

29. $k = 0.14$; $y = \dfrac{0.14}{x}$ **31.** 54 mph   **33.** 72 amps   **35.** divided by 4   **37.** $x = kyz$ **39.** $r = kst^3$ **41.** $k = \dfrac{1}{3}$; $y = \dfrac{1}{3}x^3$ **43.** $k = 0.2$; $y = 0.2\sqrt{x}$

45. $k = 1.3$; $y = \dfrac{1.3}{x^2}$ **47.** $k = 3$; $y = 3xz^3$ **49.** 22.5 tons   **51.** 15π cu. in.   **53.** 90 hp   **55.** 800 millibars   **57.** $C = 12\pi$ cm; $A = 36\pi$ sq. cm

59. $C = 14\pi$ m; $A = 49\pi$ sq. m   **61.** -3 **63.** 0   **65.** $-\dfrac{1}{5}$ **67.** $\dfrac{25}{121}$ **69.** multiplied by 2   **71.** multiplied by 4

73.   **75.**

Chapter 8 Review   1.   **3.**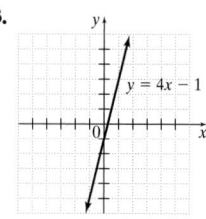

5. C   **7.** B   **9.** $m = \dfrac{2}{5}$, y-inter. $\left(0, -\dfrac{4}{3}\right)$ **11.** $f(x) = -\dfrac{2}{3}x + 4$

13. $f(x) = -2x - 2$ **15.** $f(x) = \dfrac{3}{4}x + \dfrac{7}{2}$ **17. a.** $f(x) = \dfrac{4}{3}x + 65$

b. 90%   **19.** neither   **21.** $87   **23.** 0   **25.** $-2, 4$

27. linear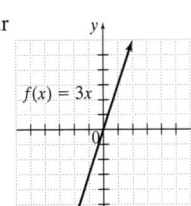

$f(x) = 3x$

29. nonlinear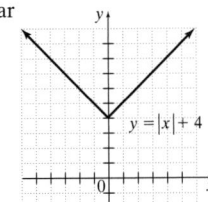

$y = |x| + 4$

31. linear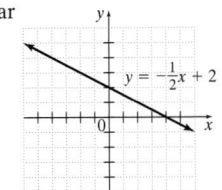

$y = -\frac{1}{2}x + 2$

33. linear $y = -1.36x$

35.

$f(x) = (x - 2)^2$

37. $3x - 4$ **39.** $2x^2 - 9x - 5$ **41.** $x^2 + 2x - 1$ **43.** 18   **45.** -2 **47.** 9   **49.** 3.125 cu. ft

Chapter 8 Test   **1.** 3   **2.** -5 **3.** 2, -2 **4.** 0   **5.**

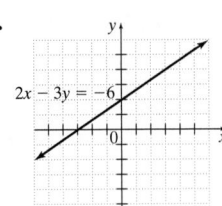

$2x - 3y = -6$

6.

$y = \frac{2}{3}x$

7.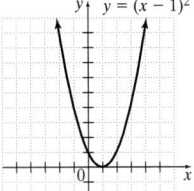

$y = (x - 1)^2$

8.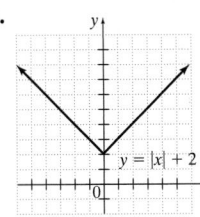

$y = |x| + 2$

9. $f(x) = -\frac{1}{2}x$ **10.** $f(x) = -\frac{1}{3}x + \frac{5}{3}$ **11.** $f(x) = -\frac{1}{2}x - \frac{1}{2}$ **12. a.** \$22,892   **b.** \$28,016   **c.** 2008

d. The average yearly earnings for high school graduates increases \$732 per year.

e. The average yearly earnings for a high school graduate in 1996 was \$21,428.   **13.** $x^2 - x + 7$ **14.** $x^3 - 2x^2 + 5x - 10$

15. 5   **16.** $x - 7$ **17.** $x^2 - 6x - 2$ **18.** 16   **19.** 9   **20.** 256 ft

Chapter 8 Cumulative Review   **1.** 66; Sec. 1.4, Ex. 4   **3. a.** $\frac{1}{2}$ **b.** 9; Sec. 1.6, Ex. 6   **5.** -11; Sec. 2.2, Ex. 6   **7.** $\frac{16}{3}$; Sec. 2.3, Ex. 2

9. $x = \dfrac{y - b}{m}$; Sec. 2.5, Ex. 6   **11.** $(-\infty, 4]$; Sec. 2.7, Ex. 5   **13.** $\frac{3}{5}$; Sec. 3.4, Ex. 6   **15. a.** 1   **b.** 1   **c.** -3; Sec. 3.7, Ex. 7   **17.** $(4, 2)$; Sec. 4.2, Ex. 1

19. $\left(-\dfrac{5}{4}, -\dfrac{5}{2}\right)$; Sec. 4.3, Ex. 6   **21.** $x + 4$; Sec. 5.6, Ex. 4   **23. a.** $6(t + 3)$ **b.** $y^5(1 - y^2)$; Sec. 6.1, Ex. 4   **25.** $(x - 2)(x + 6)$; Sec. 6.2, Ex. 3

27. $(2y - 3y)(5x + y)$; Sec. 6.3, Ex. 4   **29.** $(x + 2)(x^2 - 2x + 4)$; Sec. 6.4, Ex. 6   **31.** 4, 5; Sec. 6.5, Ex. 2   **33. a.** $\dfrac{3y^4}{x}$ **b.** $\dfrac{1}{5x - 1}$; Sec. 7.1, Ex. 2

35. $-\dfrac{3(x + 1)}{5x(2x - 3)}$; Sec. 7.2, Ex. 3   **37.** $3x - 5$; Sec. 7.3, Ex. 3   **39.** $-3, -2$; Sec. 7.5, Ex. 3   **41.** $f(x) = \dfrac{5}{8}x - \dfrac{5}{2}$; Sec. 8.1, Ex. 5

CHAPTER 9   INEQUALITIES AND ABSOLUTE VALUE

Exercise Set 9.1   **1.** $\{2, 3, 4, 5, 6, 7\}$ **3.** $\{4, 6\}$ **5.** $\{\ldots, -2, -1, 0, 1, \ldots\}$ **7.** $\{5, 7\}$ **9.** $\{x \mid x$ is an odd integer or $x = 2$ or $x = 4\}$

11. $\{2, 4\}$ **13.** $\xleftrightarrow{\quad(\underset{-2}{\quad}\overline{\quad}\underset{5}{\quad})\quad}$; $(-2, 5)$ **15.** $\xleftrightarrow{\quad[\underset{6}{\quad}\quad}$; $[6, \infty)$ **17.** $\xleftrightarrow{\quad]\underset{-3}{\quad}\quad}$; $(-\infty, -3]$ **19.** $\xleftrightarrow{\quad(\underset{11}{\quad}\underset{17}{\quad})\quad}$; $(11, 17)$

21. $\xleftrightarrow{\quad[\underset{1}{\quad}\underset{4}{\quad}]\quad}$; $[1, 4]$ **23.** $\xleftrightarrow{\quad[\underset{-3}{\quad}\underset{\frac{3}{2}}{\quad}]\quad}$; $\left[-3, \dfrac{3}{2}\right]$ **25.** $\xleftrightarrow{\quad[\underset{-21}{\quad}\underset{-9}{\quad}]\quad}$; $[-21, -9]$ **27.** $\xleftrightarrow{\quad)\underset{-1}{\quad}\quad(\underset{0}{\quad}\quad}$; $(-\infty, -1) \cup (0, \infty)$

29. $\xleftrightarrow{\quad[\underset{2}{\quad}\quad}$; $[2, \infty)$ **31.** $\xleftrightarrow{\quad\underset{0}{\quad}\quad}$; $(-\infty, \infty)$ **33.** answers may vary   **35.** $\xleftrightarrow{\quad(\underset{-1}{\quad}\underset{2}{\quad})\quad}$; $(-1, 2)$

37. $\xleftrightarrow{\quad\underset{0}{\quad}\quad}$; $(-\infty, \infty)$ **39.** $\xleftrightarrow{\quad[\underset{-1}{\quad}\quad}$; $[-1, \infty)$ **41.** $\xleftrightarrow{\quad[\underset{-5}{\quad}\quad}$; $[-5, \infty)$ **43.** $\xleftrightarrow{\quad[\underset{\frac{3}{2}}{\quad}\underset{6}{\quad}]\quad}$; $\left[\dfrac{3}{2}, 6\right]$

45. $;\left(\dfrac{5}{4},\dfrac{11}{4}\right)$ **47.** $;\varnothing$ **49.** $;(-7,\infty)$ **51.** $;\left(-5,\dfrac{5}{2}\right)$

53. $;\left(0,\dfrac{14}{3}\right]$ **55.** $;(-\infty,-3]$ **57.** $;(-\infty,1]\cup\left(\dfrac{29}{7},\infty\right)$ **59.** $;\varnothing$

61. $;\left[-\dfrac{1}{2},\dfrac{3}{2}\right)$ **63.** $;\left(-\dfrac{4}{3},\dfrac{7}{3}\right)$ **65.** $;(6,12)$ **67.** -12 **69.** -4 **71.** $-7,7$ **73.** 0

75. $-20.2^\circ\le F\le 95^\circ$ **77.** $67\le \text{final score}\le 94$ **79.** $1994\text{--}1995;1998\text{--}1999$ **81.** $;(6,\infty)$ **83.** $;[3,7]$

85. $;(-\infty,-1)$

Mental Math **1.** 7 **3.** -5 **5.** -6 **7.** 12

Exercise Set 9.2 **1.** $7,-7$ **3.** $4.2,-4.2$ **5.** $7,-2$ **7.** $8,4$ **9.** $5,-5$ **11.** $3,-3$ **13.** 0 **15.** $\varnothing$ **17.** $\dfrac{1}{5}$ **19.** $|x|=5$ **21.** $9,-\dfrac{1}{2}$ **23.** $-\dfrac{5}{2}$

25. answers may vary **27.** $4,-4$ **29.** 0 **31.** $\varnothing$ **33.** $0,\dfrac{14}{3}$ **35.** $2,-2$ **37.** $\varnothing$ **39.** $7,-1$ **41.** $\varnothing$ **43.** $\varnothing$ **45.** $-\dfrac{1}{8}$ **47.** $\dfrac{1}{2},-\dfrac{5}{6}$ **49.** $2,-\dfrac{12}{5}$

51. $3,-2$ **53.** $-8,\dfrac{2}{3}$ **55.** $\varnothing$ **57.** 4 **59.** $13,-8$ **61.** $3,-3$ **63.** $8,-7$ **65.** $2,3$ **67.** $2,-\dfrac{10}{3}$ **69.** $\dfrac{3}{2}$ **71.** $\varnothing$ **73.** answers may vary **75.** 33%

77. 38.4 lb **79.** answers may vary **81.** no solution **83.** $|x|=2$ **85.** $|2x-1|=4$ **87. a.** $c=0$ **b.** c is a negative number

c. c is a positive number

Mental Math **1.** D **3.** C **5.** A

Exercise Set 9.3 **1.** $;[-4,4]$ **3.** $;(1,5)$ **5.** $;(-5,-1)$ **7.** $;[-10,3]$

9. $;[-5,5]$ **11.** $;\varnothing$ **13.** $;[0,12]$ **15.** $;(-\infty,-3)\cup(3,\infty)$

17. $;(-\infty,-24]\cup[4,\infty)$ **19.** $;(-\infty,-4)\cup(4,\infty)$ **21.** $;(-\infty,\infty)$

23. $;\left(-\infty,\dfrac{2}{3}\right)\cup(2,\infty)$ **25.** $;\{0\}$ **27.** $;\left(-\infty,-\dfrac{3}{8}\right)\cup\left(-\dfrac{3}{8},\infty\right)$

29. $;[-2,2]$ **31.** $;(-\infty,-1)\cup(1,\infty)$ **33.** $;(-5,11)$

35. $;(-\infty,4)\cup(6,\infty)$ **37.** $;\varnothing$ **39.** $;(-\infty,\infty)$ **41.** $;[-2,9]$

43. $;(-\infty,-11]\cup[1,\infty)$ **45.** $;(-\infty,0)\cup(0,\infty)$ **47.** $;(-\infty,\infty)$

49. $;\left[-\dfrac{1}{2},1\right]$ **51.** $;(-\infty,-3)\cup(0,\infty)$ **53.** $;\varnothing$ **55.** $;(-\infty,\infty)$

57. $;\left(-\dfrac{2}{3},0\right)$ **59.** $;(-\infty,-12)\cup(0,\infty)$ **61.** $;[-1,8]$ **63.** $;\left[-\dfrac{23}{8},\dfrac{17}{8}\right]$

65. $(-2,5)$ **67.** $5,-2$ **69.** $(-\infty,-7]\cup[17,\infty)$ **71.** $-\dfrac{9}{4}$ **73.** $(-2,1)$ **75.** $2,\dfrac{4}{3}$ **77.** $\varnothing$ **79.** $\dfrac{19}{2},-\dfrac{17}{2}$

81. $\left(-\infty,-\dfrac{25}{3}\right)\cup\left(\dfrac{35}{3},\infty\right)$ **83.** $\dfrac{1}{6}$ **85.** 0 **87.** $\dfrac{1}{3}$ **89.** -1.5 **91.** 0 **93.** $|x|<7$ **95.** $|x|\le 5$ **97.** answers may vary

99. $3.45<x<3.55$

Integrated Review **1.** $(-5,7)$ **2.** $(-\infty,\infty)$ **3.** $1,\dfrac{1}{2}$ **4.** $(-3,2)$ **5.** $(-\infty,-1]\cup[1,\infty)$

6. $-18,-\dfrac{4}{3}$ **7.** $\left[-\dfrac{2}{3},4\right]$ **8.** $\varnothing$ **9.** $(-\infty,1]$ **10.** $(-\infty,0]$ **11.** $[0,6]$ **12.** $-\dfrac{3}{2},1$ **13.** B **14.** E **15.** A

16. C **17.** D

Mental Math **1.** yes **3.** yes **5.** yes **7.** no

Exercise Set 9.4   1. no; yes   **3.** no; no   **5.** no; yes

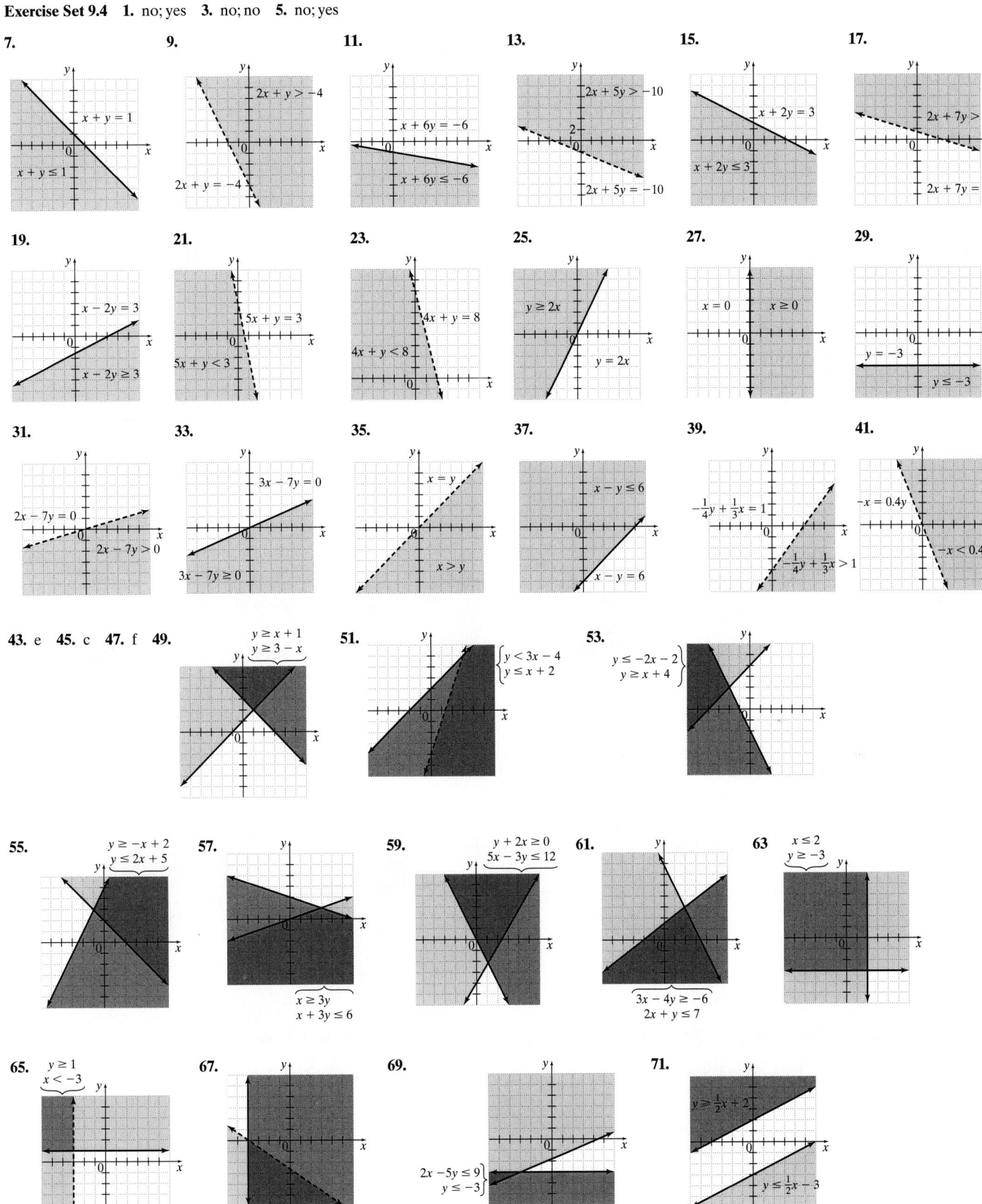

7. $x + y = 1$ $x + y \le 1$

9. $2x + y > -4$ $2x + y = -4$

11. $x + 6y = -6$ $x + 6y \le -6$

13. $2x + 5y > -10$ $2x + 5y = -10$

15. $x + 2y = 3$ $x + 2y \le 3$

17. $2x + 7y > 5$ $2x + 7y = 5$

19. $x - 2y = 3$ $x - 2y \ge 3$

21. $5x + y = 3$ $5x + y < 3$

23. $4x + y = 8$ $4x + y < 8$

25. $y \ge 2x$ $y = 2x$

27. $x = 0$ $x \ge 0$

29. $y = -3$ $y \le -3$

31. $2x - 7y = 0$ $2x - 7y > 0$

33. $3x - 7y = 0$ $3x - 7y \ge 0$

35. $x = y$ $x > y$

37. $x - y \le 6$ $x - y = 6$

39. $-\frac{1}{4}y + \frac{1}{3}x = 1$ $-\frac{1}{4}y + \frac{1}{3}x > 1$

41. $-x = 0.4y$ $-x < 0.4y$

43. e   **45.** c   **47.** f   **49.** $\begin{cases} y \ge x + 1 \\ y \ge 3 - x \end{cases}$

51. $\begin{cases} y < 3x - 4 \\ y \le x + 2 \end{cases}$

53. $\begin{cases} y \le -2x - 2 \\ y \ge x + 4 \end{cases}$

55. $\begin{cases} y \ge -x + 2 \\ y \le 2x + 5 \end{cases}$

57. $x \ge 3y$ $x + 3y \le 6$

59. $\begin{cases} y + 2x \ge 0 \\ 5x - 3y \le 12 \end{cases}$

61. $3x - 4y \ge -6$ $2x + y \le 7$

63. $\begin{cases} x \le 2 \\ y \ge -3 \end{cases}$

65. $\begin{cases} y \ge 1 \\ x < -3 \end{cases}$

67. $2x + 3y < -8$ $x \ge -4$

69. $\begin{cases} 2x - 5y \le 9 \\ y \le -3 \end{cases}$

71. $y > \frac{1}{2}x + 2$ $y \le \frac{1}{2}x - 3$

73. 16 **75.** $36x^2$ **77.** $100y^6$ **79.** C **81.** D **83.** answers may vary **85.** $x + y \geq 13$ **87.** answers may vary

89.

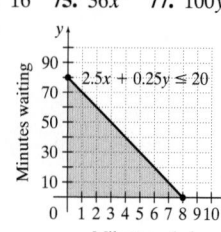

91. answers may vary **93.**

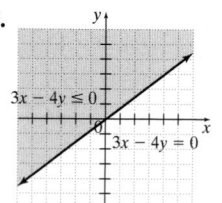

Chapter 9 Review **1.** $\left[2, \frac{5}{2}\right]$ **3.** $\left(\frac{1}{8}, 2\right)$ **5.** $\left(\frac{7}{8}, \frac{27}{20}\right]$ **7.** $(-5, 2]$ **9.** $\left(\frac{11}{3}, \infty\right)$ **11.** $16, -2$ **13.** $0, -9$ **15.** $2, -\frac{2}{3}$ **17.** $\varnothing$ **19.** $3, -3$

21. $5, -\frac{1}{3}$ **23.** $7, -\frac{8}{5}$ **25.** , $\left(-\frac{8}{5}, 2\right)$ **27.** , $(-\infty, -3) \cup (3, \infty)$ **29.** ⟵⟶, $\varnothing$

31. , $\left(-\infty, -\frac{22}{15}\right] \cup \left[\frac{6}{5}, \infty\right)$ **33.** , $(-\infty, -27) \cup (-9, \infty)$ **35.**

37.

39.

41.

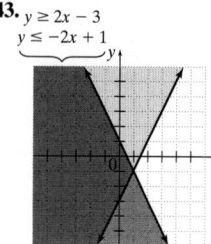

43. $y \geq 2x - 3$ $y \leq -2x + 1$

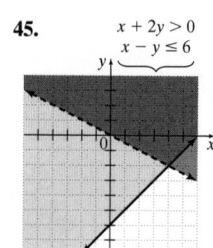

45. $x + 2y > 0$ $x - y \leq 6$

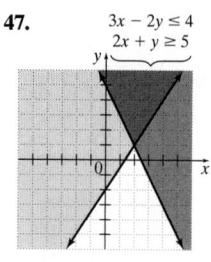

47. $3x - 2y \leq 4$ $2x + y \geq 5$

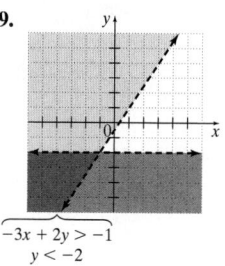

49.

$-3x + 2y > -1$ $y < -2$

Chapter 9 Test **1.** $1, \frac{2}{3}$ **2.** $\varnothing$ **3.** $\frac{3}{2}$ **4.** $\left(\frac{3}{2}, 5\right]$ **5.** $(-\infty, -2) \cup \left(\frac{4}{3}, \infty\right)$ **6.** $(3, 7)$ **7.** $(\infty, 5]$ **8.** $(-\infty, -2]$ **9.** $[-3, -1)$ **10.** $(-\infty, \infty)$

11. $3, -\frac{1}{5}$ **12.** $(-\infty, \infty)$ **13.**

14.

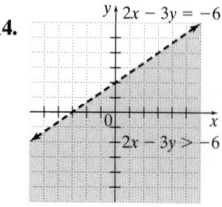

15. $y + 2x \leq 4$ $y \geq 2$

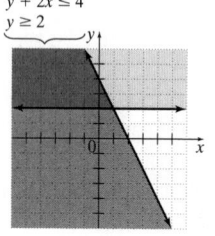

16. $2y - x \geq 1$ $x + y \geq -4$

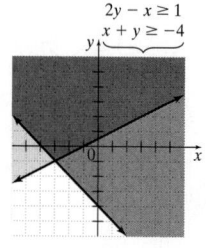

Chapter 9 Cumulative Review **1. a.** $\frac{1}{2}$ **b.** 9; Sec. 1.6, Ex. 6 **3. a.** $-\frac{39}{5}$ **b.** 2; Sec. 1.7, Ex. 9 **5. a.** $5x + 7$ **b.** $-4a - 1$ **c.** $4y - 3y^2$

d. $7.3x - 6$ **e.** $\frac{1}{2}b$; Sec. 2.1, Ex. 4 **7.** -3; Sec. 2.2, Ex. 3 **9.**

x	y
-1	-3
0	0
-3	-9

; Sec. 3.1, Ex. 5 **11. a.** x-int: $(-3, 0)$; y-int: $(0, 2)$

b. x-int: $(-4, 0)$, $(-1, 0)$; y-int: $(0, 1)$ **c.** x-int and y-int: $(0, 0)$ **d.** x-int: $(2, 0)$; y-int: none **e.** x-int: $(-1, 0)$, $(3, 0)$; y-int: $(0, 1)$, $(0, 2)$; Sec. 3.3, Ex. 1

13. parallel; Sec. 3.5, Ex. 4 **15.** $y = \frac{1}{4}x - 3$; Sec. 3.5, Ex. 5 **17.** $y = -3$; Sec. 3.6, Ex. 4 **19.** a, b, c; Sec. 3.7, Ex. 5 **21. a.** solution

b. not a solution; Sec. 4.1, Ex. 1 **23.** $12^3 - 12x^2 - 9x + 2$; Sec. 5.2, Ex. 8 **25.** $(x + 2y)(x + 3y)$; Sec. 6.2, Ex. 6 **27.** $\frac{3y^9}{160}$; Sec. 7.2, Ex. 4

29. 1; Sec. 7.3, Ex. 2 **31.** $\dfrac{x(3x-1)}{(x+1)^2(x-1)}$; Sec. 7.4, Ex. 7 **33.** -5; Sec. 7.5, Ex. 1 **35.** no solution; Sec. 4.1, Ex. 3 **37.** $(-2, 0)$; Sec. 4.2, Ex. 3

39. infinite number of solutions; Sec. 4.3, Ex. 4 **41.**

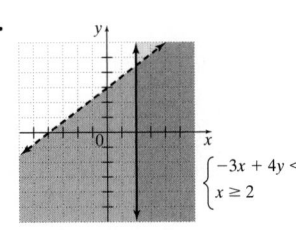

; Sec. 9.4, Ex. 8

$$\begin{cases} -3x + 4y < 12 \\ x \geq 2 \end{cases}$$

43. a. $\dfrac{25x^4}{y^6}$ **b.** x^6 **c.** $32x^2$ **d.** a^3b; Sec. 5.1, Ex. 10

45. $-6, -\dfrac{3}{2}, \dfrac{1}{5}$; Sec. 6.5, Ex. 6 **47.** 63; Sec. 7.6, Ex. 1

CHAPTER 10 RATIONAL EXPONENTS, RADICALS, AND COMPLEX NUMBERS

Mental Math **1.** d **3.** d

Exercise Set 10.1 **1.** 10 **3.** $\dfrac{1}{2}$ **5.** 0.01 **7.** -6 **9.** x^5 **11.** $4y^3$ **13.** 2.646 **15.** 6.164 **17.** 14.142 **19.** 4 **21.** $\dfrac{1}{2}$ **23.** -1 **25.** x^4 **27.** $-3x^3$

29. -2 **31.** not a real number **33.** -2 **35.** x^4 **37.** $2x^2$ **39.** $9x^2$ **41.** $4x^2$ **43.** 8 **45.** -8 **47.** $2|x|$ **49.** x **51.** $|x-5|$ **53.** $|x+2|$

55. -11 **57.** $2x$ **59.** y^6 **61.** $5ab^{10}$ **63.** $-3x^4y^3$ **65.** a^4b **67.** $-2x^2y$ **69.** $\dfrac{5}{7}$ **71.** $\dfrac{x}{2y}$ **73.** $-\dfrac{z^7}{3x}$ **75.** $\dfrac{x}{2}$ **77.** $\sqrt{3}$ **79.** -1 **81.** -3

83. $\sqrt{7}$ **85.** $[0, \infty)$; **87.** $[3, \infty)$; 0, 1, 2, 3 **89.** $(-\infty, \infty)$; **91.** $(-\infty, \infty)$; 0, 1, -1, 2, -2

   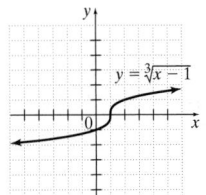

93. $-32x^{15}y^{10}$ **95.** $-60x^7y^{10}z^5$ **97.** $\dfrac{x^9y^5}{2}$ **99.** answers may vary **101.** 13 **103.** 18 **105.** 1.69 sq. m **107.** answers may vary

109. **111.**

Mental Math **1.** A **3.** C **5.** B **7.** B **9.** B

Exercise Set 10.2 **1.** 7 **3.** 3 **5.** $\dfrac{1}{2}$ **7.** 13 **9.** $2\sqrt[3]{m}$ **11.** $3x^2$ **13.** -3 **15.** -2 **17.** 8 **19.** 16 **21.** not a real number **23.** $\sqrt[5]{(2x)^3}$

25. $\sqrt[3]{(7x+2)^2}$ **27.** $\dfrac{64}{27}$ **29.** $\dfrac{1}{16}$ **31.** $\dfrac{1}{16}$ **33.** not a real number **35.** $\dfrac{1}{x^{1/4}}$ **37.** $a^{2/3}$ **39.** $\dfrac{5x^{3/4}}{7}$ **41.** answers may vary **43.** $a^{7/3}$ **45.** x **47.** $3^{5/8}$

49. $y^{1/6}$ **51.** $8u^3$ **53.** $-b$ **55.** $27x^{2/3}$ **57.** $y - y^{7/6}$ **59.** $2x^{5/3} - 2x^{2/3}$ **61.** $4x^{2/3} - 9$ **63.** $x^{8/3}(1 + x^{2/3})$ **65.** $x^{1/5}(x^{1/5} - 3)$ **67.** $x^{-1/3}(5 + x)$

69. $\sqrt{x}$ **71.** $\sqrt[3]{2}$ **73.** $2\sqrt{x}$ **75.** $\sqrt{xy}$ **77.** $\sqrt[15]{y^{11}}$ **79.** $\sqrt[12]{b^5}$ **81.** $\sqrt{a}$ **83.** $\sqrt[6]{432}$ **85.** $\sqrt[15]{343y^5}$ **87.** $25 \cdot 3$ **89.** $16 \cdot 3$ or $4 \cdot 12$ **91.** $8 \cdot 2$

93. $27 \cdot 2$ **95.** 1509 calories **97.** 97.2 million **99.** $a^{1/3}$ **101.** $x^{1/5}$ **103.** 1.6818 **105.** 5.6645 **107.** $\dfrac{t^{1/2}}{u^{1/2}}$

Exercise Set 10.3 **1.** $\sqrt{14}$ **3.** 2 **5.** $\sqrt[3]{36}$ **7.** $\sqrt{6x}$ **9.** $\sqrt{\dfrac{14}{xy}}$ **11.** $\sqrt[4]{20x^3}$ **13.** $\dfrac{\sqrt{6}}{7}$ **15.** $\dfrac{\sqrt{2}}{7}$ **17.** $\dfrac{\sqrt[4]{x^3}}{2}$ **19.** $\dfrac{\sqrt[3]{4}}{3}$ **21.** $\dfrac{\sqrt[4]{8}}{x^2}$ **23.** $\dfrac{\sqrt[3]{2x}}{3y^4\sqrt[3]{3}}$

25. $\dfrac{x\sqrt{y}}{10}$ **27.** $\dfrac{\sqrt{5x}}{2y}$ **29.** $-\dfrac{z^2\sqrt[3]{z}}{3x}$ **31.** $4\sqrt{2}$ **33.** $4\sqrt[3]{3}$ **35.** $25\sqrt{3}$ **37.** $2\sqrt{6}$ **39.** $10x^2\sqrt{x}$ **41.** $2y^2\sqrt[3]{2y}$ **43.** $a^2b\sqrt[4]{b^3}$ **45.** $y^2\sqrt{y}$

47. $5ab\sqrt{b}$ **49.** $-2x^2\sqrt[5]{y}$ **51.** $x^4\sqrt[3]{50x^2}$ **53.** $-4a^4b^3\sqrt{2b}$ **55.** $3x^3y^4\sqrt{xy}$ **57.** $5r^3s^4$ **59.** $\sqrt{2}$ **61.** 2 **63.** 10 **65.** x^2y **67.** $24m^2$

69. $\dfrac{15x\sqrt{2x}}{2}$ or $\dfrac{15x}{2}\sqrt{2x}$ **71.** $2a^2\sqrt[4]{2}$ **73.** $14x$ **75.** $2x^2 - 7x - 15$ **77.** y^2 **79.** $-3x - 15$ **81.** $x^2 - 8x - 16$ **83. a.** 20π sq. cm

b. 211.57 sq. ft **85. a.** 3.8 times **b.** 2.9 times **c.** answers may vary

Mental Math 1. $6\sqrt{3}$ **3.** $3\sqrt{x}$ **5.** $12\sqrt[3]{x}$ **7.** $2\sqrt{11}$ **9.** $8\sqrt{13}$ **11.** $10\sqrt[3]{2x}$

Exercise Set 10.4 1. $-2\sqrt{2}$ **3.** $10x\sqrt{2x}$ **5.** $17\sqrt{2} - 15\sqrt{5}$ **7.** $-\sqrt[3]{2x}$ **9.** $5b\sqrt{b}$ **11.** $\dfrac{31\sqrt{2}}{15}$ **13.** $\dfrac{\sqrt[3]{11}}{3}$ **15.** $\dfrac{5\sqrt{5x}}{9}$ **17.** $14 + \sqrt{3}$

19. $7 - 3y$ **21.** $6\sqrt{3} - 6\sqrt{2}$ **23.** $-23\sqrt[3]{5}$ **25.** $2b\sqrt{b}$ **27.** $20y\sqrt{2y}$ **29.** $2y\sqrt[3]{2x}$ **31.** $6\sqrt[3]{11} - 4\sqrt{11}$ **33.** $4x\sqrt[4]{x^3}$ **35.** $\dfrac{2\sqrt{3}}{3}$ **37.** $\dfrac{5x\sqrt[3]{x}}{7}$

39. $\dfrac{5\sqrt{7}}{2x}$ **41.** $\dfrac{\sqrt[3]{2}}{6}$ **43.** $\dfrac{14x\sqrt[3]{2x}}{9}$ **45.** $15\sqrt{3}$ in. **47.** $\sqrt{35} + \sqrt{21}$ **49.** $7 - 2\sqrt{10}$ **51.** $3\sqrt{x} - x\sqrt{3}$ **53.** $6x - 13\sqrt{x} - 5$

55. $\sqrt[3]{a^2} + \sqrt[3]{a} - 20$ **57.** $6\sqrt{2} - 12$ **59.** $2 + 2x\sqrt{3}$ **61.** $-16 - \sqrt{35}$ **63.** $x - y^2$ **65.** $3 + 2x\sqrt{3} + x^2$ **67.** $5x - 3\sqrt{15x} - 3\sqrt{10x} + 9\sqrt{6}$

69. $2\sqrt[3]{2} - \sqrt[3]{4}$ **71.** $-4\sqrt[6]{x^5} + \sqrt[3]{x^2} + 8\sqrt[3]{x} - 4\sqrt{x} + 7$ **73.** $x + 24 + 10\sqrt{x - 1}$ **75.** $2x + 6 - 2\sqrt{2x + 5}$ **77.** $x - 7$ **79.** $\dfrac{7}{x + y}$

81. $2a - 3$ **83.** $\dfrac{-2 + \sqrt{3}}{3}$ **85.** $22\sqrt{5}$ ft; 150 sq. ft **87. a.** $2\sqrt{3}$ **b.** 3 **c.** answers may vary **89.** answers may vary

Mental Math 1. $\sqrt{2} - x$ **3.** $5 + \sqrt{a}$ **5.** $7\sqrt{5} - 8\sqrt{x}$

Exercise Set 10.5 1. $\dfrac{\sqrt{14}}{7}$ **3.** $\dfrac{\sqrt{5}}{5}$ **5.** $\dfrac{\sqrt[3]{6}}{2}$ **7.** $\dfrac{4\sqrt[3]{9}}{3}$ **9.** $\dfrac{3\sqrt{2x}}{4x}$ **11.** $\dfrac{3\sqrt[3]{2x}}{2x}$ **13.** $\dfrac{2\sqrt{x}}{x}$ **15.** $\dfrac{3\sqrt{3a}}{a}$ **17.** $\dfrac{3\sqrt[3]{4}}{2}$ **19.** $\dfrac{2\sqrt{21}}{7}$ **21.** $\dfrac{\sqrt{10xy}}{5y}$

23. $\dfrac{3\sqrt[3]{2}}{2}$ **25.** $\dfrac{2\sqrt[4]{9x}}{3x^2}$ **27.** $\dfrac{5a\sqrt[3]{4ab^4}}{2a^2b^3}$ **29.** $-2(2 + \sqrt{7})$ **31.** $\dfrac{7(3 + \sqrt{x})}{9 - x}$ **33.** $-5 + 2\sqrt{6}$ **35.** $\dfrac{2a + 2\sqrt{a} + \sqrt{ab} + \sqrt{b}}{4a - b}$ **37.** $-\dfrac{8(1 - \sqrt{10})}{9}$

39. $\dfrac{x - \sqrt{xy}}{x - y}$ **41.** $\dfrac{5 + 3\sqrt{2}}{7}$ **43.** $\dfrac{5}{\sqrt{15}}$ **45.** $\dfrac{6}{\sqrt{10}}$ **47.** $\dfrac{2x}{7\sqrt{x}}$ **49.** $\dfrac{5y}{\sqrt[3]{100xy}}$ **51.** $\dfrac{2}{\sqrt{10}}$ **53.** $\dfrac{2x}{11\sqrt{2x}}$ **55.** $\dfrac{7}{2\sqrt[3]{49}}$ **57.** $\dfrac{3x^2}{10\sqrt[3]{9x}}$ **59.** $\dfrac{6x^2y^3}{\sqrt{6z}}$

61. answers may vary **63.** $\dfrac{-7}{12 + 6\sqrt{11}}$ **65.** $\dfrac{3}{10 + 5\sqrt{7}}$ **67.** $\dfrac{x - 9}{x - 3\sqrt{x}}$ **69.** $\dfrac{1}{3 + 2\sqrt{2}}$ **71.** $\dfrac{x - 1}{x - 2\sqrt{x} + 1}$ **73.** 5 **75.** $-\dfrac{1}{2}, 6$ **77.** 2, 6

79. $r = \dfrac{\sqrt{A\pi}}{2\pi}$ **81.** answers may vary

Integrated Review 1. 9 **2.** -2 **3.** $\dfrac{1}{2}$ **4.** x^3 **5.** y^3 **6.** $2y^5$ **7.** $-2y$ **8.** $3b^3$ **9.** 6 **10.** $\sqrt[4]{3y}$ **11.** $\dfrac{1}{16}$ **12.** $\sqrt[5]{(x + 1)^3}$ **13.** y **14.** $16x^{1/2}$

15. $x^{5/4}$ **16.** $4^{11/15}$ **17.** $2x^2$ **18.** $\sqrt[4]{a^3b^2}$ **19.** $\sqrt[4]{x^3}$ **20.** $\sqrt[6]{500}$ **21.** $2\sqrt{10}$ **22.** $2xy^2\sqrt[4]{x^3y^2}$ **23.** $3x\sqrt[3]{2x}$ **24.** $-2b^2\sqrt[5]{2}$ **25.** $\sqrt{5x}$ **26.** $4x$

27. $7y^2\sqrt{y}$ **28.** $2a^2\sqrt[4]{3}$ **29.** $2\sqrt{5} - 5\sqrt{3} + 5\sqrt{7}$ **30.** $y\sqrt[3]{2y}$ **31.** $\sqrt{15} - \sqrt{6}$ **32.** $10 + 2\sqrt{21}$ **33.** $4x^2 - 5$ **34.** $x + 2 - 2\sqrt{x + 1}$

35. $\dfrac{\sqrt{21}}{3}$ **36.** $\dfrac{5\sqrt[3]{4x}}{2x}$ **37.** $\dfrac{13 - 3\sqrt{21}}{5}$ **38.** $\dfrac{7}{\sqrt{21}}$ **39.** $\dfrac{3y}{\sqrt[3]{33y^2}}$ **40.** $\dfrac{x - 4}{x + 2\sqrt{x}}$

Graphing Calculator Explorations 1. 3.19 **3.** $\varnothing$ **5.** 3.23

Exercise Set 10.6 1. 8 **3.** 7 **5.** $\varnothing$ **7.** 7 **9.** 6 **11.** $-\dfrac{9}{2}$ **13.** 29 **15.** 4 **17.** -4 **19.** $\varnothing$ **21.** 7 **23.** 9 **25.** 50 **27.** $\varnothing$ **29.** $\dfrac{15}{4}$ **31.** 13 **33.** 5

35. -12 **37.** 9 **39.** -3 **41.** 1 **43.** 1 **45.** $\dfrac{1}{2}$ **47.** 0, 4 **49.** $\dfrac{37}{4}$ **51.** $3\sqrt{5}$ ft **53.** $2\sqrt{10}$ m **55.** $2\sqrt{131}$ m $\approx$ 22.9 m

57. $\sqrt{100.84}$ mm $\approx$ 10.0 mm **59.** 17 ft **61.** 13 ft **63.** 14,657,415 sq. mi **65.** 100 ft **67.** 100 **69.** $\dfrac{\pi}{2}$ sec $\approx$ 1.57 sec **71.** 12.97 ft

73. answers may vary **75.** $15\sqrt{3}$ sq. mi $\approx$ 25.98 sq. mi **77.** answers may vary **79.** 0.51 km **81.** function **83.** function **85.** not a function

87. $\dfrac{x}{4x + 3}$ **89.** $-\dfrac{4z + 2}{3z}$ **91.** 1 **93.** 2743 deliveries **95.** $-1, 0, 8, 9$ **97.** $-1, 4$

Mental Math 1. $9i$ **3.** $i\sqrt{7}$ **5.** -4 **7.** $8i$

Exercise Set 10.7 1. $2i\sqrt{6}$ **3.** $-6i$ **5.** $24i\sqrt{7}$ **7.** $-3\sqrt{6}$ **9.** $-\sqrt{14}$ **11.** $-5\sqrt{2}$ **13.** $4i$ **15.** $i\sqrt{3}$ **17.** $2\sqrt{2}$ **19.** $6 - 4i$ **21.** $-2 + 6i$

23. $-2 - 4i$ **25.** $18 + 12i$ **27.** 7 **29.** $12 - 16i$ **31.** $-4i$ **33.** $\dfrac{28}{25} - \dfrac{21}{25}i$ **35.** $4 + i$ **37.** $\dfrac{17}{13} + \dfrac{7}{13}i$ **39.** 63 **41.** $2 - i$ **43.** 20 **45.** 10 **47.** 2

49. $-5 + \dfrac{16}{3}i$ **51.** $17 + 144i$ **53.** $\dfrac{3}{5} - \dfrac{1}{5}i$ **55.** $5 - 10i$ **57.** $\dfrac{1}{5} - \dfrac{8}{5}i$ **59.** $8 - i$ **61.** 1 **63.** i **65.** $-i$ **67.** -1 **69.** -64 **71.** $-243i$ **73.** $40°$

75. $x^2 - 5x - 2 - \dfrac{6}{x - 1}$ **77.** 5 people **79.** 14 people **81.** 16.7% **83.** $1 - i$ **85.** 0 **87.** $2 + 3i$ **89.** $2 + i\sqrt{2}$ **91.** $\dfrac{1}{2} - \dfrac{\sqrt{3}}{2}i$

93. answers may vary **95.** $6 - 6i$ **97.** yes

Chapter 10 Review **1.** 9 **3.** -2 **5.** $-\dfrac{1}{7}$ **7.** -6 **9.** $-a^2b^3$ **11.** $2ab^2$ **13.** $\dfrac{x^6}{6y}$ **15.** $|-x|$ **17.** -27 **19.** $-x$ **21.** $5|(x-y)^5|$ **23.** $-x$

25. $(-\infty, \infty); -2, -1, 0, 1, 2$ **27.** $-\dfrac{1}{3}$ **29.** $-\dfrac{1}{4}$ **31.** $\dfrac{1}{4}$ **33.** $\dfrac{343}{125}$ **35.** not a real number **37.** $5^{1/5}x^{2/5}y^{3/5}$ **39.** $5\sqrt[3]{xy^2z^5}$ **41.** $a^{13/6}$ **43.** $\dfrac{1}{a^{9/2}}$

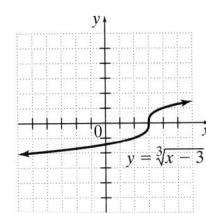
$y = \sqrt[3]{x} - 3$

45. a^4b^6 **47.** $\dfrac{b^{5/6}}{49a^{1/4}c^{5/3}}$ **49.** 4.472 **51.** 5.191 **53.** -26.246 **55.** $\sqrt[6]{1372}$ **57.** $2\sqrt{6}$ **59.** $2x$ **61.** $2\sqrt{15}$

63. $3\sqrt[3]{6}$ **65.** $6x^3\sqrt{x}$ **67.** $\dfrac{p^8\sqrt{p}}{11}$ **69.** $\dfrac{y\sqrt[4]{xy^2}}{3}$ **71. a.** $\dfrac{5}{\sqrt{\pi}}$m or $\dfrac{5\sqrt{\pi}}{\pi}$m **b.** 5.75 in. **73.** $xy\sqrt{2y}$

75. $3a\sqrt[4]{2a}$ **77.** $\dfrac{3\sqrt{2}}{4x}$ **79.** $-4ab\sqrt[4]{2b}$ **81.** $x - 6\sqrt{x} + 9$ **83.** $4x - 9y$ **85.** $\sqrt[3]{a^2} + 4\sqrt[3]{a} + 4$ **87.** $a + 64$

89. $\dfrac{\sqrt{3x}}{6}$ **91.** $\dfrac{2x^2\sqrt{2x}}{y}$ **93.** $-\dfrac{10 + 5\sqrt{7}}{3}$ **95.** $-5 + 2\sqrt{6}$ **97.** $\dfrac{6}{\sqrt{2y}}$ **99.** $\dfrac{4x^3}{y\sqrt{2x}}$ **101.** $\dfrac{x-25}{-3\sqrt{x} + 15}$ **103.** $\varnothing$

105. $\varnothing$ **107.** 16 **109.** $\sqrt{241}$ **111.** 4.24 ft **113.** $-i\sqrt{6}$ **115.** $-\sqrt{10}$ **117.** $-13 - 3i$ **119.** $10 + 4i$ **121.** $1 + 5i$ **123.** 87 **125.** $-\dfrac{1}{3} + \dfrac{1}{3}i$

Chapter 10 Test **1.** $6\sqrt{6}$ **2.** $-x^{16}$ **3.** $\dfrac{1}{5}$ **4.** 5 **5.** $\dfrac{4x^2}{9}$ **6.** $-a^6b^3$ **7.** $\dfrac{8a^{1/3}c^{2/3}}{b^{5/12}}$ **8.** $a^{7/12} - a^{7/3}$ **9.** $|4xy|$ or $4|xy|$ **10.** -27 **11.** $\dfrac{3\sqrt{y}}{y}$

12. $\dfrac{8 - 6\sqrt{x} + x}{8 - 2x}$ **13.** $\dfrac{\sqrt[3]{b^2}}{b}$ **14.** $\dfrac{6 - x^2}{8(\sqrt{6} - x)}$ **15.** $-x\sqrt{5x}$ **16.** $4\sqrt{3} - \sqrt{6}$ **17.** $x + 2\sqrt{x} + 1$ **18.** $\sqrt{6} - 4\sqrt{3} + \sqrt{2} - 4$ **19.** -20

20. 23.685 **21.** 0.019 **22.** 2, 3 **23.** $\varnothing$ **24.** 6 **25.** $i\sqrt{2}$ **26.** $-2i\sqrt{2}$ **27.** $-3i$ **28.** 40 **29.** $7 + 24i$ **30.** $-\dfrac{3}{2} + \dfrac{5}{2}i$ **31.** $\dfrac{5\sqrt{2}}{2}$

32. $[-2, \infty)$;   ; 0, 1, 2, 3 **33.** 27 mph **34.** 360 ft

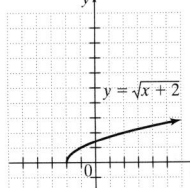
$y = \sqrt{x + 2}$

Chapter 10 Cumulative Review **1. a.** $-\dfrac{39}{5}$ **b.** 2; Sec. 1.7, Ex. 9 **3.** 12; Sec. 2.3, Ex. 3 **5.** 2 ft, 8 ft; Sec. 2.4, Ex. 3 **7.** one; Sec. 4.1, Ex. 6

9. $\left(6, \dfrac{1}{2}\right)$; Sec. 4.2, Ex. 1 **11.** no solution; Sec. 4.3, Ex. 3 **13.** 5% saline solution: 2.5L; 25% saline solution: 7.5 L; Sec. 4.5, Ex. 4 **15. a.** -4

b. 11; Sec. 5.2, Ex. 4 **17.** $2x^2 - x + 4$; Sec. 5.6, Ex. 1 **19.** $2x^2 + 5x + 2 + \dfrac{6}{x - 3}$; Sec. 5.7, Ex. 1 **21.** $(x + 2)(x + 3)$; Sec. 6.2, Ex. 6

23. a. $x^2 - 2x + 4$ **b.** $\dfrac{2}{y - 5}$; Sec. 7.1, Ex. 6 **25.** $\dfrac{3y^9}{160}$; Sec. 7.2, Ex. 4 **27.** 1; Sec. 7.3, Ex. 2 **29.** $\dfrac{x(3x - 1)}{(x + 1)^2(x - 1)}$; Sec. 7.4, Ex. 7 **31. a.** $\dfrac{x(x - 2)}{2(x + 2)}$

b. $\dfrac{x^2}{y^2}$; Sec. 7.7, Ex. 2 **33.** -5; Sec. 7.5, Ex. 1 **35.** 15 yd; Sec. 7.6, Ex. 4 **37. a.** 1 **b.** -4 **c.** $\dfrac{2}{5}$ **d.** x^2 **e.** $-3x^3$; Sec. 10.1, Ex. 3 **39. a.** $\dfrac{1}{8}$

b. $\dfrac{1}{9}$; Sec. 10.2, Ex. 3 **41.** $\dfrac{x - 4}{5(\sqrt{x} - 2)}$; Sec. 10.5, Ex. 7 **43.** Constant of variation: 15, $u = \dfrac{15}{w}$; Sec. 8.4, Ex. 3

CHAPTER 11 QUADRATIC EQUATIONS AND FUNCTIONS

Graphing Calculator Explorations **1.** $-1.27, 6.27$ **3.** $-1.10, 0.90$ **5.** $\varnothing$

Exercise Set 11.1 **1.** $-4, 4$ **3.** $-\sqrt{7}, \sqrt{7}$ **5.** $-3\sqrt{2}, 3\sqrt{2}$ **7.** $-\sqrt{10}, \sqrt{10}$ **9.** $-8, -2$ **11.** $6 - 3\sqrt{2}, 6 + 3\sqrt{2}$ **13.** $\dfrac{3 - 2\sqrt{2}}{2}, \dfrac{3 + 2\sqrt{2}}{2}$

15. $-3i, 3i$ **17.** $-\sqrt{6}, \sqrt{6}$ **19.** $-2i\sqrt{2}, 2i\sqrt{2}$ **21.** $1 - 4i, 1 + 4i$ **23.** $-7 - \sqrt{5}, -7 + \sqrt{5}$ **25.** $-3 - 2i\sqrt{2}, -3 + 2i\sqrt{2}$

27. $x^2 + 16x + 64 = (x + 8)^2$ **29.** $z^2 - 12z + 36 = (z - 6)^2$ **31.** $p^2 + 9p + \dfrac{81}{4} = \left(p + \dfrac{9}{2}\right)^2$ **33.** $x^2 + x + \dfrac{1}{4} = \left(x + \dfrac{1}{2}\right)^2$ **35.** $-5, -3$

37. $-3 - \sqrt{7}, -3 + \sqrt{7}$ **39.** $\dfrac{-1 - \sqrt{5}}{2}, \dfrac{-1 + \sqrt{5}}{2}$ **41.** $-1 - \sqrt{6}, -1 + \sqrt{6}$ **43.** $\dfrac{6 - \sqrt{30}}{3}, \dfrac{6 + \sqrt{30}}{3}$ **45.** $\dfrac{3 - \sqrt{11}}{2}, \dfrac{3 + \sqrt{11}}{2}$ **47.** $-4, \dfrac{1}{2}$

49. $-1, 5$ **51.** $-4 - \sqrt{15}, -4 + \sqrt{15}$ **53.** $\dfrac{-3 - \sqrt{21}}{3}, \dfrac{-3 + \sqrt{21}}{3}$ **55.** $-1, \dfrac{5}{2}$ **57.** $-1 - i, -1 + i$ **59.** $3 - \sqrt{6}, 3 + \sqrt{6}$

61. $-2 - i\sqrt{2}, -2 + i\sqrt{2}$ **63.** $\dfrac{-15 - 7\sqrt{5}}{10}, \dfrac{-15 + 7\sqrt{5}}{10}$ **65.** $\dfrac{1 - i\sqrt{47}}{4}, \dfrac{1 + i\sqrt{47}}{4}$ **67.** $-5 - i\sqrt{3}, -5 + i\sqrt{3}$ **69.** $-4, 1$

71. $\dfrac{2 - i\sqrt{2}}{2}, \dfrac{2 + i\sqrt{2}}{2}$ **73.** $\dfrac{-3 - \sqrt{69}}{6}, \dfrac{-3 + \sqrt{69}}{6}$ **75.** 20% **77.** $11\dfrac{1}{9}\%$ **79.** answers may vary **81.** simple **83.** $\dfrac{7}{5}$ **85.** $\dfrac{1}{5}$ **87.** $5 - 10\sqrt{3}$

89. $\dfrac{3 - 2\sqrt{7}}{4}$ **91.** $2\sqrt{7}$ **93.** $\sqrt{13}$ **95.** $-6y, 6y$ **97.** $-x, x$ **99.** 8.11 sec **101.** 6.73 sec **103.** 6 in. **105.** 16.2 × 21.6 in.

107. 2.828 thousand units or 2828 units

Mental Math 1. $a = 1, b = 3, c = 1$ **3.** $a = 7, b = 0, c = -4$ **5.** $a = 6, b = -1, c = 0$

Exercise Set 11.2 1. $-6, 1$ **3.** $-\dfrac{3}{5}, 1$ **5.** 3 **7.** $\dfrac{-7 - \sqrt{33}}{2}, \dfrac{-7 + \sqrt{33}}{2}$ **9.** $\dfrac{1 - \sqrt{57}}{8}, \dfrac{1 + \sqrt{57}}{8}$ **11.** $\dfrac{7 - \sqrt{85}}{6}, \dfrac{7 + \sqrt{85}}{6}$

13. $1 - \sqrt{3}, 1 + \sqrt{3}$ **15.** $-\dfrac{3}{2}, 1$ **17.** $\dfrac{3 - \sqrt{11}}{2}, \dfrac{3 + \sqrt{11}}{2}$ **19.** answers may vary **21.** $\dfrac{3 - i\sqrt{87}}{8}, \dfrac{3 + i\sqrt{87}}{8}$ **23.** $-2 - \sqrt{11}, -2 + \sqrt{11}$

25. $\dfrac{-5 - i\sqrt{5}}{10}, \dfrac{-5 + i\sqrt{5}}{10}$ **27.** two real solutions **29.** one real solution **31.** two real solutions **33.** two complex but not real solutions

35. $\dfrac{-5 - \sqrt{17}}{2}, \dfrac{-5 + \sqrt{17}}{2}$ **37.** $\dfrac{5}{2}, 1$ **39.** $\dfrac{3 - \sqrt{29}}{2}, \dfrac{3 + \sqrt{29}}{2}$ **41.** $\dfrac{-1 - \sqrt{19}}{6}, \dfrac{-1 + \sqrt{19}}{6}$ **43.** $-3 - 2i, -3 + 2i$

45. $\dfrac{-1 - i\sqrt{23}}{4}, \dfrac{-1 + i\sqrt{23}}{4}$ **47.** 1 **49.** $\dfrac{19 - \sqrt{345}}{2}, \dfrac{19 + \sqrt{345}}{2}$ **51.** 14 ft **53.** $2 + 2\sqrt{2}$ cm, $2 + 2\sqrt{2}$ cm, $4 + 2\sqrt{2}$ cm

55. width: $-5 + 5\sqrt{17}$ ft; length: $5 + 5\sqrt{17}$ ft **57. a.** $50\sqrt{2}$ m **b.** 5000 sq. m **59.** 37.4 ft by 38.5 ft **61.** $\dfrac{1 + \sqrt{5}}{2}$ **63.** 8.9 sec **65.** 2.8 sec

67. $\dfrac{11}{5}$ **69.** 15 **71.** $(x^2 + 5)(x + 2)(x - 2)$ **73.** $(z + 3)(z - 3)(z + 2)(z - 2)$ **75.** 0.6, 2.4 **77.** Sunday to Monday. **79.** Wednesday

81. 32; yes **83. a.** \$6901.4 million **b.** 2007 **85.** answers may vary **87.** $\dfrac{\sqrt{3}}{3}$ **89.** $\dfrac{-\sqrt{2} - i\sqrt{2}}{2}, \dfrac{-\sqrt{2} + i\sqrt{2}}{2}$ **91.** $\dfrac{\sqrt{3} - \sqrt{11}}{4}, \dfrac{\sqrt{3} + \sqrt{11}}{4}$

93. 8.9 sec: 2.8 sec: **95.** two real solutions

Exercise Set 11.3 1. 2 **3.** 16 **5.** 1, 4 **7.** $3 - \sqrt{7}, 3 + \sqrt{7}$ **9.** $\dfrac{3 - \sqrt{57}}{4}, \dfrac{3 + \sqrt{57}}{4}$ **11.** $\dfrac{1 - \sqrt{29}}{2}, \dfrac{1 + \sqrt{29}}{2}$

13. $-2, 2, -2i, 2i$ **15.** $-\dfrac{1}{2}, \dfrac{1}{2}, -i\sqrt{3}, i\sqrt{3}$ **17.** $-3, 3, -2, 2$ **19.** $125, -8$ **21.** $-\dfrac{4}{5}, 0$ **23.** $-\dfrac{1}{8}, 27$ **25.** $-\dfrac{2}{3}, \dfrac{4}{3}$ **27.** $-\dfrac{1}{125}, \dfrac{1}{8}$

29. $-\sqrt{2}, \sqrt{2}, -\sqrt{3}, \sqrt{3}$ **31.** $\dfrac{-9 - \sqrt{201}}{6}, \dfrac{-9 + \sqrt{201}}{6}$ **33.** 2, 3 **35.** 3 **37.** 27, 125 **39.** $1, -3i, 3i$ **41.** $\dfrac{1}{8}, -8$ **43.** $-\dfrac{1}{2}, \dfrac{1}{3}$

45. 4 **47.** -3 **49.** $-\sqrt{5}, \sqrt{5}, -2i, 2i$ **51.** $-3, \dfrac{3 - 3i\sqrt{3}}{2}, \dfrac{3 + 3i\sqrt{3}}{2}$ **53.** 6, 12 **55.** $-\dfrac{1}{3}, \dfrac{1}{3}, -\dfrac{i\sqrt{6}}{3}, \dfrac{i\sqrt{6}}{3}$ **57.** 5 mph, then 4 mph

59. inlet pipe, 15.5 hr; hose, 16.5 hr **61.** 55 mph, 66 mph **63.** 8.5 hr. **65.** 12 or -8 **67. a.** $(x - 6)$ in. **b.** $300 = (x - 6) \cdot (x - 6) \cdot 3$

c. 16 cm by 16 cm **69.** 22 feet **71.** $(-\infty, 3]$ **73.** $(-5, \infty)$ **75.** domain: $\{x \mid x \text{ is a real number}\}$ or $(-\infty, \infty)$; range: $\{y \mid y \text{ is a real number}\}$ or

$(-\infty, \infty)$; function **77.** domain: $\{x \mid x \text{ is a real number}\}$ or $(-\infty, \infty)$; range: $\{y \mid y \geq -1\}$ or $[-1, \infty)$; function

79. answers may vary **81. a.** 134.05 ft per sec **b.** 134.93 ft per sec **c.** Fernandez: 92.0 mph; Dominguez: 91.4 mph

Integrated Review 1. $\left\{-\sqrt{10}, \sqrt{10}\right\}$ **2.** $\left\{-\sqrt{14}, \sqrt{14}\right\}$ **3.** $\left\{1 - 2\sqrt{2}, 1 + 2\sqrt{2}\right\}$ **4.** $\left\{-5 - 2\sqrt{3}, -5 + 2\sqrt{3}\right\}$

5. $\left\{-1 - \sqrt{13}, -1 + \sqrt{13}\right\}$ **6.** $\{1, 11\}$ **7.** $\left\{\dfrac{-3 - \sqrt{69}}{6}, \dfrac{-3 + \sqrt{69}}{6}\right\}$ **8.** $\left\{\dfrac{-2 - \sqrt{5}}{4}, \dfrac{-2 + \sqrt{5}}{4}\right\}$ **9.** $\left\{\dfrac{2 - \sqrt{2}}{2}, \dfrac{2 + \sqrt{2}}{2}\right\}$

10. $\left\{-3 - \sqrt{5}, -3 + \sqrt{5}\right\}$ **11.** $\left\{-2 + i\sqrt{3}, -2 - i\sqrt{3}\right\}$ **12.** $\left\{\dfrac{-1 - i\sqrt{11}}{2}, \dfrac{-1 + i\sqrt{11}}{2}\right\}$ **13.** $\left\{\dfrac{-3 + i\sqrt{15}}{2}, \dfrac{-3 - i\sqrt{15}}{2}\right\}$

14. $\{3i, -3i\}$ **15.** $\{0, -17\}$ **16.** $\left\{\dfrac{1 + \sqrt{13}}{4}, \dfrac{1 - \sqrt{13}}{4}\right\}$ **17.** $\left\{2 + 3\sqrt{3}, 2 - 3\sqrt{3}\right\}$ **18.** $\left\{2 + \sqrt{3}, 2 - \sqrt{3}\right\}$ **19.** $\left\{-2, \dfrac{4}{3}\right\}$

20. $\left\{\dfrac{-5 + \sqrt{17}}{4}, \dfrac{-5 - \sqrt{17}}{4}\right\}$ **21.** $\left\{1 - \sqrt{6}, 1 + \sqrt{6}\right\}$ **22.** $\left\{-\sqrt{31}, \sqrt{31}\right\}$ **23.** $\left\{-\sqrt{11}, \sqrt{11}\right\}$ **24.** $\left\{-i\sqrt{11}, i\sqrt{11}\right\}$

25. $\{-11, 6\}$ **26.** $\left\{\dfrac{-3 + \sqrt{19}}{5}, \dfrac{-3 - \sqrt{19}}{5}\right\}$ **27.** $\left\{\dfrac{-3 + \sqrt{17}}{4}, \dfrac{-3 - \sqrt{17}}{4}\right\}$ **28.** $10\sqrt{2}$ ft ≈ 14.1 ft   **29.** Jack: 9.1 hr; Lucy: 7.1 hr

30. 5 mph during the first part, then 6 mph

Exercise Set 11.4   1. $(-\infty, -5) \cup (-1, \infty)$ **3.** $[-4, 3]$ **5.** $[2, 5]$

7. $\left(-5, -\dfrac{1}{3}\right)$ **9.** $(2, 4) \cup (6, \infty)$ **11.** $(-\infty, -4] \cup [0, 1]$

13. $(-\infty, -3) \cup (-2, 2) \cup (3, \infty)$ **15.** $(-7, 2)$ **17.** $(-1, \infty)$

19. $(-\infty, -1] \cup (4, \infty)$ **21.** $(-\infty, 2) \cup \left(\dfrac{11}{4}, \infty\right)$ **23.** $(0, 2] \cup [3, \infty)$

25. $(-\infty, -7) \cup (8, \infty)$ **27.** $\left[-\dfrac{5}{4}, \dfrac{3}{2}\right]$ **29.** $(-\infty, 0) \cup (1, \infty)$ **31.** $(-\infty, -4] \cup [4, 6]$

33. $\left(-\infty, -\dfrac{2}{3}\right] \cup \left[\dfrac{3}{2}, \infty\right)$ **35.** $\left(-4, -\dfrac{3}{2}\right) \cup \left(\dfrac{3}{2}, \infty\right)$

37. $(-\infty, -5] \cup [-1, 1] \cup [5, \infty)$ **39.** $\left(-\infty, -\dfrac{5}{3}\right) \cup \left(\dfrac{7}{2}, \infty\right)$ **41.** $(0, 10)$

43. $(-\infty, -4) \cup [5, \infty)$ **45.** $(-\infty, -6] \cup (-1, 0] \cup (7, \infty)$ **47.** $(-\infty, 1) \cup (2, \infty)$

49. $(-\infty, -8] \cup (-4, \infty)$ **51.** $(-\infty, 0] \cup \left(5, \dfrac{11}{2}\right]$ **53.** $(0, \infty)$

55.   **57.**   **59.**   **61.**
   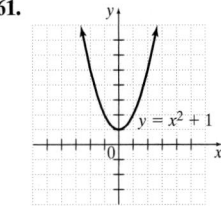

$y = |x| + 2$ $y = |x| - 1$ $y = x^2 - 3$ $y = x^2 + 1$

63. answers may vary   **65.** any number less than -1 or between 0 and 1   **67.** x is between 2 and 11

69.   **71.**
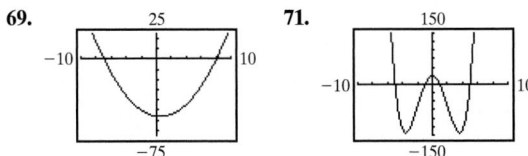

Graphing Calculator Explorations   1.   **3.**   **5.**
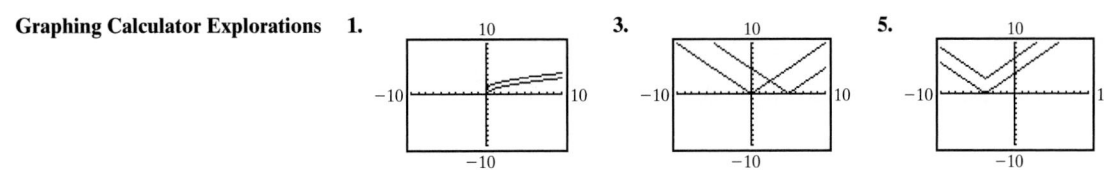

Mental Math   1. $(0, 0)$ **3.** $(2, 0)$ **5.** $(0, 3)$ **7.** $(-1, 5)$

Exercise Set 11.5   1.   **3.**   **5.**   **7.**
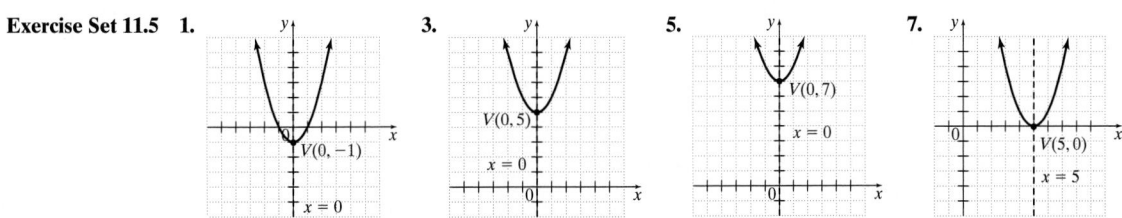

$V(0, -1)$, $x = 0$ $V(0, 5)$, $x = 0$ $V(0, 7)$, $x = 0$ $V(5, 0)$, $x = 5$

9.

11.

13.

15.

17.

19.

21.

23.

25.

27.

29.

31.

33.

35.

37.

39.

41.

43.

45.

47.

49.

51.

53.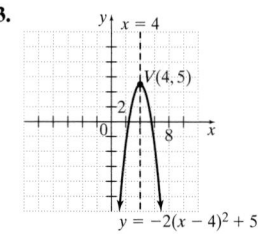

55. $x^2 + 8x + 16$ **57.** $z^2 - 16z + 64$ **59.** $y^2 + y + \dfrac{1}{4}$ **61.** $-6, 2$ **63.** $-5 - \sqrt{26}, -5 + \sqrt{26}$ **65.** $4 - 3\sqrt{2}, 4 + 3\sqrt{2}$

67. $f(x) = 5(x - 2)^2 + 3$ **69.** $f(x) = 5(x + 3)^2 + 6$ **71.**

73.

75.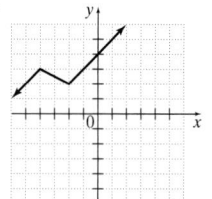

Exercise Set 11.6 **1.** $(-4, -9)$ **3.** $(5, 30)$ **5.** $(1, -2)$ **7.** $\left(\dfrac{1}{2}, \dfrac{5}{4}\right)$ **9.** D **11.** B

13. $y = x^2 + 4x - 5$
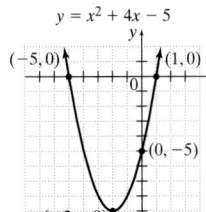

15. $y = -x^2 + 2x - 1$
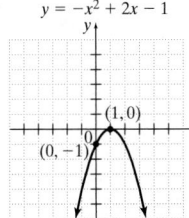

17. $y = x^2 - 4$
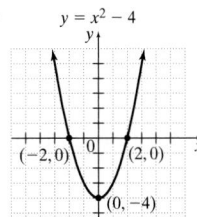

19. $y = 4x^2 + 4x - 3$
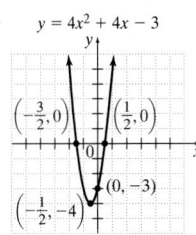

21. $y = x^2 + 8x + 15$
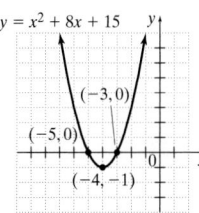

23. $y = x^2 - 6x + 5$
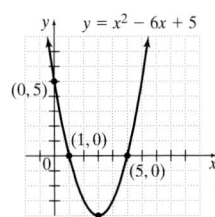

25. $y = x^2 - 4x + 5$

27. $y = 2x^2 + 4x + 5$

29. $y = -2x^2 + 12x$

31.
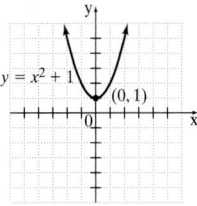
$y = x^2 + 1$

33.
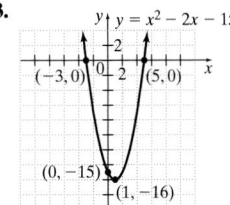
$y = x^2 - 2x - 15$

35.
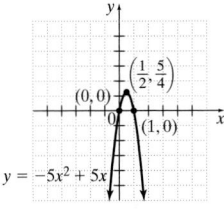
$y = -5x^2 + 5x$

37.
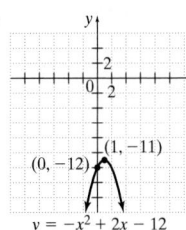
$y = -x^2 + 2x - 12$

39.
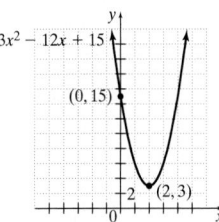
$y = 3x^2 - 12x + 15$

41. $y = x^2 + x - 6$

43.

$y = -2x^2 - 3x + 35$

45. 144 ft   **47.** 16 ft   **49.** 30 and 30   **51.** 5, −5

53. length, 20 units; width, 20 units

55.
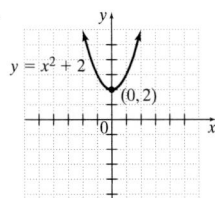
$y = x^2 + 2$

57.
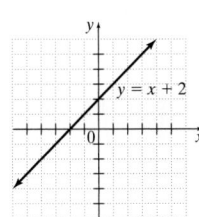
$y = x + 2$

59. $y = (x + 5)^2 + 2$

61.
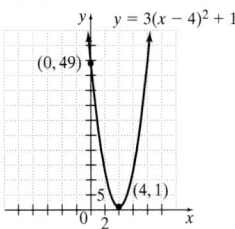
$y = 3(x - 4)^2 + 1$

63.
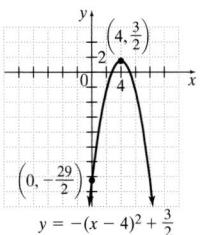
$y = -(x - 4)^2 + \frac{3}{2}$

65. $y = x^2 + 10x + 15$

67.
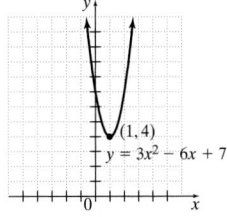
$y = 3x^2 - 6x + 7$

69. −0.84   **71.** 1.43   **73. a.** maximum; answers may vary   **b.** 2036   **c.** 32,902,500 or about 3290.3 thousands

75.

77.

Chapter 11 Review   1. 14, 1   **3.** $\frac{4}{5}, -\frac{1}{2}$ **5.** −7, 7   **7.** $-\frac{4}{9}, \frac{2}{9}$ **9.** $\dfrac{-3 - \sqrt{5}}{2}, \dfrac{-3 + \sqrt{5}}{2}$

11. $\dfrac{-3 - i\sqrt{7}}{8}, \dfrac{-3 + i\sqrt{7}}{8}$ **13.** 4.25% **15.** two complex but not real solutions **17.** two real solutions **19.** 8 **21.** $-i\sqrt{11}, i\sqrt{11}$

23. $\dfrac{5 - i\sqrt{143}}{12}, \dfrac{5 + i\sqrt{143}}{12}$ **25.** $\dfrac{21 - \sqrt{41}}{50}, \dfrac{21 + \sqrt{41}}{50}$ **27. a.** 20 ft **b.** $\dfrac{15 + \sqrt{321}}{16}$ sec; 2.1 sec **29.** 3, $\dfrac{-3 + 3i\sqrt{3}}{2}, \dfrac{-3 - 3i\sqrt{3}}{2}$

31. $\dfrac{2}{3}, 5$ **33.** $-5, 5, -2i, 2i$ **35.** 1, 125 **37.** $-1, 1, -i, i$ **39.** Jerome: 10.5 hr; Tim: 9.5 hr **41.** ← [————] → $[-5, 5]$

43. ← —[ ]— → $\left(-\infty, -\dfrac{5}{4}\right] \cup \left[\dfrac{3}{2}, \infty\right)$ **45.** ← (———) → ; (5, 6) **47.** ← —( )—(—→ $(-\infty, -6) \cup \left(-\dfrac{3}{4}, 0\right) \cup (5, \infty)$

49. ← —( )—(—→ $(-5, -3) \cup (5, \infty)$ **51.** ← —( )—(—→ ; $\left(-\dfrac{6}{5}, 0\right) \cup \left(\dfrac{5}{6}, 3\right)$

53.   **55.**   **57.**   **59.**   **61.**

63.   **65.** 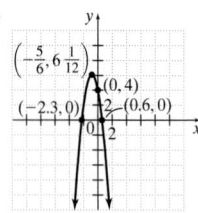   **67.** The numbers are both 210.

Chapter 11 Test **1.** $\dfrac{7}{5}, -1$ **2.** $-1 - \sqrt{10}, -1 + \sqrt{10}$ **3.** $\dfrac{1 + i\sqrt{31}}{2}, \dfrac{1 - i\sqrt{31}}{2}$ **4.** $\dfrac{3 + \sqrt{29}}{2}, \dfrac{3 - \sqrt{29}}{2}$

5. $-2 - \sqrt{11}, -2 + \sqrt{11}$ **6.** $-1, 1, -i, i, -3$ **7.** 6, 7 **8.** $3 - \sqrt{7}, 3 + \sqrt{7}$ **9.** $\dfrac{2 - i\sqrt{6}}{2}, \dfrac{2 + i\sqrt{6}}{2}$

10. ← —[ ]— → ; $\left(-\infty, -\dfrac{3}{2}\right) \cup (5, \infty)$ **11.** ← —[ ]—[ ]— → ; $(-\infty, -5] \cup [-4, 4] \cup [5, \infty)$

12. ← —( )—(—→ ; $(-\infty, -3) \cup (2, \infty)$ **13.** ← —( )—[ )— → ; $(-\infty, -3) \cup [2, 3)$

14.   **15.**   **16.**   **17.** 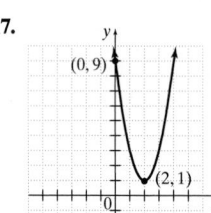   **18.** $\left(5 + \sqrt{17}\right)$ hr ≈ 9.12 hr

19. a. 272 ft **b.** 5.12 sec **20.** 7 ft

Chapter 11 Cumulative Review **1. a.** $\dfrac{1}{2}$ **b.** 9; Sec. 1.4, Ex. 3 **3. a.** $5x + 7$ **b.** $-4a - 1$ **c.** $4y - 3y^2$ **d.** $7.3x - 6$ **e.** $\dfrac{1}{2}b$; Sec. 2.1, Ex. 4

5. no solution; Sec. 4.1, Ex. 3 **7.** $(-2, 0)$; Sec. 4.2, Ex. 3 **9. a.** $\dfrac{25x^4}{y^6}$ **b.** x^6 **c.** $32x^2$ **d.** a^3b; Sec. 5.1, Ex. 10 **11. a.** 5 **b.** 5; Sec. 5.7, Ex. 3

13. $-6, -\dfrac{3}{2}, \dfrac{1}{5}$; Sec. 6.5, Ex. 6 **15.** $\dfrac{1}{5x - 1}$; Sec. 7.1, Ex. 2b **17.** $\dfrac{xy + 2x^3}{y - 1}$; Sec. 7.7, Ex. 3 **19.** $(2m - 1)^2$; Sec. 6.3, Ex. 9

21. -10 and 7; Sec. 6.6, Ex. 2 **23. a.** $5x\sqrt{x}$ **b.** $3x^2y^2\sqrt[3]{2y^2}$ **c.** $3z^2\sqrt[4]{z^3}$; Sec. 10.3, Ex. 4 **25. a.** $\dfrac{2\sqrt{5}}{5}$ **b.** $\dfrac{8\sqrt{x}}{3x}$ **c.** $\dfrac{\sqrt[3]{4}}{2}$; Sec. 10.5, Ex. 1

27. $\dfrac{2}{9}$; Sec. 10.6, Ex. 5 **29.** -5; Sec. 7.5, Ex. 1 **31.** -5; Sec. 7.6, Ex. 4 **33.** $\dfrac{1}{6}$; $y = \dfrac{1}{6}x$; Sec. 6.8, Ex. 1 **35. a.** 3 **b.** $|x|$ **c.** $|x - 2|$ **d.** -5

e. $2x - 7$ **f.** $5|x|$ **g.** $|x + 1|$; Sec. 10.1, Ex. 5 **37. a.** $\sqrt{x}$ **b.** $\sqrt[3]{5}$ **c.** $\sqrt{rs^3}$; Sec. 10.2, Ex. 7 **39. a.** $\dfrac{1}{2} + \dfrac{3}{2}i$ **b.** $-\dfrac{7}{3}i$; Sec. 10.7, Ex. 5

41. $-1 + 2\sqrt{3}, -1 - 2\sqrt{3}$; Sec. 11.1, Ex. 3 **43.** 9; Sec. 11.3, Ex. 1

Integrated Review   **1.** one-to-one; $\{(6, -2), (8, 4), (-6, 2), (3, 3)\}$ **2.** not one-to-one   **3.** not one-to-one   **4.** one-to-one

5. not one-to-one   **6.** $f^{-1}(x) = \dfrac{x}{3}$ **7.** $f^{-1}(x) = x - 4$ **8.** $f^{-1}(x) = \dfrac{x+1}{5}$ **9.** $f^{-1}(x) = \dfrac{x-2}{3}$

10.

$y = \left(\tfrac{1}{2}\right)^x$

11.
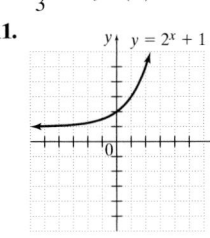
$y = 2^x + 1$

12.
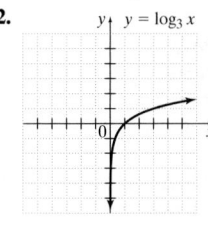
$y = \log_3 x$

13.

$y = \log_{1/3} x$

14. 3   **15.** 7   **16.** -8 **17.** 3
18. 2   **19.** $\dfrac{1}{2}$ **20.** 32   **21.** 4
22. 5   **23.** $\dfrac{1}{9}$ **24.** $\log_2 x^5$
25. $\log_2 5^x$ **26.** $\log_5 \dfrac{x^3}{y^5}$

27. $\log_5 x^9 y^3$ **28.** $\log_2 \dfrac{x^2 - 3x}{x^2 + 4}$ **29.** $\log_3 \dfrac{y^4 + 11y}{y + 2}$ **30.** $\log_7 9 + 2\log_7 x - \log_7 y$ **31.** $\log_6 5 + \log_6 y - 2\log_6 z$

Exercise Set 12.5   **1.** 0.9031   **3.** 0.3636   **5.** 0.6931   **7.** -2.6367 **9.** 1.1004   **11.** 1.6094   **13.** 1.6180   **15.** answers may vary

17. 2   **19.** -3 **21.** 2   **23.** $\dfrac{1}{4}$ **25.** 3   **27.** 2   **29.** -4 **31.** $\dfrac{1}{2}$ **33.** answers may vary   **35.** $10^{1.3} \approx 19.9526$

37. $\dfrac{10^{1.1}}{2} \approx 6.2946$ **39.** $e^{1.4} \approx 4.0552$ **41.** $\dfrac{4 + e^{2.3}}{3} \approx 4.6581$ **43.** $10^{2.3} \approx 199.5262$ **45.** $e^{-2.3} \approx 0.1003$ **47.** $\dfrac{10^{-0.5} - 1}{2} \approx -0.3419$

49. $\dfrac{e^{0.18}}{4} \approx 0.2993$ **51.** 1.5850   **53.** -2.3219 **55.** 1.5850   **57.** -1.6309 **59.** 0.8617   **61.** 4.2   **63.** 5.3

65. \$3656.38   **67.** \$2542.50   **69.** $\dfrac{4}{7}$ **71.** $x = \dfrac{3y}{4}$ **73.** $-6, -1$ **75.** $(2, -3)$ **77.** answers may vary

79.

$f(x) = e^x$

81.

$f(x) = e^{-3x}$

83.
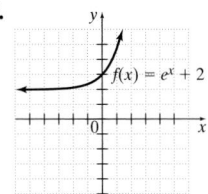
$f(x) = e^x + 2$

85.

$f(x) = e^{x-1}$

87.

$f(x) = 3e^x$

89.

$f(x) = \ln x$

91.
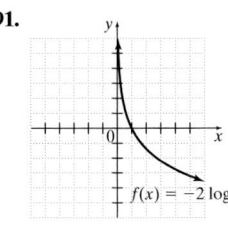
$f(x) = -2\log x$

93.
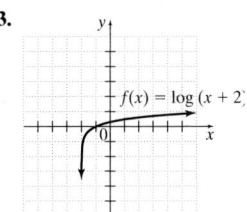
$f(x) = \log(x + 2)$

95.
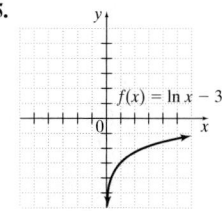
$f(x) = \ln x - 3$

97. answers may vary
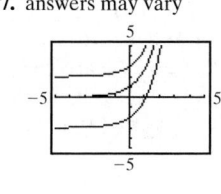

Graphing Calculator Explorations   **1.** 3.67 years, or 3 years and 8 months   **3.** 23.16 years, or 23 years and 2 months

Exercise Set 12.6   **1.** $\dfrac{\log 6}{\log 3}$; 1.6309   **3.** $\dfrac{\log 3.8}{2\log 3}$; 0.6076   **5.** $3 + \dfrac{\log 5}{\log 2}$; 5.3219   **7.** $\dfrac{\log 5}{\log 9}$; 0.7325   **9.** $\dfrac{\log 3}{\log 4} - 7$; -6.2075 **11.** $\dfrac{1}{3}\left(4 + \dfrac{\log 11}{\log 7}\right)$; 1.7441

13. $\dfrac{\ln 5}{6}$; 0.2682   **15.** 11   **17.** $9, -9$ **19.** $\dfrac{1}{2}$ **21.** $\dfrac{3}{4}$ **23.** 2   **25.** $\dfrac{1}{8}$ **27.** 11   **29.** $4, -1$ **31.** $\dfrac{1}{5}$ **33.** 100   **35.** $\dfrac{-5 + \sqrt{33}}{2}$ **37.** $\dfrac{192}{127}$ **39.** $\dfrac{2}{3}$

41. 103 wolves   **43.** 12,380,000 inhabitants   **45.** 10.5 yr   **47.** 9.9 yr   **49.** 1.7 yr   **51.** 8.8 yr   **53.** 24.5 lb   **55.** 55.7 in.   **57.** 11.9 lb/sq. in.   **59.** 3.2 mi

61. 12 weeks   **63.** 18 weeks   **65.** $-\dfrac{5}{3}$ **67.** $\dfrac{17}{4}$ **69.** $f^{-1}(x) = \dfrac{x-2}{5}$ **71.** 3.4%   **73.** answers may vary   **75.** 6.93   **77.** -3.68 **79.** 1.74   **81.** 0.2

Chapter 12 Review   **1.** one-to-one; $h^{-1} = \{(14, -9), (8, 6), (12, -11), (15, 15)\}$

3. one-to-one;

Rank in Automobile Thefts (Input)	2	4	1	3
US Region (Output)	W	Midwest	S	NE

5. a. 3   **b.** 7

7. not one-to-one

9. not one-to-one   **11.** $f^{-1}(x) = x + 9$ **13.** $f^{-1}(x) = \dfrac{x - 11}{6}$ **15.** $f^{-1}(x) = \sqrt[3]{x + 5}$ **17.** $g^{-1}(x) = \dfrac{6x + 7}{12}$

19.
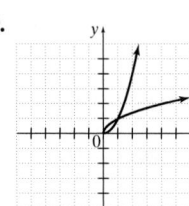

21. $f^{-1}(x) = \dfrac{x + 3}{2}$;   **23.** -2 **25.** $\dfrac{3}{2}$ **27.** $\dfrac{8}{9}$

29.

31.
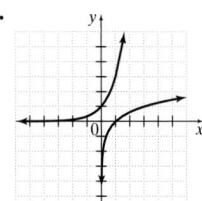

33. \$1131.82 **35.** $\log_7 49 = 2$ **37.** $\left(\dfrac{1}{2}\right)^{-4} = 16$ **39.** $\dfrac{1}{64}$ **41.** 0 **43.** 8 **45.** 5 **47.** 4

49. $\dfrac{17}{3}$ **51.** $-1, 4$ **53.**

55. $\log_3 32$ **57.** $\log_7 \dfrac{3}{4}$ **59.** $\log_{11} 4$

61. $\log_5 \dfrac{x^3}{(x + 1)^2}$ **63.** $3 \log_3 x - \log_3(x + 2)$

65. $\log_2 3 + 2 \log_2 x + \log_2 y - \log_2 z$ **67.** 2.02

69. 0.5563 **71.** 0.2231 **73.** 3 **75.** -1

77. $\dfrac{e^2}{2}$ **79.** $\dfrac{e^{-1} + 3}{2}$ **81.** 1.67 mm **83.** 0.2920

85. \$1957.30 **87.** $\dfrac{\log 7}{2 \log 3}$; 0.8856 **89.** $\dfrac{1}{2}\left(\dfrac{\log 6}{\log 3} - 1\right)$; 0.3155 **91.** $\dfrac{1}{3}\left(\dfrac{\log 4}{\log 5} + 5\right)$; 1.9538 **93.** $-\dfrac{\log 2}{\log 5} + 1$; 0.5693 **95.** $\dfrac{25}{2}$ **97.** $\varnothing$ **99.** $2\sqrt{2}$

101. 197,044 ducks **103.** 24 yr **105.** 43 yr **107.** 8.5 yr **109.** 2.82

Chapter 12 Test

1.

2. one-to-one **3.** not one-to-one **4.** one-to-one; $f^{-1}(x) = \dfrac{-x + 6}{2}$ **5** one-to-one; $f^{-1} = \{(0, 0), (3, 2), (5, -1)\}$

6. not one-to-one **7.** $\log_3 24$ **8.** $\log_5 \dfrac{x^4}{x + 1}$ **9.** $\log_6 2 + \log_6 x - 3 \log_6 y$ **10.** -1.53 **11.** 1.0686

12. -1 **13.** $\dfrac{1}{2}\left(\dfrac{\log 4}{\log 3} - 5\right)$; -1.8691 **14.** $\dfrac{1}{9}$ **15.** $\dfrac{1}{2}$ **16.** 22 **17.** $\dfrac{25}{3}$ **18.** $\dfrac{43}{21}$ **19.** -1.0979

20.

21.

22. \$5234.58 **23.** 6 yr **24.** 64,913 prairie dogs **25.** 15 yr **26.** 1.2%

Chapter 12 Cumulative Review **1. a.** $\dfrac{64}{25}$ **b.** $\dfrac{1}{20}$ **c.** $\dfrac{5}{4}$; Sec. 1.3, Ex. 4 **3.**   ; Sec. 8.2, Ex. 6

5. $\{(x, y, z) | x - 5y - 2z = 6\}$; Sec. 4.4, Ex. 4 **7. a.** -12

b. -3; Sec. 1.6, Ex. 5 **9. a.** \$102.60 **b.** \$12.60; Sec. 6.1, Ex. 9

11. -5; Sec. 2.2, Ex. 8 **13.** $2xy - 4 + \dfrac{1}{2y}$; Sec. 5.6, Ex. 3

$y = x^2$

15. $3(m + 2)(m - 10)$; Sec. 6.2, Ex. 7 **17.** $3x - 5$; Sec. 7.3, Ex. 3

19. a. 3 **b.** -3 **c.** -5 **d.** not a real number **e.** $4x$; Sec. 11.1, Ex. 4

21. a. $\sqrt[4]{x^3}$ **b.** $\sqrt[6]{x}$ **c.** $\sqrt[6]{72}$; Sec. 10.2, Ex. 8 **23. a.** $5\sqrt{3} + 3\sqrt{10}$ **b.** $\sqrt{35} + \sqrt{5} - \sqrt{42} - \sqrt{6}$ **c.** $21x - 7\sqrt{5x} + 15\sqrt{x} - 5\sqrt{5}$

d. $49 - 8\sqrt{3}$ **e.** $2x - 25$ **f.** $x + 22 + 10\sqrt{x - 3}$; Sec. 10.4, Ex. 4 **25.** $\dfrac{\sqrt[4]{xy^3}}{3y^2}$; Sec. 10.5, Ex. 3 **27.** 3; Sec. 10.6, Ex. 4

29. $\dfrac{9 + i\sqrt{15}}{6}, \dfrac{9 - i\sqrt{15}}{6}$; Sec. 11.1, Ex. 8 **31.** $\dfrac{-1 + \sqrt{33}}{4}, \dfrac{-1 - \sqrt{33}}{4}$; Sec. 11.3, Ex. 2 **33.** $[0, 4]$; Sec. 8.4, Ex. 2

35. 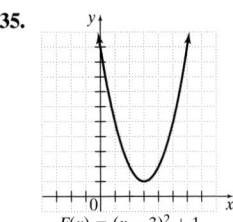 ; Sec. 11.5, Ex. 5 **37.** 63; Sec. 7.6, Ex. 1 **39.** $f^{-1}(x) = x - 3$; Sec. 12.1, Ex. 4 **41. a.** 2 **b.** -1 **c.** $\frac{1}{2}$; Sec. 12.3, Ex. 3

$F(x) = (x - 3)^2 + 1$

CHAPTER 13 CONIC SECTIONS

Graphing Calculator Explorations

1.   **3.**   **5.**   **7.**

Mental Math **1.** upward **3.** to the left **5.** downward

Exercise Set 13.1 **1.**   **3.**   **5.**   **7.**

9.   **11.** 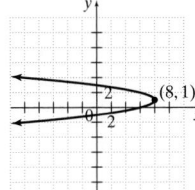 **13.** 5 units **15.** $\sqrt{41}$ units **17.** $\sqrt{10}$ units **19.** $\sqrt{5}$ units **21.** 13.88 units

23. 9 units **25.** $(4, -2)$ **27.** $\left(-5, \frac{5}{2}\right)$ **29.** $(3, 0)$ **31.** $\left(-\frac{1}{2}, \frac{1}{2}\right)$

33. $\left(\sqrt{2}, \frac{\sqrt{5}}{2}\right)$ **35.** $(6.2, -6.65)$

37.   **39.**   **41.**   **43.**   **45.**

47. 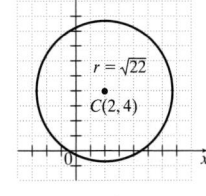 **49.** $(x - 2)^2 + (y - 3)^2 = 36$ **51.** $x^2 + y^2 = 3$ **53.** $(x + 5)^2 + (y - 4)^2 = 45$ **55.** answers may vary

57.   **59.**   **61.**   **63.**

65.   **67.**   **69.**   **71.**   **73.**

75.   **77.**   **79.**   **81.**   **83.**

85.   **87.**   **89.**   **91.**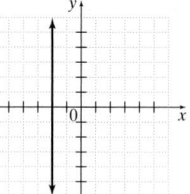

93. $\dfrac{\sqrt{10}}{4}$ **95.** $2\sqrt{5}$ **97. a.** 125 ft   **b.** 14 ft   **c.** 139 ft   **d.** $(0, 139)$ **e.** $x^2 + (y - 139)^2 = 125^2$

99. 20m   **101.** $y = -\dfrac{2}{125}x^2 + 40$ **103.**   **105.**

Graphing Calculator Explorations   1.   **3.**   **5.**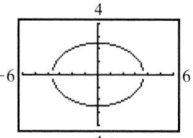

Mental Math   1. ellipse   **3.** hyperbola   **5.** hyperbola

Exercise Set 13.2   1.   **3.**   **5.**   **7.**

9.   **11.**   **13.**   **15.**   **17.**

19.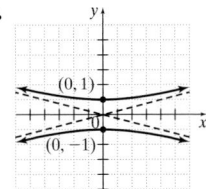

21. answers may vary   **23.** parabola   **25.** ellipse

27. hyperbola   **29.** circle   **31.** parabola   **33.** ellipse

35. hyperbola   **37.** parabola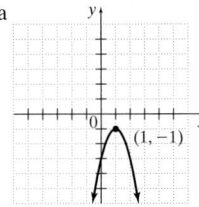

39. $(-\infty, 5)$ **41.** $[4, \infty)$ **43.** $-2x^3$ **45.** $-5x^4$

47. ellipses: C, E, H; circles: B, F; hyperbolas: A, D, G

49. A: 49, 7; B: 0, 0; C: 9, 3; D: 64, 8; E: 64, 8; F: 0, 0; G: 81, 9; H: 4, 2

51. A: $\dfrac{7}{6}$; B: 0; C: $\dfrac{3}{5}$; D: $\dfrac{8}{5}$ E: $\dfrac{8}{9}$; F: 0; G: $\dfrac{9}{4}$; H: $\dfrac{1}{6}$ **53.** equal to zero

55. answers may vary   **57.** $\dfrac{x^2}{1.69 \cdot 10^{16}} + \dfrac{y^2}{1.5625 \cdot 10^{16}} = 1$

59. $9x^2 + 4y^2 = 36$ **61.**   **63.**   **65.**

Integrated Review   **1.**   **2.**   **3.**   **4.**

5.   **6.**   **7.**   **8.**   **9.**

10.   **11.**   **12.**   **13.**   **14.**

15.

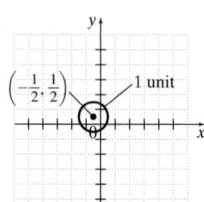

$\left(-\frac{1}{2}, \frac{1}{2}\right)$ 1 unit

Exercise Set 13.3   **1.** $(3, -4), (-3, 4)$ **3.** $\left(\sqrt{2}, \sqrt{2}\right), \left(-\sqrt{2}, -\sqrt{2}\right)$ **5.** $(4, 0), (0, -2)$ **7.** $\left(-\sqrt{5}, -2\right), \left(-\sqrt{5}, 2\right), \left(\sqrt{5}, -2\right), \left(\sqrt{5}, 2\right)$

9. ∅   **11.** $(1, -2), (3, 6)$ **13.** $(2, 4), (-5, 25)$ **15.** ∅   **17.** $(1, -3)$ **19.** $(-1, -2), (-1, 2), (1, -2), (1, 2)$ **21.** $(0, -1)$ **23.** $(-1, 3), (1, 3)$

25. $\left(\sqrt{3}, 0\right), \left(-\sqrt{3}, 0\right)$ **27.** ∅   **29.** $(-6, 0), (6, 0), (0, -6)$

31.

33.

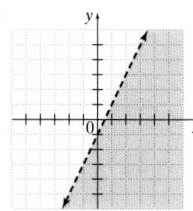

35. $(8x - 25)$ in.   **37.** $(4x^2 + 6x + 2)$ m   **39.** $0, 1, 2, 3,$ or 4 **41.** 9 and 7; 9 and -7; -9 and 7; -9 and -7 **43.** 15 cm by 19 cm   **45.** 15 thousand compact discs; price: \$3.75   **47.**

49.

Exercise Set 13.4   **1.**

3.

5.

7.

9.

11.

13.

15.

17.

19.

21.

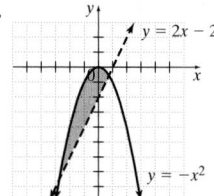

$y = 2x - 2$

$y = -x^2$

23.

25.

27.

29.

31.

33.

35.

37.

39.

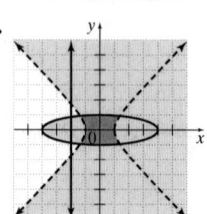

41. not a function   **43.** function   **45.** 1   **47.** $3a^2 - 2$ **49.** answers may vary

51.

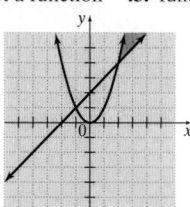

Chapter 13 Review   **1.** $\sqrt{197}$ units   **3.** $\sqrt{130}$ units   **5.** $7\sqrt{2}$ units   **7.** 16.60 units   **9.** $(-5, 5)$ **11.** $\left(-\dfrac{15}{2}, 1\right)$

13. $\left(\dfrac{1}{20}, -\dfrac{3}{16}\right)$ **15.** $\left(\sqrt{3}, -3\sqrt{6}\right)$ **17.** $(x + 4)^2 + (y - 4)^2 = 9$ **19.** $(x + 7)^2 + (y + 9)^2 = 11$

21.   **23.**   **25.**   **27.**   **29.**

31.   **33.** 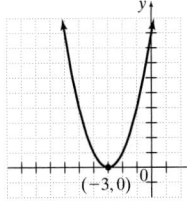   **35.** $(x - 5.6)^2 + (y + 2.4)^2 = 9.61$ **37.**

39.   **41.**   **43.**   **45.**   **47.**

49.   **51.** parabola   **53.** circle   **55.** circle   **57.** hyperbola

59. ellipse   **61.** hyperbola   **63.**   **65.** 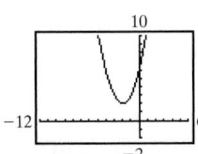   **67.** $(1, -2), (4, 4)$

69. $(-1, 1), (2, 4)$

71. $\left(2, 2\sqrt{2}\right), \left(2, -2\sqrt{2}\right)$

73. $(-1, 3), (-1, -3), (1, 3), (1, -3)$ **75.** $(1, 4)$ **77.** 15 ft by 10 ft

79.   **81.**   **83.**   **85.**   **87.**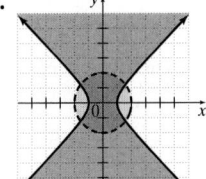

Chapter 13 Test   **1.** $2\sqrt{26}$ units   **2.** $\sqrt{95}$ units   **3.** $\left(-4, \dfrac{7}{2}\right)$

4.   **5.**   **6.**   **7.**   **8.**

9.   **10.**   **11.** 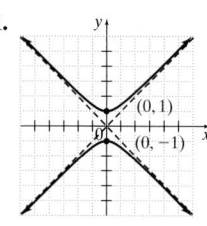   **12.** $(-5, -1), (-5, 1), (5, -1), (5, 1)$ **13.** $(6, 12), (1, 2)$

14.   **15.**   **16.** 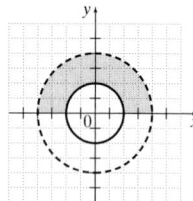   **17.** height: 10 ft; width: 30 ft

Chapter 13 Cumulative Review   1. $\varnothing$; Sec. 9.1, Ex. 3   **3.** $\left(-\infty, \dfrac{13}{5}\right] \cup [4, \ \infty \ )$; Sec. 9.1, Ex. 7   **5.** $-2, \dfrac{4}{5}$; Sec., Ex. 2   **7.** $24, -20$; Sec. 9.2, Ex. 3

9. $\dfrac{3}{4}, 5$; Sec. 9.2, Ex. 8   **11.** $(4, 8)$; Sec. 9.3, Ex. 2   **13.** $\dfrac{xy + 2x^3}{y - 1}$; Sec. 7.7, Ex. 3   **15.** $(-\infty, -\infty]$; Sec. 9.3, Ex.6   **17.** 16; Sec. 5.7, Ex. 4

19. a. 1   **b.** -4 **c.** $\dfrac{2}{5}$ **d.** x^2 **e.** $-3x^3$; Sec. 10.1, Ex. 3   **21. a.** $z - z^{17/3}$ **b.** $x^{2/3} - 3x^{1/3} - 10$; Sec. 10.2, Ex. 5   **b.**   **23. a.** 2   **b.** $\dfrac{5}{2}\sqrt{x}$

c. $14xy^2 \sqrt[3]{x}$ **d.** $4a^2b \sqrt[4]{2b}$; Sec. 10.3, Ex. 5   **25. a.** $\dfrac{5\sqrt{5}}{12}$ **b.** $\dfrac{3\sqrt[3]{7x}}{2}$; Sec. 10.4, Ex. 3   **27.** $\dfrac{\sqrt{21xy}}{3y}$; Sec. 10.5, Ex. 2   **29.** 42; Sec. 10.6, Ex. 1

31. a. $-i$ **b.** 1   **c.** -1 **d.** 1; Sec. 10.7, Ex. 6   **33.** $-1 + \sqrt{5}, -i - \sqrt{5}$; Sec. 11.1, Ex. 5   **35.** $2 + \sqrt{2}, 2 - \sqrt{2}$; Sec. 11.2, Ex. 3

37. $2, -2, i, -i$; Sec. 11.3, Ex. 3   **39.** $[-2, 3)$; Sec. 11.4, Ex. 4   **41.** 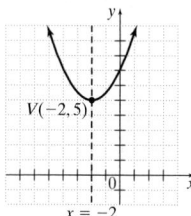   ; Sec. 11.5, Ex. 8   **43.** $(2, -16)$; Sec. 11.6, Ex. 4

45. $\sqrt{2} \approx 1.414$; Sec. 13.1, Ex. 5

CHAPTER 14   SEQUENCES, SERIES, AND THE BINOMIAL THEOREM

Exercise Set 14.1   1. $5, 6, 7, 8, 9$ **3.** $-1, 1, -1, 1, -1$ **5.** $\dfrac{1}{4}, \dfrac{1}{5}, \dfrac{1}{6}, \dfrac{1}{7}, \dfrac{1}{8}$ **7.** $2, 4, 6, 8, 10$ **9.** $-1, -4, -9, -16, -25$

11. $2, 4, 8, 16, 32$ **13.** $7, 9, 11, 13, 15$ **15.** $-1, 4, -9, 16, -25$ **17.** 75   **19.** 118   **21.** $\dfrac{6}{5}$ **23.** 729   **25.** $\dfrac{4}{7}$ **27.** $\dfrac{1}{8}$ **29.** -95 **31.** $-\dfrac{1}{25}$

33. $a_n = 4n - 1$ **35.** $a_n = -2^n$ **37.** $a_n = \dfrac{1}{3^n}$ **39.** 48 ft, 80 ft, and 112 ft   **41.** $a_n = 0.10(2)^{n-1}$; \$819.20   **43.** 2400 cases; 75 cases

45. 50 sparrows in 2004: extinct in 2010   **47.**   **49.** 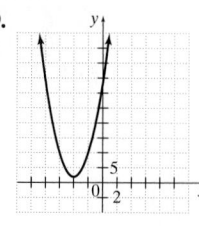   **51.** $\sqrt{13}$ units   **53.** $\sqrt{41}$ units
55. $1, 0.7071, 0.5774, 0.5, 0.4472$
57. $2, 2.25, 2.3704, 2.4414, 2.4883$

Exercise Set 14.2   **1.** $4, 6, 8, 10, 12$ **3.** $6, 4, 2, 0, -2$ **5.** $1, 3, 9, 27, 81$ **7.** $48, 24, 12, 6, 3$ **9.** 33 **11.** -875 **13.** -60 **15.** 96 **17.** -28

19. 1250 **21.** 31 **23.** 20 **25.** $a_1 = \dfrac{2}{3}; r = -2$ **27.** answers may vary   **29.** $a_1 = 2; d = 2$ **31.** $a_1 = 5; r = 2$ **33.** $a_1 = \dfrac{1}{2}; r = \dfrac{1}{5}$

35. $a_1 = x; r = 5$ **37.** $a_1 = p; d = 4$ **39.** 19 **41.** $-\dfrac{8}{9}$ **43.** $\dfrac{17}{2}$ **45.** $\dfrac{8}{81}$ **47.** -19 **49.** $a_n = 4n + 50; 130$ seats   **51.** $a_n = 6(3)^{n-1}$

53. $486, 162, 54, 18, 6; a_n = \dfrac{486}{3^{n-1}}; 6$ bounces   **55.** $a_n = 4000 + 125(n-1)$ or $a_n = 3875 + 125n; \$5375$ **57.** 25 g   **59.** $\dfrac{11}{18}$ **61.** 40 **63.** $\dfrac{907}{495}$

65. $\$11,782.40, \$5891.20, \$2945.60, \$1472.80$ **67.** $19.652, 19.618, 19.584, 19.55$ **69.** answers may vary

Exercise Set 14.3   **1.** -2 **3.** 60 **5.** 20 **7.** $\dfrac{73}{168}$ **9.** $\dfrac{11}{36}$ **11.** 60 **13.** 74 **15.** 62 **17.** $\dfrac{241}{35}$ **19.** $\displaystyle\sum_{i=1}^{5}(2i - 1)$ **21.** $\displaystyle\sum_{i=1}^{4}4(3)^{i-1}$

23. $\displaystyle\sum_{i=1}^{6}(-3i + 15)$ **25.** $\displaystyle\sum_{i=1}^{4}\dfrac{4}{3^{i-2}}$ **27.** $\displaystyle\sum_{i=1}^{7}i^2$ **29.** -24 **31.** -13 **33.** 82 **35.** -20 **37.** -2 **39.** $1, 2, 3, \ldots, 10; 55$ trees

41. $a_n = 6(2)^{n-1}; 96$ units   **43.** $a_n = 50(2)^n; n$ represents the number of 12-hour periods; 800 bacteria   **45.** 30 opossums; 68 opossums

47. 6.25 lb; 93.75 lb   **49.** 16.4 in.; 134.5 in.   **51.** 10 **53.** $\dfrac{10}{27}$ **55.** 45 **57.** 90 **59. a.** $2 + 6 + 12 + 20 + 30 + 42 + 56$

b. $1 + 2 + 3 + 4 + 5 + 6 + 7 + 1 + 4 + 9 + 16 + 25 + 36 + 49$ **c.** Answers may vary.   **d.** True; answers may vary

Integrated Review   **1.** $-2, -1, 0, 1, 2$ **2.** $\dfrac{7}{2}, \dfrac{7}{3}, \dfrac{7}{4}, \dfrac{7}{5}, \dfrac{7}{6}$ **3.** $1, 3, 9, 27, 81$ **4.** $-4, -1, 4, 11, 20$ **5.** 64 **6.** -14 **7.** $\dfrac{1}{40}$ **8.** $-\dfrac{1}{82}$ **9.** $7, 4, 1, -2, -5$

10. $-3, -15, -75, -375, -1875$ **11.** $45, 15, 5, \dfrac{5}{3}, \dfrac{5}{9}$ **12.** $-12, -2, 8, 18, 28$ **13.** 101 **14.** $\dfrac{243}{16}$ **15.** 384 **16.** 185 **17.** -10 **18.** $\dfrac{1}{5}$

Exercise Set 14.4   **1.** 36 **3.** 484 **5.** 63 **7.** 2.496 **9.** 55 **11.** 16 **13.** 24 **15.** $\dfrac{1}{9}$ **17.** -20 **19.** $\dfrac{16}{9}$ **21.** $\dfrac{4}{9}$ **23.** 185

25. $\dfrac{381}{64}$ **27.** $-\dfrac{33}{4}$, or -8.25 **29.** $-\dfrac{75}{2}$ **31.** $\dfrac{56}{9}$ **33.** $4000, 3950, 3900, 3850, 3800; 3450$ cars; $44,700$ cars   **35.** Firm A (Firm A, $\$265,000$;

Firm B, $\$254,000$)   **37.** $\$39,930; \$139,230$ **39.** 20 min; 123 min   **41.** 180 ft   **43.** Player A, 45 points; Player B, 75 points   **45.** $\$3050$

47. $\$10,737,418.23$ **49.** 720 **51.** 3 **53.** $x^2 + 10x + 25$ **57.** $\dfrac{8}{10} + \dfrac{8}{100} + \dfrac{8}{1000} + \cdots; \dfrac{8}{9}$ **59.** Answers may vary.   **65.** $8x^3 - 12x^2 + 6x - 1$

Exercise Set 14.5   **1.** $m^3 + 3m^2n + 3mn^2 + n^3$ **3.** $c^5 + 5c^4d + 10c^3d^2 + 10c^2d^3 + 5cd^4 + d^5$ **5.** $y^5 - 5y^4x + 10y^3x^2 - 10y^2x^3 + 5yx^4 - x^5$
7. Answers may vary.   **9.** 8 **11.** 42 **13.** 360 **15.** 56 **17.** $a^7 + 7a^6b + 21a^5b^2 + 35a^4b^3 + 35a^3b^4 + 21a^2b^5 + 7ab^6 + b^7$
19. $a^5 + 10a^4b + 40a^3b^2 + 80a^2b^3 + 80ab^4 + 32b^5$ **21.** $q^9 + 9q^8r + 36q^7r^2 + 84q^6r^3 + 126q^5r^4 + 126q^4r^5 + 84q^3r^6 + 36q^2r^7 + 9qr^8 + r^9$
23. $1024a^5 + 1280a^4b + 640a^3b^2 + 160a^2b^3 + 20ab^4 + b^5$ **25.** $625a^4 - 1000a^3b + 600a^2b^2 - 160ab^3 + 16b^4$

27. $8a^3 + 36a^2b + 54ab^2 + 27b^3$ **29.** $x^5 + 10x^4 + 40x^3 + 80x^2 + 80x + 32$ **31.** $5cd^4$ **33.** d^7 **35.** $-40r^2s^3$ **37.** $6x^2y^2$ **39.** $30a^9b$

41.   **43.**   **45.**   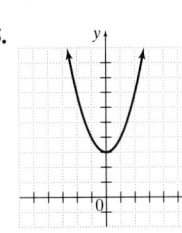   **47.** $x\sqrt[2]{x} + 5\sqrt{3x^2} + 30x\sqrt{x} + 30\sqrt{3x} + 45\sqrt{x} + 9\sqrt{3}$

49. 126 **51.** 28 **53.** answers may vary

Chapter 14 Review   **1.** $-3, -12, -27, -48, -75$ **3.** $\dfrac{1}{100}$ **5.** $a_n = \dfrac{1}{6n}$ **7.** 144 ft, 176 ft, 208 ft

9. $450, 1350, 4050, 12,150, 36,450; 36,450$ infected people in 2007   **11.** $-2, -\dfrac{4}{3}, -\dfrac{8}{9}, -\dfrac{16}{27}, -\dfrac{32}{81}$ **13.** 111 **15.** -83 **17.** $a_1 = 3; d = 5$

19. $a_n = \dfrac{3}{10^n}$ **21.** $a_1 = \dfrac{8}{3}, r = \dfrac{3}{2}$ **23.** $a_1 = 7x, r = -2$ **25.** $8, 6, 4.5, 3.4, 2.5, 1.9;$ good   **27.** $a_n = 2^{n-1}, \$512, \$536,870,912$

29. $a_n = 900 + (n-1)150$ or $a_n = 150n + 750; \$1650$/month   **31.** $1 + 3 + 5 + 7 + 9 = 25$ **33.** $\dfrac{1}{4} - \dfrac{1}{6} + \dfrac{1}{8} = \dfrac{5}{24}$ **35.** -4

37. -10 **39.** $\displaystyle\sum_{i=1}^{6}3^{i-1}$ **41.** $\displaystyle\sum_{i=1}^{4}\dfrac{1}{4^i}$ **43.** $a_n = 20(2)^n; n$ represents the number of 8-hour periods; 1280 yeast

45. Job A, $\$48,300$; Job B, $\$46,600$ **47.** 150 **49.** 900 **51.** -410 **53.** 936 **55.** 10 **57.** -25 **59.** $\$30,418; \$99,868$ **61.** $\$58; \$553$

63. 2696 mosquitoes   **65.** $\dfrac{5}{9}$ **67.** $x^5 + 5x^4z + 10x^3z^2 + 10x^2z^3 + 5xz^4 + z^5$ **69.** $16x^4 + 32x^3y + 24x^2y^2 + 8xy^3 + y^4$

71. $b^8 + 8b^7c + 28b^6c^2 + 56b^5c^3 + 70b^4c^4 + 56b^3c^5 + 28b^2c^6 + 8bc^7 + c^8$ **73.** $256m^4 - 256m^3n + 96m^2n^2 - 16mn^3 + n^4$ **75.** $35a^4b^3$

Chapter 14 Test   **1.** $-\dfrac{1}{5}, \dfrac{1}{6}, -\dfrac{1}{7}, \dfrac{1}{8}, -\dfrac{1}{9}$ **2.** 247 **3.** $a_n = \dfrac{2}{5}\left(\dfrac{1}{5}\right)^{n-1}$ **4.** $a_n = (-1)^n 9n$ **5.** 155 **6.** -330 **7.** $\dfrac{144}{5}$

8. 1 **9.** 10 **10.** -60 **11.** $a^6 - 6a^5b + 15a^4b^2 - 20a^3b^3 + 15a^2b^4 - 6ab^5 + b^6$ **12.** $32x^5 + 80x^4y + 80x^3y^2 + 40x^2y^3 + 10xy^4 + y^5$ **13.** 925

people; 250 people initially   **14.** $1 + 3 + 5 + 7 + 9 + 11 + 13 + 15$; 64 shrubs   **15.** 33.75 cm, 218.75 cm   **16.** 320 cm

17. 304 ft; 1600 ft   **18.** $\dfrac{14}{33}$

Chapter 14 Cumulative Review   **1. a.** -8 **b.** -8 **c.** 9   **d.** -9; Sec. 1.7, Ex. 4   **3.** $-2x - 1$; Sec. 2.1, Ex. 7   **5.** $y = \dfrac{1}{4}x - 3$; Sec. 3.5, Ex. 5

7. $-x + 5y = 23$ or $x - 5y = -23$; Sec. 3.6, Ex. 2   **9.** $x^3 - 4x^2 - 3x + 11 + \dfrac{12}{x + 2}$; Sec. 5.7, Ex. 2   **11. a.** $5\sqrt{2}$ **b.** $2\sqrt[3]{3}$ **c.** $\sqrt{26}$

d. $2\sqrt[4]{2}$; Sec. 10.3, Ex. 3   **13.** 10%; Sec. 11.1, Ex. 9   **15.** 2, 7; Sec. 11.3, Ex. 4   **17.** $\left(-\dfrac{7}{2}, -1\right)$; Sec. 11.4, Ex. 5   **19.** $\dfrac{25}{4}$ ft; $\dfrac{5}{8}$ sec; Sec. 11.6, Ex. 5

21. a. 25; 7   **b.** $x^2 + 6x + 9$; $x^2 + 3$; Sec. 8.3, Ex. 2   **23.** $f^{-1} = \{(1, 0), (7, -2), (-6, 3), (4, 4)\}$; Sec. 12.1, Ex. 3   **25. a.** 4   **b.** $\dfrac{3}{2}$ **c.** 6; Sec. 12.2, Ex. 4

27. a. 2   **b.** -1 **c.** 3   **d.** 6; Sec. 12.3, Ex. 5   **29. a.** $\log_{11}30$ **b.** $\log_3 6$ **c.** $\log_2 (x^2 + 2x)$; Sec. 12.4, Ex. 1   **31.** $2509.30; Sec. 12.5, Ex. 8

33. $\dfrac{\log 7}{\log 3} \approx 1.7712$; Sec. 12.6, Ex. 1   **35.** 18; Sec. 12.6, Ex. 2   **37.** 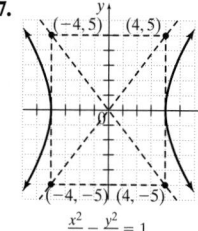 ; Sec. 13.2, Ex. 4   **39.** $(2, \sqrt{2})$; Sec. 13.3, Ex. 2

41. 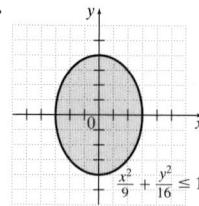 ; Sec. 13.4, Ex. 1

43. 0, 3, 8, 15, 24; Sec. 14.1, Ex. 1   **45.** 72; Sec. 14.2, Ex. 3
47. a. $\dfrac{7}{2}$ **b.** 56; Sec. 14.l3, Ex. 1   **49.** 465; Sec. 14.4, Ex. 2

APPENDIX A   SOLVING EQUATIONS AND INEQUALITIES—STUDY SKILLS PRACTICE

Section 2.3   **1.** $-\dfrac{8}{3}$ **2.** 6   **3.** 0   **4.** $\varnothing$

Section 2.7   **1.** $(-1, \infty)$ **2.** $\left(-\infty, \dfrac{3}{2}\right]$ **3.** 3   **4.** all real numbers or $(-\infty, \infty)$

Section 6.5   **1.** $9, -\dfrac{1}{2}$ **2.** 0   **3.** $4, -9$ **4.** $(3, \infty)$ **5.** $(-\infty, 2)$ **6.** $\left(-\infty, \dfrac{5}{4}\right]$

Section 7.5   **1.** $7, -2$ **2.** $-\dfrac{8}{5}$ **3.** $7, -\dfrac{3}{2}$ **4.** all real numbers or $(-\infty, \infty)$ **5.** $(-\infty, 5]$ **6.** $0, 5, -5$

Section 9.1   **1.** $[-1, 3]$ **2.** $(-1, 6)$ **3.** 5.1   **4.** $(-\infty, -4)$ **5.** $(-\infty, -3]$

Section 9.2   **1.** $\dfrac{5}{3}, -3$ **2.** $(-\infty, -2) \cup (2, \infty)$ **3.** 4   **4.** $\dfrac{1}{2}$ **5.** $[19, \infty)$ **6.** $7, -\dfrac{21}{2}$

Section 9.3   **1.** $(-\infty, 4] \cup [18, \infty)$ **2.** 4, 18   **3.** $(-3, 2)$ **4.** $-\dfrac{7}{9}, \dfrac{7}{9}$ **5.** 10

Section 9.6   **1.** 3   **2.** $\dfrac{\log 5 - 7 \log 2}{\log 2}$ **3.** $(-\infty, 0) \cup (9, \infty)$ **4.** $\dfrac{7}{6}$ **5.** $4, -1$ **6.** $\dfrac{-1 \pm \sqrt{10}}{3}$ **7.** $(2, 4]$ **8.** $\dfrac{61}{16}$ **9.** 1   **10.** $\left(-\infty, \dfrac{2}{3}\right) \cup (1, \infty)$

Section 10.6   **1.** $\dfrac{1}{3}, -5$ **2.** $\dfrac{9}{5}$ **3.** $5, -\dfrac{1}{3}$ **4.** $\dfrac{25}{3}$ **5.** $(-\infty, -4]$ **6.** $(-\infty, -2) \cup (8, \infty)$

Section 11.2   **1.** $2 \pm \sqrt{17}$ **2.** $\dfrac{5 \pm \sqrt{17}}{2}$ **3.** 3, 2   **4.** $(10, \infty)$ **5.** 3   **6.** $0, -\dfrac{22}{3}$ **7.** 2   **8.** $(-26, 34)$

Section 11.4   **1.** $5, -2$ **2.** $(-2, 5)$ **3.** -4 **4.** $(-\infty, -4] \cup (10, \infty)$ **5.** 23   **6.** $\dfrac{2 \pm \sqrt{10}}{5}$ **7.** $\varnothing$ **8.** 15   **9.** $(1, \infty)$ **10.** $(-\infty, 1]$

APPENDIX B   OPERATIONS ON DECIMALS

Appendix B Exercise Set   **1.** 17.08   **3.** 12.804   **5.** 110.96   **7.** 2.4   **9.** 28.43   **11.** 227.5   **13.** 2.7   **15.** 49.2339   **17.** 80   **19.** 0.07612   **21.** 4.56
23. 648.46   **25.** 767.83   **27.** 12.062   **29.** 61.48   **31.** 7.7   **33.** 863.37   **35.** 22.579   **37.** 363.15   **39.** 7.007

APPENDIX C   REVIEW OF ALGEBRA TOPICS

C-1 Equations   **1.** $-3, -8$ **3.** -2 **5.** $\dfrac{1}{4}, -\dfrac{2}{3}$ **7.** 1, 9   **9.** 0   **11.** 5   **13.** $\dfrac{3}{5}, -1$ **15.** no solution   **17.** $\dfrac{1}{8}$ **19.** 0   **21.** $6, -3$ **23.** 0   **25.** $\dfrac{2}{5}, -\dfrac{1}{2}$

27. $\dfrac{3}{4}, -\dfrac{1}{2}$ **29.** 29   **31.** -8 **33.** $-\dfrac{1}{3}, 0$ **35.** $-\dfrac{7}{8}$ **37.** $\dfrac{31}{4}$ **39.** 1   **41.** $-7, 4$ **43.** 4, 6   **45.** $-\dfrac{1}{2}$ **47. a.** incorrect   **b.** correct   **c.** correct

d. incorrect **49.** $K = -11$ **51.** $K = 24$ **53.** -4.86 **55.** 1.53 **57.** $\dfrac{8}{x}$ **59.** $8x$ **61.** $3x + 2$

C-2 Problem Solving **1.** $4y$ **3.** $3z + 3$ **5.** $(15x + 30)$ cents **7.** $10x + 3$ **9.** -5 **11.** 45,225 **13.** 78 **15.** 1.92 **17.** 1612.41 million acres
19. 1991 earthquakes **21.** 860 shoppers **23.** 17% **25.** 6750 users **27.** Los Angeles: 61.6 million; Atlanta: 76.9 million; Chicago: 67.4 million
29. $737 - 200$: 113 seats; $737 - 300$: 134 seats; $757 - 200$: 190 seats **31.** $430.00 **33.** Mile High Stadium: 76,125 seats; Heinz Field Stadium: 64,450
seats **35.** AOL: 23,700,000; Earthlink: 4,700,000; MSN: 4,000,000 **37.** 28.6 million **39. a.** 214,866 operators **b.** answers may vary **41.** 75, 76, 77
43. $64°, 32°, 84°$ **45.** height: 48 in.; width: 108 in. **47.** length: 14 cm; width: 6 cm **49.** $80°, 100°$ **51.** $15°, 75°$ **53.** $40°, 140°$ **55.** width: 8.4 m;
height: 47 m **57.** incandescent: 1500 bulb hours; fluorescent: 100,000 bulb hours; halogen: 4000 bulb hours

C-3 Graphing **1.** (5.2) **3.** $(3,0)$ **5.** $(-5, -2)$ **7.** $(-1, 0)$ **9.** QI **11.** QII **13.** QIII **15.** y-axis **17.** QIII **19.** x-axis **21.** QIV **23.** x-axis

25. QIII **27.** $y = -x - 2$ **29.** $3x - 4y = 8$ **31.** $y = \dfrac{1}{3}x$ **33.** $y + 4 = 0$ **35.** 1 **37.** -4 **39.** 1, 3

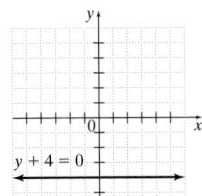

41. $q(-1) = -2$

C-4 Polynomials and Factoring **1.** $2y^2 + 2y - 11$ **3.** $x^2 - 7x + 7$ **5.** $25x^2 - 30x + 9$ **7.** $2x^3 - 4x^2 + 5x - 5 + \dfrac{8}{x + 2}$
9. $(x - 4 + y)(x - 4 - y)$ **11.** $x(x - 1)(x^2 + x + 1)$ **13.** $2xy(7x - 1)$ **15.** $4(x + 2)(x - 2)$ **17.** $(3x - 11)(x + 1)$ **19.** $4(x + 3)(x - 1)$
21. $(2x + 9)^2$ **23.** $(2x + 5y)(4x^2 - 10xy + 25y^2)$ **25.** $8x^2(2y - 1)(4y^2 + 2y + 1)$ **27.** $(x + 5 + y)(x^2 + 10x - xy - 5y + y^2 + 25)$
29. $(5a - 6)^2$ **31.** $7x(x - 9)$ **33.** $(a + 7)(b - 6)$ **35.** $(x^2 + 1)(x - 1)(x + 1)$ **37.** $(5x - 11)(2x + 3)$ **39.** $5a^3b(b^2 - 10)$
41. prime **43.** $10x(x - 10)(x - 11)$ **45.** $a^3b(4b - 3)(16b^2 + 12b + 9)$ **47.** $2(x - 3)(x^2 + 3x + 9)$ **49.** $(y^4 + 2)(3y - 5)$
51. $100(z + 1)(z^2 - z + 1)$ **53.** $(2b - 9)^2$ **55.** $(y - 4)(y - 5)$ **57.** $A = 9 - 4x^2 = (3 + 2x)(3 - 2x)$

C-5 Rational Expressions **1.** $\dfrac{1}{2}$ **2.** 10 **3.** $\dfrac{1 + 2x}{8}$ **5.** $\dfrac{2(x - 4)}{(x + 2)(x - 1)}$ **7.** 4 **9.** -5 **11.** $\dfrac{2x + 5}{x(x - 3)}$ **13.** -2 **15.** $\dfrac{(a + 3)(a + 1)}{a + 2}$

17. $-\dfrac{1}{5}$ **19.** $\dfrac{4a + 1}{(3a + 1)(3a - 1)}$ **21.** $-1, \dfrac{3}{2}$ **23.** $\dfrac{3}{x + 1}$ **25.** -1 **27. a.** $\dfrac{x}{5} - \dfrac{x}{4} + \dfrac{1}{10}$ **b.** Write each rational expression term so that the

denominator is the LCD, 20. **c.** $\dfrac{-x + 2}{20}$ **29.** b **31.** d **33.** d

APPENDIX D   AN INTRODUCTION TO USING A GRAPHING UTILITY

Viewing Window and Interpreting Window Settings Exercise Set

1. yes **3.** no **5.** answers may vary **7.** answers may vary **9.** answers may vary

11. Xmin = −12 Ymin = −12 **13.** Xmin = −9 Ymin = −12 **15.** Xmin = −10 Ymin = −25 **17.** Xmin = −10 Ymin = −30
Xmax = 12 Ymax = 12 Xmax = 9 Ymax = 12 Xmax = 10 Ymax = 25 Xmax = 10 Ymax = 30
Xscl = 3 Yscl = 3 Xscl = 1 Yscl = 2 Xscl = 2 Yscl = 5 Xscl = 1 Yscl = 3

19. Xmin = −20 Ymin = −30
Xmax = 30 Ymax = 50
Xscl = 5 Yscl = 10

Graphing Equations and Square Viewing Window Exercise Set   **1.** Setting B   **3.** Setting B   **5.** Setting B

7.   **9.**   **11.**   **13.**

15. **17.** **19.** **21.**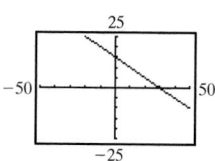

APPENDIX E SOLVING SYSTEMS OF EQUATIONS BY MATRICES

1. $(2, -1)$ **3.** $(-4, 2)$ **5.** $\varnothing$ **7.** $\{(x, y)|3x - 3y = 9\}$ **9.** $(-2, 5, -2)$ **11.** $(1, -2, 3)$ **13.** $(4, -3)$ **15.** $(2, 1, -1)$ **17.** $(9, 9)$ **19.** $\varnothing$ **21.** $\varnothing$
23. $(1, -4, 3)$ **25. a.** end of 1984 **b.** black-and-white sets; microwave ovens; The percent of households owning black-and-white television sets is decreasing and the percent of households owning microwave overs is increasing; answers may vary **c.** in 2002 **d.** no; answers may vary **27.** answers may vary

APPENDIX F SOLVING SYSTEMS OF EQUATIONS BY DETERMINANTS

1. 26 **3.** -19 **5.** 0 **7.** $(1, 2)$ **9.** $\{(x, y)|3x + y = 1\}$ **11.** $(9, 9)$ **13.** 8 **15.** 0 **17.** 54 **19.** $(-2, 0, 5)$ **21.** $(6, -2, 4)$ **23.** 16 **25.** 15

27. $\dfrac{13}{6}$ **29.** 0 **31.** 56 **33.** $(-3, -2)$ **35.** $\varnothing$ **37.** $(-2, 3, -1)$ **39.** $(3, 4)$ **41.** $(-2, 1)$ **43.** $\{(x, y, z)|x - 2y + z = -3\}$ **45.** $(0, 2, -1)$ **47.** 5

49. 0 **51.**
$$\begin{array}{cccc} + & - & + & - \\ - & + & - & + \\ + & - & + & - \\ - & + & - & + \end{array}$$
53. -125 **55.** 24

APPENDIX G MEAN, MEDIAN, AND MODE

Appendix D Exercise Set
1. mean: 29, median: 28, no mode **3.** mean: 8.1, median: 8.2, mode: 8.2 **5.** mean: 0.6, median: 0.6, mode: 0.2 and 0.6
7. mean: 370.9, median: 313.5, no mode **9.** 1214.8 ft **11.** 1117 ft **13.** 6.8 **15.** 6.9 **17.** 85.5 **19.** 73 **21.** 70 and 71 **23.** 9
25. 21, 21, 24

APPENDIX H REVIEW OF ANGLES, LINES, AND SPECIAL TRIANGLES

Appendix B Exercise Set

1. $71°$ **3.** $19.2°$ **5.** $78\frac{3}{4}°$ **7.** $30°$ **9.** $149.8°$ **11.** $100\frac{1}{2}°$ **13.** $m\angle 1 = m\angle 5 = m\angle 7 = 110°; m\angle 2 = m\angle 3 = m\angle 4 = m\angle 6 = 70°$ **15.** $90°$

17. $90°$ **19.** $90°$ **21.** $45°, 90°$ **23.** $73°, 90°$ **25.** $50\frac{1}{4}°, 90°$ **27.** $x = 6$ **29.** $x = 4.5$ **31.** 10 **33.** 12

PHOTO CREDITS

Chapter 1 CO © Garry Black/Masterfile, (p. 2) Wagne Sepp Seitz/Woodfin Camp & Associates, (p. 4) © Rachel Epstein/Photo Edit, (p. 10) Tim Davis/Photo Researchers, Inc., (p. 15) Photo Researchers, Inc., (p. 27) Joseph McBride/Stone/Getty Images, (p. 39) © Freeman Patterson/Masterfile, (p. 39) © ZEFA/Masterfile, (p. 46) Bishop Airport, (p. 46) © Sharna Balfour/Gallo Images/Corbis

Chapter 2 CO © MTPA/Masterfile, (p. 89) Laurence R. Lowry/Stock Boston, (p. 89) Hugh Sitton/Stone/Getty Images, (p. 89) Klaus G. Hinkelmann, (p. 101) Rolf Bruderer/Masterfile, (p. 103) Getty Images, Inc., (p. 104) AP/Wide World Photos, (p. 104) © Rick Gomez/Masterfile, (p. 108) NASA/Science Source/Photo Researchers, Inc., (p. 109) Bill Bachman/Stock Boston, (p. 111) Sean Reid/Alaska Stock, (p. 119) Colin Braley/Hulton Archive/Getty Images, (p. 120) © Norbert Wu/www.norbertwu.com, (p. 120) John Elk III/Stock Boston, (p. 121) © Jonathan Nourok/Photo Edit, (p. 122) ImageState/International Stock Photography Ltd., (p. 127) AP/ Wide World Photos, (p. 136) Jeff Greenberg/The Image Works, (p. 146) Jean-Claude LeJeune/Stock Boston

Chapter 3 CO © Royalty Free/CORBIS, (p. 159) David Young-Wolff/Stone/Getty Images, (p. 170) © Michael Newman/Photo Edit, (p. 172) © Tony Freeman/Photo Edit, (p. 173) © Amy C. Etra/Photo Edit, (p. 181) © Tony Freeman/Photo Edit, (p 181) © Tony Freeman/Photo Edit, (p. 188) © John Neubaur/Photo Edit, (p. 203) © Amy Etra/Photo Edit, (p. 203) © Cathy Melloan Resources/Photo Edit, (p. 207) Photodisc/Getty Images, (p. 209) © Spencer Grant/Photo Edit, (p. 209) © Mary Kate Denny/Photo Edit, (p. 210) © Ann States/Corbis/SABA Press Photos, Inc.

Chapter 4 CO © George Disario/CORBIS, (p. 248) © Michael Newman/Photo Edit, (p. 254) © Peter Hvizdak/The Image Works, (p. 254) Michael Kevin Daly/CORBIS, (p. 255) © David Young Wolff/Photo Edit Inc., (p. 273) SuperStock, Inc., (p. 274) © James Leynse/CORBIS SABA, (p. 275) © Armando Arorizo/ZUMA/Corbis, (p. 275) AP/Wide World Photos, (p. 276) Pierre Tremblay/Masterfile Corporation, (p. 284) AP/Wide World Photos

Chapter 5 CO © Tom Stewart/CORBIS, (p. 328) Jim Pickerell/Stock Boston, (p. 328) G. Brad Lewis/SeaPics.com, (p. 347) NASA IMAGE/POETRY Education and Public Outreach Program, (p. 347) Alvis Upitis/Image Bank/Getty Images, (p. 347) National Space Science Data Center/NASA

Chapter 6 CO © Ariel Skelley/CORBIS, (p. 359) Photodisc/Getty Images, (p. 378) R. Ian Lloyd/Masterfile Corporation, (p. 379) John Freeman/Stone/Getty Images

Chapter 7 CO © Spencer Grant/Photo Edit, (p. 417) Chuck Keeler/Frozen Images, (p. 423) Tebo Photography, (p. 425) © Bettmann/CORBIS, (p. 425) AP/Wide World Photos, (p. 452) © Charles O'Rear/CORBIS, (p. 465) Kelly-Mooney Photography/Corbis/Bettmann, (p. 465) Pictor/ImageState/International Stock Photography Ltd., (p. 475) Photodisc/Getty Images

Chapter 8 CO Getty Images, Inc, (p. 486) Masterfile Corporation, (p. 487) AGE Fotostock America, Inc., (p. 487) Mark Richards/Photo Edit, (p. 506) Richard A. Cooke III/Getty Images Inc.-Stone Allstock, (p. 510) © Ken Walsh/Age Fotostock, (p. 516) Wolff/Photo Edit, Inc.

Chapter 9 CO © Kevin Fleming/CORBIS

Chapter 10 CO Frank Whitney/Getty Images Inc.-Image Bank, (p. 579) © Susan Findlay/Masterfile, (p. 608) Steve Gottlieb/Getty Images, Inc.-Taxi

Chapter 11 CO © Gibson Stock Photography, (p. 639) Tony Freeman/Photo Edit, (p. 642) David Parker/Science Photo Library/Photo Researchers, Inc., (p. 652) Dr. Chikaraishi/National Institute for Fusion Science, (p. 652) © Shane Pederson/Wildlight, (p. 658) © ZEFA/Masterfile, (p. 661) Tim Flach/Getty Images Inc. Stone, (p. 661) Arthur S. Aubry Photography/Getty Images, Inc.-Photodisc, (p. 690) Robert Harding/Robert Harding World Imagery, (p. 696) AP/Wide World Photos

Chapter 12 CO © David R. Frazier Photolibrary, Inc. www.drfphoto.com, (p. 719) © I and I/Masterfile, (p. 719) Peter Arnold, Inc., (p. 744) © Weiss/Sunset/Animals Animals

Chapter 13 CO Superstock Royalty Free

Chapter 14 CO © David Young Wolff/Photo Edit, Inc.

Appendix (p. 849) AP/Wide World Photos, (p. 852) Michael Newman/Photo Edit, (p. 854) Jeff Robbins World Photos, (p. 854) © David Jennings/The Image Works

Single PC Site License

1.   GRANT OF LICENSE and OWNERSHIP: The enclosed computer programs and any data ("Software") are licensed, not sold, to you by Pearson Education, Inc. publishing as Pearson Prentice Hall ("We" or the "Company") in consideration of your adoption of the accompanying Company textbooks and/or other materials, and your agreement to these terms. You own only the disk(s) but we and/or our licensors own the Software itself.   This license allows instructors and students enrolled in the course using the Company textbook that accompanies this Software (the "Course") to use and display the enclosed copy of the Software on an unlimited number of computers, for academic use only, so long as you comply with the terms of this Agreement. You may make one copy for back up only. We reserve any rights not granted to you.

2.   USE RESTRICTIONS: You may <u>not</u> sell or license copies of the Software or the Documentation to others. You may <u>not</u> transfer, distribute or make available the Software or the Documentation. You may <u>not</u> reverse engineer, disassemble, decompile, modify, adapt, translate or create derivative works based on the Software or the Documentation. You may be held legally responsible for any copying or copyright infringement that is caused by your failure to abide by the terms of these restrictions.

3.   TERMINATION: This license is effective until terminated. This license will terminate automatically without notice from the Company if you fail to comply with any provisions or limitations of this license. Upon termination, you shall destroy the Documentation and all copies of the Software. All provisions of this Agreement as to limitation and disclaimer of warranties, limitation of liability, remedies or damages, and our ownership rights shall survive termination.

4.   DISCLAIMER OF WARRANTY: THE COMPANY AND ITS LICENSORS MAKE <u>NO</u> WARRANTIES ABOUT THE SOFTWARE, WHICH IS PROVIDED "<u>AS-IS</u>." IF THE DISK IS DEFECTIVE IN MATERIALS OR WORKMANSHIP, YOUR ONLY REMEDY IS TO RETURN IT TO THE COMPANY WITHIN 30 DAYS FOR REPLACEMENT UNLESS THE COMPANY DETERMINES IN GOOD FAITH THAT THE DISK HAS BEEN MISUSED OR IMPROPERLY INSTALLED, REPAIRED, ALTERED OR DAMAGED.   THE COMPANY DISCLAIMS ALL WARRANTIES, EXPRESS OR IMPLIED, INCLUDING WITHOUT LIMITATION, THE IMPLIED WARRANTIES OF MERCHANTABILITY AND FITNESS FOR A PARTICULAR PURPOSE. THE COMPANY DOES NOT WARRANT, GUARANTEE OR MAKE ANY REPRESENTATION REGARDING THE ACCURACY, RELIABILITY, CURRENTNESS, USE, OR RESULTS OF USE, OF THE SOFTWARE.

5.   LIMITATION OF REMEDIES AND DAMAGES: IN NO EVENT, SHALL THE COMPANY OR ITS EMPLOYEES, AGENTS, LICENSORS OR CONTRACTORS BE LIABLE FOR ANY INCIDENTAL, INDIRECT, SPECIAL OR CONSEQUENTIAL DAMAGES ARISING OUT OF OR IN CONNECTION WITH THIS LICENSE OR THE SOFTWARE, INCLUDING, WITHOUT LIMITATION, LOSS OF USE, LOSS OF DATA, LOSS OF INCOME OR PROFIT, OR OTHER LOSSES SUSTAINED AS A RESULT OF INJURY TO ANY PERSON, OR LOSS OF OR DAMAGE TO PROPERTY, OR CLAIMS OF THIRD PARTIES, EVEN IF THE COMPANY OR AN AUTHORIZED REPRESENTATIVE OF THE COMPANY HAS BEEN ADVISED OF THE POSSIBILITY OF SUCH DAMAGES. SOME JURISDICTIONS DO NOT ALLOW THE LIMITATION OF DAMAGES IN CERTAIN CIRCUMSTANCES, SO THE ABOVE LIMITATIONS MAY NOT ALWAYS APPLY.

6.   GENERAL: THIS AGREEMENT SHALL BE CONSTRUED IN ACCORDANCE WITH THE LAWS OF THE UNITED STATES OF AMERICA AND THE STATE OF NEW YORK, APPLICABLE TO CONTRACTS MADE IN NEW YORK, AND SHALL BENEFIT THE COMPANY, ITS AFFILIATES AND ASSIGNEES. This Agreement is the complete and exclusive statement of the agreement between you and the Company and supersedes all proposals, prior agreements, oral or written, and any other communications between you and the company or any of its representatives relating to the subject matter. If you are a U.S. Government user, this Software is licensed with "restricted rights" as set forth in subparagraphs (a)-(d) of the Commercial Computer-Restricted Rights clause at FAR 52.227-19 or in subparagraphs (c)(1)(ii) of the Rights in Technical Data and Computer Software clause at DFARS 252.227-7013, and similar clauses, as applicable.

Should you have any questions concerning this agreement or if you wish to contact the Company for any reason, please contact in writing: Customer Service Pearson Prentice Hall, 200 Old Tappan Road, Old Tappan NJ 07675.

Minimum System Requirements	Macintosh
Windows	Power PC G3 233 MHz or better
Pentium II 300 MHz processor	Mac OS 9.x or 10.x
Windows 98 or later	64 MB RAM
64 MB RAM	800 x 600 resolution
800 x 600 resolution	8x or faster CD-ROM drive
8x or faster CD-ROM drive	QuickTime 6.0 or later
QuickTime 6.0 or later	